Costa Rican Ecosystems

Costa Rican Ecosystems

Edited by Maarten Kappelle

The University of Chicago Press
Chicago and London

Maarten Kappelle is currently coordinator for the United Nations Environment Programme's global Chemicals and Waste Subprogramme and has previously held science and leadership roles in the World Wide Fund for Nature (WWF), The Nature Conservancy (TNC), Costa Rica's Instituto Nacional de Biodiversidad (INBio), and several universities in the Netherlands and abroad. He is author, editor, or coeditor of many scientific books in Spanish and English, including *Ecology and Conservation of Neotropical Montane Oak Forests*, *Biodiversity of the Oak Forests of Tropical America*, *Páramos de Costa Rica*, and *Diccionario de Biodiversidad*. He lives and works in Nairobi, Kenya.

The University of Chicago Press, Chicago 60637
The University of Chicago Press, Ltd., London

Printed in the United States of America

24 23 22 21 20 19 18 17 16 1 2 3 4 5

ISBN-13: 978-0-226-12150-5 (cloth)
ISBN-13: 978-0-226-27893-3 (paper)
ISBN-13: 978-0-226-12164-2 (e-book)
DOI: 10.7208/chicago/9780226121642.001.0001

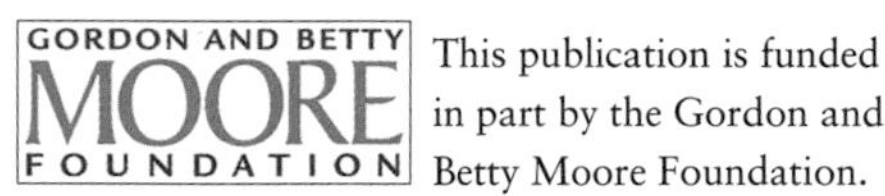

This publication is funded in part by the Gordon and Betty Moore Foundation.

Front cover photograph taken by Yamil Sáenz.

LIBRARY OF CONGRESS CATALOGING-IN-PUBLICATION DATA

Costa Rican ecosystems / edited by Maarten Kappelle.
pages ; cm
Includes bibliographical references and index.
ISBN 978-0-226-12150-5 (cloth : alk. paper)—ISBN 978-0-226-27893-3 (pbk. : alk. paper)—ISBN 978-0-226-12164-2 (e-book)
1. Ecology—Costa Rica. I. Kappelle, Maarten, editor.
QH108.C6C664 2015
577.097286—dc23
2015011315

∞ This paper meets the requirements of ANSI/NISO Z39.48-1992 (Permanence of Paper).

Contents

Dedication ix

List of Contributors xi

Foreword xv
Thomas E. Lovejoy

Presentation xvii
Rodrigo Gámez Lobo

Preface xix
Maarten Kappelle

Part I. Introduction

1 Costa Rica's Ecosystems: Setting the Stage 3
Maarten Kappelle

Part II. The Physical Environment

2 Climate of Costa Rica 19
Wilberth Herrera

3 Geology, Tectonics, and Geomorphology of Costa Rica: A Natural History Approach 30
Guillermo E. Alvarado and Guaria Cárdenes

4 Soils of Costa Rica: An Agroecological Approach 64
Alfredo Alvarado and Rafael Mata

Part III. The Pacific Ocean and Isla del Coco

5 The Pacific Coastal and Marine Ecosystems 97
Jorge Cortés

6 The Gulf of Nicoya Estuarine Ecosystem 139
José A. Vargas

7 Isla del Coco: Coastal and Marine Ecosystems 162
Jorge Cortés

8 Isla del Coco: Terrestrial Ecosystems 192
Michel Montoya

Part IV. The Northern Pacific Dry Lowlands

9 The Northern Pacific Lowland Seasonal Dry Forests of Guanacaste and the Nicoya Peninsula 247
Quírico Jiménez M., Eduardo Carrillo J., and Maarten Kappelle

10 Biodiversity Conservation History and Future in Costa Rica: The Case of Área de Conservación Guanacaste (ACG) 290
Daniel H. Janzen and Winnie Hallwachs

Part V. The Central and Southern Pacific Seasonally Moist Lowlands and Central Valley

11 The Central Pacific Seasonal Forests of Puntarenas and the Central Valley 345
Quírico Jiménez M. and Eduardo Carrillo J.

12 The Southern Pacific Lowland Evergreen Moist Forest of the Osa Region 360
Lawrence E. Gilbert, Catherine A. Christen, Mariana Altrichter, John T. Longino, Peter M. Sherman, Rob Plowes, Monica B. Swartz, Kirk O. Winemiller, Jennifer A. Weghorst, Andres Vega, Pamela Phillips, Christopher Vaughan, and Maarten Kappelle

Part VI. The Moist and Clouded Highlands

13 The Montane Cloud Forests of the Volcanic Cordilleras 415
Robert O. Lawton, Marcy F. Lawton, R. Michael Lawton, and James D. Daniels

14 The Montane Cloud Forests of the Cordillera de Talamanca 451
Maarten Kappelle

15 The *Páramo* Ecosystem of Costa Rica's Highlands 492
Maarten Kappelle and Sally P. Horn

Part VII. The Wet Caribbean Lowlands

16 The Caribbean Lowland Evergreen Moist and Wet Forests 527
Deedra McClearn, J. Pablo Arroyo-Mora, Enrique Castro, Ronald C. Coleman, Javier F. Espeleta, Carlos García-Robledo, Alex Gilman, José González, Armond T. Joyce, Erin Kuprewicz, John T. Longino, Nicole L. Michel, Carlos Manuel Rodríguez, Andrea Romero, Carlomagno Soto, Orlando Vargas, Amanda Wendt, Steven Whitfield, and Robert M. Timm

Part VIII. The Caribbean Sea and Shore

17 The Caribbean Coastal and Marine Ecosystems 591
Jorge Cortés

Part IX. The Rivers, Lakes, and Wetlands

18 Rivers of Costa Rica 621
Catherine M. Pringle, Elizabeth P. Anderson, Marcelo Ardón, Rebecca J. Bixby, Scott Connelly, John H. Duff, Alan P. Jackman, Pia Paaby, Alonso Ramírez, Gaston E. Small, Marcia N. Snyder, Carissa N. Ganong, and Frank J. Triska

19 Lakes of Costa Rica 656
Sally P. Horn and Kurt A. Haberyan

20 Bogs, Marshes, and Swamps of Costa Rica 683
Jorge A. Jiménez

Part X. Conclusion

21 Costa Rican Ecosystems: A Brief Summary 709
Maarten Kappelle

Acronyms 723

Subject Index 727

Systematic Index of Common Names 747

Systematic Index of Scientific Names 757

In Memory of Dr. Luis Diego Gómez Pignataro (1944–2009)

The editor wishes to dedicate this book on Costa Rican ecosystems to Dr. Luis Diego Gómez Pignataro. His untimely death due to leukemia over five years ago occurred as this publication was being written and edited.

Luis Diego Gómez P., Costa Rican biologist, botanist, geologist, and palaeontologist, was in his life a great naturalist, an unstoppable explorer, and a true *uomo universale* who mastered Latin—a neo-Renaissance person *pur sang* in the community of tropical scientists in Central America. Having learnt from *generosos maestros* like Paul W. Richards and Leslie R. Holdridge, he held a PhD from the Universidad de Loyola in Spain and a PhD *honoris causa* from the European Academy of Sciences.

Luis Diego discovered the flora of the most remote parts of the country, from the inaccessible sectors of Parque Internacional La Amistad and Parque Nacional Chirripó, to the botanically unexplored core of Isla del Coco. And, as Jorge Arturo Jiménez pointed out, Luis "was capable of moving easily from an organismic perspective to an ecosystemic viewpoint, . . . as a true son of Humboldtian science"—something not often seen among scholars today.

His vast knowledge of the arts and sciences went way beyond his in-depth understanding of the taxonomy and ecology of Neotropical ferns and mushrooms for which he was globally renowned as an outstanding eminence. For instance, he was an excellent piano player and cofounder of the Costa Rican music institution Fundación Ars Musica. Moreover, he published on fossils and early indigenous cultures.

From 1970 to 1985, Luis served as General Director of the Museo Nacional de Costa Rica in San José. Under his leadership the Herbario Nacional flourished and attracted dozens of international botanists—including the editor of this book—who contributed enormously to the national plant collection by donating their duplicates. During his administration he was able to form a dream team of Costa Rican plant scientists including Luis Poveda, Nelson Zamora, Pablo Sánchez, and Quírico Jiménez, among many others. To ensure the sharing of scientific knowledge among Costa Rican scientists he launched the biological journal *Brenesia*, in honor of Alberto Brenes Mora, the famous botanist from San Ramón.

Later on, from 1986 to 2005 Luis headed the Organization for Tropical Studies (OTS)–administered Estación Biológica Las Cruces near San Vito de Java, Coto Brus, in southern Costa Rica. Here, he turned the Wilson Botanical Garden, founded in 1962 by Catherine and Robert Wilson, into a world-class research facility for plant taxonomists and ecologists alike. After the 1994 fire that destroyed the station's library, herbarium, and other facilities he made sure the station was rebuilt and the library collection restored. After his retirement from Las Cruces, a new herbarium was established in his honor (Herbario Luis Diego Gómez [HLDG]) for his distinguished contributions to research, knowledge sharing, and land acquisition around the station. From 2005 to 2007 he served as Director of the OTS-administered Estación Biológica La Selva near Puerto Viejo de Sarapiquí, in northern Costa Rica. For more than twenty years, Luis worked for OTS as a professor of many courses, including ecology, ethnobiology, and tropical medicine.

Luis' visionary approach made him cofound Costa Rican organizations like Fundación de Parques Nacionales (FPN), the Academia Nacional de Ciencias, and the Instituto Nacional de Biodiversidad (INBio). During his prolific career he published over two hundred scientific articles and several books on, amongst others, Costa Rican plant taxonomy, ecology, and palaeontology, as well as a number of maps of the country's vegetation and biotic units. In fact, he was the first national palaeontologist in Costa Rica doing research in the country and publishing his findings in scientific journals and magazines, including the description of new fossils.

Some 30 years ago I first met Luis. It was in November 1985, when I walked into the herbarium of the Museo Nacional where he served as its director. He was sitting at his desk reviewing and identifying a number of ferns that he and others had collected in remote sectors of southern Costa Rica. I asked him if he could help me identify the dozens of ferns I had just collected for my master studies on the plant diversity and ecology of the montane oak forests of the Cordillera de Talamanca. He immediately started to put names on the newspapers that contained the specimens I had brought, explaining to me how I could recognize the different genera. Wow, how blessed I was to count on his taxonomic brilliance! We spent all afternoon going through the never-ending pile; he even came across a specimen of *Thelypteris gomeziana*, named after him by renowned fern taxonomist David Lellinger! At that moment we became friends forever and worked closely for years on the pteridophytic flora of the tall oak forests and *páramos* of the high Talamancas.

Twenty years after our first encounter in 1985, I started to develop the idea for the current volume and decided to discuss the plan with Luis in order to hear his viewpoint and invite him to serve as coeditor. Luis was enthusiastic from the first moment and provided helpful suggestions. Unfortunately, his progressing disease and the fatigue that immediately came with it did not allow him to play a role as coeditor, nor to write any piece of text for this volume. However, he continually encouraged me to work on the book and make sure I would finish it. When I learnt about his sudden departure on November 13, 2009, I was in shock. His death affected my energy levels and drive, and caused a setback in the development of the book. However, to this day I remember the last words he said to me only two weeks before he passed away: "Martín, asegúrete que se publique este libro! El pais lo necesita!"—"Martin, make sure this book gets published! The country needs it!"

Although Luis Diego's departure has left us with an immense void, his life and works remain to inspire young Costa Rican and visiting researchers, students, and naturalists. His first-ever studies on the classification and mapping of Costa Rica's vegetation and biotic units, his papers on the Costa Rican ferns and fungi, as well as his numerous contributions on paleontological features and fossils, will continue to serve the many people who have an interest in the knowledge and conservation of the unique biodiversity that characterizes Costa Rica as a pioneer and leader in making sustainability work for people and nature alike in the twenty-first century. Science will remember him through the twenty-seven newly described species that bear his name today.

Luis Diego enjoys a restful moment on a downed log in a Costa Rican forest. *Photo courtesy: Rebeca Brenes.*

We will remember "Luigi" (as he used to sign off his letters) or "Ludovicus" (as he wrote under his Latin comments on collected plant specimens) as an extraordinarily knowledgeable man, a sensible friend, and a wonderful colleague. Someone who always took the time to listen and share his knowledge and experience with others—always with a great sense of humour. We will dearly miss him.

Maarten Kappelle, editor

Contributors

Mariana Altrichter
Prescott College
Prescott, AZ 86301
USA

Alfredo Alvarado
Centro de Investigaciones Agronómicas
Facultad de Ciencias Agroalimentarias
Universidad de Costa Rica (UCR)
San Pedro de Montes de Oca
Costa Rica

Guillermo E. Alvarado
Área de Amenazas y Auscultación Sísmica y Volcánica
Instituto Costarricense de Electricidad (ICE)
San José
Costa Rica

Elizabeth P. Anderson
Department on Earth & the Environment
Florida International University
Miami, FL
USA

Marcelo Ardón
Department of Biology
East Carolina University
Greenville, NC
USA

J. Pablo Arroyo-Mora
Department of Geography
McGill University
Montreal, Quebec
Canada

Rebecca J. Bixby
Department of Biology
University of New Mexico
Albuquerque, NM
USA

Guaria Cárdenes
Escuela Centroamericana de Geología
Centro de Investigaciones en Ciencias Geológicas (CICG)
Universidad de Costa Rica (UCR)
San Pedro de Montes de Oca
Costa Rica
and
Biology Department
Florida Institute of Technology
Florida
USA

Eduardo Carrillo J.
Instituto Internacional en Conservación y Manejo en Vida Silvestre (ICOMVIS)
Universidad Nacional
Heredia
Costa Rica

Enrique Castro
La Selva Biological Station
Organization for Tropical Studies
San Pedro de Montes de Oca
Costa Rica

Catherine A. Christen
Smithsonian Conservation Biology Institute
Front Royal, VA 22630
USA

Ronald C. Coleman
Department of Biological Sciences
California State University Sacramento
Sacramento, CA 95819-6077
USA

Scott Connelly
Odum School of Ecology
University of Georgia
Athens, GA 30602
USA

Jorge Cortés
Centro de Investigación en Ciencias del Mar y Limnología (CIMAR)
and Escuela de Biología
Universidad de Costa Rica (UCR)
San Pedro de Montes de Oca
Costa Rica

James D. Daniels
Department of Biology and Cell Biology
Huntingdon College
Montgomery, AL 36106
USA

John H. Duff
US Geological Survey
Menlo Park, CA
USA

Javier F. Espeleta
Tropical Science Center
San José
Costa Rica

Rodrigo Gámez Lobo
Instituto Nacional de Biodiversidad (INBio)
Santo Domingo de Heredia
Costa Rica

Carlos García-Robledo
Laboratory of Interactions and Global Change
Department of Multitrophic Interactions
Institute of Ecology (INECOL)
Mexico
and
National Museum of Natural History
Departments of Botany and Entomology
Smithsonian Institution
Washington, DC 20013-7012
USA

Lawrence E. Gilbert
Section of Integrative Biology and Brackenridge Field Laboratory
University of Texas
Austin, TX 78712
USA

Alex Gilman
La Selva Biological Station
Organization for Tropical Studies
San Pedro de Montes de Oca
Costa Rica

Carissa N. Ganong
Odum School of Ecology
University of Georgia
Athens, GA 30602
USA

José González
La Selva Biological Station
Organization for Tropical Studies
San Pedro de Montes de Oca
Costa Rica

Kurt A. Haberyan
Department of Natural Sciences
Northwest Missouri State University
Maryville, MO 64468
USA

Winnie Hallwachs
Department of Biology
University of Pennsylvania
Philadelphia, PA 19104
USA

Wilberth Herrera
Apartado 2183-4050
Alajuela
Costa Rica

Sally P. Horn
Department of Geography
University of Tennessee
Knoxville, TN 37996
USA

Alan P. Jackman
Department of Chemical Engineering
University of California
Davis, CA
USA

Daniel H. Janzen
Department of Biology
University of Pennsylvania
Philadelphia, PA 19104
USA

Jorge A. Jiménez
Fundación MarViva
Santa Ana
Costa Rica

Quírico Jiménez
Empresa de Servicios Públicos de Heredia (ESPH)
Heredia
Costa Rica

Armand T. Joyce
1408 Eastwood Drive
Slidell, LA 70458
USA

Maarten Kappelle
World Wide Fund for Nature (WWF International)
Gland
Switzerland
and
Department of Geography
University of Tennessee
Knoxville, TN 37996
USA

Erin Kuprewicz
National Museum of Natural History
Department of Botany
Smithsonian Institution
Washington, DC 20013
USA

Marcy F. Lawton
Monte Sano Learning Center
Huntsville, AL 35801
USA

Robert O. Lawton
Department of Biological Sciences
University of Alabama
Huntsville, AL 35899
USA

R. Michael Lawton
Department of Ecology and Evolutionary Biology
University of Tennessee
Knoxville, TN 37996
USA

John T. (Jack) Longino
Department of Biology
University of Utah
Salt Lake City, UT 84112
USA

Thomas E. Lovejoy
Environmental Science and Policy
George Mason University
Fairfax, VA 22030
USA

Rafael Mata
Centro de Investigaciones Agronómicas
Facultad de Ciencias Agroalimentarias
Universidad de Costa Rica (UCR)
San Pedro de Montes de Oca
Costa Rica

Deedra McClearn
National Museum of Natural History
Department of Botany
Smithsonian Institution
Washington, DC 20013
USA
and
DKU Program Office
Duke University
Durham, NC 27708
USA

Nicole L. Michel
Department of Animal and Poultry Science
University of Saskatchewan
Saskatoon, SK S7N 5A6
Canada

Michel Montoya
Fundación Amigos de la Isla del Coco (FAICO)
San Pedro de Montes de Oca
Costa Rica

Pia Paaby
Organization for Tropical Studies (OTS)
San Pedro de Montes de Oca
Costa Rica

Pamela Phillips
Livestock Insects Research Laboratory
Agricultural Research Service
United States Department of Agriculture
Kerrville, TX 78028
USA

Rob Plowes
Section of Integrative Biology
Brackenridge Field Laboratory
University of Texas
Austin, TX 78712
USA

Catherine M. Pringle
Odum School of Ecology
University of Georgia
Athens, GA 30602
USA

Alonso Ramírez
Institute for Tropical Ecosystem Studies
University of Puerto Rico
Rio Piedras
Puerto Rico

Carlos Manuel Rodríguez
Conservation International
San José
Costa Rica

Andrea Romero
University of Wisconsin-Whitewater
Whitewater, WI 53190
USA

Peter M. Sherman
Prescott College
Prescott, AZ 86301
USA

Gaston E. Small
University of St. Thomas
Saint Paul, MN
USA

Marcia N. Snyder
Odum School of Ecology
University of Georgia
Athens, GA 30602
USA

Carlomagno Soto
La Selva Biological Station
Organization for Tropical Studies
San Pedro de Montes de Oca
Costa Rica

Monica B. Swartz
Section of Integrative Biology
Brackenridge Field Laboratory
University of Texas
Austin, TX 78712
USA

Robert M. Timm
Department of Ecology and Evolutionary Biology & Biodiversity Institute
University of Kansas
Lawrence, KS 66045
USA

Frank J. Triska
US Geological Survey
Menlo Park, CA
USA

José A. Vargas
Centro de Investigación en Ciencias del Mar y Limnología (CIMAR)
Universidad de Costa Rica (UCR)
San Pedro de Montes de Oca
Costa Rica

Orlando Vargas
La Selva Biological Station
Organization for Tropical Studies
San Pedro de Montes de Oca
Costa Rica

Christopher Vaughan
ICOMVIS
Universidad Nacional
Heredia
Costa Rica
and
Department of Forest and Wildlife Ecology
University of Wisconsin
Madison, WI 53706
USA

Andres Vega
AMBICOR
Tibas
Costa Rica

Jennifer A. Weghorst
Natural History Museum and Biodiversity Research Center
University of Kansas
Lawrence, KS 66045
USA

Amanda Wendt
Department of Ecology and Evolutionary Biology
University of Connecticut
Storrs, CT 06269-3043
USA

Steven Whitfield
Zoo Miami
Conservation and Research Department
Miami, FL 33177
USA

Kirk O. Winemiller
Department of Wildlife and Fisheries Sciences
Texas A&M University
College Station, TX 77843
USA

Foreword

It was inevitable that my path would lead to Costa Rica and its dazzling array of ecosystems. For decades the main entry point for North American students to tropical biology was through the Organization for Tropical Studies (OTS) and its field stations, most especially La Selva. I was a major exception in that I did my dissertation in the Brazilian Amazon, but I almost immediately (1971) made a pilgrimage to Central America to broaden my perspective. Just to get to La Selva involved a lengthy trip from San José: along the Cordillera, then up through the cloud forest and over a pass, and then down into the lowlands with the last leg of the journey up the Sarapiqui River by boat. It was a splendid partial introduction to Costa Rica's ecosystems.

In a sense I knew them already, but only in the abstract. The logical framework of Leslie R. Holdridge's Life Zone System was fundamental for every tropical biology student, with its elegant simplicity constructed from gradients of temperature, moisture, and altitude. As important as that construct was and is, it paled in contrast with the biological reality of the ecosystems in question.

Rich as the scientific knowledge of Costa Rica's ecosystems was 40 years ago it is dwarfed by that of today. That stems from important scientific institutions in addition to the Organization for Tropical Studies: among others, the Tropical Science Center (TSC) of Holdridge and Joseph Tosi in San Pedro, the Tropical Agricultural Research and Higher Education Center (CATIE) at Turrialba, the epicenter of tropical dry forest ecology at Santa Rosa in Guanacaste, continually sparked by Dan Janzen, and the visionary Instituto Nacional de Biodiversidad (INBio). As a consequence this book on Costa Rica's ecosystems—really the first for any tropical country—is long overdue but has a lot to build from. There have been extremely significant works on individual ecosystems (La Selva's rainforests, Monteverde's cloud forests, and the tropical alpine treeless *páramos*), but no comprehensive overview—other than the visitor experience on the grounds at INBioparque.

In the same four decades Costa Rica has gone from being the Central American country with the most national parks to being the self-styled "Green Republic"—one that is often globally in the forefront as a major innovator in environment and sustainability. At the Earth Summit in Rio de Janeiro in 1992 Costa Rica was very much the showcase country in both biodiversity science and in conservation. Costa Rica wasn't perfect, though. Much needed to be done in protecting marine and freshwater ecosystems, but Costa Rica was in every sense the leading nation in the tropics and to a major extent globally.

Now with Rio+20 behind us—the United Nations Conference on Sustainable Development held in Rio de Janeiro, Brazil, on June 20–22, 2012—and as the nations of the planet struggle with defining Sustainable Development Goals (SDGs), the environmental horizon is very dark indeed. The Stockholm Environment Institute's diagnosis of Planetary Boundaries shows three major transgressions: in the carbon cycle (climate change), the nitrogen cycle (and proliferating dead zones in coastal waters), and, above all, in biodiversity.

What is needed is a transition to planetary management, where this planet that we call "home" is managed as the biophysical system that it actually is. That, in turn, means conservation, management, Sustainable Development Goals that explicitly incorporate ecosystems and biodiversity, *and*, indeed, restoration of ecosystems at scale. That is the only way in which humanity can avoid a train wreck with climate change and the distorted nitrogen cycle with consequent staggering additional loss of biodiversity. It is also the only way that Latin America and the Caribbean can realize their potential as what the United Nations Development Program (UNDP) terms the Biodiversity Superpower.

So, the book *Costa Rican Ecosystems*, edited by Dr. Maarten Kappelle, comes at an extraordinarily opportune time. The culmination of decades of scientific achievement and experience, it provides an intellectual template upon which sustainability can be built for Costa Rica. It also serves as a model for an ecosystems overview that all nations should aspire to and emulate. When—not if—global sustainability is achieved, it will be recognized that one of the places that it started was in this remarkable country and with this equally remarkable book.

Dr. Thomas Eugene (Tom) Lovejoy
Senior Fellow, United Nations Foundation
Professor of Environmental Science and Policy at George Mason University
Member of the Council of the World Wildlife Fund (WWF-US)
Washington, DC
USA

Dr. Thomas Eugene (Tom) Lovejoy.

Presentation

During the second half of the twentieth century and the first decade of the twenty-first century, Costa Rica's natural environment underwent profound changes due to the socioeconomic development processes that took place. Costa Rica's increasing hunger for food, water, energy, infrastructure, and urban development has had a considerable, negative impact on its ecosystems and the environmental services on which its population depends—though the magnitude of this impact may be less severe when compared to other countries or regions. This has become clear from global environmental evaluations like the 2006 Millennium Ecosystem Assessment (MA), a multi-stakeholder effort endorsed by the United Nations (UN). Moreover, on top of these negative pressures occurred a series of environmental problems triggered by modern climate change.

In the environmental arena, and especially in the field of biodiversity, Costa Rica has made great strides to conserve and use its extraordinary natural richness in a smart way, thanks to its particular social and political conditions. Consequently, today more than 26% of its surface is managed under some kind of formally established public protection. Additionally, an approximate 2% is safeguarded in private reserves, while another 14% of privately owned land is maintained as forest through innovative funding schemes based on the concept of payments for environmental services (PES). Thanks to such incentives, Costa Rica has recovered a large portion of the dense forest cover that it harbored in previous decades, either through natural regeneration or by planting trees. In fact, the Costa Rican landscape has recently changed favorably due to this novel instrument.

Simultaneously, scientific knowledge about biodiversity—especially at the species level—has increased significantly during the past two decades, thanks to efforts of national scientific institutions with support from external parties. The generation of information and knowledge on the environment in general and biodiversity in particular, as well as their use in public and informal education, has increased in an equally important manner in the public domain. As a result, Costa Rica's society is now much more aware of the enormous biological richness that inhabits the country. Costa Ricans have become truly conscious about the need to protect the nation's biodiversity and use it sustainably. These particular conditions have enabled the simultaneous development of the natural tourism sector as a key industry at national level. In fact, nature-based tourism has now turned into an important source of income to many Costa Ricans and has positioned the country as a truly green destiny in the world.

At the same time the corporate sector is gradually becoming more involved in conservation efforts. Awareness about the need for corporate social responsibility (CSR) has increased among entrepreneurs. Now, they begin to understand that taking such responsibility is also in their own, for-profit interest. This change in attitude leads to innovative opportunities for action in the field of biodiversity conservation—action that had previously not been foreseen. Simultaneously, a modern political and legal framework for implementation is being developed, allowing the country to establish solid rules that guarantee the conservation, use, and equal distribution of benefits derived from its biodiversity.

However, as Costa Rica's program on the State of the Nation in Human Sustainable Development (Programa Estado de la Nación en Desarrollo Humano Sostenible) clearly states, at present the country is starting to face important internal and political conflicts related to the environment; the unsustainable use of land for urban, coastal, or touristic purposes; the generation of electricity; mining and oil exploration; and extensive export-driven agriculture, among others. The availability, quality, and quantity of suitable drinking water as well as the management of solid waste are also prime topics that attract considerable attention from civil society today.

Never before has Costa Rica so much felt the need for a broad, knowledge-based, integrated vision on the use of its territory and natural resources. Above all, it is now much more aware of the need to develop and implement a fully integrated management approach, beyond the establishment of protected wildlife areas. Adopting such an approach is the key to managing conservation areas in a way that will guarantee the delivery of essential ecosystem services for human benefits in the coming future.

That is exactly why the book *Costa Rican Ecosystems* appears at the right moment. Although some authors previously published essential works about the ecology and natural history of individual sectors of Costa Rica, no other volume has yet dealt with terrestrial, freshwater, and marine ecosystems in a more holistic and integrated manner than this book, bringing together and analyzing the existing wealth of ecological information and knowledge now available—a feature that makes this volume a unique contribution to our shared knowledge.

The editor of the book and author of various chapters therein, Dr. Maarten Kappelle, is highly qualified to lead and conduct the immense task of assembling a book of this proportion. He possesses a broad and in-depth knowledge of the country, both at the ecological and cultural level. Highly motivated, he lived and worked for long periods in Costa Rica and has shown a special appreciation for our country and its biodiversity. These characteristics allowed him to bring together a group of outstanding authors, environmental scientists and specialists alike, very familiar with this particular nation and knowledgeable about a wide range of its biodiversity aspects.

This book will, without doubt, serve as a key piece of study and reference for all who are interested to learn about Costa Rican biodiversity and its conservation from a scientific angle. The volume has enormous educational value as it focuses on ecosystems on the basis of characteristics that are easily visible and understandable by any student, any individual citizen, or tourist. It will notably contribute to our capability to value the magnitude of Costa Rica's natural richness, since its ecological perspective will allow the reader to visualize, understand, and better appreciate the extraordinary variety of landscapes and seascapes that thrive in this small corner of the world.

It will be, for sure, an invaluable book that will serve much-needed land use planning efforts, taking into account social, economic, and ecological dimensions. It will help ensure that we choose the appropriate type of resource use for each place, allowing current and future generations to enjoy Costa Rica's biodiversity and essential ecosystem services on which we depend so strongly. Such are choices that ultimately will allow Earth's evolutionary life processes to continue progressively in our beloved country and beyond.

Dr. Rodrigo Gámez Lobo
President and Former Director General of Instituto Nacional de Biodiversidad (INBio)
Former Advisor to the President of the Republic of Costa Rica
Former Director of the Escuela de Fitotécnia and Vice Rector of Research of the Universidad de Costa Rica
Santo Domingo de Heredia
Costa Rica

Dr. Rodrigo Gámez Lobo.

Preface

More than thirty years ago, in 1983, Daniel H. Janzen published his now classic book *Costa Rican Natural History* at the University of Chicago Press. In that magnum opus a total of 174 contributors shared with us their knowledge on Costa Rica's biophysical aspects and the natural history of the cornucopia of plants and animals that inhabit this tiny tropical country. Until that crucial moment, only Leslie Holdridge's *Forest Environments in Tropical Life Zones: A Pilot Study* (Holdridge et al. 1971) provided a national overview of diversity, though looking at the country's forest types from a forester's perspective.

At the time Dan Janzen's colossal volume appeared, remote areas of the country's hinterland, such as the Atlantic slopes of the Cordillera de Talamanca, had not yet been inventoried in detail. Specific fields of study such as the marine realm or soil microfauna had received very little attention as yet. Innovative techniques ranging from the interpretation of satellite imagery and the use of radio collars to track vertebrates, to DNA sequencing and bar-coding were just being developed and far from being "common business practice."

Since the publication of Janzen's monumental work hundreds if not thousands of scientific articles and technical reports on every aspect of Costa Rica's biodiversity have seen the light, including the description of hundreds of new species, genera, and even new families of flowering plants such as the Ticodendraceae—literally a family of *tico* (i.e., Costa Rican) trees (Gómez-Laurito and Gómez 1991). In this context, it is worthwhile to mention the 1986 book *Vegetación de Costa Rica* by Luis Diego Gómez, to whom this book is dedicated. The publication of that work now almost thirty years ago represented another milestone in biological sciences in the country. It dealt with Costa Rica's plant communities from the viewpoint of a taxonomist and biogeographer.

However, most biological and ecological research conducted in Costa Rica during the last quarter of the twentieth century focused on selected species and their biotic and/or physical environment, often from a reductionist, organismic, or population biological viewpoint rather than from a holistic, ecosystem-based approach. These studies were mainly carried out at well-equipped facilities (see, e.g. Hartshorn 1983), including the OTS-administered biological stations at La Selva (Atlantic lowland rain forest in the northeast), Palo Verde (Pacific wetlands along the Tempisque river), and Las Cruces (Pacific premontane forest in the south), the TSC-administered Monteverde research facility (montane forests in the Cordillera de Tilarán), the ACG-managed Santa Rosa station (dry forests along Guanacaste's northern Pacific coast), the UNA- and CATIE-managed site facilities in the Cordillera de Talamanca (montane oak forests near La Esperanza, San Gerardo de Dota, and Villa Mills, respectively), and marine biological stations such as the establishment at Punta Morales north of the city of Puntarenas. In the recent past research results from several of these places have been assembled in single volumes (McDade et al. 1994, on lowland rainforests at La Selva; Kappelle 1996, on montane oak forests in the Cordillera de Talamanca; Nadkarni and Wheelwright 2000, on cloud forests at Monteverde; Frankie et al. 2004, on seasonal dry forests in Guanacaste; Kappelle and Horn 2005, on alpine *páramos*; Wehrtmann and Cortés 2009, on marine biodiversity).

Still, no single account had treated Costa Rica's biodiversity in an ecosystemic and integrated manner or made an attempt to summarize the huge body of ecological knowledge that has appeared over the past decades throughout the country. Hence, twenty-five years after Janzen's milestone book, I felt the need to produce a comprehensive volume that would look at Costa Rica's biodiversity from an ecological perspective, paying extensive attention to the full range of the country's major ecosystems in a holistic way: from lowland to highland, from dryland to wetland, from Atlantic to Pacific, from freshwater to brackish and salty, from land to sea, and from continent to oceanic island—covering the wide variety of ecosystems that are the home of Costa Rica's extraordinary, alpha, beta, and gamma diversity.

Thus, in 2005 I contacted the late Luis Diego Gómez Pignataro to discuss the idea. As Costa Rica's most experienced, all-round vegetation scientist familiar with every corner of the country, he would be best positioned to serve as sparring partner and discuss the plan with me. I invited him to join the project and he gladly accepted. Immediately after, we proposed the idea to the University of Chicago Press, which enthusiastically joined the cause in 2006. Very sadly, shortly after, Luis became terminally ill. His worsened health condition did not allow him to become actively involved in the development of the book itself (see the dedicatory section "In Memory of Dr. Luis Diego Gómez Pignataro (1944–2009)" in this volume).

Since then it took nine years to come to the final product that is lying in front of us: the first scientific volume on Costa Rican ecosystems. Although the book is not meant to provide a definitive statement about Costa Rican ecosystems, it is meant to serve as a good start to the reader who wants to get a grasp of the unique ecological variety of one of the planet's most species-dense countries, if not the densest. Dan Janzen, in the preface of his 1983 book *Costa Rican Natural History*, wrote: "I hope this book will be out of date in ten or twenty years; some sections were out of date as they were being written. Those who read it will make it obsolete. . . . I have been impressed with how fragmentary is the knowledge each of us has even of our own areas of specialization and of the organisms we are supposed to be familiar with. Rather than being scornful of this sorry state of tropical biology, however, I encourage the reader to work doubly hard to rectify it." These same thoughts and words apply to this volume today. Hence, the editor hopes that this book will encourage young students to further investigate Costa Rica's ecosystems and the species of which they are composed, their relations with other species and the environment, and the interaction with human society.

The objective of this book is to introduce to the reader the entire range of principal ecosystems that dominate Costa Rica's landscapes and seascapes: its dry, seasonally-moist, rain and cloud forests; its lowland and highland woodlands, and alpine *páramos*; its rivers and lakes; its bogs, marshes, and swamps; and its estuaries, mangroves, sandy and rocky beaches, coral reefs, pelagic seas, and unique oceanic island, Isla del Coco. These chapters are geophysically put in perspective by providing an overview of the country's climate, geology, geomorphology, and soils. Each of the sixteen ecological chapters explores a particular type of ecosystem, discussing its physical environment, biogeography, species diversity, community composition, species interactions, structure and functioning, land use history, and conservation.

By analyzing, integrating, and synthesizing data from literature, museum collections, observational measurements, and field experiments–from tissues to plots to landscapes–the authors provide a vast array of state-of-the-art information and knowledge necessary to understand the diversity and complexity of these ecological systems. The editor hopes that the readers will be able to visualize the continuum that exists among these ecosystems. Moreover, it is hoped that the book will trigger the curious among the readers to conduct much-needed innovative research that will fill the many gaps that remain in our understanding of Costa Rica's ecosystems. Similarly, the book is intended to motivate people in general to make an extra effort to conserve this country's extraordinary biological riches in the long run, for people and nature alike.

Without the active help and encouragement of many individuals and organizations the present book would not have seen the light. I thank the 65 contributors who collaborated with their chapters without charging any fee for their efforts. In particular I express my gratitude to the lead authors for their huge effort to develop the chapters as well as for their unparalleled patience awaiting final publication of the book: Alfredo Alvarado, Bob Lawton, Cathy Pringle, Dan Janzen, Deedra McClearn, Guillermo Alvarado, Jorge Cortés, Jorge Jiménez, José Vargas, Larry Gilbert, Michel Montoya, Quírico Jiménez, Sally Horn, and Wilberth Herrera.

Numerous anonymous specialists are kindly thanked for their help in reviewing the twenty chapters on a voluntary basis. Costa Rican institutions like CATIE, INBio, Fundación MarViva, MINAET (and its predecessors SPN and MIRENEM), OTS, TSC, UCR, and UNA: all provided technical support to the project. INBio, TNC, and WWF—my employers over the past years—continually encouraged me to finish the book. I owe a special acknowledgment to the late Luis Diego Gómez for the many inspiring discussions we had about the idea that led to this book.

I also would like to acknowledge the University of Chicago Press for its great interest in publishing the book. Particularly, I am thankful to Christie Henry, editorial director for sciences and social sciences at the press, who showed interest in the development of the book since the very beginning of this journey in 2006 and helped to make sure we got to the final stages of publishing in 2015. Amy Krynak assisted energetically in formatting the volume for publication at the press, while Mary Corrado was instrumental in copyediting all chapters in a very professional manner. Gary S. Hartshorn and an anonymous reviewer were so kind to take the time and effort to review the entire volume and provided many constructive comments that helped improve the manuscript as a whole.

The Gordon and Betty Moore Foundation is much thanked for its generous financial contribution that made it possible to include a large series of color illustrations. A special thanks goes to Steve McCormick, Guillermo Castilleja, Jennifer Rae, and Cathy Manovi, who facilitated the Moore Foundation's financial support.

Finally, I am very grateful to Antoine Cleef and the late Adelaida Chaverri who first brought me to Costa Rica in 1985 to study its montane oak forests and *páramos*. I am indebted to the Dutch research funding organization NWO and Amsterdam-based UvA university that together funded part of my studies in Costa Rica. Rodrigo Gámez Lobo at INBio gave me in 1998 the opportunity to study Costa Rica's ecosystems at a national scale. At the same time, my parents, Mary E. Mohr and the late Dirk Kappelle, supported me in every step along the road of life, enabling me to get to today's result. Ultimately, it was my wife, Marta E. Juárez Ruiz, who encouraged me unconditionally during the past twenty-five years to make sure that the knowledge we generated on Costa Rica's extraordinary nature was shared more broadly and made available to the people in the country. *Costa Rican Ecosystems* would not have come to fruition without her never-ending optimism, drive, and motivation. Now, it is up to the next generation—the generation of our two sons, Derk Frederik and Bernard Floris—to ensure that Costa Rica's ecological systems will thrive well into the future, healthy and abundant, serving humankind in the twenty-first century.

Dr. Maarten Kappelle
Director of Programme Office Conservation Performance at the World Wide Fund for Nature (WWF International), Gland, Switzerland, outposted in Nairobi, Kenya
Adjunct Associate Professor in Geography at the University of Tennessee, Knoxville, TN, USA
Visiting Professor at the UN-mandated University for Peace, Ciudad Colon, Costa Rica

Dr. Maarten Kappelle.

Abbreviations

ACG: Área de Conservación Guanacaste (Guanacaste Conservation Area). Management unit of SINAC (Sistema Nacional de Áreas de Conservación), Costa Rica's National System of Conservation Areas.

CATIE: Centro Agronómico Tropical de Investigación y Enseñanza (Tropical Agronomical Center for Research and Education), Turrialba, Costa Rica.

INBio: Instituto Nacional de Biodiversidad (National Institute for Biodiversity), Santo Domingo de Heredia, Costa Rica.

MINAET: Ministerio de Ambiente, Energía y Tecnología (Ministry of Environment, Energy and Technology), San José, Costa Rica.

MIRENEM: Ministerio de Recursos Naturales, Energía y Minas (Ministry for Natural Resources, Energy and Mines), San José, Costa Rica.

NWO: Nederlandse Organisatie voor Wetenschappelijk Onderzoek (Netherlands Organisation for Scientific Research), The Hague, the Netherlands.

OTS: Organization for Tropical Studies, San Pedro de Montes de Oca, Costa Rica.

SPN: Servicio de Parques Nacionales (National Park Service), San José, Costa Rica.

TNC: The Nature Conservancy, Washington, DC, USA.

TSC: Tropical Science Center, San José, Costa Rica.

UCR: Universidad de Costa Rica (University of Costa Rica), San Pedro de Montes de Oca, Costa Rica,

UNA: Universidad Nacional (National University), Heredia, Costa Rica.

UvA: Universiteit van Amsterdam (University of Amsterdam), Amsterdam, the Netherlands.

WWF: World Wide Fund for Nature, Gland, Switzerland.

References

Frankie, G.W., A. Mata-Jiménez, and S.B. Vinson, eds. 2004. *Biodiversity Conservation in Costa Rica: Learning the Lessons in a Seasonal Dry Forest*. Berkeley: University of California Press. 341 pp.

Gómez, L.D. 1986. *Vegetación de Costa Rica*. Vol. 1 of L.D. Gómez, ed., *Vegetación y Clima de Costa Rica*. With 10 maps (scale 1:200,000). San José: EUNED.

Gómez-Laurito, J., and L.D. Gómez. 1991. Ticodendraceae: a new family of flowering plants. *Annals of the Missouri Botanical Garden* 78: 87–88.

Hartshorn, G.S. 1983. Plants: introduction. In D.H. Janzen, ed., *Costa Rican Natural History*, 118–57. Chicago: University of Chicago Press.

Holdridge, L.R., W.C. Grenke, W.H. Hatheway, T. Liang, and J.A. Tosi. 1971. *Forest Environments in Tropical Life Zones: A Pilot Study*. Oxford: Pergamon Press. 735 pp.

Janzen, D.H., ed. 1983. *Costa Rican Natural History*. Chicago: University of Chicago Press.

Kappelle, M. 1996. *Los Bosques de Roble (*Quercus*) de la Cordillera de Talamanca, Costa Rica: Biodiversidad, Ecología, Conservación y Desarrollo*. Amsterdam: University of Amsterdam. 336 pp.

Kappelle, M., and S.P. Horn, eds. 2005. *Páramos de Costa Rica*. Instituto Nacional de Biodiversidad (INBio)—The Nature Conservancy—WOTRO Foundation. Santo Domingo de Heredia, Costa Rica: INBio Press. 767 pp.

McDade, L.A., K.S. Bawa, H.A. Hespenheide, and G.S. Hartshorn, eds.

1994. *La Selva: Ecology and Natural History of a Neotropical Rain Forest*. Chicago: University of Chicago Press.
Nadkarni, N.M., and N. Wheelwright, eds. 2000. *Monteverde: Ecology and Conservation of a Tropical Cloud Forest*. New York: Oxford University Press.
Wehrtmann, I.S., and J. Cortés, eds. 2009. *Marine Biodiversity of Costa Rica, Central America*. Berlin: Springer 538 pp.

Part I

Introduction

Chapter 1 Costa Rica's Ecosystems: Setting the Stage

Maarten Kappelle[1]

We should preserve every scrap of biodiversity as priceless while we learn to use it and come to understand what it means to humanity.
—Edward O. Wilson, Professor Emeritus, Harvard University

No other area of equal size anywhere in America possesses so rich and varied flora, and none in North America is at all comparable in these respects. It is improbable that in any part of the Earth there can be found an equal area of greater botanical interest. . . . In few countries of the world, I believe, would it be possible to travel so much and find only pleasant and ever varied scenes, and be received everywhere with simple and sincere hospitality.
—Paul C. Standley, in *Flora of Costa Rica*, October 12, 1937

Ecosystem Discovery, Exploitation, Conservation, and Sustainability

Some twenty years after Christopher Columbus visited in 1502 the coast of today's Puerto Limón on his fourth and final voyage to the New World, the Spanish conquistador Gil González D'Ávila, while on a royal expedition sailing from Panama to Nicaragua, named the country *Costa Rica*, or Rich Coast. He did so because of the golden objects that were used by pre-Columbian indigenous tribes for body decoration and rank distinction, including necklaces, nose plugs, ear plugs, bracelets, and bells (Quilter and Hoopes 2003). However, ultimately it was not the golden treasures that justified the name of Costa Rica, but rather its biological richness: its huge variety of life, piled up in a small corner of the world (Gómez and Savage 1983). Ever since foreign naturalists like Anders Sandoe Ørsted, William More Gabb, Karl Sapper, Karl Hoffmann, Alexander von Frantzius, Karl Wercklé, and Henri François Pittier visited the country and were astonished by its rich flora and fauna, Costa Rica and its ecosystems have been considered by specialists and laymen a true Valhalla of biotic diversity in all its senses (Pittier 1908, Gómez and Savage 1983, Hartshorn 1983, Gómez 1986).

However, over the past 150 years Costa Rica's lush ecosystems have become more and more threatened, pricipally as a result of land conversion for cattle ranching, coffee growing, and large-scale banana production (Hall 1985). Particularly since World War II when the interest in precious hardwoods increased and construction of highways flourished (Merker et al. 1943), accelerated deforestation became the prime driver of biodiversity loss in the country (Sader and Joyce 1988). During the past few decades forest conversion together with other stress factors—such as climate change, overfishing, the introduction of aggressive invasive alien species, the construction of large infrastructure features such as roads and dams, sedimentation, environmental pollution, urban sprawl, and coastal encroachment—have begun to put ever-increasing pressures

[1] World Wide Fund for Nature (WWF International), Gland, Switzerland; Department of Geography, University of Tennessee, Knoxville, Tennessee

on the fragile cornucopia of Costa Rica's ecological systems, impoverishing and reshaping them in already fragmented landscapes and seascapes.

While during the second half of the twentieth century Costa Rica lost almost half of its forest cover, since the early 1970s to date (2015) the country has been able to save millions of hectares in 169 protected areas, ranging from absolute reserves and national parks to forest reserves and protective zones (Gámez and Ugalde 1988, Boza 1992, Wallace 1992, SINAC 2009, Obando 2011). Together, Costa Rica's protected areas cover 26.2% of the country's territory today. The development of the national park system initially occurred simultaneously with massive deforestation in unprotected areas, a phenomenon now known as the "Grand Contradiction" (Evans 1999).

The first wildlife area that received formal protection was created in 1945. It concerned the montane oak forest zone just south of Cartago, along both sides of the Inter-American Highway (Kappelle 1996). From 1969 to the late 1970s this and other early protected areas, including the Reserva Natural Absoluta Cabo Blanco and the Volcán Turrialba and Volcán Irazú national parks, were formally administered by the Departamento de Parques Nacionales. Then in 1977, protected area management passed on to the Servicio de Parques Nacionales (SPN), which was formally created as a specialized unit under direction of the Ministerio de Agricultura y Ganadería (MAG). Key protected areas like the now famous Cahuita, Chirripó, Corcovado, Santa Rosa, and Tortuguero national parks were created during that decade, as the environmental movement became stronger and focused hard on safeguarding the country's last remaining wild places (Gámez and Ugalde 1988, Wallace 1992, Boza 1993).

At the end of the next decade, in 1988, SPN was incorporated into the new Ministerio de Recursos Naturales, Energía y Minas (MIRENEM). Then, in 1995, new responsibilities were added while MIRENEM was restructured. It became the Ministerio del Ambiente y Energía (MINAE). In that same year, SPN was merged with both the Dirección General Forestal (DGF) and the Dirección General de Vida Silvestre (DGVS) into the innovative Sistema Nacional de Áreas de Conservación (SINAC), a subdivision of the young MINAE (Evans 1999). During the second administration of President Oscar Arias Sánchez (2006 to 2010) the telecommunications sector was added to MINAE, to become the Ministerio del Ambiente, Energía y Telecomunicaciones (MINAET). In 2013 MINAET became again MINAE, as the telecommunication department was moved to another ministry. As of 2014, MINAE is also referred to as the Ministerio del Ambiente, Energía, Aguas y Mares, recognizing the growing importance of the freshwater and marine resources for the country.

SINAC was foremost created to serve as a facilitating mechanism necessary to administer all protected areas in an integrated manner at regional level (SINAC 2009). In total, eleven Áreas de Conservación (ACs) were established as part of SINAC, covering the full territory of the country. Costa Rica's 1998 biodiversity law (Ley de Biodiversidad) legally formalized and strengthened this organizational structure and its holistic, decentralized, and inclusive approach.

Thanks to extraordinary efforts in the past, Costa Rica has now been able to devote nearly a third of its territory to the conservation into perpetuity of its rich biological diversity, spread over eleven conservation areas (for a more detailed historical account, see Wallace 1992, García 1997, Evans 1999, and Gámez 2003). Therefore, today Costa Rica serves as a successful model of biodiversity research and conservation. It is a country in which many innovative ideas were first conceptualized, tested, and implemented (Fournier 1991, Wallace 1992, Evans 1999). These ideas range from all-taxa biodiversity inventories (ATBI) (Janzen and Gámez 1997) and biological prospecting meant to discover wild species with medicinal properties (Tamayo et al. 2004) to avant-garde bar-coding of plant and insect specimens in ex situ collections (Gámez 1999); from an out-of-the-box means of linking debt reduction with environmental protection measures through Debt-for-Nature Swaps proposed by Tom Lovejoy in 1984 (Thapa 1998) to revolutionary Payments for Environmental Services (PES; Pagiola 2008, and see Arriagada et al. 2012); and from ecosystem-based sustainable tourism models (Aylward et al. 1996) to leadership in climate change discussions (Castro et al. 2000), mostly recently about Reduced Emissions from Deforestation and forest Degradation (REDD and REDD+; Karousakis 2007). Hence, since the early 1970s Costa Rica has been at the forefront in developing and implementing new and bold ideas to study and safeguard its extraordinary biodiversity.

Over the coming decades, such novel approaches will allow the country to catalyze its human sustainable development model based on twenty-first-century principles of a truly green economy. In this context it is important to mention that Costa Rica has announced its intention to become the first carbon dioxide–neutral country in 2030. Another very hopeful sign is the fact that Costa Rica has recently recorded a change from having a net loss of forests to having a net gain in forest area (UNEP/GRID-Arendal 2009): while in 1991, 29 percent (ca 14,000 km^2) of Costa Rica's land cover qualified as closed forest (Sánchez-Azofeifa et al. 2006), by 2010 thanks to forest-related interventions up to

51 percent of its land area could be classified again as closed forest (UNEP/GRID-Arendal 2009, Stone 2011)—a unique success story at global level!

History of Costa Rica's Biogeography

To understand the diversity and complexity of Costa Rica's ecosystems today, it is essential to get a grasp of their biogeographic history. When did these ecosystems actually originate and how did they develop over time? As Coates and Obando (1996) discussed in their treatment of the geologic evolution of the Central American Isthmus, its formation has indeed been a complex and extended process that stretched over the last 15 million years and had huge consequences to ocean circulation, global climatic patterns, biogeography, ecology, and the evolution of both terrestrial and marine organisms in the region (also, see Jones and Hasson 1985, and Stehli and Webb 1985). Today the rise of the isthmus is considered to be the culmination of an extended geologic history involving the growth and migration of the Central American volcanic arc, at the junction of the Pacific and Caribbean Plates, and its collision with South America (Coates and Obando 1996).

The formation and closure of the Central American land bridge about 2.7 million years (Ma) triggered the migration of plants and animals from North America (the Nearctic region) into South America (the Neotropical region) and vice versa, contributing to today's extraordinary isthmian biodiversity at all levels, from genes to landscapes (Rich and Rich 1983, Stehli and Webb 1985, Webb 2006). Recently, some authors concluded that the Central American closure took place some ten million years earlier (Montes et al. 2015). As a result land mammal faunas from North and South America mingled on a continental scale, including North American ungulates that found their way to South America (tapirs, deer, horses, pumas, canids, bears, and a number of rodents), while glyptodonts—more heavily armored relatives of modern armadillos—and giant anteaters (*Myrmecophaga* spp.), among others, migrated along the inverse route, from South to North America (Rich and Rich 1983, Webb 2006). Some families of northern land mammals diversified at moderate rates (Procyonidae, Felidae, Tayassuidae, and Camelidae), while others such as Canidae, Mustelidae, Cervidae, and especially Muridae, evolved explosively (Webb 2006). Today this hemispherical process of species migration is known as the *Great American Biotic Interchange* (GABI), which in fact was first observed by the nineteenth-century English naturalist Alfred Russell Wallace (1876).

At the same time, the Pliocene closure of the isthmus separated the Pacific and Atlantic Oceans that had been connected since the Mesozoic by an interoceanic seaway through the Central American volcanic island arc. As a result, two marine floras and faunas became disconnected, allowing evolution to take place among the now-separated Atlantic and Pacific species populations (Cronin and Dowsett 1996). On top of that, the occurrence of past glaciations on the highest mountains (Kappelle and Horn, chapter 15 of this volume), differences in seasonal patterns of rainfall superimposed on discontinuous mountain chains, temperature gradients changing over short altitudinal ranges (Herrera, chapter 2 of this volume), and the development of rich mineral soils on rugged terrain and lowland plains (Alvarado and Mata, chapter 4 of this volume), led to even higher levels of biotic diversity and ecosystem complexity (Burger 1980, Stehli and Webb 1985, Gómez 1986, Kappelle et al. 1992, Alvarado and Cárdenes, chapter 3 of this volume).

Costa Rican Biodiversity at the Species Level

Over the past three decades, biodiversity inventories in Costa Rica and the world have increased and improved considerably, allowing us to make relatively good estimates of current species diversity (Groombridge 1992, Obando 2002). A recent review of Costa Rican species data shows that out of about 2 million species that have been discovered on Earth, around 95,000 are found in Costa Rica (Obando 2011). That is about 5% of all species officially known to exist on our planet.

Similarly, it is expected that around half a million species thrive in Costa Rica, including all the species unknown until date. That is about 3.6% of the 14 million species that have been estimated to live on Earth (Obando 2002, 2011). Thus, it is believed that so far only 19% of all species living in Costa Rica have been formally discovered and scientifically described. This small percentage underlines the need to continue to invest in species inventories and—in the case of species new to science—prepare and publish formal species descriptions. However, the lack of trained taxonomists and curators needed to conduct the correct identification of biological specimens—a problem known as the *Taxonomic Impediment* and first observed by the International Union of Biological Sciences (IUBS) and its DIVERSITAS Programme—withholds the country from quickly raising the number of formally known native species, while limiting its ability to conserve, use, and share the benefits of its biological diversity in a socially just manner.

Today, Costa Rica is home to at least 125 species of viruses, 213 Monera (among others, bacteria), 2,300 fungi (with ca. 700 Ascomycota and 1,300 Basidiomycota), 564 algae (including 205 microalgae), 11,467 plants including some 2,000 tree species, 670 Protozoa, 88 nematodes, 66,000 insects (including 16,000 Lepidoptera or butterflies), 1,550 mollusks, 916 fish (781 marine species and 135 freshwater fish), 189 amphibians, 234 reptiles, 854 birds, and 237 mammals of which 107 are bats and 20 are marine mammals (Herrera and Obando 2009). These data reveal that 75% of the 87,000 species known from Costa Rica actually are insects. It also demonstrates that almost 10% of all bird species in the world are indeed found in Costa Rica.

Furthermore, Mug et al. (2001) report that a total of about 5,000 species (that is 5.7% of 87,000 known species) are coastal-marine in distribution, with at least 3,650 species known from the Pacific Ocean and coast, and over 1,325 species known from the Caribbean Sea and shores. These data demonstrate that the Pacific waters of Costa Rica are almost three times as rich in species compared with the country's portion of the Caribbean Sea. In this regard, the reader should be aware that the 1,100 km Pacific coast line is five times as long as the Caribbean coast line (ca. 200 km), and that the Economic Exclusive Zone (EEZ) of Costa Rica amounts to ca. 570,000 km^2 in the Pacific Ocean but extends over only approximately 24,000 km^2 in the waters of the Caribbean Sea (Mug et al. 2001, Obando 2002). All together, Costa Rica's marine territory is eleven times bigger than its land surface.

On a global scale, Costa Rica is certainly among the twenty most species-rich countries (Groombridge 1992). At the same time, Costa Rica's land surface covers a mere 51,100 km^2, which corresponds to only 0.03% of the global land surface. Hence the notion, that Costa Rica is the world's number one country in terms of species density (i.e., the number of species present per 1,000 km^2; see García 1996, and Obando 2002). Although other so-called megadiverse countries do contain more species in absolute numbers (e.g., Australia, Brazil, Colombia, Indonesia, and Mexico), they are 22 (Colombia) to 150 (Australia) times bigger in size (measured as land surface).

When looking at the potentially unique presence of these many species, it turns out Costa Rica is a country with moderate levels of endemism. This is due to the fact that important Costa Rican ecosystems are shared with neighboring countries: for instance, the dry forest of Guanacaste extends into northern Pacific Nicaragua, the northern Caribbean lowland rainforests of Sarapiquí and San Carlos are similar to forest types found across the border in the Atlantic sector of Nicaragua, and the montane forests and *páramos* of the southern highlands extend into the Panamanian sector of the *Cordillera de Talamanca*. However, still some 1,200 species of plants (that is, 12% of all plant species), including some 177 tree species, are unique to Costa Rica (Obando 2002). Similarly, around 14% of all freshwater fish species, 20% of amphibians, 16% of reptiles, 0.8% of birds, and 2.5% of mammals, are endemic to Costa Rica (Obando 2002).

Areas richest in endemic plant species are the volcanic *cordilleras* (Guanacaste, Tilarán, and Central) and the Osa Peninsula. Other important endemic regions include the high Talamancas, the central Pacific sector, and the Pacific oceanic island Isla del Coco (Marco V. Castro C., pers. com.; Obando 2002). On the other hand, levels of endemicity—and species distributions in general—may be due to some kind of collection bias: historically, not all parts of the country have been sampled with the same level of intensity, and sites along easily accessible roads and near biological field stations may suffer from oversampling by taxonomists and other collectors (Marco V. Castro C., pers. com.).

Unfortunately, many species thriving in Costa Rica are becoming threatened by a variety of stress factors mentioned earlier in this chapter. Obando (2011) estimates, that populations of about 2% of the species known from Costa Rica are severely under pressure, some of which are actually threatened with extinction. Amphibians are perhaps the species group that is most threatened; some species such as the golden toad (*Bufo periglenes*) are believed to have gone extinct over the past 25 years (Pounds 2001, Pounds and Crump 1994).

Costa Rican Biodiversity at the Ecosystem Level

The thousands of native species (*alpha diversity*) that thrive in Costa Rica compose a wide array of complex ecological systems. In fact, all main terrestrial Central American ecosystems—that is, tropical rainforest, seasonally dry tropical forest, tropical cloud forest, temperate (oak) forests, and high-elevation ecosystems such as *páramo*—are found in tiny Costa Rica (Dirzo 2001). This country is therefore perhaps the best representative of Central America's land-based ecodiversity. Its complex geological history, variety of climates, and topographic heterogeneity make it a kaleidoscope of terrestrial ecosystems found in the region (Burger 1980, Gómez 1986).

On the salty side, the presence of estuaries, mangroves, coastal channels, sandy beaches, rocky intertidal zones, intertidal mud flats, seagrass beds, coral reefs, rodolith beds, pelagic systems, deep benthic zones, various kinds of islets, and a tropical fjord turn Costa Rica into a coastal-marine biodiversity paradise, a mecca for both oceanographers and marine biologists (Mug et al. 2001, Wehrtmann and Cortés 2009; and see various chapters by Cortés, this volume). Freshwater ecosystems are no less numerous, though less species-rich. Throughout the country, one can observe wetlands such as rivers and rivulets, lakes and lagoons, as well as swamps, marshes, and peatbogs. They occur in alpine *páramos,* in mountainous areas, along hilly terrains, in lowland regions, and in coastal zones.

Environmental heterogeneity along gradients (*beta diversity*) and in and among broad landscapes and seascapes (*gamma diversity*) is overwhelming. One can take an imaginary walk from coast to coast, along a cross-section, starting at the dry Peninsula de Santa Elena along the northern Pacific coast where cacti locally dominate rocky beaches, through the seasonally dry forests of Guanacaste, on to the steep volcanic cordilleras with their epiphyte-laden montane cloud forests near Monteverde and subalpine dwarf forests at the summit of Volcán Barva; then hike towards the *páramo* patches and volcanic lakes at elevations over 3,000 m on top of Volcán Irazú and V. Turrialba, or further south at Cerro Chirripó, and walk down into the Atlantic zone with its lush premontane and lowland tropical rainforests and broad rivers meandering through the plains of San Carlos, Sarapiqui, and Limón; and then continue to the southeast through the dense, humid and hot forests along the lower slopes of the Cordillera de Talamanca, and end near Cahuita and Gandoca-Manzanillo where sandy palm beaches and coral reefs dot the coast line. Hence, over a length of a few hundred kilometers, one traverses the full array of tropical ecosystems, over land from ocean to ocean. Nowhere else in the tropics one can do such a complete "ecological triathlon," while observing a dozen magnificent ecosystems and thousands of unique biological species!

Brief Overview of Existing Ecosystem Classifications in Costa Rica

It is well known that species communities occur as continua, some species decreasing in importance as others become more common along environmental gradients. On the other hand, scientists and practitioners find it useful to classify assemblages of plants, animals, and even microorganisms into species communities and—at higher levels of organization—ecosystems. Graphic visualizations of community or ecosystem-level units as expressed in maps are essential to understand and delineate the geographic distribution of species combinations and landscape patches. Such spatial tools and related geographic information systems (GIS) are vital to revealing landscape patterns, assessing land cover change, conducting land use zoning, and conserving biodiversity along ecological gradients, among other actions (Savitsky and Lacher 1998).

Edward O. Wilson defines biodiversity as "the totality of the inherited variation of all forms of life across all levels of variation, from ecosystem to species to gene." In the case of Costa Rica, biodiversity at the species level has received relatively considerable attention from scientists (e.g., Janzen 1983, Gámez 1991, Obando 2002, Hammel et al. 2004, Wehrtmann and Cortés 2009, and see the website of the Costa Rican Instituto Nacional de Biodiversidad, INBio, at www.inbio.ac.cr). Not so the diversity at the level of ecosystems. Those few scholars who actually did study Costa Rican species communities or ecosystem-level units, however, did not necessarily agree on how they should be defined or classified (Gómez 1986). Hence, the historic development of ecosystem classifications and country-wide maps that use different hierarchical systems and distinct nomenclatures. Below, the most important ecological classifications and maps that exist for Costa Rica at a national level—in particular regarding its land-based vegetation—are briefly discussed.

Although foreign scholars like Anders Ørsted, Karl Hoffmann, Helmuth Polakowsky, Karl Wercklé (1909), Otto Porsch, and Paul H. Allen each made an attempt to classify and describe (part of) Costa Rica's vegetation and flora (Gómez 1986), it was not until Leslie R. Holdridge and his coworkers published their classic field study of *Forest Environments in Tropical Life Zones* (1971) that a full overview of all Costa Rica's vegetation formations and their structure and composition was first assembled (see fig. 1.1.). Holdridge's overview of Costa Rican vegetation was based on forest stand data collected in 1964–1966. It had its roots in his earlier, more methodological studies (Holdridge 1947, 1967), which detailed the theoretical fundamentals of a global bioclimatic scheme for the classification of land areas known as *life zones* (Spanish: *zonas de vida*). The three climatic axes of the triangular system with its 30+ hexagons are defined by annual precipitation, mean annual biotemperature (based on temperature and the length of the growing season), and potential evapotranspiration ratio (PET) (Holdridge 1967). By incorporating biotemperature changes as they occur along an altitudinal gradient Holdridge followed in the footsteps of geographer Alexander von Humboldt: the German nineteenth-century naturalist was the first to notice, while traveling through the South American Andes, that tropical temperatures drop with increasing elevation above sea level and that vegetation changes accordingly (von Humboldt and Bonpland 1814).

According to Holdridge's assessment, Costa Rica's small land surface is home to twelve life zones that include dry, moist, wet, and rain forests, altitudinally distributed over

Fig. 1.1 Maarten Kappelle (left) and Leslie (Les) R. Holdridge (1907–1999) in the garden behind Holdridge's house in San Isidro de Heredia, Costa Rica. *Photo taken by Mrs. Holdridge in April 1986.*

tropical (i.e., lowland), premontane, lower montane, and montane belts (Holdridge et al. 1971, and see Hartshorn 1983). Additionally, six climatic transitions occur in the country, ranging from warm to cool-dry and cool-wet transitions. At elevations over 3,000 m (e.g, at the summit of Cerro Chirripó; see Kappelle and Horn, chapter 15 of this volume), Costa Rica's only non-forest life zone is found: the sub-alpine rain *páramo*. To visualize the potential distribution of Holdridge's life zones and transitions in Costa Rica, their extent was mapped by Holdridge's colleague Joseph Tosi (1969), and thirty years later remapped with more precision and geographic detail by Rafael Angel Bolaños et al. (1999).

Some ten years ago, descriptions of Holdridge's life zones together with Tosi's map served as key input data to determine the carbon absorption capacity of Costa Rican forests with the aim to facilitate negotiations for joint implementation projects within the framework of the Kyoto Protocol (Obando 2002), a key international agreement that focuses at stabilizing greenhouse gas (GHG) concentrations in the atmosphere at a level that would prevent dangerous anthropogenic interference with the climate system.

Growing empirical evidence, however, has called into question the predictive power of the climate-based Holdridge model (Cornell 1998). As a reply to the life zone system, Luis Diego Gómez (1986) proposed a classification of 53 vegetation macrotypes (Spanish: *macrotipos de vegetación*) that are principally defined by vegetation morphology (i.e., physiognomy), stand structure, species composition, and soil characteristics. An essential climate feature that the Holdridge system lacked but that was incorporated by Gómez, concerned the seasonal distribution of rainfall as expressed in the length of the dry season measured in months (Gómez 1986, and see Herrera 1986 and chapter 2 of this volume).

Gómez' approach was inspired by the kind of vegetation analyses conducted previously in other parts of the tropics by scholars like Beard (1944) and Grubb et al. (1963). It followed a floristic-ecological method to classify vegetation types and associations in the European plant sociological tradition of Braun-Blanquet (1932; and see Westhoff and Van der Maarel 1973). When possible, Gómez adopted the associated plant community nomenclature that was originally ratified during the Sixth International Botanical Congress held in 1935 in Amsterdam, the Netherlands (Gómez 1986).

During the mid-1990s Gómez' map of vegetation macrotypes (scale 1:200,000) proved extremely useful in assessing gaps in Costa Rica's protected area system (García 1996)—much more than, for example, Savitsky et al.'s (1995) coarse map of Costa Rican habitats (scale 1:500,000). That gap analysis—known as the GRUAS I project—together with additional fieldwork in remote areas, led to the establishment of new nature reserves such as Parque Nacional Maquenque in northeastern Costa Rica, a park that attempts to protect the habitat of the endangered Great Green Macaw (*Ara ambigua*) (Chassot and Monge 2002). Complementary to Gómez' national vegetation classification, Zamora et al. (2004) described a total of 24 geographically defined botanical or floristic regions present in Costa Rica. These floristic regions were mapped by Marco V. Castro Campos and colorfully published in Hammel et al. (2004). The result is very useful when locating plant species populations or herbarium collection sites.

Seven years after Gómez' classification of vegetation macrotypes he merged his plant ecological scheme with Wilberth Herrera's (1986) elaborate climate types (also, see Herrera, chapter 2 of this volume). This integration led to the definition, classification, and mapping of Costa Rica's biotic units (Spanish: *unidades bióticas*; Herrera and Gómez 1993). More than in the case of the vegetation macrotypes, the newly combined system took into account physiographic and climatic data, as well as distributional data on flora and fauna. Since the biotic unit itself was defined by its ecological and other biotic features it could easily be distinguished in the field and spatially visualized in a distribution map. During the late 1990s, the biotic units system was used to select novel collection sites necessary to complete Costa Rica's national biodiversity inventory under the auspices of the Instituto Nacional de Biodiversidad (INBio) (Obando 2002).

Following the creation of Costa Rica's biodiversity law, INBio and SINAC/MINAE joined forces in 1998 and created the ECOMAPAS Project with Dutch governmental aid. Its purpose was to establish a decision-support tool that could monitor changes in terrestrial ecosystem cover at a national level and guide management for ecosystem health in protected areas and surrounding buffer zones.

The project's first step was a GIS-based mapping of ecosystems at a scale of 1:50,000 (Kappelle et al. 2003a). A set of four out of eleven Áreas de Conservación (ACs) were prioritized for mapping: Osa (ACOSA), La Amistad-Pacífico (ACLA-P), La Amistad-Caribe (ACLA-C), and Pacífico Central (ACOPAC). Mapping was done on the basis of the interpretation of 1995–1996 aerial photographs at scales of 1:25,000 to 1:40,000 and produced by the German company Hansa-Luftbild using funding provided by the Global Environmental Facility (GEF). Applying the well-established Rapid Ecological Assessment (REA) methodology presented in Sayre et al. (2000), map data were verified through extensive ground-truthing in the field, which included hundreds of sample points in situ (Kappelle et al. 2003a). Ecosystem categorization followed UNESCO's (1973) physiognomic vegetation classification system, which is grounded in the methodology proposed by Mueller-Dombois and Ellenberg (1974)—two scholars who built upon the legacy inherited from the plant sociologist Braun-Blanquet (1932). Then, with the aim to develop a thorough landscape-ecological classification sensu Troll (1939) and Zonneveld (1995), the ECOMAPAS team crossed vegetational data with a series of thematic layers covering geological units, land forms, soil types, temperature provinces, and rainfall regimes (Kappelle et al. 2003a, 2003b).

So far, two printed volumes have been published within the ECOMAPAS framework: one covering the ecosystems of ACOSA (Kappelle et al. 2003b) and one focusing on systems found in a portion of ACOPAC—that is, the Rio Savegre watershed (Acevedo et al. 2002). Owing to limited resources, further hard-copy publication has been put on hold, but distributional data on ecosystems—now available for about half of the country and its ACs—can be easily accessed in digital formats through INBio's website and SINAC/MINAET.

Only a few years ago (2007), a second national gap assessment was conducted to identify remaining weaknesses in Costa Rica's protected area system. This process, known as GRUAS II, included terrestrial, freshwater, and marine components (SINAC 2007, Arias et al. 2008). The land-based analysis (including the land portion of Isla del Coco) was performed on the basis of a newly assembled map that displayed the distribution of 33 so-called phytogeographic units (Spanish: *mapa de unidades fitogeográficas*) (Zamora 2008). These "coarse filter" units were the result of a GIS exercise in which the 53 vegetation macrotypes of Gómez (1986) were overlain on top of the 24 floristic regions of Zamora et al. (2004) (for details, see SINAC 2007, and Arias et al. 2008). Currently, the GRUAS II map of phytogeographic units is informing policy decision making to establish new protected areas and, above all, biological corridors that connect core areas (B. Herrera, Tropical Agricultural Research and Higher Education Center, CATIE, pers. com., and see Arias et al. 2008). An example of the latter is the Amistosa corridor initiative that links Parque Internacional La Amistad with the Peninsula de Osa (F. Carazo, The Nature Conservancy, pers. comm.).

All of the above-mentioned classifications dealt with land-based systems, ranging from life zones, to vegetation (macro)types, biotic units, botanical regions, landscape-ecological types, and ultimately, phytogeographical units. Almost none of them, however, addressed marine ecosystems, and only a few provided major details on freshwater systems. An example of the latter is Kappelle et al.'s (2003b) classification, mapping, and description of Osa's freshwater and coastal-marine plant-dominated ecosystems. These included floating vegetation in freshwater lakes (e.g., Laguna Corcovado), *Symphonia globulifera* forest swamps, brackish *Raphia taedigera* palm swamps, flooded *Acrostichum aureum* fernlands, and *Rhizophora mangle* mangrove forests. Nomenclature of these aquatic ecosystem types followed a combination of UNESCO's (1973) vegetation classification system and Cowardin et al.'s (1979) hierarchy of wetland types.

Fortunately, there are a few more accounts that actually discuss aquatic ecosystems, though still in a concise fashion. For instance, Gómez (1984) provided an overview of aquatic plants and their habitats, while Jiménez (1994) described the mangrove ecosystems of the Pacific coast. Obando (2002), in turn, briefly listed the most important aquatic systems in the country. She reported that Costa Rica harbors some 350 wetlands that all in all make up for 7% of its territory. The most significant freshwater wetland types are flooded forests, swamps, marshes, bogs, lakes, and lagoons. Many of these wetland systems have been evaluated at the ecoregional level by The Nature Conservancy (TNC; 2009) with the aim to identify priority sites for freshwater conservation in Costa Rica and neighboring countries.

The most significant coastal-marine ecosystems, on the other hand, have been briefly described by Mug et al. (2001). They cover mangrove systems, sandy beaches, coastal rock formations, estuaries, coastal channels and brackish lagoons (e.g., such as found in Parque Nacional Tortuguero), coral reefs, rocky reefs, island coasts, mud flats and rocky banks, pelagic and oceanic environments, deep sea benthic environments, and a kind of tropical, anoxic fjord (i.e., the Golfo Dulce in southern Costa Rica). Complementing the ecosystem-level outline of Mug and coworkers, Wehrtmann and Cortés (2009) provided a comprehensive overview of the species that inhabit the ecological systems of the seas and shores of Costa Rica (also, see several chapters by Cortés, this volume). Additionally, ecological data for Costa Rican portions of four coastal-marine ecoregions (Nicoya, Panama Bight, Isla del Coco, and Southwestern Caribbean) were recently analyzed by TNC (2008) to help select key areas for marine conservation purposes.

The Ecosystem Concept within the Context of this Book

Strangely enough, so far no scientific book has dealt in an integrated manner with the full range of terrestrial, freshwater, and marine ecosystems that occur in Costa Rica. At the same time, none of the above-described classifications of forests, vegetation, land cover types, wetlands, and coastal-marine ecosystems have been formalized and officially adopted by the Costa Rican government or other national entities. Hence, there is a clear need to publish a book that provides an overview of Costa Rica's main ecosystems by compiling what we know today about their physical setting, biogeography, species diversity, community composition, species interactions, structure, functioning, land use history, and conservation. Ecosystems have been defined by the Convention on Biological Diversity (CBD) as "dynamic complexes of plant, animal and micro-organism communities and their non-living environment interacting as a functional unit." This definition is valid within the context of this book, applies well to all its chapters, and is in line with Eugene P. Odum's key fundamentals of ecology (Odum 1971).

In order to keep it simple, I decided to avoid the use of existing national ecosystem classification systems like those developed by Leslie R. Holdridge, Luis Diego Gómez, or Nelson Zamora. The main reason is that their classifications are too detailed within the scope of this book. At the same time, they do not treat marine ecosystems or discuss any freshwater system in an in-depth manner. Rather, I tried to use a higher level modus operandi, more or less close to the concept of the "biome" (Campbell 1996) and "ecoregion" (Brunckhorst 2000).

However, following such an approach in its strict sense probably would have led to a classification too coarse for the purpose of this book. There are indeed only three terrestrial biomes occurring in Costa Rica: tropical rain forest, tropical dry forest, and tundra—the latter in a tropical fashion (that is, with a diurnal climate) and regionally known as *páramo*.

Similarly, only five terrestrial ecoregions occur in the country if one excludes the mangroves (Olson et al. 2001). One of these is the "Talamancan montane forest," which ranges from the volcanoes of southern Nicaragua southeastward across Costa Rica into central Panama (Olson et al. 2001). Hence, Olson et al. (2001) erroneously classify the volcanic *cordilleras* of northern and central Costa Rica as "Talamancan" and wrongly incorporate, under "montane forest," the biome-level, tundra-like treeless *páramos*.

The other four terrestrial ecoregions (sensu Olson et al. 2001) that are observed in Costa Rica and have been evaluated for conservation purposes by Calderón et al. (2004) are the Central American Dry Forest in Guanacaste (dealt with in this volume by Q. Jiménez et al. [chapter 9] and by D.H. Janzen and W. Hallwachs [chapter 10]), the Costa Rican Seasonal Moist Forest in the southern sector of Peninsula de Nicoya and in the Valle Central (see Q. Jiménez and E. Carrillo, chapter 11 of this volume), the Isthmian Pacific Moist Forest in the south including the Peninsula de Osa (L. Gilbert et al., chapter 12 of this volume), and the Isthmian Atlantic Moist Forest in the Caribbean lowlands that range from Upala and Los Chiles to Sixaola (D. McClearn et al., chapter 16 of this volume).

Therefore, in this book I have included some fifteen large natural ecosystems that cover all of Costa Rica's territorial lands and waters. Eight are terrestrial in distribution and correspond mostly to forests. The land-based ecosystem of the oceanic island Isla del Coco is treated as a distinct system (M. Montoya, chapter 8 of this volume), whereas montane forests along the volcanic *cordilleras* in the North (R. Lawton et al., chapter 13 of this volume) are considered distinctively from the montane forests found in the Cordillera de Talamanca (M. Kappelle, chapter 14 of this volume), as they clearly differ in structure and composition. Freshwater ecosystems, in turn, are separated in rivers (C. Pringle et al., chapter 18 of this volume), lakes (S.P. Horn and K. Haberyan, chapter 19 of this volume), and other wetlands like swamps, marshes, and bogs (J. Jiménez, chapter 20 of this volume). Coastal-marine ecosystems, on the other hand, are defined and discussed for the Pacific Ocean and Caribbean Sea (J. Cortés, chapters 5 and 17 of this volume), as well as more specifically for the estuary of the Golfo de Nicoya (J.A. Vargas, chapter 6 of this volume) and the oceanic waters around Isla del Coco (J. Cortés, chapter 7 of this volume). In summary, the natural ecosystems dealt with in this book are as follows:

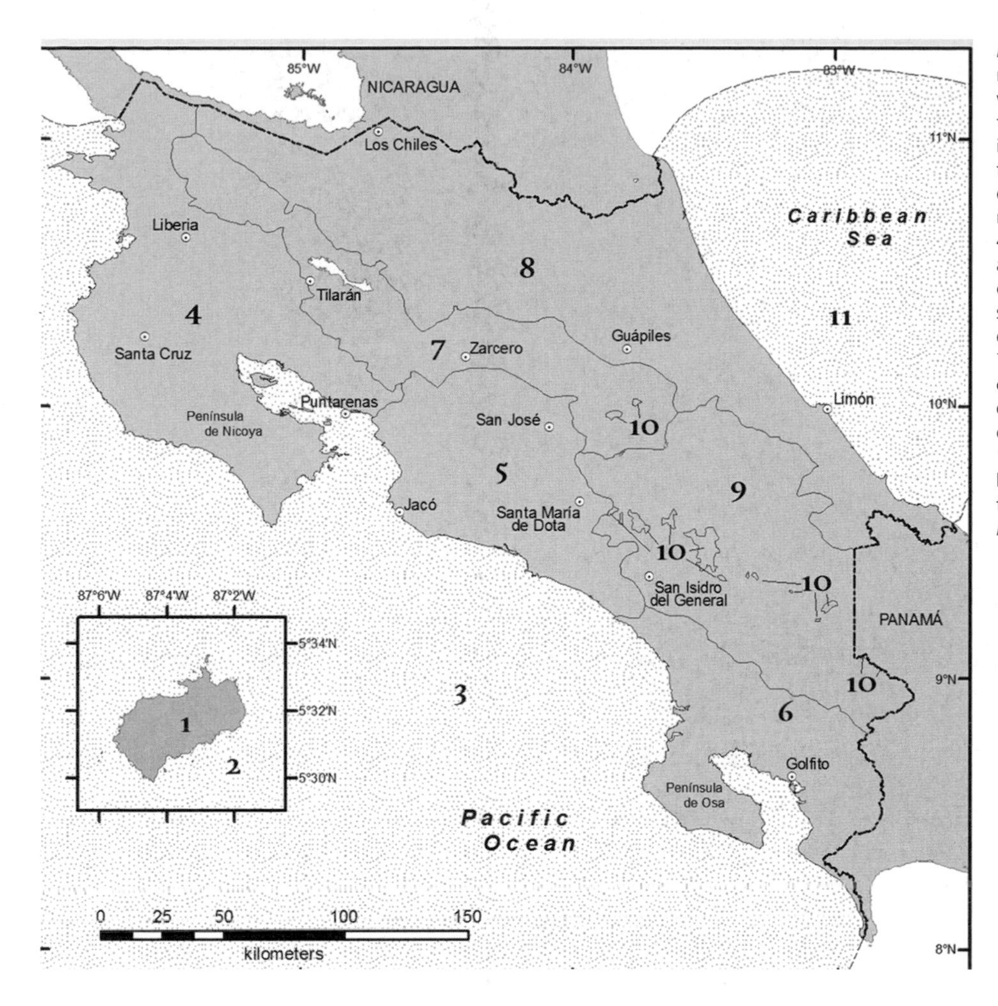

Fig. 1.2 Map of the main terrestrial and marine ecosystems treated in this book. Freshwater ecosystems are embedded within the terrestrial units and have not been specifically indicated. Legend: 1. The terrestrial ecosystem of Isla del Coco. 2. The coastal-marine ecosystem of Isla del Coco. 3. The coastal-marine ecosystem of the Pacific Ocean. 4. The seasonal dry forests of Guanacaste and Nicoya. 5. The seasonal forests of the central Pacific zone and Central Valley. 6. The southern Pacific lowland evergreen moist forest of the Peninsula de Osa and surroundings. 7. The montane cloud forests of the volcanic *cordilleras*. 8. The evergreen moist forests of the Caribbean lowlands. 9. The montane cloud forests of the Cordillera de Talamanca. 10. The alpine *páramos* of Costa Rica's highlands. 11. The coastal-marine ecosystem of the Caribbean Sea.
Map prepared by Marco V. Castro.

A) Terrestrial ecosystems:
   i) The terrestrial ecosystem of Isla del Coco;
   ii) The seasonal dry forests of Guanacaste and Nicoya in the northern Pacific lowlands;
   iii) The seasonal forests of the central Pacific zone and Central Valley;
   iv) The southern Pacific lowland evergreen moist forest of the Peninsula de Osa;
   v) The evergreen moist forests of the Caribbean lowlands;
   vi) The montane cloud forests of the volcanic *cordilleras* in the North;
   vii) The montane cloud forests of the Cordillera de Talamanca in the south;
   viii) The alpine *páramos* of Costa Rica's highlands;

B) Freshwater ecosystems:
   i) Costa Rican rivers;
   ii) Costa Rican lakes;
   iii) Costa Rican bogs, marshes, and swamps;

C) Coastal-marine ecosystems:
   i) The coastal-marine ecosystem of Isla del Coco;
   ii) The coastal-marine ecosystem of the Pacific Ocean;
   iii) The estuary of the Gulf of Nicoya; and
   iv) The coastal-marine ecosystem of the Caribbean Sea.

The map provided in Fig. 1.2 clearly shows the distribution of the terrestrial and marine natural ecosystems at a national scale. Each ecosystem chapter of this book deals with one of these systems, evaluating its diversity and complexity, and discussing its resource use and conservation. The sequence in which they are presented reflects a journey over land and sea: a fly-over from the southwest to the northeast, from the land and waters of Isla del Coco towards the Pacific coast of continental Costa Rica, into the high *cordilleras* and down to the Atlantic forests, then into to the Caribbean lowlands and coast. Fig. 1.3 gives a good photographic impression of the variety of the ecological systems

one may encounter when making such a Grand Tour across the seas, shores, and inlands of *Tiquicia*, as Costa Rica is often amically called by its inhabitants.

This journey-wise sequence will allow the reader to get familiar with the ecological systems she or he may find in the field, when visiting rural regions and coastal zones, from west to east, from south to north, and vice versa. It will help her or him to understand the full range of varied seascapes and landscapes as they intergrade with each other. This touring approach was selected over a merely technical or even clinical overview of hierarchically categorized ecosystems, for such a framework would suggest the use of a thoroughly developed classification system—which is neither the case nor pretended by the editor.

Before each of Costa Rica's ecosystems is presented, three introductory chapters treat the non-biotic environment of the country: its climate, geology, and soils. Learning about Costa Rica's geophysical factors, patterns, and processes is essential to understand the *medium* or *entorno* in which its ecosystems have developed over time, and how they function. All three chapters have been prepared by outstanding national scholars in physical sciences: Wilberth Herrera (climate), Guillermo Alvarado and Guaria Cárdenes (geology and geomorphology), and Alfredo Alvarado and Rafael Mata (soils).

As the reader may have noted while interpreting the deliberations made in this chapter, this book's emphasis is definitely not on man-made ecosystems or anthropic "biomes" such as croplands, rangelands, tree plantations, urban parks, or villages and dense settlements. However, such non-natural, semi-natural, and sometimes truly cultural ecosystems are occasionally treated within the scope of the predominant natural ecosystems that provide the matrix of human-influenced mosaics of landscapes or seascapes. Hence, human-influenced forest restoration, for instance, is discussed in detail in the chapters on seasonal dry forests of Guanacaste and Nicoya (Janzen and Hallwachs, chapter 10 of this volume). Similarly, human threats like deforestation or overfishing are treated extensively in chapters on montane forests (Lawton et al. [chapter 13] and Kappelle [chapter 14]) and coastal-marine ecosystems (chapters 5, 7, and 17 by Cortés and chapter 6 by Vargas), respectively. Future works, however, should make an attempt to assess, describe, and discuss much more specifically the man-made and often novel ecosystems that are spreading throughout the country (e.g., coffee and banana plantations, grasslands for dairy cattle, central parks in villages, dam reservoir waters, and fish ponds), and their impact on biodiversity (e.g., see Banks et al. 2013). An interesting example of a novel ecosystem is the young forest that is currently developing in the core area of INBioparque in Santo Domingo de Heredia. It is the result of human-assisted, accelerated restoration on a former coffee plantation. Many similar restoration efforts are underway elsewhere in the country.

On a final note, it is important to mention that the contributors who prepared the chapters of this book include national and foreign scientists who are outstanding experts on the ecosystems they present and analyze. Their contributions are often based on many decades of intense field study that includes both observational and experimental research. Each of them has done a major job in compiling the current knowledge that has been generated over the past fifty years or so. Owing to their extraordinary efforts that knowledge is now made available to academic scientists, students, natural history guides, conservationists, educators, park guards, and visitors alike. The editor expresses the hope that this vast scientific knowledge, now integrated and available in a single, comprehensive volume, will serve all who are interested in understanding the diversity and complexity of these marvelous natural ecosystems—knowledge necessary to be able to safeguard and sustainably use Costa Rica's natural riches. In this way, societies like Costa Rica's will be able to work towards true sustainable development in sync with frameworks such as provided by *The Future We Want*[2] and the *Sustainable Development Goals* currently being implemented.

Fig. 1.3 Photographic impressions of Costa Rican ecosystems. First row (upper row): left: *Cyathea* treeferns at Isla del Coco; center: *Porites*-dominated coral reef at Isla del Coco; right: Punta Uvita south of Dominical along the Pacific shore showing the "tombolo," a narrow sandbar that connects the reef to the coast. Second row: left: vegetated sedimentary rocks at Isla del Caño, Pacific Ocean; center: *Pelliciera* mangroves at Río Estero, close to Bahía Drake, Península de Osa; right: *Raphia* palm swamp in the Térraba-Sierpe national wetland, north of the Península de Osa. Third row: left: large buttresses at the base of a moist forest tree in Corcovado National Park, Península de Osa; center: wetlands along the Río Tempisque in Palo Verde National Park; right: dense moist forest at Carara National Park. Fourth row: left: mid-elevation cloud forest along the ridges of the Pacific south; center: *Blechnum*-dominated peatbog in the upper montane belt of La Amistad International Park; right: *Chusquea*-dominated *páramo* near Cerro Urán, Chirripo National Park. Fifth row (bottom row): far left: seasonal dry forest in April at Santa Rosa, Guanacaste; center left: Laguna Hule in the northern volcanic highlands; center right: lowland creek at La Selva in the Caribbean moist forest region; far right: *Quercus*-dominated montane forest near San Gerardo de Dota in the Cordillera de Talamanca.
Photo credits—First row (upper row): left: Bárbara Sperl; center: Jorge Cortés; right: Yamil Sáenz. Second row: left: Jorge Cortés; center: Luis González; right: Luis González. Third row: left: Larry Gilbert; center: Garret Crow; right: Quírico Jiménez. Fourth row: left: Yamil Sáenz; center: Luis González; right: Felipe Carazo. Fifth row (bottom row): far left: Dan Janzen; center left: Kurt Haberyan; center right: Cathy Pringle; far right: Maarten Kappelle.

[2] *The Future We Want* is the outcome document that was adopted by world leaders at *Rio+20*, the United Nations Conference on Sustainable Development (UNCSD) that was held in Rio de Janeiro, Brazil, 20–22 June 2012. This global meeting took place precisely twenty years after the 1992 Rio Conference during which the *Convention on Biological Diversity* (CBD) was established.

References

Acevedo, H., J. Bustamante, L. Paniagua, R. Chaves, and F. Quesada. 2002. Ecosistemas de la Cuenca Hidrográfica del Río Savegre, Costa Rica. Santo Domingo de Heredia, Costa Rica: Instituto Nacional de Biodiversidad (INBio) and Ministerio de Ambiente y Energía (MINAE). 352 pp.

Arias, E., O. Chacón, G. Induni, B. Herrera-F., H. Acevedo, L. Corrales, J.R. Barborak, M. Coto, J. Cubero, and P. Paaby. 2008. Identificación de vacíos en la representatividad de ecosistemas terrestres en el Sistema Nacional de Áreas Protegidas de Costa Rica. *Recursos Naturales y Ambiente* 54: 21–27.

Arriagada, R.A., P.J. Ferraro, E.O. Sills, K.P. Subhrendu, and S. Cordero-Sancho. 2012. Do payments for environmental services affect forest cover? A farm-level evaluation from Costa Rica. *Land Economics* 88: 382–99.

Aylward, B., K. Allen, J. Echeverría, and J. Tosi. 1996. Sustainable ecotourism in Costa Rica: the Monteverde Cloud Forest Preserve. *Biodiversity and Conservation* 5: 315–43.

Banks, J. E., L. Hannon, P. Hanson, T. Dietsch, S. Castro, N. Ureña, and M. Chandler. 2013. Effects of proximity to forest habitat on hymenoptera diversity in a Costa Rican coffee agroecosystem. *Pan-Pacific Entomologist* 89: 60–68.

Beard, J.S. 1944. Climax vegetation in tropical America. *Ecology* 25: 127–58.

Bolaños, R., V. Watson, and J.A. Tosi. 1999. Mapa Ecológico de Costa Rica (Zonas de Vida) según el Sistema de Clasificación de Zonas de Vida del Mundo de L.R. Holdridge. San Pedro de Montes de Oca, Costa Rica: Tropical Science Center (TSC).

Boza, M.A. 1992. *Costa Rica National Parks*. Madrid, Spain: INCAFO. 91 pp.

Boza, M.A. 1993. Conservation in action: past, present, and future of the National Park System of Costa Rica. *Conservation Biology* 7(2): 239–47.

Braun-Blanquet, J. 1932. *Plant Sociology: The Study of Plant Communities*. (Transl. from German by G.D. Fuller and H.S. Conard. Original title: *Pflanzensoziologie: Grundzüge der Vegetationskunde*). New York: McGraw-Hill. 458 pp.

Brunckhorst, D. 2000. Bioregional planning: resource management beyond the new millennium. Sydney, Australia: Harwood Academic Publishers.

Burger, W.C. 1980. Why are there so many kinds of flowering plants in Costa Rica? *Brenesia* 17: 371–88.

Calderón, R., T. Boucher, M. Bryer, L. Sotomayor, and M. Kappelle. 2004. *Setting Biodiversity Conservation Priorities in Central America: Action Site Selection for the Development of a First Portfolio*. San José, Costa Rica: The Nature Conservancy (TNC). 32 pp.

Campbell, N.A. 1996. *Biology*. 4th ed. Menlo Park, CA: Benjamin/Cummings.

Castro, R., F. Tattenbach, L. Gámez, and N. Olson. 2000. The Costa Rican experience with market instruments to mitigate climate change and conserve biodiversity. *Environmental Monitoring and Assessment* 61(1): 75–92.

Chassot, O., and G. Monge. 2002. Great Green Macaw: flagship species of Costa Rica. *PsittaScene* 53: 6–7.

Coates, A.G., and J.A. Obando. 1996. The geologic evolution of the Central American Isthmus. In J.B.C. Jackson, A.F. Budd, and A.G. Coates, eds., *Evolution and Environment in Tropical America*, 21–56. Chicago: University of Chicago Press.

Cornell, J.D. 1998. The status of the Holdridge life zone model on its 50th anniversary. In Abstracts of the Annual Meeting (August 2–6, 1998) of the Association for Tropical Biology (ATB) and the American Institute of Biological Sciences (AIBS). Published as a supplement to *Biotropica* 30(2). Baltimore, MD.

Cowardin, L.M., V. Carter, F.C. Golet, and E.T. LaRoe. 1979. *Classification of Wetlands and Deepwater Habitat of the United States*. Publication No. FWS/OBS-79/31. Washington, DC: Biological Service Program, US Fish and Wildlife Service.

Cronin, T.M., and H.J. Dowsett. 1996. Biotic and oceanographic response to the Pliocene closing of the Central American Isthmus. In J.B.C. Jackson, A.F. Budd, and A.G. Coates, eds., *Evolution and Environment in Tropical America*, 76–104. Chicago: University of Chicago Press.

Dirzo, R. 2001. Ecosystems of Central America. In S.A. Levin, ed., *Encyclopedia of Biodiversity*. Vol. I, 665–76. San Diego, CA: Academic Press.

Evans, S. 1999. *The Green Republic: A Conservation History of Costa Rica*. Austin: University of Texas Press. 317 pp.

Fournier, L.A. 1991. *Desarrollo y Perspectivas del Movimiento Conservacionista Costarricense*. San José, Costa Rica: Editorial de la Universidad de Costa Rica (EUCR). 113 pp.

Gámez, R. 1991. Biodiversity conservation through the facilitation of its sustainable use: Costa Rica's National Biodiversity Institute. *Trends in Ecology and Evolution* 6: 377–78.

Gámez, R. 1999. *De Biodiversidad, Gentes y Utopías: Reflexiones en los 10 Años del INBio*. Santo Domingo de Heredia, Costa Rica: Instituto Nacional de Biodiversidad (INBio).

Gámez, R. 2003.*The Link between Biodiversity and Sustainable Development: Lessons from INBio's Bioprospecting Program in Costa Rica*. Santo Domingo de Heredia, Costa Rica: Instituto Nacional de Biodiversidad (INBio).

Gámez, R., and A. Ugalde. 1988. Costa Rica's National Park System and the preservation of biological diversity: linking conservation with socio-economic development. In F. Almeda and C.M. Pringle, eds., *Tropical Rainforests: Diversity and Conservation*, 131–42. San Francisco: California Academy of Sciences.

García, R. 1996. *Propuesta Técnica de Ordenamiento Territorial con Fines de Conservación de Biodiversidad en Costa Rica: Proyecto GRUAS*. San José, Costa Rica: SINAC, MINAE. 114 pp.

García, R. 1997. *Biología de la Conservación y Áreas Silvestres Protegidas: Situación Actual y Perspectivas en Costa Rica*. Santo Domingo de Heredia, Costa Rica: Instituto Nacional de Biodiversidad (INBio).

Gómez, L.D. 1984. Las Plantas Acuáticas y Anfibias de Costa Rica y Centroamérica. San José, Costa Rica: Editorial Universidad Estatal a Distancia (EUNED).

Gómez, L.D. 1986. Vegetación de Costa Rica: Apuntes para una Biogeografía Costarricense. San José, Costa Rica: Editorial Universidad Estatal a Distancia (EUNED).

Gómez, L.D., and J.M. Savage. 1983. Searchers on that rich coast: Costa Rican field biology, 1400–1980. In D.H. Janzen, ed., *Costa Rican Natural History*, 1–11. Chicago: University of Chicago Press.

Groombridge, B. 1992. *Global Biodiversity: Status of the Earth's Living Resources*. World Conservation Monitoring Centre (WCMC). New York: Chapman and Hall. 585 pp.

Grubb, P.J., J.R. Lloyd, T.D. Pennington, and T.C. Whitmore. 1963. A comparison of montane and lowland rain forest in Ecuador. I. The forest structure, physiognomy, and floristics. *Journal of Ecology* 51: 567–601.

Hall, C. 1985. *Costa Rica, a Geographical Interpretation in Historical Perspective*. Boulder, CO: Westview. 348 pp.

Hammel, B.E., M.H. Grayum, C. Herrera, and N. Zamora. 2004. *Manual de Plantas de Costa Rica*. Vol. 1: *Introducción. Monographs in Systematic Botany from the Missouri Botanical Garden* 97: 1–300.

Hartshorn, G.S. 1983. Plants: introduction. In D.H. Janzen, ed., *Costa Rican Natural History*. Chicago: University of Chicago Press.

Herrera, A., and V. Obando. 2009. Algunos datos sobre biodiversidad de Costa Rica. In SINAC (Sistema Nacional de Áreas de Conservación). IV. Informe de País sobre la implementación del Convenio sobre la Diversidad Biológica. SINAC. San José, Costa Rica. Accessed October 22, 2011: http://www.inbio.ac.cr/estrategia/default.html

Herrera, W. 1986. Clima de Costa Rica. San José, Costa Rica: Editorial Universidad Estatal a Distancia (EUNED).

Herrera, W., and L.D. Gómez. 1993. Mapa de Unidades Bióticas de Costa Rica. Scale 1:685,000. US Fish and Wildlife Service—TNC—INCAFO—CBCCR—INBio—Fundación Gómez-Dueñas. San José, Costa Rica.

Holdridge, L.R. 1947. Determination of world plant formations from simple climatic data. *Science* 105(2727): 367–68.

Holdridge, L.R. 1967. *Life Zone Ecology*. San José, Costa Rica: Tropical Science Center.

Holdridge, L.R., W.C. Grenke, W.H. Hatheway, T. Liang, and J.A. Tosi. 1971. *Forest Environments in Tropical Life Zones: A Pilot Study*. Oxford: Pergamon. 735 pp.

von Humboldt, A., and A. Bonpland. 1814. Relation Historique du Voyage aux Régions Équinoxiales du Nouveau Continent fait en 1799, 1800, 1801, 1802, 1803 et 1804. Paris: Dufour.

Janzen, D.H., ed. 1983. *Costa Rican Natural History*. Chicago: University of Chicago Press.

Janzen, D.H., and R. Gámez. 1997. Assessing information needs for sustainable use and conservation of biodiversity. In P.M. Hawksworth and S. Dextre Clarke, eds., *Biodiversity Information: Needs and Options*, 21–29. Wallingford, Oxon, UK: CAB International.

Jiménez, J.A. 1994. Los Manglares del Pacífico de Centroamérica. Heredia, Costa Rica: Editorial Fundación Universidad Nacional. 352 pp.

Jones, D.S., and P.F. Hasson. 1985. History and development of the marine invertebrate faunas separated by the Central American Isthmus. In F.G. Stehli and S.D. Webb, eds., *The Great American Biotic Interchange*, 325–55. New York: Plenum.

Kappelle, M. 1996. *Los Bosques de Roble (*Quercus*) de la Cordillera de Talamanca, Costa Rica: Biodiversidad, Ecología, Conservación y Desarrollo*. Amsterdam: Universidad de Amsterdam. 336 pp.

Kappelle, M., A.M. Cleef, and A. Chaverri. 1992. Phytogeography of Talamanca montane *Quercus* forests, Costa Rica. *Journal of Biogeography* 19(3): 299–315.

Kappelle, M., M. Castro, H. Acevedo, P. Cordero, L. González, E. Méndez, and H. Monge. 2003a. A rapid method in ecosystem mapping and monitoring as a tool for managing Costa Rican ecosystem health. In D.J. Rapport, W.L. Lasley, D.E. Rolston, N.O. Nielsen, C.O. Qualset, and A.B. Damania, eds., *Managing for Healthy Ecosystems*, 449–58. Boca Raton, FL: Lewis.

Kappelle, M., M. Castro, H. Acevedo, L. González, and H. Monge. 2003b. Ecosystems of the Osa Conservation Area, Costa Rica. Bilingual ed. Santo Domingo de Heredia, Costa Rica: Instituto Nacional de Biodiversidad (INBio) and Ministerio de Ambiente y Energía (MINAE). 496 pp.

Karousakis, K. 2007. *Incentives to Reduce GHG Emissions from Deforestation: Lessons Learned from Costa Rica and Mexico*. Paris: Organisation for Economic Co-operation and Development (OECD). 50 pp.

Merker, C.A., W.R. Barbour, J.A. Scholten, and W.A. Dayton. 1943. *The Forests of Costa Rica: A General Report on the Forest Resources of Costa Rica*. Washington, DC: Forest Service of the US Department of Agriculture (USDA) and Office for the Coordinator of Inter-American Affairs. 84 + 49 pp.

Montes, C., A. Cardona, C. Jaramillo, A. Pardo, J.C. Silva, V. Valencia, C. Ayala, I.C. Pérez-Angel, L.A. Rodríguez-Parra, V. Ramírez, and H. Niño. 2015. Middle Miocene closure of the Central American Seaway. *Science* 348(6231): 226–29.

Mueller-Dombois, D., and H. Ellenberg. 1974. *Aims and Methods of Vegetation Ecology*. New York: John Wiley and Sons.

Mug, M., M.A. Bolaños, J. Sheffield, and F. Liebinger. 2001. Diagnóstico sobre la Investigación Marino-Costera en Costa Rica: Informe final. Santo Domingo de Heredia, Costa Rica: Instituto Nacional de Biodiversidad (INBio) and Ministerio de Ambiente y Energía (MINAE). 31 pp.

Obando, V. 2002. Biodiversidad en Costa Rica: Estado del Conocimiento y Gestión. Santo Domingo de Heredia, Costa Rica: Instituto Nacional de Biodiversidad (INBio). 81 pp.

Obando, V. 2011. Algunos datos sobre la biodiversidad en Costa Rica. Lecture held at Instituto Nacional de Biodiversidad (INBio). Santo Domingo de Heredia, Costa Rica.

Odum, E.P. 1971. *Fundamentals of Ecology*. 3rd ed. Philadelphia: Saunders. 574 pp.

Olson, D.M., E. Dinerstein, E.D. Wikramanayake, N.D. Burgess, G.V.N. Powell, E.C. Underwood, J.A. D'Amico, I. Itoua, H.E. Strand, J.C. Morrison, C.J. Loucks, T.F. Allnutt, T.H. Ricketts, Y. Kura, J.F. Lamoreux, W.W. Wettengel, P. Hedao, and K.R. Kassem. 2001. Terrestrial ecoregions of the world: a new map of life on Earth. *BioScience* 51(11): 933–38.

Pagiola, S. 2008. Payments for environmental services in Costa Rica. *Ecological Economics* 65: 712–24.

Pittier, H. 1908. Ensayo sobre las Plantas Usuales de Costa Rica. Washington, DC: H.L. and J.B. McQueen.

Pounds, J.A. 2001. Climate and amphibian declines. *Nature* 410: 639–40.

Pounds, J.A., and M.L. Crump. 1994. Amphibian declines and climate disturbance: the case of the golden toad and the harlequin frog. *Conservation Biology* 8: 72–85.

Quilter, J., and J.W. Hoopes. 2003. *Gold and Power in Ancient Costa Rica, Panama, and Colombia*. Washington, DC: Dumbarton Oaks Publications. 428 pp.

Rich, P.V., and T.H. Rich. 1983. The Central American dispersal route: biotic history and paleogeography. In D.H. Janzen, ed., *Costa Rican Natural History*, 12–34. Chicago: University of Chicago Press.

Sader, S., and A. Joyce. 1988. Deforestation rates and trends in Costa Rica, 1940–1983. *Biotropica* 20: 11–19.

Sánchez-Azofeifa, G.A., R.C. Harris, and D.L. Skole. 2006. Deforestation in Costa Rica: a quantitative analysis using remote sensing imagery. *Biotropica* 33(3): 378–84.

Savitsky, B., and T.E. Lacher Jr., eds. 1998. *GIS Methodologies for Developing Conservation Strategies: Tropical Forest Recovery and Wildlife Management in Costa Rica.* New York: Columbia University Press.

Savitsky, B.G., D. Tarbox, D. van Blaricom, T.E. Lacher, Jr., and J. Fallas. 1995. Habitats of Costa Rica: an annotated map. Scale 1:500,000. Strom Thurmond Institute and Archbold Tropical Research Center, Clemson University. Clemson, SC.

Sayre, R., E. Roca, G. Sedaghatkish, B. Young, S. Keel, R. Roca, and S. Sheppard. 2000. *Nature in Focus: Rapid Ecological Assessment*. Washington, DC: The Nature Conservancy (TNC) and Island Press. 182 pp.

SINAC (Sistema Nacional de Áreas de Conservación). 2007. GRUAS II: Propuesta de Ordenamiento Territorial para la Conservación de la Biodiversidad de Costa Rica. Volumen 1: Análisis de Vacíos en la Representatividad e Integridad de la Biodiversidad Terrestre. San José, Costa Rica: SINAC, Ministerio de Ambiente y Energía (MINAE). 100 pp.

SINAC (Sistema Nacional de Áreas de Conservación). 2009. IV Informe de País sobre la implementación del Convenio sobre la Diversidad Biológica. San José, Costa Rica: SINAC. Accessed October 22, 2011: http://www.inbio.ac.cr/estrategia/default.html

Standley, P.C. 1937. *Flora of Costa Rica*, part I. Chicago: Field Museum of Natural History.

Stehli, F.G., and S.D. Webb, eds. 1985. *The Great American Biotic Interchange*. New York: Plenum.

Stone, S. 2011. Forests in a Green Economy. Presentation on World Environment Day, June 5, 2011. Economics and Trade Branch, United Nations Environment Programme (UNEP). Geneva, Switzerland.

Tamayo, G., L. Guevara, and R. Gámez. 2004. Biodiversity prospecting: the INBio experience. In A.T. Bull, ed., *Microbial Diversity and Bioprospecting*, 445–49. Washington, DC: ASM.

Thapa, B. 1998. Debt-for-nature swaps: an overview. *International Journal of Sustainable Development & World Ecology* 5(4): 249–62.

TNC (The Nature Conservancy). 2008. *Evaluación de Ecorregiones Marinas en Mesoamérica: Sitios Prioritarios para la Conservación en las Ecorregiones Bahía de Panamá, Isla del Coco y Nicoya del Pacífico Tropical Oriental, y en el Caribe Suroccidental de Costa Rica y Panamá*. San José, Costa Rica: The Nature Conservancy (TNC). 165 pp.

TNC (The Nature Conservancy). 2009. *Evaluación de Ecorregiones de Agua Dulce de Mesoamérica: Sitios Prioritarios para la Conservación en las Ecorregiones de Chiápas a Darién*. San José, Costa Rica: The Nature Conservancy (TNC). 515 pp.

Tosi, J.A. 1969. Mapa Ecológico de Costa Rica, Basado en la Clasificación Vegetal Mundial de L.R. Holdridge. Scale 1: 750,000. San Pedro de Montes de Oca, Costa Rica: Tropical Science Center (TSC).

Troll, C. 1939. Luftbildplan and ökologische Bodenforschung. *Zeitschrift der Gesellschaft für Erdkunde zu Berlin* 7–8: 1–58.

UNEP/GRID-Arendal. 2009. Change Forest Cover Costa Rica. UNEP/GRID-Arendal Maps and Graphics Library. Accessed August 18, 2011: http://maps.grida.no/go/graphic/change-forest-cover-costa-rica

UNESCO (United Nations Educational, Scientific and Cultural Organization). 1973. International Classification and Mapping of Vegetation. Paris: UNESCO.

Wallace, A.R. 1876. *The Geographical Distribution of Animals, with a Study of the Relations of Living and Extinct Faunas as Elucidating the Past Changes of the Earth's Surface*. New York: Harper and Brothers.

Wallace, D.R. 1992. *The Quetzal and the Macaw: The Story of Costa Rica's National Parks*. San Francisco: Sierra Club Books. 222 pp.

Webb, S.D. 2006. The great American biotic interchange: patterns and processes. *Annals of the Missouri Botanical Garden* 93(2): 245–57.

Wehrtmann, I.S., and J. Cortés, eds. 2009. *Marine Biodiversity of Costa Rica, Central America*. Berlin: Springer. 538 pp.

Wercklé, C. 1909. *La Subregión Fitogeográfica Costarricense*. San José, Costa Rica: Tipografia Nacional. 55 pp.

Westhoff, V., and E. van der Maarel. 1973. The Braun-Blanquet approach of phytosociology. In R.H. Whittaker, ed., *Manual of Vegetation Science*. Vol. 5: *Junk*, 619–726. The Netherlands: The Hague.

Zamora, N. 2008. Unidades fitogeográficas para la clasificación de ecosistemas terrestres en Costa Rica. *Recursos Naturales y Ambiente* 54: 14–20.

Zamora, N., B.E. Hammel, and M.H. Grayum. 2004. Vegetation. In B.E. Hammel, M.H. Grayum, C. Herrera, and N. Zamora, eds. *Manual de Plantas de Costa Rica*. Vol. 1: *Introducción*. *Monographs in Systematic Botany from the Missouri Botanical Garden* 97: 91–216.

Zonneveld, I.S. 1995. *Land Ecology*. Amsterdam: SPB Academic. 199 pp.

Part II

The Physical Environment

Chapter 2 Climate of Costa Rica

Wilberth Herrera[1]

Introduction

Costa Rica is a territory located in the Central American isthmus, between 8° 22' 26" and 11° 13' 12" North latitude and 82° 33' 48"and 85° 57' 57" West longitude. It is bounded on the north by Nicaragua (300 km), on the southeast by Panama (363 km), on the east by the Caribbean Sea (212 km), and on the south and west by the Pacific Ocean (1,016 km). The country covers a total area of 51,100 square kilometers, of which 50,980 km^2 are continental territory, and 120 km^2 are insular territories, notably Chira Island (43 km^2) located in the Gulf of Nicoya and Cocos Island, a 24-km^2 territory of volcanic mountain origin in the Pacific Ocean, 498 kilometers off the coast of the Osa Peninsula (Punta Llorona), at 5° 33′ North latitude and 87° 03′ West longitude.

Despite its small size, 95 Climate Groups (Herrera 1986) and 55 Biotic Units (Herrera and Gómez 1993) have been identified in the country, ranging from the sub-moist dry and very warm climate of the lowlands of Guanacaste Province to the very wet, cold climate of the Talamanca Mountain Range. The three main factors that determine Costa Rica's climate are its orography, its latitude, and the fact that it is an isthmus or land bridge.

Climate and Orography

The country is divided into two parts of almost equal size by a central mountain range that runs from northwest to southeast and rises to an elevation of 3,819 meters. To the northeast and east of this system is the Caribbean slope, with an area of 24,395 km^2, and to the west and southwest is the Pacific slope with a coverage of 26,585 km^2. In the case of both slopes, 51% of the territory corresponds to warm lands below 300 meters elevation. This mountainous feature, together with the other orographic systems that run parallel to the coastline and intercept air masses coming from both oceans, produces very different climatic characteristics in a territory as small as Costa Rica. Within a single slope, along a stretch of only 14 kilometers but with an altitudinal difference of 700 meters, moisture gradients vary from humid to pluvial; the dry period ranges from zero to five months; the sheets of rain vary from 1,900 to 7,600 mm, and there is a gradient of 180 to 318 rainy days.

Other climatic elements, such as solar radiation and hours of sunlight, are highly influenced by the configuration of the orographic system. Windward slopes and mountain passes are generally very cloudy sites, where hours of sunlight are less than 50% of what they would normally be on the basis of their latitudinal position. Considerable amounts of cloud form above 2,200 meters and water deficit conditions and high radiation rates may occur, along with strong winds, mainly during the period of January, February, March, and April.

Climate and Latitudinal Position

Costa Rica is located in the Equatorial Tropical belt with a median latitude of 10° North. This means that the sun's rays attain zenith positions twice a year (April and August), with twelve hours of sunlight per day. Monthly thermal variation is less than 4°C and the coasts are affected by the

[1] Apartado 2183-4050, Alajuela, Costa Rica

Costa Rican Coastal Current in the Pacific, and by a branch of the union of the North Equatorial Current and the South Equatorial Current in the Caribbean Sea (Martínez 1970). Owing to its latitudinal position, the atmospheric systems that determine the country's thermal and pluvial regimes are the Intertropical Convergence Zone (ITCZ), cold fronts, the easterly waves, and marine and terrestrial breezes.

Intertropical Convergence Zone

The Intertropical Convergence Zone (ITCZ) is the area of interaction where the northeastern Trade Winds of the Northern Hemisphere meet the southeastern Trade Winds of the Southern Hemisphere (Zárate 1978). The ITCZ is a convergence zone of up to 200 km in width where cloud systems develop that are composed of layers or tiers of cumulonimbus, cumulus, altostratus, or other clouds, which generate strong rains, storms, turbulence, and winds predominately from the southwest (Equatorial Westerlies or "anti-trades" winds). The ITCZ shifts from north to south and vice versa, depending on the zenith position of the sun. It affects Costa Rica, particularly the Pacific slope, with copious and strong rains in the afternoons during the wet season from April to December. During the months that the country is influenced by the ITCZ, winds from the southwest blow along the entire Pacific slope, especially in the afternoons. These winds are weak, but they are heavily laden with clouds. As the ITCZ weakens or is displaced farther to the south of Costa Rica, the trade winds gather strength, blowing from the east and northeast, generating a water deficit on the Pacific slope and persistent rains on the Caribbean slope, mainly in the months of November, December, July, and August.

Cold Fronts

At the beginning of the Northern Hemisphere winter, and as the Bermuda Anticyclone (Bermuda High) is displaced farther to the south of its usual position (30° N), the ITCZ weakens, and trade winds from the northeast affect the whole of Central America, pushing modified cold front air masses that generate strong precipitation and floods on Central America's Caribbean slope. According to Martínez (1970), these polar air masses become warmer and acquire greater moisture when they move through the Gulf of Mexico, but maintain some of their original conditions when they cross the Mexican *Meseta*. If they enter Costa Rica, they can cause rains for nearly 24 hours a day during three consecutive days, saturating the soil and causing floods with substantial losses in the productive sector. As these air masses rise they shower the hills, mountain passes, and continental divides with light rains and drizzle, but produce clear, rain-free skies in the North Pacific, the Central Pacific, and the Central Valley regions. By contrast, the South Pacific region may experience isolated storms coming from the Pacific Ocean, even when the Caribbean is under the influence of a cold front.

Easterly Waves

Easterly waves are atmospheric disturbances that are characteristic of the tropical region and result from the elongation of cyclones, which sporadically generate convergence, instability, and bad weather, especially from June to October. These are disturbances of the trade winds that do not cross the Intertropical Convergence line (Andrade 1968). According to Chorley and Barry (1972), the easterly waves are closed low pressure zones in the middle troposphere. Behind the trough or *thalweg* (the deepest parts of a valley or watercourse), convective systems and storms develop that affect large areas of the Caribbean slope for one or two days. The rest of the territory is affected by the same phenomenon but with less intensity in temporal, quantitative, and spatial terms.

The Isthmian Effect

The Costa Rican territory is influenced by both the Atlantic and the Pacific Oceans and, due to the narrowness of its territory (minimum 119 km, maximum 464 km), the atmospheric systems that originate in the Atlantic affect the Pacific region within a matter of hours, and the systems that station themselves in the Pacific can alter weather conditions on the Caribbean slope; however, the central mountainous system exerts a barrier effect on wind flow, cloudiness, and rains, giving rise to a mosaic of climates and ecosystems. For example, according to Costa Rica's National Meteorological Institute (IMN), in July 1979 the country was influenced 61% of the time by an atmospheric condition originating in the Pacific Ocean, 29% by one in the Atlantic Ocean, and the rest of the time (10%) by a mixed condition. The degree of modification exerted by the continental territory on the air masses that cross it can be expressed using an index of continentality (terrestrial coefficient; +) or oceanicity (maritime coefficient; –). The continental index ranges from 0 to 100% and the oceanicity index from 0 to –20%. Herrera (1986) reports an extreme continentality index of +4.7% in Alajuela, a town located in the center of the country between the mountain chains, and an oceanicity index of –10.6% in Sanatorio Durán, a site on the southern slope of the Cordillera Central (Costa Rica's central mountain range).

Temporal and Spatial Climate Variations in Costa Rica

A general outline follows on the temporal and spatial variations in temperature, solar radiation and hours of sunlight, precipitation, and relative humidity in Costa Rica. The presented analysis includes all available data up to 2004, from more than 600 stations of the National Meteorological Institute (IMN), the Costa Rican Electricity Institute (ICE), and the National Aqueduct and Sewerage System (AyA).

Temperature

The mean annual temperature varies from 27.6° C on the North Pacific coast and 26°C on the Caribbean coast, to 6° C on Cerro Chirripó, the highest peak in the country (3,819 m). The annual thermal gradient is 5.7° C for every 1,000-meter increase in altitude on the Pacific slope, and 5.2° C for every 1,000-meter increase in altitude on the Caribbean slope. Minimum temperature extremes of –3° C have been recorded in areas above 3,000 meters, and maximum extremes of 41°C in the hot, dry lowlands of the Tempisque River Basin. The mean monthly temperature oscillation does not exceed 4°C between the hottest months—usually March or April with zenith sun positions—and November, the coolest month. Figure 2.1 shows the hourly temperature variations on mountain peaks, on plains, and in valleys.

Hours of Sunlight and Solar Radiation

At 10° mean latitude for Costa Rica, and in the absence of cloud cover, the sun is above the horizon for 11.41 hours on the winter solstice (December 22) and 12.58 hours on the summer solstice (June 21). However, abundant cloud cover, particularly from 600 to 2,200 meters, reduces hours of sunlight by at least 50% throughout the Caribbean slope, and in some areas the daily mean is less than 2.5 hours during periods of strong rain activity. If there is less sunshine due to the presence of clouds, the radiation is also reduced. At the top of the atmosphere above the 10th parallel, the rate of solar radiation varies from 37.5 megajoules (MJ) per square meter in April to 30.7 MJ/m^2 in November; however, the total daily average recorded is 23 MJ/m^2 in the dry season and the minimum is 13 MJ/m^2 at very cloudy sites. Radiation rates are also affected by the inclination and orientation of hillsides. For example, on the summer solstice, a 30° north-facing slope receives 29 MJ/m^2 (clear sky) while a south-facing slope would receive 16 MJ/m^2. This situation gives rise to hillside microclimates that are particularly important for biodiversity, especially above 3,000 meters elevation. Figures 2.2 and 2.3 show monthly patterns of solar radiation and hours of sunlight at sites with opposite climates.

Precipitation

Annual sheets of rain vary from 1,300 millimeters in the dry climates of Guanacaste Province to 7,467 mm in the watershed of the Río Grande de Orosí on the Caribbean slope, where there is no dry season and precipitation exceeds potential evapotranspiration during every month of the year.

The monthly and hourly rainfall patterns are very variable, depending on the location—the Caribbean slope or Pacific slope, mountain passes, continental divides, sites that are windward or leeward to the trade winds, and the position of the Intertropical Convergence Zone. In continental and insular areas, there are three precipitation regimes (Fig. 2.4): the Pacific, the Atlantic, and the Coastal Atlantic. The Coastal Atlantic regime comprises the coastal strip and the plains. A dry season does not occur in this regime; however, rainfall decreases considerably in March, April, September, and October.

Precipitation occurs during the night and the morning, but does not have a defined pattern during the diurnal period. The Atlantic Regime, characteristic of areas from 500 to 2,700 meters elevation on the Caribbean slope, receives

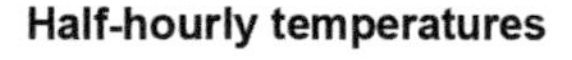

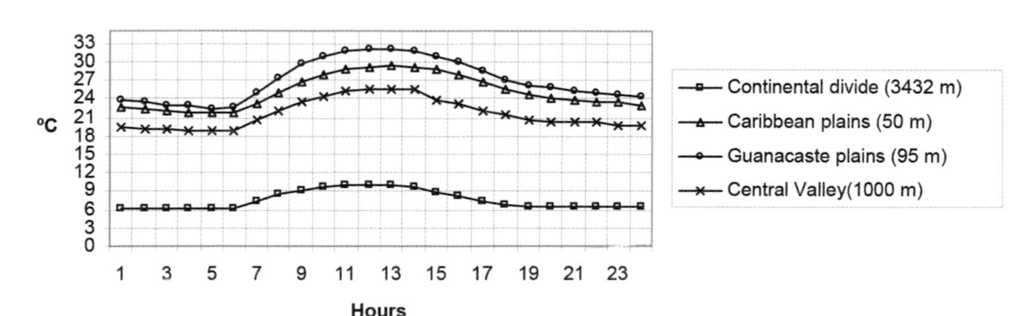

Fig. 2.1 Mean hourly temperatures at continental divides, plains, and valleys.

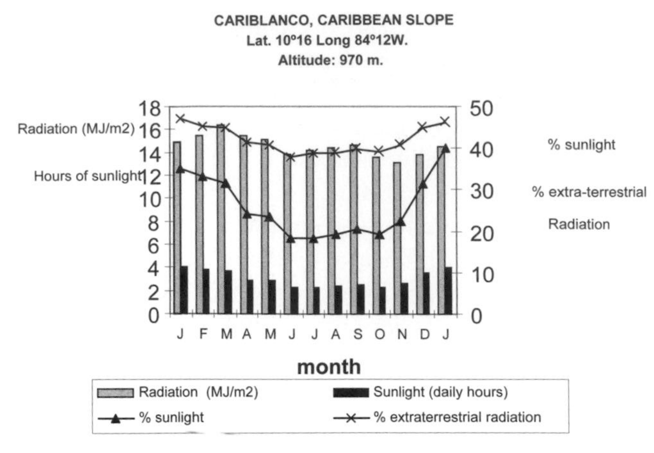

Fig. 2.2 Monthly variation in solar radiation and hours of sunlight at the cloudiest site in Costa Rica.

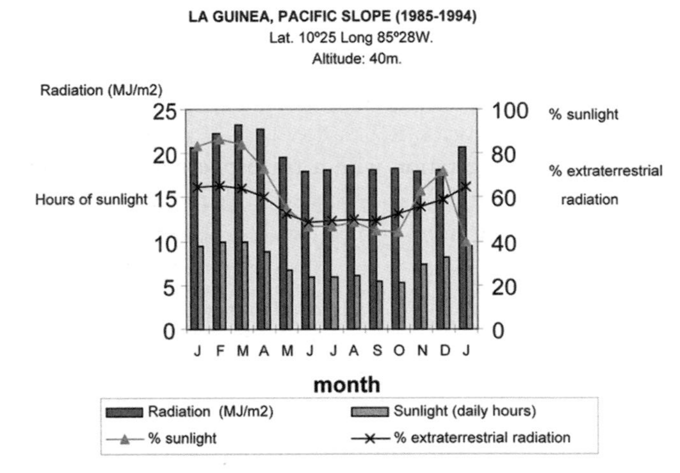

Fig. 2.3 Monthly variation in solar radiation and hours of sunlight at the sunniest site of Costa Rica.

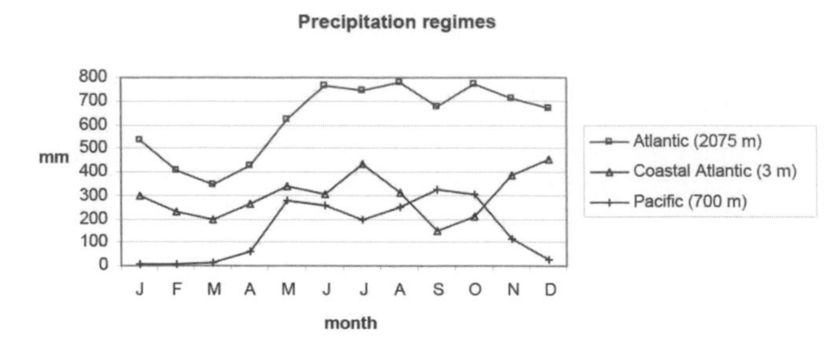

Fig. 2.4 Pluviometric regimes of Costa Rica.

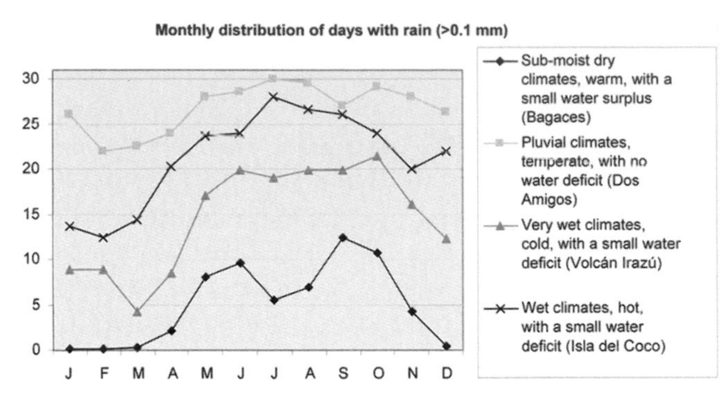

Fig. 2.5 Rainy days (precipitation >0.1 mm).

maximum rainfall in the afternoon and the early hours of the evening in the months of May, June, July, August, September, October, and November. From December to April, the rains are distributed over 24 hours of the day, with a maximum at night. The Pacific regime is characterized by rainfall during the afternoons and the early hours of the evening in the months of April, May, June, July, August, September, October, November, and December. The entire Pacific slope and the plains of Los Guatusos are affected. On the mountain summits of both slopes, the pluvial regime is a mixture of both regimes.

In Costa Rica there are very marked contrasts in the monthly and annual rainfall levels and in the number of rainy days (precipitation ≥ 0.1 mm). At the rainiest site (Río Grande de Orosí basin) there is activity on 325 days of the year, with a minimum of 22 days in January and 31 days in September. At the least rainy site in the Guanacaste lowlands, it rains only 60 days per year, with a minimum of one day in January to 27 days in October. Some of the rainiest sites are located on the continental divides below 2,500 meters, some in mountain passes, and others on hillsides that are perpendicular to the trade wind flow (Fig. 2.5).

Relative Humidity

Average monthly relative humidity in these climates varies between 65 and 90%, and average hourly relative humidity varies from 52 to 98% (Fig. 2.6). In sites above 3,000 meters, in the period February–April, with little cloud cover, hourly relative humidity drops to 20 or 25%, favoring desiccation and the propagation of forest fires. The same situation occurs in the Central Valley, between Turrúcares and Alajuela, during the same period. In the narrow valleys and lowlands of the South Pacific, leeward to the trade winds, morning cloud banks form, an indication of the high relative humidity rates above 90%. The morning fog phenomenon in the valleys and plains is partly explained by the accumulation of cold air in the *thalwegs*, weak winds, soil humidity that is always near field capacity, and the presence of vegetative cover.

In Costa Rica's *páramos*—areas above 3,200 m—in calm wind conditions, the necessary hygrothermic conditions are generated for white frosts (hoarfrosts) and black frosts on more than 34 days from December to April. A

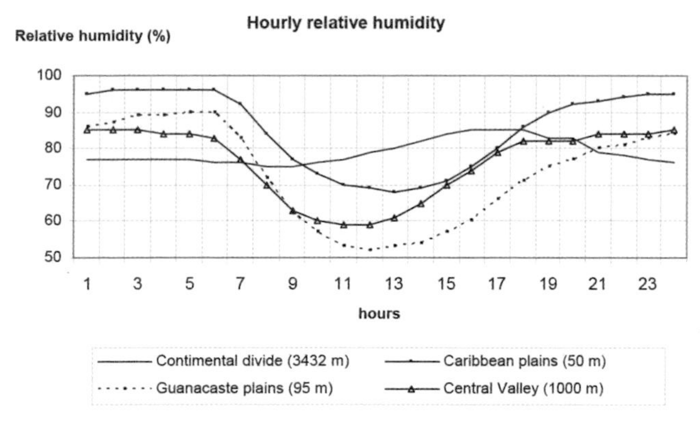

Fig. 2.6 Hourly relative humidity at continental divides, valleys, and plains.

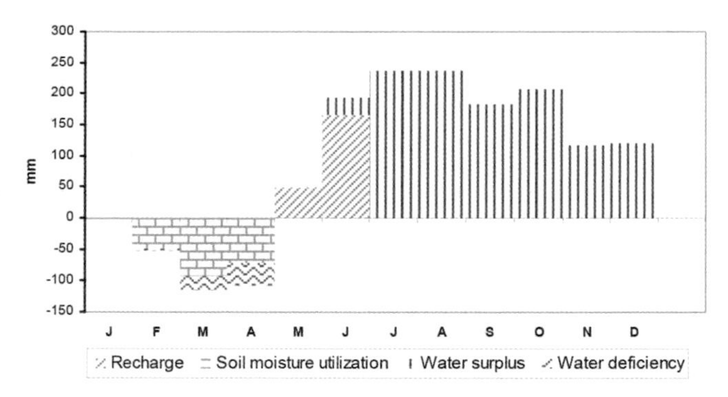

Fig. 2.7 Monthly water balance for the San Jorge de Los Chiles station (1980–2004).

black frost causes the death of a plant or its sensitive parts, due to cold exposure. This limits the growth of arboreal vegetation above 3,300 meters elevation (Herrera 2005).

Hydric Balance

The comprehensive relationship between land-atmosphere systems is expressed using soil moisture balance, an approach that allows us to determine potential and actual evapotranspiration rates, levels of water deficits and surpluses, periods of field capacity, periods of moisture extraction, and hydric recharge seasons. Using this method, it is possible to integrate and quantify the behavior of the climatic elements and their effects on different ecosystems. In different regions of Costa Rica, soil climatology conditions are generally analyzed to a depth of one meter, using the model developed by Thornthwaite and Mather (1955, 1957) and the graphic modeling of Rolim et al. (1998) to examine climatic hydric balance. In this study, a conditional drought period is when there is a water deficit in the soil but the permanent wilting point has not been reached, and absolute drought is the period when water reserves have fallen below the permanent wilting point value. The water deficit is the difference between the water that is actually evapotranspirating and the maximum that could evapotranspirate. Water deficits occur when rains and water reserves are not sufficient to satisfy the climatic demand (potential evapotranspiration).

Climate Regions

San Carlos Plains and Los Guatusos Plains

Annual precipitation in the San Carlos and Los Guatusos plains ranges from 2,500 to 4,000 mm, with a considerable decline in the rains during the months of March, April, and May. The need for water (potential evapotranspiration), according to climate, ranges from 116 millimeters in January to 165 mm in May, for a potential sheet of rain of 1,618 mm per year. One balance for San Jorge (Fig. 2.7) indicates that vegetation with a rhizosphere at one meter depth suffers a water deficit after the sixth day without rain, a situation that occurs in February. In March and April, the deficit is so great that water reserves fall below the wilting point, giving rise to an absolute drought. For May and June the soil recharges moisture, and by the end of June there may already be a small water surplus, which is maintained for a period of seven consecutive months (June–December).

Tortuguero Plains

The plains of Sarapiquí, Tortuguero, Guápiles and Matina have abundant rainfall, which ranges from 3,000 to 6,000 mm annually. Precipitation occurs on more than 170 days of the year; however, in March and April the amount of rain and the number of rainy days decline considerably, forcing the vegetation to extract moisture from the soil, which is then recharged in the next month. Although the monthly balances do not indicate the presence of absolute drought, this phenomenon occurs with some recurrence. For example, in the decade 1993–2003, absolute droughts were recorded in March, April, and May 1994; and in January and March of 1998, a year that was under the influence of the warm phase of the El Niño Southern Oscillation (ENSO), also referred to as *El Niño*.

On the hot and wet lands of the plains, beginning with the soil at field capacity, the vegetation experiences deficit following the fifth day without rain. The extraction of moisture occurs in March; recharge occurs in April, and the rest of the year there is a water surplus in the soil (Fig. 2.8). Occasionally deficits may occur in September and/or October.

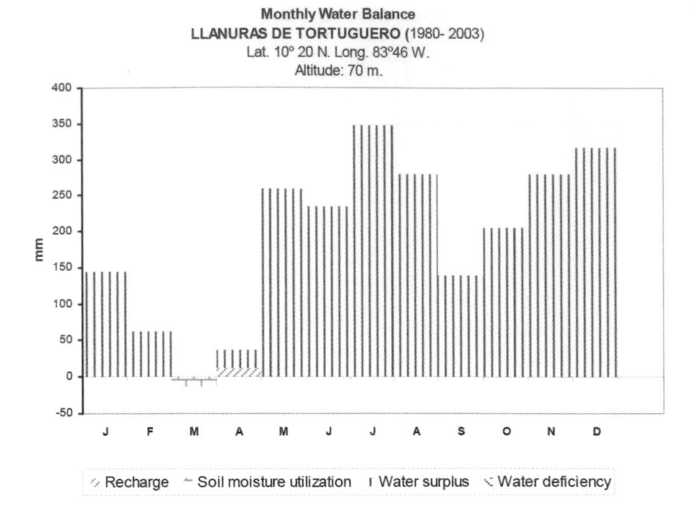

Fig. 2.8 Water balance of the Tortuguero Plains.

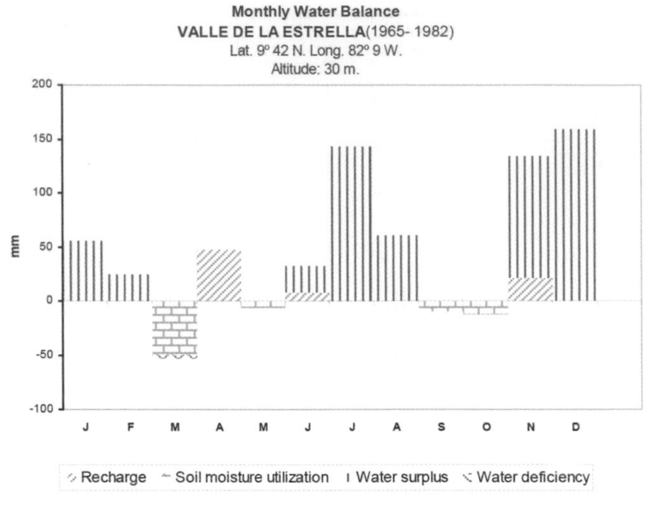

Fig. 2.9 Water balance of Valle de la Estrella.

From December to February, the soil can remain saturated with pools or puddles with the passage of modified cold fronts that contribute copious and persistent rains for three consecutive days.

Southern Caribbean

The small coastal plains of the Southern Caribbean, located between the 10th parallel and the border with Panama, have a climatic and edapho-climatic regime that is different from the rest of the Caribbean slope. Here there are two peaks of pluvial activity, one in July and the other in November, altered by three water deficit periods. Mean annual precipitation ranges from 1,800 to 3,500 mm. The water balance (Fig. 2.9) indicates that the rains are insufficient to cover potential evapotranspiration in the months of March, May, September, and October, and that there is a water deficit, mainly in March. Average water reserves are always maintained well above the wilting point; however, the absence of rains can provoke an absolute drought phenomenon in this region, of the same frequency and magnitude as occurs on the Tortuguero Plains—in other words, with a probability of twice every ten years. The amount and distribution of the monthly rains is not sufficient to recharge soils to field capacity during seven months of the year. If the Caribbean slope were to be affected by droughts as a result of global climate change, the soil climatology of the southern Caribbean could suffer a major impact.

South Pacific Plains

The Coto 47 station, located in the lower watershed of the Río Coto Colorado, represents in part the soil climatology conditions that prevail on most of the western coast

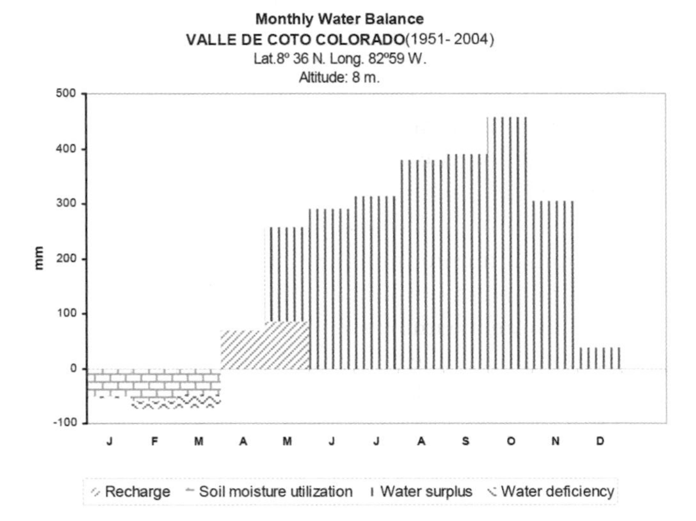

Fig. 2.10 Water balance of Valle de Coto Colorado.

of Golfo Dulce; a very warm, wet territory with precipitation from 3,000 to 6,000 mm, with potential needs below 1,800 mm. During the months of January, February, and March, intense evapotranspiration reduces water levels below the wilting point, and the entire region is affected by the absolute drought phenomenon. In April and May, the rains exceed needs and recharge occurs until mid-May, a period when the water surplus begins to increase, with peaks of 458 mm in October. The surplus infiltrates to deep layers and/or runs off via the drainage network. Excess water in the soil occurs over nine months: from May to December, with a total amount of 2,349 mm (Fig. 2.10).

Some areas in the South Pacific region—on the plains between the north coast and the Brunqueña Range—have very

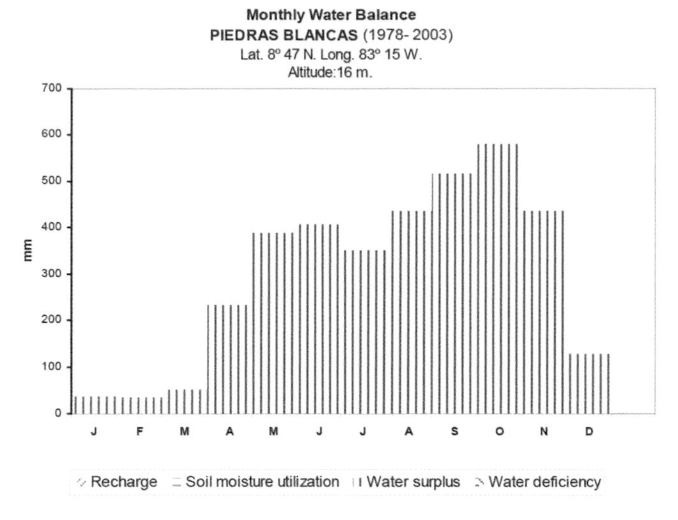

Fig. 2.11 Water balance of Piedras Blancas, South Pacific.

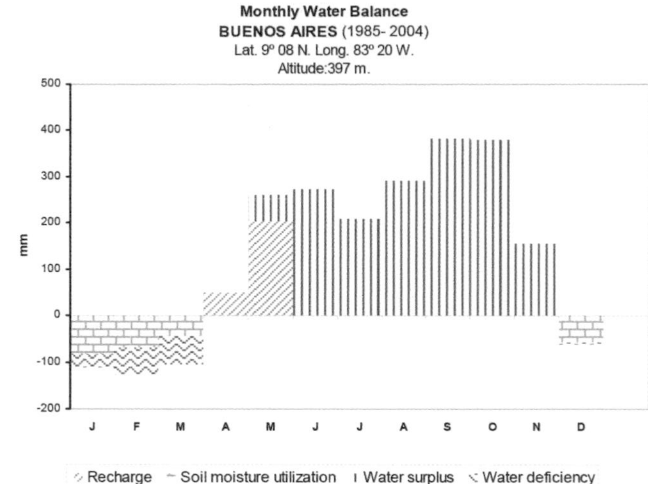

Fig. 2.12 Water balance at Buenos Aires, Río Grande de Térraba Basin.

special climates without dry seasons. This situation occurs because of the sheltered conditions provided by the high Talamanca Mountains and the Brunqueña Range against the trade winds from the north and northeast. This favors the incursion of a moist breeze from the Pacific throughout the year, although with less frequency from January to April, which generates convective and orographic rains year-round as it tries to rise up the Brunqueña Range. One station that is typical of this condition is Piedras Blancas (Fig. 2.11). Mean monthly values indicate that a deficit does not occur, but the sequential balances reveal that once every three years water deficits occur in January, February, or March.

The Térraba Basin

The Térraba Basin covers an area of 5,079.4 km^2 and contains the El General Valley and the Valley of Coto Brus. It is bounded on the north by the Talamanca Mountain Range, with maximum elevations of 3,820 meters; to the south, the Cal Range and the Brunqueña Range separate it from the Pacific Ocean. This geographic situation produces a Pacific-type precipitation regime in the high, middle, and lowland areas of the basin. The Pindeco station in the Buenos Aires district is a very good example of the functioning of soil climatology, principally at 200 to 1,000 meters elevation (Fig. 2.12). Mean precipitation in the basin ranges from 1,800 mm at Maíz de Boruca on the leeward side of the Cal Range, to 7,000 mm on the southern Talamanca slopes. At Buenos Aires, annual mean precipitation is 3,317 mm; water needs reach 1,721 mm, varying from 119 mm in November to 170 mm in March, a month of high temperatures, elevated sun positions, and low relative humidity. By mid-December, soil moisture falls rapidly and gives rise to a conditional drought, which is accentuated in the months of January, February, and March, attaining levels of absolute drought. The rains start in April and the soils begin to recharge moisture in mid-May, when they begin to reach field capacity and there is excess water for the next seven months, for a total of 1,750 mm. According to the classification system of Thornthwaite and Mather (1955, 1957), the aridity at Buenos Aires is low (9% of the evapotranspiration potential) and the surplus is very large (109% of annual evapotranspiration).

Continental Divides

Above 3,000 meters elevation, climate conditions can be very changeable within a period of hours. Between the months of December and March, Irazú Volcano, Turrialba Volcano, and all the continental divide regions of the Talamanca Cordillera are affected by strong trade winds from the north and northeast, and by abundant fog, rain, and drizzle. However, for several days or hours, the weather can become sunny and windy, with relative humidity below 35%, high evapotranspiration rates, and no rain. These are times of soil water deficits (Fig. 2.13), very low temperatures, black and white frosts, and possibly forest fire propagation. By contrast, during the rest of the year, cloudy and very cloudy skies predominate with low rates of evapotranspiration and cold, moist winds.

Central Pacific Plains

The Central Pacific Plains, between the Tulín and Térraba rivers, are slightly or moderately arid. The drought here is not as severe as on the plains and hills of Guanacaste

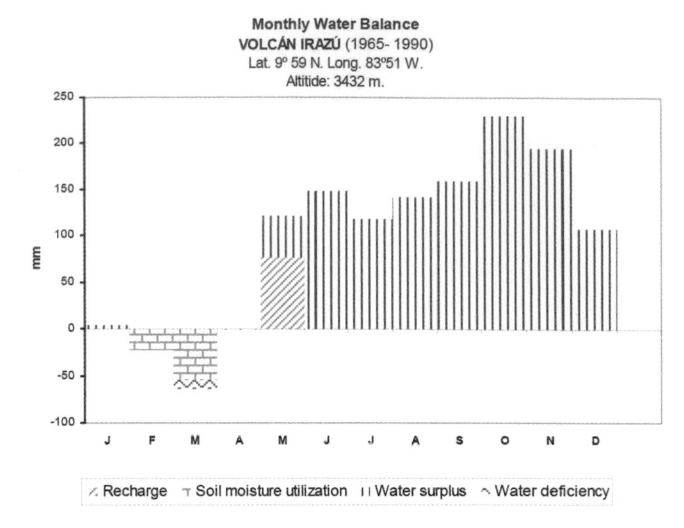

Fig. 2.13 Water balance of Irazú Volcano, continental divide.

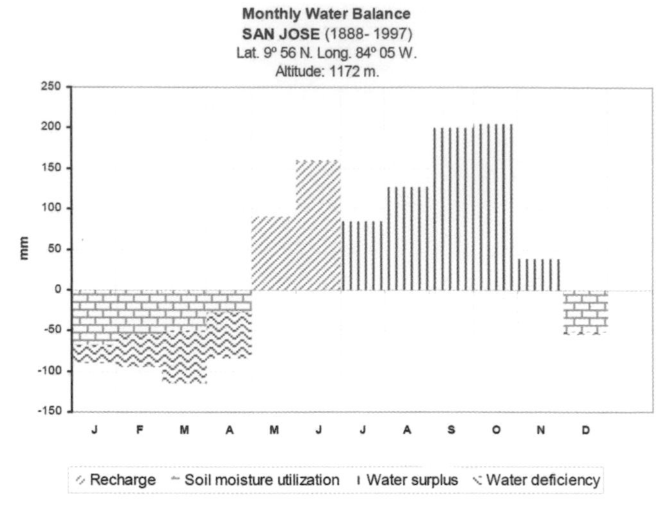

Fig. 2.15 Water balance at San José, Río Grande de Tárcoles Basin.

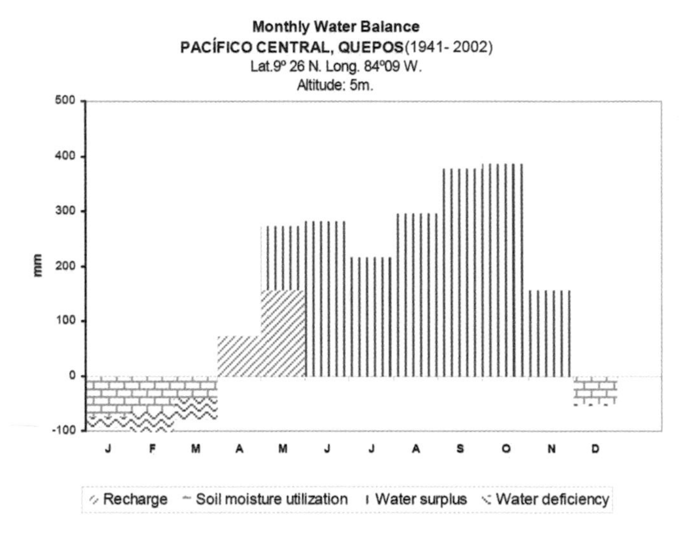

Fig. 2.14 Water balance at Puerto Quepos, Central Pacific.

province, nor is it as slight as the drought that occurs in the Coto Colorado valley. The hydric balance at Puerto Quepos shows seven months of water surplus and four months of absolute drought. According to the classification system of Thornthwaite and Mather (1955, 1957), the climate is very wet, with a small water deficit and a surplus that exceeds needs by 116% (Fig. 2.14). In the Central Pacific region the mountain chains run parallel to the coast and at a distance of less than 15 km from the shore.

These cordilleras provoke intense orographic rains, mainly when a hurricane is present in the Caribbean. Precipitation can exceed 300 mm in one day and, as a consequence, the soils become saturated with pools or puddles for periods of two to three days, mainly in September and October. At Naranjillo, at 700 m elevation, the most intensive rainfall for the entire Pacific slope of Costa Rica has been recorded, with an annual precipitation of 6,500 mm. Monthly precipitation varies from 109 (March) to 1,027 mm (October). Conditions similar to this also occur in the hills of the western slope of the Corcovado Peninsula in the South Pacific region.

Río Grande de Tárcoles Basin

The Río Grande de Tárcoles Basin encompasses an area of 2,116 square kilometers, with the Central Mountain Range to the north, the Talamanca Range to the south, and the Aguacate Hills on the west. Nearly the entire area has a Pacific regime climate; however in the mountain passes and on the peaks of the Central Mountain Range, the climatic condition is a mixture of the Atlantic mountain regime and that of the Pacific, with rains throughout the year and at any hour. Precipitation in the Tárcoles Basin varies from 1,600 mm in Santa Ana, to 4,650 mm in Gallito, on the continental divide of the Cordillera Central. The thermic, pluviometric, and soil climatology gradients vary considerably from north to south and vice versa. For example, in the country's capital city, San José, there is a pronounced dry season from December to April (Fig. 2.15), with an absolute drought during four consecutive months. Surplus water in the soil occurs over six consecutive months, from June to November. If the country is affected by the warm phase of the El Niño or ENSO phenomenon, soil moisture can decrease until it attains the conditional drought category in July or August, during the short dry period known as the Veranillo de San Juan. In the case of the Gallito Station (2,120 m elevation) located at a distance of 18 km northwest of San José, there is no dry season and a surplus, on average, is present every month.

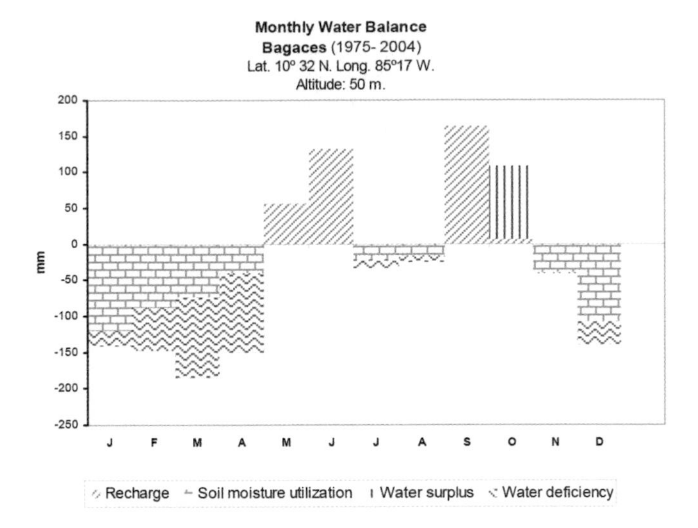

Fig. 2.16 Water balance at Bagaces, Baja del Río Tempisque Basin.

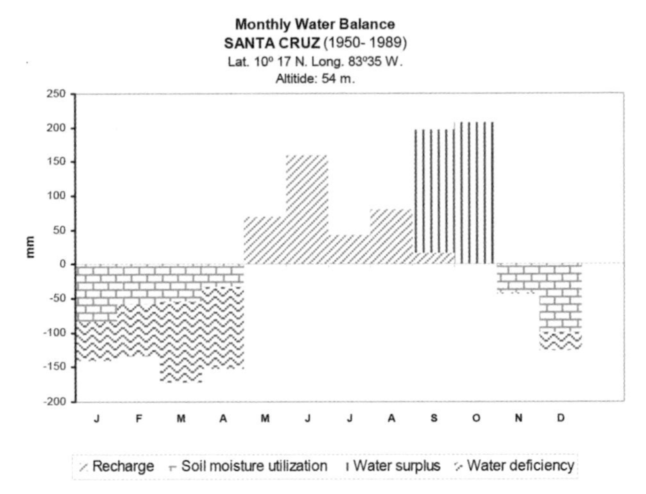

Fig. 2.17 Water balance at Santa Cruz, the boundary between dry and moist climates of the North Pacific.

North Pacific

The North Pacific Region includes the Río Tempisque basin (5,460.2 km^2), the Nicoya Peninsula, and part of the dry coastal strip to the east of the Gulf of Nicoya. The Pacific precipitation regime predominates in nearly the entire region, with afternoon rains in the period from May to November. However, in the mountain passes, and on the continental divide of the Guanacaste Mountain Range and the Tilarán or Minera Range, where maximum elevations do not exceed 2,000 meters, a mixed rain regime occurs, in-between a mountain Atlantic and a Pacific regime. From November to February, rains and drizzle fall on these peaks and mountain passes from air masses transported by the trade winds.

The Föhn effect leaves copious rains on the windward side (Caribbean slope) and near the divide; however, below 600 meters on the leeward side, clear skies predominate with strong winds, low relative humidity, high temperatures, and dry soils. From May to November, the entire region receives moderate rains, interrupted by the normal occurrence of the Veranillo de San Juan or Canícula in July and August. Annual precipitation varies from 1,200 to 4,000 mm, from zero in January to 700 mm in October. In September and October, the presence of low pressure systems in the Pacific or hurricanes in the Caribbean gives rise to rains that persist for three days, known as *temporales*, which are often accompanied by flooding episodes.

In the month of April, with zenith sun positions, moderate to weak winds, and low relative humidity, potential evapotranspiration reaches 190 mm per month. The annual amount of this same element—potential evapotranspiration—exceeds 1,900 mm in some sites such as Filadelfia and Nicoya. The influence of El Niño can trigger periods of water deficit throughout the lower and middle watershed for 33 consecutive months.

In the southern part of the peninsula the rains are more abundant and the dry episodes are less severe than in the rest of the Northern Region. The monthly humidity balance of Bagaces, a site at 90 meters elevation, reflects the soil-related climatology of areas with less pluvial activity in Costa Rica (Fig. 2.16). These soil-related climatic conditions are (1) uninterrupted water deficit for eight months (November, December, January, February, March, April, July, August); (2) absolute drought for five consecutive months: December, January, February, March, and April; (3) uninterrupted water recharge for four months: May and June, September and October; and (4) only 75% of annual water needs (1,883 mm) are satisfied.

Soil-related climatology conditions in the moist and submoist climate regions of the North Pacific are differentiated from those that prevail at Bagaces (sub-moist dry climate). The absolute drought begins in December and ends in April; the soil recharges in May, June, July, and August. There is a surplus in only two months: September and October (Fig. 2.17).

Perspectives on Climate Change and Ecosystems in Costa Rica[2]

Past climate change in Costa Rica has been documented by Islebe and Hooghiemstra (1997) who studied the presence and abundance of fossil pollen in soil cores from montane

[2] This chapter's section on *Perspectives on Climate Change and Ecosystems in Costa Rica* was prepared and contributed by Maarten Kappelle.

(2,300 m) peatbog locations in the Cordillera de Talamanca (Trinidad, La Chonta). These palynologists noted that Central American climate change during the mid-Holocene seems more affected by changes in humidity than temperature. They reconstructed distribution maps of *paramó* and montane vegetation in Costa Rica for periods between 10,000 and 18,000 years before present. Their data indicate that during the Last Glacial Maximum a *paramó* vegetation corridor existed between Costa Rica and Panama.

At the same time, it seems that future climate change will be significant in Costa Rica. Recent studies on modern climate change in the country suggest the possibility of a two-degree increase in temperature by the middle of the twenty-first century in tropical forest regions, and an increase of 1°C in tropical cloud forests (Pounds et al. 1997, 1999, Karmalkar et al. 2008). Should this trend be consolidated, the existing cloud forest patches in Costa Rica would be located 200 meters higher than their current positions and tropical forests would tend to occupy more territory, between 3,100 and 3,400 meters elevation (Lawton et al. 2001, Nair et al. 2003). As a result, the current area of *páramo* vegetation would be reduced to a small, restricted area between 3,400 and 3,820 meters elevation; by the middle of the century, only the area of Cerro Chirripó would offer suitable climate conditions that would allow the survival of *páramo* vegetation in Costa Rica (Herrera 2005).

In a detailed simulation for doubled atmospheric carbon dioxide concentration (2 x CO_2) conditions in tropical montane cloud forests like those in Costa Rica, Still et al. (1999) found that the relative humidity surface is shifted upwards by hundreds of meters during the dry winter season when these forests typically rely mostly on moisture from cloud contact. At the same time, these authors noted an increase in the warmth index that could imply an increased level of evapotranspiration. According to these authors, this combination of reduced cloud contact and increased evapotranspiration could have serious conservation implications, given that these ecosystems typically harbor a high proportion of endemic species and in Costa Rica are often situated on mountain tops or ridge lines.

Enquist (2002) predicted regional impacts of climate change on the geographical distribution and diversity of tropical forests in Costa Rica. On basis of climate change scenarios she concluded that elevation-associated life zones and ecosystems in Costa Rica may be particularly vulnerable to future climatic changes. This is also true for lowland seasonally dry forest. Geographical regions in Costa Rica that contain these life zones are likely to warrant special management and conservation attention in the event of predicted climate change, she pointed out.

The small patches of middle-elevation cloud forests that are currently found on mountain chains or on isolated peaks, on both the Caribbean and the Pacific slopes, will be replaced by tropical lowland vegetation. This is the case of the Nicoya Peninsula, the Central Pacific, and the South Pacific. Other mountain chains with cloud forests at lower elevations today, such as the Sierra Minera de Tilarán, are already reporting the loss of amphibian species due to global warming and to the altitudinal migration of the level of the condensation line (Pounds & Crump 1994, Pounds et al. 1997, 1999, Lawton et al. 2001, Ray et al. 2006). The degree of impact that climate change is having on ecosystems is highly variable, both spatially and temporally, and depends on the location of the ecosystem with respect to the mountain chains, directions of the dominant winds, land use, proximity to the sea, elevation, and soil moisture retention capacity, among other factors.

Future changes in temperature and precipitation could alter cloud cover at the vegetation level and seriously affect mountain ecosystems (Lawton et al. 2001, Ray et al. 2006, Karmalkar et al. 2008). The disappearance or reduction of cloud forest would also cause a reduction in the water volumes of rivers on the Pacific, which benefit from the cloud forests' sponge effect, capturing passing clouds and/or favoring condensation, mainly during low water periods (Lawton et al. 2001). It is assumed that in the lowlands, where moist and wet forests predominate, and to a lesser extent, in the sub-moist dry forests, evapotranspiration will increase and thus the current boundaries could be displaced to one side or the other, depending on the variation in annual precipitation. However, the variation could be of lower impact than in the sub-tropical thermal, temperate, cold temperate and boreal provinces, where altitudinal variation implies major climate changes in reduced spaces.

Much more integrated climatological research is needed to better understand the full impact of modern climate change on Costa Rican ecosystems and human society. A holistic, interdisciplinary approach will be essential to get a full picture of human-driven climate change in the twenty-first century. Only by understanding the full scope and impact of current and potential climate change on our lives, in every sense, will we be able to mitigate and manage adaptively, thus ensuring a sustainable future for both mankind and nature in biodiverse tropical countries like Costa Rica.

To mitigate modern and future climate change in Costa Rica—and hence safeguard the country's rich biological diversity—it will be crucial to develop and apply science-based, integrated systems and tools. One key example is presented by Castro et al. (2000) who discuss Costa Rica's policy framework that provides an appropriate context for the actual and proposed development of market-based instruments designed to attract capital investments for carbon sequestration and biodiversity conservation. Such a framework allows the establishment of mechanisms to use those funds to compensate owners for the environmental services provided by their land to the society. As a developing economy, they point out, Costa Rica is striving to internalize the benefits from the environmental services it offers, as a cornerstone of its sustainable development strategy. It is this kind of win-win tool that will help Costa Rica get a grip on climate change while protecting its lush nature.

References

Andrade, E. 1968. *Introducción a la Meteorología en Honduras*. Tegucigalpa.

Castro, R., F. Tattenbach, L. Gámez, and N. Olson. 2000. The Costa Rican experience with market instruments to mitigate climate change and conserve biodiversity. *Environmental Monitoring and Assessment* 61(1): 75–92.

Chorley, R., and R. Barry. 1972. *Atmósfera, Tiempo y Clima*. Barcelona: Ediciones Omega.

Enquist, C.A.F. 2002. Predicted regional impacts of climate change on the geographical distribution and diversity of tropical forests in Costa Rica. *Journal of Biogeography* 29(4): 519–34.

Herrera, W. 1986. Clima de Costa Rica. In L. D. Gómez, ed., *Vegetación y Clima de Costa Rica*. San José, Costa Rica: Editorial EUNED. 118 pp.

Herrera, W. 2005. El clima de los *Páramos* de Costa Rica. In M. Kappelle and S. Horn, eds., *Páramos de Costa Rica*, 113–28. Costa Rica: Editorial INBio.

Herrera, W., and L. Gómez. 1993. Mapa de Unidades Bióticas de Costa Rica. US Fish and Wildlife Service, The Nature Conservancy, INBio. San José, Costa Rica: INCAFO.

Islebe, G.A., and H. Hooghiemstra. 1997. Vegetation and climate history of montane Costa Rica since the last glacial Quaternary. *Science Reviews* 16(6): 589–604.

Karmalkar, A.V., R.S. Bradley, and H.F. Diaz. 2008. Climate change scenario for Costa Rican montane forests. *Geophysical Research Letters* 35: L11702. doi:10.1029/2008GL033940

Lawton, R.O., U.S. Nair, R.A. Pielke, Sr., and R.M. Welch. 2001. Climatic impact of tropical lowland deforestation on nearby montane cloud forests. *Science* 294: 584–87.

Martínez, A. 1970. Anexo A. Meteorología e Hidrología Istmo Centroamericano. Naciones Unidas, Consejo Económico y Social, Costa Rica.

Nair, U.S., R.O. Lawton, R.M. Welch, and R.A. Pielke. 2003. Impact of land use on Costa Rican tropical montane cloud forests: sensitivity of cumulus cloud field characteristics to lowland deforestation. *Journal of Geophysical Research–Atmospheres* 108: 193.

Pounds, J.A., and M.L. Crump. 1994. Amphibian declines and climate disturbance: the case of the golden toad and the harlequin frog. *Conservation Biology* 8: 72–85.

Pounds, J.A., M.P.L. Fogden, and J.H. Campbell. 1999. Biological response to climate change on a tropical mountain. *Nature* 398: 611–15.

Pounds, J.A., M.P.L. Fogden, J.M. Savage, and G.C. Gorman. 1997. Tests of null models for amphibian declines on a tropical mountain. *Conservation Biology* 11: 1307–22.

Ray, D.K., U.S. Nair, R.O. Lawton, R.M. Welch, and R.A. Pielke. 2006. Impact of land use on Costa Rican tropical montane cloud forests: sensitivity of orographic cloud formation to deforestation in the plains. *Journal of Geophysical Research–Atmospheres* 204: 111.

Rolim, G., P. Sentelhas, and V. Barbieri. 1998. Planilhas no ambiente excel para os cálculos de balanços hídricos: normal, sequencial, de cultura e de produtividade real e potencial. *Revista Brasileira de Agrometeorologia, Santa Maria* 6(1): 133–37.

Still, C.J., P.N. Foster, and S.H. Schneider. 1999. Simulating the effects of climate change on tropical montane cloud forests. *Nature* 398: 608–10.

Thornthwaite, C.W., and J.R. Mather. 1955. The water balance. Centerton, NJ: Laboratory of Climatology, Publications in Climatology, vol. 8, no. 1, p. 1–104.

Thornthwaite, C.W., and J.R. Mather. 1957. Instructions and tables for computing potential evapotranspiration and the water balance. Centerton, NJ: Laboratory of Climatology, Publications in Climatology, vol. 10, no. 3, p. 185–311.

Zárate, E. 1978. Comportamiento del viento en Costa Rica. Instituto Meteorológico Nacional. Nota Técnica de investigación No. 2. San José, Costa Rica.

Chapter 3 Geology, Tectonics, and Geomorphology of Costa Rica: A Natural History Approach

Guillermo E. Alvarado[1,*] and Guaria Cárdenes[2,3]

Introduction

Costa Rica occupies an interoceanic and intercontinental position at the narrow Central American isthmus, which separates North America from South America and the Atlantic Ocean from the Pacific Ocean. The unique location of Costa Rica along this land-bridge and the country's long geological, biological, and climatological history have motivated researchers to conduct studies in these fields using the country as a natural, long-term laboratory. The resulting laboratory experiment has led to an intricate mosaic of dynamic landscapes shaped by a wide range of processes, such as volcanism, tectonics, fluvial and marine erosion and deposition, weathering, and hydrothermal, karst, glacial, and periglacial processes, and its consequent deposits.The result is a physiography characterized by a heterogeneous array of geomorphic and tectonic provinces, each featuring a distinctive assemblage of landforms that contains a unique history of landscape evolution.

Costa Rica also hosts a wide variety of climatic and ecological zones, ranging from humid tropical rainforests in the Caribbean and southern Pacific lowlands, with >4,000 mm/yr of rainfall (see McClearn et al., chapter 16 of this volume, and Gilbert et al., chapter 12), to the dry tropical vegetation of the northern Pacific coastal plains, with <1,500 mm/yr of highly seasonal precipitation (see Jiménez et al., chapter 9, and Janzen and Hallwachs, chapter 10 of this volume). Similarly, vegetation zones within mountainous regions range from the humid cloud forests in the highlands (Lawton et al., chapter 13, and Kappelle, chapter 14 of this volume) to the dwarf scrublands of the high-altitude *páramo* (Kappelle and Horn, chapter 15). Dramatic climatic and topographic gradients juxtapose 3,400–3,800 m peaks that were glaciated in the Pleistocene in proximity to humid lowland basins mantled by thick lateritic Oxisols. Topographic extremes coupled with variations in slope, wind direction, and orographic precipitation result in extraordinarily diverse microclimates, vegetation cover, and soil types within a single mountain range (Gómez 1986, Kappelle et al. 1995, Marshall 2007, and references therein).

The history contained in the geology and its relationship with the landscapes, climate, and vegetation of Costa Rica have long interested geographers, biologists, archaeologists, civil engineers, and of course, geologists. Systematic regional geological investigations in Costa Rica began in the late nineteenth century (Denyer and Alvarado 2000).

A synopsis of the geology of Costa Rica with particular emphasis on the most recent geological history, the Quaternary (the last 2.59 million years, or 2.59 Ma), and particularly the Holocene (last 11,500 years or 11.5 ka) is presented here. It has been written and illustrated in a way that is more accessible to the non-geologist. Nevertheless, in this chapter we cannot avoid complex terminology. Therefore and to facilitate reading we include a glossary for non-specialists in geology, which can be found at the end. Readers interested in learning more about specific geological topics, including lithology and geochronological periods, may want to consult original papers and references cited in this chapter (see also Denyer and Alvarado

[1] Área de Amenazas y Auscultación Sísmica y Volcánica, Instituto Costarricense de Electricidad (ICE), San José, Costa Rica

[2] Escuela Centroamericana de Geología, Centro de Investigaciones en Ciencias Geológicas (CICG), Universidad de Costa Rica (UCR), San Pedro de Montes de Oca, Costa Rica

[3] Biology Department, Florida Institute of Technology, Florida, USA

* Corresponding author

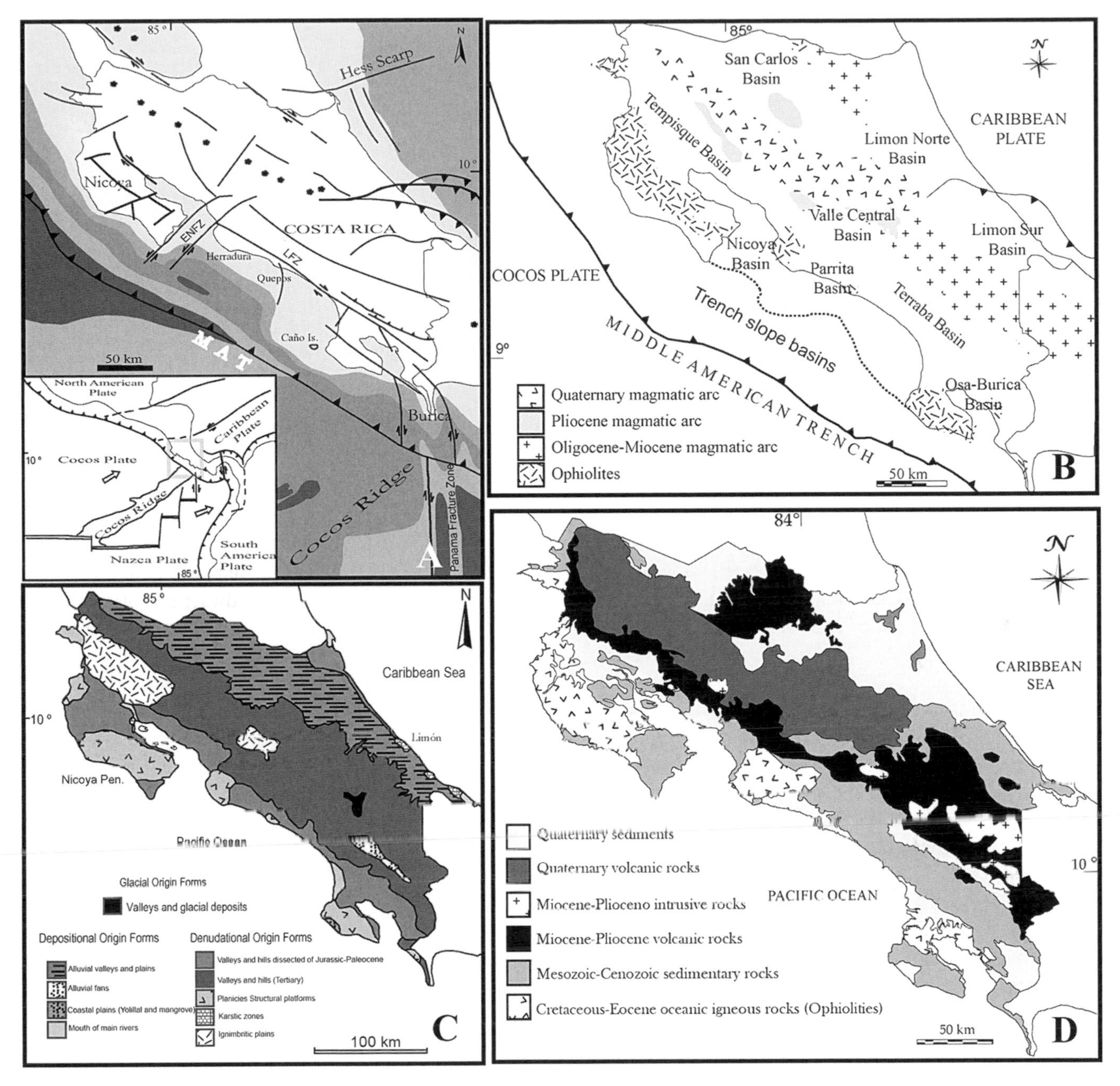

Fig. 3.1 Different simplified types of geological maps of Costa Rica: (a) major tectonic features, (b) marine basins, (c) geomorphologic map, (d) geological map.

2007, Bundschuh and Alvarado 2007, Alvarado and Gans 2007).

Tectonics

Costa Rica maintains a complex geology, directly related to the presence of three plates (Caribbean, Cocos, and Nazca, Fig. 3.1), the Panama microplate, and an uncertain number of tectonic *terranes* (exotic megablocks from a faraway source). The Costa Rican crust itself (about 40 km thick) is quasi-continental. This type of crust is not as thick, old, or crystalline as typical continental crust, but rather thickened compared to normal oceanic crusts. It shows an acidic evolution through time and Vp velocities similar to typical, true continental crust, indicating an increase in maturity of the arc (Pichler and Weyl 1975, Matumoto et al. 1977, Vogel et al. 2004, Lücke et al. 2010).

From the tectonic viewpoint, Costa Rica is located at the southern *terminus* of the Middle America Trench (MAT)

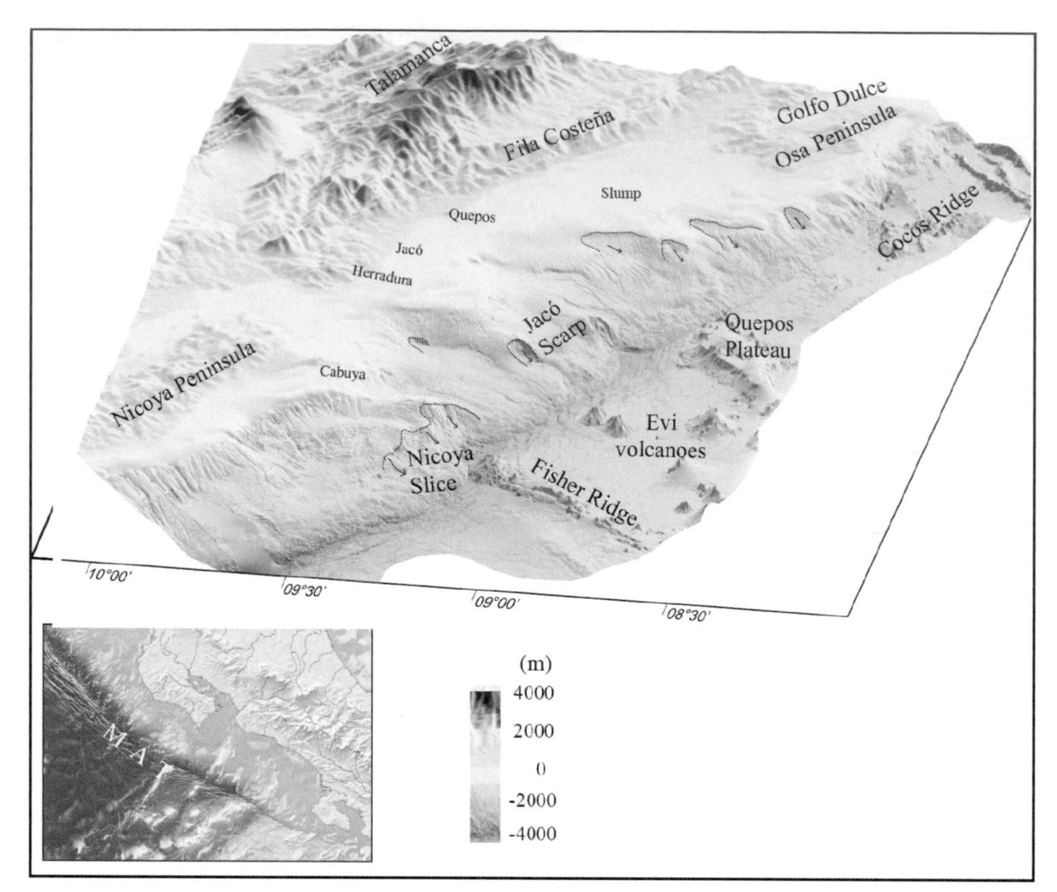

Fig. 3.2 Overview perspective of the Costa Rican continental margin morphology and adjacent land. Hydrosweep swath bathymetry off Costa Rica from SONNE cruises. Along the MAT it is possible to see the grooves or tracks, and multiple slide scars produced by subducted seamounts. Note the fault scarps in the subducted Cocos plate and at the mid- to upper-slope break of the margin. Note also the relationship between the Talamanca and Fila Costeña ranges and the subducted Cocos Ridge, a volcanic submarine range.
Figure courtesy of University of Kiel-SFB574-IFM-GEOMAR.

(Fig. 3.1a). The most prominent geotectonic features in the area are (a) the tectonic trench, a product of the subduction of the Cocos plate under the Caribbean plate; (b) subduction of a major topographic seafloor feature called the Cordillera del Coco (or Cocos Ridge, in English), a submarine volcanic range, extending along much of southern Costa Rica, where the trench apparently has been plugged by the anomalously buoyant oceanic crust; (c) the Panama Fracture Zone, which defines a triple junction between Cocos, Nazca, and Caribbean plates off the coast of southern Costa Rica; (d) Costa Rica's territory, formed by well-defined magmatic and sedimentary chains, its older oceanic igneous complexes (also known as ophiolites), the back-arc plains, and a passive marine margin, which has a listric fault system in the northeast, and the back-arc thrusting of the Caribbean sea floor under western Costa Rica (Figs. 3.1 and 3.2).

The northwestern part of the seafloor is relatively smooth and it underthrusts the continent without much disruption. The dip of the slab under the Caribbean plate is steep, more than 50°, with seismicity up to 200 km depth in a well-defined Wadati-Benioff zone (Protti et al. 1995). On the other hand, to the southeast, the seismically defined slab dips at about 60° and is only 80 km deep (Arroyo 2001).

The pattern of faulting in Neogene sediments in Costa Rica appears to reflect a predominantly horizontal shortening perpendicular to the trench. It is consistent with the width, indentation, and subsequent subduction of the submarine buoyant Cocos plate and the Cordillera del Coco (Montero 1994, Kolarsky et al. 1995), but is also due to the compressive stress generated by the Panama microplate or block (de Boer et al. 1995). The maximum horizontal compressive stress direction in Costa Rica tends towards N 10° E, and overthrusts are nearly perpendicular to that direction (Montero 1994).

Geology and Geomorphology by Region

In Costa Rica, morphotectonic units were first established by Weyl (1971) and subsequently by Mora (1981). Later, Weyl (1980) as well as Madrigal and Rojas (1980) expanded these geomorphological studies in Costa Rica with various levels of detail. More recently, several geomorphological topics were included in a volume edited by Bundschuh and Alvarado (2007). Here we present an up-to-date account of the main morphotectonic regions in Costa Rica, and include data from several recent surveys in marine geology (Fig. 3.1).

From the Abyssal Plain to the Continental Slope in the Marine Pacific Margin

The Abyssal Plain and the Middle America Subduction Trench
At the Costa Rica Pacific margin, the Cocos Plate is subducting beneath the Caribbean Plate at a rate of 78 mm/yr (Protti et al. 2012). On the oceanic plate, three domains can be clearly delineated using seafloor morphology (von Huene et al. 1985, 2000; see Fig. 3.2): (1) Northwest of Fisher Seamount; here, the oceanic crust is 22–23 Ma old, has a relatively smooth surface with normal faulting, and is presently being subducted opposite the Nicoya Peninsula, where a simple continental-slope morphology occurs. (2) The adjacent area to the southeast; its oceanic crust is approximately 13–19 Ma old and is covered for about 40% with seamounts (inactive volcanoes, some still with a crater) that range in size from 1 to 2 km high and 10 to 20 km wide. Here the continental slope is much more rugged and shows scars along the slope, which mark the places where seamounts have subducted in the past. (3) Further to the southeast, the Cordillera del Coco lifts up the margin at the Osa Peninsula.

According to available seismic reflection data, a 300–600 m thick pelagic sequence of calcareous and siliceous biogenic sediments enters the subduction zone. The presence of a sedimentary cover on the Cocos Plate has been verified several times by drilling and dredging (McIntosh et al. 1993, Vannucchi and Tobin 2000). Its sedimentary section can be subdivided into three major units, comprising from top to bottom: (a) diatomaceous ooze, (b) silty clay, and (c) calcareous ooze, of early Middle Miocene age (about 16.4 Ma) to Holocene.

The margin wedge is composed of igneous rocks covered by 1–2 km thick slope sediments (Ranero and von Huene 2000). The middle continental slope is characterized by a dominant subsidence from upper bathyal (200–600 m) and even nearshore limestone breccia of Middle Miocene (16.4 Ma) to its modern abyssal depths of more than 2,000 m (Vannucchi et al. 2003). Thick Pliocene and Pleistocene marine sediments are reported in the seismic reflection images and drilling of the margin, with highest sedimentation rates occuring in the Late Pleistocene (125–164 m/Ma) (von Huene et al. 1985, Kimura et al. 1997).

Detailed morphological investigations have revealed regions of sliding in the submarine Costa Rica margin (Fig. 3.2), among which the 50 km wide prehistoric Nicoya slide had a tsunamogenic potential, with an estimated maximum wave height above the slide of 27 m, among several other smaller slides like the Quepos and Jacó scarps, each of which had the potential to generate a 6–7 m high wave (von Huene et al. 2003). This highly tsunamogenic potential is the result of seamounts being transported on the rapidly subducting Cocos plate and have led to a characteristic pattern of upper Caribbean plate deformation. Each seamount created a broad furrow from the deformation front to its current position, which is clearly indicated by a circular uplift. The uplifted domes are associated seaward with landslides and steep carps. Seamount subduction is important for other geological processes such as earthquakes (Bilek et al. 2003) or subduction erosion (Ranero and von Huene 2000).

Bioturbation and fecal material are common throughout the sedimentary sequence; in fact, numerous fluid escape structures show episodes of rapid sedimentation (Vannucchi and Tobin 2000). Mud diapirs or mud volcanoes have also been detected at the Costa Rica margin between water depths of 1,850 and 3,100 m. Some have a diameter of approximately 800–1,700 m and are as much as 14–115 m high above the surrounding seafloor, displaying slope angles of 3–10°. These structures present acumulations of authigenic carbonates and show a typical abundance of cold vent chemosynthetic organisms: *Pogonophora* colonies and bivalve clusters (vesicomyid clams, solemyid and mytilid mussels). Scattered *Calyptogena* clams and shell debris were observed at 1,800 to 2,300 m water depth (Fig. 3.2 and 3.3). Futhermore, widespread bacterial mats were observed around 400 m water depth (Bohrmann et al. 2002, Mörz et al. 2004).

Gas hydrates (crystalline structures of water and methane molecules) are common and are mostly disseminated through pore spaces or as thin sheets of pure gas hydrates that fill microfractures. They can also be disseminated as massive nodules and have been detected as cement in ash layers. They are of biogenic origin and could be supplied from below the hydrate stability zone or could form in situ. The base of hydrate-bearing sediments off Costa Rica's coast is commonly found between 100–400 m below the seafloor (mbsf) (Pecher et al. 1998).

The trench along the northern Costa Rican margin is ~4,000–5,000 m deep while off the southern coast the trench is locally less 1,000 m deep. The buoyant Cordillera del Coco and large flanking seamounts—the southern reaches of the MAT—have been proposed as sites of accelerated subduction erosion, due to faulting and mass slides (Ranero and von Huene 2000, von Huene et al. 2000).

Subduction along the MAT produces most of the seismicity recorded in Costa Rica, including large earthquakes (M_w 7.0–7.7). On the basis of the changing nature of the Wadati-Benioff zone alongside the MAT, Protti et al. (1995) divided the western Costa Rican margin into northern (Nicoya), central, and southern (Osa) components. The recent events (i.e., the 1983 event of Golfito, M_w 7.4) exhibited a complex

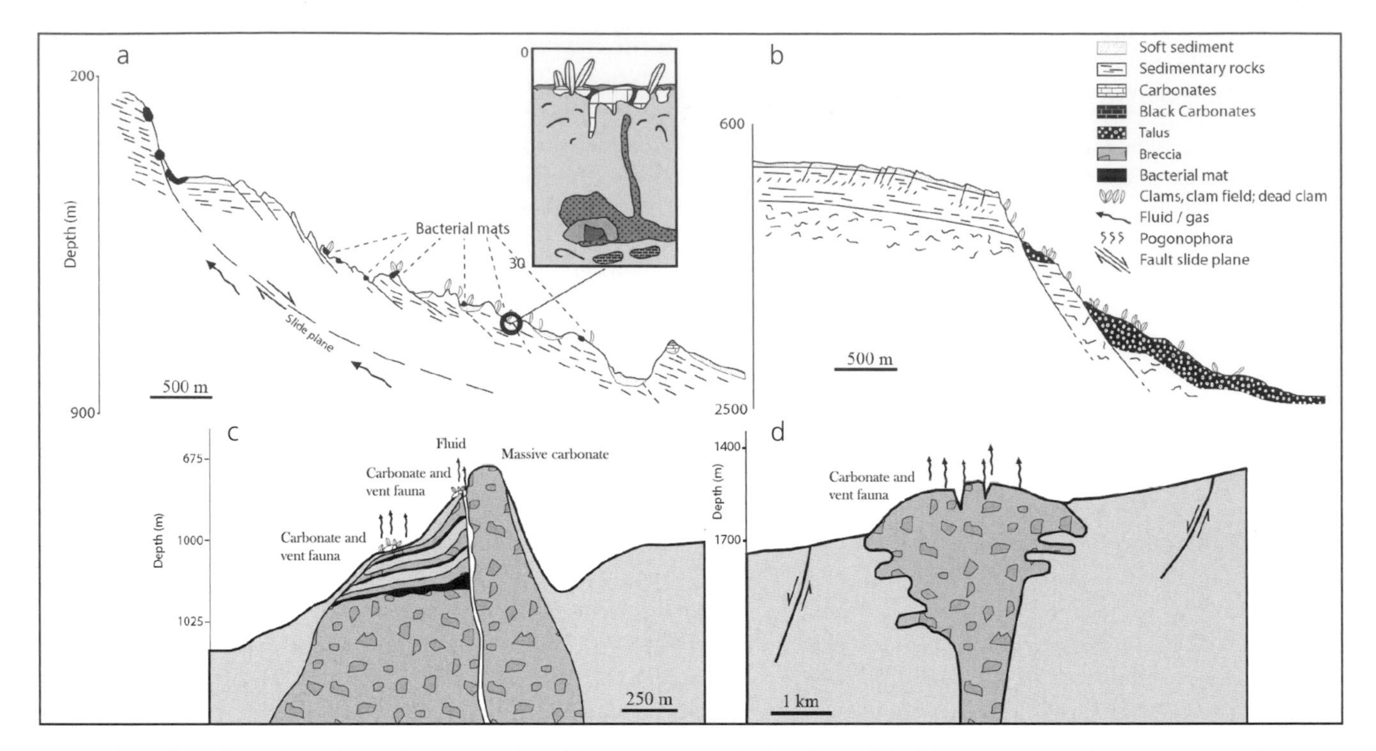

Fig. 3.3 (a) Seafloor observation and geological interpretation of the Quepos submarine landslide, and the injection of methane between 350 and 400 m water depth, (b) Geological interpretation of the uplift area above the subducting seamount over the erosive scar down to the bottom of Jacó Scar, (c) Schematic cross-section through the fault controlled the Mound 12 mud volcano, with failure of the seaward half, following by the formation of a basal carbonate and shell debris, subsequently covered by mud flows and hemi-pelagic sediments, (d) Schematic cross-section through the fault-controlled breccia cone of the diapir of Mound Culebra. Insert, sketch of carbonate-bearing core showing structures like the sandy conduit that are thought to provide fluids and methane for the above-vent community and authigenic carbonate production.
(a and b simplified after Bohrmann et al. 2002, c and d simplified after Mörz et al. 2004.)

rupture pattern indicative of increased plate coupling due to ridge subduction, and possibly reflecting the complex bathymetry (Tajima and Kikuchi 1995). Large earthquakes up to M_w 7.6 have occurred along the Osa segment with an estimated recurrence interval of about 40 years (Montero 1986).

The Nicoya Peninsula segment of the MAT has been recognized as a seismic gap (Montero 1986, Protti et al. 1995) with a potential for large (M_w 7.4–7.9) earthquakes over a recurrence interval of 50–80 years. In fact, on September 5, 2012, a strong earthquake (M_w 7.6) happened in front of Sámara beach on the Nicoya Peninsula (Linkimer et al. 2013). Additionally, off the coast of the Nicoya Peninsula large normal faults have formed along the outer rise.

The central segment of the MAT—historically the most seismically active region of the margin—has generated earthquakes up to M_w 7.0 over a short recurrence interval. This region is bounded to the south by the subducting flank of the submarine volcanic range called Cordillera del Coco (defined by the 2,000 m bathymetric contour) and to the north by the Fisher Ridge, the seamount chain, and the abrupt jog in the MAT (Protti et al. 1995).

Cordillera del Coco

The Cordillera del Coco, located on the Cocos Plate, extends from southern Costa Rica to the Galápagos Archipelago. It is the longest mountain range in Costa Rica, extending about 1,200 km as a submarine volcanic range (about 780 km in the marine territory) (Alvarado 2000). The section of the Cordillera del Coco closest to Costa Rica is a 180 km wide, shallow, bathymetric feature that rises 2–2.5 km above the ocean floor (Figs. 3.1 and 3.2). The oceanic crust is up to 25 km thick beneath its crest (Walther 2002). It is the product of hot spot volcanism at the Galápagos plume, which close to the Costa Rica margin has been dated between 13 to 14.5 Ma (Werner et al. 1999).

Outside of central Costa Rica, seamounts (mainly extinct volcanoes) cover about 40% of the seafloor and range between 0.1–2.5 km high and 10–20 km wide. North of the seamount domain, the smoother, older oceanic crust (ca. 25–30 Ma) is derived from both the East Pacific Rise and the Cocos-Nazca Spreading Center subducts under the Nicoya Peninsula (von Huene et al. 1985, Werner et al. 1999).

The Isla del Coco or Cocos Island (surface: 24 km^2),

located 496 km southwest of Cabo Blanco, is the only subaerial outcrop of the Cordillera del Coco. It is the result of an extinct basaltic shield volcano and trachytic domes (Castillo et al. 1988). Their ages lie between 2.2 and 1.5 Ma (Bellon et al. 1983, O'Connor et al. 2007) and are therefore much younger that the surrounding volcanic basement (Hey 1977). Thus, this volcano could be the product of either a hotspot (Burke and Wilson 1976), or a hot line (Alvarado et al. 1992), or an anomalous and extinct oceanic rift (Meschede et al. 1998), or it could even be the result of the reactivation of a larger Galápagos mantle plume (O'Connor et al. 2007; see the discussion in Rojas and Alvarado 2012).

Forearc "Ophiolitic" Promontories

The space between the trench—particularly at the continental slope—and the arc is characterized by principal peninsulas and promontories made up of Costa Rica's oldest rocks ("ophiolites" or "basic-ultrabasic oceanic igneous complexes" with ages between 200 and 40 Ma), and relatively recent sedimentary sequences (Fig. 3.1). All these conspicuous geomorphic features on the Central American land-bridge seem to be the result of the uplift that was caused by long-term subduction of submarine seamounts and volcanic ranges taking place at least since 40 Ma (Hoernle et al. 2004).

Uplifted terraces are very common along the Costa Rican Pacific coast, and show a northward decreasing rate of Quaternary uplift from 4.7–7 m/ka in the south to rates of less than 2 m/ka in the north (Marshall and Vannuchi 2003, Sak et al. 2003, Marshall 2007, Denyer et al. 2014).

Nicoya and Santa Elena Peninsulas

The Santa Elena Peninsula (containing ultramafic rocks) follows a linear E-W trending fault that bisects the Peninsula, forming several prominent bays. The persistant fault-bounded basement outcrops determine the morphology of the rugged shoreline extending southward from the Santa Elena Peninsula to the northern Nicoya Peninsula, as well as the steep cliffs formed of overlying Guanacaste ignimbrites. The rugged Pacific coastline of the Nicoya Peninsula features abundant pocket bays and sandy beaches, bounded by steep, rocky headlands. Uplifted marine terraces and paleobeach deposits indicate active emergence throughout the late Quaternary (Gardner et al. 2001, Marshall 2007). In contrast, the Peninsula's gulf coast follows a low-relief alluvial plain with extensive mangrove estuaries. The coastal piedmont along all sides of the Nicoya Peninsula rises steeply into a mountainous interior highland that reaches over 900 m in elevation.

Nowadays, the basement rocks exposed on the Nicoya Peninsula consist of an intensely deformed oceanic sequence of Cretaceous pillow basalts, intrusive rocks, and pelagic sediments. Along the margins of the Peninsula, a sequence of Upper Cretaceous to Quaternary marine sediments unconformably overlies the Nicoya Complex basement. These sediments include Cretaceous-Paleocene turbidites, Paleocene to Eocene deep- to shallow-water carbonates, Miocene shelf clastics, and a shallowing upward sequence of Plio-Pleistocene shelf sandstones and conglomerates (Jaccard et al. 2001, Denyer et al. 2014).

At Punta Descartes (an anticline) and adjacent bays (synclines), the west-northwest grain of the ridges and valleys is controlled by a system of parasitic folds on a broader anticline. Uplifted Holocene shore platforms, beach ridges, and stream terraces record active coastal emergence on the Descartes headland. Valley-fill terraces within a coastal embayment extend several kilometers inland reaching 15–20 m elevation. Streams draining towards the Bahía de Salinas are incised into these deposits, exposing an uplifted shallow bay to intertidal muds overlain by beach ridge sands and fluvial gravels. Radiocarbon dating of Holocene deposits indicates uplift rates of 2.0–3.5 m/ka (Marshall and Vannuchi 2003).

The Nicoya Peninsula, located only 60 km inboard of the Middle America trench, lies directly above the seismogenic zone. This unique location results in pronounced seismic cycle deformation, which is readily observed along the Peninsula's shorelines. Co-seismic uplift of >1 m affected the Nicoya Peninsula's central Pacific coastline during the M_w 7.7 subduction earthquake of 1950 (Marshall and Anderson 1995). The Nicoya Peninsula coastline has a particular morphology; the western part is an emerging coast, typified by an intercalation of cliffs and sandy beaches, and the eastern one is a submerging coast, along which there are well-developed mangroves (e.g., the Nicoya Gulf-Tempisque estuarine system; see Vargas, chapter 6 of this volume).

The continuing uplift has been recorded by a sequence of Quaternary marine and fluvial terraces at the Nicoya Peninsula. High-elevation remnants (400–1,000 m above sea level [m a.s.l.]) of a Pliocene-Pleistocene marine erosion surface (Cerro Azul surface) are preserved at the mountain block of the Peninsula. Deformation of this surface manifests a differential uplift across a series of mountain block faults. The lower elevation alluvial terrace (La Mansión surface) occupies interior river valleys at 4–10 m above local base level. Also, two laterally extensive, Holocene marine terraces are well developed along 40 km of nearly perpendicular coastlines at the tip of the Nicoya Peninsula, which extend nearly 1 km inland and are locally covered with up to 2 m of fossiliferous, intertidal sand and beach rocks (Marshall 1991, Gardner et al. 2001, Denyer et al. 2014).

Thus, several paleo-beach ridges and beach rock layers are evident in the topography of the continuous blanket of unconsolidated Holocene beach deposits that cover the wavecut platform from Cabuya to Montezuma. The upper Cabuya surface is a 1.5 m thick unconsolidated deposit of shell-rich beach sand and gravels overlying the paleoplatform at an elevation of 12.5–15.0 m above the highest high tide; the sand and gravel yielded calibrated radiocarbon ages of 4,190 ± 100 yr before present (BP) and 4,690 ± 220 yr BP, respectively. In the case of the lower Cabuya surface, the most prominent of these reaches a height of over 1.5 m, and is continuous for over 500 m. The calibrated radiocarbon ages varied from 2,330 ± 70 yr BP and 700 ± 90 yr BP, for elevations of 8.3 and 2.5 m above the highest high tide, respectively (Marshall 1991).

A small coastal bluff at the Cabuya-Montezuma coast, ranging in height from 0.5 m to over 6 m, is composed of high-energy, shell-rich, pebble-cobble beach deposits—an upward fining sequence of gravels and sand, overlain by alluvial gravels, sands, and muds. The estimated age of these deposits is ca. 650 yr BP. Similarly, the Isla Cabuya is surrounded at high tide by water, but as the tide falls, it exposes the modern intertidal platform and the island becomes connected to the mainland and can be reached by foot. The upper part of the platform has a 0.5 m thick unconsolidated beach deposit predominantly composed of marine shells and coral fragments of a ^{14}C calibrated age of 490 ± 60 yr BP. The uplifted wavecut platform beneath this deposit is about 2 m elevation above the present highest high tide (Marshall 1991).

This uplift and tilting of Holocene terraces to the southeastof the Nicoya Peninsula is occurring in response to subduction of seamounts along the projected trend of the Fisher seamount chain, onto the Pacific margin of the Caribbean plate (Fig. 3.4). Uplift rates decrease linearly from a maximum of about 6.0 m/ka near the tip of Cabo Blanco to less 1.0 ka/m along both coastlines (Gardner et al. 2001).

Turrubares Block and Quepos Promontory

The basement is composed of basic igneous rocks (basalts, gabbros, volcaniclastic sediments) of Cretaceous to Early Eocene age (ca. 65–45 Ma), exposed mainly in the Herradura block but also in the Quepos promontory (Hauff et al. 2000, Arias 2003).

The Herradura headland exhibits the highest topographic relief within the Chorotega forearc (>1,700 m). This fault-bounded block exposes Late Cretaceous oceanic basalts, which have been stripped of their sedimentary cover by rapid Quaternary uplift and erosion. The differential uplift between the Herradura block and adjacent lower-relief blocks is accommodated by dip slip along steep margin-perpendicular faults. Holocene river terraces and wavecut benches attest to rapid uplift along the Herradura headland.

The Turrubares-Quepos-Sierpe segment of the Middle America trench is a known source of large ($M_w \leq 7.0$) subduction earthquakes (Montero 1986, Protti et al. 1995).

Osa Peninsula and Punta Burica

The Osa Peninsula is a 62 km long, northwest trending, outer forearc, high inboard of the subducting submarine volcanic Cordillera del Coco. The Peninsula consists of a narrow coastal piedmont surrounding a northwest trending mountainous core that locally exceeds 700 m elevation.

The rocks cropping out along the Osa Peninsula are a Middle Eocene-Middle Miocene (45–15 Ma) *mélange* (mixture) dominated by basalt, cherts, and limestone resulting from accretion of seamounts. At the end of Miocene time, the subduction of a paleo-Cordillera del Coco caused uplift and severe tectonic erosion of the accretionary edifice allowing exhumation of the *mélange*. The *mélange* is overlain by a clastic sequence of Pliocene-Quaternary age, represented by conglomerates, mudstone ,and siltstone with fine-grained volcanoclastic turbiditic layers. Locally there are megabreccias formed by slumps and calcareous greywacke turbidities (Corrigan et al. 1990, Berrangé 1992, di Marco et al. 1995, Vannucchi et al. 2006).

Late Pleistocene fossiliferous marine sands unconformably overly beveled surfaces that cut across the competent rocks of the *mélange* basement. Nowadays, exposures of these rocks are found more than 75 m a.s.l., requiring uplift rates in excess of 6 m/ka. Furthermore, analysis suggests that the arrival of the blunt-tipped leading edge of the Cordillera del Coco likely resulted in a short-lived (ca. 42 ka) interval of an initial period of very rapid (ca. 30 m/ka) surface uplift (Sak et al. 2004).

The elongate Burica Peninsula juts southward into the Pacific Ocean forming a 25-km-long promontory at the Costa Rica-Panama border (Fig. 3.1 and 3.4). This emergent fragment of the Cretaceous-Paleogene oceanic basalts basement is overlaid unconformably by a Plio-Pleistocene sequence of marine sands, conglomerates, and turbidities beds. The relationships of facies and faunal assemblages indicate that the Pliocence subsidence was interrupted by a rapid Pleistocene uplift. The Pio-Pleistocene sediments exhibit significant folding and vertical displacement along a prominent north-trending fault valley that bisects the Peninsula. Uplifted wavecut platforms along the Osa Peninsula coast attest to ongoing deformation (Corrigan et al. 1990, Morell et al. 2011).

The Osa Peninsula segment of the Middle America

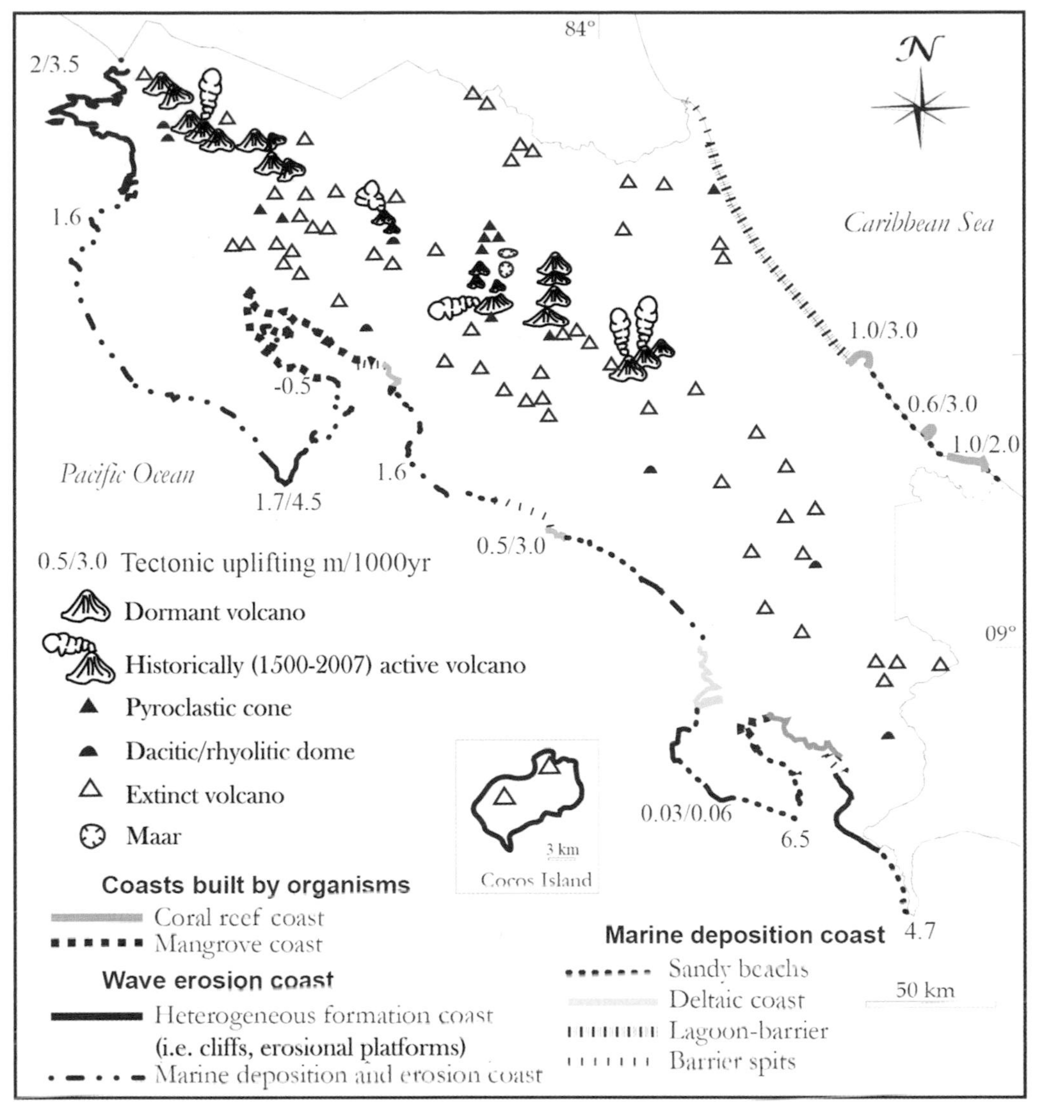

Fig. 3.4 Volcanic edifices and marine coast morphology of Costa Rica.
Volcanic edifices after Alvarado 2000, and marine coast morphology after Denyer and Cárdenes 2000.

trench is a known source of large ($M_w \geq 7.0$) subduction earthquakes associated with underthrusting of the buoyant Cordillera del Coco (Montero 1986, Tajima and Kikuchi 1995). The elongate N-S form of the Burica Peninsula is a consequence of the Panama Fracture Zone, also known as a source of large (M_w 7) strike-slip earthquakes. Several N-S faults in the southern inner forearc (Cordillera Costeña and San Vito plain) are the response of the subduction of the Panama Fracture Zone (Arroyo 2001, Morell et al. 2008, 2011).

The Golfo Dulce goldfield has been mined since pre-Columbian times and has produced at least twice as much gold as the entire Tilarán-Aguacate gold province in northern Costa Rica. The gold occurs in alluvial, colluvial, and beach placers (mainly gravels or conglomerates) of Pliocene to Holocene age, overlying the basic igneous complexes of Late Cretaceous to Eocene age (Berrangé 1992; Alvarado and Gans 2012).

Marine Basins between the Forearc and the Magmatic Arc

Tempisque, Nicoya, and Orotina Basins

At the Nicoya Peninsula, deep water sedimentary rocks (calciturbidites and mass flow deposits) of Late Cretaceous origin rest unconformably on basalts of the Nicoya Complex. Pelagic limestone clasts with planktonic foraminifera and abundant rudist fragments.

The Barra Honda limestone, of Late Paleocene-Lower Eocene age, represents remnants of a formerly continuous carbonate platform from a restricted to open marine environment (Jaccard et al. 2001). Barra Honda limestone

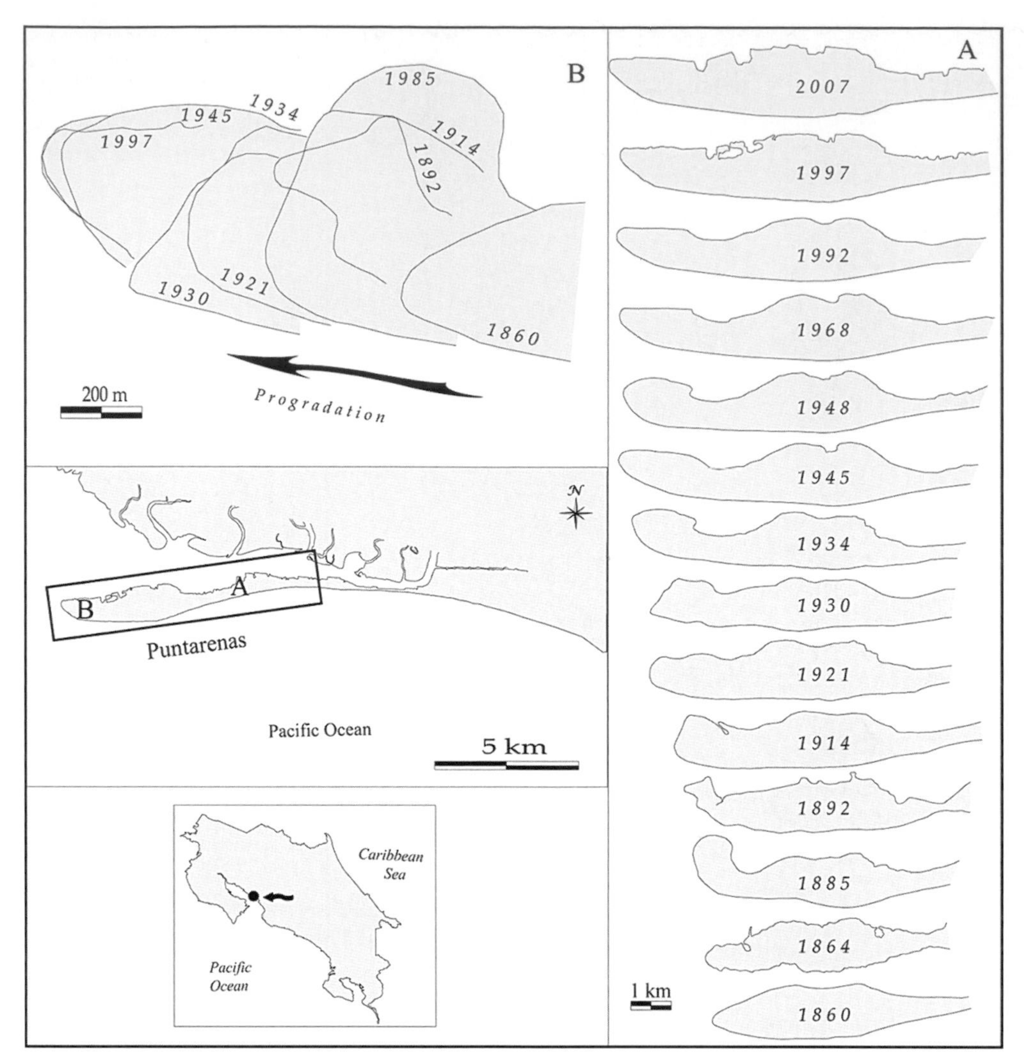

Fig. 3.5 Geomorphological evolution of Puntarenas sand bar. *Modified after Denyer et al. 2005.*

exposures in the hill flanks do not exceed 90 m in thickness; however, caves extend to depths of 200 m, suggesting the development of syn-sedimentary loading and deformation in the central part of the platform, where continuous deposition caused subsequent weight-subsidence in the center of the hills (Mora 1979). Although there are some extensive cave systems and peripheral springs, the karst landforms are restricted to sinkholes and dry valleys (Day 2007, Ulloa et al. 2011).

In the offshore region of the Nicoya Gulf there is about 3.5 km of Cretaceous-Quaternary sediment extending toward the northeast on the on-shore region, limited in the south by a shallow uplift of the basement. The sediments represent a typical shallowing, upward prograding succession of sediments that grade from pelagic sediments to slope and continental deposits; alluvial and lahar sediments represent the most recent ones (Barboza et al. 1995).

Along a stretch of 150 km of coastline south of the Nicoya Peninsula, major trunk rivers draining the inner forearc flow along a system of active, coast-orthogonal faults. These steep faults segment the inner forearc coastline into several fault-bounded blocks with sharply differing Quaternary uplift rates as determined from elevated marine and fluvial terraces (Fig. 3.1 and 3.4). The rivers follow fault-controlled valleys incised within Neogene-Quaternary nearshore sediments, volcaniclastic debris, and pyroclastic deposits. The Barranca, Jesús María, and Tárcoles faults form the boundaries of the Esparza and Orotina fault blocks (Marshall et al. 2003, Marshall 2007, Denyer et al. 2010).

The low-lying Orotina fault block between the Jesús María and Tárcoles rivers is covered by a >100 m thick Quaternary sequence of lahar-derived debris avalanche deposits, ignimbrites, and their fluvial gravels equivalents. During the early Quaternary (about 0.6 Ma), a series of

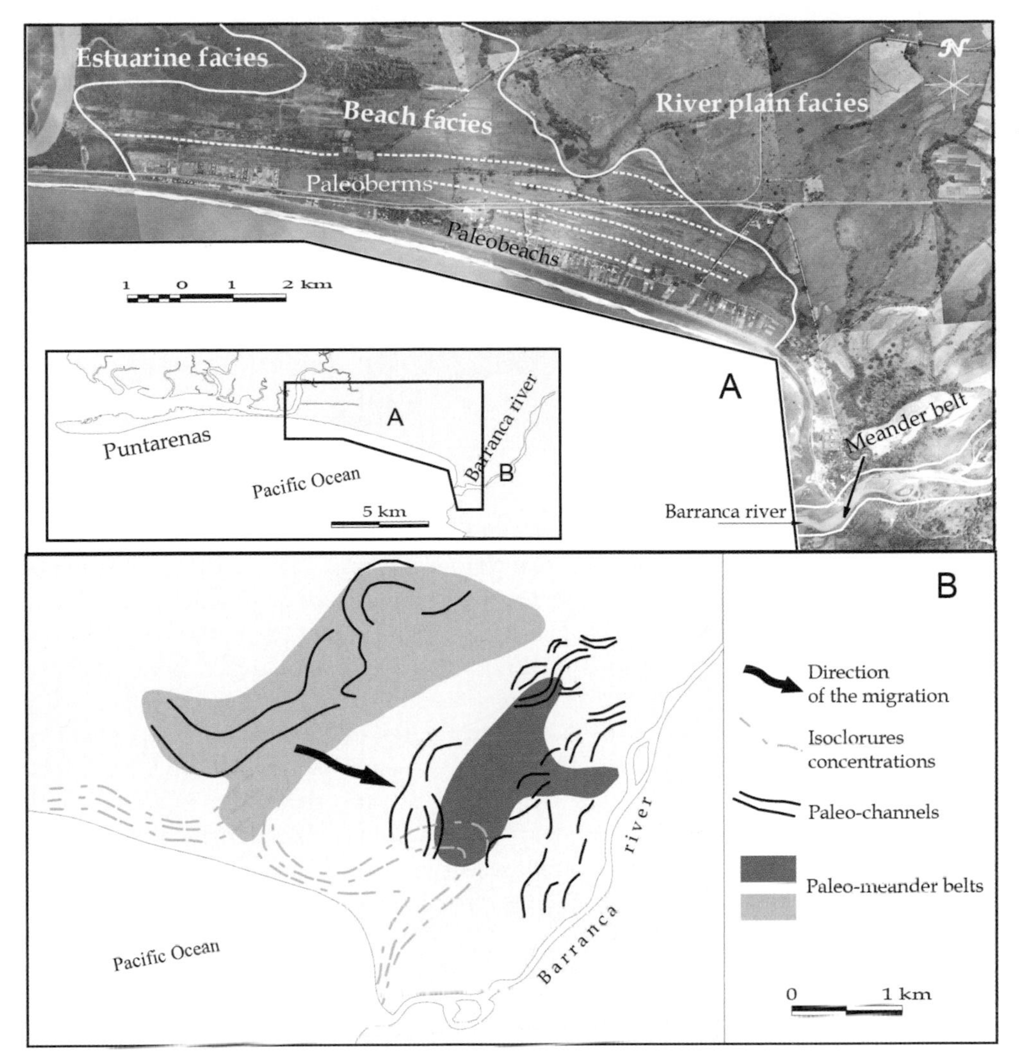

Fig. 3.6 (a) Distribution of sedimentary facies at part of Barranca river, (b) Migration of the main channel of the Barranca river during the last 40 years. *After Denyer et al. 2005.*

eruption-generated debris avalanches, diluted by water incorporation into lahars, descended from the volcanic front onto the coastal plain, forming the framework of a 25-km-wide debris fan (Orotina fan). Meandering paleochannels of the Tárcoles River are preserved across the fan surface as inverted topographic ridges of welded tuff overlying river gravels (Marshall et al. 2003, Denyer et al. 2010, Alvarado and Gans 2010).

Up to five late Quaternary alluvial fill terraces (10–260 m elevation) occur along the lower reaches of the fault-controlled Barranca and Tárcoles rivers. The total number of terraces, and the vertical spacing between them, varies along the coast with respect to the magnitude of local tectonic uplift rates. This relationship suggests that terrace generation along this coastline is strongly controlled by the interaction of rock uplift and eustatic sea level fluctuation (Marshall et al. 2000, 2003).

The Puntarenas sand spit (part of the Gulf of Nicoya-Tempisque estuarine system), which is the foundation of the capital of the province, appears in maps dating from the eighteenth century. It is 600 m wide, 7 km long, and has an average elevation of 3 m above sea level. Based on historical maps, aerial photographs (since 1860), and geophysical data, Denyer et al. (2004) concluded that Puntarenas is part of an estuarine system growing southward. The sand bar shows lateral growth, primarily at the end ("La Punta") where the lateral southwards growing rate is 14 m/year (prior to human control), so the origin of the spit could be extrapolated from about 500 years ago, and it is related to the sediment transportation for the marine currents from the Barranca estuarine system (see Fig. 3.5 and 3.6 for actualized analysis). Denyer et al. (2004) conclude that the origin of this sand bar is the NW to SE migration of the Barranca channel, driven by neotectonic activity on

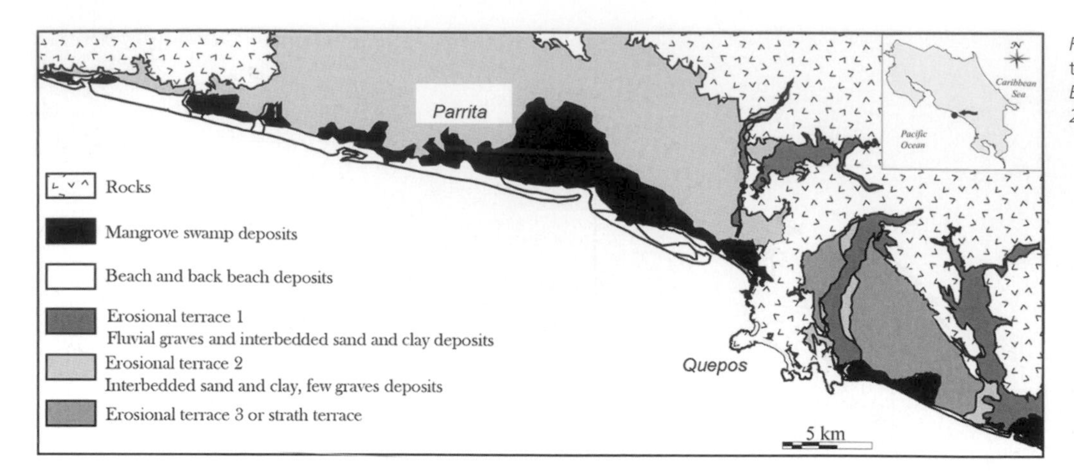

Fig. 3.7 Distribution of sedimentary deposits at Parrita-Quepos area. *Based on Drake 1989 and Cárdenes 2003.*

the Barranca fault. The distribution of recent sedimentary facies at Puntarenas documents the extraordinary preservation of extended paleo-beaches, flood plains, and estuary sediments, which could indicate that the estuarine system is actually prograding.

Parrita-Térraba-Golfo Dulce basins

The Late Cretaceous to Quaternary sequence in the Parrita, Térraba, and Golfo Dulce basins of southwestern Costa Rica includes more than 4.5 km of strata ranging from pelagic limestones and turbidites of Late Cretaceous to Paleocene rocks, shallow-marine carbonate sedimentation of Middle Eocene, deep water Oligocene to Miocene sedimentation (mudstone and sandstone), and shallow water to subaerial Miocene to Pliocene mudstone and sandstone, followed by Pliocene to Lower Quaternary volcaniclastic rocks and non-marine sands and gravels of Quaternary age. These basins are located off-shore and on-shore, limited in the south by a shallow uplift of the basement. The sedimentary rocks represent a typical shallowing, upward prograding succession of sediments that grade from pelagics to slope and continental deposits; and alluvial and deltaic paralic sediments representing the most recent ones (Coates et al. 1992, Corrigan et al. 1990, Barboza et al. 1995, di Marco et al. 1995). The Golfo Dulce is a tropical fjord, 215 m deep (Hebbeln and Cortés 2001, and see Cortés, chapter 5 of this volume).

Several recent sedimentary environments have been recognized at the central Pacific coast of Costa Rica. Alluvial deposits including alluvial plains, active and inactive meandering channels, active braided river channels, and colluvial and small lacustrine deposits represent a continental environment. The main rivers like Tárcoles, Parrita, Savegre, and others, which are meandering systems with flood plains limited at the north by the ranges, correspond with steep dip-slip fault zones. For this reason, the analysis of the morphological and sedimentological changes associated with the local tectonic setting suggests a SE inclination of the Parrita area. The alluvial system presents a meandering belt (Parrita river), which migrated SE to the actual alluvial plain. This abnormal situation would have been generated by the SE inclination of the area (Cárdenes 2003).

The drainage networks on the Parrita coastal plain are deflected around the Quepos highland and exhibit at least four late Quaternary terraces that attest to active uplifting (see Marshall 2007, and references therein). The oldest unit is composed of laminated, fine-grained red and grey clay containing foraminifera and poorly preserved bivalves. The deposit is mottled extensively with purple mottles in a blue-grey matrix, and a minimum thickness of 10 m, and typically forms low flat-topped hills with elevations of 25–35 m. The most recent units are composed of coarse-grained alluvium, highly weathered fluvial deposits, 2.5 to 30 m thick, distal fine-grained deposits, and buried soils, with local fluvial channels. From regional correlation and local ^{14}C dating, the ages of the terraces are estimated from less than 34,000 years old to about only 400 yrs BP. Average incision rates (0.5–3.0 m/ka) are estimated from the terrace sequence and apparently reflect regional uplift rates (Drake 1989) (Fig. 3.7).

The coastal environment presents a series of sandy beaches including gravel, sand bars, and tombolos. A common feature is the built-bar-estuaries system, which is colonized by mangrove vegetation (Fig. 3.8 and 3.9). The sand bars and beaches are the result of NW-SE littoral currents, and a mesotidal range. Additionally, there are erosional platforms made up of sedimentary rocks, which are affected by faults and synsedimentary deformation (Fig. 3.10). The most striking change at the coastal zone over the last 50 years is the Damas bar migration. These dramatic changes have caused the destruction of about 15 houses and one aquaculture business. The possible causes of

Fig. 3.8 (a) and (b) Bioerosion of sedimentary rocks at Punta Judas erosive platform, (c) Sandy beach at central Pacific coast, (d) Rocky beach at south Pacific coast, (e) Erosive platform at Dominical, (f) Erosive platform at Punta Judas (e and f at central Pacific coast).

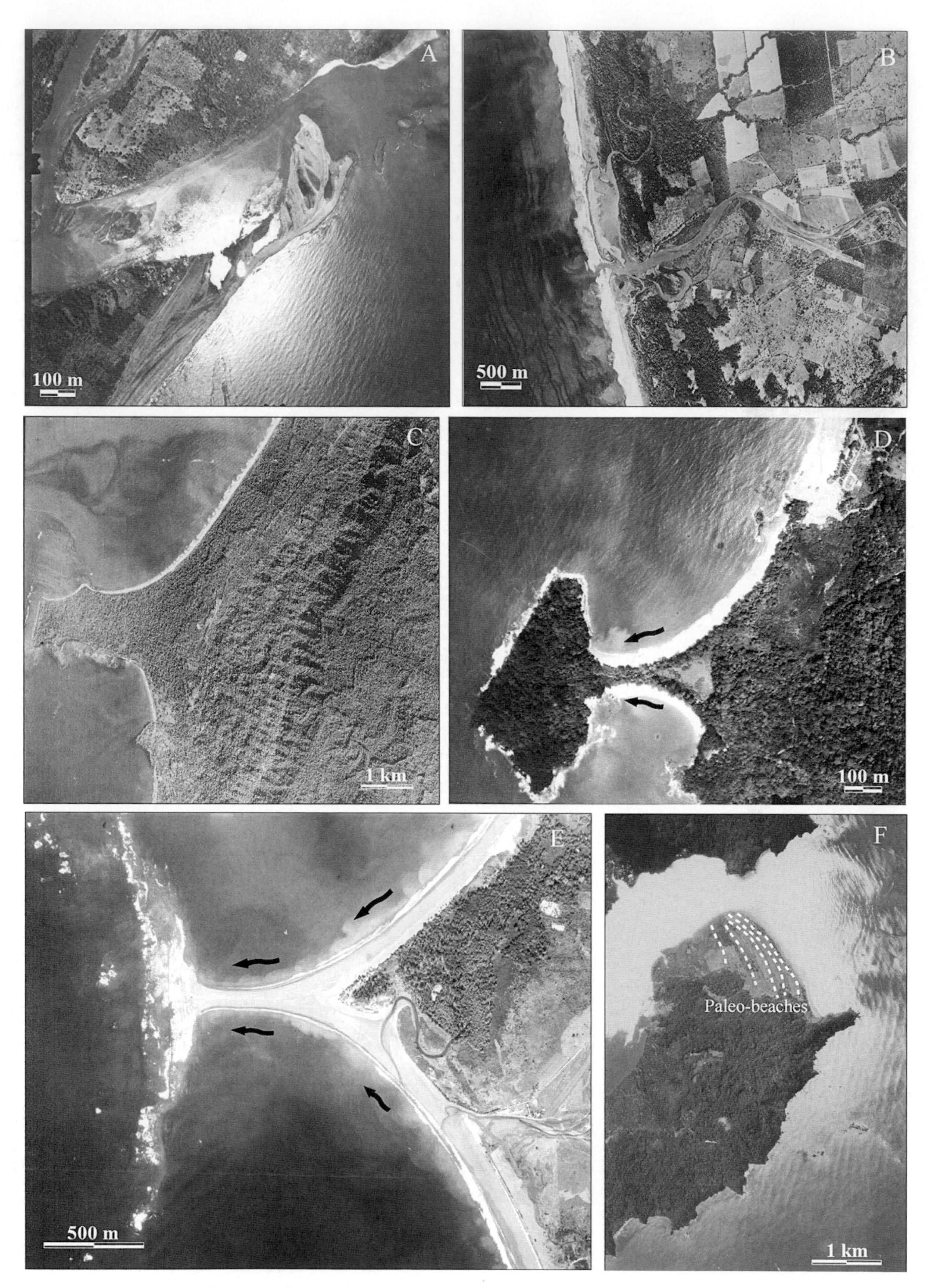

Fig. 3.9 Coastal geomorphology: (a) San Juan river deltaic system at north Caribbean coast, (b) Parrita river estuarine system at central Pacific coast, (c) Cahuita coral reef at Limón, (d) Punta Catedral tombolo, (e) Punta Uvita tombolo (d and e at central Pacific coast), (f) Paleo-beaches at Violines island at south Pacific coast.

Fig. 3.10 (a) Erosive platform at Dominical, (b) Marine erosion through a fault in the basaltic platform of Herradura, Playa Hermosa, (c) At least three different erosive levels in the marine platform due to tectonic uplift during the Quaternary.

this migration are a combination of local agents (e.g., tectonic setting), the increase of volume of water transported by the principal stream channels, and an increase of wave, tides, and currents' energy (Cárdenes 2003). Drake (1989) reports the presence of several terraces, which are the sedimentary response to the neotectonic activity of the area (Fig. 3.7 and 3.10). According to our observations, it is possible to count 2 to 5 raised paleo-beach ridges, which are locally visible as part of the topography characterized by a continuous blanket of unconsolidated Holocene beach deposits that occur from Savegre to the Playa Ballena beaches. One of these has been dated, using radiocarbon techniques, at 5,540 ± 70 yr (Fisher et al. 2004). Main geomorphic features are the natural tunnels and arcs that appear at the Ventanas beach, between the sandy beach area and the rock promontory that is affected by the intertidal area. These natural tunnels and arcs were produced by marine erosion occurring along tectonic faults and fractures.

Fig. 3.11 (a) Laguna Hule explosion caldera (maar) formed 6,200 years ago, (b) Cones of Barva andesitic shield volcano, (c) Turrialba volcanic graben with craters, (d) The twin cones of Arenal volcano in March 1987, (e) Cave in lahar created when a tree was putrefied after burial, (f) Cave in a lava flow.
(a) photo by G.E. Alvarado, (b) photo courtesy of Raúl Mora, (c) photo courtesy of Raúl Mora-Amador, (d) photo by G.E. Alvarado, (e) photo courtesy of Leonel Rojas, (f) photo by Leonel Rojas.

Quaternary marine sands, beach ridges, and alluvial gravels along the Osa Peninsula shorelines show high rates of tectonic uplift (6.5–2.1 m/ka) that decrease along an arc-ward trend from the Peninsula's interior, northeastward toward the Dulce Gulf. Along the northeastern coastal piedmont, a sequence of uplifted beach ridges yield radiocarbon ages ranging from <1 ka near the modern shoreline to >30 ka at an elevation of 25 m. Rivers draining the coastal piedmont exhibit two extensive Pleistocene gravel terraces that form a thick alluvial apron across the fault-bounded mountain front. These deposits overlie nearshore marine sediments dated at >30 ka. Two lower terraces with late Holocene radiocarbon ages occur adjacent to active channels attesting to continued uplift (see Marshall 2007 and references therein).

The Holocene development history of the fringing reef at Punta Islotes (Golfo Dulce) was reconstructed. It can be divided into four different stages: (a) an initial stage (5,500–4,000 yr BP), which includes the settling of *Pocillopora damicornis* and formation of a small fringing reef; (b) a reef establishment stage (4,000–1,500 yr BP), showing continuous growth of a branching, massive coral facies, and a drop in accumulation rates of the fore-reef talus facies; (c) a stage characterized by rapid vertical growth (1,500–500 yr BP), with accumulation rates of 5–8.3 m/1,000 yr, and growth of most of the reef's framework; and (d) a final stage (500 yr BP to present), in which accumulation rates decline, owing first to an increase in freshwater, and then to a presence of terrigenous sediments related to deforestation on adjacent shores (Cortés 1991, and see Cortés, chapter 5 of this volume).

Magmatic Arc

Truly speaking, the magmatic (volcanic and intrusive) chain in all of Central America is no arcuate form, but for convenience and tradition, it is classified as an island arc. The magmatic arc is the axis of Costa Rica formed by two active volcanic ranges, the Cordillera Central and Cordillera de Guanacaste, and by two extinct ranges, the Cordillera de Tilarán-Aguacate, and the Cordillera de Talamanca. The latter two ranges do not have huge stratovolcanoes because they are mostly extinct and deeply eroded. There is isolated and local evidence of relatively recent volcanic activity during the Quaternary, as shown by the huge Monteverde volcanic plateau (2.1–1.1 Ma), the Perdidos-Chato-Arenal volcanoes (0.1–0 Ma), and the volcanic domes in Talamanca (4–0.1 Ma). For details, see Alvarado (2000) and Alvarado and Gans (2012).

The major volcanic centers in the Quaternary ranges are more or less regularly spaced, 22 km apart, with one exception between Barva and Irazú, which is filled by the Zurquí hills. Volcán Arenal is also isolated, being only separated from the Cordillera de Guanacaste (Volcán Tenorio) and the Cordillera Central (Volcán Platanar) by recent volcanic gaps of 40 and 42 km each, respectively (see Figs. 3.1, 3.4, 3.8).

Cordillera de Guanacaste and Santa Rosa Ignimbrite Plateau

The Cordillera de Guanacaste is located in northern Costa Rica. Geologically, it comprises an 110 km long chain of four major stratovolcanic complexes (Orosí-Cacao, Rincón de la Vieja-Santa María, Miravalles-Zapote, and Tenorio-Montezuma) oriented NW-SE. Rincón de la Vieja is the only volcano active in historical time. The Quaternary volcanoes (basalts to andesites, with rare dacites) of Guanacaste grew about 0.6 Ma over a regional basaltic to dacitic basement of Late Miocene to Lower Pleistocene age. Two remnant calderas occur along the Cordillera de Guanacaste, the Alcántaro-Guachipelín and the Guayabo calderas. An extended fluvio-lacustrine sequence (ca. 100 km^2) with a strong volcanic influence (volcanic sandstones and siltstones) and diatomite deposits represent the sediments of an ancient lake that originated within the Alcántaro-Guachipelín caldera (Zamora et al. 2004). Other local diatomite deposits, interbedded with fluvial and ash flow deposits, are present around Montano and La Ese localities, representing a lacustrine sequence filling in the small local lake basins.

The oldest volcanic rocks exposed in this area consist of a pile of pyroclastic flow deposits (ignimbrites) with minor interbedded lava flows and terrigenous and paralic sediments of Upper Miocene (8 Ma) to Middle Pleistocene age, about 0.6 Ma (Alvarado et al. 1992, Carr et al. 2007, Alvarado and Gans 2012). These units form a broad plateau (2,000 km^2) that extends seaward from the base of the volcanic chain, consisting of ignimbrites emitted from old stratovolcanoes. They form a gently undulating plain that ends in an abrupt 100–150 m high escarpment near the modern Pacific coast. Rivers draining the volcanic range have incised deep valleys and canyons (i.e., Río Liberia) into the plateau. Owing to tectonic uplift since the Pliocene, several rivers near the Pacific coast drained to the Tempisque River instead of taking a more straighfoward path towards the ocean (Madrigal and Rojas 1980).

Cordillera de Tilarán and Montes del Aguacate

The extinct Cordillera de Tilarán and Montes del Aguacate (105 km long) consist of heavily dissected remnants of stratovolcanoes, andesitic shield volcanoes, and old volcanic calderas composed of Miocene-Lower Pleistocene basaltic to andesitic lavas, rare dacites, and volcaniclastic rocks (breccias, conglomerates, and tuffs). Hydrothermal alteration and deep tropical weathering have destabilized the steep slopes of these ranges, resulting in pervasive landsliding (Alvarado et al. 1992, 2007; Alvarado and Gans 2012).

Throughout the central Montes del Aguacate, an extinct volcanic range, deeply incised linear canyons have developed along active, northwest-and-northeast-trending faults of the Central Costa Rica deformed belt. The Río Grande de Tárcoles cuts a deep gorge through the Montes del Aguacate, connecting rivers of the Valle Central basin with the Pacific coastal plain to the southwest. Along the Tárcoles canyon and many of its tributaries, resistant ignimbrite deposits form level benches and isolated hilltops 50–100 m above the valley floor. On the basis of late Quaternary isotopic ages for the ignimbrites, the bedrock incision rates range is 0.1 to 0.5 mm/yr (Marshall et al. 2003).

A large former lake basin (>50 km^2) developed during the Middle Pleistocene in the Palmares and San Ramón area, and persists today as a depression within the Pliocene volcanic terrain. Within this basin is a thick sequence (up to 90 m) of lacustrine and fluvial sediments with beds of diatomites, pumiceous conglomerates, tuffaceous siltstones, and sandstones. The lake drained via the impressive canyon of the Río Grande that flows eastwards (Rojas 2013). Other diatomite deposits are also known to fill smaller lake basins in Turrúcares and Agua Caliente. Small lenticular (5–30 m thick) diatomite and fluvial outcrops are also present along the Santa Rosa and San José rivers, south of Líbano, and near Peñas Blancas (Mathers 1989).

The Los Perdidos domes (active about 90 ka), and the Chato (40–3.4 ka) and Arenal (7–0 ka) volcanoes (Alvarado 2000, Alvarado and Gans 2012), are isolated Upper Quaternary vents located on the northern slope of the Cordillera de Tilarán, showing the same trend as the other volcano alignments (Cordillera de Guanacaste and Central). Arenal was one of the 16 most active volcanoes in the world between 1968 and 2010.

Cordillera Central

The stratovolcanoes and complex andesitic shield volcanoes of the Cordillera Central form a 130 km-long volcanic chain including Platanar-Porvenir, Poás, Barva, Irazú and Turrialba volcanoes. Minor but important volcanic centers include Congo, Hule, Río Cuarto, and Cacho Negro, along with a dozen parasitic small pyroclastic cones (i.e., Sabana Redonda, Monte de la Cruz, Pasquí, and Armado cones). The Cordillera Central contains the largest volcanoes (elevations ranging from 2,000–3,400 m), in both area and volume, of the entire Central American volcanic front (Fig. 3.11). Their summits exhibit multiple craters and transverse alignments of parasitic cones, and collapsed scarps formed extensive volcanic slides that generated volcanic debris avalanche deposits. Poás, Irazú, and Turrialba have been active in historical time; Barva and Hule were active in Holocene time (Alvarado 2000). The Turrialba Volcano erupted several times between 2010 and 2015, particularly on March 12, 2015, causing explosions of gas and spreading gray ash across parts of Costa Rica, in Coronado, San José, Alajuela, and Heredia. The international airport Juan Santamaría had to be temporarily closed during the fall of Turrialba's volcanic ash.

A strong climatic gradient across the range results in greater weathering and erosion including fluvial canyons (i.e., La Vieja, Aguas Zarcas, Toro, Sarapiquí, Río Sucio, Toro Amarillo, etc.), waterfalls, and more frequent landslides on the humid Caribbean slope (Alvarado 2000, Marshall 2007). Along both flanks of the Cordillera Central, gravitational spreading of the volcanic massif generates prominent fault-propagation-fold scarps along the base of the mountains (Borgia et al. 1990).

Rare, small, non-carbonate caves are found on the volcanic range of Cordillera Central (Fig. 3.11). Some are related to marine erosion on the rock (lahar/debris avalanche deposits) cliffs (i.e., Guacalillo beach), or due to lateral fluvial erosion of lavas by mountain rivers (i.e., Toro Amarillo river), or volcanic caves in lava flows (i.e., Cervantes and Ángeles lava fields), or more strangely, due to natural putrefaction of large trees in lahar deposits (i.e., El General Hydroelectrical Plant).

Cordillera de Talamanca

The Cordillera de Talamanca (200 km long in Costa Rica) is conspicuous and unique within Central America. It is the highest mountain range in Costa Rica, with maximum elevations over 3,500 m, and generally above 2,000 m, and represents an uplifted inactive segment of the Central American Magmatic arc (Fig. 3.1). The topography of the Cordillera de Talamanca is asymmetric in cross-section. The SW flank is steep, whereas the NE slope is moderately inclined but with profound large scarps. It constitutes the magmatic axis in the southeast part of Costa Rica, and is composed of intrusive batholiths and stocks of quartz-diorites and monzonites. Subordinate granites and gabbros are pervasive intruding sedimentary and volcanic rocks. The sedimentary rocks are predominantly volcarenites, breccias, fossiliferous calcarenites, and sandy black shales.

Most of the intrusive rocks of Talamanca are Miocene

Fig. 3.12 (a–c) Glaciar morphology at the Chirripó National Park, (d) The Puente de Piedra, a natural bridge and arch of ignimbrite, Tacares, Grecia.

in age between 8 and 12 Ma. There are only a few intrusive rocks with questionable ages as old as the Early Miocene to Late Eocene (19–35 Ma) or as recent as the Pliocene age in Dota-Candelaria and Guacimal, in the Cordillera de Tilarán (23.5–6.3 Ma) (de Boer et al. 1995; Alvarado and Gans 2012). Dacites and andesites produced by the partial melting of hydrated oceanic crust (called adakites) erupted as the latest phase of magmatic activity at the location of previously voluminous calc-alkaline arc magmatism (Abrattis and Wörner 2001).

One of the most conspicuous characteristics of the Cordillera de Talamanca is the glacial and periglacial geomorphology, which is best expressed in the Chirripó National Park (Weyl 1971, Hastenrath 1973). The largest scale erosional landforms are U-shaped valleys excavated by Pleistocene glaciers in preexisting fluvial valleys, the cirques, horns, and minor features such as whalebacks, striated, grooved, and polished bedrock (Fig. 3.12). Lateral, terminal, and medial moraines (glacial and subglacial tills) are the most prominent depositional feature, but there are also fluvioglacial terraces and glacial lakes. The area of glaciers in the Cordillera de Talamanca was estimated to be about 35 km^2 in the Chirripó National Park, but there is an additional 14 km^2 in neighboring areas (Cerro de la Muerte, 5 km^2; Cerro Urán, 5 km^2; Kamuk, 2 km^2; and Cuericí, 2 km^2).

The uplift rate for the Talamanca has been estimated at 1.4 ± 0.5 km/my, with the beginning of exhumation at 3.5 Ma (Grafe et al. 2002). The thick accumulation (up to 2 km) and growth of alluvial fans in the Valle del General indicate uplift of the Cordillera de Talamanca during the glacial and interglacial time (Kesel 1983). The presence of striated boulders at lower elevations in the Valle de El General, and the rectangular shaped valleys some 1,000 m below clear glacial features (Protti 1996), are products of erosive debris flows instead of glacial erosion.

Intra-Magmatic Axis Basins

Arenal Depression

The Arenal lake with a WNW-ESE trend (ca. 80 x 5 km, 600 m elevation) is a complex tectonic depression limited by active strike slip and normal faults. It is situated between the Arenal volcano and the eroding remnants of Monteverde volcanism (2.1–1.1 Ma). Several earthquakes of intermediate magnitude ($5 < M_w < 6.5$) have affected the region (Montero 1986).

Valle Central

The elongate Valle Central of Costa Rica consists of an east–west trending (ca. 70 x 10 km) basin (600–1,500 m elevation) situated between the active volcanoes of the Cordillera Central and the eroding volcanic remnants of the Montes del Aguacate and Cordillera de Talamanca. Throughout the Quaternary, this highland basin filled with a thick accumulation (>1 km) of volcanic products (andesitic to dacitic lavas, pyroclastic rocks, lahar and debris avalanche deposits). The Valle Central consists of a low-relief upland surface with deeply incised river canyons (i.e., Río Virilla) cut into the underlying Quaternary volcanic rocks and less deeply into the Miocene sedimentary sequence.

The Quaternary migration of the magmatic front shifted volcanism northeastward to the Caribbean slope, creating a new topographic divide and forming the Valle Central basin (Marshall et al. 2003). The fault-controlled drainage networks of the Valle Central feed into the Tárcoles gorge, a prominent canyon cut through the eroded highlands of the Montes del Aguacate, which provides a link between the Valle Central rivers and the Pacific coastal plain downstream. During the Middle-Late Pleistocene, the Tárcoles river breached the Montes del Aguacate drainage system, leading to progressive capture and re-routing of Valle Central drainage networks toward the Pacific slope (Marshall et al. 2000, 2003).

Neotectonic and seismic data show that the Central Costa Rica Deformed Belt is a wide, diffuse, active fault system through the central part of the country (Montero 2001, Marshall et al. 2000). Paleoseismicity and historical studies include the occurrence of intermediate magnitude earthquakes ($M_w < 6.5$) and seismic swarms, related mostly to NW dextral strike slip faults and ENE to NE sinestral strike slip faults.

General and Coto Brus Valleys

The General and Coto Brus valleys occupy an elongate structural basin (100–1000 m elevation) that stretches over 90 km in length and 10 km in width and is located between the Pacific slopes of the Cordillera de Talamanca and the Cordillera Costeña. A series of broad alluvial fans, which coalesce along the foot of the Cordillera de Talamanca, form an extensive piedmont surface on the valley bottom. Tributaries of the General and Coto Brus rivers, which drain the Talamanca highlands, have deeply incised this fan complex, leaving a sequence of terrace remnants along canyon margins. These alluvial surfaces are distinguished from one another on the basis of geomorphic settings, sediment texture, and morphologic and chemical characteristics of the soils. The oldest geomorphic surfaces coincide with the extensive piedmont upland in the northwestern portion of the General valley. These well-drained upland surfaces exhibit dark-red, deeply weathered lateritic Oxisols. A series of lower fan surfaces with less-developed soils yield Late Pleistocene to Holocene radiocarbon ages. The youngest alluvial surfaces consist of low elevation agradational terraces in-set along river canyons and abandoned braided channel bars of the General and Coto Brus rivers (Mora 1979, Kesel 1983).

In the Río Corredor basin, in southeastern Costa Rica, karst has developed in limestones of Late Eocene age, outcropping the Fila de Cal (part of the Cordillera Costeña), particularly where surface drainage is directed onto it from overlying sandstones and siltstones. Karst features like dolines and dry valleys are well developed in the basin, particularly in fault locations, but are relatively young in age and maturity. Where surface streams encounter the clastic/limestone contacts or faults, they are captured via large insurgences that ultimately supply resurgence 100–200 m lower in the Quebrada Seca or Corredor river valleys. Caves are largely fault-controlled, and two main levels separated by an elevational difference of about 25 m may reflect rapid uplift during the Quaternary (see Day 2007, Ulloa et al. 2011, and references therein).

Thrust-Fold Deformation Belts

A complex tectonic system of reverse faults (thrust faults) and folds forms a range that in literature is referred to as the thrust-fold deformation belt. In Costa Rica this system is present at both the Pacific and Caribbean sides of the Talamanca range. Along the Pacific it forms part of the forearc system and is called the Cordillera Costeña and Fila Bustamante. On the Caribbean side it is called Baja Talamanca and forms part of the back-arc system. Both are grouped here in the thrust-fold deformation belts for its similar tectonic origin and structures (Fig. 3.1 and 3.2).

Cordillera Costeña and Fila Bustamente

At the forearc, the Cordillera Costeña is a 150-km-long and 15–25-km-wide mountain range with peak elevations of 1,100–1,400 m. It runs NW-SE parallel to the south Pacific coast. The Fila Bustamante is a morphologically less defined, 64-km-long and 25–35-km-wide mountain range with peak elevations of 1,400–2,500 m. Morphometric analysis of mountain front sinuosity and facet development indicate rapid uplift of the Cordillera Costeña in response to compressional tectonism (Weyl 1971, Wells et al. 1998, Fisher et al. 2004).

Major thrust faults generally bound the exterior and interior of the mountain front that imbricate the Eocene-Miocene forearc basin sequence of the Térraba basin. Another distinct strike slip fault (mainly in the eastern part) can be observed perpendicular to the Cordillera Costeña. Bedrock has a predominant NW strike and NE dip of 20°–45° (Mora 1979, Fisher et al. 2004). The sedimentary sequence includes Middle to Late Eocene bioclastic carbonate rocks and a transitional zone to the turbiditic sequence (mudstones and volcaniclastic rocks) of Early to Late Oligocene, indicating a deeper marine environment. The existence of a very local Pliocene marine mudstone (Kesel 1983) is one indicator that thrusting within the Térraba basin began after the Pliocene, sometime before 5.3 Ma. The sedimentary rocks are intruded by gabbroic and hypoabysal (dolerite, basalt, andesite) dikes of Middle Miocene and Pliocene age (McMillan et al. 2004).

The coalescence of the General and Coto Brus rivers forms the Río Grande de Térraba, which cuts the Cordillera Costeña, indicating an antecedent river cut during the uplift of the range (Henningsen 1966). Significant Quaternary deposits (mainly debris and hypoconcentrated flows) within the Cordillera Costeña are limited to four discontinuous fluvial terraces up to 100 m above river levels, but their ages are poorly constrained (Bullard 2002).

Baja Talamanca

The Panama block moves independently from both the Nazca and Caribbean Plates and is bounded to the north by subduction of the Caribbean Plate along a series of fold-and-thrust belts called the Northern Panama Thrust Belt. These faults currently propagate west toward the coast, generating thrust faulting and earthquakes. Some authors include the Baja Talamanca as part of the back-arc area.

Fossiliferous and non-fossiliferous sandstones and limestones are composed of coral fragments in a sandy matrix, indicating a shallow marine enviroment (littoral to sublittoral with patch reefs), in which growth was interrupted by the fast sedimentation and/or uplifting of alluvial delta fan deposits (Aguilar and Denyer 1994). In the back-arc, the faulting and the thrust and folds system shows a NW trend and near Turrialba a combining dextral strike slip and reverse faulting (Denyer et al. 2003).

Uplift was clearly observed during the April 22, 1991, Limón earthquake, reaching 4.46 m inland and between 0.5 and 1.85 m in the Caribbean coast (Denyer et al. 2003). As a consequence of the presence of coral material in the ancient uplifted platform, it was possible to study the Holocene tectonic history in the area. On the basis of ^{14}C calibrated ages and the current elevation, corrected with eustatic sea level curves, it was possible to determine an average uplift of about 1.75 m/ka (Denyer 2007).

Sediments and the marine currents' system at Gandoca-Manzanillo define two sedimentary environments. One is between Punta Uva and Manzanillo, where sediments were derived from the coral reefs and local geological formations (Miocene to Pliocene clastic sedimentary rocks), and the other lies between Punta Mona and Río Sixaola, where sediments arrive primarily from outside the area (Fig. 3.9).

Back-Arc Region

The back-arc region of Costa Rica is not completely a passive margin, because the Limón basin can be divided into northern and southern sub-basins, which are characterized by different structural settings. Firstly, very low magnitude historical earthquakes (typical of passive continental margin) and an extension tectonic regime (normal listric faults) dominate the North Limón basin and the Llanuras de Tortuguero. Secondly, the South Limón basin has a strong compressional regime with several large earthquakes, a topic that is treated in the previous Baja Talamanca section (Fig. 3.1).

The San Carlos-Caño Negro-Tortuguero Plain and Low Hills

The regional basement contains serpentinized peridotites, Albian siliceous pelagites, and Paleocene to Middle Eocene turbidites, covered by a thick Miocene to Pliocene volcanic sequence. Quaternary fine alluvial and palustrine deposits represent the recent deposits (Gazel et al. 2005, Alvarado et al. 2007, and references therein).

The lowlands encompass an extensive alluvial plain (usually less than 100 m in elevation) and a low volcanic relief (less than 300 m) that reaches 35–150 km seaward from the base of the Quaternary volcanic range. A series of major rivers draining the volcanic range traverse the alluvial lowlands, transporting a high sediment load for deposition across broad inland flood plains and a coalescing delta complex at the coast. A sequence of massive alluvial fans has developed along the foot of the volcanic range where the major rivers exit the mountain front. In many cases, modern rivers have incised below extensive upland fan surfaces comprising thick accumulations of Pleistocene fluvial gravels, capped by well-developed, deep-red, clay-rich soils.

At several locations, the low-relief landscape is interrupted by abrupt hills generated by Lower Miocene to Pliocene volcanism (the Sarapiquí arc, a former remnant volcanic arc), and isolated Quaternary basaltic volcanism. The volcanic rocks usually are deeply eroded and covered by deep-red, clay-rich soils (up to 50 m thick). In some geological and geomorphological maps these features are incorrectly mapped as alluvial terraces, instead of lithological terraces.

The north Caribbean coast of Costa Rica is receiving abundant fluvial discharge from the Colorado, San Juan, and Tortuguero rivers (Fig. 3.9), but despite abundant sediment discharge, the regional wave climate has been sufficient to inhibit delta-plain progradation into the Caribbean Sea. Instead, a relatively straight coastal plain, consisting of fluvial-deltaic sands capped by wet forest and palm swamp detritus, has formed along a passive continental margin (Parkinson et al. 1998). Thus, along the Tortuguero coast, the low-relief, sediment-laden shoreline traces a broad, continuous arc for over 120 km between the San Juan River in the north and the Limón headland in the south. This coastline consists of a 10–15 km wide band of prograding, shore-parallel, beach ridges that stretch along the margin of the vast alluvial plain. The lower reaches of rivers approaching the coastline are often deflected between the shore-parallel beach ridges, resulting in a coastal morphology of elongate lagoons and narrow barrier islands (Marshall 2007).

North Limón Basin

The North Limón basin on the Caribbean Sea has a wide and structurally homogeneous depocenter developed in a relative tectonic quiescent area, however, with normal listric faults, particularly active during the Quaternary (Brandes et al. 2007). Some small earthquakes ($M_w < 5$) are reported by the Parismina nest and along the Hess Scarp (Fernández et al. 1994).

South Limón Basin

Compressional tectonics have created a different structural history in the South Limón basin at the Caribbean Sea. It is defined by a shortening due to a heterogeneous pattern of accommodation space and sediment thickness, where a number of small depocenters have been active since the Middle Miocene and their location changed continually through time. The hinges are most obvious in the Pliocene-Pleistocene, where a succession of piggy-back basins evolved in response to off-shore activity of the Limón fold-and-thrust belt. This pattern of tectonics causes topographic breaks at the sea-floor that control the position of recent submarine channels (Barboza et al. 1995, Denyer et al. 2003, Brandes et al. 2007).

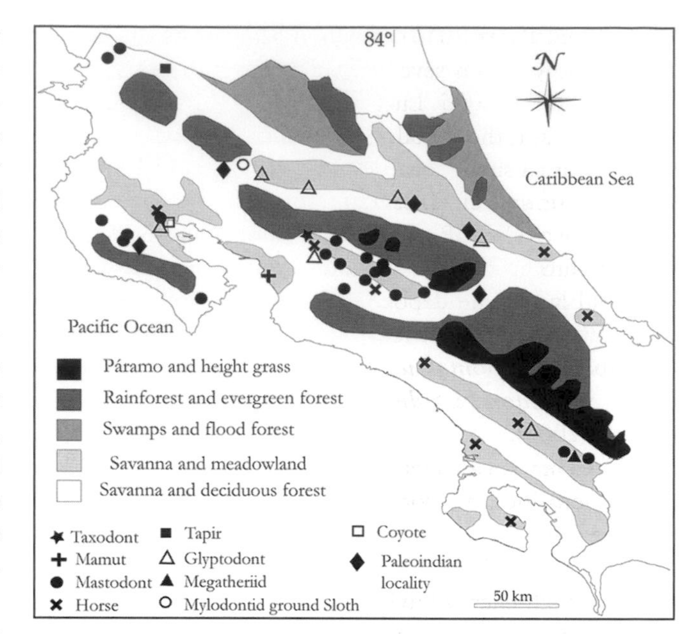

Fig. 3.13 Map of the Pleistocene fossil mega-mammal localities of Costa Rica, distribution of Paleoindian spear point, and distribution of the vegetation. No reconstruction of the coastline was done considering that the sea level varied at the end of the Last Glacial Event while tectonic uplift took place. Map based on and modified after Alvarado 1986, Gómez 1986, Lucas et al. 1997, Laurito et al. 2005.

Pleistocene Fossil Mammals

Fossil mammals are known from more than 45 localities of Pleistocene age in Costa Rica (Fig. 3.13). Most of these mammals are proboscideans referred to as the gomphothere *Cuvieronius hyodon* (Laurito 1988). One occurrence of *Mammuthus columbi* is known from Costa Rica (Lucas et al. 1997), being the southernmost record of *Mammuthus* in Central America (Alvarado 1994, Lucas and Alvarado 2010).

Less well documented are occurrences of a pampathere, megatheriid (*Eremotherium* sp.) and mylodontid (*Glossotherium* aff. *tropicorum*) ground sloths and glyptodonts (*Glyptotherium* aff. *texanum*, *Glyptotherium* cf. *arizonae*, *Pachyarmatherium*) (Gómez 1986, Lucas et al. 1997, Mead et al. 2006, Pérez 2013). *Equus* (i.e., *E. conversidens*) is poorly known from several localities (Gómez 1986, Laurito et al. 1993, 2005, Lucas et al. 1997, Valerio and Laurito 2004), and the toxodont *Mixotoxodon larensis* is well known from a single locality (Laurito 1993). *Canis latrans* and *Tapirus* sp., cf. *T. terrestris* is also recorded for the Pleistocene of Costa Rica (Lucas et al. 1997). A fossil record of the family Camelidae (*Palaeolama mirifica*) is described from old lacustrine deposits found near the Río Grande (Pérez 2013). A rodent fossil fauna is also described with four species: *Tylomys watsoni, Reithrodontomys mexicanus, Sigmodon hispidus*, and *Proechimys semispinosus* (Laurito 2003).

The Costa Rican Pleistocene fossil record is from numerous localities, but consists of one or a few taxa of large mammals and little research on small mammals at each site has been done. It suggests a probable bias towards preservation in high-energy fluvial deposits, alluvial, ignimbrites, and lahars, and a collecting and/or preservation bias toward fossils of large size. None of the Costa Rican Pleistocene mammals is directly associated with human artifacts or remains (Lucas et al. 1997) (Figs. 3.13 and 3.14).

Central America has acted as a filter to dispersal and/or as a center of evolution during the great American interchange, but there is no evidence of this in its Pleistocene mammal record. Most Pleistocene mammal taxa from Central America are of North American origin (leporids, felids, canids, gomphotheriids, mammutids, elephantids, tapirids, equids, tayassuids, camelids, cervids, and bovids); the remainder belongs to families of South American origin (dasypodids, glyptodontids, megalonychids, megatheriids, mylodontids, hydrochoerids, and toxodontids). If Central America acted as a filter, then that filter prevented strongly the dispersal of mammals from South America to North America than the reverse. No evidence of an endemic center of evolution is evident in the Central American record of Pleistocene mammals. The Pleistocene record of mammals from Costa Rica documented here reinforces these conclusions. Most Costa Rican Pleistocene taxa were of North American origin; the others are of South American origin, and there are no endemic taxa (Lucas et al. 1997, 2007) (Fig. 3.14).

The Late Pleistocene record of *Mammuthus columbi* may be part of a late Pleistocene maximum range of this

Fig. 3.14 A Pleistocene scene in the present San Carlos plains, based on some of the fossils found in Costa Rica. Horses, mastodon, and glyptodont dominate the landscape. A megatheriid and mylodontid were also present. In the foreground is the eastern part of the Nicaragua Lake during the Pleistocene, in today's Costa Rican territory (before drying and being restricted to the present Nicaraguan territory), as well as the active volcanoes of Guanacaste. The vegetation consists of savanna and sparse deciduous forest. The present coastline is represented by discontinuous lines.

species from North America far into Central America. The Early Pleistocene (or Late Pliocene) record at Bajo Barrantes consists only of taxa of South American origin (*Mixotoxodon larensis)* and may predate the arrival in Costa Rica of North American immigrants (Lucas et al. 1997).

Fossil Plants

In this section we will present a summary of the fossil plants that have been reported, and the locality where they were found.

Paleobotanical research in Costa Rica began in 1921, when William Berry did his study on "Tertiary fossil plants from Costa Rica." Berry described a Talamancan fossil flora found at Sixaola. The species he described are *Annona costaricana, Ficus talamanca, Heliconia* sp., *Nectandra aerolata, Nectandra woodringi, Phyllites costaricensis, Piperites cordatus, Piperites quinquecostatus*, and *Inga sheroliensis.*

Lohmann and Brinkman (1931) reported *Lithothamnion* and *Lithophorella* from the Fila de Cal Formation, at Las Animas, Turrialba. Also, Gómez (1973) reported calcareous fossils of Cryptonemiales: *Archaeolithophyllum, Goniolithon*, and *Lithothamnion.*

Gómez (1970) studied several fragments of rocks that have prints of fern-like Pteropsida. The samples were collected in Río General valley. The specimens were described as resembling the *Pecopteris* and *Mixonera* types of Paleozoic era and some actual species of *Thelypteris.*

In 1971, Gómez analysed some specimens found at the junction of Parrita and Candelaria rivers at the central Pacific coast. He reported a new species under the name *Palmacites berryanum*, which could be related to the modern genera of the *Palmae.*

The first fossil record of Bromeliaceae in Costa Rica was actually a new species, *Karatophyllum bromeliodes*. The specimen was found in "Tertiary" rocks in the central part of the country, at San Ramón de Alajuela (Gómez 1972). These rocks are probably Pliocene or Quaternary in age. Also, the fossil impressions of leaves of *Ficus padifolia* were found in diatomitic rocks in Guanacaste (Gómez 1974), and in 1978 Luis Diego Gómez reported *Equisetum* aff. *giganteum* in calacareous rocks at Navarro, Cartago.

During the 1980s and 1990s paleo-research focused especially on fossil pollen collected from soils, lake sediments, and peat bogs. This paleobotanical research provided important information on paleoclimatic conditions during the late Pleistocene and early Holocene (Hooghiemstra et al. 1992, Islebe and Hooghiemstra 1997, Islebe et al. 2005, Kappelle et al. 2005, Horn 2007). Palynological evidence from peat bogs near La Chonta close to El Empalme in the northwestern portion of the Cordillera de Talamanca demonstrated how plant species of mountain forests and *páramo* vegetation types moved up and down mountain slopes as a result of iterative warming and cooling that took place during the glacial-interglacial cycles of the Quaternary (Hooghiemstra et al. 1992). These studies reported a series of plant genera that were recorded as fossilized pollen in peat and clayey soils: *Alchornea, Alnus, Drimys, Grammitis, Hedyosmum, Hymenophyllum, Hypericum, Ilex, Jamesonia, Myrica, Podocarpus, Puya, Quercus, Viburnum,* and *Weinmannia.* Plant families present with unidentified genera that were also reported as fossils from this area include Apiaceae, Araliaceae, Asteraceae, Cyatheaceae, Ericaceae, Gentianaceae, Poaceae, Solanaceae, and Urticaceae (Hooghiemstra et al. 1992, Islebe and Hooghiemstra 1997, Islebe et al. 2005, Kappelle et al. 2005, Horn 2007). All these genera and families are still found in today's highland vegetation (e.g, Kappelle and Horn 2005; and see of this volume: Kappelle, chapter 14, and Kappelle and Horn, chapter 15).

The fossil plants record of Costa Rica is limited. The studies and reports on the fossil flora are very few and most of them concern palynological studies, which provide paleoclimatic information (see Islebe and Hooghiemstra 1997, Horn 2007).

There are a few other reports on the fossil flora of Costa Rica, especially on black shales, or from petrified wood. Most of these studies indicate the presence of leaves, wood, or other plant materials, but none of them report any classified specimen (see Obando 1986 and others). Woody species in the Valle del General suggest that the woodland there used to be a dry forest type, in which *Byrsonima*, a major component of most savannas (grasslands) since the Late Pleistocene, was once present (Kesel 1983). Later, Pérez (1998) reported in detail on the fossil plants found at La Palmera, Alajuela, the locality with the most diverse and abundant flora of Costa Rica. He found five families: Bromeliaceae, Araliaceae, Lauraceae, Piperaceae, and Moraceae.

Pérez and Laurito (2003) described twenty-one impressions of acorns of *Quercus corrugata*, from the La Palmera locality in San Carlos. They confirm an important decrease in temperature during the Pleistocene, which corresponds with the maximum glacial between 50,000 to 13,000 years BP. Pérez and Laurito (2003) also found and described *Juglans olanchana* for the first time in southern Central America. The material they studied at La Palmera was composed of two nuts and an impressed leaf of Pleistocene age.

Paleoenvironments and Climate Change during the Pleistocene and Holocene

The paleoenvironment and climate have been little addressed in geological studies on Costa Rica, and much of the existing information to date comprises puzzling and incidental fragments and isolated pieces of evidence that often lack a good geochronological control (see a summary in Gómez 1986). For a more complete, though still preliminary, summary of the last 100,000 years in Costa Rica and 36,000 years for the entire Central America region, we refer to Lachniet and Asmerom (2007) and Horn (2007).

Several periods in the last 100 ka were as wet as the Holocene (11–0 ka), with a minimum at ca. 60 ka, and several rainfall maxima between 100–65 ka and at around 37 ka, with a lesser extent during the Holocene (Lachniet and Asmerom 2007).

During Pleistocene glacial epochs, snow accumulated and glaciers formed on top of the Talamanca range (i.e., the peaks of Chirripó, Kámuk, and in some instances Cerro de la Muerte) and perhaps on the highest volcanoes as well (i.e., Irazú and Turrialba). The equilibrium line altitudes (ELAs) of permanent ice (the Pleistocene snowline) were about 3,500 m a.s.l. (Weyl 1971, Orvis and Horn 2000, Kappelle and Horn 2005). The lowest reconstructed extent of the glacier tongues at Chirripó was about 3,100 m high (Orvis and Horn 2000), with a maximum ice thickness of 150–175 m (Lachniet 2007). Pollen in the La Chonta bog indicates that the last glacier interval (ca. 50,000–15,600 yr BP) was 7–8° C cooler than present, and the treeless *páramo* extended down to 2,100 m elevation (Hooghiemstra et al. 1992, Islebe and Hooghiemstra 1997, Horn 2007).

About 16 ka after the Last Glacial Maximum, the climate was so dry that it permitted the formation of sand dunes at Islas Murciélago (Denyer et al. 2005), at the moment in which the ocean level was about 100–120 m lower than the present-day level (i.e., Pinter and Garner 1989). During the beginning of the last deglaciation (15,600–13,000 yr) the upper forest limit rose as high as 2,700–2,800 m, indicating a temperature increase of up to 4.6°C, and precipitation may have increased. The upper forest limit dropped 300–400 m during the Younger Dryas from 13,100–12,300 yr, indicating a temperature decline of 2–3°C. From 12,300–11,200 yr, the glaciers retreated above 3,500 m, and subalpine rain forest was gradually replaced by mountain rain forest as both the forest limit and temperatures rose toward present-day values (Hooghiemstra et al. 1992, Islebe and Hooghiemstra 1997, Horn 2007).

Also, at least a limited distribution of savanna has existed as is indicated by fossils of grazing mammals, such as horses and mammoths, during the Late Pleistocene (Lucas et al. 1997). Unfortunately, the Pleistocene record of mammals in Costa Rica is too diffuse and biased (see above) to allow many conclusions to be drawn from this rough temporal organization. Cleary, glyptodonts, *Cuvieronius*, and *Equus* were present throughout the Middle-Late Pleistocene, at altitudes at least up to 300, 1,200, and 1,850 m a.s.l., respectively, precisely in proposed savanna-like areas, and in correspondence with the isolated and rare occurrence of Paleo-Indian (about 13,500–9,000 yr BP) spear points found in Costa Rica (Fig. 3.12 and 3.13).

Analyses of pollen, diatoms, and microscopic and macroscopic charcoal in sediment cores from lakes and bogs in the Cordillera de Talamanca suggest wetter conditions during the early Holocene, especially between 7,700–4,800 yr BP, with minor changes in temperatures (Islebe and Hooghiemstra 1997, Horn 2007). A brief early-Holocene dry period (8,300–8,000 yr BP) is evident in a speleothem record from northern Costa Rica and may correspond with the high-latitude 8,200 yr BP cold event, but afterwards a stable, relatively wetter monsoon climate was established ca. 7,600 yr BP. In the late Holocene a trend toward distinctively drier climates began about 3,200 yr BP throughout much of the circum-Caribbean region, including Costa Rica (see Horn 2007).

Brief Summary of Costa Rica's Geological History

The early Caribbean basin began as a narrow seaway between the Pacific and Proto-Caribbean, when the last physical connection between North and South America ended about 170 Ma. The next intercontinental connection occurred ca. 75 Ma ago and lasted for about 4–6 Ma (Bonaparte 1984 a, b), when the proto-Antilles formed a land-bridge between the Americas in the current position of Central America. Another period of ephemeral land-bridge formation in the ancestral Antilles also occurred during the Paleocene (Alvarado 1994, Lucas and Alvarado 1994 and references therein). The emplacement of the oceanic plateaus (the Caribbean Large Igneous Province and other mafic igneous events around 69–139 Ma) between the Americas and the interaction of the Galápagos hotspot track with the Central American volcanic front (between 70 Ma to present), played a fundamental role in the formation of land bridges (ancestral Antilles) between the Americas during the Campanian, Paleocene, and Pliocene (Central America isthmus) (Hoernle et al. 2002, 2004, Alvarado and Gans 2012).

In the Late Cretaceous, an intraoceanic arc formed along

the western margin of the Caribbean Plate. This arc marked the beginning of the subduction of the Farallón Plate in the region. The present territory of Costa Rica emerged above sea-level several times, as a consequence of tectonic events, mostly related to subduction. Andesitic calc-alkaline clasts in submarine sediments give evidence that arc volcanism dates back to the Campanian (Kuijpers 1980).

During the Middle Eocene, the areas now occupied by the Tempisque, Nicoya, Parrita, and Térraba basins were part of a major forearc system that spanned along the Pacific margin of Costa Rica. The present structural configuration of the Parrita and Nicoya basins began in the Middle Eocene, related to shear stresses caused by the rotation in clockwise direction of the southern portion of the Costa Rican territory along a main transcurrent fault zone. Such movement displaced the Tempisque and Térraba in the south and north, respectively, developing the half-graben configuration and the system of strike-slip faults of the transtensive basins of Parrita and Nicoya (Barboza et al. 1995).

The northern part of Costa Rica emerged and was subjected to erosion during the Oligocene, whereas in the southern part, sedimentation continued. The modern tectonic configuration began during the Late Oligocene to Early Miocene (27–25 Ma), when the Farallón Plate broke up into the Cocos and Nazca plates separated by the newly formed Galápagos rift zone (Hey 1977, Wortel and Cloetingh 1981). Since the Miocene, shallow marine deposition has been widespread in all of Costa Rica.

There are few Oligocene intrusions in Costa Rica; it was not until the Miocene that magmatism began on a wide scale. Subduction of the young lithosphere 14–25 Ma (von Huene et al. 2000) could have caused migration of volcanism landward in central and north Costa Rica. The Late Oligocene-Miocene (29–8 Ma) was a period of high volcanic activity throughout Costa Rica (Alvarado et al. 1992, Alvarado and Gans 2012). In Costa Rica there has been a 30° counterclockwise rotation of the arc from its Middle Miocene position close to the modern volcanic front. This occurred from 15 to 8 Ma and is attributed to deformation in the overriding plate (shortening in the south coeval with extension in the NW), accompanied by a trench retreat in the north (McMillan et al. 2004) (Fig. 3.15).

Tectonic and magmatic activity during the Middle Miocene to Pliocene in Costa Rica was characterized by an uplift, intrusions along the inner arc, extensive volcanism, and folding with reverse and thrust faulting. Since 8 Ma the arc has been parallel to the modern volcanic front but progressively retreated to the northeast in Costa Rica. Adakite restrictive volcanism in Talamanca, represented by small volcanic domes (4.4–0.9 Ma), corresponds in space and time with the subduction of a large scarp associated with a tectonic boundary in southern Panama (McMillan et al. 2004).

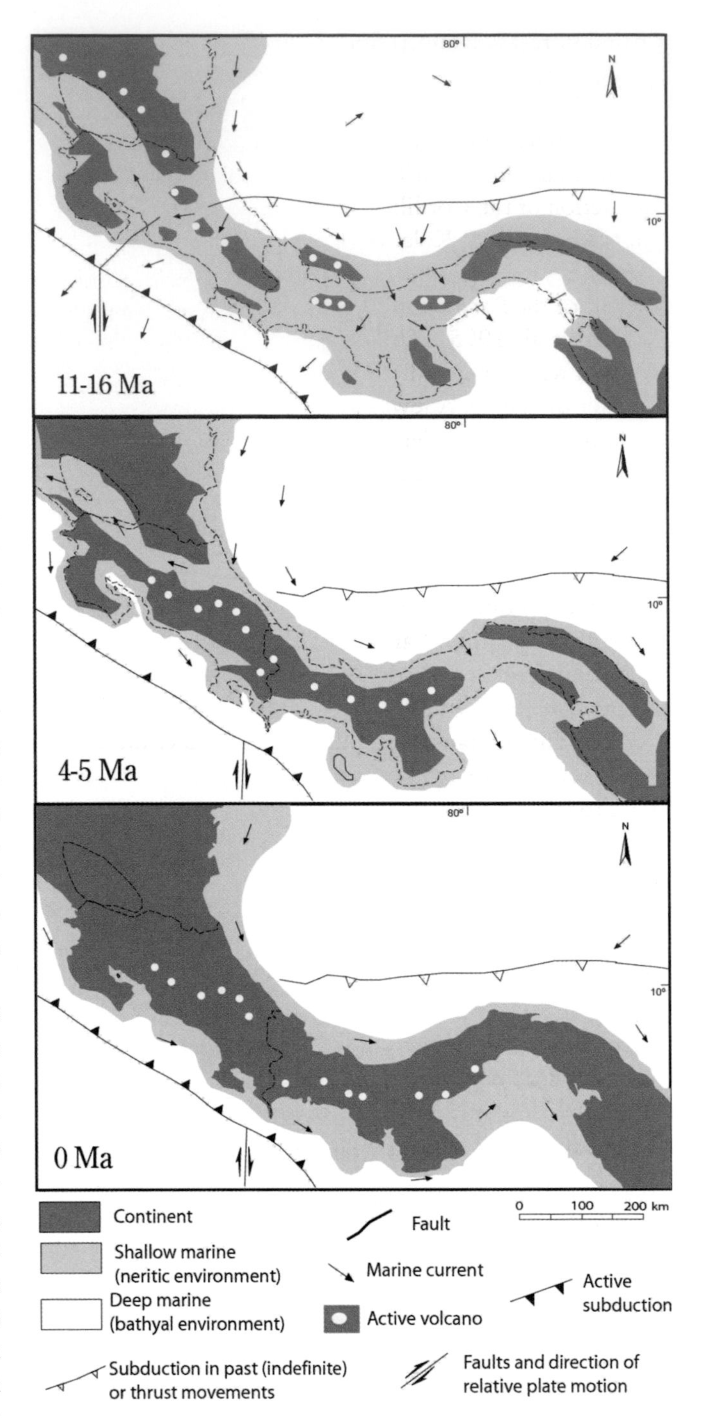

Fig. 3.15 Paleographic reconstruction of Costa Rica during the past 16 m.y. *Modified after Alvarado et al. 2007.*

Subduction of the Cordillera del Coco produced a shallow ridge indentation (Montero 1994), affected the southern part of Costa Rica by basement-rooted thrusting, back arc and intra-arc compression (Rivier 1985, Kolarsky et al. 1995), volcanic arc extinction (McGeary et al. 1985), and uplift of the costal zone (Cross and Pilger 1982, Madrigal and Rojas 1980; see also Marshall 2007 and references therein). The rise of the Talamanca range was caused by the subduction of the Cordillera del Coco beneath the volcanic front (Rivier 1985, Kolarsky et al. 1995), and the collision of the arc with the South American plate led to the development of the Panama microplate (Mann and Burke 1984, de Boer et al. 1995). The age of the Cordillera del Coco collision with the Middle America Trench is still, however, a matter of debate. A few authors claim the age of collision of a older Cordillera del Coco as early as Middle Miocene, on the basis of geological evidence (Rivier 1985), although the most accepted age at present is around 5–6.5 Ma (see Vannucchi et al. 2003, Alvarado and Gans 2012). Where the crustal shortening is larger in the Cordillera Costeña, the highest mountains of Talamanca (*cerros* Buenavista, Chirripó, and Kámuk) are present (Fig. 3.1 and 3.2).

During the Pliocene the Panamanian basins filled with marine and continental clastic sediments (fan and braided delta to alluvium fans) (Fig. 3.15). In the higher mountains, glaciers and snowcaps were present. Melting of the ice and snow together with volcanic activity in a tropical environment contributed to generate very thick lahars and alluvial deposits.

The Isthmus of Panama began to close 15.2–13 Ma ago (Duque-Caro 1990, Haug and Tiedemann 1998, Montes et al. 2015). In Costa Rica, the coast consisted of a shallow-marine embayment during the Neogene. Vertebrate paleontology strongly suggests that Central America became a peninsula of North America in the Miocene, as land vertebrates of North American affinity are known from Miocene fossils recovered from Guatemala to Panama. The arrival of South American mammals like Xenarthrans in southern Central America (today's Costa Rica and Panama) around 8.5 to 6.5 Ma resulted from their capacity to swim and hop from island to island at the time Panamanian territory was close to South America (Laurito and Valerio 2012, Montes et al. 2015). This Central American peninsula clearly became connected to South America circa 3.5 Ma ago (see Lucas 2014, and references therein). In fact, fossils indicate that the closure of the Isthmus of Panama was almost complete at 3.7–3.6 Ma (Coates et al. 1992). The final closure allowed a land mammal exchange at 2.7 Ma (Marshall 1988), coinciding with the glacial-induced sea-level fall during the peak of the northern hemisphere ice-sheet growth (Haug and Tiedemann 1998).

Most authors (i.e., Coates 1997) assume that the closure took place in the Panama-Costa Rica region ("Panamanian isthmus" [Fig. 3.15]), but Gartner et al. (1987) argue that the last interchange of waters between the tropical Atlantic and Pacific was across the isthmus of Tehuantepec, from the Gulf of Tehuantepec into the Gulf of Mexico (Fig. 3.14). The last significant change in the planktonic foraminiferal assemblage occurs around 1.9 Ma and is related to the last closure of the isthmus of Panama (Keller et al. 1989). The Panama land bridge permitted the great American biotic interchange of terrestrial animals (mastodons, saber-tooth cats, tapirs, ground sloths, armadillos, etc.) and plants between North and South America. In several cases it also worked as a biological filter (or biogeographic frontier) owing to differences in topography, vegetation, climate, and geological evolution along the isthmus (see Gómez 1986, Alvarado 1994, Lucas and Alvarado 1994, Kohlmann and Wilkinson 2007).

The beginning of rapid (0.55–0.6 km/Ma) marine subsidence of the Pacific margin of Costa Rica at 6.5–5 Ma is coincident with the arrival of the Cordillera del Coco crest at the Costa Rica trench (Vanucchi et al. 2003); the arc volcanism extended along the length of Costa Rica more or less coincident with the modern volcanic front (Alvarado and Gans 2012). The Pliocene subduction of the Cordillera del Coco about 5.4 Ma led to uplift and increased exhumation of the Cordillera de Talamanca in the southeast, such that stratovolcanoes are preserved and active to the north and south of the Cordillera de Talamanca, whereas in the area of major uplift they were eroded to deeper crustal levels. This exhumation was characterized by a sudden increase in uplift rates between 5.5 and 3.5 Ma (Gräfe et al. 2002).

There is a clear biogeographic frontier formed by the border between the Pacific dry forest (Mexican Pacific Coast Province) and the Pacific rain forest (Western Panamanian Isthmus Province) at the Río Grande de Tárcoles river, possibly related to the Talamanca cordillera (Kohlmann and Wilkinson 2007).

The glaciations in Costa Rica would have occurred when the Cordillera de Talamanca was uplifted sufficiently high (>3,100 m high) and some volcanoes grew above this altitude; the precise ages of the first glaciations that took place in the modern territory of Costa Rica are still unclear, but it is clear that the last glaciation affected the country between ca. 18 and 11 ka. At the end of the Pleistocene, an intensification of continental glaciation resulted in sea-level changes with a stronger amplitude (Haug and Tiedemann 1998). At Isla del Coco, submarine erosive arcs and platforms (90–110 m and 183 m depth) are probably the result of erosion that occurred during the last two glacial maxima (sea levels fell to about 130 to 160 m depth around 18 ka, and 130–155 m depth about 130 ka), besides slow subsidence events of Coco Island due to thermal cooling of the volcanic shield and oceanic crust (Rojas and Alvarado 2012).

Marine sediments deposited from 11,300 to 9,600 years

ago contain evidence of an apparent downslope migration of some mountain forest taxa, interpreted to reflect cooler climatic conditions during this period, especially since deglaciation about 10,000 yr BP (Horn 1993).

The second dramatic subsidence of the Pacific margin was very close to the Pliocene-Pleistocene transition (1.8 Ma; Vannucchi et al. 2003), just contemporaneous with the initiation of Monteverde volcanism (2.1–1.1 Ma old). It was followed by the present Middle-Late Quaternary volcanic activity, the Guanacaste and Central ranges, which grew mainly during the last 1 Ma, with major episodes of cone/shield building at 1.61–0.85 Ma (Proto-Cordillera), 0.74–0.20 Ma (Paleo-Cordillera), and 0.25 Ma (Neo-Cordillera) separated by 0.1–0.3 Ma intervals of dormancy (erosion) and/or explosive silicic volcanism (Alvarado and Gans 2012).

Finally, as a note of interest, both the oldest and the most recent volcanic rocks of Costa Rica are found in the magmatic arc. One of the oldest rocks found in Costa Rica corresponds to a very exotic rock, which is the case of sighting and recovering a meteorite in Costa Rica that happened on April 1, 1857, in Heredia. It consists of an H5 Chondrite that probably originated in asteroid 6 Hebe (Soto 2004). In opposition, the most recent rocks in Costa Rica were produced by persistent eruption (lava flows and pyroclastic rocks) at Arenal Volcano, active for 42 years, until the start of its new dormancy phase at the end of 2010, and more recently by the eruption at Turrialba Volcano in 2014–2015.

Acknowledgments

The authors thank David Szymanski, Thomal Vogel, and Roland von Huene for their valuable comments at the early preliminary draft of this chapter, and Maarten Kappelle for review of the final versions. Matthew Lachniet, Sally Horn, and Ken Orvis provided valuable comments on the glacial and paleoclimate aspects. Our studies have been supported by the Escuela Centroamericana de Geología, and Centro de Investigaciones en Ciencias Geológicas, both at the University of Costa Rica (UCR), and also by the Instituto Costarricense de Electricidad (ICE). Cristian Corrales, Maikol Rojas, and Krista Thiele helped with figures. We thank the project SFB-574 of the University of Kiel and the IFM-GEOMAR (Germany) for their continuous support to our research.

References

Abrattis, M., and G. Wörner. 2001. Ridge collision, slab-window formation, and the flux of Pacific asthenosphere into the Caribbean realm. *Geology* 29: 127–30.

Aguilar, T., and P. Denyer. 1994. Bioestratigrafía del parche arrecifal de la Quebrada Brazo Seco, Plio-Pleistoceno, Limón, Costa Rica. *Revista Geológica de América Central* 17: 55–66.

Alvarado, G.E. 1994. *Historia Natural Antigua: Los Intercambios Biológicos Interamericanos*. San José: Ed. Tecnológica de Costa Rica.

Alvarado, G.E. 2000. *Los Volcanes de Costa Rica: Geología, Historia y Riqueza Natural*. 2nd ed. San José: EUNED.

Alvarado, G.E., and P.B. Gans. 2012. Síntesis geocronológica del magmatismo, metamorfismo y metalogenia de Costa Rica, América Central. *Revista Geológica de América Central* 46: 7–122.

Alvarado, G.E., C. Dengo, U. Martens, J. Bundschuh, T. Aguilar, and S.B. Bonis. 2007. Stratigraphy and geologic history. In J. Bundschuh and G.E. Alvarado, eds., *Central America: Geology, Resources and Hazards*, 1: 345–94. London: Taylor and Francis.

Alvarado, G.E., S. Kussmaul, S. Chiesa, P.-Y. Gillot, G. Wörner, and C. Rundle. 1992. Cuadro cronoestratigráfico de las rocas ígneas de Costa Rica basado en dataciones radiométricas. *Journal of South American Earth Sciences* 6(3): 151–68.

Arias, O. 2003. Redefinición de la Formación Tulín (Maastrichtiano-Eoceno Inferior) del pacífico central de Costa Rica. *Revista Geológica de América Central* 28: 47–68.

Arroyo, I.G. 2001. Sismicidad y Neotectónica en la region de influencia del proyecto Boruca: hacia una major definición sismogénica del sureste de Costa Rica. San Pedro de Montes de Oca, Costa Rica: Escuela Centroamericana de Geología, University of Costa Rica. 162 pp. [Lic. thesis].

Barboza, G., J. Barrientos, and A. Astorga. 1995. Tectonic evolution and sequence stratigraphy of the Central Pacific margin of Costa Rica. *Revista Geológica de América Central* 18: 43–63.

Bellon, H., R. Sáenz, and J. Tournon. 1983. K/Ar radiometric ages of lavas from Cocos Island (Eastern Pacific). *Marine Geology* 54: M17-M23.

Berrangé, J.P. 1992. Gold in Costa Rica. *Mining Magazine* 402–7.

Berry, E.W. 1921. Tertiary fossil plants from Costa Rica. *Proceedings of the United States National Museum* 59: 169–85.

Bilek, S.L., S.Y. Schwartz, and H.R. de Shon. 2003. Control of seafloor roughness on earthquake rupture behaviour. *Geology* 31: 455–58.

Bohrmann, G., C. Jung, K. Heeschen, W. Weinrebe, B. Baranov, B. Cailleux, R. Heath, V. Huehnerbach, M. Hort, T. Kath, D. Masson, and I. Trummer. 2002. Widespread fluid expulsion along the seafloor of the Costa Rica convergent margin. *Terra Nova* 14: 69–79.

Bonaparte, J.F. 1984a. Nuevas pruebas de la conexión física entre Sudamérica y Norteamérica en el Cretácico Tardío (Campaniano). *Actas III Congreso Argentino de Paleontologia y Bioestratigrafía* 141–49.

Bonaparte, J.F. 1984b. El intercambio faunístico de vertebrados continentales entre América del Sur y del Norte a fines del Cretácico. *Mememorias III Congreso Latinoamericano de Palaontología, Oaxtepec* 438–50.

Borgia, A., J. Burr, W. Montero, L.D. Morales, and G. Alvarado. 1990.

Fault-propagation folds induced by gravitational failure and slumping of the Costa Rica volcanic range: implications for large terrestrial and Martian edifices. *Journal of Geophysical Research* 95(B9): 14357–82.

Brandes, C., A. Astorga, S. Back, and R. Littke. 2007. Fault controls on sediment distribution patterns, Limón Basin, Costa Rica. *Journal of Petroleum Geology* 20(1): 25–40.

Bullard, T.F. 2002. Geomorphic history and fluvial responses to active tectonics and climate change in an uplifted forearc region, Pacific coast of southern Costa Rica, Central America. *Zeitschrift für Geomorphologie* 129: 1–29.

Bundschuh, J., and G.E. Alvarado. 2007. *Central America: Geology, Resources and Hazards*. 2 vol. London: Taylor and Francis.

Burke, K.C., and J.T. Wilson. 1976. Hot spots on the Earth's surface. *Scientific American* 235: 46–57.

Cárdenes, G. 2003. Evolución de los sistemas sedimentarios costero y aluvial de la región de Parrita, Pacífico Central de Costa Rica. *Revista Geológica de América Central* 29: 69–76.

Carr, M.J., I. Saginor, G.E. Alvarado, L.L. Bolge, F.N. Lindsay, K. Milidakis, B.D. Turrin, M.D. Feigenson, and C.C. Swisher III. 2007. Element fluxes from the volcanic front of Nicaragua and Costa Rica. *Geochemistry Geophysics Geosystems* 8: Q06001. doi:10.1029/2006GC001396

Castillo, P., R. Batiza, D. Vanko, R. Malavassi, J. Barquero, and E. Fernández. 1988. Anomalously young volcanoes and old hot-spot traces, I: geology and petrology of Cocos Island. *Geological Society of America Bulletin* 100: 1400–1414.

Coates, A.G., ed. 1997. *Central America: A Natural and Cultural History*. New Haven: Yale University Press.

Coates, A.G., J.B.C. Jackson, L.S. Collins, T.M. Cronin, H.J. Dowsett, L.M. Bybell, P. Jung, and J.A. Obando. 1992. Closure of the Isthmus of Panama: the near-shore marine record of Costa Rica and western Panama. *Geological Society of America Bulletin* 104: 814–28.

Corrigan, J.D., P. Mann, and J.C. Ingle. 1990. Forearc response to subduction of the Cocos ridge, Panama-Costa Rica. *Geological Society of America Bulletin* 102: 628–52.

Cortés, J. 1991. Los arrecifes coralinos de Golfo Dulce, Costa Rica: aspectos geológicos. *Revista Geológica de América Central* 13: 15–24.

Cross, T.A., and R.H. Pilger. 1982. Controls of subduction geometry, location of magmatic arcs, and tectonics of arc and back-arc regions. *Geological Society of America Bulletin* 93: 545–62.

Day, M.J. 2007. Karst landscapes. In J. Bundschuh and G.E. Alvarado, eds., *Central America: Geology, Resources and Hazards*, 1: 154–70. London: Taylor and Francis.

de Boer, J.Z., M.S. Drummond, M. Bordelon, M.J. Defant, H. Bellon, and R.C. Maury. 1995. Cenozoic magmatic phases of the Costa Rica island arc (Cordillera de Talamanca). In P. Mann, ed., Geologic and tectonic development of the Caribbean plate boundary in Southern Central America. *Geological Society of America Special Papers* 295: 35–55.

Denyer, P. 2007. Tectonically controlled uplifting of the Caribbean coast of Costa Rica. Abstracts of 20th Colloquium on Latin American Earth Sciences, Kiel, Germany, April 11–13, pp. 99–100.

Denyer, P., and G.E. Alvarado. 2007. Mapa Geológico de Costa Rica (1: 400,000). Librería Francesa, San José.

Denyer, P., A.T. Aguilar, and G.E. Alvarado. 2000. Geología y estratigrafía de la hoja Barranca, Costa Rica. *Revista Geológica de América Central* 29: 105–25.

Denyer, P., and G.E. Alvarado. 2000. Desarrollo y evolución de la geología. In P. Denyer and S. Kussmaul, eds., *Geología de Costa Rica*, 471–92. Cartago: Ed. Tecnológica de Costa Rica.

Denyer, P., and G. Cárdenes. 2000. Costas marinas. In P. Denyer and S. Kussmaul, eds., *Geología de Costa Rica*, 185–218. Cartago: Ed. Tecnológica de Costa Rica.

Denyer, P., J. Cortés, and G. Cárdenes. 2005. Hallazgo de dunas fósiles de final del Pleistoceno en las islas Murciélago, Costa Rica. *Revista Geológica de América Central* 33: 29–44.

Denyer, P., S. Kruse, and G. Cárdenes. 2004. Registro histórico y evolución de la barra arenosa de Puntarenas, golfo de Nicoya, Costa Rica. *Revista Geológica de América Central* 31: 45–59.

Denyer, P., W. Montero, and G.E. Alvarado. 2003. *Atlas Tectónico de Costa Rica*. San José: Ed. Univ. Costa Rica.

Denyer, P., A. Teresita, and W. Montero. 2014. *Cartografía geológica de la península de Nicoya, Costa Rica: estratigrafía y tectónica*. San José: Ed. Univ. Costa Rica.

di Marco, J., P.O. Baumgartner, and J.E. Channel. 1995. Late Cretaceous-early Tertiary paleomagnetic data and a revised tectonostratigraphic subdivision of Costa Rica and western Panama. In P. Mann, ed., Geologic and tectonic development of the Caribbean plate boundary in Southern Central America. *Geological Society of America Special Papers* 295: 1–27.

Drake, P.G. 1989. Quaternary geology and tectonic geomorphology of the coastal piedmont and range, Puerto Quepos area, Costa Rica. University of New Mexico, p. 183. [M.Sc. thesis].

Duque-Caro, H. 1990. Neogene stratigraphy, paleoceanography and paleobiogeography in northwest South America and the evolution of the Panama Seaway. *Palaeogeography, Palaeoclimatology, Palaeoecology* 77: 203–34.

Fernández, J.A., G. Bottazzi, G. Barboza, and A. Astorga. 1994. Tectónica y estratigrafía de la Cuenca Limón Sur. *Revista Geológica de América Central*, vol. esp. *Terremoto de Limón*, 15–28.

Fisher, D.M., T.W.Gardner, P.B. Sak, J.D. Sanchez, K. Murphy, and P. Vannucchi. 2004. Active thrusting in the inner forearc of an erosive convergent margin, Pacific coast, Costa Rica. *Tectonics* 23. doi:10.1029/2002TC001464

Gardner, T., J. Marshall, D. Merritts, B. Bee, R. Burgette, E. Burton, J. Cooke, N. Kehrwald, and M. Protti. 2001. Holocene forearc block rotation in response to seamount subduction, southeastern Península de Nicoya, Costa Rica. *Geology* 29(2):151–54.

Gartner, S., J. Chow, and R.J. Stanton. 1987. Late Neogene paleooceanography of the eastern Caribbean, Gulf of Mexico and eastern equatorial Pacific. *Marine Micropaleontology* 12: 255–304.

Gazel, E., G.E. Alvarado, J. Obando, and A. Alfaro. 2005. Geología y evolución magmática del arco de Sarapiquí, Costa Rica. *Revista Geológica de América Central* 32: 13–31.

Gómez, L.D. 1970. A first report of fossil fern-like Pteropsida from Costa Rica. *Revista de Biologia Tropical* 16(2): 255–58.

Gómez, L.D. 1971. *Palmacites berryanum*, a new palm fossil from the Costa Rican Tertiary. *Revista de Biologia Tropical* 19(1,2): 121–32.

Gómez, L.D. 1972. *Karatophyllum bromelioides* (Bromeliaceae), nov. gen. et sp., del Terciario Medio de Costa Rica. *Revista de Biologia Tropical* 20(2): 221–29.

Gómez, L.D. 1973. *Criptonemiales* calcáreas fósiles en las calizas terciarias de Patarrá, Costa Rica. *Revista de Biologia Tropical* 21(1): 107–10.

Gómez, L.D. 1974. *Ficus padifolia* H.B.K. en la diatomite pliocena/pleistocena de la Formación Bagaces, Guanacaste, Costa Rica. *Veröff. Uberseemuseum Bremen, Reihe A* 4(15): 141–48.

Gómez, L.D. 1978. Preliminary note on a fossil *Equisetum* from Costa Rica. *Fern Gazette (London)* 11(6): 401–4.

Gómez, L.D. 1986. *Vegetación de Costa Rica*. San José: EUNED.

Gräfe, K., W. Frisch, I.M. Villa, and M. Meschede. 2002. Geodynamic evolution of southern Costa Rica related to low-angle subduction of the Cocos Ridge: constraints from thermochronology. *Tectonophysics* 348: 187–204.

Hastenrath, S. 1973. On the Pleistocene glaciation of the Cordillera de Talamanca, Costa Rica. *Zeitschrift fur Gletscherkunde und Glazial* 9: 105–12.

Hauff, F., K. Hoernle, P. van der Bogaard, G. Alvarado, and D. Garbe-Schönberg. 2000. Age and geochemistry of basaltic complexes in western Costa Rica: contributions to the geotectonic evolution of Central America. *Geochemistry, Geophysics, Geosystems* 1: 5. doi:10.1029/1999GC000020

Haug, G.H., and R. Tiedemann. 1998. Effect of the formation of the Isthmus of Panama on Atlantic Ocean thermohaline circulation. *Nature* 393: 673–76.

Hebbeln, D., and J. Cortés. 2001. Sedimentation in a tropical fjord: Golfo Dulce, Costa Rica. *Geo-Marine Letters* 20: 142–48.

Henningsen, D. 1966. Notes on stratigraphy and paleontology of Upper Cretaceous and Tertiary sediments in Southern Costa Rica. *Bulletin of the American Association of Petroleum Geologists* 50: 562–66.

Hey, R. 1977. Tectonic evolution of the Cocos-Nazca spreading center. *Geological Society of America Bulletin* 88: 1404–20.

Hoernle, K., P. van der Bogaard, R. Werner, B. Lissina, F. Hauff, G. Alvarado, and D. Garbe-Schönberg. 2002. Missing history (16–71 Ma) of the Galápagos hotspot: implications for the tectonic and biological evolution of the Americas. *Geology* 30: 795–98.

Hoernle, K., F. Hauff, and P. van der Bogaard. 2004. A 70 m.y. history (139–69 Ma) for the Caribbean large igneous province. *Geology* 32: 697–700.

Hooghiemstra, H., A.M. Cleef, G. Noldus, and M. Kappelle. 1992. Upper Quaternary vegetation dynamics and palaeoclimatology of the La Chonta bog area (Cordillera de Talamanca, Costa Rica). *Journal of Quaternary Science* 7(3): 205–225.

Horn, S.P. 1993. Postglacial vegetation and fire history in the Chirripó Páramo of Costa Rica. *Quaternary Research* 40: 107–16.

Horn, S.P. 2007. Late Quaternary lake and swamp sediments: recorders of climate and environment. In J. Bundschuh and G.E. Alvarado, eds., *Central America: Geology, Resources and Hazards*, 423–41. London: Taylor and Francis.

Islebe, G.A., and H. Hooghiemstra. 1997. Vegetation and climate history of montane Costa Rica since the Last Glacial. *Quaternary Science Reviews* 16: 589–604.

Islebe, G.A., H. Hooghiemstra, and R. van 't Veer. 2005. Historia holocénica de la vegetación y del nivel de agua en dos turberas de la cordillera de Talamanca, Costa Rica. In M. Kappelle and S.P. Horn, eds., *Páramos de Costa* Rica, 237–52. Santo Domingo de Heredia, Costa Rica: Ed. Instituto Nacional de Biodiversidad, INBio.

Jaccard, S., M. Münster, P.O. Baumgartner, C. Baumgartner-Mora, and P. Denyer. 2001. Barra Honda (Upper Paleocene-Lower Eocene) and El Viejo (Campanian-Maastrichtian) carbonate platforms in the Tempisque area (Guanacaste, Costa Rica). *Revista Geológica de América Central* 24: 9–28.

Kappelle, M., ed. 2006. Ecology and Conservation of Neotropical Montane Oak Forests. *Ecological Studies Series*, Vol. 185. Berlin—Heidelberg—New York: Springer. 483 pp.

Kappelle, M., and S.P. Horn, eds. 2005. *Páramos de Costa Rica*. Santo Domingo de Heredia, Costa Rica: Instituto Nacional de Biodiversidad, INBio. 767 pp.

Kappelle, M., K.A. Haberyan, and S.P. Horn. 2005. Algas fósiles y recientes de los páramos de Costa Rica (Fossil and recent algae from the Costa Rican páramos). In M. Kappelle and S. Horn, eds., *Páramos de Costa Rica*, 331–41. Santo Domingo de Heredia, Costa Rica: Instituto Nacional de Biodiversidad, INBio.

Kappelle, M., J.G. van Uffelen, and A.M. Cleef. 1995. Altitudinal zonation of montane *Quercus* forests along two transects in the Chirripó National Park, Costa Rica. *Vegetatio* 119: 119–153.

Keller, G., C.E. Zenker, and S.M. Stone. 1989. Late Neogene history of the Pacific-Caribbean gateway. *Journal of South American Earth Sciences* 2: 73–108.

Kesel, R.H. 1983. Quaternary history of the Río General valley, Costa Rica. *National Geographic Society Research Report* 15: 339–58.

Kimura, G., E.A. Silver, and P. Blum, et al. 1997. Introduction. *Proceedings of the Ocean Drilling Program, Initial Reports* 170: 1–17.

Kohlmann, B., and M.J. Wilkinson. 2007. The Tárcoles Line: biogeographic effects of the Talamanca Range in lower Central America. *Giornale Italiano di Entomologia* 54(12): 1–30.

Kolarsky, R.A., P. Mann, and S. Monechi. 1995. Stratigraphic development of southwestern Panama as determined from integration of marine seismic data and onshore geology. In P. Mann, ed., Geologic and tectonic development of the Caribbean plate boundary in Southern Central America. *Geological Society of America Special Paper* 295.

Kuijpers, E. 1980. The geologic history of the Nicoya ophiolite complex, Costa Rica and its geotectonic significance. *Tectonophysics* 68: 233–55.

Lachniet, M.S. 2007. Glacial geology and geomorphology. In J. Bundschuh J, and G.E. Alvarado, eds., *Central America: Geology, Resources and Hazards*, 1: 171–84. London: Taylor and Francis.

Lachniet, M.S., and Y. Asmerom. 2007. Central American rainfall variations over the past 100,000 yr from Costa Rican stalagmites: a proxy for cross-isthmian water vapor transport. AGU Joint Assembly, Acapulco, Mexico.

Laurito, C. 1988. Los proboscídeos fósiles de Costa Rica y su contexto en la América Central. *Vínculos* 14(1–2): 29–58.

Laurito, C. 1993. Análisis topológico y sistemático del toxodonte de Bajo de los Barrantes, Provincia de Alajuela, Costa Rica. *Revista Geológica de América Central* 16: 61–68.

Laurito, C. 2003. Roedores fósiles del Pleistoceno Superior de la localidad La Palmera de San Carlos, Provincia de Alajuela, Costa Rica. *Revista Geológica de América Central* 29: 43–52.

Laurito, C., and A.L. Valerio. 2012. Paleobiogeografía del arribo de mamíferos suramericanos al sur de América Central de previo algran intercambio biótico Americano: un vistazo al GABI en América Central. *Revista Geológica de América Central* 46: 123–44.

Laurito, C., A.L. Valerio, and E.A. Pérez. 2005. Los xenarthras fósiles de la localidad de Buenos Aires de Palmares (Blancano Tardío-

Irvingtoniano Temprano), Provincia de Alajuela, Costa Rica. *Revista Geológica de América Central* 33: 83–90.

Laurito, C., W. Valerio, and E. Vega. 1993. Nuevos hallazgos paleovertebradológicos en la Península de Nicoya: implicaciones paleoambientales y culturales de la fauna de Nacaome. *Revista Geológica de América Central* 16: 113–15.

Linkimer, L., I. Arroyo, M.M. Mora, A. Vargas, G.J. Soto, R. Barquero, W. Rojas, W. Taylor, and M. Taylor. 2013. El terremoto de Sámara (Costa Rica) del 5 de setiembre del 2012 (Mw 7,6). *Revista Geológica de América Central* 49: 73–82.

Lohmann, W., and M. Brinkmann. 1931. Uber Obereocäne Kalke, Gabbros und Andesite von Costa Rica. *Central Journal of Mineralogy, Geology and Paleontology* in connection with the New Year Book of Mineralogy, Geology and Paleontology 1931: 553–59.

Lucas, S.G. 2014. Vertebrate paleontology in Central America: 30 years of progress. *Revista Geológica de América Central*, Número Especial 2014: 30 Aniversario: 139–55. doi:10.15517/rgac.v0.16576

Lucas, S.G., and G.E. Alvarado. 1994. Role of Central America in land-vertebrate dispersal during the Late Cretaceous and Cenozoic. *Profil* 7: 401–12.

Lucas, S.G., and G.E. Alvarado. 2010. Fossil Proboscidea from the Upper Cenozoic of Central America: taxonomy, evolutinary and paleobiogeographic significance. *Revista Geológica de América Central* 42: 9–16.

Lucas, S.G., G.E. Alvarado, R. García, E. Espinoza, J.C. Cisneros, and U. Martens. 2007. Vertebrate paleontology. In J. Bundschuh and G.E. Alvarado, eds., *Central America: Geology, Resources and Hazards*, 1: 443–51. London: Taylor and Francis.

Lucas, S.G., G.E. Alvarado, and E. Vega. 1997. The Pleistocene mammals of Costa Rica. *Journal of Vertebrate Paleontology* 17(2): 413–27.

Lücke, O.H., H.-J. Götze, and G.E. Alvarado, 2010. A constrained 3D density model of the upper crust from gravity data interpretation for Central Costa Rica. *International Journal Geophysics* 2010: 860902. doi:10.1155/2010/860902

Madrigal, R., and E. Rojas. 1980. *Manual Descriptivo del Mapa Geomorfológico de Costa Rica* (escala 1: 20,000). San José: SEPSA Imprenta Nacional.

Mann, P., and K. Burke. 1984. Neotectonics of the Caribbean. *Reviews of Geophysics and Space Physics* 22(4): 309–62.

Marshall, J.S. 1991. Neotectonics of the Nicoya Peninsula, Costa Rica: a look at forearc response to subduction at the Middle America trench. University of California, Santa Cruz. 196 pp. [M.Sc. thesis].

Marshall, J.S. 2007. Geomorphology and physiographic provinces. In J. Bundschuh and G.E. Alvarado, eds., *Central America: Geology, Resources and Hazards*, 1: 75–122. London: Taylor and Francis.

Marshall, J.S., and R.S. Anderson. 1995. Quaternary uplift and seismic cycle deformation, Península de Nicoya, Costa Rica. *Geological Society of America Bulletin* 107: 463–73.

Marshall, J.S., D.M. Fisher, and T.W. Gardner. 2000. Central Costa Rica deformed belt: kinematics of diffuse faulting across the western Panama block. *Tectonics* 19(3): 468–92.

Marshall, J.S., B.D. Idleman, T.W. Gardner, and D.M. Fisher. 2003. Landscape evolution within a retreating volcanic arc, Costa Rica, Central America. *Geology* 31(5): 419–22.

Marshall, J.S., and P. Vannuchi. 2003. Forearc deformation influenced by subducting plate morphology and upper plate discontinuity, Pacific Coast, Costa Rica. *Geological Society of America Abstracts with Programs* 35: 74.

Marshall, L.G. 1988. Land mammals and the Great Interchange. *American Scientist* 76: 380–88.

Mathers, S. 1989. Costa Rican diatomite: a review of existing knowledge and future potential. *Revista Geológica de América Central* 10: 3–17.

Matumoto, T., M. Ohtake, G. Latham, and J.E. Umaña. 1977. Crustal structure in Southern Central America. *Bulletin of the Seismological Society of America* 67: 121–34.

McGeary, S., A. Nur, and Z. Ben-Avraham. 1985. Spatial gaps in arc volcanism: the effect of collision or subduction of oceanic plateaus. *Tectonophysics* 119: 195–221.

McIntosh, K., E.A. Silver, and T. Shipley. 1993. Evidence and mechanisms for forearc extension at the accretionary Costa Rica convergent margin. *Tectonics* 12: 1380–92.

McMillan, I., P. Gans, and G. Alvarado. 2004. Middle Miocene to present plate tectonic history of the southern Central American Volcanic Arc. *Tectonophysics* 392: 325–48.

Mead, J.I., R. Cubero, A.L. Valerio Zamora, S.L. Swift, C. Laurito, and L.D. Gómez. 2006. Plio-Pleistocene *Crocodylus* (Crocodylia) from southwestern Costa Rica. *Studies on Neotropical Fauna and Environment* 41: 1–7.

Meschede, M., U. Barckahusen, and H.-U. Worm. 1998. Extinct spreading on the Cocos Ridge. *Terra Nova* 10: 211–16.

Montero, W. 1986. Períodos de recurrencia y tipos de secuencias sísmicas de los temblores interplaca e intraplaca en la región de Costa Rica. *Revista Geológica de América Central* 5: 35–72.

Montero, W. 1994. Neotectonics and related stress distribution in a subduction-collisional zone: Costa Rica. *Profil* 7: 125–41.

Montero, W. 2001. Neotectónica de la región central de Costa Rica: frontera oeste de la microplaca de Panamá. *Revista Geológica de América Central* 24: 29–56.

Montes, C., A. Cardona, C. Jaramillo, A. Pardo, J.C. Silva, V. Valencia, C. Ayala, I.C. Pérez-Angel, L.A. Rodríguez-Parra, V. Ramírez, and H. Niño. 2015. Middle Miocene closure of the Central American Seaway. *Science* 348(6231): 226–29.

Mora, S. 1979. Estudio Geológico de una parte de la región Sureste del Valle del General, Provincia de Puntarenas, Costa Rica. San Pedro de Montes de Oca, Costa Rica: Escuela Centroamericana de Geología, Universidad de Costa Rica (UCR). 188 pp. [Lic. thesis].

Mora, S. 1981. Clasificación morfotectónica de Costa Rica. Informe Julio-Diciembre 1981. *Instituto Geografico Nacional* 33–55.

Morell, K.D., D.M. Fisher, and T.W. Gardner. 2008. Inner forearc response to subduction of the Panama Fracture Zone, southern Central America. *Earth and Planetary Science Letters* 265: 82–95.

Morell, K.D., D.M. Fisher, T.W. Gardner, P. La Femina, D. Davidson, and A. Telezke. 2011. Quaternary outer fore-arc deformation and uplift inboard of the Panama Triple Junction, Burica Peninsula. *Journal Geophysical Research* 116: B05402. doi:10.1029/2010JB007979

Mörz, T., N. Fekete, A. Kopf, W. Brückmann, S. Kreiter, V. Huehnerbach, D. Masson, D.A. Hepp, M. Schmidt, S. Kutterolf, H. Sahling, F. Abegg, V. Spiess, E. Suwss, and C.R. Ranero. 2004. Styles and Productivity of Mud Diapirism along the Middle American Margin, 2: Mound Culebra and Mounds 11 and 12. NATO Science Series. Dordrecht, Netherlands: Kluwer Academic.

Obando, L.G. 1986. Estratigrafía de la Formación Venado y rocas sobreyacentes (Mioceno-Reciente). Provincia de Alajuela, Costa Rica. *Revista Geológica de América Central* 5: 73–104.

O'Connor, J.M., P. Stoffers, J.R. Wijbrans, and T.R. Worthington. 2007. Migration of widespread long-lived volcanism across the Galapagos Volcanic Province: evidence for a broad hotspot melting anomaly? *Earth and Planetary Science Letters* 263: 339–54.

Orvis, K.H., and S.P. Horn. 2000. Quaternary glaciers and climate on Cerro Chirripó, Costa Rica. *Quaternary Research* 54: 24–37.

Parkinson, R.W., J. Cortés, and P. Denyer. 1998. Passive margin sedimentation on Costa Rica's north Caribbean coastal plain, Río Colorado. International Coastal Symposium 1998 Proceedings. *Journal of Coastal Research Special Issue* 26: 110–22.

Pecher, I.A., C.R. Ranero, R. von Huene, T.A. Minshull, and S.C. Singh. 1998. The nature and distribution of bottom simulating reflectors at the Costa Rican convergent margin. *Geophysical Journal International* 133: 219–29.

Pérez, E.A. 1998. Evaluación paleoambiental de una localidad con flora fósil del Pleistoceno, La Palmera de San Carlos, Provincia de Alajuela. Final report, Bachiller en Manejo de Recursos Naturales, Universidad Estatal a Distancia.

Pérez, E.A. 2003. Presencia de *Juglans olanchana* standley & LO Williams (Juglandaceae) en el territorio costarricense durante el pleistoceno. *Revista Geológica de América Central* 28: 77–81.

Pérez, E.A. 2003. Los mamíferos fósiles del distrito de Puente de Piedra (Xenarthra, Glypodontidae; Artiodactlyla, Camelidae, Lamini), Grecia. *Revista Geológica de América Central* 49: 33–44.

Pérez, E.A., and C.A. Laurito. 2003. *Quercus corrugata* Hooker (Fagaceae) como indicador paleoclimático del Pleistoceno de Costa Rica. *Revista Geológica de América Central* 28: 83–90.

Pichler, H., and R. Weyl. 1975. Magmatism and crustal evolution in Costa Rica (Central America). *Geologische Rundschau* 64: 457–75.

Pinter, N., and T.W. Gardner. 1989. Construction of a polynomial model of glacio-eustatic fluctuation: estimating paleo-sea levels continuously through time. *Geology* 7: 295–98.

Protti, M., F. Güendel, and E. Malavassi. 2001. Evaluación del Potencial Sísmico de la Península de Nicoya. Heredia: Editorial Fundación UNA.

Protti, M., F. Güendel, and K. McNally. 1995. Correlation between the age of the subducting Cocos plate and the geometry of the Wadati-Benioff zone under Nicaragua and Costa Rica. In P. Mann, ed., Geologic and tectonic development of the Caribbean plate boundary in Southern Central America. *Special Paper–Geological Society of America* 295: 309–26.

Protti, M., V. González, J. Freymueller, and S. Doelger. 2012. Isla del Coco, on Cocos Plate, converges with Isla de San Andrés, on the Caribbean Plate, at 78 mm/year. *Revista de Biología Tropical 60* (Suppl. 3): 33–41.

Protti, R. 1996. Evidencias de glaciación en el Valle del General (Costa Rica) durante el Pleistoceno Tardío. *Revista Geológica de América Central* 19/20:75–85.

Ranero, C.R., and R. von Huene. 2000. Subduction erosion along the Middle America convergent margin. *Nature* 404: 748–52.

Rivier, F. 1985. Sección geológica del Pacífico al Atlántico a través de Costa Rica. *Revista Geológica de América Central* 2: 23–31.

Rojas, K.V. 2013. Relación entre los procesos volcano-sedimentarios y el neotectonismo de la Cuenca lacustrina de Palmares y San Ramón, Costa Rica. San Pedro de Montes de Oca, Costa Rica: Escuela Centroamericana de Geología, University of Costa Rica. 150 pp.[Lic. thesis].

Rojas, W. and G.E. Alvarado. 2012. Marco geológico y tectónico de la Isla del Coco y la región maritima circunvecina, Costa Rica. *Revista de Biología Tropical 60* (Suppl. 3): 15–33.

Sak, P.B., D.M. Fisher, and T.W. Gardner. 2004. Effects of subducting seafloor roughness on upper plate vertical tectonism: Osa Península, Costa Rica. *Tectonics* 23: TC1017. doi:10.10290/2002TC001474

Soto, G.J. 2004. Meteoritos y meteoros en Costa Rica (verdaderos, posibles, falsos). *Revista Geológica de América Central* 31: 7–23.

Tajima, F., and M. Kikuchi. 1995. Tectonic implications of the seismic ruptures associated with the 1983 and 1991 Costa Rica earthquakes. In P. Mann, ed., Geologic and tectonic development of the Caribbean plate boundary in South America. *Special Paper–Geological Society of America* 295: 327–40.

Ulloa, A., T. Aguilar, C. Goicoechea, and R. Ramírez. 2011. Descripción, clasificación y aspectos geológicos de las zonas kársticas de Costa Rica. *Revista Geológica de América Central* 45: 53–74.

Valerio, A., and C. Laurito. 2004. Paleofauna de Aguacaliente de Cartago, Costa Rica, I: *Equus* cf. *E. conversidens* Owen, 1869. *Revista Geológica de América Central* 31: 87–92.

Vannucchi, P., D.M. Fisher, S. Bier, and T.W. Gardner. 2006. From seamount accretion to tectonic erosion: formation of Osa Mélange and the effects of Cocos Ridge subduction in southern Costa Rica. *Tectonics*. doi:10.1029/2005TC001855

Vannucchi, P., C. Ranero, S. Galeotti, S.M. Straub, D.W. Scholl, and K. McDougall-Ried. 2003. Fast rates of subduction erosion along the Costa Rica Pacific margin: implications for nonsteady rates of crustal recycling at subduction zones. *Journal of Geophysical Research* 108(B11): 2511. doi:10.1029/2002JB002207

Vannucchi, P., and H. Tobin. 2000. Deformation structures and implications for fluid flow at the Costa Rica convergent margin, ODP Sites 1040 and 1043, Leg 170. *Journal of Structural Geology* 22: 1087–103.

Vogel, T.A., L.C. Patino, G.E. Alvarado, and P.B. Gans. 2004. Silicic ignimbrites within the Costa Rican volcanic front: evidence for the formation of continental crust. *Earth Planetary Science Letters* 226: 149–59.

von Huene, R., J. Azéma, W.T. Coulbourn, D.S. Cowan, J.A. Curiale, C.A. Dengo, R.W. Faas, R. Harrison, R. Hesse, J.W. Ladd, N. Muzilev, T. Shiki, P.R. Thompson, and J. Westberg. 1985. *Initial Reports of the Deep Sea Drilling Project.* Washington, DC: Government Printing Office, vol. 67.

von Huene, R., C. Ranero, and P. Watts. 2003. Tsunamigenic slope failure along the Middle America Trench in two tectonic settings. *Marine Geology* 3415: 1–15.

von Huene, R., C.R. Ranero, W. Weinrebe, and K. Hinz. 2000. Quaternary convergent margin tectonics of Costa Rica, segmentation of the Cocos Plate and Central American volcanism. *Tectonics* 19 (2): 314–34.

Walther, C. 2002. Crustal structure of the Cocos Ridge off Costa Rica. *Journal of Geophysical Research* 108. doi:10.1029/2001JB000888

Wells, S., T.F. Bullard, C.M. Menges, P.G. Drake, P.A. Karas, K.I. Kelson, J.B. Rutter, and J.R. Wesling. 1998. Regional variations in tectonic geomorphology along segmented convergent plate boundary, Pacific coast of Costa Rica. *Geomorphology* 1: 239–65.

Werner, R., K. Hoernle, P. van der Bogaard, C. Ranero, R. von Huene, and D. Korivh. 1999. Drowned 14-m.y. old Galápagos archipelago off the coast of Costa Rica: implications for tectonic and evolutionary models. *Geology* 27 6: 499–502.

Weyl, R. 1971. La clasificación morfotectónica de Costa Rica (América Central). Informe Julio-Diciembre. *Instituto Geografico Nacional* 107–25.

Weyl, R. 1980. Geology of Central America. Berlin: Gebrüder Borntraeger.

Wortel, R., and S. Cloetingh. 1981. On the origin of the Cocos-Nazca spreading center. *Geology* 9: 425–30.

Zamora, N., J. Méndez, M. Barahona, and L. Sjöbohm. 2004. Volcano-estratigrafía asociada al campo de domos de Cañas Dulces, Guanacaste, Costa Rica. *Revista Geológica de América Central* 30: 41–58.

Glossary

Andesite: A dark-colored, fine-grained extrusive rock (SiO_2: 52–63%) that could be composed primarily of plagioclase crysts (phenocrysts) and one or more of the mafic minerals, with a groundmass composed generally of the same minerals.

Anticline: A fold, generally convex upward, whose core contains the stratigraphically older rocks.

Arch: Natural arch. A basement doming.

Aseismic: Area that is not subject to large or even minor earthquakes.

Basalt: A general term for dark-colored mafic igneous rocks (SiO_2 < 52%), commonly extrusive, composed chiefly of calcic plagioclase, clinopyroxene, and olivine.

Basement: The undifferentiated complex of rocks that underlies the rocks of interest in an area.

Basic: Said of an igneous rock having relatively low silica content, sometimes delimited arbitrarily as 45 to 52%.

Beachrock: A friable to well-cemented sedimentary rock, formed in the intertidal zone in a tropical or subtropical region (beach line zone), consisting of sand or gravel (detrital and/or skeletal, including sandy coral beach or rich in shell fragments) cemented with calcium carbonate (aragonite).

Bioturbation: The churning and stirring of sediment by organisms.

BP: Before present (that is, before the year 1950).

Breccia: A coarse-grained clastic rock composed of angular broken rock fragments held together by cement or in a fine-grained matrix; it differs from conglomerate in that the fragments have sharp edges and unworn corners. Breccia may originate as a result of talus accumulation, igneous processes, disturbance during sedimentation, collapse of rock material, or tectonic processes.

Calciturbidites: Type of rock formed by carbonate talud sediments.

Caldera: A large, basin-shaped volcanic depression, more or less circular or rectangualar in form, the diameter of which is more than 1.5 km.

Chondrite: A stony meteorite containing spheroidal granules embedded in a fine-grained matrix of pyroxene, olivine, and Fe-Ti minerals, with or without glass.

Cirques: A deep steep-walled half-bowl-like recess or hollow, variously described as horseshoe-or crescent-shaped or semicircular in plan, situated high on the side of a mountain and commonly at the head of a glacial valley, and produced by the erosive activity of a glacier. It often contains a small round lake, and it may or may not be occupied by ice or snow.

Colluvial: Pertaining to colluvium or colluvial deposits.

Colluvium: General term applied to loose and incoherent deposits, usually at the foot of a slope or cliff as a result from gravity, chiefly.

Conglomerate: A coarse-grained clastic sedimentary rock, composed of rounded to subangular fragments larger than 2 mm in diameter set in a fined-grained matrix of sand or silt, and commonly cemented by calcium carbonate, iron oxide, silica, or hardened clay.

Cretaceous: The final period of the Mesozoic era, thought to have covered the span of time between 135 and 65 million years ago.

Debris Avalanche: The very rapid and usually sudden megasliding and flowage of incoherent, unsorted mixtures of soil and weathered bedrock.

Debris Flow: A high-density moving mass of rock fragments, soil, mud, and water; more than half of the particles being larger than sand size.

Delta: The low, nearly flat, alluvial tract of land at or near the mouth of a river, commonly forming a triangular or fan-shaped plain of considerable area, crossed by many distributaries of the main river, perhaps extending beyond the general trend of the coast, and resulting from the accumulation of sediment supplied by the river in such quantities that it is not removed by tides, waves, and currents. Most deltas are partly subaerial and partly submerged.

Diatomite: A light-colored soft friable siliceous sedimentary rock, consisting chiefly of opaline frustules of the diatom, a unicellular aquatic plant related to the algae.

Dolerite: In British usage, the preferred term for what is called diabase in reference to the fine-grained character of the rock between a microgabbro and a basalt.

Doline: A karst depression.

Estuarine: Pertaining to or formed or living in an estuary; esp. said of deposits and of the sedimentary or biological environment of an estuary.

Eustatic: Pertaining to worldwide changes of sea level that affect all the oceans. Eustatic changes may have various causes, but the changes dominant in the last few million years were caused by additions of water to, or removing water from, the continental icecaps.

Fault-propagation Fold: A fold formed in front of a fault surface, commonly associated with the upward termination of a thrust fault.

Fold: A curve or bend of a planar structure such as rock strata, bedding planes, foliation, or cleavage. A fold is usually a product of deformation, although its definition is descriptive and not generic and may include primary structures.

Foraminifera: Any protozoan belonging to the subclass Sarcodina, order Foraminifera, characterized by the presence of a test of one to many chambers composed of secreted calcite (rarely silica or aragonite) or of agglutinated particles. Most foraminifers are marine but freshwater forms are known. Range from Cambrian to the present.

Fracture Zone: On the deep-sea floor, an elongate zone of unusually irregular topography that often separates regions of different depths. Such zones commonly cross and apparently displace the mid-oceanic ridge by faulting.

Gabbro: A group of dark-colored, mafic intrusive igneous rock composed principally of basic plagioclase (commonly labradorite or bytownite) and clinopyroxene (augite), with or without olivine and orthopyroxene.

Geomorphology: The science that treats the general configuration of the Earth's surface; specifically, the study of the classification, description, nature, origin, and development of present landforms and their relationship to underlying structures, and of the history of geological changes as recorded by these surface features. The term is applied to the general interpretation of landforms, but has also been restricted to features produced only by erosion or sedimentation.

Graben: An elongate, relatively depressed crustal unit or block that is bounded by faults on its long sides. It is a structural form that may or may not be geomorphologically expressed as a valley.

Greywacke: An old rock name that has been variously defined but is now generally applied to a dark gray firmly indurate coarse-grained sandstone that consists of poorly sorted angular to subangular grains of quartz and feldspar, with a variety of dark rock and mineral fragments embedded in a compact clayey matrix having the general composition of slate and containing an abundance of clay minerals. The original German word *grauwacke* means a grey, earthy rock.

Holocene: An epoch of the Quaternary period, from the end of the Pleistocene, at 11,500 years ago, to the present time.

Horn: A high, rocky, sharp-pointed mountain peak with prominent faces and ridges, bounded by the intersecting walls of three or more cirques that have been cut back into the mountain by the headward erosion of glaciers.

Hypobysal: Subvolcanic rock or igneous intrusion emplaced at shallow level between plutonic and volcanic (subaerial) manifestations.

Ignimbrite: The rock formed by widespread deposition and consolidation of pyroclastic ash flow deposits. The term originally implied dense welding but there is no longer such a restriction, so that the term includes rock types such as welded to non-welded pyroclastic pumice flow deposits.

Intrusion: The process of emplacement of magma in preexisting rock; magmatic activity; also the igneous rock mass formed within the surrounding rock.

Karst: A type of topography that is formed on limestone, gypsum, and other rocks by dissolution and that is characterized by sinkholes, caves, and underground drainage. The word is of German-Slovenian origin and refers to the Karst Plateau—a region in Slovenia partially extending into Italy.

Lahar: A mudflow (debris flow to hypoconcentrated flow) composed chiefly of volcaniclastic materials, water, and other detritus on the flank of a volcano.

Listric: A curvilinear, usually concave-upward surface of fracture that curves, at first gently and then more steeply, from a horizontal position. Listric surfaces bound wedge-shaped masses, appearing to be thrust against or along each other.

Lithosphere: A layer of strength relative to the underlying asthenosphere for deformation at geologic rates. It includes the crust and part of the upper mantle and is of the order of 100 km in thickness.

Margin: A continent-ocean basin transition marked by an active plate boundary (active continental margin), in most of the cases a subduction zone, or not marked by an active plate boundary (passive margin).

Mass Flow: A unit movement of a portion of the land surface; specif. mass wasting or the gravitative transfer of material down a slope.

Matrix: The finer-grained material enclosing, or filling the interstices between, the larger grains of particles of a sediment or sedimentary rock; the natural material in which a sedimentary particle is embedded.

mbsf: Meters below sea floor. It is a convention for depths below the seabed used in geology.

Meander: One of a series of regular, freely developing sinuous curves, bends, loops, turns, or windings in the course of a stream.

Mélange: A mappable body of rock characterized by the inclusion of fragments and blocks of all sizes, both exotic and native, embedded in a fragmented and generally sheared matrix of more tractable material. Mode or origin could be tectonic, sedimentary, or a mix.

Moraine: A mound, ridge, or other distinct accumulation of unsorted, unstratified glacial drift, predominantly till, deposited chiefly by direct action of glacier ice, in a variety of topographic landforms that are independent of control by the surface on which the drift lies.

Mudstone: An indurated mud having the texture and composition of shale, but lacking its fine lamination or fissility; a blocky or massive, fine-grained sedimentary rock in which the proportions of clay and silt are approximately equal, a non-fissile mud shale, or when it is desirable to characterize the whole family of fined-grained sedimentary rocks (as distinguished from sandstones, conglomerates, and limestones).

Normal Fault: A fault in which the handing wall appears to have moved downward relative to the footwall. The angle of the fault is usually 45–90°. There is dip separation but there may or may not be dip slip.

Olivine: An olive-green, grayish-green, or brown orthorhombic mineral: $(Mg, Fe)_2\ SiO_4$. Olivine is a common rock-forming mineral of basic, ultrabasic, and low-silica igneous rock (gabbro, basalt, peridotite, dunite).

Ophiolite: A group of mafic and ultramafic igneous rocks ranging from basalt to gabbro and peridotite, including serpentinite derived from them by later metamorphism, whose origin is usually associated with tectonic emplaced oceanic lithosphere.

Overthrust: A low-angle thrust fault of large scale, with displacement generally measured in kilometers.

Paralic: Said of sedimentary deposits formed along the margin of the sea, in shallow water subject to marine invasion, and of environments (such as lagoonal or littoral) of the marine borders. Also said of basins, platforms, marshes, swamps, and other features marked by thick terrigenous deposits intimately associated with estuarine and continental deposits, such as deltas formed on the heavily alluviated continental shelves.

Parasitic Cone: Said of a volcanic cone, crater that occurs on the side of a larger cone; it is a subsidiary form.

Periglacial: Said of the processes, conditions, areas, climates, and topographic features at the immediate margins of former and existing glaciers and ice sheets, and influenced by the cold temperature of the ice. By extension, said of the environment in which frost action is an important factor, or of phenomena induced by a periglacial climate beyond the periphery of the ice.

Planktonic: Said of that type of pelagic organism that floats.

Plate: A torsionally rigid, thin segment of the earth's lithosphere, which may be assumed to move horizontally and adjoins other plates along zones of seismic activity. Ca. 100 km thick.

Pleistocene: An epoch of the Quaternary period that began 2.59 million years ago and lasted until the start of the Holocene (11,500 years ago). If the Quaternary is designated as an era, then the Pleistocene is considered to be a period.

Prograding: The building forward or outward toward the sea of a shoreline or coastline (as of a beach, delta, or fan) by nearshore deposition of river-borne sediments or by continuous accumulation of beach material thrown up by water or moved by longshore drifting.

Pyroclastic Cone: Cone made of particles ejected during a volcanic eruption, usually of small size.

Pyroclastic Flow: Fluidized mass of hot ash. Term often used in a more general sense.

Quaternary: Geological era that began 2.59 million years ago and extends to the present.

Rift: A long, narrow continental trough that is bounded by normal faults; a graben of regional extent. It marks a zone along which the entire thickness of the lithosphere has ruptured under extension.

Rudist: Any bivalve mollusk belonging to the superfamily Hippuritacea, characterized by an inequivalve shell, usually attached to a substrate, and either solitary or gregarious in reeflike masses. They are frequently found in association with corals. Range: Upper Jurassic to Upper Cretaceous, possibly Paleocene.

Sand Spit: A spit consisting chiefly of sand.

Sandstone: A medium-grained clastic sedimentary rock composed of abundant rounded or angular fragments of sand size set in a fine-grained matrix (silt or clay) and more or less firmly united by a cementing material (commonly silica, iron, oxide, or calcium carbonate); the consolidated equivalent of sand, intermediate in texture between conglomerate and shale.

Seamount: An elevation of the sea floor, 1,000 m or higher, either flat-topped (called a guyot) or peaked (called a seapeak). Seamounts may be discrete, arranged in a linear or random grouping, or connected at their bases and aligned along a ridge or rise.

Seismic Gap: A segment of an active fault zone that has not experienced a principal earthquake during a time interval when most other segments of the zone have. Seismologists commonly consider seismic gaps to have a high future-earthquake potential.

Shield Volcano: A volcano in the shape of a flattened dome, broad and low, built by lava flows or even ignimbrites.

Siltstone: A rock whose composition is intermediate between those of sandstone and shale and of which at least two-thirds is material of silt size; it tends to be flaggy, containing hard, durable, generally thin layers, and often showing various primary current structures.

Sinkhole: A circular depression in a karst area. Its drainage is subterranean, its size is measured in meters or tens of meters, and it is commonly funnel-shaped.

Stratovolcano: A conical volcano that is constructed of alternating layers of lava and pyroclastic deposits.

Subduction: The process of one lithospheric plate descending beneath another.

Subduction Erosion: Erosion caused in the narrow belt in which subduction takes place.

Sublittoral: Said of that part of the littoral zone that is between low tide and a depth of about 100 m.

Syncline: A fold of which the core contains the stratigraphically younger rocks; it is generally concave upward.

Terrace: Any long, narrow, relatively level or gently inclined surface, generally less broad than a plain, bounded along one edge by a steeper descending slope and along the other by a steeper ascending slope; a large bench or steplike ledge breaking the continuity of a slope. The term is usually applied to both the lower or front slope (the riser) and the flattish surface (the tread), and it commonly denotes a valley-contained, aggradational form composed of unconsolidated material as contrasted with a bench eroded in solid rock.

Terranes: An exotic megablock (several km^2) of different rocks tectonically accreted. The term is used in a general sense and does not imply a specific rock unit.

Tombolo: A bar or barrier that connects an island with the mainland.

Tuff: A general term for all consolidated ash.

Turbidite: Sediment of rock deposited from, or inferred to have been deposited from, a turbidity current. It is characterized by graded bedding, moderate sorting, and well-developed primary structures in the sequence.

Ultramafic: Said of an igneous rock (SiO_2 < 45%) composed chiefly of mafic minerals, e.g., monomineralic rocks composed of olivine and augite.

Unconformity: A substantial break or gap in the geologic record where a rock unit is overlain by another that is not next in stratigraphic succession, such as an interruption in the continuity of a depositional sequence of sedimentary rocks or a break between eroded igneous rocks and younger sedimentary strata. It results from a change that caused deposition to cease for a considerable span of time, and it normally implies uplift and erosion with loss of the previously formed record.

U-Shaped Valley: A valley having a pronounced parabolic cross-profile suggesting the form of a broad letter "U," with steep walls and a broad, nearly flat floor; specifically, a valley carved by glacial erosion, such as a glacial trough.

Volcanic Dome: A steep-sided, rounded extrusion of highly viscous lava squeezed out from a volcano, and forming a dome-shaped or bulbous mass of congealed lava above and around the volcanic vent. The structure generally develops inside a volcanic crater or on the flank of a large volcano, and is usually much fissured and brecciated.

Volcanic Gap: Sector without volcanic activity in a specific time.

Volcanic Plateau: Surface formed by extensive lava or ash flows that cover topographic irregularities.

Wadati-Benioff: A narrow zone of earthquake foci that seismically illuminates a subduction zone.

Whaleback: A large mound or hill having the general shape of a whale's back, especially a smooth elongated ridge of a glacier plain having a rounded crest and ranging widely in size, about 300 km long, 1–3 km wide, and perhaps 50 m high. It forms a coarse-grained clastic pedestal build up and left behind by a succession of longitudinal glacial deposits along the same path.

Chapter 4 Soils of Costa Rica: An Agroecological Approach

Alfredo Alvarado[1,*] and Rafael Mata[1]

Introduction

While investigating the geology of Central America, Sapper (1903) depicts the first soil map of the region, including the territory of Costa Rica. However, soil science in the country begins a little later when Prescott (1918) and Bennett (1926) selected lands to plant lowland crops in the Caribbean Region. During the 1940s, the University of Costa Rica (UCR) reopens and the Inter American Institute of Agricultural Sciences (IICA, Turrialba) starts operations in Costa Rica. This was a remarkable landmark for the country's agricultural research, since many areas of knowledge were housed in Faculties or Departments at UCR, and medium-term agricultural (soil) research projects were carried out at IICA. During these years the first works on soil analysis for soil fertility purposes were conducted in a small laboratory at UCR in collaboration with the Departamento Nacional de Agricultura de la Secretaría de Agricultura and Ganadería (Ramírez 2001).

At the beginning of the 1950s, with the expansion of medium- and high-altitude crops (mainly coffee, sugarcane, and vegetables), soil studies were carried out in the Central Valley (Dondoli and Torres 1954, COSTA RICA-MAI 1958). Government-sponsored colonization programs for the northern region in the 1950s and 1960s led to increased soil knowledge for this region (COSTA RICA-ITCO 1964, Sandner et al. 1966). In December 1955 the Agricultural Research Center of UCR was established as an institution that produced important research papers on Costa Rican soils until date.

It is during the 1960s that a group of soil scientists at IICA, Turrialba (now CATIE), promoted soil science knowledge in Latin America; as a result of this effort, other areas of soil science developed, and books in Spanish were prepared on soil microbiology (Blasco 1970), soil chemistry (Fassbender 1975), soil physics (Gavande 1973, Forsythe 1975), and later on soil genesis and classification (Alvarado 1985), soil clay mineralogy (Besoain 1985), and soil and forest ecosystems (de las Salas 1987). Most of the knowledge generated until 1960 allowed the drafting of the first maps of potential land use in Costa Rica at a scale of 1:750,000 (Plath and van der Sluis 1965, Coto and Torres 1970), as well as the first semi-detailed taxonomic soil map of the country (USAID 1965) and the characterization of Costa Rican soils on the Soil Map of the World (FAO-UNESCO 1976). In 1960, the Programa Cooperativo Oficina de Café-MAG initiated activities related to nutrition and fertilization of coffee plantations. At the end of the decade Fertilizantes de Centroamérica (Costa Rica) S.A. (FERTICA) began operations, a company that for more than 30 years dominated the market of this type of products and promoted its use exponentially.

By the mid-1960s, the Tropical Science Center (TSC; a.k.a. Centro Científico Tropical, CCT, in Spanish) and the Organization for Tropical Studies (OTS, or OET in Spanish) initiated their activities in Costa Rica. Until the mid-1980s TSC focused its attention mostly on life zone ecology and land use capability / suitability studies throughout the country. Examples of their land use assessments are the evaluations conducted in the Salitre indigenous reserve (Tosi

[1] Centro de Investigaciones Agronómicas, Facultad de Ciencias Agroalimentarias, Universidad de Costa Rica (UCR), San Pedro de Montes de Oca, Costa Rica

* Corresponding author

1967a), in Guanacaste (Tosi 1967b), in the Boruca area south of San Isidro del General (Tosi and Zadroga 1975), and in the northern zone (Gow et al. 1988). These land use–oriented environmental assessments evaluated the potential capability and suitability of the land for future agricultural and forestry use. Interestingly, some 40 years ago, foresters and soil experts Joseph Tosi, Leslie Holdridge, and Gary Hartshorn, together with other TSC and OTS colleagues, already stressed the need to maintain natural vegetation in those areas where slopes were steep and the terrain was rugged. In the particular case of Salitre, for instance, they pleaded for the establishment of a Forestry Reserve that should be preserved in the long run, in order to avoid loss of vegetation and soil erosion. Quite an advanced viewpoint, at that time in history.

Since then, many studies on the relationship between soil and vegetation have been published by visiting professors and their students. Soil science developed quickly in Costa Rica and soil maps of the entire country were prepared at a detailed scale (1:200,000 by Pérez et al. 1978 and Vásquez 1979). The soil laboratory of MAG improved its facilities at the end of the decade, allowing it to contribute significantly to soil analysis for agricultural uses.

During the 1980s the Soil Science Society of Costa Rica was created with a mandate to organize the VII Latin American Soil Science Congress, and by the 1990s published 12 books on soil science topics. In the late 1980s and early 1990s soils of Talamancan montane cloud forests and *páramo* vegetation received particular attention (van Uffelen 1991, Kappelle and van Uffelen 2005, 2006, and see Kappelle and Horn, chapter 15 of this volume). More recently a new group of authors contributed to the knowledge of the soils in the Talamanca Range (Winowiecki 2008, Schembre 2009, Chinchilla et al. 2011a,b,c,d, Salazar 2014). In 1991, Servicios Eléctricos Potosí S.A. (SEPSA) published a soil map of the country, incorporating detailed studies prepared for two large hydroelectric development projects in mountainous regions, irrigation projects in Guanacaste, banana expansion in the Atlantic region, and rural development at the border with Panama. Furthermore, several studies have already integrated geographic information systems (GIS) with modern concepts of sustainable development (e.g., see Stoorvogel and Eppink 1995, Arroyo 1996, Ugalde 1996).

Description of Major Soil Orders

This section describes the usefulness and importance of Costa Rican soils for agriculture, following soil taxonomy standards. The analysis includes a summary of the type of areas occupied by each major soil type; the crops planted there; its geographic distribution within the country and position in the landscape; its probable means of origin; the principal mineralogical, physical, and nutritional characteristics of each group; and the management practices that could be applied to each to achieve best its productive potential. In this way, this section can be considered an update to the findings presented by Bertsch et al. (2000).

It is possible to find in Costa Rica all 12 major soil orders recognized by soil taxonomists except desert soils (Aridisols) and frost soils (Gelisols). Six of these 10 have major agricultural relevance: Inceptisols (38.6%), Ultisols (21%), Andisols (14.4%), Entisols (12.4%), Alfisols (9.6%), and Vertisols (1.6%); the percentages mentioned represent the relative area each soil order covers (Mata 1991).

Entisols

Entisols are very recent soils that exhibit little development. Therefore it is not possible to distinguish any defined horizon sequence in the profile. The most common Entisol suborders in the country are fluvents, aquents, orthents, and psamments (i.e., soils derived from recent alluvial deposits, under stagnant water, shallow soils on hard rock, and sandy textured materials, respectively). The presence of psamments in coastal beach fronts or elevated coastal structures is a consequence of continental uplift due to plate tectonics activity (Fig. 4.1).

Recent alluvial deposits lead to the formation of fluvents in areas where frequent flooding does not allow soils to remain undisturbed long enough to permit the development of horizons. Under these conditions, a sequence of layers of contrasting particle sizes occurs. In the same geomorphic surfaces, these soils turn into aquents when the water table remains near or above the soil surface and restricts soil development for long periods of time. Orthents, being the most abundant Entisols, prevail on rocky hillsides with low temperatures and/or very erosive rains, recent volcanic depositions such as ashes or lava, and the presence of parental material resistant to weathering. Other Orthents are found in low relief portions in regions of rhyolithic origin with less than 1500 mm of precipitation in the dry Pacific areas.

Most Entisols are of limited agricultural potential due to their high flood risk, restricted soil rooting depth, low fertility status, or location on steep slopes. Their use by humans should be restricted to forestry or conservation activities. Nevertheless, in Costa Rica these soils are frequently used for annual crops, and extensive cattle operations on both

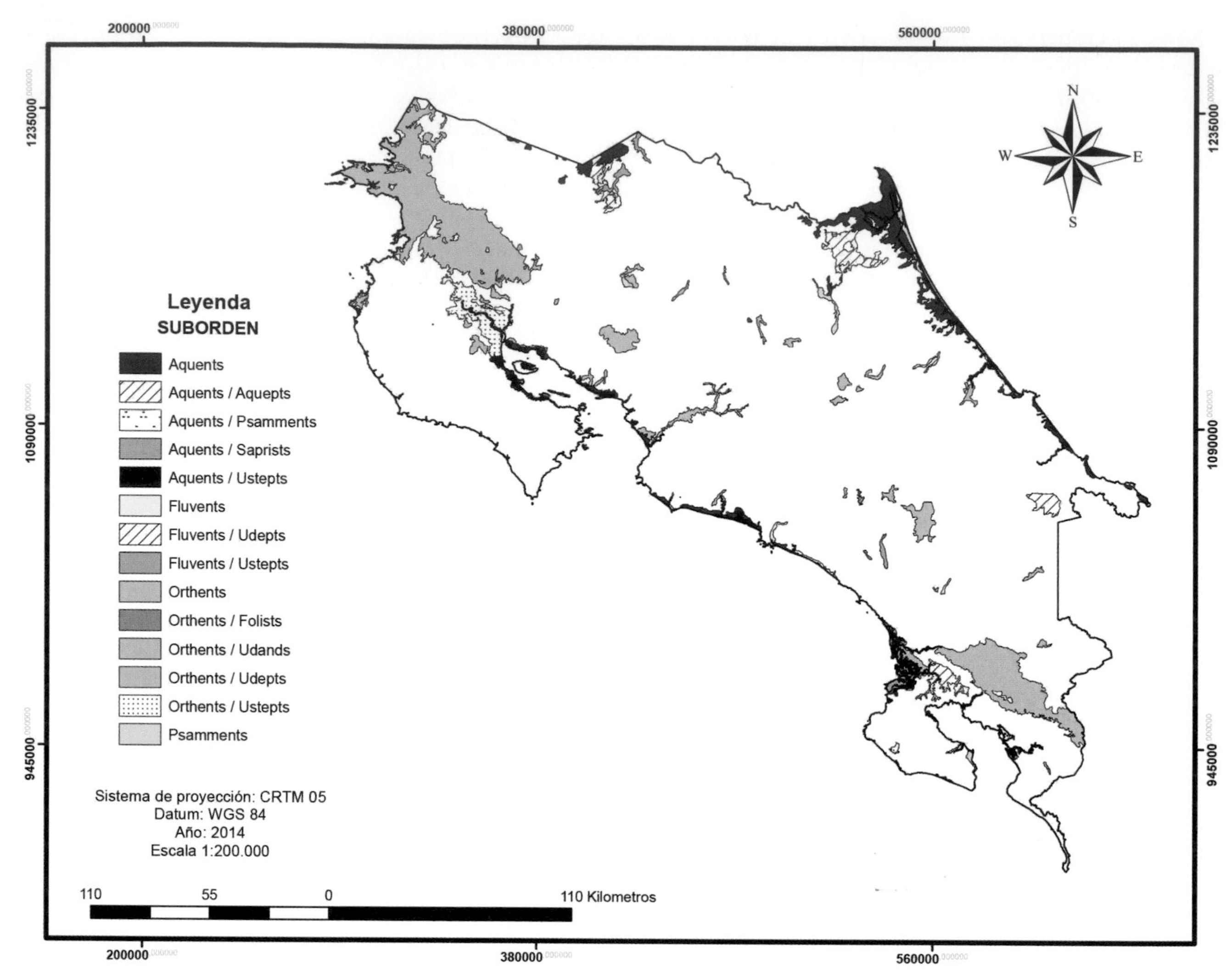

Fig. 4.1 Distribution and characteristics of Entisols of Costa Rica.

flat and steep lands. Owing to their minimal development, these soils reflect the properties of the parent material out of which they were formed, which is why they display a very varied mineralogy. In general, they are not very suitable for agricultural purposes because of shallow rooting depth, reduced conditions, and, as noted, frequent flooding and high susceptibility to hydric and aeolian erosion.

In wetlands, aquents and fluvents are associated with aquepts (Inceptisols) at the Caribbean side covered by natural vegetation dominated by yolillo palms (*Raphia* spp.) and cativo trees (*Prioria copaifera*) and mangrove vegetation in the Pacific coast, recently exploited for banana and oil palm production, later for charcoal production and today in danger by a possible large sediment deposition if the proposed hydroelectrical plant at Diquís becomes a reality (Torrealba et al. 2011).

Once drained, these soils are used to plant bananas, cocoa, and oil palm. Orthents form on thin volcanic ash deposits over lava flows (e.g., in Cervantes, Cartago province, and Paso Canoas, Puntarenas province). Although they are not very productive, they may be heavily fertilized and planted with vegetables serving nearby markets. In other regions, where the exposed rock is not really hard (e.g., as in the case of rhyolithic materials in Guanacaste) they are used for ranching purposes and, recently (and with very little success), for forestry. At the hillsides in the Southern and Central Pacific regions orthents are commonly used for low technology bean planting ("frijol tapado").

Inceptisols

Inceptisols are widely distributed in Costa Rica. Because the country is geologically and geomorphologically relatively young, Inceptisols cover about 39 percent of the country's territory. The young age of the soils also implies that they strongly reflect their parent material, including Lithic (rock), Fluventic (riverine), Andic (volcanic), Vertic (clay mineralogy), or Oxic (trivalent cation accumulation) properties. They are common on hillsides where erosion due to a combination of earthquakes and heavy rainstorms induces landslides, which limit soil formation to a profile with very weak horizon development. Under these conditions, a typical Udepts toposequence includes Vitrandic Dystrudepts (high volcanic glass, coarse textured soils) in the upper positions, Humic Dystrudepts (high organic matter content soil) in the middle positions, and Typic Dystrudepts (soils with low bases status) in the lower positions of the landscape.

The Ustepts are found in rolling and flat geomorphic surfaces. Among these, Dystrustepts (low base saturation) and Haplustepts (little soil development) form from the weathering of relatively old alluvial and/or colluvial fans. In the same environment, Inceptisols classified as Aquepts are found when there is a perched water table. These are the most important soils for agriculture of the lowlands less than 100 meters above sea level. A sulfidic horizon forms where brackish water and mangrove vegetation occurs along coastal back swamps. These are classified as Sulfaquepts. The most important Inceptisols are found in alluvial valleys in the coastal plains. These soils have the highest agricultural potential of the country and can be found in the valleys of the Tempisque, Bebedero, Tárcoles, Parrita, Térraba, Sierpe, and Coto rivers on the Pacific side, and the Matina, Reventazón, Parismina, Pacuare, Estrella, and Sixaola rivers on the Caribbean side (Fig. 4.2). In many cases these soils develop from basic parent materials such as limestone, from which they inherited their high base saturation, adequate texture, and moisture retention.

Inceptisols (except those with poor drainage) generally

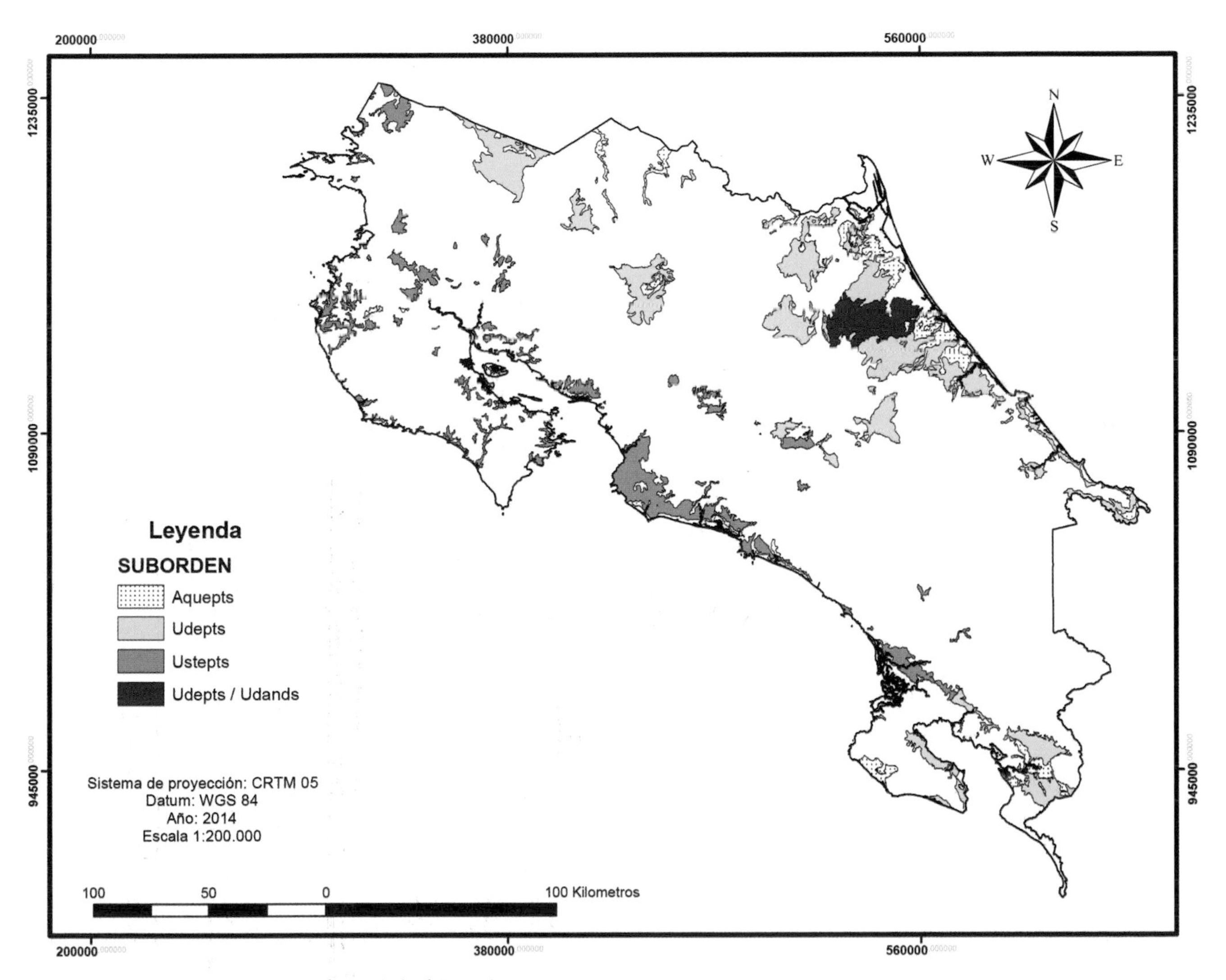

Fig. 4.2 Distribution and characteristics of Inceptisols of Costa Rica.

have good characteristics for management, since they do not possess the properties of more developed soils, such as cation depletion, that affect management adversely. For this reason, they can be used for a large range of agricultural production activities, including banana, oil palm, sugarcane, cocoa, coffee, staple crops, livestock, forestry and, recently, non-traditional crops such as mango, avocado, cantaloupe, pepper, roots and tubers, tropical flowers, etc. Even the Sulfaquepts along the coast are important for mangrove forestry, shrimp aquaculture, and extraction of salt.

Chemical and mineralogical properties of Inceptisols vary according to their origin. Therefore, their range of chemical characteristics is broad. Each soil tends to include many clay mineral types mixed together, including smectites, allophane, kaolinites, and organic and oxidic coatings (Alvarado et al. 2014a,b). When there is a preponderance of volcanic ash materials some amorphous clay develops. In the alluvial valleys of both the Caribbean and the Pacific sides, montmorillonite is found. The extreme weathering conditions in tropical environments of the El General River Valley result in the formation of 1:1 fractions of clays and oxides in red soils of very high acidity values, and cation depletion. These are the most infertile Inceptisols of the country.

Those Inceptisols used for commercial plantations in the poorly drained lowlands require drainage (Epi- and Endoaquepts). For example, as Eastern banana plantations spread from the slopes toward Limon extensive networks of 1- to 2-meter-deep ditches are required. Such ditches are economically viable only when flood frequency remains low. The fertility of Inceptisols in the North Atlantic Zone is much higher than in the South Atlantic Region because they are developed from volcanic materials spread downslope by rivers. In the South Atlantic Region of the country Inceptisols were formed from much less productive calcareous materials, and are also subject to much greater frequency of flooding.

The fertility of Inceptisols in the Guanacaste lowlands can be greatly enhanced with applications of S and Zn, especially in rice plantations (Bornemisza 1990, Cordero 1994). Moisture availability is critical on these Inceptisols in ustic (long dry season) environments. These properties have been mapped, and used for categories of crop insurance (IICA 1979).

Rice cultivars on Inceptisols in the South Pacific Valleys have been subject to Cu toxicity generated by massive applications of copper-containing Bordeaux fungicides to banana plantations in the 1940s and 1950s (Cordero and Ramírez 1979). Much of this very fertile land had to be abandoned and owing to silt deposition by river flooding they have been rehabilitated for annual crops. During the early days, these lowlands were planted with banana and cocoa without any fertilizer application; at present, improved varieties and higher productivity of oil palm and banana plantations require large amounts of complete fertilizer formulas and drainage systems.

Small farmers, living in government settlements in the Northern and Atlantic regions of the country, plant cereals and roots and tubers using low inputs. Because of the predominantly perudic environments in these regions, the traditional slash and burn system is not practiced since the slashed vegetation does not dry and cannot be burned. This particular problem does not allow for obtaining beneficial effects from liming and fertilizing with added ashes. Normally, the accumulated biomass slowly decomposes with time, releasing nutrients only gradually (Bertsch and Vega 1991).

Andisols

A comprehensive summary of Andisols of Costa Rica is presented by Alvarado et al. (2001). Andisols are formed from volcanic ash deposits and occupy: (a) the Central Valley and surrounding mountains; (b) hillsides of the Guanacaste Mountain Range, (c) the region between Coto Brus and the border with Panama influenced by the Barú Volcano's ashes, and (d) some regions of the Northern and Atlantic zones where fluvio-volcanic depositions occur (Fig. 4.3). Volcanoes are still quite active in Costa Rica, and their activity influences agricultural potential directly as well as indirectly through soil building and acid rain depositions (Alvarado and Cárdenes, chapter 3 of this volume). The emissions of acidic clouds from volcanoes turn into acid rain in nearby zones, which leads to an intensive weathering of the land system, enhancing basic cation leaching and causing considerable loss of crop yields. Although Andisols cover only 14 percent of the nation's territory, many major agricultural products like coffee, sugarcane, vegetables, non-traditional export crops (flowers, ferns, strawberry), and dairy products like milk and cheese are indeed produced on lands dominated by these soils. Part of the latest large banana boom of the 1990s was settled on volcanic soils of the Northern Zone and parts of the Atlantic Region. In the lowlands, in terms of non-traditional crops, Andisols can produce very good roots and tubers, heart of palm, and a huge range of tropical ornamental plants.

The frequent rejuvenation of these soils by andesitic volcanic ash additions constantly enriched the environment with nutrients. Large depositions of debris, particularly near the craters, allows the formation of Vitrands, while Udands form under repetitive deposition of thin volcanic layers in the middle positions of the landscape in udic environments.

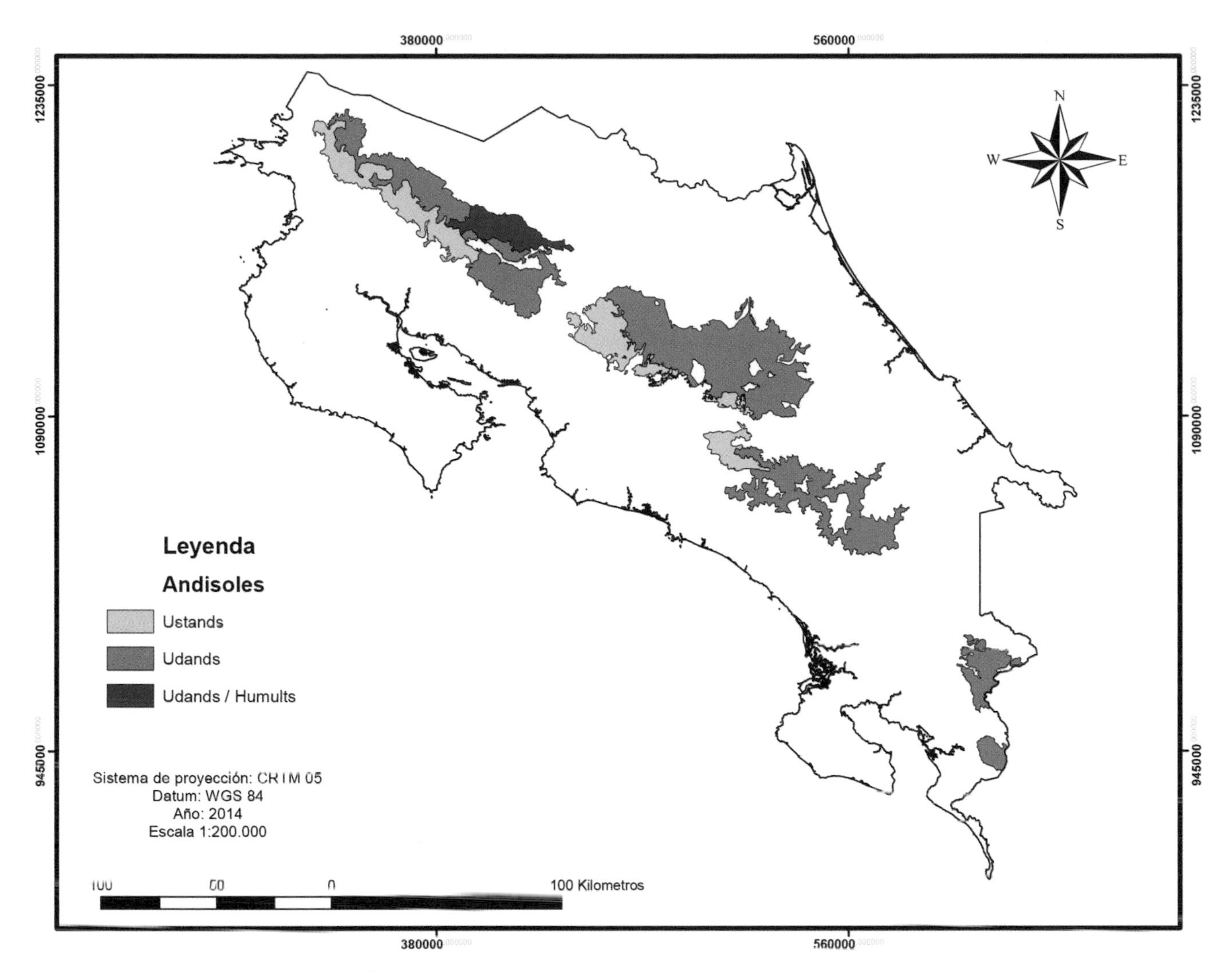

Fig. 4.3 Distribution and characteristics of Andisols of Costa Rica.

In the lower parts of the landscape, where a distinctive dry season occurs, Ustands are predominant. Andisols of lighter color are found along the Guanacaste Mountain Range, in the north of the country, and originated from the deposition of rhyolithic/dacitic ashes. Andesitic basaltic ashes predominate in the central and southern parts of the country, and give rise to dark-colored soils.

The effective depth of the top soil layer of Andisols generally depends on the magnitude of the volcanic deposition that formed that layer. Deep soils tend to be formed from the deposition of many small layers of ashes, while thin Andisols are formed by one event, which can be of small or large magnitude. It is possible to observe the ash deposition frequency and magnitude in deep road cuts, as well as the presence of paleosols with a different degree of weathering.

Soil particles are generated and distributed initially by the nature of the original volcanic activity, and then sorted by prevailing winds according to particle size and density, creating a textural gradient along the hillsides of volcanic craters. Coarser material is deposited in the vicinity of craters, resulting in sandy to sandy-loam materials. Further away silty loam or loam textures are predominant. Finer textures are found farther away from the volcano, particularly in the B horizons of well-developed soils. This textural gradient notoriously affects nutrient availability and irrigation needs.

Once the original deposition has taken place climate forces predominate. For example, if a moist and cold environment ensues near the volcano, this allows for a weak weathering process of volcanic glass, releasing small amounts of Si, Al, and Fe oxides and hydroxides. If long periods of volcanic inactivity follow, the translocation of the oxides will form a cemented layer (called a placic horizon) wherever an abrupt textural change is present. Farther from the crater, allophane becomes predominant. This type of clay is an amorphous and hydrated colloid, which forms organomineral compounds and represents required products of volcanic ash decomposition in humid zones (Alvarado et al. 2014a). Allophane is an unstable clay-size particle of high reactivity that gives a peculiar behavior to these soils. Secondary organomineral compounds possess a very large hydration capacity that enables them to enlarge their total surface, therefore increasing their capacity to retain or exchange cations and anions.

In the Central Valley, farther away from the volcanoes, rainfall decreases and a long dry period permits the formation of a 1:1 crystalline clay named halloisite. This type of clay has shrink and swell properties, low water retention characteristics, and less nutrient retention than allophanes. They are predominant in the brown-yellowish soils of the coffee and sugarcane plantations of the Central Valley. Each mineral type gives a characteristic color to the soils that are formed from them. Dark-colored Andisols are associated with a high allophane content; brown yellowish Andisols are dominated by halloisite; while brown reddish Andisols are related to kaolinite. White-colored Andisols are associated with the presence of gibbsite (Colmet Daage et al. 1973, Besoain 1985).

Owing to the presence of highly stable organomineral compounds, especially in the A horizon, Andisols tend to be very well structured. This results in a high infiltration capacity leading to both good drainage, and good moisture retention characteristics. One unfortunate consequence of these properties is that these soils promote the leaching of nitrate from agricultural systems and human sewage down to underground waters, contaminating ground waters and reducing their value for human use (Reynolds 1991), although these relationships have a variety of applications (Radulovich et al. 1992) that are currently being studied.

These soils have low bulk density and low resistance to tangential forces, making them easy to plow. In Costa Rica this task should be done by animal traction in order to prevent erosion, instead of using heavy machinery that tends to compact the soil (RELACO 1996). Overgrazing causes a similar effect. During periods of high volcanic activity, large amounts of very unstable ashes are deposited as blankets that cover the landscape. This material partially dissolves when subject to alternating dry and moist periods, inducing redistribution of soluble elements at the surface, cementing small pores, and reducing infiltration by crusting. This phenomenon develops into massive erosion, which encourages the formation of colluvio-alluvial fans at the bottom of the landscape. This is the main factor generating catastrophic events when deposited as "lahars" or "debris avalanches" (Alvarado, Vega, et al. 2004) in populated areas. Also, these soils are intensively used for agricultural activities that greatly trigger their erosion and cause silting of hydroelectric dams.

Most Andisols have a moderate fertility depending on the composition of their parent material. In general, soils formed from the ashes of the Irazú Volcano are richer in bases than those formed from Poás Volcano materials (Alvarado 1975); Andisols around Barú Volcano in the Southern Region are even poorer than those of the Central Valley. Nutrient leaching in volcanic areas is counterbalanced by new additions of volcanic ash; this process enables nature to maintain the base saturation of the ecosystem. Generally, Andisols have pH values near neutral except in agricultural areas with poor management or where the decomposition of abundant organic matter content of Andisols gradually acidify soils, particularly when large amounts of N are applied. When this happens they do respond to liming with calcitic (Ca carbonate) or dolomitic (Ca and Mg carbonates) products.

The soil fertility potential of Andisols can be estimated by the sum of cations (Ca, Mg, K, Na). Higher values indicate a better condition for crop development and imply that other nutrients are also abundant. In Andisols of the Southern Region, the predominance of plagioclases over orthoclases creates a pronounced K deficiency (Molina et al. 1986, Henriquez and Bertsch 1994). In recent volcanic ashes, N is the most limiting factor for crop production. But P, although abundant in total, creates difficulties for farmers too. P is held tightly by the clay lattices of Andisols so that it is not available to plants. Retention is generally over 70 percent, which is very high, and it can easily reach values of 95 percent. This problem constitutes by far the major limitation for crop development on these soils (Alvarado 1982, Canessa et al. 1987). In addition, B and S can also be tightly held as anions. The application of these two elements is essential for coffee production all over Costa Rica. Andisols formed in the lowlands of fluviovolcanic origin of the Northern Zone and part of the Atlantic Region, along the Sarapiquí, Sucio, Chirripó, Tortuguero, and Destierro rivers, are poorly understood. Under very high temperature and rainfall conditions, they seem to weather to form soils with more nutritional problems than those of the highlands. In addition, the low relief of these areas enables water to accumulate on the surface, thus enhancing soil compaction, particularly in pasture lands.

Owing to the high P retention of Andisols, most crops require large fertilizer applications with soluble P. The exact location and granule size of such fertilizers are important; it should be applied along with light applications of lime that increase the availability of P retained in organic materials. N is also a limiting factor for crop production, except when legume species are planted to fix N, such as when white clover is planted with kikuyo grass. Large applications of ammonium N result in the release of hydrogen ions, enhancing acidification of extensive areas, particularly in grasslands and coffee plantations. To correct for this condition, frequent applications of lime are required to get good yields, as has been used for sugarcane (Chaves and Alvarado 1994). Other elements, such as Mg, can limit crop performance if the parent material is low in Mg (Poás slopes, primarily) or when large K applications induce nutrient antagonisms. As with B and Zn, foliar and soil analyses should be run on a regular base to correct for deficiencies.

Tree-shaded coffee plantations require smaller additions of fertilizer than new full-sun varieties since shading reduces photosynthesis of coffee plants. If the shading tree is a legume species, such as *Erythrina* and *Inga*, biologically fixed N is added to the system. Coffee plantations also are associated with contour planting, use of tills, windbreak barriers, and hedge rows, practices which are necessary to reduce erosion, particularly during crop establishment. The use of agrochemical products on these types of soils have different effects over long times. In the case of potato fields many years of fertilization generates P accumulation. In areas where potatoes have been cropped for more than 25 years, concentrations of more than 80 ppm of available P have been found. In the case of cupric fungicides, used as disease protectors in intensively managed coffee plantations, however, Cu accumulates at a rate of approximately 1 ppm/year, which might become a long-term problem because plant toxicity begins at 100 ppm (Cabalceta et al. 1996).

Vertisols

Vertisols are found mainly in the Northwest Dry Pacific region of Costa Rica (see chapter 9 on Nicoya and Guanacaste by Jiménez, Carrillo, and Kappelle, this volume), on either plains or depressions where the dry season extends from 4 to 6 months, and are often associated with small patches of similar Mollisols. Although most Vertisols have a neutral or basic status, a few of them located near the border of Nicaragua are acid. Vertisols occupy only 2 percent of the country's area, and are restricted to depressional areas in the most important alluvial valleys of the Dry Pacific, and to similar locations in the western part of the Central Valley (Santa Ana, Pozos, Lindora, Ciruelas) (Fig. 4.4).

Vertisols are used intensively for both agricultural practices and—in the Central Valley—for urban development. During the rainy season the main crop on these soils is rice, either flooded or rain-fed. With irrigation and adequate soil water management sugarcane, soybean, melon, cotton, or even hot chili pepper and sauce tomato can be grown. Trees grow poorly on these soils, owing to root damage caused by alternate seasonal periods of dryness and water excess. Thus, commercial forests are neither abundant nor recommended on Vertisols. Even though pastures are found there, their management is very difficult and beef production remains very poor.

Vertisols in Costa Rica originate mostly from rhyolithic tuffs high in biotitic micas, with some recent additions of very fine volcanic ash. The confluence of several factors is necessary for Vertisols to form: the presence of a depressional zone, which prevents a good drainage; the occurrence of materials rich in Si, Ca, and Mg that accumulate in dry alluvial and/or fluvio-lacustrine deposits; and a well-defined season. An exception occurs in the Central Valley, where climatic and tectonic dynamics first created and then eliminated lakes, which served the same functions as depressional areas.

The conditions necessary for Vertisols also favor the formation of 2:1 montmorillonite type clays, which have very high Si content and the strongest colloidal properties of all clays (Alvarado et al. 2014b). This generates Si films interlayered between, but very poorly bonded with, montmorillonite particles. The result is very small, highly individual particles with unlimited and reversible water absorption capacity. The resulting soil is highly expandable, and has high specific surface activity, cohesiveness, adhesiveness, plasticity, and water retention capacity. What this means is that these soils can get both very wet and very dry, change their shape greatly, crack when they get dry, and become extremely slippery and sticky when wet. Even with their large water holding capacity, the difference between their field capacity and the permanent wilting point is rather low, so they dry out easily from a plant's perspective. Overall Vertisols are poorly suited for most agricultural and engineering operations owing to their contractions and expansions in response to seasonal fluctuations of rainfall.

Most Vertisols (usually Usterts) are less than 1 meter deep, are dark colored, and have little horizon differentiation and a clayish texture. Shallower Vertisols tend to lay over low permeability tuffs, and become water saturated and anoxic during the rainy season. Because of the reduced conditions grayish subsurface horizons (aquerts) are present. When the dry season arrives, Vertisols dry very drastically, forming massive blocks with open cracks in between that affect irrigation, electric poles, and engineering operations. At the onset of the rains, a vertical water flow runs down the cracks, causing the subsoil clays to expand rapidly. This process effectively seals the whole system, causing heavy floods as the rains increase. The use of mechanized cultivation on these soils is difficult and expensive.

Vertisols are potentially quite fertile soils, with high

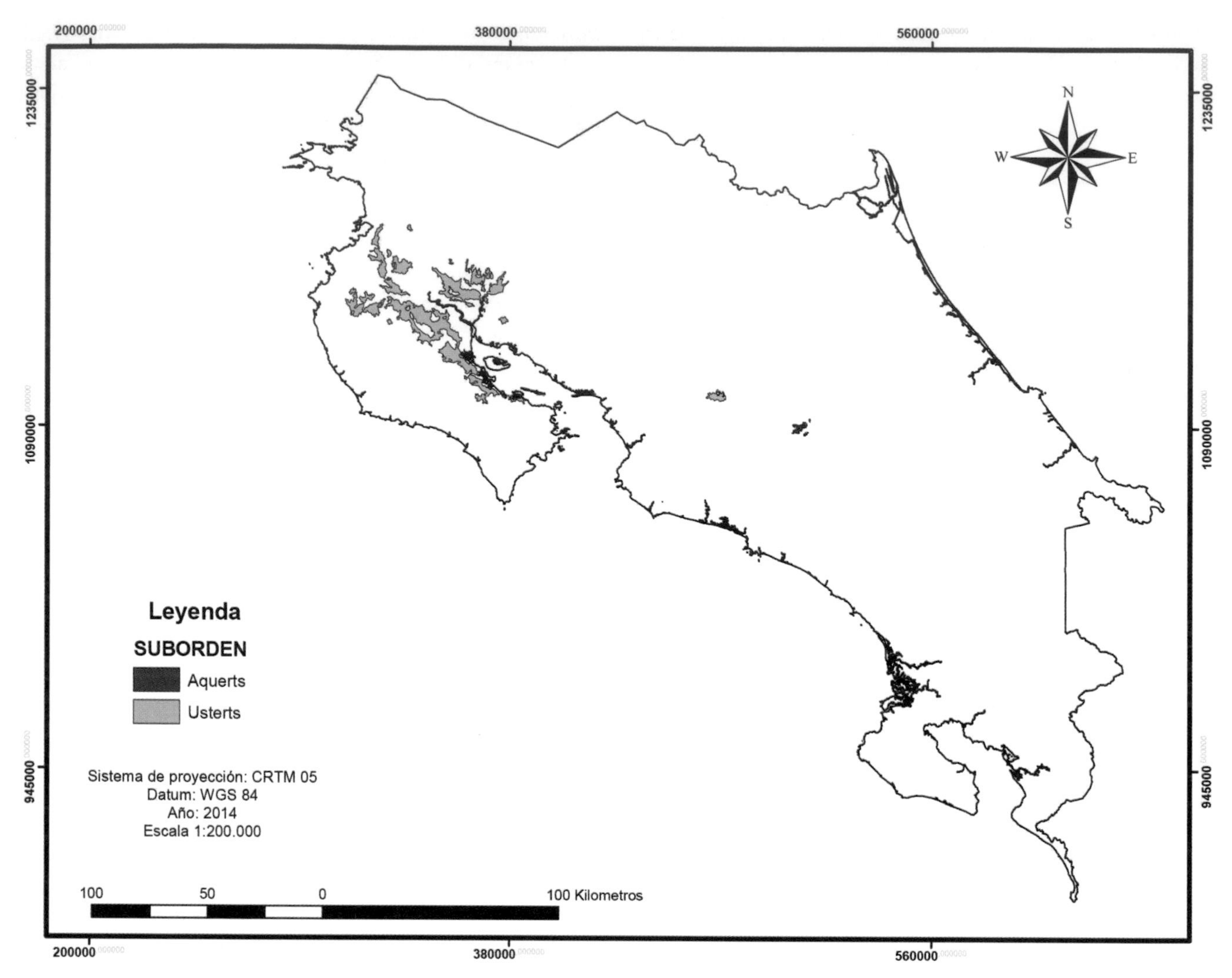

Fig. 4.4 Distribution and characteristics of Vertisols of Costa Rica.

pH, Ca, and Mg contents. Thus, the constraints for high productivity on Vertisols are mainly physical rather than nutritional. Nevertheless there are many nutritional problems that get in the way of its high potential fertility being expressed by high plant growth. Organic matter additions under flooding conditions may induce the reduction of Fe and Mn to toxic levels for most crops. The high Ca and Mg concentrations generate additional problems leading to difficult uptake of other nutrients by plants, and hence poor plant growth, especially when the K content is low. Even though Ca phosphate complexes are the most soluble among all phosphates, a plant's ability to use P is limited owing to its binding to Ca. Additionally, the content of minor cations is low in response to the high pH. All of these lead to serious limitations on plant growth.

The basic nutrient management strategy for Vertisols is maintenance fertilization with particular consideration of the levels of K and Zn (Sancho et al.1984). Sulfur fertilization may also be useful (Bornemisza 1990). The 2:1 clays display a high cation retention capacity, especially for K and NH_4, on both the external and the internal surfaces, resulting in peculiar behaviors of these cations. To reduce K-induced deficiencies, this element needs to be applied, particularly for annual crops. The use of pesticides must be planned carefully when crop rotation is practiced since the active ingredients can be trapped in clay particles during the first culture cycle to be released later when irrigation is applied during the second cycle. Irrigation of Vertisols in Costa Rica will be possible soon because of the hydropower plant projects taking place in Guanacaste Province. Significant investments in infrastructure, such as canals, need to be done in order to achieve a sustainable and profitable use of such irrigation. Research and technology adaptation programs are needed to ensure, for example, that the expandable clays will not destroy the new infrastructure.

Alfisols and Ultisols

The oldest and most weathered soils of Costa Rica belong to these orders, the differences being chemical and found in the subhorizon. Alfisols have more basic subhorizons and, particularly in Costa Rica, occur in dryer environments. In agronomic terms, both types of soils have a very similar "plow layer." The real differences arise after intensive use, when Ultisols start exhibiting more marked fertility problems. In Costa Rica, these soils occupy a large area: about 31 percent of the territory (21 percent Ultisols, 10 percent Alfisols).

In older times, and in other regions of the tropics today, the prevalent land use for these soils was slash-and-burn agriculture. This is not relevant in Costa Rica today because of the high input agricultural system, the wet climatic conditions where natural vegetation cannot be burned, and the high quality requirements for agricultural products. In general, they are considered marginal for contemporary agriculture because of their low and rapidly declining fertility, and only some of them are in use, particularly for roots, tubers, and pineapple. During the beef cattle boom of the 1970s, these soils were most preferred for grazing purposes. However, the cattle degraded these soils rapidly. Most of these pastures were abandoned, leading to abandoned grasslands, secondary shrublands, and, eventually, to secondary forests.

These soils, however, have some good functions when properly managed. Virtually all the pineapple produced in Costa Rica is grown on these soils, as well as significant amounts of citrus, mango, avocado, palm heart (*palmito*), sugarcane, roots and tubers, etc. In the Southern Pacific Region, large coffee plantations and *Gmelina arborea* plantations for pulp production are being established, although both face severe nutritional constraints. The acidity problems of many Ultisols might be reduced by liming, which decreases Al and increases fertility, or through the selection of species, varieties, or strains tolerant to acid soils and low P contents (Acuña and Uribe 1996, Uribe 1994).

Ultisols are found in the Northern Zone of Costa Rica in Sarapiquí, San Carlos, and Cutris districts; in areas of the Southern portion of the country in Pérez Zeledón and Buenos Aires district and in the proximity of the border with Panama, as well as the foothills (Atlantic and Pacific) of the Talamanca Mountain Range (Fig. 4.5).The main areas containing Alfisols are located on the Nicoya Peninsula and, associated with Vertisols, on the flood plains of the Tempisque River. In these areas, commercial plantation forests of *Tectona grandis*, *Bombacopsis quinata*, and *Gmelina arborea* have been successfully established, along with small coffee plantations. Alfisols also occur in the Central Pacific Zone in Grecia, Atenas, Orotina, and San Mateo districts, where small-scale fruit plantations (mango, tamarind, cashew, caimito) and recreational villas are the main forms of land use. Wherever they are found, these soils occupy the highest positions in the watersheds and along the slopes; that is, Alfisols are not subject to frequent addition of fresh materials and/or, are exposed to mild leaching conditions with consequent base accumulation at the subsoil (Fig. 4.6).

These soils originate from the downward flow of water through the soil profile over long periods of time, under high temperature conditions and from practically any parent material. Their main feature is the presence of an argillic clay horizon formed from the water-borne migration of clay particles from the superficial horizons to the deepest layers of the soil. For this movement to occur, precipitation must be higher than potential evapotranspiration under free-drainage conditions, that is, the water table must extend very deep into the soil and be separated from the surface. This process involves the loss of Na, K, Ca, and Mg from the soil profile, leaving behind high concentrations of Al, Fe, and Si (tri and tetravalent cations) in greater extent in Ultisols than Alfisols.

The high concentration of hydrated iron accounts for the reddish color of these soils. More specifically, they are brownish red to reddish in the concave parts of the relief and brownish yellow to yellow in the convex depressions when Fe is bound with water molecules. The most relevant criterion considered for classification of Ultisols and Alfisols is the presence of an argillic and/or kandic subsurface horizon, which is acid in Ultisols (humid tropics) and neutral or basic in Alfisols (dry/humid tropics).

Kaolinites (1:1 clays) as well as Fe and Al oxides predominate in these soils (Alvarado et al. 2014b). Even when composed of fine materials, the formation of H bonds in 1:1 clays fosters particle aggregation and, therefore, a more developed structure. As these aggregates get coated by oxides a larger particle known as "pseudosand" is formed. In some regions, Fe and Al accumulation is so high as to allow their exploitation as bauxite. Such deposits, also known as plinthite, display a white mottled surface over a red matrix in which gibbsite can be found.

The presence of stable aggregates in granular structures gives these soils excellent physical properties for agriculture, especially with respect to drainage. Overgrazing and intensive mechanization, however, can deteriorate their favorable physical properties irreversibly. Liming improves fertility, but when excessive, increases erosion by favoring clay deflocculation. Such effects influence productivity much more drastically in Ultisols, because of their low fertility. Unfortunately the good aggregation qualities of these soils also result in ideal conditions for nutrient losses, especially bases (Ca, Mg, K). This in turn, brings about severe acidity problems, including toxicity caused by Al and to a lesser extent Mn, and P availability problems due to its fixation on Fe and Al oxides and hydroxides surfaces. The rapid leaching also results in poor Effective Cation Exchange Capacity (ECEC) owing to a restricted specific surface of clay particle aggregation. Because no favorable conditions for organic matter accumulation are present, nitrates are easily lost by leaching and N availability is always limited (Schwartz 1998). Leaching of micronutrients due to acidity results in deficiencies more commonly observed in even older soils highly exposed to run-off. All of these properties in turn account for the low fertility of Alfisols and especially Ultisols. The priority in managing these soils is replacing the lost Ca and Mg by liming, along with the selection of acid tolerant germplasm. Agriculture is possible in these soils with an intense and well-balanced N-P-K fertilization program if an adequate supply of minor elements is included. The use of organic fertilizers, along with liming, can be an important source of nutrients while at the same time improving the physical properties altered by soil mismanagement.

Environmental Relationships between Soils, Litter, and Organisms

Soil as a Habitat

Many different types of animals live in and around the soil, leaving imprints in soil formation, nutrient recycling, and environmental biodegradation of organic residues (including pesticides), which have long been reported in tropical environments. Among the types of effects that animals

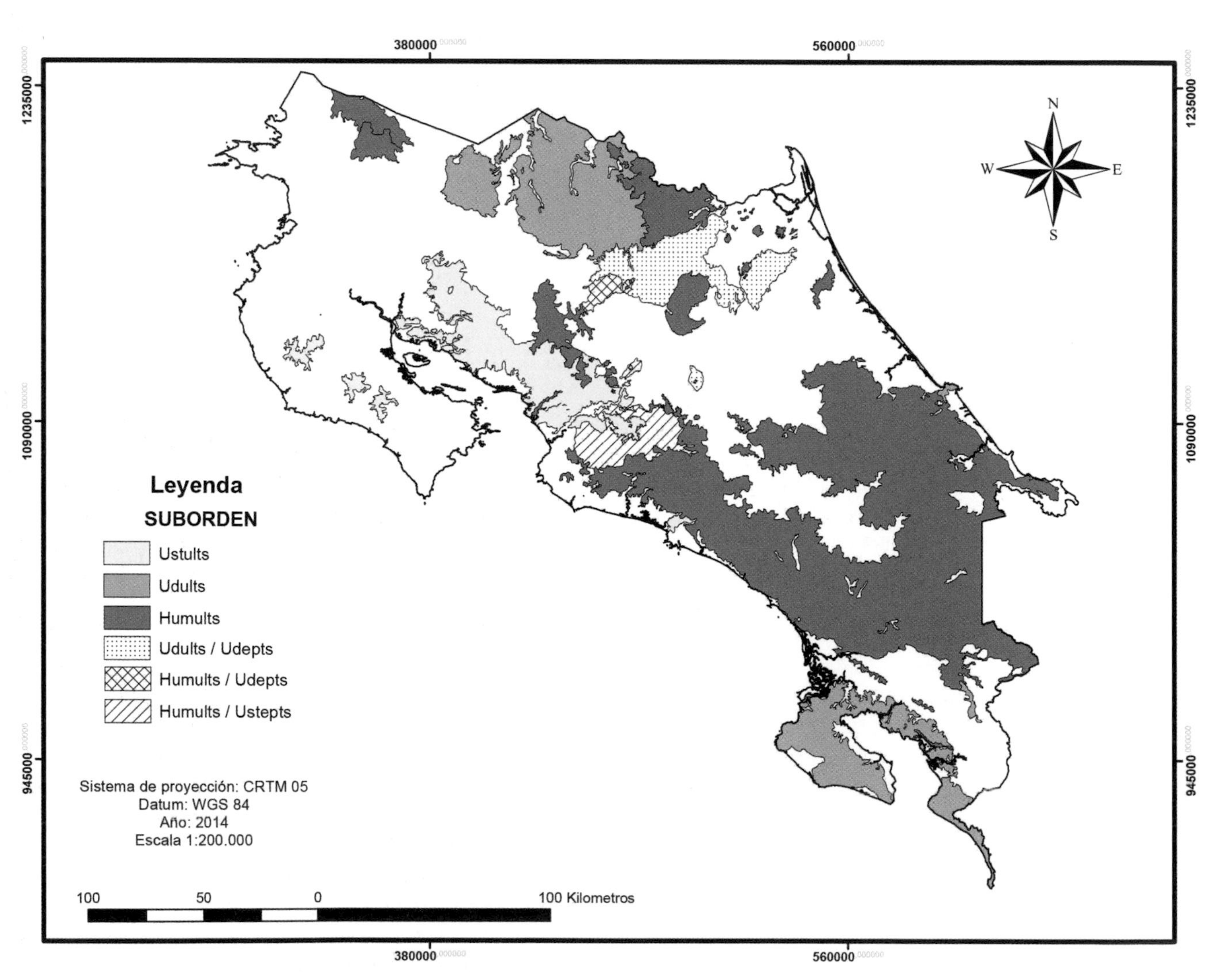

Fig. 4.5 Distribution and characteristics of Ultisols of Costa Rica.

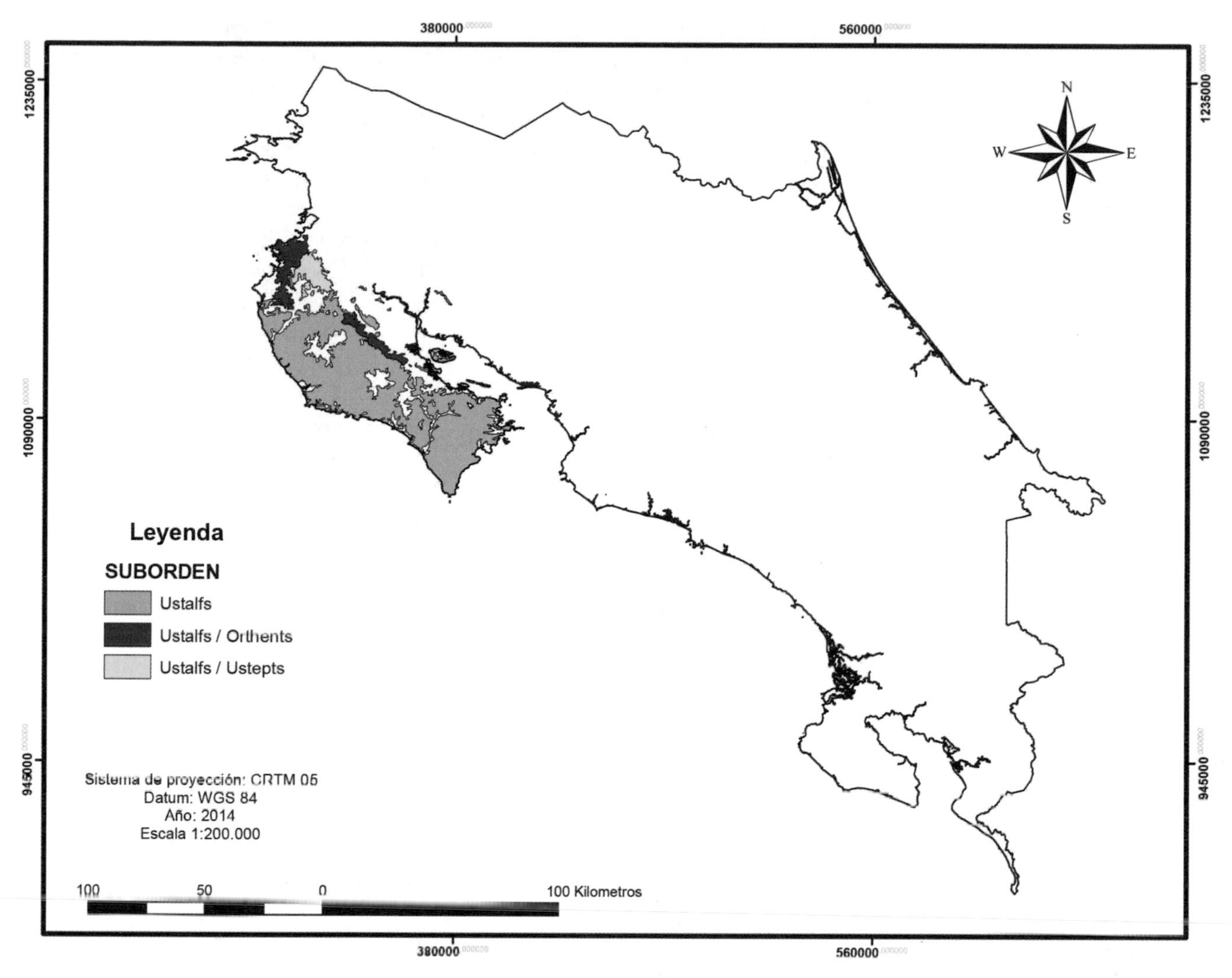

Fig. 4.6 Distribution and characteristics of Alfisols of Costa Rica.

leave in the soil, secretion of binding substances, burrows, and soil particle transportation (including organic matter/plant residues, clay, silt and sand particles), are mostly mentioned. Earthworms, termites, and ants are most visible, but other animals, like rodents, birds, and crabs, contribute to mixing soil material in specific environments as well. There is a group of animals that just live on the soil, causing little effect on its properties, including snails, snakes, and deer (Fig. 4.7).The various ways animals affect soil properties vary among them, mainly because of their size and number in the soils. The following sections intend to document these effects on Costa Rican soils.

Bacteria and Fungi

Bacteria and fungi in soil and soil litter are the most abundant organisms. From the agricultural perspective most of these microorganisms are harmful and well-studied, particularly when they turn into diseases for agricultural crops. Here, we describe several free N-fixing microorganisms and their host plants: *Rhizobium/Phaseolus leguminosarum* (bean), *Rhizobium/Erythrina* (*poró*), and *Frankia/Alnus acuminata* (alder). Mycorrhizae are discussed when related to forestry species (Fig. 4.8). Numerous publications address the beneficial effects of *Rhizobium* to the production of common *Phaseolus* beans, as well as the various soil properties that negatively affect the symbiotic relationship between the two partners (Ramírez and Alexander 1980a,b, Araya et al. 1986, Uribe et al. 1990, Acuña and Ramírez 1992a, Uribe 1993a,b, Castro et al. 1993, Acuña and Uribe 1996); similarly other studies emphasize the relationship between *Rhizobium/Glycine max* (Acuña and Ramírez1992b, Acuña et al. 1987, Ortiz et al. 1986, 1990), and others explain the relationship *Rhizobium/Erythrina* (Escobar et al.

Fig. 4.7 Snails in gardens and snakes in teak plantations in Costa Rica. Both are animals that live on top of the soil but have little effect on its properties.

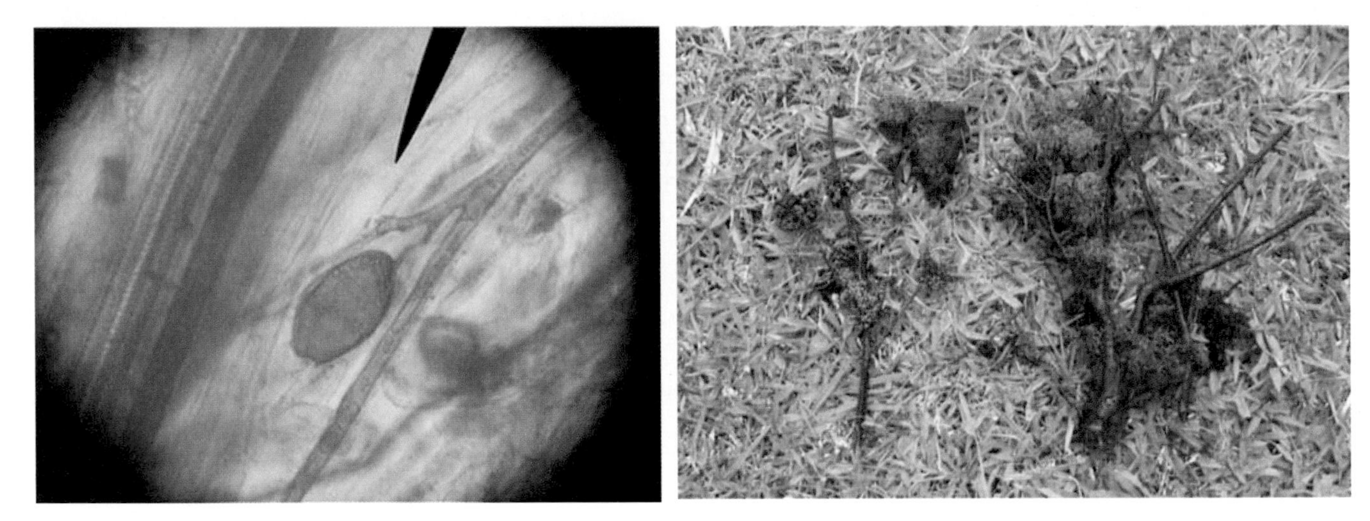

Fig. 4.8 Arbuscular mycorrhizae in teak roots and *Frankia* nodules in *Alnus acuminata* roots growing in Costa Rican Alfisols and Andisols, respectively.

1994, Ramírez and Flores 1994, Gross et al. 1993, Nygren and Ramírez 1993, 1995, Nygren et al. 1993). The symbiosis between *Frankia* and *A. acuminata* is also documented in various papers (Álvarez 1956, Russo 1989, Meza 1994, Segura et al. 2006). In the case of the mycorrhizae the available information mainly discusses the effects of its interaction with seedlings and trees in forestry nurseries and forest plantations (Vega 1964, Rojas 1992, Gadea et al. 2004, Alvarado et al. 2004).

Nematodes

Most nematodes are plant-parasitic species (e.g., see López and Salazar 1987), although a few free living nematodes exist in Costa Rica (Zullini et al. 2002). Various authors describe how soil texture, humidity, cation exchange capacity, pH, and organic matter content affect nematode population dynamics, directly and indirectly, by affecting the growth of living vegetation. In general, plant-parasitic nematode populations decrease with soil depth owing to a reduction in root biomass in the least aerated subsoil horizons; this has been proven for *Meloidogyne incognita* and *Rotylenchulus reniformis* on papaya plantations (Jiménez and López 1987), and rice fields all over Costa Rica (López and Salazar 1987, López 1988), with the exception of *Longidorus* sp. and *Criconemella palustris*. However, the population of nematodes depends to a major degree on the parasitic species they feed upon (López 1981), the crop phenology (Esquivel 1994), and the crop distribution in the field (Meneses et al. 2003). In order to control nematodes in banana plantations, nematicides were used in the past, causing human infertility to employees spraying the chemical, as reported by Ramírez and Ramírez (1980).

Arthropods

McGlynn et al. (2007) tested the effects of soil and litter nutrient stoichiometry on the invertebrate litter fauna of a Costa Rican tropical rain forest. Animal densities were estimated from 15 sites across a phosphorus gradient. The density of the invertebrate litter fauna varied, and was strongly tied to soil and litter phosphorus concentrations. An increase in phosphorus concentration corresponded with an equally proportionate increase in animal densities. Natural variation in nutrient levels can thus serve as a predictor of density in a highly diverse tropical animal community. Haggar and Ewel (1994) found that on fertile soils in the humid lowlands at La Selva Biological Station, phosphorus is abundant and preliminary data indicate that much of the organically bound P is under microbial control. Under the dry-wet climatic conditions of the Central Valley of Costa Rica, Herrera and Fournier (1977) and Fraile and Serafino (1978) found that soil and litter moisture availability plays an important role in determining the density and vertical distribution of soil microarthropods; Collembola, Protura, Symphyla, and Acarina were those most adversely affected by rainfall, while Coccoidea and other groups were unaffected by it (Fraile and Serafino 1978). Also, the invertebrate population was more diverse and stable in the oldest forest, but populations increased when litter moisture content increased with the first rains (Herrera and Fournier 1977). Populations of arthropods decrease with elevation above sea level (Bruhl et al.1999). Atkin and Proctor (1988) studied the litter and soil fauna in1 ha plots at six altitudes along a transect ranging from 100 to 2,600 m a.s.l. on the Caribbean slope of Barva Volcano. They showed that the invertebrate biomass in soils at 100 and 500 m a.s.l. seem to be the highest ever recorded for tropical rainforests. This is partially attributed to the presence of a clear soil temperature gradient (Fig. 4.9).

Springtails

Guillén et al. (2006a) studied the diversity and abundance of soil springtails (Collembola) in a primary forest, a secondary forest, and a coffee plantation, in Tapantí National Park. Each month, eight soil samples were taken in each ecosystem, totaling 360 samples. A total of 23,751 springtails were found, belonging to 9 families and 16 species. Of the three ecosystems, the primary forest was the most diverse (H' = 2.406), followed by the secondary forest (H' = 2.174), and the coffee plantation presented the lowest diversity (H' = 1.651). In contrast with diversity, the greatest abundance of springtails was found in the coffee plantation with 10,111 individuals. Guillén et al. (2006b) also found that the largest springtail biomass is associated with the highest organic matter content, lower penetration resistance, and lower pH values of the primary forest. The results showed an association between these variables and some collembolan species, which indicates that changes in the structure of collembolan communities can be used as biological indicators of soil quality and management of ecosystems.

Ants and Termites

In Costa Rica, 85 genera and at least 620 species of ant have been identified so far (Longino and Hanson 1995,

Fig. 4.9 Examples of Quilopoda and Diplopoda, common in soil litter of Costa Rican soils at mid-elevation.

Hölldobler and Wilson 1990). However, the number of genera and species present in different agroecosystems varies between 9–16 genera and 13–23 species in coffee plantations (Benítez and Perfecto 1989, Perfecto and Vandermeer1994, Barbera 2001), and 10–19 genera and 16–26 species in cacao plantations (Young 1986). The influence of some of these species on Costa Rican soil properties is also being recognized. Araya and Alvarado (1978) and Araya (1980) determined the influence of leaf-cutter ants (*Atta cephalotes*) on the chemical and morphological properties of soil of the Premontane Wet forest (Fabio Baudrit Experimental Station), the Premontane Wet forest transition to lowland (Santa Rosa National Park), and the Tropical Wet forest transition to per-humid conditions (La Lola farm). Results shows that vertical translocation of materials from the upper to the lower soil horizons and vice versa is large, and varies in different ecosystems; values of cation exchange capacity in ant-affected and undisturbed soil were 30 and 50 cmol(+) $100g^{-1}$, respectively, while Ca, Mg, and K contents were 2 and 23, 1 and 28, and 0.36 and 1.6 cmol(+) $100g^{-1}$, respectively. Other soil properties, like organic matter content, available P, pH, and sand, silt, and clay content, were also affected significantly.

Alvarado et al. (1981) studied the influence of leaf-cutter ants (*A. cephalotes*) on the morphology of twenty-seven soil profiles of Andisols distributed within a 2.5-ha site in Turrialba, Costa Rica. Leaf-cutter ant influence on each profile was noted in 85% of the soil profiles or *pedons*. The influence on each profile was estimated, and out of all profiles, 37% had low, 26% medium, and 22% high disturbance. The surface area covered by leaf-cutter mounds was 38.9% of the study area; only 1% of the aboveground disturbed area was active, however. Leaf-cutter ants transport material from the AB and B horizons to the soil surface, producing a new A1 horizon. In addition, some subsoil chambers are filled with plant material. Knowledge of termites that are active at soil level is still scanty, however (Fig. 4.10).

Fig. 4.10 Leaf-cutter ant (*Atta cephalotes*) (top) and termite mounts (bottom) in grasslands.

Worms

Monge and Alfaro (1995) studied the geographical variation of habitats in Costa Rican velvet worms (Onychophora: Peripatidae), comparing 20 onychophoran localities from the seasonally dry western Pacific forest to the rainforests of the Caribbean. The authors found that the Costa Rican species *Epiperipatus biolleyi* Bouvier, compared with the Brazilian species *Peripatus acacioi*, was found (1) in sandy, not clay-rich, soil, (2) closer to the surface, and (3) in burrows whose temperature is more similar to the external air temperature. For both species the sod humidity (mean, 35%) and acidity (pH = 5.2–6.2) were similar. The *E. biolleyi* population density was 0.25 individuals m^{-2}, and no clear-cut trends in associated flora and fauna were found but the animals preferred rotten to non-rotten wood, and water-soaked soil to oven-dried soil, during periods of inactivity.

Earthworms

The knowledge of soil-born native earthworm species of Costa Rica is limited. Esquivel (1997) found no relationship between density of earthworms and different agroecosystem properties in the wet humid lowlands of the Atlantic Zone; the author found an average of 194 individuals m^{-2}, with *Pontoscolex corethrurus* being the dominant species owing to its high adaptability to disturbed ecosystems. López and Kass (1996) found that *Erythrina* mulch and mucuna green manure treatments resulted in better phosphorus balances and higher earthworm populations and increased the yield of common bean (*Phaseolus vulgaris*).While sampling earthworm communities at eight sites of the Caribbean Coast to assess the distribution of the peregrine pantropical species *P. corethrurus* and its

Fig. 4.11 Earthworms and earthworm casts in Inceptisol grasslands of Guanacaste, Costa Rica.

relationships with native species depending on the type of land use, Lapied and Lavelle (2003) found that this species is largely dominant in almost all habitat types with a density range of 143 to 182 individuals m^{-2}.The species became dominant even in remaining plots of primary forests. In contrast, the species has not yet penetrated the large primary forest of the northeast of the country, where only native species could be found, and reached a maximum density of ca. 361 individuals m^{-2} in banana plantation sites. In all sites, a density increase of this species corresponds significantly with a reduction of the rest of the earthworm fauna except for *Dichogaster* sp. Where *P. corethrurus* was absent, density of other species reached 34.4 individuals m^{-2}. In southern Costa Rica, human immigration and sustained activities probably favored the establishment of *P. corethrurus*. León, Bolaños, and Fraile (1993) studied the relationship between edaphic conditions and the abundance and biomass of earthworms at eight sites in Costa Rica where organic waste accumulates, finding that the abundance and the biomass of *P. corethrurus* follows models that depend on soil carbon percentages (Fig. 4.11).

Crabs

The land crab *Gecarcinus quadratus* (Gecarcinidae) lives in densities exceeding 10,000 adults ha^{-1} in the coastal forests of Corcovado National Park, Costa Rica (Fig. 4.12). Crabs, living solitarily in half-meter deep burrows, forage nocturnally, transporting plant propagules and fallen leaves to subterranean chambers. The influence of *G. quadratus* on the distribution of soil organic carbon and root distribution in a Costa Rican rain forest was studied by Sherman (2006). Percent organic carbon in the crabzone (CZ) soils decreased with depth. The carbonless zone (CLZ) soils contained significantly more carbon at the topsoil and 32 cm depths but significantly less carbon at 72 cm. Carbon values at these depths, however, differed in regards to season. Vertical root profiles taken from the adjacent zones all indicated greater densities at the surface soils than below. The CZ had relatively lower root densities in the top 15 cm of the soil than the nearby CLZ. Surface densities of very fine and fine roots were 50% and 72% lower in the CZ than in the CLZ respectively.

Rodents

Many rodent species spend their lives in and around the soil, where they mix large amounts of material while digging the ground to build their nests and tunnels. According to Reid (1997) rodents of Costa Rica include 2 suborders, 8 families, and 48 species. Among the families, 33 species of rats and mice are most abundant in soil ecosystems. They belong to the families Muridae, Heteromyidae, and Echimyidae; the remaining rodents belong to Sciuridae (5 species of squirrels), Geomyidae (4 species of pocket gophers or *taltuzas*), and one species each in the Erethizontidae (porcupine), Dasyproctidae (*guatusa*), and Agoutidae (agouti, *tepezcuintle*) (Fig. 4.13).

Rodents can be grouped by size, according to Javier Monge (University of Costa Rica, pers. comm.), as follows: (1) large rodents (>2 kg/adult) such as the agouti or tepezcuintle (*Agouti paca*), the guatusa (*Dasyprocta punctata*), and the porcupine (*Coendou mexicanus*); (2) medium-sized rodents (0.1–1 kg/adult), including squirrels (Sciuridae), taltuzas (Geomyidae), and large rats (Echimyidae), and (3) small rodents (< 0.1 kg/adult) that includes other rats and mice

Fig. 4.12 Crabs on a sandy beach close to a mangrove forest at Montezuma, Guanacaste, Costa Rica.

Fig. 4.13 An agouti (*Agouti paca*, tepezcuintle) in captivity, Buenos Aires, Puntarenas, Costa Rica.

(Muridae and Heteromyidae). The pocket gophers occupy many ecological niches, except those with clayey soils. To give an idea of the amount of soil material reworked by this species: their nesting chamber might be up to 60 x 110 x 30 cm in width, length, and height, respectively; their tunnels can be as long as 192 m but average values lie between 30–80 m, disturbing an area of around 200–325 m^2 per individual (Monge 2006). According to Rodríguez and Vaughan (1985), a female agouti (*D. punctata*) has a home range of 3.9 ha, and maximum and minimum distances travelled daily reach 1,800 and 727 m, respectively. When roaming around in the La Selva tropical rain forest, agoutis not only mix soil materials but also help transport *Carapa guianensis* seeds (Arias 2001). The species is also the main secondary disperser of large seeds in the cloud forest of Monteverde (Wenny 2002).

Cattle

In the dry tropical ecosystems of Guanacaste, various authors have found that cattle grazing: (1) improves seed

Fig. 4.14 Soil compaction under shading trees (left) and hillside erosion (right) caused by cattle trampling in Guanacaste grasslands, Costa Rica.

dispersion via excreta of *Enterolobium cyclocarpum* and *Crescentia alata* (Janzen 1982a,b, Alvarado et al. 1982), (2) increases soil P availability, (3) reduces bulk density under dung droppings from 1.05 to 0.93 g cm^{-3}, (4) increases soil porosity from 13 to 21% (Herrick and Lal 1995, 1996), and (5) decreases aboveground residue accumulation, thus reducing fire frequency in dry tropical environments (Barbosa 1994). Changes in chemical properties of the soils are not as clear (Daubenmire 1972, Johnson and Wedin 1997). While evaluating the compaction and compactability of agricultural and cattle-raising soils in Guanacaste, Aguero and Alvarado (1983) found in cattle areas compaction average values of 62 kg cm^{-2}, values that doubled the ones found in cropping lands (average 30 kg cm^{-2}). These effects on the soil physical characteristics delay the regeneration of natural forest. The effect of cattle dung on the physical properties of the dry tropical pasture lands of Guanacaste was studied by Herrick and Lal (1995, 1996); these authors found an increase of soil porosity from 13 to 21% in the soil without dung, and under dung additions, respectively (Fig. 4.14).

Soils and Vegetation

As previously stated, to a certain extent soils can help determine the best land use. However, land use naturally affects soil properties, effects enhanced by human activities. As an example, the distribution of the cloud forest (wet tropical montane forest) and *páramo* vegetation at Cerro de la Muerte and Cerro Chirripó is associated with climatic and edaphic altitudinal gradients (Kappelle and van Uffelen 2005, 2006, Kappelle and Horn, chapter 15 of this volume). Inversely, in the same region, Blaser and Camacho (1990) described the occurrence of two types of soils developed from volcanic ash under different vegetation cover. Below the cover of mixed oak forest (*Chusquea talamancensis*, *Quercus costaricensis*, *Grammadenia myricoides*, *Prunus cornifolia*, and *Vaccinium consanguineum*), Placudands are dominant. However, below the white oak forest cover (*Chusquea tomentosa*, *Ardisia glandulosa-marginata*, *Quercus copeyensis*, *Weinmannia pinnata*, *Ocotea* spp., *Nectandra* spp., *Styrax argenteus*, and *Ilex* spp.), Dystrudands are dominant. The difference between the two soils is considerable, since Placudands present a thin hard layer of iron known as "placic horizon" formed by preferential movement of Fe, chelated by organic substances produced under the specific oak trees, thus impeding good drainage.

Among the various reasons explaining shifting cultivation in tropical environments, Sánchez (1985) mentions soil fertility decline, weed invasion, and the impact of insects and diseases on crop yield over time. The issue of soil fertility decline is strongly related to low-input agricultural systems, and usually associated with soil organic matter depletion, nutrient leaching and retention in highly weathered soils, and nutrient extraction via agricultural and forestry products (Hartemink 2003). Recently, studies on soil carbon sequestration tend to provide an alternative to identify proper land uses, with the purpose of recycling residues, planting crops that enhance soil organic matter accumulation, and applying environmentally friendly soil practices (Lal et al. 2006). The next paragraphs describe the main factors that affect Soil Organic Matter (SOM) in Costa Rica.

Soil Organic Matter

Differences in Soil Organic Matter (SOM) by Life Zone

Data collected by Alvarado (2006) show that Soil Organic Matter (SOM) content in Costa Rica increases from the tropical (warm) to the montane (cool) belt, both in the topsoil and subsoil (Table 4.1). The amount of residues added to the ecosystem in natural forests of Costa Rica decreases with altitude above sea level (Heaney and Proctor1989), 9.0, 6.6, 5.8, and 5.3 t ha^{-1} $year^{-1}$ at elevations of 100, 1,000, 2,000, and 2,600 m a.s.l., respectively. This is due to lower photosynthesis rates and exposition of vegetation to strong winds. However, the accumulation of residues under cool mountainous weather ecosystems is explained in terms of having low N content (Heaney and Proctor 1989), being hard and waxy (Holdridge et al.1971), containing high amounts of phenols (Montagnini and Jordan 2002), and having a low population of arthropods (Bruhl et al. 1999), all of them contributing to a reduction of residues' mineralization rates (Tanner et al. 1998). Powers and Schlesinger (2002a) also attribute this accumulation to the occurrence of amorphous clay minerals in mid-elevation Andisols in Costa Rica. These minerals make it difficult for organic materials to easily decompose.

As precipitation increases along vegetation gradients, soil organic matter increases, although the effect of temperature is more acute than that of moisture. Residue in lowland humid tropical forests undergoes a fast and efficient recycling process, hence its short life span in the ecosystem (Montagnini 2002). In tropical forests, the production of residue increases with average annual rainfall (Bernhard and Loumeto 2002) and its accumulation lasts only for the length of the dry season period.

On the basis of Holdridge's Life Zone distribution and area (ha) in Costa Rica, a total of 1,348.2 Tg of SOM was calculated for the country (Table 4.2). The life zones that contribute the most to SOM in Costa Rica are Premontane Wet Forest (391.0 Tg), Tropical Moist Forest (203.2 Tg), and Premontane Moist Forest (172.9 Tg), respectively.

Differences in Soil Organic Matter (SOM) by Soil Type

Small areas of peat deposits (Histisols) have been found in Costa Rica as the following: (1) thin blanket deposits about 1 m thick in the highlands of the Talamanca Range (e.g., at La Chonta close to El Empalme along the Inter-American Highway), such as in Sphagnofibrists (Kappelle and van Uffelen 2005), (2) peat layers interbedded with alluvium layers

Table 4.1. Soil Organic Matter (SOM %[a]) Content as Related to Holdridge's Life Zones in Costa Rica

Life Zone	Dry	Moist	Wet	Rain forest	AVERAGE
Tropical	3.3/1.0 (7)	3.3/0.9 (7)	4.3/1.3 (13)		3.6/1.1 (27)
Premontane		6.3/2.8 (2)	6.9/2.20 (6)	6.6/2.0 (5)	6.6/2.3 (13)
L. Montane		4.9/1.6 (1)	19.1/4.9 (1)	20.4/6.7 (3)	14.8/4.4 (5)
Montane				19.8/Rock (1)	19.8/Rock (1)
AVERAGE	3.3/1.0 (7)	4.8/1.8 (10)	10.1/2.8 (20)	15.6/4.3 (9)	

[a] SOM (0–0.30 m) / SOM (0.31–1.00 m) (Number of samples).
Source: Alvarado 2006.

Table 4.2. Soil Organic Matter (SOM) Stock in Costa Rica Calculated by Holdridge's Life Zones

LIFE ZONE (LZ)	Extension (Ha x 1,000)	Mg SOM to a depth of (m) x = 0.3	Mg SOM to a depth of (m) y = 1	SOM / LZ (Tg)	No. of samples	Regression equation	Regression coefficient
Tropical dry forest	150.3	82.7	142.9	21.5	7	y = 1.2697x + 37.852	0.955
Tropical moist forest	1,058.2	119.9	192.0	203.2	7	y = 1.4993x + 49.034	0.5945
Tropical wet forest	1,083.6	86.9	144.4	156.5	13	y = 1.1008x + 4.7945	0.9127
Premontane moist forest	556.7	193.5	310.5	172.9	2	nd	nd
Premontane wet forest	1,217.7	191.6	321.1	391.0	6	y = 2.6939x − 59.582	0.9598
Premontane rain forest	445.3	86.1	291.2	129.7	5	y = 6.283x − 249.47	0.9803
Lower montane rain forest	137.6	309.6	586.2	80.7	3	y = 1.7414x + 220.37	0.9994
Montane moist forest	335.5	188.2	342.0	114.7	1	nd	nd
Montane wet forest	1.9	228.3	420.8	0.8	1	nd	nd
Montane rain forest	118.7	244.3	651.6	77.3	6	y = 1.4338x − 24.441	0.9372
COSTA RICA	5,105.6			1348.2	51		

NOTE. nd = no data are available.
Source: Alvarado 2006.

south of Lake Nicaragua (Cocibolca), and (3) thick layers of peat at the Parismina River Basin. The accumulation of organic materials is the result of tectonism and volcanism (Cohen et al. 1986), including the burial of large amounts of vegetation by volcanic debris, and low temperature in the highlands. Apart from catastrophic events, above- and belowground factors that decrease or increase carbon pools in soils are those presented in Lal and Kimble (2000) and Buurman et al. (2004).

A study conducted in Costa Rica by Cabalceta (1993) to evaluate different methods for extracting soil-available nutrients also included SOM determinations as a part of the soils' characterization. In this study, 25 topsoils of each of the 4 major soil orders of the country were sampled, and results are presented in Table 4.3. The author reported a sequence for SOM content in the A soil horizon, in the following order: Vertisols < Inceptisols < Ultisols < Andisols. It is noteworthy that the larger the average of SOM of a particular soil order, the larger the range of its Soil Organic Content (SOC) is.

Under natural conditions, the sequestration of SOC in each soil order can be increased up to the maximum value of its range. However, unpublished results in organic farming systems show that in spite of the large amounts of compost or organic residues applied (10–30 Mg/ha), the total amount of C in soils rarely increases. According to Schlesinger (2000), only a small sink for SOC in soils may derive from the adoption of conservation tillage and the regrowth of native vegetation on abandoned agricultural land, but no net sink for SOC is likely to occur through application of manure to agricultural lands.

Soil organic matter contents for the country calculated by soil order (Alvarado 2006), came to a total of 1,445.7 Tg (Table 4.4), which is in fact a larger amount than found when using the Life Zone approach (Table 4.2), though within reasonable assumptions. SOM depends on soil order, reflecting their genesis: Entisols << Ultisols = Inceptisols < Mollisols = Alfisols << Andisols << Histisols. However, the soil orders that contribute most (owing to the size of the area they cover) were Inceptisols (492.8 Tg), Andisols (301.3 Tg), and Ultisols (261.5 Tg), respectively. In this case, correlations between SOM at 0.3 m vs. 1.0 m depth for each order were quite significant, with the exception of Entisols.

Soil Organic Matter (SOM) Turnover

Sauerbeck and González' (1977) study on the decomposition of ^{14}C-labeled wheat (*Triticum aestivum*) straw in 12 representative soils of Costa Rica showed that, under field conditions, after one year, 23 to 36% of the ^{14}C added in the wheat straw remained in the soil. However, four years later the residual ^{14}C ranged from 11 to 23%. The asymptotic model best fitted the turnover of the residues. Similar results were obtained while describing the turnover of banana (*Musa* spp.) (Vargas and Flores 1995) and peach palm (*Bactris gasipaes*) fresh residues and stems under field conditions (Soto et al. 2002), using decomposition litterbags. In these studies, fruits were harvested on a weekly basis, applying residues after each harvest. A linear model that relates to different stages of decomposed residues, like the one proposed in Heal et al. (1997), is more appropriate.

Table 4.3. Soil Organic Matter (SOM %) Content in 100 Samples of Vertisols, Inceptisols, Ultisols, and Andisols (25 A Horizons of Each Order) from Costa Rica

Soil Order	Minimum	Maximum	Average
Vertisols	1.6	5.9	3.5
Inceptisols	1.0	9.9	4.2
Ultisols	1.9	9.7	5.7
Andisols	4.8	24.0	10.9

Adapted from Cabalceta 1993.

Table 4.4. Soil Organic Matter (SOM) in Costa Rica Calculated by Soil Order

SOIL ORDER (SO)	Extension (Ha x 1,000)	Mg SOM to a depth of (m) x = 0.3	y = 1	SOM / SO (Tg)	No. of samples	Regression equation	Regression coefficient
Inceptisols	1,976.0	127.5	249.4	492.8	27	y = 2.1565x – 17.388	0.8411
Ultisols	1,069.0	138.4	244.6	261.5	15	y = 2.0729x – 42.378	0.7764
Andisols	750.0	222.0	401.8	301.3	14	y = 1.8351x – 5.6576	0.9133
Entisols	627.0	116.7	286.1	179.4	8	y = 1.3424x + 129.44	0.5315
Alfisols	487.2	156.7	260.7	127.0	8	y = 1.4003x + 41.212	0.8051
Vertisols	78.4	111.2	223.8	17.6	20	y = 1.6993x + 34.82	0.7232
Mollisols	69.0	144.0	269.2	18.6	16	y = 1.8974x – 4.1355	0.8006
Histosols	49.3	362.6	967.3	47.7	3	y = 8.2793x – 2035.1	0.9619
COSTA RICA	5,106.0			1,445.7	111		

Source: Alvarado 2006.

Other decomposition studies, using rain forest species residues, show similar results (Babbar and Ewel 1989, Byard et al.1996). Land management practices such as fire, grazing, tillage, and fertilizer application, among others, affect the distribution or SOC (Townsend et al. 2002). The accumulation of SOC underground is strongly related to the following: (1) fine roots decomposing naturally (humification) in the soil; (2) self-pruning of root during dry season in deciduous species (Ordóñez 2003); (3) root chopping by plowing the land (Veldkamp 1994); (4) root decay after pruning the crop (coffee); and (5) decay of microorganism's biomass. A special case of C build-up in the soil is that of so-called black carbon. This fraction is the result of burning lands for cropping purposes, and often due to common savannah and forest fires that may convert up to 2% of the standing biomass into charred material. Char production due to burning of deciduous wood is normally higher; at the same time, the char fraction in soils may last for long periods.

Soil Organic Matter (SOM) and Ecosystem Management

According to Gichuru et al. (2003), alteration of forest cover due to human intervention ranges from marginal modification to fundamental transformation; selective extraction of wood represents one extreme, while deforestation represents the opposite. After deforestation, land use changes will have a profound influence on soil properties (i.e., organic matter content and bulk density), since both the residue addition and decomposition rates are considerably affected (Ewel et al. 1981, Raich 1983, Veldkamp 1994, Johnson and Wedin 1997, Guggenberger and Zech 1999, Powers 2001, 2004, Powers and Schlesinger 2002a,b).

The recovery of C in the deeper layers of Andisols is higher than that of the A soil horizons, which is related to a lower presence of plant residues with increased depth and to an accumulation of fulvic acids in these layers (Alvarado 1974). Van Dam et al. (1997) observed that decomposition rates decrease strongly with depth and that diffusional transport alone is insufficient for the simulation of SOC movements into the soil; it had to be augmented by depth-dependent decomposition rates to explain the dynamics of SOC, delta ^{13}C, and delta ^{14}C. Cleveland et al. (2004) concluded that, with regards to the presence of physicochemical reactions with soil surfaces, humic (hydrophobic) fractions of dissolved organic matter (DOM) become more abundant than non-humic (hydrophilic) fractions over time. The latter fraction is the one that migrates into deeper layers of soil profiles. Both Cleveland et al. (2004) and Powers and Schlesinger (2002b) reported that neither the changes in delta ^{13}C isotopic fraction during DOM uptake by soil organisms, nor the difference in composition of litter and roots, explained the variation of delta ^{13}C values with soil depth.

To evaluate the impact of the change in vegetation cover on soil properties, three different data types are used to assess soil changes: (1) expert knowledge, (2) nutrient balances, and (3) monitoring of soil properties over time (Hartemink 2003). Independently of how the impact was measured, available information should be used to reduce data variability and improve data interpretation, while looking at (1) ecological differences (e.g., dry vs. wet tropics, or Andisols vs. Ultisols), (2) the kind of a change in land use (e.g., forest to grassland vs. grassland to secondary forest), and (3) type of data collected (e.g., changes monitored over time in samples taken at one site or chronosequential sampling vs. changes estimated for samples taken in different adjacent land-use systems sampled at the same time and known as false-time series). The effect of the ecological impact on soil organic matter accumulation was covered in previous sections, while the impact of the change in vegetation cover will be discussed in the next paragraphs.

Changes in Soil Organic Content (SOC)

According to Veldkamp (1994) deforestation, followed by 25 years of pasture, caused a net loss of 21.8 Mg/ha of SOC for Eutric Hapludands and 1.5 Mg/ha for Oxic Humitropepts in the Atlantic lowlands of Costa Rica. In an Andic Humitropept the author found that the decomposition of tree roots caused an extra input of SOC during the first year after deforestation and a strong stabilization of SOC by forming Al-organic matter complexes.

Similarly, slash and burning of a 8–9-yr-old evergreen forest around Turrialba, Costa Rica, volatilized 31% of the initial amount of SOC, 22% of N, and 49% of S. Soil CO_2 evolution was higher beneath an 11-week-old slash field (3.6 g/m^2 daily of C) than from beneath an evergreen forest (2.5 g/m^2), probably because slash conserved soil moisture better than actively transpiring forest (Ewel et al. 1981). Particle-sized separation of SOM, where particulate SOM (light fraction and sand-associated SOM) is separated from mineral-bond SOM (silt- and clay-associated SOM), revealed that under agricultural use of a soil formerly under primary forest in the country's northern lowland Huetar region, a depletion of the particulate SOM occurred, whereas clay- and silt-bond SOM was less affected (Guggenberger and Zech 1999).

These authors also observed that abandonment of pastures and growth of secondary forests raised SOC content in all separates to a pre-cultivation level within 18 years, and sand-associated SOC was even higher compared with values for the primary forest. Results suggest that land use primarily influences the balance across the light fraction and the size separates, with the particulate SOM pool being the most significant component in the context of management impacts on these soils.

Powers (2001, 2004) looked for changes in total SOC concentration (Mg C/ha) at a depth of 0.30 cm that occur during different land-use transitions in northeastern Costa Rica. The study included 12 sites where old-growth forest was converted into banana plantations, 15 sites that suffered a conversion of pasture to cash crops, and four sites demonstrating the change of pastureland into *Vochysia guatemalensis* forest. At the sites converted to banana plantations, the top soil SOC concentration decreased from 37% to 16.5%. Similar results were obtained for sites where pastures were replaced by crops. However, sites with soils now under *V. guatemalensis* did not show an increase in SOC storage, at least not during the first decade. In conclusion, reduced C input to the soil may be a key reason that explains the loss of SOC pools during land-use changes in ecosystems replaced by pastures or crops, since SOC restoration rates appear to be slow.

To determine how the conversion in a Tropical Premontane Wet Forest and a nearby secondary forest affects the SOC budget, major soil C storages, inputs, and CO_2 evolution from a tropical Inceptisol were measured by Raich (1983) over a six month period. Total C storage in and on the mature forest soil comprised 9,330 g C m^{-2} in SOM, 1,850 g C m^{-2} in litter, and 340 g C m^{-2} in small roots (diameter of 5 mm); larger roots were not measured. Average daily inputs to the mature forest soil include 1.3 g C m^{-2} in litterfall and 0.10 g dissolved organic carbon (DOC) m^{-2} in precipitation (throughfall + stem flow). The evolution of CO_2 from the mature forest averaged 3.4 g C m^{-2} d^{-1} or 2.6 times the average rate of litterfall. Total C storage in and on the secondary growth soil was composed of 8,600 g C m^{-2} in SOM, 700 g C m^{-2} in litter and 157 g C m^{-2} in small roots, or 2,060 g C m^{-2} less than in the mature forest. Average daily inputs to the mature forest soil include 0.7 g C m^{-2} in litterfall and 0.12 g DOC m^{-2} in precipitation (throughfall + stem flow). The evolution of CO_2 from mature forest averaged 4.6 g C m^{-2} d^{-1} or 1.4 times the average rate in the mature forest. Measured inputs of C to soils were considerably less than soil-CO_2 evolution rates at both sites.

Johnson and Wedin (1997) reported a loss of SOM in the Guanacaste region while comparing mature forest and grassland soils—an effect attributed to a larger rate of mineralization of residues due to higher temperatures in the later ecosystem. In a study to determine soil changes associated with the conversion of grasslands to 13-year-old and 17-year-old secondary forests in tropical dry forest of Guanacaste on Andic and Typic Haplustepts, Alfaro et al. (2001) found that plant cover did not affect SOM, pH, soil acidity, and K content; however, Ca and Mg contents were higher under the 13-year-old forest cover (nutrients associated to litter added to 7.03%) than the 17-year-old forest cover (nutrients associated to litter added to 4.51%). In the same environment, Leiva (2007) did not find any differences in SOC due to changes in vegetation cover after pasture abandonment in Entisols of the ignimbritic plateau of Guanacaste; however, the author found nutritional differences related to forest age (an initial decrease and a further increase of nutrient availability with forest age).

Oelbermann et al. (2004) mention that the potential to sequester C in aboveground components in agroforestry systems is estimated to be 2.1×10^9 Mg C $year^{-1}$ in tropical and 1.9×10^9 Mg C $year^{-1}$ in temperate biomes. Studies from Costa Rica have shown that a 10-year-old system with *Erythrina poeppigiana* sequestered C at a rate of 0.4MgC ha^{-1} $year^{-1}$ in coarse roots and 0.3 Mg C ha^{-1} $year^{-1}$ in tree trunks. Tree branches and leaves are added to the soil as mulch, contributing 1.4 Mg C ha^{-1} $year^{-1}$ in addition to 3.0 Mg ha^{-1} $year^{-1}$ from crop residues. This resulted in an annual increase of the SOC pool by 0.6 Mg ha^{-1} $year^{-1}$. Oelbermann et al. (2005) quantified the C stock of tree roots and C input from tree prunings and crop residues in 19-, 10- and 4-year-old *E. poeppigiana* and *Glyricidia sepium* alley cropping systems in Costa Rica. The 19-year-old alley cropping system was studied at two fertilizer levels (tree prunings only [–N], and tree prunings plus chicken manure [+N]), and was compared with a sole crop. The 10- and 4-year-old systems were also studied at two fertilizer levels (tree prunings only [–A], and tree prunings plus *Arachis pintoi* as a groundcover [+A]), and compared with a sole crop. In the 19-year-old system C input from *G. sepium* was significantly greater compared with *E. poeppigiana*, but for both tree species there was no significant difference between +N and –N treatments. For the 10- and 4-year-old systems, *E. poeppigiana* had a significantly higher C input from prunings compared to *G. sepium*, and the presence of *A. pintoi* increased pruning biomass productivity significantly in these systems. Tree roots of 10- (4,527 kg C ha^{-1}) and 4-year-old (3,667 kg C ha^{-1}) *E. poeppigiana* represented 16 and 28% of the total C allocation. Carbon input from maize (*Zea mays* L.) and bean (*Phaseolus vulgaris* L.) residues were not significantly different between alley crops and sole crops in the 19-year-old system per unit of cropped land. In this system, +N treatments had a significantly greater C input from bean residue than in –N treatments, but no such trend was observed for maize residues. Carbon input from maize and bean residues were significantly greater in alley crops than the sole crops, but not significantly different between +A and –A treatments in the younger system. The greatest input of organic material occurred in the 19-year-old alley crop followed by the 10- and 4-year-old alley crops.

Nutrient Availability

The soils of Costa Rica are, in overview, moderately fertile by global standards, but very fertile compared to other tropical conditions. The old soils of much of Africa and Amazonia present far more severe problems, but the young soils of Northern glaciated areas in the Midwest of the USA are much more fertile. According to Bertsch (1986), the most important chemical qualities of soils with respect to their nutritional value for plants relate to the abundance and availability of the major plant nutrients, nitrogen, phosphorus, and potassium (NPK). By these criteria specific tracks of land in Costa Rica present some problems for economic agricultural activities. As in other tropical areas, all Costa Rican soils are deficient in N. In addition, 74 percent are deficient in P, and 22 percent of the country is deficient in K. Ca and Mg are low in 35 percent of the soils, a problem considered more relevant than Al toxicity that occupied a larger area (20–30 percent). There is less information on micronutrients. Boron deficiency is probably most important, followed by Zn (26 percent) and Mn (23 percent). Very few parts of the country present Fe deficiencies (6 percent), and only some areas show Cu toxicity problems (Cordero and Ramírez 1979). All of these nutritionally related problems can easily be overcome with the use of soil amendments like lime and fertilizers.

More recent literature seems to be showing that the physical properties of soils can restrict agricultural activities at a national level. Most important are shallowness, steep relief, erosion susceptibility (see section 2.3), flooding risk, drought-inducing sandy textures or heavy textures (generated by swamps), and compacted soils (Beets 1990). Animal nutrition as well as human can be influenced by the soils via mineral concentration in feeding material. The quantity of most nutrients in pasture soils and hence forages is adequate (Bertsch 1986) except that 63 percent of soil pasture samples had insufficient levels of Co or Cu for ruminant growth (Vargas et al. 1992, Vargas and McDowell1993). In past years Costa Ricans have suffered from iodine deficiency, a typical problem related to volcanic ash-derived soils. When salt was iodized this deficiency disappeared except for some remote coastal regions where people consume non-iodized salt.

Soil Compaction and Erosion

Land deterioration via soil compaction is reported by different authors in Costa Rica. An example of minor impact is presented by Radulovich and Sollins (1985) who described the effect of foot traffic at La Selva Biological Station where penetration resistance (kg cm^{-2}) increased from 0.41 to 0.96, while bulk density increased from 0.63 to 0.68. A major impact case is presented by Agüero and Alvarado (1983) who measured the effect of cattle trampling on grasslands in Guanacaste dry lowlands, finding penetration resistance (kg cm^{-2}) values of 62 in cattle areas, a value that more than doubles the agricultural land average of 30. On the hillsides of the Poás Volcano, wood volume of 15-year-old plantations of *Cedrela odorata* decreased from 110 to 11 m^3 ha^{-1} and was associated with changes in soil bulk density between 0.96–1.12 and 1.18–1.34 g cm^{-3}, respectively (Castaing 1982). In the same area but on kikuyu grasslands (*Pennisetum clandestinum*) compaction is being caused by cattle stepping on top of poorly aerated soils (Mora 1988). Lang (2000) attributed large soil erosion rates while logging wet tropical forest in the Península de Osa, at least partially, to the compaction of road surfaces. Tafur and Forsythe (1988) discovered that a slight increase in penetration resistance in sweet potato (*Ipomoea batatas*) plantations reduces the negative effect of *Rhissomatus subcostatus* (Coleoptera: Curculionidae) to the tubers; however, in the case of maize, disking the soils at field capacity reduced corn yield by 48% compared with disking the dry soil.

The main factors causing erosion under tropical humid conditions are when frequent rainfalls exceed the soil infiltration rate, hence increasing run-off, a problem aggravated by loss of soil cover. Also, catastrophic climatic events, including tropical storms, hurricanes, floods, and drought, induce soil erosion, and geologic phenomena, including volcanic ash deposition and earthquakes, accelerate soil erosion. Overall, however, it is important to emphasize that many of these processes also form soils. The problem of erosion occurs when the natural factors interact with loss of vegetation cover, such as is generally the result of human activities on the land. In particular, the level of erosion is high in areas with greater road and human settlement density. This can be seen in a comparison of the Pacific Slope, which was settled first, with the Atlantic Slope. Even though the Atlantic Slope gets more rain the Pacific Slope suffers from greater erosion impact.

Erosion in Costa Rica, according to the Tropical Science Center (TSC; 1991) is larger in areas planted with annual crops than in grasslands and relatively small when the land is planted with perennial crops. Even though estimated erosion values for the country steadily increased from 1973 to 1983, after this period soil losses have been stable, reaching an estimated value of 190,000 MT/year.

Conclusions

The variability of the soils of Costa Rica is almost as large as the number of different agroecosystems developed to match the varied ecological niches. Farmers and scholars (both national and international) have contributed to the development of more environmentally sound agricultural practices. Soil management errors of the past are lessons for the present. Therefore, new agricultural practices, including the use of soil inoculants (*Rhizobium* and Mycorrhizae), compost, minimum tillage, crop associations, and greenhouse operations, are now *en vogue*. It is expected that these approaches will lead to the conservation of soils, in terms of controlling both chemical contamination and physical erosion.

The considerations made here about soil organisms and land use changes on organic matter changes in the soil are pertinent in an environment where this type of knowledge is misused or little used. International studies on issues like soil carbon sequestration, recuperation of degraded lands, organic farming, use of imported inoculants, and recognition of bioindicators are examples of soil research that need less emphasis; on the contrary, soil research is needed to enhance nutrient use by crops, participatory approaches to involve more farmers into the generation of new knowledge, and identification of new agricultural possibilities with added value. Among the few alternatives to improve land (soil) use, ecotourism seems important. Additionally, an assessment of the carrying capacity of soils in Costa Rica's main protected areas visited by ecotourists is also urgently needed.

This chapter shows that modern national and international soil research has contributed equally to elucidate soil-related issues and understand links among soils, ecosystems, and people in Costa Rica. The authors hope that this positive trend continues, not only in Costa Rica, but also in other countries on the American continent.

References

Acuña, O., and C. Ramírez. 1992a. Contenidos de nitrato y amonio en dos suelos tratados con diciadiamida y urea y su efecto sobre el crecimiento y nodulación de la soya (*Glycine max* Merr.). *Turrialba* 42: 127–32.

Acuña, O., and C. Ramírez. 1992b. Efecto de los protozoarios sobre la colonización y nodulación en el frijol (*Phaseolus vulgaris*) por *Rhizobium leguminosarum* biovar *phaseoli*. *Turrialba* 42: 411–14.

Acuña, O., C. Ramírez, R. Montero, and E. Mata. 1987. Respuesta de la soya (*Glycine max* Merr.) en Liberia, Guanacaste, a la inoculación con diversas cepas de *Rhizobium japonicum*. *Agronomía Costarricense* 11: 33–38.

Acuña, O., and L. Uribe. 1996. Inoculación con 3 cepas seleccionadas de *Rhizobium leguminosarum* bv. *phaseoli*. *Agronomía Mesoamericana* 7(1): 35–40.

Agüero, J.M., and A. Alvarado. 1983. Compactación y compactabilidad de suelos agrícolas y ganaderos de Guanacaste, Costa Rica. *Agronomía Costarricense* 7(1/2): 27–33.

Alfaro, E.A., A. Alvarado, and A. Chaverri. 2001. Cambios edáficos asociados a tres etapas sucesionales de bosque tropical seco en Guanacaste, Costa Rica. *Agronomía Costarricense* 25(1): 7–20.

Alvarado, A. 1974. A volcanic ash soil toposequence in Costa Rica, Central America. M.Sc. Thesis, North Carolina State University. Raleigh, NC. 89 pp.

Alvarado, A. 1975. Fertilidad de algunos andepts dedicados a potreros en Costa Rica. *Turrialba* 25(3): 265–70.

Alvarado, A.1982. Phosphate retention in andepts from Guatemala and Costa Rica as related to other soil properties. PhD Thesis. North Carolina State University. Raleigh, NC. 82 pp.

Alvarado, A. 1985. *El origen de los suelos*. *Turrialba*. Costa Rica: CATIE. 52 pp.

Alvarado, A. 2006. Potential soil carbon sequestration in Costa Rica. In R. Lal et al., eds.,*Carbon Sequestration in Soils of Latin America*, 147–68. Binghamton, NY: The Harworth Press.

Alvarado, A., C.W. Berish, and F. Peralta. 1981. Leaf-cutter ant (*Atta cephalotes*) influence on the morphology of andepts in Costa Rica. *Soil Science Society of America Journal* 45(4): 790–94.

Alvarado, A., F. Bertsch, E. Bornemisza, G. Cabalceta, W. Forsythe, C. Henríquez, R. Mata, E. Molina, and R. Salas. 2001. *Suelos Derivados de Cenizas Volcánicas (Andisoles) de Costa Rica*. San José, Costa Rica: Asociación de la Ciencia del Suelo. 111 pp.

Alvarado, A., M. Chavarría, R. Guerrero, J. Boniche, and J.R. Navarro. 2004. Características edáficas y presencia de micorrizas en plantaciones de teca (*Tectona grandis* L.f.) en Costa Rica. *Agronomia Costarricense* 28(1): 89–100.

Alvarado, A., R. Mata, and M. Chinchilla. 2014a. Arcillas identificadas en suelos de Costa Rica a nivel generalizado durante el período 1931–2014: I. Historia, metodología de análisis y mineralogía de arcillas en suelos derivados de cenizas volcánicas. *Agronomía Costarricense* 38(1): 76–106.

Alvarado, A., R. Mata, and M. Chinchilla. 2014b. Arcillas identificadas en suelos de Costa Rica a nivel generalizado durante el período 1931–2014: II. mineralogía de arcillas en suelos con características vérticas y oxídico-caoliníticas. *Agronomía Costarricense* 38(1): 107–131.

Alvarado, A., A. Stam, and S. Readhead. 1982. *Crescentia alata* (jícaro) and Vertisol distribution in burned savannah in Guanacaste, Costa Rica. In B. Williams, ed., *OTS Tropical Biology: an Ecological Approach*, 245–47. Dept. of Biology, University of Miami, Florida.

Alvarado, G.E., E. Vega, J. Chaves, and M. Vásquez. 2004. Los grandes deslizamientos (volcánicos y no volcánicos) de tipo debris avalanche en Costa Rica. *Revista Geológica de América Central* 30: 83–100.

Álvarez, H. 1956. Estudio forestal del jaúl (*Alnus jorullensis* HBK) en Costa Rica. Tesis maestría, Instituto Interamericano de Ciencias Agrícolas. Turrialba, Costa Rica. 87 pp.

Araya, L.M. 1980. Influencia de la hormiga *Atta* spp. en la génesis de suelos en tres ecosistemas de Costa Rica. Thesis Ing. Agr., Universidad de Costa Rica. San José, Costa Rica. 90 pp.

Araya, L.M., and A. Alvarado.1978. Influencia de la hormiga *Atta* spp. en la génesis de suelos. III Congreso Agronómico Nacional, Resúmenes 1: 86. San José, Costa Rica: Universidad de Costa Rica.

Araya, R., O. Acuña, and C. Ramírez. 1986. Efecto del fósforo y del *Rhizobium phaseoli* en frijol común intercalado con cafeto. *Agronomía Costarricense* 12(1): 81–86.

Arias, H. 2001. Remoción y germinación de semillas de *Dipteryx panamensisy* y *Carapa guianensis* en bosques fragmentados de Sarapiquí, Costa Rica. *Revista Forestal Centroamericana* 34: 42–46.

Arroyo, L.A. 1996. Método de evaluación de tierras para cultivos anuales, por medio del sistema de información geográfica: estudio de caso, distrito de Upala. In F. Bertsch et al., eds., *Congreso Nacional Agronómico y de Recursos Naturales*, 10th ed. (San José, Costa Rica, 1996), 29–37. Memorias. San José, Costa Rica: EUNED, EUNA.

Atkin, L., and J. Proctor. 1988. Invertebrates in the litter and soil on Volcán Barva, Costa Rica. *Journal of Tropical Ecology* 4(3): 307–10.

Babbar, L.I., and J.J. Ewel. 1989. Descomposición del follaje en diversos ecosistemas sucesionales tropicales. *Biotropica* 21(1): 20–29.

Barbera, N. 2001. Diversidad de especies de hormigas en sistemas agroforestales contrastantes de café, en Turrialba, Costa Rica. Thesis Mg. Sc., Centro Agronómico de Investigación and Enseñanza. Turrialba, Costa Rica. 78 pp.

Barbosa, G. 1994. Restauración ecológica y control de incendios a través del pastoreo en el bosque seco de Palo Verde, Costa Rica. San José, Costa Rica: Organización de Estudios Tropicales. 7 pp.

Beets, W.C. 1990. Raising and sustaining productivity of small-holder farming systems in the tropics. Holland, AgBe Publishing. 738 pp.

Benítez, J., and I. Perfecto. 1989. Efecto de diferentes tipos de manejo de café sobre las comunidades de hormigas. *Agroecología Neotropical* 1(1): 11–15.

Bennett, H.H.1926. General soil regions of Eastern and Northern Costa Rica. In J.C. Treadvell, *Possibilities to Rubber Production in Northern Tropical America*, 66–83.Washington, D.C.: Bureau of Foreign and Domestic Commerce.

Bernhard, F., and J.J. Loumeto. 2002. The litter systems in African forest-tree plantations. In M.V. Reddy, ed., Management of Tropical Plantation-forests and Their Soil-litter System, 11–39. Enfield, NH: Science Publishers, Inc.

Bertsch, F. 1986. Manual para Interpretar la Fertilidad de los Suelos de Costa Rica. San José: Oficina de Publicaciones, Universidad de Costa Rica. 81 pp.

Bertsch, F., A. Alvarado, C. Henríquez, and R. Mata. 2000. Properties, geographic distribution, and major soil orders of Costa Rica. In C.A.S. Hall, ed., *Quantifying Sustainable Development*, 265–94. San Diego, CA: Academic Press.

Bertsch, F., and V. Vega. 1991. Dinámica de nutrimentos en un sistema de producción con bajos insumos en un Typic Dystropept del trópico muy húmedo, Río Frío, Heredia, Costa Rica. In T.J. Smyth, W.R. Raun, and F. Bertsch, eds., *Manejo de Suelos Tropicales en Latinoamérica*, 28–32.Raleigh, NC: North Carolina State University.

Besoain, E. 1985. *Mineralogía de Arcillas de Suelos*. San José, IICA Serie de libros y materiales educativos no. 60. 1205 pp.

Blasco, M. 1970. *Microbiología de Suelos* (curso). Turrialba, Costa Rica: Instituto Interamericano de Ciencias Agrícolas. Mimeografiado. 246 pp.

Blaser, J., and M. Camacho. 1990. *Estructura, Composición y Aspectos Silviculturales de un Bosque de Roble (Quercus spp.) del Piso Montano de Costa Rica*. Turrialba, Costa Rica: Proyecto CATIE/COSUDE, Silvicultura de Bosques Naturales. 241 pp.

Bornemisza, E. 1990. *Problemas del Azufre en Suelos y Cultivos de Mesoamérica*. San José: Editorial de la Universidad de Costa Rica. 101 pp.

Bruhl, C.A., M. Mohamed, and K.L. Linsenmair. 1999. Altitudinal distribution of leaf litter ants along a transect in primary forests on Mount Kinabalu, Sabah, Malaysia. *Journal of Tropical Ecology* 15(3): 265–77.

Buurman, P., M. Ibrahim, and M.C. Amézquita. 2004. Mitigation of greenhouse gas emissions by tropical silvopastoral systems: optimism and facts. In L. Mannetje, L. Ramírez, M., C. Ibrahim, J. Sandoval, M. Ojeda, and J. Ku, eds., *The Importance of Silvopastoral Systems in Rural Livelihoods to Provide Ecosystem Services*, 61–72. Mérida, México.

Byard, R., K.C. Lewis, and F. Montagnini. 1996. Leaf litter decomposition and mulch performance from mixed and mono specific plantations of native tree species in Costa Rica. *Agriculture, Ecosystems & Environment* 58: 145–55.

Cabalceta, G. 1993. Niveles críticos de fósforo, azufre y correlación de soluciones extractoras en Ultisoles, Vertisoles, Inceptisoles y Andisoles de Costa Rica. Thesis Mg. Sc., Universidad de Costa Rica. San José, Costa Rica. 167 pp.

Cabalceta, G., A. D'Ambrosio, and E. Bornemisza. 1996. Evaluación de cobre disponible en Andisoles e Inceptisoles de Costa Rica plantados de café. *Agronomía Costarricense* 20(2): 125–34.

Canessa, J., F. Sancho, and A. Alvarado. 1987. Retención de fosfatos en andepts de Costa Rica. II. Respuesta a la fertilización fosfórica. *Turrialba* 37(2): 211–18.

Castaing, A. 1982. Algunos factores edáficos y dasométricos relacionados con el crecimiento y comportamiento de *Cedrela odorata* L. Thesis Mg. Sc., Programa Recursos Naturales, Programa Universidad de Costa Rica–Centro Agronómico Tropical de Investigación y Enseñanza. Turrialba, Costa Rica. 123 pp.

Castro, L., L. Uribe, and A. Alvarado 1993. Efecto del enriquecimiento del inoculante de *Rhizobium* con dosis crecientes de fósforo sobre el crecimiento del frijol común (*Phaseolus vulgaris*). *Agronomía Costarricense* 17(1): 55–59.

Centro Científico Tropical and World Resources Institute. 1991. *La Depreciación de los Recursos Naturales en Costa Rica y su Relación con el Sistema de Cuentas Nacionales*. San José, Costa Rica.

Chaves, M., and A. Alvarado. 1994. Manejo de la fertilización en plantaciones de caña de azúcar (*Saccharum* spp.) en Andisoles de ladera de Costa Rica. In *World Congress of Soil Science*, 15th ed.(Mexico, 1994). Vol. 7a: Commission VI Symposia, 353–72. Acapulco, Mexico: ISSS.

Chinchilla, M., A. Alvarado, and R.A. Mata. 2011a. Factores formadores y distribución de suelos de la subcuenca del río Pirris, Talamanca, Costa Rica. *Agronomía Costarricense* 35(1): 33–57.

Chinchilla, M., A. Alvarado, and R.A. Mata. 2011b. Caracterización y clasificación de algunos Ultisoles de la región de Los Santos, Talamanca, Costa Rica. *Agronomía Costarricense* 35(1): 59–81.

Chinchilla, M., A. Alvarado, and R.A. Mata. 2011c. Andisoles, Inceptisoles y Entisoles de la región de Los Santos, Talamanca, Costa Rica. *Agronomía Costarricense* 35(1): 83–107.

Chinchilla, M., A. Alvarado, and R.A. Mata. 2011d. Capacidad de las tierras para uso agrícola en la subcuenca media-alta del río Pirrís, Los Santos, Costa Rica. *Agronomía Costarricense* 35(1): 109–130.

Cleveland, C.C., J.C. Neff, A.R. Townsend, and E. Hood. 2004. Composition, dynamics, and fate of leached dissolved organic matter in terrestrial ecosystems: results from a decomposition experiment. *Ecosystems* 7: 275–85.

Cohen, A.D., R. Raymond, S. Mora, A. Alvarado, and L. Malavassi. 1986. Características geológicas de los depósitos de turba en Costa Rica (preliminary report). *Revista de Geología América Central* 4: 47–67.

Colmet-Daage, F., F. Maldonado, C. de Kimpe, M. Trichet, and G. Fusil. 1973. Caractéristiques de quelques sols dérives de cendre volcanique de la Cordillère Central du Costa Rica. Ed. Prov. Office de la Recherche Scientifique et Technique OutreMer, Centre des Antilles, Bureau des Sols, Guadaloupe. 32 pp.

Cordero, A. 1994. *Fertilización y Nutrición Mineral del Arroz*. San José, Editorial UCR. 100 pp.

Cordero, A., and G.F. Ramírez. 1979. Acumulamiento de cobre en los suelos del Pacífico Sur de Costa Rica y sus efectos detrimentales en la agricultura. *Agronomía Costarricense* 3(1): 63–78.

Costa Rican Instituto de Tierras y Colonización (ITCO). 1964. *Estudio de la Región de Upala*. San José, Costa Rica.

Costa Rican Ministerio de Agricultura e Industria (MAI). 1958. Estudio preliminar de suelos de la región occidental de la Meseta Central. Boletín Técnico no. 22. San José, Costa Rica. 64 pp.

Coto, J.A., and J.E. Torres. 1970. Mapa de uso potencial de la tierra de Costa Rica. San José, Costa Rica: Ministerio de Agricultura y Ganadería. Scale 1:750.000.

Daubenmire, R. 1972. Ecology of *Hyparrhenia rufa* (Ness) in derived savanna in north-western Costa Rica. *Journal of Applied Ecology* 9: 11–23.

de las Salas, G. 1987. *Suelos y Ecosistemas Forestales con Énfasis en América Tropical*. San José, Costa Rica: Servicio Editorial IICA. 447 pp.

Dóndoli, C., and J.A. Torres. 1954. *Estudio Geoagronómico de la Región Oriental de la Meseta Central*. San José, Costa Rica: Ministerio de Agricultura e Industrias. 180 pp.

Escobar, M., C. Ramírez, and D. Kass.1994. Nitrógeno en un cultivo de callejones de poró (*Erythrina poeppigiana*), madero negro (*Gliricidia sepium*) con frijol común (*Phaseolus vulgaris*). In H. D. Thurston, M. Smith, G.Abawi and S. Kearl, eds., *Tapado, Los Sistemas de Siembra con Cobertura*, 140–56. Ithaca, NY: CIIFAD.

Esquivel, A. 1994. Variación estacional de la distribución espacial de *Meloidogyne incognita* (Nemata: Heteroderidae) en fincas tabacaleras de Pérez Zeledón. Thesis Mg. Sc., Universidad de Costa Rica, Facultad de Agronomía. San José, Costa Rica.

Esquivel, J.O. 1997. Efecto del componente arbóreo de un sistema silvopastoril sobre la distribución espacial de nutrientes, biomasa microbial y densidad de lombrices en un suelo bajo pastoreo, en la Zona Atlántica de Costa Rica. Tesis Mag. Sc, Centro Agronómico Tropical de Investigación y Enseñanza. Turrialba, Costa Rica. 65 pp.

Ewel, J.J., C.W. Berish, B.J. Brown, N. Price, and J.W. Raich. 1981. Slash and burn impacts on a Costa Rican wet forest site. *Ecology* 62(3): 816–29.

FAO/UNESCO. 1976. Mapa mundial de suelos. Vol. III. México y Centro América. Place de la Fontenoy, París, France. 104 pp.

Fassbender, H.W. 1975. *Química de Suelos con Énfasis en Suelos de América Latina*. San José, Costa Rica: IICA. 398 pp.

Forsythe, W. 1975. *Manual de Laboratorio de Física de Suelos*. San José, Costa Rica: IICA. 212 pp.

Fraile, J., and A. Serafino. 1978. Variaciones mensuales en la densidad de microartrópodos edáficos en un cafetal de Costa Rica. *Biología Tropical* 26(2): 291–301.

Gadea, P., R.O. Russo, M.P. Bertoli, and J.L. Sosa. 2004. Respuesta de 5 especies forestales en etapa de vivero al tratamiento con micorrizas arbusculares. In *Memorias del 1er. Congreso sobre Suelos Forestales*, Heredia, Costa Rica: INISEFOR/UNA.

Gavande, S.A. 1973. *Física de Suelos: Principios and Aplicaciones*. México, D.F., México: Limusa and Wiley. 351 pp.

Gichuru, M.P., A. Bationo, M.A. Bekunda, H.C. Goma, P.L. Mafongonya, D.N. Megendi, H.M. Murwira, S.M. Nadwa, P. Nyathi, and M.J. Swift, eds. 2003. *Soil Fertility Management in Africa: A Regional Perspective*. Nairobi, Kenya: Academy Science Publishers. 306 pp.

Gow, D., R. Bolaños, L. Bonnefil, C. Donato, G. Hartshorn, J. Laarman, D. Norman, J. Tolisano, and J. Tosi. 1988. *Environmental Assessment for the Northern Zone Consolidation Project in Costa Rica*. Vol. 1: Environmental assessment. Vol. 2: Technical reports. Washington, DC: Development Strategies for Fragile Lands.

Gross, L., C. Ramírez, and D. Kass. 1993. Selection of efficient *Bradyrhizobium* strains for *Erythrina poeppigiana* in soils with high aluminum saturation. In S. Westley and M. Powell, eds., Proceedings of International Conference on *Erythrina* in the New and Old Worlds (October 19–23, 1992), 283–91. Turrialba, Costa Rica: CATIE.

Guggenberger, G., and W. Zech. 1999. Soil organic matter composition under primary forest, pasture, and secondary succession, Region Huetar Norte, Costa Rica. *Forest Ecology and Management* 124(1): 93–104.

Guillén, C., F. Soto, and M. Springer. 2006a. Diversidad y abundancia de colémbolos edáficos en un bosque primario, un bosque secundario y un cafetal en Costa Rica. *Agronomía Costarricense* 30(2): 7–11.

Guillén, C., F. Soto, and M. Springer. 2006b. Variables físicas, químicas y biológicas del suelo sobre las poblaciones de colémbolos en Costa Rica. *Agronomía Costarricense* 30(2): 13–22.

Haggar, J.P., and J.J. Ewel. 1994. Experiments on the ecological basis of sustainability: early findings on nitrogen, phosphorus, and root systems. *Interciencia* 19(6): 347–51.

Hartemink, A.E. 2003. *Soil Fertility Decline in the Tropics with Case Studies on Plantations*. The Netherlands: CABI Publishing. 360 pp.

Heal, O.W., J.M. Anderson, and M.J. Swift. 1997. Plant litter quality and decomposition: an historical overview. In G. Cadish and K.E. Giller, eds., *Driven by Nature: Plant Litter Quality and Decomposition*, 3–29. Oxon, UK: CAB International.

Heaney, A., and J. Proctor. 1989. Chemical elements in litter in forests on Volcán Barba, Costa Rica. In J. Proctor, ed., *Mineral Nutrients in Tropical Forest and Savanna Ecosystems*, 255–71. The British Ecological Society, Special Publication No. 9. Oxford, England: Blackwell Scientific Publications.

Henriquez, C., and F. Bertsch. 1994. Efecto de la aplicación fraccionada del fertilizante potásico en un Andisol bajo cultivo de maíz and frijol en Coto Brus, Costa Rica. *Agronomía Costarricense* 18(1): 53–59.

Herrera, M.E., and L.A. Fournier. 1977. Producción, descomposición e invertebrados del mantillo en varias etapas de la sucesión en Ciudad Colón, Costa Rica. *Biología Tropical* 25(2): 275–88.

Herrick, J.E., and R. Lal. 1995. Soil physical property changes during dung decomposition in a tropical pasture. *Soil Science Society of America Journal* 59(3): 908–12.

Herrick, J.E., and R. Lal. 1996. Dung decomposition and pedoturbation in a seasonally dry tropical pasture. *Biology and Fertility of Soils* 23(2): 177–81.

Holdridge, L.R. 1987. *Ecología Basada en Zonas de Vida*. San José, Costa Rica: IICA. Serie de libros y materiales educativos no. 34. 216 pp.

Holdridge, L.R., W.C. Grenke, W.H. Hatheway, T. Liang, and J.A. Tosi. 1971. *Forest Environments in Tropical Life Zones: A Pilot Study*. Great Britain: Pergamon Press. 741 pp.

Hölldobler, B., E.O. Wilson. 1990. *The Ants*. Cambridge, MA: The Belknap Press of Harvard University Press. 571 pp.

Instituto Interamericano de Ciencias Agrícolas (IICA). 1979. Estudio de planificación agrícola del Pacífico Seco. San José, Costa Rica: IICA. Scale 1:200.000.

Janzen, D.H. 1982a. Removal of seeds from horse dung by tropical rodents: influence of habitat and amount of dung. *Ecology* 63(6): 1887–1900.

Janzen, D.H.1982b. How and why horses open *Crescentia alata* fruits. *Biotropica* 14(2): 149–52.

Jiménez, G.G., and R. López. 1987. Fluctuación estacional de la distribución espacial de *Meloidogyne incognita* y *Rotylenchulus reniformis* en papaya (*Carica papaya* L.). *Turrialba* 37(2): 165–70.

Johnson, N.C., and D.A. Wedin. 1997. Soil carbon, nutrients, and mycorrizhae during conversion of dry tropical forest to grassland. *Ecological Applications* 7(1): 171–82.

Kappelle, M., and J.G. van Uffelen. 2005. Los suelos de los páramos de Costa Rica. In M. Kappelle and S.P. Horn, eds., *Páramos de Costa Rica*, 147–59. Instituto Nacional de Biodiversidad (INBio)—The Nature Conservancy (TNC)—WOTRO Foundation. Santo Domingo de Heredia, Costa Rica: INBio Press.

Kappelle, M., and J.G. van Uffelen. 2006. Altitudinal zonation of montane oak forest along climate and soil gradients in Costa Rica. *Ecological Studies* 185: 39–54.

Lal, R., C.C. Cerri, M. Bernoux, J. Etchevers, and J. Cerri, eds. 2006. *Carbon Sequestration in Soils of Latin America*. Binghamton, NY: The Harworth Press. 554 pp.

Lal, R., and J.M. Kimble. 2000.Tropical ecosystems and the global C cycle. In: R. Lal, J.M. Kimble, and B.A. Stewart, eds., *Global Climate Change and Tropical Ecosystems, Advances in Soil Science*, 3–32.

Lang, S.B. 2000. Effects of logging roads on erosion in a wet tropical forest in the río Chiquito watershed, Península de Osa, Costa Rica. Thesis M. Sc., Colorado State University. Fort Collins, CO. 87 pp.

Lapied, E., and P. Lavelle. 2003. The peregrine earthworm *Pontoscolex corethrurus* in the East coast of Costa Rica. *Pedobiologia* 47(5/6): 471–74.

Leiva, J.A. 2007. Regeneración arbórea y características edáficas en bosques secos tropicales desarrollados sobre la meseta ignimbrítica de Santa Rosa, Noroeste de Costa Rica. Thesis Mg. Sc., Universidad de Costa Rica. San José, Costa Rica. 148 pp.

León, S., J. Bolaños, and J. Fraile. 1993. Características edáficas, densidad y biomasa de poblaciones de lombrices de tierra. *Brenesia* (39/40): 125–30.

Longino, J.I., and P.E. Hanson. 1995. The ants (Formicidae). In P.E. Hanson and I.D. Gauld, *The Hymenoptera of Costa Rica*. New York: Oxford University Press and The Natural History Museum. 893 pp.

López, F.L., and D.L. Kass. 1996. Efecto de enmiendas orgánicas en la dinámica del fósforo e indicadores de actividad biológica sobre el rendimiento del frijol en un suelo Acrudoxic Melanudand. *Agroforestería en las Américas* 3(11–12): 12–15.

López, R. 1981. Distribución espacial de nematodos del arroz después de la cosecha en el sureste de Costa Rica. *Agronomía Costarricense* 5(1/2): 49–54.

López, R. 1988. Nuevas observaciones sobre la distribución espacial de nematodos parásitos del arroz (*Oryza sativa* L.) en Costa Rica. *Turrialba* 38(1): 39–44.

López, R., and L. Salazar. 1987. Observaciones sobre la distribución espacial de nematodos fitoparásitos en árboles frutales. *Agronomía Costarricense* 11(2): 141–48.

Mata, R. 1991. Los órdenes de suelos de Costa Rica. In *Memoria Taller de Erosión de Suelos*, 28–32. Heredia, Costa Rica: MADE, UNA.

McGlynn, T.P., D.J. Salinas, R.R. Dunn, T.E. Wood, D.A. Lawrence, and D.A. Clark. 2007. Phosphorus limits tropical rain forest litter fauna. *Biotropica* 39(1): 50–53.

Meneses, A., L.E. Pocasangre, E. Somarriba, A.S. Riveros, and F.E. Rosales. 2003. Diversidad de hongos endofíticos y abundancia de nematodos en plantaciones de banano y plátano de la parte baja de los territorios indígenas de Talamanca. *Agroforestería en las Américas* 10 (37/38): 59–62.

Meza, E. 1994. Efecto de las condiciones climáticas y edáficas sobre la interacción entre el actinomicete *Frankia* sp. y el jaúl (*Alnus acuminata*). Tesis Licenciatura,Universidad de Costa Rica. San José, Costa Rica. 198 pp.

Molina, E., A. Cordero, and F. Bertsch. 1986. Potasio en andepts de Costa Rica. II. Respuesta a la fertilización con P and K en invernadero. *Turrialba* 36(3): 289–98.

Monge, J. 2006. Las taltuzas de Costa Rica: historia natural, impacto y manejo. Laboratorio Plagas Invertebrados, CIPROC, Escuela Agronomía, Facultad Ciencias Agroalimentarias, Universidad de Costa Rica. San José, Costa Rica. 45 pp.

Monge, J., and J.P. Alfaro. 1995. Geographic variation of habitats in Costa Rican velvet worms (Onychophora: Peripatidae). *Biogeographica* 71(3): 97–108.

Montagnini, F. 2002. Tropical plantations with native trees: their function in ecosystem restoration. In M.V. Reddy, ed., *Management of Tropical Plantation-forests and Their Soil-litter Systems*, chapter 4, 73–94. Enfield, NH: Science Publishers, Inc.

Montagnini, F., and C.F. Jordan. 2002. Reciclaje de nutrientes en bosques lluviosos neotropicales. In M.R. Guariguata and G. Kattan, eds., *Ecología de Conservación de Bosques Lluviosos Neotropicales*, 167–91. Cartago, Costa Rica: Ediciones LUR.

Mora, A. 1988. Estudio preliminar sobre el efecto de la compactación del suelo sobre el rendimiento and calidad de pasto kikuyo (*Pennisetum clandestinum*) durante la época de verano. Thesis B.S., Facultad de Agronomía. San José, Costa Rica. 58 pp.

Nygren, P., and C. Ramírez.1993. Phenology of N_2 fixing nodules in pruned clones of *Erythrina poeppigiana*, pp. 297–305. In S.N. Westley and M. Powell, eds., Proceedings of International Conference on *Erythrina* in the New and Old Worlds (October 19–23, 1992). Turrialba, Costa Rica: CATIE.

Nygren, P., and C. Ramírez.1995. Production and turnover of N_2 fixing nodules in relation to foliage development in periodically pruned *Erythrina poeppigiana* (Fabaceae) trees. *Forest Ecology & Management* 73: 59–73.

Nygren, P.,C. Ramírez, and G. Sanchez.1993. Growth of seedlings of five half-sib families of *Erythrina poeppigiana* to inoculation with a selected *Bradyrhizobium* sp. Strain. In S.N. Westley and M. Powell, eds., Proceedings of International Conference on *Erythrina* in the New and Old Worlds (October 19–23, 1992), 278–82.Turrialba, Costa Rica: CATIE.

Oelbermann, M., R.P. Voroney, and A.M. Gordon. 2004. Carbon sequestration in tropical and temperate agroforestry systems: a review with examples from Costa Rica and southern Canada. *Agriculture, Ecosystems and Environment* 104: 359–77.

Oelbermann, M., R.P. Voroney, D.C.L. Kass, and A.M. Schlönvoigt. 2005. Above- and below-ground carbon inputs in 19-, 10- and 4-year-old Costa Rican alley cropping systems. *Agriculture, Ecosystems and Environment* 105: 163–72.

Ordóñez, H. 2003. Fenología de la copa y de las raíces finas de *Simarouba glauca* y *Dalbergia retusa* (cocobolo) con riego en una plantación mixta de Guanacaste. Thesis Mg. Sc., Universidad de Costa Rica. San José, Costa Rica. 134 pp.

Ortiz, R., O. Acuña, and C. Ramírez. 1986. Necesidades nutricionales de la soya inoculada con *Rhizobium japonicum* en un suelo de Pijije, Guanacaste, mediante la prueba biológica de invernadero. In Memorias, Séptimo Congreso Agronómico Nacional (28 de Julio–1 de Agosto), 34–35. San José, Costa Rica.

Ortiz, R., O. Acuña, C. Ramírez, and A. Cordero.1990. Determinación de las necesidades nutricionales de la soya inoculada con *Rhizobium japonicum* en un humitropept de Guanacaste, en invernadero. *Agronomía Costarricense* 14: 115–20.

Pérez, S., A. Alvarado, and E. Ramírez. 1978. Asociaciones de subgrupos de suelos de Costa Rica (mapa preliminar). San José, Costa Rica: OPSA. Scale 1:200.000. 9 h.

Perfecto, I., and J. Vandermeer. 1994. Understanding biodiversity loss in agroecosystems: reduction of ant diversity resulting from transformation of the ecosystem in Costa Rica. *Entomology (Trends in Agricultural Sciences)* 2: 7–13.

Plath, C.V., and A.J. van der Sluis. 1965. Mapa de uso potencial de la tierra. Turrialba, Costa Rica: IICA.

Powers, J.S. 2001. Geographic variation in soil organic carbon dynamics following land-use change in Costa Rica. Thesis Ph. D., Duke University, Graduate School and Management of Biology. Durham, NC. 281 pp.

Powers, J.S. 2004. Changes in soil carbon and nitrogen after contrasting land-use transitions in northeastern Costa Rica. *Ecosystems* 7(2): 134–46.

Powers, J.S., and W.H. Schlesinger. 2002a. Geographic and vertical patterns of stable carbon isotopes in tropical rain forest soils of Costa Rica. *Geoderma* 109(1–2): 140–60.

Powers, J.S., and W.H. Schlesinger. 2002b. Relationships among soil carbon distributions and biophysical factors at nested spatial scales in rain forest of northern Costa Rica. *Geoderma* 109(3–4): 165–90.

Prescott, S.C. 1918. Examination of tropical soils. United Fruit Company. The Research Laboratory. Bulletin no. 3. 594 pp.

Radulovich, R., and P. Sollins. 1985. Compactación de un suelo aluvial de origen volcánico por tráfico de personas. *Agronomía Costarricense* 9(2): 143–48.

Radulovich, R., P. Sollins, P. Baveye, and E. Solórzano. 1992. Bypass water flow through unsaturated microaggregated tropical soil. *Soil Science Society of America Journal* 56(3): 721–26.

Raich, J.W. 1983. Effects of forest conversion on the carbon budget of a tropical soil. *Biotropica* 15(3): 177–84.

Ramírez, A.L., and C. Ramírez.1980. Esterilidad masculina causada por la exposición laboral con el nematicida 1,2-dibromo-3-cloropropano. *Acta Médica Costarricense* 23: 219–22.

Ramírez, C., and M. Alexander. 1980a. Evidence suggesting protozoan predation on *Rhizobium* associated with germinating seeds and in the rhizosphere of beans (*Phaseolus vulgaris* L.). *Applied and Environmental Microbiology* 40: 492–99.

Ramírez, C., and M. Alexander. 1980b. Increased bacterial colonization of the rhizosphere by controlling the soil protozoan population in the *Rhizobium* legume system. In: P. O. Vose and A. P. Rushell, eds., *Associative Dinitrogen Fixation*, chapter 8. Boca Raton, FL: CRC Press.

Ramírez, C., and D. Flores. 1994. Structure of dinitrogen-fixing nodules in *Erythrina poeppigiana*. *Revista de Biología Tropical* 42(supl.2): 15–28.

Ramírez, G.F. 2001. Contribución del ingeniero agrónomo al desarrollo de la fertilidad de suelos y de la nutrición mineral de las plantas. In A. Jiménez, ed., *Medio Siglo de Contribución al Progreso Nacional 1941 1991*, capítulo 11, 253–61. San José, Costa Rica: Colegio de Ingenieros Agrónomos.

Reid, F.A. 1997. A Field Guide to the Mammals of Central America and Southeast Mexico. New York: Oxford University Press. 334 pp.

RELACO. 1996. El uso sostenible del suelo en zonas de ladera: el papel esencial de los sistemas de labranza conservacionista. In F. Bertsch and C. Monreal, eds., Reunión Bienal de la Red Latinoamericana de Labranza Conservacionista, 3rd ed. (San José, Costa Rica, 1995). Memorias. San José, Costa Rica: ACCS. 307 pp.

Reynolds, J.N. 1991. Soil nitrogen dynamics in relation to groundwater contamination in the Valle Central, Costa Rica. PhD Thesis, University of Michigan.

Rodríguez, J.M., and C. Vaughan. 1985. Notas sobre la ecología de la guatuza (*Dasyprocta punctata* Gray) en el bosque seco tropical de Costa Rica. *Brenesia* 24: 353–60.

Rojas, I. 1992. Efecto de la micorrización sobre el crecimiento de tres especies forestales en dos suelos de Guanacaste, Costa Rica. Tesis de Maestría, Universidad de Costa Rica, San José, Costa Rica. 73 p.

Russo, R.O. 1989. Evaluating alder endophyte (*Alnus acuminata*-*Frankia*-Mycorrhizae) interactions. *Plant and Soil* 118: 151–55.

Salazar, C. 2014. Caracterización de los suelos principales de la cuenca

Part III

The Pacific Ocean and Isla del Coco

Chapter 5 The Pacific Coastal and Marine Ecosystems

Jorge Cortés[1]

Introduction

Background

Costa Rica is located in the southern section of the Central American Isthmus, which emerged around 3 million years ago and separates the two largest oceans of the world, the Atlantic and the Pacific (Coates et al. 1992, Coates 1997, Denyer et al. 2003). For an alternative closure time of the isthmus see Montes et al. (2015). The isthmus created a land bridge between two large continental masses, the North American and the South American plates, but at the same time, erected a barrier to marine organisms and changed the oceanographic conditions on both sides of the isthmus (Stehli and Webb 1985, Alvarado-Induni 1994, D'Croz and Robertson 1997, Cortés 2003, 2007). For some groups of marine organisms the separation resulted in the evolution of transisthmian sister species (Lessios 1990, Knowlton 1993, Knowlton et al. 1993, Wehrtmann and Albornoz 2002), while in other cases, such as in reef-building corals, the development of the land bridge completely changed species compositions (Cortés 1986, 2003).

The Pacific waters of Costa Rica cover an area of 565,683 km^2 (INCOPESCA 2006), which is more than 10 times the terrestrial area (Fig. 5.1), and represents 96% of the marine area of the country. The Pacific mainland coastline (1,254 km; Fig. 5.2) is about six times the length of the Caribbean coast and harbors a great diversity of coastal and marine ecosystems. The most extensive mangrove forests are along the Pacific coast, as well as rocky intertidal zones of diverse rock types, beaches consisting of a great variety of sediment composition and grain size, large and small estuaries, a tropical fjord, islets and islands of various sizes and proximity to the coast, plus one oceanic island, seagrasses, coral reefs, and many gulfs and bays (Jiménez and Soto 1985, Denyer and Cárdenas 2000, Cortés and Jiménez 2003, Cortés and Wehrtmann 2009).

The northern coastal zone along the Pacific is characterized by a marked dry season, with very strong winds, the presence of dry tropical forest (Mata and Echeverría 2004, Alfaro et al. 2012, and see this volume's chapters 9 and 10 by Jiménez et al. and Janzen and Hallwachs, respectively), and seasonal upwelling (Cortés 1996/1997, Jiménez 2001, Alfaro and Cortés 2012, Cortés et al. 2014). The dry conditions extend southward to the Golfo de Nicoya. The central Pacific is a transitional zone towards a more humid climate in the south (see e.g., Jiménez and Carrillo, chapter 11 of this volume). The south Pacific coast is covered with tropical rain forest, where it rains year-round (Herrera 1985, Jiménez and Soto 1985, Kappelle et al. 2002, and Gilbert et al., chapter 12 of this volume).

In this chapter I will describe the main coastal and marine ecosystems found along the Pacific coast of Costa Rica. [Please note that coastal and marine systems of Isla del Coco are presented in chapter 7 in this book.] The first section concerns a brief history of marine research, followed by some notes on the state of scientific knowledge and descriptions of the physical conditions of the coast. I will describe the coastal and marine ecosystems and provide comments on the main natural and anthropogenic impacts that influence those ecosystems. Also, several conservation initiatives are presented, while it ends with research, management, and conservation perspectives for the future.

[1] Centro de Investigación en Ciencias del Mar y Limnología (CIMAR), and Escuela de Biología, Universidad de Costa Rica (UCR), San Pedro de Montes de Oca, 11501-2060 San José, Costa Rica

Historical overview

Cortés (2009) presented a history of marine biodiversity research in Costa Rica, from which most of the information for this chapter is derived, but complemented here with data from other areas of marine research. In this chapter I comment on the most important expeditions and collections that have been carried out, and that significantly increased our knowlegde of the marine organisms, ecosystems, and processes in Costa Rica's marine waters.

Marine scientific research in Costa Rica started in the early nineteenth century, with collections of specimens made in the Golfo de Nicoya by ship captains or professional collectors, such as Hugh Cuming, and later described by

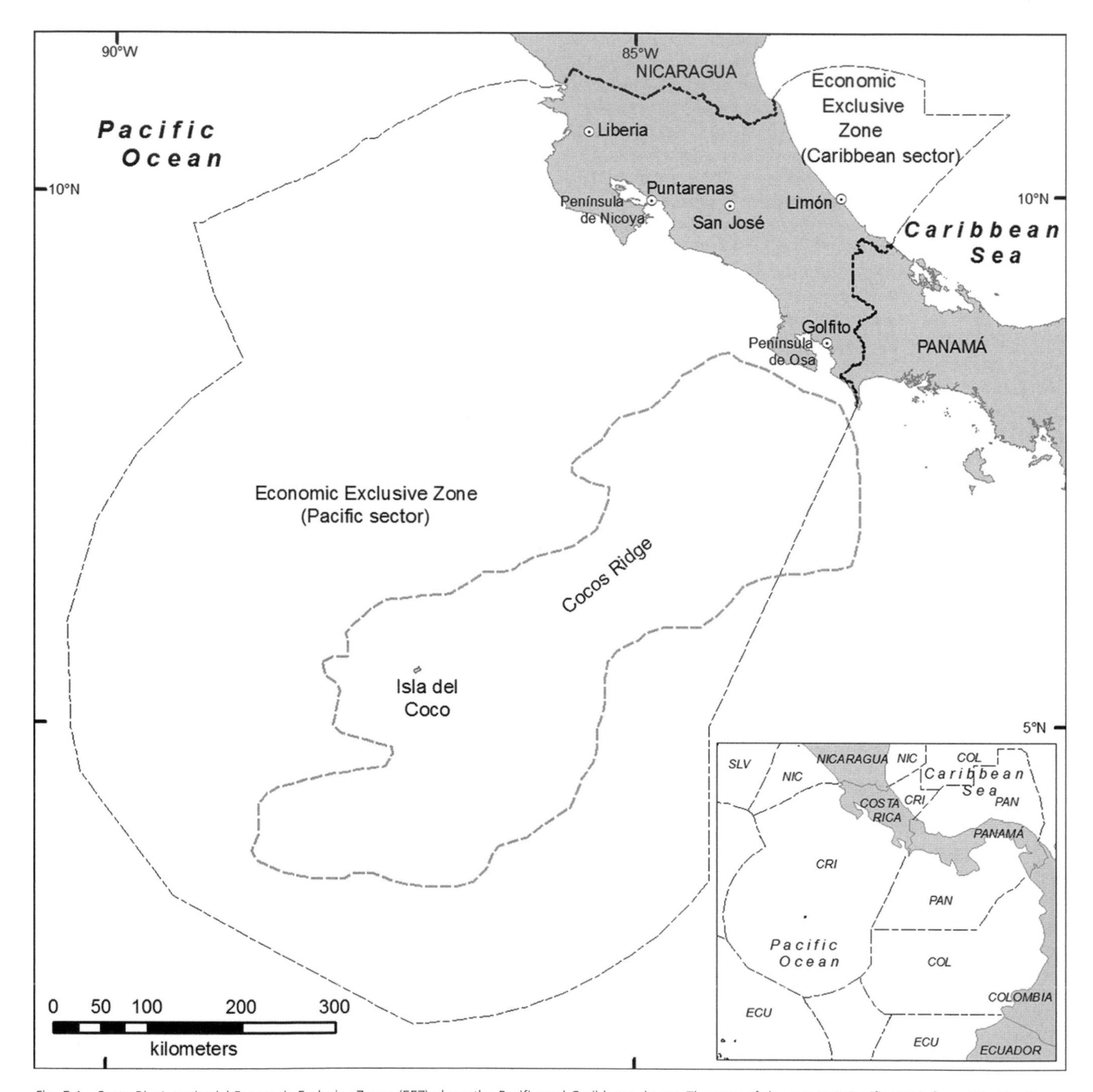

Fig. 5.1 Costa Rica's territorial Economic Exclusive Zones (EEZ) along the Pacific and Caribbean shores. The core of the country's Pacific EEZ is formed by the Cocos Ridge, a submarine mountain range. Its highest peak is formed by Isla del Coco.
Map prepared by Marco V. Castro.

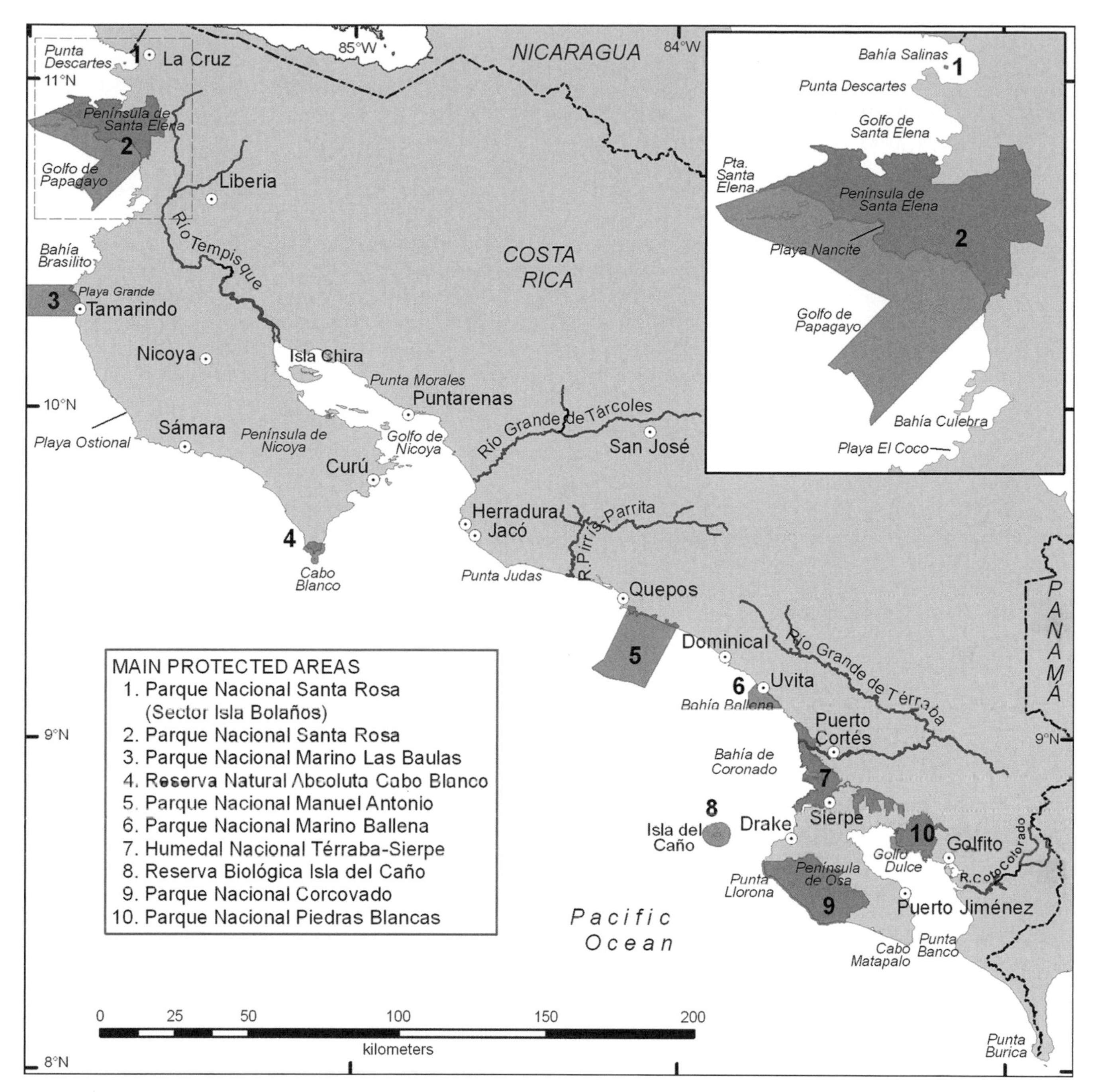

Fig. 5.2 Map of the Pacific coast of Costa Rica showing main cities and ports, topographic features, and protected areas. *Map prepared by Marco V. Castro.*

scientists—for example, George B. Sowerby (1832) on the mollusks, and Addison E. Verrill (1870) on the corals and octocorals. Towards the end of the nineteenth century large scientific expeditions were organized and extensive collections made in the Eastern Pacific, but none of them reached the Pacific coast of Costa Rica, only Isla del Coco—for example, the *Albatross* expeditions (Agassiz 1892, Townsend 1901), or the Hopkins Stanford Galapagos Expedition of 1898–1899 (Snodgrass and Heller 1905). During the first quarter of the twentieth century there were no expeditions made in the Pacifc waters of Costa Rica, possibly owing to impacts of the First World War.

In the 1920s and 1930s there was a new surge of expeditions to the Eastern Tropical Pacific, most of them funded by philantropists—for example, the expeditions by William K. Vanderbilt aboard his yachts *Eagle* and *Ara* (Boone

1930), which conducted collections along the Pacific coast of Costa Rica in 1928; the 1932 California Academy of Sciences Expedition funded by Templeton Crocker, which visited Costa Rica from June 22 to July 4 in that year, first at Puntarenas and then, after returning from Isla del Coco, along the northern section of the coast (Crocker 1933); or the Eastern Pacific Expedition of the Zoological Society of New York on board the *Zaca*, which sampled many places along the Pacific coast in 1938 (Beebe 1938, 1942). This latter expedition was led by one of the most outstanding naturalist of the twentieth century, William Beebe (Gould 2004). Possibly the scientifically most productive series of expeditions were the ones conducted by the Allan Hancock Foundation on the *Velero III*, with Allan Hancock himself serving as the ship's Captain (Fraser 1943).

During the Second World War there was a gap in scientific exploration in the region, but in the late 1950s there was a number of new expeditions and visits by individual scientists. The Beaudette Foundation funded, in 1959, an expedition by the *Stella Polaris* to the Eastern Pacific (Dawson and Beaudette 1959). During that expedition Elmer Yale Dawson collected algae and listed several species for Costa Rica. He described a new species of coralline algae from Isla del Caño (Dawson 1960). Previously, he had published on the algae of the Golfo de Nicoya and Golfo Dulce (Dawson 1957). In 1964, William C. Banta collected and described 13 bryozoan species from Playas del Coco (Banta and Carson 1977). During the Stanford Oceanographic Expedition #18 in 1968, aboard the RV (Research Vessel) *Te Vega*, intertidal organisms were collected in the Golfo de Nicoya and Bahía Brasilito (Ball 1972, Ball and Haig 1974). Oceanographic research was done in the Golfo Dulce in 1969 aboard the RV *T.G. Thompson*, and Francis A. Richards, James J. Anderson, and Joel D. Cline published on the chemical and physical characteristics of the gulf (Richards et al. 1971). Jean Nicholls-Driscoll (1976) worked on the benthic fauna in samples collected during that cruise.

Playas del Coco has been an important collection site along the Pacific coast. In 1970, Charles Birkeland and Tom M. Spight (Smithsonian Tropical Research Institute, STRI) collected ascidians that were later worked on by Takasi Tokioka (1971, 1972). Deborah M. Dexter (1974) published the first paper on beach fauna of Costa Rica, on the basis of collections she made in 1971 at several beaches along the Caribbean and Pacific coasts. Again from Playas del Coco, Allan Child (1979) reported on sea spiders collected in 1972 by C.E. Dawson and himself. The RV *Searcher* visited Costa Rica in 1972, and collections were made along the central and southern Pacific coast, including Isla del Caño and Golfo Dulce; Raymond B. Manning and Marjorie L. Reaka described the stomatopod Crustacea (Manning and Reaka 1979, Reaka and Manning 1980). In 1973, aboard the RV *Agassiz*, and again in 1978, with the RV *Alpha Helix*, the Scripps Institution of Oceanography collected along the coast from the Península de Santa Elena down to Punta Judas, including the Golfo de Nicoya. Spencer R. Luke lists the specimens of decapod crustaceans and stomatopods (Luke 1977), echinoderms (Luke 1982), mollusks (Luke 1995), and cnidarians (Luke 1998) in the Benthic Invertebrate Collections of the Scripps Institution of Oceanography.

A study on the chemical characteristics of the Costa Rican Thermal Dome (Domo de Costa Rica, DCR) was published by Broenkow (1965). Biological studies of the DCR were carried out by the Instituto de Ciencias del Mar y Limnología of the Universidad Nacional Autónoma de México (UNAM) during the DOMO Expeditions. They collected plankton samples with the RV *Mariano Matamoros*, twice in 1979, and with the RV *Puma* in 1981 and 1982 (Vicencio-Aguilar and Fernández-Alamo 1996). Several theses and papers were written—for example, Sánchez-Navas and Segura Puertas (1987) on pelagic mollusks, Segura-Puertas (1991) on medusae, Gasca and Suárez (1992) on siphonophores, and Vicencio-Aguilar and Fernández-Alamo (1996) presented a summary on the taxonomy and biogeography of the zooplankton of the Costa Rican Thermal Dome.

One paper (Bianchi 1991) was published on basis of the 1987 collections of the RV *F. Nansen*, in the Golfo de Papagayo. In 1990, the RV *Gyre* visited Isla del Caño; some coral cores were taken (Macintyre et al. 1992) to study the growth history of the coral reefs; the ship continued on to Isla del Coco.

During the 1950s there was a great interest in the physical oceanography of the Eastern Pacific to understand the distribution of tuna. Klaus Wrytki started a series of studies in the region in 1961 and published classical papers on the physical oceanography of the Eastern Tropical Pacific (Wyrtki 1966). Recently, an issue of *Progress in Oceanography* was dedicated to the Eastern Pacific. It includes excellent papers on different oceanographic aspects of the region (Fiedler and Lavín 2006).

In the past two decades some large expeditions have been dedicated to geochemical and geophysical research, because of the interesting condition of the subducting Cocos Plate and the deep abyssal plains just off the Pacific coast of Costa Rica (e.g., McAdoo et al. 1996, Linke et al. 2005, Mau et al. 2006, Ranero et al. 2007). As part of research in the cold seeps with the submersible *Alvin* operated by the Woods Hole Oceanographic Institution, many organisms were collected (L. Levin pers. comm. 2009, Opresko and Breedy 2010, Levin et al. 2012).

The creation of the Centro de Investigación en Ciencias

del Mar y Limnología (Marine Sciences and Limnology Research Center, CIMAR) at the Universidad de Costa Rica (UCR), in 1979, is an important milestone in the development of marine science in Costa Rica. For the first time Costa Rican scientists started to study the marine ecosystems of their own country. During the subsequent years numerous papers have been published on a wide variety of topics, mainly covering topics in marine biodiversity and ecology (see: http://cimar.ucr.ac.cr/).

The RV *Skimmer*, of the University of Delaware, was operated and used by CIMAR in the Golfo de Nicoya from 1979 to 1981 (Cortés 2009), which generated many theses and publications (Vargas 1995). This was the first large-scale project developed by CIMAR. It was followed by many individual and group projects, as well as other expeditions. Little by little the coasts of the country have been explored, and several sites are now continuously being monitored (Cortés 2009, 2012). Between 1993 and 1994, the RV *Victor Hensen* was used to sample along the Pacific coast, especially in the Golfo Dulce area (Vargas and Wolff 1996).

Current State of Knowledge

The Golfo de Nicoya is probably one of the best-known tropical estuaries in the world (Vargas 1995, and chapter 6 of this volume). The first publications in which marine organisms of Costa Rica are mentioned are from the Golfo de Nicoya. Gonzalo Fernández de Oviedo's book "Historia general y natural de las Indias, islas y tierra firme del mar océano," published in 1535, mentions the extraction of mollusks from Golfo de Nicoya (Meléndez 1974). The first scientific papers in specialized journals are based on specimens collected in the gulf (Sowerby 1832, Mörch 1859, Verrill 1870, Stebbing 1906), and the first paper on physical oceanography was from Golfo de Nicoya (Peterson 1960).

In the 1950s, Dawson collected and published results on the algae of the Golfo de Nicoya, among other areas of the country (Dawson 1957, 1960, Dawson and Beaudette 1959). In the 1970s and especially in the late 1970s and early 1980s, the Golfo de Nicoya was intensively studied. One of the first papers was on fish communities (León 1973), followed by detailed studies on benthic communities (Maurer and Vargas 1984, Maurer et al. 1984, Vargas et al. 1985, Dittel 1991), macro- and meiofauna of intertidal mudflats (De la Cruz and Vargas 1987, Vargas 1987, 1988a, b, 1989a, b), physico-chemical characteristics and pollution (Epifanio et al. 1983, Voorhis et al. 1983, Dean et al. 1986, Fuller et al. 1990), fish taxonomy, behavior, distribution and abundance (López and Bussing 1982, Bartels et al. 1983, 1984, López 1983, Araya 1984, Duncan and Szelistowski 1998, Ramírez et al. 1989, 1990, Szelistowski 1989, 1990, Thorne et al. 1989, 1990, Bussing and López 1993, 1996), phytoplankton (Hargraves and Víquez 1985), primary productivity (Córdoba-Muñoz 1998, Gocke et al. 2001a, b), larval ecology (Epifanio and Dittel 1984, Dittel and Epifanio 1990, Dittel et al. 1991), and fisheries (Campos 1983, Campos et al. 1984, 1993, Dittel et al. 1985, Campos and Corrales 1986, Thorne et al. 1989).

In the 1990s as well as in the past two decades, the study of the Golfo de Nicoya continued, covering topics in physical oceanography (Lizano and Vargas 1993, Brenes et al. 1996, 2001, Cháves and Birkicht 1996, Lizano 1998, Kress et al. 2002, Lizano and Alfaro 2004, Palter et al. 2007), pollution (Acuña et al. 1998, Loría et al. 2002, Acuña-González et al. 2004, Spongberg 2004a, García-Céspedes et al. 2006, Lizano et al. 2008, Spongberg et al. 2011), plankton (Morales and Vargas 1995, Víquez and Hargraves 1995, Molina-Ureña 1996, Morales-Ramírez 1996, Wangelin and Wolff 1996, Brugnoli-Olivera and Morales-Ramírez 2001, Morales-Ramírez and Brugnoli-Olivera 2001, Brugnoli-Olivera et al. 2004, Vargas-Montero and Freer 2004a, b, c), fisheries and aquaculture (Mug-Villanueva et al. 1994, Rostad and Hansen 2001, Fischer and Wolff 2006, Stern-Pirlot and Wolff 2006, Tabash and Chávez 2006, Tabash 2007, Herrera-Ulloa et al. 2009, López-Garro et al. 2009, Zanella et al. 2009, Zanella-Cesarotto et al. 2010, Hernáez and Wehrtmann 2014), biodiversity (Emig and Vargas 1990, Dean 1996a, 1998, 2009, Mielke 1997, Dean and Blake 2007, Acuña et al. 2013, Morales Ramírez et al. 2014, Murase et al. 2014), observation of fur seals (Montero-Cordero et al. 2010), the effectiveness of a marine reserve on fish diversity and abundance (Myers et al. 2010), description of a new genus and species of gastropod (Høisæter 2012), and on the natural history of several soft bottom organisms: a polychaete worm (Rojas-Figueroa and Vargas-Zamora 2008), a cephalochordate (Vargas and Dean 2010), two species of echinoderms (Vargas & Solano 2011), mollusks (Vargas-Zamora and Sibaja-Cordero 2011), crustaceans (Vargas-Zamora et al. 2012) and a symbiotic crustacean (Salas-Mota et al. 2014). The rocky intertidal zones were studied as a whole (Sibaja-Cordero 2005, Sibaja-Cordero and Vargas-Zamora 2006), as well as some of their organisms (Villalobos 1980a, b, Wehrtmann et al. 2012). In mangrove areas, the ecology of several animals have been studied—for example, mollusks (Castaing et al. 1980, Cruz and Jiménez 1994 and references therein) and crustaceans (Hernáez et al. 2012). A trophic model was developed as well (Wolff et al. 1998), while Vargas and Mata (2004) published a summary of research done in the Golfo de Nicoya.

Another region along Costa Rica's Pacific coast that has

been studied with a certain level of detail is Golfo Dulce, located in the south of the country. The first publication on this sector that became available dealt with the chemical and physical characteristics of the gulf itself (Richards et al. 1971), while the second one was on benthic invertebrate communities (Nicholls-Driscoll 1976). Golfo Dulce is an anoxic basin and, because of its oceanographic characteristics, has been called a tropical fjord (Hebbeln and Cortés 2001). In the late 1980s, studies were initiated about the coral reefs of the gulf; a dissertation was published on the Holocene growth history of one of its reefs (Cortés 1990a). Subsequently, a series of papers on the reefs of the gulf were published (Cortés 1990b, 1991, 1992, Cortés et al. 1994). Later, a thesis on bioerosion and bioaccretion in Golfo Dulce was completed (Fonseca 1999) and two more papers were published (Fonseca and Cortés 1998, Fonseca et al. 2006).

Our knowledge on Golfo Dulce increased significantly after the 1993–1994 *Victor Hensen* expedition (Wolff and Vargas 1994, Vargas and Wolff 1996). Many fields of study were addressed, ranging from geology (Hebbeln et al. 1996) to biogeochemistry (Thamdrup et al. 1996), including microbiology (Kuever et al. 1996), taxonomy of several groups (polychaetes: Dean 1996b; mollusks: Cruz 1996, Høisæter 1998; crustaceans: Jesse 1996, Morales-Ramírez 1996, Vargas et al. 1996; chaetognaths: Hossfeld 1996; fishes: Bussing and López 1996; and benthic macrofauna: León-Morales and Vargas 1998), zooplankton (Wangelin and Wolff 1996), and a trophic analysis (Wolff et al. 1996). More papers were produced in subsequent years covering the physical oceanography of the gulf (Svendsen et al. 2006), its biogeochemical cycles and the distribution of parameters related to the water column and bottom (Dallsgaard et al. 2003, Acuña-González et al. 2006, Ferdelman et al. 2006, Silva and Acuña-González 2006), pollution (Spongberg and Davis 1998, Spongberg 2004b), plankton and El Niño effect on species composition (Morales-Ramírez and Nowaczyk 2006, Quesada-Alpízar and Morales-Ramírez 2006, Morales-Ramírez and Jakob 2008), and on the possible effects of tuna feed-lots on cetaceans (Oviedo-Correa et al. 2009). Feutry et al. (2010) published results on the ichthyofauna of a mangrove forest in Golfo Dulce, Breedy and Guzman (2012) and Breedy et al. (2013) described new species of octocoral from the gulf, and Samper-Villareal et al. (2014) reported on the presence of seagrasses in the gulf. A synthesis of published marine research in the region was published by Quesada-Alpizar and Cortés (2006), while Morales-Ramírez (2011) published on the high marine biodiversity of Golfo Dulce (1,022 species). Alvarado et al. (2012b) point out that the Golfo Dulce represents one of the largest marine conservation gaps in Costa Rica.

A third region along the Pacific shores that has received some attention is Bahía Culebra, in the northern sector of Pacific Costa Rica. There, mainly the coral reef ecosystem has been studied (Glynn et al. 1983, Cortés and Murillo 1985, Jiménez 1997, 1998, Jiménez et al. 2001, Jiménez and Cortés 2003a, Dominici-Arosemena et al. 2005, Fernández and Cortés 2005, Sunagawa 2005, Bezy et al. 2006, Fernández-García 2007, Sunagawa et al. 2008, Bezy 2009, Alvarado and Vargas-Castillo 2012, Alvarado et al. 2012a, Fernández-García et al. 2012) as well as its plankton (Morales-Ramírez et al. 2001, Bednarski and Morales-Ramírez 2004, Vargas-Montero 2004, Rodríguez-Sáenz 2005, Rodríguez-Sáenz and Morales-Ramírez 2012, Rodríguez-Sáenz et al. 2012). The rocky intertidal zone has been investigated by Madrigal-Castro et al. (1984) and by Sibaja-Cordero and García-Méndez (2014) while the fauna of sandy beaches was assessed by Sibaja-Cordero et al. (2014). Pollution in the bay has been evaluated as well (Acuña-González et al. 2004, García-Céspedes et al. 2004, 2006, Spongberg et al. 2011). Additionally, sightings of the spotted dolphin, *Stenella attenuata*, have been reported (Rodríguez-Sáenz and Rodríguez-Fonseca 2004), and its ecology has been studied and described (May-Collado and Forcada 2012). The presence of radionuclides and other compounds in the bay's sediments has been assessed by Lizano et al. (2008, 2012). In turn, Vargas-Castillo (2012) has added 15 new species of crustaceans to the known biodiversity of Bahía Culebra (Cortés et al. 2012).

Isla del Caño off the coast of the Osa Peninsula has been visited and sampled on several occasions. Species from this island were described by, for example, Dawson (1960) who published on species of algae, and Manning and Reaka (1979) and Reaka and Manning (1980) who reported on stomatopods. However, our marine knowledge of this island, especially with regards to its coral reefs, has increased significantly after Héctor M. Guzman started to study them back in 1984 (Guzman 1986). He published about the meiofauna (Guzman et al. 1987a), the impact of the 1982/83 El Niño (Guzman et al. 1987b), the corallivores (Guzman 1988a, b, Guzman and Robertson 1989), plankton (Guzman and Obando 1988, Guzman et al. 1988a), its coral reefs and how they changed through time (Guzman and Cortés 1989a, 2001), coral growth rates (Guzman and Cortés 1989b), coral mortality due to a dinoflagellate algal bloom (Guzman et al. 1990), and the restoration of reefs through coral transplant activities (Guzman 1991).

A series of papers was published on the reproduction of the main coral species in the Eastern Tropical Pacific, including Isla del Caño (Glynn et al. 1991, 1994, 1996, 2000, 2008, 2011, 2012). Oil pollution levels were measured and resulted to be very low (Acuña and Murillo 1987). Moreover, Fonseca et al. (2010) published a marine habitat map of the island, from the shore to the border of the reserve, covering a distance of 3 km. Ornelas-Gatdula et al. (2012) studied sea slugs and found that they were comprised of three species, *Navanax aenigmaticus* being the species found along the Pacific coast. Samples from the island were used in a study on cryptic nudibranch species (Pola et al. 2012). A species of chimaeroid fishes was collected off Isla del Caño (Angulo et al. 2014).

From the rest of Costa Rica's coast there is little information available (Cortés and Wehrtmann 2009). A project is now underway to study the biological components and the physico-chemical conditions of the seasonal upwelling at Bahía Salinas in the extreme northern sector along the Pacific coast (Sibaja-Cordero and Cortés 2008, Cortés et al. 2014).

Regarding ecosystems, some have been studied at system level, for example, the mangrove forests, intertidal mud flats in the Golfo de Nicoya, the rocky intertidal zones of some coastal sectors, the seagrass beds found in Bahía Culebra, and coral communities and reefs (see below). Other ecosystems, such as beaches and deep waters, have received very little attention so far. In terms of taxonomic groups some have been studied in detail, for example, phytoplankton, zooplankton, octocorals, stony corals, polychaetes, mollusks, crustaceans, fishes, and turtles (Wehrtmann and Cortés 2009).

The Physical Environment

Geology and Geomorphology

The geology and geomorphology of the Pacific coast and ocean bottoms of Costa Rica are treated elsewhere in this book (Alvarado and Cárdenas, chapter 3 of this volume). Here I just include a few notes on this matter as they are particularly relevant to the marine ecosystems. For further details, reference is made to chapter 3.

The Pacific coastline is irregular, containing two main, closed gulfs, the Nicoya and Dulce gulfs, and two open gulfs, Papagayo and Coronado (Fig. 5.2). The width of the continental shelf (down to the 200 m isobath) ranges from less than 5 km to over 70 km, and has an area of about 15,000 km^2. It drops rapidly down to depths of 4,000 m in the Mesoamerican Trench (Figs. 5.1 and 5.3). The open, deep sea (more than 200 m depth) covers a surface of about 550,000 km^2 and is divided into two very distinctive regions, one smooth, with almost no seamounts, and another

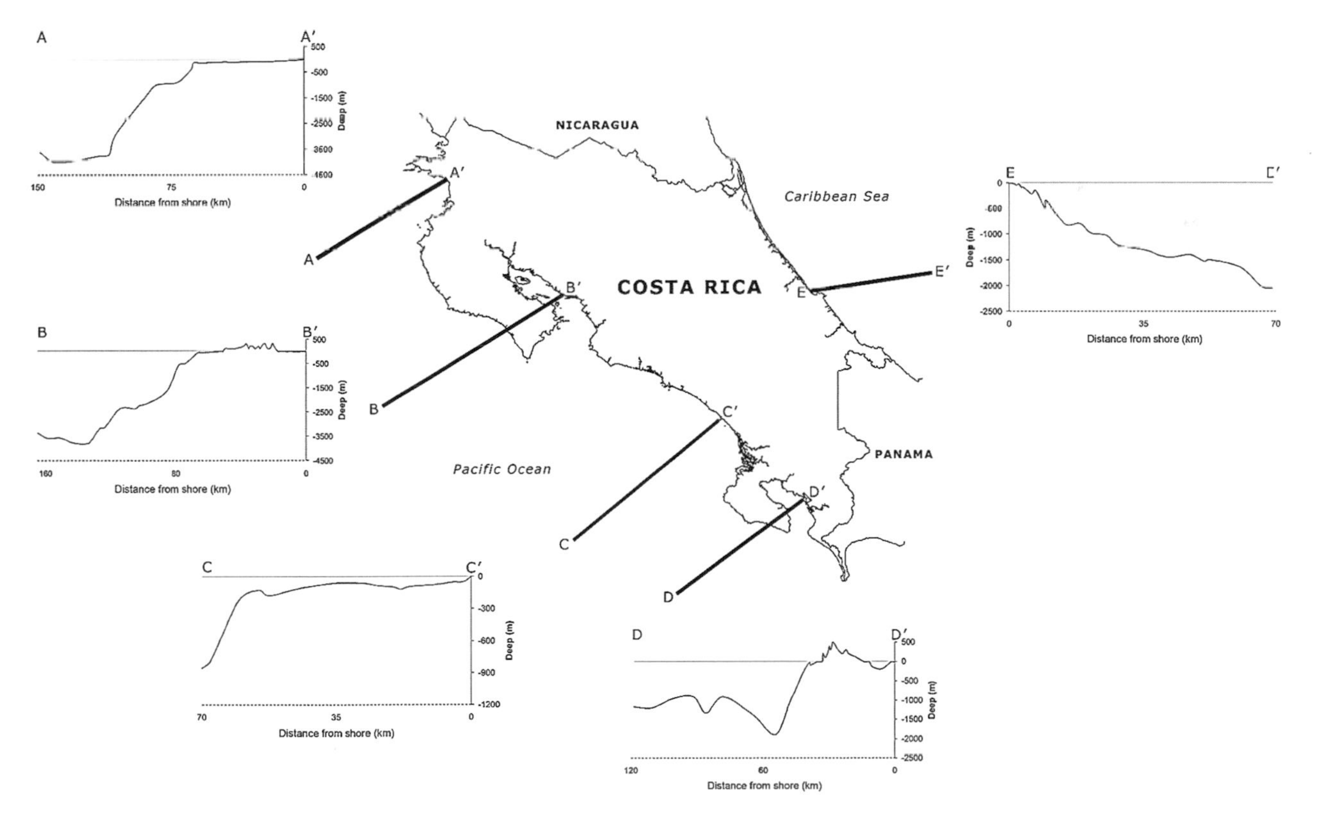

Fig. 5.3 Bottom profiles of four sections along Costa Rica's Pacific shore and one at the Caribbean coast. *Profiles prepared by Alexander Rodriguez.*

one with extensive areas covered by sea mounts and by the Cocos Ridge or Cocos Cordillera (Fig. 5.1).

Climate

The north Pacific coast of Costa Rica has well-defined periods of rain (summer) and drought (winter), while the south is rainy year-round; the central section of the coast is transitional between these two extremes (see also Herrera, chapter 2 of this volume). The dry period extends from December to April-May and occurs as a result of the southward displacement of the Intertropical Convergence Zone (ITCZ) also associated with a more stable atmosphere (Alfaro 2002, Taylor and Alfaro 2005, Amador et al. 2006, Herrera, chapter 2 of this volume). The northern region has in general smaller rivers and, therefore, its sediment input is smaller than in the southern sector. The local difference in climate has a profound effect on mangrove forest growth, as the largest trees are found in the south.

Hydrography

The waters off Costa Rica's Pacific coast are located just south of the Eastern Pacific Warm Pool as defined in Fiedler and Talley (2006), while sea water temperature is around 27–28°C year-round, except in the northern section along the coast, which experiences seasonal upwelling (see below). The Eastern Tropical Pacific is characterized by a shallow and very well-defined thermocline (<40 m), a thick Oxygen-Minimum Zone (OMZ) with significant levels of hypoxia (Fiedler and Talley 2006, Stramma et al. 2008), and lower pH compared to other tropical regions (Feely et al. 2004, Orr et al. 2005, Kleypas et al. 2006). This is a region of eddy formation between the boreal autumn and spring. They can range in size from 180 to 500 km in diameter. These eddies are formed by strong wind jets that flow across the low altitude sections of Mesoamerica (Tehuantepec, México; Gulf of Papagayo, Costa Rica-Nicaragua; and Gulf of Panama, Panamá) (Amador et al. 2006, Willet et al. 2006, Alfaro and Cortés 2012, Alfaro et al. 2012).

Another important feature offshore is the Costa Rican Thermal Dome (*Domo de Costa Rica*) (Fiedler 2002, Ballestero 2006). The dome is a depression on the ocean surface where the thermocline is shallower than surrounding waters; it is forced by the Papagayo wind jet, which forms the previously mentioned eddies, during the Pacific dry season in this region (Fiedler 2002, Amador et al. 2006). This feature is important as it generates an upwelling of nutrients.

The Intertropical Convergence Zone is a low-pressure belt just north of the equator, and moves north around mid-year, and south towards the end of the year and beginning of the following year (Amador et al. 2006, Herrera, chapter 2 of this volume). The dry season along the Pacific coast of Costa Rica extends from December to April-May, when the ITCZ is located farther south and the northern section of the coast experiences seasonal upwelling (Alfaro et al. 2012).

The important El Niño-Southern Oscillation (ENSO) event is a coupled ocean-atmosphere phenomenon generated in the Eastern Pacific, and has climatic and economic impacts worldwide (Glantz 2001). During an ENSO event the waters of the Eastern Tropical Pacific warm beyond the yearly variability, resulting in some cases, especially in the past two decades, in mass mortality of corals and other marine organisms (Enfield 1989, Glynn et al. 1988, Glynn 1990, Wang and Fiedler 2006). The impact of recent ENSO events in marine ecosystems in Costa Rica on coral reefs and plankton communities has received attention from several scientists. The 1982–1983 and the 1997–1998 events have been extremely devastating phenomena that affected the coral reefs along the Pacific coast of Costa Rica, causing mortalities of over 50% among the main reef-building corals (Cortés et al. 1984, Guzman et al. 1987b, Guzman and Cortés 1989a, Jiménez and Cortés 2001, 2003b, Jiménez et al. 2001). The effect of the 1997–1998 ENSO event was studied in plankton communities of the Golfo de Nicoya and Golfo Dulce. Some consequences observed were changes in the phytoplankton community structure and effects on the phyto-zooplankton coupling patterns in the Golfo de Nicoya, whereas in the Golfo Dulce there was an increase in gelatinous zooplankton, and a decrease in biomass and abundance of non-gelatinous zooplankton (Morales-Ramírez and Brugnoli-Oliveira 2001, Quesada-Alpízar and Morales-Ramírez 2006, Morales-Ramírez and Nowaczyk 2006, Brugnoli-Olivera and Morales-Ramírez 2008).

In summary, ocean dynamics in the Eastern Tropical Pacific off the coast of Costa Rica change during the year as a result of seasonal wind variations and the displacement of the Intertropical Convergence Zone. The same happens on larger time scales during El Niño-Southern Oscillation (ENSO) events (Wang and Fiedler 2006, Lizano 2008).

Circulation, Tides, and Waves

The main ocean current along the Pacific coast is the Costa Rican Coastal Current that flows from the southeast to the northwest, parallel to the Pacific coastline of Central America (Kessler 2006). Closer to the shore, eddies move in an opposite direction, while subsurface flows move towards the northeast.

The tides along the Pacific coast are semidiurnal (i.e., two high and two low water levels each day). The average tidal range expands from 207 to 234 cm at Quepos and Golfito, respectively, with extreme ranges of 256 and 289 cm at these sites (Lizano 2006). The difference is probably due to the level of openness of the sites: Quepos is open to the Pacific Ocean, while Golfito is an enclosed area within Golfo Dulce.

Waves along the Pacific coast are predominantly from the southwest and south, indicating that they originate far away. This is corroborated by the long period of the waves most of the time. Average wave height (1–1.3 m) is lower between November and April, with minimum values all along the coast in January and February (Amador et al. 2006, Lizano 2007). Average wave height values are higher (1.5–1.7 m) between May and October (Lizano 2007) when the *westerlies* are active along the Pacific coast.

Upwelling

The northern section of Costa Rica is characterized by lowlands and isolated volcanoes. Hence, the NE Trade Winds (*vientos alisios*) blow from the Caribbean across the isthmus during the dry season (December to April-May). Strong winds reach the Pacific coast during that period; the wind jets fan out and produce a wind velocity variation to the left of the axis generating considerable upwelling, the so-called Ekman pumping. The upwelled waters are cold, nutrient-rich, low in oxygen, and with a low pH (Rixen et al. 2012). This phenomenon is known as the Papagayo upwelling, one of three seasonal coastal upwellings that occur along the Pacific coast of Meso-America (Legeckis 1988, McCreary et al. 1989, Alfaro and Lizano 2001, Fiedler 2002, Willet et al. 2006, Alfaro and Cortés 2012, Alfaro et al. 2012). Temperatures as low as 15.5°C have been recorded in Bahía Culebra (Jiménez 2001) and 14.7°C in Bahía Salinas (J. Cortés unpubl. data). During the upwelling period, the alga *Sargassum liebmanni* grows extremely fast (0.68 cm per day) (Cortés et al. 2014).

Ecosystems

Beaches

The beaches of the Pacific coast of Costa Rica are one of the main tourist attractions of the country, but have not been studied much. The composition of the beaches is very variable (Fig. 5.4) and is made up of material ranging from fine grain sediments to cobbles, and from basaltic to carbonate sands (Denyer and Cárdenas 2000). Beaches are the result of the interphase between air, land, and sea water, where environmental parameters fluctuate widely. The highly variable combinations of these environmental factors in relation with shifting sands, make these beaches a tough habitat to live in. Only a few organisms can actually live within the sand itself—for example, polychaetes, amphipods, crabs, and irregular sea urchins (Dexter 1974). Some beaches are used by sea turtles for laying their eggs (see below).

Three papers have been published on the infauna of sandy beaches in Costa Rica. Dexter (1974) examined the macroscopic infaunas of seven beaches along the Pacific coast (Playita Blanca in Playas del Coco, Tamarindo, Sámara, La Punta in Puntarenas, Boca de Barranca, Playa Cocal in Quepos, Playa Espadilla in Manuel Antonio), during the spring of 1971. She also sampled the infauna found along the Caribbean coast and observed that Pacific beaches averaged seven times the density of individuals, had significantly larger numbers of species, and had significantly finer sand. From December 1997 to November 1998 Díaz-Ferguson and Vargas-Zamora (2001) studied the abundance of the

Fig. 5.4 Various beach types found along the Pacific coast of Costa Rica. (A) Sandy beach at Isla del Caño; (B) cobble beach on Península de Osa; and (C) a combination of sandy beach (upper section) and mud flat (lower section) at Bahía Salinas (forefront).
Photographs by Jorge Cortés.

porcellanid crab, *Petrolisthes armatus*, at a rocky beach in the Golfo de Nicoya, and recorded observations on its larvae (Díaz-Ferguson et al. 2008). Years later Wehrtmann et al. (2012) studied the reproductive plasticity of this species. Ultimately, Sibaja-Cordero et al. (2014) described the species richness of the invertebrates of sandy beaches along the north Pacific coast of Costa Rica.

As mentioned above, sea turtles are important users of the sandy beaches found along the Pacific coast of Costa Rica. Four species nest on those beaches: the leatherback or *tortuga baula*, *Dermochelys coriacea* (Vandelli, 1761); hawksbill or carey, *Eretmochelys imbricata* (Linné, 1766); black turtle, *Chelonia agassizii* Bocourt, 1868; and the olive ridley turtle or *tortuga lora*, *Lepidochelys olivacea* (Eschscholtz, 1829); see for instance, Sasa et al. (2009) and Q. Jiménez et al. in chapter 9 of this volume. Three beaches are especially important for turtle nesting: Nancite, Ostional, and Playa Grande. At Nancite and Ostional the olive ridley turtles nest during "*arribadas*" or massive synchronized nesting periods, when several thousand females come out over two or three days (Fig. 5.5A). The other beach, Playa Grande, which is located in the Parque Nacional Marino Las Baulas, is one of the last significant nesting sites for leatherback turtles (Fig. 5.5B), but numbers of nesting turtles have been dropping here lately (Spotila et al. 2000, Pihen et al. 2006).

Two papers have been published on parasites of marine turtles, one on the cloacal and nasal bacterial flora (Santoro et al. 2006), and another one on the presence of digenetic trematode parasites that thrive in olive ridley sea turtles (Santoro and Morales 2007). Recently, Santoro and Mattiucci (2009) published a chapter on parasites found in Costa Rican marine turtles.

Rocky Intertidal Zone

Rocky shores make up for more than half of the coast line of the Pacific zone of Costa Rica (Denyer and Cárdenas 2000). For most organisms the rocky intertidal zone is quite a harsh environment to thrive in (Fig. 5.6). During parts of the day this habitat is situated underwater, while at other moments it is exposed to direct sunlight, causing wide temperature fluctuations, leading to possible dessication or exposure to freshwater during heavy rains. Additionally, there might be a strong wave impact as well. Organisms in this environment showed a discernible zonation across this particular habitat (Sibaja-Cordero 2005, Sibaja-Cordero and Vargas-Zamora 2006, Sibaja-Cordero and Cortés 2008). The upper zone (supralittoral and high littoral) turned out to be occupied by species in the families Littorinidae and Neritidae. Just next to it, there was a strip with a dense cover of barnacles of the genus *Chthamalus*, whose resulting patterns of distribution and abundance are probably a function of past variation in recruitment (Sutherland 1987). In this zone also another barnacle species was observed: *Tetraclita stalactifera*. Villalobos (1980a) found that this latter species reproduced only once a year at Bahía Ballena, and that its relative density was fairly constant throughout the year. He also determined that growth was constant on an annual basis, with a slight decline just before the release of the nauplii (Villalobos 1980b). The lower zone was most diverse in terms of both animal and algal species. Here, assemblages between places were more variable than in the higher parts of the intertidal zone (Bakus 1968, Villalobos 1980c, Sibaja-Cordero and Vargas-Zamora 2006, Sibaja-Cordero and Cortés 2008). In some areas that were studied, about half of the gastropods were carnivores while the other

Fig. 5.5 (A) "*Arribada*" or massive synchronized nesting at Refugio Nacional de Vida Silvestre Ostional; (B) Leatherback turtle nesting in Playa Grande (now Parque Nacional Las Baulas) in 1983.
(A) photograph by Dirk Kappelle, (B) photograph by Jorge Cortés.

Fig. 5.6 Different types of rocky intertidal zones: (A) Metasedimentary rocks on the southern side of Isla del Caño; (B) Fossil dunes at Isla Murciélago. *Photographs by Jorge Cortés.*

half were herbivores (Paine 1966). In general, populations living in this zone resulted to be highly dynamic (Spight 1978, Ortega 1987a).

In his study on the zonation of gastropods in Costa Rica, Bakus (1968) also observed that about half of the intertidal gastropods are carnivores and another 50% are herbivores, while the number of snails found at the Pacific side was about twice the amount found at the Caribbean coast. Spight (1976) compared gastropod diversity between Costa Rica and Washington State in the USA. He observed that Costa Rica had five times more species than found in Washington, although the individual tropical sample quadrats had as many species as the temperate quadrats, indicating a more patchy distribution among the tropical species. In fact, most tropical species thrived in fewer habitats than the temperate species. Later, Ortega (1986) compared fish predation of gastropods in Costa Rica, Panama, and temperate zones. Her studies show that fish predation intensity in Costa Rica was variable in space and time, comparable to temperate zones, and lower than in the Bay of Panama. She indicates that local variation within each region should be considered when temperate-tropical comparisons are made.

Ortega (1987b) found that the mean and maximum sizes of *Siphonaria gigas* (local name: *casco de burro*) were smaller at a site accessible to fishermen, as compared to a site in Parque Nacional Manuel Antonio where extraction is prohibited. Interestingly, Bakus (1968) did not report this species from Manuel Antonio before it became a National Park (Willis and Cortés 2001). *Siphonaria gigas* was sometimes surrounded and imprisoned by the barnacle *Chthamalus fissus*. The limpets lost weight when imprisoned, a situation that persisted until the barnacles were removed following *Thais brevidentata* predation (Sutherland and Ortega 1986).

Siphonaria gigas has been one of the most studied intertidal gastropods in Costa Rica. Lahmann and González (1982) marked shells of this mollusk and found that 84% returned to the same home scar, and that the size of the individuals decreased higher on the rocks and farther from the low tide level. Crisp et al. (1990) found that *S. gigas* deposited one microgrowth band per tidal cycle, and determined a growth rate between 0.0305 and 0.0401 mm day^{-1}. Sibaja Cordero (2008) at Isla del Coco, recorded the largest specimens of *S. gigas* ever measured in Costa Rica.

The chiton fauna at the Sámara rocky intertidal site (exposed rocks and tide pools) consisted of nine species, with *Ischnochiton dispar* (Sowerby 1832) being the predominant species. Zones with intermediate environmental conditions and high algal diversity had the highest abundance and diversity of chitons (Jörger et al. 2008).

Sibaja-Cordero and Cortés (2008) published a paper on the rocky intertidal zone of Bahía Salinas, which is located in the northwest sector of the country. As mentioned earlier, it is an area of important seasonal upwelling. Sibaja-Cordero and Cortés (2008) found that during the upwelling season, the algal cover increased significantly as well as the abundance of herbivores, whereas sessile invertebrate presence and abundance did not change between seasons, but were rather different between sheltered and protected sites in the area. In another study, Nova-Bustos et al. (2010) investigated the abundance and distribution of ascidians in Bahía Cuajiniquil and very recently Quesada et al. (2014) studied the diet of the sea anemone, *Anthopleura nigrescens*, which varies between sites depending on the time of the

Fig. 5.8 (A) Intertidal mud flat in Golfo de Nicoya; (B) Leaf-covered tubes of polychaete worms in a mudflat at Bahía Salinas. *(A) photograph by Ulrich Saint-Paul, (B) photograph by Jorge Cortés.*

a 500 µm mesh sieve) of the Punta Morales mudflat (more than 30% silt+clay) was numerically dominated by deposit feeders (Vargas 1987), including the polychaetes *Paraprionospio pinnata* (Spionidae) and *Mediomastus californiensis* (Capitellidae), the ostracod *Cyprideis pacifica*, and the cumacean *Coricuma nicoyensis*. Mean density of the macrofauna was 13,827 ± 10,185 individuals per m^2, while diversity (*H'*) ranged from 1.75 to 3.36 and equitability (Evenness) ranged from 0.48 to 0.87. A seasonal pattern of the community (dry vs. rainy seasons) was detected using statistical multivariate techniques. Seasonality of the population of the cumacean *Coricuma nicoyensis* was evident with higher abundance levels during the dry season (December to April) and lower levels during the rainy season (May to November), salinity being the most obvious environmental factor influencing the abundance of this microcrustacean (Vargas 1989a). The importance of deposit feeding invertebrates, the types of feeding modes and habitat utilization, and the existence of seasonal patterns, make this community similar to certain temperate zone communities (Vargas 1988a).

There are very few studies of yearly fluctuations of tropical intertidal mudflat communities even though there can be large changes over relatively short periods (Vargas 1989b). Some of the changes are related to seasons (Vargas 1996) but others are not. Vargas (1988a) observed significant oscillations in the dominant species at the same site over the course of a few years.

Vargas (1988b) also studied the meiofauna (organisms passing through a 500 µm mesh sieve but retained on a 62 µm mesh sieve) at Punta Morales, and found densities of approximately 2.4x10^6 individuals per m^2, with the nematodes representing 82% of the total, followed by foraminifers (6.1%), copepods (4.2%), and ostracods (3.7%). There was a positive correlation between the percentage of silt+clay and the abundance of organisms.

Dittmann and Vargas (2001) compared the tropical tidal flat benthos from Central America with the one found in Australia. It turned out that the organisms found at these sites are similar in terms of their ecological roles and demonstrate high levels of taxonomic diversity. The dominant groups in both biogeographic regions were polychaetes (Fig. 5.8B), crustaceans, and mollusks, while most are deposit feeders. In fact, related species had similar ecological roles in both regions.

In an intertidal sandflat, also found along the eastern shore of Golfo de Nicoya, Rojas-Figueroa and Vargas-Zamora (2008) studied the world´s largest polychaete, *Americonuphis reesei*. This polychaete is used as bait in the area, and was most abundant in medium sand with low contents of silts and clays. It had a patchy distribution, with an average density of 9.3 individuals per m^2 and a maximum of 134 individuals per m^2. José A. Vargas has recently re-analyzed samples and data collected from Golfo de Nicoya in 1980 and from 1984 to 1986. Vargas and Dean (2010) published a paper on the distribution and abundance of the cephalochordate *Branchiostoma californiense*. With samples collected between 1984 and 1996, Vargas and Solano (2011) presented the population fluctations of two echinoderms from an intertidal flat, the sand dollar, *Mellitella stokesii*, and the burrowing brittle star, *Amphipholis geminata*. Vargas-Zamora and Sibaja-Cordero (2011) re-analyzed molluscan data from the 1980s to determine fluctuation patterns, which were more significant during the rainy season. Vargas-Zamora et al. (2012) re-analyzed crustacean diversity and population fluctuations from cores collected in the 1980s. They found that populations did not have a yearly fluctuation but responded to events like El Niño and to strong harmful algal blooms.

Seagrass Beds

Seagrass beds in the tropics are formed by marine flowering plants (angiosperms) but live their entire cycle underwater (Fig. 5.9). They have roots and can take up nutrients both from sediments and the water column itself. The roots, which are part of a rhizome system, help anchor the plants underwater so they can thrive in nutrient-poor environments (Björk et al. 2008). Although seagrass beds have a high productivity and serve as habitat to many organisms, they are now being impacted worldwide (Waycott et al. 2009).

Seagrass beds along the Pacific coast of Costa Rica (and all of Central America) are rare and small compared to those along the Caribbean coast (Cortés and Salas 2009, and see Cortés, chapter 17 in this book). Seagrasses have been collected at several points along the Pacific coast (Gómez 1984, Davidse et al. 1994), but patches have been observed only at Bahía Culebra, Herradura, and the mouth of the Sierpe River (Cortés 2001). Only two species of seagrass have been reported from the Pacific coast of Costa Rica: *Ruppia maritima* Linnaeus, 1753, Family Potamogetonaceae (Fig. 5.9) and *Halophila baillonii* Ascherson, 1874, Family Hydrocharitaceae. Only the largest patch at Bahía Culebra was studied, but it disappeared in 1996 after a severe storm (Cortés 2001). The patch, made up mainly by *R. maritima* and isolated plants of *H. baillonii* in deeper waters, was a band parallel to the coast about 500 m long by 10 m wide. At least 44 invertebrate species and three species of fishes were associated with that seagrass bed. Surveys in 2011 and 2012 in this same area found only a few isolated plants (M. Loría pers. comm., pers. obs.). A seagrass field of *Halophila baillonii* and its associated fauna was described for the first time from the Golfo Dulce (Samper-Villareal et al. 2014). The dominant associated fauna was composed of polychaetes, which included several new records for Costa Rica. Seagrasses were not known from the Golfo Dulce until recently.

The main stressor to seagrass beds is a change in water quality. Sedimentation and euthrophication have significant impacts on seagrasses. Others are changes in currents that will uproot plants or changes in pollutants that may affect them. Climate change will result in changes in water quality, water level regime, and storm intensity. A few other types of impacts that may happen result from changes in pH and CO_2 concentrations. However, the effect that these factors will have on the ecosystem is just now being studied and understood (Björk et al. 2006).

Coral Reefs

Eastern Tropical Pacific coral reefs are characterized by their relatively small size. They cover only a few hectares, have a low coral diversity, and are discontinuous. This is why Cortés (1997) called them the minimum expression of a coral reef (Fig. 5.10A). The Eastern Tropical Pacific is an area exposed to extreme temperatures, from cold upwelled waters (Glynn and Stewart 1973, Cortés 2003) to warm waters during El Niño years (Glynn 1990, Guzman and Cortés 2007). It is also a region of low salinity (Fiedler and Talley 2006) and low pH (Kleypas et al. 2006), especially in the areas of seasonal upwelling (Manzello et al. 2008, Manzello 2009, 2010, Rixen et al. 2012). For these reasons

Fig. 5.9 The seagrass, *Ruppia maritima*, as it grew in Bahía Culebra in the mid-1990s.
Photograph by Jorge Cortés.

microatolls (round colonies with dead centers that mark the sea level) of *Porites lobata* (Fig. 5.10D). The reef slope and base are the most diverse areas and have the highest coverage of live coral. The shallow sections of the reef were structured mainly by physical factors: wave action, temperature and salinity fluctuations, and low tidal exposure. In contrast, deeper sections were controlled by biological interactions: bioerosion, damselfish algal lawns, and corallivores (Guzman 1988a, Guzman and Cortés 1989a).

The corals of Isla del Caño, as in most parts of the Eastern Tropical Pacific, have been severely impacted by the 1982–1983 El Niño event, resulting in losses of up to 50% of live coral coverage (Guzman et al. 1987b, Glynn et al. 1988). During the 1990s, other El Niño events impacted the reefs but mortality was much lower (Guzman and Cortés 2001). Reefs have also been impacted by phytoplankton blooms (Guzman et al. 1990).

Twenty species of octocorals, two black corals, seventeen reef-building corals (hermatypic corals), and four ahermatypic coral species have been identified from Isla del Caño so far (Cortés et al. 2009), making it the richest coral reef area of the country. The growth rates of the predominant species, *Porites lobata,* and pocilloporids, were greater during the dry season, with light (i.e., turbidity, cloud cover) and other physical factors probably controlling seasonal growth, instead of temperature, which is the limiting factor in other regions (Guzman and Cortés 1989b).

The reproductive ecology of eight species of reef-building corals and one non-zooxanthellate coral have been studied at Isla del Caño as part of a larger Eastern Pacific project (Glynn et al. 1991, 1994, 1996, 2000, 2008, 2011, 2012). All species exhibited year-round gonadal development with peaks at certain times, but only at Isla del Caño significant recruitment has been observed (Guzman and Cortés 2001). Asynchronous release of male and female gonads may be an explanation for the lack of sexual recruits (Bezy 2009). In some species, fragmentation may be the most important form of reproduction (Guzman 1988b, 1991, Guzman and Cortés 1989a). Boulay et al. (2012) in their study on genotypic diversity of the reef-building coral, *Porites lobata*, found that some populations at Isla del Caño reproduce mainly asexually. They also found the cryptic species *Porites evermanni* at the island (Boulay et al. 2014).

Coral reefs of the Pacific coast of Costa Rica have been impacted by natural and anthropogenic factors (Cortés and Jiménez 2003), and, combined with expected climate change expressed in higher temperatures and ocean acidification, more perturbations are expected. However, recent observations indicate that recovery is occurring faster at reefs exposed to minimum levels of human impacts (e.g., at Isla del Coco), compared to Isla del Caño and more impacted areas such as Bahía Culebra. In other regions, reef resilience is apparently higher among less stressed reefs (Hughes et al. 2005). Mora et al. (2011) demonstrated that reef fish assemblages, especially the most diverse, are very sensitive to biodiversity loss due to anthropogenic stressors.

Rhodolith Beds

Rhodoliths, equivalent to maërl, are free-living forms of non-geniculate coralline red macroalgae that form extensive beds worldwide (Fig. 5.11) (Foster 2001, Harvey and Woelkerling 2007, Steller et al. 2007). Individual rhodoliths might be composed of single species or a combination of species, and beds are also composed by a combination of a single or several species (Steller et al. 2009). The depth to which these beds can be found ranges from the lowermost intertidal zone to depths of over 286 m (Foster 2001, Amado-Filho et al. 2007). These beds are relevant because (1) its complex three-dimensional structure provides microhabitats and shelter, which sustain a rich biodiversity of sea weeds, invertebrates, and fishes, some of which are commercially important and endemic species (Barbera et al. 2003, Steller et al. 2003, Hinojosa-Arango and Riosmena-Rodríguez 2004, Figueiredo et al 2007, Riul et al. 2009, pers. obs.); (2) they function as recruitment areas for many species (Steller et al. 2003, Kamenos et al. 2004, pers. obs.); (3) its structures have the potential to provide information on past oceanic conditions (Foster et al. 1997, Frantz et al. 2000); (4) they contribute to keeping pH levels of seawater steady (Canals and Ballesteros 1997); and (5) they are exploited as a source of calcium carbonate for a wide variety of human uses, including as fertilizer and soil improving substance, biological denitrificator, drinking water improver, toxin eliminator, animal fodder additive, and as a ingredient in pharmaceutics, cosmetics, bone surgery, and even in nuclear industry (Barbera et al. 2003, Grall and Hall-Spencer 2003, Riul et al. 2008).

In several parts of the world rhodolith beds and rhodolith-forming species groups are listed as Special Areas of Conservation (Europe, e.g. in the UK) or are considered in the development of Marine Parks (BIOMAERL 1998). In the northeastern Atlantic, two representative species, *Phymatolithon calcareum* and *Lithothamnion corallioides*, belong to the few algal species requiring appropriate management measures under the European Community Habitats Directive, as species of community interest whose collection in the wild and exploitation may be subject to management measures (Wilson et al. 2004). Maërl beds have also been included in the UK Biodiversity Action Plan (Birkett et al.

Fig. 5.11 (A) Individual rodolith, a non-geniculate coralline red macroalga. Rodoliths can grow (B) between rocks and corals, San Juanillo, or (C) in extensive beds as on Isla del Caño.
Photographs by Jorge Cortés.

1998), in the Natura 2000 sites, in the OSPAR Commission list, and in the Mediterranean Red Book of threatened habitats (http://www.ospar.org).

These management measures have led to the prohibition of their commercial exploitation across Europe, and ultimately they have been proposed as critical marine habitats for conservation (Birkett et al. 1998, De Grave and Whitaker 1999, Hall-Spencer 2005). Likewise, in New Zealand and Australia, rhodolith beds have also been recognized as important biogenic habitats to fisheries productivity, and different monitoring programs are currently being developed to integrate these habitats into management systems, which will help resource users and managers find a better balance between resource extraction and ecosystem integrity and resilience (Morrison et al. 2008).

Furthermore, rhodolith-forming species have been included in the New Zealand Threat Classification System list (http://www.doc.govt.nz). In México, rhodolith beds were used to design marine protected areas and are now considered in the management plan of Natural Protected Areas (NPAs). Because of their influence on ecological processes and economic activities (fisheries), monitoring is necessary (Riosmena-Rodriguez 2001).

Along the Pacific and Atlantic coasts of Central America, rhodolith beds are known mostly from Panama (Glynn et al. 1972, Littler and Littler 2008). The first record of relevant communities in Costa Rica was done by Fernández (2008) at Isla del Coco. Most of the beds along the Pacific coast are known from areas with low terrigenous sedimentation (J. Cortés in prep.). Species composition and distribution is currently being assessed (C. Fernandez unpubl. data). Also, in a recent study more than 100 species have been recognized associated with a rhodolith bed (Solano-Barquero 2011). It seems that rhodoliths are important nursing grounds for polychaetes and ophiuroids since a considerable number of organisms found are juveniles. In particular, this habitat should be studied in much more detail to understand the full scope of its ecological role and function in coastal areas.

Pelagic ecosystem

The largest area of Costa Rica's geographic jurisdiction is the open ocean (pelagic and deep benthos). Although its size is more than half a million square kilometers (more than 10 times its land area), it has been studied very little, with the exception of the Costa Rica Thermal Dome (Sameoto 1986, Vicencio-Aguilar and Fernández-Alamo 1996, Fiedler 2002, Ballestero 2006). The pelagic ecosystem (the water column of the oceans) can be divided into a neritic or coastal zone (the water column on top of the continental shelf) and an oceanic zone (the water that covers the rest of the ocean at depths over 200 m).

Some of the topics that have been studied in the neritic zone of Costa Rica include plankton diversity and abundance (Morales-Ramírez 1996, Vicencio-Aguilar and Fernández-Alamo 1996), fish diversity (Bussing and López 1996), cetacean presence (May-Collado and Morales-Ramírez 2005, Martínez-Fernández et al. 2011, 2014), the use of modified circle hooks to reduce the impact of fisheries (Swimmer et al. 2011), and the potential impact of the mahi-mahi fishery (Whoriskey et al. 2011). Morales-Ramírez (1996) presented a list of copepods collected during the Victor Hensen Expedition in 1993–1994, while Bussing and López (1996) presented a list of fishes collected during that expedition. Some sampling stations were established in the neritic zone during the DOMO III and IV Expeditions aboard the RV *Puma* (Vicencio-Aguilar and Fernández-Alamo 1996). May-Collado and Morales-Ramírez (2005) studied the presence and behavior of the spotted coastal dolphin, *Stenella attenuata*. In their study of the coastal cetaceans of Costa Rica, Martínez and co-workers (2011) found that the three most sighted species were *Stenella attenuata* (68%), *Megaptera novaeangliae* (13%), and *Tursiops truncatus* (10%). In collaboration with the Costa Rican Non-Govermental Organization, PRETOMA, Swimmer et al. (2011) demonstrated that the use of modified circle hooks can improve fishing selectivity and reduce incidental capture of non-target species in longline fisheries. Also, Whoriskey et al. (2011) presented the potential impact on sharks and turtles of the mahi-mahi fishery.

Studies of the oceanic zone include plankton (Vicencio-Aguilar and Fernández-Alamo 1996), cetacean sightings (May-Collado et al. 2005), and studies by Swimmer et al. (2011) and Whoriskey et al. (2011) that extend into this zone. During the four DOMO expeditions, held between 1979 and 1982, a large number of plankton stations were put in the oceanic zone (Vicencio-Aguilar and Fernández-Alamo 1996), including the Costa Rica Thermal Dome. May-Collado et al. (2005) reported 19 species of cetaceans (families Balaenopteridae, Kogiidae, Physeteridae, Ziphiidae and Delphinidae) from the pelagic zone of Costa Rica, with the majority of species being mainly oceanic in distribution.

Deep Benthos

A few samples were collected from the deep sea between 1888 and 1905 during the *Albatross* expeditions, while recently several papers have been published on mollusks,

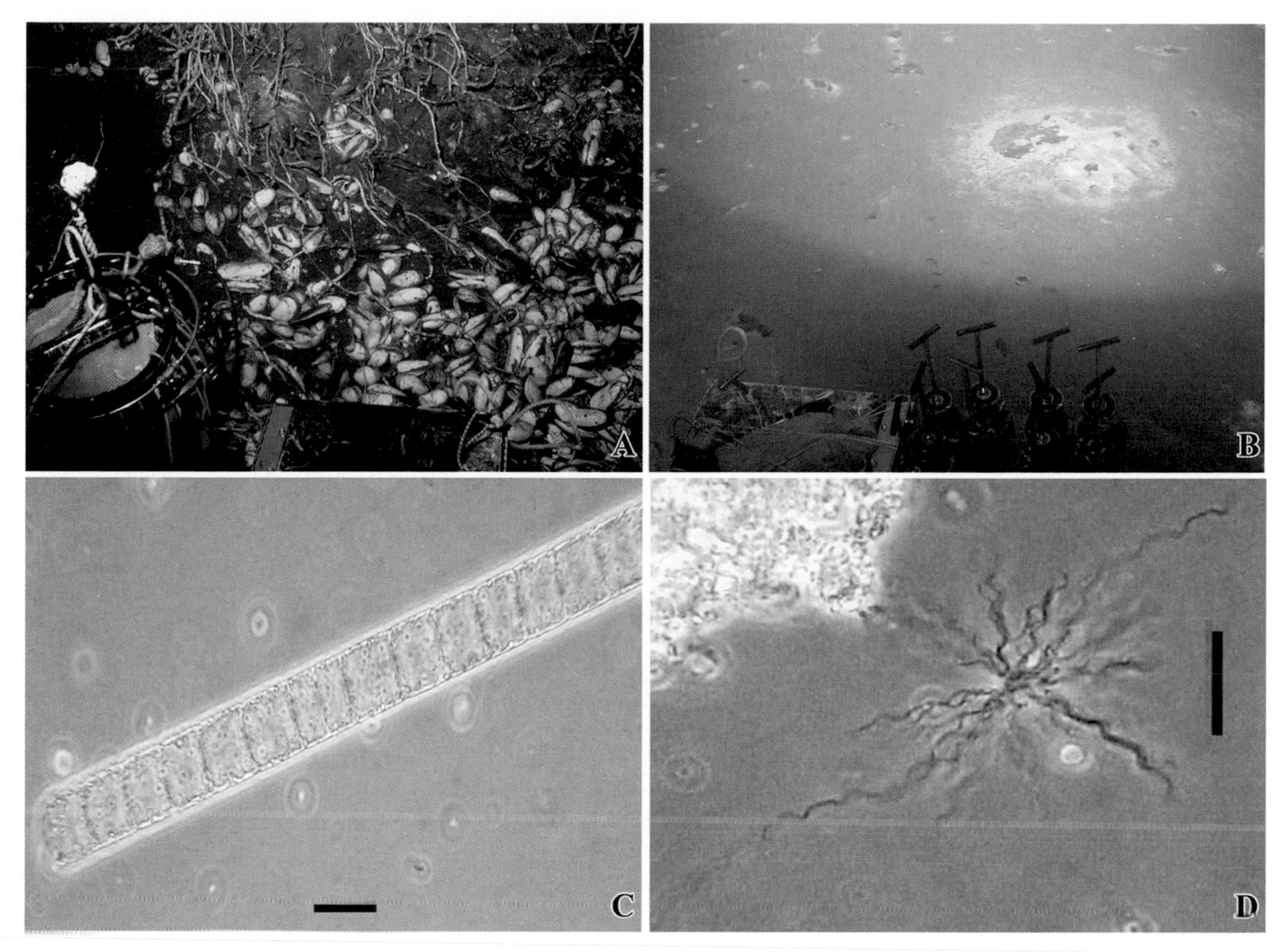

Fig. 5.12 (A) Macroinvertebrates (the mussels are *Bathymodiolus* sp. nov.; the clams are *Calyptogena* sp.; the pogonophorans [tubes] are mainly *Lamellibrachia* sp.) from 1,792 m deep, off the Pacific coast of Costa Rica. (B) Bacterial mat over soft sediments at a depth of 400 m. C and D show macrobacteria from Golfo Dulce, collected and prepared by Victor A. Gallardo and Carola Espinoza: (C) *Thioploca* cf. *chileae*, (D) spirochaetes. Scale bar 10 μm.
(A) and (B) photographs by CRROCKS/L.Levin, taken from the DSRV Alvin*; (C) and (D) photographs by Victor A. Gallardo and Carola Espinoza.*

decapods, and echinoderms from this habitat (Cortés 2009). In the past decade, most of the cruises in the open ocean have been geophysical in scope, but organisms have been photographed using Remote Operated Vehicles (ROVs) and some have also been collected with the *Alvin* submarine (Fig. 5.12A) (Cortés 2009, Cortés and Wehrtmann 2009, L.A. Levin pers. comm. 2009, Opresko and Breedy 2010, Levin et al. 2012).

Bacterial mats have been observed in some deep areas (Fig. 5.12B), and cold (methane) seeps that are distributed along the continental slope (Bohrmann et al. 2002, Han et al. 2004, Sahling et al. 2008). More than 100 cold seeps have been located along the coast. On average there is one every four kilometers but with variable density, and most are in a band about 28 ± 8 km landward of the trench (Sahling et al. 2008). Associated with the fluid expulsion sites are chemosynthetic vent organisms such as bacterial mats; vesicomyid, solemyid, and mytilid bivalves; and siboglinid tubeworms (Bohrmann et al. 2002). A species of vesicomyid bivalve, *Calyptogena costaricana*, was described from material collected from this area. These bivalves live in symbiosis with sulphide-oxidizing bacteria in reducing habitats (Krylova and Sahling 2006). Other organisms have been observed in cold seeps and probably represent new species. Deep water samples from the Cocos Cordillera and from the continental slope have yielded new species of kinorhynchs (Neuhaus 2004, Neuhaus and Blasche 2006). These deep areas are in need of much more research.

In some parts of the oceans at midwater depth (100 to 900 m) levels of oxygen are very low (<0.5 ml l^{-1} dissolved oxygen, or less than 7.5% saturation). These are called Oxygen-Minimum Zones (OMZ), and are widespread beneath the most productive regions of the world oceans (Levin 2002, 2003, Helly and Levin 2004, Karstensen et al. 2008). They can cover about 8% of the total oceanic area (Paulmier and Ruiz-Pino 2009) but are more abundant in some parts around the globe, for example in the Eastern Pacific (Helly and Levin 2004, Karstensen et al. 2008). Off the coast of South America the OMZ can extend for thousands of square kilometers representing millions of cubic meters of low oxygen waters (Fuenzalida et al. 2009). Although biodiversity is low in the OMZ, there are very interesting communities of organisms that do tolerate low oxygen levels (Helly and Levin 2004). The OMZs are characterized by the presence of large filamentous bacteria (Fig. 5.12C) (Gallardo 1977, Gallardo and Espinoza 2007), and influence the distribution of the fauna, from the smallest to the largest (Quiroga et al. 2009, Veit-Köhler et al. 2009).

There is an OMZ several hundreds of meters thick along the continental slope of Costa Rica (L. Levin pers. comm. 2009) and within the Golfo Dulce (Córdoba-Muñoz and Vargas 1996, Acuña-González et al. 2006). Samples from Golfo Dulce and the Quepos Landslide at a depth of 400 m contained macrobacterias (Gallardo and Espinoza 2007, Levin et al. unpubl. data); many samples from the continental slope were collected in February-March 2009 and January 2010, and are currently being analyzed (L.A. Levin pers. comm. 2009). Levin et al. (2012) found "hydrothermal seeps," a combination of hydrothermal vent and cold seep faunas. A new species of black coral was collected during the 2009 expedition and described by Opresko and Breedy (2010). Goffredi (2010) reports on ectosymbiont bacteria on the Yeti crab described by Thurber et al. (see below). Aguado and Rouse (2011) described three new species of polychaetes *Laubierus alvini, Natsushima sashai,* and *Shinkai fontefridae*, and a new report for the region, *Shinkai longipedata*. Thurber et al. (2011) described a new species of Yeti crab, *Kiwa puravida*, that feeds on epibiotic bacteria that grow on its claw. A new species of clam, *Pliocardia krylovata*, was described from another cold seep on the Pacific coast of Costa Rica (Martin and Goffredi 2012).

Bianchi (1991) analyzed bottom trawl samples collected from the continental shelf and slope edge between Gulf of Tehuantepec, Mexico, and Gulf of Papagayo, Costa Rica. She found that the most abundant organism in the Gulf of Papagayo was *Pleuroncodes monodon*. Ingo S. Wehrtmann of the Universidad de Costa Rica and his collaborators have started to publish a series of papers on deep water fisheries (Wehrtmann and Nielsen-Muñoz 2009), and organisms collected during commercial trawling operations. There are papers on organisms of the deep water fishery off the Pacific coast of Costa Rica: crustaceans (Stomatopoda: Decapoda) (Wehrtmann and Echeverría-Sáenz 2007, Echeverría-Sáenz and Wehrtmann 2011), stomach content of the threadfin anglerfish (Espinoza and Wehrtmann 2008), lithodid crabs (MacPherson and Wehrtmann 2010), the species diversity, spatial and bathymetric distribution of squat lobsters (Wehrtmann et al. 2010), and on gonad development of a deep water shrimp (Villalobos-Rojas and Wehrtmann 2011). Patricio Hernáez-Bové (2010), also with deep water samples, completed a thesis on demography and spatial distribution of the red squat lobster (*Pleuroncodes planipes* Stimpson), and published a paper on sexual maturity and egg production (Hernáez and Wehrtmann 2011). He also published on the population demography and spatial distribution of the mantis shrimp, *Squilla biformes* (Hernáez et al. 2011). A paper on the feeding ecology of two elasmobranchs has also been published (Espinoza et al. 2012).

Fig. 5.13 Islands and islets of the Pacific of Costa Rica: (A) Isla del Caño (from the southwest) and (B) one of the Catalina Islands off the Guanacastecan coast in the northern Pacific region of Costa Rica.
Photographs by Jorge Cortés.

Islands and Islets

There are many islands and islets along the Pacific coast of Costa Rica that range in size from just pinnacles to tens of square kilometers (Fig. 5.13), for example Isla Chira (43 km^2). Apart from Isla del Coco, which is an oceanic island that has never been in contact with the continent (see chapters 7 and 8 of this volume on Isla del Coco by J. Cortés and M. Montoya, respectively), all other islands are remants of rocky outcrops that used to be part of the continent (Denyer and Cárdenas 2000). Some of those islets and islands, like Isla Bolaños (which is a National Wildlife Refuge), are important nesting areas for seabirds, while others might harbor an interesting flora and fauna as well. Unfortunately, there are no systematic studies of the islands and islets, and most are greatly affected by human occupation.

Species Diversity

The marine flora and fauna of the Pacific coast of Costa Rica is part of the central core of the Panamic biogeographic province, the most diverse province of the Eastern Pacific (Robertson and Cramer 2009). Almost 5,000 species (Table 5.1) have been reported from marine and coastal environments of the Pacific coast of Costa Rica (excluding species found only at Isla del Coco or in the Costa Rica Thermal Dome), representing about 73% of the total number of species recorded from both coasts of Costa Rica (6,805 spp.). Of those species, five are bacterias, 3,996 are invertebrates, and 964 are vertebrates (Table 5.1). Sixty-two of the reported species are endemic to the Pacific coast of Costa Rica, mostly belonging to the arthopods and annelids (Table 5.1). However, these numbers represent an underestimation of the marine biodiversity of the Pacific coast of Costa Rica because several environments have never been sampled or are heavily undersampled—for example, seamounts, caves, the pelagic environment, and deep regions. Also, several taxonomic groups have not been studied yet, or published on, just to mention a few groups: cyanobacteria, non-parasitic flatworms, and nemerteans (Table 5.1) (Wehrtmann et al. 2009).

The groups with most species are Mollusca (1,495 spp.), Chordata (986 spp., of which 838 are fishes), Arthropoda (915 spp.), Annelida (322 spp.), and Cnidaria (223 spp.) (Table 5.1). Micro- and macroalgae diversity is relatively high, with the most specious groups being Chrysophyta (151 spp.) and Rhodophyta (158 spp.) (Table 5.1). Several groups have been observed but there are no published accounts, for example, on Ctenophora and Nemertea (Table 5.1).

Table 5.1. Marine Biodiversity of Costa Rica's Pacific Coast

TAXON	No. of species	Endemic species
Virus	n.a.[a]	n.a.
Bacteria	5 genera[b]	n.a.
Cyanophyta/Cyanobacteria	23	0
Granuloreticulosa (Foraminifera)	94	0
Ciliophora	1	0
Chrysophyta	151	0
Dinophyta	97	0
Haptophyta	1	0
Ochrophyta	3	0
Neomonada	2	0
Chlorophyta	46	0
Ochrophyta	26	0
Rhodophyta	158	6
Fungi	5	0
Plantae	10	0
Porifera	62	0
Cnidaria	223	7
Ctenophora	n.a.[c]	n.a.
Platyhelminthes	42[d]	0
Nemertea	n.ae	n.a.
Kinorhyncha	2	0
Nematoda	1[f]	0
Acanthocephala	1	0
Sipuncula	20	0
Echiura	1	0
Annelida	322	16
Arthropoda	915	30
Mollusca	1,495	3
Bryozoa	48	0
Brachiopoda	8	0
Echinodermata	191	0
Chaetognatha	25	0
Hemichordata	1	n.a.
Chordata	986[g]	0
Urochordata	21[h]	n.a.
Cephalochordata	1	n.a.
Pisces	838	0
Reptilia	5	0
Aves	93	0
Mammalia	28	0
TOTAL	**4,965**	**62**

NOTE. Data do not include species from Parque Nacional Isla del Coco or the Costa Rica Thermal Dome. n.a. = information not available.

[a] Virus are probably present but have not been assessed so far.

[b] Many more genera and species of bacteria are to be expected.

[c] Ctenophores have been observed but no published records are available.

[d] More species of Platyhelminthes are expected to be listed in the future since a number of free-living flatworms have been observed though not recorded in the past.

[e] Nemerteans have been observed, but published records are lacking.

[f] More species of Nematoda are expected to be listed formally, but have not been identified so far.

[g] Total number of chordates.

[h] Salps have been observed, but not reported formally.

Source: Data are based on Wehrtmann et al. 2009, and have been expanded and updated with additions from Vargas et al. 1985, Kuever et al. 1996, Kelmo and Vargas 2002, Gallardo and Espinoza 2007, Breedy and Guzman 2008, Fernández and Alvarado 2008, Bernecker and Wehrtmann 2009, Breedy et al. 2009, Excoffon et al. 2009, Alvarado et al. 2010, Goffredi 2010, MacPherson and Wehrtmann 2010, Opresko and Breedy 2010, Aguado and Rouse 2011, Thurber et al. 2011, and Martin and Goffredi 2012.

People and Nature

Human Population and Demography

The Pacific coast of Costa Rica has a population of over 600,000 inhabitants in two provinces, Guanacaste and Puntarenas. The fertility rate for Guanacaste is 2.2 and for Puntarenas 2.3, while unemployment is 7.7% and 6.1%, respectively. Both values are above the national average of 2 and 4.6%, respectively (http://www.inec.go.cr/; Instituto Nacional de Estadísticas y Censos—National Institute of Statistics and Census). The largest population centers are Puntarenas, Quepos, and Golfito, and together with Playas del Coco, they are the most important fishing towns of the coast. The most important economic activity at present is tourism, but in the past these were fisheries and coastal agriculture.

Threats

Natural and Anthropogenic Disturbances

Beaches are dynamic systems which can be severely eroded during periods of storms and strong wave action. The impact of beach erosion has been reported as a source of sea turtles' nest destruction (Fish et al. 2005) but there are no reports of impacts on other beach organisms. Human impacts on beaches can be triggered by altering currents and sediment regimes, resulting in excess deposition or erosion, pollution, and soil compaction by vehicles and walking people. So far, very little research has been done on this topic.

Rocky intertidal organisms are subject to predation (Ortega 1986) and may be dislodged by extreme wave action. The main anthropogenic impact they suffer is deliberate destruction—for example, by building over them and by their extraction. Mollusks of commercial value, such as *Siphonaria gigas*, have almost completely disappeared from many rocky shores, and now are found only in protected areas (Ortega 1987b, Willis and Cortés 2001).

Mangroves are affected by changes in sedimentation regimes that are caused by either natural or human-induced processes. Sections of mangrove forests in the Térraba-Sierpe National Wetland have been impacted by changes in sedimentation rates (M. Silva, pers. comm. 2008). They can also be impacted by strong winds during storms or hurricanes, but there are no records of this sort of impact in Costa Rica.

In the past an extensive seagrass bed and several smaller ones were devastated by powerful surge waves in Bahía Culebra (Cortés 2001). All the seagrass beds disappeared and after more than 13 years recovery has still been minimal (J. Cortés pers obs).

The main natural perturbation to the Pacific coral reefs has been the warming impact during El Niño-Southern Oscillation (ENSO) events. Reef-building coral mortalities of up to 50% were recorded at several reef sites after bleaching during (Fig. 5.14A) the 1982–1983 and 1997–1998 ENSO events (Cortés et al. 1984, Guzman et al. 1987b, Guzmán and Cortés 2001, Jiménez and Cortés 2001, 2003b, Jiménez et al. 2001). Recovery has been observed but after each ENSO event the percentage of live coral and the number of colonies was reduced. The end result is a slow recovery process (Guzman and Cortés 2001).

La Niña events are characterized by cooler water and usually with higher levels of nutrients, which can result in

Fig. 5.14 (A) Bleached and dead corals at Los Mogos, Golfo Dulce; (B) Corals killed by a dense and persistant phytoplankton bloom in 2007. *(A) photograph by Jaime Nivia, (B) photograph by Jorge Cortés.*

Fig. 5.15 Partially dissolved coral, *Pavona gigantea*, possibly due to the seasonal upwelling of low pH waters in the northern section of the Pacific coast of Costa Rica.
Photograph by Jorge Cortés.

harmful plankton blooms. Such an algal bloom impacted the shallow coral reefs of Isla del Caño in 1985, causing 100% coral mortality between 0 and 3 m depth (Guzman et al. 1990). Phytoplankton blooms have been observed in Guanacaste and cause extensive coral death (Fig. 5.14B) (Jiménez et al. in prep). Between 2007 and 2009 these blooms became more intense and extensive and caused higher levels of coral death along Guanacaste's coasts in the northwest (C.E. Jiménez unpubl. data, J Cortés pers. obs.).

Rhodolith beds are constantly subject to a variety of natural (bioturbation and high levels of sedimentation, turbidity, temperature, and water movement) and anthropogenic (exploitation through extraction, trawling harvesting, reduced water quality caused by shellfish and fish farm waste) disturbances (Barbera et al. 2003, Steller et al. 2003, 2007, Wilson et al. 2004, Hall-Spencer et al. 2006). Given its relatively slow growth rates (< 1 mm/year), these habitat-forming species have been considered a non-renewable resource (Bruno and Bertness 2001, Blake and Maggs 2003). Thus, the slow rhodolith growth combined with negative impacts of the above mentioned disturbances makes recovery predictably slow (Steller et al. 2003, Rivera et al. 2004).

The main threat to the open ocean within the country's Exclusive Economic Zone (EEZ) is the lack of knowledge of what is living in the water and thriving on the sea bottom. But worst of all is the inability of Costa Rica's government to patrol the waters and control what is happening in this vast part of the country.

Climate Change

Seawater warming; changes in the inflow of freshwater, sediments, and nutrients; more and stronger storms; and increased dissolution of CO_2, are all having a major impact on the marine and coastal ecosystems of Costa Rica. As mentioned above, El Niño warming events have resulted in significant coral decline, higher seawater temperatures will only exacerbate these impacts. The influx of freshwater loaded with terrigenous sediments and nutrients are impacting coastal ecosystems; predictions indicate that they will increase. During extreme high tides, strong wave action, and storms, some areas of the coastal zone get flooded. With a higher sealevel these areas will increase in size, affecting not only the coastal ecosystems but also the coastal human population that depend on these ecosystems. Finally, as dissolved CO_2 levels increase in seawater, pH values go down, dissolving carbonate skeletons (Fig. 5.15). Knowing that it is difficult to disentangle the effects of climate change from those of other factors, we need to develop and implement adequate monitoring programs to observe and evaluate these phenomena and impacts as soon as possible.

Conservation

Marine Protected Areas

Six National Parks, one Absolute Reserve, one Biological Reserve, five National Widlife Refuges, and three National Wetlands along the Pacific coast of Costa Rica contain marine areas that are protected by law (see Fig. 5.2, for the most important protected areas; for further details, see Mora et al. 2006, Alvarado et al. 2012b). Unfortunately, most are so-called paper parks. For example, Parque Nacional Manuel Antonio has a marine area of 42,016 ha, but most of the time, either the small patrolling boat is broken, or its outboard motor is damaged, or there is just no fuel available to navigate the park's waters. Even when the boat is functioning well and fuel is in place, there is often no qualified person with experience to navigate the boat.

Alvarado et al. (2012b) found that the average area of Marine Protected Areas (MPAs) along the Pacific coast of Costa Rica (not including Isla del Coco) was close to 55 km^2 (ranging from 0.05 km^2 [Humedal Nacional Marino Playa Blanca] to 46,391 ha [Parque Nacional Santa Rosa]). At the same time, the average distance between MPAs along the coast was about 21 km. These values are above the world average MPA size published by Halpern (2003), and distances are shorter than those proposed by Shanks et al. (2003) as ideal for good interconnections between MPAs. Costa Rica seems to offer a good network of MPAs along the Pacific coast, but as Alvarado et al. (2012b) warn, more study is needed to prove the effectiveness of any connectivity among areas.

A multi-stakeholder ecoregional assessment of coastal and marine ecosystems found along the Pacific and Caribbean coasts of Costa Rica, among other countries, evaluates diversity and threats of habitats (TNC 2008). The study, developed with the participation of numerous marine specialists, provides an initial blueprint for the establishment of new marine protected areas in the Nicoya, Panama Bight, and Isla del Coco marine ecoregions. It is expected the assessment will serve as a key tool for guiding conservation strategies and action in this part of the Eastern Tropical Pacific region.

Legislation

Cajiao-Jiménez (2003) indicated that Costa Rica's legislation on coastal zones and marine areas provides almost full protection, though pieces of legislation are often dispersed and sometimes overlapping. For instance, there is an important law on the use and protection of the coastal zone (Ley Marítimo Terrestre), which guarantees public access to all beaches. There are other laws on the protection and sustainable use of biodiversity, including marine organisms (Ley de Conservación de Vida Silvestre, Ley de Biodiversidad), and on the regulation of fishing activities and marine turtles (Ley de Pesca y Caza Marítima, Ley de Protección, Conservación y Recuperación de las Poblaciones de Tortugas Marinas). Costa Rica is a signatory party to many UN conventions including CITES (traffic of endangered species), UNCLOS (Law of the Sea), RAMSAR (wetland protection), CBD (biodiversity), and UNFCCC (climate change).

Conservation efforts and policies can have positive impacts. For example, the reduction of turtle egg poaching and decreased juvenile and adult mortality of the green turtle in Tortuguero have contributed to an increase in nesting since 1963 (Troëng and Rankin 2005). This example demonstrates that long-term conservation efforts can reverse negative impacts and offer us hope that adequate management actually can result in recovery of marine species populations.

Conservation Organizations

There are a number of national non-governmental organizations (NGOs) dedicated to the study, protection, and conservation of marine organisms, mainly charismatic species such as sea turtles and cetaceans. Some of the most active local NGOs are Keto, MarViva, PRETOMA, and PROMAR. There are also several big international non-governmental organizations (BINGOs) such as The Nature Conservancy (TNC), the International Union for Conservation of Nature (IUCN), Conservation International (CI), and the World Wildlife Fund (WWF). All of them have been contributing in some way or another to the study, protection, conservation, and awareness of the marine richness of Costa Rica.

Conservation Strategies and Action

Threat Abatement

At present there are no projects in progress dealing specifically with threat abatement. At most there are some campaigns to raise awareness about the problems that affect coastal and marine environments along the Pacific shores of Costa Rica.

Ecological Restoration

There are also no coastal or marine ecological restoration projects happening at this moment. Several artificial reefs have been constructed and some were maintained for a while (Guzman et al. 1988b), showing initially positive results. However, none have been studied for periods long enough to draw any long-term conclusions.

Economic Valuation of Environmental Goods and Services

The economic valuation of environmental goods, and especially of the services of marine and coastal ecosystems, is very urgent. However, so far only one thesis has addressed this issue. Ibarra (1996) found that fish have a higher value for tourism when they are alive than when they are dead.

Many more studies along this line are urgently needed. Recent studies conducted in other important marine parts of the world such as Hawai'i in the Pacific Ocean, and the Great Barrier Reef in Australia have valued these areas at amounts between US $100,000 and $1,000,000 per km^2 (Jones 2007).

Future Perspectives

Scenarios for the Twenty-First Century: Towards Sustainability?

The world's oceans have been understudied, unappreciated, and at present taken to their limits through overexploitation of resources and increased pollution. The oceans were seen as inexhaustable and with a limitless capacity to absorb all our insults (Earle 2009). Now we know this is not so and if we don't act now it will be too late for the oceans and seas, and as such for our own species.

Costa Rica is no exception to this. Most fisheries have been overexploited, several coastal areas are highly polluted, there is illegal fishing going on, sea turtle eggs are being poached, and legal and illegal dumping of waste materials is continuing. There is an important network of marine protected areas (MPAs), but we need more and bigger areas, not only along the coast but also in the open seas.

There are some positive signs in Costa Rica, all happening since 2009. Costa Rica now has an office related to marine issues at the Ministry of the Environment and Energy (MINAE), a National Marine Strategy, a Marine Program as part of the public educational system, and initiaves funded by the afore-mentioned BINGOs meant to develop national monitoring programs of key coastal ecosystems, among others.

Importance of Ecosystem Health for Human Survival

Sylvia A. Earle states it very clearly in her book, *The World Is BLUE*: "If there is no 'blue', referring to healthy oceans, there is no 'green', referring to life on land. The oceans are the blue heart of our planet" (Earle 2009).

Moreover, healthy coastal and marine ecosystems are needed to face extreme events, which we are already witnessing and which will intensify with climate change. As ecosystems deteriorate negative social and environmental impacts will arise. Our society needs resilient social-ecological systems that incorporate the concepts of care for the environment, mechanisms for living with disasters, and know-how necessary to act when disasters occur (Adger et al. 2005).

Research Needs

Much more research is needed on the biodiversity, natural history, and especially the environmental services of marine and coastal ecosystems of the Pacific shores of Costa Rica. Some regions will need more work, for example, the continental slope, the abyssal plains, and the pelagic environment, which cover also the largest parts of Costa Rica's national area. As a country Costa Rica should demand equal time from large research vessels of developed countries that carry out their research projects in its national waters. With few exceptions collaboration has been extremely limited.

Another area of research is the re-evaluation of coastal ecosystems that were systematically studied in the past. Such projects should go back to historical sampling by William Beebe (Beebe 1938) or the Allan Hancock Foundation (Fraser 1943), who took detailed notes at sampling sites that should be revisited. Additionally, visits should be made to more recent sampling sites, for example, those at Playas del Coco that were visited by Spight (1976) in 1970 and 1971. Here a census of intertidal gastropods could be repeated and compared to Spight's data. Even more recent surveys can be repeated as well, as Carlos Jiménez and Juan José Alvarado have been doing in Bahía Culebra, and elsewhere along the coast.

Retrospective analyses of coral or sediment cores are also an excellent way of looking at the past in order to understand the present. Research along this line has been done in Costa Rica, for example, by looking at the Holocene growth history of a coral reef in Golfo Dulce (Cortés et al. 1994) or at the recent sedimentation in Golfo Dulce (Hebbeln and Cortés 2001). Analyses of coral cores is being done by Carlos Jiménez with the aim to reconstruct past conditions at several localities in Costa Rica.

Conclusion

Protection and conservation of the marine and coastal ecosystems are not only an obligation of the government, but a need and responsibility of everyone. It doesn't matter where one is living: we all depend on the oceans' resources and more importantly, on the environmental services they offer to our society.

Acknowledgments

I thank Maarten Kappelle and the late Luis Diego Gómez, to whom I dedicate this chapter, for the invitation to write this chapter. Funding over the years to study the marine ecosystems of Costa Rica has been provided mainly by the Vicerrectoría de Investigación of the Universidad de Costa Rica (UCR), CONICIT, Ecodesarrollo Papagayo, The Nature Conservancy (TNC), Conservation International (CI), USAID-Israel, and M. Tupper. Omar Lizano and Eric Alfaro reviewed the sections on climate and physical oceanography, José A. Vargas on Intertidal Flats, Jeffrey Sibaja Cordero on Rocky Intertidal, Rafael Riosmena on Rhodoliths, Lisa A. Levin on Oxygen Minimum Zones, and Juan José Alvarado on Marine Protected Areas; Maarten Kappelle reviewed the entire chapter twice. The following persons helped with information on various groups of marine organisms: Cindy Fernández (algae), Jeffrey Sibaja-Cordero (polychaetes), Rita Vargas (crustaceans), Juan José Alvarado (echinoderms), William Bussing and Ana Ramírez (fishes), and Gilbert Barrantes (birds). The DSRV *Alvin* deep water images were obtained through the US National Science Foundation (NSF) grant OCE-0826254 to L. Levin. I thank Marco Castro (TNC) for the preparation of the maps. Logistic support has been provided by the governmental Conservation Areas, Hotel Ecoplaya, private funders, and CIMAR. Many mentors, students, and colleagues have helped me get to know the coastal region of Costa Rica, which I greatly appreciate.

References

Acuña, F.H., A. Garese, A.C. Excoffon, and J. Cortés. 2013. New records of sea anemones (Anthozoa: Actiniaria) from Costa Rica. *Revista de Biología Marina y Oceanografía* 48: 177–84.

Acuña, J., V. García, and J. Mondragón. 1998. Comparación de algunos aspectos físico-químicos y calidad sanitaria del Estero de Puntarenas, Costa Rica. *Revista de Biología Tropical* 46(Suppl. 6): 1–10.

Acuña, J.A., and M.M. Murillo. 1987. La contaminación por hidrocarburos de petróleo en la Isla del Caño. *Ingeniería y Ciencia Química* 11: 95–98.

Acuña-González, J., J. Vargas-Zamora, and R. Córdoba-Muñoz. 2006. A snapshot view of some vertical distributions of water parameters at a deep (200 m) station in the fjord-like Golfo Dulce embayment, Costa Rica. *Revista de Biología Tropical* 54(Suppl. 1): 193–200.

Acuña-González, J., J. Vargas-Zamora, E. Gómez-Ramírez, and J. García-Céspedes. 2004. Hidrocarburos de petróleo, disueltos y dispersos, en cuatro ambientes costeros de Costa Rica. *Revista de Biología Tropical* 52(Suppl. 2): 43–50.

Adger, W.N., T.P. Hughes, C. Folke, S.R. Carpenter, and J. Rockström. 2005. Social-ecological resilience to coastal disasters. *Science* 309: 1036–39.

Agassiz, A. 1892. Reports on the dredging operations off the west coast of Central America to the Galápagos, to the west coast of Mexico, and in the Gulf of California, in charge of Alexander Agassiz, carried on by the U.S. Fish Commission steamer "Albatross," Lieut. Commander Z.L. Tanner U.S.N., commanding. II. General sketch of the expedition of the "Albatross," from February to May, 1891. *Bulletin of the Museum of Comparative Zoölogy* 23: 1–89.

Aguado, M.T., and G.W. Rouse. 2011. Nautiliniellidae (Annelida) from Costa Rican cold seeps and a western Pacific hydrothermal vent, with description of four new species. *Systematics and Biodiversity* 9: 109–31.

Alfaro, E.J. 2002. Some characteristics of the annual precipitation cycle in Central America and their relationships with its surrounding tropical oceans. *Tópicos Meteorológicos y Oceanográficos* 9: 88–103.

Alfaro, E.J., and J. Cortés. 2012. Atmospheric forcing of cold subsurface water events in Bahía Culebra, Costa Rica. *Revista de Biología Tropical* 60(Suppl. 2): 173–86.

Alfaro, E.J., J. Cortés, J.J. Alvarado, C. Jiménez, A. León, C. Sánchez-Noguera, J. Nivia-Ruiz, and E. Ruiz-Campos. 2012. Clima y variabilidad climática de la temperatura subsuperficial del mar en Bahía Culebra, Guanacaste, Costa Rica. *Revista de Biología Tropical* 60(Suppl. 2): 159–71.

Alfaro, E.J., and O.G. Lizano. 2001. Algunas relaciones entre las zonas de surgencia del Pacífico Centroamericano y los océanos Pacífico y Atlántico tropicales. *Revista de Biología Tropical* 49(Suppl. 2): 185–93.

Alongi, D.M. 2002. Present state and future of the world's mangrove forests. *Environmental Conservation* 29: 331–49.

Alvarado, J.J. 2006. Factores físico-químicos y biológicos que median en el desarrollo de los arrecifes y comunidades coralinas del Parque Nacional Marino Ballena, Pacífico sur, Costa Rica. M.Sc. Thesis, Universidad de Costa Rica. San Pedro, Costa Rica.

Alvarado, J.J., and J.F. Aguilar. 2009. Batimetría, salinidad, temperatura y oxígeno disuelto en aguas del Parque Nacional Marino Ballena, Pacífico, Costa Rica. *Revista de Biología Tropical* 57(Suppl. 1): 19–29.

Alvarado, J.J., J. Cortés, C. Fernández, and J. Nivia. 2005. Coral communities and reefs of Ballena Marine National Park, Pacific coast of Costa Rica. *Ciencias Marinas* 31: 641–51.

Alvarado, J.J., J. Cortés, and H. Reyes-Bonilla. 2012a. Bioerosion impact model for the sea urchin *Diadema mexicanum* on three Costa Rican Pacific coral reefs. *Revista de Biología Tropical* 60(Suppl. 2): 121–32.

Alvarado, J.J., J. Cortés, M.F. Esquivel, and E. Salas. 2012b. Costa Rica's Marine Protected Areas: review and perspectives. *Revista de Biología Tropical* 60: 129–42.

Alvarado, J.J., C. Fernández, and J. Cortés. 2009. Water quality conditions on coral reefs at the Marino Ballena National Park, Pacific Costa Rica. *Bulletin of Marine Science* 84: 137–52.

Alvarado, J.J., F.A. Solís-Marín, and C.G. Ahearn. 2010. Echinoderms (Echinodermata) diversity off Central American Pacific. *Marine Biodiversity* 41: 261–85.

Alvarado, J.J., and R. Vargas-Castillo. 2012. Invertebrados asociados al coral constructor de arrecifes *Pocillopora damicornis* en Playa Blanca, Bahía Culebra, Costa Rica. *Revista de Biología Tropical* 60(Suppl. 2): 77–92.

Alvarado-Induni, G.E. 1994. *Historia Natural Antigua*. Cartago, Costa Rica: Editorial Tecnológica de Costa Rica.

Amado-Filho, G.M., G. Maneveldt, R.C.C. Manso, B.V. Marins-Rosa, M.R. Pacheco, and S.M.P.B. Guimarães. 2007. Structure of rhodolith beds from 4 to 55 meters deep along the southern coast of Espírito Santo State, Brazil. *Ciencias Marinas* 33: 399–410.

Amador, J.A., E.J. Alfaro, O.G. Lizano, and V.O. Magaña. 2006. Atmospheric forcing of the eastern tropical Pacific: a review. *Progress in Oceanography* 69: 101–42.

Angulo, A., M.I. López, W.A. Bussing, and A. Murase. 2014. Records of chimaeroid fishes (Holocephali: Chimaeriformes) from the Pacific coast of Costa Rica, with the description of a new species of *Chimera* (Chimaeridae) from the eastern Pacific Ocean. *Zootaxa* 3861: 554–74.

Araya, H.A. 1984. Los sciaenidos (corvinas) del Golfo de Nicoya, Costa Rica. *Revista de Biología Tropical* 32: 179–96.

Bakus, G.J. 1968. Zonation in marine gastropods of Costa Rica and species diversity. *Veliger* 10: 207–11.

Ball, E.E. 1972. Observations on the biology of the hermit crab, *Coenobita compresus* H. Milne Edwards (Decapoda, Anomura) on the west coast of the Americas. *Revista de Biología Tropical* 20: 265–73.

Ball, E.E., and J. Haig. 1974. Hermit crabs from the tropical eastern Pacific. I. Distribution, color and natural history of some common shallow-water species. *Bulletin of the Southern California Academy of Sciences* 73: 95–104.

Ballestero, D. 2006. El Domo Térmico de Costa Rica. In V. Nielsen-Muñoz and M. Quesada-Alpizar, eds., *Informe Técnico: Ambientes Marinos de Costa Rica*, 69–85. San José, Costa Rica: Comisión ZEE Costa Rica, CIMAR, CI, TNC.

Banta, W.C., and R.J.M. Carson. 1977. Bryozoa from Costa Rica. *Pacific Science* 31: 381–424.

Barbera, C., C. Bordehore, J.A. Borg, M. Glemarec, J. Grall, J.M. Hall-Spencer, C. De La Huz, E. Lanfranco, M. Lastra, P.G. Moore, J. Mora, M.E. Pita, A.A. Ramos-Espla, M. Rizzo, A. Sánchez-Mata, A. Seva, P.J. Schembri, and C. Valle. 2003. Conservation and management of Northeast Atlantic and Mediterranean maërl beds. *Aquatic Conservation: Marine and Freshwater Ecosystems* 13: 65–76.

Bartels, C., K. Price, M. López, and W.A. Bussing. 1983. Occurrence, distribution, abundance and diversity of fishes in the Gulf of Nicoya, Costa Rica. *Revista de Biología Tropical* 31: 75–101.

Bartels, C., K.S. Price, M. López-Bussing, and W.A. Bussing. 1984. Ecological assessment of finfish as indicators of habitats in the Gulf of Nicoya, Costa Rica. *Hydrobiologia* 112: 197–207.

Bednarski, M., and A. Morales-Ramírez. 2004. Composition, abundance and distribution of macrozooplankton in Culebra Bay, Gulf of Papagayo, Pacific coast of Costa Rica and its value as bioindicator of pollution. *Revista de Biología Tropical* 52(Suppl. 2): 105–18.

Beebe, W. 1938. Eastern Pacific expeditions of the New York Zoological Society, XIV. Introduction, itinerary, list of stations, nets and dredges of the eastern Pacific *Zaca* expedition, 1937–1938. *Zoologica* 23: 287–98.

Beebe, W. 1942. *Book of Bays*. New York: Harcourt, Brace and Company.

Bernecker, A., and I.S. Wehrtmann. 2009. New records of benthic marine algae and Cyanobacteria for Costa Rica, and a comparison with other Central American countries. *Helgoland Marine Research* 63: 219–29.

Bezy, M.B. 2009. Reproducción sexual y reclutamiento del coral masivo, *Pavona clavus*, en Bahía Culebra, Golfo de Papagayo, Costa Rica. M.Sc. Thesis, Universidad de Costa Rica. San Pedro, Costa Rica.

Bezy, M.B., C. Jiménez, J. Cortés, A. Segura, A. León, J.J. Alvarado, C. Gillén, and E. Mejía. 2006. Contrasting *Psammocora*-dominated coral communities in Costa Rica, tropical eastern Pacific. In Proceedings of the 10th International Coral Reef Sympsoium, Okinawa: 376–81.

Bianchi, G. 1991. Demersal assemblages of the continental shelf and slope edge between the Gulf of Tehuantepec (Mexico) and the Gulf of Papagayo (Costa Rica). *Marine Ecology Progress Series* 73: 121–40.

BIOMAERL. 1998. Maerl grounds: habitats of high biodiversity in European waters. In Proceedings of the 3rd European Marine Science and Technology Conference, I. Marine Ecosystems, Lisbon: 169–78.

Birkett, D.A., C.A. Maggs, and M.J. Dring. 1998. *An Overview of Dynamic and Sensitivity Characteristics for Conservation Management of Marine SACs, 5 Maerl*. Scotland: Scottish Association of Marine Science. 116 pp.

Björk, M., F. Short, E. McLeod, and S. Beer. 2008. *Managing Seagrasses for Resilience to Climate Change*. Gland, Switzerland: IUCN.

Blake, C., and C.A. Maggs. 2003. Comparative growth rates and internal banding periodicity of maërl species (Corallinales, Rhodophyta) from northern Europe. *Phycologia* 42: 606–12.

Bohrmann, G., K. Heeschen, C. Jung, W. Weinrebe, B. Baranov, B. Cailleau, R. Heath, V. Hühnerbach, M. Hort, D. Masson, and I. Trummer. 2002. Widespread fluid expulsion along the seafloor of the Costa Rica convergent margin. *Terra Nova* 14: 69–79.

Boone, L. 1930. Scientific results of the cruises of the yachts "Eagle" and "Ara", 1921–1928, William K. Vanderbilt, commanding. Crustacea: Stomatopoda and Brachyura. *Bulletin of the Vanderbilt Marine Museum* 2: 1–228.

Boulay, J.N., J. Cortés, J. Nivia-Ruiz, and I.B. Baums. 2012. High genotypic diversity of the reef-building coral *Porites lobata* (Scleractinia: Poritidae) in Isla del Coco National Park, Costa Rica. *Revista de Biología Tropical* 60(Suppl. 3): 279–92.

Boulay, J.N., M.E. Hellberg, J. Cortés, and I.B. Baums. 2014. Unrecognized coral species diversity masks differences in functional ecology. *Proceedings of the Royal Society B* 281: 20131580.

Breedy, O., and H.M. Guzman. 2008. *Leptogorgia ignita*, a new shallow-water coral species (Octocorallia: Gorgoniidae) from the tropical eastern Pacific. *Journal of the Marine Biological Association of the United Kingdom* 88: 893–99.

Breedy, O., and H.M. Guzman. 2012. A new species of *Leptogorgia* (Cnidaria: Anthozoa: Octocorallia) from Golfo Dulce, Pacific, Costa Rica. *Zootaxa* 3182: 65–68.

Breedy, O., H.M. Guzman, and S. Vargas. 2009. A revision of the genus *Eugorgia* Verrill, 1868 (Coelenterata: Octocorallia: Gorgoniidae). *Zootaxa* 2151: 1–46.

Breedy, O., G.C. Williams, and H.M. Guzman. 2013. Two new species of gorgonian octocorals from the Tropical Eastern Pacific Biogeographic Region (Cnidaria, Anthozoa, Gorgoniidae). *ZooKeys* 350: 75–90.

Brenes, C., S. León, and G. Arroyo. 1996. Influence of coastal waters on some physical and chemical oceanographic characteristic of Gulf of Nicoya, Costa Rica. *Tópicos Meteorológicos y Oceanográficos* 3: 65–72.

Brenes, C., S. León, and J. Cháves. 2001. Variaciones en las propiedades termohalinas en el Golfo de Nicoya, Costa Rica. *Revista de Biología Tropical* 49(Suppl. 2): 145–52.

Broenkow, W.W. 1965. The distribution of nutrients in the Costa Rica Dome in the Eastern Tropical Pacific Ocean. *Limnology and Oceanography* 10: 40–52.

Brugnoli-Olivera, E., E. Díaz-Ferguson, M. Delfino-Machin, A. Morales-Ramírez, and A. Dominici-Arosemena. 2004. Composition of the zooplankton community, with emphasis in copepods, in Punta Morales, Golfo de Nicoya, Costa Rica. *Revista de Biología Tropical* 52: 897–902.

Brugnoli-Olivera, E., and A. Morales Ramírez. 2001. La comunidad fitoplánctica de Punta Morales, Golfo de Nicoya, Costa Rica. *Revista de Biología Tropical* 49(Suppl. 2): 11–17.

Brugnoli-Olivera, E., and A. Morales-Ramírez. 2008. Trophic planktonic dynamics in a tropical estuary, Gulf of Nicoya, Pacific coast of Costa Rica during El Niño 1997 event. *Revista de Biología Marina y Oceanografía* 43: 75–89.

Bruno, J., and M. Bertness. 2001. Habitat modification and facilitation in benthic marine communities. In M. Bertness, M. Hay, and S. Gaines, eds., *Marine Community Ecology*, 201–18. Sunderland, MA: Sinauer Associates, Inc.

Bussing, W.A., and M.I. López. 1993. Peces demersales y pelágicos costeros del Pacífico de Centro América Meridional. Guía Ilustrada. *Revista de Biología Tropical* (Special Pub.): 1–164.

Bussing, W.A., and M.I. López. 1996. Fishes collected during the R.V. Victor Hensen Costa Rica Expedition (1993/1994). *Revista de Biología Tropical* 44(Suppl. 3): 183–86.

Cahoon, D.R., P.F. Hensel, J. Rybczyk, K.L. McKee, C.E. Proffitt, and B.C. Perez. 2003. Mass tree mortality leads to mangrove peat collapse at Bay Islands, Honduras after Hurricane Mitch. *Journal of Ecology* 91: 1093–105.

Cajiao-Jiménez, M.V. 2003. *Régimen Legal de los Recursos Marinos y Costeros en Costa Rica*. San José, Costa Rica: Editorial IPECA. 192 pp.

Campos, J. 1983. Estudio sobre la fauna de acompañamiento del camarón en Costa Rica. *Revista de Biología Tropical* 31: 291–96.

Campos, J., B. Burgos, and C. Gamboa. 1984. Effect of shrimp trawling on the commercial ichthyofauna of the Gulf of Nicoya, Costa Rica. *Revista de Biología Tropical* 32: 203–7.

Campos, J., and A. Corrales. 1986. Preliminary results on the trophic dynamics of the Gulf of Nicoya, Costa Rica. *Anales del Instituto de Ciencias del Mar y Limnología, UNAM* 13: 329–34.

Campos, J., A. Segura, O. Lizano, and E. Madrigal. 1993. Ecología básica de *Coryphaena hippurus* (Pisces, Coryphaenidae) y abundancia de otros grandes pelágicos en el Pacífico de Costa Rica. *Revista de Biología Tropical* 41: 783–90.

Canals, M., and E. Ballesteros. 1997. Production of carbonate particles by phytobenthic communities on the Mallorca-Menorca shelf, northwestern Mediterranean Sea. *Deep Sea Research* 44: 611–29.

Castaing, A., J.M. Jiménez, and C.R. Villalobos. 1980. Observaciones sobre la ecología de manglares de Costa Rica y su relación con la distribución del molusco *Geloina inflata* (Philippi) (Pelecypoda: Corbiculidae). *Revista de Biología Tropical* 28: 323–40.

Cháves, J., and M. Birkicht. 1996. Equatorial subsurface water and the nutrient seasonality distribution of the Gulf of Nicoya, Costa Rica. *Revista de Biología Tropical* 44(Suppl. 3): 41–48.

Child, C.A. 1979. Shallow-water Pycnogonida of the Isthmus of Panama and the coast of Middle America. *Smithsonian Contributions to Zoology* 293: 1–86.

Coates, A.G., ed. 1997. *Central America: A Natural and Cultural History*. New Haven and London: Yale University Press.

Coates, A.G., J.B.C. Jackson, L.S. Collins, T.M. Cronin, H.J. Dowsett, L.M. Bybell, P. Jung, and J.A. Obando. 1992. Closure of the Isthmus of Panama: the near-shore marine records of Costa Rica and western Panama. *Geological Society of America Bulletin* 104: 14–28.

Córdoba-Muñoz, R. 1998. Primary productivity in the water column of Estero Morales, a mangrove system in the Gulf of Nicoya, Costa Rica. *Revista de Biología Tropical* 46(Suppl. 6): 257–62.

Córdoba-Muñoz, R., and J.A. Vargas. 1996. Temperature, salinity, oxygen, and nutrient profiles at a 200 m deep station in Golfo Dulce, Pacific coast of Costa Rica. *Revista de Biología Tropical* 44(Suppl. 3): 233–36.

Cortés, J. 1986. Biogeografía de corales hermatípicos: el istmo centroamericano. *Anales del Instituto de Ciencias del Mar y Limnología, UNAM* 13: 297–304.

Cortés, J. 1990a. Coral reef decline in Golfo Dulce, Costa Rica, eastern Pacific: anthropogenic and natural disturbances. PhD diss., University of Miami. Miami, FL.

Cortés, J. 1990b. The coral reefs of Golfo Dulce, Costa Rica: distribution and community structure. *Atoll Research Bulletin* 344: 1–37.

Cortés, J. 1991. Los arrecifes coralinos de Golfo Dulce, Costa Rica: aspectos geológicos. *Revista Geológica de América Central* 13: 15–24.

Cortés, J. 1992. Los arrecifes coralinos de Golfo Dulce, Costa Rica: aspectos ecológicos. *Revista de Biología Tropical* 40: 19–26.

Cortés, J. 1996/1997. Comunidades coralinas y arrecifes del Área de Conservación Guanacaste, Costa Rica. *Revista de Biología Tropical* 44(3)/45(1): 623–25.

Cortés, J. 1997. Biology and geology of coral reefs of the eastern Pacific. *Coral Reefs* 16(Suppl.): S39–S46.

Cortés, J. 2001. Requiem for an eastern Pacific seagrass bed, Bahía Culebra, Costa Rica. *Revista de Biología Tropical* 49(Suppl. 2): 273–78.

Cortés, J. 2003. Coral reefs of the Americas: an introduction to Latin American Coral Reefs. In J. Cortés, ed., *Latin American Coral Reefs*, 1–7. Amsterdam: Elsevier Science.

Cortés, J. 2007. Coastal morphology and coral reefs. In J. Bundschuh and G.E. Alvarado, eds., *Central America: Geology, Resources and Hazards*, Vol. 1, 185–200. London: Taylor & Francis.

Cortés, J. 2009. A history of marine biodiversity scientific research in Costa Rica. In I.S. Wehrtmann and J. Cortés, eds., *Marine Biodiversity of Costa Rica, Central America*, 47–80. Berlin: Springer.

Cortés, J. 2012. Historia de la investigación marina en Bahía Culebra, Guanacaste, Costa Rica. *Revista de Biología Tropical* 60(Suppl. 2): 19–37.

Cortés, J., A.C. Fonseca, and D. Hebbeln. 1996. Bottom topography and sediments around Isla del Caño, Pacific of Costa Rica. *Revista de Biología Tropical* 44(Suppl. 3): 11–17.

Cortés, J., H.M. Guzmán, A.C. Fonseca, J.J. Alvarado, O. Breedy, C. Fernández, A. Segura, and E. Ruiz. 2009. Ambientes y organismos marinos de la Reserva Biológica Isla del Caño, Área de Conservación Osa, Costa Rica. Serie Técnica: Apoyando los esfuerzos en el manejo y protección de la biodiversidad tropical, No. 13. San José, Costa Rica: TNC.

Cortés, J., and C.E. Jiménez. 1996. Coastal-marine environments of Parque Nacional Corcovado, Puntarenas, Costa Rica. *Revista de Biología Tropical* 44(Suppl. 3): 35–40.

Cortés, J., and C.E. Jiménez. 2003. Corals and coral reefs of the Pacific of Costa Rica: history, research and status. In J. Cortés, ed., *Latin American Coral Reefs*, 361–85. Amsterdam: Elsevier Science.

Cortés, J., I.G. Macintyre, and P.W. Glynn. 1994. Holocene growth history of an eastern Pacific fringing reef, Punta Islotes, Costa Rica. *Coral Reefs* 13: 65–73.

Cortés, J., and M.M. Murillo. 1985. Comunidades coralinas y arrecifes del Pacífico de Costa Rica. *Revista de Biología Tropical* 33: 197–202.

Cortés, J., M.M. Murillo, H.M. Guzmán, and J. Acuña. 1984. Pérdida de zooxantelas y muerte de corales y otros organismos arrecifales en el Caribe y Pacífico de Costa Rica. *Revista de Biología Tropical* 32: 227–32.

Cortés, J., and E. Salas. 2009. Seagrasses. In I.S. Wehrtmann and J. Cortés, eds., *Marine Biodiversity of Costa Rica, Central America*. Berlin, Springer. Text: 119–22, Species List CD: 71–72.

Cortés, J., R. Vargas-Castillo, and J. Nivia-Ruiz. 2012. Marine biodiversity of Bahía Culebra, Guanacaste, Costa Rica: published records. *Revista de Biología Tropical* 60(Suppl. 2): 39–71.

Cortés, J., and I.S. Wehrtmann. 2009. Diversity of marine habitats of the Caribbean and Pacific of Costa Rica. In I.S. Wehrtmann and J. Cortés, eds., *Marine Biodiversity of Costa Rica, Central America*, 1–45. Berlin: Springer.

Cortés, J., J. Samper-Villareal, and A. Bernecker. 2014. Seasonal phenology of *Sargassum liebmannii* J. Agardh (Fucales, Heterokontophyta) in an upwelling area of the Eastern Tropical Pacific. *Aquatic Botany* 119: 105–10.

Crisp, D.J., J.G. Wieghell, and C.A. Richardson. 1990. Tidal microgrowth bands in *Siphonaria gigas* (Gastropoda, Pulmonata) from the coast of Costa Rica. *Malacologia* 31: 229–36.

Crocker, T. 1933. The Templeton Crocker Expedition of the California Academy of Sciences, 1932, No. 2: Introductory statement. Proceedings of the California Academy of Sciences 4th Series 21: 3–9.

Cruz, R.A. 1996. Annotated checklist of marine mollusks collected during the R.V. Victor Hensen Costa Rica Expedition 1993/1994. *Revista de Biología Tropical* 44(Suppl. 3): 59–68.

Cruz, R.A., and J.A. Jiménez. 1994. *Moluscos Asociados a las Áreas de Manglar de la Costa Pacífica de América Central: Guía*. Heredia, Costa Rica: Editorial Fundación UNA.

Dallsgaard, T., D.E. Canfield, J. Peterson, B. Thamdrup, and J. Acuña-González. 2003. N_2 production by the anammox reaction in the anoxic water column of Golfo Dulce, Costa Rica. *Nature* 422: 606–8.

Danielsen, F., M.K. Sørensen, M.F. Olwig, V. Selvam, F. Parish, N.D. Burgess, T. Hiraishi, V.M. Karunagaran, M.S. Rasmussen, L.B. Hansen, A. Quarto, and N. Suryadiputra. 2005. The Asian tsunami: a protective role for coastal vegetation. *Science* 310: 643.

Darwin, C.R. 1842. *The Structure and Distribution of Coral Reefs*. Tuscon, AZ: University of Arizona Press, 1984.

Davidse, G., M. Sousa, and A.O. Chater. 1994. *Flora Mesoamericana: Vol. 6. Alismataceae a Cyperaceae*. México DF, México: Instituto de Biología, UNAM.

Dawson, E.Y. 1957. Marine algae from the Pacific Costa Rican gulfs. Los Angeles County Museum *Contributions in Science* 15: 1–28.

Dawson, E.Y. 1960. New records of marine algae from Pacific Mexico and Central America. *Pacific Naturalist* 1(20): 31–52.

Dawson, E.Y., and P.T. Beaudette. 1959. Field notes from the 1959 eastern Pacific cruise of the Stella Polaris. *Pacific Naturalist* 1(13): 1–24.

D'Croz, L., and D.R. Robertson. 1997. Coastal oceanographic conditions affecting coral reefs on both sides of the Isthmus of Panama. Proceedings of the 8th International Coral Reef Symposium, Panama 2: 2053–58.

Dean, H.K. 1996a. Subtidal benthic polychaetes (Annelida) of the Gulf of Nicoya, Costa Rica. *Revista de Biología Tropical* 44(Suppl. 3): 69–80.

Dean, H.K. 1996b. Polychaete worms (Annelida) collected in Golfo Dulce, during the Victor Hensen Costa Rica expedition (1993/1994). *Revista de Biología Tropical* 44(Suppl. 3): 81–86.

Dean, H.K. 1998. A new species of Hesionidae, *Glyphohesione nicoyensis* (Annelida: Polychaeta), from the Gulf of Nicoya, Costa Rica. *Proceedings of the Biological Society of Washington* 111: 257–62.

Dean, H.K. 2009. Polychaetes and echiurans. In I.S. Wehrtmann and J. Cortés, eds., *Marine Biodiversity of Costa Rica, Central America*. Berlin: Springer + Business Media BV. Text: 181–91, Species List CD: 122–59.

Dean, H.K., and J.A. Blake. 2007. *Chaetozone* and *Caulleriella* (Polychaeta: Cirratulidae) from the Pacific coast of Costa Rica, with description of eight new species. *Zootaxa* 1451: 41–68.

Dean, H.K., D. Maurer, J.A. Vargas, and C.H. Tinsman. 1986. Trace metal concentrations in sediment and invertebrates from the Gulf of Nicoya, Costa Rica. *Marine Pollution Bulletin* 17: 128–31.

De Grave, S., and A. Whitaker. 1999. Benthic community re-adjustment following dredging of a muddy-maërl matrix. *Marine Pollution Bulletin* 38: 102–8.

De la Cruz, E., and J.A. Vargas. 1987. Abundancia y distribución vertical de la meiofauna en la playa fangosa de Punta Morales, Golfo de Nicoya, Costa Rica. *Revista de Biología Tropical* 35: 363–67.

Denyer, P., and G. Cárdenas. 2000. Costas marinas. In P. Denyer and S. Kussmaul, eds., *Geología de Costa Rica*, 185–218. Cartago, Costa Rica: Editorial Tecnológica de Costa Rica.

Denyer, P., W. Montero, and A.E. Alvarado. 2003. Atlas tectónico de Costa Rica. Editorial Universidad de Costa Rica. San Pedro, Costa Rica.

Dexter, D.M. 1974. Sandy-beach fauna of the Pacific and Atlantic coasts of Costa Rica and Colombia. *Revista de Biología Tropical* 22: 51–66.

Díaz-Ferguson, E., and J.A. Vargas-Zamora. 2001. Abundancia de *Petrolisthes armatus* (Crustacea: Porcellanidae) on a tropical estuarine intertidal rocky beach, Gulf of Nicoya estuary, Costa Rica. *Revista de Biología Tropical* 49(Suppl. 2): 97–102.

Díaz-Ferguson, E., D. Arroyo, A. Morales, and J.A. Vargas. 2008. Observaciones sobre la larva del cangrejo marino tropical (Decapoda: Porcellanidae) *Petrolisthes armatus* en el Golfo de Nicoya, Costa Rica. *Revista de Biología Tropical* 56: 1209–23.

Dittel, A.I. 1991. Distribution, abundance and sexual composition of stomatopod Crustacea in the Gulf of Nicoya, Costa Rica. *Journal of Crustacean Biology* 11: 269–76.

Dittel, A.I., and C.E. Epifanio. 1990. Seasonal and tidal abundance of crab larvae in a tropical mangrove system, Gulf of Nicoya, Costa Rica. *Marine Ecology Progress Series* 65: 25–34.

Dittel, A.I., C.E. Epifanio, and J.B. Chavarría. 1985. Population biology of the portunid crab *Callinectes arcuatus* Ordway in the Gulf of Nicoya, Costa Rica. *Estuarine, Coastal and Shelf Science* 20: 593–602.

Dittel, A.I., C.E. Epifanio, and O. Lizano. 1991. Flux of crab larvae in a mangrove creek in the Gulf of Nicoya, Costa Rica. *Estuarine, Coastal and Shelf Science* 32: 129–40.

Dittmann, S., and J.A. Vargas. 2001. Tropical tidal flat benthos compared between Australia and Central America. In K. Reise, ed., Ecological Comparisons of Sedimentary Shores. *Ecological Studies* 151: 275–293.

Dominici-Arosemena, A., E. Brugnoli-Olivera, J. Cortés-Núñez,

H. Molina-Ureña, and M. Quesada-Alpízar. 2005. Community structure of eastern Pacific reef fishes (Gulf of Papagayo, Costa Rica). *Tecnociencia* 7: 19–41.

Duke, N.C., J.O. Meynecke, S. Dittman, A.M. Ellison, K. Anger, U. Berger, S. Cannicci, K. Diele, K.C. Ewel, C.D. Field, N. Koedam, S.Y. Lee, C. Marchand, I. Nordhaus, and F. Dahdough-Guebas. 2007. A world without mangroves? *Science* 317: 41.

Duncan, R.S., and W.A. Szelistowski. 1998. Influence of puffer predation on vertical distribution of mangrove littorinids in the Gulf of Nicoya, Costa Rica. *Oecologia* 117: 433–42.

Earle, S.A. 2009. *The World is Blue: How our Fate and the Ocean's are One*. Washington, D.C.: National Geographic. 303 pp.

Echeverría-Sáenz, S., and I.S. Wehrtmann. 2011. Egg production of the commercially exploited deepwater shrimp, *Heterocarpus vicarius* (Decapoda: Pandalidae), Pacific Costa Rica. *Journal of Crustacean Biology* 31: 434–40.

Ellison, J.C. 2000. How South Pacific mangroves may respond to predicted climate change and sealevel rise. In A. Gillespie and W. Burns, eds., *Climate Change in the South Pacific: Impacts and Responses in Australia, New Zealand, and Small Islands Status*, 289–301. Dordrecht: Kluwer Academic Publishers.

Emig, C., and J.A. Vargas. 1990. *Glottidia audebarti* (Broderip) (Brachiopoda, Lingulidae) from the Gulf of Nicoya, Costa Rica. *Revista de Biología Tropical* 38: 251–58.

Enfield, D.B. 1989. El Niño, past and present. *Reviews of Geophysics* 27: 159–87.

Epifanio, C.E., and A.I. Dittel. 1984. Seasonal abundance of Brachyuran crab larvae in a tropical estuary: Gulf of Nicoya, Costa Rica, Central America. *Estuaries* 7: 501–5.

Epifanio, C.E., D. Maurer, and A.I. Dittel. 1983. Seasonal changes in nutrients and dissolved oxygen in the Gulf of Nicoya, a tropical estuary on the Pacific coast of Central America. *Hydrobiologia* 101: 231–38.

Espinoza, M., T.M. Clarke, F. Villalobos-Rojas, and I.S. Wehrtmann. 2012. Ontogenetic dietary shifts and feeding ecology of the rasptail skate *Raja velezi* and the brown smoothhound shark *Mustelus henlei* along the Pacific coast of Costa Rica, Central America. *Journal of Fish Biology* 81: 1578–95.

Espinoza, M., and I.S. Wehrtmann. 2008. Stomach content analyses of the threadfin anglerfish *Lophiodes spilurus* (Lophiiformes: Lophiidae) associated with deepwater shrimp fisheries from the central Pacific of Costa Rica. *Revista de Biología Tropical* 56: 1959–70.

Excoffon, A.C., F.H. Acuña, and J. Cortés. 2009. The sea anemone *Nemanthus californicus* (Cnidaria, Actiniaria, Nemanthidae) from Costa Rica: re-description and first record outside the type locality. *Marine Biodiversity Records* 2: 1–5. doi: 10.1017/S1755267209990601

Feely, R.A., C.L. Sabine, K. Lee, W. Berelson, J. Kleypas, V.J. Fabry, and F.J. Millero. 2004. Impact of anthropogenic CO_2 on the $CaCO_3$ system in the oceans. *Science* 305: 362–66.

Ferdelman, T.G., B. Thamdrup, D.E. Canfield, R.N. Glud, J. Kuever, R. Lillebaek, N.B. Ramsing, and C. Wawer. 2006. Biogeochemical controls on the oxygen, nitrogen and sulfur distritributions in the water column of Golfo Dulce: an anoxic basin on the Pacific coast of Costa Rica revisited. *Revista de Biología Tropical* 54(Suppl. 1): 171–91.

Fernández, C. 2008. Flora marina del Parque Nacional Isla del Coco, Costa Rica, Pacífico Tropical Este. *Revista de Biología Tropical* 56(Suppl. 3): 57–69.

Fernández, C., and J.J. Alvarado. 2008. Chlorophyta de la costa Pacífica de Costa Rica. *Revista de Biología Tropical* 56(Suppl. 4): 149–62.

Fernández, C., and J. Cortés. 2005. Reef Site: *Caulerpa sertularioides*, a green alga spreading aggressively over coral reef communities in Culebra Bay, North Pacific of Costa Rica. *Coral Reefs* 24: 10.

Fernández-García, C. 2007. Propagación del alga *Caulerpa sertularioides* (Chlorophyta) en Bahía Culebra, Golfo de Papagayo, Pacífico norte de Costa Rica. Thesis M.Sc., Universidad de Costa Rica. San Pedro, Costa Rica.

Fernández-García, C., J. Cortés, J.J. Alvarado, and J. Nivia-Ruiz. 2012. Physical factors contributing to the benthic dominance of the alga *Caulerpa sertularioides* (Caulerpaceae, Chlorophyta) in the upwelling Bahía Culebra, north Pacific of Costa Rica. *Revista de Biología Tropical* 60(Suppl. 2): 93–107.

Feutry, P., H.J. Hartmann, H. Casabonnet, and G. Umaña. 2010. Preliminary analysis of the fish species of the Pacific Central American mangrove of Zancudo, Golfo Dulce, Costa Rica. *Wetlands Ecology and Management* 18: 637–50.

Fiedler, P.C. 2002. The annual cycle and biological effects of the Costa Rica Dome. *Deep Sea Research I* 49: 321–38.

Fiedler, P.C., and M.F. Lavín. 2006. Introduction: a review of eastern tropical Pacific oceanography. *Progress in Oceanography* 69: 94–100.

Fiedler, P.C., and L.D. Talley. 2006. Hydrography of the eastern tropical Pacific: a review. *Progress in Oceanography* 69: 143–80.

Figueiredo, M.A. de O., K. Santos de Menezes, E.M. Costa-Paiva, P.C. Paiva, and C.R.R. Ventura. 2007. Experimental evaluation of rhodoliths as living substrate for infauna at the Abrolhos Bank, Brazil. *Ciencias Marinas* 33: 427–40.

Fischer, S., and M. Wolff. 2006. Fisheries assessment of *Callinectes arcuatus* (Brachyura, Portunidae) in the Gulf of Nicoya, Costa Rica. *Fisheries Research* 77: 301–11.

Fish, M.R., I.M. Côté, J.A. Gill, A.P. Jones, S. Renshoff, and A.R. Watkinson. 2005. Predicting the impact of sea-level rise on Caribbean sea turtle nesting habitat. *Conservation Biology* 19: 482–91.

Fonseca, A.C. 1999. Bioerosión y bioacreción en arrecifes coralinos del Pacífico sur de Costa Rica. Thesis M.Sc., Universidad de Costa Rica. San Pedro, Costa Rica.

Fonseca, A.C., and J. Cortés. 1998. Coral borers of the eastern Pacific: the sipunculan *Aspidosiphon* (*A.*) *elegans* and the crustacean *Pomatogebia rugosa*. Pacific Science 52: 170–75.

Fonseca, A.C., H.K. Dean, and J. Cortés. 2006. Non-colonial macroborers as indicators of coral reef stress in the south Pacific of Costa Rica. *Revista de Biología Tropical* 54: 101–15.

Fonseca, A.C., H.M. Guzmán, J. Cortés, and C. Soto. 2010. Marine habitat map of "Isla del Caño", Costa Rica, comparing Quickbird and Hymap images classification results. *Revista de Biología Tropical* 58: 373–81.

Foster, M.S. 2001. Rhodoliths: between rocks and soft places. *Journal of Phycology* 37: 659–67.

Foster, M.S., R. Riosmena-Rodríguez, D.L. Steller, and W.M.J. Woelkerling. 1997. Living rhodolith beds in the Gulf of California and their implications for paleoenvironmental interpretation. Geological Society of America Special Paper 318: 127–39.

Frantz, B.R., M.K. Kenneth, H. Coale, and M.S. Foster. 2000. Growth rate and potential climate record from a rhodolith using ^{14}C Accelerator Mass Spectrometry. *Limnology and Oceanography* 45: 1773–77.

Fraser, C.M. 1943. General account of the scientific work of the *Velero III* in the eastern Pacific, 1931–1941, Part III: a ten-year list of the *Velero III* collecting stations (Charts 1–115). With an appendix of collecting stations of the Allan Hancock Foundation for the year 1942. *Allan Hancock Pacific Expeditions* 1: 259–431.

Fuenzalida, R., W. Schneider, J. Garcés-Vargas, L. Bravo, and C. Lange. 2009. Vertical and horizontal extension of the oxygen minimum zone in the eastern South Pacific Ocean. *Deep Sea Research II* 56: 992–1003.

Fuller, C.C., J.A. Davis, D.J. Cain, P.J. Lamothe, T.L. Fries, G. Fernández, J.A. Vargas, and M.M. Murillo. 1990. Distribution and transport of sediment-bound metal contaminants in the Río Grande de Tárcoles, Costa Rica (Central America). *Water Research* 24: 805–12.

Gallardo, V.A. 1977. Large benthic microbial communities in sulphide biota under Peru-Chile subsurface countercurrent. *Nature* 268: 331–32.

Gallardo, V.A., and C. Espinoza. 2007. New communities of large filamentous sulfur bacteria in the eastern South Pacific. *International Microbiology* 10: 97–102.

García-Céspedes, J., J.A. Acuña-González, and J.A. Vargas-Zamora. 2004. Metales traza en sedimentos de cuatro ambientes costeros de Costa Rica. *Revista de Biología Tropical* 52(Suppl. 2): 51–60.

García-Céspedes, V., J. Acuña-González, J.A. Vargas-Zamora, and J. García-Céspedes. 2006. Calidad bacteriológica y desechos sólidos en cinco ambientes costeros de Costa Rica. *Revista de Biología Tropical* 54(Suppl. 1): 35–48.

Gasca, R., and E. Suárez. 1992. Sifonóforos (Cnidaria: Siphonophora) del Domo de Costa Rica. *Revista de Biología Tropical* 40: 125–30.

Gateño, D., A. León, Y. Barki, J. Cortés, and B. Rinkevich. 2003. Skeletal tumor formations in the massive coral *Pavona clavus*. *Marine Ecology Progress Series* 258: 97–108.

Gilman, E.L., J. Ellison, N.C. Duke, and C. Field. 2008. Threats to mangroves from climate change and adaptation options: a review. *Aquatic Botany* 89: 237–50.

Glantz, M. 2001. Currents of Change: Impacts of El Niño and La Niña on Climate and Society, 2nd ed. Cambridge, UK: Cambridge University Press.

Glynn, P.W. 1990. Coral mortality and disturbances to coral reefs in the tropical eastern Pacific. In P.W. Glynn, ed., *Global Ecological Consequences of the 1982–83 El Niño-Southern Oscillation*, 55–126. Netherlands: Elsevier.

Glynn, P.W. 1997. Eastern Pacific reef coral biogeography and faunal flux: Durham's dilemma revisited. Proceedings of the 8th International Coral Reef Symposium, Panamá 1: 371–78.

Glynn, P.W. 2003. Coral communities and coral reefs of Ecuador. In J. Cortés, ed., *Latin American Coral Reefs*, 449–72. Amsterdam: Elsevier Science.

Glynn, P.W., S.B. Colley, C.M. Eakin, D.B. Smith, J. Cortés, N.J. Gassman, H.M. Guzmán, J.B. del Rosario, and J. Feingold. 1994. Reef coral reproduction in the eastern Pacific: Costa Rica, Panama, and Galapagos Islands (Ecuador). II. Poritidae. *Marine Biology* 118: 191–208.

Glynn, P.W., S.B. Colley, N.J. Gassman, K. Black, J. Cortés, and J.L. Maté. 1996. Reef coral reproduction in the eastern Pacific: Costa Rica, Panamá, and Galápagos Islands (Ecuador). III. Agariciidae (*Pavona gigantea* and *Gardineroseris planulata*). *Marine Biology* 125: 579–601.

Glynn, P.W., S.B. Colley, H.M. Guzman, I.C. Enochs, J. Cortés, J.L. Maté, and J.S. Feingold. 2011. Reef coral reproduction in the eastern Pacific: Costa Rica, Panamá and the Galápagos Islands (Ecuador). VI. Agariciidae, *Pavona clavus*. *Marine Biology* 158: 1601–17.

Glynn, P.W., S.B. Colley, J.L. Maté, I.B. Baums, J.S. Feingold, J. Cortés, H.M. Guzmán, J.C. Afflerbach, V.W. Brandtneris, and J.S. Ault. 2012. Reef coral reproduction in the equatorial eastern Pacific: Costa Rica, Panamá, and the Galápagos Islands (Ecuador). VII. Siderastreidae, *Psammocora stellata* and *Psammocora profundacella*. *Marine Biology* 159: 1917–32.

Glynn, P.W., S.B. Colley, J.L. Maté, J. Cortés, H.M. Guzmán, R.L. Bailey, J.S. Feingold, and I.C. Enochs. 2008. Reproductive ecology of the azooxanthellate coral *Tubastrea coccinea* in the Equatorial Eastern Pacific. Part V. Dendrophylliidae. *Marine Biology* 153: 529–44.

Glynn, P.W., S.B. Colley, J.H. Ting, J.L. Maté, and H.M. Guzmán. 2000. Reef coral reproduction in the eastern Pacific: Costa Rica, Panama, and Galápagos Islands (Ecuador). IV. Agariciidae, recruitment and recovery of *Pavona varians* and *Pavona* sp.a. *Marine Biology* 136: 785–805.

Glynn, P.W., J. Cortés, H.M. Guzmán, and R.H. Richmond. 1988. El Niño (1982–83) associated coral mortality and relationship to sea surface temperature deviations in the tropical eastern Pacific. Proceedings of the 6th International Coral Reef Symposium, Australia 3: 237–43.

Glynn, P.W., E.M. Druffel, and R.B. Dunbar. 1983. A dead Central American coral reef tract: possible link with the Little Ice Age. *Journal of Marine Research* 41: 605–37.

Glynn, P.W., N.J. Gassman, C.M. Eakin, J. Cortés, D.B. Smith, and H.M. Guzmán. 1991. Reef coral reproduction in the eastern Pacific: Costa Rica, Panama, and Galapagos Islands (Ecuador), Part I: Pocilloporidae. *Marine Biology* 109: 355–68.

Glynn, P.W., and R.H. Stewart. 1973. Distribution of coral reefs in the Pearl islands (Gulf of Panama) in relation to thermal conditions. *Limnology and Oceanography* 18: 367–79.

Glynn, P.W., R.H. Stewart, and J.E. McCosker. 1972. Pacific coral reefs of Panamá: structure, distribution and predators. *Geologische Rundschau* 61: 483–519.

Glynn, P.W., G.M. Wellington, E.A. Wieters, and S.A. Navarrete. 2003. Reef-building coral communities of Easter Island (Rapa Nui), Chile. In J. Cortés, ed., *Latin American Coral Reefs*, 473–94. Amsterdam: Elsevier Science.

Gocke, K., J. Cortés, and M.M. Murillo. 2001a. Planktonic primary production in a tidally influenced mangrove forest on the Pacific coast of Costa Rica. *Revista de Biología Tropical* 49(Suppl. 2): 279–88.

Gocke, K., J. Cortés, and M.M. Murillo. 2001b. The annual cycle of primary productivity in a tropical estuary: The inner regions of the Golfo de Nicoya, Costa Rica. *Revista de Biología Tropical* 49(Suppl. 2): 289–306.

Goffredi, S.K. 2010. Indigenous ectosymbiotic bacteria associated with diverse hydrothermal vent invertebrates. *Environmental Microbiology Reports* 2: 479–88.

Gómez, L.D. 1984. Las plantas acuáticas y anfibias de Costa Rica y Centroamérica: I. Liliopsida. San José, Costa Rica: Editorial UNED.

Gould, C.G. 2004. *The Remarkable Life of William Beebe: Explorer and Naturalist*. Washington: Island Press.

Grall, J., and J.M. Hall-Spencer. 2003. Problems facing maerl conservation in Brittany. *Aquatic Conservation: Marine and Freshwater Ecosystems* 13: 55–64.

Guzman, H.M. 1986. Estructura de la comunidad arrecifal de la Isla del

Caño, Costa Rica, y el efecto de perturbaciones naturales severas. Thesis M.Sc., Universidad de Costa Rica. San Pedro, Costa Rica.

Guzman, H.M. 1988a. Distribución y abundancia de organismos coralívoros en los arrecifes coralinos de la Isla del Caño, Costa Rica. *Revista de Biología Tropical* 36: 191–207.

Guzman, H.M. 1988b. Feeding behavior of the gastropod corallivore *Quoyula monodonta* (Blainville). *Revista de Biología Tropical* 36: 209–12.

Guzman, H.M. 1991. Restoration of coral reefs in Pacific Costa Rica. *Conservation Biology* 5: 189–95.

Guzman, H.M., J. Campos, C. Gamboa, and W.A. Bussing. 1988b. Un arrecife artificial de llantas: su potencial para el manejo de pesquerías. *Anales del Instituto de Ciencias del Mar y Limnología, UNAM* 15: 249–54.

Guzman, H.M., and J. Cortés. 1989a. Coral reef community structure at Caño Island, Pacific Costa Rica. PSZNI: *Marine Ecology* 10: 23–41.

Guzman, H.M., and J. Cortés. 1989b. Growth rates of eight species of scleractinian corals in the eastern Pacific (Costa Rica). *Bulletin of Marine Science* 44: 1194–86.

Guzman, H.M., and J. Cortés. 2001. Changes in reef community structure after fifteen years of natural disturbances in the eastern Pacific (Costa Rica). *Bulletin of Marine Science* 69: 133–49.

Guzman, H.M., and J. Cortés. 2007. Reef recovery 20-yr after the 1982–83 El Niño massive mortality. *Marine Biology* 151: 401–11.

Guzman, H.M., J. Cortés, P.W. Glynn, and R.H. Richmond. 1990. Coral mortality associated with dinoflagellate blooms in the eastern Pacific (Costa Rica and Panama). *Marine Ecology Progress Series* 60: 299–303.

Guzman, H.M., J. Cortés, R.H. Richmond, and P.W. Glynn. 1987b. Efectos del fenómeno de "El Niño-Oscilación Sureña" 1982/83 en los arrecifes de la Isla del Caño, Costa Rica. *Revista de Biología Tropical* 35: 325–32.

Guzman, H.M., and V.L. Obando. 1988. Diversidad y abundancia diaria y estacional del zooplancton marino de la Isla del Caño, Costa Rica. *Revista de Biología Tropical* 36: 139–50.

Guzman, H.M., V.L. Obando, R.C. Brusca, and P.M. Delaney. 1988a. Aspects of the population biology of the marine isopod *Excorallana tricornis occidentalis* Richardson, 1905 (Crustacea: Isopoda: Corallanidae) at Caño Island, Pacific Costa Rica. *Bulletin of Marine Science* 43: 77–87.

Guzman, H.M., V.L. Obando, and J. Cortés. 1987a. Meiofauna associated with a Pacific coral reef in Costa Rica. *Coral Reefs* 6: 107–12.

Guzman, H.M., and D.R. Robertson. 1989. Population and feeding responses of the corallivorous pufferfish *Arothron meleagris* to coral mortality in the eastern Pacific. *Marine Ecology Progress Series* 55: 121–31.

Hall-Spencer, J. 2005. Ban on maërl extraction. *Marine Pollution Bulletin* 50: 121.

Hall-Spencer, J., N. White, E. Gillespie, K. Gillham, and A. Foggo. 2006. Impact of fish farms on maerl beds in strongly tidal areas. *Marine Ecology Progress Series* 326: 1–9.

Halpern, B.S. 2003. The impact of marine reserves: do reserves work and does reserve size matter? *Ecological Applications* 13: S117–S137.

Han, X., E. Suess, H. Sahling, and K. Wallmann. 2004. Fluid venting activity on the Costa Rica margin: new results from authigenic carbonates. *International Journal of Earth Sciences* 93: 596–611.

Hargraves, P.E., and R. Víquez. 1985. Spatial and temporal distribution of phytoplankton in Gulf of Nicoya, Costa Rica. *Bulletin of Marine Science* 37: 577–85.

Harvey, A.S., and W.J. Woelkerling. 2007. A guide to nongeniculate coralline red algal (Corallinales, Rhodophyta) rhodolith identification. *Ciencias Marinas* 33: 411–26.

Hebbeln, D., D. Beese, and J. Cortés. 1996. Morphology and sediment structures in Golfo Dulce, Costa Rica. *Revista de Biología Tropical* 44(Suppl. 3): 1–10.

Hebbeln, D., and J. Cortés. 2001. Sedimentation in a tropical fjord: Golfo Dulce, Costa Rica. *Geo-Marine Letters* 20: 142–48.

Helly, J.J., and L.A. Levin. 2004. Global distribution of naturally occurring marine hypoxia on continental margins. *Deep Sea Research I* 51: 1159–68.

Hernáez, P., T.M. Clarke, C. Benavides-Varela, F. Villalobos-Rojas, J. Nívia-Ruiz, and I.S. Wehrtmann. 2011. Population demography and spatial distribution of the mantis shrimp *Squilla biformes* (Stomatopoda, Squillidae) from Pacific Costa Rica. *Marine Ecology Progress Series* 424: 157–68.

Hernáez, P., E. Villegas-Jiménez, F. Villalobos-Rojas, and I.S. Wehrtmann. 2012. Reproductive biology of the ghost shrimp *Lepidophthalmus bocourti* (A. Milne-Edwards, 1870) (Decapoda: Axiidea: Callianassidae): A tropical species with a seasonal reproduction. *Marine Biology Research* 8: 635–43.

Hernáez, P., and I.S. Wehrtmann. 2011. Sexual maturity and egg production in an unexploited population of the red squat lobster *Pleuroncodes monodon* (Decapoda, Galatheidae) from Central America. *Fisheries Research* 107: 276–82.

Hernáez, P., and I.S. Wehrtmann. 2014. Breeding cycle of the red squat lobster *Pleuroncodes monodon* H. Milne Edwards, 1837 (Decapoda, Muninidae) from deepwater Pacific of Costa Rica. *Marine Ecology* 35: 204–11.

Hernáez-Bové, P. 2010. Aspectos demográficos y distribución espacial de *Pleuroncodes planipes* Simpson (Decapoda, Anomura, Galatheidae) en el Pacífico de Costa Rica. Thesis M.Sc., Universidad de Costa Rica. San Pedro, Costa Rica.

Herrera, W. 1985. *Clima de Costa Rica*. San José, Costa Rica: Editorial UNED.

Herrera-Ulloa, A., J. Chacón-Guzmán, G. Zúñiga-Calero, O. Fajardo, and R. Jiménez-Montealegre. 2009. Acuicultura de pargo La Mancha *Lutjanus guttatus* (Steindachner, 1869) en Costa Rica dentro de un enfoque ecosistémico. *Revista Ciencias Marinas y Costeras* 1: 197–213.

Hinojosa-Arango, G., and R. Riosmena-Rodríguez. 2004. Influence of rhodolith-forming species and growth-form on associated fauna of rhodolith beds in the central-west Gulf of California, México. PSZN: *Marine Ecology* 25: 109–27.

Høisæter, T. 1998. Preliminary check-list of the marine, shelled gastropods (Mollusca) of Golfo Dulce, on the Pacific coast of Costa Rica. *Revista de Biología Tropical* 46(Suppl. 6): 263–70.

Høisæter, T. 2012. *Cimaria vargasi* n. gen, n. sp. (Gastropoda: Pyramidellidae: Odostomiinae) from the Pacific Coast of Costa Rica, Central America. *Zootaxa* 3178: 63–67.

Hossfeld, B. 1996. Distribution and biomass of arrow worms (Chaetognatha) in Golfo de Nicoya and Golfo Dulce, Costa Rica. *Revista de Biología Tropical* 44(Suppl. 3): 157–72.

Hughes, T.P., D.R. Bellwood, C. Folke, R.S. Steneck, and J. Wilson. 2005. New paradigms for supporting the resilience of marine ecosystems. *Trends in Ecology and Evolution* 20: 380–86.

Ibarra, E. 1996. El valor de uso del paisaje submarino en el Golfo de Papagayo: comparación de la industria de buceo recreativo con la industria de extracción de peces para acuario. Licentiate thesis, Universidad de Costa Rica. San Pedro, Costa Rica.

INCOPESCA. 2006. Memoria Institucional 2002–2006: Instituto Costarricense de Pesca y Acuicultura. San José, Costa Rica: Imprenta Nacional.

Jesse, S. 1996. Demersal crustacean assemblages along the Pacific coast of Costa Rica: a quantitative and multivariate assessment based on the Victor Hensen Costa Rica Expedition (1993/1994). *Revista de Biología Tropical* 44(Suppl. 3): 103–14.

Jiménez, C.E. 1997. Corals and coral reefs of Culebra Bay, Pacific coast of Costa Rica: anarchy in the reef. Proceedings of the 8th International Coral Reef Symposium, Panamá 1: 329–34.

Jiménez, C.E. 1998. Arrecifes y comunidades coralinas de Bahía Culebra, Pacífico norte de Costa Rica (Golfo de Papagayo). Thesis M.Sc., Universidad de Costa Rica. San Pedro, Costa Rica.

Jiménez, C. 2001. Seawater temperature measured at the surface and at two depths (7 and 12 m) in one coral reef at Culebra Bay, Gulf of Papagayo, Costa Rica. *Revista de Biología Tropical* 49(Suppl. 2): 153–61.

Jiménez, C., G. Bassey, A. Segura, and J. Cortés. 2010. Characterization of the coral communities and reefs of two previously undescribed locations in the upwelling region of Gulf of Papagayo (Costa Rica). *Revista Ciencias Marinas y Costeras* 2: 95–108.

Jiménez, C.E., and J. Cortés. 2001. Effects of the 1991–92 El Niño on scleractinian corals of the Costa Rican central Pacific coast. *Revista de Biología Tropical* 49(Suppl. 2): 239–50.

Jiménez, C.E., and J. Cortés. 2003a. Growth of seven species of scleractinian corals in an upwelling environment of the eastern Pacific (Golfo de Papagayo, Costa Rica). *Bulletin of Marine Science* 72: 187–98.

Jiménez, C.E., and J. Cortés. 2003b. Coral cover change associated to El Niño, eastern Pacific, Costa Rica, 1992–2001. PSZN: *Marine Ecology* 24: 179–92.

Jiménez, C.E., J. Cortés, A. León, and E. Ruiz. 2001. Coral bleaching and mortality associated with El Niño 1997/98 event in an upwelling environment in the eastern Pacific (Gulf of Papagayo, Costa Rica). *Bulletin of Marine Science* 69: 151–69.

Jiménez, J.A. 1981. The mangroves of Costa Rica: a physiognomic characterization. Thesis M.Sc., University of Miami. Miami, FL. 130 pp.

Jiménez, J.A. 1999a. Ambiente, distribución y características estructurales de los manglares del Pacífico de Centro América: contrastes climáticos. In A. Yáñez-Arancibia and A.L. Lara-Domínguez, eds., *Ecosistemas de Manglar en América Tropical*, 51–70. México: Instituto de Ecología; Costa Rica: UICN/ORMA; Siliver Spring, MD: NOAA/NMFS.

Jiménez, J.A. 1999b. El manejo de los manglares del Pacífico de Centro América: usos tradicionales y potenciales. In A. Yáñez-Arancibia and A.L. Lara-Domínguez, eds., *Ecosistemas de Manglar en América Tropical*, 275–90. México: Instituto de Ecología; Costa Rica: UICN/ORMA; Siliver Spring, MD: NOAA/NMFS.

Jiménez, J., and R. Soto. 1985. Patrones regionales en la estructura y composición florística de los manglares de la costa Pacífica de Costa Rica. *Revista de Biología Tropical* 33: 25–37.

Jones, S. 2007. *Coral: A Pessimist in Paradise*. London: Little, Brown Book.

Jörger, K., R. Meyer, and I.S. Wehrtmann. 2008. Species composition and vertical distribution of chitons (Mollusca: Polyplacophora) in a rocky intertidal zone of the Pacific coast of Costa Rica. *Journal of the Marine Biological Association of the United Kingdom* 88: 807–16.

Kamenos, N.A., P.G. Moore, and J.M. Hall-Spencer. 2004. Nursery-area function of maerl grounds for juvenile queen scallops *Aequipecten opercularis* and other invertebrates. *Marine Ecology Progress Series* 274: 183–89.

Kappelle, M., M. Castro, H. Acevedo, L. González, and H. Monge. 2002. *Ecosistemas del Área de Conservación Osa (ACOSA)*. Sto. Domingo, Heredia, Costa Rica: Editorial INBio. 496 pp.

Karstensen, J., L. Stramma, and M. Visbeck. 2008. Oxygen minimum zones in the eastern tropical Atlantic and Pacific oceans. *Progress in Oceanography* 77: 331–50.

Kelmo, F., and R. Vargas. 2002. Anthoathecatae and Leptothecatae hydroids from Costa Rica (Cnidaria: Hydrozoa). *Revista de Biología Tropical* 50: 599–627.

Kessler, W.S. 2006. The circulation of the eastern tropical Pacific: a review. *Progress in Oceanography* 69: 181–217.

Kleypas, J.A., R.A. Feely, V.J. Fabry, C. Langdon, C.L. Sabine, and L.L. Robbins. 2006. *Impacts of Ocean Acidification on Coral Reefs and Other Marine Calcifiers: A Guide for Future Research*. NSF/NOAA/USGS.

Knowlton, N. 1993. Sibling species in the sea. *Annual Review of Ecology and Systematics* 24: 189–216.

Knowlton, N., L.A. Weigt, L.A. Solórzano, D.K. Mills, and E. Bermingham. 1993. Divergence in proteins, mitocondrial DNA and reproduction compatibility across the Isthmus of Panama. *Science* 260: 1629–32.

Kress, N., S. León-Coto, C.L. Brenes, S. Brenner, and G. Arroyo. 2002. Horizontal transport and seasonal distribution of nutrients, dissolved oxygen and chlrophyll-*a* in the Gulf of Nicoya, Costa Rica: a tropical estuary. *Continental Shelf Research* 22: 51–66.

Krylova, E.M., and H. Sahling. 2006. Recent bivalve mollusks of the genus *Calyptogena* (Vesicomyidae). *Journal of Molluscan Studies* 72: 359–95.

Kuever, J., C. Wawer, and R. Lillebæk. 1996. Microbiological observations in the anoxic basin Golfo Dulce, Costa Rica. *Revista de Biología Tropical* 44(Suppl. 3): 49–58.

Lacerda, L.D., J.E. Conde, B. Kjerfve, R. Álvarez-León, C. Alarcón, and J. Polanía. 2002. American mangroves. In L.D. Lacerda, ed., *Mangrove Ecosystems: Function and Management*, 1–62. Berlin: Springer.

Lahmann, E.J., and W. González. 1982. Observaciones sobre la distribución espacial y el comportamiento de *Siphonaria gigas* Sowerby, 1825 en la costa Pacífica de Costa Rica (Gastropoda: Siphonariidae). *Anales del Instituto de Ciencias del Mar y Limnología, UNAM* 9: 101–10.

Legeckis, R. 1988. Upwelling off the gulfs of Panamá and Papagayo in the tropical Pacific during March 1985. *Journal of Geophysical Research* 93:15485–15489.

León, P. 1973. Ecología de la ictiofauna del Golfo de Nicoya, un estuario tropical. *Revista de Biología Tropical* 21: 5–30.

León-Morales, R., and J.A. Vargas. 1998. Macroinfauna of a tropical fjord-like embayment: Golfo Dulce, Costa Rica. *Revista de Biología Tropical* 46(Suppl. 6): 81–90.

Lessios, H.A. 1990. Adaptation and phylogeny as determinants from the two sides of the Isthmus of Panamá. *American Naturalist* 135: 1–13.

Levin, L.A. 2002. Deep-ocean life where oxygen is scarce. *American Scientist* 90: 436–44.

Levin, L.A. 2003. Oxygen minimum zone benthos: adaptation and community response to hypoxia. *Oceanography and Marine Biology: An Annual Review* 41: 1–45.

Levin, L.A., V.J. Orphan, G.W. Rouse, A.E. Rathburn, W. Ussler III, G.S. Cook, S.K. Goffredi, E.M. Perez, A. Waren, B.M. Grupe, G. Chadwick, and B. Strickrott. 2012. A hydrothermal seep on the Costa Rica margin: middle ground in a continuum of reducing ecosystems. *Proceedings of the Royal Society B* 279: 2580–88.

Linke, P., K. Wallmann, E. Suess, C. Hensen, and G. Rehder. 2005. In situ benthic fluxes from an intermittently active mud volcano at the Costa Rica convergent margin. *Earth and Planetary Science Letters* 235: 79–95.

Littler, M.M., and D.S. Littler. 2008. Coralline algal rhodoliths form extensive benthic communities in the Gulf of Chiriqui, Pacific Panama. *Coral Reefs* 27: 553.

Lizano, O. 1998. Dinámica de las aguas en la parte interna del Golfo de Nicoya ante altas descargas del Río Tempisque. *Revista de Biología Tropical* 46(Suppl. 6): 11–20.

Lizano, O.G. 2006. Algunas características de las mareas en las costas Pacífica y Caribe de Centroamérica. *Ciencia y Tecnología* 24: 51–64.

Lizano, O.G. 2007. Climatología del viento y oleaje frente a las costas de Costa Rica. *Ciencia y Tecnología* 25: 43–56.

Lizano, O.G. 2008. Dinámica de aguas alrededor de la Isla del Coco, Costa Rica. *Revista de Biología Tropical* 56(Suppl. 2): 31–48.

Lizano, O.G., and E. Alfaro. 2004. Algunas características de las corrientes marinas en el Golfo de Nicoya, Costa Rica. *Revista de Biología Tropical* 52(Suppl. 2): 77–94.

Lizano, O.G., E.J. Alfaro, and A. Salazar-Matarrita. 2012. Un método para evaluar el enriquecimiento de metales en sedimentos marinos en Costa Rica. *Revista de Biología Tropical* 60(Suppl. 2): 197–211.

Lizano, O.G., L.G. Loría, E.J. Alfaro, and M. Badilla. 2008. Distribución espacial de radionucleídos en sedimentos marinos de Bahía Culebra y el Golfo de Nicoya, Pacífico, Costa Rica. *Revista de Biología Tropical* 56(Suppl. 4): 83–90.

Lizano, O., and J.A. Vargas. 1993. Distribución espacio-temporal de la salinidad y la temperatura en la parte interna del Golfo de Nicoya. *Tecnología en Marcha* 12(2): 3–16.

Loaire, S.R., P.B. Duffy, H. Hamilton, G.P. Asner, C.B. Field, and D.D. Ackerly. 2009. The velocity of climate change. *Nature* 462: 1052–55.

López, M. 1983. *Lycodontis verrilli* (Pisces: Muraenidae) descripción de su larva leptocéfala del Golfo de Nicoya, Costa Rica. *Revista de Biología Tropical* 31: 343–44.

Lopéz, M.I., and W.A. Bussing. 1982. Lista provisional de los peces marinos de la costa Pacífica de Costa Rica. *Revista de Biología Tropical* 30: 5–26.

López-Garro, A., R. Arauz-Vargas, I. Zanella, and L. Le Foulgo. 2009. Análisis de las capturas de tiburones y rayas en las pesquerías artesanales de Tárcoles, Pacífico Central de Costa Rica. *Revista Ciencias Marinas y Costeras* 1: 145–57.

Loría, L.G., R. Jiménez, and O. Lizano. 2002. Radionucleídos naturales y antropogénicos en el estuario del Golfo de Nicoya, Costa Rica. *Tópicos Meteorológicos y Oceanográficos* 9: 74–78.

Loría-Naranjo, M., J. Samper-Villarreal, and J. Cortés. 2014. Potrero Grande and Santa Elena mangrove forest structure, Santa Rosa National Park, North Pacific, Costa Rica. *Revista de Biología Tropical* 62(Suppl. 4): 33–41.

Luke, S.R. 1977. Catalog of the benthic invertebrate collections. I. Decapod Crustacea and Stomatopoda. *Scripps Institution of Oceanography Reference Series* 77–9. 72 pp.

Luke, S.R. 1982. Catalog of the benthic invertebrate collections. Echinodermata. *Scripps Institution of Oceanography Reference Series* 82–5. 66 pp.

Luke, S.R. 1995. Catalog of the Benthic Invertebrate Collections of the Scripps Institution of Oceanography. Mollusca. *Scripps Institution of Oceanography Reference Series* 95–24. 477 pp.

Luke, S.R. 1998. Catalog of the Benthic Invertebrate Collections of the Scripps Institution of Oceanography. Coelenterata. *Scripps Institution of Oceanography Reference Series* 98–02. 58 pp.

Macintyre, I.G., P.W. Glynn, and J. Cortés. 1992. Holocene reef history in the eastern Pacific: mainland Costa Rica, Caño Island, Cocos Island, and Galápagos Islands. *Proceedings of the 7th International Coral Reef Symposium, Guam* 2: 1174–84.

MacPherson, E., and I.S. Wehrtmann. 2010. Occurrence of lithodid crabs (Decapoda, Lithodidae) on the Pacific coast of Costa Rica, Central America. *Crustaceana* 83: 143–51.

Madrigal-Castro, E., J. Cabrera-Peña, J. Monge-Esquivel, and F. Pérez-Acuña. 1984. Comparación entre dos poblaciones de *Acanthina brevidentata* (Gastropoda: Mollusca) en dos zonas rocosas de Playa Panamá, Guanacaste, Costa Rica. *Revista de Biología Tropical* 32: 11–15.

Manning, R.B., and M.L. Reaka. 1979. Three new stomatopod crustaceans from the Pacific coast of Costa Rica. *Proceedings of the Biological Society of Washington* 92: 634–39.

Manzello, D.P. 2009. Reef development and resilience to acute (El Niño warming) and chronic (high-CO_2) disturbances in the eastern tropical Pacific: a real-world climate change model. Proceedings of the 11th International Coral Reef Symposium, Fort Lauderdale 2: 1299–304.

Manzello, D.P. 2010. Ocean acidification hot spots: Spatiotemporal dynamics of the seawater CO_2 system of eastern Pacific coral reefs. *Limnology and Oceanography* 55: 239–48.

Manzello, D.P., J.A. Kleypas, D.A. Budd, C.M. Eakin, P.W. Glynn, and C. Langdon. 2008. Poorly cemented coral reefs of the eastern tropical Pacific: Possible insights into reef development in a high-CO_2 world. *Proceedings of the National Academy of Sciences, USA* 105: 10450–55.

Martin, A.M., and S.K. Goffredi. 2012. *"Pliocardia" krylovata*, a new species of vesicomyid clam from cold seeps along the Costa Rica Margin. *Journal of the Marine Biological Association of the United Kingdom* 92: 1127–37.

Martínez-Fernández, D., A. Montero-Cordero, and L. May-Collado. 2011. Cetáceos de las aguas costeras del Pacífico norte y sur de Costa Rica. *Revista de Biología Tropical* 59: 283–90.

Martínez-Fernández, D., Montero-Cordero, A., and Palacios-Alfaro, D. 2014. Áreas de congregación de cetáceos en el Pacífico norte de Costa Rica: propuestas para su manejo. *Revista de Biología Tropical* 62(Suppl. 4): 99–108.

Mata, A., and J. Echeverría. 2004. Introduction. In G.W. Frankie,

A. Mata, and S.B. Vinson, *Biodiversity Conservation in Costa Rica: Learning the Lesson in a Seasonal Dry Forest*, 1–12. Berkeley: University of California Press.

Maté, J.L. 2003. Corals and coral reefs of the Pacific coast of Panamá. In J. Cortés, ed., *Latin American Coral Reefs*, 387–417. Amsterdam: Elsevier Science.

Mau, S., H. Sahling, G. Rehder, E. Suess, P. Linke, and E. Soeding. 2006. Estimates of methane output from mud extrusions at the erosive convergent margin off Costa Rica. *Marine Geology* 225: 129–44.

Maurer, D., C. Epifanio, H. Dean, S. Howe, J. Vargas, A. Dittel, and M. Murillo. 1984. Benthic invertebrates of a tropical estuary: Gulf of Nicoya, Costa Rica. *Journal of Natural History* 18: 47–61.

Maurer, D., and J.A. Vargas. 1984. Diversity of soft bottom benthos in a tropical estuary: Gulf of Nicoya, Costa Rica. *Marine Biology* 81: 97–106.

May-Collado, L.J., and J. Forcada. 2012. Small-scale estimation of relative abundance for the coastal spotted dolphins (*Stenella attenuata*) in Costa Rica: the effect of habitat and seasonality. *Revista de Biología Tropical* 60(Suppl. 2): 133–42.

May-Collado, L., T. Gerrodette, J. Calambokidis, K. Rasmussen, and I. Sereg. 2005. Patterns of cetacean sighting distribution in the Pacific Exclusive Economic Zone of Costa Rica based on data collected from 1979–2001. *Revista de Biología Tropical* 53: 249–63.

May-Collado, L., and A. Morales-Ramírez. 2005. Presencia y patrones de comportamiento del delfín manchado costero, *Stenella attenuata* (Cetacea: Delphinidae) en el Golfo de Papagayo, Costa Rica. *Revista de Biología Tropical* 53: 265–76.

McAdoo, B.G., D.L. Orange, E.A. Silver, K. McIntosh, L. Abbott, J. Galewsky, L. Kahn, and M. Protti. 1996. Seafloor structural observations, Costa Rica accretionary prism. *Geophysical Research Letters* 23: 883–86.

McCreary, J.P., H.S. Lee, and D.B. Enfield. 1989. The response of the coastal ocean to strong offshore winds: with application to circulation in the gulfs of Tehuantepec and Papagayo. *Journal of Marine Research* 47: 81–109.

McLeod, E., and R.V. Salm. 2006. *Managing Mangroves for Resilience to Climate Change*. Gland, Switzerland: IUCN. 64 pp.

Meléndez, C. 1974. Viajeros por Guanacaste. Ministerio de Cultura, Juventud y Deportes, Costa Rica. *Serie Nos Ven* 4. 92 pp.

Mielke, W. 1997. New findings of interstitial Copepoda from Punta Morales, Pacific coast of Costa Rica. *Microfauna Marina* 11: 271–80.

Molina-Ureña, H. 1996. Ichthyoplancton assemblages in the Gulf of Nicoya and Golfo Dulce embayments, Pacific coast of Costa Rica. *Revista de Biología Tropical* 44(Suppl. 3): 173–82.

Montero-Cordero, A., D. Martínez-Fernández, and G. Hernández-Mora. 2010. Mammalia, Carnivora, Otariidae, *Arctocephalus galapagoensis* Heller, 1904: First continental record for Costa Rica. *Check List* 6: 630–32.

Montes, C., A. Cardona, C. Jaramillo, A. Pardo, J.C. Silva, V. Valencia, C. Ayala, I.C. Pérez-Angel, L.A. Rodríguez-Parra, V. Ramírez, and H. Niño. 2015. Middle Miocene closure of the Central American Seaway. *Science* 348(6231): 226–29.

Mora, A., C. Fernández, and A.G. Guzmán. 2006. *Áreas Marinas Protegidas y Áreas Marinas de Uso Múltiple de Costa Rica: Notas para una Discusión*. San José, Costa Rica: MarViva. 104 pp.

Mora, C., O. Aburto-Oropeza, A. Ayala Bocos, et al. 2011. Global human footprint on the linkage between biodiversity and ecosystem functioning in reef fishes. *PLOS Biology* 9(4): e1000606. doi:10.1371/journal.pbio.1000606

Morales, A., and J.A. Vargas. 1995. Especies comunes de copépodos (Crustacea: Copepoda) pelágicos del Golfo de Nicoya, Costa Rica. *Revista de Biología Tropical* 43: 207–18.

Morales-Ramírez, A. 1996. Checklist of copepods from Gulf of Nicoya, Coronado Bay and Golfo Dulce, Pacific coast of Costa Rica, with comments on their distribution. *Revista de Biología Tropical* 44(Suppl. 3): 103–14.

Morales-Ramírez, A. 2011. La diversidad marina del Golfo Dulce, Pacífico sur de Costa Rica: amenazas a su conservación. *Biocenosis* 24: 9–20.

Morales-Ramírez, A., and E. Brugnoli-Oliveira. 2001. El Niño 1997–1998 impact on the plankton dynamics in the Gulf of Nicoya, Pacific coast of Costa Rica. *Revista de Biología Tropical* 49(Suppl. 2): 103–14.

Morales-Ramírez, A., and J. Jakob. 2008. Seasonal vertical distribution, abundance, biomass, and biometrical relationships of ostracods in Golfo Dulce, Pacific coast of Costa Rica. *Revista Biología Tropical* 56(Suppl. 4): 125–47.

Morales-Ramírez, A., and J. Nowaczyk. 2006. Gelatinous zooplankton in the Golfo Dulce, Pacific coast of Costa Rica during transition from rainy season to dry season 1997–1998. *Revista de Biología Tropical* 54(Suppl. 1): 201–23.

Morales-Ramírez, A., E. Suárez-Morales, M. Corrales, and O. Esquivel-Garrote. 2014. Diversity of the free-living marine and freshwater Copepoda (Crustacea) in Costa Rica: a review. *ZooKeys* 457: 15–33.

Morales-Ramírez, A., R. Víquez, K. Rodríguez, and M. Vargas. 2001. Marea roja producida por *Lingulodinium polyedrum* (Peridiniales, Dinophyceae) en Bahía Culebra, Golfo de Papagayo, Costa Rica. *Revista de Biología Tropical* 49(Suppl. 2): 19–23.

Mörch, O.A.L. 1859. Beiträge zur Molluskenfauna Central Amerika's. *Malakozool. Blät* 6: 102–26.

Morrison, M., M. Consalvey, K. Berkenbusch, and E. Jones. 2008. Biogenic habitats and their value to New Zealand fisheries. *Water and Atmosphere* 16: 20–21.

Mug Villanueva, M., V.F. Gallucci, and H.L. Lai. 1994. Age determination of corvina reina (*Cynoscion albus*) in the Gulf of Nicoya, Costa Rica, based on examination and analysis of hyaline zones, morphology and microstructure of otoliths. *Journal of Fish Biology* 45: 177–91.

Mumby, P.J., A.J. Edwards, J.E. Arias-Gonzalez, K.C. Lindeman, P.G. Blackwell, A. Gall, M.I. Gorczynska, A.R. Harborne, C.L. Pescod, H. Renten, C.C.C. Wabnitz, and G. Llewellyn. 2004. Mangroves enhance the biomass of coral reef fish communities in the Caribbean. *Nature* 427: 533–36.

Murase, A., A. Angulo, Y. Miyazaki, W.A. Bussing, and M.I. López. 2014. Marine and estuarine fish diversity in the inner Gulf of Nicoya, Pacific coast of Costa Rica, Central America. *Check List* 10: 1401–13.

Myers, M.C., J. Wagner, and C. Vaughan. 2010. Long-term comparison of the fish community in a Costa Rican rocky shore marine reserve. *Revista de Biología Tropical* 59: 233–46.

Neuhaus, B. 2004. Description of *Campyloderes* cf. *vanhoeffeni* (Kinorhyncha, Cyclorhagida) from the Central American East Pacific Deep with a review of the genus. *Meiofauna Marina* 13: 3–20.

Neuhaus, B., and T. Blasche. 2006. *Fissuroderes*, a new genus of Kinorhyncha (Cyclorhagida) from the deep sea and continental shelf of New Zealand and from the continental shelf of Costa Rica. *Zoologischer Anzeiger* 245: 19–52.

Nichols-Driscoll, J. 1976. Benthic invertebrate communities in Golfo Dulce, Costa Rica, an anoxic basin. *Revista de Biología Tropical* 24: 281–98.

Nova-Bustos, N., A.C. Hernández-Zanuy, and R. Viquez-Portuguez. 2010. Distribución y abundancia de las ascidias de los fondos rocosos de la Bahía de Cuajiniquil, Costa Rica. *Boletín de Investigaciones Marinas y Costeras* 39: 57–66.

Opresko, D.M., and O. Breedy. 2010. A new species of antipatharian coral (Cnidaria: Anthozoa: Antipatharia: Schizopathidae) from the Pacific coast of Costa Rica. *Proceedings of the Biological Society of Washington* 123: 234–41.

Ornelas-Gatdula, E., Y. Camacho-García, M. Schrödl, V. Padula, Y. Hooker, T.M. Gosliner, and A. Valdés. 2012. Molecular systematics of the "*Navanax aenigmaticus*" species complex (Mollusca, Opisthobranchia): Coming full circle. *Zoologica Scripta* 41: 374–85.

Orr, J.C., V.F. Fabry, O. Aumont, et al. 2005. Anthropogenic ocean acidification over the twenty-first century and its impact on calcifying organisms. *Nature* 437: 681–86.

Ortega, S. 1986. Fish predation on gastropods on the Pacific coast of Costa Rica. *Journal of Experimental Marine Biology and Ecology* 97: 181–91.

Ortega, S. 1987a. Habitat segregation and temporal variation in some tropical intertidal populations. *Journal of Experimental Marine Biology and Ecology* 113: 247–65.

Ortega, S. 1987b. The effects of human predation on the size distribution of *Siphonaria gigas* (Mollusca: Pulmonata) on the Pacific coast of Costa Rica. *Veliger* 29: 251–55.

Oviedo-Correa, L.E., J.D. Pacheco, and D. Herra-Miranda. 2009. Evalución de los riesgos de afectación por el establecimiento de granjas atuneras en relación con la distribución espacial de cetáceos en el Golfo Dulce, Costa Rica. *Revista Ciencias Marinas y Costeras* 1: 159–74.

Paine, R.T. 1966. Food web complexity and species diversity. *The American Naturalist* 100: 65–75.

Palter, J., S. León-Coto, and D. Ballestero. 2007. The distribution of nutrients, disolved oxygen and chlorophyl *a* in the upper Gulf of Nicoya, Costa Rica, a tropical estuary. *Revista de Biología Tropical* 55: 427–36.

Paulmier, A., and D. Ruiz-Pino. 2009. Oxygen minimum zones (OMZs) in the modern ocean. *Progress in Oceanography* 80: 113–28.

Peterson, C.L. 1960. The physical oceanography of the Gulf of Nicoya, Costa Rica, a tropical estuary. *Bulletin Inter-American Tropical Tuna Commission* 4: 139–216.

Pihen, E., V. Nielsen, and M. Espinoza. 2006. Tortugas marinas. In V. Nielsen-Muñoz, and M. Quesada-Alpizar, eds., *Informe Técnico: Ambientes Marinos de Costa Rica*, 149–65. San José, Costa Rica: Comisión ZEE Costa Rica, CIMAR, CI, TNC.

Pola, M., Y. Camacho-García, and T. Gosliner. 2012. Molecular data illuminate cryptic nudibranch species: The evolution of the Scyllaeidae (Nudibranchia: Dendronotina) with a revision of *Notobryon*. *Zoological Journal of the Linnean Society* 165: 311–36.

Polanía, J. 1993. Mangroves of Costa Rica. In L.D. Lacerda, ed., *Conservation and Sustainable Utilization of Mangrove Forests in Latin America and Africa Regions*, 129–37. ITTO/ISME Project PD 114/90 (F), Part I. Okinawa, Japan.

Polidoro, B.A., K.E. Carpenter, L. Collins, et al. 2010. The loss of species: mangrove extinction risk and geographic areas of global concern. *PLOS ONE* 5(4): e10095. doi:10.1371/journal.pone.0010095

Quesada, A.J., F.H. Acuña, and J. Cortés. 2014. Diet of the sea anemone *Anthopleura nigrescens*: composition and variation between daytime and nighttime high tides. *Zoological Studies* 53: 26. doi:10.1186/s40555-014-0026-2

Quesada-Alpízar, M.A., and J. Cortés. 2006. Los ecosistemas marinos del Pacífico sur de Costa Rica: estado del conocimiento y perspectivas de manejo. *Revista de Biología Tropical* 54(Suppl. 1): 101–45.

Quesada-Alpízar, M.A., and A. Morales-Ramírez. 2006. The possible effect of "El Niño" in the non-gelatinous zooplancton in Golfo Dulce, Pacific coast of Costa Rica during the period of 1997–1998. *Revista de Biología Tropical* 54(Suppl. 1): 225–40.

Quiroga, E., J. Sellanes, W.E. Arntz, D. Gerdes, V.A. Gallardo, and D. Hebbeln. 2009. Benthic megafaunal and demersal fish assemblages on the Chilean continental margin: The influence of the oxygen minimum zone on bathymetric distribution. *Deep Sea Research II* 56: 1112–23.

Ramírez, A.R., M.I. López, and W.A. Szelistowski. 1990. Composition and abundance of ichthyoplancton in a Gulf of Nicoya mangrove estuary. *Revista de Biología Tropical* 38: 463–66.

Ramírez, A.R., W.A. Szelistowski, and M. López. 1989. Spawning pattern and larval recruitment in Gulf of Nicoya anchovies (Pisces: Engraulidae). *Revista de Biología Tropical* 37: 55–62.

Ranero, C.R., R. von Huene, W. Weinrebe, and U. Barckhausen. 2007. Convergent margin tectonics: a marine perspective. In J. Bundschuh and G.E. Alvarado, eds., *Central America: Geology, Resources and Hazards*, Volume 1, 239–65. London: Taylor & Francis.

Reaka, M.L., and R.B. Manning. 1980. The distributional ecology and zoogeographical relationships of stomatopod crustacea from Pacific Costa Rica. *Smithsonian Contributions to Marine Sciences* 7: 1–29.

Reyes-Bonilla, H. 2003. Coral reefs of the Pacific coast of México. In J. Cortés, ed., *Latin American Coral Reefs*, 331–49. Amsterdam: Elsevier Science.

Reyes-Bonilla, H., and E. Barraza. 2003. Corals and associated marine communities from El Salvador. In J. Cortés, ed., *Latin American Coral Reefs*, 351–60. Amsterdam: Elsevier Science.

Richards, F.A., J.J. Anderson, and J.D. Cline. 1971. Chemical and physical observations in Golfo Dulce, an anoxic basin on the Pacific coast of Costa Rica. *Limnology and Oceanography* 16: 43–50.

Riosmena-Rodríguez, R. 2001. Mantos de rodolitos en el Golfo de California: implicaciones en la biodiversidad y el manejo de las zonas costeras. *Biodiversitas* 36: 12–14.

Riul, P., P. Lacouth, P.R. Pagliosa, M.L. Christoffersen, and P.A. Horta. 2009. Rhodolith beds at the easternmost extreme of South America: Community structure of an endangered environment. *Aquatic Botany* 90: 315–20.

Riul, P., C.H. Targino, J.N. Farias, P.T. Visscher, and P.A. Horta. 2008. Decrease in *Lithothamnion* sp. (Rhodophyta) primary production due to the deposition of a thin sediment layer. *Journal of the Marine Biological Association of the United Kingdom* 88: 17–19.

Rivera, M.G., R. Riosmena-Rogríquez, and M.S. Foster. 2004. Age and growth of *Lithothamnion muelleri* (Corallinales, Rhodophta) in the southwestern Gulf of California, Mexico. *Ciencias Marinas* 30: 235–49.

Rixen, T., C. Jiménez, and J. Cortés. 2012. Impact of upwelling events on the sea water chemistry in the of Gulf of Papagayo (Culebra Bay), Costa Rica. *Revista de Biología Tropical* 60(Suppl. 2): 187–95.

Robertson, D.R., and K.L. Cramer. 2009. Shore fishes and biogeographic subdivisions of the Tropical Eastern Pacific. *Marine Ecology Progress Series* 380: 1–17.

Rodríguez-Sáenz, K.E. 2005. Distribución espacial y temporal de la biomasa, composición y abundancia del zooplancton, con énfasis en hidromedusas de Bahía Culebra, durante La Niña 1999 y el 2000. Thesis M.Sc., Universidad de Costa Rica. San Pedro, Costa Rica.

Rodríguez-Sáenz, K., and A. Morales-Ramírez. 2012. Composición y distribución del mesozooplancton en una zona de afloramiento costero (Bahía Culebra, Costa Rica) durante La Niña 1999 y el 2000. *Revista de Biología Tropical* 60(Suppl. 2): 143–57.

Rodríguez-Sáenz, K., and J. Rodríguez-Fonseca. 2004. Avistamientos del delfín manchado, *Stenella attenuata* (Cetacea: Delphinidae) en Bahía Culebra, Costa Rica, 1999–2000. *Revista de Biología Tropical* 52(Suppl. 2): 189–93.

Rodríguez-Sáenz, K., J.A. Vargas-Zamora, and L. Segura-Puertas. 2012. Medusas (Cnidaria: Hydrozoa) de una zona de afloramiento costero, Bahía Culebra, Pacífico, Costa Rica. *Revista de Biología Tropical* 60: 1731–48.

Rojas-Figueroa, R.E., and J.A. Vargas-Zamora. 2008. Abundancia, biomasa y relaciones sedimentarias de *Americonuphis reesei* (Polychaeta: Onuphidae) en el Golfo de Nicoya, Costa Rica. *Revista de Biología Tropical* 56(Suppl. 4): 59–82.

Rostad, T., and K.L. Hansen. 2001. The effects of trawling on the benthic fauna of the Gulf of Nicoya, Costa Rica. *Revista de Biología Tropical* 49(Suppl. 2): 91–95.

Sahling, H., D.G. Masson, C.R. Ranero, V. Hühnerbach, W. Weinrebe, I. Klaucke, D. Bürk, W. Brückmann, and E. Suess. 2008. Fluid seepage at the continental margin offshore Costa Rica and southern Nicaragua. *Geochemistry, Geophysics, Geosystems* 9: doi:10.1029/2008GC001978.

Salas-Moya, C., S. Mena, and I.S. Wehrtmann. 2014. Reproductive traits of the symbiotic pea crab *Austinotheres angelicus* (Crustacea, Pinnotheridae) living in *Saccostrea palmula* (Bivalvia, Ostreidae), Pacific coast of Costa Rica. *ZooKeys* 457: 239–52.

Sameoto, D.D. 1986. Influence of the biological and physical environment on the vertical distribution of mesozooplankton and micronekton in the eastern tropical Pacific. *Marine Biology* 93: 263–79.

Samper-Villarreal, J., A. Bourg, J.A. Sibaja-Cordero, and J. Cortés. 2014. Presence of a *Halophila baillonii* Asch. (Hydrocharitaceae) seagrass meadow and associated macrofauna on the Pacific coast of Costa Rica. *Pacific Science* 68: 435–44.

Samper-Villareal, J., J. Cortés, and C. Benavides. 2012. Description of the Panamá and Iguanita mangrove stands within Bahía Culebra, north Pacific coast of Costa Rica. *Revista de Biología Tropical* 60(Suppl. 2): 109–20.

Sánchez-Nava, S., and L. Segura-Puertas. 1987. Los moluscos pelágicos (Gastropoda: Heteropoda y Pteropoda) recolectados en el Domo de Costa Rica y regiones adyacentes. *Memorias de la Sociedad Méxicana de Malacología* 3: 232–40.

Santoro, M., and S. Mattiucci. 2009. Sea turtle parasites. In I.S. Wehrtmann and J. Cortés, eds., *Marine Biodiversity of Costa Rica, Central America*. Berlin: Springer. Text: 507–19, Species List CD: 497–500.

Santoro, M., and J.A. Morales. 2007. Some digenetic trematodes of the olive ridley sea turtle, *Lepidochelys olivacea* (Testudines, Cheloniidae) in Costa Rica. *Helminthologia* 44: 25–28.

Santoro, M., C.M. Orrego, and G. Hernández-Gómez. 2006. Flora bacteriana clocal y nasal de *Lepidochelys olivacea* (Testudines, Cheloniidae) en el Pacífico norte de Costa Rica. *Revista de Biología Tropical* 54: 43–48.

Sasa, M., G.A. Chaves, and L.D. Patrick. 2009. Marine reptiles and amphibians. In I.S. Wehrtmann and J. Cortés, eds., *Marine Biodiversity of Costa Rica, Central America*. Berlin: Springer. Text: 459–68, Species List CD: 474–78.

Segura-Puertas, L. 1991. Medusas (Cnidaria: Hydrozoa y Scyphozoa) de la región del Domo de Costa Rica. *Revista de Biología Tropical* 39: 159–63.

Shanks, A.L., B.A. Grantham, and M.H. Carr. 2003. Propagule dispersal distances and the size and spacing of marine reserves. *Ecological Applications* 13: S159–S169.

Sibaja-Cordero, J.A. 2005. Distribución vertical de la epifauna en zonas rocosas de entre mareas, Golfo de Nicoya, Costa Rica. Licentiate thesis, Universidad de Costa Rica. San Pedro, Costa Rica.

Sibaja-Cordero, J.A. 2008. Vertical zonation in the rocky intertidal at Cocos Island (Isla del Coco), Costa Rica: a comparison with other tropical locations. *Revista de Biología Tropical* 56(Suppl. 2): 171–87.

Sibaja-Cordero, J.A., Y.E. Camacho-García, and R. Vargas-Castillo. 2014. Riqueza de especies de invertebrados en playas de arena y costas rocosas del Pacífico Norte de Costa Rica. *Revista de Biología Tropical* 62(Suppl. 4): 63–84.

Sibaja-Cordero, J.A., and J. Cortés. 2008. Vertical zonation of rocky intertidal organisms in a seasonal upwelling area (eastern tropical Pacific). *Revista de Biología Tropical* 56(Suppl. 2): 171–87.

Sibaja-Cordero, J.A., and K. García-Méndez. 2014. Variación espacial y temporal de los organismos de un intermareal rocoso, Bahía Panamá, Pacífico Norte, Costa Rica. *Revista de Biología Tropical* 62(Suppl. 4): 85–97.

Sibaja-Cordero, J.A., and J.A. Vargas-Zamora. 2006. Zonación vertical de epifauna y algas en litorales rocosos del Golfo de Nicoya, Costa Rica. *Revista de Biología Tropical* 54(Suppl. 1): 49–67.

Silva, A.M., and J. Acuña-González. 2006. Caracterización físico-química de dos estuarios en la bahía de Golfito, Golfo Dulce, Pacífico de Costa Rica. *Revista de Biología Tropical* 54(Suppl. 1): 241–56.

Silva-Benavides, A.M. 2009. Mangroves. In I.S. Wehrtmann and J. Cortés, eds., *Marine Biodiversity of Costa Rica, Central America*. Berlin: Springer. Text: 123–30, Species List CD: 73–78.

Snodgrass, R.E., and E. Heller. 1905. Papers from the Hopkins-Stanford Galapagos Expedition, 1898–1899. XVII. Shore fishes of the Revillagigedo, Clipperton, Cocos and Galapagos islands. *Proceedings of the Washington Academy of Science* 6: 333–427.

Solano-Barquero, A. 2011. Macrofauna asociada a rodolitos en el Parque Nacional Isla del Coco, Costa Rica. Lic. Thesis, Universidad de Costa Rica. San Pedro, Costa Rica.

Soto, R., and J. Jiménez. 1982. Análisis fisonómico estructural del manglar de Puerto Soley, La Cruz, Guanacaste, Costa Rica. *Revista de Biología Tropical* 30: 161–68.

Sowerby, G.B. 1832. Characters of new species of Mollusca and Conchifera, collected by Hugh Cuming. *Proceedings of the Zoological Society of London Part II* 1832: 25–33.

Spight, T.M. 1976. Censuses of rocky shore prosobranchs from Washington and Costa Rica. *Veliger* 18: 309–17.

Spight, T.M. 1978. Temporal changes in a tropical rocky shore snail community. *Veliger* 21: 137–43.

Spongberg, A.L. 2004a. PCB concentrations in sediments from the Gulf of Nicoya estuary, Pacific coast of Costa Rica. *Revista de Biología Tropical* 52(Suppl. 2): 11–22.

Spongberg, A.L. 2004b. PCB contamination in marine sediments from Golfo Dulce, Pacific coast of Costa Rica. *Revista de Biología Tropical* 52(Suppl. 2): 23–32.

Spongberg, A.L., and P. Davis. 1998. Organochlorinated pesticide contaminants in Golfo Dulce, Costa Rica. *Revista de Biología Tropical* 46(Suppl. 6): 111–24.

Spongberg, A.L., J.D. Witter, J. Acuña, J. Vargas, M. Murillo, G. Umaña, E. Gómez, and G. Perez. 2011. Reconnaissance of selected PPCP compounds in Costa Rican surface waters. *Water Research* 45: 6709–17.

Spotila, J.R., R.D. Reina, A.C. Steyermark, P.T. Plotkin, and F.V. Paladino. 2000. Pacific leatherback turtles face extinction. *Nature* 405: 529–30.

Stebbing, T.R.R. 1906. A new Costa Rica amphipod. *Proceedings of the United States National Museum* 31: 501–4.

Stehli, F., and S. Webb. 1985. *The Great American Biotic Interchange*. New York: Plenum Press.

Steller, D.L., M.S. Foster, and R. Riosmena-Rogríguez. 2007. Sampling and monitoring rhodollith beds: rhodolith distribution and taxonomy, biodiversity and long-term sampling. In K.I.P.R. Rigy and Y. Shirayama, eds., *Sampling Biodiversity in Coastal Communities: NaGISA Protocols for Seagrass and Macroalgal Habitats*, 93–97. Kyoto: Kyoto University Press.

Steller, D.L., M.S. Foster, and R. Riosmena-Rodríguez. 2009. Rhodolith banks. In M.E. Johnson and J. Ledesma-Vázquez, eds., *Atlas of Coastal Ecosystems in the Gulf of California: Past and Present*, 72–82. Tucson, AZ: University of Arizona Press.

Steller, D.L., R. Riosmena-Rodríguez, M.S. Foster, and C. Roberts. 2003. Rhodolith bed diversity in the Gulf of California: the importance of rhodolith structure and consequences of anthropogenic disturbances. *Aquatic Conservation: Marine and Freshwater Ecosystems* 13: 5–20.

Stern-Pirlot, A., and M. Wolff. 2006. Population dynamics and fisheries potential of *Anadara tuberculosa* (Bivalvia: Arcidae) along the Pacific coast of Costa Rica. *Revista de Biología Tropical* 54(Suppl. 1): 87–99.

Stramma, L., G.C. Jonson, J. Sprintall, and V. Mohrholz. 2008. Expanding oxygen-minimum zones in the Tropical oceans. *Science* 320: 655–58.

Sunagawa, S. 2005. Seasonal variation in symbiont densities and skeletal phosphorus concentrations in the eastern Pacific (Costa Rica) coral *Pavona clavus*. Thesis M.Sc., Universität Bremen. Bremen, Germany.

Sunagawa, S., J. Cortés, C.E. Jiménez, and R. Lara. 2008. Variation in cell densities and pigment concentrations of symbiotic dinoflagellates in the coral *Pavona clavus* in the eastern Pacific (Costa Rica). *Ciencias Marinas* 34: 113–23.

Sutherland, J.P. 1987. Recruitment limitation in a tropical intertidal barnacle: *Tetraclita panamensis* (Pilsbry) on the Pacific coast of Costa Rica. *Journal of Experimental Marine Biology and Ecology* 113: 267–82.

Sutherland, J.P., and S. Ortega. 1986. Competition conditional on recruitment and temporary escape from predators on a tropical rocky shore. *Journal of Experimental Marine Biology and Ecology* 95: 155–66.

Svendsen, H., R. Rosland, S. Myking, J.A. Vargas, O. Lizano, and E. Alfaro. 2006. A physical-oceanographic study of Golfo Dulce, Costa Rica. *Revista de Biología Tropical* 54(Suppl. 1): 147–70.

Swimmer, Y., J. Suter, R. Arauz, K. Bigelow, A. López, I. Zanela, A. Bolaños, J. Ballestero, R. Suárez, J. Wang, and C. Boggs. 2011. Sustainable fishing gear: the case of modified circle hooks in a Costa Rican longline fishery. *Marine Biology* 158: 757–67.

Szelistowski, W.A. 1989. Scale-feeding in juvenile marine catfishes (Pisces: Ariidae). *Copeia* 1989: 517–19.

Szelistowski, W.A. 1990. A new clingfish (Teleostei: Gobiesocidae) from the mangroves of Costa Rica, with notes on its ecology and early development. *Copeia* 1990: 500–507.

Tabash, F.A. 2007. Exploración de la pesquería de arrastre de camarón durante el período 1991–1999 en el Golfo de Nicoya, Costa Rica. *Revista de Biología Tropical* 55: 207–18.

Tabash, F.A., and E.A. Chávez. 2006. Optimizing harvesting strategies of the white shrimp fisheries in the Gulf of Nicoya, Costa Rica. *Crustaceana* 79: 327–43.

Taylor, M.A., and E.J. Alfaro. 2005. Climate of Central America and the Caribbean. In J.E. Oliver, ed., *Encyclopedia of World Climatology*, 183–89. Netherlands: Springer.

Thamdrup, B., D.E. Canfield, T.G. Ferdelman, R.N. Glud, and J.K. Gundersen. 1996. A biogeochemical survey of the anoxic basin Golfo Dulce, Costa Rica. *Revista de Biología Tropical* 44(Suppl. 3): 19–34.

Thorne, R.E., J.B. Hedgepeth, and J. Campos. 1989. Hydroacoustic observations of fish abundance and behavior around an artificial reef in Costa Rica. *Bulletin of Marine Science* 44: 1058–64.

Thorne, R.E., J.B. Hedgepeth, and J.A. Campos. 1990. The use of stationary hydroacoustic transducers to study dial and tidal influences of fish behaviours. *Rapports et Procés-verbaux des Reúnions / Conseil Permanent International pour l'Exploration de la Mer* 189: 167–75.

Thurber, A.R., W.J. Jones, and K. Schnabel. 2011. Dancing for food in the deep sea: Bacterial farming by a new species of Yeti crab. *PLOS ONE* 6(11): e26243. doi:10.1371/journal.pone.0026243

TNC (The Nature Conservancy). 2008. Evaluación de ecorregiones marinas en Mesoamérica. Sitios prioritarios para la conservación en las ecorregiones Bahía de Panamá, Isla del Coco y Nicoya del Pacífico Tropical Oriental, y en el Caribe Suroccidental de Costa Rica y Panamá. San José, Costa Rica: TNC. 165 pp.

Tokioka, T. 1971. A new species of *Rhopalaea* from the Pacific coast of Costa Rica (Tunicata, Ascidiacea). *Publications of the Seto Marine Biological Laboratory XIX* (2/3): 119–22.

Tokioka, T. 1972. On a small collection of ascidians from the Pacific coast of Costa Rica. *Publications of the Seto Marine Biological Laboratory XIX* (6): 383–408.

Tomlinson, P.B. 1986. *The Botany of Mangroves*. Cambridge, UK: Cambridge University Press.

Townsend, C.H. 1901. Dredging and other records of the United States Fish Commission Steamer Albatross, with bibliography relative to the work of the vessel. US Fish Commission Report for 1900: 387–562.

Troëng, S., and E. Rankin. 2005. Long-term conservation efforts contribute to positive green turtle *Chelonia mydas* nesting trend at Tortuguero, Costa Rica. *Biological Conservation* 121: 111–16.

Valiela, I., J.L. Bowen, and J.K. Cork. 2001. Mangrove forests: one of the world's threatened major tropical environmentes. *BioScience* 51: 807–15.

Vargas, J.A. 1986. A description of the structure of a tropical intertidal mud flat community. PhD diss., University of Rhode Island. Narragansett, RI. 180 pp.

Vargas, J.A. 1987. The benthic community of an intertidal mud flat in the Gulf of Nicoya, Costa Rica. Description of the community. *Revista de Biología Tropical* 35: 299–316.

Vargas, J.A. 1988a. Community structure of macrobenthos and the result of macropredator exclusion on a tropical intertidal mud flat. *Revista de Biología Tropical* 36: 287–308.

Vargas, J.A. 1988b. A survey of the meiofauna of an eastern Pacific intertidal mudflat. *Revista de Biología Tropical* 36: 541–44.

Vargas, J.A. 1989a. Seasonal abundance of *Coricuma nicoyensis* Watling and Breedy, 1988 (Crustacea: Cumacea) on a tropical intertidal mudflat. *Revista de Biología Tropical* 37: 207–12.

Vargas, J.A. 1989b. A three year survey of the macrofauna of an intertidal mud flat in the Gulf of Nicoya, Costa Rica. Proceedings of the 6th Symposium on Coastal and Ocean Management 2: 1905–91.

Vargas, J.A. 1995. The Gulf of Nicoya estuary: past, present and future cooperative research. *Helgoländer Meeresuntersuchungen* 49: 821–28.

Vargas, J.A. 1996. Ecological dynamics of a tropical intertidal mudflat community. In K.F. Nordstrom and C.T. Roman, eds., *Estuarine Shores: Evolution, Environments and Human Alterations*, 355–71. London: John Wiley and Sons Ltd.

Vargas, J.A., and H.K. Dean. 2010. On *Branchiostoma californiense* (Cephalochordata) from the Gulf of Nicoya estuary, Costa Rica. *Revista de Biología Tropical* 58: 1143–48.

Vargas, J.A., H.K. Dean, D. Maurer, and P. Orellana. 1985. Lista preliminar de los invertebrados asociados a los sedimentos del Golfo de Nicoya, Costa Rica. *Brenesia* 24: 327–42.

Vargas, J.A., and A. Mata. 2004. Where the dry forest feeds the sea: the Gulf of Nicoya Estuary. In G.W. Frankie, A. Mata, and S.B. Vinson, eds., *Biodiversity Conservation in Costa Rica: Learning the Lessons in a Seasonal Dry Forest*, 126–35. Berkeley: University of California Press.

Vargas, J.A., and S. Solano. 2011. On *Mellitella stokesii* and *Amphipholis geminata* (Echinodermata), from an intertidal flat in the upper Gulf of Nicoya estuary, Pacific, Costa Rica. *Revista de Biología Tropical* 59: 193–98.

Vargas, J.A., and M. Wolff. 1996. Preface. Pacific coastal ecosystems of Costa Rica with emphasis on the Golfo Dulce and adjacent areas: a synoptic view based on the R.V. Victor Hensen-expedition 1993/94 and previous studies. *Revista de Biología Tropical* 44(Suppl. 3): iii–vi.

Vargas, R., S. Jesse, and M. Castro. 1996. Checklist of crustaceans (Decapoda and Stomatopoda), collected during the Victor Hensen Costa Rica expedition (1993/1994). *Revista de Biología Tropical* 44(Suppl. 3): 97–102.

Vargas-Castillo, R. 2012. Nuevas adiciones a la fauna de crustáceos decápodos de Bahía Culebra, Guanacaste, Costa Rica. *Revista de Biología Tropical* 60(Suppl. 2): 73–76.

Vargas-Montero, M. 2004. Floraciones algales en Costa Rica y su relación con algunos factores meteorológicos y consideraciones sobre sus efectos socioeconómicos. Thesis M.Sc., Universidad de Costa Rica. San Pedro, Costa Rica.

Vargas-Montero, M., and E. Freer. 2004a. Presencia de los dinoflagelados *Ceratium dens*, *C. fusus* y *C. furca* (Gonyaulacales: Ceratiaceae) en el Golfo de Nicoya, Costa Rica. *Revista de Biología Tropical* 52(Suppl. 1): 115–20.

Vargas-Montero, M., and E. Freer. 2004b. Proliferaciones algales nocivas de cianobacterias (Oscillatoriaceae) y dinoflagelados (Gymnodiniaceae) en el Golfo de Nicoya, Costa Rica. *Revista de Biología Tropical* 52(Suppl. 1): 121–25.

Vargas-Montero, M., and E. Freer. 2004c. Proliferaciones algales de la diatomea toxigénica *Pseudo-Nitzschia* (Bacillariophyceae) en el Golfo de Nicoya, Costa Rica. *Revista de Biología Tropical* 52(Suppl. 1): 127–32.

Vargas-Zamora, J.A., and J.A. Sibaja-Cordero. 2011. Molluscan assemblage from a tropical intertidal estuarine sand-mud flat, Gulf of Nicoya, Pacific, Costa Rica (1984–1987). *Revista de Biología Tropical* 59: 1135–48.

Vargas-Zamora, J.A., J.A. Sibaja-Cordero, and R. Vargas-Castillo. 2012. Crustaceans from a tropical estuarine sand-mud flat, Pacific, Costa Rica, (1984–1988) revisited. *Revista de Biología Tropical* 60: 1763–81.

Veit-Köhler, G., D. Gerdes, E. Quiroga, D. Hebbeln, and J. Sellanes. 2009. Metazoan meiofauna within the oxygen-minimum zone off Chile: Results of the 2001-PUCK expedition. *Deep Sea Research II* 56: 1105–11.

Verrill, A.E. 1870. Notes on Radiata in the Museum of Yale College, Number 6: Review of the corals and polyps of the west coast of America. *Transactions of the Connecticut Academy of Arts and Sciences* 1: 377–558.

Vicencio-Aguilar, M.E., and M.A. Fernández-Alamo. 1996. Zooplancton del Domo de Costa Rica: taxonomía y biogeografía. *Revista de Biología Tropical* 44: 631–42.

Villalobos, C.R. 1980a. Variations in population structure in the genus *Tetraclita* (Crustacea: Cirripedia) between temperate and tropical populations. III. *I. stalactifera* in Costa Rica. *Revista de Biología Tropical* 28: 193–201.

Villalobos, C.R. 1980b. Variations in population structure in the genus *Tetraclita* (Crustacea: Cirripedia) between temperate and tropical populations. IV. The age structure of *T. stalactifera* and concluding remarks. *Revista de Biología Tropical* 28: 353–59.

Villalobos, C.R. 1980c. Algunas consideraciones sobre el efecto de los factores físicos y biológicos en la estructura de una comunidad de algas en el Pacífico de Costa Rica. *Brenesia* 18: 289–300.

Villalobos-Rojas, F., and I.S. Wehrtmann. 2011. Gonad development in the commercially exploited deepwater shrimp *Solenocera agassizii* (Decapoda: Solenoceridae) from Pacific Costa Rica, Central America. *Fisheries Research* 109: 150–56.

Víquez, R., and P.E. Hargraves. 1995. Annual cycle of potentially toxic dinoflagellates in the Golfo de Nicoya, Costa Rica. *Bulletin of Marine Science* 57: 467–75.

Voorhis, A.D., C.E. Epifanio, D. Maurer, A.I. Dittel, and J.A. Vargas. 1983. The estuarine character of the Gulf of Nicoya, an embayment on the Pacific coast of Central America. *Hydrobiologia* 99: 225–37.

Wang, C., and P. Fiedler. 2006. ENSO variability in the eastern tropical Pacific: a review. *Progress in Oceanography* 69: 239–66.

Wangelin, M. von, and M. Wolff. 1996. Comparative biomass spectra and species composition of the zooplancton communities in Golfo

Dulce and Golfo de Nicoya, Pacific coast of Costa Rica. *Revista de Biología Tropical* 44(Suppl. 3): 135–56.

Waycott, M., C.M. Duarte, T.J.B. Carruthers, R.J. Orth, W.C. Dennison, S. Olyarnik, A. Calladine, J.W. Fourqurean, K.L. Heck, A.R. Hughes, G.A. Kendrick, W.J. Kenworthy, F.T. Short, and S.L. Williams. 2009. Accelerating loss of seagrasses across the globe threatens coastal ecosystems. *Proceedings of the National Academy of Sciences* 106: 12377–81.

Wehrtmann, I.S., and L. Albornoz. 2002. Evidence of different reproductive traits in the transisthmian sister species, *Alpheus saxidomus* and *A. simus* (Decapoda, Caridea, Alpheidae): description of the first postembryonic stage. *Marine Biology* 140: 605–12.

Wehrtmann, I.S., and J. Cortés, eds. 2009. *Marine Biodiversity of Costa Rica, Central America*. Berlin: Springer. Text: 538 pp, Species List CD: 500 pp.

Wehrtmann, I.S., J. Cortés, and S. Echeverría-Sáenz. 2009. Marine biodiversity of Costa Rica: perspectives and conclusions. In I.S. Wehrtmann and J. Cortés, eds., *Marine Biodiversity of Costa Rica, Central America*, 521–33. Berlin: Springer + Business Media BV.

Wehrtmann, I.S., and S. Echeverría-Saénz. 2007. Crustacean fauna (Stomatopoda: Decapoda) associated with the deepwater fisheries of *Heterocarpus vicarius* (Decapoda: Pandalidae) along the Pacific coast of Costa Rica. *Revista de Biología Tropical* 55(Suppl. 1): 121–30.

Wehrtmann, I.S., J. Herrera-Correal, R. Vargas, and P. Hernáez. 2010. Squat lobsters (Decapoda: Anomura: Galatheidae) from deepwater Pacific Costa Rica: species diversity, spatial and bathymetric distribution. *Nauplius* 18: 69–77.

Wehrtmann, I.S., I. Miranda, C.A. Lizana-Moreno, P. Hernáez, V. Barrantes-Echandi, and F.L. Mantelatto. 2012. Reproductive plasticity in *Petrolisthes armatus* (Anomura, Porcellanidae): a comparison between a Pacific and an Atlantic population. *Helgoland Marine Research* 66: 87–96.

Wehrtmann, I.S., and V. Nielsen-Muñoz. 2009. The deepwater fishery along the Pacific coast of Costa Rica, Central America. *Latin American Journal of Aquatic Research* 37: 543–54.

Whoriskey, S., R. Arauz, and J.K. Baum. 2011. Potential impacts of emerging mahi-mahi fisheries on sea turtle and elasmobranch bycatch species. *Biological Conservation*: 144(6): 1841–49.

Willet, S.W., R.R. Leben, and M.F. Lavín. 2006. Eddies and tropical instability waves in the eastern tropical Pacific: a review. *Progress in Oceanography* 69: 218–38.

Willis, S., and J. Cortés. 2001. Mollusks of Manuel Antonio National Park, Pacific Costa Rica. *Revista de Biología Tropical* 49(Suppl. 2): 25–36.

Wilson, S., C. Blake, J.A. Berges, and C.A. Maggs. 2004. Environmental tolerances of free-living coralline algae (maerl): implications for European marine conservation. *Biological Conservation* 120: 279–89.

Wolff, M., J. Chavarría, V. Koch, and J.A. Vargas. 1998. A trophic flow model of the Golfo de Nicoya, Costa Rica. *Revista de Biología Tropical* 46(Suppl. 6): 63–79.

Wolff, M., H.J. Hartmann, and V. Koch. 1996. A pilot trophic model for Golfo Dulce, a fjord-like tropical embayment, Costa Rica. *Revista de Biología Tropical* 44(Suppl. 3): 215–31.

Wolff, M., and J.A. Vargas, eds. 1994. RV Victor Hensen Costa Rica Expedition 1993/1994 Cruise Report. *ZMT Contribution 2*. Bremen, Germany: Centre for Tropical Marine Ecology.

Wyrtki, K. 1966. Oceanography of the eastern equatorial Pacific Ocean. *Oceanography and Marine Biology: An Annual Review* 4: 33–68.

Zamora-Trejos, P. 2006. Manglares. In V. Nielsen-Muñoz and M.A. Quesada-Alpízar, eds., *Ambientes Marino Costeros de Costa Rica*, 23–39. Comisión Interdisciplinaria Marino Costera de la Zona Económica Exclusiva de Costa Rica, Informe Técnico. San José, Costa Rica: CIMAR, CI, TNC.

Zamora-Trejos, P., and J. Cortés. 2009. Los manglares de Costa Rica: Pacífico norte. *Revista de Biología Tropical* 57: 473–88.

Zanella, I., A. López, and R. Arauz. 2009. Caracterización de la pesca del tiburón martillo, *Sphyrna lewini*, en la parte externa del Golfo de Nicoya, Costa Rica. *Revista Ciencias Marinas y Costeras* 1: 175–95.

Zanella-Cesarotto, I., A. López-Garro, and R. Arauz-Vargas. 2010. La alimentación de tiburones martillo jóvenes (*Sphyrna lewini*) capturados en el Golfo de Nicoya, Costa Rica. *Boletín de Investigaciones Marinas y Costeras* 39: 457–64.

Zapata, F., and B. Vargas-Ángel. 2003. Corals and coral reefs of the Pacific coast of Colombia. In J. Cortés, ed., *Latin American Coral Reefs*, 419–47. Amsterdam: Elsevier Science.

Chapter 6 The Gulf of Nicoya Estuarine Ecosystem

José A. Vargas[1]

Introduction

Ecosystem Boundaries

The Gulf of Nicoya (*Golfo de Nicoya*) is located in the northwestern region of the Pacific coast of Costa Rica. The extensive hydrological connection between the Gulf and watersheds draining into it gives origin to one of the most prominent ecological and geographical systems of Costa Rica. If the Tempisque and Tárcoles River basins are considered as contributing directly to the estuarine system via freshwater discharges and associated pollutants and sediments, then this estuarine ecosystem comprises more than 25% of the total mainland area (51,000 km^2) of the country (Vargas and Mata 2004). However, traditionally the Gulf has been viewed as a separate ecosystem whose limits start at the water edge. If for convenience we follow this approach, then the Gulf could be centered at 10° N–85° W, having a length of about 100 km at its longitudinal axis, and comprising an area of about 1,500 km^2. Depths along the axis range from less than a meter at low tide near the mouth of the Tempisque River, to more than 500 m at the mouth of the Gulf (Fig. 6.1). The shape of the Gulf and its bathymetry allow it to be divided into three main regions: the shallow inner region, extending from a line between the tip of the Puntarenas peninsula and San Lucas island to the mouth of the Tempisque River; the middle gulf region, limited to the south by a line from the Negritos Islands on the western shore to the Caldera Port on the eastern shore; and the lower region, from the Negritos Islands south to a line crossing the Gulf from Ballena Bay to Herradura Bay (Fig. 6.1), which deepens and broadens towards the open ocean. The inner gulf is bordered mostly by mangrove forests, sandy beaches, and tidal flats. The middle gulf is limited by sandy beaches, and the lower gulf includes sandy beaches and rocky cliffs. Tides in the Gulf are semidiurnal, with a mean spring tide range of 2.8 m at the port of Puntarenas (Voorhis et al. 1983).

Two of the photographs included in Fig. 6.1 illustrate the main pier at the port of Puntarenas in 1901 and in 2007, leaving the false impression that not much has changed there in more than a century, since what is visible now and then looks nearly the same. However, there have been drastic changes in the waters of this ecosystem that are not as visible, as it may be a deforested patch on one of its many islands. Contrary to land-based ecosystems, nobody owns the resources in the waters, in the sediments, and along the intertidal zone of the Gulf of Nicoya. Fish, crustaceans, mollusks, and other species of commercial interest are common property, or *commons*. These resources do not belong to anyone, but can be used by everyone, and no fisherman takes personal responsibility for the sustainable use of its resources, a condition known as the *Tragedy of the Commons* (Hardin 1968).

What is notorious in the Gulf today is the development of the coastline, with new housing complexes being built every year. More deforested patches appear on the very few and already disturbed surviving coastal forests, including mangroves. Cargo ships unload containers at the new port facilities of Caldera, and cruise ships dock at the upgraded Puntarenas pier, where fast sail clippers did so more than a century ago. Further upstream the new port of Punta Morales serves foreign cargo ships loading ethylic alcohol and cane sugar. The third photograph, taken in 1900, shows a

[1] Centro de Investigación en Ciencias del Mar y Limnología (CIMAR), Universidad de Costa Rica (UCR), San Pedro de Montes de Oca, 11501-2060 San José, Costa Rica

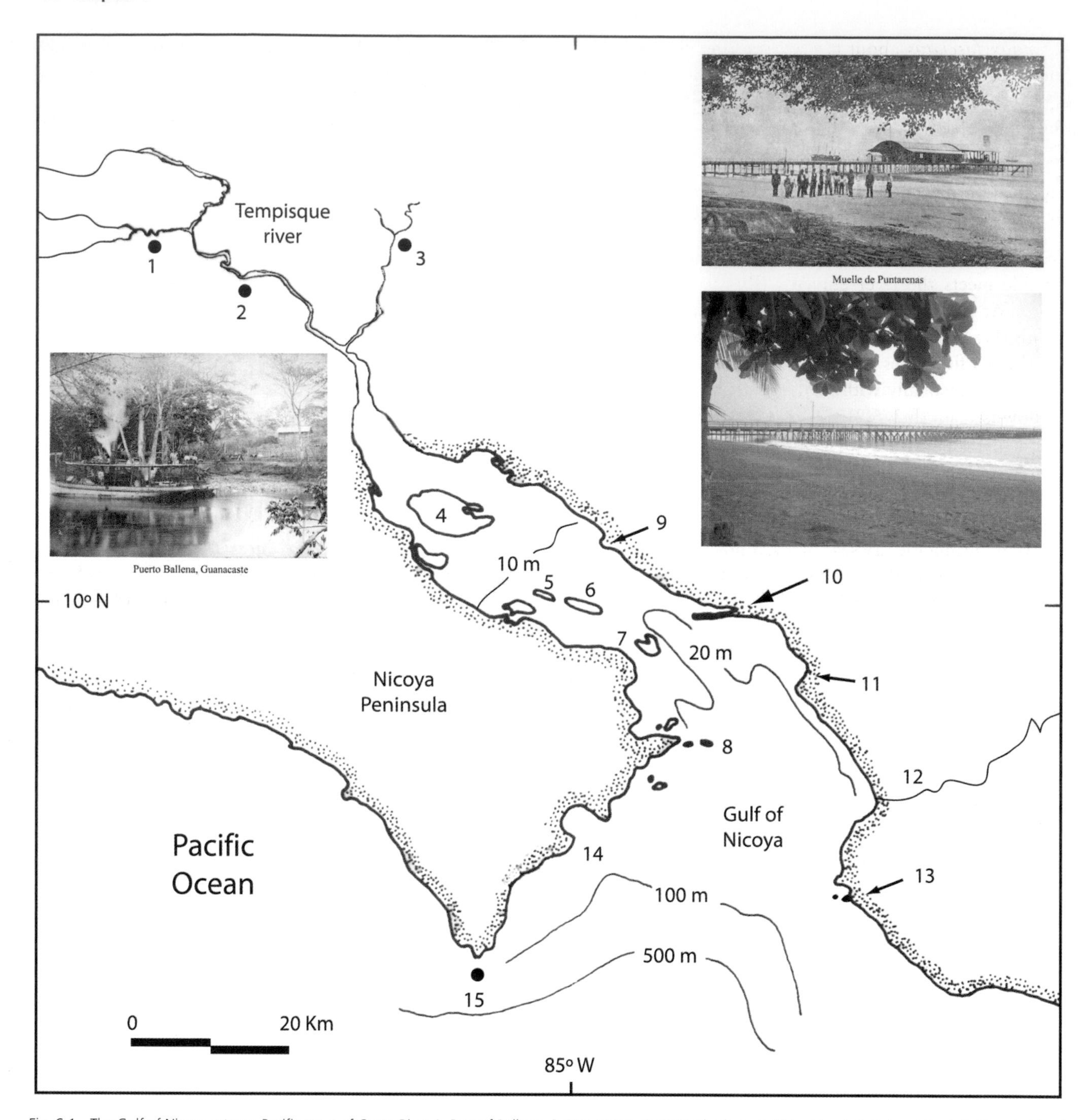

Fig. 6.1 The Gulf of Nicoya estuary, Pacific coast of Costa Rica: 1. Port of Ballena, 2. Port Humo, 3. Port Bebedero, 4. Chira Island, 5. Bejuco Island, 6. Venado Island, 7. San Lucas Island, 8. Negritos Islands, 9. Port of Punta Morales, 10. Port city of Puntarenas, 11. Port of Caldera, 12. Tárcoles river mouth, 13. Herradura Bay, 14. Ballena Bay, 15. Cape Blanco. Depth contours in meters. Top right: the Puntarenas pier in 1901. Bottom: same pier site, 2007. Left: a steam boat at the Ballena river port, 1900. *Photos taken in 1900 and 1901 by P. Baixench (*Revista de Costa Rica*, 1902. Tipografía Nacional, San José); photo taken in 2007 by José A. Vargas.*

steam boat at the former fluvial port of Ballena, upstream the Tempisque River at a time when most of the access to the core of the Guanacaste Province was by boat to ports like Ballena, Humo, and Bebedero (Fig. 6.1), and boats such as shrimp trawlers were not known in the estuary. New roads, and a new bridge crossing the river at its mouth, make river navigation obsolete. Today, the Tempisque River flow is negligible at the former Ballena port site owing to the use of freshwater further upstream for crop irrigation and industrial use, and to increased sedimentation (Jiménez and González 2001). The main seasonal biological event in the Tempisque River is the upstream migration of the sardine

Astyanax fasciatus about two weeks after the end of the rainy season (López 1978). The photographs (López 1978) of the 1971 massive fish migration just remind us of better times in the past.

Physical Characteristics, or Why the Gulf of Nicoya Is an Estuary

An estuary is the mouth of a river or arm of the sea where the tide meets the river currents (Emery and Stevenson 1957). At a river mouth, such as the Tempisque River mouth (Fig. 6.1), the salinity gradient between freshwater and the sea fluctuates daily with the level of the tide, and also varies seasonally with the amount of freshwater coming downstream (Fig. 6.2). In this and other estuaries there is also a surface-to-bottom salinity gradient, with lighter freshwater forming a layer above the heavier saline water column. This layer is more prominent during the rainy season (May to November), when the Gulf is vertically stratified, than during the dry season (December to April), when strong winds break the stratification and increase mixing.

Vertical stratification keeps nutrient-rich Equatorial Sub-surface Water at deeper levels below the thermocline over most of the lower section of the Gulf, and higher tidal energy dissipation on the mid-gulf appears to be mixing these nutrients over the water column (Chaves and Birkicht 1996). The vertical distribution of temperature and salinity in the Gulf indicates that the mid to upper estuary regions are highly stratified during the rainy season, and much less stratified, or even well mixed, during the dry season. The saline field has a maximum during April, and minimum salinity values are found during the peak of the rainy season (Fig. 6.2) in October-November (Brenes et al. 2001). River water often carries large amounts of silt and clay that become deposited at the river mouth, and this input of fine sediments and associated compounds increases during the rainy season.

In July 1981, the concentration of sediments within the surface and near-surface waters on the upper gulf was found to range from 20 mg/L at the mouth of the Tempisque River, to near 3 mg/L near the port of Puntarenas (Klemas et al. 1983). The lower gulf shores differ on the input of nutrients and sediments carried by the Tárcoles and Barranca rivers on the eastern shores compared with the western shores, where no major rivers are found. The sediment plume of the Tárcoles River, which drains the heavily populated central Metropolitan Area of the country, was found to extend an average of 10 km off-shore, and had sediment concentrations ranging from 1.0 mg/L off-shore to more than 10 mg/L at the mouth (Klemas et al. 1983). The freshened surface water from the upper gulf combines with the discharges of these rivers and flows towards the ocean in

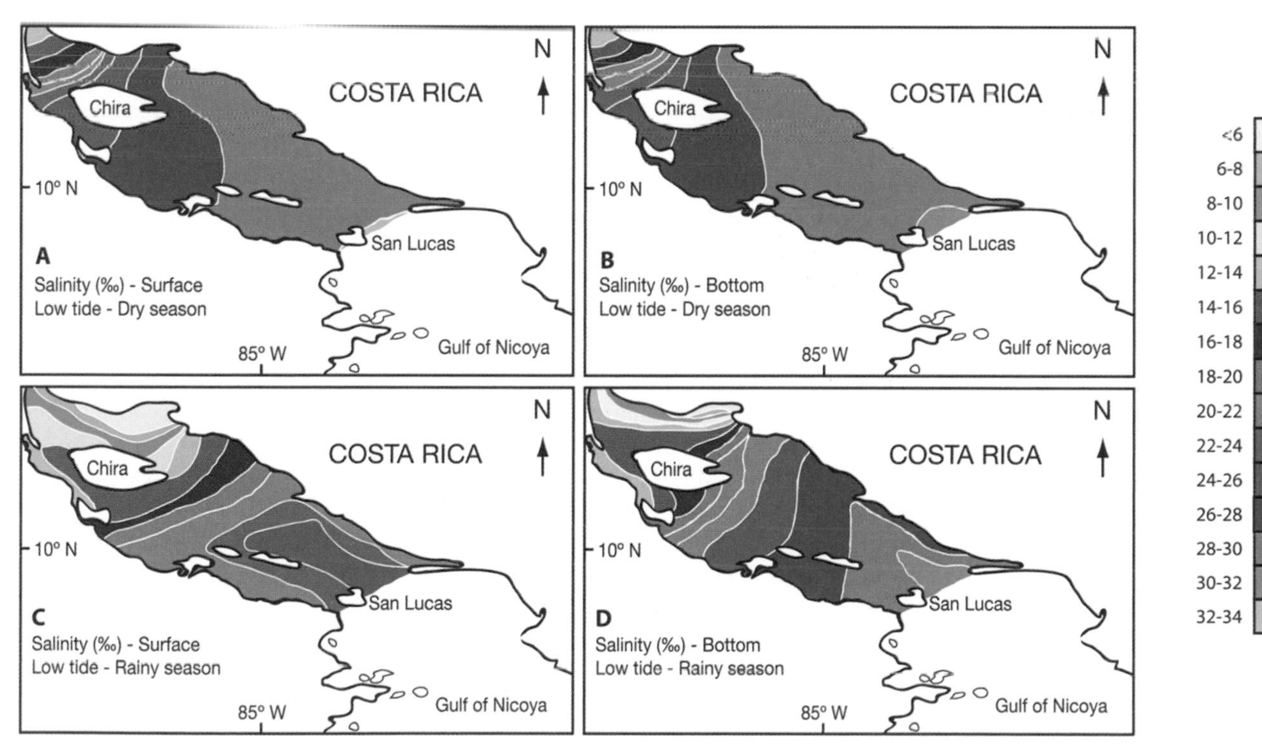

Fig. 6.2 Salinity (parts per thousand) gradients along the axis of the upper region of the Gulf of Nicoya during the dry and rainy season. *Modified from Lizano and Vargas (1993).*

the surface layer along the eastern shore (top 20 m). In a typical estuarine circulation, this outflow is compensated by an inflow of more saline waters on the western side at all depths, and on the eastern side by waters deeper than 20 m (Voorhis et al. 1983). This is why debris from the Tárcoles River is carried towards the ocean when floating on the surface layer, but when it sinks to the bottom layer it may be brought back into the system by bottom currents.

Salinity values close to those of brackish water are found north of Chira Island during high discharges of the Tempisque River in the rainy season (Figs. 6.1, 6.2). A salinity gradient is developed during low tide, while a high freshwater river discharge is coupled with an upstream bottom flow (salt wedge) that reaches the head of the estuary (Lizano 1998). Moreover, a study by Lizano and Alfaro (2004) supports the view that the spatial and vertical variation of water currents in the Gulf of Nicoya is not only related to the temperature-salinity structure, but also to the tidal cycles and tidal ranges characteristic of this estuary, where currents with magnitudes over 1m/sec were measured at a point between San Lucas Island and the port of Puntarenas (Fig. 6.1), during spring tides.

Water Chemistry and Primary Productivity: What Fuels the System?

The Gulf of Nicoya differs from most temperate estuaries in that seasonality is more related to salinity changes due to seasonal rainfall, than to seasonal changes in water temperature. Seasonal changes in concentrations of nutrients (nitrate, nitrite, ammonium, and silicate) have been described by Epifanio et al. (1983), who found that the Gulf differs from most temperate estuaries in that much of the nitrogen entering the ecosystem is from off-shore deep Pacific waters transported across the shelf into the Gulf. More recently, Tabash-Blanco (2007a) indicated that the Gulf is a net source of dissolved inorganic nitrogen that varies seasonally.

There is a balance between surface waters descending to the bottom and upwelling waters that carry nutrients and other chemical elements to the surface. An important source of nitrogen into the system is due to the input by the Tárcoles and Tempisque rivers. These rivers contribute nitrogen in the form of nitrate ions having great spatial and temporal variability owing to different oxidation stages. This author argues that this variability explains the temporal and spatial changes of primary production in the Gulf (Tabash-Blanco 2007a). Palter et al. (2007) provide data in support of the hypothesis that discharges from the Tempisque River play a major role in the spatial and temporal variability of nutrients in the upper estuary.

Primary production is among the highest reported for tropical estuaries worldwide. Córdoba-Muñoz (1998) measured net primary production at a mangrove creek near the port of Punta Morales (Fig. 6.1), and found values close to 450 g C/m^2/year, while Gocke et al. (2001a) found a net primary production of 278 g C/m^2/year at a nearby site. They also studied the annual cycle of primary productivity near the tip of the Punta Morales pier, and found that this region of the Gulf is an extremely productive, phytoplankton-dominated system, with an annual gross primary production of 1,037 and a net primary production of 610 g C/m^2/year. Respiration was estimated at 427 g C/m^2/year. Peaks of primary production coincided with blooms of red tide–forming algae. On an annual basis they estimated that 41% of the organic carbon produced in the system is already consumed in the euphotic layer, and close to 80% is consumed in the less-than-10 m deep water column in the upper gulf (Fig. 6.1), with the remaining production probably reaching the benthos (Gocke et al. 2001b). The early work by Ryther (1969) provides reference values for worldwide marine primary production: for the open ocean 50 g C/m^2/year, for the coastal zones 100 g C/m^2/year, and for upwelling zones 300 g C/m^2/year. In an unpublished survey (Fig. 6.3), the Gulf stands as a unique ecosystem when compared with three other coastal embayments of Costa Rica: the Golfo Dulce estuary and Culebra Bay, both on the Pacific coast, and the Moín Bay on the Caribbean coast. Incubations of phytoplankton in light and dark bottles at these environments yielded net productivity estimates as high as 1,500 g C/m^2/year for the Gulf of Nicoya (Fig. 6.3). The Gulf was the only system where all incubations yielded positive balances between production and respiration. The high productivity of the Gulf of Nicoya, transferred to other trophic levels, explains in part its high fisheries production. In this tropical estuary photosynthesis is possible year-round, since there is no seasonal low irradiance or low temperature limitations, as in temperate estuaries. Nixon (1995) classified estuaries as eutrophic if primary productivity levels were around 400 g C/m^2/year. On the basis of Nixon (1995) and of a recent review of primary productivity in estuaries done by Cloern et al. (2014), several of the values included in Fig. 6.3 for the Gulf of Nicoya may be considered as indicative of eutrophic to hypertrophic conditions.

The importance of the relative contribution of intertidal microalgae to the total primary production of the Gulf of Nicoya was put forward by Vargas (1996), but it remains as yet unaccounted for. Therefore, the productivity of this tropical ecosystem might be even greater than the available information indicates. The contribution of benthic microalgae such as pennate diatoms to the primary productivity of estuaries has been a research area of recent development,

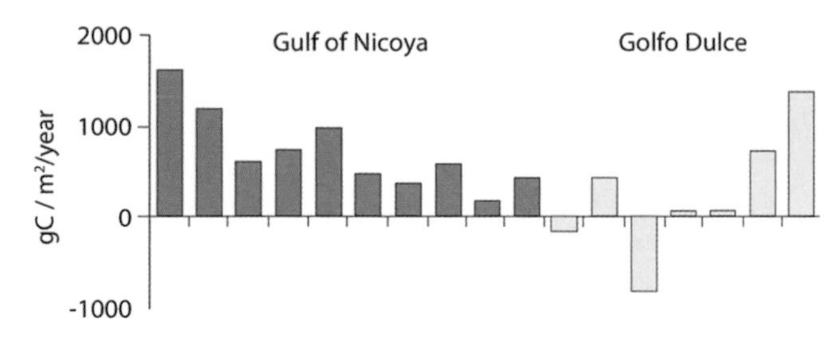

Fig. 6.3 Primary net productivity (positive values), and respiraton (negative values), at two estuarine environments on the Pacific coast (Gulf of Nicoya: Punta Morales - Golfo Dulce: Golfito Bay mouth) of Costa Rica. Data obtained using the light-dark bottle method. Productivity values extrapolated to g C/m^2/year. At least four of the estimates for the Gulf of Nicoya may be indicative of eutrophic to hypertrophic conditions. Unpublished results of CIMAR-CoCosRi survey (March-November, 2000).

and field methods are still being developed (Kromkamp and Forster 2006). As pointed out by Cahoon (2006), microphytobenthic production can equal or exceed phytoplankton production, supports significant secondary production, alters the properties of shallow sediments, and plays a key role in nutrient cycling in these ecosystems. An evaluation of several studies on phytobenthos production in the tropical regions yielded a mean of 527 g C/m^2/year (Cahoon 2006).

There is only one study in the Gulf of Nicoya focusing on production-consumption processes in the sediments. It was conducted by Gocke et al. (1981), who found that decomposition of organic matter in mangrove sediments was near 400 mg C/m^2/day. More studies are needed to estimate benthic production-decomposition rates on intertidal muddy and sandy flats of the Gulf of Nicoya. Much-needed updated estimates could be included in future runs of the trophic model already available (Wolff et al. 1998) for this estuary.

The Gulf of Nicoya as a Single Ecosystem

The term *ecosystem* is frequently used to include all relationships between organisms of a given area and their interactions with the physical environment (Bush 2000). This definition poses a problem when the spatial limits of the ecosystem need to be clearly defined for management purposes. Moreover, the hierarchical components of an ecosystem operate at different temporal scales, which make management goals even more difficult to achieve for the system as a whole (Wolff 2009). In the case of the Gulf of Nicoya estuary, its operational limits are clearly marked at the water edge and the upper intertidal zone, where an ecotone or transition zone separates terrestrial ecosystems from estuarine waters. However, since the ecosystem definition implies interaction with the physical environment such limits become less practical at river mouths, where freshwater inputs play a major role on the functioning of the estuary through the influx of nutrients, sediments, and pollutants. The connection of the Gulf of Nicoya with dry forest ecosystems via the Tempisque River (see Jiménez and Carrillo, chapter 11 of this volume) has been emphasized by Vargas and Mata (2004).

Ecosystems are considered as hierarchically organized, with communities as the next lower level (Wolff 2009). In the Gulf of Nicoya we could refer to benthic and planktonic communities. The methods used for the study of plankton communities (plankton nets) differ from those of benthic ecologists (grabs, dredges, cores), and from those of fish biologists (trawls, nets). In the following section I will mention, without being exhaustive, published results on the most important communities of the Gulf of Nicoya ecosystem. In this context the recent book on the marine biodiversity of Costa Rica (Wehrtmann and Cortés 2009) provides updated information on early and recent taxonomic and ecological research conducted in the Gulf of Nicoya in many groups of invertebrates, vertebrates, benthic algae, phytoplankton, and coastal vegetation (see also Cortés, chapter 5 of this volume).

Communities and Ecotones

Mangroves

The description of mangrove systems has been covered in more detail by J. Cortés (this volume; see chapter 5 on Pacific coastal and marine ecosystems; and see J. Jiménez, chapter 20 in this book). Mangrove forests have been the center of more research efforts than any other estuarine ecosystem in the Gulf of Nicoya. These forests are relatively easy to access, and the standard sampling methods do not require heavy gear or its operation from a research vessel. In addition, the objects of study (the trees) are visible to the researcher, contrary to the blind approach when sampling subtidal benthos, plankton, and fish.

Jiménez and Soto (1985) described the regional patterns in the structure and floral composition of the Pacific coast mangroves of Costa Rica. The core vegetation of the mangroves is composed of several species of trees, among these: *Rhizophora mangle, R. harrisonii, Avicennia bicolor, A. germinans, Laguncularia racemosa,* and *Pelliciera rhizophorae. R. mangle* (red mangrove) grows mostly at the edges of tidal channels and the roots are colonized by a diverse fauna of which boring isopods, like *Sphaeroma peruvianum*, and the encrusting barnacle, *Balanus* spp., exert a certain influence on the growth and productivity of this tree species (Perry 1988). Soto (1988) focused on the geometry, biomass, and leaf life-span of the black mangrove (*A. germinans*), at the small Salinas mangrove swamp near the port of Caldera (Fig. 6.1). Here he found 9 cm tall, dwarfish reproductive plants, and trees of 6 m height growing along a salinity gradient. The dwarfish mature trees were found in high-salinity areas.

The linkages between the Gulf's mangrove forests and other components of the ecosystem are just beginning to be understood. Preliminary evidence points to the importance of mangroves as a refuge, or as nursing grounds, for several species of fish (Ramírez et al. 1990), to the use of floating mangrove leaves as a transport mechanism of estuarine organisms like decapod crustaceans (Wehrtmann and Dittel 1990), to the flux of crab larvae in mangrove creeks (Dittel et al. 1991), as well as to the influence of pufferfish predation on the vertical distribution of mangrove littorinid snails (Duncan and Szelistowski 1998).

An attempt has been made to integrate in a trophic model suitable information on mangrove species and the ecology known from the Gulf of Nicoya. The resulting model (Wolff et al. 1998) is the only one available for the Gulf and one of the first developed for a tropical estuarine ecosystem. The model emphasizes that detritus matter exported from the mangroves to the estuary feeds an aquatic food web, with penaeid shrimps and other detritivores in the center of the web.

Intertidal Sediments

In the Gulf of Nicoya there is usually a sharp ecotone between the edge of the mangrove forest and other environments, and the edge of intertidal flats. Most of the information on intertidal soft-sediment communities from the Gulf is based on studies conducted at the sand-mud sediments near the Punta Morales port (Fig. 6.1), and summarized in Vargas (1995, 1996). Intertidal benthic research has focused primarily on the description of structural aspects of communities (variation in space and time of the numbers of individuals and species), while much work remains to be done on functional aspects (energy flow). Statistical multivariate methods applied to a three-year (1984–1987) data set revealed a seasonal (dry vs. rainy) pattern of the macrofaunal (= organisms retained on a 500 micron mesh) community, with several species showing marked peaks of abundance (Vargas 1987, 1988b, 1989b). The community was numerically dominated by deposit-feeding polychaete worms, with characteristic species like *Paraprionospio pinnata* and *Mediomastus californiensis*. The density of individuals per square meter ranged from about 4,000 to more than 40,000, and diversity ranged from 1.75 to 3.36 as estimated, using Shannon-Wiener's diversity function H (Vargas 1987). More than a hundred species of benthic macrofaunal organisms were identified belonging to the Polychaeta, Crustacea, Mollusca, and other groups. The reviews by Dean (2008, 2009) on the polychaete species so far reported for Costa Rica and on their use as indicators of pollution have clarified many taxonomic problems. Most polychaete records from Costa Rica come from the Gulf of Nicoya. Population fluctuations of some species of polychaetes from the Gulf were as high as those reported from temperate estuaries (Vargas 1987,1988a,1989b). Recently, Vargas and Dean (2010) described long-term abundance patterns of the lancelet *Branchiostoma californiense* in subtidal and intertidal habitats in the Gulf of Nicoya, and Vargas and Solano (2011) included data on population fluctuations of two species of intertidal echinoderms collected during 1984–1987 and 1994–1996. More recently, Vargas-Zamora and Sibaja-Cordero (2011) and Vargas-Zamora et al. (2012) provided evidence for long-term fluctuation patterns of intertidal mollusks and crustaceans collected during the same period in the Gulf. These types of long-term data sets are rare for tropical estuaries.

A survey of the hard (formalin-preserved) meiofaunal component (= organisms retained on a 62 micron mesh) conducted at the sand-mud Punta Morales intertidal flat revealed that Nematoda, Foraminifera, Copepoda, and Ostracoda accounted for more than 90% of the number of individuals and that most of the meiofauna was restricted to the top 2 cm in the sediments. Kinorhynchs (Phylum Kinorhyncha) were reported for the first time for this tropical estuary (De la Cruz and Vargas 1987, Vargas 1988b). No attempt was made to identify the meiofaunal organisms at the species level, as meiofaunal studies in the tropics are at an early stage, and taxonomic literature deals mostly with temperate faunas. However, some species of benthic copepods have been described by Mielke (1997), from intertidal sediments of the Gulf of Nicoya.

Caging experiments aimed at evaluating the role of predators in regulating community structures were performed at the Punta Morales sand-mud flat by Vargas (1988a). Caging experiments designed to exclude predators in temperate and subtropical latitudes have resulted, with few exceptions, in an increase in the number of individuals and species inside the protected areas. The observed increases have been related to a diminished role of predation, thus allowing prey species to reach higher abundances inside the caged areas. This expected result was not observed in the Gulf of Nicoya when total number of individuals and species were considered. Vargas (1988a) concluded that the role of disturbance in community structure might be more important than the role of predators on a community where most of the species are sediment re-workers, such as many polychaetes, crustaceans, mollusks, and rays that feed on or in the sediments. Much remains to be done to better understand the roles of the macrofaunal, meiofaunal, and microfaunal (= organisms smaller than 62 microns), components on the functioning of the benthic communities, and their role in the estuarine ecosystem. Moreover, Vargas (1996) compared the impact of caging on benthos at Punta Morales (10° N, Costa Rica) and Rehoboth (38° N, eastern USA), and showed that the number of individuals (polychaetes, crustaceans, mollusks) inside cages at the end of the experiments was 1,414 at the tropical site and 8,791 at the temperate site, while numbers at the beginning of the experiment were 1,201 and 1,924, respectively.

Very little research has been conducted in sandy sediments of the Gulf. The report by Dexter (1974) is indeed a pioneer study describing the fauna of a high-energy sandy beach near the tip of the Puntarenas peninsula (Fig. 6.1). Dexter found that the intertidal fauna was characterized by the isopod *Cirolana salvadorensis* (600 individuals/m^2), the sand crab *Emerita rathbunae* (300/m^2), the bivalve *Donax panamensis,* the snail *Olivella semistriata*, and the polychaete worm *Scolelepis agilis* (250/m^2).

Pereira (1996) assessed the impact of migratory shorebirds preying on the fauna of an intertidal sand flat located halfway between the ports of Puntarenas and Punta Morales (Fig. 6.1). A total of 27 species of charadriiform birds were observed in the study area, with as many as 14 species and 713 individuals at a single count.

A study from a sandy shore was conducted at the Cocorocas sand flat (Dittmann and Vargas 2001), near the port of Punta Morales. Here, the fauna included several species of polychaete worms shared with a nearby sand-mud flat. The fauna of the sedimentary environments in the Gulf was compared by Dittmann and Vargas (2001), who noted a similar fauna in comparable environments in Australia. This comparison revealed a high degree of similarity despite the different biogeographic regions. Species numbers were moderately high, and the polychaete worms, mollusks, and crustaceans dominated the macrofauna at both sides of the Pacific. Most of the species were deposit feeders. An interesting finding is that related species play similar ecological roles at both sides, and only a few cases were identified in which an ecological counterpart did not exist in both the Gulf of Nicoya and the Australian tidal flat communities. For instance, the counterpart for the brachiopod *Glottidia audebarti* (Emig and Vargas 1990) is the brachiopod *Lingula* sp. in Australia. The predatory role played in the Gulf by the snail *Natica unifasciata* is assumed by *Natica* spp. snails in Australia. Related species of the fiddler crab, *Uca* spp., were found performing similar roles at both sites. The same is true for several species of hemichordates, echinoderms, and benthic feeding fish, like rays. These comparisons are useful to understand how benthic ecosystems function worldwide, and to understand the status of ecosystems along environmental gradients of human impact. The relative importance of predation on the benthos itself, and the similarity of trophic roles played by invertebrate species on a mud flat in the Gulf in comparison with similar species present in sandy and muddy sedimentary environments along the eastern shores of the United States, has been discussed by Vargas (1996).

Subtidal Sediments

Subtidal sediments comprise the whole area that is permanently submerged in the Gulf of Nicoya. Therefore, the benthic communities here cover most of the area of the Gulf and grade into each other. Information available on these communities comes from surveys conducted aboard the *RV Skimmer* in the early 1980s (Maurer and Vargas 1984, Maurer et al. 1984, 1988). Infaunal (organisms living *in* the sediments) communities were sampled using grabs taken at 42 stations distributed over most of the Gulf of Nicoya. The survey yielded about 5,000 individuals belonging to 205 species, of which the polychaete worms were represented by 120 species (58%), the crustaceans by 46 (22%), and the mollusks by 22 (11%). The maximum density found was 8,744 ind/m^2, and maximum diversity measured using the Shannon-Wiener function H was 3.09, which was considered not as high as expected for a tropical estuary, probably due to the numerical dominance of certain species like the polychaete worm *Mediomastus californiensis*. The subtidal sediment fauna of the Gulf of Nicoya was characterized by broadly occurring species fluctuating in abundance with change in depth and sediment type.

To provide some measure of seasonal trends, four stations located near the Tárcoles River mouth were sampled over five times during the period running from October 1980 to August 1981 (Maurer et al. 1988). Twenty numerically dominant polychaete species indicated three basic abundance patterns: species with a marked seasonal peak and subsequent stable densities, species with a moderate seasonal peak and subsequent stable densities, and species with no discernible seasonal peak. These seasonal patterns were similar to those found in intertidal sediments (Vargas 1987, 1988a).

The soft bottom epifaunal (organisms living *on* the sediments) communities were evaluated using a modified shrimp-trawl net dragged over the bottom (Maurer et al. 1984). A total of 98 species comprising more than 56,000 individuals were collected at 20 stations. Decapod crustaceans were the dominant taxon (55%), followed by mollusks (30%), and echinoderms (9%). Regardless of region or season there was considerable station-to-station variability in abundance and biomass. The most common species in the lower Gulf were the penaeid shrimps, *Penaeus brevirostris, P. stylirostris*, and *Trachypenaeus byrdi*. Thirty-four species of crabs and lobsters were collected as well as five species of stomatopods.

Among the decapods, the portunid crab *Callinectes arcuatus* was of particular interest owing to its potential for fisheries (DeVries et al. 1983, Dittel et al. 1985). This crab plays a role that is ecologically very similar to that of its Atlantic counterpart *C. sapidus*. Moreover, a comparison of the dispersal of crab larvae in the Delaware bay (USA) and the Gulf of Nicoya found that *Callinectes* spp. crabs spawn in the lower reaches of estuaries in both regions, and their larvae are exposed to predation and transport out of the estuary (Epifanio and Dittel 1982). These findings provide useful information for the management of *C. arcuatus* in the Gulf. A study on *C. arcuatus* in the upper Gulf was performed by Fischer and Wolff (2006), who found that a proposed maximum fishing effort of 1,600 traps is unlikely to be sustained by this species, since a decrease in the proportion of large males in the catches has already been observed in the past years under a lower fishing regime. Another important crustacean group concerns the predatory mantis shrimps (Stomatopoda), whose distribution and sexual composition in the Gulf has been evaluated by Dittel (1991).

About a decade after the *RV Skimmer*'s surveys, a new benthic study was conducted in the Gulf of Nicoya aboard the *RV Victor Hensen* (Vargas and Wolff 1996). The benthic communities were studied with particular emphasis on mollusks (Cruz 1996), polychaete worms (Dean 1996), crustaceans (Jesse 1996), and fish (Bussing and López 1996,Wolff 1996). Most of the information gathered during this survey was used as input for the development of a trophic model for the Gulf of Nicoya (Wolff et al. 1998).

Rocky Intertidal Substrates

Rocky intertidal shores are found mostly on the western side of the Gulf of Nicoya, usually bordering small islands (Fig. 6.1), or forming rocky cliffs on the mainland shores. The study of the fauna associated with hard substrates in the Gulf of Nicoya was initiated by Paine (1966), who described the food web at a rocky site near the port of Caldera. Villalobos (1980) studied the population structure of the intertidal barnacles *Tetraclita rubescens* and *T. stalactifera*. A study by Cutler et al. (1992) focused on these hard substrates as habitats preferred by the peanut worms *Phascolosoma perlucens* (in rocks) and *Antillesoma antillarum* (between rocks). The list of sipunculid species found in Costa Rica has been updated by Vargas and Dean (2009). The patterns of abundance of the porcellanid crab *Petrolisthes armatus* was described by Diaz-Fergusson and Vargas-Zamora (2001), and by Diaz-Fergusson et al. (2008) from a rocky beach near Punta Morales in the mid-upper gulf. This crab is found in high numbers under rocks at low tide, and reproduce continuously year-round, with some peaks of abundance during the dry and rainy seasons. A study of the vertical distribution (vertical zonation) of the communities at four rocky intertidal localities in the Gulf of Nicoya was performed by Sibaja-Cordero and Vargas-Zamora (2006), and data were analyzed using multivariate statistical methods. Zonation patterns were found at the four sites, in which the lower zone was characterized by the presence of several species of algae, and the upper band was occupied by several species of barnacles and small snails. Cluster analyses summarizing the vertical zonation of the biota at two localities on opposite coasts of the lower Gulf of Nicoya (Caldera Bay and Montezuma Beach) are presented in Fig. 6.4A and 6.4B.

Plankton Communities

Phytoplankton

The first survey oriented to identify the main components of the phytoplankton communities in the Gulf of Nicoya was published by Hargraves and Víquez (1985). Their survey included seven stations distributed from Chira Island to the mouth of the Tárcoles River (Fig. 6.1). These authors sampled during approximately weekly intervals from January to November 1982 and found that planktonic microalgae were dominated by diatoms, of which *Chaetoceros* (23 spp.), and *Rhizosolenia* (10 spp.) were most diverse, and by dinoflagellates that form red tides (*marea roja*). The most abundant diatom was *Skeletonema costatum*, followed by *Cylindrotheca closterium*, *Nitzschia pungens*, and *R. stolterfothii*. Total diatom species abundance was highest during the rainy season. There was a conspicuous similarity between diatom floras of the upper and lower sections of the estuary, when compared with other eastern Pacific coastal waters.

Red tides observed during 1979–1984 shared the dinoflagellate *Cochlodinium catenatum*, being the most common species, while species like *Gonyaulax digitale*, *Ceratium furca*, and *Noctiluca scintillans* were found accompanying the dominant *C. catenatum*. The annual cycle of potentially harmful dinoflagellates in the Gulf of Nicoya was described by Víquez and Hargraves (1995). A one-year evaluation of the phytoplankton community at a station near the port of Punta Morales (Fig. 6.1) by Brugnoli-Olivera and Morales-

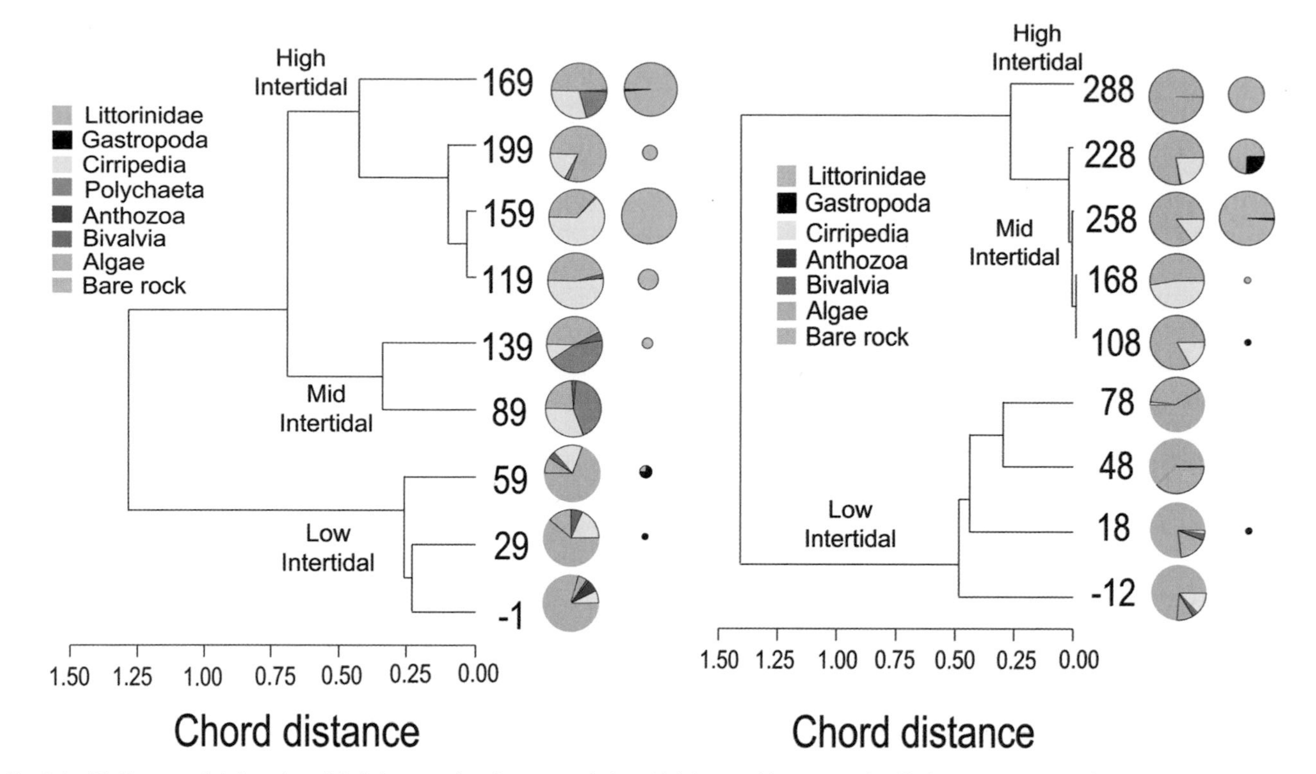

Fig. 6.4 (A) Cluster analysis based on digital photographs taken at a rocky intertidal site at Caldera Bay, mid Gulf of Nicoya. Numbers refer to the tidal levels referred to low tide (0.0 m chart datum) at the port of Puntarenas. The high intertidal zone was characterized by a high proportion of littorinid gastropods. The mid intertidal zone was dominated by barnacles (Cirripedia), and the reef-building sabellariid polychaete *Phragmatopoma attenuata*. The low intertidal zone was characterized mainly by algae (mainly Rhodophyta). (B) Cluster analysis based on digital photographs taken at a rocky intertidal site at Montezuma, near Ballena Bay, lower Gulf of Nicoya. Numbers refer to the tidal levels referred to low tide (0.0 m chart datum) at the port of Puntarenas. The high intertidal zone was dominated by littorinid gastropods, the mid intertidal zone by barnacles, and the lower intertidal zone by algae (mainly Phaeophyta and Chlorophyta).
Modified from Sibaja-Cordero and Vargas-Zamora (2006).

Ramírez (2001) yielded a total of 43 taxa, with centric and pennate diatoms and flagellates representing 90% of the total abundance and *S. costatum* represented by 2%. In the nano-phytoplankton fraction (cells less than 30 microns diameter), *Chaetoceros* spp. cells were most abundant, followed by flagellates (23%). The number of nanophytoplanktonic cells varied seasonally, probably owing to nutrient availability.

Zooplankton

The study of zooplankton communities in the Gulf of Nicoya started as part of the evaluation of populations of the blue crab *Callinectes arcuatus* (Epifanio and Dittel 1982), and brachyuran crabs (Epifanio and Dittel 1984). The importance of mangrove creeks as sites for the development of certain stages of the life cycle of estuarine crustaceans was evaluated by Dittel and Epifanio (1990), and Dittel et al. (1991). A study of copepods—the most common microcrustaceans in plankton worldwide—was initiated by Morales and Vargas (1995).

Because of their high abundance in estuarine waters, copepods are key components of the food web, and include many species that feed on small phytoplankton algae like diatoms. Other copepod species behave more like carnivores and feed on invertebrate larvae and other invertebrates, like arrow worms (Chaetognatha). The copepods in turn are eaten mainly by planktivorous fish. A great amount of knowledge on the dynamics of the plankton communities of the Gulf of Nicoya was obtained during the *RV Victor Hensen* expedition in 1993/1994 (Vargas and Wolff 1996). As part of this expedition, Morales-Ramírez (1996) conducted a study of the copepod fauna of the Gulf using a 200 micron mesh net. Results showed that the Gulf waters were dominated in its upper part by neritic estuarine copepod species like *Acartia lilljenborgii*, *Paracalanus parvus*, and *Hemicyclops thalassius*, as well as species in the genus *Pseudodiaptomus*. An oceanic group of species was observed in the lower gulf where small *Oncaea venusta* as well as larger species like *Pleuromanna robusta*, *Eucalanus attenuatus*, *E. elongatus*, and *Rhincalanus nasutus* were common. Other plankton components were studied by Von Wangelin and Wolff (1996), who found that bivalve larvae, foraminifers, ostracods, and mysid crustaceans as well as nauplii larvae increase in abundance in comparison with more open-water conditions. In the Gulf, inshore plankton is neritic. The dominance of small calanoid copepods and the high abundance of fish eggs and invertebrate larvae suggest that this estuary is indeed an important spawning site.

The first study on the distribution and abundance of carnivorous arrow worms (Chaetognatha) in the Gulf of Nicoya was conducted by Hossfeld (1996), who was able to identify nine species in the genus *Sagitta* and one in the genus *Kronitta*, with *S. enflata* having the highest frequency of occurrence in the middle and lower Gulf of Nicoya regions. She found that only *S. friderici* was associated with lower salinity and near-shore waters and occupies the inshore part of the estuary as a unique species. The dominance of juveniles and small individuals at the inshore stations reflects high reproduction rates, and a high energy flow through chaetognaths preying on plankton. Hossfeld (1996) also found an increase of chaetognath biomass during the change from the rainy season to the dry season conditions and points to high zooplankton production during this time of the year. This is another evidence for strong seasonality in the estuarine ecosystem.

The study of the fish larval stages (ichthyoplankton) started in the Gulf of Nicoya by López and Arias (1987) who identified 20 fish larval types from a mangrove creek in Ballena Bay, and by Ramírez et al. (1989,1990) who described larval recruitment of anchovies, and the composition and abundance of fish larval stages at a mangrove estuary. Molina-Ureña (1996) as part of the *RV Victor Hensen* expedition, was able to identify in the estuary a neritic assemblage of larval fish characterized by engraulids, sciaenids, and gobiids, as well as an oceanic group dominated by myctophids, bregmacerotids, ophiidids, and trichiurids. The influence of the El Niño Southern Oscillation (ENSO) on the 1997–1998 plankton dynamics of the Gulf has been evaluated by Morales-Ramírez and Brugnoli-Oliveira (2001). They found that ENSO affected little the zooplankton communities, but influenced species composition and nutrient assimilation of phytoplankton during months with higher surface water temperatures (up to 31.5°C).

Fish Communities

The Gulf of Nicoya is the main fishing ground of Costa Rica and (over)fishing has heavily impacted its estuarine ecosystem since the mid-twentieth century, when shrimp trawling started. In fact, the removal of shrimps and associated fauna including fish predators from the ecosystem has had a direct impact on the tropho-dynamics of this estuary, as the trophic model predicts (Wolff et al. 1998). Fish harvested in the Gulf are sold in most of the main cities in the country, and species of high commercial value include top predators like corvina (Sciaenidae) and red snapper (Lutjanidae), but also sharks.

The first detailed evaluation of the ichthyofauna of the Gulf of Nicoya was conducted by León (1973), and updated using trawl fishing by Bartels et al. (1983, 1984). However, a new survey is urgently needed to evaluate changes in recent years. Bartels et al. (1983, 1984) collected a total of 29,812 individuals in the estuary, representing 214 species found in trawl hauls during three cruises conducted by the ship *RV Skimmer*. The sciaenids and sea catfishes were most important in terms of biomass while *Stellifer* sp., *Porichthys* sp., and several flounder and catfish species contributed to the greatest number of individuals.

Two major types of fish distributional patterns were observed: several species were ubiquitous and were found throughout the gulf in varying abundances, while other species were restricted to either the upper or the lower Gulf. The upper Gulf was characterized by sciaenids (corvinas), sea catfishes (Ariidae), and flatfishes (several families). The deeper regions of the lower estuary included flounders, gobies, morays and congers, and several other species.

In his early study, León (1973) found that fish diversity (as measured by *H*) increased from the Tempisque River mouth (2.3) to the mid Gulf (2.8), while Bartels (1984) found H values ranging from 2.9 to 3.2 in the upper Gulf. These values were considered higher than those found in non-tropical environments. The identification of fish aboard the *RV Victor Hensen* yielded a total of 118 species of fish caught by dredges and benthic trawls (Bussing and López 1996). Wolff (1996) performed a multivariate analysis of fish data collected during the *RV Victor Hensen* survey and found that the estuary could be divided into three areas: (1) an interior shallow area above the thermocline (less than 50 m deep) characterized by sciaenids, sea catfishes, stingrays, flatfishes, and sea robins; (2) an outer area (deeper than 100 m) characterized by cods, scorpion fishes, gobies, cutlass fishes, serranids, anglerfishes, and flatfishes; and (3) a transition zone composed of the central and lateral parts of the estuary, with a mixed species assemblage with carangids, pufferfish, snappers, several flatfish species, and lizardfish.

As in the studies done by Bartels et al. (1983, 1984), the survey aboard the *RV Victor Hensen* did not include any sample from rocky areas; thus, fishes associated with reefs were not collected, nor were those found in very shallow sedimentary shores, or in tide pools. Consequently, the fish fauna of the Gulf may be more diverse than the species so far reported by both ship surveys. A study by Phillips (1983) on the diversity of fish at a littoral site describes the diurnal and monthly variation within the fish fauna near the port of Punta Morales (Fig. 6.1). The survey by Rojas et al. (1996), on the other hand, describes the fish diversity in mangrove areas in the upper section of the Gulf. In general, the earlier findings by León (1973), Phillips (1983), and Bartels et al. (1983, 1984) regarding the dominant groups in the upper gulf (sciaenids, catfishes, and flatfishes), and those of the lower gulf (flounders, gobies, morays and congers among others), were confirmed by these studies. It is important to mention that in recent years new species of fish have been described from the estuary, like the ray *Urotrygon cimar* (López and Bussing 1998), as well as several new stingray parasites like *Acanthobotrium nicoyaense*, which is named after this ecosystem (Brooks and McCorquodale 1995).

The first attempt to understand the trophic relationships among the fishes of the Gulf of Nicoya was made by Campos and Corrales (1986). On the basis of data collected monthly with a bottom trawl at eight stations in the Gulf of Nicoya, and the analysis of the stomach contents of 21 species of fish, the authors found that crustacean prey items represented 58% of the total prey organisms (46% were shrimps), followed by other species of fish (18%), ophiuroid echinoderms (7.8%), polychaete worms (7.4%), and mollusks (3.6%). From the 21 species of fish, 52% were classified as omnivores, and 48% carnivores. A recent study along this line of research is by Espinoza and Wehrtmann (2008) who analyzed the stomach contents of *Lophiodes spilurus*, an anglerfish associated with the deep water shrimp (*Heterocarpus vicarius*) fishery near the entrance (100 to 240 m depth) to the Gulf of Nicoya. They found that this fish preys only on crustaceans (30%) and teleost fish (70%).

Trophic Modeling

There are many ecosystems in Costa Rica, but attempts to integrate all the useful information into models are rare, in spite of the potential of modeling as a tool for management purposes. The ECOPATH methodology (Wolff et al. 1998, Wolff 2009) was applied to available data from the Gulf of Nicoya, and the resulting trophic model (Fig. 6.5) provides a useful management tool to understand how this estuarine ecosystem works. The model is of the steady-state type, in which biomass production of—and imports to—the system compartments is balanced by consumption and exports. The size of each box is proportional to the square root of the compartment biomass. According to the model the ingestion by detritus and phytoplankton feeders alone represents 82% of the total Gulf of Nicoya ecosystem consumption.

Mangroves, although covering a small relative area of the Gulf, contribute 75% to the system biomass, but only 1% to the primary production. Epibenthos, shrimps, and small demersal fish are the most important groups in terms of biomass, while plankton, small pelagic fish, carangids, and squids dominate the pelagic biomass. Shrimps have a central role in the model as converters of much of the system detritus and other bentho-pelagic food into food biomass for several shrimp predators. Thus, their overexploitation means a significant reduction of the food stock for these predators (Wolff et al. 1998).

According to Wolff (2006), it is improbable that other species can compensate for the removal of shrimps from the Gulf, and the decline of many commercially important populations of shrimp-feeding species seems a logical consequence of this overexploitation. The removal of shrimps and other components (cnidarians, mollusks, crabs, echinoderms, and fish) from the system (known as *by-catch*), due to shrimp harvesting, will slowly bring the ecosystem to a state where most of the energy fixed by the primary producers will settle to the sediments. Then, recycling of this energy will have to be done mostly by an enhanced microbial food web.

The Gulf of Nicoya's trophic model was compared with a similar model (Wolff et al. 1996) available for the Golfo Dulce, the second largest estuarine embayment of Costa Rica on the Pacific coast. In the deep Golfo Dulce (200 m in its core basin, and therefore a "tropical fjord"), phytoplankton and benthic production is low and the pelagic community seems to be more tightly structured, leading to a larger fractional throughput in the higher trophic levels. The Gulf of Nicoya model was also compared with a trophic model developed for a mangrove estuary in Brazil, also using the ECOPATH II methodology. This latter model predicts that most of the energy remains in the benthic system of the Brazilian mangrove forest, where it is transferred to leaf-consuming mangrove crabs (Wolff 2006), a group having a negligible impact in the Gulf of Nicoya. The Brazilian ecosystem also differs from the Gulf of Nicoya in the importance of the mangrove biomass, which is two orders of magnitude larger than that in the Gulf, while pointing at the relatively little importance of the Brazilian epibenthos compartment. These comparisons are useful to understand how estuarine ecosystems work at different latitudes.

The model developed for the Gulf of Nicoya by Wolff et al. (1998) needs to be updated with recent data. The construction of two seasonal models (rainy and dry season) was recommended by Wolff et al. (1998), considering the strong seasonality detected in most ecosystem studies in the Gulf. The contribution of phytobenthos to the primary

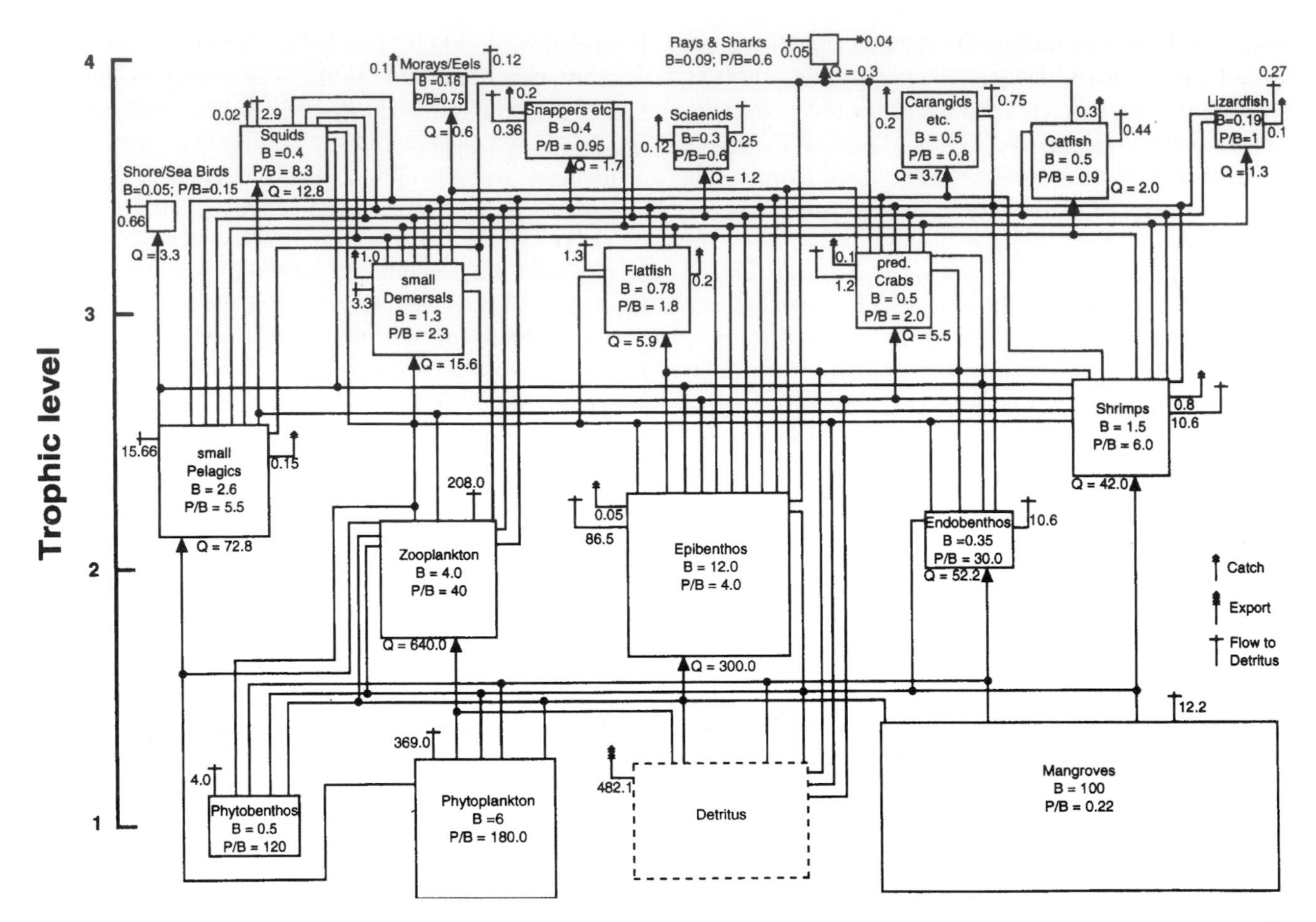

Fig. 6.5 The Gulf of Nicoya trophic model of Wolff et al. (1998). Box size is proportional to the square root of the compartment biomass. Q represents the total consumption of each box and the flows are given in grams of wet weight per m^2.
Reprinted from Revista de Biología Tropical *(RBT) 46 (Suppl. 6), page 68, Figure 2, by permission of the editor of RBT.*

productivity pool is now under evaluation, as well as the importance of riverine imports into the system. The latter topic has been addressed recently in a biogeochemical model developed by Tabash-Blanco (2007).

Worldwide Estuarine Trends

The review of paleo-ecological, archaeological, historical, and ecological data allowed Jackson et al. (2001) to identify different human impacts on coastal ecosystems over time. They found that temperate estuaries worldwide are undergoing drastic changes in their water characteristics and ecology owing to human exploitation and pollution, rendering them the most degraded of marine ecosystems. Their list of threats and their trends includes the following: (1) loss of seagrasses and suspension feeders; (2) increased sedimentation and water turbidity; (3) enhanced episodes of hypoxia and anoxia; (4) higher frequency and duration of algal and toxic dinoflagellate blooms; (5) outbreaks of medusae; (6) fish kills; (7) eutrophication; (8) enhanced microbial production; and (9) shifts from benthic primary-production ecosystems to plankton primary-production ecosystems.

The Gulf of Nicoya estuary has been used by humans as a source of food since pre-Columbian times. Since its discovery in 1519 by Spanish navigators, the human impact on its waters and watershed has been on the rise. The port of Puntarenas became very important for the export of coffee to Europe beginning in the mid-nineteenth century, and fishing activities became a key source of income for local people towards the mid-twentieth century, with the advent of industrial fishing of penaeid shrimp, and with the demand for more types of sea food being offered at the local markets as alternatives to traditional red meat consumption. The human population in the Central Valley, which is drained by the Tárcoles River (Fig. 6.1), has increased drastically over the past 50 years, posing a greater demand on resources from the Gulf, but also turning this river into an open sewer that gathers sediments, detergents, trace metals, untreated sewage, and thousands of organic and inorganic chemical pollutants including fertilizers, hydrocarbons, pesticides, antibiotics, and hormones (Spongberg et al. 2011).

The Gulf of Nicoya in View of Worldwide Estuarine Trends

Loss of Suspension Feeders

There are many old anecdotic accounts on the use of certain species of shellfish for human consumption in the Gulf of Nicoya. The islands of the Gulf were described for the first time in 1529 by Gonzalo Fernández de Oviedo (Meléndez 1974), who sketched the first known map of the estuary, and mentioned the presence of pearl-producing oysters near Chira Island (Fig. 6.1). He also mentioned the use for food of very thick oyster-like bivalves that "Christians call *pie de burro*" (donkey hoof), a name probably applied to the thick *Spondylus calcifer* found attached to hard subtidal substrates. In muddy intertidal sediments the thick ark clam *Grandiarca grandis*, known in Costa Rica as *chucheca*, was heavily overfished in the past (Stern-Pirlot and Wolff 2006). The specimen from the Gulf of Nicoya shown in Fig. 6.6 is among the largest known, with valves measuring 14 cm in length, and weighing 1.7 kg.

The British engineer Richard Trevithick, inventor of the steam rail locomotive, arrived at the port of Puntarenas in 1824, and requested permission for harvesting oysters with machinery he had brought aboard his ship. He offered to pay the Costa Rica government in return 5% of all the pearls found, and also offered to teach local fishermen on the use of the machinery. Gold mining in the nearby Aguacate Mountains proved more attractive, and the fishing operation was aborted (Fernández-Guardia 1938). Trevithick also sold to the government 2,000 pounds of mercury for use in the Aguacate gold mines (Fernández Guardia 1938). This is the first account on the use of mercury in Costa Rica, and traces of the metal might have reached the Gulf of Nicoya. The pearl-producing mollusks were probably the clams *Pinctada mazatlanica* or *Crassostrea prismatica* (= *Ostrea iridescens*), or both. These species are still found in low numbers in the less turbid waters of the lower gulf, such as the oyster bed of *O. iridescens* described by Campos

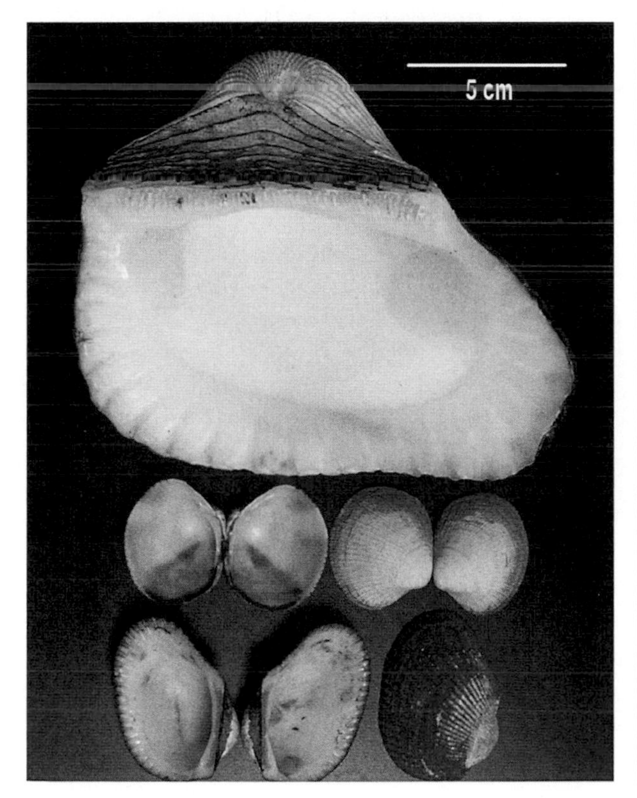

Fig. 6.6 Vanishing estuarine resources. a gigantic specimen of the ark clam, *Grandiarca grandis* (local name: chucheca), collected near the Negritos Islands in the Gulf of Nicoya in 1990. Valve length is 14 cm and the shells weigh 1,724 g. Middle left: two shells of the smaller, *Polymesoda radiata* (local name: almeja blanca) found in sandy sediments. Bottom left: two shells of the ark clam, *Anadara tuberculosa* (local name: piangua), found in mangrove muddy sediments. The shellfish fishery shifted from the large *G. grandis* to the smaller species, once *G. grandis* became rare due to overfishing.

and Fournier (1989) at Curú Bay, near the Negritos Islands (Fig. 6.1). Pearl and mother-of-pearl were part of an important artisanal fishery in the Gulf of Nicoya during the first half of the nineteenth century. In a brochure printed in London (Molina 1849), which aimed at describing Costa Rica to attract potential colonizers, the following marine products are listed among the miscellaneous exports of the country: pearls, mother-of-pearl, tortoise shell, and purple (murex). Mother-of-pearl was used mostly to make shirt buttons and other objects, and the tortoise shell refers to the *carey* turtle, *Eretmochelys imbricata*. The shell of this turtle was used locally for the fabrication of jewelry objects like rings, necklaces, combs, and others. This artisanal tradition was typical for the port of Puntarenas until the mid-twentieth century, when conservation measures were put in place.

The fishery of the murex snails in the Gulf of Nicoya was already mentioned by Thomas Gage, a traveling Irish priest, as early as 1636. The Native Americans living on the shores of the upper Gulf of Nicoya collected the snails when the water was clear, and the mucus released by the mollusk was used to stain fabrics with a purple color. The snails were carefully returned alive to their habitat, a famous early example of sound fishery management (Meléndez 1974). This murex snail (Fig. 6.7) is known today as *Plicopurpura patula*. Gage also mentioned the great diversity and abundance of other mollusks found in the upper region of the Gulf. Today the above mentioned species of mollusks are becoming very scarce in the estuary, and overfishing of *Grandiarca grandis* (known as *chucheca*) has almost led to extinction of this species. Attempts have been made to cultivate this ark clam in the laboratory, or to recolonize

Fig. 6.7 Estuarine rocky shore fauna: a shell of the muricid snail, *Plicopurpura patula pansa*, commonly known as murex or purpura, and formerly abundant in shallow rocky areas in the upper estuary. When disturbed, it produces a mucus secretion that was used by the native Americans in the upper Gulf of Nicoya for staining cotton strings with a purple color. Once the mucus was released, the snail was returned unharmed to its original collecting site. This is perhaps the first known management strategy for a marine resource in the Gulf, and reported by an European traveler early in the seventeenth century.

former beds with juveniles—with limited success in both cases (Villalobos and Báez 1983, Fournier and De la Cruz 1987).

The *Anadara* spp. fishery has now moved to the exploitation of a smaller sister species, *Anadara tuberculosa*, known locally as *piangua* (Fig. 6.6). The most recent estimate of its fishery potential was provided by Stern-Pirlot and Wolff (2006). They concluded that heavy overfishing in the Gulf of Nicoya and other estuaries, also mirrored by the low average harvested size of the clams, is due to the relatively small surviving populations in each fishing area, the easy access of this intertidal resource, and its high economic importance. They warn that if conservation measures are not implemented it is likely that *A. tuberculosa* may follow *G. grandis* on its way to local extinction. Other bivalves are now being harvested heavily in the Gulf of Nicoya, such as several species of razor clams, *Tagelus* spp. (Rojas et al. 1988), and *Polymesoda radiata* (Cruz and Jiménez 1994); see Fig. 6.6. The mussel *Mytella guyanensis* forms mats over the muddy sediments of the upper Gulf. Its small size and the high water content of its meat have saved this species from overfishing (Arroyo and Marín-Alpízar 1998).

The above-mentioned data support the idea that removal of filter feeding bivalves was and still is a serious problem in the Gulf of Nicoya estuary. The impact of the removal of other components mentioned, such as marine turtles, and several species of snails like the murex and the big *Strombus galetus* (known as *cambute*), whose main populations are today mainly restricted to the Absolute Natural Reserve Cabo Blanco (Fig. 6.1) (Arroyo and Mena 1998), is unknown. Other important snails being harvested are the intertidal pulmonate *Siphonaria gigas* found on rocky shores (Ortega 1987), *Hexaplex radix* found on mangrove creeks, and the subtidal *black cambute, Melongena* spp., found on subtidal muddy sediments. In addition, many species of herbivorous intertidal chitons (Polyplacophora) are also being driven to local extinction due to overfishing. The book by Wehrtmann and Cortés (2009) provides updated lists of all mollusks reported from Costa Rica. However, much work remains to be done on the taxonomy and ecological role of micro-mollusks. The description of *Cimaria vargasi*, a minute snail from the upper Gulf of Nicoya, is a recent example.

Shrimp Trawling and Bottom Damage

Estuaries in many regions around the world support important fisheries. The blue crab *Callinectes sapidus* fishery in the Chesapeake estuary (eastern shore of the USA) is a good example from a temperate region (Epifanio and Dittel 1982). In the subtropical and tropical regions some estuaries are important fishing grounds for benthic fish and penaeid shrimps. These fisheries usually involve the use of trawl gear to catch the epibenthic fauna, of which penaeids usually make a fraction of the total catch. Other invertebrates and fish caught in the nets are by-catch, and those of low commercial value are frequently discarded dead overboard (Bush 2000). In addition to the removal of benthic fish, decapod and stomatopod crustaceans, and other invertebrates, trawling on the sea floor causes direct impact on the benthos by the mechanical action of the trawl doors, chains, and net on the sediment. This impact has been compared to that of forest clearcutting (Watling and Norse 1998).The impact of bottom trawling depends on factors like the weight of the gear, the towing speed of the ship, the grain size of the sediments, and the speed of the bottom currents (Jones 1992, Warnken et al. 2003).

The upper Gulf of Nicoya has been closed to trawling for decades, but the mid and lower sections of the Gulf have been open to this activity since the mid-1950s, when shrimp fisheries became an important source of income for industrial fishermen. There are several studies on shrimp by-catch from the Gulf (Campos 1983a, Campos et al. 1984, Campos 1986). An early study by Campos (1983a) indicates that in 1980 at least 27 species of fish were discarded by white shrimp trawlers, and only five species of fish were represented by individuals of more than 20 cm in length, which means that most fish were non-reproductive juveniles. More recent data (Tabash-Blanco 2007b) on *Farfantepenaeus californiense* (brown shrimp), *F. brevirostris* (pink shrimp), and *Solenocera agassizi* (fidel shrimp) indicate that these species reached their Maximum Sustainable Yield (MSY; Wolff 2009) in the decades of 1970 and 1980, and are now found at overexploitation levels. Rostad and Hansen (2001) conducted a preliminary survey to test for differences in the benthic fauna of trawled vs. untrawled stations in the Gulf of Nicoya. In spite of the small number of stations sampled, they found that the samples from trawled stations were composed of more than 50% polychaetes, 30% arthropods (mainly amphipods), some mollusks, a few amphioxus (Cephalochordates), sipunculans, and ophiuroid echinoderms. In contrast, the samples from the non-trawled stations had a similar percentage of worms, but the arthropods represented less than 20%, and ophiuroids reached 15%, a result similar to that reported by Ball et al. (2000). The decline of shallower penaeid shrimp populations has shifted part of the shrimp fishing pressure to resources found in deeper waters at the entrance to the Gulf of Nicoya. Wehrtmann and Echeverría-Sáenz (2007) report on the fauna associated with the fishery of the pandalid shrimp *Heterocarpus vicarius* at depths from 200 to

350 meters. They found that 75% of the catch comprised a total of 28 decapod and two stomatopod species. The remaining 25% included several species of fish.

Overfishing of Top Fish Predators and Herbivorous Fish

The German traveler Wilhelm Marr (1819–1904) provides an interesting description of the life at the port of Puntarenas in 1853, then a city of only 1,200 inhabitants. Shortly after his arrival at the port, he is warned by local sailors not to swim there because of the danger of being killed by sharks that, according to him, are found in such high numbers that they are a nuisance to whale-fishing by North American ships in the Gulf of Nicoya. Once a cetacean was left dead overnight in the water, nothing could be used of its body the following morning. He further indicated that sharks are found at high tide in the Estero (a small secondary estuary behind the port), where they are seen in schools near the site where the slaughterhouse was located (Marr 2004).

The probabilities of being bitten by a shark when swimming in the Estero are extremely small today owing probably to shark overfishing and pollution. As for the whales, in 1980 during the *RV Skimmer* cruises, I remember having seen a whale's right jaw bone stranded on the beach at Jesusita Island, between San Lucas and Negritos Islands (Fig. 6.1). The bone was so big that three people could barely move it.

Rodríguez-Fonseca (2001) reports four species of dolphins from the outer Gulf of Nicoya: *Peponecephala electra, Tursiops truncates, Stenella attenuata,* and *S. longirostris.* The whale *Balaenoptera musculus* is reported from off-shore Pacific waters, and *B. edeni* and *Megaptera novaeangliae* are reported from the Golfo Dulce region, but no whales have been observed in the Gulf of Nicoya.

Other predator species are being removed by fishing from the ecosystem at a high rate. Rojas (2006) points out that 13 species of penaeid shrimp, five species of corvinas (*Cynoscion* spp.), three species of snappers (*Lutjanus* spp.), and the shark *Mustelus dorsalis* make the most important source of fish protein for thousands of people living in the Gulf of Nicoya. Rojas (2006) also reports that the stomach contents of *M. dorsalis* include fish (*Anchoa* sp., *Caranx* sp., *Lutjanus* sp., *Engraulis* sp., and *Ophistonema* sp.), crustaceans like the stomatopods *Squilla hancocki* and *S. parva,* and the shrimp *Farfantepenaeus* sp., as well as cephalopods. *S. hancocki* was found in 65% of the stomach contents examined.

Rays and sharks are included in the Gulf of Nicoya trophic model of Wolff et al. (1998). However, the size of the box (Fig. 6.5) is probably underestimated owing to the lack of published data on the feeding ecology of these fish. In the Gulf, the best known sites for shark fishing are near Chira Island (Fig. 6.1), where fishermen capture *M. dorsalis*, *M. lunulatus*, and *Sphyrna lewini* at the beginning of the rainy season when the sharks spawn (Rojas et al. 2000). The number of invertebrates and fish species captured as part of shrimp trawling operations has been mentioned already, while studies by Campos (1983a,b, 1986) provide information on the magnitude of this problem, which continues until today. An updated assessment of the shrimp by-catch is urgently needed, as well as an update on the status of the corvinas-snappers-sharks fisheries in the Gulf of Nicoya.

Red Tides and Fish Kills

The first scientific description of a red tide in the Gulf of Nicoya was prepared by Hargraves and Víquez (1981), who found that the dinoflagellate causing the red tide was *Cochlodinium catenatum,* which reached concentrations of 80 million cells per liter. Gocke et al. (1990) reported that near the Punta Morales port (Fig. 6.1) the massive occurrence of red tides in distinct patches strongly influenced the spatial and temporal concentrations of dissolved oxygen. The depth at which photosynthesis is possible was restricted to the upper two meters due to high concentrations of microalgae near the surface and intensive shelf-shading. Their observations on red tides were accompanied by boat trips to more distant regions of the inner Gulf, while sometimes dead fishes were found floating on the water. Gocke et al. (1990) argued that red tides contribute significantly to the primary productivity of the Gulf, but whether the algae are used directly by herbivorous organisms or only after their partial degradation to detritus via the food web remains to be investigated. Vargas-Montero and Freer (2004a) report that massive mortality of cultured fish occurred near Negritos Islands (Fig. 6.1), where the dinoflagellates *Ceratium dens*, *C. fusca*, and *C. fusus* were identified in water samples. They also report (Vargas-Montero and Freer 2004b) that in May 2002 a harmful algal bloom (HAB, the term commonly used nowadays) left a large number of dead fish in the Gulf of Nicoya, and water samples were found to contain the cyanobacterium *Trichodesmium erythraeum* as well as the dinoflagellate *Cochlodinium* cf. *polykrikoides*. The latter species is associated with fish kills in other regions.

Vargas-Montero and Freer (2004c) have also reported the presence of the toxic diatom *Pseudo-nitzschia pungens* in water samples taken from a HAB near San Lucas Island (Fig. 6.1). Unusual fish mortality, nearly specific for the Sciaenidae fish (corvinas, croakers, drums) occurred in the inner Gulf of Nicoya in 1985, and was reported by Szelistowski and Garita (1989). Fishes comprising a total of 15 sciaenid species, of which *Cynoscion squamipinnis* was the most common, were found dead upon collection. The combination of observed tissue irregularities was compatible with pollutant-induced damage. However, the possibility that this mass mortality was caused by red tide toxins could not be excluded, as HAB occurred in this region of the Gulf in 1985 (Szelistowski and Garita 1989).

Toxic algal blooms have been reported to have also a significant impact on benthic communities elsewhere. Olsgard (1993) found evidence for a clear switch in the faunal composition in a study area near Oslo, Norway, subsequent to a bloom of a toxic flagellate, *Chrysochromulina polylepis*. The impacted fauna was observed to have a tendency to return to pre-bloom communities only after two years. The main species affected were moderately abundant oligochaetes, polychaetes, crustaceans, mollusks, and sipunculans. Recently, red tide outbreaks were considered important by Vargas-Zamora and Sibaja-Cordero (2011) and by Vargas-Zamora et al. (2012) to explain changes in the abundance of mollusks and crustaceans at an intertidal flat in the Gulf of Nicoya. In the Gulf of Nicoya, the high organic input to the bottom material provided by settling dead bloom cells might have an impact on the detritus-feeding benthic communities found in most of the gulf. These topics deserve further research in the future.

Pollution (Trace Metals, Organics, Sewage)

Water pollution has been blamed as the most likely cause of changes in the biota of the Gulf of Nicoya, even before quantitative data on pollution were available for this estuary. The reason for this issue is that a great portion of the pollution load coming into the Gulf of Nicoya is visible in the form of beach debris, or as sewage discharges from the port city of Puntarenas (Fig. 6.1), into the Estero, a subsystem behaving like a partially mixed small estuary. The Estero is limited at one side by the city of Puntarenas and at the other side by mangrove forests. It receives a heavy load of fecal coliform bacteria, as well as contaminants carried by run-off, including detergents, hydrocarbons, and metals (Acuña et al. 1998, Acuña-González et al. 2004, García et al. 2006, García-Céspedes et al. 2004).

Research on pollution in the Gulf of Nicoya formed an important part of the expeditions aboard the *RV Skimmer* (1979–1981). Trace metals in sediments and a few invertebrates were evaluated with emphasis on four locations located near the mouth of the Tárcoles River (Fig. 6.1), which drains the highly populated Central Valley of Costa Rica. Metal concentrations of Cu, Pb, Cr, Fe, and Zn were found to be characteristic of non-industrialized estuaries (Dean et al. 1986). A more detailed assessment of trace metals at the mouth of the Tárcoles River and at stations further upstream was conducted by Fuller et al. (1990). Their study identified Cr as a river pollutant, probably originating from the operation of leather tanneries that were common in the Central Valley towards the mid-twentieth century.

Recent information on sediment and water pollution for the Gulf of Nicoya comes from the 2000–2003 UCR-CIMAR-Coastal Pollution of Costa Rica (CoCosRi) research project, which has produced several reports on PCB (polychlorinated biphenyl) concentrations (Spongberg 2004a,b, Spongberg 2006), petroleum hydrocarbons (Acuña-Gonzalez et al. 2004), trace metals (García-Céspedes et al. 2004), and bacteriological contamination and beach litter (García et al. 2006). These studies were conducted in four different coastal ecosystems of Costa Rica, allowing useful comparisons on the relative impact of pollution at each site. In their review Spongberg and Witter (2008) point out that values of PCB concentrations from the Gulf of Nicoya and other ecosystems are in general low when compared to sites in temperate ecosystems. A greater spatial coverage of sampling stations for trace metal evaluation has been provided by Salazar et al. (2004), using a different analytical technique. More recently, Lizano et al. (2008) studied concentrations of ten natural and artificial radionuclides in marine sediments of the Gulf of Nicoya. The range of values for these elements was considered normal when compared to global reports.

The above-mentioned studies point to the fact that levels of most pollutants in the gulf are still below those reported for industrialized estuaries. However, in recent years there has been growing concern about the disrupting impact of very low concentrations of pollutants on endocrine systems of estuarine organisms (Oberdorster and Cheek 2000). Actually, the Gulf of Nicoya is among the first tropical estuarine ecosystems where the disrupting impact of pollutants on the endocrine system of selected species was evaluated. A review by Cheek (2006) and the detection (Gravel et al. 2006) of imposex (a disorder in sea snails) in the intertidal snail *Thais brevidentata* at the port of Caldera (Fig. 6.1) points towards the importance of this novel research topic. As pointed out by Cheek (2006), endocrine disruption may not lead to extinction of a species over its entire range, but, in combination with other stressors such as habitat loss, overfishing, and global climate change, may contribute to local extinctions. Among the many endocrine-disrupting compounds are pesticides and herbicides such as DDT, as well as synthetic hormones. A study by Humbert et al. (2007), which focused on the assessment of toxicity of the main pesticides applied in Costa Rica, supports the need for urgent monitoring of pesticide residues in tissues of estuarine species of commercial and non-commercial value in the Gulf of Nicoya. A significant step towards the identification of other sources of pollutants entering coastal waters in Costa Rica represents a recent study by Spongberg et al. (2011) who found a wide diversity of pharmaceutical and personal care products (PPCPs), including antibiotics, that had reached the waters of the Gulf of Nicoya estuary.

Slowing the Trends: The Need for Ecosystem-Based Management

The information mentioned in this chapter makes the Gulf of Nicoya one of the better-known tropical estuaries worldwide. However, the ecosystem trends mentioned above indicate that the gulf is shifting its operational status in such a way that the services provided by the estuary (fisheries, tourism, navigation, and biodiversity) are being drastically impacted in a negative manner. If these trends continue, this ecosystem will be driven to a status where the number of trophic levels will be reduced, and most of the carbon fixed at the photic layer will end up buried in sediments.

Of course, more research is still needed to understand in detail the functioning of certain compartments of the ecosystem. In the meantime, the development of a model would be a significant step towards the implementation of the *ecosystem management approach*, rather than a continuation of focusing mostly on the management of individual species—a management strategy that has proved unsuccessful over the past years. As indicated by Link (2000), single species management strategies do not consider species interactions, allocation of biomass, changes in ecosystem structure or function, biodiversity, non-fishing ecosystem services, protected or rare species, non-target species, discarded by-catch, and gear impacts.

However, in the Gulf of Nicoya the beginning of new fisheries is likely to increase in the future. A few years ago fishing pressure was directed to the giant onuphid polychaete worm, *Americonuphis reesei*, an important species used as food in shrimp culture (Rojas-Figueroa and Vargas-Zamora 2008). Polychaetes are a group for which stock-assessment methods still have to be designed from scratch. This is not the case for tropical fish, since a great deal of effort was done to develop sampling and stock assessment methods applicable to small-scale fishers in Costa Rica, like those published by Conquest et al. (1996).

According to Jennings (2004), the *ecosystem approach* to fishery management puts emphasis on a management regime that maintains the health of the ecosystem, alongside appropriate human use of the marine environment. From an ecological perspective, such an ecosystem approach takes into account, and aims to remedy, the unwanted impacts of fishing on non-target species, habitats, and ecological interactions. This approach also recognizes that ecosystems provide many goods and services other than just fish, crustaceans, and shellfish (Jennings 2004).

As indicated by Browman and Stergiou (2004), a central role in the concept of ecosystem management is played by Marine Protected Areas (MPAs), which are intended to help recover overexploited fish stocks, preserve habitats and biodiversity, maintain ecosystem structure, and serve as control sites for comparison purposes. The Gulf of Nicoya includes several islands and coastal sites under some kind of management regime, and shrimp trawling is prohibited in the upper Gulf region west of the port of Puntarenas. These areas could serve as starting sites for the implementation of the ecosystem approach. Significant efforts have been undertaken to provide alternative uses for the Gulf of Nicoya, such as the pioneer deployment of artificial reefs (Campos and Gamboa 1989, Thorne et al. 1989).

Jiménez and González (2001) as well as Vargas and Mata (2004) have emphasized that the implementation of new management strategies for the Gulf of Nicoya should first bring to the table representatives from the private sectors and the main governmental agencies involved, like the Ministry of the Environment and Energy (MINAE), the Costa Rican Institute of Tourism (ICT), the Fisheries and Aquaculture Institute (INCOPESCA), the port authority (INCOP), and the municipal governments. This is of particular relevance when conflict-of-interest issues are to be addressed and jurisdictions overlap, such as in the case of artisanal vs. industrial fishing, or the construction of new housing and/or tourism complexes vs. setting aside these areas to promote nature-oriented management programs. Research institutions and international cooperation must continue to play a key role (Vargas 1995), as updated and new scientific information and knowledge should serve as the solid foundation on which management strategies must be grounded.

As a closing remark I quote a statement made in 1978 by the marine ecologist J.W. Hedgpeth:

> Whatever the ecosystem may be, or how complicated, or whether it is simply another word for the natural world we are part of, there are obviously too many things going on to study all of them or gather data on everything at once and ask the computer to tell us what it means. Our concern is to understand the environment well enough to make predictions and hope to manage it, or at least control ourselves and/or actions so that we will not find ourselves living on a vast dung heap beside a vast cesspool. Such understanding and ultimate management is incompatible with political exigencies and the need for a "quick fix."

References

Acuña, J., V. García, and J. Mondragón. 1998. Comparación de algunos aspectos físico-químicos y calidad sanitaria del Estero de Puntarenas, Costa Rica. *Revista de Biología Tropical* 46(Suppl. 6): 1–10.

Acuña-González, J.A., J.A. Vargas, E. Gómez-Ramírez, and J. García-Céspedes. 2004. Hidrocarburos de petróleo, disueltos y dispersos, en cuatro ambientes costeros de Costa Rica. *Revista de Biología Tropical* 52(Suppl. 2): 43–50.

Arroyo, D., and B. Marín-Alpízar. 1998. Crecimiento de *Mytella guyanensis* (Bivalvia: Mytilidae) en balsas flotantes. *Revista de Biología Tropical* 46(Suppl. 6): 21–26.

Arroyo, D., and L. Mena. 1998. Estructura de la población del cambute *Strombus galeatus* (Gastropoda: Strombidae) en Cabo Blanco, Costa Rica. *Revista de Biología Tropical* 46(Suppl. 6): 37–46.

Ball, B., B. Mundy, and I. Tuck. 2000. Effect of otter-trawling on the benthos and environments in muddy sediments. In M.J. Kaiser and S.J. de Groot, eds., *The Effects of Fishing on Non-target Species and Habitats*. Oxford: Blackwell Science Ltd.

Bartels, C., K. Price, M. López, and W.A. Bussing. 1983. Occurrence, distribution, abundance and diversity of fishes in the Gulf of Nicoya, Costa Rica. *Revista de Biología Tropical* 31: 75–101.

Bartels, C., K.S. Price, M. López-Bussing, and W. Bussing. 1984. Ecological assessment of finfish as indicators of habitats in the Gulf of Nicoya, Costa Rica. *Hydrobiologia* 112: 197–207.

Brenes, C.L., S. León, and J. Chaves. 2001. Variación de las propiedades termohalinas en el Golfo de Nicoya, Costa Rica. *Revista de Biología Tropical* 49(Suppl. 2): 145–52.

Brooks, D.R., and S. McCorquodale. 1995. *Acanthobothrium nicoyaense* n. sp. (Eucestoda: Tetraphyllidea: Onchobothriidae) in *Aetobatus narinari* (Euphrasen), (Chondrichthyes: Myliobatiformes: Myliobatidae) from the Gulf of Nicoya, Costa Rica. *Journal of Parasitology* 81: 244–46.

Browman, H.L., and K.I. Stergiou. 2004. Marine Protected Areas as a central element of ecosystem-based management: defining their locations, size and number. *Marine Ecology Progress Series* 274: 271–303.

Brugnoli-Olivera, E., and A. Morales-Ramírez. 2001. La comunidad fitoplánctica de Punta Morales, Golfo de Nicoya, Costa Rica. *Revista de Biología Tropical* 49(Suppl. 2): 11–17.

Bush, M.B. 2000. *Ecology of a Changing Planet*. New Jersey: Prentice Hall. 498 pp.

Bussing, W.A., and M.I. López. 1996. Fishes collected during the R.V. Victor Hensen Expedition (1993 1994). *Revista de Biología Tropical* 44(Suppl. 3): 183–86.

Cahoon, L.B. 2006. Upscaling primary production estimates: regional and global scale estimates of microphytobenthos production. In J.C. Kromkamp, J.F.C. de Brouwer, G.F. Blanchard, R.M. Foster, and V. Creach, eds., *Functioning of Microphytobenthos in Estuaries*, 99–108. Amsterdam: Royal Netherlands Academy of Arts and Sciences (KNAW).

Campos, J. 1983. Estudio sobre la fauna de acompañamiento del camarón en Costa Rica. *Revista de Biología Tropical* 31: 291–96.

Campos, J. 1986. Fauna de acompañamiento del camarón en el Pacífico de Costa Rica. *Revista de Biología Tropical* 34: 185–97.

Campos, J., B. Burgos, and C. Gamboa. 1984. Effect of shrimp trawling on the commercial ichthyofauna of the Gulf of Nicoya, Costa Rica. *Revista de Biología Tropical* 32: 203–7.

Campos, J., and A. Corrales. 1986. Preliminary results on the trophic dynamics of the Gulf of Nicoya, Costa Rica. *Anales del Instituto de Ciencias del Mar y Limnología, UNAM* 13: 329–34.

Campos, J., and M.L. Fournier. 1989. El banco de *Ostrea iridescens* (Pterioidea: Ostreidae) en Bahía Curú, Costa Rica. *Revista de Biología Tropical* 38: 331–33.

Campos, J., and C. Gamboa. 1989. An artificial tire reef in a tropical marine system: a management tool. *Bulletin of Marine Science* 44: 757–66.

Chaves, J., and M. Birkicht. 1996. Equatorial subsurface water and nutrient seasonality distribution of the Gulf of Nicoya, Costa Rica. *Revista de Biología Tropical* 44(Suppl. 3): 41–47.

Cheek, A.O. 2006. Subtle sabotage: endocrine disruption in wild populations. *Revista de Biología Tropical* 54(Suppl. 1): 1–19.

Cloern, J.E., S.Q. Foster, and A.F. Kleckner. 2014. Phytoplankton primary production in the world's estuarine ecosystems. *Biogeosciences* 11: 2477–501.

Conquest, L., R. Burr, R. Donnelly, J. Chavarría, and V. Gallucci. 1996. Sampling methods for stock-assessment for small-scale fisheries in developing countries. In V. Gallucci, S.B. Saila, D.J. Gustafson, and B.J. Rothschild, eds., *Stock Assessment: Quantitative Methods and Applications for Small-scale Fisheries*, 271–353. New York: Lewis Publishers.

Córdoba-Muñoz, R. 1998. Primary productivity in the water column of Estero Morales, a mangrove system in the Gulf of Nicoya, Costa Rica. *Revista de Biología Tropical* 46(Suppl. 6): 257–62.

Cruz, R.A. 1996. Annotated checklist of marine molluscs collected during the R.V. Victor Hensen Costa Rica expedition 1993/1994. *Revista de Biología Tropical* 44(Suppl. 3): 59–67.

Cruz, R.A., and J.A. Jiménez. 1994. *Moluscos Asociados a las Áreas de Manglar de la Costa Pacífica de América Central: Guía*. Heredia, Costa Rica: Editorial Fundación UNA. 180 pp.

Cutler, N., E. Cutler, and J.A. Vargas. 1992. Peanut worms (Phylum Sipuncula) from Costa Rica. *Revista de Biología Tropical* 40: 153–158.

Dean, H.K. 1996. Subtidal benthic polychaetes (Annelida) of the Gulf of Nicoya, Costa Rica. *Revista de Biologia Tropical* 44 (Suppl. 3): 69–80.

Dean, H.K. 2008. The use of polychaetes (Annelida) as indicator species of marine pollution: a review. *Revista de Biología Tropical* 56(Suppl. 4): 11–38.

Dean, H.K. 2009. Polychaetes and Echiurans. In I. Wehrtmann and J. Cortés, eds., *Marine Biodiversity of Costa Rica*, 181–91. *Monographiae Biologicae* 86. Dordrecht, Netherlands: Springer.

Dean, H.K., D. Maurer, J.A. Vargas, and C.H. Tinsman. 1986. Trace metal concentrations in sediment and invertebrates from the Gulf of Nicoya, Costa Rica. *Marine Pollution Bulletin* 17: 128–31.

De la Cruz, E., and J.A. Vargas. 1987. Abundancia y distribución vertical de la meiofauna en la playa fangosa de Punta Morales, Golfo de Nicoya, Costa Rica. *Revista de Biología Tropical* 35: 363–67.

DeVries, M., C.E. Epifanio, and A.I. Dittel. 1983. Reproductive periodicity of the tropical crab *Callinectes arcuatus* Ordway in the Gulf of Nicoya, Costa Rica. *Estuarine, Coastal and Shelf Science* 17: 709–16.

Dexter, D.M. 1974. Sandy beach fauna of the Pacific and Atlantic coast of Costa Rica and Colombia. *Revista de Biología Tropical* 22: 51–66.

Diaz-Fergusson, E., D. Arroyo, A. Morales, and J.A. Vargas. 2008. Observaciones sobre la larva del cangrejo marino tropical (Decapoda: Porcellanidae) *Petrolisthes armatus* en el Golfo de Nicoya, Costa Rica. *Revista de Biología Tropical* 56(3): 1209–23.

Díaz-Fergusson, E., and J.A. Vargas. 2001. Abundance of *Petrolisthes armatus* (Crustacea: Porcellanidae) on a tropical estuarine intertidal rocky beach, Gulf of Nicoya estuary, Costa Rica. *Revista de Biología Tropical* 49(Suppl. 2): 97–101.

Dittel, A.I. 1991. Distribution, abundance and sexual composition of stomatopod Crustacea in the Gulf of Nicoya, Costa Rica. *Journal of Crustacean Biology* 11: 269–76.

Dittel, A.I., and C.E. Epifanio. 1990. Seasonal and tidal abundance of crab larvae in a tropical mangrove system, Gulf of Nicoya, Costa Rica. *Marine Ecology Progress Series* 65: 25–34.

Dittel, A., C.E. Epifanio, and J.B. Chavarría. 1985. Population biology of the portunid crab *Callinectes arcuatus* Ordway in the Gulf of Nicoya, Costa Rica; Central America. *Estuarine, Coastal and Shelf Science* 20: 593–602.

Dittel, A.I., C.E. Epifanio, and O. Lizano. 1991. Flux of crab larvae in a mangrove creek in the Gulf of Nicoya, Costa Rica. *Estuarine, Coastal and Shelf Science* 32:129–40.

Dittmann, S., and J.A. Vargas. 2001. Tropical tidal flat benthos compared between Australia and Central America. In K. Reise, ed., *Ecological Comparisons of Sedimentary Shores*, 275–93. *Ecological Studies* 151.

Duncan, R.S., and W.A. Szelistowsky. 1998. Influence of puffer predation on vertical distribution of mangrove littorinids in the Gulf of Nicoya, Costa Rica. *Oecologia* 117: 433–42.

Emery, K.O., and R.R. Stevenson. 1957. Estuaries and lagoons. In J.W. Hedgpeth, ed., *Treatise on Marine Ecology and Paleoecology*, 673–750. *Geological Society of America Memoir* 67.

Emig, C., and J.A. Vargas. 1990. *Glottidia audebarti* (Broderip), (Brachiopoda, Lingulidae) from the Gulf of Nicoya, Costa Rica. *Revista de Biología Tropical* 38: 251–58.

Epifanio, C.E., and A. Dittel. 1982. Comparison of dispersal of crab larvae in Delaware Bay, (USA) and the Gulf of Nicoya, Central America. In V. Kennedy, ed., *Estuarine Comparisons*, 447–87. New York: Academic Press.

Epifanio, C.E., and A.I. Dittel. 1984. Seasonal abundance of Brachyuran crab larvae in a tropical estuary: Gulf of Nicoya, Costa Rica, Central America. *Estuaries* 7: 501–5.

Epifanio, C.E., D. Maurer, and A.I. Dittel. 1983. Seasonal changes in nutrients and dissolved oxygen in the Gulf of Nicoya, a tropical estuary on the Pacific coast of Central America. *Hydrobiologia* 101: 231–38.

Espinoza Mendiola, M., and I. Werhtmann. 2008. Stomach content analyses of the treadfin anglerfish *Lophiodes spilurus* (Lophiiformes: Lophiidae) associated with deepwater shrimp fisheries from the central Pacific of Costa Rica. *Revista de Biología Tropical* 56: 1971–90.

Fernández-Guardia, R. 1938. Documentos relativos a Mr. Richard Trevithick. *Revista de los Archivos Nacionales* 9 / 10: 508–21.

Fischer, S., and M. Wolff. 2006. Fisheries assessment of *Callinectes arcuatus* (Brachyura, Portunidae) in the Gulf of Nicoya, Costa Rica. *Fisheries Research* 77: 301–11.

Fournier, M.L., and E. De la Cruz. 1987. Reproduction of the cockle *Anadara grandis*, in Costa Rica. *NAGA, ICLARM Quarterly* 10(1): 6.

Fuller, C.C., J.A. Davis, D.J. Cain, P.J. Lamothe, T.L. Fries, G. Fernández, J.A. Vargas, and M.M. Murillo. 1990. Distribution and transport of sediment-bound metal contaminants in the Río Grande de Tárcoles, Costa Rica (Central America). *Water Research* 24: 805–12.

García, V., J. Acuña-González, J.A. Vargas, and J. García-Céspedes. 2006. Calidad bacteriológica y desechos sólidos en cinco ambientes costeros de Costa Rica. *Revista de Biología Tropical* 54(Suppl. 1): 35–48.

García-Céspedes, J., J.A. Acuña-González, and J.A. Vargas. 2004. Metales traza en sedimentos de cuatro ambientes costeros de Costa Rica. *Revista de Biología Tropical* 52(Suppl. 2): 51–60.

Gocke, K., J. Cortés, and M.M. Murillo. 2001a. Planktonic primary production in a tidally influenced mangrove forest on the Pacific coast of Costa Rica. *Revista de Biología Tropical* 49(Suppl. 2): 279–88.

Gocke, K., J. Cortés, and M.M. Murillo. 2001b. The annual cycle of primary productivity in a tropical estuary: The inner regions of the Golfo de Nicoya, Costa Rica. *Revista de Biología Tropical* 49(Suppl. 2): 289–306.

Gocke, K., J. Cortés, and C. Villalobos. 1990. Effects of red tides on oxygen concentration and distribution in the Gulf of Nicoya, Costa Rica. *Revista de Biología Tropical* 38: 401–7.

Gocke, K., M. Vitola, and G. Rojas. 1981. Oxygen consumption patterns in a mangrove swamp on the Pacific coast of Costa Rica. *Revista de Biología Tropical* 29:143–54.

Gravel, P., K. Johanning, J. McLachlan, J.A. Vargas, and E. Oberdörster. 2006. Imposex in the intertidal snail *Thais brevidentata* (Gastropoda: Muricidae) from the Pacific coast of Costa Rica. *Revista de Biología Tropical* 52(Suppl. 2): 21–26.

Hardin, G. 1968. The tragedy of the commons. *Science* 162: 1243–48.

Hargraves, P., and R. Víquez. 1981. The dinoflagellate red tide in Golfo de Nicoya, Costa Rica. *Revista de Biología Tropical* 29: 31–38.

Hargraves, P., and R. Víquez. 1985. Spatial and temporal distribution of phytoplankton in the Gulf of Nicoya, Costa Rica. *Bulletin of Marine Science* 37: 577–85.

Hedgpeth, J.W. 1978. As blind men see the elephant: the dilemma of marine ecosystem research. In M.L. Wiley, ed., *Estuarine Interactions*, 3–15. New York: Academic Press.

Hoisaeter, T. 2012. *Cimaria vargasi* n. gen., n. sp. (Gastropoda: Pyramidellidae: Odostomidae) from the Pacific coast of Costa Rica, Central America. *Zootaxa* 3178: 63–67.

Hossfeld, B. 1996. Distribution and biomass of arrow worms (Chaetognatha) in Golfo de Nicoya and Golfo Dulce, Costa Rica. *Revista de Biología Tropical* 44(Suppl. 3): 157–72.

Humbert, S. 2007. Toxicity assessment of the main pesticides used in Costa Rica. *Agriculture, Ecosystems and Environment* 118: 183–90.

Jackson, J.B., M.X. Kirby, W.H. Berger, K.A. Bjorndal, L.W. Botsford, B.C. Bourque, R.H. Bradbury, R. Cooke, J. Erlandson, J.A. Estes, T.P. Hughes, S. Kidwell, C.B. Lange, H.S. Lenihan, J.M. Pandolfi, C.H. Peterson, R.S. Steneck, M.J. Tegner, and R.R. Warner. 2001. Historical overfishing and the recent collapse of coastal ecosystems. *Science* 293: 629–38.

Jennings, S. 2004. The ecosystem approach to fishery management: a significant step towards sustainable use of the marine environment? *Marine Ecology Progress Series* 274: 269–303.

Jesse, S. 1996. Demersal crustacean assemblages along the Pacific coast of Costa Rica: a quantitative and multivariate assessment based on the Víctor Hensen Costa Rica Expedition (1993/1994). *Revista de Biología Tropical* 44(Suppl. 3): 115–34.

Jiménez, J., and E. González, eds. 2001. *La Cuenca del Río Tempisque: Perspectivas para un Manejo Integrado*. San José, Costa Rica: Organización para Estudios Tropicales. 137 pp.

Jiménez, J., and R. Soto. 1985. Patrones regionales en la estructura y composición florística de los manglares de la costa Pacífica de Costa Rica. *Revista de Biología Tropical* 33: 25–37.

Jones, J.B. 1992. Environmental impact of trawling on the sea bed: a review. *New Zealand Journal of Marine and Freshwater Research* 26: 59–67.

Klemas, V., S. Ackleson, M.M. Murillo, and J.A. Vargas. 1983. Water quality assessment of the Golfo de Nicoya, Costa Rica. Phase 1 of the Remote Sensing Task. Progress Report of the 1980–1981 International Sea Grant Program. Newark, DE: University of Delaware, College of Marine Studies. DEL-SG-04–83. 96 pp.

Kromkamp, J.C., and R.M. Forster. 2006. Developments in microphytobenthos primary productivity studies. In J.C. Kromkamp, J.F.C. de Brouwer, G. F. Blanchard, R.M. Forster, and V. Creach, eds., *Functioning of Microphytobenthos in Estuaries*, 9–30. Amsterdam: Royal Netherlands Academy of Arts and Sciences (KNAW).

León, P. 1973. Ecología de la ictiofauna del Golfo de Nicoya, un estuario tropical. *Revista de Biología Tropical* 21: 5–30.

Link, J.S. 2000. What does ecosystem-based fisheries management mean? *Fisheries* 27: 18–21.

Lizano, O. 1998. Dinámica de las aguas en la parte interna del Golfo de Nicoya ante altas descargas del Río Tempisque. *Revista de Biología Tropical* 46(Suppl. 6): 11–20.

Lizano, O.G., and E. Alfaro. 2004. Algunas características de las corrientes marinas en el Golfo de Nicoya, Costa Rica. *Revista de Biología Tropical* 52(Suppl. 2). 77–94.

Lizano, O., L.G. Loria, F.J. Alfaro, and M. Badilla. 2008. Distribución espacial de radionucleidos en sedimentos marinos de Bahia Culebra y el Golfo de Nicoya, Pacifico, Costa Rica. *Revista de Biología Tropical* 56(Suppl. 4): 83–90.

Lizano, O.G., and J.A. Vargas. 1993. Distribución espacio-temporal de la salinidad y la temperatura en la parte interna del Golfo de Nicoya. *Tecnología en Marcha* 12(2): 3–16.

López, M.I. 1978. Migración de la sardina *Astyanax fasciatus* (Characidae) en el río Tempisque, Guanacaste, Costa Rica. *Revista de Biología Tropical* 26: 261–75.

López, M.I., and C. Arias. 1987. Distribución temporal y espacial del ictioplancton en el estuario de Pochote, Bahía Ballena, Pacífico de Costa Rica. *Revista de Biología Tropical* 35: 121–26.

López, M.I. and W.A. Bussing. 1998. *Urotrygon cimar*, a new Eastern Pacific stingray (Pisces: Urolophidae). *Revista de Biologia Tropical* 46(Suppl. 6): 271–77.

Loría, L.G., R. Jiménez, and O. Lizano. 2002. Radionucleídos naturales y antropogénicos en el estuario del Golfo de Nicoya, Costa Rica. *Tópicos Meteorológicos y Oceanográficos* (2):74–78.

Marr, W. 2004. Viaje a Centroamérica. Translated by I. Reinhold, from *Reise nach Central America* (Marr, 1863). San José, Costa Rica: Editorial de la Universidad de Costa Rica. 472 pp.

Maurer, D., C.E. Epifanio, H.K. Dean, S. Howe, J.A. Vargas, A.I. Dittel, and M.M. Murillo. 1984. Benthic invertebrates of a tropical estuary: Gulf of Nicoya, Costa Rica. *Journal of Natural History* 18: 47–61.

Maurer, D., and J.A. Vargas. 1984. Diversity of soft-bottom benthos in a tropical estuary: Gulf of Nicoya, Costa Rica. *Marine Biology* 81: 97–106.

Maurer, D., J.A. Vargas, and H.K. Dean. 1988. Polychaetous annelids from the Gulf of Nicoya, Costa Rica. *Internationale Revue der Ges. Hydrobiologie und Hydrographie* 73: 43–59.

Meléndez, C. 1974. *Viajeros por Guanacaste*. Serie Nos Ven. No. 4. San José, Costa Rica: Ministerio de Cultura, Juventud y Deportes, Departamento de Publicaciones. 557 pp.

Mielke, W. 1997. New findings of interstitial Copepoda from Punta Morales, Pacific coast of Costa Rica. *Microfauna Marina* 11: 271–80.

Molina, F. 1849. Brief sketch of the Republic of Costa Rica. London: P.P. Thoms. 15 pp.

Molina-Ureña, H. 1996. Ichthyoplankton assemblages in the Gulf of Nicoya and Golfo Dulce embayments, Pacific coast of Costa Rica. *Revista de Biología Tropical* 44(Suppl. 3): 173–82.

Morales, A., and J.A. Vargas. 1995. Especies comunes de copépodos (Crustacea: Copepoda) pelágicos del Golfo de Nicoya, Costa Rica. *Revista de Biología Tropical* 43: 207–18.

Morales-Ramírez, A. 1996. Checklist of copepods from Gulf of Nicoya, Coronado Bay and Golfo Dulce, Pacific coast of Costa Rica, with comments on their distribution. *Revista de Biología Tropical* 44(Suppl. 3): 103–13.

Morales-Ramírez, A., and E. Brugnoli-Oliveira. 2001. El Niño 1997–1998 impact on the plankton dynamics in the Gulf of Nicoya, Pacific coast of Costa Rica. *Revista de Biología Tropical* 49(Suppl. 2): 103–14.

Nixon, S.W. 1995. Coastal marine eutrophication—a definition, social causes and future concerns. *Ophelia* 41: 199–219.

Oberdorster, E, and A.O. Cheek. 2000. Gender benders at the beach: endocrine disruption in marine and estuarine organisms. *Environmental Toxicology and Chemistry* 20: 21–26.

Olsqard, F. 1993. Do toxic algal blooms affect subtidal soft-bottom communities? *Marine Ecology Progress Series* 102: 269–86.

Ortega, S. 1987. The effect of human predation of the size distribution of *Siphonaria gigas* (Mollusca: Pulmonada) on the Pacific coast of Costa Rica. *Veliger* 29: 251–55.

Paine, R.T. 1966. Food web complexity and species diversity. *American Naturalist* 100: 66–75.

Palter, J., S. León, and D. Ballestero. 2007. The distribution of nutrients, dissolved oxygen and chlorophyll a in the upper Gulf of Nicoya, Costa Rica, a tropical estuary. *Revista de Biología Tropical* 55: 427–36.

Pereira, A.I. 1996. The impact of foraging by sandpipers (Scolopacidae) on populations of invertebrates in the intertidal zone of Chomes beach, Gulf of Nicoya, Costa Rica. In P. Hicklin, ed., *Shorebird Ecology and Conservation in the Western Hemisphere*, 44–51. *International Water Studies* 8.

Perry, D. 1988. Effects of associated fauna on growth and productivity in the red mangrove. *Ecology* 69: 1064–75.

Phillips, P.C. 1983. Diel and monthly variation in abundance, diversity, and composition of littoral fish populations in the Gulf of Nicoya. *Revista de Biología Tropical* 31: 297–306.

Ramírez, A.R., M.I. López, and W.A. Szelistowski. 1990. Composition and abundance of ichthyoplankton in a Gulf of Nicoya mangrove estuary. *Revista de Biología Tropical* 38: 463–66.

Ramírez, A.R., W.A. Szelistowski, and M. López. 1989. Spawning pattern and larval recruitment in Gulf of Nicoya anchovies (Pisces: Engraulidae). *Revista de Biología Tropical* 37: 55–62.

Rodríguez-Fonseca, J. 2001. Diversidad y distribución de los cetáceos de Costa Rica (Cetacea: Delphinidae, Physeteridae, Ziphiidae y Balaenopteridae). *Revista de Biología Tropical* 49: 135–43.

Rojas, R. 2006. Reproducción y alimentación del tiburón enano, *Mustelus dorsalis* (Pisces: Triakidae) en el Golfo de Nicoya, Costa Rica: elementos para un manejo sostenible. In M. Rojas and I. Zanella, eds., *Memoria del Primer Seminario Taller sobre el Estado del Conocimiento de la Condrictiofauna de Costa Rica*, 29–32. Santo Domingo de Heredia: Instituto Nacional de Biodiversidad.

Rojas, J.R., J. Campos, A. Segura, M. Mug, R. Campos, and O. Rodríguez. 2000. Shark fisheries in Central America: a review and update. *Uniciencia* 17: 49–56.

Rojas, J.R., J.F. Pizarro, and M. Castro. 1996. Diversidad y abundancia íctica en tres áreas de manglar en el Golfo de Nicoya, Costa Rica. *Revista de Biología Tropical* 42: 663–72.

Rojas, J., C.E. Villalobos, F. Chartier, and C.R. Villalobos. 1988. Tamaño, densidad y reproducción de la barba de hacha *Tagelus peruvianus* (Bivalvia: Solecurtidae) en el estero de Puntarenas, Costa Rica. *Revista de Biología Tropical* 36: 479–83.

Rojas-Figueroa, R., and J.A. Vargas-Zamora. 2008. Abundancia, biomasa y relaciones sedimentarias de *Americonuphis reesei* (Polychaeta: Onuphidae) en el Golfo de Nicoya, Pacifico, Costa Rica. *Revista de Biología Tropical* 56(Suppl. 4): 59–82.

Rostad, T., and K.L. Hansen. 2001. The effects of trawling on the benthic fauna of the Gulf of Nicoya, Costa Rica. *Revista de Biología Tropical* 49(Suppl. 2): 91–95.

Ryther, J.M. 1969. Photosynthesis and fish production in the sea. *Science* 166: 72–76.

Salazar, A., O.G. Lizano, and E.J. Alfaro. 2004. Composición de sedimentos en las zonas costeras de Costa Rica utilizando Fluorescencia de Rayos-x (FRX). *Revista de Biología Tropical* 52(Suppl. 2): 61–75.

Sibaja-Cordero, J.A., and J.A. Vargas-Zamora. 2006. Zonación vertical de epifauna y algas en litorales rocosos del Golfo de Nicoya, Costa Rica. *Revista de Biología Tropical* 54(Suppl. 1): 49–67.

Soto, R. 1988. Geometry, biomass allocation and leaf demography of *Avicennia germinans* (L.)L. (Avicenniaceae) along a salinity gradient in Salinas, Puntarenas, Costa Rica. *Revista de Biología Tropical* 36: 309–23.

Spongberg, A.L. 2004a. PCB concentrations in sediments from the Gulf of Nicoya estuary, Pacific coast of Costa Rica. *Revista de Biología Tropical* 52(Suppl. 2): 11–22.

Spongberg, A.L. 2004b. PCB contamination in surface sediments in the coastal waters of Costa Rica. *Revista de Biología Tropical* 52(Suppl. 2): 1–10.

Spongberg, A.L. 2006. PCB concentrations in intertidal sipunculan (Phylum Sipuncula) marine worms from the Pacific coast of Costa Rica. *Revista de Biología Tropical* 52(Suppl. 1): 27–33.

Spongberg, A.L., and J. Witter. 2008. A review of PCB concentrations in tropical media, 1996–2007. *Revista de Biología Tropical* 56(Suppl. 4): 1–9.

Spongberg, A. L., J. D. Witter, J. Acuña, J. A. Vargas, M. Murillo, G. Umaña, E. Gómez, and G. Pérez. 2011. Reconnaissance of selected PPCP compounds in Costa Rican surface waters. *Water Research* 45: 6709–17.

Stern-Pirlot, A., and M. Wolff. 2006. Population dynamics and fisheries potential of *Anadara tuberculosa* (Bivalvia: Arcidae) along the Pacific coast of Costa Rica. *Revista de Biología Tropical* 54(Suppl. 1): 87–99.

Szelistowski, W.A., and J. Garita. 1989. Mass mortality of sciaenid fishes in the Gulf of Nicoya, Costa Rica. *Fishery Bulletin* 87: 363–65.

Tabash-Blanco, F. 2007a. A biogeochemical model for the Gulf of Nicoya, Costa Rica. *Revista de Biología Tropical* 55: 33–42.

Tabash-Blanco, F. 2007b.Explotación de la pesquería de arrastre del camarón durante el periodo 1991–1999 en el Golfo de Nicoya, Costa Rica. *Revista de Biología Tropical* 55: 207–18.

Thorne, R.E., J.B. Hedgepeth, and J. Campos. 1989. Hydroacoustic observations of fish abundance and behavior around an artificial reef in Costa Rica. *Bulletin of Marine Science* 44:1058–64.

Vargas, J.A. 1987. The benthic community of an intertidal mud flat in the Gulf of Nicoya, Costa Rica. Description of the community. *Revista de Biología Tropical* 35: 229–316.

Vargas, J.A. 1988a. Community structure of macrobenthos and the results of macropredator exclusion on a tropical mud flat. *Revista de Biología Tropical* 36: 287–308.

Vargas, J.A. 1988b. A survey of the meiofauna of an eastern tropical Pacific intertidal mud flat. *Revista de Biología Tropical* 36: 541–44.

Vargas, J.A. 1989a. A three year survey of the macrofauna of an intertidal mud flat in the Gulf of Nicoya, Costa Rica. In O. Magoon, M. Converse, D. Miner, L.T. Tobin, and D. Clark, eds., *Proceedings of the 6th Symposium on Coastal and Ocean Management*,1905–19. Vol. 2. New York: American Society of Civil Engineers.

Vargas, J.A. 1989b. Seasonal abundance of *Coricuma nicoyensis* Watling and Breedy (Crustacea, Cumacea), on an intertidal mud flat in the Gulf of Nicoya, Costa Rica. *Revista de Biología Tropical* 37: 207–11.

Vargas, J.A. 1995. The Gulf of Nicoya estuary, Costa Rica: past, present, and future cooperative research. *Helgolander Meeresunters.* 49: 821–28.

Vargas, J.A. 1996. Ecological dynamics of a tropical intertidal mudflat community. In K.F. Nordstrom and C.T. Roman, eds., *Estuarine Shores: Evolution, Environments and Human Alterations*, 355–71. London: John Wiley and Sons Ltd.

Vargas, J. A. and H. K. Dean. 2009. Sipunculans. In I. Wehrtmann and J. Cortés, eds., *Marine Biodiversity of Costa Rica*, 175–80. *Monographiae Biologicae* 86. Dordrecht, Netherlands: Springer.

Vargas, J.A., and H.K. Dean. 2010. On *Branchiostoma californiense* (Cephalochordata) from Costa Rica. *Revista de Biología Tropical* 58: 1143–48.

Vargas, J.A., and A. Mata. 2004. Where the dry forest feeds the sea: the Gulf of Nicoya Estuary. In G.W. Frankie, A. Mata, and S.B. Vinson, eds., *Biodiversity Conservation in Costa Rica: Learning the Lessons in a Seasonal Dry Forest*, 126–35. Berkeley: University of California Press.

Vargas J.A., and S. Solano. 2011. On *Mellitella stokesii* and *Amphipholis geminata* (Echinodermata), from a tropical intertidal flat in the upper Gulf of Nicoya estuary, Pacific, Costa Rica. *Revista de Biología Tropical* 59: 193–98.

Vargas, J.A., and M. Wolff, eds. 1996. Pacific ecosystems of Costa Rica with emphasis on the Golfo Dulce and adjacent areas: a synoptic view based on the RV Victor Hensen expedition 1993/1994 and previous studies. *Revista de Biología Tropical* 44(Suppl. 3): 1–238.

Vargas-Montero, M., and E. Freer. 2004a. Presencia de los dinoflagelados, *Ceratium dens*, *C. fusus* y *C. furca* (Gonyaulacales: Ceratiaceae) en el Golfo de Nicoya, Costa Rica. *Revista de Biología Tropical* 52(Supl. 1): 115–20.

Vargas-Montero, M., and E. Freer. 2004b. Proliferaciones algales nocivas de cianobacterias (Oscillatoriaceae) y dinoflagelados (Gymnodiniaceae) en el Golfo de Nicoya, Costa Rica. *Revista de Biología Tropical* 52(Suppl. 1): 121–25.

Vargas-Montero, M., and E. Freer. 2004c. Proliferaciones algales de la diatomea toxifenica, *Pseudo-Nitzschia* (Bacillarophyceae) en el Golfo de Nicoya, Costa Rica. *Revista de Biología Tropical*. 52(Suppl. 1): 127–32.

Vargas-Zamora, J.A., and J.A. Sibaja-Cordero. 2011. A molluscan assemblage from a tropical intertidal estuarine sand-mud flat, Gulf of Nicoya, Pacific, Costa Rica (1984–1987). *Revista de Biologia Tropical* 59(3): 1135–48.

Vargas-Zamora, J.A., J.A. Sibaja-Cordero, and R. Vargas-Castillo. 2012. Crustaceans from a tropical estuarine sand-mud flat, Pacific, Costa Rica, (1984–1988) revisited. *Revista de Biologia Tropical* 60(4): 1763–81.

Villalobos, C.R. 1980. Variations in population structure in the genus *Tetraclita* (Crustacea: Cirripedia) between temperate and tropical populations. IV. The age structure of *T. stalactifera*, and concluding remarks. *Revista de Biología Tropical* 28: 353–59.

Villalobos, C.R., and A.L. Báez. 1983. Tasa de crecimiento y mortalidad en *Anadara tuberculosa* (Bivalvia: Arcidae) bajo dos sistemas de cultivo. *Revista Latinoamerica de Acuicultura* 17: 9–18.

Víquez, R., and P. Hargraves. 1995. Annual cycle of potentially harmful dinoflagellates in the Gulf of Nicoya, Costa Rica. *Bulletin of Marine Science* 57(2): 467–75.

Von Wangelin, M., and M. Wolff. 1996. Comparative biomass spectra and species composition of the zooplankton communities in Golfo Dulce and Golfo de Nicoya, Pacific coast of Costa Rica. *Revista de Biología Tropical* 44(Suppl.3): 135–55.

Voorhis, A., C.E. Epifanio, D. Maurer, A.I. Dittel, and J.A. Vargas. 1983. The estuarine character of the Gulf of Nicoya, an embayment on the Pacific coast of Central America. *Hydrobiologia* 99: 225–37.

Warnken, K.W., G.A. Gill, T.M. Dellapenna, R.D. Lehman, D.F. Harper, and M.A. Alison. 2003. The effects of shrimp trawling on sediment oxygen consumption and the fluxes of trace metals and nutrients from estuarine sediments. *Estuarine, Coastal and Shelf Science* 57: 25–42.

Watling, L., and E.A. Norse. 1998. Disturbance of the sea-bed by mobile fishing gear: a comparison to forest clearcutting. *Conservation Biology* 12: 1180–97.

Wehrtmann, I., and J. Cortés, eds. 2009. Marine Biodiversity of Costa Rica. *Monographiae Biologicae* 86. Dordrecht, Netherlands: Springer. 538 pp. and CD.

Wehrtmann, I., and A.I. Dittel. 1990. Utilization of floating mangrove leaves as a transport mechanism of estuarine organisms, with emphasis on Decapod Crustacea. *Marine Ecology Progress Series* 60: 67–73.

Wehrtmann, I.S., and S. Echeverría-Sáenz. 2007. Crustacean fauna (Stomatopoda, Decapoda), associated with the deepwater fishery of *Heterocarpus vicarius* (Decapoda: Pandalidae) along the Pacific coast of Costa Rica. *Revista de Biología Tropical* 55(Suppl. 1): 141–152.

Wolff, M. 1996. Demersal fish assemblages along the Pacific coast of Costa Rica: a quantitative and multivariate assessment based on the RV Victor Hensen Costa Rica expedition (1993–1994). *Revista de Biología Tropical* 44(Suppl. 3): 187–214.

Wolff, M. 2006. Biomass flow structure and resource potential of two mangrove estuaries: insights from comparative modelling in Costa Rica and Brazil. *Revista de Biología Tropical* 54(Suppl. 1): 69–86.

Wolff, M., ed. 2009. *Tropical Waters and their Living Resources: Ecology, Assessment and Management*. Bremen, Germany: H.M. Hauschild. 343 pp.

Wolff, M., I. Chavarría, V. Koch, and J.A. Vargas. 1998. A trophic flow model of the Golfo de Nicoya, Costa Rica. *Revista de Biologia Tropical* 46(Suppl. 6): 63–79.

Wolff, M., H.J. Hartmann, and V. Koch. 1996. A pilot trophic model for Golfo Dulce, a fjord-like tropical embayment, Costa Rica. *Revista de Biología Tropical* 44(Suppl.3). 215–31.

Chapter 7 Isla del Coco: Coastal and Marine Ecosystems

Jorge Cortés[1]

Introduction

Background

Isla del Coco (Cocos Island) is an oceanic island located between 5°30'–5°34'N and 87°01'–87°06'W in the Eastern Tropical Pacific (ETP) portion of the Pacific Ocean, approximately 500 km southwest from Costa Rica and over 600 km northeast from the Galápagos Islands, Ecuador (Fig. 7.1). The island is the only subaerial exposed portion of the Coco Volcanic Cordillera, which extends from the Galápagos Islands to the southern part of Costa Rica (Fig. 7.1). It is the nucleus of the Área de Conservación Marina Isla del Coco (ACMIC) (Cocos Island Marine Conservation Area), which includes 9,640 km^2 of the Sea Mounts Marine Management Area (created in March 2011). The island measures 4.4 km by 7.6 km, has a surface area of approximately 24 km^2 and a perimeter of 23.3 km. Its highest point (575.5 m) is at Cerro Iglesias (Figs. 7.2 and 7.3). At present, the marine protected area (MPA) (which is a no-take area) extends 22.2 km (12 nautical miles) around the island, covering an area of about 2,000 km^2, making it the largest national park in Costa Rica. A characteristic that makes Isla del Coco very different from all other oceanic islands in the ETP is the abundance of rain, with a yearly average around 5,000 mm and a maximum of 7,000 mm (Garrison 2005, Alfaro 2008). For this reason the island is covered in dense tropical rain forest (although comprising only a few species) and harbors an abundance of waterfalls (Fig. 7.4, and see Montoya, chapter 8 of this volume).

[1] Centro de Investigación en Ciencias del Mar y Limnología (CIMAR), and Escuela de Biología, Universidad de Costa Rica (UCR), San Pedro de Montes de Oca, 11501-2060 San José, Costa Rica

The discovery of Isla del Coco has been attributed to the Spanish pilot Joan Cabeças de Grado in 1526, but other dates and names are mentioned and contended (Montoya 2007a, and see Montoya, chapter 8 of this volume). The first representation of the island was in the world map (*mapamundi*) of Enrique II, in 1542, with the name "Ysle des Coques" (Lièvre 1893). It appeared again in the Andreas Homo Atlas of 1559 and in the planisphere of Nicolas Desliens printed in 1559 (Lièvre 1893, Montoya 2007a, and see Montoya, chapter 8 of this volume). From the seventeenth to the nineteenth centuries the island was used by pirates. William Dampier mentioned Isla del Coco in his 1697 book, *A New Voyage around the World.* Three famous legends of treasures buried on the island exist, which is why Isla del Coco has been called Treasure Island and may have been the inspiration for Robert Louis Stevenson's book of the same name. One of the treasures was supposedly buried by Edward Davies between 1683 and 1702, another was taken to the island in 1819 by Benito "Bloody Sword" Bonito, and the largest of all, the Treasure of Lima, was left there in 1821 by Captain William Thompson (Weston 1992). Apart from pirates, whalers were also common visitors to the island. Starting in the late seventeenth century and continuing into the early twentieth century, the island was an obligate stop for whalers (Weston 1992, Montoya 2007a, Montoya chapter 8 of this volume).

Costa Rica, 48 years after its separation from Spain (September 15, 1821), took possession of the island on Independence Day 1869, when Jesús Jiménez Zamora was President of the Republic of Costa Rica. It was used as a prison from 1879 to 1881, but because of the cost of reaching the island it was abandoned. The German sailor August Gissler was

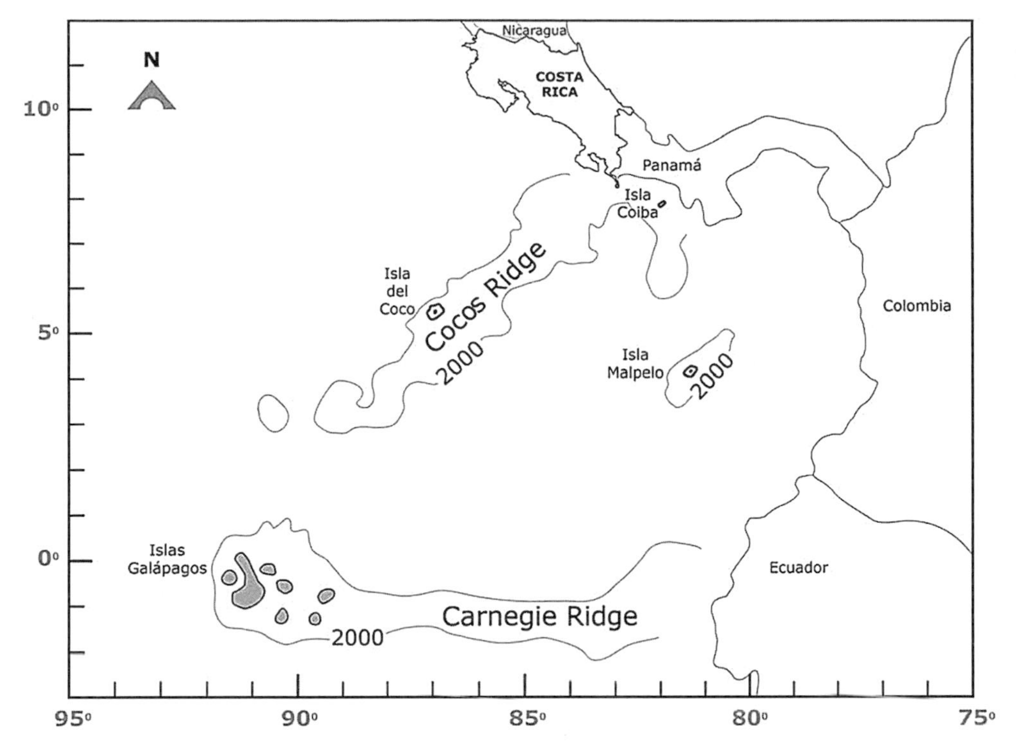

Fig. 7.1 Location of Isla del Coco on the Cocos Ridge (Cocos Cordillera) in the Eastern Tropical Pacific.
Modified and corrected from Castillo et al. (1988).

Fig. 7.2 Early morning view of Isla del Coco, with Cerro Iglesias on the right.
Photograph by Jorge Cortés, January 2007.

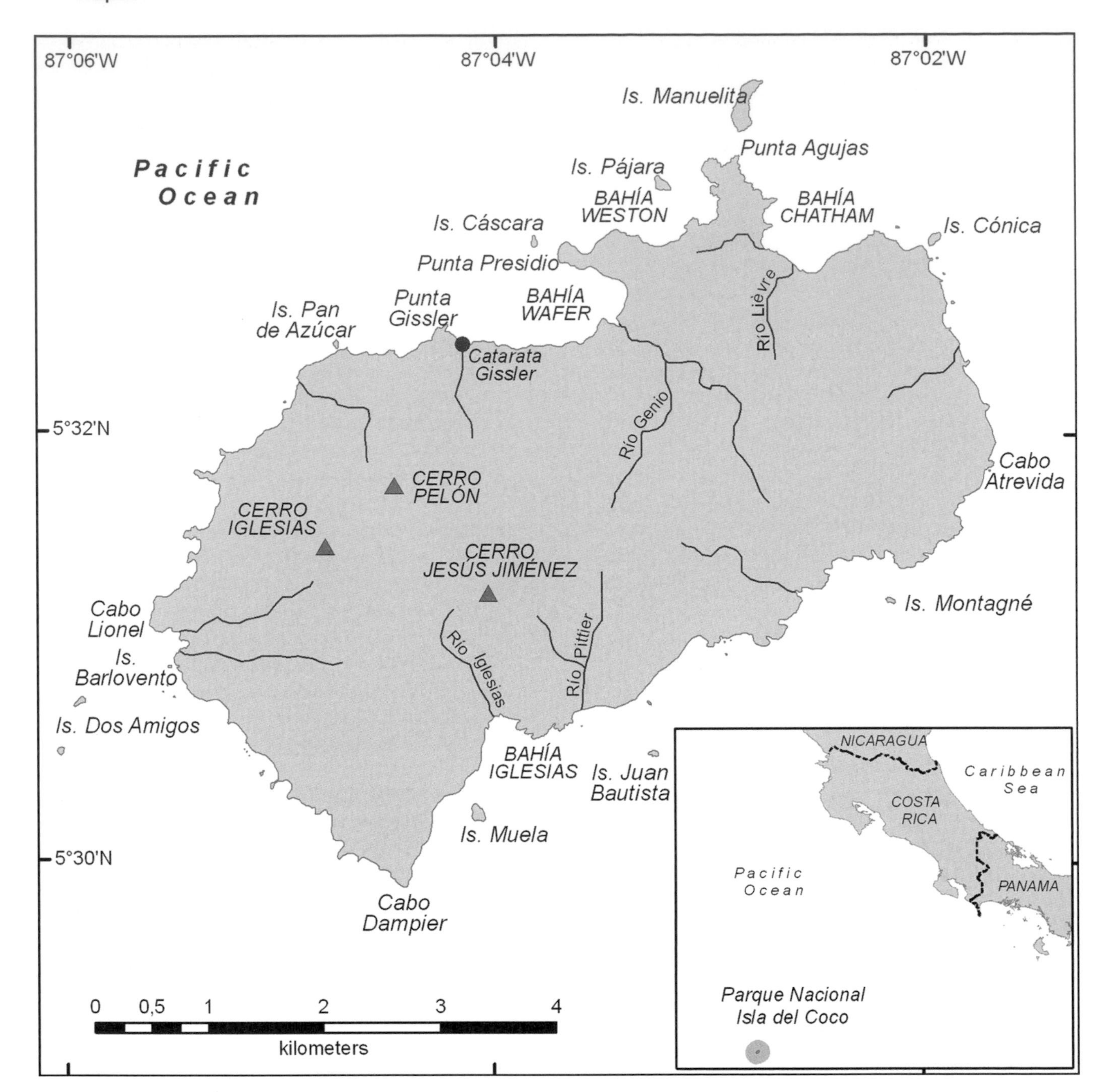

Fig. 7.3 Map of Isla del Coco with the names of topographic features as they are known today. Note that many of these names have changed once or several times over the past centuries. Is = Islet.
Map prepared by Marco V. Castro.

named Governor of the Island by the Government of Costa Rica in 1887, and lived on it from 1894 to 1908 (Beebe 1926a, Montoya 2007a). He tried along with other settlers to establish an agricultural colony, known as The Cocos-Island Plantation Company. Philippe Cousteau visited the island in 1976 and on the basis of his images, his father, Jean Jacques Cousteau, wrote that Isla del Coco was "the most beautiful island in the world," which apparently has been important in terms of receiving worldwide attention for its conservation.

In 1978, during the Rodrigo Carazo Odio administration, Isla del Coco was declared a National Park. Within the Law of the Sea of the United Nations (UNCLOS), with the declaration of Isla del Coco as a National Park, Costa

Rica had the right to claim 200 miles of sea around the island, an area of 290,000 km^2. This area around the island, together with the 200 miles from the coast, results in a Pacific marine area for Costa Rica covering 565,683 km^2 (INCOPESCA 2006), which is over 10 times its terrestrial area (51,100 km^2). The most important declaration was in 1997, when Parque Nacional Isla del Coco was proclaimed a World Heritage Site by the United Nations Educational, Scientific and Cultural Organization (UNESCO), and in 1998, the island was nominated an internationally important wetland within the Ramsar Convention. Parque Nacional Isla del Coco is part of the Eastern Tropical Pacific Marine Corridor that includes the Galápagos Islands (Ecuador), Malpelo and Gorgona islands (Colombia), and Coiba Island (Panamá) (Fig. 7.1).

Isla del Coco is the first shallow area touched by the North Equatorial Countercurrent, the main west-east current near to the equator (Montoya and Kaiser 1988, Guzmán and Cortés 1993, Glynn et al. 1996, Cortés 1997, Kessler 2006). This current transports larvae and juveniles of marine organisms from the Central Pacific to the ETP—for example, corals (Dana 1975, Cortés 1986, 2011, Richmond 1990, Glynn and Ault 2000), mollusks (Montoya and Kaiser 1988, Scheltema 1988, Shasky 1988), echinoderms (Lessios et al. 1996, 1998) and fishes (Robertson and Allen 1996, Robertson et al. 2004, Lessios and Robertson 2006). All reef-building corals of the Eastern Tropical Pacific originated in the Indo-Pacific, with the exception of a few endemic species—for example, *Pavona gigantea* Verrill, 1869 and *Porites panamensis* Verrill, 1866; in fact, one species, *Leptoseris scabra* Vaughan, 1907, is possibly a recent arrival (Cortés 1997, 2011, Cortés and Guzmán 1998).

Isla del Coco is the entrance gate for Indo-Pacific organisms to the Eastern Tropical Pacific. Currents bring larvae, juveniles, and possibly adults from the other side of the Pacific Ocean, crossing the largest and deepest gap in the world (Grigg and Hey 1992, Lessios et al. 1998, Lessios and Robertson 2006). The island provides the first shallow water encountered by the North Equatorial Counter Current (NECC); organisms can settle, and then move on to other oceanic islands and to the mainland coasts of the Americas. Some marine organisms of Parque Nacional Isla del Coco have a close affinity with Indo-Pacific species—for example, 7.5% of the mollusks of this park are of Indo-Pacific origin (Montoya 1983). Manning (1972) described several ETP species of stomatopods with Indo-West Pacific affinity. Lessios et al. (1996) reported four species of Indo-Pacific sea urchin from oceanic islands in the ETP, while Lessios et al. (1998) found a high degree of genetic identity between central-western Pacific and ETP populations of the sea urchin *Echinothrix diadema* (Linnaeus, 1758). The shore-fish fauna of the island has a high percentage (32.3%) of trans-Pacific (from the Western Pacific Ocean) species (Robertson et al. 2004). About 75% of zooxanthellate corals are recent eastward-moving migrants (Cortés 1986, Glynn and Ault 2000).

Isla del Coco has attracted the attention not only of treasure hunters. Owing to its unique and abundant flora and fauna, numerous scientists and scientific expeditions have visited it, contributing significantly to our present knowledge of its terrestrial and marine biodiversity. A brief history of marine research at Isla del Coco is presented in the following section. The island has several coastal and marine habitats, including sandy and cobble beaches, rocky intertidal zones, sandy and rocky bottoms at various depths, coral communities and reefs, open marine waters and deep benthic habitats. These ecosystems will be described below.

Historical Overview

The following historical account of marine research at Isla del Coco is based on a larger publication on marine research in Costa Rica (Cortés 2008, 2009a); only publications in scientific journals are considered here. Previous chronicles—some of which mention marine organisms—can be found in Lièvre (1893), Weston (1992), and Montoya (2007a). Towards the end of the nineteenth and early in the twentieth century there were six scientific expeditions that visited Isla del Coco. The first was in 1888 by the US Fish Commission Steamer *Albatross* under the leadership of Alexander Agassiz. On that occasion four stations south of Isla del Coco were established on April 1 and 2. There, hydrographic data were recorded in one of the stations, and a few samples were collected using scoop nets and electric lights (Townsend 1901). Apparently the scientists landed on the island, because two species of snails collected from the beaches of the island are reported in the single existing reference that reports specimens from Costa Rica collected during the expedition (Dall 1900).

A second major expedition made significant contributions to the knowledge of the marine biodiversity in the region: it was conducted by the US Fish Commission Steamer *Albatross* 1891 Expedition and visited the west coast of Central America and the Galápagos. Again it was led by Alexander Agassiz (Agassiz 1892, Townsend 1901). The *Albatross* 1891 Expedition took place in Costa Rican waters from February 26 to March 2 in that year, and most of the 15 stations sampled were located around Isla del Coco. Dredging and trawling were conducted at 12 stations; hydrographic soundings were done at three more stations; additionally, six surface tow-net stations (where dredging was also carried out) and six serial-temperature stations,

Fig. 7.4 High vertical sea cliffs in Bahía Iglesias, on the south side of the island. *Both photographs (pages 166–67) by Jorge Cortés, January 2007.*

with records from the surface to the bottom (being the deepest station: 3,616 m) ranging from 29.9 to 2.5°C, were completed (Agassiz 1892, Townsend 1901). Consequently, many papers were published and new species described (Cortés 2012a), including the first paper on Crustacea from Costa Rica (Faxon 1895), and the first paper on sponges from the Pacific of Costa Rica (Wilson 1904).

The first Costa Rican scientific expedition to Isla del Coco was conducted in 1898, on the Steamer *Poás*. It was organized by the then-recently-created Instituto Físico-Geográfico Nacional (National Physical Geographical Institute) of Costa Rica (Pittier 1899, Eakin 1999). The expedition was lead by Henri Pittier, Director of the Institute. In his report he presents a summary of the state of knowledge on the island's history, geography, and flora (Pittier 1899). Anastasio Alfaro, of the Museo Nacional of Costa Rica, commented on the birds and mammals of the island, including seabirds (Alfaro 1899).

The fourth major expedition to visit Isla del Coco was the Hopkins Stanford Galápagos Expedition of 1898–1899 (Heller 1903). This expedition was organized by the Zoology Department at Stanford University under the patronage of Timothy Hopkins, who deployed a sailing schooner. Although the primary interest was to collect vertebrates, ultimately many groups of animals and plants were collected, including coastal marine organisms. On the way back to the United States from the Galápagos Islands, the expedition stopped at Isla del Coco on June 29, 1899, and stayed there for four days. As a result of this expedition several papers were published, mainly on terrestrial vertebrates and plants, as well as on marine organisms (mollusks: Pilsbry and Vanatta 1902, crustaceans: Rathbun 1902; fishes: Snodgrass and Heller 1905). Papers based on those collections were published only until the early twentieth century (e.g., Dall 1920).

In January 1902, another expedition organized by the

Instituto Físico-Geográfico Nacional, and again led by Henri Pittier, visited Isla del Coco, aboard the Steamer *Turrialba*. Paul Biolley participated in this expedition as naturalist of this institute. He collected and described nine species of terrestrial mollusks, two species from fresh or brackish waters, and 23 marine species. He also mentioned five species previously reported by other authors (Biolley 1907). During that expedition Biolley collected a new species of amphipod that was described by Stebbing (1903).

From 1904 to 1905, the US Fish Commission Steamer *Albatross* was deployed again in the Eastern Pacific. No information on this cruise has been found, but several papers have been located in which samples from Costa Rica are mentioned: one containing species of mollusks (Dall 1908), one on medusae (Bigelow 1909), and another one on siphonophorans (Bigelow 1911).

The sixth and last expedition of this first period of exploration of Isla del Coco, conducted at the beginning of the twentieth century, was the California Academy of Sciences Expedition to the Galápagos Islands aboard the Schooner *Academy*. This expedition visited the island from September 3 to 13, 1905, and collected insects, birds, lizards, and freshwater animals. All the marine animals mentioned were large vertebrates: fishes, turtles, sharks, devilfish, and whales, and most of the marine animals they collected were eaten (Slevin 1931).

The next intense phase of exploration occurred between 1924 and 1939 when numerous expeditions visited Isla del Coco. In 1924 there was a British expedition, one of only three European Expeditions to Costa Rica during the early twentieth century. It was led by Cyril Crossland aboard the *St. George*. Mollusks from Isla del Coco were reported by Tomlin (1927) and medusae, collected at two tow-netting stations south of the island, by Kramp (1956). An important expedition during this period was the Arcturus Oceanographic Expedition in 1925, the first oceanographic expedition organized by the New York Zoological Society, and led by William Beebe (Beebe 1926a, b, Tee-Van 1926). It undertook plankton tows south and north of Isla del Coco, and made collections on and around the island for ten days in May 1925 (Beebe 1926c). At least eight papers were published which included information on Costa Rica (Cortés 2012a)—for example, polychaetes (Treadwell 1928), siphonophorans (Bigelow 1931), and cephalopods (Robson 1948).

The next expedition to the island was the cruise with the yacht *Eagle* in 1926, owned and commanded by William Kissam Vanderbilt. It collected specimens at Isla del Coco in 1926, and Lee Boone reported and described new species in many crustacean groups, coelenterates, echinoderms, and mollusks (Boone 1930a, b, 1933). In 1929, Gifford Pinchot traveled to the South Seas, across the Caribbean, the Panama Canal, then Isla del Coco and the Galápagos, before heading to the Marquesas, Tahiti, and Society islands, aboard the MY *Mary Pinchot*. During this so-called Pinchot South Seas Expedition, A.K. Fisher and Henry A. Pilsbry collected fishes, including five species of marine fishes and one freshwater species from Isla del Coco (Fowler 1932).

In 1932, the Templeton Crocker Expedition of the California Academy of Sciences set sail for the Eastern Pacific. During this expedition they sampled in transit from Puntarenas to Isla del Coco and then at the island from June 26 to June 30 (Crocker 1933). Many papers were published, ranging from fungi to birds; at least nine contained information on marine organisms collected in Costa Rica (Cortés 2012a)—for example, Hertlein (1935) on pectinidid bivalves, Burkenroad (1938) on crustaceans, and Bigelow (1940) on medusae.

The Allan Hancock Foundation *Velero III* expeditions to the Eastern Pacific, in collaboration with the University of Southern California, visited Isla del Coco between 1932 and 1938 (Fraser 1943a, b). Samples were collected over seven years: Bahía Chatham in February 1932; Chatham and Wafer bays in February-March 1933; and these two bays plus Nuez Island (now Isla Manuelita) in January 1938. At least 58 papers have been published on the collected material (Cortés 2012a), ranging from algae (Taylor 1945) to echinoderms (Ziesenhenne 1942, Diechmann 1958), and including corals (Durham and Barnard 1952), bryozoans (Osburn 1950), and crustaceans (Haig and Provenzano 1965), just to mention a few. Specimens from these collections are still being analyzed and papers published (e.g., Barnard 1980, Castro 1996).

Another expedition in the 1930s was the Presidential Cruise (Franklin D. Roosevelt, US President, on board the USS *Houston*) from July to August 1938, under the scientific leadership of Waldo L. Schmitt of the Smithsonian Institution. He collected samples on August 3 at Isla del Coco (Schmitt 1939). At least five papers (Cortés 2012a) covering five phyla were published on species from Isla del Coco—for example, Bartsch and Rehder (1939) on mollusks, Hartman (1939) on polychaetes, and Shoemaker (1942) on amphipods.

During the next 15 years, which included the Second World War, only the US Navy Galápagos Expedition in 1941 visited Isla del Coco. Waldo L. Schmitt collected some specimens (Smithsonian Institution Archives). Two publications have been found in which specimens collected during that expedition are mentioned: the first by Roth and Coan (1971) in which they described a new species of gastropod, and another by Manning (1972) who described a new species of stomatopod. The next expeditions took place in the 1950s. First, the Woodrow G. Krieger Expedition to the eastern Pacific collected fishes around the Galápagos Islands and Isla del Coco in 1952–1953. Halstead and Schall (1955) published on the toxicity of venomous fish of Isla del Coco. In 1953 and 1954 the RV *Xarifa*, led by Hans Hass (the third European expedition to Costa Rica), visited the Eastern Pacific. Corals were collected from Isla del Coco by Georg Scheer in 1954 and were described by Durham (1962). Apparently the RV *Vema*, of the Lamont-Doherty Earth Observatory at Columbia University, visited Isla del Coco on November 30, 1958 (Child 1992), but no information about this expedition, nor any other publication in which it is mentioned, has been found.

The Stanford Oceanographic Expedition #20 (September-December 1968) occupied two stations off Isla del Coco where cephalopods were collected (Fields and Gauley 1972). During that expedition Isla del Coco was visited on October 22 but weather conditions were not very good.

Trawls were conducted with no success due to clumps of coral that snagged the net. John S. Pearse collected *Diadema mexicanum* (Agassiz, 1863) by diving (Stanford University Archives), but no publication has been found in which samples collected are mentioned.

The Janss Foundation vessel, RV *Searcher*, was in Costa Rica in 1972, and the chief scientist was William A. Bussing from the Universidad de Costa Rica. It was at the island between March 29 and April 8. Later, Gerald Joseph Bakus published a paper on the toxicity of holothurians (Bakus 1974), and was the first to report on the intertidal and shallow subtidal zonation of organisms (Bakus 1975). Bussing (1983, 1997) described species on the basis of specimens collected during that specific expedition. Then, from May 5 to July 1, 1973, the RV *Velero IV*, cruising primarily off the Pacific coast of Costa Rica, collected a few samples off Panama and Isla del Coco (J. Crampon and K. Fauchald, pers. comm., 2005); no publication from this expedition has been found.

Between 1982 and 1989, malacological collection trips were carried out to Parque Nacional Isla del Coco aboard the Schooner *Victoria af Carlstat*, resulting in more than 50 papers published (Cortés 2012a). These papers significantly increased the number of mollusk species reported for the island, including a large number of new records and species (e.g., Montoya and Bertsch 1983, Shasky 1983, 1989, D'Attilio et al. 1987, Ferreira 1987, Montoya and Kaiser 1988, Reid and Kaiser 2001, Kaiser 2002). In 1986, the Harbor Branch Oceanographic Institution at Fort Pierce, Florida, with the RV *Seward Johnson,* used its submarine, the *Johnson-Sea-Link I*, to collect around the Galápagos Islands and Parque Nacional Isla del Coco, for pharmaceutical prospecting (Pomponi and van Hoek 1987). Only three papers resulted from this expedition; one on azooxanthellate corals (Cairns 1991a), another one on stylasters (Cairns 1991b), and a description of a new species of glass sponge (Reiswig 2010). The island was visited by scientists from the Smithsonian Tropical Research Institute (STRI) in Panamá, on December 2–7, 1987, aboard the RV *Benjamin*, and again on November 7–9, 1990, on the RV *Gyre*. Papers on the distribution, gene flow, and phylogenetics of echinoderms in the eastern Pacific were published (Lessios et al. 1996, 1998, 1999). In 1988 an expedition was organized by the Los Angeles County Museum of Natural History to collect invertebrates and fishes at Parque Nacional Isla del Coco. The sailboat *Victoria af Carlstad* was used (Camp and Kuck 1990, Zimmerman and Martin 1999). During the 1990 RV *Gyre* cruise, 7.5 cm-diameter cores were obtained from six large colonies of the coral *Porites lobata* along the north coast of the island to reconstruct the Holocene growth history of the island's coral reefs (Macintyre et al. 1992). In 1997, aboard the RV *Urracá*, scientists from the Smithsonian Tropical Research Institution, the Bishop Museum, and the University of California, Santa Barbara collected fishes at Parque Nacional Isla del Coco as part of a basin-wide study of fish biogeography (D.R. Robertson, pers. comm., 2007). In 2001, the SSV *Robert Seamans* was in Costa Rican waters between November 9 and 16, and close to Parque Nacional Isla del Coco between November 10 and 14. Plankton and neuston samples were collected, dredging and trawling was done close to the island, and oceanographic data were collected (Graziano 2001). Papers based on the zooplankton samples were published (Morales-Ramírez 2008, Castellanos-Osorio et al. 2012, Gasca and Morales-Ramírez 2012, Jiménez-Cueto et al. 2012). In 2004, during the Global Ocean Sampling Expedition, the J. Craig Venter Institute scientists, aboard *Sorcerer II*, collected two surface samples in Costa Rican waters, one 57 km from Parque Nacional Isla del Coco and the other at Roca Sucia, an islet close to the island. They did metagenomic analyses and found a complex mixture of microorganisms (Rusch et al. 2007).

Some organisms that have been collected have been described as new species of marine organisms found at Parque Nacional Isla del Coco. For instance, a new species of shrimp collected by Michel Montoya in 1992 from Bajo Alcyone at a depth of 35 m was described by Wicksten and Vargas (2001). Curiously, the species described is not intertidal and it is not associated with other organisms like most other members of the genus *Thor* (Kingsley, 1878). In 1993, Ginger Garrison of the US Geological Survey began a study of fish diversity at Parque Nacional Isla del Coco in support of park conservation management efforts. Subsequently, in 2000 a first edition of a guide to the fishes of the island was published and in 2005 a second revised edition appeared (Garrison 2005). In 1999, Rita Vargas-Castillo of the Museo de Zoología, Universidad de Costa Rica, collected a species of brachyuran crab that was later described as a new species (Thoma et al. 2005).

In the early 1990s Alejandro Acevedo-Gutiérrez carried out research on marine mammals at Parque Nacional Isla del Coco. He reported that the most common marine mammal was the bottlenose dolphin, *Tursiops truncatus* Montagu, 1821, which was frequently observed around the island; the Cuvier's beaked whale, *Ziphius cavirostris* Cuvier, 1823, and the false killer whale, *Pseudorca crassidens* Owen, 1846, were occasionally seen; while the humpback whale, *Megaptera novaeangliae* Borowski, 1781, and the Galápagos sea lion, *Zalophus wollebaeki* Sivertsen, 1953 (reported as *Z. californianus*), were rarely encountered (Acevedo-Gutiérrez 1994, 1996, Acevedo-Gutiérrez et al. 1997).

During a 14-month study, Acevedo-Gutiérrez identified 756 individuals of *T. truncatus* from dorsal fin photographs; 89% were sighted just once and most of the other individuals were sighted twice (A. Acevedo-Gutiérrez pers. comm. 2007). *Tursiops truncatus* was observed feeding in waters deeper than 100 m about 500 m off the shore at Parque Nacional Isla del Coco, in groups of about 11 dolphins, with a maximum of 70 individuals (Acevedo-Gutiérrez and Würsig 1991, A. Acevedo-Gutiérrez, pers. comm., 2007). Individuals of this species of dolphin moved faster, jumped more (Acevedo-Gutiérrez 1999), and increased their number of whistles during feeding (Acevedo-Gutiérrez and Stienessen 2004). The spatial arrangement and movement of the dolphins were related to the spatial arrangement of the prey (Acevedo-Gutiérrez and Parker 2000). Acevedo-Gutiérrez and Würsig (1991) and Acevedo-Gutiérrez and Parker (2000) observed several species of seabirds feeding with the dolphins: the brown, *Sula leucogaster* (Boddaert, 1783) and the red-footed, *Sula sula* (Linnaeus, 1766) boobies, and the frigatebird, *Fregata minor* (J.F. Gmelin, 1789). When sharks are present *T. truncatus* stops feeding in the fish clump, possibly to avoid accidental wounds from the sharks. As the number of dolphins increased the number of feeding sharks decreased (Acevedo-Gutiérrez 2002).

Acevedo-Gutiérrez (1994) reported for the first time at Parque Nacional Isla del Coco, the sea lion, *Zalophus wollebaecki* (reported as *Z. californianus*), a species that lives at the Galápagos Islands. Montoya (2008a) analyzed historical records of otariids observed at Isla del Coco and concluded that their occurrence is accidental but recurrent, possibly due to climatic and oceanographic conditions.

Acevedo-Gutiérrez (1994) reported the presence of humpback whales, *M. novaeangliae*, at Parque Nacional Isla del Coco and Golfo Dulce. This last observation suggests the possibility of an overlap of northern and southern populations (Acevedo-Gutiérrez and Smultea 1995). At Parque Nacional Isla del Coco, 22 individuals of false killer whales (*P. crassidens*) were re-sighted at the island 1–4 times, spanning periods of 1 to 730 days (Acevedo-Gutiérrez et al. 1997).

Michel Montoya has been studying the Parque Nacional Isla del Coco for many years (Montoya 2007a and see chapter 8 by Montoya of this volume), especially the island's mollusks (Montoya 1983) and the birds (Montoya 2007b). In his paper on marine birds (Montoya 2008b), he reported 32 species of which eight are reproductive residents. The rest are visitors from the Nearctic (8 species), other tropical regions (11 spp.) and the Eastern Pacific region (13 spp.). Young et al. (2009) reported the presence of the swallow-tailed gull around Parque Nacional Isla del Coco as well as in other areas of the Exclusive Economic Zone (EEZ) of Costa Rica. Montoya (2008b) recommends to study the human impact and the effect of introduced species on the avifauna, as well as the establishment of a monitoring program.

In 2004, Randall Arauz of PRETOMA, a Costa Rica marine conservation advocacy group, in collaboration with the NOAA Southwest Fisheries Science Center, started a shark tagging program using Pop-off Archival Tags and SPOT Smart Position Tags. Data were retrieved from six silky sharks, *Carcharhinus falciformis* (Bibron, 1839), and four pelagic thresher, *Alopias pelagicus* (Nakamura, 1935) (reported as the bigeye thresher, *A. superciliosus*). The migratory routes, covering the Exclusive Economic Zone (EEZ) of six countries and International Waters, point to the need of management at a multinational level (Kohin et al. 2006). Since July 2005, Arauz has been directing a hammerhead shark, *Sphyrna lewini* (Griffith and Smith, 1834), tagging program, in collaboration with the Shark Research Institute, based in Boston, Massachusetts.

Up to 2009, these organizations have jointly carried out five shark tagging expeditions to Parque Nacional Isla del Coco. So far, twenty-five sharks have been tagged with acoustic transmitters (VEMCO V16 Coded Multi-Purpose Transmitters), and four receivers have been deployed around the Island. Additionally, four sharks have been tagged with PAT satellite tags. Preliminary results indicate hammerheads school around different seamounts during the day, while during the night they travel to deep waters (over 300 m) where they feed (Arauz and Antoniou 2006). Unfortunately, tag loss rate is high, limiting observations to the first three weeks, although two sharks tagged in July of 2005 with acoustic transmitters returned to the Island over a year later (R. Arauz, pers. comm., 2007). Another shark study was published by Whitney et al. (2004). On the basis of videos, these auhtors described courtship and mating events of white-tip reef sharks, *Triaenodon obesus* (Rüppell, 1837), at Parque Nacional Isla del Coco.

They observed some types of behavior never before recorded for that species and some never observed in any elasmobranch. Whitney and Motta (2008) observed that the white-tip reef shark, at Isla Manuelita, assumed a cleaning (mouths widely agape) pose when tactilely stimulated by swarming hyperiid amphipods, but are not cleaned. Four recent papers have been published on sharks from Isla del Coco: two are on deep water observations of sharks and rays (Cortés et al. 2012, Starr et al. 2012a), another is a report of a species of shark that was not known from the Eastern Tropical Pacific (López-Garro et al. 2012), *Carcharhinus melanopterus*, and finally, there is a paper on the abundance and population structure of the white-tip reef shark (Zanella et al. 2012), *Triaenodon obesus*.

Starting in the 1980s and continuing until today, Parque Nacional Isla del Coco has been visited by scientists from the Centro de Investigación en Ciencias del Mar y Limnología (Marine Science and Limnology Research Center, CIMAR) of the Universidad de Costa Rica. William A. Bussing has described new species of fish from Parque Nacional Isla del Coco (e.g., Bussing 1997, Bussing and Lavenberg 2003), including endemic species (Bussing 1983, 1991, 2010). Bussing and López (2005) published a book on marine fishes of Parque Nacional Isla del Coco, and recently, Angulo et al. (2014) reported the presence of manefish that are associated to siphonophores.

Two papers on the impact of the El Niño-Southern Oscillation (ENSO) on the island's corals reefs have been published (Guzmán and Cortés 1992, 2007) and a list of 17 species of zooxanthellate scleractinians have been reported (Cortés and Guzmán 1998). A detailed bathymetric map of the insular platform around Parque Nacional Isla del Coco was produced by Lizano (2001), and updated including previously unknown seamounts (Lizano 2012). The platform is less than 180 m deep and extends 18 km in a northeast-southwest direction, being widest on the east side.

Two expeditions conducted in 2006–2007, aboard the MV *Proteus* of MarViva, resulted in at least five new records of echinoderms (Alvarado and Chiriboga 2008, Alvarado 2010), a new species of octocoral (Breedy and Cortés 2008, 2011), new records of crustaceans (Vargas-Castillo and Werhtmann 2008), and possibly other new species and records of marine organisms. The rocky intertidal has been studied in detail (Sibaja-Cordero 2008, Sibaja-Cordero and Cortés 2010), and environments between 50 and 450 m are described for the first time based on photographs with a submersible (Cortés and Blum 2008).

Since 2007 scientists from CIMAR have been visiting the island to study the marine environments, with funding from the French Fund for the World Environment (FFEM), and since 2010 with funds of the Consejo Nacional de Rectores (CONARE: Council of Rectors of the Public Universities of Costa Rica). Consequently, more species have been discovered resulting in new records as well as new species (Fernández 2008, Suárez-Morales and Morales-Ramírez 2009, Alvarado 2010, Dean et al. 2010a, b, 2012, Breedy and Cortés 2011, Acuña et al. 2012, 2014, Breedy et al. 2012, Gasca and Morales-Ramírez 2012, Sibaja-Cordero et al. 2012c, Suárez-Morales and Gasca 2012, Vargas-Montero et al. 2012, Esquete et al. 2013, Sibaja-Cordero et al. 2013, Angulo et al. 2014), new environments have been explored (Sibaja-Cordero and Cortés 2010, Cortés et al 2012, Sibaja-Cordero et al. 2012a,b,c, 2014, Starr et al. 2012b), many samples have been taken, and several environmental processes are being studied.

The large-scale circulation around Isla del Coco has now been described (Lizano 2008), as well as the climatic conditions and how they will probably change in the future (Alfaro 2008, Quirós-Badilla and Alfaro 2009, Maldonado and Alfaro 2010, 2012, Hidalgo and Alfaro 2012). A bioerosion model on the impact of *Diadema mexicanum* was developed as well (Alvarado et al. 2012), and the genetic structure of the main reef-building coral, *Porites lobata*, was determined recently (Boulay et al. 2012).

In September 2009, the National Geographic Society organized an expedition to Isla del Coco, addressing two objectives. The first was to estimate the abundance and biomass of reef fishes and sharks in the shallow reefs around the Island, a relatively pristine ecosystem, and contribute to the understanding of the migratory movements of sharks. The second objective was to explore the deep water environments of Parque Nacional Isla del Coco and the seamount southwest of the Island, Los Picos de la Isla del Coco (also known as Las Gemelas), down to a depth of 400 m. Results of the fish study indicate that Parque Nacional Isla del Coco has the highest biomass of fish of any place globally studied so far, the most complete assemblage of trophic levels, and the highest densities of apex carnivores (Sala 2009, Friedlander et al. 2012). The seamount, whose shallowest point was 165 m deep, has been heavily overfished, but there are still populations that could recover if the exploitation stops. During this expedition the mechanical arm of the *DeepSee* submersible was used and collections of numerous species were made, many of which are new reports for the area and some are species new to science (Cortés et al. 2009, 2012, Breedy et al. 2012, Starr et al. 2012a,b, Angulo et al. 2014).

Current State of Knowledge

The first synthesis on the state of knowledge of marine biodiversity of Isla del Coco was published by Leo George Hertlein (1963). In his excellent compilation, which includes an extensive bibliography, he included 319 marine species. By 2009, more than 1,100 marine species had been identified from Parque Nacional Isla del Coco, of which over 30, that is 3% of the total, are endemic species (Wehrtmann et al. 2009). Cortés (2012b) reported 1,688 marine species from Isla del Coco. The species richness of Parque Nacional Isla del Coco is very high in view of its relatively small size (Cortés 2012b). Zimmerman and Martin (1999) reported ten species of brachyuran crabs from Parque Nacional Isla del Coco, including two new species; they wrote: "The increase in species richness suggested by these collections may make Cocos the most specious single island in the eastern Pacific, after the remaining crab are studied." Allen (2007) reported that Parque Nacional Isla del Coco is in the list of top-ranking areas on the basis of levels of endemism and in number of endemics (per unit area) of coral reef fishes in the Indo-Pacific.

Some taxonomic groups—for example, fishes (Bussing and López 2005, Garrison 2005, Cortés et al. 2012, López-Garro et al. 2012, Angulo et al. 2014), corals (Marenzeller 1904, Durham 1962, 1966, Durham and Barnard 1952, Cairns 1991a, Cortés and Guzmán 1998, Lattig and Cairns 2000) and mollusks, especially gastropods (Montoya 1983, Shasky 1988, Rodríguez et al. 2009), have been studied for many years and in great detail, while other important components of the marine biodiversity have been neglected—for example, benthic macroalgae and plankton—until recently (Fernández 2008, Morales-Ramírez 2008, Bernecker and Wehrtmann 2009, Castellanos-Osorio et al. 2012, Gasca and Morales-Ramírez 2012, Suárez-Morales and Gasca 2012, Vargas-Montero et al. 2012). There are groups that have never been studied even though they are living on and around the island—for example, cyanobacteria and flatworms. New records and species are found in practically every expedition to the island.

The number of endemic species is low, but that is common in marine environments (Cortés 2012b). Most endemics are fishes, a very well-studied group, and most have been described during the last 30 years (15 of the 23 species). Of the other endemics most were described before 1955, although there are a few exceptions—for example, the stony corals (Hydrozoa and Scleractinia) described in 1991. A new wave of species descriptions is taking place with a glass sponge in 2010, two octocorals, one in 2011 and one in 2012, and more species descriptions currently being prepared.

The coral reefs of the island have been studied during several occasions (Bakus 1975, Macintyre et al. 1992, Guzmán and Cortés 1992, 2007), and a summary of those studies is presented below in the coral reef section. With the CIMAR-FFEM project and the CONARE funding new information on the coral reef ecology and history, the diversity of marine organisms, plankton diversity and ecology, physical and chemical oceanography, and climatology of Isla del Coco is currently being generated.

The Physical Environment

Geology and Geomorphology

Isla del Coco is the only subaerial exposed section of the aseismic Cocos Cordillera that extends from the Galápagos hotspot to the Osa Peninsula in southern Costa Rica, a total of 1,200 km. The Coco Volcanic Cordillera is the longest cordillera (mountain range) of Costa Rica with 700 km within its EEZ (Alvarado-Induni 2000, Rojas and Alvarado 2012; Alvarado and Cárdenas, chapter 3 of this volume). On the basis of GPS data, the island, which is sitting on the Cocos Plate, is converging toward Costa Rica at the NE between 72 (Kellog and Vega 1995) and 90 mm/yr (Bohrmann et al. 2002), with the most recent estimate as 78 mm/yr (Protti et al. 2012), one of the highest rates in the world. Isla del Coco is the summit of a seamount, with waters shallower than 180 m extending 18 km in a southwest-northeast orientation (Lizano 2001, 2012), with 111 km^2 of substrate less than 100 m deep (Robertson et al. 2004). From the edge of the insular platform the depth gradient is very steep down to 1,000 m to the crest of the cordillera and continuing down to over 3,000 m at the base. Submarine arches have been observed around the island. The cordillera below the island is thought to be middle Miocene (10–15 million years). The island is much younger, about 2 million years old (Dalrymple and Cox 1968, Bellon et al. 1983), and is an example of a shield volcano superimposed on early seamount (hot-spot) volcanism (Rojas and Alvarado 2012).

On the basis of field work and petrologic analyses, Castillo et al. (1988) proposed the following geological history of Isla del Coco: the first stage corresponds to a shield volcano formation, followed by explosive volcanism (pyroclastic and trachyte series) in a caldera, in what is now the northeast of the island, around Wafer Bay. In the third and youngest volcanic stage, volcanism (massive lava flows) moved to the southwest, characterized by fissural eruptions and centered at Cerro Iglesias (Malavassi 1982, Castillo et al. 1988, Alvarado-Induni 2000, and see Alvarado and Cárdenas, chapter 3 of this volume). The last period is the post–explosive volcanism stage, when erosion takes over, resulting in the present geomorphology. The island is surrounded by very steep and high sea cliffs, with high waterfalls in some areas (Fig. 7.4), rivers and creeks, deeply eroded ravines, caves, and only two sandy beaches (Bahía Wafer and Bahía Chatham) (Rojas and Alvarado 2012, and see Montoya, chapter 8 of this volume).

Climate

The climate at Isla del Coco was first studied by Henri Pittier (Pittier 1899), and later by Protti (1964) and Rojas-Acuña (1964). A semi-automatic weather station was installed in Bahía Chatham, and was operational from May 1979 to June 1980 (Reyes and Vogel 1981, Fernández 1984, Alfaro 2008). During that period minimum average air temperature was 22.7°C and the maximum average was 25.1°C (Fernández 1984). Precipitation is between 5,000 and

5,500 mm at Bahía Chatham, 5,500 to 6,000 mm at Bahía Wafer, and between 6,000 and 7,000 mm at Cerro Iglesias (Herrera 1985 and chapter 2 of this volume, Alfaro 2008). Isla del Coco is in the band of heavy rains of the Eastern tropical Pacific (Amador et al. 2006). The island's weather depends on the north-south movement of the Intertropical Convergence Zone (ITCZ). Between February and March the ITCZ is at its maximum southern position (Taylor and Alfaro 2005, Amador et al. 2006) and this is the period of minimum rainfall at Isla del Coco. The rainy season takes place during the northern summer months with the highest peak in October. In late July-early August there is the mid-summer drought ("el veranillo" in Spanish), similar to the one observed on the mainland Pacific coast, when the amount of rain diminishes (Amador et al. 2006). Winds are predominantly coming from the southwest (Fernández 1984, Herrera 1985, Amador et al. 2006). Winds are stronger during May-December (mainly in October) with relatively low wind speed occuring from January to April (Fernández 1984). The available data are from short periods, thus much longer time series are necessary to refine the climatology of Isla del Coco (Alfaro 2008).

Water Circulation

Isla del Coco is located within the Eastern Pacific Warm Pool (Fiedler and Talley 2006, Pennington et al. 2006). The waters around the island are influenced by the North Equatorial Counter Current (NECC) (Fig. 7.5) and the fresh water plume from the Panama Bight (Fiedler and Talley 2006, Kessler 2006, Lizano 2008). As with the climate of the island, the position of the NECC varies seasonally in accordance with the position of the Intertropical Convergence Zone (ITCZ). Between February and April, when the ITCZ is located more to the south, the NECC does not reach Isla del Coco, and a cyclonic (counterclockwise) wind-driven gyre centered on the island is established. Between August and September, when the ITCZ migrates north, the island is reached by a strong NECC (Kessler 2006, Lizano 2008). At this time of the year, larvae and juveniles of marine organisms may be transported from the Central Pacific to Isla del Coco, and especially during ENSO conditions in which the easterly flow of the current intensifies (Scheltema 1988, Montoya and Kaiser 1988, Richmond 1990, Guzmán and Cortés 1993, Cortés 1997). Waves are generally running from the southwest but during the boreal winter waves generated in the North Pacific and by the Papagayo wind jet do reach the island (Amador et al. 2006, Alfaro 2008). The thermocline is relatively shallow, around 50 m or less (Fiedler and Talley 2006).

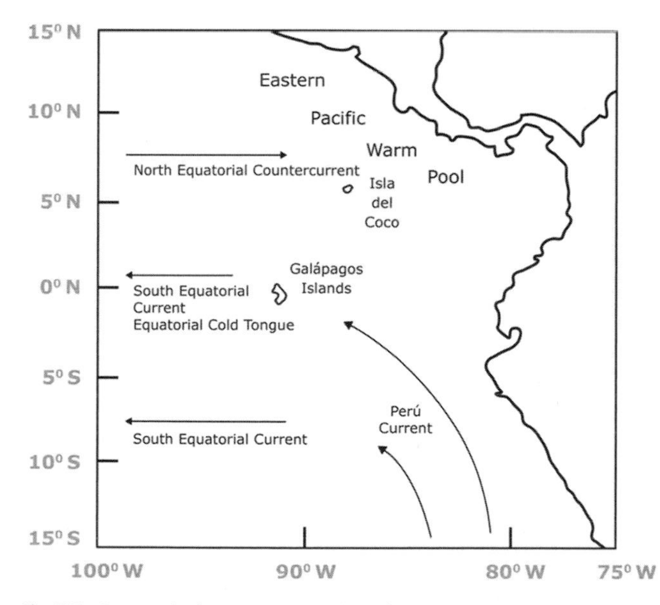

Fig. 7.5 Currents in the Eastern Tropical Pacific. *Based on Kessler (2006).*

Ecosystems

Beaches

Isla del Coco has two sandy beaches, both on the north side of the island: Bahía Wafer and Bahía Chatham. These are the two main landing sites for boats. There are also several cobble beaches around the island (Fig. 7.6). There are no publications on the organisms, dynamics, or sediment composition of the beaches.

Rocky Intertidal Zones

The rocky intertidal zones of the island were first described by Bakus (1975) on the basis of his observations at the Chatham and Wafer bays in 1972. In 2007, Jeffrey Sibaja-Cordero studied the rocky intertidal zones around the island, as well as around the adjacent islands and islets (Sibaja-Cordero 2008). He reported a distinct zonation of organisms on the rocks and species composition varied according to rock type (Fig. 7.7). Sibaja-Cordero (2008) found that the lowest intertidal zone is covered by pink and black incrusting algae that grow over the barnacles *Tetraclita stalactifera* (Lamarck, 1818) and *Megabalanus coccopoma* (Darwin, 1854). The barnacles extend above the algal band, and were more abundant over continuous rock outcrops than on large blocks in the bays. Before the barnacles, gastropods were the predominant organisms, with *Siphonaria gigas* Sowerby, 1825 in the mid-intertidal zone, and *Nodilittorina modesta* (Philippi, 1846), *Nerita scabricosta* Lamarck, 1822, and *Nerita funiculata* Menke,

Fig. 7.6 Beach at Bahía Chatham, one of the two sandy beaches at Isla del Coco. *Photograph by Mariana Cortés Kandler, January 2007.*

1851 in the upper intertidal zone. Bakus (1975) indicated that *N. modesta* (listed as *Littorina modesta*) was below *S. gigas*, and that *N. scabricosta* (and hundreds of their egg capsules) was the predominant organism in the upper intertidal zone. Sibaja-Cordero (2008) also noted that *S. gigas* was very abundant at some outcrops, while absent in other sections of the island. Above the *S. gigas* band the crab *Grapsus grapsus* (Linnaeus, 1758) was observed with a high variation in abundance (Sibaja-Cordero 2008). Bakus (1975) indicated that the rock louse (isopod) *Ligia* sp. ranged throughout most of the intertidal zone.

The temporal change of the rocky intertidal zone was studied by Sibaja-Cordero and Cortés (2010). Changes were observed but they were small compared to a seasonal upwelling area (Sibaja-Cordero and Cortés 2008). Even so, when the temperature was lower the abundance of green algae increased in the infralittoral; when the temperature was higher red algae predominated. Temporal changes increased from the supralittoral to the infralittoral (Sibaja-Cordero and Cortés 2010).

In the intertidal zone of Isla del Coco there are 35 caves, which range in size at the entrance from 0.5 to 6.5 m high, and 1 to 12 m wide. They have a water depth at high tide from 0 to 11 m, and the length of those caves ranges from 2 to 42 m (Fig. 7.8). There are also three arches: 0.5–2.5 m high, 1–2 m wide, 0.5–4 m deep at high tide, and from 2.5 to 10 m long. Finally, there are two tunnels (more than 10 m long), both 5 m high and 3 to 4 m wide, with a water depth between 4 and 7.5 m, and lengths of 40 and 91 m. The latter one is at Punta Presidio (Fig. 7.9). On Isla Manuelita there are 6 caves, ranging from 0.5 to 11 m high and 0.5 to 13.5 m wide, with a water depth of 0 to 10 m, and length from 8 to 35 m (Benumof and Lockwood 2000, Rojas and Alvarado 2012). Additionally, there is an arc on Isla Muela, which is about 8 m high, 2 m wide, and 3 to 4 m long (Fig. 7.10). The flora and fauna of these caves, arches, and tunnels have not yet been studied.

Coral Reefs

Isla del Coco has extensive coral reefs and coral communities (Bakus 1975, Guzmán and Cortés 1992). Eighteen species of zooxanthellate corals (Cortés and Guzmán 1998) and 15 of azooxanthelates (Cairns 1991a, Cortés 2012b) have been reported from the island (Table 7.1). This is the highest number of coral species reported from any site in Costa Rica (Cortés et al. 2010). The reefs range from less than one hectare to more than 50 and most are on the northern side of the island (Bakus 1975, Guzmán and Cortés 1992). The main reef builder is *Porites lobata* Dana, 1846 (Fig. 7.11), but there are some areas with a predominance of agariciids: *Pavona clavus* (Dana, 1846), *Pavona varians* Verrill, 1864, *Leptoseris scabra* and *Gardineroseris planulata* (Dana, 1846). Senn and Glasstetter (1989) indicated the presence of several barnacle reefs, but mentioned only the one at Roca Sumergida on the southern side of the island. They also say, erroneously, that corals are sparse, are less than one meter in size, and that they are probably killed by decreased salinity due to rainfall and to the cold waters of the Humboldt Current. Corals are abundant and form reefs, low salinity does not seem to be an important limiting factor, and the Humboldt Current does not reach Parque Nacional Isla del Coco.

Macintyre et al. (1992) collected six 7.5 cm (diameter) cores from large colonies of *P. lobata* from the reefs in Bahía Chatham and Isla Pájara. The longest cores were 3.1 m and 3.0 m, and their bases were dated at 250 ± 50 and 430 ± 80 years BP, respectively. They indicated that the bases of large colonies are subject to intense bioerosion, so the dates found are an underestimation of the age of the largest colonies and the reefs themselves. In January 2008, as part of the FFEM-CIMAR project, Carlos E. Jiménez collected

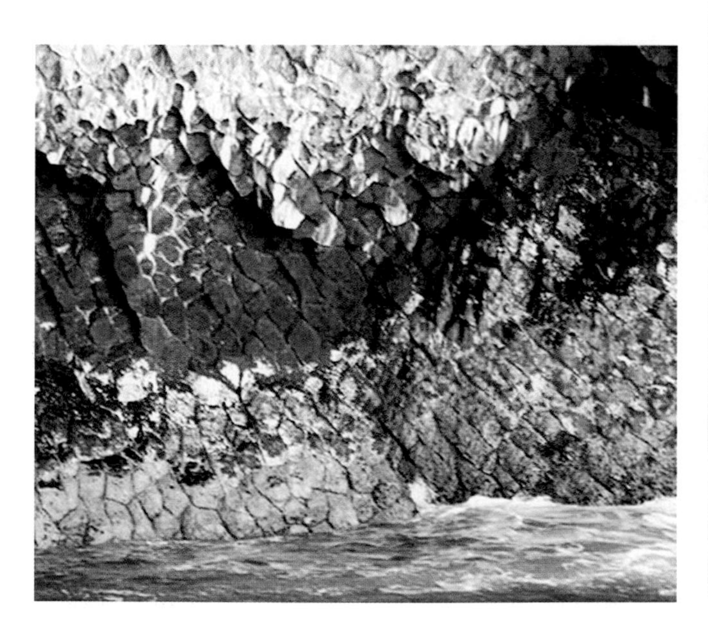

Fig. 7.7 (a) Isla Juan Bautista on the south side of the island. The rocks are columnar basalts. (b) Rocky intertidal zone at Isla Muela on the south side of the island. In both photographs the pink and black bands on the lower intertidal zone are incrusting algae. No *Siphonaria gigas* is observed on the left, while several individuals are observed in the photograph on the right.
(a) photograph by Jorge Cortés, January 2007, (b) photograph by Jeffrey Sibaja-Cordero, January 2007.

Fig. 7.8 Caves in the intertidal zone at Isla del Coco.
Photographs (top and bottom) by Mariana Cortés Kandler, January 2007.

coral cores ranging from 0.7 to 1.5 m, with possible ages from 90 to 190 years.

The coral reefs at Isla del Coco were severely impacted by the 1982–1983 ENSO event, with around 90% of the corals dying (Guzmán and Cortés 1992). In 1987, live coral cover was 2.99 ± 0.25% (Guzmán and Cortés 1992), but in 2002 the same reefs had a live coral cover of 14.87 ± 6.78% (Guzmán and Cortés 2007).

There are organisms that impact corals—for example, the corallivores that feed on them or the bioeroders that convert them to rubble and sand. Among the corallivores the crown-of-thorns starfish *Acanthaster planci* (Linnaeus, 1758) and the pufferfish *Arothron meleagris* (Bloch and Schneider, 1801) are the most important ones. *Acanthaster planci* had densities of 10.4 ± 3.1 ind/m^2 in 1987 that dropped to 7.4 ± 2.3 ind/m^2 in 2002, although this difference was statistically not significant. Conversely, during the same period densities of *A. meleagris* increased significantly from 8.2 ± 0.5 to 152.6 ± 37.5 ind/m^2. This phenomenon could affect the recovery of some coral reefs on the island (Guzmán and Cortés 2007).

Among the important bioeroders one finds the sea urchin *D. mexicanum*, whose densities decreased from 14.5 ± 1.6 ind/m^2 in 1987 to 0.84 ± 0.3 ind/m^2 in 2002. Decreases in bioerosion are important in reducing the biological destruction of the reef structure at Isla del Coco (Guzmán and Cortés 2007, Alvarado et al. 2012). In 2007, average densities of *D. mexicanum* around Isla del Coco ranged from 0.03 to 6.53 ind/m^2 (Alvarado and Chiriboga 2008, Alvarado et al. 2012).

Fig. 7.9 The tunnel at Punta Presidio, which is 91 m long.
Photograph by Jorge Cortés, January 2007.

Open Ocean and Deep Waters

Isla del Coco is surrounded by very deep waters, but there have been very few studies on the organisms living in the water column or as part of the benthos. Agassiz (1892), in his report on the 1891 *Albatross* Expedition, indicates that samples were collected from as deep as 3,566 m, and from surface tows. The deep sea bottom consisted mainly of globigerina ooze. Papers have been published on some of the animals collected—for example, Faxon (1895) on crustaceans, Agassiz (1898) on deep water echinoids, Wilson (1904) on deep water sponges, and Dall (1908) on mollusks, but no publication has been found on the surface tow-net samples that were collected at that time.

Since 2006, a submersible (*DeepSee*, operated by the Undersea Hunter Group), with capability of immersions down to 450 m, has been used around Parque Nacional Isla del Coco. Pelagic and benthic organisms have been photographed, mainly on the northern side of the island and in 2007 in the southern sector. What is surprising is that there is no gradual change from a shallow (< 50 m) to a deep fauna, but an abrupt change is observed around 50 m.

Fig. 7.10 Isla Muela and its arch, Bahía Iglesias, south side of Isla del Coco. *Photographs by Jorge Cortés, January 2007.*

Below that depth many species, including entire groups of organisms never seen in the shallow waters, are being observed (Fig. 7.12) (Cortés et al. 2012, Starr et al. 2012a, b). Another surprise is that some organisms at the southern side of the island are actually different to those found in the northern sector (Cortés and Blum 2008).

People and Nature

Human Population

Today the island is inhabited only by park personnel, volunteers, and other temporary visitors—for example, scientists. At this point in time there are no permanent human residents on the island.

Threats

The main natural threat to the marine habitats of Parque Nacional Isla del Coco is the recurrent ENSO warming event, which has been shown to be devastating (Guzmán and Cortes 2007). As climate change intensifies it is predicted that seawaters will get warmer and more acidic (Kleypas et al. 2006, IPCC 2007), which will lead to a greater negative impact on the marine environment and reduced calcification by organisms. Additionally, future warming may have an adverse effect on biodiversity (Mayhew et al. 2007).

In 2007, fishing pressure was the leading human impact affecting Parque Nacional Isla del Coco (Fig. 7.13). According to reports by park personnel, MarViva, and the Costa Rican Coast Guard, 1,214 boats were observed around Parque Nacional Isla del Coco in 2006, most between August and November. That year, 393 lines (most of them several kilometers long) were retrieved (Fig. 7.14). Attached to those lines were 605 tunas, 190 sharks, 34 manta rays, 16 turtles, and 12 marlins, among other species. During the first half of 2007, 62 lines were collected with thousands of hooks and many animals. Some estimates indicate a 60% decline of relative abundance of sharks in the EEZ of Costa Rica during 1991 to 2001 (Arauz et al. 2004, 2006). For now the island's marine resources are being protected but at a very high economic cost.

Mora et al. (2011) analyzed human impact on coral reef

Table 7.1. Scleractinian Corals Reported for Isla del Coco, Costa Rica

	Zooxanthellate	Azooxanthellate
FAMILY POCILLOPORIDAE		
Pocillopora damicornis Linnaeus, 1758	X	
Pocillopora elegans Dana, 1846	X	
Pocillopora eydouxi Milne Edwards & Haime, 1860	X	
Pocillopora meandrina Dana, 1846	X	
FAMILY PORITIDAE		
Porites lobata Dana, 1840	X	
FAMILY SIDERASTREIDAE		
Psammocora stellata Verrill, 1866	X	
Psammocora superficiales Gardiner, 1898	X	
FAMILY AGARICIIDAE		
Gardineroseris planulata Dana, 1846	X	
Leptoseris papyracea Dana, 1846	X	
Leptoseris scabra Vaughan, 1907	X	
Pavona chiriquensis Glynn, Maté & Stemann, 2001	X	
Pavona clavus Dana, 1846	X	
Pavona gigantea Verrill, 1869	X	
Pavona maldivensis Gardiner, 1905	X	
Pavona varians Verrill, 1864	X	
Pavona xarifae Scheer & Pillai, 1974	X	
FAMILY FUNGIDAE		
Fungia (Cycloseris) curvata Hoeksema, 1989	X	
Fungia (Cycloseris) distorta Michelin, 1842	X	
FAMILY RHIZANGIIDAE		
Astrangia dentata Verrill, 1866		X
Cladocora debilis Milne Edwards & Haime, 1849		X
Cladocora pacifica Cairns, 1991		X
Culicia stellata Dana 1846		X
FAMILY CARYOPHYLLIIDAE		
Anomocora carinata Cairns, 1991		X
Caryophyllia diomedeae Marenzeller, 1904		X
Caryophyllia perculta Cairns, 1991		X
Coenocyathus bowersi Vaughan, 1906		X
Desmophyllum dianthus Esper, 1794		X
Polycyathus hondaensis Durham & Barnard, 1952		X
Tethocyathus prahli Lattig & Cairns 2000		
FAMILY FLABELLIDAE		
Javania cailleti Duchassaing & Michelotti, 1864		X
FAMILY DENDROPHYLIIDAE		
Dendrophyllia oldroydae Oldroyd, 1924		X
Endopachys grayi Milne Edwards & Haime, 1848		X
Rhizopsammia verrilli Van der Horst, 1922		X
Tubastrea coccinea Lesson, 1829		X

Source: Cairns 1991a, Cortés and Guzmán 1998, Cortés and Jiménez 2003, and Cortés 2009b, 2012.

biodiversity around the world, including the Eastern Tropical Pacific (ETP). They found a negative correlation between human stressors and biodiversity. Edgar et al. (2011) compared "no-take" with limited protected and openly fished areas at 190 sites in the ETP, 18 of them around Parque Nacional Isla del Coco. The "no-take" areas had higher biomass of carnivorous fishes and higher coral cover than limited protected or fished areas.

Another threat to the island ecosystems is the insufficient control of marine tourist activity by the park's authorities. There has been no underwater monitoring of diving activity, so the effect of voluntary and involuntary contact with the marine fauna is unknown. However, it is being studied at present. Other types of human impact, such as water pollution, remain relatively low in terms of severity (Acuña-González et al. 2008).

Conservation

Protected Area

Isla del Coco has been a National Park since 1978, but it was not until 1992 that the National Park Service took full control of the island, led by Joaquín Alvarado García (El Indio) (Fig. 7.15) as its Director. This is a turning point in the island's history, because at that moment more effective protection took place, especially of the marine environments. Other significant turning points are the total restriction of access of fishing boats to the protected area in 2002, and the creation of a buffer zone around Parque Nacional Isla del Coco, the Seamounts Marine Management Area, in 2011.

Conservation Organizations

Three organizations have been assisting the National Park. The first is the Fundación Amigos de la Isla del Coco (Friends of Cocos Island Foundation, FAICO), created in 1994, a non-governmental organization devoted to seeking funds and promoting conservation of the island worldwide. Asociación MarViva, a non-profit, non-governmental organization, was created in 2002. Its mission is to collaborate with local authorities in the conservation and protection of marine environments, especially at Parque Nacional Isla del Coco. Conservation International (CI) has funded several expeditions to the island.

Conservation Strategies and Action

To further protect the island from overfishing, a buffer zone beyond the current boundaries at 22.2 km has been created (March 2011). Within this zone non-industrial exploitation of fishes would be allowed, and industrial exploitation banned. The aim is to reduce the pressure on commercial fish stocks because every year on average, the

Fig. 7.11 Coral reef at Isla Pájara. The predominant coral species is *Porites lobata*. *Photograph by Jorge Cortés, January 2007.*

Fig. 7.12 Unknown species of soft coral of the order Alcyonacea, family Alcyoniidae (identified by Leen van Ofwegen, Leiden Museum, the Netherlands), photographed from the submersible DeepSee (UnderSea Hunter Group) on the north side of Isla del Coco, at a depth between 280 and 330 m.
Photograph by Nico Ghersinich, October 2005.

international industrial fishing fleets capture more than ten times what the combined national non-industrial fleets capture (INCOPESCA 2006). Patrolling operations of the near-shore waters of the island will continue, as will monitoring of the shallow benthic environments.

An important effort in the region is the Eastern Pacific Marine Corridor Initiative for the conservation of marine habitats and organisms in the region, especially in relation to the UNESCO World Heritage Site islands. The agreement was signed in 2004 by Ecuador (Galápagos Islands), Colombia (Malpelo and Gorgona), Panamá (Coiba), and Costa Rica (Isla del Coco). The Marine Corridor covers an area of 211 million hectares and its main objective is the protection, conservation, and wise use of marine resources in the region to improve the quality of life of its inhabitants. Many organizations are involved: several United Nations bodies, government institutions of the four countries, nongovernmental organizations, especially Conservation International (CI), academic institutions, tourist organizations, and local organizations and communities.

In close cooperation with CI and specialists of other conservation and research institutions—for example, CIMAR and the Smithsonian Tropical Research Institute (STRI)—The Nature Conservancy (TNC) recently published an ecoregional assessment that makes use of benthic modeling and identifies coastal and marine priority sites of important conservation value in the Nicoya, Panama Bight, and Cocos Island marine ecoregions (TNC 2008). This study can serve as a key tool for setting conservation priorities around Parque Nacional Isla del Coco and further guide conservation

Fig. 7.13 Fishing boats at 22.2 km (12 miles) from the island—that is, at the border of the protected marine area.
Photograph by Jorge Cortés, January 2007.

strategies and action in this National Park and other marine protected areas in the Eastern Tropical Pacific (ETP).

A Costa Rican government project for Parque Nacional Isla del Coco, funded by the United Nations Development Program (UNDP) and the French Fund for the World Environment (FFEM), has started to monitor diving and visitation intensities, and the biological, physical, and chemical conditions of the surrounding waters of the island, by enhancing basic scientific research and establishing a monitoring system. The monitoring program will contribute to ACMIC's decision-making regarding marine tourism management, and it will be valuable in establishing carrying capacity levels for visitors and vessels entering the waters around the island. Scientific research has continued with funds from the Consejo Nacional de Rectores (CONARE) of Costa Rica.

Economic Valuation of Environmental Goods and Services
Funded by CONARE, Mary Luz Moreno-Díaz has applied a series of methodologies to value goods and services of Parque Nacional Isla del Coco (Moreno-Díaz 2012). She calculated an overall gross income derived from activities related to Isla del Coco at approximately $8.3 million, mainly as a result of tourist visitation. She indicates that if the natural resources being visited by tourists degrade owing to the effects of climate variability these benefits could decrease by 30%. To this degradation factor we must add the issue of illegal fishing, which is exerting a heavy toll on marine organisms of Parque Nacional Isla del Coco.

Research Needs

Most of the research related to marine ecosystems of Parque Nacional Isla del Coco has been biological in nature, with a focus on the biodiversity of some particular taxonomic groups—for example, fishes, mollusks, and corals. And as mentioned above, some taxa have never been studied, even though they are known from the island—for example, cyanobacteria and flatworms. Until 2012, no investigation has looked at primary productivity of the water column or its benthos, or on interactions among species, with a few exceptions concerning reef organisms (Guzmán and Cortés 2007). There are no studies on population dynamics of most reef organisms or on the reproductive biology of corals or other marine species. There are just very few, if any, publications on the physical, chemical, or geological oceanography.

Even though the pelagic fishes and resident sharks are one of the main attractions of the island, if not the most important one, until 2012 there were no studies on their species populations, dynamics, interactions, seasonality, or any other aspect of their ecology or fisheries status. And almost nothing is known about the behavior, habitat preferences, and migrations of pelagic sharks, many of which are being caught incidentally by major fisheries in the ETP. This type of information on pelagic fish is strongly needed to facilitate the development of effective management strategies, at local, national, and international levels.

Conclusions

Why should Costa Rica or, for that matter, the world be concerned about Parque Nacional Isla del Coco? This is a valid question because the island is remote and expensive to protect, and very few people will ever visit it. The island is a real biological, ecological, and scenic treasure. In fact, Parque Nacional Isla del Coco is an important and increasingly rare natural laboratory. Because of its isolation, we can study phenomena that are relatively unaffected by human activity. The island is an excellent experimental control site for comparative studies on processes that transform the world we live in. We should also remember that Isla del

Fig. 7.14 Fishing lines retrieved by park personnel, MarViva, and the Coast Guard, within the park boundaries. *Photograph by Odalisca Breedy, January 2007.*

Fig. 7.15 Joaquín Alvarado, El Indio, Director of Parque Nacional Isla del Coco from 1992 to 1997.
Photograph by Yanina Rovinsky, 1993.

Coco acts as the entrance gate for Indo-Pacific organisms into the ETP (Cortés and Jiménez 2003), which was one of the reasons why it has formally received the status of Natural World Heritage Site.

Acknowledgments

I thank Maarten Kappelle and the late Luis Diego Gómez for their invitation to contribute to this book. Economic support to write this chapter was provided by the Universidad de Costa Rica through the Escuela de Biología, and for field work by the Vicerrectoría de Investigación and the Centro de Investigación en Ciencias del Mar y Limnología (CIMAR), both at the Universidad de Costa Rica (UCR). Funding to visit the island has been provided by the Universidad de Costa Rica, Conservation International, the French Fund for the World Environment (FFEM), and the Consejo Nacional de Rectores de las Universidades Públicas de Costa Rica (CONARE). Michel Montoya reviewed the manuscript and provided a huge amount of bibliography, and many comments about the islands' history, for which I am grateful. Eric Alfaro reviewed the section on climate, Guillermo Alvarado-Induni the section on geology, Omar Lizano the section on circulation, Gary Williams the paragraphs on the expeditions of the California Academy of Sciences, and Randall Arauz the shark tagging section. Kifah Sasa checked the chapter and contributed valuable information on conservation strategies for the island. Ginger Garrison, Yanina Rovinsky, Pablo Madriz, Alejandro Acevedo, and Peter J. Mumby did detailed reviews of the chapter, greatly improving it. Jeffrey Sibaja-Cordero provided unpublished information and advice regarding the rocky intertidal zones. Xochilt Lezama Cáceres helped with the maps. I appreciate the help and support of ACMIC, SINAC, MarViva, Conservation International, the French Fund for the World Environment, and the Consejo Nacional de Rectores de las Universidades Públicas de Costa Rica (CONARE).

References

Acevedo-Gutiérrez, A. 1994. First record of a sea lion, *Zalophus californianus*, at Isla del Coco, Costa Rica. *Marine Mammal Science* 10: 484–85.

Acevedo-Gutiérrez, A. 1996. Lista de mamíferos marinos en Golfo Dulce e Isla del Coco. *Revista de Biología Tropical* 44: 933–34.

Acevedo-Gutiérrez, A. 1999. Aerial behavior is not a social facilitator in bottlenose dolphins hunting in small groups. *Journal of Mammology* 80: 768–76.

Acevedo-Gutiérrez, A. 2002. Interactions between marine predators: dolphin food intake is related to number of sharks. *Marine Ecology Progress Series* 240: 267–71.

Acevedo-Gutiérrez, A., and N. Parker. 2000. Surface behavior of bottlenose dolphins is related to spatial arrangement of prey. *Marine Mammal Science* 16: 287–98.

Acevedo-Gutiérrez, A., and M.A. Smultea. 1995. First records of humpback whales including calves at Golfo Dulce and Isla del Coco, Costa Rica, suggesting geographical overlap of northern and southern hemisphere populations. *Marine Mammal Science* 11: 554–60.

Acevedo-Gutiérrez, A., and S.C. Stienessen. 2004. Bottlenose dolphins (*Tursiops truncatus*) increase number of whistles when feeding. *Aquatic Mammals* 30: 357–62.

Acevedo-Gutiérrez, A., and B. Würsig. 1991. Preliminary observations on bottlenose dolphins, *Tursiops truncatus*, at Isla del Coco, Costa Rica. *Aquatic Mammals* 17: 148–51.

Acevedo-Gutiérrez, A., B. Brennan, P. Rodríguez, and M. Thomas. 1997. Resightings and behavior of false killer whales (*Pseudorca crassidens*) in Costa Rica. *Marine Mammal Science* 13: 307–14.

Acuña, F.H., J. Cortés, and A. Garese. 2012. Occurrence of the sea anemone *Telmatactis panamensis* (Verrill, 1869) (Cnidaria: Anthozoa: Actiniaria) at Isla del Coco National Park, Costa Rica. *Revista de Biología Tropical* 60(Suppl. 3): 201–5.

Acuña, F.H., A. Garese, A.C. Excoffon, and J. Cortés. 2013. New records of sea anemones (Anthozoa: Actiniaria) from Costa Rica. *Revista de Biología Marina y Oceanografía* 48: 177–84.

Acuña-González, J., J. García-Céspedes, E. Gómez-Ramírez, J.A. Vargas-Zamora, and J. Cortés. 2008. Parámetros físico-químicos en aguas costeras de la Isla del Coco, Costa Rica (2001–2007). *Revista de Biología Tropical* 56(Suppl. 2): 49–56.

Agassiz, A. 1892. Reports on the dredging operations off the west coast of Central America to the Galápagos, to the west coast of Mexico, and in the Gulf of California, in charge of Alexander Agassiz, carried on by the U.S. Fish Commission steamer "Albatross", Lieut. Commander Z.L. Tanner U.S.N., commanding. II. General sketch of the expedition of the "Albatross," from February to May, 1891. *Bulletin of the Museum of Comparative Zoölogy* 23: 1–89.

Agassiz, A. 1898. Reports on the dredging operations off the west coast of Central America to the Galápagos, to the west coast of Mexico, and in the Gulf of California, in charge of Alexander Agassiz, carried on by the U.S. Fish Commission steamer "Albatross", Lieut. Commander Z.L. Tanner U.S.N., commanding. XXIII. Preliminary report on the echini. *Bulletin of the Museum of Comparative Zoölogy* 32: 71–86.

Alfaro, A. 1899. Flora y fauna de la Isla del Coco. Memoria Secretaría de Fomento, San José, Costa Rica 1899: 31–36.

Alfaro, E.J. 2008. Ciclo diario y anual de variables troposféricas y oceánicas en la Isla del Coco. *Revista de Biología Tropical* 56(Suppl. 2): 19–29.

Allen, G.R. 2007. Conservation hotspots of biodiversity and endemisms for Indo-Pacific coral reef fishes. *Aquatic Conservation: Marine and Freshwater Ecosystems*. Published online at Wiley InterScience (http://www.interscience.wiley.com). DOI: 10.1002/aqc.880

Alvarado, J.J. 2010. Isla del Coco (Costa Rica) echinoderms: state of knowledge. In L.G. Harris, S.A. Böttger, C.W. Walter, and M.P. Lesser, eds., *Echinoderms: Durham*, 103–13. *Proceedings of the 12th International Echinoderm Conference*. Leiden, Netherlands: CRC Press, Taylor & Francis Group, Balkema.

Alvarado, J.J., and A. Chiriboga. 2008. Distribución y abundancia de equinodermos de las aguas someras en la Isla del Coco, Pacifico Oriental, Costa Rica. *Revista de Biología Tropical* 56(Suppl. 2): 99–111.

Alvarado, J.J., J. Cortés, and H. Reyes-Bonilla. 2012. Bioerosion impact model for the sea urchin *Diadema mexicanum* on three Costa Rican Pacific coral reefs. *Revista de Biología Tropical* 60(Suppl. 2): 121–32.

Alvarado-Induni, G.E. 2000. *Los Volcanes de Costa Rica: Geología, Historia y Riqueza Natural*. San José, Costa Rica: Editorial UNED.

Amador, J.A., E.J. Alfaro, O.G. Lizano, and V.O. Magaña. 2006. Atmospheric forcing of the eastern tropical Pacific: a review. *Progress in Oceanography* 69: 101–42.

Angulo, A., B. Naranjo-Elizondo, M. Corrales-Ugalde, and J. Cortés. 2014. First record of the genus *Paracaristius* (Perciformes: Caristiidae) from the Pacific of Central America, with comments on their association with the siphonophore *Praya reticulata* (Siphonophorae: Prayidae). *Marine Biodiversity Records* 7: e132. doi:10.1017/S1755267214001262

Arauz, R., and A. Antoniou. 2006. Preliminary results: movement of scalloped hammerhead shark (*Sphyrna lewini*) tagged in Cocos Island National Park, Costa Rica, 2005. In R. Rojas and I. Zanella, eds., *Memoria: Primer Seminario Taller sobre el Estado del Conocimiento de la Condrictiofauna de Costa Rica*, 8–9. Santo Domingo, Heredia, Costa Rica: INBio.

Arauz, R., Y. Cohen, J. Ballestero, A. Bolaños, and M. Pérez. 2004. Decline of shark populations in the Exclusive Economic Zone of Costa Rica. International Symposium on Marine Biological Indicators for Fisheries Management. Paris, France: UNESCO, FAO.

Arauz, R., A. López, J. Ballestero, and A. Bolaños. 2006. Estimación de la abundancia relativa de tiburones en la Zona Económica Exclusiva de Costa Rica a partir de observaciones a bordo de la flota de palangre de Playas del Coco, Guanacaste, Costa Rica. In R. Rojas and I. Zanella, eds., *Memoria: Primer Seminario Taller sobre el Estado del Conocimiento de la Condrictiofauna de Costa Rica*, 10–12. Santo Domingo, Heredia, Costa Rica: INBio.

Bakus, G.J. 1974. Toxicity in holothurians: A geographical pattern. *Biotropica* 6: 229–36

Bakus, G.J. 1975. Marine zonation and ecology of Cocos Island, off Central America. *Atoll Research Bulletin* 179: 1–9.

Barnard, J.L. 1980. Revision of *Metharpinia* and *Microphoxus* (marine Phoxocephalid Amphipoda from the Americas). *Proceedings of the Biological Society of Washington* 93: 104–35.

Bartsch, P., and H.A. Rehder. 1939. Mollusks collected on the Presidential Cruise of 1938. *Smithsonian Miscellaneous Collections* 98(10): 1–18.

Beebe, W. 1926a. *The Arcturus Adventure: An Account of the New York Zoological Society's First Oceanographic Expedition*. New York: G.P. Putman's Sons.

Beebe, W. 1926b. Cocos-The isle of pirates. *New York Zoological Society Bulletin* 29: 55–67.

Beebe, W. 1926c. The Arcturus Oceanographic Expedition. *Zoologica* 8: 1–45.

Bellon, H., R. Sáenz, and J. Tournon. 1983. K/Ar radiometric ages of lavas from Cocos Island (eastern Pacific). *Marine Geology* 54: M17–M23.

Benumof, B.T., and J.P. Lockwood. 2000. Investigación de las cuevas marinas de la Isla del Coco: datos de campo. Volcano, HI: Geohazards Consultant International and Puffin Investment Company.

Bernecker, A., and I.S. Wehrtmann. 2009. New records of benthic marine algae and Cyanobacteria for Costa Rica, and a comparison with other Central American countries. *Helgoland Marine Research* 63: 219–29.

Bigelow, H.B. 1909. Reports of the scientific results of the Expedition to the Eastern Tropical Pacific, in charge of Alexander Agassiz, by the U.S. Fish Commission Steamer "Albatross," from October, 1904, to March, 1905, Lieut. Commander L.M. Garrett, U.S.N., commanding. XVI. The Medusae. *Memoirs of the Museum of Comparative Zoology* 37: 1–243.

Bigelow, H.B. 1911. Reports of the scientific results of the Expedition to the Eastern Tropical Pacific, in charge of Alexander Agassiz, by the U.S. Fish Commission Steamer "Albatross," from October, 1904, to March, 1905, Lieut. Commander L.M. Garrett, U.S.N., commanding. XXIII. The Siphonophorae. *Memoirs of the Museum of Comparative Zoology* 38: 369–408.

Bigelow, H.B. 1931. Siphonophorae from the Arcturus Oceanographic Expedition. *Zoologica* 8: 525–92.

Bigelow, H.B. 1940. Eastern Pacific Expeditions of the New York Zoological Society. XX. Medusae of the Templeton Crocker and Eastern Pacific *Zaca* Expeditions, 1936–1938. *Zoologica* 25: 281–321.

Biolley, P. 1907. Mollusques de l'Isle del Coco. San José, Costa Rica: Museo Nacional de Costa Rica, Tipografía Nacional. (Published in Spanish: Biolley, P. 1935. Moluscos de la Isla del Coco. *Revista del Colegio de Señoritas* (San José) 2(6): 2–18.)

Bohrmann, G., K. Heeschen, C. Jung, W. Weinrebe, B. Baranov, B. Cailleau, R. Heath, V. Hühnerbach, M. Hort, D. Masson, and I. Trummer. 2002. Widespread fluid expulsion along the seafloor of the Costa Rica convergent margin. *Terra Nova* 14: 69–79.

Boone, L. 1930a. Scientific results of the cruises of the yachts "Eagle" and "Ara", 1921–1928, William K. Vanderbilt, commanding. Crustacea: Stomatopoda and Brachyura. *Bulletin of the Vanderbilt Marine Museum* 2: 1–228.

Boone, L. 1930b. Scientific results of the cruises of the yachts "Eagle" and "Ara", 1921–1928, William K. Vanderbilt, Commanding. Crustacea: Anomura, Macrura, Schizopoda, Isopoda, Amphipoda, Mysidacea, Cirripedia, and Copepoda. *Bulletin of the Vanderbilt Marine Museum* 3: 1–221.

Boone, L. 1933. Scientific results of cruises of the yachts "Eagle" and "Ara", 1921–1928, William K. Vanderbilt, commanding. Coelenterata, Echinodermata and Mollusca. *Bulletin of the Vanderbilt Marine Museum* 4: 1–217.

Boulay, J.N., J. Cortés, J. Nivia-Ruiz, and I.B. Baums. 2012. High genotypic diversity of the reef-building coral *Porites lobata* (Scleractinia: Poritidae) in Isla del Coco National Park, Costa Rica. *Revista de Biología Tropical* 60(Suppl. 3): 279–92.

Breedy, O., and J. Cortés. 2008. Octocorals (Coelenterata: Anthozoa: Octocorallia) of Isla del Coco, Costa Rica. *Revista de Biología Tropical* 56(Suppl. 2): 71–77.

Breedy, O., and J. Cortés. 2011. Morphology and taxonomy of a new species of *Leptogorgia* (Cnidaria: Octocorallia: Gorgoniidae) in Cocos Island National Park, Pacific Costa Rica. *Proceedings of the Biological Society of Washington* 124: 62–69.

Breedy, O., L.P. Van Ofwegen, and S. Vargas. 2012. A new family of soft corals (Anthozoa, Octocorallia, Alcyonacea) from the aphotic tropical eastern Pacific waters revealed by integrative taxonomy. *Systematics and Biodiversity* 10: 351–59.

Burkenroad, M.D. 1938. The Templeton Crocker Expedition. XII. Sergestidae (Crustacea: Decapoda) from the Lower California region, with descriptions of two new species and some remarks on the Organ of Pesta in *Sergestes*. *Zoologica* 22: 315–29.

Bussing, W.A. 1983. A new tropical eastern Pacific labrid fish, *Halichoeres discolor* endemic to Isla del Coco. *Revista de Biología Tropical* 31: 19–23.

Bussing, W.A. 1991. A new genus and two new species of tripterygiid fishes from Costa Rica. *Revista de Biología Tropical* 39: 77–85.

Bussing, W.A. 1997. *Chriolepis atrimelum* (Gobiidae) a new species of gobiid fish from Isla del Coco, Costa Rica. *Revista de Biología Tropical* 45: 1547–52.

Bussing, W.A. 2010. A new fish, *Peristedion nesium* (Scorpaeniformes: Peristediidae) from Isla del Coco, Costa Rica. *Revista de Biología Tropical* 58: 1149–56.

Bussing, W.A., and R.J. Lavenberg. 2003. Four new species of eastern tropical Pacific jawfishes (*Opistongathus*: Opistognathidae). *Revista de Biología Tropical* 51: 529–50.

Bussing, W.A., and M.I. López. 2005. Peces de la Isla del Coco y peces arrecifales de la costa Pacífica de América Central meridional/Fishes of Cocos Island and reef fishes of the Pacific coast of lower Central America. *Revista de Biología Tropical* 53(Suppl. 2). 192 pp.

Cairns, S.D. 1991a. A revision of the ahermatypic Scleractinia of the Galápagos and Cocos Islands. *Smithsonian Contributions to Zoology* 504: 1–33.

Cairns, S.D. 1991b. New records of Stylasteridae (Hydrozoa: Hydroida) from the Galápagos and Cocos Islands. *Proceedings of the Biological Society of Washington* 104: 209–28.

Camp, D.K., and H.G. Kuck. 1990. Additional records of stomatopod crustaceans from Isla del Coco and Golfo de Papagayo, east Pacific Ocean. *Proceedings of the Biological Society of Washington* 103: 847–53.

Castellanos-Osorio, I., R.M. Hernández-Flores, Á. Morales-Ramírez, and M. Corrales-Ugalde. 2012. Apendicularias (Urochordata) y quetognatos (Chaetognatha) del Parque Nacional Isla del Coco, Costa Rica. *Revista de Biología Tropical* 60(Suppl. 3): 243–55.

Castillo, P., R. Batiza, D. Vanko, E. Malavassi, J. Barquero, and E. Fernández. 1988. Anomalously young volcanoes on hot-spot traces: I. Geology and petrology of Cocos Island. *Geological Society of America Bulletin* 100: 1400–14.

Castro, P. 1996. Eastern Pacific species of *Trapezia* (Crustacea, Brachyura: Trapeziidae), sibling species symbiotic with reef corals. *Bulletin of Marine Science* 58: 531–54.

Child, C.A. 1992. Pycnogonida of the Southeast Pacific Biological Oceanographic Project (SEPBOP). *Smithsonian Contributions to Zoology* 526: 1–43.

Cortés, J. 1986. Biogeografía de corales hermatípicos: el istmo centroamericano. *Anales del Instituto de Ciencias del Mar y Limnología, UNAM* 13: 297–304.

Cortés, J. 1997. Biology and geology of coral reefs of the eastern Pacific. *Coral Reefs* 16(Suppl.): S39–S46.

Cortés, J. 2008. Historia de la investigación marina de la Isla del Coco, Costa Rica. *Revista de Biología Tropical* 56(Suppl. 2): 1–18.

Cortés, J. 2009a. A history of marine biodiversity scientific research in Costa Rica. In I.S. Wehrtmann and J. Cortés, eds., *Marine Biodiversity of Costa Rica, Central America*, 47–80. *Monographiae Biologicae* 86. Berlin: Springer.

Cortés, J. 2009b. Stony corals. In I.S. Wehrtmann and J. Cortés, eds., *Marine Biodiversity of Costa Rica, Central America*. *Monographiae Biologicae* 86. Berlin: Springer. Text: 169–73, Species list CD: 112–18.

Cortés, J. 2011. Eastern Tropical Pacific coral reefs. In D. Hopley, ed., *The Encyclopedia of Modern Coral Reefs: Structure, Form and Process*, 351–58. Berlin: Springer.

Cortés, J. 2012a. Bibliografía sobre investigaciones marinas, oceanográficas, geológicas y atmosféricas en el Parque Nacional Isla del Coco y aguas adyacentes, Pacífico Costa Rica. *Revista de Biología Tropical* 60(Suppl. 3): 363–92.

Cortés, J. 2012b. Marine biodiversity of an Eastern Tropical Pacific oceanic island, Isla del Coco, Costa Rica. *Revista de Biología Tropical* 60(Suppl. 3): 131–85.

Cortés, J., and S. Blum. 2008. Life down to 450 m at Isla del Coco, Costa Rica. *Revista de Biología Tropical* 56(Suppl. 2): 189–206.

Cortés, J., O. Breedy, S. Blum, S. Earle, K. Green, A. Klapfer, B. Robison, R. Starr, and E. Widder. 2009. Investigaciones con submarino de la Isla del Coco y el Monte Submarino Las Gemelas: Informe de la expedición. Presented to the Ministry of the Environment of Costa Rica.

Cortés, J., and H.M. Guzmán. 1998. Organismos de los arrecifes coralinos de Costa Rica: descripción, distribución geográfica e historia natural de los corales zooxantelados (Anthozoa: Scleractinia) del Pacífico. *Revista de Biología Tropical* 46: 55–92.

Cortés, J., and C.E. Jiménez. 2003. Corals and coral reefs of the Pacific of Costa Rica: history, research and status. In J. Cortés, ed., *Latin American Coral Reefs*, 361–85. Amsterdam: Elsevier Science B.V.

Cortés, J., C.E. Jiménez, A.C. Fonseca, J.J. Alvarado. 2010. Status and conservation of coral reefs in Costa Rica. *Revista de Biología Tropical* 58(Suppl. 1): 33–50.

Cortés, J., A. Sánchez-Jiménez, J.A. Rodríguez-Arrieta, G. Quirós-Barrantes, P.C. González, and S. Blum. 2012. Elasmobranchs observed in deepwaters (45–330m) at Isla del Coco National Park, Costa Rica (Eastern Tropical Pacific). *Revista de Biología Tropical* 60(Suppl. 3): 257–73.

Crocker, T. 1933. The Templeton Crocker Expedition of the California Academy of Sciences, 1932, No. 2: Introductory statement. *Proceedings of the California Academy of Sciences, 4th Series* 21: 3–9.

Dall, W.H. 1900. Additions to the insular land-shell faunas of the Pacific Coast, especially of the Galápagos and Cocos islands. *Proceedings of the Academy of Natural Sciences of Philadelphia* 52: 88–106.

Dall, W.H. 1908. Reports on the dredging operations off the west coasts of Central America to the Galápagos, to the west coast of Mexico, and in the Gulf of California, in charge of Alexander Agassiz, carried on by the U.S. Fish Commission steamer "Albatross", during 1891, Lieut. Commander Z.L. Tanner, U.S.N., commanding. XXXVII, and Reports on the scientific results of the expedition to the eastern tropical Pacific, in charge of Alexander Agassiz, by the U.S. Fish Commission steamer "Albatross," from October, 1904, to March, 1905, Lieut. Commander L.M. Garrett, U.S.N., commanding. XIV. The Mollusca and Brachiopoda. *Bulletin of the Museum of Comparative Zoölogy* 43: 205–487.

Dall, W.H. 1920. Annotated list of the recent Brachiopoda in the collection of the United States National Museum, with descriptions of thirty-three new forms. *Proceedings of the United States National Museum* 57(2314): 261–377.

Dalrymple, G.B., and A. Cox. 1968. Paleomagnetism, potassium-argon ages and petrology of some volcanic rocks. *Nature* 217: 323–26.

Dana, T.F. 1975. Development of contemporary eastern Pacific coral reefs. *Marine Biology* 33: 355–74.

D'Attilio, A., B.W. Myers, and D.R. Shasky. 1987. A new species of *Phyllonotus* (Muricidae: Muricinae) from Isla del Coco, Costa Rica. *Nautilus* 101: 62–65.

Dean, H.K., J.A. Sibaja-Cordero, and J. Cortés. 2010a. Occurrence of the phoronid *Phoronopsis albomaculata* in Cocos Island, Costa Rica. *Pacific Science* 64: 459–62.

Dean, H.K., J.A. Sibaja-Cordero, J. Cortés, R. Vargas, and G.Y. Kawauchi. 2010b. Sipunculids and Echiurans of Isla del Coco (Cocos Island), Costa Rica. *Zootaxa* 2557: 60–68.

Dean, H.K., J.A. Sibaja-Cordero, and J. Cortés. 2012. Polychaetes (Annelida: Polychaeta) of Cocos Island National Park, Pacific Costa Rica. *Pacific Science* 66: 347–86.

Deichmann, E. 1958. The Holothurioidea collected by the Velero III and IV during the years 1932 to 1954. Part II. Aspidochirota. *Allan Hancock Pacific Expedition* 11: 253–349.

Durham, J.W. 1962. Corals from the Galápagos and Cocos islands. *Proceedings of the California Academy of Sciences, 4th Series* 32: 41–56.

Durham, J.W. 1966. Coelenterates, especially stony corals, from the Galápagos and Cocos Islands. In R.I. Bowman, ed., *The Galápagos. Proceedings of the Symposia of the Galápagos International Scientific Project* 15: 123–35.

Durham, J.W., and J.L. Barnard. 1952. Stony corals of the eastern Pacific collected by the Velero III and Velero IV. *Allan Hancock Pacific Expedition* 16: 1–110.

Eakin, M.C. 1999. The origins of modern science in Costa Rica: The Instituto Físico-Geográfico Nacional, 1887–1904. *Latin American Research Review* 34: 123–50.

Edgar, G.J., S.A. Banks, S. Bessudo, J. Cortés, H.M. Guzman, S. Henderson, C. Martínez, F. Rivera, G. Soler, D. Ruiz, and F. Zapata. 2011. Variation in reef fish and invertebrate communities with level of protection from fishing across the Eastern Tropical Pacific seascape. *Global Ecology and Biogeography*. DOI: 10.1111/j.1466-8238.2010.00642.x

Esquete, P., J.A. Sibaja Cordero, and J.S. Troncoso. 2013. A new genus and species of Leptocheliidae (Crustacea: Peracarida: Tanaidacea) from Isla del Coco (Costa Rica). *Zootaxa* 3741: 228–42.

Faxon, W. 1895. Reports on an exploration off the west coasts of Mexico, Central and South America, and off the Galápagos Islands, in charge of Alexander Agassiz, by the U.S. Fish Commission steamer Albatross, during 1891, Lieut. Commander Z.L. Tanner U.S.N., commanding. XV. The stalk-eyed Crustacea. *Memoirs of the Museum of Comparative Zoology* 18: 1–292.

Fernández, C. 2008. Flora marina del Parque Nacional Isla del Coco, Costa Rica, Pacífico Tropical Oriental. *Revista de Biología Tropical* 56(Suppl. 2): 57–69.

Fernández, W. 1984. Comments on meteorological and climatological observations on Cocos Island. *Revista de Geofísica* 20: 9–19.

Ferreira, A.J. 1987. The chiton fauna of Cocos Island, Costa Rica (Mollusca: Polyplacophora) with the description of two new species. *Bulletin of the Southern California Academy of Sciences* 86: 41–53.

Fiedler, P.C., and L.D. Talley. 2006. Hydrography of the eastern tropical Pacific: a review. *Progress in Oceanography* 69: 143–80.

Fields, W.G., and V.A. Gauley. 1972. A report on the cephalopods collected by Stanford Oceanographic Expedition 20 to the eastern tropical Pacific Ocean, September to November, 1968. *Veliger* 15: 113–18.

Fowler, H.W. 1932. The fishes obtained by the Pinchot South Seas Expedition of 1929, with description of one new genus and three

sent in the Americas—was introduced from Europe to the island, as well as to other eastern Pacific islands and coastal zones of the Americas. Now, almost five centuries later, the black rat still continues to pose a threat to the conservation of the island's plants, animals and ecosystems. Also, during this period a first few cultivated plants were imported from Peru, but it is not clear which species actually were introduced to the island (Jinesta 1940).

The Period Dominated by Pirates, Corsairs, and Filibusters

From the end of the sixteenth century until the beginning of the nineteenth century, pirates, corsairs, and filibusters frequently visited Isla del Coco. Some of them established camps that were maintained for a while. Chronicles from the time that they stayed at the island provide descriptions on the nature of the island and the activities that these pirates and filibusters undertook to extract and exploit its biodiversity (Wafer 1699, Duret 1720, Betagh 1728, Dampier 1697, Kerr 1814, Burney 1816). The main human activities that were reported from those times are the clearing of forest meant to provide room for their temporary camps in the bays at the north, the cutting of coconut palms to gather their nuts, the extraction of timber to repair their ships, fishing in the bays, the collection of marine mollusks and crustaceans, and the capturing of marine birds and collection of their eggs.

The Time of the Whalers

According to Epler (1987) and Kasteleijn (1987) the whalers' period in the eastern tropical Pacific extended from 1790 to 1840. During that time Isla del Coco was frequently visited by whalers who were mainly attracted by the good drinking quality of its water. In fact, the island's water could be stored and kept safe for long times due to its purity and chemical composition. The negative effect the whalers caused to the island's biodiversity is documented in the logbook of these ships and in the diaries of a few columnists writing at the time (Colnett 1798, Belcher 1843, Coulter 1847, Davis 1874, Lièvre 1893, 1962, Epler 1987). They report on impacts like the intentional introduction of domesticated pigs (*Sus scrofa*) and goats (*Capra aegagrus hircus*) and some plants cultivated in 1793, by the English whaling ship *Rattler*, commanded by Captain James Colnett (1798).

Another kind of exploitation concerned the extraction of firewood and especially that of the endemic *palo de hierro* or iron stick tree (*Sacoglottis holdridgei*, Humiriaceae), the wood of which can be set on fire when it is still green. It was used to melt whale fat and turn it into oil. Such extractions were especially done in the bays and at the more accessible cliffs around the island. Coconuts were also gathered in an uncontrolled manner; the indiscriminate cutting of coconut palm trees by whalers to harvest the fruits took such forms that at some point coconut palms were no longer found on the island named after them (Belcher 1843). Moreover, whalers hunted wild pigs and goats on the island, at the same time when they went after whales, sperm whales, and otariid seals in the waters around the island (Coulter 1847, Montoya 2008b).

The Time of Scientific Explorations and Research

This period is characterized by two phases. The first corresponds to the occurrence of large expeditions directed at exploring the island geographically and scientifically—expeditions that were carried out from the end of the eighteenth century until the last part of the nineteenth century and were sponsored by the world powers of that time. The second one is characterized by expeditions that visited specific sites at and around the island. The latter were sponsored by scientific institutions such as museums and universities. They took place from the end of the nineteenth century and continued throughout the twentieth century.

Large expeditions that took place during the first of these historical phases include the voyage of Malaspina in January 1791 (Malaspina 1885, Purrua 2001), the visit that George Vancouver made in January 1795 (Vancouver 1798, 1801), an expedition by Edward Belcher in April 1838 (Belcher 1843), and the visit by Count Henri Louis de Gueydon in November 1846 (Gueydon 1948). The main legacy of these larger expeditions is the considerable array of pioneer narrations they produced on a variety of aspects of the island's nature and ecosystems, including geographic measurements of the island's terrain and hydrography, cartographic studies of the territory and coastal zone, and first scientific collections of animals and plants. Vancouver's expedition also reported that he and his crew had planted domestic species in Wafer Bay during their visit to the island (Vancouver 1798).

Scientific expeditions during the second phase were generally facilitated by universities, museums, foundations, or individual scholars. They included a fair number that had as final destination the Galápagos Islands, but that made sure to allow for a stop-over at Isla del Coco. The research activities carried out by these expeditions—some fifty in total—resulted in a large number of scientific reports that contributed to the rich and specialized bibliography that exists today on Isla del Coco (Fuentes et al. 2005). In several historical accounts on the island's nature, ecosystems,

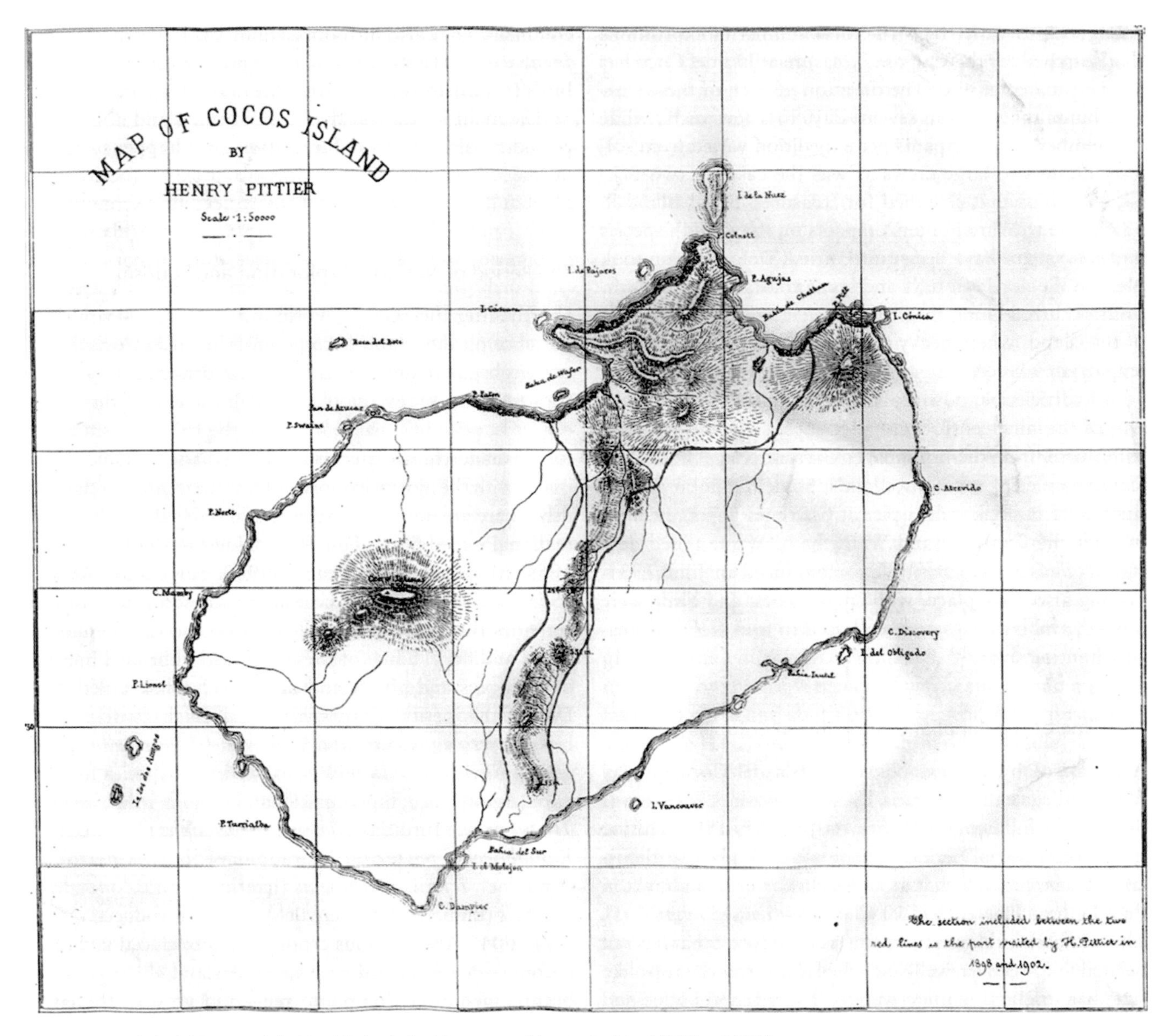

Fig. 8.1 Map of Isla del Coco made by Henri Pittier during his two expeditions to the island (1898 and 1902).

and species one can read about the arrival and activities that many of these expeditions undertook (Hertlein 1963, Hogue and Miller 1981, Montoya 1990, 2004, 2007, Montoya and Pascal 2005, Cortés 2008, Cortés (chapter 7 in this volume).

The first Costa Rican scientific expedition was carried out in June 1898, when the naturalists Anastasio Alfaro and Henri Pittier visited the island (Alfaro 1899, Pittier 1899, 1935). The island's land portion that was explored by Pittier is shown in Fig. 8.1. At the same time, an overview map of Isla del Coco and its geographic features can be found in Cortés (chapter 7 of this volume).

The Period of Treasure Hunters

A range of authors has discussed the historical period of the treasure hunters who visited Isla del Coco and the impact they had on its local biodiversity (Hancock and Weston 1960, Vergnes 1978, Montoya 1990, Weston 1992, and Arias 1997). According to Sinergia 69 (2002b), the treasure hunters' period began after 1830. Treasure hunting activities concentrated fundamentally in the two bays to the north (Chatham and Wafer), where visiting vessels anchored and enjoyed a certain level of protection, and in the small valley of the Minute Stream (*Quebrada Minuto*)

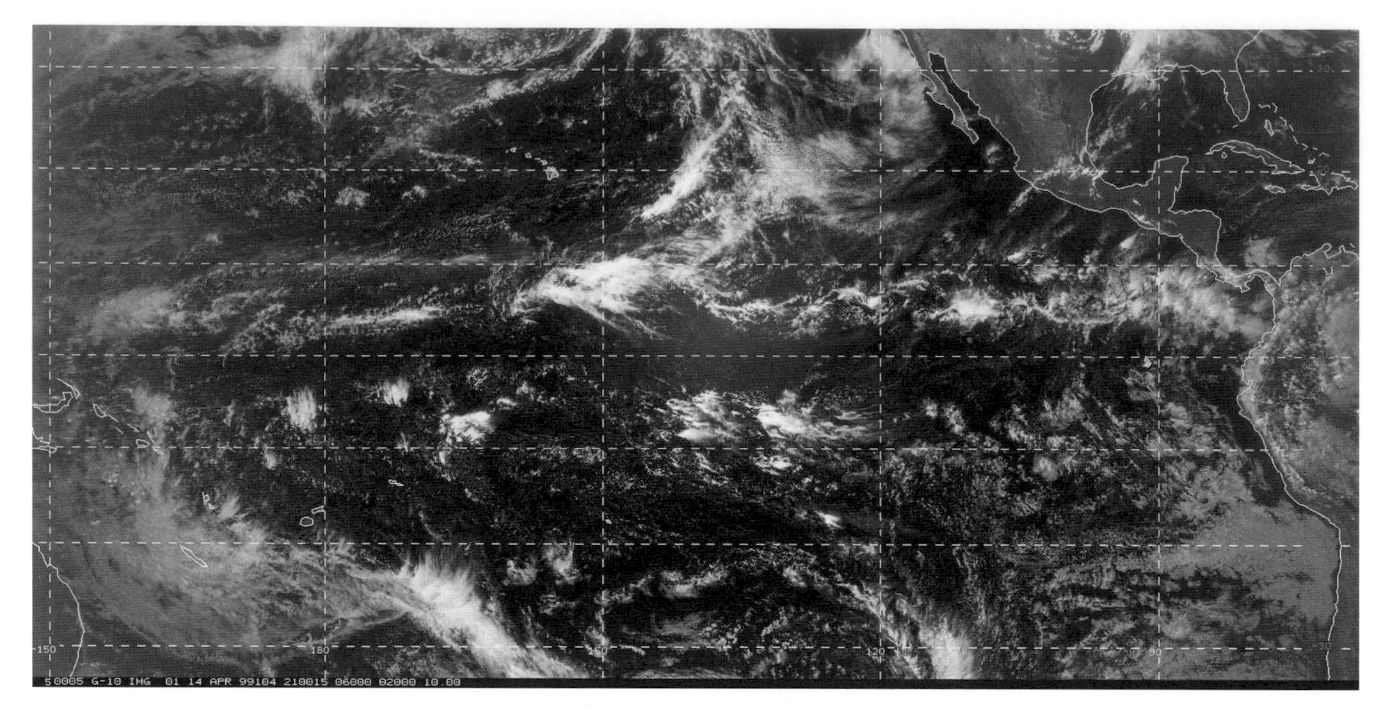

Fig. 8.2 The cloudiness over the Equator corresponds to the Intertropical Convergence Zone (ITCZ). *Satellite image by NASA.*

reliability. For this reason it is almost impossible to conduct a deep quantitative analysis of the various climatic parameters that would allow us to define accurately the island's climate. Fortunately, Herrera (1984, and chapter 2 of this volume) presents information based on climate data gathered at the island during 1979–1982 and in 1984. He concludes that, above all, the governing climate at Isla del Coco depends exclusively on the north-south and south-north movements of the ITCZ. This is evidenced in terms of abundant cloud cover (cumulus and cumulonimbus, etc.), heavy rainfall throughout the year, and predominantly southwestern winds.

Herrera (1984) points out that the average annual rainfall in Chatham Bay oscillates between 5,000 and 5,500 mm, while in Wafer Bay and near Cerro Iglesias precipitation levels reach between 6,000 and 6,500 mm annually. On the other hand, this author states that the rainiest period ranges from May to July, when monthly amounts of rainfall fluctuate between 900 and 1,000 mm. The months that have the lowest precipitation levels are January, February, and March (with a monthly registered minimum of 150 mm).

According to the same author, the number of rainy days—days with 0.1 mm rain or more—varies per month from 11 rainy days in March to up to 30 rainy days a month in July and August (Herrera 1984). Owing to such a strong pluvial regime in combination with a short and dense hydrographic system, no negative water balance occurs on the island. Therefore, no ecologically dry season exists, while soils provide sufficient water to ensure the vegetation remains evergreen and lush throughout the year.

Precipitation at Isla del Coco is usually torrential and linked to convection systems that produce intense and highly erosive rains. Such a pluvial condition associated with a steep topography causes any human intervention on the island to have serious consequences. An increase in soil erosion will lead to a rise in downslope sedimentation, affecting the fragile coastal and marine communities that surround the island, hence reducing the chances of survival of the organisms that compose them.

The island's annual average temperature at sea level is 25.5°C, with a thermal gradient of 0.4°C for each 100 m increase in elevation (Herrera 1984). The minimum annual average is 23.1°C while the annual maximum is 27.6°C. More recently Sinergia69 (2000a) presented rainfall data for Isla del Coco, starting with an analysis of existing information such as generated by the Atlas of Surface Marine Data (NOAA/PEML-MAP-FERRET, version 4.91). The values of monthly precipitation (interpolated averages of several years) that they presented are as follows: 200–240 mm in January, 120 mm in February, 80–100 mm in March, 220–240 mm in April, 420–450 mm in May, 480–500 mm in June, 440–460 mm in July, 520 mm in August, 560 mm in September, 520 mm in October, 400–420 mm in November, and 280–300 mm in December. According to these data the months with lowest precipitation would be January, February, and March. This period coincides with the core of the dry season on Costa Rica's mainland.

Recently a study has been carried out that focused on the island's diurnal and annual cycles of troposphere and oceanic variables (Alfaro 2008). On the basis of data gathered between 1997 and 2005 and some existing, older data series this author confirmed that seasonal variations at Isla del Coco are associated with the movement of the ITCZ, since the island is under its influence from the boreal spring (March-June) until the northern hemisphere's fall (September-December).

Another study on climate change variability at Isla del Coco analyzed recent climate records gathered at the island's weather stations in the recent past (1979–2005) (Quirós-Badilla and Alfaro 2009). Results from this analysis showed that during January, February, and March levels of precipitation on the island are lowest, while heavy rains occur mainly from April through December, with July and August being slightly drier than the other months of the rainy season. Temperatures, on the other hand, are highest from January to March and somewhat lower during the period from June through December. Data demonstrate that warmer events at the island caused by the El Niño Southern Oscillation (ENSO) correlate to rainy periods and surface temperatures above normal values, while ENSO-related colder events correlate to temperatures below normal levels. This kind of data seems to be useful in predicting the future climate at Isla del Coco in relation to ENSO events.

The Oceanographic Environment

Considering that Isla del Coco is an oceanic island, knowledge on the system dynamics of its marine currents can help us to understand the origin and evolution of its terrestrial and marine species and ecosystems (see also Cortés, chapter 7 of this volume). The island is located in the center of the Eastern Tropical Pacific region where a complex series of marine currents converge (Montoya 1990, The Nature Conservancy 2008).

To understand the conditions that determine this insular territory it is essential to explain the regional oceanographic context. As a matter of fact, the system of oceanic circulation that surrounds the island is influenced by three main factors: (a) a complex of strong currents that circulate warm tropical waters towards the east, just slightly north of the equatorial line; (b) two currents of cold water, more or less well defined, that circulate towards the west, parallel to the warm currents mentioned above; and (c) a smaller complex of currents that originate in front of the Pacific coasts of Central America and northern South America and have influence on the 800-km-long coastline.

The complex of warm currents is composed of two well-defined currents: the North Equatorial Countercurrent (NECC) and the Pacific Equatorial Undercurrent (EUC) or Cromwell Current.

The NECC originates in the central and western part of the Pacific Ocean and circulates towards the east along the parallels of 4° and 10° northern latitude, starting from 110° western longitude. The force and penetration of this current varies considerably during the course of the year, and from year to year. This current responds to patterns of atmospheric circulation and in particular to the north-south-north movement of the ITCZ.

The NECC gets closer to the Central American coast during the year reaching higher levels of intensity between August and December (see also Cortés, chapter 7 of this volume). From February to April this current seems not or only weakly developed, and it is during those months when it occurs around 125° western longitude, reaching its most eastern position. During the rest of the year this countercurrent can penetrate the zone between 95° and 85° western longitude from where it starts to fork. A portion of this flow of warm water takes a northeast direction and later moves northward to merge with the waters of the Costa Rican Coastal Current (CRCC), in this way contributing to the generation of a cyclonal flow around the Costa Rican Dome (CRD). The other part of the NECC flows towards the southeast and further on towards the south, turning into an anticyclonic movement to finally merge with the South Equatorial Current (SEC).

The Pacific Equatorial Undercurrent (EUC), on the other hand, first appears in the central part of the Pacific Ocean (approximately at 160° western longitude). It is a straight but well-defined current of warm water that flows towards the east along the equatorial line. This current is considered submarine as it circulates to depths over 20 m below sea level and is covered by a thin water layer that flows in a western direction. Its width is about 300 km; its thickness is approximately 200 m, when it circulates down to 100 m deep in its extreme western end; and it is only 40 m thick when it arrives at its oriental-most limits near the Galápagos Islands where the current disappears (Abbott 1966). To date there is no information available about possible transport of plankton by this current, neither is there any clear evidence of a potential direct influence of this current on Isla del Coco itself.

The main currents of the Eastern Tropical Pacific (ETP) region that flow towards the west are two wide currents that transport cold waters. First, there is the oceanic Peruvian Current—also called Humboldt Current—that circulates northward along the South American coast and changes its direction towards the west at 10° southern latitude, where it becomes the Equatorial Current of the South (ECS). At its northern limit close to 4° northern latitude this current borders the NECC while it extends down south to the subtropical region with falling current speeds (Montoya 1990, Kessler 2006).

The second current of cold waters is the Californian Current (CC), which flows towards the south along the western coast of the United States and Mexico, rotating west to 10° northern latitude, where it becomes the North Equatorial Current (NEC). This last current also captures waters of the Costa Rican Coastal Current (CRCC) during the months of December to March, as well as waters of the NECC from July through December, between 10° and 20° northern latitude.

These two cold currents are too far from Isla del Coco to have a direct influence during normal years. However, an indirect influence is possible owing to the penetration of waters from the SEC into the NECC at its southern limit.

The third component of the system of oceanic currents that exercises influence on Isla del Coco corresponds to a complex of flows that develop and evolve in front of the coast of mainland Central America and north of South America. This complex has several elements, among which the CRCC stands out. It flows in a northwestern to western direction in front of the Central American coast starting from Costa Rica. It seems that this current brings part of the waters accumulated by the NECC to the Gulf of Panama and to the zone in front of the southern coast of Costa Rica. From June to July it stays parallel to the coast of Central America and moves fast. From August onwards, this current moves west until 10° northern latitude, where it merges with the NECC. This pattern remains in place until the end of December. Then, from January to March the CRCC flows towards the west between the parallels of 9° and 12° northern latitude, merging its waters with a cyclonal circulation flow whose center is the Costa Rican Dome (CRD).

The CRD is an area located in front of the northwest coast of Costa Rica, precisely where a strong tropical thermocline exists that causes a difference of up to 10 m deep with respect to sea surface level (Montoya 1990, Kessler 2006). The Dome has a diameter of 150 to 300 km and is located between 8° and 10° northern latitude and 88° and 90° western longitude. The cyclonal circulation around the Dome is determined by the NECC in the south, by the CRCC in the east, and partly by the NEC in the north (Wyrtki 1964).

In front of the western coast of Colombia a cyclonic circulation eddy—a large-scale system of rotating ocean currents—develops in an elliptical shape. Its northern circulation component along the coast corresponds to the coastal Colombia Current (CCC), while its southern circulation component leaves the Gulf of Panama towards the south and southwest, having its maximum development during the months from December to April. During this period most of these waters come from the Gulf of Panama and move westward until they merge into an anticyclonic flow that has its center in the proximity of Isla del Coco, close to 5° northern latitude and 88° western longitude (Wyrtki 1964, 1965).

This latter anticyclonic eddy that develops best between December and April is called the Cocos Anticyclonic Eddy (CAE), for its nearness to Isla del Coco and the influence it has on this particular oceanic island. In general, at the end of every year, the waters of the Eastern Tropical Pacific region experience some kind of thermal increase. This phenomenon is the result of a strong flow of warm water in front of the coasts of tropical America. This warm flow moves westward owing to the effect of the NECC. It occurs when the Trade Winds (*Vientos Alisios*) that come from South America and blow towards the central Pacific region get weaker while the El Niño Southern Oscillation (ENSO) event occurs. ENSO originates at a gradient between water levels of the western and eastern sectors of the Pacific Ocean, as a result of differences between the high atmospheric pressure zone along the South American coast and the low pressure zone near Indonesia. With certain frequency this annual warming-up of surface waters gets stronger and causes a catastrophic lack of balance in the region's terrestrial and marine environments. Such an oceanic and atmospheric fluctuation occurs periodically around Isla del Coco and is an important factor that influences the origin and evolution of the island's biodiversity. In addition to that, a recent study on the water circulation dynamics of the sea around Isla del Coco analyzes some of the related chemical and physical aspects of these marine currents (Lizano 2008).

Geology

Cocos Ridge—a submarine volcanic mountain range or backbone—runs over the Cocos Plate in a southwest-northeast direction, from the Galápagos towards the Costa Rican mainland. It is built up by a series of submarine volcanoes that presumably originated following the displacement of an anomalous *hot spot* on the sea floor, which dates back to the Middle Miocene (Castillo et al. 1988). Isla del Coco is located in the central area of the northwestern part of this seismic submarine mountain range. It is the only volcanic structure along the entire mountain range that emerges from the waters, from a depth of approximately 3,000 m (Malavassi 1982; Alvarado and Cárdenes, chapter 3 of this volume).

On the basis of K-Ar radiometric measurements made in alkaline lavas, the age of the island is estimated to be between 1.91 and 2.44 M years (Bellon et al. 1983). In this regard it has been suggested that Isla del Coco is the product of a localized kind of volcanism along the Cocos Ridge that took place during the Late Plioceno (Castillo et al. 1988). For matters of comparison, the neighboring Galápagos Islands have an average age of 1.8 to 4.2 M years according to K-Ar dating done in volcanic rocks (Bailey 1976). The Galápagos archipelago is located about 700 km southwest of Isla del Coco at the intersection of the Cocos Ridge and Galápagos Fracture Zone, close to the Carnegie Ridge.

According to Castillo et al. (1988), the oldest exposed rocks on the island are thin (1 to 5 m thick) flows of aa-type basalt with well-developed paleo-soils at the surface. This rock type is widely distributed across the island and locally may be overlaid by alkaline rocks and other, differentiated lavas or by pyroclastic, trachytic rocks. Such layer sequences are generally thin (less than 100 m thick) and cut by dikes, intrusions, or trachytic plugs. The oldest series are normally overlaid by younger ones. The latter are composed of olivine types of alkali basalt and their corresponding, differentiated subtypes. The younger series are relatively thinner in the eastern part of the island and a bit thicker in the western portion (up to 300 m thick). This difference results from the fact that the most recent volcanic activity happened in the western part of the island.

In Wafer Bay near Punta Presidio (Prison Point) heavy, more than 100-m-thick lava flows and well-developed columnar structures have been observed (Malavassi 1982). According to this author, Punta Presidio is built up of a series of aa-type flows (*coladas*) formed by a 10-m-thick, upper portion of scoria and breccias and an even denser, lower portion composed of 1- to 30-m-thick, sheet-like lava flows. The difference in thickness of these lava flows seems to result from an increase of the viscosity of the magma itself in the upper portion.

Castillo et al. (1988) propose the following sequence of volcanic events that occurred during the geologic development of the island: (a) volcanism of the shield volcano type produced by the lower part of the older series; (b) the collapse of the caldera together with an eruption of pyroclastic rocks and localized trachytic intrusives; and (c) fissure eruptions observed in the younger series and that occurred at the end of, or immediately after, the formation of the calderas (Fig. 8.3).

From the petrochemical point of view most of these lavas are associated with the alkaline series, while only a few show a tendency towards potassic, shoshonitic, and tholeitic series (Alvarado 1984a). In a lithostratigraphic context, Alvarado (1984b) defined the formation *Alkali*

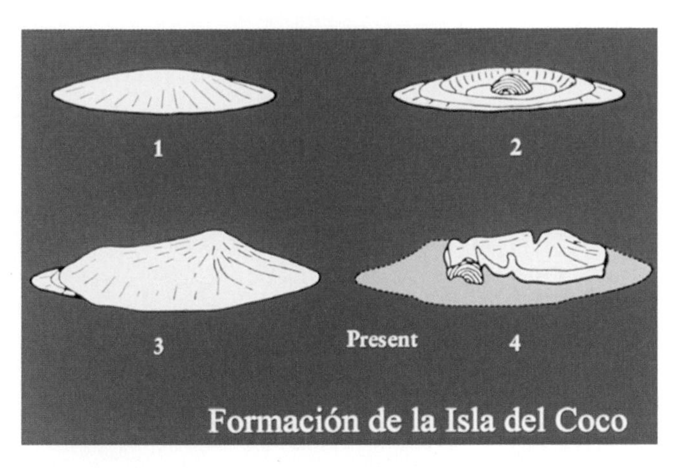

Fig. 8.3 The geological evolution of Isla del Coco.
Source: Castillo et al. (1984).

Basalts of Isla del Coco (type locality: Wafer Bay), which is characterized by olivine basalts, with or without hornblendes, hawaiites, mugerites, and trachytics with tuffs and breccias, all forming part of an alkaline sequence (Alvarado and Cárdenes, chapter 3 of this volume).

More recent geologic research done at Isla del Coco seems to contradict the observations from some previous studies (Lockwood and Benumof 2000). These authors developed a new reconnaissance map of the geology of Isla del Coco and described the following lithological units that allow us to comprehend the island's geomorphological development, its topography, its soils, and the specific location of its ecosystems. In reverse chronological order—starting with the youngest—these units are presented below.

Sedimentary Deposits from the Holocene (Less than 10,000 Years Old)

BEACH DEPOSITS. Beach deposits are the youngest sediments on the island and occur in the intertidal zone along the beaches and in small pockets. They consist of unconsolidated deposits of sand, gravel, and cobbles that are transported and deposited during each tidal cycle. It also includes deposits of pebbles that occur along the tidal line and are moved around by storm waves or tsunamis. A good example of beach deposits is found in the Chatham Bay (Fig. 8.4).

ALLUVIUMS. Alluviums consist of narrow strips of gravel and unconsolidated sediments found along all water courses (rivers and creeks), and in broader areas with unconsolidated deposits of sediments with sand and mud in the Chatham and Wafer bays. Here they may contain cobbles and pebbles behind the beach berm.

COLLUVIUMS. Colluviums are formed by a mixture of fragments of subangular rocks and unconsolidated soil material.

Fig. 8.4 The beach at Bahía Chatham with unconsolidated Holocene deposits (sand, gravel, and rocks). *Photograph by Michel Montoya, April 2005.*

They are observed along low slopes and in all the canyons of the island's water courses. They are particularly extensive in the Chatham, Wafer, and Iglesias bays. At these places the colluvial material has been partially weathered by the water, but comes mostly from the accumulation of rock and soil fragments proceeding from the contiguous slopes. It includes deposits from small landslides and pieces of vegetation. Wild pigs dig into the surface of this unit, which is the source of most sediment dragged by the streams.

LANDSLIDE DEPOSITS. Collapses of the terrain, or landslides, are common features along the entire island, particularly near the cliffs along the coast and in the canyons of the streams. Landslides are particularly abundant at Isla del Coco and can be observed in the geomorphological sketch of the island prepared by Bergoeing (1994; see Fig. 8.5). Most of these big landslides happened in prehistoric times. The deposits that form after landslides have occurred consist predominantly of mixed materials and angular blocks of rock that overlay more resistant volcanic material.

For example, the deposits caused by an important landslide in Chatham Bay have had big impacts on sediment distribution along the coast. In fact, they originated the rocks that hold engravings by visitors from the eighteenth century. Some rather important landslides have happened in the past on top of volcanic upper series. This is the case along the south and southeast coast of the island.After intense and persistent rains landslides and other small collapses of the terrain are frequently observed in the island's interior. They contain volcanic material and plant remains. Likewise, following intense rainfall it is easy to observe landslides along the island's entire coast line. The areas that Bergoeing (1994) marked with yellow and an arrow in his geomorphological sketch of Isla del Coco (Fig. 8.5) correspond to landslides. This phenomenon clearly demonstrates the intense dynamics that shape the island's terrain. It is a typical feature of young oceanic islands with land forms that are highly impacted by their specific climate conditions. More rounded land forms, on the other hand, characterize a more mature geomorphology.

Volcanic Rocks (over 2.0 M Years Old)

UPPER VOLCANIC SERIES (UVS). According to Castillo et al. (1988) all volcanic rocks developed on top of the lower volcanic rock layers belong to the Upper Volcanic Series. It is made up of massive lava flows with inserted breccias that form coastal cliffs almost everywhere around the island. Volcaniclastic material and siliceous lavas found below, do not belong to these upper volcanic rocks.

The lavas of the Upper Volcanic Series are mostly made up of hawaiites (Castillo et al. 1988). In many places the thickness of these upper lavas gets over 100 m, especially where it has filled previously existing canyons. Such thick flows constitute sheet flow units and form the many cliffs along the coast line of Isla del Coco.

The Upper Volcanic Series extends over older rocks and usually cuts the hillsides by stratifying the underlying volcaniclastic rocks. Some exposed upper volcanic rocks develop on top of underlying, soft, and deformed sedimentary rocks where they form cushion-like structures. Some of these flows are located below sea level. Some of the thick sheets of the Upper Volcanic Series suggest a movement as if they were intrusive cornices. The Upper Volcanic unit represents the youngest volcanic unit on the island and their upper surface forms the island's original shield. Although the initial surface has been eroded since volcanism ceased

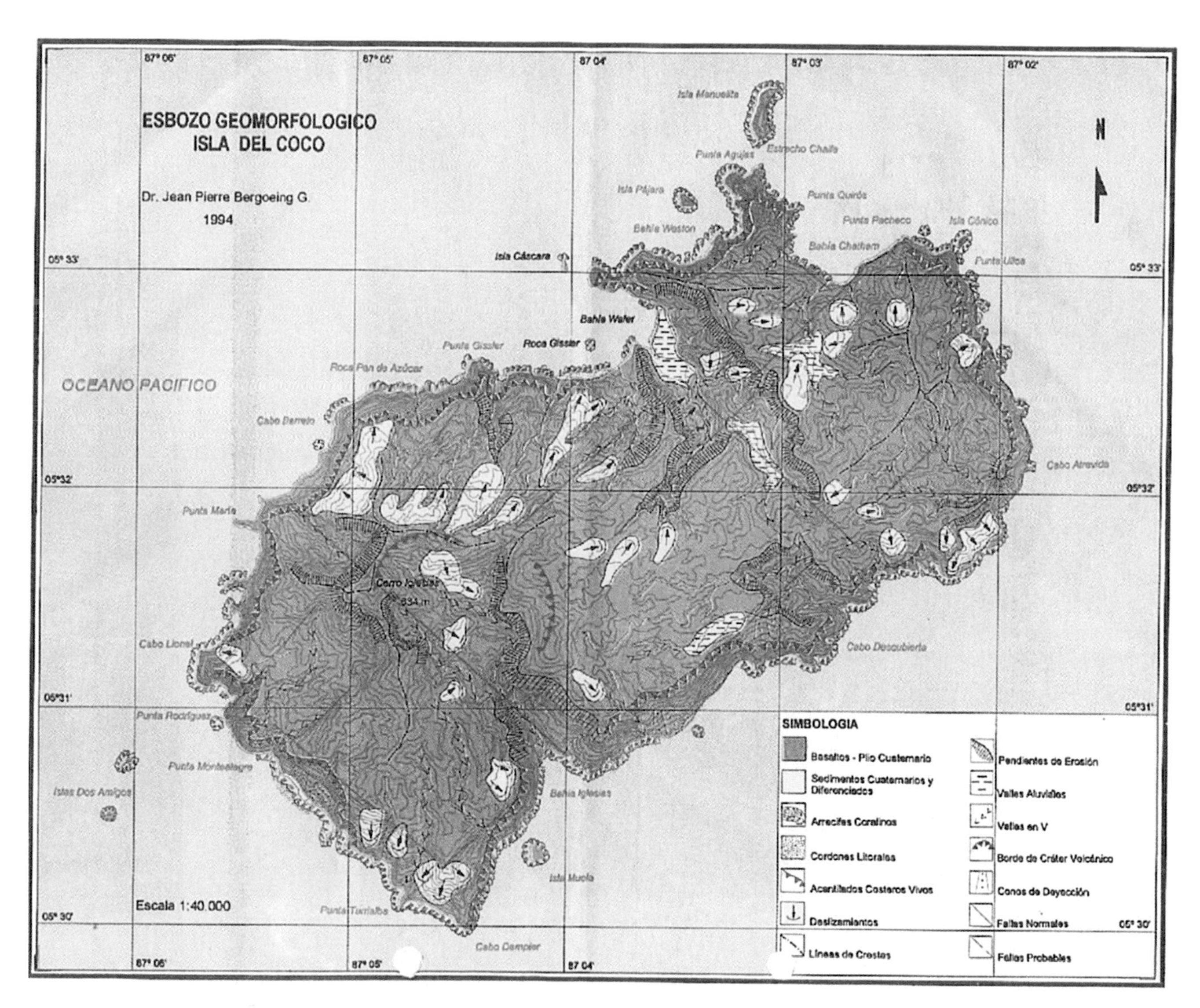

Fig. 8.5 Geomorphological map of Isla del Coco.
Source: Bergoeing (1994).

two million years ago, extensive superficial remnants have been preserved in the higher parts of the island. Still, the original surface has been strongly weathered to form deep, 15-m-thick lateritic soils. In a lot of places they conserve the structure of the original lavas, as a past witness of the original porphyritic and vesicular structures, now converted into clays.

SILICEOUS ROCKS. The pale-colored, siliceous volcanic rocks that possibly are made up of trachytics occur as protruding intrusives in three of the outer islets around Isla del Coco, and in a slightly sloping wide plain located between the Chatham and Wafer bays and in front of Weston Bay. The intrusive bodies in the sea correspond to small islands located near the south, west, and north coast (Isla Juan Bautista, Islas Dos Amigos, and Isla Pájara, respectively). These islands (Fig. 8.6) are characterized by gnarled columnar structures, which give the impression that they are part of domes or sheets that have been eroded previously at other places. The stratigraphic relationship between these flows and the lavas of the Upper Volcanic Series (UVS) is not known. However, the presence of ashy siliceous horizons in volcaniclastic rocks adjacent to the UVS suggests that siliceous volcanism occurs before the UVS lavas are deposited.

VOLCANICLASTIC ROCKS. Volcaniclastic rocks are broadly distributed on Isla del Coco and the surrounding islets. They are largely responsible for the geomorphological evolution of the island and influenced its current landforms. They vary largely in texture, from poorly stratified rocks and chaotic breccias, to vitrified and finely stratified tuffs. The most common rocks correspond to relatively well-stratified rocks, poorly distributed breccias that contain angular and subangular basalt fragments derived from UVS material. These rocks are exposed to the sea's wave dynamics and are

Fig. 8.6 Geological features at a rocky islet off the nearby coast of Isla del Coco. *Photograph by Vivian Araya, April 2005.*

usually very hard and erosion resistant, forming prominent protrusives like Punta Agujas. In the island's interior they are covered by muddy soils, constantly water saturated and strongly weathered, while they form soft and friable (i.e., easily reduced to powder) rock outcrops. At Wafer Bay treasure hunters have been digging a lot into these rocks. They are less resistant to climatic forces than other volcanic rock deposits and are continuously subject to erosion. In fact, they are responsible for the formation of the Wafer and Iglesias bays and, to a lesser extent, of Chatham Bay. They are in control of the development and shaping of the Genio and Iglesias rivers.

The rocks exposed in Wafer Bay are characterized by thick volcanic breccias with individual subangular basalt fragments of less than 1 m diameter in their northernmost extension. The texture of these thick breccias is much finer in the southwestern sector. Most of it has been deposited immediately under the cliffs. This phenomenon can be interpreted like a series of cuts of an active fault that was formed at the northern border of an older version of today's Wafer Bay. The water has been pushing the volcaniclastic sediments and weathered most of it, as is evidenced by the dispersed presence of rounded fragments in the area. Here fossils have not been observed—an indication of very fast rates of accumulation.

LOWER VOLCANIC SERIES (LVS). Castillo et al. (1988) describe the lava flows of the Lower Volcanic Series as the oldest exposed rocks on Isla del Coco. Predominantly they are made up of aa-type basalt flows that are commonly mixed with ashy, yellowish brown horizons and, in some cases, with volcanic breccias.

Apparently, these volcanic lava flows have developed under open air (that is, subaerially), since no cushion-like lavas or other evidence of marine deposition have been observed. These lava flows are best exposed at Punta Dampier and north of Wafer Bay. Although they have not been mapped so far, small protrusives are observed in the intertidal zone along the coastline of the entire island. In the Weston Bay area massive breccias accompany these basaltic flows. They are cut at numerous places by almost vertical volcanic dikes of up to 3 m thick, and in some places by narrow dikes of sandy rock derived from volcaniclastic sediments.

Topography and Hydrology

As a consequence of the island's geological development its topography is very irregular. However, the terrain is quite rolling in the central and eastern parts of the island between 200 and 260 m elevation. Here, erosion patterns with V-shaped valleys are found that represent a transition between juvenile and submature phases of the geomorphologic cycle (Malavassi 1982, Alvarado 1984a,b, and see Alvarado and Cárdenes, chapter 3 of this volume). The southwestern part of the island is so steep that it is almost impossible to reach the sea from the cliffs. The southern part of the island, on the other hand, is like a rolling plain that continues down to sea level.

The peaks and hills in the interior are built up of lava flows with small quantities of breccias and tuffs. They are all easily eroded, leading to irregular topographies. Cerro Iglesias is the highest point on the island. According to the Costa Rican Electricity Institute (ICE) it reaches 575.5 m altitude. Malavassi (1982) reports that its core is a volcano without any present activity near its cone.

The hydrological system of the island is made up of three main watershed basins. One is dominated by the Genio River, which flows in a south-north direction towards the Wafer Bay. A second watershed is controlled by the Pittier River, which flows in a north-south direction and abruptly ends in the Iglesias Bay. The third basin is drained by the Lièvre River, which streams eastwards into the Chatham Bay. All three watersheds are small—ca. 5 km long in the case of the two bigger ones—and have limited water-capturing surfaces.

Nevertheless, waterfalls occur frequently because of the large amounts of rainfall they capture and the abruptness of the terrain and the river flows themselves. The most important one is the Gissler waterfall that is located a few hundreds meters west of the Genio River mouth at Wafer Bay. The fall itself is approximately 55 m tall. In turn, the waterfall of the Iglesias River that also drains into the Iglesias Bay is about 110 m tall. The island's hydrological system also counts on a series of small water courses, many of which are temporary and drain from the cliffs directly into the sea.

Isla del Coco is blessed with four bays, three of which are located along the northern coast: Chatham, Wafer (Fig. 8.7), and Weston. The first two are the only roadsteads on the island that serve as natural harbors. In the southwestern sector of the island a fourth bay exists: Iglesias. This one is too exposed to serve as a secure natural anchor place. The Chatham and Wafer bays are the only ones with sandy beaches on the island. They are small and their development is incipient. Wafer Bay seems more evolved and is broader. The almost total absence of sandy beaches on the island is due to the strong activity of the waves and tides that don't allow any accumulation of material deposits. A study of the sand types found at the beaches of the Chatham and Wafer bays demonstrate that granulometric curves follow a sigmoidal pattern characteristic of coastal zones. They include fine- to medium-grained sands (Brenes and González 1995), which are basically coralline (80%). Their organic

Fig. 8.7 Bahía Wafer, Isla del Coco.
Photograph by Cristian Cotonas, April 2005.

origin and granular diameters demonstrate that the current beach "stock" is continuously being eroded, with little lateral displacement. In fact, this beach drift is produced by the waves during the tides. Therefore, the stability of these beaches depends on the growth of the contiguous coral reefs that contribute most of the beach material.

The other coasts around the island are mostly steep cliffs of volcanic rock that can reach heights of 180 m. They were formed by the columnar disjunction of powerful lava flows (Fig. 8.8). Some of these cliffs have banks of landslides at their bases and are the product of marine erosion. Others, however, are directly exposed to the wave dynamics. It is here, at these cliffs, that a large number of intermittent cascades can be observed after intense rains.

These cliffs continue below the current sea level, forming caverns, canyons, and submarine banks. According to Chubb and Richardson (1933) this is due to the fact that, at some point in its geologic history, the island sunk about 60 to 90 m with respect to its sea level. This phenomenon would also explain the submarine continuation of the sandy beaches at the Chatham and Wafer bays. On the other hand, many of these cliffs have been weathered at their base as a result of erosion caused by pounding waves and tidal oscillations.

Recently (Lockwood and Benumof 2000) did a reconnaissance survey in the intertidal caverns or caves at Isla del Coco. Through their assessment they were able to assign to each of the caves and caverns an identification code, collect georeferential data, measure the mouth dimensions, identify the type of parent rock in which each cave was developed, evaluate the morphology of the cave bottoms, and make other general observations. Of a total of 46 identified caverns, forty occur along the coast of the main island and six on the outer islets, many of which have not yet been explored. However, recently Sibaja-Cordero (2008) conducted a study on the organisms present in the rocky intertidal zones of the island, which were compared with other intertidal areas in the Tropics.

Detailed studies on the hydrogeology of the island have been carried out in Bahía Wafer, whereas at other places on the island more generalized data have been collected. In Bahía Wafer a geophysical study was conducted that made use of electric vertical cores and an electric pseudosection. This was done with the aim to get insight in the below-ground hydrology along the coastal zone of this bay (Arias-Salguero 2000, 2002). Results determined that an aquifer had developed in unconsolidated sediments belonging to recent deposits favoring infiltration. The study also indicated the presence of some clayey lenses that may serve as sealing layers and that influence the generation of both free and confined aquifers. Furthermore, an assessment was made of the profile's freshwater–salty water interface. Information from this assessment is essential in developing a water resource management that takes into account the possibility of water salinization and pollution. Such elements should be considered in environmental impact assessments (EIAs) when human settlements are to be extended in the Bahía Wafer area.

Soils

According to Brenes and González (1995), the soils of Isla del Coco show characteristics that strongly relate to the island's topography. At the lava plateaus of the interfluvial zone soils are clayey and soft, have a well-developed organic layer, and demonstrate leaching of clay matter into the B horizon. These soils are mostly Alfisols and associated Inceptisols. At the sides of the terrain where Inceptisols display a strong clayey or loamy matrix with fine- to coarse-grained sands, they appear in association with Entisols. Here, fine matter has been washed away by natural weathering processes and the erosive disturbance that large numbers of wild pigs cause when they search for food and

Fig. 8.8 Column-shaped vertical geological structures at Isla del Coco.
Photograph by Andrea Ferris, April 2009.

especially with the floras of the Pacific slopes of Central America and northwestern South America. The arrival of plants at the island has been principally facilitated by dispersal by wind, marine currents, floating material, and visiting birds, among other mechanisms.

The biggest proportion of plant species on the island has affinities with floras of the Guyanan-Amazonian phytogeographic province in the Neotropical region. This condition is illustrated by the arrival of seeds and other reproductive and vegetative materials, which came from the mainland or neighboring oceanic islands. This dispersal process started during the Pliocene-Pleistocene, before the Central American isthmus closed in the area where Panama lies today. It took place when South American rivers like the Magdalena and Orinoco drained into a common sea with westbound currents that transported plant elements towards Isla del Coco.

According to a recent floristic inventory conducted at Isla del Coco (Trusty et al. 2006), a total of 263 terrestrial vascular plants have been recorded on the island. Of these, 81 are pteridophytes (ferns and allies) representing 30.8% of its flora. Another 182 are spermatophytes (phanerogams: plants with flowers and seeds) that make up 69.2% of the island's vascular flora.

Out of the 81 ferns, 16 species are endemic (19.75%) and 65 are native to the island (80.25%). It is noteworthy to report the abundant presence of three endemic arborescent *Cyathea* ferns across the entire island: *C. alfonsiana, C. nesiotica*, and *C. notabilis* (Fig. 8.10). Note that these arborescent ferns occur at lower altitudes than at the continent. This difference is explained by the island's microclimate that is typical for islands with mountainous areas in the proximity of their coasts (Rojas-Alvarado and Trusty 2004), known as the telescope effect.

When analyzing the entire vascular flora of Isla del Coco, one will note that 71 species (27%) correspond to species

Fig. 8.10 Treeferns in the genus *Cyathea* at Isla del Coco. *Photograph by Bárbara Sperl, April 2008.*

introduced by man, either by accident or intentionally. The remaining 192 species (73%) are reported to be native species and characteristic of the island. Among the native species 37 are endemic to the island, resulting in an endemism level of 19.27% for native plants.

Trusty and Blanco (2005) and Trusty et al. (2006) determined that five orchid species occur on the island. Three belong to the genus *Epidendrum* (*E. cocoense*, *E. insularum*, and *E. jimenezii*) and two to *Maxillaria* (*M. adendrobium* and *M. parviflora*). The three *Epidendrum* species are endemic to Isla del Coco. This confirms the high proportion of endemic species on oceanic islands. Recently, Bogarín et al. (2011) has put the two *Maxillaria* species in other genera and species: *Ormilthidium adendrobium* and *Camaridium micracanthum*. Similary, these authors present an identification key to the island's orchids, on the basis of morphological characters, and discuss their biogeography, ecology, taxonomy, evolution, and conservation.

The vascular plants introduced by man include nutritious, fruit-bearing, aromatic, ornamental, and medicinal plants, as well as grasses and weeds. Recent studies executed within the framework of the FFEM-GEF-UNDP project for management and conservation of the biodiversity of Isla del Coco, reveal the presence of another 20 species that have been introduced to the island (Pablo Madriz Masís, pers. comm., February 2007, Madriz Masís 2009), increasing the total number of introduced plant species to 103. This situation is worrisome since some of these species have already behaved as aggressive invasive species on other islands in the world. These introduced species occur mostly in the surroundings of existing settlements like those found in the Chatham and Wafer bays and should be eradicated as soon as possible, since they represent a true threat to the island's genetic variety and may affect the success of species and ecosystem conservation.

According to a study by Dauphin (1995, 1999), the epiphytic and terrestrial bryophyte flora on the island is composed of 153 species, 98 of which are liverworts (Hepaticae), belonging to 43 genera in 10 families, while the remainder (55) are mosses (Musci) occurring in 33 genera in 17 families (Fig. 8.11). In general, these species are common on the continent and have affinities with the Neotropics (60%), the Pantropics (10.4%), northern South America (4.5%), and the Caribbean (1.3%). Levels of endemism reach 3.2% while sub-endemism is around 1.3%.

In turn, Isla del Coco's epiphyllous bryophytes—those that grow on top of living leaves of higher plants—have been studied by Bernecker and Lücking (2000), who studied collections made in March 1992. This inventory included 45 epiphyllous bryophyte species that belonged to four families and 22 genera, of which only two were mosses and the rest liverworts. Of a total of 43 hepatic species, 42 belonged to the Lejeuneaceae and only one to the Radulaceae.The floristic affinities of Isla del Coco's epiphyllous bryophytes are mainly Neotropical (68%), Neotropical-African (12%, trans-Atlantic), Central American (10%), Neotropical-Australasian (5%, trans-Pacific), and Pantropical (5%).

Owing to the aforementioned reasons, Isla del Coco can be considered, from the floristic viewpoint, a phytogeographically complex and disjunct unit, separated from the continent. It constitutes a true botanical garden of tropical floras that includes elements from diverse geographical regions. An analysis of the current knowledge on its floristic diversity, its biogeographic relationships, and its endemism demonstrates the island's exceptional conditions and its great importance for science.

There are also a number of studies that deal with specific taxa found at Isla del Coco, such as liverworts (Clark 1953, Morales 1991), mosses (Williams 1924, Bartram 1933), ferns (Svenson 1938, Gómez 1971, 1975, 1976, Rojas-Alvarado 2001, Rojas-Alvarado and Trusty 2004), the endemic clubmoss *Lycopodium brachiatum* (a synonym of *Huperzia brachiata*; Maxon 1913), the endemic fern *Thelypteris cocos* (Smith and Lellinger 1985), the endemic treefern *Alsophila notabilis* (a synonym of *Cyathea notabilis*; Maxon 1922), the endemic *Acalypha pittieri* (Pax and Hoffman 1924), the endemic vine *Marcgravia waferi* (Standley 1937), the endemic cyperous *Killinga nudiceps* (Tucker 1984), the lauraceous *Ocotea insularis* (Burger and Van der Werff 1990), and the palm *Rooseveltia franckianiana* (a synonym of *Euterpe precatoria* var. *longevaginata*; Cook 1939, 1940). Additionally, several preliminary lists of plant species have been elaborated (Rose 1892, Pittier 1899, 1935, Steward 1912, Svenson 1935, Fosberg and Klawe 1966, Montoya 1991b, Soto 1995).

Vegetation and Plant Communities

The most abundant vegetation on Isla del Coco is found in its Tropical Rain Forest (Fig. 8.12), its Tropical Cloud Forest (Figs. 8.13 and 8.14), and in a series of very local associations or plant communities defined by edaphic and climatic factors. Given that the island is characterized by a dynamic geomorphology with natural landslides and human intervention, across the island all these vegetation types can be observed in diverse states of succession or natural regeneration.

A number of authors have proposed all kinds of vegetation classification schemes for the island, including Fournier (1966, 1968), Gómez (1975a, 1975b), Montoya (1990, 1991b, 2008), Lücking and Lücking (1995b), Dauphin (1995, 1999), and Bernecker and Lücking (2000). However, to describe the vegetation types found on the island this

Fig. 8.11 Bryophytes in the tropical cloud forest at Isla del Coco.
Photograph by Bárbara Sperl, April 2008.

Fig. 8.12 The tropical rain forest at Isla del Coco.
Aerial photograph by Jorge Rodríguez, April 2000.

author has selected the general scheme proposed by Trusty et al. (2006)—though with some modifications—since it represents the most up-to-date classification and takes into account the vegetation units previously defined by other authors, while it mentions the most common species for most vegetation types. Thus, according to Trusty and colleagues, seven main vegetation types can be identified at Isla del Coco. They are shortly described in the next paragraphs.

TROPICAL RAIN FOREST. The Tropical Rain Forest corresponds to Trusty et al.'s (2006) *Low Elevation Humid Forest*. In this forest type, the diversity of tree species is extremely low. For example, their study of trees with diameters at breast height (DBH) over 10 cm found in ten plots of 400 m^2 located between 30 m elevation and the island's summit, didn't record more than five species: *Sacoglottis holdridgei, Ocotea insularis, Clusia rosea, Henriettella fascicularis*, and *Miconia dodecandra*. The tree species *Ficus pertusa, Eugenia cocoensis*, and *Brosimum* sp. were found only sporadically along the northern part of the island. The shrub layer of this vegetation type is dense and diverse, with many Melastomataceae and two common and endemic species of ferns: *Cyathea alfonsiana* and *C. notabilis*. The only palm in this forest is *Euterpe precatoria* var. *longevaginata*. The herb layer is formed by ferns and the cyperaceous *Hypolytrum amplissimum* (Alves and Thomas 2002). Thick lianas of *Schlegelia brachyanta* and *Entada gigas* hang from the tree branches. A remarkable aspect of this vegetation type is the density and diversity of epiphytic species. The forest looks like a mixture of red and green because of the abundance of the bromeliad *Guzmania sanguinea* that grows on almost every tree (Fig. 8.15). In addition, four species of orchids are commonly observed, including the endemic *Epidendrum cocoense* and *E. insularum*. There are also several epiphytic ferns as well as two endemic hanging epiphytic clubmosses, *Huperzia branchiata* and *H. pittieri*.

TROPICAL CLOUD FOREST. The Tropical Cloud Forest coincides with Trusty et al.'s (2006) *High Elevation Cloud Forest*. This forest occurs above 450 m elevation and is mainly found at the two highest peaks, Cerro Iglesias and Cerro Pelón. The canopy of this forest type is dominated almost exclusively by the tree *Sacoglottis holdridgei*, locally known as Palo de Hierro, while the understory is dominated

Fig. 8.13 Tropical cloud forest at Isla del Coco, dominated by *Sacoglottis holdridgei* and *Cyathea* treeferns. *Photograph by Juán José Pucci, April 2005.*

by the treefern *Cyathea alfonsiana* (Fig. 8.14). These big, arborescent ferns are almost completely covered by a moss layer. When fog ascends to the highest parts on the island the whole forest starts to drip. The best known species of the island's Cloud Forest are *Freziera calophylla* (Theaceae), *Hedyosmum racemosum* (Chloranthaceae), and the endemic fern *Elaphoglossum reptans*.

This forest type responds well to the ecological definition of cloud forests proposed by Stadtmüller (1987). According to this author, all the forests of the humid Tropics that are frequently covered by clouds or fog receive an additional amount of humidity—additional to normal rainfall—when fine droplets of water (horizontal precipitation) condense on trunks, branches, and leaves, hence having an impact on the forest's hydrological regime and radiation balance and therefore affecting its climatic, edaphic, and ecological parameters.

BAY COMMUNITIES. The vegetation type known as Bay Communities corresponds to Trusty et al.'s (2006) *Bayshore Communities*. These are plant communities that historically have suffered major alterations due to the presence of man. They are located in the Chatham and Wafer bays, close to their beaches and the forests behind their berms. At Wafer Bay trees and shrubs belong to the species *Talipariti tiliaceum* var. *pernambucense* (locally known as *majagua*), *Annona glabra*, *Terminalia catappa*, *Erythrina fusca*, and *Ochroma pyramidale*. Here, along the coast, the climbers *Mucuma sloanei*, *M. mutisiana*, and *Canavalia maritima* grow in and on top of trees and bushes. At all places where this vegetation type is found and dominated by *majagua* it is called "majagual."

A small beach area occurs next to the old construction site at Bahía Wafer. Here some coconut palm trees (*Cocos nucifera*) grow while underneath, at ground level, herbs like *Setaria geniculata*, *Sphagneticola trilobata*, and *Hydrocotyle umbellata* thrive. The beach at Chatham Bay is flooded during each tide. The only place that is not flooded is located near the mouth of the Lièvre River and is covered with *majagua* and *Annona glabra* plants (Fig. 8.16).

Contrary to its name, there are only few coconut trees on Isla del Coco. They occur in small and isolated pockets along the beach of Bahía Iglesias, and in the southern sector of the island near the lower parts of the cliffs. The most extensive coconut groves on the island, however, are found at Bahía Iglesias (Fig. 8.17), just behind the beach's berm, which is composed of pebbles. Here, they are accompanied by *Ipomoea pes-caprae* and the fern *Blechnum occidentale*. According to Trusty et al. (2006), no true mangrove forest occurs at the island.

The island's most altered area is found at Bahía Wafer, where human settlements occurred historically. It used to be the landing place for most navigators who arrived at the island in search of water and wood necessary to repair their vessels and serve as fuel to melt the fat of the whales they caught. Also, in this area, humans redirected the water currents, hence eliminating the environments suitable for the development of mangroves.

Near the mouth of the Genio River just behind the berm of the beach, a tidally flooded zone existed that originated conditions for the development of a mangrove forest. Currently, in this area some specific mangrove species are found—for example, *Conocarpus erectus*, *Annona glabra*, *Talipariti tiliaceum* var. *pernambucense*, and the fern *Acrostichum aureum*, as well as the crabs *Cardisoma crassum*, and *Caenobita compressus*. These are all species that can be considered pioneers of an incipient mangrove forest, or are rather remnants of an old mangrove stand that has disappeared.

COASTAL CLIFF COMMUNITIES. The here-described vegetation type called Coastal Cliff Communities is the same as Trusty et al.'s (2006) *Coastal Cliff Communities*. It is typically found along most cliffs of the island's shores. Here, the endemic *Cecropia pittieri*—locally known as the island's *guarumo*—and *Clusia rosea* (common name: *copey*) dominate the landscape with their open crowns. A blend of climbers belonging to various *Ipomoea* species covers the trees and shrubs, forming green vertical walls (Fig. 8.17).

Fig. 8.14 Tropical cloud forest at Isla del Coco, dominated by *Sacoglottis holdridgei* and *Cyathea* treeferns. *Photograph by Michel Montoya, April 2005.*

Near the tidal line of these cliffs several herbaceous species are commonly observed: the sharp and irritating rush *Rhynchospora polyphylla* (Cyperaceae), *Laportea aestuans* (Urticaceae), and *Kohleria spicata* (Gentianaceae) with its red flowers. In less sloping areas, the closed canopy forest reaches the seashore and may become dominated by *Sacoglottis holdridgei* and tree species as *Ocotea insularis* and *Clusia rosea*. The understory is made up of shrubby melastomes like *Ossaea macrophylla* and *O. bracteata*. One of the particular species of this vegetation type is the endemic treefern *Cyathea nesiotica*, which is restricted to places with moderate slopes that are exposed to a more intense solar radiation.

RIPARIAN COMMUNITIES. Another vegetation type known as Riparian Communities corresponds to Trusty et al.'s (2006) *Riparian Areas*. Its presence along the banks of the Chatham, Iglesias, and Wafer rivers that drain into the bays is determined by high rainfall at sloping areas.

The canopy of the riparian forests is made up of tree species that also occur elsewhere on the island but abound here: *Sacoglottis holdridgei*, *Ocotea insularis*, and *Clusia rosea*. The rush *Calyptrocarya glomerulata*, the aroid *Spathiphyllum laeve*, and the fern *Danaea nodosa* are abundant along the banks of the streams throughout the entire island. In the lower part of the Wafer Bay, the Genio River and its tributaries serve as habitat for *Rustia occidentalis* and *Pilea gomeziana*. The endemic species *Hoffmannia piratarum* (Rubiaceae) is wholly restricted to the riparian zone along this river.

The Pittier River that ends in the Iglesias Bay in the southwestern part of the island and is accessible over only a short

Fig. 8.15 Guzmania sanguinea bromeliads in Isla del Coco's tropical rain forest. *Photograph by Guillermo Blanco, April 2008.*

Fig. 8.16 Coastal vegetation with *Annona glabra* and *Talipariti tiliaceum* var. *pernambucense* (locally known as *majagua*). *Photograph by Michel Pascal and Michel Montoya, March 2004.*

section between the waterfall and the sea serves as habitat for a small forest of endemic *Eugenia cocosensis* trees (Myrtaceae) with an understory in which the shrub *Psychotria gracilenta* (Rubiaceae) thrives. The Lièvre River that drains into Chatham Bay and passes through a rocky and sloping terrain is characterized by the common shrub *Ardisia cuspidata* (Myrsinaceae), which flourishes along its banks.

LANDSLIDE VEGETATION. The vegetation type called Landslide Vegetation corresponds to Trusty et al.'s (2006) *Landslides*. As has been explained previously, owing to the abundance of rains and the irregularity of the terrain many landslides occur on the island, both in the coastal areas and in its interior (see Fig. 8.5, the geomorphological map developed by Bergoeing in 1994).

These open areas are often quickly covered by creeping ferns that require support from other plants. They include *Dicranopteris flexuosa, D. pectinata, Sticherus remotus*, and other plant forms like climbers in the genus *Ipomoea*. The endemic *guarumo* tree species *Cecropia pittieri* also accompanies these species. This vegetation type is a successional community that grows in zones where landslides are common. In fact, it is a characteristic element of the vegetation found at Isla del Coco.

VEGETATION OF THE ISLETS. The vegetation type called Vegetation of the Islets corresponds to Trusty et al.'s (2006) *Islets*. In fact, it includes two distinct vegetation communities: one thriving on the islets north of Isla del Coco (Islotes de Cónica, Manuelita, Pájara, Cáscara, and Pan de Azúcar), which are more protected from beating by the sea and winds and are scarcely vegetated with *Clusia rosea* trees and

Fig. 8.17 Cliff communities and landslide vegetion at Bahía Iglesias, Isla del Coco.
Photograph by Elvis Porras, May 2007.

the endemic grass *Chloris paniculata* (Fig. 8.18); and another one growing on the islets located at the southern tip of the island (Islotes de Barlovento, Dos Amigos Grande, Dos Amigos Pequeña, Muela, Juan Bautista, and Montagne). The latter are strongly impacted by the wind and waters, and do not have any arborescent or shrubby vegetation. These southern islands are partly covered by algae, mosses, lichens, and other lower plants that give them a greenish color during the rainy season.

Montoya and López Pozuelo (2012) presented the first estimation of land cover percentages for vegetation types at Isla del Coco. Their vegetation assessment happened within the framework of a project to define Important Areas for Bird Conservation (AICA) in Costa Rica, in which Isla del Coco corresponded to AICA CR-21. According to their study, the vegetation cover percentages for each plant community type are as follows: 65% Tropical Rain Forest, 20% Tropical Cloud Forest, 4% Landslide Vegetation, 4% Coastal Cliff Communities, 3% Riparian Communities, 1% Bay Communities, 1% Wind-driven Shrub Vegetation, and 2% deforested areas.

Animals

Vertebrates

Recently Montoya (2004) and Montoya and Pascal (2005) conducted studies on the terrestrial resident vertebrate fauna of Isla del Coco. They presented a synthesis of the vertebrate fauna that evolved on the island during the past 500 years and started with the discovery of the island by European navigators somewhere between 1531 and 1542 (see also previous sections in this chapter).

The terrestrial vertebrates are represented by 26 species of which 19 are native (autochthonous) and 7 are introduced (allochthonous). The native portion constitutes 73% of today's diversity. Out of the 21 species of vertebrates (11 mammals and 10 birds) that have been introduced in a certain manner during the course of the past 500 years, seven (six mammals and one bird)—that is 33% of the total—still feed and reproduce on the island as of today. The other ones have not been able to survive and became extinct at the island. This percentage is almost three times higher than the predicted value according to the 10% rule that is used in estimatations for oceanic islands. Nineteen of the 21 exotic species have been introduced intentionally, whereas only two species were introduced in an accidental way: the rats *Rattus rattus* and *R. norvegicus* (Montoya and Pascal 2005).

The island's terrestrial fauna is characterized by the absence of native mammals. However, vertebrate endemism fluctuates around 42%. On the other hand, the available literature doesn't allow us to draw well-grounded conclusions about the potential disappearance of native species that lived on the island at the moment of its discovery by the Europeans and do not occur there anymore today. In spite of its geographical isolation, its distant location with respect to the main navigation routes, and the absence of permanent human settlements and port facilities, over time the composition and population size of the island's terrestrial vertebrate fauna has been considerably modified by man, including the deliberate introduction of a number of exotic species (Montoya and Pascal 2005).

Furthermore, the presence on the island of two pinnipeds (fin-footed mammals) has already been mentioned. These marine mammals spend part of their life on the mainland. Recent records report the recurrent arrival of sea lions (*Zalophus wollebaeki*) that proceed from the Galápagos Islands and come to Isla del Coco either as solitary animals or in very small groups. Other reports from the mid-nineteenth century highlight the presence of a small fur seal (*Arctocephalus galapagoensis*), also originating from the Galápagos Islands. The presence of these otariids (eared seals) native to the Galápagos Islands may be related to the warming of the sea waters due to the El Niño Southern Oscillation (ENSO) phenomenon. For a full account and analysis of the historical presence of otariids (Pinnipedia) on Isla del Coco, please see Montoya (2008c).

Fishes

The ichthyofauna of Isla del Coco has been studied recently by Lavenberg and Bussing (2000), Bussing and López (2005), and Garrison (2005). Their studies allowed for analyzing and summarizing in the best way possible the diversity of marine and freshwater fishes on the island.

According to the list of fish species available for Isla del Coco and prepared by Lavenberg and Bussing (2000),

Fig. 8.18 Pájara Islet with vegetation dominated by *Clusia rosea* and *Chloris paniculata*. *Photograph by Juan José Pucci, April 2005.*

which was subsequently updated in May 2005 by Garrison (2005), the ichthyofauna of this insular territory is composed of a total of 274 species (including 5 species of freshwater fish) that belong to 19 orders and 79 families.

Most of the fish species of Isla del Coco can be attributed to the biogeographical region or province known as Panamic, which extends from northern Baja California down to the northern sector of Peru, including the oceanic islands of the Eastern Tropical Pacific (Isla del Coco, Galápagos Archipelago, and Isla Malpelo) that belong to Costa Rica, Ecuador, and Colombia, respectively (see also Cortés, chapter 7 of this volume).

In order to understand the freshwater ichthyofauna, it is necessary to realize that the fluvial system of Isla del Coco is rather particular. It is characterized by a small geographical land area (24 km^2) with an extremely uneven terrain and a seasonal precipitation regime with over 6,500 mm of annual rainfall. Under such conditions rivulets and rivers drain quickly and have very variable flows, with numerous waterfalls and puddles, and hard and stable substrates especially formed by volcanic rocks and pebbles. Only the Genio, Lièvre, and Pittier rivers and some other small streams have permanent waters. Generally, such freshwater courses form at their mouths different levels of salinity (brackish waters).

It is in these aquatic environments that one finds an interesting freshwater ichthyofauna assemblage, including five species, three of which are endemic to the island: the Cocos Island goby (*Sicydium cocoensis*), the Cocos Island clingfish (*Gobiesox fulvus*), and the Cocos Island sleeper (*Eleotris tubularis*). The other two, non-endemic fish species are the mountain mullet (*Agonostomus monticola*) and the spotted sleeper (*Eleotris picta*), whose geographical distribution areas also extend to continental freshwaters.

On the other hand, since they have flat bodies, a ventral ventose, and benthic feeding habits, four of those species are perfectly adapted to the aforementioned freshwater habitats that are most common on the island: quick currents, variable flows, waterfalls, hard and stable substrates, etc. The fifth species (mountain mullet) is adapted to very quick currents of variable salinity of the river mouths.

The presence on the island of continental freshwater fishes and the precursors of today's endemics is explained by the complex system of superficial marine currents that flow around the island and facilitated the fishes' past dispersal and arrival. Some continental freshwater species have pelagic larvae that can be transported passively by superficial marine currents, which consequently make their arrival at the island feasible. This has been the case for the Mountain Mullet whose larvae have been found in ichthyoplankton from the open seas. Other freshwater fish genera like *Sicydium* and *Eleotris* live part of their life in the sea, a feature that would allow them to be transported by marine currents to Isla del Coco.

Another possible explanation for the arrival of freshwater fish at the island is the dispersal of fish eggs and larvae by occasionally visiting water birds that fly with eggs and larvae stuck to their extremities (e.g., legs, feet) (Montoya 2006). On the other hand, the arrival, colonization, adaptation, and evolution of these continental fish species didn't suffer from any competition, since Isla del Coco is an oceanic island of recent formation and did not have any freshwater fishes that could compete with the recently arrived colonists. For example, it is well known that species in the genus *Eleotris* prosper only in freshwater environments where ichtyological diversity is limited (Bussing and López, 1977). The lack of interspecific competition on recent oceanic islands would also explain the possible colonization of freshwater environments by some marine species at Isla del Coco. Practically all the freshwater fish species on the island have some kind of affinity with today's marine species and complete part of their life cycle in marine waters. Therefore, they easily migrate between the island's water courses in the higher parts and the sea.

The diversity and abundance of the present freshwater ichthyofauna on the island can serve as a biological indicator of the health state of its freshwater ecosystems. To do so, it will be necessary to define appropriate protocols to systematically monitor these systems.

Reptiles

The herpetofauna of Isla del Coco is numerically poor. There are only two existent terrestrial reptiles on the island: the small Coco Island lizard (*Norops townsendi*) (Fig. 8.19) and the Coco Island gecko (*Sphaerodactylus pacificus*) (Fig. 8.20). Both species are endemic to the island. They have been studied by some authors (Stejnerger 1900, 1903, Carpenter 1965, Jennsen and Rothblum 1977, Montoya 2005), who mainly focused on their behavior. At this moment these two reptiles suffer from predation by introduced cats and some visiting birds like the cattle egret (*Bubulcus ibis*).

Furthermore, marine turtles have been observed in the waters surrounding the island: the leatherback sea turtle (*Dermochelys coriacea*), the olive ridley sea turtle (*Lepidochelys olivacea*), the Pacific black sea turtle (*Chelonia agassizii*), and the hawksbill sea turtle (*Eretmochelys imbricata*). Although prints of turtles are only occasionally observed on the few sandy beaches of the island, nesting seems very rare. At the end of December 2007 a nest of a pacific black sea turtle was found and subsequently monitored at the beach along Wafer Bay. Eggs eclosion happened on January 22, 2008, and was documented with digital images (Geiner

Fig. 8.19 The endemic, small Coco Island lizard (*Norops townsendi*). *Photograph by Alvaro Saborío, May 2008.*

Golfín, ACMIC park ranger, pers. comm., April 28, 2008). Like on other oceanic islands around the world, the nesting success of these turtles depends in part on the predatory behavior of wild pigs and rats, which have inhabited the island for about two centuries now.

Additionally, there are some rare records of the marine Pacific snake (*Pelamis platurus*), which enriches the island's herpetological fauna. This species possibly arrives at the island by means of floating plant remains that originate on the continent.

Birds

Recently several studies have been carried out to better define and characterize the avifauna of Isla del Coco (Montoya 2003a, 2003b, 2004, 2006, 2007, 2008a, 2009b, Montoya and López Pozuelo 2007, 2012, Montoya and Pascal 2004, 2005, Dean and Montoya 2005, Easley and Montoya 2006, López Pozuelo and Montoya 2009, Huertas and Sandoval, 2012, Obando-Calderón et al. 2012). The island's bird fauna consists of 128 recorded species in 79 genera belonging to 31 families. Thirty-four species are marine-oceanic-pelagic in behavior (17 genera), 44 are coastal-lacustrine-estuarine (23 genera), and 50 are truly terrestrial (34 genera). One of the terrestrial birds was introduced to the island in 1965 (*Icterus pectoralis*) while another one is a domestic pigeon (*Columbia livia*).

A total of 116 bird species are visitors from elsewhere, while thirteen are native or resident species and regularly nest and reproduce on the island. Ninety-three of the visiting bird species are Nearctic migrants distributed over several categories: 44 reproduce exclusively in the Nearctics, 35 reproduce both in the Nearctics and in the Neotropics, and 14 are Nearctic species with congeners in the Neotropics. Five of the Neotropical species occur accidentally

Fig. 8.20 The endemic Cocos gecko (*Sphaerodactylus pacificus*). *Photograph by Michel Pascal and Michel Montoya, June 2002.*

on Isla del Coco (*Laterallus ambigularis, Crotophaga ani, Phaeothlypis fulvicauda, Cyanocompsa pallerina,* and the introduced *Icterus pectoralis*).

The marine avifauna known at Isla del Coco includes 34 species in 9 families and 17 genera. Eight do nest on the island and its peripheral islets, while 27 are only visiting. Out of the 34 species, eight are Nearctic migrants, 12 are circumtropical, and 14 have a distribution that corresponds to the Eastern Pacific region (Montoya 2008b).

Five of the resident species are terrestrial, three of which are endemic to the island: the Cocos flycatcher (*Nesotriccus ridgwayi*), the Cocos cuckoo (*Coccyzus ferrugineus*) (Fig. 8.21), and the Cocos Island finch or Cocos finch (*Pinaroloxias inornata*). A fourth species is the yellow warbler (*Dendroica petechia aureola*) (Fig. 8.22), which is endemic to Isla del Coco and the Galápagos islands. Finally, the fifth residing species known at this moment is the spot-breasted oriole (*Icterus pectoralis*), a species that was introduced intentionally, and whose population is currently very limited. The eight remaining resident species are marine birds that reproduce on the island and its outer islets: the masked booby (*Sula dactylatra*), the brown booby (*S. leucogaster*), the red-footed booby (*S. sula*), the great frigatebird (*Fregata minor*), and the brown noddy or common noddy (*Anous stolidus*) (Fig. 8.23), the black noddy or white-capped noddy (*A. minutus*), the sooty tern (*Onychoprion fuscatus*), and the white tern (*Gygis alba*). For the latter species the island is the most important nesting site in the whole Eastern Tropical Pacific region.

The Cocos finch (*Pinaroloxias inornata*) (Fig. 8.24), together with the 13 finches inhabiting the Galápagos Islands, form the group of Darwin's finches that served this outstanding scientist to establish his theory on the origin and evolution of species. Recent microsatellite studies and analysis of mitochondrial DNA sequences have determined that the predecessors of the Cocos finch flew in from the

Galápagos Islands (Sato et al. 1999, 2004, and Petren et al. 1999).

With regards to bird species presence-abundance data for Isla del Coco (Montoya 2009b), the following values have been determined: eight species are abundant (6.2%), three are common (2.3%), four are not very common (3.1%), ten are rare (7.8%), 39 are casual (30.2%), and 64 occur accidentally (49.6%).

In summary, the following are the main characteristics of the island's avifauna: (a) most of the species that have been observed on the island (77%) display a casual or accidental presence, since long periods of time may pass by without any bird being observed, and when they are seen again, they occur either solitary or in very small numbers; (b) a large portion of the birds that are observed on the island are visitors (90%), most of which are Nearctic migratory species (71%), while only a minority originates from the Neotropics (4.4%) and the remainder is mainly marine-circumtropical in its distribution throughout the Eastern Pacific; (c) thirteen species are native and nest and reproduce on the island, three of which are endemic to the island; and, (d) there is no information available to make sound decisions on bird management and conservation since no permanent biological monitoring program has been in place until date.

Historical information and references on the island's avifauna can be found in Montoya (2007). Following Slud's (1967) synthesis, other recent studies on the birds of Isla del Coco have been conducted by Acevedo-Gutiérrez (1994), who reported three new records for the island; Anderson et al. (1980) who studied a nematode in the endemic Cocos flycatcher; Dudzik (1996) with a first record [*sic*] of *Pelecanus occidentalis* and *Phaeton lepturus*; Grant and Grant (1997), who extracted and analyzed blood samples of the island's finch for genetic studies and reported a first sight of *Cyanocompsa pallerina*; Kroodsma et al. (1984), who

Fig. 8.21 The endemic Cocos cuckoo (*Coccyzus ferrugineus*). *Photograph by Michel Pascal and Michel Montoya, June 2002.*

analyzed the songs of the island's flycatcher; Layton (1984), on the systematic position of the island's flycatcher; Lücking and Lücking (1995a), who were the first to report *Passarina cyanea*; Mora and Barrantes (1995), with a first record of *Phaeothlypis fulvicauda*; Petren et al. (1999) and Sato et al. (1999) on the phylogeny of Darwin's finches including Isla del Coco's finch; Schluter (1984), who studied the morphological and phylogenetic relationships of Darwin's finches including the endemic species on the island; Sherry (1985, 1986), who analyzed the adaptations of the island's flycatcher with respect to its feeding and foraging behavior and morphology, and studied its nests, eggs, and reproductive behavior; Sherry and Werner (1984), who reported the first sightings of 19 bird species; Smith and Sweatman (1976) and Werner and Sherry (1987), who all studied the feeding behavior of Cocos Island's Darwin finch; and Stiles and Skutch (1989), who report a total of 26 bird species for the island in their classic book on the birds of Costa Rica; and Obando-Calderón et al. (2012) and Huertas and Sandoval (2012),who wrote about the occasional presence of ten Nearctic migrant bird species seen on the island in October 2010.

Mammals

As mentioned above, on Isla del Coco no native or autochthonous terrestrial mammals are found. The island houses only exotic and introduced mammals. In a study on the exotic vertebrates of the island, Montoya (2004) tells the story of mammal introductions by man—introductions that occurred intentionally or accidentally. The six exotic mammals that currently thrive on the island are the ship rat or black rat (*Rattus rattus*), the brown rat or Norwegian rat (*Rattus norvegicus*), the feral pig (*Sus scrofa*), the domestic cat (*Felis silvestris*), the wild goat (*Capra aegagrus hircus*), and the white-tailed deer (*Odocoileus virginianus*). The current presence of these species constitutes a true threat to the

Fig. 8.22 The yellow warbler (*Dendroica petechia aureola*), which is common at Isla del Coco and the Galápagos Islands. *Photograph by Kevin Easley, May 2006.*

Fig. 8.23 The brown noddy or common noddy (*Anous stolidus*), which nests on the islets without vegetation. *Photograph by Elvis Porras, May 2007.*

species and ecosystems on the island, and poses challenges to their conservation.

Without doubt the rats of Isla del Coco have been introduced over several times. The first accidental introduction of the ship rat happened somewhere during the midst of the sixteenth century, when the first European vessels arrived at the island. The brown rat was probably first introduced before the end of the nineteenth century. The June 1898 Costa Rican–led scientific expedition to Isla del Coco was the first one that reported the presence of brown rats on the island (Alfaro 1899, Pittier 1899, Montoya 2004, Montoya and Pascal 2005).

The single estimate of the rat population on the island was done by Gómez (2004). He confirmed the presence of the two species of rat on the entire island, including the surrounding islets. Actual rat densities rise as one gets closer to the settlements in the Chatham and Wafer bays. Gómez reports an average density of rats on the island of 87.5 individuals per hectare, ranging from 63 rats per ha at Cerro Pelón to 156 per ha in the *Talipariti*-dominated *majaguales* vegetation of Wafer Bay. These ranges for rat densities are much higher than those observed in wild populations and are comparable to those of large urban centers.

The first introduction of goats and pigs to the island is well recorded and occurred in 1793 during the arrival of the English whaler *Rattler*, which was commanded by Captain James Colnett, who wanted to assure the provisioning of fresh meat for future whaling expeditions.

The island's wild pigs (Fig. 8.25) have been the subject of studies by Sierra and colleagues (Sierra 2000, 2001a, 2001b, 2001c, 2002). They conducted research on their diet, reproductive state, and genetic makeup. They also assessed the erosive impact that their behavior had on the soil, calculating erosion rates of 23.6 to 200.4 kg/ha/year. The wild pigs' population appeared to oscillate between 400 and 500 individuals.

Fig. 8.24 The endemic Cocos finch (*Pinaroloxias inornata*), the single Darwin's finch outside the Galápagos Islands.
Photograph by Guillermo Blanco, April 2007.

No studies have addressed the goats on the island, and therefore little is known about their biology, natural history, and local behavior. At the moment their population is small, but individuals still seem to have an impact on the vegetation along the forest borders and in areas altered by landslides or human influence, in this way affecting processes of natural regeneration of the island's vegetation.

From historical accounts it is not clear when cats (*Felis silvestris*) were introduced to the island. Some authors point out that the introduction of cats took place when pigs and goats were brought to the island by Captain Colnett (Vergnes 1978), but the accounts on the 1794 *Rattler* expedition do not mention this. On the other hand, the first written record on the presence of cats on the island was provided by Captain Luis de Goeris (1879), who visited the island in July 1879 on board the ship *Liberia* bringing provisions and staff to cover for the recently established prison on the island (Montoya 2004). Although nothing is known about the population and behavior of the cats, we do know they predate mainly on resident birds (e.g., *Sula sula* and *Gygis alba*) and on other vertebrates such as the small lizard and gecko, both endemics at Isla del Coco.

A male and two female white-tailed deer (Fig. 8.26) were introduced in June 1935 by the crew of the *Veracity* who came to the island for treasure hunting during an expedition of *Treasure Recovery Limited* (Hancock and Weston 1960). Its population size, ecology, and distribution at the island are unknown. However, it has been revealed that it is mostly present in forest clearings and along forest margins altered by man or natural landslides, where it feeds upon pioneer plants in successional vegetation, hence altering natural regeneration processes.

Generally speaking, the introduced vertebrate fauna poses a real threat to the ecological integrity of the island, owing to the impact that these species have on the environment in general and the biological diversity of the ecosystems in particular. Monitoring and management actions are required to control or even eradicate them—if that is justified.

Insects and Other Terrestrial Invertebrates

A study on the entomological fauna of Isla del Coco was carried out by Hogue and Miller (1981). These authors identified 362 species, among which some 65 are probably endemic, 47 native, and 12 accidental species. Data analysis shows that the island's entomofauna is rather disharmonious, which is very normal for oceanic islands. Many insects from continental lands are represented by one or only a few species while others are completely absent. Aquatic insects, for instance, are poorly represented, although freshwater environments on the island are quite varied. The main groups of terrestrial arthropods are all present, although in small numbers. Bark-perforating insects (wood borers), leaf miners, and epiphyte-dwelling species are proportionally abundant. This is a common phenomenon at oceanic islands and reveals the importance of transport of plant remains from the continent to the island through drift (rafting), which serves as a key dispersal mechanism and triggers colonization of the island.

Twenty years ago, Charles L. Hogue of the Los Angeles Museum of Natural History in California and Scott E. Miller of the B.P. Bishop Museum in Honolulu, Hawai'i, had studied the island's arthropods over many years and reported at that time (Hogue and Miller, pers. comm., May 22, 1990), that the island's entomofauna collected and recorded until then possibly reached up to 800 species, 450 of which had been identified only at the family level (400 insects and 50 species belonging to other arthropods). At the time, there were 58 species of endemic insects indicating levels of endemism to be around 14.5%. It was expected that these levels would double or even triple if all collected material was ultimately identified.

Hogue and Miller pointed out that Isla del Coco's entomofauna is of particular importance to the zoogeography of the islands in the Eastern Pacific Ocean, as well as to the paleography of Central America and the Galápagos Islands. This importance results from the very humid environment and exuberant vegetation present at the island, which allows for more founding species than any other oceanic island in the region.

The island's arthropod fauna has several particular species, including the island's endemic scorpion (*Opisthacanthus valerioi*) (Fig. 8.27), which belongs to a very old genus with a disjunctive, Gondwanian distribution, that includes species spread over Africa, South America, and the Caribbean.

Another particular aspect of the insects at Isla del Coco is that only one single species of diurnal butterfly, *Historis*

Fig. 8.25 A feral pig at Isla del Coco (*Sus scrofa*).
Photograph by Guillermo Blanco, April 2007.

odius (Fig. 8.28), seems to reproduce on the island. It uses the endemic *guarumo* (*Cecropia pittieri*) as its host plant, but may also be observed feeding on mature fruits of other plants.

ANTS. At the beginning of the twentieth century, Isla del Coco's ant fauna (Hymenoptera, Formicidae) was studied by Forel (1902), Wheeler (1919), and Emery (1919). More recently, Solomon and Mikheyev (2005) retook the study of the island's ants, after recognizing that its myrmecological fauna had not received a lot of attention in spite of its relevance for biogeography. In their detailed, systematic inventory carried out during three weeks in June 2003 they listed 19 ant species in four subfamilies and 14 genera. Their main conclusions suggest first of all, that in spite of low human presence on the island, the ant fauna is dominated by introduced species. Secondly, they concluded that the current ant fauna differs substantially from the one that was reported during expeditions at the beginning of the twentieth century. For instance, one of the endemic ant species, *Camponotus biolleyi*, which was described about a century ago by Forel (1902), had not been recorded during the 2003 study.

It was also determined that *Wasmannia auropunctata*, a well-known invasive species at many places around the world, occurred in extremely high numbers at altered places but was absent in unaltered natural sites. According to these entomologists, the population of this ant species behaves aggressively, both interspecifically and intraspecifically—that is, it is aggressive towards populations of other species as well as towards individuals of the same species, a phenomenon not frequently observed among invasive species (Solomon and Mikheyev 2005).

BEETLES. Bark beetles and *Ambrosia* beetles (Curculionidae, Scolytidae) at Isla del Coco also have received increased

Fig. 8.26 The white-tailed deer (*Odocoileus virginianus*), which was introduced to Isla del Coco in 1935. *Photograph by Guillermo Blanco, March 2007.*

Fig. 8.27 Isla del Coco's endemic scorpion (*Opisthacanthus valerioi*).
Photograph by Jonathan Rawlinson, February 2001.

attention from scientists lately. Kirkendall and Bjarte (2006), for instance, found 19 species in five different genera, of which seven are endemic to the island. It makes the Scolytidae the most numerous beetle group here. These authors described three species that are considered new to science while being endemic to Isla del Coco: *Pycnarthrum pseudoinsularis*, *Xyleborinus cocoensis*, and *X. sparsegranularum*. Another species, *Xyleborus bispinatus*, was again recognized as a distinct species by separating it from *X. ferruginosus*. These authors also mention six other species that are known from the larger region, but had not been recorded for the island in the past. Additionally, they report ten species that were already known from Isla del Coco.

On the basis of this and other studies we can confirm that the flora and fauna thriving on Isla del Coco are the result of processes of natural dispersion and human-induced transport. After analyzing patterns of species colonization and distribution in relation to endemism, we may conclude that the largest numbers of taxa have emigrated from the American mainland. However, there are also elements that have their origin at the Galápagos Islands or have migrated all the way from the Caribbean region.

SPHINGIDS OR HAWK MOTHS. It is common to observe some big nocturnal butterflies on Isla del Coco that are attracted by the light of vessels that anchor in the island's bays. These butterflies or moths belong to the sphingids (Lepidoptera, Macrolepidoptera, Sphingidae), which are characterized by a frenulum that connects the two forewings with the two hindwings and controls movement when they fly, and which have a conical body profile with a rounded head and a long and pointed abdomen. During their larval stage these butterflies generally feed on one or a few species of plants in the same genus. Adults are considered excellent

Fig. 8.29 Leucauge argyra, one of the arachnids at Isla del Coco.
Photograph by Guillermo Blanco, April 2007.

In the island's intertidal zone a total of eight terrestrial species of crab have been identified. Five belong to the section Brachyura: *Johngarthia cocoensis* (Figs. 8.30 and 8.31) and *Cardisoma crassum* (Fig. 8.32), both in the Gecarcinidae family, and *Uca panamensis* (Panamian fiddler crab), *U. zacae* (Zaca's fiddler crab), and *Ocypode gaudichaudii* (ghost crab) in the Ocypodidae family. The species *J. cocoensis* is endemic to the island and was recently described by Perger et al. (2011).

Three land crabs that are hermits in the section Anamura thrive on the island: *Coenobita compresssus* in the Coenobitidae and, according to Gómez (1977, two species in the family Paguridae: *Pagurus californiensis* and *Pylopagurus longimanus*.

It is important to point out the presence of the fly *Pterogramma cardiosomi* (Diptera: Sphaeroceridae), which is an endemic species of Isla del Coco that lives in association with the blue crab (*Cardisoma crassum*) (Fig. 8.32) (Gómez 1977, Norrbom et al. 1984).

The blue crab has been observed in the area close to the outlets of the Genio and Pittier rivers at the Wafer and Iglesias bays, respectively. The Cocos Island orange crab (*Johngarthia cocoensis*) has been observed across the entire island though it seems to be more frequent on the outer islets that surround the island.The ghost crab (*Ocypode gaudichaudii*) has been observed only at the sandy beach contiguous to the mouth of the Genio River at Wafer Bay. The Panamanian fiddler crab, in turn, maintains a small colony in the sandy and gravel-containing intertidal zone at the Genio River mouth. Furthermore, Zaca's fiddler crab has not been observed since it was recorded in 1932.

The terrestrial hermit crabs of the island thrive mainly at the island's scarce sandy beaches (Chatham and Wafer). *Caenobita compressus* is the most common hermit crab on the island. The survival of these hermit crabs depends on the availability of empty shells of marine snails. Therefore, any collection of shells of these gastropods should be strictly forbidden at the island.

Besides the previously mentioned crabs the presence of *Grapsus grapsus* at the island should be reported. This jumpy crab, known as the Sally Lightfoot crab, thrives in every rocky coastal area that is wetted by the sea. It has a wide geographical distribution and extends its range from Baja California in Mexico down south to Talcahuano in Chile, inhabiting the islands of the Eastern Tropical Pacific.

Freshwater Invertebrates

The invertebrate fauna of the freshwater environment on Isla del Coco is not very well known. The best-studied group is formed by the crustaceans. In the island's abundant streams the presence of five different species of freshwater shrimps has been recorded. Four belong to the genus *Macrobrachium* (Palaemonidae) and one to the genus *Archaeatya* (Atyidae). Species in the family Palaemonidae are characterized by a well-developed and generally jagged face, with the first two pairs of pereopods (i.e., the walking limbs of a crustacean) ending in pincers, the second pair being bigger and more robust. In the central part of the Eastern Pacific this family is represented by 48 species belonging to 21 genera (Hendrickx 1995). Species in this genus mostly inhabit freshwater streams, but some of them occasionally and regularly migrate to brackish waters of coastal lagoons and estuary systems.

The Cauque River prawn (*Macrobrachium americanum*) ranges from Isla de Cedros in Baja California, Mexico, down to the Chira River in the Piura Department of northern Peru. It has also been collected at the Galápagos Islands and at Isla de Gorgona, Colombia. It is a large-sized

Fig. 8.30 The Cocos orange land crab (a new species new to science, currently being described) at Isla del Coco. *Photograph by Michel Pascal and Michel Montoya (March 2004).*

species with a grayish brown body and abdominal segments that have three, reddish brown to black, longitudinal fringes (one dorsal and two dorsolateral), sometimes presenting blue stains in the lower side of the shell. Usually, during the driest time of the year (December to April at Isla del Coco) the adults of this species are more easily observed in the higher parts of the streams, from which they descend to lower areas and estuaries in the rainy season when levels rise and the water becomes turbid. This pattern relates to reproductive periods when the species migrates toward the mouths of the streams where larvae are liberated that obligatorily need to pass their first life stages in the sea. At Isla del Coco this species has been collected from permanent rivers like the Genio, Lièvre, and Iglesias.

The prawn or shrimp *Macrobrachium hancocki* is found in coastal waters along the Pacific slope between Costa Rica and Márcona in the Department of Piura, Peru. Its geographical distribution includes Isla del Coco and some islands of the Galápagos Archipelago. It is medium-sized and blue with a brown toned back. This coloration can vary and become more brownish with blue patches on the chelipeds (pincers). The habitat of these shrimps is similar to that of *M. americanum*. In the freshwater streams they prefer puddles with a sandy bottom, some vegetation, and calm waters with abundant litter. Although these shrimps live as adults in freshwater, during their larval stage they require a marine environment. Therefore they can easily be transported and dispersed by marine currents. This would explain their presence in the freshwaters of Isla del Coco where it has been collected in the Genio and Iglesias rivers.

The Cocos Island prawn (*Macrobrachium cocoense*) was described in 1984 on the basis of a collection made in the Genio River that drains into the Wafer Bay, which is now the type locality for this endemic species (Abele and Kim 1984). This freshwater crustacean has a shell and abdomen that are predominantly clear brown with black spots. Later

Fig. 8.31 The purple crab (*Gecarcinus quadratus*) at Isla del Coco. *Photograph by Jonathan Rawlinson, February 2002.*

Fig. 8.32 The blue crab (*Cardisoma crassum*) at Isla del Coco. *Photograph by Michel Pascal and Michel Montoya, June 2002.*

in its life, its abdomen becomes bronzed while chelipeds (pincers) turn black. The biology and ecology of this species has not been the subject of any study yet.

In 1959, the genus and species of the Cocos Island atyid (*Archaeatya chacei*) were originally described on the basis of specimens collected at the island, hence the island serves as the type locality for both taxonomic levels. Later the species was gathered from water courses of the Pacific slope of Costa Rica, as well as on the islands of the Perlas Archipelago in Panama (Abele and Kim 1984). Villalobos (1959) indicates that the insular form of this shrimp is of continental origin and has followed an atyine evolution. Because of its morphological features and systematic position the species is located between the genera *Potimirin* and *Atyia* in the Atyidae family. Also, it is characterized by its small size and measures only 12 to 15 mm in length. It is common at Isla del Coco and has been found in the main streams of the island, from 250 m elevation down to the mouths that drain into the sea. It occupies practically all freshwater microhabitats found on the island.

Other Arthropods

According to Brölemann (1903, 1905) and Hertlein (1963), seven species of myriapods have been collected on the island (3 chilopods and 4 diplopods). They report the following endemic diplopods for the island: *Epinannolene pittieri*, *Leptodesmus folium*, and *Orthomorpha coactata*. Likewise, Hoffman (1979) described a new myriapod for Isla del Coco: *Trichomorpha hyla* (Polydesmida: Chelodesmidae).

People and Nature

Comparing the island's conditions in 2009 with those found in 1992 (Montoya 1992), one notes that the human presence on Isla del Coco has increased considerably. This implies a serious threat to the future of the island's species and ecosystems, and has implications for its conservation.

First, it is important to highlight that the Costa Rican government has maintained a permanent presence on the island since 1978. In the beginning it was characterized by a small Costa Rican naval base that was replaced in 1982 by staff working for the National Parks Service (SNP) and later by officials of the National System of Conservation Areas (SINAC) among other governmental institutions. This presence generated a progressive growth in residential infrastructures, an increase in introduced plants for human consumption (vegetables) and ornamental purposes, an increase in human-produced waste—mainly around the two settlements located in the Chatham and Wafer bays—as well as a rise in the use of fossil fuels, among other impacts that negatively affected the island's biodiversity.

Secondly, since the beginning of the 1980s there has been a constant presence of organized, international tourism, particularly dedicated to diving. Although diving norms and regulations do exist on the island, they are not fully applied and its impact on marine species and ecosystems is not monitored, owing to the lack of specialized personnel and operational resources.

Thirdly, there is a sustained presence of illegal, national and international fishermen in the island's protected waters, who apply unethical fishing arts to catch all sorts of fish, though mainly tuna and shark.

Fourthly, the number of occasional visits by pleasure yachts and other ships has increased lately, including those of cruise ships that organize trips specifically to visit the island itself.

In summary, although in the past few years protection of the island has improved and new norms and regulations for conservation and use have been established and put in place, the impact caused by increasing human presence on the island continues to affect its ecosystems, its biological communities, and the species they are composed of.

Negative impacts result from inadequate management of liquid and solid waste, produced by both residents and visitors. An example of such an impact is the recent demographic explosion in two species of rats (*Rattus rattus* and *R. norvegicus*) at the turn of the new century, with the highest population densities closer to the human settlements at the Wafer and Chatham bays (Gómez 2004). Another example concerns the loss of docility of some birds, characteristic of oceanic islands, that in the past nested close to the northern bays and now nest in the island's interior as a consequence of human-induced pollution with sound and lights in the northern bays (Montoya 2007).

In the past few years, the impact on the island's marine biodiversity caused by illegal fishing in its protected area (i.e., the Isla del Coco National Park) and adjacent zone has been controlled to some extent. It is, however, still a factor of concern that affects its species diversity. The management of introduced and invasive species, including both plants and animals, continues to be the main conservation issue to be addressed on the island.

So far, very little has been done to develop sound management instruments needed for successful conservation. No biodiversity management program is in place, nor is there any personnel to execute such a program if it existed. Unfortunately, no systematic and permanent biodiversity monitoring on the island is being done, and consequently there are no criteria to make any good decision on the possible eradication of invasive plant and animal species or on the need to monitor the impact that today's human presence (tourists, residents, visitors, etc.) has on the island's environment.

Future Perspectives

Scenarios for the Twenty-First Century: Towards Sustainability?

Any future scenario that is proposed for Isla del Coco should put at its center the island's extraordinary conditions to achieve successful conservation of its species and ecosystems.

The island has been declared a natural World Heritage Site by UNESCO, a Wetland of International Importance by the Ramsar Convention, and a Marine National Park and Marine Conservation Area by the Government of Costa Rica. These global and national recognitions confirm and ratify its importance for conservation while stressing the responsibility that the Costa Rican government has in guaranteeing, in an effective and unavoidable manner, its management for biodiversity conservation purposes, as the only way to ensure its future sustainability.

For these reasons, the only realistic scenario for the island would be natural resource management necessary to enable its permanent conservation. If such a goal is not achieved, the loss of species and ecosystem integrity will be imminent. Sustainable natural resource management approaches at the island should include the wise and environmentally healthy management of human presence, infrastructures and impacts, increasing amounts of waste, and rising numbers of tourists, both on the land and in the waters, as well as the successful eradication of invasive species and the restoration of degraded vegetation where necessary.

The establishment of a permanent biodiversity management program, with adequate staff who have the right skills and tools to monitor biodiversity and are able to apply best management practices, is essential to ensure effective conservation at the island in the long run.

In this regard, during the past few years, organizations like the MarViva Foundation, the Foundation of Friends of Isla del Coco (FAICO), UNESCO's World Heritage Program, the Global Environmental Facility (GEF), and the French Fund for the Global Environment (FFEM), as well as a few other organizations, have offered considerable support to the Government of Costa Rica in order to sustain conservation efforts at the island. However, to date, it has not been possible to install and institutionalize a single, integrated, and well-staffed management program with a clear biodiversity conservation purpose and adequate equipment necessary to protect and conserve the island effectively.

On a final note, it should be said that any conservation effort at the island should adequately address the issue of modern climate change and take into account any impact that results from human-induced climate change and will affect the island well before the end of the twenty-first century. Climate changes like sea level rise will indeed have a major negative effect on the coastal and inland ecosystems of Isla del Coco, which depend so much on the specific conditions of its fragile physical and geographical environment.

Research Needs

In 1999 a research strategy was developed for Isla del Coco under the auspices of the National System for Conservation Areas (SINAC). As part of this strategy, a series of key criteria were defined that any research project at the island should take into account (Montoya 2001b). Some of the most important criteria were: (a) the oceanic character of the island and its implications for the natural environment; (b) the geographic location of the island and its interaction with the particular local geological, climatic, and oceanographic conditions; (c) the current state of knowledge on its biodiversity; and (d) research needs to support and guide biodiversity monitoring and management. It was suggested that these criteria should serve as the basis for defining and prioritizing any future research interest and project.

An overall strategy for the island that was elaborated subsequently (MINAE, SINAC, and ACMIC 1999) did not take into account any of these criteria and led to a long list of research topics suggested by participants during workshops. In fact, the proposed research themes did not coincide much with the reality and needs of the island. In the following period, the strategy was neither implemented nor evaluated again.

Later in 2006, the Isla del Coco Marine Conservation Area (ACMIC) prepared a new research program for the island (ACMIC 2007). This program also lists a series of research activities that should be implemented by external organizations, but did not provide any prioritization or indication about follow-up, staff training, internalization, and/or institutionalization of research results and conclusions.

The most important research needs in fields related to the island's terrestrial biodiversity are today as follows:

- The continued inventory of species and study of their ecology and biogeography, putting emphasis on synthesizing knowledge on taxonomic groups that have received little or no attention so far;
- The development of tools, systems, and protocols to monitor native and introduced species as bioindicators of the state of the environment, especially focusing on endemic and invasive species;
- The study of ecosystem functioning and dynamics, particularly with respect to species interactions (behavior, symbiosis, mutualism, competition, parasitism, etc.), aspects of predation (herbivory, carnivory), pollination, frugivory and seed dispersal;
- The study of succession and natural regeneration, especially in relation to vegetation structure, productivity, biogeochemical cycles, and ecosystem seasonality.

Any research project that focuses on the topics mentioned above should take place within the framework of: (a) a specific biodiversity management program; (b) a geo-referenced database on biodiversity; (c) existing studies on the geomorphology and macro/micro-climatology of the island; and (d) available personnel to ensure research results feed into the furthering of a biodiversity management program.

Conclusions

After having assessed the current state of knowledge on the biodiversity of Isla del Coco and its geographical surroundings, and having evaluated the institutional and programmatic framework within which biodiversity management and conservation should take place, we can draw the following conclusions:

a. The historical threats that affect the biodiversity of the island continue to persist and even increase in some cases. The strongest threat concerns the increasing human presence on the island, of which the impact has been neither monitored nor fully controlled. The second most important threat is the presence of introduced and invasive plant and animal species that have been little managed so far.

b. The biodiversity knowledge that has been generated is still insufficient and fragmented. Many fields of study haven't received any attention. No basic information is available that would help establish protocols to monitor species that would require management (e.g., invasive species eradication, or other management practices). Therefore it is necessary to establish a well-designed management program for biodiversity conservation. Such a program should be the central axis of a general management plan for the island and should come with adequate personnel and funds to ensure its successful functioning. It should establish the activities, procedures, and protocols to implement a kind of integrated management and conservation of the island's species and ecosystems with the aim to guarantee a sustainable future for its biodiversity.
c. The national and international recognition of Isla del Coco as national park, conservation area, World Heritage Site, and international wetland indicate that Isla del Coco's future is in the conservation arena and hence, that the Costa Rican government and international organizations have a responsibility to work together to conserve the island's natural riches for the future. Past experiences, however, demonstrate that the government-administered Isla del Coco Marine Conservation Area (ACMIC) must improve its capacity to receive international cooperation and assume concrete responsibilities to manage the island's biodiversity to achieve successful conservation in the long term.

Acknowledgments

This chapter is dedicated, in memoriam, to the late Luis Diego Goméz, explorer, botanist, and researcher of Isla del Coco and many other places in the Neotropics. Together with Maarten Kappelle, he invited me to write this comprehensive chapter. I am also thankful to Gregorio Dauphin, Pablo Madriz Masís, Carlos Rojas, and Carlos Víquez who studied Isla del Coco and provided me with scientific information on specific topics dealt with in this chapter. Furthermore, I am very grateful to the participants of the various Isla del Coco BioCourses that have been organized annually since 2004 by the Organization for Tropical Studies (OTS). They provided me with information and a number of photographs that are included in this publication. Ultimately, I am thankful to the numerous researchers, students, park guards, and other observers who during the past 29 years supported my fieldwork and shared with me their insights and knowledge on this unique oceanic island.

References

Abbott, D.P. 1966. Factors influencing the zoogeographic affinities of Galápagos inshore marine fauna. In R.I. Bowman, ed., *The Galapagos Proceeding Symposia International Scientific Project*, 108–22. Berkeley: University of California Press.

Abele, L.G., and W. Kim. 1984. Notes on the freshwater shrimps of Isla del Coco with description of *Macrobrachium cocoense*, a new species. *Proceedings of the Biological Society of Washington* 97(4): 951–60.

Acevedo-Gutiérrez, A. 1994. First records of occurrence and nesting of three birds species at Isla del Coco, Costa Rica. *Revista de Biología Tropical* 42(3):762.

ACMIC. 2007. Programa de Investigación. Área de Conservación Marina Isla del Coco 2007. San José, Costa Rica: Área de Conservación Marina Isla del Coco. 30 pp.

Adams, P.A. 1983. A new subspecies of *Chrysoperla externa* from Cocos Island (Neuroptera: Chrysopidae). *Bulletin of the Southern California Academy of Sciences* 82(1): 42–45

Alfaro, A. 1899. Informe sobre una expedición a la Isla del Coco. Fauna. San José, Memoria de la Secretaría de Fomento 1898/1899: 159–97. Also in *Reproducciones científicas, una expedición y legislación de la Isla del Coco*, San José, Instituto Geográfico Costa Rica, 31–36.

Alfaro, E.J. 2008. Ciclo diario y anual de variables troposféricas y oceánicas en la Isla del Coco, Costa Rica. *Revista de Biología Tropical* 56(Suppl. 2): 19–29.

Alvarado, G.E. 1984a.Distribución de las unidades litoestratigráficas igneas del Neogeno y Cuaternario de Costa Rica. In P. Sprechmann, ed., *Manual de Geología de Costa Rica*, 301–7. Vol. 1. San José, Costa Rica: Editorial Universidad de Costa Rica.

Alvarado, G.E. 1984b. Aspectos petrológicos-geológicos de los volcanes y unidades lávicas del Cenozoico Superior de Costa Rica. Thesis, Escuela Centroamericana Geología, Universidad de Costa Rica. San José, Costa Rica. 187 pp.

Alves, M.V., and W. Thomas. 2002. Four new species of *Hypolytrum* Rich. (Cyperaceae) from Costa Rica and Brazil. *Feddes Repertorium* 113(2–4): 261–70.

Anderson, R.C., L.P. Wong, and T.W. Sherry. 1980. *Diplotriaena muscisaxicola* Shuurmans-Stekhoven, 1952 (Nematoda: Diplotriaenoidea) from *Nesotriccus ridgwayi* Townsend (Tyranidae) of Cocos Island, Costa Rica. *Canadian Journal of Zoology* 58:1923–26.

Anderson, R.S. 2002. The Dryiphthoridae of Costa Rica and Panama: checklist with keys, new synonymy and description of new species of *Cactofagus, Mesocordylus, Metamasius*, and *Rhodobaenus* (Coleoptera; Curculionidae). *Zootaxa* 80:1–94.

Anderson, R.S., and A.A. Lanteri. 2000. New genera and species of weevils from the Galápagos, Ecuador and Cocos Island, Costa Rica (Coleoptera, Curculionidae; Entiminae; Entimini). *American Museum Novitates* 3299:1–15.

Arias, R. 1997. Isla del Coco. Historia y leyenda. Thesis, Universidad de Costa Rica. San José, Costa Rica. 261 pp.

Arias-Salguero, M.E. 2000. Informe de prospección geoeléctrica en la Isla del Coco. Sección Sismología, Vulcanología y Exploración Geofísica, Universidad de Costa Rica. 12 pp.

Arias-Salguero, M.E. 2002. Aplicaciones biofísicas a la hidrogeología en Costa Rica (Aplicación en Isla del Coco). *Revista Geológica de América Central* 27:11–20.

Atkinson, I.A.E. 1985. The spread of commensal species of *Rattus* to oceanic islands and their effects on island avifauna. In P.J. Moors, *Conservation of Birds*, 35–81. International Council for Bird Preservation Technical Publication No. 3. Cambridge, England.

Baert, L. 1995. The Anyphaenidae of the Galápagos Archipelago and Cocos Island, with a redescription of *Anyphaenoides pluridentata* Berland, 1913. *Bulletin of the British Arachnological Society* 10(1):10–14.

Bailey, C.F. 1976. The potassium-argon ages from Galapagos Islands. *Science* 192:465–67.

Banks, N. 1905. Arachnids from Cocos Island. *Proceedings of the Entomological Society of Washington* 7(1):20–23.

Bartram, E.C. 1933. Mosses of the Templeton Crocker Expedition collected by John T. Howell and list of mosses known from the Galapagos Islands and from Cocos Island. *Proceedings of the California Academy of Sciences, 4th Series* 21(8):75–76.

Belcher, E.1843. *Narrative of a voyage round the World, performed in the Majesty's Ship Sulphur. During the years 1836–1842; including details of the naval operations in China, from Dec.1840, to Nov. 1841*. London: Henry Colburn. 387 pp.

Bellamy, C.L. 1986. A new species of *Halecia* from Cocos Island, Costa Rica, with review of the Neotropical genera of the tribe Chalcophorini (Coleptera: Buprestidae). *Coleopterists Bulletin* 40(4): 381–87.

Bellon, H., R. Sáenz, and J. Tournon. 1983. K-Ar radiometric ages of lavas from Cocos Island (Eastern Pacific). *Marine Geology* 54:17–23.

Bentham, G. 1844–1846. The botany of the voyage of H.M.S. Sulphur, under command of Capitain Sir Edward Belcher, during the years 1836–42. Vol. 12. Published under the authority of the Lords Commissioners of the Almiralty, London. 196 pp.

Bergoeing, J.P. 1994. Esbozo geomorfológico. Isla del Coco. Map. San José, Costa Rica.

Bernecker, A., and R. Lücking. 2000. Epiphyllous bryophytes from Cocos Island, Costa Rica. A floristic and phytogeographic study. *Ecotropica* 6: 55–69.

Betagh, W. 1728. A voyage round the word, being an account of remarkable enterprise, begun the year 1719, chiefly to the cruise on the Spaniards in the great South Ocean, relating the true historical facts of that whole affair . . . , by William Betagh, captain of marines in the expedition. Printed for T. Combes, J. Lacy, and J. Clarke. London: H. Colburn.

Bickel, D.J., and B.J. Sinclair. 1997. The Dolichopodidae (Diptera) of the Galapagos Islands, with notes on the New World fauna. *Entomologica Scandinavica* 28(3):241–70.

Bogarín, D., J. Warner, M. Powell, andV. Savolainen. 2011. The orchid flora of Cocos Island National Park, Puntarenas, Costa Rica. *Botanical Journal of the Linnean Society*166(1):20–39.

Brenes, C., and C. González. 1995. Geología, hidrología, clima y suelos. In *Plan de Manejo. Isla del Coco*, Anexo 1, Sondeo Ecológico Rápido, 22–39. San José, Costa Rica: Fundación de la Universidad de Costa Rica para la Investigación (FUNDEVI), Fundación Pro-Ambiente (PROAMBI), Instituto Costarricense de Turismo (ICT), y Servicio de Parques Nacionales (SPN).

Bright, D.E. 1982. Scolytidae (Coleoptera) from Cocos Islands, Costa Rica, with description of one new species. *Coleopterists Bulletin* 36(1):127–30.

Brölemann, E.W. 1903. Myriapodes recueillis a l'Isla de Coco par M. le Professeur P. Biolley. *Annales de la Société Entomologique de France* 72:128–42.

Brölemann, E.W. 1905. Myriapodes du Costa Rica recueillis par M. le Professeur P. Biolley. *Annales de la Société Entomologique de France* 74:335–80.

Brown, J.W. 1990. Sphingidae (Lepidoptera) of Isla del Coco, Costa Rica, with remarks on the macrolepidoptera fauna. *Brenesia* 33:81–84.

Brown, J.W., J.P. Donahue, and S.E. Miller. 1991. Two new species of geometrid moths (Lepidoptera: Geometridae: Ennominae) from Cocos Island, Costa Rica. Natural History Museum of Los Angeles. *Contributions in Science* 423:11–18.

Brown, J.W., and S.E. Miller. 1999. A new species of *Coelostathma* Clemens (Lepidoptera: Tortricidae) from Cocos Island, Costa Rica, with comments on the phylogenetic significance of abdominal dorsal pits in Sparganothini. *Proceedings of the Entomological Society of Washington* 101(4):701–7.

Burney, J. 1816. History of the buccaneers of America. Luke Hansan's & Sons, near Lincoln's Inn, for Payne and Foss, Pall Mall, XII, 326 pp.

Bussing, W.A., and M.I. López. 1977. Distribución y aspectos ecológicos de los peces de las cuencas hidrográficas del Arenal y Tempisque, Costa Rica *Revista de Biología Tropical* 25(1):13–37.

Bussing, W.A., and M.I. López. 2005 Peces de la Isla del Coco y peces arrecifales de la costa Pacífica de América Central meridional. Guía ilustrada / Fishes of Cocos Island and reef fishes of the Pacific coast of lower Central America. *Revista de Biología Tropical* 23(suppl. 1):1–192.

Byers, G.W. 1981. The crane flies (Diptera: Tipulidae) of Cocos Island, Costa Rica, with descriptions of four new species. Natural History Museum of Los Angeles, *Contributions in Science* 335:1–8

Carpenter, C.C.1965.The display of the Cocos Island Anole. *Herpetologica* 21(4):256–60.

Castillo, P.R., R. Batiza, D. Vanko, E. Malavassi, E. Barquero, and E. Fernández. 1988. Anomalously young volcanoes on old hot-spot traces. I. Geology and petrology of Cocos Island. *Geological Society of America Bulletin* 100: 1400–1414.

Chavarría Mora, N. 1899. Observaciones sobre la Isla del Coco. In *Reproducciones Científicas: Una Expedición y Legislación de la Isla del Coco* (1963), 39–43. San José, Costa Rica: Instituto Geográfico Costa Rica.

Chemsak, J.A., and E.G. Linsley. 1982. *Checklist of the Cerambycidae and Disteniidae of North, Central America, and the West Indies (Coleoptera)*. Medford: Plexus Publishing Inc. 138 pp.

Choe, J.C. 1992. Zoraptera of Panama with a review of the morphology, systematics, and biology of the Order. In D. Quintero and A. Aiello, eds., *Insects of Panama and Mesoamerica: Selected Studies*, 249–56. New York: Oxford University Press.

Chubb, L.J., and R. Richardson. 1933.Geology of Galapagos, Cocos and Easter Islands. Honolulu, *Bernice P. Bishop Museum Bulletin* 110:1–67.

Clark, L. 1953. Some hepaticae from Galapagos, Cocos and other Pacific coast Islands. *Proceedings of the California Academy of Sciences, 4th Series* 27:593–624.

Colnett, J. 1798. Voyage to the South Atlantic and round Cape Horn into the Pacific Ocean, for the purpose of extending the spermaceti whale fisheries, and other objects of commerce, by ascertaining the ports, bays, harbours, and anchoring births, in certain islands and coasts in those seas at which the ships of the British merchants might be refitted. Undertaken and performed by Captain James Colnett, of the Royal Navy, in the ship Rattler. London: printed for the author by W. Bennet, Marsham Street, SOHO; Stockdale, Strand; and White, Fleet Street.

Cook, O.F. 1939.A new palm from Cocos Island collected on the presidential cruise of 1938. *Smithsonian Miscellaneous Collections* 98(7):1–26.

Cook, O.F. 1940. An endemic palm on Cocos Island near Panama mistaken for the coconut palm. *Science* 9:140–42.

Coolidge, K.R. 1909. The Arachnida of the Galapagos Islands. *Psyche* 16(5): 112.

Cortés, J. 2008. Historia de la investigación marina de la Isla del Coco, Costa Rica. *Revista de Biología Tropical* 56(Suppl. 2): 1–18.

Coulter, J. 1847. *Adventures on the West Coast of South America and the interior of California*. Vol. 2. London: Longman, Brown, Green and Longmans.

Dampier, W. 1697. The new voyage round the World. Describing particularly the Isthmus of America, several coasts and islands in the West Indies, the isles of Cape Verde, the passage by Terra del Fuego, the South Sea Coasts of Chili, Peru, and Mexico; the isle of Guam, one of the Ladrones, Mindanao, and other Philippine and East India, islands near Cambodia, China, Formosa, Luconia, Celebes, &c. New Holland, Sumatra, Nicobar Isles; Cape of Good Hope, and Santa Helena. Their Soils, Rivers, Harbours, Plants, Fruits, Animals, and Inhabitants. Their Customs, Religion, Government, Trade, &c. London: printed for James Knapton at the Crown in St Paul's Church-yard.

Dauphin, G. 1995. Briófitos de la Isla de Cocos: diversidad ecológica. Thesis, Ciudad Universitaria, Universidad de Costa Rica.San Pedro, Costa Rica. 63 pp.

Dauphin, G. 1999. Bryophytes of Coco Island, Costa Rica: biodiversity, biogeography and ecology. *Revista de Biología Tropical* 47(3): 309–28.

Davis, D.R. 1994. Neotropical Tinidae. V: The Tineidae of Cocos Island, Costa Rica (Lepidoptera: Tineoidea). *Proceedings of the Entomological Society of Washington* 94(4):735–48.

Davis, W.D. 1874. Nimrod of the sea; or, the American whaleman. New York: Harper & Brother Publisher. 403 pp. Also in: Boston, Charles Lauriat Company, 406 p. 1926. and North Quincy, MA, Christopher Publishing House, 405 p. 1972.

Dean, R., and M. Montoya. 2005. Ornithological observations from Cocos Island, Costa Rica (April 2005). *Zeledonia* 9(1):62–69.

Dudzik, K.J. 1996. First record of *Pelecanus occidentalis* (Aves: Pelecanidae) and *Phaeton lepturus* (Aves: Phaethontidae) at Isla del Coco, Costa Rica. *Revista de Biología Tropical* 44: 303–4.

Duret (Sieur D.). 1720. Voyage de Marseille a Lima, et dans autres lieux des Indes Occidentales. Avec une exacte description de ce qu'il y a de plus remarquable tant pour la géographie, que pour les moeurs, le coûtumes, le commerce, le gouvernement et la religion des peuples; avec notes et des figures en taille-douce. Par Sieur D., Paris, Jean Baptiste Coignard. 282 pp.

Easley, K., and M. Montoya. 2006. Observaciones ornitológicas en la Isla del Coco, Costa Rica (Mayo 2006). *Zeledonia* 10(2):31–41.

Eberhard, G.W. 1989. Niche expansion in the spinder *Wendilgarda galapagensis* (Araneae, Theridiosomatidae) on Cocos Island. *Revista de Biología Tropical* 37(2): 163–68

Eibl-Eibesfeldt, I. 1986. Una excursión a la Isla del Coco. In I. Eibl-Eibesfeldt, *Las Islas Galápagos: Un Arca de Noé en el Pacífico*, 174–82. Madrid: Alianza Editorial.

Emery, C. 1919. Formiche dell' Isola Cocos. Rendiconto delle sessioni della R. Accademia della Scienze dell' Instituto di Bologna. *Classe de Scienze Fisiche, Nouva Serie* 23: 36–40.

Epler, B.C. 1987. Whalers, whales and tortoises. *Oceanus* 39(2): 86–92.

Fau, F. 1960. Diario de a bordo. Expedición a la Isla del Coco. La República I. 15 abril, p.10; II. 22 abril, p.10; III. 13 mayo, p. 10; IV. 20 mayo, p.10, 12; V. 3 junio, p.10,18; VI. 10 junio, p.12; VII. 17 junio, p. 11; VIII. 24 junio, p.11; IX. 1 julio, p.10; X. 8 julio, 10,15; XI. 29 julio, p.10.

Fernández de Oviedo y Valdés, G. 1535–1547. *Historia General y Natural de las Indias: Islas y Tierra Firme del Mar Océano*. Asunción, Paraguay: Editorial Guaranía (1944–1945). 14 Tomos (Primera edición americana).

Forel, A. 1902. *Quatre Notices Myrmecologiques*. II. Fourmis de l' Île des Cocos, etc . . . *Annales de la Societé Royale Entomologique de Belgique* 46:170–182.

Fosberg, F.R., and W.L. Klawe. 1966. Preliminary list of plants from Cocos Island. In R.I. Bowman, ed., *Proceedings of the Symposia of the Galapagos International Scientific Project*, 187–89. Berkeley: University of California Press.

Fournier, L.A. 1966. Botany of Cocos Island. In R.I. Bowman, ed., *Proceedings of the Symposia of the Galapagos International Scientific Project*, 183–86. Berkeley, California: University of California Press.

Fournier, L.A. 1968. Descripción de la vegetación de la Isla del Coco. Instituto Geográfico Nacional, *Informe Semestral* (enero–junio 1968): 49–64.

Francke, O.F. 1974. Nota sobre los géneros *Opisthacanthus* Peters y *Nepabellus* nom. nov. (Scorpionida, Scorpionidae) e informe sobre el hallazgo de *O. lepturus* en la Isla del Coco. *Brenesia* 4:31–35.

Fuentes, G., A.B. Azofeifa, and S. Aguilar. 2005. *Bibliografía sobre el Parque Nacional Isla del Coco*. Serie Bibliografias de la OET 3. San José, Costa Rica: Organización Estudios Tropicales (OET). 238 pp.

Garrison, G. 2005. Peces de la Isla del Coco / Isla del Coco fishes. Santo Domingo de Heredia, Costa Rica: Instituto Nacional de Biodiversidad. 430 pp.

Goeris de, L. 1879. Informe y mapa del viaje en el Pailebot Liberia a mi mando, de Puntarenas a la Isla del Coco y regreso. *La Gaceta* 436: 1.

Gómez, J.R. 2004. Estudio denso poblacional de los roedores introducidos y su impacto sobre la fauna y flora nativa en la Isla del Coco. Thesis, Universidad Nacional, Programa Regional en Manejo de Vida Silvestre, p.v. Heredia, Costa Rica.

Gómez, L.D. 1971. Two new tree ferns from Costa Rica. *American Fern Journal* 61(4): 166–70.

Gómez, L.D. 1975a. Contribuciones a la pteridología costarricense. VII. Pteridofitos de la Isla del Coco.*Brenesia* 6:33–48.

Gómez, L.D. 1975b. The ferns and fern-allies of Cocos Island, Costa Rica. *American Ferns Journal* 65:102–4.

Gómez, L.D. 1977. Contribuciones a la pteridología costarricense. X. Nuevos pteridófitos de la Isla del Coco. *Brenesia* 8:97–101.

Gómez, L.D. 1983. The fungi of Cocos Island, Costa Rica. *Brenesia* 21: 355–64.

Goodnight, J.C., and M.L. Goodnight. 1947. Studies on the phalangid fauna of Central America. *American Museum Novitates* 1340:1–21.

Grant, P.R., and R. Grant.1997. Expedición del Instituto Smithsonian de Investigaciones Tropicales a la Isla del Coco. Informe del estudio sobre pinzones. Panamá: Instituto Smithsonian de Investigaciones Tropicales (STRI). 2 pp.

Güenchor, L.A. 1975. Los náufragos del Coco. *Excélsior* I. Abandonados 4 meses en el paraíso, 10 noviembre p.1, 3; II. Riqueza fabulosa de la Isla del Coco, 11 noviembre, p. 5; III. Guerra con cerdos y extrañas enfermedades, 13 noviembre p. 4; IV. La odisea lleva ya 15 años, 14 noviembre, p.8; V. Historia no termina al regresar, 16 noviembre, p. 2.

Gueydon, H.L (Compte de). 1948. Report on Cocos (Traduction d'un rapport addressé à M. le Contre-Amiral Hamelin, par M. le Comte de Gueydon, Capitaine de Corvette, Commandant du brick Le Génie sur la relâche à l'île des Cocos en Novembre 1846). *Pacific Discovery* 1(6):8–14.

Hancock, R., and J.A. Weston. 1960. The lost treasure of Cocos Island. New York: Thomas Nelson & Sons. 325 pp.

Hendrickx, M.E. 1995.Camarones. In W. Fisher, F. Krupp, W. Schneider, C. Sommer, K.E. Carpenter, and V.H. Niem, *Guía de FAO para la Identificación de Especies para los Fines de Pesca: Pacífico Centro-oriental*, Vol. 1: Plantas e invertebrados, 417–69. Roma: UN Food and Agriculture Organization (FAO).

Heppner, J.B. 1981. A new *Tortyra* from Cocos Island, Costa Rica (Lepidoptera: Choreutidae). *Journal of Research on the Lepidoptera* 19(4):196–98.

Herrera, W.1984. Informe de campo. Gira realizada a la Isla del Coco con el objetivo de recabar información climatológica. San José, Costa Rica: Servicio de Parques Nacionales. 6 pp.

Hertlein, L.G. 1963.Contribution to the biogeography of Cocos Island, including a bibliography. *Proceedings of the California Academy of Sciences, 4th Series* 32(8):219–89.

Hoffman, R.L. 1979. Chelodesmil Studies VIII. A new milliped of the genus *Trichomorpha* from Cocos Island, with notes on related species and proposed species of the new tribe Trichomorphini (Polydesmida: Chelodesmidae). Natural History Museum of Los Angeles. *Contributions in Science* 305:1–7.

Hogue, C.L., and S.E. Miller. 1981. Entomofauna of Cocos Island, Costa Rica. Smithsonian Institution, *Atoll Research Bulletin* 250:1–29.

Huber, B. 1998. The pholcid spiders of Costa Rica (Araneae: Pholcidae). *Revista de Biología Tropical* 45: 1583–634.

Huertas, J.A., and L. Sandoval. 2012. Ten new bird species for Isla del Coco, Costa Rica. *Check List: Journal of Species Lists and Distribution* 8(3):568–71.

Huertas, V. 1959. Cuatro días en la Isla del Coco. La República I. 7 junio, p.19; II. 13 junio, p.9; III. 17 junio, p. 4; IV. 28 junio, p. 29; V. 11 julio, p.6.

Jennsen, T.A., and L.M. Rothblum. 1977. Display repertoire analysis of *Anolis townsendi* (Sauria: Iguanidae) from Cocos Island. *Copeia* 1:103–9.

Jinesta, R. 1937. *La Isla del Coco*. San José, Costa Rica: Imprenta Falco Hermanos. 24 pp.

Jinesta, R. 1939. La isla del Coco. *Elevación* 1(5):4–31. Versión revisada de R. Jinesta, 1937.

Jinesta, R. 1940. Piezas arqueológicas de la Isla del Coco. *Prensa Libre* (14 mayo): 4.

Johnson, C. 1982. Ptillidae (Coleoptera) from the Galápagos and Cocos Island. *Brenesia* 19/20:189–99.

Jordal, B.H. 1998. A review of *Scolytodes* Ferrari (Coleoptera: Scolytidae) associated with *Cecropia* (Cecropiaceae) in northern Neotropics. *Journal of Natural History* 32:31–85.

Kasteleijn, H.W. 1987. Marine biological research in the Galapagos: past, present and future. *Oceanus* 30(2):33–41.

Kerr, R. 1814. General history and collection of voyages and travels, arranged in systematic order: Forming a complete history of the origin and progress of navigation discovery, and commerce, by sea and land, from the earliest ages to the present time. Vol. X. Edinburgh, UK: Ballantyne and Company.

Kessler, W.S. 2006. The circulation of the eastern tropical Pacific: a review. *Progress in Oceanography* 69: 181–217.

Kirkendall, L.R., and H.J. Bjarte. 2006. The bark and ambrosia beetles (Curculionidae, Scolytinae) of Cocos Island, Costa Rica and the role of mating systems in island zoogeography. *Biological Journal of the Linnean Society* 89(4):728–42.

Kroodsma, D.E., V.A. Ingalls, T.W. Sherry, and T.K. Werner. 1984. Songs of the Cocos Flycatcher: vocal behavior of a suboscine on an isolate oceanic island. *Condor*: 75–84.

Kury, A.B. 2003. Annotated catalogue of the Laniatores of the New World (Arachnida, Opiliones). *Revista Ibérica de Aracnología*, Volumen Especial 1:1–337.

Lavenberg, R., and W.A. Bussing. 2000. Peces: lista de especies (actualizada a marzo del 2005). In G. Garrison (2005), *Peces de la Isla del Coco*, 357–79. Santo Domingo de Heredia, Costa Rica: Instituto Nacional de Biodiversidad (INBio).

Layton, W.E. 1984. The systematic position of the Cocos Flycatcher. *Condor* 86:42–47.

Levi, H.W. 1963. American spiders of the genus *Theridion* (Araneae, Theridiidae). *Bulletin of the Museum of Comparative Zoölogy* 129(10): 481–589.

Lièvre, D. 1893. Une île déserte du Pacifique. L'île de Cocos (Amérique). *Bulletin de la Societé de Géographie Commerciale du Hávre* 1893: 233–58. Revue de Géographie (Paris), Tome 32, livraison de mai: 349–357, 2 illustrations; livraison de juin: 416–422; Tome 33, livraison de juilet: 34–41.

Lièvre, D. 1962. Una isla desierta en el Pacífico, Isla del Coco (América). In *Los Viajes de Cockburn y Lièvre por Costa Rica*, 105–34. Editorial Costa Rica. Traducción de Jorge y Maruja León, de Lièvre, D. Une île déserte du Pacifique, l'île des Cocos (Amérique). *Bulletin Société de Géographie Commerciale du Hávre* 1893: 233–58.

Lizano, O.G. 2008. Dinámica de aguas alrededor de la Isla del Coco, Costa Rica. *Revista de Biología Tropical* 56(Suppl. 2):31–48.

Lockwood, J.P., and B.J. Benumof. 2000. *Initial Isla del Coco Geologic Investigation*. Field Report (based on January 31–February 14, 2000 fieldwork). Hawaii: Geohazards Consultants International Inc. 14 pp. Map.

López Pozuelo, F., and M. Montoya. 2009. Observaciones ornitológicas en la Isla del Coco, Costa Rica. IV. Enero–Mayo 2008. *Zeledonia* 13(2):55–60.

Lourenço, W.R. 1980. A propósito de duas novas espécies de *Opisthacanthus* para a região Neotropical. *Opisthacanthus valerioi* da "Isla del Coco", Costa Rica e *Opisthacanthus heurtaultae* da Guiana Francesa (Scorpiones, Scorpionidae). *Revista Nordestina de Biologia* 3(2):179–94.

Lücking, A., and R. Lücking. 1995a. *Passerina cyanea* (Passeriformes: Emberizidae), nuevo informe ornitológico para la Isla del Coco, Costa Rica. *Revista de Biología Tropical* 41(3): 928–29.

Lücking, R., and A. Lücking. 1995b. Folliicolous lichens and bryophytes from Cocos Island, Costa Rica. A taxonomical and ecogeographical study. I. Lichens. *Herzogia* 11: 143–74.

Madrigal, M. 1954. El tesoro de la Isla del Coco es su caza y su pesca. La Nación I. 27 junio, p 28; II. 4 julio, p.18; III. 6 julio, p.4; IV. 11 julio, p.35; V. 13 julio, p.10. Also in *Centroamericana, Revista Cultural del Istmo*1956; 3(9): 80–84.

Madriz Masís, J.P. 2009. El Parque Nacional Isla del Coco (PNIC) una isla invadida. *Biocenosis* 22(1–2):61–72.

Malaspina, A.1885.Viaje político-científico alrededor del Mundo por las corbetas Descubierta y Atrevida al mando de los capitanes de navío Don Alejandro Malaspina y Don José de Bustamante y Guerra desde 1788 a 1794. Publicado con una introducción por Pedro de Novo y Colson. Madrid: Imprenta Viuda e hijos de Avianzú.

Malavassi, E. 1982.Visita al Parque Nacional Isla del Coco. *Revista Geográfica de América Central* (15–16): 211–16.

Mathis, W.N., and W.W. Wirth. 1978. A new genus near *Canaceoides* Gresson, three new species and notes on their classification (Diptera: Canacidae). *Proceedings of the Entomological Society of Washington* 80:524–37.

Maxon, W.R. 1913. *Lycopodium brachiatum*. *Contributions from the United States National Herbarium* 17:176.

Maxon, W.R. 1922. *Alsophila notabilis*. *Contributions from the United States National Herbarium* 24:39.

MINAE, SINAC, and ACMIC. 1999. *Isla del Coco: Estrategia de Investigación*. San José, Costa Rica: Ministerio del Ambiente y Energía (MINAE), Sistema Nacional de Áreas de Conservación (SINAC) y Área de Conservación Marina Isla del Coco (ACMIC). 44 pp.

Modestel, Y. 1960. El paraíso está en el Océano Pacífico. *Prensa Libre* I. 26 abril, p.2; II. 27 abril, p.2; III. 28 abril, p.2; IV.29 abril, p.2; V. 30 abril, p.2; VI. 3 mayo, p.2; VII. 4 mayo, p.2; VIII. 5 mayo, p.2.

Montoya, M. 1990. *Plan de manejo Parque Nacional Isla del Coco*. San José, Costa Rica: Sistema de Parques y Reservas Marinas (SIPAREMA), Servicio de Parques Nacionales (SPN), Ministerio de Recursos Naturales, Energía y Minas (MIRENEM). 104 pp.

Montoya, M.1991a. Isla del Coco, un ejemplo de isla oceánica. In J. Monge-Nájera, *Introducción al Estudio de la Naturaleza: Una Visión desde el Trópico*, 86–87. San José, Costa Rica: Editorial Universidad a Distancia.

Montoya, M. 1991b. La flora y vegetación de la Isla del Coco, Costa Rica (1991). Escuela de Biología, Universidad de Costa Rica. 6 pp. Anexos. Manuscrito.

Montoya, M. 1992. Bases para la aplicación del concepto de capacidad de carga en el Parque Nacional Isla del Coco. In T. Maldonado, L.A. Hurtado de Mendoza, and O. Saborio,eds., *Análisis de Capacidad de Carga para Visitación en las Áreas Silvestres de Costa Rica*, 78–85.San José, Costa Rica: Fundación Neotrópica, Centro de Estudios Ambientales y Políticas.

Montoya, M. 2001a. La biota en una isla oceánica como la Isla del Coco. *Ambientico* 88:11–12.

Montoya, M. 2001b. La excepcionalidad de la Isla del Coco: criterios para una estrategia de investigación. *Ambientico* 88:16–17.

Montoya, M. 2002. La Isla del Coco en la Historia General y Natural de la Indias de Fernández de Oviedo. Sus Implicaciones. San José. Manuscrito.

Montoya, M. 2003a. *Isla del Coco: Una Introducción a Su Historia Natural*. San José, Costa Rica: Fundación Amigos de la Isla del Coco (FAICO). 48 pp.

Montoya, M. 2003b. Sobre la formación de una colonia de *Sula dactylatra* (Pelecaniformes: Sulidae) en la Isla del Coco, Costa Rica. *Zeledonia* 7(2):24–28.

Montoya, M. 2004. Vertebrados alóctonos de la isla del Coco, Costa Rica. San José, Costa Rica: Fundación Amigos de la Isla del Coco (FAICO) y Área de Conservación Marina Isla del Coco (ACMIC). 27 pp.

Montoya, M. 2005. Salamanquesa de la Isla del Coco. *El Financiero* 508(4–10 abril 2005):30. Pausa ecológica. Organización para Estudios Tropicales.

Montoya, M. 2006. Las aves acuáticas de la Isla del Coco. Humedal de importancia internacional de la Convención de Ramsar. *Zeledonia* 10(2):42–52.

Montoya, M. 2007. Notas históricas sobre la ornitología de la Isla del Coco, Costa Rica. *Brenesia* 68: 37–57. (Publicado noviembre 2008)

Montoya, M. 2008a. Conozca la Isla del Coco: una guía para su visitación. In Organización para Estudios Tropicales (OET), BioCurso "Conozca el Parque Nacional Isla del Coco" (26 abril al 3 de mayo 2008), 36–177. 180 pp.

Montoya, M. 2008b. Aves marinas de la Isla del Coco, Costa Rica y su conservación. *Revista de Biología Tropical* 56(suppl. 2): 133–49.

Montoya, M. 2008c. La presencia de otáridos (Carnivora : Otariidae) en la Isla del Coco, Costa Rica. *Revista de Biología Tropical* 56(suppl. 2): 151–58.

Montoya, M. 2009a. Insecta, Lepidoptera, Sphingidae, *Cocytius anteus* Drury: First record for Isla del Coco, Costa Rica and notes on its hostplant. *Check List: Journal of Species Lists and Distribution* 5(2):151–53.

Montoya, M. 2009b. Relaciones biogeográficas del avifauna de la Isla del Coco, Costa Rica.20 pp. Manuscript.

Montoya, M., and F. López Pozuelo. 2007. Observaciones ornitológicas en la Isla del Coco, Costa Rica (marzo–agosto 2007). *Zeledonia* 11(2):1–11.

Montoya, M., and F. López Pozuelo. 2012. Isla del Coco (CR-21). In L. Sandoval and C. Sánchez, eds., *Áreas Importantes para la Conservación de Aves (AICA) en Costa Rica*, 161–66. San José, Costa Rica: Unión de Ornitólogos de Costa Rica.

Montoya, M., and M. Pascal. 2004. Dos nuevos registros para la avifauna de la Isla del Coco. *Zeledonia* 8(2): 7–11.

Montoya, M., and M. Pascal. 2005. Un demi millénaire d'évolution de la faune de vertébrés de l'Ile Cocos, (Costa Rica-Patrimoine Mondial). *Revue d' Écologie, La terre et la vie* (France) 60(4): 211–22.

Mora, J.M., and G. Barrantes. 1995. 4. Fauna. In *Plan general de manejo de la Isla del Coco*, Anexo 1, Sondeo Ecológico Rápido, 61–73. San José, Costa Rica: PROAMBI-FUNDEVI, Escuela de Biología, UCR.

Morales, M.I. 1991. Las hepáticas comunicadas para Costa Rica. *Tropical Bryology* 4:25–51.

Nickle, D.A. 1983. A new species of pseudophylline katydid *Parascopioricus binoditergus* n.s. from Cocos Island, Costa Rica (Orthoptera: Tettigoniidae). *Entomological News* 96(1):1–6.

Norrbom, A.L., K.C. Kin, and F.D. Fee. 1984. *Pterogramma cardisomi*, sp. n. (Diptera: Sphaeroceridae) from Cocos Island, Costa Rica: description of adults and immatures. *Brenesia* 22:285–91.

Obando-Calderón, G., J. Chavez-Campos, R. Garrigue, A. Martínez-Salinas, M. Montoya, O. Ramirez, and J. Zook. 2012. Lista oficial de las aves de Costa Rica. Actualización 2012. *Zeledonia* 16(2): 70–84.

Pax, F., and K. Hoffman. 1924. *Acalipha pittieri*. In A. Engle, ed., *Das Pflanzenreich IV*, 18. Leipzig, Germany: Engelmann.

Penelas, A.G. 1963. Isla del Coco, paraíso de los deportistas. *Prensa Libre*(11 junio): 4b.

Perger, R., R. Vargas, and A. Wall 2011. *Johngarthia cocoensis* (Crustacea, Decapoda, Brachyura) from Cocos Island, Costa Rica. *Zootaxa* 2911:57–68.

Petren, K., B.R. Grant, and P.R. Grant. 1999. A phylogeny of Darwin's finches, based on microsatellite DNA length variations. *Proceedings of the Royal Society of London B* 266: 321–29.

Pittier, H. 1899. Apuntamientos preliminares sobre la Isla de Cocos, posesión costarricense en el Océano Pacífico. San José, Memoria de la Secretaría de Fomento 1998/1999: 15–28.

Pittier, H. 1935. Apuntamientos preliminares sobre la Isla de Cocos, posesión costarricense en el Océano Pacífico. *Revista del Colegio de Señoritas* 2(4–5):2–11.

Platnick, N.I. 2008. The world spider catalog, version 9.0. American Museum of Natural History. http://research.amnh.org/entomology /spiders/catalog/index.html.

Purrua, E.J., ed. 2001. *The Diary of Antonio de la Tova on the Malaspina Expedition, 1789–1794.* Lewiston, NY: Edwin Mellen Press.

Quiros-Badilla, E., and E.J. Alfaro. 2009. Algunos aspectos relacionados con la variabilidad climática en la Isla del Coco, Costa Rica. *Revista de Climatología* 9: 33–44.

Reyes-Castillo, P., and G. Halffter-Salas. 1978. Análisis geográfico de la distribución de la tribu Proculini (Coleoptera: Passalidae). *Folia Entomológica Mexicana* 39/40:222–24.

Rojas, C. 2005. Cocos Island trip, april 23–30, 2005. Ciudad Universitaria, Universidad de Costa Rica, Escuela de Biología. 4 pp.

Rojas, C., and S.L. Stephenson. 2008. Myxomycete ecology along an elevation gradient on Cocos Island, Costa Rica. *Fungal Diversity* 29:117–27.

Rojas-Alvarado, A.F. 2001. Seis especies nuevas y dos nuevos registros de helechos (Pteridophyta) para Costa Rica. *Revista de Biología Tropical* 49(2): 435–52.

Rojas-Alvarado, A.F., and J.L.Trusty. 2004. Diversidad de pteridofitas en la Isla del Coco. *Brenesia* 62:1–14.

Rose, J.N. 1892. List of plants from Cocos Island. p 135. In Lists of plantas collected by the USS Albatross in 1887–1891 along the western coast of America. *Contributions from the United States National Herbarium* 1.

Salguero, M. 1965. Isla del Coco, huerto en el océano. La Nación, 24 abril, p. 33–40; 1 mayo p. 27, 29, 30, 35, 37, 38.

Salguero, M. 1975. Isla del Coco, huerto en el océano. In M. Salguero, *Crónicas de Tierra Adentro*, 166–79. San José, Costa Rica: Editorial Costa Rica.

Sato, A., W.E. Mayer, H. Tichy, P.R. Grant, B.R. Grant, and J. Klein. 2004. Evolution of Mhc class genes II B in Darwin's finches and their closest relatives: birth of a new gene. *Immunogenetics* 53:792–801.

Sato A., C. O'Huigin, P. Figueroa, P.R. Grant, B.R. Grant, H. Tichy, and J. Klein. 1999. Phylogeny of Darwin's finches as revealed by nt-NDA sequences. *Proceedings of the National Academy of Sciences* 96:5101–6.

Schluter, D. 1984. Morphological and phylogenetic relations among the Darwin's finches. *Evolution* 38:921–30.

Sherry, T.W. 1985. Adaptations to a novel environment: food, forging and morphology of the Cocos Island Flycatcher. In M.S. Buckley, ed., *Neotropical Ornithology. Ornithological Monographs* 36: 908–20. Washington: American Ornithologists' Union.

Sherry, T.W. 1986. Nest, eggs, and reproductive behavior of the Cocos Island Flycatcher (*Nesotriccus ridgwayi*). *Condor* 88:531–32.

Sherry, T.W., and T.K. Werner. 1984. List of birds species new to Cocos Island, Costa Rica, since the report of Paul Slud. San José, Costa Rica: Blue Scorpion Expedition. 1 p.

Sibaja-Cordero, J.A. 2008. Vertical zonation in the rocky intertidal at Cocos Island (Isla del Coco), Costa Rica: comparison with other tropical locations. *Revista de Biología Tropical* 56(Suppl. 2): 171–87.

Sierra, C. 2000. Ecología, impactos y métodos de erradicación del cerdo feral (*Sus scrofa*) en la Isla del Coco, Costa Rica. Heredia, Costa Rica: Universidad Nacional, Programa Regional en manejo de vida silvestre. 49 pp.

Sierra, C. 2001a. El cerdo cimarrón (*Sus scrofa*, Suidae) en la Isla del Coco: composición de su dieta, estado reproductivo y genética. *Revista de Biología Tropical* 49(3–4):1147–57.

Sierra, C. 2001b. El cerdo cimarrón (*Sus scrofa*, Suidae) en la Isla del Coco: escarbaduras, alteraciones al suelo y erosión. *Revista de Biología Tropical* 49 (3–4):1159–70.

Sierra, C. 2001c. La vida silvestre y asilvestrada en la Isla del Coco. *Ambientico* 88:13–15.

Sierra, C. 2002. Estrategia para la erradicación de cuatro vertebrados introducidos en la Isla del Coco, Costa Rica: cerdos cimarrones (*Sus scrofa*), gatos cimarrones (*Felis catus*), rata negra (*Rattus rattus*), rata de alcantarilla (*Rattus norvegicus*). San José, Costa Rica: ACMIC, MINAE y UNESCO. 310 pp.

Sinergia 69. 2000a. Vol. 2. Aspectos meteorológicos y climatológicos del ACMIC y su área de influencia. San José, Costa Rica: Proyecto GEF/PNUD Conocimiento y uso de la biodiversidad del ACMIC. 184 pp. Anexos.

Sinergia 69. 2000b. Vol. 5. Diagnóstico preliminar de las alteraciones, degradaciones y amenazas en el ACMIC. San José, Costa Rica: Proyecto GEF/PNUD Conocimiento y uso de la biodiversidad del ACMIC. 72 pp. Anexos.

Slater, J.A. 1981. A new species of *Ozophora* from Cocos Island (Hemiptera: Lygaeidae). *Journal of the Kansas Entomological Society* 54(1):22–26.

Slud, P. 1967. The birds of Cocos Island (Costa Rica). *Bulletin of the American Museum of Natural History* 134:261–96.

Smith, A.R., and D.B. Lellinger. 1985. *Thelypteris cocos*. *Proceedings of the Biological Society of Washington* 98(4):918.

Chapter 9 The Northern Pacific Lowland Seasonal Dry Forests of Guanacaste and the Nicoya Peninsula

Quírico Jiménez M.,[1] Eduardo Carrillo J.,[2] and Maarten Kappelle[3,4,*]

Dedication

We dedicate this chapter to our friend and one of our most prominent teachers, the late Alfonso Mata Jiménez (1939–2010), an outstanding Costa Rican pioneer in environmental chemistry, ecology, and the conservation of natural resources. Dr. Mata received his PhD in organic chemistry from the University of Detroit (1970) and conducted postdoctoral studies in Liverpool, UK, and Delft, the Netherlands. Following service as full professor for several decades at the Universidad de Costa Rica (UCR) he served from 1984 through 1988 as dean at the university's Faculty of Sciences. He was past president and executive director of the Tropical Science Center (CCT) in San Pedro de Montes de Oca. As a prolific scholar and active environmentalist he helped develop environmental legislation at national levels and authored more than 70 publications on organic chemistry, environmental pollution, and natural resources, as well as several books including the *Dictionary of Ecology* and *The Golfo de Nicoya* (national science award). As a pioneer at the Club de Montañismo based at the Universidad de Costa Rica he explored the Chirripó Massif in the early 1970s, accompanied by Roger Bourillon, Jorge Moya, Chris Vaughan, and the late Adelaida Chaverri. Their scientific visits to the highest parts of the country raised awareness about the importance of this unique *páramo* environment and led to the establishment of Parque Nacional Chirripó in 1975. More recently, *don Alfonso* co-edited, together with Gordon W. Frankie and S. Bradleigh Vinson, a very comprehensive volume on the ecology, biodiversity, and conservation status of seasonal dry forests in Costa Rica, based on more than thirty years of study (Frankie et al. 2004a). The papers gathered in that book have served as key references for the current chapter.

Introduction to Costa Rica's Dry Forests

Hundreds of years ago the tropical dry forest covered ca. 17% of the planet's surface (Murphy and Lugo 1986, 1995). Today tropical dry forests in the Western Hemisphere extend for 519,597 km² across North and South America with Mexico, Brazil, and Bolivia harboring the largest and best-preserved dry forest fragments (Portillo Quintero and Sánchez Azofeifa 2010).

When the Spaniards arrived in the Americas in the fifteenth and sixteenth centuries, this fragile forest system covered about 550,000 km² in Mesoamerica and stretched from the Península de Nicoya and northern part of the Central Pacific Region of Costa Rica all the way north to the foothills of the mountains west of Mazatlán in Mexico (Janzen 1986b, 1988b, and 2004, Martin et al. 1998, Stoner and Timm 2004). A low extent and high level of fragmentation of Central American dry forests in countries like Guatemala, Nicaragua, and Costa Rica means that these forests are at a higher risk from human disturbance and deforestation than those of the larger Latin American countries today.

At the middle of the 1980s less than 2% of it remained

[1] Empresa de Servicios Públicos de Heredia (ESPH), Apartado 26-3000, Heredia, Costa Rica
[2] Instituto Internacional en Conservación y Manejo en Vida Silvestre (ICOMVIS), Universidad Nacional, Apartado 1350-3000, Heredia, Costa Rica
[3] World Wide Fund for Nature (WWF International), Avenue du Mont-Blanc 1196, Gland, Switzerland
[4] Department of Geography, University of Tennessee, Knoxville, TN
* Corresponding author

relatively intact, while only 0.08% has been placed under some category of protection (Janzen 1986, Gradwohl and Greenberg 1988). According to Janzen (1988b, and see Janzen and Hallwachs, chapter 10 of this volume) this forest type is perhaps the most endangered large and widespread ecosystem in the tropics.

It is estimated that prior to 1940 but long after much of it had been cleared, the total dry forest area in Costa Rica covered approximately 400,000 ha, which equals almost 8% of the national territory. By 1950 this area was already reduced to only 40,200 hectares, mainly due to logging and clearing for pastures to raise cattle in Guanacaste (Fig. 9.1) and the Golfo de Nicoya Basin (GNB), the vast region around the Golfo de Nicoya (Sader and Joyce 1988, Mata and Echeverría 2004) (see Fig. 9.2, for a map of Costa Rica's northern dry forest lowland region and its protected areas).

According to Murphy and Lugo (1995), the Costa Rican dry forests, as in the rest of Mesoamerica, suffered from uncontrolled deforestation mainly due to 400 years of logging and forest conversion for pastures and the extraction of precious woods (also, see Sader and Joyce 1988, Sánchez Azofeifa 2000, Sánchez Azofeifa et al. 2001, Quesada Avendaño and Stoner 2004). Deforestation through clear-cutting was stimulated by successive governments during the first half of the twentieth century in order to promote agriculture and extensive cattle ranching (Myers 1981, Parsons 1983, Thrupp 1988, Harrison 1991, Sánchez Azofeifa et al. 2001).

Although forests started to recover in northwestern Costa Rica beginning in 1980 (Janzen 1986a and 2002, Allen 2001, Calvo Alvarado et al. 2008), regular anthropogenic fires in the forests and pastures in the dry season continue to be set in Guanacaste. This has been reported

Fig. 9.1 Typical Guanacaste pasture with isolated trees in the plains surrounding the forested hills close to Hatillo de Santa Cruz. *Photograph by Maarten Kappelle.*

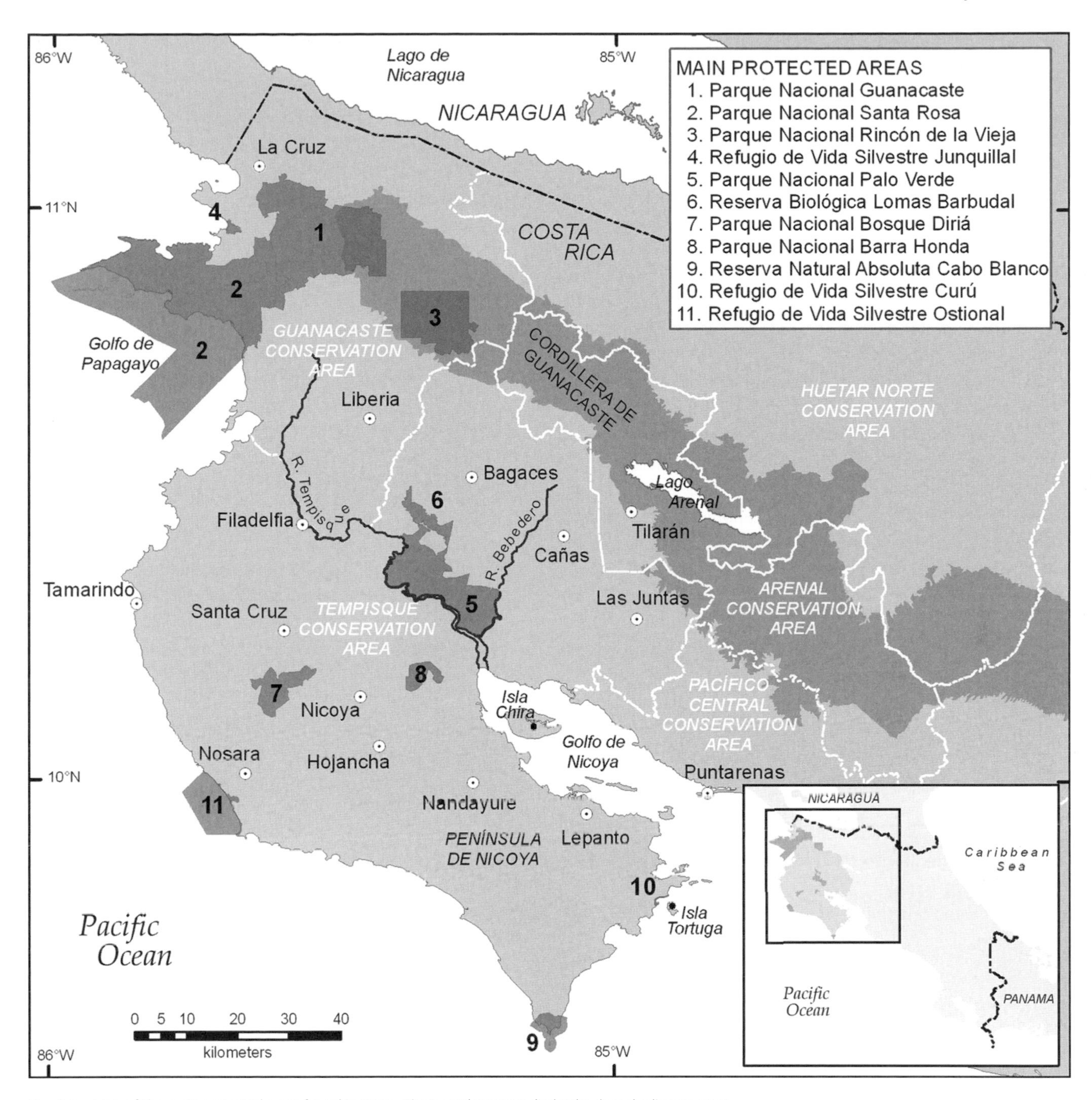

Fig. 9.2 Map of the main protected areas found in Costa Rica's northwestern dry lowlands and adjacent areas.
Map prepared by Marco V. Castro.

by Alfaro et al. (1999) on the basis of the analysis of remotely sensed GOES-8 satellite images in the late 1990s. Fortunately, since about 2000 the size, intensity, and frequency of these fires have greatly diminished in the region, although up to today human-set fires still do occur. In April 2010, for instance, a total area of at least 1,500 ha—mainly covered by *Typha* and grasses—was burnt in the Parque Nacional Palo Verde. Similarly, in the same month fires were developing in hundreds of hectares near Estación Experimental Horizontes, in the Nacascolo district close to Liberia, in the Parque Nacional Rincón de la Vieja, and in the Parque Nacional Guanacaste near La Cruz.

In the past, land colonization and commercial logging in Central America seem to have started in the dry forests of most of the region (Janzen 1988, Janzen and Hallwachs, chapter 10 of this volume). Practically all of Costa Rica's dry forests disappeared in a few decades, with exception of some remaining, protected fragments in a few areas in Provincia de Guanacaste (e.g., Parque Nacional Santa Rosa, including its Sector Santa Elena and Sector Murcielago, as well as Sector Poco Sol and Sector Horizontes in the eastern part of the Área de Conservación Guanacaste, and the Parque Nacional Palo Verde and Reserva Biológica Lomas de Barbudal, both in the Área de Conservación Tempisque). These forest islands were the refuges of the marvelous dry forest flora and fauna (Wallace 1989), and contribute to the regeneration of the largest patches of remaining dry forest in the WWF-defined Central American Dry Forest Ecoregion (Olson et al. 2001).

In general, dry forests are less complex than moist forests in terms of floristic composition and structure (Holdridge 1967, Murphy and Lugo 1986). However, there is a range of differences among dry forests mainly due to changing climatic and edaphic conditions, as well as atmospheric and soil moisture conditions. Furthermore, the gamut of species interactions in these forests seems to be exceptionally varied (Janzen 1988). Apparently, some 300,000 species (65% of the estimated number of species in Costa Rica) call the Área de Conservación Guanacaste (ACG) their home (D.H. Janzen, pers. comm.), though the ACG also contains cloud forest, rain forest, and their intergrades. Since the inventory of species—collection, preparation, identification, storage, and listing or description of specimens—is often a very slow process (Hanson 2004), in the 1990s Dan Janzen and colleagues came up with an All Taxa Biodiversity Inventory (ATBI) for the ACG (Anderson and Erwin 1995, Rodriguez and Gauld 1995, Brooks 1996, Janzen 1996). It was meant to speed up the collection and identification of Costa Rica's species and especially triggered the inventory of insect and plant species in northern Guanacaste (Janzen 1996, and pers. comm.).

Climate

The climate of the northern Pacific lowlands is mostly dry to sub-humid but becomes somewhat moister going south into the Península de Nicoya (Herrera 1985, and chapter 2 of this volume; Sánchez Azofeifa et al. 2003). In these lowlands, more than 95% of the annual precipitation falls during the rainy season (May to December) while the remaining 5% of the rain falls during the dry season (December to mid-May). This marked seasonality is principally determined by the annual north-south movement of the thermal equator, and the first half of the dry season is characterized by winds that can reach speeds between 10 and 80 km per hour (Asch et al. 2000, Mata and Echeverría 2004; Herrera, chapter 2 of this volume).

Accurate long-term meteorological records are scarce in lowland Guanacaste and consist mainly of monthly precipitation values for Liberia, La Pacífica, and the Parque Nacional Santa Rosa in the ACG (Enquist and Leffler 2001). Rainfall in Costa Rica's dry Pacific lowlands shows a marked seasonal variation with an average of 1,800 mm annually (Mata and Echeverría 2004). The dry season starts in December and ends in the middle of May, ranging from 6 months in the northwest of Guanacaste (Santa Rosa) to 5 months in the southeast along the Río Tempisque (Palo Verde) (Gillespie et al. 2000). Owing to its very permeable serpentine soils, the Península de Santa Elena in the northwest is the driest part in the entire country (Fig. 9.3). Along the Pacific coast just south of Santa Elena the annual precipitation in the 108 km^2 Parque Nacional Santa Rosa is 1,528 mm (Janzen 1988, 1998). Towards the east, in the Río Tempisque Basin north of the Golfo de Nicoya, precipitation is slightly lower and reaches annual amounts of 1,000 to 1,500 mm (Asch et al. 2000, Mateo-Vega 2001), with an average yearly precipitation of 1,267 to 1,717 mm in the Parque Nacional Palo Verde (Gillespie et al. 2000). Further south, on the more humid Península de Nicoya, the average yearly rainfall fluctuates around 2,000 mm (Herrera 1985, and chapter 2 of this volume).

Average monthly temperatures in the northern dry lowlands are among the highest in the country (Herrera 1985, and chapter 2 of this volume) and range from 24.6 to 30°C (Asch et al. 2000), but may reach a monthly average of 36°C at some places like Santa Elena or Santa Rosa along the Pacific coast (Janzen 2004). Santa Rosa itself has a mean annual temperature of 25°C (Gillespie et al. 2000). On basis of data gathered from twelve meteorological stations over the period 1954–1999 Mateo-Vega (2001) has been able to report a yearly average temperature of 27.4°C for the Río Tempisque Basin. In the core of the lower basin, Parque Nacional Palo Verde has a mean annual temperature of 25°C (Gillespie et al. 2000).

Geology, Geomorphology, and Soils

The dry Tempisque depression in Costa Rica's northern dry Pacific lowlands—the culturally rich Chorotega region—extends northward from the shores and head of the Golfo de Nicoya almost to Nicaragua. On the northeast rise the volcanoes of the Cordillera de Guanacaste, while to the west the lowlands stretch to the Pacific Ocean, and south to the hilly core of the Península de Nicoya (Slud 1980).

Fig. 9.3 The very dry Cerros de Santa Elena at the Península de Santa Elena, as seen from the Pacific Ocean. *Photograph by Barry Hammel.*

It includes the Península de Santa Elena, the Península de Nicoya, and the Río Tempisque Basin (RTB). The latter has an average annual flow of 308 m³ per second (Mata 2004).

Throughout the wider Área de Conservación Guanacaste (ACG), the underlying physical substrate ranges from 85-million-year-old serpentine barrens to marine limestone and volcanic deposits from tens of millions of years to only a few days old to the east (Janzen 2004). A common rock type on the Península de Nicoya is unique serpentine peridotite (Azéma and Tournon 1980), which is underlain by even older marine sediments and volcanic rocks, among which appear the oldest exposed rocks known from Costa Rica, the Nicoya complex (Tournon and Alvarado 1997, Mata and Echeverría 2004).

In the northwestern part of the dry forest region of Guanacaste the Península de Santa Elena is found, which represents the oldest piece of land mass in all of Central America. It constituted an isolated oceanic island before it became part of the Central American Isthmus (Janzen 1988, 1998). During the past 85 million years its rocks have been weathered, giving rise to poor soils. Here, the Rivas Formation stands out (Dengo 1962). It contains lutites and is made up of tuff sheets and thick sandstones (little stratified Grauwacke and iimolites). Geomorphologically, the peninsula's terrain is formed of coastal hills with an advanced stage of development due to its old age and consequently long process of erosion. Its valleys are determined by geological structures including faults.

Gómez (1986) reports ignimbrites, pumice tuffs, and derivates from this region, which forms part of the Río Tempisque Basin, the most important watershed area in the Provincia de Guanacaste. In fact, this river watershed and its surrounding areas are composed of a series of stratigraphic units that correspond to different geological periods, the principal ones being the Nicoya Complex, the Bagaces and Liberia Formations, the Aguacate Group, undifferentiated lahars and volcanic edifices, as well as fluvial, alluvial, and coastal deposits. They originated at different moments during the period ranging from the Cretaceous to the Quaternary (Bergoeing 1998, Dengo 1962, Mateo-Vega 2001).

The Parque Nacional Barra Honda in the eastern part of the Península de Nicoya west of the Río Tempisque mouth protects caves characterized by stalactites in calcareous formations that were shaped by rain filtering through the limestone over 70 million years or more. Northwest of Barra Honda but south of the town of Santa Cruz is the 983 m tall Cerro Vista al Mar, the highest peak of the Península de Nicoya.

The Parque Nacional Palo Verde in the Río Tempisque Basin has the unique feature of having both 200 m tall hills with steep gradients formed by limestone rocks and a lagoon and freshwater swamps that used to be bathed by the high tide of the Rio Tempisque mouth, especially in the rainy season (see chapter 20 by Jorge Jiménez in this volume).

The limestone hills around the Río Tempisque Basin, such as found at Cerro Guayacán (Fig. 9.4), consist of three types of landforms (Jiménez 1999b): (1) plains at the foothills that have a flat to undulating topography (0–10%), are more humid, have soils with few rocks (mostly occasional small isolated rocks), and display a large species diversity; (2) slopes dominated by rocks, either in groups or isolated, with pronounced gradients between 10 and 30%; and (3) higher part of the foothills where the ground is generally flat, and mainly covered by groups of rocks of different sizes.

Further south, across the Península de Nicoya the topography is rather irregular. Here, the terrain is mainly dominated by denudational landforms, with 700 m tall mountains and narrow valleys that are often as wide as their streams (Gómez 1986).

Differences in soils at the Península de Nicoya and in the Río Tempisque Basin—which includes the Tempisque and Bebedero sub-basins—are large. Soils vary from volcanic types in the upper parts of the Cordillera de Guanacaste to flooded alluvial soils in the Tempisque valley (Mata and Echeverría 2004). Here, a complex mosaic of edaphically determined vegetation types is the rule (Mateo-Vega 2001). South of Liberia, for instance, soils are derived from rhyolithic ash while black montmorillonite clay soils locally dominate in the Tempisque basin (Hartshorn 1983). According to Vaughan et al. (1996) more than 20 soil subgroups belonging to five taxonomic orders have been identified here, including Alfisols, Entisols, Inceptisols, Mollisols, and Vertisols (also, see chapter 4 by Alvarado and Mata, this volume). In the edaphically young Río Tempisque Basin, 38% of the soils are Inceptisols, while 26% are Entisols and 13% Alfisols (TSC 1999).

Vegetation and Flora

Tropical Dry Forest Structure

The northern Pacific lowlands of Costa Rica are characterized by dry forests and other seasonal vegetation types that range from sea level up to average elevations of 400 m, and up to 700 m at some local peaks. Janzen and Hallwachs (chapter 10, this volume) list a total of 10 different Holdridge life zones for the Área de Conservación Guanacaste, including the typical Dry Tropical Forest life zone (*bosque seco tropical,* bs-T). Gómez (1986) reports a total of nine vegetation macrotypes, which include mangrove forests, swamps and reed beds, as well as seasonal deciduous, semideciduous and evergreen forests—the latter occurs at the extreme southern end of the Península de Nicoya. Over 30 years ago Janzen and Liesner (1980) listed 975 species of broad-leaved plants for the lowlands of Guanacaste exclusive of the Península de Nicoya.

The northern dry lowland forests are dominated by deciduous, semi-deciduous, and evergreen trees that annually withstand six dry months (Fig. 9.5). This multilayered

Fig. 9.4 Dry forest dominated by the cactus *Stenocereus aragonii* in the higher part of Cerro Guayacán, Parque Nacional Palo Verde. *Photograph by Quírico Jiménez.*

Fig. 9.5 Dry forest with deciduous and semi-deciduous trees in the hills near Hatillo de Santa Cruz, Guanacaste.
Photograph by Maarten Kappelle.

forest type has originally a 20 to 30 m tall canopy, with short, stout trunks and large, often spreading, flat-topped crowns, usually not in lateral contact with each other, especially when it has been logged, burned, and otherwise perturbed for four centuries (Hartshorn 1983). Tree species density is generally low with an average of 30 species per ha. Canopy cover in mature forest may reach 70 to 100%, which is 85% on average (Gómez 1986). Dry-season deciduous compound leaves dominate the foliage in the canopy of the secondary successional forest as exemplified by pinnately leaved mimosoid and caesalpinioid leguminose trees. Many leaves are mesophyllous or have a coriaceous texture. About 20% of the trees show some kind of buttresses.

The subcanopy layer is 10 to 20 m tall and contains more evergreen species that have slender, crooked or leaning trunks and small open crowns. A 2–5 m tall shrub layer is characterized by multiple-stemmed scrubby plants often covered with spines or thorns (Hartshorn 1983). Low branching of trees at heights of 3 to 6 m is frequent (Gómez 1986). The ground layer is poor in species presence and abundance. Gómez (1986), however, reports at least 250 species of vascular plants that thrive on the ground in Costa Rican deciduous and semi-deciduous forests. Epiphytes are few (see elsewhere in this chapter), while up to 20% of the dry forest flora corresponds to vines and lianas (e.g., Hartshorn 1983), many of which are dispersed by wind. In fact, in young successional dry forest communities wind-dispersed species dominate the floristic composition, whereas animal-dispersed species predominate in tropical dry oak forests (Powers et al. 2009). Table 9.1 shows the most abundant plant families of the Costa Rican dry forest.

In a study of stems over 2.5 cm diameter at breast height in seven 1,000 m^2 plots established in variously aged secondary successional tropical dry forests in Costa Rica and

Table 9.1. Most Abundant Plant Families of Costa Rica's Northwestern Lowland Dry Forests

Families with greatest abundance of individuals	Families with greatest number of species
Asteraceae	Acanthaceae
Bignoniaceae	Apocynaceae
Convolvulaceae	Boraginaceae
Cyperaceae	Capparaceae
Euphorbiaceae	Malpighiaceae
Fabaceae	Polygonaceae
Malvaceae	Sapindaceae
Moraceae	Sterculiaceae
Poaceae	Verbenaceae
Rubiaceae	
Solanaceae	

Nicaragua, Gillespie et al. (2000) found that the Parque Nacional Santa Rosa was the richest site with the highest family (33), genera (69), and species (75) diversity of all sites, largely due to the extremely diverse disturbance history of the site, and that it is the only easily accessible area in Provincia de Guanacaste containing anything approximating original forest. Species richness and forest structure were significantly different between sites. Fabaceae was the dominant tree and shrub family at most sites, but no species was repeatedly dominant based on number of stems in all fragments of secondary successional tropical dry forest (Gillespie et al. 2000). Previously, Burnham (1997) had done a similar analysis of stand characteristics in a one hectare plot of secondary successional dry semideciduous forest at Santa Rosa. She inventoried, mapped, and measured all stems >10 cm DBH and counted 354 stems in 56 species. Basal area totaled 20.40 m^2 per ha and total leaf litter mass was 666.3 g. Eight of the 56 species accounted for 74.8% of the total basal area: *Quercus oleoides* (25.9%), *Hymenaea courbaril* (18.49%), *Manilkara zapota* (12.16%), *Sloanea terniflora* (5.87%), *Luehea* spp. (5.20%), *Ficus* spp. (2.91%), *Dilodendron costaricense* (2.15%), and *Calycophyllum candidissimum* (2.12%) (Burnham 1997).

Vargas Ulate (2001) observed that plant community distribution at the ignimbrite meseta of ACG is intimately related to the type of relief, soils, and humidity. In the upper parts of the meseta, characterized by soils that are stony, sandy, and acidic, herbaceous vegetation such as savanna and edaphic steppe is dominant in the old fields. By contrast, secondary successional forest is found on the deep and organically rich soils of the valley floors (Vargas Ulate 2001). Dwarf varieties of *Byrsonima crassifolia*, *Curatella americana*, and *Quercus oleoides* form part of the vegetation growing on acid and infertile soils.

In a 4 ha oak forest plot 10 km NW of Bagaces in Provincia de Guanacaste, Hartshorn (1983) found a stem density of 204 stems per ha, with stems in 44 species. Basal area reached 12.7 m^2 per ha. The dominant tree species was *Quercus oleoides* (oak) which made up for 34% of the counted stems and 58% of the plot's basal area. For this reason, Gómez (1986) called this forest association a *Quercetum*, following the phytosociological nomenclature proposed by European authors like Braun-Blanquet (1979; and see Kappelle 1996, chapter 14 of this volume). Other important secondary successional species were *Byrsonima crassifolia*, *Apeiba tibourbou*, *Spondias mombin*, *Cordia alliodora*, *Guazuma ulmifolia*, *Luehea candida*, *Annona reticulata*, *Luehea speciosa*, and *Cochlospermum vitifolium*. In another study also conducted near Bagaces, Frankie et al. (2004b) assessed the vascular flora of a 10 x 10 km area, and reported a total of 487 flowering plants species spread over 175 herbs (36% of all species), 144 trees, 78 shrubs, 56 vines, 23 lianas, 8 epiphytes, 2 parasites, and 1 terrestrial cactus.

At Palo Verde, along the Calle Apiary, Hartshorn (1983) inventoried another 4 ha secondary successional forest plot which was co-dominated by tree species like *Calycophyllum candidissimum*, *Licania arborea*, *Brosimum alicastrum*, *Spondias mombin*, *Guazuma ulmifolia*, *Thouinidium decandrum*, *Caesalpinia eriostachys*, *Luehea candida*, *Tabebuia ochracea*, and *Pachira quinata* (synonymous with *Bombacopsis quinata*, the pochote tree). A total of 68 species were recorded. None of these were as dominant as the oak species near Bagaces. Stem density at Palo Verde was similar as near Bagaces, with 219 stems per ha, while basal area resulted slightly higher with 19.9 m^2 per ha (Hartshorn 1983).

Below, an overview is given of the different vegetation types and their characteristic plant species for each of the three main sub-regions in the northern Pacific dry lowlands: (1) Northern Guanacaste and the Península de Santa Elena; (2) Central Guanacaste, the Río Tempisque Basin, and Palo Verde Wetland; and (3) the Península de Nicoya.

Floristic Composition in Guanacaste, Península de Nicoya and the Río Tempisque Basin

1. Northern Guanacaste and the Península de Santa Elena

The heavily disturbed Península de Santa Elena and particularly Punta Descartes include one of the most important areas of dry forest in the country when plant diversity is considered (Fig. 9.6). According to Zamora et al. (2004) the dry forest flora is largely composed of elements of Mexican and Guatemalan origin, and includes species that have their southernmost distribution limit in Panama, Colombia, Venezuela, or the Guyanas. A number of species have a Caribbean (Antillean) origin while the remainder is tropical

Fig. 9.6 Dry forest during the growing season near Potrero Grande at the Península de Santa Elena. In the back the Cerros de Santa Elena are observed.
Photograph by Barry Hammel.

in distribution, some of them being pan-tropical. Some dry forest species that have their southernmost distribution limit in northern Guanacaste are *Bursera graveolens*, *Capparis incana*, *Cordia gerascanthus*, *Jatropha costaricensis*, *Karwinskia calderoni*, *Agave seemanniana*, *Quercus oleoides*, and *Thouinia serrata*.

The hot and dry climate in combination with the old, eroded, serpentine soils at the Península de Santa Elena have allowed for the establishment of a very distinct, semi-arid, herbaceous to shrubby vegetation type characterized by thorny scrub and small, scattered trees (Elizondo and Jiménez 1988), but largely generated by being anthropogenically burned every few years for at least four centuries. At places where trees abound, a dry deciduous forest occurs interspersed by patches of more evergreen forest along the seasonal streams. Until now at least some 3,350 plant specimens have been collected at a number of different spots along the Península de Santa Elena and surroundings—for example, at the Hacienda Murciélago, Laguna La Calavera, Cerro El Inglés, Cerro Calera, Playa Potrero Grande, Punta El Respingue, Cerro Murciélago, and Isla San José (Grayum 2004).

Conspicuous dry forest trees often seen at the serpentine barrens along the hilltops and ridges at Santa Elena are *Byrsonima crassifolia*, *Cochlospermum vitifolium*, *Curatella americana*, *Gliricidia sepium*, and *Roupala montana*. Some other woody species common in this habitat are *Aeschynomene* sp., *Agave seemanniana*, *Calliandra tergemina*, *Coursetia elliptica*, *Curatella americana*, *Diphysa humilis*, *Euphorbia schlechtendalii*, *Haematoxylum brasiletto*, *Krameria ixine*, *Melanthera nivea*, *Mimosa tricephala*, *Russelia sarmentosa*, *Trixis inula*, *Turnera diffusa*, *Vachellia villosa* (= *Acacia villosa*), and the endemic *Simsia santarosensis* (Grayum 2004). Herbs that are most characteristic for this community are, among others, *Alophia silvestris*, *Asclepias woodsoniana*, *Axonopus aureus*, *Buchnera pusilla*, *Bulbostylis paradoxa*, *Cipura campanulata*, *Declieuxia fruticosa*, *Evolvulus alsinoides*, *Jacquemontia mexicana*, *Oxalis frutescens*, *Paspalum pectinatum*, *Sauvagesia pulchella*, *Schoenocaulon officinale*, *Schwenckia americana*, *Trachypogon plumosus*, and *Waltheria indica* (Grayum 2004).

The pastures were burned at 1–3 year intervals until the mid-1980s and relatively common species are *Paspalum pectinatum*, *Trachypogon plumosus*, and the fire-adapted, exotic star grass, *Hyparrhenia rufa*. This latter grass species was originally introduced from Africa in the 1920s to feed beef cattle in the Guanacaste pastures. It is locally known as *jaragua*. All these grasses are often accompanied by a variety of herbs and small shrubs (Zamora et al. 2004). During the past decade, cattle have been being used as a "management tool" to reduce the amount of combustible material in *jaragua* dominated pastures in the Parque Nacional Palo Verde and Reserva Biológica Lomas de Barbudal, with the aim to reduce the risk of fires in these fragmented dry forest areas (Stern et al. 2002, and see Castro 1993 for initial thoughts on management of Palo Verde).

Along the streams that traverse the serpentine terrain a

Fig. 9.7 The Palo Verde marshes with the *palo verde* tree (*Parkinsonia aculeata*) and dense floating mats of the water hyacinth (*Eichhornia crassipes*). *Photograph by Garret Crow.*

Fig. 9.8 The Palo Verde Wetland. The reddish-gray color is due to the recent burning of the vegetation. The Cerros del Rosario appear in the background. *Photograph by Quírico Jiménez.*

Fig. 9.9 The Palo Verde Wetland with the Río Tempisque flowing in the background.
Photograph by Garret Crow.

Fig. 9.10 Forested crests along the hills in Parque Nacional Palo Verde. The blooming trees with white flowers are individuals of the species *Calycophyllum candidissimum* (Rubiaceae).
Photograph by Quírico Jiménez.

as well as *Bauhinia glabra* (Fabaceae) and *Combretum decandrum* (Combretaceae).

One of the most attractive species found on the limestone hills along the Río Tempisque is the well-known guayacán real (*Guaiacum sanctum*; Hernández Garzón 2001, González Jiménez and Hernández Garzón 2006). In the nearby area of Lomas de Barbudal one still finds Mahogany trees (*Swietenia humilis*, locally known as *caoba*) as well as Cristóbal (*Platymiscium parviflorum*), both of which are heavily endangered species (Jiménez 1999a).

In the area's foothill plains tree species like *Enterolobium cyclocarpum* (the guanacaste tree), *Samanea saman*, *Spondias mombin*, *Astronium graveolens*, *Sideroxylon capiri*, and *Calycophyllum candidissimum* are frequent. Dominant along the slopes are species such as *Bursera simaruba*, *Astronium graveolens*, *Erythroxylum havanense*, *Lonchocarpus hughesii*, *Pachira quinata* (= *Bombacopsis quinata*), and the columnar cactus *Stenocereus aragonii*. Furthermore, there is clear evidence of natural regeneration of guayacán real (*Guaiacum sanctum*) in this part of the northern dry forest region.

and *Pachira quinata*. In the forests' understory *Aphelandra scabra*, *Palicourea guianensis*, *Piper reticulatum*, and *Vachellia collinsi* are common. This vegetation type may extend right down to the shoreline at places where land conversion for agriculture, cattle ranching, or development hasn't occurred yet. Here, species as *Plumeria rubra*, *Bursera simaruba*, and *Stenocereus aragonii* are typical dwellers at the rocky ridges and crevices.

Similar patches of forests are still found in the Curú Wildlife Refuge at the southern end of the Península de Nicoya (Puntarenas province), close to the Golfo de Nicoya. Species like *Platymiscium curuense*, a local species of *Cristóbal* named after the Curú Wildlife Refuge, are clear evidence of this pattern (Zamora and Klitgaard 1997). Other more scarce species that grow in this part of the country are *Sterculia apetala*, *Pseudobombax septenatum*, *Ceiba pentandra*, *Schizolobium parahyba*, and *Terminalia oblonga*. The forest of the Reserva Natural Absoluta Cabo Blanco, located at the southernmost tip of the Península de Nicoya, is characterized by evergreen species such as *Anacardium excelsum*, and a small population of *Copaifera aromatica*.

Finally, east of the Península de Nicoya in the Golfo de Nicoya, several islands such as Chira, Venado, Caballo, and San Lucas emerge as a continuity of the Peninsula's vegetation and flora. Some of these islands, including the latter two, retain more than 95% of their vegetation cover including mangrove forests (and see Soto Soto and Jiménez Ramón 1982, for similar mangrove forests at Puerto Soley in northern Guanacaste). The majority is inhabited with the exception of the Refugio de Vida Silvestre Isla San Lucas. To date these islands haven't received much formal attention from scientists and require thorough studies in the fields of ecology, vegetation science, and botanical inventory.

Vertebrates

Mammals

Although the vertebrate fauna of Costa Rica's northern dry deciduous and semi-deciduous forests is truly enormous (Fleming 1981, Gill et al. 1988, Bussing et al. 1995, Carillo et al. 1999), the mammals that thrive in these forests are among the most poorly known of any of Holdridge's bioclimatic life zones in the country (Stoner and Timm 2004). According to these authors, of a total of 114 species of mammals that were originally present in Guanacaste's dry forests, probably 110 are still found in this habitat. Bats are by far the most diverse group with more than 66 species, followed by rodents (11 species), marsupials (7), weasels (6), cats (5), raccoons (3), primates (3), artiodactyls (3), canids (2), edentates (2), one species of rabbit and one species of tapir (Stoner and Timm 2004), many of which reach their southernmost limit in the Costa Rican dry forests (*Didelphis virginiana*, the Virginia opossum; *Liomys salvini*, Salvin's spiny pocket mouse; *Reithrodontomys gracilis*, slender harvest mouse; *R. paradoxus*, a harvest mouse endemic to the Pacific lowlands of Costa Rica and Nicaragua; and *Mephitis macoura*, the hooded skunk; plus a number of bat species).

Beginning in the late 1500s, and then more intensively from the 1940s until the late 1970s, large-scale deforestation for cattle ranching and beef production has led to major habitat loss in the dry lowlands of Guanacaste, the Río Tempisque Basin, and the Península de Nicoya (Myers 1981, Parsons 1983, Sader and Joyce 1988, Sánchez Azofeifa 2000, Quesada Avendaño and Stoner 2004, Calvo Alvarado et al. 2009, Castillo et al 2012, and see Janzen and Hallwachs, chapter 10 of this volume). This large-scale habitat conversion in combination with hunting and poaching has significantly affected the population viability of many big mammals that used to be common in this part of the country—for example, the jaguar (*Panthera onca*), puma (*Puma concolor*, also known as cougar or mountain lion; Fig. 9.13), Baird's tapir (*Tapirus bairdii*, Fig. 9.14), and peccary (*Tayassu pecari*) (Rodríguez et al. 2002, and for peccaries, see McCoy and Vaughan 1990). Populations of small- to medium-sized cats like the ocelot or ocelote (*Leopardus pardalis*), margay or tigrillo (*Leopardus wiedii*), and jaguarundi (*Puma yagouaroundi*) suffered as well, though to a lesser extent. These species mostly disappeared from the area during the second half of the twentieth century—especially outside the protected areas (Vaughan et al. 1996), though ACG has apparently healthy populations of everything but the now extinct giant anteater (D.H. Janzen, pers. comm. 2010, and Janzen and Hallwachs, chapter 10 of this volume).

During the 1990s abandonment of unproductive, extensive pastures in many parts of northern dry lowlands triggered forest regrowth, a process that allowed many mammals to increase their population sizes toward healthier levels. The jaguar, for instance, had disappeared from most of the dry forests in the past, but over the past five years several sightings of this large cat have been reported from the Península de Nicoya and Río Tempisque Basin and it ranges throughout ACG. Similarly, populations of the spider monkey (*Ateles geoffroyi*) that were declining in Guanacaste, Nicoya, and the Tempisque valley during the past decades (Massey 1987) are now benefiting from forest recovery and further protection of dry deciduous and other seasonal lowland forests in the area. Other mammal species that have taken advantage of recent land abandonment and

Fig. 9.13 The puma or mountain lion (*Puma concolor*).
Photograph by Eduardo Carrillo.

Fig. 9.14 The tapir (*Tapirus bairdii*).
Photograph by Eduardo Carrillo.

subsequent dry forest recovery are the puma, the ocelot, Baird's tapir, and two species of peccary (*Pecari tajacu* and *Tayassu pecari*). Populations of the latter have mostly recovered in the higher parts of the area—for example, in the Cordillera de Guanacaste, and have been observed migrating along altitudinal corridors from the upper zones down to the dry lowlands of the ACG.

Another monkey that is common in northern Guanacaste, the Península de Nicoya, and the Río Tempisque Basin is *Cebus capucinus* ssp. *imitator*, the white-faced capuchin or *mono cariblanca* (for Palo Verde, see Massey 1987; for Santa Rosa, see Fedigan et al. 1985, O'Malley and Fedigan 2005, Fedigan and Jack 2012). Chevalier-Skolnikoff (1990) studied tool use by capuchins during some 300 hr of observation in the Parque Nacional Santa Rosa and reported 31 incidents of tool use, including eight different types of tool-use behavior. This author concluded that tool use to get and prepare food, among other activities, is a notable behavior pattern in the studied *Cebus* troop, suggesting high levels of behavioral adaptability. These findings would also support the notion that *Cebus* and the African and Asian great apes have followed a parallel evolutionary development of tool-using capacity (Chevalier-Skolnikoff 1990). Studies of food habits of these white-faced capuchins in the Parque Nacional Santa Rosa and elsewhere (Moscow and Vaughan 1987, O'Malley and Fedigan 2005) have revealed that vertebrate predation in *Cebus capucinus* is considerable (Freese 1977, Fedigan 1990). Lizards, birds, bats, squirrels, and even coatis all form part of the diet of this monkey (Fedigan 1990, O'Malley and Fedigan 2005).

A third monkey frequently observed and heard in Provincia de Guanacaste is the mantled howler monkey (*Alouatta palliata*, locally known as *mono aullador* or *mono congo*) (e.g., for Palo Verde, see Massey 1987; for Santa Rosa, see Fedigan et al. 1985). Apparently, in protected places like La Pacifica in Provincia de Guanacaste, the population size of the howler monkey did not change a lot between the early 1970s and the early 1980s when deforestation peaked in many parts of the northern dry forest region. In July 1984 Clarke et al. (2005) located some 257 howlers at La Pacifica, representing 16 different social groupings and nine solitary animals. The total number of howlers, the number and location of groups, and the age-sex composition they reported were very similar to a 1972–1976 survey of the same population. Habitat fragmentation, though, does affect the genetic variation of the mantled howler monkey in the Área de Conservación Tempisque and causes animals to move towards more favorable—that is, forested—habitats (Quan Rodas 2002).

Comparing densities of howler monkeys and white-faced capuchins in the Parque Nacional Santa Rosa, Fedigan et al. (1985) observed 25 *Alouatta palliata* groups totaling 342 individuals in 1983–1984, giving a park density of 3.4 howler monkeys per square kilometer. Similarly, they recorded 28 *Cebus capucinus* groups totaling 393 individuals, which give a crude density of 3.9 cebus monkeys per square kilometer. As a result, cebus monkey density seems slightly higher than howler density in this protected area during the mid-1980s.

Other important mammals in the dry forest region of Costa Rica are the porcupine (*Sphiggurus mexicanus*), the cotton-tail rabbit (*Sylvilagus floridanus*), and the coati (*Nasua narica*), which lives in social groups of more than 30 individuals, and is abundant in the area probably due to a lack of predators. Furthermore, in northern Pacific Costa Rica one may observe the mapache or raccoon (*Procyon lotor*), the coyote (*Canis latrans*), and the collared peccary (*Pecari tajacu*), which lives in packs of 2 to 15 individuals (Fig. 9.15).

Among the bats, the most conspicuous species in the Pacific dry forests are the vampire bat (*Desmodus rotundus*; see Wohlgenant 1994) and the false vampire or *vampiro falso* (*Vampyrum spectrum*), one of the largest bats in the world (Bradbury 1983) and still found in reduced populations in the country (Stoner and Timm 2004). The false vampire lives in tree holes in groups of four or five individuals,

Fig. 9.15 The collared peccary (*Pecari tajacu*).
Photograph by Eduardo Carrillo.

but has very low population densities. Important pollinating bats are the nectarivorous *Glossophaga soricina* and *Phyllostomus discolor* (e.g., see Wohlgenant 1994, Stoner 2005). Both species were captured with *Ceiba pentandra* pollen in northern coastal Guanacaste (Lobo et al. 2005). Examples of insectivorous bats in these forests are *Natalus stramineus* and *Pteronotus* spp., while typical frugivorous species are *Phyllostomus discolor* and *Sturnira lilium*. Bats that have their southernmost distribution limit in Costa Rican dry forests are *Carollia subrufa* and *Glossophaga leachii* (Stoner 2001, 2005, Stoner and Timm 2004, and see LaVal 2004). Three of 47 bat species recorded during the mid-1990s at Palo Verde make up 60% of the 1,245 captured individuals: *Carollia perspicillata*, *Artibeus jamaicensis*, and *Sturnira lilium*. Here at Palo Verde, the abundance of several species of bats changes significantly over seasons (Stoner and Timm 2004).

Rats, mice, and other small terrestrial rodents abound in the Guanacaste and Nicoya drylands (Bonoff and Janzen 1980). Salvin's spiny pocket mouse, for instance, is very common and acts as a major predator of dry forest seeds such as those from *Enterolobium cyclocarpum*, the Guanacaste tree (Janzen 1982a, 2004), as well as of moth pupae in the litter (Janzen 1986c). Furthermore, the Central American agouti (*Dasyprocta punctata*) is also frequently observed in Costa Rican dry forest and plays a key role as one of the ecosystem's most important scatterhoarders (Hallwachs 1994).

Less common—now probably extinct—are the giant anteater (*Myrmecophaga tridactyla*), the silky anteater (*Cyclopes didactylus*), the rare squirrel monkey (*Saimiri oerstedii*), the tepezcuintle or paca (*Agouti paca*), and perhaps the manatee (*Trichechus manatus*).

The Río Tempisque Basin is still one of the most mammal-rich areas of the northern Pacific dry lowlands. Mateo-Vega et al. (2001) report a total of 111 mammal species for this wetland area. All of these species occur in ACG today. To the west, in the Refugio Nacional de Vida Silvestre Bosque Nacional Diriá (now a National Park) south of the town of Santa Cruz, at least 32 mammal species are found in premontane forest (Villarreal Orias 2003). This is about 15% of the total number of mammal species for Costa Rica. The Diriá fauna includes two species of Didelphimorphia, one species of Xenarthra, one primate species, four carnivores, two species of Artiodactyla, five rodent species, and 17 bat species (Chiroptera) (Villarreal Orias 2003). Probably the most conspicuous species observed in this wildlife refuge are the Virginia opossum, the howler monkey, the gray fox (*Urocyon cinereoargenteus*), the tayra (*Eira barbara*), and the white-tailed deer (*Odocoileus virginianus*).

The known mammal fauna of the Reserva Natural Absoluta Cabo Blanco at the southernmost tip of the Península de Nicoya included at least 37 species of non-flying mammals and 39 species of bats in 1963 when the reserve was established. However, 45 years later at least six species of mammals have disappeared from the reserve as a result of human pressures on the environment (Timm et al. 2009). Today, northern raccoons are one of the most frequently seen mammals here.

Birds

Birds have been formally studied in Guanacaste's dry lands at least since World War II (Wetmore 1944). Slud (1980) did a major assessment of bird diversity in the lower Tempisque valley and reported some 272 species of bird from localities like Palo Verde, Catalina, Bolsón, and Ballena. Further avifaunal inventories throughout the years (e.g., see Stiles and Skutch 1991, Campos Ramírez 1990, and Langen 1994), have led to a long list of 345 bird species for the dry forest of Costa Rica, many of which are associated with aquatic environments like brackish mangrove forests and marshes (Barrantes and Sánchez 2004). Examples are the many herons, egrets, ibises, jacanas, ducks, and a species of large stork. About 58% of the 345 species correspond to resident birds, 36% are latitudinal migrants, and around 6% have resident and migratory populations (Barrantes and Sánchez 2004).

Over the years it has become well known that the avifauna of Central America's dry forests has its southernmost distribution limit in Costa Rica's northern dry region (Slud 1980, Stiles 1983). Here the Península de Nicoya has levels of bird diversity that are characteristic of Central America's Pacific lowlands (Howell and Webb 1995). Some of the species observed in the drylands of Nicoya and the Río Tempisque Basin also occur in moister but still seasonal forests such as found in Costa Rica's Central Pacific region southward. This is especially true for bird species that prefer open or altered areas within forests, or seek forest edges in zones that suffer from deforestation.

A recent assessment of forest birds that are rare in small fragments of Central American dry forest demonstrates that ACG and the Parque Nacional Palo deserve a high priority for regional conservation of dry forest avifauna (Gillespie 2000). Here, rare species like the spot-breasted oriole (*Icterus pectoralis*) merit special attention from environmentalists and should be considered a top priority for conservation of the regional avifauna (Gillespie 2000).

Although Stiles and Skutch (1991) describe the Península de Nicoya as a good bird-watching area at the national level, few ornithological studies have been carried out in this part of the country. Villarreal Orias (2003) reports a total of 94 bird species for the premontane forest of the Bosque Nacional Diriá in the northern sector of the Península de Nicoya, the most diverse zone here, being above 600 m elevation. Among the species recorded at the Bosque Nacional Diriá are the short-tailed hawk (*Buteo brachyurus*, locally known as *gavilán colicorto*), the crested guan (*Penelope purpurascens* or *pava moñuda*) and the long-tailed manakin (*Chiroxiphia linearis* or *saltarín colilargo*), which is one of the most abundant species in the area. Big game bird species like the great currasow (*Crax rubra*) are still frequent in northern Guanacaste as well (Vaughan and Weis 1999).

Bird diversity in dry forests at Lomas Barbudal is equally rich and includes species like the turquoise-browed motmot (*Eumomota superciliosa*), the yellow-naped parrot (*Amazona auropalliata*), the black vulture (*Coragyps atratus*) and turkey vulture (*Cathartes aura*), the black-headed trogon (*Trogon melanocephalus*), the white-throated magpie-jay (*Calocitta formosa*), as well as many egrets, various herons and kingfishers, and numerous species of hummingbirds, tinamous, and woodcreepers (Campos Ramírez 1990). Black-headed trogons have been observed consuming both fruits and insects, depending on the seasonal availability of fruits (Riehl and Adelson 2009). Apparently, adult birds consume more fruit and fewer large insects (Phasmatodea and Mantodea) than nestlings and eat more types of arthropods and fruit. However, adults preferentially consumed small caterpillars and delivered large ones to their nestlings. Riehl and Adelson (2009) conclude that black-headed trogons time reproduction to coincide with arthropod rather than fruit abundance, a pattern that may be more common among omnivorous forest birds than previously recognized.

The Río Tempisque Basin with its complex system of wetlands, particularly Parque Nacional Palo Verde, contains the richest freshwater avifauna in all of Central America (Stiles 1983). Here, the protected wetland areas of the Tempisque valley alone harbor a total of 306 bird species, sixty of which are aquatic birds (Mateo-Vega et al. 2001). This watershed basin includes the most important freshwater and brackish marsh areas in the country (see chapter 20 by Jorge Jiménez, this volume) and provides a refuge for a large number of aquatic birds, including the large wood stork, *Jabiru mycteria* (locally called *cigueñón* or *jabirú*), the beautiful pink colored roseate spoonbill (*Ajaia ajaja*, or *espatula rosada*, Fig. 9.16) and the colorful scarlet macaw (*Ara macao*, locally known as *guacamayo rojo*).

Although the jabiru wood stork—the biggest of three Neotropical stork species—is also known to inhabit the Guatuso plains in the northern part of the Alajuela Province, the Río Tempisque Basin is its single nesting site in Costa Rica. Unfortunately, its population size and nesting area have declined considerably in recent years, so it is now considered highly endangered (Villarreal Orias 2000). Over ten years ago, its population in the Tempisque valley reached only 52 individuals with very few nests found in trees between the wetland and the flooded forest (Villarreal Orias 1997).

A detailed study of the behavior of the jabiru wood stork by Gamboa Poveda (2001, 2003) included 2,607 individual bird observations in the Tempisque valley. Analysis revealed that most activity takes place during the dry season (February-May) when prey is harder to locate. The wood stork seems particularly active during morning hours. In the Río Tempisque Basin, this large bird is most frequently observed in the Palo Verde Lagoon and Corral de Piedra Lagoon. During the study period (1999–2000) foraging behavior was more common in the Palo Verde Lagoon (79% of all observations).

Also common is the blue-winged teal (*Anas discors*), a species of migratory duck that prefers freshwater swamps for feeding and resting. Its populations also decreased considerably from the 1960s throughout 1980s in areas such as the Parque Nacional Palo Verde, as the pools of water became clogged up with vegetation. Fortunately, in recent years there has been a noticeable increase in the number of individuals visiting the marshes and open waters of the area, thanks to efforts to manage and maintain the Tempisque lagoons. For instance, Vaughan et al. (1996) reported the presence of up to 15,000 individuals of blue-winged teal, 25,000 black-bellied whistling ducks (*Dendrocygna autumnalis*, locally known as *pijije común*), and 4,000 jabiru

Fig. 9.16 The roseate spoonbill (*Ajaia ajaja*). *Photograph by Eduardo Carrillo.*

wood storks in the restored wetland areas, along with other less abundant species such as the roseate spoonbill. Unfortunately, agricultural businesses consider the black-bellied whistling duck a pest because it feeds inside their rice plantations, which are typical in many of the lower, wetter areas. Some years ago thousands of ducks were poisoned, causing a significant decrease in their populations.

Many of the birds that inhabit the Río Tempisque Basin have to deal with seasonal changes determined by a six- to seven-month-long dry season when the lagoons dry up, their water flows diminish considerably—along with their food resources—and the trees lose their leaves completely, except in riverine environments (e.g., gallery forests along the main streams). To deal with these changes the birds adjust their diet temporarily or seasonally migrate to other, more hospitable places (Stiles 1983).

At the mouth of the Río Tempisque at the northern side of the Golfo de Nicoya, Isla de Pájaros is located, precisely opposite the national park. Here, numerous nesting sites of species such as the great egret (*Casmerodius albus*, the *garceta grande*), the cattle egret (*Bubulcus ibis* or *garcilla bueyera*), the white ibis (*Eudocimus albus* or *ibis blanco*), the roseate spoonbill, the water turkey or anhinga (*Anhinga anhinga* or *pato aguja*), and the wood stork have been observed in the mid-1990s (Maldonado et al. 1995, Vaughan et al. 1996). Other species found on Isla de Pájaros include the black-bellied whistling duck, the threatened scarlet macaw, and the Muscovy duck (*Cairina moschata* or *pato real*), a species that lives in streams within the forests.

Recently, abandonment of agricultural lands in parts of the lower Río Tempisque Basin such as the Palo Verde wetland has led to recovery of seasonal flooding regimes, which has allowed for the re-establishment of many resident and migratory bird species that in the past used to be common in this wetland refuge—before they locally disappeared due to land conversion for rice fields and pastures (Calvo Alvarado and Arias 2004, Trama 2005).

Reptiles and Amphibians

Little is known about the diversity, natural history, and ecology of amphibians and reptiles of Costa Rica's northern Pacific dry lowlands (Savage and Villa 1986, Savage 2002) and neighboring Cordillera de Guanacaste (Stafford 1998). Authors like Sasa and Solórzano (1995), however, examined some aspects of the natural history, habitats, periods of activity, relative abundance, diet, and reproduction of specific amphibians and reptiles in this part of the country.

Herptofaunal diversity in the northern Pacific lowlands is particularly high in the protected dry forests of northwestern Guanacaste. Sasa and Solórzano (1995), for instance, report a total of 18 amphibian species and 59 reptiles for the Parque Nacional Santa Rosa. In general, the Costa Rican dry forest serves as habitat to some 76 herpetofaunal species, comprising 34 species of snake (Serpentes), 19 frogs and toads (Anurans), 17 lizards (Sauria), 5 turtles and crocodiles, and 1 salamander (Sasa and Bolaños 2004).

Fig. 9.17 The *castellana* or cantil snake (*Agkistrodon bilineatus*). *Photograph by Eduardo Carrillo.*

To the west, in the Río Tempisque Basin some seven families of amphibians and fourteen of reptiles have been recorded (Alpízar et al. 1998, Vaughan et al. 1996). Its protected core wetlands harbor at least 22 amphibian and 15 reptile species (Mateo-Vega et al. 2001). Here along the Río Tempisque, most of the species reported by Sasa and Solórzano (1995) for Santa Rosa are found, with the exception of the castellana or cantil snake (*Agkistrodon bilineatus* or *serpiente moccasin*, Fig. 9.17), a pit viper that in Costa Rica is restricted to the drier habitats of the extreme northwestern part of the country. It can reach up to 1.35 m in length.

The smaller number of amphibians found in Santa Rosa is due to the drastic seasonality of the local climate, when compared with the central and southern Península de Nicoya, where the climate is often more humid. In this regard, Villarreal Orias (2003) reports from the Bosque Nacional Diriá in north-central Nicoya close to Santa Cruz, the presence of 20 species of amphibians and reptiles, of which six are frog species, 10 are lizards, and four are snakes. Of particular importance are the yellow toad (*Bufo luetkenii* or *sapo amarillo*) and the anole lizard (*Norops pentaprion*, locally known as a *lagartija*) species with much-reduced pop-

eating dry forest weevils have to wait one and sometimes two years as hidden and reproductively dormant adults between (dry-season) fruit crops (Janzen 2004).

Lepidoptera—Butterflies and Moths

Over twenty years ago Janzen (1987a) reported 3,140 species of moths and butterflies from the dry forest of the Parque Nacional Santa Rosa. About 2,800 of these are moths (Janzen 1987a, 1988c). It is expected that these numbers will continue to rise as research continues. The modern use of DNA barcoding to supplement and strengthen the taxonomic platform underpinning the inventory of thousands of sympatric species of caterpillars in tropical dry forest, for instance, will be particularly helpful in this regard and may reveal latent species diversity among moths and butterflies like skippers (e.g., Janzen et al. 2005, Smith et al. 2007, Burns and Janzen 2001, Burns et al. 2008).

During the dry season butterfly diversity and abundance are much reduced in Guanacaste's dry deciduous lowland forest with many resident species such as *Eurema daira* passing the dry season in river bottoms (DeVries 1987). Four weeks after the first rains have fallen butterflies abound everywhere. According to DeVries (1987) some genera are most abundant the first two weeks of the rainy season (*Eurytides*), whereas others are most abundant much later (*Archaeoprepona*, *Zaretis*, and *Memphis*). Reproductive diapause among resident butterflies occurs in the dry season whereas a fair amount of migrants seasonally move to wetter places. Some of the most typical butterflies of the dry deciduous forest are *Eurytides philolaus*, *E. epidaus*, *Itaballia demophile*, *Kricogonia lyside*, *Marpesia petreus*, *Microtia elva*, *Memphis forreri*, *Taygetis kerea*, *Eunica monima*, and *Caligo memnon* (DeVries 1987). Species like *Eurytides epidaus*, *Kricogonia lyside*, and *Memphis forreri* reach their southernmost limit in this part of Central America.

DeVries (1987, 1997)—who collected a lot of specimens in Parque Nacional Santa Rosa—estimates known butterfly diversity in the Pacific deciduous region of Costa Rica at 178 species, spread over 14 Papilionidae (*swallowtail butterflies*), 31 Pieridae (*sulphurs and whites*; medium-sized mostly pale-colored butterflies), 101 Nymphalidae (*brush-footed butterflies* or *four-footed butterflies*), and 32 Riodinidae (known as *metalmarks*). In the Parque Nacional Santa Rosa, butterfly species diversity is around 486 species with 13 papilionids, 21 pierids, 89 nymphalids, 32 riodinids, ca. 50 lycaenids, and some 281 hesperiids (Haber and Stevenson 2004). None of the dry forest butterflies in Costa Rica are endemic to the country (DeVries 1987, 1997). Certain butterflies like *Anartia fatima* (Nymphalidae), the banded peacock, show clear aggregative behavior in Guanacaste during the dry season (Young 1979).

Like many other Lepidoptera (Janzen 1987a, 1988c), every year a skipper butterfly, *Aguna asander* (Hesperiidae), whose caterpillars feed on *Bauhinia ungulata* (Fabaceae) leaves in the rainy season in the dry forest lowlands, abandons its nests in the dry forests and flies up into the cold cloud forests on the volcano tops. There these butterflies roost individually in the cloud forest understory before they migrate back to the dry forest lowlands and establish their nests when the rain comes (Janzen 1987, 1988c, 2004). Janzen (2004) explains that migrating moth species can occur in huge numbers at lights in passes at 600 to 2,500 m elevation in Costa Rica's mountain ranges that separate the rain forest from the dry forests.

According to Janzen (1987a, 1988c) moths pass the six month rain-free dry season and some other portions of the year, by (a) remaining dormant in the egg stage (one species only), (b) remaining dormant in the pupal or prepupal stage (many species), (c) undergoing larval development (a few species of particular life forms), (d) remaining in the Santa Rosa sector as a potentially active but non-reproductive adult (many species), and (e) migrating out of Santa Rosa after one to two generations and then returning at the beginning of the following rainy season, like in the case of *Aguna asander* (a few species of particular life forms).

Janzen (1986b) reported from the Santa Rosa sector a total of 30 breeding saturniid moth species (and 5 waif species) and a sphingid moth fauna (hawk moths, sphinx moths, and hornworms) of 64 regularly breeding species, 10 occasional breeding species, and 9 waifs (83 species in total). Saturniidae are among the largest and most spectacular of the moths. To given an example, the saturniid *Hylesia lineata* (Saturniidae: Hemileucinae) passes the dry season in the egg stage, lays all its eggs in one nest, has strongly urticating hairs on the adult, and has caterpillars that feed on at least 46 species of plants in 17 families in Parque Nacional Santa Rosa (Janzen 1984).

In the Santa Rosa area only one endemic saturniid thrives, and no endemic sphingids. At the same time, nearly all of the saturniids and sphingids of Santa Rosa have very broad geographic and ecological ranges (Janzen 1986b, 1993). The saturniids are dormant during the six month dry season and all survive the dry season within the Park. More than half of the sphingids migrate out of the Park during the dry season or even the second half of the rainy season. Santa Rosa supports a formidable array of predators and parasitoids that eat saturniid and sphingid caterpillars (Janzen 1986b, 1993). In the mean time, researchers Dan Janzen and Winnie Hallwachs (2009) continue to publish information on the natural history of Costa Rican dry forest caterpillars, pupae, butterflies, and moths (http://janzen.sas.upenn.edu).

Hymenoptera—Bees, Wasps, and Ants.
Costa Rica harbors at least 785 species of bees (Apidae) and 620 species of ants (Formicidae), many of which thrive in the dry forests of Guanacaste (Vinson et al. 2004). Major bee groups are Halictinae, Anthophorinae, Megachilinae, Euglossinae, Meliponinae, and Xylopinae. Frankie et al. (2004b) and Vinson et al. (2004) highlight the importance of large bees in the genera *Centris*, *Epicharis*, *Mesoplia*, and *Mesocheira* as pollinators of many flowering trees in Guanacaste. Melipones and introduced honeybees are common as well. Nests of *Centris aethyctera* and *C. flavifrons* (Hymenoptera: Apoidea: Anthophoridae) in the dry forest were studied by Vinson and Frankie (1977). *C. flavifrons* nests in the ground and prefers soil, while others (*C. flavofasciata*) prefer sand, and still others (*C. bicornuta*) nest in preexisting wood cavities (Vinson et al. 2004).

In a dry deciduous forest at least 15 species of male euglossine bees were attracted that apparently do not have breeding populations in this forest and do not visit orchids here. Probably, they travel long distances to find females and chemical resources elsewhere in distant habitats or forest types (Janzen 1971, Janzen et al. 1982).

As part of their annual phenology social wasps such as the paper wasp *Polistes variabilis* (Hymenoptera: Vespidae) seem to migrate, once the dry season starts, out from the dry forests of Guanacaste to other, more distant and colder, high elevation ecosystems (Hunt et al. 1999), to be dormant there to wait for the next year's rain (Janzen 2004). The brood of such highly social wasps serves as prey to many tropical dry forest bird species (Windsor 1975). Leaf-cutter ants (*Atta colombica;* Hymenoptera: Formicidae) constitute an important element of the dry forest ecosystem. Apparently they may harvest higher quality food in the dry season (Wirth et al. 1997), though the dry season air is too dry to permit foraging in the daytime (Janzen 2004). To adapt to changing environmental and seasonal conditions *Atta colombica* colonies are known to relocate their nests (Rockwood 1973, 1975). Other important dry-forest ants are several species in the genus *Pseudomyrmex* that are known to develop mutualistic relations with dry forest *Vachellia* (= *Acacia*) plants (Janzen 1966 and 1983, Vinson et al. 2004, and Janzen and Hallwachs, chapter 10 of this volume).

Crustaceans: Freshwater Shrimps

The freshwater prawn *Macrobrachium americanum*, also known as the Cauque River prawn (Crustacea: Palaemonidae), has been studied in the San Andrés River, Santa Cruz, Guanacaste. Álvarez et al. (1996) collected a total of 290 specimens and did some morphometrical calculations: minimum commercial size was 55 to 60 mm total length, between 3.6 and 4.8 g abdominal weight, and between 6.9 and 9.8 g total weight. The presence and abundance of this and other macroinvertebrates are excellent bioindicators of water quality and levels of chemical pollution in Neotropical streams and rivers, and hence of ecosystem integrity and species population health (William McLarney, pers. comm.).

Species Interactions

Flowering, Pollination, and Breeding Patterns

Frankie et al. (2004b) provide a thorough overview of flowering phenology and pollination systems diversity in the seasonal dry forest of northern Pacific Costa Rica. Their account includes a nice overview of early and recent phenological studies in this part of the country. Many of the more recent data they present are from a study site near Bagaces in Guanacaste. Generally, they found that community-level periodicity patterns of leaf fall, leaf flushing, flowering, and fruiting are closely associated with the highly seasonal dry and wet periods. Of a total of 140 tree species observed, almost half bloom in February and to a lesser extent in March, both being dry season months. In November the fewest tree species are flowering (less than 20). Referring to unpublished observations by G. Frankie, Bawa (2004) wrote that in 1998 following the El Niño year of 1997, the intensity of tree flowering was reduced in Guanacaste. Apparently, reduction in food resources influenced the abundance of bees almost throughout the dry season.

The most prominent pollination systems in the Bagaces dry forest are large-bee, small-bee/generalist, moth (Haber and Frankie 1989), and bat pollination types (Table 9.3). Nectarivorous bats that pollinate dry forest tree species are known to forage over distances of several kilometers in a single night (Heithaus et al. 1975). Similarly, in the dry season tree-pollinating moths migrate from Guanacaste's deciduous and semideciduous forests to adjacent wet seasonal forest types (Janzen 1987b).

Large and small bees appear to pollinate most of the trees and non-tree plants that were recorded (Frankie et al. 1983). Large bees like those in the genera *Centris*, *Epicharis*, *Mesoplia*, and *Mesocheira* visit mainly large, brightly colored flowers that are bilaterally symmetrical and bloom during the dry season (Frankie et al. 2004b). Frankie and coworkers estimated that at peak foraging periods there could be up to 50,000 bees on a single large flowering crown of *Andira inermis*—a "large-bee flower" tree—and most of these belonged to the genus *Centris*. Occasional visitors of "large-bee flowers" are *Xylocopa* species and euglossine bees, also called orchid bees (Frankie et al. 1983).

out weevils), *guapinol* trees have shorter intervals between seed crops, an earlier age at first reproduction, smaller seed crops, and less resin in the pod walls in the absence of these weevils (Janzen 1975). Results suggest selective pressures on *H. courbaril* exerted by Costa Rican *Rhinochenus* weevils.

If *guapinol* seeds survive predation by insects they become available for dispersion by mammals. Hallwachs (1986) has suggested that the agouti (*Dasyprocta punctata*) is perhaps the only effective *guapinol* seed dispersal agent present today in dry deciduous forests of northern Costa Rica. It seems to have inherited the role of disperser of *H. courbaril* seeds from large Pleistocene mammals that are now extinct (Janzen 1975b). Only 10,000 years ago, some 15 genera of Central American large herbivores became extinct, including frugivorous horses, gomphotheres (extinct elephant-like animals), ground sloths, and other Pleistocene megafauna (Janzen and Martin 1982). This has resulted in an alteration of seed dispersal and subsequent distributions of many plant species like *H. courbaril* and trees like *Crescentia alata* (known as calabash or *jícaro*) and *Enterolobium cyclocarpum*, the *guanacaste*, *caro caro*, or elephant ear tree (Janzen and Martin 1982). Apparently, the introduction of horses and cattle may have in part restored the local ranges of these tree species previously seed-dispersed by now extinct Pleistocene mammals (Janzen 1980, 1981b, Janzen and Martin 1982, and see Janzen and Hallwachs, chapter 10 of this volume).

In disturbed dry forest sites (e.g., pastures), today *E. cyclocarpum* is primarily dispersed by cattle and horses, whose movements are restricted by pasture boundaries (Janzen 1980, 1981a, 1981b, Gonzales et al. 2009). Often *E. cyclocarpum* seeds do not accumulate below the parent tree over the years, even when large potential dispersal agents are missing from its habitats. By placing fruits containing known numbers of seeds below a parent *guanacaste* tree in the Parque Nacional Santa Rosa, Costa Rica, and monitoring their disappearance, Janzen (1982a) determined that almost 94 percent of the seeds were removed by a small native forest-floor rodent, the spiny pocket mouse *Liomys salvini*. Hence, only a few percent of the seed crop seemed to accumulate below the parent tree.

Since the arrival of the Spaniards in northern Costa Rica and the introduction of the horse, free-ranging horses in its forested grasslands have been swallowing about half of the seeds in the *E. cyclocarpum* fruits that they eat, while some of them have been observed defecating at least 9–56% of the seeds alive (Janzen 1981a). Hence, horses seem to serve as key dispersal agents for seeds of the *guanacaste* tree today (Janzen 1981b). The horse-seed interaction suggests that Pleistocene horses may have contributed to both local and long-distance population recruitment by *E. cyclocarpum*, and contemporary horses certainly have the potential to do so (Janzen 1981a, 1981b). An additional study on variations in *E. cyclocarpum* seed size suggests that different parts of the *guanacaste* seed crop may end up in different dispersers, and dispersers with different preferences for fruit seediness and tolerances for seed size may remove different portions of the seed crop (Janzen 1982d).

A study of in vitro fermentation of *guanacaste* seeds in rumen vs. in caecal inoculum to test their germination response to the fermentation processes of cattle and horses demonstrated that large, non-ruminant mammalian herbivores function as seed predators while ruminants are better disposal agents for guanacaste seeds (Janzen et al. 1985).

The hard, ripe fruits of the previously mentioned species of *Crescentia alata* (*jícaro*) are broken by range horses with their incisors; they swallow the small seeds embedded in the sugar-rich fruit pulp. The seeds survive the trip through the horse and germinate in large numbers where horses have defecated (Janzen 1982b). In this regard, horses are considered surrogate Pleistocene dispersal agents (Janzen 1982c). *Jícaro* seeds were also found to be highly edible to *Liomys salvini*—just like in the case of the *guanacaste* seeds (see previous paragraphs)—which avidly harvested the *jícaro* seeds from horse dung but could not open the hard fruits itself. Janzen (1982c) believes that this kind of interaction among horses, *jícaro* fruits, and *Liomys* mice is probably representative of that which used to occur between Pleistocene megafauna and a number of contemporary Central American trees (Janzen 1982c).

Mutualism: Ant-Plant Relations

A well known case of plant-animal mutualism is found in the intimate relationship between *Pseudomyrmex* ants (*hormiga del cornizuelo*) and *Vachellia* plants (swollen-thorn acacia, formerly in the genus *Acacia*), each involving several species (Janzen 1966 and 1983, Vinson and Frankie 2004). Janzen (1966) discussed in detail the coevolution between *Vachellia cornigera* L. (Mimosoideae; Leguminosae), and *Pseudomyrmex ferruginea* F. Smith (Pseudomyrmecinae; Formicidae). *V. cornigera*, though, is less common in the northern Pacific lowlands than the more ubiquitous *V. collinsii*, which in Provincia de Guanacaste and northern Puntarenas is often occupied by *P. ferruginea* ants. Janzen (1966, 1983) gives a full account of the life cycle of *P. ferruginea* queens and males, its reproductive cycle in relation to its acacia habitat, its daily movements from acacia trees back and forth to surrounding trees and shrubs, and predation of social wasps on male swarms. In the end, an ant colony may occupy as many as ten to thirty acacia shoots (Janzen 1983).

Epiphytism

Epiphytes are relatively uncommon in tropical dry forest when compared to epiphyte diversity in tropical semihumid, rain, and cloud forests (Benzing 1990). Epiphytes account for an estimated 2 to 4% of the total flora of tropical dry forests, while in moist or rain forests they may reach 25% (Gentry and Dodson 1987, Bullock et al. 1995).

The few epiphytes that do occur in the northern dry forests of Guanacaste and Nicoya are adapted to withstand long dry periods. A number of them extend their distribution into the seasonal forests of the Central Pacific region (in Esparza, Orotina, and along the Candelaria River near Puriscal and Acosta). Good examples of epiphytic species adapted to these dry climates are cacti like *Hylocereus costaricensis* (perhaps the most common cactus epiphyte in the country, locally known as *pitahaya*), *Selenicereus testudo*, and *S. wercklei*, which forms large hanging masses in the tree tops of the dry forest (Morales 2000). A rather rare epiphytic cactus is *Disocactus amazonicus*, which has been reported from three collections in the Península de Nicoya and the slopes of the Cordillera de Guanacaste (Morales 2000).

In dry forest lowlands epiphytic bromeliads and orchids are more common than cacti. Morales (2000) reports at least 28 species of orchid and 15 species of bromeliad (*bromelia*) for the deciduous and semideciduous forests of the North. The most species-rich orchid genera are *Epidendrum* and *Oncidium*. Locally less diverse genera include *Cattleya*, *Elleanthus*, *Encyclia*, *Laelia*, *Myrmecophila*, *Scaphyglottis*, *Sobralia*, and *Trigonidium*. Occasionally *Vanilla planifolia*—world renowned for its vanilla flavor—is found in the wettest parts of the Península de Nicoya (Morales 2000). Yeaton and Gladstone (1982) studied the pattern of colonization of epiphytes on calabash trees (*Crescentia alata*) in Guanacaste and found that four orchids dominated its epiphyte community and colonized the tree in the following order, from pioneer to late colonizer: *Oncidium cebolleta*, *Encyclia cordigera*, *Brassavola nodosa*, and *Laelia rubescens*.

Among the bromeliads, on the other hand, one observes genera such as *Aechmea*, *Bromelia*, *Catopsis*, *Pitcairnia*, *Tillandsia*, and *Werauhia*. The genus *Tillandsia* is represented by at least six species that grow on trunks and branches of dry forest trees. *Tillandsia schiedeana* is perhaps one of the most common plant species in dry forests at mid-elevation in Guanacaste (Morales 2000).

While the presence of hygrophilic ferns and fern-allies—both epiphytic and terrestrial—is not a characteristic feature of the northern Pacific dry lowland region, two species are relatively common here: *Lygodium venustum* and *Selaginella pallescens*. Less abundant but occasionally important are *Selaginella sertata*, *Thelypteris minor*, *Cyclopeltis semicordata*, and *Adiantum concinnum*. At higher elevations and in slightly more humid areas it is even possible to find *Polypodium attenuatum*, *Polypodium wagneri*, *Cheilanthes brachypus*, and *Adiantum trapeziforme*.

Parasitism

Parasitism is a very common type of species interaction in the tropical dry forests of northern Costa Rica. Numerous parasitic species of flies, flatworms, and lianas, among others, are known to use organisms like dry forest caterpillars, mollusks, frogs, lizards, and birds as hosts. A good example is provided by parasitoid flies (Diptera), such as tachinids (Tachinidae), which abound in the Área de Conservación Guanacaste (ACG). They parasitize thousands of species of caterpillars. Also, some species of wasps such as the hyperparasitoid wasps in the Trigonalidae do parasitize caterpillars (Lepidoptera) (Smith et al. 2012).

Smith et al. (2007) encountered >400 parasitoid fly species of specialist tachinids with only a few generalists in Guanacaste. Their DNA barcode studies affirm that at least 16 species of apparently generalist tropical parasitoid tachinids are not all generalists. Similarly, parasitic Braconidae (Hymenoptera) are known to feed on larvae of the diverse Costa Rican dry forest skipper butterflies (Lepidoptera: Hesperiidae: Pyrginae) (Janzen et al. 1998, Burns and Janzen 2001 and 2005).

Also, many species of helminth parasites have been reported to parasitize vertebrate species in Costa Rican dry forest, particularly birds, lizards, and frogs. For instance, the parasitic flatworm *Halipegus eschi* (Platyhelminthes: Digenea: Hemiuridae) has been described from the esophagus of the semi-aquatic Vaillant's frog (*Rana vaillanti*) found in deeper pools of stagnant or slow-moving water in the Provincia de Guanacaste (Zelmer and Brooks 2000). This dry forest anuran also serves as host to another flatworm species, the frog lung fluke *Haematoloechus meridionalis* (Digenea: Plagiorchioidea: Haematoloechidae) (León-Règagnon et al. 2001).

Similarly, Zamparo et al. (2003a, 2003b, 2003c, 2003d, 2004, and 2005) described a number of new species of trematodes, including *Whallwachsia illuminata* (Trematoda: Digenea), which was named after Winnie Hallwachs, longtime researcher in the Parque Nacional Santa Rosa, and co-author of chapter 10 of this volume. Most of these parasitic trematodes are parasites of vertebrates when in adult stage. They have complex life cycles requiring one or more intermediate hosts like mollusks (e.g., snails or clams). Most are hermaphroditic and many capable of self-fertilization.

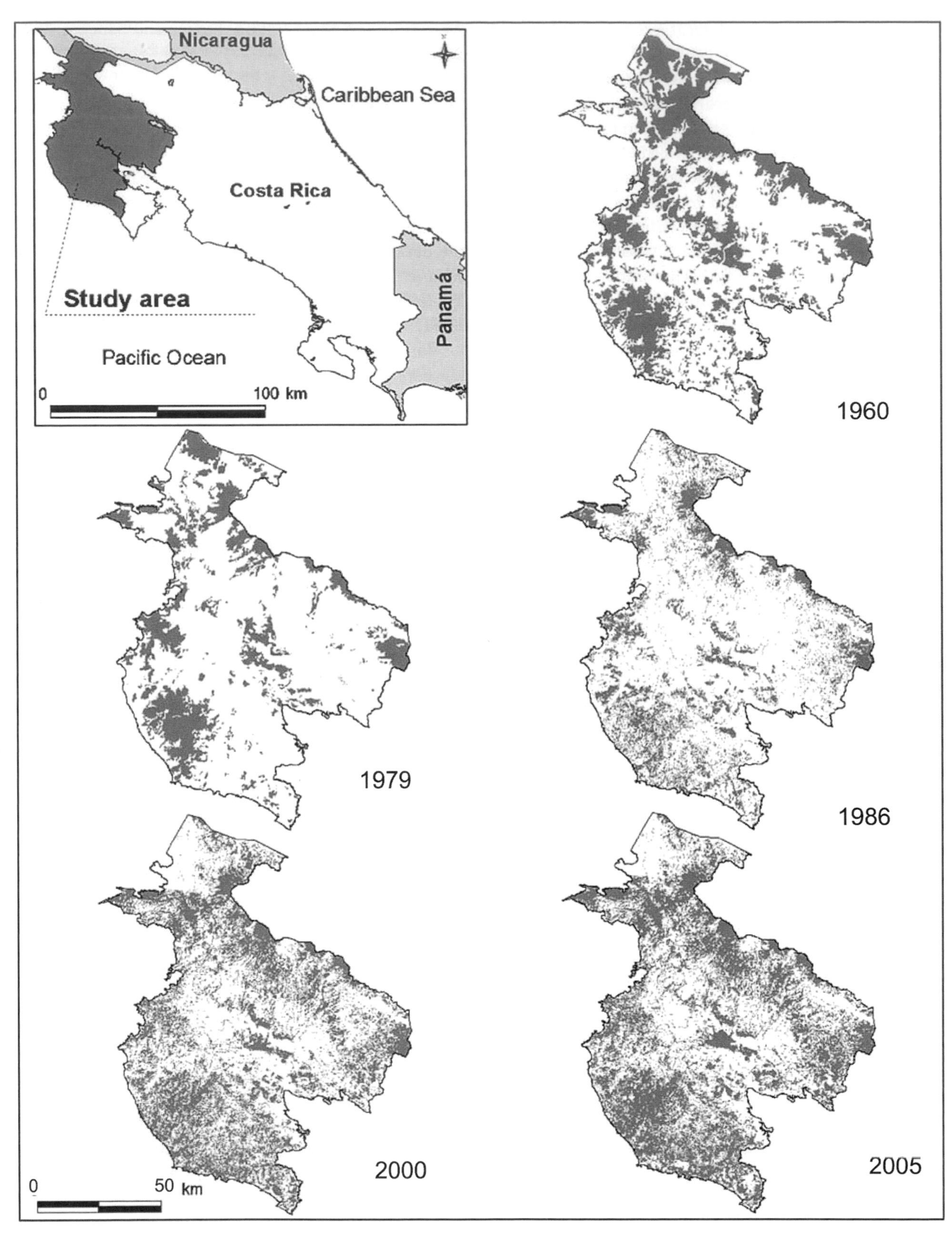

Fig. 9.18 Forest coverage maps of the Chorotega region, Costa Rica, showing the period from 1960 to 2005. Dark areas correspond to forest cover. *Source: Calvo Alvarado et al. (2009). Reproduced with kind permission from Elsevier B.V. Publishers.*

(b) 45 families in the intermediate stage (1,137 individuals in 106 species); and (c) 45 families and 751 individuals (92 species) in the late stage (Kalácska et al. 2004). Successional dry forest stand height ranged from 7.5 (early successional) to 15 m tall (late), while basal area increased from 11.7 (early) to 30.1 m^2 per ha (late), and species density augmented from 15 to 29 species per 0.1 ha. Stem density, however, was highest in the intermediate stage (130 stem per 0.1 ha) and lower in early (112) and late (107) phases of succession (Kalácska et al. 2004). Floristic descriptions of the successional stages (early, intermediate, and late) have been provided by Arroyo Mora (2002). A complementary study that integrated data on forest structure and remotely sensed data for four successional stages (pasture, early, intermediate, and late) and compared successional stages with chronological age, demonstrated there was no separability in the spectral reflectance among different age classes. In contrast, successional stages showed distinct groups with minimal overlap (Arroyo Mora et al. 2005b).

In a similar study on the diversity and structure of regenerating dry forests in northern Costa Rica, Powers et al. (2009) analyzed data from sixty 20x50 m forest plots in the Santa Rosa and Palo Verde protected areas. Using the common chronosequence approach of identifying stands that differ in age, they measured forest structure and composition, and inferred regeneration dynamics by assuming a space-for-time substitution. The study by Powers et al. (2009) differed from the study by Kalácska et al. (2004) in that they considered forest age as a quantitative variable, sampled over a wider range of forest types, and emphasized how regeneration dynamics depend upon edaphic variables. Successional dynamics as assessed from plots of different age showed that the patterns of change in indices of stand structure, species richness, and tree community composition varied with forest type (and hence soil properties), while forest structure (densities of stems in different size classes) recovered to levels found in mature forest within 4–5 decades in regenerating dry forest at Santa Rosa and Palo Verde, but increased with stand age in Guanacaste's oak dominated forests (Powers et al. 2009).

Reforestation and Forest Restoration

Before accelerated natural regeneration was considered a major option to actively restore the dry forests of northern Costa Rica by the mid-1980s, reforestation with exotic species dominated in this region. In fact, reforestation with exotic trees and, to a lesser extent, native species has happened a lot in Guanacaste over the past twenty years. Quiros Herrera (1998), for instance, reports on reforestation projects in the Nicoya region with pochote (*Pachira quinata*), a native species of high commercial value in the domestic market, and exotic teak (*Tectona grandis*). In this regard, the private company Bosque Puerto Carrillo (BPC) started a commercial plantation in 1983, establishing 12 hectares of pochote near Puerto Carrillo and Buenos Aires de Sámara, Península de Nicoya. In 1986 it tried teak (*Tectona grandis*), planting on almost 30 ha that first year. Later, in the early 1990s it concentrated its efforts on planting teak and initiated plantations at Palo Arco near Jabillos de Nandayure, expanding its efforts to the central part of the Península de Nicoya. The market success of these teak plantations led to the establishment of an industrial plant at Palo Arco, designed to produce teak parquet. The example of BPC has been copied across the northern dry lands of Guanacaste and southern Nicoya, as well as in the Atlantic lowlands of the northeastern part of the country (the plains of San Carlos, Guatuso, Sarapiquí, etc.).

During the mid-1980s a large group of Costa Rican and American scientists and volunteers set out to save the tropical dry forests in the northwestern part of Costa Rica (Janzen 1986a, 1988a, Cherfas 1987, Tenebaum 1994, Allen 2001). The team, led by the dynamic Dan Janzen and his wife, Winnie Hallwachs, from the University of Pennsylvania, moved relentlessly ahead, taking a broad array of political, ecological, and social steps necessary for restoration of the until then fire-scarred range lands in Guanacaste (for a full account see Janzen and Hallwachs, chapter 10 of this volume). The use of Costa Rican debt-for-nature swaps to save Guanacaste's dry forests has been just one of many ways in which tools have been applied to conserve and restore these ecosystems over the past 25 years (Cherfas 1986). As a result, by the year 2000 large patches of formerly degraded dry forest were being restored, allowing thousands of plant and animal species like the guanacaste tree, Baird's tapir, and the howler monkey to return (Allen 2001). Most of these patches were stitched together in the newly formed Parque Nacional Guanacaste (1989), the core area of the Área de Conservación Guanacaste (ACG) (see, e.g., Allen 1988 and 2001, Janzen 1986a, 1987c, and 2002, the section on forest conservation, management, and sustainability in this chapter, and Janzen and Hallwachs, chapter 10 of this volume). At the beginning of the new millennium over 463 square miles of dry forest land were being restored in the northwestern lowlands of Costa Rica.

Fortunately, not only in Guanacaste is the dry forest recovering. Some sectors of Nicoyan dry forest as well as the moister systems in the southern part of the Península de Nicoya are regenerating. This is in part due to the abandonment of lands previously used for cattle ranching. Only a few years ago, these forests appeared to have recovered up to 50% of their original cover (Sánchez Azofeifa et al.

Arroyo Mora, J.P., G.A. Sánchez Azofeifa, B. Rivard, J.C. Calvo, and D.H. Janzen. 2005a. Dynamics in landscape structure and composition for the Chorotega region, Costa Rica from 1960 to 2000. *Agriculture, Ecosystems and Environment* 106: 27–39.

Arroyo Mora, J.P., G.A. Sánchez Azofeifa, M. Kalácska, and B. Rivard. 2005b. Secondary forest detection in a Neotropical dry forest using Landsat 7 ETM+ imagery. *Biotropica* 37(4): 497–507.

Asch, C., G. Oconitrillo, and J.L. Rojas. 2000. Delimitación cartográfica y otras consideraciones sobre las zonas afectadas por inundaciones en la cuenca baja del río Tempisque, Guanacaste, Costa Rica. San José, Costa Rica: Instituto Geográfico Nacional, Departamento de Geografía. 28 pp.

Azéma, J., and J. Tournon. 1980. La péninsule de Santa Elena, Costa Rica: un massif ultrabasique charrié en marge pacifique de l'Amérique Centrale, Costa Rica. Paris: C. R. Acad. Sci. 290: 9–12.

Barrantes, G., and J.E. Sánchez. 2004. Geographical distribution, ecology, and conservation status of Costa Rican dry-forest avifauna. In G.W. Frankie, A. Mata-Jiménez, and S.B. Vinson, eds., *Biodiversity Conservation in Costa Rica: Learning the Lessons in a Seasonal Dry Forest*, 147–59. Berkeley: University of California Press.

Bassey, G., and G. Barboza-Jiménez. 1989. Proyecto Parque Nacional Guanacaste: apuntes sobre recursos costeros del Proyecto Parque Nacional Guanacaste. San José, Costa Rica. 17 pp.

Bawa, K.S. 1974. Breeding systems of tree species of a lowland tropical community. *Evolution* 28: 85–92.

Bawa, K.S. 2004. Impact of global changes on the reproductive biology of trees in tropical dry forests. In G.W. Frankie, A. Mata-Jiménez, and S.B. Vinson, eds., *Biodiversity Conservation in Costa Rica: Learning the Lessons in a Seasonal Dry Forest*, 38–47. Berkeley: University of California Press.

Benzing, D.H. 1990. *Vascular Epiphytes*. Cambridge, UK: Cambridge University Press.

Bergoeing, J.P. 1998. *Geomorfología de Costa Rica*. San José, Costa Rica: Instituto Geográfico Nacional. 423 pp.

Bolaños, R.A., and V. Watson. 1993. Ecological Map of Costa Rica. According to L.R.Holdridge's Life Zones Classification System. San José, Costa Rica: Tropical Science Center. Scale 1.200.000.

Bonoff, M. B., and D.H. Janzen. 1980. Small terrestrial rodents in 11 habitats in Santa Rosa National Park, Costa Rica. *Brenesia* 17: 163–74.

Bradbury, J. 1983. *Vampyrum spectrum* (vampiro falso, false vampire bat). In D.H. Janzen, ed., *Costa Rican Natural History*, 500–501. Chicago: University of Chicago Press.

Braun-Blanquet, J. 1979. Fitosociología: bases para el estudio de las comunidades vegetales. Madrid, Spain: Blume. 820 pp.

Brooks, D.R. 1996. The all taxa biodiversity inventory in the Área de Conservación Guanacaste, Provincia de Guanacaste, Costa Rica. In *Proceedings of the American Association of Zoo Veterinarians Annual Conference*, 343–49. Puerto Vallarta, Mexico. November 3–8, 1996.

Brooks, S.J. 1989. Odonata colectados en el Parque Nacional Guanacaste, Costa Rica, Julio 1988. *Notulae Odonatologicae* 3(4): 49–52.

Bullock, S.H., H.A. Mooney, and E. Medina. 1995. *Seasonally Dry Tropical Forests*. Cambridge: Cambridge University Press. 450 pp.

Burger, J., and M. Gochfeld. 1991. Burrow site selection by Black Iguana (*Ctenosaura similis*) at Palo Verde, Costa Rica. *Journal of Herpetology* 25(4): 430–35.

Burnham, R.J. 1997. Stand characteristics and leaf litter composition of a tropical dry forest hectare in Santa Rosa National Park, Costa Rica. *Biotropica* 29: 387–95.

Burns, J.M., and D.H. Janzen. 2001. Biodiversity of pyrrhopygine skipper butterflies (Hesperiidae) in the Área de Conservación Guanacaste, Costa Rica. *Journal of the Lepidopterists' Society* 55: 15–43.

Burns, J.M., and D.H. Janzen. 2005. Pan-Neotropical genus *Venada* (Hesperiidae: Pyrginae) is not monotypic: four new species occur on one volcano in the Área de Conservación Guanacaste, Costa Rica. *Journal of the Lepidopterists' Society* 59: 19–34.

Burns, J.M, D.H. Janzen, M. Hajibabaei, W. Hallwachs, and P.D.N. Hebert. 2008. DNA barcodes and cryptic species of skipper butterflies in the genus *Perichares* in Área de Conservación Guanacaste, Costa Rica. *Proceedings of the National Academy of Sciences* 105(17): 6350–55.

Bussing, W.A. 1998. *Freshwater Fishes of Costa Rica*. 2nd ed. San José, Costa Rica: Universidad de Costa Rica. 468 pp.

Bussing, W.A. 2008. A new species of poeciliid fish, *Poeciliopsis santaelena*, from Peninsula Santa Elena, Área de Conservación Guanacaste, Costa Rica. *Revista de Biología Tropical* 56(2): 829–38.

Bussing, W., K. Winemiller, M. López, F. Bolaños, M. Sasa, M. Donnelly, J. Savage, C. Guyer, A. Solorzano, M. Araya, G. Barrantes, J. Blake, D. Hernandez, F. Joyce, B. Loiselle, A. Pereira, J.M. Mora, B. Rodriguez, L.M. Collado, J. Saenz, J. Salazar, V.A. Vega, and T. Yates. 1995. An inventory of the vertebrate fauna of the Área de Conservación Guanacaste. Unpublished manuscript. 123 pp.

Calvo Alvarado, J., and O. Arias. 2004. Restauración hidrológica del humedal Palo Verde. *Ambientico* 129: 7–8.

Calvo Alvarado, J., B. McLennan, A. Sánchez Azofeifa, and T. Garvin. 2009. Deforestation and forest restoration in Guanacaste, Costa Rica: putting conservation policies in context. *Forest Ecology and Management* 258(6): 931–40.

Calvo Alvarado, J., G.A. Sánchez Azofeifa, and M. Kalácska. 2008. Deforestation and restoration of a tropical dry forest in the Chorotega region, Costa Rica. In H. Tiessen and J.W.B. Stewart, eds., *Applying Ecological Knowledge to Landuse Decisions*, 123–33. Sao Paulo, Brazil: Scientific Committee on Problems of the Environment (SCOPE)—Inter-American Institute for Global Change Research (IAI)—Inter-American Institute for Cooperation on Agriculture (IICA).

Campos Ramírez, R.G. 1990. The birds of Reserva Biológica Lomas de Barbudal Guanacaste. Bagaces County. Guanacaste, Costa Rica. *Bee Line: News and Bulletin* [*Friends of Lomas Barbudal*]: 1–4.

Carrillo, E., G. Wong, and J.C. Sáenz. 1999. *Mamíferos de Costa Rica*. Santo Domingo de Heredia, Costa Rica: Instituto Nacional de Biodiversidad (INBio). 248 pp.

Castillo, M., B. Rivard, A. Sánchez-Azofeifa, J. Calvo-Alvarado, and R. Dubayah. 2012. LIDAR remote sensing for secondary Tropical Dry Forest identification. *Remote Sensing of Environment* 121: 132–43.

Castro, G. 1993. Palo Verde: Un reto de Manejo. *Newsletter of Fundación Neotrópica* 1: 8–11.

Chavarría, U., J. González Ramírez, and N.A. Zamora Villalobos. 2001. *Árboles Comunes del Parque Nacional Palo Verde* [*Common Trees of Parque Nacional Palo Verde*]. Santo Domingo: Instituto Nacional de Biodiversidad (INBio).

Cherfas, J. 1986. Costa Rica swaps its national debt for forests in Guanacaste. *New Scientist* (August 6, 1986).
Cherfas, J. 1987. How to grow a tropical forest. *New Scientist* (October 23, 1987).
Chevalier-Skolnikoff, S. 1990. Tool use by wild *Cebus* monkeys at Santa Rosa National Park, Costa Rica. *Primates* 31(3): 375–83.
Choudhury, A., R. Daverdin, and D.R. Brooks. 2002. *Wallinia chavarriae* n. sp. (Trematoda: Macroderoididae) in *Astyanax aenaeus* (Günther, 1860) and *Bryconamericus scleroparius* (Regan, 1908) (Osteichthyes: Ostariophysi: Characidae) from the Área de Conservación Guanacaste, Costa Rica. *Journal of Parasitology* 88: 107–12.
Clarke, M.R., E.L. Zucker, and N.J. Scott, Jr. 2005. Population trends of the mantled howler groups of La Pacífica, Guanacaste, Costa Rica. *American Journal of Primatology* 11(1): 79–88.
Cordero-Montoya, R., H. Acevedo-Mairen, and J.C. Calvo Alvarado. 2008. Cambio de la cobertura de la tierra para el Área de Conservación Tempisque 1998–2003, Guanacaste, Costa Rica. *Kurú: Revista Forestal* (Instituto Tecnológica de Costa Rica) 5(15): 1–15.
Cornelius, S.E. 1986. *The Sea Turtles of Santa Rosa National Park*. Madrid, Spain: Incafo. 64 pp.
Crow, G.E. 2002. *Plantas Acuáticas del Parque Nacional Palo Verde y el Valle del Río Tempisque*. 1st ed. Santo Domingo de Heredia, Costa Rica: Instituto Nacional de Biodiversidad (INBio). 296 pp.
Crow, G.E., and D.I. Rivera. 1986. Aquatic vascular plants of Parque Nacional Palo Verde, Costa Rica. *Uniciencia* 3: 71–78.
Daniels, A.E. 2004. Protected area management in the watershed context: a case study of Parque Nacional Palo Verde, Costa Rica. M.Sc. Thesis, University of Florida.
Dauphin, G., and M. Grayum. 2005. Bryophytes of the Península de Santa Elena and Islas Murciélago, Guanacaste, Costa Rica, with special attention to Neotropical dry forest habitats. *Lankesteriana* 5(1): 53–61.
Dengo, G. 1962. *Estudio Geológico de la Región de Guanacaste, Costa Rica*. San José, Costa Rica: Instituto Geográfico Nacional (IGN). 122 pp.
DeVries, P.J. 1987. *The Butterflies of Costa Rica and Their Natural History*. Vol. I. Papilionidae, Pieridae and Nymphalidae. New Jersey: Princeton University Press. 327 pp.
DeVries, P.J. 1997. *The Butterflies of Costa Rica and Their Natural History*. Vol. II. Riodinidae. New Jersey: Princeton University Press. 288 pp.
Edelman, M. 1985. Extensive land use and the logic of the latifundio: a case study in Provincia de Guanacaste, Costa Rica. *Human Ecology* 13(2): 153–85.
Elizondo, L.H., and Q. Jiménez. 1988. La sabana arbolada "El Escobio," Liberia, Guanacaste, Costa Rica. *Revista de Biología Tropical* 36: 175–85.
Enquist, B.J., and A.J. Leffler. 2001. Long-term tree ring chronologies from sympatric tropical dry-forest trees: individualistic responses to climate variation. *Journal of Tropical Ecology* 17: 41–60.
Fedigan, L.M. 1990. Vertebrate predation in *Cebus capucinus*: meat eating in a Neotropical monkey. *Folia Primatologica* 54: 196–205.
Fedigan, L.M., L. Fedigan, and C. Chapman. 1985. A census of *Alouatta palliata* and *Cebus capucinus* monkeys in Santa Rosa National Park, Costa Rica. *Brenesia* 23: 309–22.
Fedigan, L.M., and K.M. Jack. 2012. Tracking neotropical monkeys in Santa Rosa: lessons from a regenerating Costa Rican dry forest. In P.M. Kappeler and D.P. Watts, eds., *Long-Term Field Studies of Primates*, 165–84. Berlin: Springer.
Fitch, H.S., and J. Hackforth-Jones. 1983. *Ctenosaura similis*. In D.H. Janzen, ed., *Costa Rican Natural History*. Chicago: University of Chicago Press.
Fleming, T.H. 1981. *Los Mamíferos del Parque Nacional Santa Rosa*. Serie Materiales de Enseñanza No. 1. San José, Costa Rica: Editorial Universidad Estatal a Distancia (EUNED). 13 pp.
Fleming, T.H. 1985. Coexistence of five sympatric *Piper* (Piperaceae) species in a tropical dry forest. *Ecology* 66(3): 688–700.
Flowers, R.W. 1991. Aggregations of Cassidinae (Chrysomelidae) in Santa Rosa and Parque Nacional Guanacaste, Costa Rica. *Biotropica* 23(3): 308–10.
Frankie, G.W., W.A. Haber, P.A. Opler, and K.S. Bawa. 1983. Characteristics and organization of the large bee pollination system in the Costa Rican dry forest. In C.E. Jones and R.J. Little, eds., *Handbook of Experimental Pollination Biology*, 411–47. New York: Scientific and Academic Editions.
Frankie, G.W., A. Mata-Jiménez, and S.B. Vinson, eds. 2004a. *Biodiversity Conservation in Costa Rica: Learning the Lessons in a Seasonal Dry Forest*. Berkeley: University of California Press. 341 pp.
Frankie, G.W., W.A. Haber, S.B. Vinson, K.S. Bawa, P.S. Ronchi, and N. Zamora. 2004b. Flowering phenology and pollination systems diversity in the seasonal dry forest. In G.W. Frankie, A. Mata-Jiménez, and S.B. Vinson, eds., *Biodiversity Conservation in Costa Rica: Learning the Lessons in a Seasonal Dry Forest*, 17–29. Berkeley: University of California Press.
Frankie, G.W., V. Solano, and M. McCoy. 1994. Una evaluación técnica del corredor la Mula entre la Reserva Biológica Lomas de Barbudal y el Parque Nacional Palo Verde. 16 pp. San José, Costa Rica.
Freckman, D.W., coordinator. 1995. An inventory of the nematode fauna of the Área de Conservación Guanacaste, Costa Rica. Manuscript. 44 pp.
Freese, C.H. 1977. Food habits of the whited-faced capuchins *Cebus capucinus* L. (Primates: Cebidae) in Santa Rosa National Park, Costa Rica. *Brenesia* 10/11: 43–56.
Gamboa Poveda, M.E. 2001. Comportamiento del jabirú (*Jabiru mycteria*) en tres humedales de la cuenca baja del río Tempisque, Costa Rica. Thesis, Universidad Nacional. Heredia, Costa Rica.
Gamboa Poveda, M.E. 2003. El comportamiento de *Jabirú mycteria* durante la época reproductiva en humedales de la zona norte de Costa Rica. *Zeledonia* 7(1): 4, 25–32.
Gentry, A.H., and C.H. Dodson. 1987. Diversity and biogeography of Neotropical vascular epiphytes. *Annals of the Missouri Botanic Gardens* 74: 205–33.
Gerhardt, K. 1998. Leaf defoliation of tropical dry forest tree seedlings—implications for survival and growth. *Trees* 13: 88–95.
Gill, D.E., and OTS staff. 1988. A naturalist's guide to the OTS Palo Verde Field Station, el Refugio de Fauna Silvestre "Dr. Rafael Lucas Rodriguez Caballero", and Reserva Biológica Lomas de Barbudal, Provincia de Guanacaste, Costa Rica. Maryland: Department of Zoology, University of Maryland. 107 pp.
Gillespie, T.W. 2000. Rarity and conservation of forest birds in the tropical dry forest region of Central America. *Biological Conservation* 96(2): 161–68.

Chapter 10 Biodiversity Conservation History and Future in Costa Rica: The Case of Área de Conservación Guanacaste (ACG)

Daniel H. Janzen[1,*] and Winnie Hallwachs[1]

Prologue

This essay-like chapter is directed to the audience of, in this order of importance, present and future staff of Área de Conservación Guanacaste (ACG), staff of the entire system of conserved wildlands of Costa Rica, and the people of Costa Rica (for all, to be transliterated into Spanish in another document), foreign visitors of all kinds and nations to Costa Rica, the world audience, and the scientific community.

History of Área de Conservación Guanacaste (ACG)

In the Beginning: Geology

Eighty-five million years ago and before, when dinosaurs and their associates wandered and swam the earth, today's 163,000 ha Área de Conservación Guanacaste (ACG, Fig. 10.1) was a big blue patch of a great expanse of ocean between what today we call the Pacific Ocean and the Atlantic Ocean. There was no Central America. Far out to the west there was an island, the terrain that today we call Península Santa Elena, the peninsular portion of ACG that extends to the west. There were other Pacific islands as well, islands that gradually moved eastward through what would become Costa Rica, to become islands in today's Caribbean (see Fig. 10.9–10 in Graham 2003; Fig. 10.4 in Hoernle et al. 2002; and see Alvarado and Cárdenes, chapter 3 of this volume).

Some 65 million years ago, when Central America still had not emerged from the ocean, a 20 km diameter meteor hit where today lies the tip of the Yucatan Peninsula, creating global-encompassing tidal waves that were kilometers in height. The aerial debris blocked out the sun for years. *Adiós* to almost all that depended directly on photosynthesis and the sun's daily dose of heat. *Adiós* to the dinosaurs and most of their associates—though note that not only did some small mammals survive this event, but so did "microdinosaurs," the ancestors of the animals that today we call "birds" (Zelenitsky et al. 2012). This perturbation opened the world for a great evolutionary explosion. Who survived this catastrophic event, to become the raw material for this evolutionary explosion? Those species that could live (or stay dormant) for years in nearly continuous darkness and cold, those who could feed on dead and dormant species, and those who in turn ate them. Look to today's ocean depths for some living examples, to creatures of the night for others. From these ragged refugees and their predators and parasites largely evolved the world of macro-organisms that we know today, when the sun shone again on nature's green solar panels.

About 16 million years ago, the island that was to become Península Santa Elena in ACG, the land once walked by dinosaurs and ever since above the sea, crunched into the emerging archipelago of Central America. This Central American land bridge was first an archipelago lying between the southernmost extension of the Rocky Mountains where they end today in southern Honduras, and northern Colombia-Venezuela as the northern end of the Andes. Central America eventually became a solid land bridge between North and South America, finally closing a mere 3 million years ago. With this closing, the Pacific and the Atlantic oceans became two separate puddles, with all the expected

[1] Department of Biology, University of Pennsylvania, Philadelphia, PA 19104 djanzen@sas.upenn.edu, whallwac@sas.upenn.edu

* Corresponding author

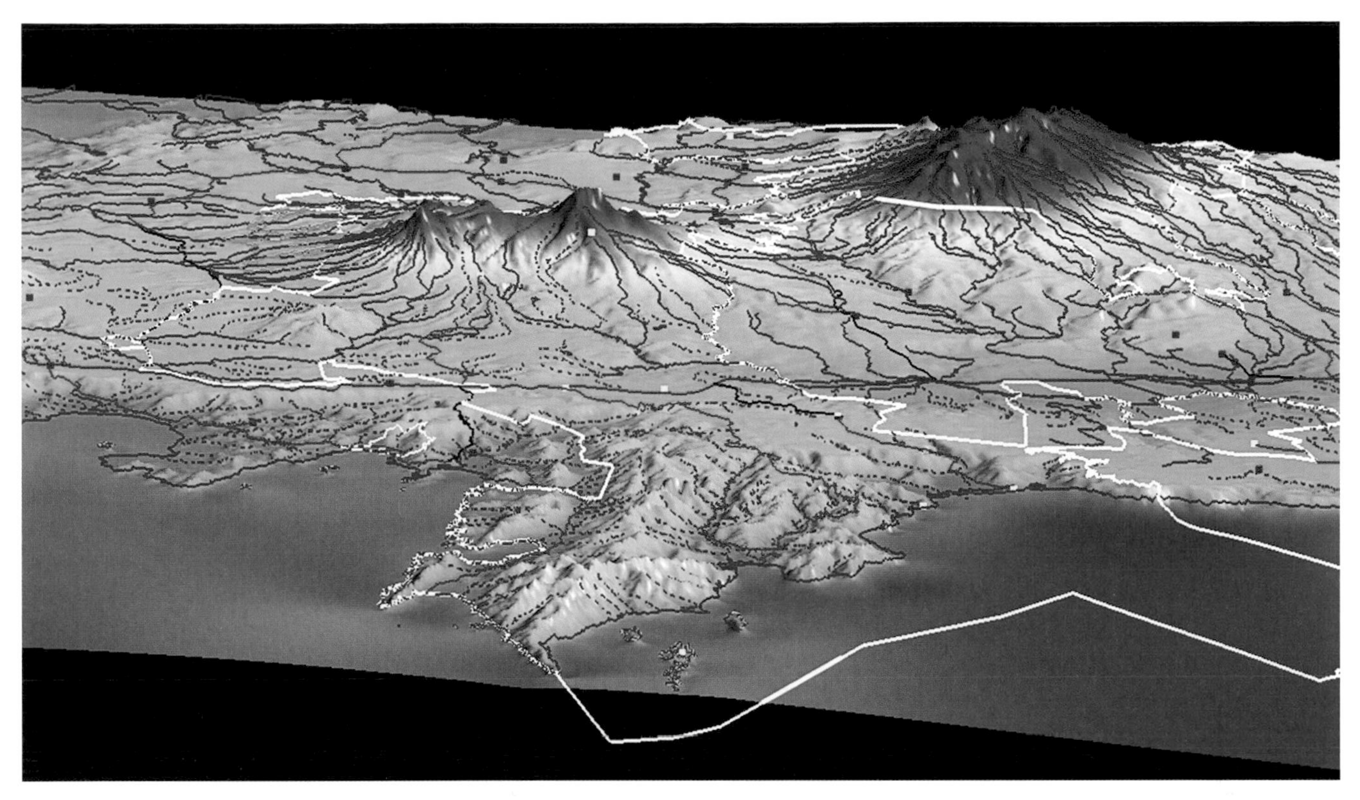

Fig. 10.1 Área de Conservación Guanacaste (ACG), enclosed by a white boundary (orange around Sector Del Oro), as viewed from the west out over the Pacific Ocean; ACG extends from 6 to 18 km into the Pacific Ocean across the dry-forested coastal lowlands up to 1,450–1,650–2,000 m cloud-forested volcano peaks (red) and down to 90 m in the rain-forested Caribbean lowlands (see Figs. 10.2, 10.3, 10.8, and 10.33 for map views).

evolutionary consequences of dividing a population into two parts, each to now live out separate evolutionary trajectories according to the different ecological worlds of the two different oceans (Jackson and D'Croz 1998). And with this closing, the land bridge that allowed extensive blending of North and South American biodiversity was complete (Cody et al. 2010, Marshall et al. 1982, Rich and Rich 1983, Iturralde-Vincent and MacPhee 1999, Jackson and D'Croz 1998, Wallace 1997).

However, for tens of millions of years before the closure there must have been substantial south-north interchange through island-hopping up and down the archipelago of Central America-to-be and between the two continents (and see Johnson and Weckstein 2011). Not only were these islands stepping stones, but also they had their isolated, and thus speciating, populations that were then thrown together after full emergence of the land bridge, to yield part of the high biodiversity encountered in Central America today.

As recently as 1.5 million years ago, there was a large lake perched on this land bridge, extending at least from today's Managua/Granada in central Nicaragua south to Liberia, in northwestern Costa Rica. The trade winds from the east/northeast brought sufficient moisture-laden air from the Caribbean, and perhaps from the lake itself, to foment rain forest all the way from the Caribbean coast to the Pacific coast across the narrow land bridge at this point. Among the emergent volcanic islands in the lake was a 6-million-year-old volcano, today a weathered 617 m hill, Cerro El Hacha, located in north central ACG (Figs. 10.2, 10.13). And on the southeastern shore of this lake was a 2,500 m volcano, roughly centered on what is today the Rincón de la Vieja complex of the Cordillera de Guanacaste that crosses ACG from southeast to northwest (Figs. 10.1 and 10.2). For reference, 2,500 m is nearly as high as the highway over Cerro de la Muerte in the Talamanca mountain range lying between San José and Panama.

And then there was a Krakatoa-scale explosion of this volcano. In 20 minutes, perhaps, the entire content of this volcano was exploded-lifted-and-dropped on most of the Costa Rican portion of Lake Nicaragua, forming what we think of today as the Mesa Santa Rosa, the 200–300 m elevation flat area crossed by the Inter-American Highway (Pan-American Highway of older terminology) as it bisects ACG from south to north (Figs. 10.2, 10.8). The pink area of the geological map of the ACG area (Fig. 10.3) outlines the western portion of this volcanic debris. When drilling for water wells in 2008 in the Área Administrativa of ACG, Sector Santa Rosa, rotting organic matter from the old lake

Fig. 10.4 Volcán Concepción in Nicaragua in full eruption on February 9, 2007. Volcano formations of this nature must have been an integral part of the birth of the volcanoes of Cordillera de Guanacaste.
Photograph by Emanuel Sferios.

Arrival of European Culture and Subsequent Land Use History

Europeans arrived in the ACG in two very different ways. While Pizarro did walk across it on his way south to Nicoya in 1523, all of the subsequent marine traffic up and down the Pacific coast passed close by the shores of ACG, and probably harvested from it. According to rumor, colonial records document that many houses in Lima, Peru, are constructed with wood harvested from the forests behind the ACG beaches, and specifically, from the shores of Península Santa Elena. Timber, dye wood (*Haematoxylum brasiletto*), tannin and medicinal-bearing woods in coastal forest would have been a ready harvest for any boat with space in the hold. Fires that escaped from camping on the beach would have followed, easily invading the lightly logged forests and quickly converting them to mixes of native grasses and secondary succession small woody plants, which are much more inflammable in the dry season than is old-growth forest. For example, the flatland forest rich in mahogany/caoba (*Swietenia macrophylla*) trees immediately northwest of the north end of Playa Potrero Grande (Fig. 10.15) appears to be untouched by "modern" society, but is suspiciously even-aged (Janzen 1998). Yes, perhaps a tsunami in the 1800s leveled it to bare mud, along with the even-aged mangrove forest behind Playa Potrero Grande, but on the other hand it may also be a case of natural restoration following thorough harvest as well. The name "Potrero Grande" (big pasture) dates to at least the 1600s and refers to the enormous long and thoroughly deforested valley bottom (and its stone corrals built with indigenous labor under early Spanish administrations) turned into an ocean-accessible pasture that backs the swamp behind Playa Potrero Grande and is the site of Oliver North's airstrip built during the Contra war orchestrated by the United States of America against the Nicaraguan *Sandinistas* in the 1980s (Fig. 10.7). It will take major biological detective work to puzzle out the details of European impact on the biodiversity of the coastal zone of ACG from the mid-1500s to the time of the World War II in the 1940s (if indeed it is even possible). However, whatever the details, it is clear that none of the ACG lowland coastal forest, including that of its islands, can be regarded as pristine old-growth, even if the loss of the megafauna is disregarded. However, above about 200 m elevation on the outer end of the Península Santa Elena there are about 2,000 ha of highly inaccessible and generally unburned dwarf forest that probably approximates old-growth status. At the uppermost elevations of this forest (about 700 m) there even survives a mossy remnant of a depauperate and very insular cloud forest (Maria Marta Chavarria, pers. comm.) that has probably never seen an axe or chain saw.

Modern Landscape-Level Changes since the 1500s

Each region of Costa Rica has its own peculiar four centuries of history of European colonization. In one sentence, the goal of the colonists and their offspring—no matter in which century—was to harvest and/or remove the natural vegetation and replace it with anthropogenic structures, croplands, plantations, and pastures that produced goods and services for direct use and sale. In the case of the dry northwestern part of Costa Rica, the region of ACG with its Pacific dry forest and wetter Cordillera de Guanacaste, this was a four-century-long and complicated process, and fortunately only variably successful and thorough.

Its nature was determined largely by transportation routes to other places, soil type, climate, and world-level markets/inventions. A brief visit to this past sheds light on some of the larger social processes that led to the region of ACG being, in 1966–1985, still populated by remnants of the wild organisms and ecosystems that could generate

Fig. 10.5 Volcán Orosí viewed from the north, with the Del Oro orange plantations in the foreground (March 7, 2001). The blue cloudless skies of the ACG rain shadow dry forest in the dry season on the right (west) contrast strongly with the heavy rain clouds of the rain forest to the left (northeast), and the tuft of clouds bathing the Volcán Orosí cloud-forested top.

Fig. 10.6 Pedregal Orosí at the western, dry forest base of Volcán Orosi (Fig. 10.5), 500 m elevation, on February 13, 1987, while still under a regime of heavy grazing and annual burning. Many hundreds of the rocks in this *pedregal* have deeply carved petroglyphs (Fig. 10.26) dating from pre-Columbian times, and the site may well have been deforested since then.

the landscape-level restoration process in motion at present. Perhaps the most important trait of this past was the combination of being a long distance from major social centers (Managua-Granada far to the north in Nicaragua, and the Puerto Viejo-Meseta Central-Puntarenas axis to the south in Costa Rica), and having soils bad for agriculture, scarce water, and severe dry seasons, all of which were inhospitable to farming and ranching. There is not a single ever-flowing river on the Mesa Santa Rosa, the Mesa is essentially a volcanic tuff-rock tableland derived from the innards of that volcano that exploded 1.5 m years ago (Fig. 10.3), and the long dry season (December-May) is largely rain-free while the short dry season (July-August) is erratic in length and intensity. These obstacles are what allowed the survival of ACG biodiversity long enough and completely enough to perform the beginnings of landscape-level restoration when allowed (Janzen 1988a, 1988c, 2002), starting with the inauguration of Monumento Nacional Santa Rosa and Parque Nacional Santa Rosa (today Sector Santa Rosa of ACG) in 1966–1971 (Government of Costa Rica 1998) and intensified with the beginning of ACG as the Proyecto Parque Nacional Guanacaste in 1986 (Allen 1988, 2001, Janzen 1988a, 1988c, 2000a, 2002).

During the 1500s and 1600s, a major route from the eastern United States and Europe to California and Chile was for an ocean-going boat to go up the Río San Juan all the way to Lake Nicaragua, sail across it, disembark at Rivas, walk and mule-train the 20 km to the Pacific, and then sail north to Mexico and California, or south to Panama, Peru, and Chile. In the mid-1700s, an earthquake raised the bed of the Río San Juan shortly after it left Lake Nicaragua, forcing large boats from the Atlantic to disembark at that point. From the date when the opening of the Panama Canal in 1914 rendered this transport route obsolete, trains of hundreds of mules carried the cargo to Lake Nicaragua for paddle-wheeler boat transport across to Rivas, or muled all the way from the Río San Juan to Rivas along the southern lake shore.

The 1848 California Gold Rush was largely populated via this route and it was very active at the time of the US

Fig. 10.7 Playa Potrero Grande (background) and Potrero Grande (center) with Oliver North's Contra airstrip remnants in the center, Península Santa Elena (January 9, 1988). The dark green forest near the beach is mangrove forest while the woodland immediately towards the inland is regenerating freshwater swamp forest. The yellow-gray hills are covered with a native (and inedible to livestock) grass, *Trachypogon plumosus*, that has spread from natural disturbance sites (cliffs and rocky ridges) following centuries of annual to semi-annual manmade burning of these serpentine barrens that were originally covered with low-stature and highly deciduous dry forest.

Civil War in 1861; William Walker was ousted from Nicaragua by the US military in 1857 for tampering with US business interests that managed this cross-Nicaraguan transport (Jamison 1909, Doubleday 1886, Wight 1860, Wells 1856). The mules for this massive transport were produced in Hacienda Santa Rosa, which stretched from the coast at Playa Naranjo and Nancite (and probably Potrero Grande) to the mid-elevational, moister slopes of Volcán Cacao (and mules were presumably also produced at other haciendas in its vicinity, such as Hacienda Inocentes and Hacienda Orosí on the drier foothills of Volcán Orosí). The implication is that the original erratic clearing (largely by fire) of the ACG dry forest for native grass pasture, beginning in the late 1500s, was to grow large herds of horses, *burros*, and mules as well as some long-horned red-furred cattle ("*ganado*"). In the 1970s, 80-year-old Santa Rosa employees said that in their youth, large herds of horses produced foals in the pastures near the coast in the early rainy season, and then moved on their own to the still-green pastures on the foothills of the Cordillera de Guanacaste during the long dry season, and then back again to the coast at the beginning of the rainy season in mid-May (see Fig. 10.8 for this transect).

In the 1600s–1800s a major cattle slaughter industry developed at the port of Puntarenas far to the south, where hides and tallow were packed into barrels and shipped around the bottom of South America to Europe. These cattle would have come from all parts of Guanacaste Province but ACG was probably too far north to have contributed heavily. However, when the indigo dye industry in Guatemala and El Salvador emerged in the late 1700s to 1800s, requiring large amounts of field labor to pick and process leaves from plantings of native *Indigofera* (a small shrub in the Fabaceae), herds of cattle were produced in northwestern Costa Rica and driven north as food for this labor. This was not a very lucrative business, at least in part because the cattle lost a lot of weight on the drive north, and because there were no easy pastures to fatten them on in El Salvador (the pastures having been converted to indigo plantations). Furthermore, when the cattle herd arrived, the buyer offered a very low price because there were no other buyers and certainly they were not going to be driven all the way back to Costa Rica if prices were low. At the same time, the growing urbanization of the Meseta Central was becoming a cattle buyer in the same way, and the business was also not lucrative for the same reasons. Hacienda Santa Rosa was very far from everywhere economically, and in combination with its very poor soils and being very far from the political centers of power to the north and south, it was apparently always a low-yield largely extensive (rather than small area intensive) cattle-ranching system. Records in the colonial archives in Granada, Nicaragua, indicate that Santa Rosa had more than 40 different owners between the late 1500s and the 1960s, suggesting repeated ranching and farming failures by a succession of owners. This in turn saved nearly its entire dry forest species from local elimination, though these ranching/burning/hunting activities certainly altered their population, community, and ecosystem biology.

On March 29, 1856, when an exploratory contingent of William Walker's army encountered General Mora and his Costa Rican troops at the Casona Santa Rosa (Wells 1856), the greater Hacienda Santa Rosa extended from the tip of Península Santa Rosa to the forests of Cordillera de Guanacaste and had been an extensive low-productivity mule and cattle ranch for two and a half centuries. Apparently it was the second hacienda to be established in Costa Rica, after Hacienda Inocentes on the northern foothills of Volcán Orosí (with a somewhat moister climate and much closer to Lake Nicaragua). Santa Rosa was paying taxes to Spain by the end of the 1500s. It appears that it was first created as a land grant from the Queen of Spain to one of her captains, who then committed some offense that caused her to take it back. The captain's son then bought it back and the record of this transaction is somewhere in the colonial archives in Granada, Nicaragua (fortunately not burned by one of the marauding pirates of the 1600s–1700s, when there was still big-boat-free access to the lake from the Atlantic to Grenada via the Río San Juan).

The traditional way of clearing land for cattle pasture was still in use in the 1900s–1960s. The "traditional way," probably in operation since the 1500s, was as follows. Wherever a natural (or anthropogenic) break in the forest canopy allowed sunlight down to ground level and therefore a patch of herbaceous vegetation developed, the cowboy encountering it in the dry season simply set a fire in it. The same process was repeated the next year and within a few years flat areas on good soil were largely covered by grasses and herbs, with the rocky (and wetter) ravines and cliffs still largely forested. Such an extensive pasture system was easily maintained relatively free of woody (shading) vegetation by annual or semiannual burning near the end of the long dry season (April-May). The indigenous population would probably not have used fire this way because the last thing that they wanted—except for ceremonial purposes—would have been a large patch of grass (not even the deer are grass-eaters; however, the indigenous people might have hunted *Sigmodon hispidus*, the hispid cotton rat, which can be extremely common in large pasture systems in Guanacaste). The African grass *jaragua* (*Hyparrhenia rufa*) was introduced to Costa Rica in the 1920s and by the 1940s the Hacienda Santa Rosa cowboys were carrying sacks of *jaragua* seed and sprinkling it into the pastures

Fig. 10.8 The transect from the north end of Playa Naranjo (foreground) to the lower slopes of Volcán Orosí (center rear) and Volcán Cacao (right rear) (January 29, 1988). The Mesa Santa Rosa is evident as a flat mosaic of yellow jaragua grass pasture and forest remnants through the center of the image. The Inter-American Highway roughly bisects the Mesa from right to left slightly inland from the dark central line of the cliffs bordering the Cafetal-Rio Cuajiniquil headwaters.

Fig. 10.9 Mosaic of ungrazed and unburned jaragua grass pasture and forest remnants on the western edge of Mesa Santa Rosa between Quebrada Vaca Blanca (to the left) and Portón de los Perros (out of sight to the right) (March 16, 1987). This is the patch of yellow jaragua on the far left in Fig. 10.8. See also Fig. 10.21.

after burning, gradually converting all the pastures to a much higher yield forage (Daubenmire 1972) (Fig. 10.9). Simultaneously, cebu cattle, which are far more resistant to the Guanacaste heat and long dry season than are European breeds (derived from *Bos taurus*), were introduced into these same pastures. Cebu cattle are genetically derived from *Bos indicus*, an Indian-African species of tropical dry forest origin, with part of their resistance prominently displayed as the large fat-containing food reservoir hump over the shoulders. By the 1960s–1970s, during the neo-flowering of the ancient Guanacaste cattle industry (Parsons 1983, Shane 1986, Janzen 1988b), the Guanacaste herds still often contained a mix of both cattle breeds (Fig. 10.10).

While many ranchers in southern Guanacaste Province were gradually converting to more intensive cattle ranching, with more thorough elimination of wild biodiversity (and replacement of *jaragua* with other, even more productive species of African grasses), ACG pastures were treated in the old-style extensive and non-intensive ranching that was basically unchanged from the 1700s except for better breeds of cattle and deliberate seeding of *jaragua* (local ranchers were still coming to Parque Nacional Santa Rosa in the 1970s to harvest truck-loads of *jaragua* to seed their pastures). In brief, a herd of several thousand or more cattle (steers, cows, and bulls) were turned loose in many thousands of hectares of brushy pasture (with watercourses) and then harvested at yearly or more intervals.

Pasture "care" consisted of setting fire to it, often in the second or third month of the dry season but later in the year if it was desired that the forest margins be more thoroughly pushed back (by the hotter fire created by the drier soil and fuel). The wind-driven fire did its job. The only pasture management was to "*hacer rondas*," which meant clearing the herbaceous material to nearly bare ground for a long meter back from the barbed-wire fences, in hopes that when the fire swept through, the wooden fence posts would not be incinerated. The fence posts came from trees cut either during the original clearing of the forest, or trees cut from forest patches that had burned less readily. Posts were therefore abundant and cost only cheap labor. By the early 1990s, when these posts needed to be replaced at a much higher cost, the somewhat higher salaries (formerly a

Fig. 10.10 Management of mixed cebu and brown European cattle (Palo Verde, Guanacaste, March 10, 1976). The brown cattle (*Bos taurus*) were being replaced by the much more heat-resistant and productive cebu (*Bos indicus*).

mere pittance) and other costs made extensive cattle ranching marginally economic, if that.

All of this is relevant to conservation today because the large extensive cattle pastures were quite rich in biodiversity that survived on the patchy sites that burned poorly because they were moist or contained low amounts of herbaceous fuel (because of thorough grazing by livestock). As vehicle-based agro-civilization developed, the large holdings were gradually partitioned into small areas of higher yield (rice fields, higher quality [often replanted] African grass, fruit orchards, sugar cane, peanuts, cotton, pineapple, etc.), sometimes irrigated, and polished quite free of their biodiversity. Other areas were abandoned, logged, burned, and used for subsistence-level micro-ranching of a few horses and a small cattle herd. Both treatments resulted in a gradual erosion of dry forest biodiversity, exacerbated by the usual effects of fragmentation and insularization. Even the strips of forest, left by tradition and some laws standing along Guanacaste Province watercourses, had gradually dissolved from their once species-rich community structure to a still-dwindling mix of living dead (Janzen 2001), ruderals, and agrochemical-resistant species. This mess is sprinkled with a few species that "do well" (whatever that means) in the presence of massive human manipulation of the system (boat-tailed grackles, magpie jays, ctenosaurs, *Guazuma ulmifolia*, *Senna pallida*, *Sida* spp., etc.).

Throughout this process, the state of the erratically damaged vegetation suggests that large-scale farming was not a part of the European land-clearing process in the tens of thousands of hectares of Hacienda Santa Rosa, the core area of ACG dry forest, until about the 1940s. At this time some of the flatter and better soil pastures to the west of the Casona Santa Rosa were plowed and seeded to dryland rice. This was a short-lived experiment, due to both the low-quality soil and the erratic rain patterns. Various rocky slopes and hills of the lowland coastal area (e.g., the hillside inland from Argelia behind central Playa Naranjo and the southern sides of the valley containing the Río Poza Salada, Fig. 10.14), and almost all of the flat coastal plain behind Playa Naranjo (5–100 m elevation, Fig. 10.8), were thoroughly cleared of their dry forest to grow corn, yucca (cassava), beans, papaya, banana, fruit trees, and rice planted in the style of colonizing small-scale farmers and squatters. These areas are today covered by secondary forests of varied ages, heights, and characteristics, such as their degree of deciduousness in the long dry season. What distinguishes them and their restoration from that of the Mesa Santa Rosa is that they were never planted with pasture grasses, they were closely surrounded by forest seed-sources, and they were on somewhat good soil. The consequence is that they quickly turned back to young forest when abandoned in the 1960s and 1970s at the time Parque Nacional Santa Rosa was decreed and the colonists relocated outward. While these lowland areas were swept by dry season fires, both set on-site and invading from the fires sweeping the pastures of the Mesa Santa Rosa in the 1940s–1960s, they have for the most part not been burned since their abandonment with the removal of the colonists in 1970. The other small-scale European clearing of ACG dry forest was the subsistence farming (which cleared most of the forest off of the lower slopes of Cerro El Hacha in the 1970s, Fig. 10.13) and rudimentary coffee plantings in a few places on the southwestern intermediate elevation slopes (Sector Pailas and Sector Santa Maria) of the Rincón de la Vieja massif. Coffee planting was one of several clearing activities of Hacienda Santa Maria (today, Sector Santa Maria of ACG) above Liberia, and along the trail from Estación Pailas to the Rincón Crater in Sector Pailas. This

was its first director. In 1977 it was increased to a size of 10,800 terrestrial ha through expropriation of a portion of a neighboring ranch that owned the land behind the southern part of Playa Naranjo. This version of PNSR contained 23,000 ha of marine area at that time as well, though this is generally not included in statements about its early size and was essentially ignored at that time.

The decree that created Costa Rica's first dry forest national park was blemished by mistakenly viewing the pastures and old fields populated by introduced grasses (largely *jaragua* or *Hyparrhenia rufa* from East Africa) as "savannas," seemingly and romantically analogous to African grasslands. In this context, the annual anthropomorphic dry season fires seemed almost natural, and besides, the fires were something so large, that a few no-budget "*guardaparques*" could not even dream of eliminating. At best they could prevent the fires from consuming buildings. Restoration was not even a management concept, though of course it began to occur by default on the prior farmland and pastures within the park close to the Casona. The PNSR staff focus was on stopping the hunting, which seemed to be as much an insult to the "King's Garden" as it was biologically undesirable, on removing the colonizing farmers/squatters, stopping further logging, and simply surviving. As these things gradually came under control, the focus shifted to expelling the 2000+ cattle still foraging throughout Parque Nacional Santa Rosa. This expulsion was both because "it is a national park" and because the cattle and horses symbolized people, something that was anathema to national park systems everywhere at that time (and that regularly led to the destruction of buildings, roads, fruit trees, etc., on the farms that were incorporated into lands of the emerging national park system).

A landmark court decision was won by Costa Rica's *Servicio de Parques Nacionales* (SPN) in 1976, telling the owner along the south boundary of Sector Santa Rosa that he had to remove his cattle from the park or they would be shot. About 1,000 cattle were removed, and the Guardia Rural came into Santa Rosa and shot 1,000+ cattle in 1977. Every vulture in Costa Rica appeared to be working

Fig. 10.13 Cerro El Hacha, the core of a 6 million year old volcano that once was an island in Lake Nicaragua, as viewed from the fire tower lookout on Cerro Pedregal on the upper slopes of Volcán Cacao. Its yellow upper portion is covered with a native grass *Trachypogon plumosus*, a population expanded by man-made fire from small populations once occupying natural disturbance sites. The yellow flatland pastures are jaragua grass (*Hyparrhenia rufa*) introduced from East Africa.

Fig. 10.14 Estación Argelia, the administration station for Playa Naranjo in Sector Santa Rosa, and originally the house for the large salt extraction works in the estuary behind it (July 1973). Note the cleared hillside dry forest behind the station, the kind of clearing that characterized all of the slopes near the ocean, slopes that were probably also cleared of their old-growth forest centuries earlier through ocean-access logging.

in Santa Rosa, and this was the only time that nesting vultures have been encountered there. At this time, there was no effective fire control program in Santa Rosa. Removal of the cattle was the removal of the biotic mowing machines. For centuries before this time, there had been a crude equilibrium between the highly fractured patches of dry forest, the closely grazed pastures, and the human-set fires (Janzen 1988c, 2002). Each annual fire burned the closely cropped pastures and trimmed the edges of the forest/secondary woody succession (Figs. 10.16 and 10.17).

Fires in dry years (or later in the dry season) burned further into the forest margin, expanding the pastures. Fires in wet years (or earlier in the dry season) did not kill the woody marginal vegetation, thereby shrinking the pastures. If an area did not burn for several years, the fire was more severe when it did arrive. When the cattle were removed, the result was instantaneous and highly predictable. The African grass grew to dense stands 1–2 m in height (Fig. 10.18), and the fuel for the annual fires was sufficient to create a wall of flame 1–4 m high that gobbled up the forest margins in huge bites (Fig. 10.19 and 10.20). Trees and patches of forest that had coexisted with the cattle-grazed grass-fires were incinerated, and fires burned downwind all the way through the understory of dense forest. By 1984 it was obvious that the multiple-aged stands of variously aged secondary dry forest and even the tiny fragments of "original" forest (old-growth) would be burned to total elimination if something was not done (Janzen 1988d, e). A fire elimination program was initiated by visiting research biologists and Santa Rosa staff. The current Santa Rosa and ACG status of being largely covered with various stages of woody succession vegetation is a direct result (compare Fig. 10.21 and 10.22). There are no natural fires in a "natural" ACG. There are no lightning strikes in the dry season; a rainy season lightning strike may set an individual *veranillo*-dried pasture or dead tree on fire, but that fire does not burn past the limit of the dry spot. The Costa Rican habit of labeling fires as "forest fires" is a misnomer, since intact living ACG forest basically does not burn; what burns is the pasture grasses and low herbs, with the heat killing tree seedlings and other young trees, thereby allowing more sunlight, which in turn generates more dry season fuel for the next fire.

Fig. 10.21 Jaragua-forest edge that was characteristic of tens of thousands of hectares of Mesa Santa Rosa of central ACG dry forest at the beginning of the restoration process (December 30, 1980) and burned every 1–3 years. This pasture is at least 200 years old and was previously populated by native grasses. The forest to the left and background is encino (*Quercus oleoides*) old succession following logging and burning. Compare with Fig. 10.22.

intensity of the late July-August short dry season is highly variable (see Fig. 10.25 for an example of a very poorly developed *veranillo* in 2006). The short dry season is a time of pupal dormancy and adult sexual inactivity among Santa Rosa moths and butterflies, a dormancy that is generally broken by the second peak of heavy rain in early September (Janzen 1987, 1988f, 2004). If the short dry season is exceptionally dry and hot, moth pupae that would normally have remained dormant until the following May may eclose with the arrival of the September rains. For example, adults of the moths *Xylophanes turbata*, *Manduca dilucida*, and *Manduca lanuginosa* (Sphingidae) may be (unusually) caught in a September light trap. They usually remain as dormant underground pupae from June-July all the way through to the following May. This is just one of many small suggestions that the drying and warming of the Sector Santa Rosa dry forest under climate change will change its overall biology.

The tapering off (December-January) of the long rainy season into the beginning of the long dry season is highly variable in intensity, depending on how much rain fell during the total rainy season, and its pattern. The duration of dry season leafiness is also impacted by the intensity of the strong tree-level December-to-mid-March trade winds that blow across Santa Rosa after having passed over the Cordillera de Guanacaste. When they abruptly stop in March, the temperature climbs rapidly to the largely wind-free and intense dry season 33–35°C daily maxima temperatures that last until the long-rainy season begins in May. However, as mentioned above, about mid-March this increase in temperature often brings 1–2 days of early afternoon strong rain showers, a "false beginning" that seems to deceive no species but can soak unwary campers expecting a rain-free dry season.

The semi-orderly seasonality described above, created by the combination of the orderly seasonal passage back and forth of the thermal equator across ACG, the proximity of the Pacific Ocean, and the rain shadow of the Cordillera de Guanacaste, is strongly disrupted by hurricanes in either ocean. A Caribbean hurricane brings 2–5 days of heavy overcast to the Santa Rosa dry forest, and frequent heavy rain intermingled with lighter showers and drizzle. If combined with either the May-June or September-October rainy peaks, the Santa Rosa seasonal watercourses can contain major temporary streams and rivers (Figs. 10.28 and 10.29). A Pacific hurricane can be even more intense, filling almost all seasonal watercourses to their brim and overflowing (Figs. 10.30 and 10.31). However, it is noteworthy that (apparently) owing to their dense and relatively undisturbed forest cover, even during a hurricane the watercourses of the upper slopes of the ACG volcanoes flow strongly but do not rise substantially, with the flooding and major erosion being more noticeable at lower elevations where there is massive lateral surface run-off from pastures and saturated young secondary forest (e.g., Figs. 10.28–10.31).

Moving away from the Santa Rosa "classical" dry forest climate on the Mesa Santa Rosa (at 250–300 m elevation) towards the ocean yields no major climate surprises or disjunctions. Owing to the high soil porosity of Península Santa Elena, the coastal portion of the ACG dry forest ecosystem is conspicuously drier, as expressed by the morphology and species of its biodiversity (see general ecosystem aspect in Fig. 10.32). To the north, the climate gradually moistens; Piedras Blancas on the border with Nicaragua had a tree species composition (prior to deforestation) that was similar to that of Cabo Blanco and Puntarenas to Carara (Fig. 10.33), complete with *Attalea rostrata* (formerly called *Scheelea rostrata*) palms and espavel trees (*Anacardium excelsum*). This great expanse of dry forest to rain forest intergrade between the Pacific and Lake Nicaragua and the La Cruz area, and then eastward towards Santa Cecilia, has been so thoroughly destroyed that it is difficult to imagine what it was originally. To the south, toward Liberia, the current species composition of insects suggests that the climate was somewhat more moist than is Sector Santa Rosa, but it is very hard to distinguish that possibility from the effects of less wind, better soil, and the more moist flood plains that are today occupied by pastures and crops.

Fig. 10.22 Exactly the same view as Fig. 10.21, after 20 years without fire and absence of livestock (November 4, 2000). The canopy of encino is still visible and Winnie Hallwachs' hand (center) is positioned at a height of 2 m. The isolated jicaro (*Crescentia alata*) tree in Fig. 10.21 is completely shaded by this new forest (and therefore dying). The bulk of this young forest is wind-dispersed yayo (*Rehdera trinervis*, Verbenaceae) intermixed with another seventy wind- and vertebrate-dispersed woody species. Such invasion by forest is characteristic of many tens of thousands of unburned hectares of ACG pasture. Though moderate grazing speeds up forest invasion (and reduces the damage by escaped fires), livestock also does considerable damage to watercourses, and cowboys and their dogs cannot resist hunting.

Fig. 10.23 Forty-year-old dry forest in the rainy season (June), Sector Santa Rosa; same view as Fig. 10.24.

Fig. 10.24 Forty-year-old dry forest in the dry season (April), Sector Santa Rosa; same view as Fig. 10.23.

the same mixing pot rather than allowing the fine-scale species-to-species contacts that must have originally occurred when all was old-growth forest and the disturbances were "natural" (Janzen 1986a, b).

The upper elevations of the Cordillera de Guanacaste were very likely to have been lightly glaciated in the last Pleistocene cool period, at its maximum about 20,000 years ago. Volcán Orosí, at 20,000 years of age now, would have been newly formed then, and the other older volcanoes would have been in a state of becoming vegetated. Narrow altitudinal bands of different kinds of tropical forest would have marched up the slopes, from the most cold-intolerant lowland dry forest (Pacific side) and rain forest (Atlantic side) up to the most cold-tolerant evergreen cloud forest at the top. However, at the peak of the last glaciation, sea level was about 100 meters below current levels, so there would have been more linear space over which to spread that gradient at least on the Pacific dry forest side.

As the climate gradually warmed, the continuous (across the Cordillera) mid-elevation bands would have moved upward and become fragmented into upper elevational (and progressively smaller) islands by the mid-elevation valleys between them. For example, one of these valleys is today the Quebrada Grande-Nueva Zelandia-Dos Ríos mountain pass. Today, these upper elevation insular cloud forests (visible as blue area in Fig. 10.2), and their biodiversity, are being literally cooked off the tops of the Cordillera as global warming heats (and dries) this part of Costa Rica (and other tropical mountains as well). Equally threatening to these ecological islands, the upper elevations are being reduced in size and quality as dry season refuges for birds and flying insects of the lowlands that spend parts of the year at cool, upper elevations (e.g., Hunt et al. 1999; Fig. 10.34). Perhaps even more unfortunate, those portions of cloud forest biodiversity that can tolerate the increased temperatures and loss of rain/cloud cover are now being subjected to invasion

Fig. 10.28 Río Tempisque at Potrerillos, Inter-American Highway, ACG margin and collecting all ACG Pacific run-off from the western side of Volcán Cacao and Volcán Orosí; normal rainy season level (July 22, 2003); compare with Fig. 10.29.

Fig. 10.29. Río Tempisque at Potrerillos, Inter-American Highway, ACG margin when collecting all ACG Pacific run-off from the western side of Volcán Cacao and Volcán Orosí; flood level during a hurricane from the Caribbean (October 1999); compare with Fig. 10.28. Photograph by Felipe Chavarría.

of novel competitor and predator/parasite regimes from the lowlands. For example, the army ant *Eciton burchelli* and other species of ants are now foraging all the way to the top of Volcán Cacao (1,650 m) and directly attacking dormant *Polistes* (Hunt et al. 1999) from the lowlands, whereas in 1985 there were no visible ants at all at 1,000 m and above on the same volcano. A particularly dramatic case of this impact is probably taking place now in the small patches of inaccessible remnant cloud forest at 700 m elevation at the mountaintops of Península Santa Elena, and along the upper elevations of the Rincón de la Vieja massif.

The general long-term climate-change trend of increasing rainfall in the tropics is concentrated well to the south of Costa Rica (Adler and Gu 2007) but apparently moving slowly northward to engulf the region ACG to El Salvador (Sachs and Myhrvold 2011). In ACG as a whole, global climate change is warming and drying ACG—all ecosystems—and reducing the total amount of water arriving by rainfall, as well as reducing the amount condensing on the vegetation in the upper parts of the Cordillera de Guanacaste. This is reducing annual and seasonal river flow from the volcanoes, especially in the dry seasons. When the dry forest rainy season is shorter, less intense, and more erratic in start and end dates, there are very visible effects on the biodiversity. At the same time second order consequences are rampant, such as litter decomposition happening less thoroughly in a dry rainy season, presumably affecting all of the organisms that depend on litter recycling, as well as leaving more fuel for a dry season anthropogenic fire.

As the dry season advances, seasonal watercourses flow less and leave temporary pools of shorter duration. This curtails the activity of aquatic species and reduces the capacity (and ranges) of water-dependent species. As ACG dries, there is a reduction in the amount and diversity of the annual rainy season invasion of the dry forest by "rain forest species," species that generationally move from east to west during the rainy season. There is also a reduction in the number of species that survive multiple years in local small moist habitats as residual fragments of once larger populations. Particular weather events that are used as cues by dry forest biota for reproduction, dormancy, migration, growth, and other periodic events, such as the timing of the beginning of the May rainy season, or cool windy weather in November, may easily be out of phase with other processes or species needed as mutualists or prey/hosts. Particular seasons needed for fat storage, fledgling-friendly learning, growth, migration, dormancy, etc., may not be as long as needed, may be de-synchronized with other events, or may even be too long. The occasional hurricane or other intense tropical storm that dumps 0.5 to 1.5 m of rainfall over a few weeks certainly does not compensate for the general de-synchronizing and drying that is taking place, even if it results in the annual average rainfall staying about the same year by year (e.g., Fig. 10.35).

In theory, and probably in practice, part of the salvation of the ACG dry forest biodiversity in the face of climate change is in having the opportunity to move from west to east, moving up the moisture gradient and to some degree, up to cooler elevations as well. This has been a major part of the rationale for expanding ACG to the wetter east during the past decade (http://janzen.sas.upenn.edu/RR/Rincón_rainforest.htm) and is currently the impetus for further eastward expansion (http://janzen.bio.upenn.edu/saveit.html; http://www.gdfcf.org). However, the warming also means that the species at the uppermost and the easternmost portions of the moisture/elevation gradient have nowhere to flee, since the upper portion is bounded by air, and the eastern section is bounded by cultivation and lower elevations. A part of future management discussion in ACG will be whether to contemplate moving species from the 1400–1600 m elevations on the tops of Volcán Orosí and Volcán Cacao to upper elevations at the Rincón de la Vieja complex (up to 2000 m), a sort of local "Noah's ark rescue operation" or whether to simply let their Orosí and Cacao populations die out (assuming that these species are already on Rincón de la Vieja), as would happen with the smaller and lower islands in an oceanic archipelago following a rise in sea level. This "*conservation* conversation" will need to be part of the larger discussion (e.g., Thomasa 2011) of what to do when some other more distant conservation area wishes to view the elevated parts of Cordillera de Guanacaste as a refuge, a conversation that belongs in the yet larger discussion of what to do about the species, habitats, and minute ecosystem fragments that will gradually disappear as the entire ACG moves over centuries through its successional stages to become once again old-growth forest—and a new one at that, in view of climate change.

Returning to the question of "what is the impact of climate change on ACG?", there is no straightforward reply—other than "perturbation." While the climate is unambiguously changing as perceived by both humans and others living there, and by weather station records, the multiple ACG dry forest biotic changes that are evident even without looking explicitly for them have at least four major causes. All four causes are intertwined and interinfluencing: (1) climate change per se, (2) climate/habitat change from the woody succession/restoration/de-fragmentation taking place throughout all ACG ecosystems (many species become common, many other species become rare or much less ubiquitous), (3) insularization of ACG as a whole (yes, the species richness of its final equilibrium will not be as great as it once was when it was part of an unbroken expanse of natural ecosystems) and occupied many of its Holdridge Life Zones (Fig. 10.33), and finally, (4) the gradual dying out of the "living dead" individuals, populations, and ecosystems (Janzen 2001) that characterize the biodiversity that seems to be surviving in the surrounding agro-scape (and seascape). Additionally, other macroinfluences will arrive (e.g., invasive alien species, overwhelming human desire for targeted resource extraction [especially water and geothermal energy], evolutionary change in the fragments of once-widespread species, sea-level rise, etc.). It is a striking and inconvenient characteristic of ACG that whenever a particular conservation topic—biofuels, introduced species, climate change, fragmentation, territorial expansion, water extraction, whatever—is put in the spotlight, it cannot be discussed and treated in isolation from the other challenges and traits. An integrated solution or action must be sought, yet simultaneously management has to resist the temptation to let all considerations tangle up each other to where there is no action. This means that solutions need to be integral—the so-called "ecosystem approach" (Janzen 2000a, 2000b), both biologically and sociologically—yet also take specific and rapid actions for a specific problem even if, because of the other considerations, the solution is not a perfect solution. When the victim is bleeding to death, you stop the bleeding, even if it leads to scars and amputations down the road.

Dry Forest Seasonality

The most visibly outstanding trait of ACG dry forest is its strong seasonality, briefly alluded to above (and see Fig. 10.23, 10.24, 10.25, 10.30, 10.31, 10.33, and see

Fig. 10.30 Cafetal, headwaters of Río Cuajiniquil, a typical ACG seasonal river bed during the dry season (March 15, 1999). Compare with Fig. 10.31.

Herrera, chapter 2 of this volume). Dry forest seasonality, and its impact on organisms, merits many books full of descriptions, analyses, and discussion on the topic. Here we touch on just a few of what have been some of the more startling seasonal aspects to us as biologists, with an extra-tropical upbringing, living in ACG in all seasons of the year.

Seasonality of Insect Migration

When the heavy rains begin in mid-May in ACG dry forest (Fig. 10.25), a large number of insects abruptly appear in any census system—moths and beetles in light traps (Fig. 10.36), wasps and flies in Malaise traps, caterpillars on foliage, defoliation by caterpillars and beetles, caterpillars brought to nestling birds, wasp nests being built, etc. These peaks were previously spoken of as the result of "hatching with the rains," but it is much more complex than that. As each species is investigated in detail, each is found to have its own complex relationship to the breaking of the long (and hot) dry season. These sets of relationships become progressively harder to summarize the more that is known.

However, all of them have to a great degree the phenology of their lives tied to the seasonality of the beginning of the end of the long dry season and the beginning of the long rainy season (Janzen 1983b, 1984a, 1987, 1988f). As an example, the large number of moths that appear in a light trap in the period from early May (1–2 weeks before the rainy season) through the end of May (two weeks after the beginning of the rains), have three principal origins. There is a set of species that have been dormant pupae suspended in silk cocoons (a few such as *Rothschildia erycina* and *R. lebeau*, *Automeris zozimanaguana* [formerly *A. zugana*], and *A. tridens*, some Megalopygidae and Limacodidae) on the

Fig. 10.31 Cafetal, headwaters of Río Cuajiniquil, a typical ACG seasonal river bed during the full rainy season strongly enhanced by the edge of Hurricane Mitch (October 30, 1998). Compare with Fig. 10.30.

foliage. There is a set that pass the dry season as dormant pupae in the litter/soil (mostly) since the previous June (e.g., *Manduca dilucida*, *M. lanuginosa*, *Xylophanes turbata*, *X. juanita*, *Schausiella santarosensis*, *Arsenura arianae*, *Eutelia furcata*, *Euscirrhopterus poeyi*, *Neotuerta sabulosa*, *Dysodia speculifera*, *Holochroa ochra*, *Protographium epidaus*, *P. philolaus*, etc.). In entomological terminology many of these species are univoltine, meaning that they have just one generation per year. These species are "migrating" into the rain-soaked dry forest at the beginning of the rains from a dry forest microecosystem—a pupal or prepupal dormancy chamber underground or in the litter. They have been hiding for 11 months, "hoping" not to be found by a predator or parasite. Their cue for eclosion ranges from extreme heat, and perhaps an internal calendar, to the cooling that comes from the first rain. Several of them are conspicuous in emerging as adults days to weeks before the actual rain (Janzen 1988f), indicating the ability of the pupa to "anticipate" the rains. Next, there are many species that have been in the forest as adults in reproductive stasis (in diapause or aestivation in entomological terminology), hiding as inactive adults in crevices in tree bark and other nooks and crannies. When the cool and moist nights of the rains come, or some other associated weather cue arrives, they "turn on" to fly, mate, and search out food plants on which to lay their eggs, thereby beginning a period of 1–2 to 3–4 generations during the rainy season. The last generation of adults to eclose at the end of the long rainy season in December-January again hides to pass the long dry season (or sometimes the adults both hide and search for food in rotting fruit). This group contains many species of small Noctuidae and Erebidae, with many species of *Eulepidotis* and *Letis* being conspicuous examples; but butterflies such as *Eunica monima*, *Myscelia pattenia*, *Eurema daira*, *Urbanus proteus*, *Callicore pitheas*, *Archaeoprepona demophon*, and *Historis acheronta* do the same (Janzen 1983b). Again, many of the species of moths of this ilk appear in a dry forest light trap 1–10 days before the rains actually begin.

And third, there is a set of species that arrive by migration from the wetter side of ACG—from the rain forest, cloud forest, and intergrades (Janzen 1988g). These species left the ACG dry forest at the end of their first generation in late June to July, or in a few cases at the end of the rainy season, apparently flying to the "other side" of Costa Rica to be sexually inactive (but feeding) adults and to return 10 months later, or to have one or more generations in the rain forest to cloud forest before their offspring return to the dry forest at the beginning of the following rainy season (Janzen 1988b, Haber and Stevenson 2004). Some members of some of these populations apparently remain in ACG dry forest feeding at flowers for a while after the caterpillar season has passed, and sometimes have a light second generation during the September- November second heavy peak of rains (e.g., the sphingid moths *Pachylia ficus*, *Xylophanes chiron*, *X. anubus*, *Aellopos fadus*, *Enyo ocypete*). This cycle of seasonal migration is being viewed here as seen from the dry forest–centric viewpoint.

Alternatively, some of these species may be viewed as a rain forest species that, at the beginning of the rainy season, expands into the dry forest to have a large generation at a time of superabundant edible foliage and minimal populations of predators and parasitoids (that are satiated by the very large synchronized peak of prey), and then "retreats" to the rain forest for subsequent generations.

However, the situation is yet more complex than it appeared in the first several decades of study of Santa Rosa dry forest insects. The insect and caterpillar inventory (Janzen 1996, Janzen et al. 2009) of ACG was very pleased to find

Fig. 10.32 On the horizon, the dark evergreen encino (lowland oak: *Quercus oleoides*) forest on the splash of white volcanic tuff (see Fig. 10.3) overlying the Santa Elena Peninsula serpentine vegetation (scattered green to yellow to brown to gray deciduous) at La Angostura (March 25, 2002).

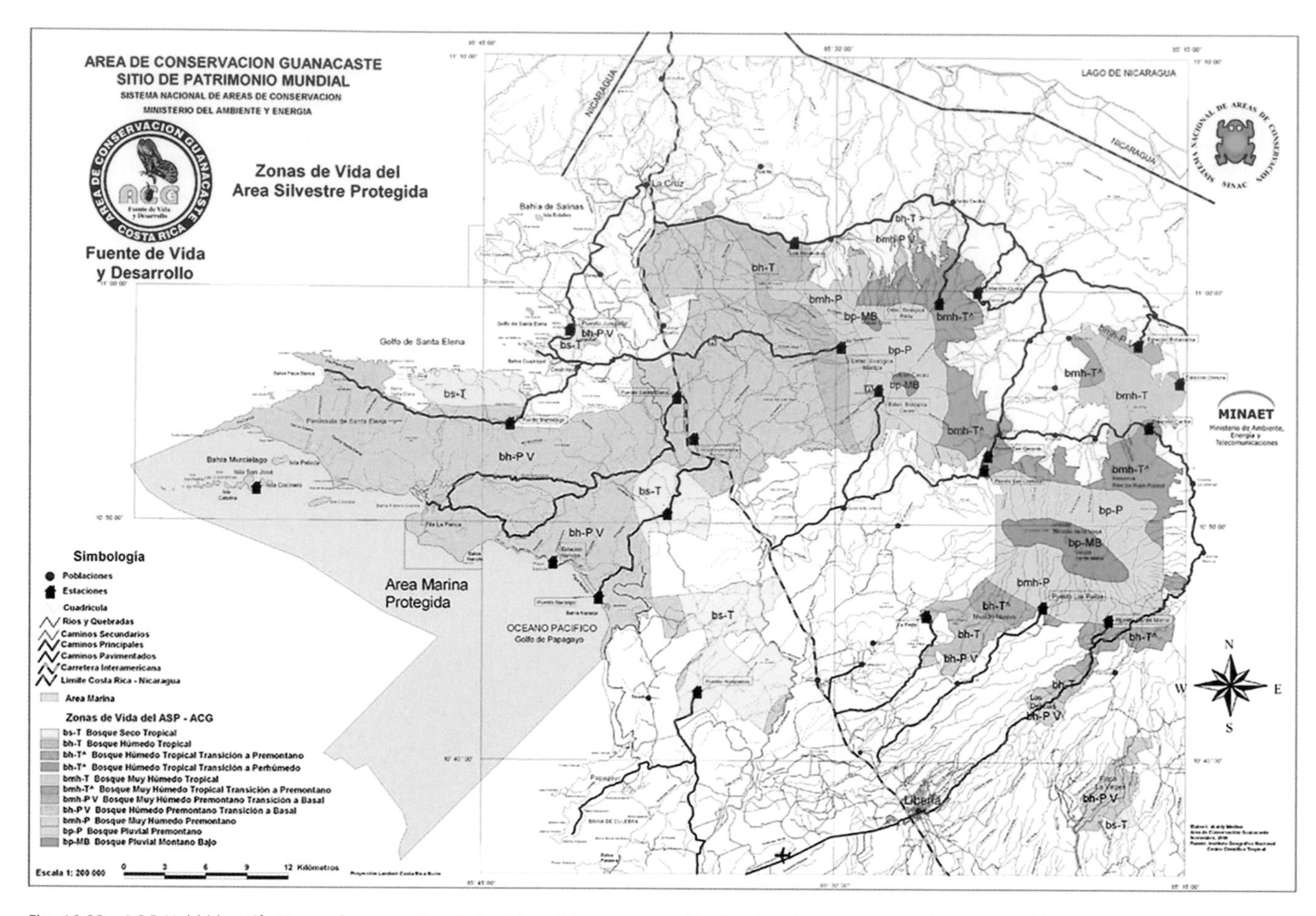

Fig. 10.33 ACG Holdridge Life Zones, demonstrating their wide variety as expressed in the changing transect running westward from the dry Pacific sector to the cloud forest and then on the Atlantic rain forest.
Source: Waldy Medina, September 2007.

many "dry forest species" of Sphingidae caterpillars in the ACG rain forest at the time of year when they have disappeared from the dry forest after their first generation in May-July. However, when adults reared from all of these species were DNA barcoded (Janzen et al. 2005, 2009, Holloway 2006), it was found that *Xylophanes porcus* and *X. libya,* for example, were in fact each two populations (= two species) (other species do appear to be just one population, migrating as previously believed). While *X. porcus* does not seem to have a morphological trait to distinguish the dry forest population from the rain forest population, *X. libya* does. Its dry forest specimens are lighter in color, a trait that would normally have been viewed as an ecophenotype, except that DNA-identified adults of the dry forest population reared from wild-caught rain forest caterpillars are just as light in color as are its conspecifics from the dry side of ACG. This means that when the dry forest *X. libya* caterpillars disappear from the dry forest for 9 months of the year, the population may in fact not be gone but rather, may be simply invisible to our trapping methods (not being attracted to lights, and not generating caterpillars). Alternatively it might be living in the moister interface between dry forest and rain forest at the western base of the volcanoes. Only more inventories will determine what is happening. However, when the adults caught at lights over the years—specimens accumulated by the INBio national biodiversity inventory and of unknown caterpillar origins—were then DNA barcoded (see Janzen et al. 2005), it was found that there are in fact three *X. libya* populations (= species) of free-flying adults—those of Pacific dry forest, those of Atlantic rain forest at intermediate elevations, and those from Atlantic rain forest low elevations. And, adults of all three have been collected from a single light trap at Estación Biológica Pitilla, which is Atlantic rain forest at an intermediate elevation (700 m), but within a few tens of kilometers of both Pacific dry forest and Atlantic lowland rain forest on the northeastern side of Volcán Orosí-Orosilito. There is much more complexity to both the species and migrations of *X. libya* than meets the eye, and this is probably the case with many ACG species of moths and butterflies.

A seasonal migratory phenomenon that is very visible to humans is the annual single generation of a set of species of large yellow and white pierid butterflies. It begins with the arrival in the dry forest of single and unnoticed females (and perhaps males, but they have not been censused) of a group of species during the 1–2 weeks bracketing the beginning of the rains, species that lay their eggs on newly expanded leaves (e.g., *Phoebis philea* and *P. sennae* on *Senna* spp.; *Anteos maerula* and *A. clorinde* on *Senna atomaria*; *Aphrissa statira* on *Callichlamys latifolia* and *Xylophragma seemannianum*). These species of butterflies are also present in low to intermediate elevation rain forest, where their caterpillars have been occasionally encountered by the ACG caterpillar inventory at all times of year (adults reared from these caterpillars have the same DNA barcode as those of the May-June generation in ACG dry forest). The dry forest generation of adults ecloses in large numbers in the last half of June and then both sexes are encountered, flying east in large numbers, back to the rain forest. Many hundreds per hour of 5–8 species may pass a single point on the Sector Santa Rosa entrance road. For the remainder of the rainy season, an occasional caterpillar of one of these species (most commonly, *Phoebis philea*) is encountered in the ACG dry forest, and the occasional adult is seen. During the long dry season, only the very occasional adult is encountered (and if they were not large and yellow, their density is so low that they would not be noticed). To complicate matters, as woody succession—also known as forest restoration—has progressed in ACG dry forest, the density of these species per year has steadily declined with the shrinking populations of their caterpillar food plants, all of which are short, shrubby, and shaded out by the growing trees. In the same dry forest, two bright orange nearly identical cryptic species, *Phoebis argante* and *Phoebis hersilia*, are having nearly year-round continuous generations on their evergreen (and somewhat riparian) food plants—*Inga vera* and *Zygia longifolia* (food plant records in Janzen and Hallwachs 2011, discussion in Janzen et al. 2009).

A conspicuous example of ACG dry forest seasonal migration and reproductive dormancy is displayed by the (previously) common social wasp *Polistes instabilis* (Hunt et al. 1999; Fig. 10.34) that is a predator on lowland caterpillars. This wasp builds open-faced *papier-maché* nests ("lengua de vaca") initiated by single females (or perhaps small groups of females as well) hanging on twigs within dense foliage 1–2 m above the ground in the lowland ACG dry forest at the beginning of the long rainy season (these nests were also commonly built under the eaves of buildings). The wasps actively seek caterpillars, chew them into pieces, and carry the pieces to their larvae. The nest size grows rapidly to where there can be 10–30 wasps at a nest by the end of June. Throughout the remainder of the rainy season, the nest grows more slowly, but may have as many as 100 wasps by December. Sometime in November-January, the nest is abandoned and the wasps fly to the cloud forest at 1,000+ m elevation on Volcán Cacao (Hunt et al. 1999). Thousands of wasps, mostly females with males scattered among them, aggregate in hollow trees (and buildings when available) in the cold foggy climate to inactively pass the long dry season. In effect, they have put themselves in a refrigerator. On hot days, some become active and fly out of the cavities—not to forage but "sunning" on foliage. When the rains begin in Santa Rosa, the wasps abruptly disappear from the cloud forest, and appear in the dry forest in the lowlands below, building their nests. As the clouds have moved up Volcán Cacao during the past 20 years of climate change, wasp-occupied hollow trees are now encountered higher up on the volcano, and with fewer wasps in them. Equally bad for the wasps, *Eciton* army ants (major predators on *Polistes* wasp nests) have now reached the (warming) very top of Volcán Cacao, thereby adding yet another dry season challenge to the wasps' survival. Over the same period, rainy season nests have become noticeably scarcer in the lowland dry forest. Very few *P. instabilis* individuals remain in the dry forest during the long dry season; while active nests have not been located, the adults are encountered drinking water from waterholes.

While this obvious reduction in *P. instabilis* wasp density is likely due to the shrinking size and quality of the volcano-top "refrigerator" in which the wasps hibernate, it may also be caused by the general decline in caterpillar density being experienced by the Santa Rosa dry forest over this same period. It may even be associated with cooler temperatures at the more strongly shaded ground-level dry forest microhabitat occupied by these wasps in the rainy season, as forest succession advances; cooler temperatures lead to slower larval growth and hence smaller population sizes. While this wasp displays unambiguous seasonal migration in ACG, exactly analogous to monarch butterflies *(Danaus plexippus)* in their Canada to Mexico annual seasonal migration, it does have a large neotropical dry forest distribution from Mexico to Panama (http://zipcodezoo.com/Animals/P/Polistes_instabilis.asp). In all of its range it is within a few tens of kilometers of higher elevations where it could be passing the long dry season.

However, it is also the case that a few active nests of *P. instabilis* survived the dry season in Sector Santa Rosa in the 1980–1990s, at a time when the overall rainy season nest density was easily 100-fold greater than at present. These could be the breeding stock for quick selection for a non-migrating population if the volcanic refrigerator disappears, and the lowland caterpillar density remains high enough for

Fig. 10.34 Caterpillar-hunting *Polistes instabilis* (Vespidae) social wasps from Sector Santa Rosa dry forest passing the dry season aestivating/hibernating in a cavity (building) in cloud forest at 1,150 m on Volcán Cacao (Estación Biológica Cacao, January 18, 2007); these wasps are mostly female individuals, which are derived from many hundreds of small nests in the lowlands in the previous rainy season. They will return to the lowlands at the start of the next rainy season to solitarily found new nests.

them to maintain a rainy-season population large enough to have enough nests for some to make it through the largely caterpillar-free long dry season (the dry season also means heavy nest predation by white-faced capuchin monkeys and *Eciton* army ants, and no free-standing drinking water over large areas). Alternatively, these dry season nests may have been aberrant, sick, or otherwise dead-end members of the population, and offer no genetic escape for a species locked into seasonal migration to the volcano cloud forest as a way of escaping the inimical dry time of the year.

Of Plant Reproduction and Plant-Animal Interactions
Reproductive dormancy by adult organisms is most vividly displayed by ACG dry forest trees. Each species flowers, and matures its seeds, at quite specific periods of the year, many in the dry season and some in the rainy season. They display strong within- and between-year synchronies as well. There are three major seasonality-derived arrays of evolutionary drivers behind this phenology: (1) competition among vegetative individuals during the rainy season selectively favors minimal use of product and reserves for flowering and seeding at this time; this is the time when vegetative growth is most feasible and necessary in crown-to-crown and root-to-root competitive interactions (e.g., Janzen 1967a, 1978, 1982d); (2) seasonally influenced (largely) co-evolutionary interactions with pollinators, seed dispersers, and seed predators evolutionarily drives reproductive phenology; and (3) the "best" physical conditions for seedling and flower/fruit growth select for timing of reproduction. These three drivers and their interactions generate an incredibly diverse array of species-specific and ecosystem-specific patterns of reproductive, as well as vegetative, dormancy. When climate change and microclimate change generated by succession is added in, the patterns and the departures from "one size fits all" become even more complex.

The reproductive dormancy of a large ACG dry forest tree, *Hymenaea courbaril* ("guapinol"; Fabaceae, the bean family) offers multiple examples. The flower buds are visible immediately following the January shedding of the 11.5-month-old leaf crop and its replacement with a new set of leaves during 1–2 weeks; in other words, despite it seeming to have leaves year-round, *H. courbaril* is not evergreen as an adult. The visible flower buds are produced in January-February (early dry season). Following massive bud predation by the larvae of three species of *Anthonomus* weevils (Clark 1992; *Anthonomus* is the genus of the cotton boll weevil and in the Curculionidae), the buds open to flowers that are visited and sometimes pollinated by *Glossophaga* bats (see Teixeira de Moraes and Sebbenn 2011) in late February to early April. On most years, no fruits are set from these (perfect) flowers. The trees are essentially acting as males, presumably with their pollen being carried by the far-flying bats (at least up to 7 km), to distant trees that will bear a fruit crop that year. On those "male" years, the tree is not investing in the cost of fruit production. At long intervals a patch of *H. courbaril* has many members that bear a fruit crop of 100–2,000 (large) fruits per individual. The last massive and synchronized *H. courbaril* fruit crop in Bosque Húmedo of Sector Santa Rosa dry forest was in 1983, over 33 years ago. In 2007 a small patch of trees on the Sendero Natural in Sector Santa Rosa had their first large crop in 24 years (and some may well have been pollinated by flowering "males" in the Bosque Húmedo 3–4 km distant). In a fruiting year, the young green fruits gradually accumulate bulk and reserves until the end of the long rainy season. The mature fruits fall from the tree in late December to early February (a full eleven months after flowering). Once on the ground, these hard large megafauna fruits (Janzen and Martin 1982) are today gnawed open by agoutis (*Dasyprocta punctata*) and their seeds both preyed upon and dispersed/cached underground (Hallwachs 1986).

Throughout tens of millions of years (Neotropical commercial amber is the 30–40 million-year-old fossil resin of *H. courbaril*) the dispersal coterie of *H. courbaril* also included animals big enough to molar-crunch the hard fruits and swallow the hard large seeds entirely, which in turn would generate a quite different seed shadow than occurs today with agoutis being the primary dispersal agent (Hallwachs 1986). While the fruits are full-sized but still in the canopy and protected by a thick resin-rich rind (fruit wall), two species of *Rhinochenus* weevils (Curculionidae) oviposit through the fruit wall and the large larvae prey on the seeds (Janzen 1974, 1975).

Each year a few trees produce 1–20 fruits, presumably "physiological accidents," and this steady low dribble of fruits is what sustains the *Rhinochenus* population as well as being one of the many species of seeds and fruits that sustain the agouti population. Throughout this cycle there are seasonal aspects. Fallen fruits (dropped to dispersal agents) are harvested (and their seeds dispersed) at very different

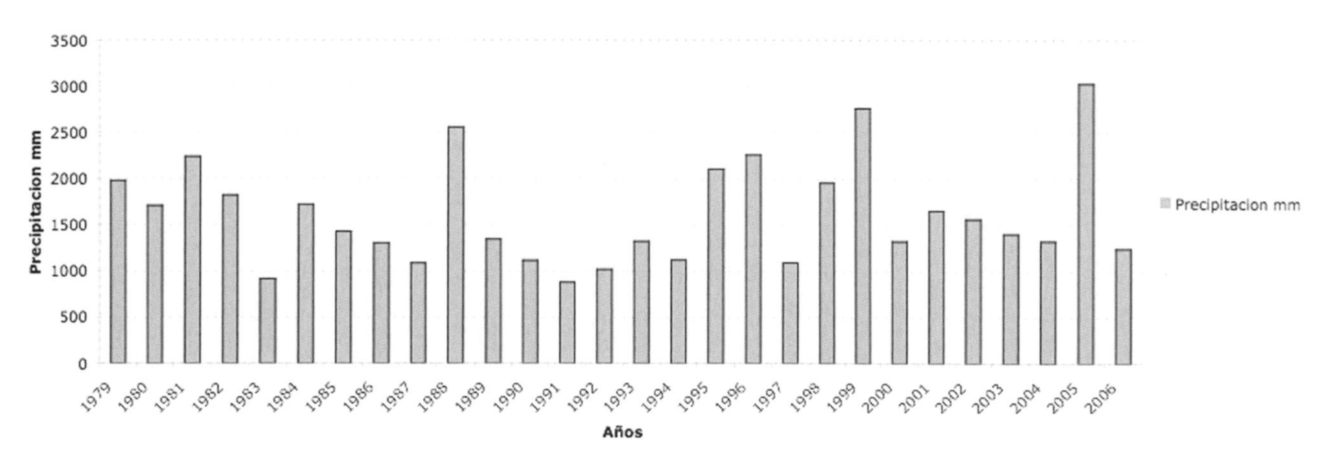

Fig. 10.35 Total rainfall recorded per year by the weather station at Administrative Area of ACG in Sector Santa Rosa dry forest. To appreciate the general dessication of this dry forest rainy season input, one needs to subtract the heavy load (0.5 to 1 m) of rainfall occurring over a few days during the event of a hurricane in the vicinity (e.g., 1988, 1999, 2005), and recognize that after the mid-1980s, the beginning of the rainy season has been characterized by sporadic rains, rather than many consecutive days of soaking rainfall as in the past

Fig. 10.36 Moths attracted to battery-powered fluorescent lights in ACG dry forest, just four days after the beginning of the rainy season (May 19, 2007). There are at least 300 species of moths captured in this photograph, some of which have migrated into the dry forest from cloud and rain forest. Some of these moths have very recently eclosed from long-dormant pupae, and some have become active after passing the dry season as quiescent adults.

rates by agoutis in the rainy season and the dry season, presumably both because of alternative food availability and because of the loud (and revealing) sound of gnawing into one of the very hard fruits in the dry season (Hallwachs 1986). The tiny *Anthonomus* weevils have just one February-March generation and then immediately emerge as adults from the aborted fallen flower buds. They then have to hide somewhere in a crevice in tree bark or in a dry rolled leaf for the remainder of the dry season and all of the six-month rainy season, and then into the beginning of the following dry season before there is again an oviposition site available in a flower bud. Flowering in the rain-free dry season obviates having physiologically rain-resistant flowers and pollen, and occurs at a time of the year when there are no other known bat-visited flowers in ACG old-growth dry forest. The tree is also investing flowering resources in the dry season, when there is no threat of crown-to-crown vegetative competition with other large trees. Avoidance of inter-crown competition may well be why *H. courbaril* has been selected to dump its old leaf crop and grow a new one in the early dry season (as opposed to some other time of the year) *before* bud expansion and flowering, though it does not explain why it drops a full crop of what appear to be perfectly normal leaves that are demonstrably capable of withstanding all the rigors of the dry season. The tree even grows its woody diameter seasonally, leading to very well defined September-October growth rings that allow a tree's age to be counted from a polished cross section.

The light defoliation by caterpillars of adult, sapling, and seedling *H. courbaril* is also highly seasonal, occurring almost entirely in the May-June first part of the rainy season (and conspicuously not at the time of new leaf production). Even the spectacular total defoliations of *H. courbaril* crowns by the large caterpillars of *Schausiella santarosensis* (Saturniidae), observed during 2003–2008 (and absent during 2010–2012) in Bosque Húmedo in Sector Santa Rosa, is restricted to the first six weeks of the rainy season. The great bulk of the host-specific monophagous *S. santarosensis* caterpillars are univoltine and remain dormant pupae in the leaf litter for 11 months. The defoliated trees gradually and weakly refoliate with smaller leaves during the second half of the rainy season. All of these seasonal events can be, and probably are, interrupted or calendar-shifted by climate change. Now multiply that across 400 species of dry forest trees in the same 10 km^2!

Dry Forest Succession, Restoration, and Conservation

Fragmentation and Restoration of the Dry Forest

Parque Nacional Santa Rosa (PNSR) was established in 1971 as a "right-now" act of conservation of dry forest remnants. Even then they were not understood to be as much remnant as they were, since they were embedded in old fields and pastures that were incorrectly called "savannah" (e.g., Daubenmire 1972). These remnants were also disappearing in the face of advancing local, national, and international agricultural society (Government of Costa Rica 1998).

The Parque Nacional Santa Rosa absorption into "Proyecto Parque Nacional Guanacaste (PPNG)" in 1985, which morphed into ACG over the following two decades (Allen 1988, 2001, Janzen 2000a), was driven by the concept of "restoration of tropical dry forest" (Janzen 1988a, 2002). Had there been some huge portion of Meso-American old-growth dry forest still intact to preserve in 1985, as there was and still are large areas of rain forest in other parts of the world, and had tropical dry forest still been relatively well conserved in other parts of the tropics, there would have been little incentive to invent the concept and strategy of PPNG through its restoration. In the simplistic terminology of those years, "since there is no dry forest conserved, large enough to biologically and sociologically survive, it is necessary to restore one." At that time, the mantra of conservation was "decree it, buy it, and conserve it now before it is cut down, because once cut, it is gone forever." While there was obvious truth in this mantra, it was also obvious from many examples of "accidental", naturally occurring restoration that a forest could be restored if the seed sources (animal and plant) were within dispersal distance (e.g., living in the woody succession in tree falls, landslides, old river courses, abandoned/failed/neglected farms and ranches, etc.). Within a few years, and stimulated by the propaganda being created by PPNG, it became generally accepted that the donor audience, and policy-maker audience, could grasp two messages—yes, save it if it is still there, but grow it back if it is not there. A large amount of farming/ranching/plantation activity is, in fact, dedicated to blocking restoration or succession (burning pastures, shooting elephants, weeding plantation understory, etc.). To quote an old-time rancher neighbor of ACG in 1986, "Of course the forest will come back if you let it. Why do you think we burn the pastures?"

The ACG dry forest landscape, and that of the more wet ecosystems to the east, are restoring themselves (Janzen 2000a, 2002) as hunting, logging, burning, grazing, and farming have been removed and the wild population fragments allowed to re-invade and re-occupy their former terrain. But there was a very human-generated underlying condition. While ACG had suffered more than four centuries of European-style intervention, and in select sites, as mentioned earlier, millennia of indigenous peoples' intervention, it is generally a quite inhospitable place for ranching and subsistence agriculture. Below much of its dry forest were truly poor soils, some of the worst in Costa Rica—both recent and unweathered volcanic soils below the dry forest in general, and mineral-rich serpentine in Santa Elena. ACG was very far from the centers of social and political power, and from markets in both Nicaragua and the Meseta Central of Costa Rica (though the production of beef for the *indigo* farms of El Salvador and mules for the cross-isthmus transport from the Río San Juan or Lago Nicaragua to the Pacific, and the Inter-American Highway in the 1940s did generate major ranching impacts). Its volcano slopes are steep and wet with rain and fog/clouds. It had no big navigable rivers, but instead many small rocky ones, and none at all in the Mesa Santa Rosa. Exploitable indigenous labor appears to have been generally minimal by the time of original European colonization. All of these things, as mentioned earlier, are reflected in the colonial archives in Granada, which indicate that Hacienda Santa Rosa had more than 40 different owners over four centuries, an ownership turnover that is fully congruent with the observation that despite the enormous impact of European uses, they were very far from thorough—far less thorough than the uses experienced by large holdings on good soil and climate, and near social centers, that would be owned by a several-generation family lineage.

As a consequence, until as recently as 1965–1985, the Mesa Santa Rosa, and its surroundings, contained fragments of populations of easily 99% of the species that were present when the Europeans arrived (only the giant anteater, *Myrmecophaga tridactyla*, is known to have been lost). These surviving species were intermingled throughout in an extremely diverse mosaic of different ages and histories of incomplete ecosystem damage (though some of the pastures and former fields were many hundreds of hectares in area and many individuals and populations were "living dead"). The dry forest area of ACG was dotted with single trees, small patches of secondary forest, wooded ravines, and rocky slopes (see Fig. 10.9). The more wet forests on the western volcano slopes were bullet-holed by 1–300 ha pastures or fields, but these were often bordered by equal-sized expanses of forest that were either unlogged, or had been only high-grade logged and then ignored. It was as though a hurricane had passed through. On the eastern and northern rain-forested volcano slopes, the same was true but even larger expanses were covered by old-growth or only high-grade logged forest without subsequent ranching or cultivation.

During the original focus in 1985 on restoring the dry forest of Sector Santa Rosa (well before we understood the importance of incorporating the wetter forests of the Cordillera de Guanacaste and beyond), the most self-evident management decision was to stop/eliminate the man-made fires (Janzen 2000a, 2002). In 1986–1988, minimalist experiments in tree planting in pastures were tried, but it became immediately obvious that when confronted with 60,000 ha of pasture with scattered dry forest remnants, the correct use of budget and administrative resources was not in planting trees, but rather in stopping the fires and letting nature do the reforestation. The policy quickly became "let the plant species mutualize with animals and fight it out among themselves, and thereby they themselves determine what species and community composition will eventually come to occupy what particular soil type, slope, moisture condition, elevation, etc." Naturally occurring woody succession into the pastures began as soon as the fires were stopped. However, it was initially discouraging that 3–5 years were required for the woody plants to be tall enough to be generally visible over the oceans of ungrazed pasture and begin to strongly shade out the introduced 1–2 m tall African grasses (Janzen 2002). After this 3–5 year delay, the ACG staff and planning felt that they were unambiguously on the right course and that all the work of eliminating the fires was worth the effort.

This invasion of dry forest pastures—"ensuciarles" (dirty them)—by fire-free woody succession was unambiguously aided by four biological factors, all four of which were generally understood prior to this very pragmatic application of anthropogenic succession ecology (and all four of which are qualitatively different in ACG rain forest pastures).

First, more than 100 species of common ACG dry forest trees, vines, and shrubs have wind-dispersed seeds. There was a continual downwind rain of these seeds from scattered adults surviving at pasture edges. ACG rain forest is characterized by having very few wind-dispersed species.

Second, dry forest vertebrates, nearly all of which are to some degree frugivorous and therefore walking or flying containers of seeds, about to be defecated, spit out, or buried, are generally exposed to heat/sun/dryness in their life cycles. They are therefore generally willing to (or forced to) cross the sun/heat/wind/exposure of pastures (and perch in or below isolated pasture trees), thereby dispersing hundreds of species of animal-dispersed seeds into pastures. Furthermore, their seed shadows are very different from those created by wind and can extend over distances of many kilometers (Janzen 1988c). In this context, some small bird or bat, or even a larger animal (coyote, coati, ctenosaur, opossum, collared peccary, white-tailed deer) that seems of minor conservation importance because it appears to be not threatened with extinction, can be of enormous importance in the restoration process. The same applies to the importance of individual woody plants—very few dry forest species (e.g., only "caoba" [or *Swietenia macrophylla*] and "cocobolo" [*Dalbergia retusa*] of the original invaders in the Santa Rosa restoration process are considered by Costa Rican legislation to be threatened). Rain forest animals in general avoid the sun and harshness of pastures.

Third, the establishing woody plant seedlings encountered the spores of mycorrhizal fungi for their roots in the treeless dry forest pastures, owing to these fungi annually producing wind-dispersed spores in the dry season. Rain forest mycorrhizal fungus spores are generally dispersed by surface ground water flow and therefore generally rare to absent in treeless pastures.

Fourth, while the dry forest juvenile attempting to establish in the "death valley" habitat of a dry forest pasture does encounter severe climate challenges, they are not nearly as different from those inside its "normal" dry forest as are the challenges encountered by a rain forest woody plant juvenile attempting to survive in the full sun and erratic drought of a rain forest pasture.

In the early years of fire control and elimination in ACG dry forest there were three key methods, each of which had their special importance and place in time. First and foremost was the decision to actually put out the fires, both in ACG and sometimes, before the fire got to ACG. This required: 1) funds for minor equipment, and 2) the willingness to realize that something as simple as a broom (better, water-soaked wet brooms, in quantity) and effort could extinguish many kinds of grass fires. Most of them were burning in litter or pasture grass, and not the forest per se. And it required: 3) team-based individual initiative to be permitted around-the-clock with full-time (specialized) staff, with a fire program both responsible for all fire extinguishing and allowed to receive appropriate applause. In other words, fire control and elimination was a "people problem," rather than a biological problem.

Second, for the first time in ACG history, fire-breaks and back-fires were set and managed, often at night. Using fire to fight fire was a reasonable and familiar tool for Guanacaste resident neighbors hired onto the fire team, and setting fire to a national park was not anathema to them.

Third, for large areas of ACG, cattle were left in the brushy pastures for 1–3 years after land was purchased; in the late 1980s there were as many as 7,000 cattle in ACG at the peak time of using cattle as biotic mowing machines (and they even disperse some seeds). They did eat broad leaf plants to some degree, but they saved far more by grass removal (reducing fuel for fire and reducing competition from grass). By far the fastest way to create young woody vegetation (also known as forest invasion) in an ACG dry forest pasture was to have moderate cattle density coupled with full elimination of fire. However, once the woody vegetation had made a firm start at forest restoration, the cattle were removed because they can also maintain the forest in a state of deflected succession through selective browsing and trampling, and because they perturb streams, ponds, marshes, and other water sources. Horses are more complex, both because they are (a) native animals (the horse was evolutionarily invented in the New World); (b) good seed dispersers (Janzen 1981a, 1981b, 1982b) (the guanacaste tree, *Enterolobium cyclocarpum*, is going extinct in ACG and throughout much of Guanacaste Province because of horse removal); (c) present in much lower numbers than cattle; and (d) less damaging to watercourses. To make this all yet more complex, however, it needs to be noted that pre-human ACG dry forest had a complement of large vegetarian mammals (and their predators and scavengers) that would have done substantial "damage" to the vegetation and watercourses in what we think of as being pristine old-growth today. No East African waterhole remains clear and clean during the dry season. While it is an easy decision to exclude cattle once the woody restoration process is well on its way, there is an argument to be made for having a sector of ACG dry forest with free range horses and some kind of artificial predation to hold their numbers at a reasonable level, just as we would do if someone were to discover a population of ground sloths today.

In general terms, the Santa Rosa dry forest invades (and invaded) the abandoned pastures by creeping in downwind from the upwind side of the pasture. The accidents of what species of what age happens to be growing on the upwind side thereby influences the nature of the succession invasion for centuries to come. The other form of invasion is the gradually enlarging patch of animal-dispersed vegetation that accumulates below and next to a single large tree in a pasture (Janzen 1988c). Again, the accidents of which particular species of dispersal agents happen to be visiting a particular perch or shade tree will determine the structure of the resulting forest for centuries.

These two invasion processes gradually merge as the woody vegetation takes over the pasture. And, as the forest becomes shadier and denser, there is more lateral movement by larger and more ground-bound vertebrates that bring larger seeds, many of which are seeds of seemingly evergreen species of trees (e.g., agoutis as seed dispersers, Hallwachs 1986). This in turn begins the centuries-long process of recovering the largely evergreen to semi-evergreen aspect of the forest that once clothed Mesa Santa Rosa. In short, the initially invading forest is almost entirely deciduous during the long dry season. The exception is when a coyote (*Canis latrans*) goes into the forest and eats fallen fruits of *Manilkara chicle* (a long-lived seemingly evergreen tree), and then defecates their seeds in the shade of an isolated pasture tree. This eventually creates a very shady spot in the regenerating deciduous forest. This secondary forest, made of species with a) fast-growing and light-demanding seedlings, b) annually-produced large crops of small seeds, and c) relatively short-lived trees, is the first owner of the terrain previously occupied by a semi-evergreen forest (removed by logging or simply clearing). The final old-growth forest, almost unknown to the residents of Guanacaste Province but easily accessible in the Bosque Húmedo of Sector Santa Rosa, contains all these species but at much lower density and intermingled with 5–15 species of apparently evergreen canopy trees (they really are not evergreen, but drop and replace their leaves over just a few weeks at a species-idiosyncratic time of year, and therefore seem to be evergreen, as described earlier for *Hymenaea courbaril*).

As a pasture fills in with rapidly growing juvenile woody plants (large numbers of individuals of the same species), there are several very visible consequences. First, these juveniles are largely non-reproductive, which means that the site is very poor in food for the many species of ACG seed predators and fruit feeders—an ocean of young forest can be a desert for them. Second, the growing vegetative parts that are potential food for browsers are literally moved too high for the ground-bound animals to get to them, and rendered much harder for the ground-bound observing human being to see them when they are being eaten. The consequence is that the biologist, tourist, or manager may feel that a species of animal is becoming rare when it may have simply moved up and out of sight (as well as its population becoming "diluted" among the ocean of juvenile plants). Third, many species of herbs, low shrubs, vines, and even stunted larger woody plants are no longer insolated near ground level (as they were before when living on the pasture edge). They either go locally extinct or persist as the few and relatively invisible individuals that make it into the generally rising canopy (where they may continue to reproduce, but be nearly invisible to the ground-based observer). Fourth, for both shade and disease causes that are not understood, the wave of surviving juveniles (which largely later die through competition and herbivory) that appear in the pasture when the fires are halted does not continue to annually appear in the understory of the successful adults now reproducing in the canopy. For example, it is very striking to encounter a large and somewhat widely spaced stand of adult guacimo (*Guazuma ulmifolia*) trees filling an old pasture and now generating billions of viable seeds annually, yet having no juvenile guacimo growing in the mottled shade below or even in the clearings created by adult deaths or other species in between the adult *G. ulmifolia*. It is likely that this is because the soil saturated with germinating seedlings of this species quickly becomes a killing zone due to a population explosion of lethal fungi specific to that species of seedling. This phemenon is very visible below large-seeded guanacaste trees, where no seedling survives past about 50 cm in height, and surviving seedlings require a seed disperser to drop them relatively far from a seeding adult.

The initial ACG fire-control/elimination program was and still is not perfect. While most of ACG dry forest and accompanying pastures/fields has not been burned since 1984–1986, a few areas have been burned at 2–10 year intervals, and yet fewer more frequently.

The relationship of the occasional fire to the succession process is highly variable, but in general the longer the site goes without a fire, the more shade there is and therefore the less herbaceous vegetation there is at ground level to fuel a dry season fire. Very low quality soils (e.g., volcanic tuff, serpentine) restore their forests the slowest of all, with the consequence that they can remain as "open vegetation" when impacted by a fire only every few decades, while high quality and more moist soils can re-vegetate with woody plants so rapidly that now after two decades without a fire they are almost unburnable (and certainly a litter fire can be rapidly extinguished in them). In the management goal of not letting fire deflect the process of woody succession, perhaps the most important strategy is to avoid a second fire in the next year in a burned early succession forest. This is because the first fire opens the canopy enough (by the heat killing woody stems, more than by direct consumption of woody plants) to allow substantial herbaceous development low to the ground, which means that the fire in the next year burns much more intensely because it is drier and has more fuel. Second, the first year fire may wound the base of, but not fell, a large tree. The second-year fire then burns in and through the dead wounded area, felling the tree and creating a large tree fall that fills with herbaceous vegetation. Third, the root reserves of a woody plant are often enough to substantially replace the shade-generating leafy crown after a single fire, but the plant is often not able to do this if burned again the next year (Janzen 2002).

The only natural "grasslands" in ACG are those growing on the old volcanic mud flows on the western slopes of Volcán Rincón de la Vieja, and those in swamps such as Punta Respingue at Península Santa Elena. These open areas will eventually be reinvaded by the same kinds of forest as currently occupy the other slopes of the same volcanic complex, if they are allowed to do so by management and climate change over the centuries. All of the grasslands (actually old pastures, sometimes incorrectly called "savannas") on ACG good soils are anthropogenic (and largely occupied by introduced African grasses) and are rapidly filling with many kinds of secondary forest now that the contemporary assault by fire has ceased (though climate change may cause any given hectare to fill to a different equilibrium state than it bore in pre-European times, for obvious reasons).

Those "pastures" on poor soils—for example, those on Santa Elena serpentine, volcanic tuff from Volcán Cacao to Santa Maria to Liberia, or the volcanic core of Cerro El Hacha (Fig. 10.13)—are largely covered with a single species of native grass (*Trachypogon plumosus*) with hundreds of species of herbaceous plants mixed in. These grasslands are being reoccupied by forest much more slowly than are the pastures on high-quality soil. Even with the elimination of man-made fires they will use centuries to shrink back to their former, small and fractured habitat of cliffs, rocky ridges, landslide scars, ravine margins, etc., embedded in closed-canopy forest. It is worth noting that the only "natural" fires in ACG occur when a rainy season lightning strike hits a tree (lightning does not occur in the long dry season). If that tree is in a pasture (especially an ungrazed one), it may well start a pasture grass fire that burns to the forest edge and stops. This event was occasional in the 1970–1980s in Sector Santa Rosa's abandoned pastures. If the struck tree is dead and in forest, the tree may well burn, leaving a pile of white ash where it lay, but the litter (and surrounding vegetation) is too moist to burn past this point.

The ACG pastures in cloud forest, rain forest, and ecosystem intergrades display forest invasion and restoration very differently than has been the experience in the ACG dry forest. If a well-established and clean rain forest pasture is simply abandoned, it often grows a dense waist-deep stand of whichever of tens of species of introduced Old World grasses occupy it, and decades later it is still largely in that condition. Intense or light grazing, plowing, burning, and seedling planting does little to hasten forest invasion, though it eventually can help. The reasons are the four listed earlier: rain forest pastures have almost no wind-dispersed woody plants in their vicinity, are much disliked by rain forest potential seed dispersers, are sprinkled very little with wind-dispersed mycorrhizal spores, and are an extreme microclimate for rainforest juvenile plants. However, by noticing that a rain forest understory develops in the strong shade of abandoned gmelina plantations (*Gmelina arborea* is a verbenaceous southeast Asian tree grown for cardboard fiber and cheap lumber), ACG has successfully removed well-established rain forest pastures by planting them with gmelina as if for a timber plantation, which effectively shades out the grasses, but "abandoning" the plantation 1–2 years after planting. The plantation dies of old age after 10–20 years, and the juveniles do not survive in the plantation's own shade.

All of the problems with mid-elevation rain forest restoration on old ACG pastures are exacerbated with increasing elevation on the western, drier sides of the Cordillera de Guanacaste, but with a new complication. Because the upper elevations are less species-rich, and becoming even more so with climate change, there are yet fewer species to contribute to the invasion of old pastures. The restoration toolbox has fewer tools.

Additionally, on the western slopes of the Cordillera, as the forests were cleared, their more xeric nature and yet more abundant moisture made them prime habitat for lowland Guanacaste dry forest plants and animals to invade. This invasion gives the wildland-plus-agro-scape mix a drier aspect, a more "dry forest aspect," than the site "should" have. These "weed" species in themselves will also increase the competitive blockage of return by the native rain forest species. On the eastern/northern, and therefore wetter slopes of the Cordillera within ACG, the same process occurs, albeit involving those yet fewer species of dry forest invaders that have moved to the wet side.

For a variety of sociological reasons, the decision to fully protect and restore the marine portion of ACG got a very slow start, and today is approximately where the terrestrial effort was about 1985–1990—there is removal of sport fishing and large commercial fishers, and now only light poaching by artisanal fishers (= marine hunters). Owing to difficult access, the dry forest beaches themselves, at least of the most remote parts of Península Santa Elena, remained about as pristine in their rock formations and contemporary back beach vegetation as any on the Pacific coast of Costa Rica. This became apparent only during the detailed examination (Janzen 1998) associated with the court case for the Hacienda Santa Elena expropriation for inclusion in ACG. Below the surface of the ocean, the exceptionally high biodiversity of Sector Marino survived as fractured and human-hunted populations largely because of the unfriendliness of the area (windy, rocky, current-rich). However, the area was overrun by the desperate fishers pushed south by the wars in Nicaragua in the 1960–1990 period. Since the damage was largely through underwater hunting/poaching, and since marine populations that are only partly impacted seem to display good natural restoration on their own when unmolested, the primary ACG challenge is in continuing the gradual-to-abrupt removal of artisanal hunting from Sector Marino, and letting nature take its course in restoration.

By far the most damaging invasive species in ACG is *Homo sapiens*. For ACG to survive as a conserved wildland (and wild marine area) into the indefinite future it will have to withstand and tolerate a light level of this impact indefinitely (roads, power lines, people presence, trails, buildings, lights, sounds, pesticides, smells, invasive other species, insularization, etc.). This impact is an inevitable cost of ACG being accepted as a full-status social institution within Costa Rica. The next most visible invader, *Hyparrhenia rufa*, or "*jaragua*," the East African pasture grass introduced to Costa Rica in the 1920s, has been found to be a trivial challenge as woody vegetation restores itself throughout ACG dry forest on the "good" soils occupied by *jaragua*. This is simply because *jaragua* is very susceptible to shade. Poor soils are "invaded" by *Trachypogon plumosus*, which is apparently native to natural disturbance sites in ACG dry forest. However, this species also dwindles to trivial presence as woody vegetation reappears (more slowly) on poor soils. The rain forest and cloud forest portions of ACG are strongly invaded by an array of African pasture grasses and several hundred species of Guanacaste dry forest plants and animals, but these also dwindle to zero or very close to it as the forest restores itself or is restored with gmelina, except in roadsides, buildings, and other sites that are deliberately kept open.

The generally observed inability of introduced garden plants and animals to invade undisturbed mainland tropical ecosystems is in full play in ACG, with the dry forest itself showing many examples of human-associates hurtling into local extinction in ACG dry forest wildlands now that their nurturing humans have been removed (e.g., *Musa, Citrus, Cassia grandis, Enterolobium cyclocarpum, Spondias purpurea, Persea americana, Hyparrhenia rufa, Vachellia farnesiana, Crescentia alata, Acrocomia aculeata*, etc.).

ACG island habitats are a different matter. The super abundant *Rattus norvegicus* population on Isla San José of the Islas Murciélagos in Sector Marino, yet absent from the mainland as a wild animal, is a conspicuous example. However, the mainland ACG is probably due for multiple invaders that are not so easily removed as was *jaragua*. For the past ten years, ACG staff have noticed a growing population of the introduced African terrestrial orchid *Oeceoclades maculata* in the understory of lightly disturbed vegetation on the dry forest—rain forest intergrade in Sector Orosí at 400–600 m elevation and hence towards the Pacific in the dry forest. This plant could become a major ACG weed since it is a widespread invader, apparently tolerant of many ecological circumstances, not removed by shade, and with no visible herbivory.

The Africanized honeybee (*Apis mellifera*) invasion swept through ACG dry forest in 1984, built up to high density (1986–1988), and has now shrunk to be almost

synergism (or the lack of it) among these major forces is unclear. Furthermore, the amount of on-site, detailed and multi-year observations and experiments that it will take to tease out the "why and how" is so large that it will not be done during the lifespan of the major successional changes taking place now.

Perhaps the most easily observed is the conspicuous decline in density and ecosystem occupation of many species of dry forest plants as human impacts are strongly reduced. For example, in the 1960s to 1980s, and presumably for several centuries before this, the ACG dry forest ant-acacia (*Acacia collinsii*, now known as *Vachellia collinsii*) with its three species of obligate mutualistic *Pseudomyrmex* acacia ants ("*hormiga del cornizuelo*") (Belt 1985, Janzen 1983c), were very common plants in pastures, pasture-forest interfaces, roadsides, and "naturally disturbed" sites such as tree falls, landslides, ravines, and cliffs. Today the overall density of this plant (and therefore its ant colonies) in ACG is not more than 5% of what it was 20 years ago. The surviving population of ant-acacias has a very high proportion of senescent old individuals, and juveniles are very rare even in both natural and man-made disturbance sites (many of which were colonized previously by seed flow from anthropogenically disturbed sites; e.g., Janzen 1983d). It will not be surprising for the plant to be extinct within-ACG, or nearly so, fifty years from now. Heavy shade created by the closed and rising canopy severely weakens the few *V. collinsii* that germinate and are fortunate enough to acquire a healthy *Pseudomyrmex* colony (just as was found to be the case with the ant-acacia *Vachellia cornigera* in remnants of Guanacaste-like dry forest and pasture in Veracruz, Mexico; Janzen 1967b). There may also be a major decline in the density of fruit-eating bats (*Corollia perspicillata*, *Glossophaga soricina*) and magpie jays (*Calocitta formosa*), all of which were major seed dispersers for ant-acacias (as well as many other species of plants), when there was a massive community of frequently disturbed sites filled with early succession fruit-bearing herbs, vines, shrubs, and trees in ACG dry forest.

Many other ACG dry forest early successional species are "suffering" the same decline in density as are *Vachellia collinsii*—*Casearia nitida* (known as *Casearia corymbosa* in earlier literature), *Cochlospermum vitifolium*, *Senna pallida* (known as *Cassia biflora* in earlier literature, e.g., Janzen 1980), *Senna atomaria*, *Bauhinia ungulata*, *Malvaviscus arborea*, *Cayaponia racemosa*, *Ipomoea digitata*, *Sida acuminata*, *Byrsonima crassifolia*, *Sida* spp., *Vachellia farnesiana*, and even *Guazuma ulmifolia*. However, guacimo (*G. ulmifolia*) can grow to a tree as much as 20 m in height and persist in at least some pasture-to-forest succession for 100+ years. Some of the disappearing species were very directly dependent on cattle, horses, and people for their maintenance as a "population" in ACG dry forest (e.g., *Enterolobium cyclocarpum* [the introduced national tree of Costa Rica and the namesake for Guanacaste Province], *Crescentia alata*, *Spondias purpurea*, *Cassia grandis*, *Vachellia farnesiana*, and *Vachellia cornigera* [an ant-acacia of wetter areas, such as the margins of Palo Verde swamp in Área de Conservación Tempisque, and even there may have been introduced by cattle brought from Mexico centuries ago]). These pasture associates are/were so omnipresent throughout Costa Rica's dry forest agro-scape that they are thought of as characteristic members of the dry forest of Guanacaste Province. However, they may largely or totally disappear from dry forest in the absence of manmade disturbance (just as they probably did with the initial disturbance of extinction of the megafauna).

Future generations will discover what are the predominant ACG dry forest species after several centuries of succession, and which species drop out entirely or are restricted to a few scattered peculiar and perpetually disturbed sites (probably on steep slopes and ravines). However, it is even now abundantly clear that the density and community structure components of animal-plant interactions will be very different once the ACG dry forest is again a relatively "natural" community. For example, in the 1980s and earlier it was the norm to encounter healthy, large acacia-ant colonies of all three species in any ACG dry forest habitat, though there was a slight bias towards *Pseudomyrmex flavicornis* being more abundant in very insolated sites and *P. nigrocinctus* and *P. spinicola* being more abundant within shady young forest. Each of these three species was probably the competitive dominant in some particular kind of naturally occurring disturbed site (e.g., tree falls, versus cliffs and landslides, versus watercourse margins), but anthropogenic disturbance had leveled the playing field by offering the brushy pasture to all. Today, the distinctive black *P. flavicornis*, the most sun- and heat-tolerant species, is nearly extinct following three decades of succession.

A particularly noticeable aspect of dry forest succession following massive manmade disturbance is the re-emergence of interdigitating tongues or peninsulas of wetter, more evergreen, forest extending out from the evergreen rainforests bordering ACG dry forest on its eastern margins (Janzen 1986a, b). However, this phenomenon is simultaneously being extinguished by the general drying of ACG dry forest as expressed in such things as the post-1980 disappearance of "rain forest and sometimes dry forest species" from the Bosque Húmedo in Sector Santa Rosa. For example, the following animals and plants were there but now appear to be missing or almost so: *Anolis biporcatus*, *Ochroma lagopus*, *Ceiba pentandra*, *Banara guianensis*, *Piper marginatum*, *Piper psilorhachis* [formerly *P. amalago*], *Cecropia obtusifolia*, *Trema micrantha*, *Muntingia calabura*, *Vismia baccifera*, *Phaenostictus mcleannani*, *Herpetotheres cachinnans*, *Chromacris colorata*, *Biblis hyperia*, *Nasutitermes* sp., *Iguana iguana*, *Nyctomys nyctomys*, *Dasypus novemcinctus*, and *Didelphis virginiana*.

Conclusion

On that sad note we end this discourse on some salient features of ACG dry forest and its immediately adjacent wetter and cooler ecosystems. The biology of the ACG dry forest ecosystem deserves a massive book-length treatment, and a massive website of images and description; one has germinated (http://www.acguanacaste.ac.cr, and see also http://www.gdfcf.org). However, in the urgency of taking the conservation steps to ensure that a maximum amount of it is still with us a millennium from now, most of that documentation and analysis has had to wait. If we expend our energy and effort on that more enjoyable and more academic task, we will then turn around one day and find that what we documented is gone. Yet, if we fail to portray it and make it available to all of society, it will likewise be gone.

Acknowledgments

This essay and the information in its contents have been strongly supported by the staff and parataxonomists of Área de Conservación Guanacaste (ACG) and by ideas and information from many hundreds of natural historians, ecologists, and taxonomists, as well as by more than 9,500 private and government donors to ACG, and by 45 years of generous support of our research and these conservation efforts by the US National Science Foundation (NSF), the Wege Foundation, the Government of Canada, the Biodiversity Institute of Ontario, the JRS Biodiversity Foundation, Permian Global, the University of Pennsylvania, and major private donors.

References

Adler, R.F., and G. Gu. 2007. Long term increase in rainfall seen in tropics. http://www.nasa.gov/centers/goddard/news/topstory/2007/rainfall_increase.html.

Allen, W.H. 1988. Biocultural restoration of a tropical forest. *BioScience* 38(3): 156–61.

Allen, W.H. 2001. *Green Phoenix: Restoring the tropical forests of Guanacaste, Costa Rica*. New York: Oxford University Press. 301 pp.

Belt, T. 1985. *The Naturalist in Nicaragua*. Chicago: University of Chicago Press. 403 pp. Reprint of 1874 publication.

Bullock, S.H., H.A. Mooney, and E. Medina, eds. 1995. *Seasonally Dry Tropical Forests*. Cambridge: Cambridge University Press. 450 pp.

Clark, W.E. 1992. The *Anthonomus marmoratus* species group (Coleoptera: Curculionidae). *Transactions of the American Entomological Society* 118: 129–45.

Cody, S., J.E. Richardson, V. Rull, C. Ellis, and R.T. Pennington. 2010. The Great American Biotic Interchange revisited. *Ecogeography* 33: 326–32.

Daubenmire, R. 1972. Ecology of *Hyparrhenia rufa* (Mees.) in derived savannah in northwestern Costa Rica. *Journal of Applied Ecology* 9: 11–13.

Dirzo, R., H.S. Young, H.A. Mooney, and G. Ceballos, eds. 2011. *Seasonally Dry Tropical Forests*. Washington, DC: Island Press. 392 pp.

Doubleday, C.W. 1886. *Reminiscences of the Filibuster War in Nicaragua*. New York: G.P. Putnam's Sons.

Durham, W.H. 1979. *Scarcity and Survival in Central America: Ecological Origins of the Soccer War*. Stanford: Stanford University Press. 232 pp.

Government of Costa Rica. 1998. Área de Conservación Guanacaste: Nomination for inclusion in the World Heritage List of natural properties. Submitted to UNESCO, July 1, 1998. 38 pp.

Graham, A. 2003. Historical phytogeography of the Greater Antilles. *Brittonia* 55: 357–83.

Haber, W.A., and R.D. Stevenson. 2004. Diversity, migration, and conservation of butterflies in northern Costa Rica. In G.W. Frankie, A. Mata, and S.B. Vinson, eds., *Biodiversity Conservation in Costa Rica: Learning the Lessons in a Seasonal Dry Forest*, 99–114. Berkeley: University of California Press.

Hallwachs, W. 1986. Agoutis (*Dasyprocta punctata*): the inheritors of guapinol (*Hymenaea courbaril*: Leguminosae). In A. Estrada and T. Fleming, eds., *Frugivores and Seed Dispersal*, 285–304. Dordrecht: Dr. W. Junk Publishers.

Hays, S.T., and S. Conant. 2007. Biology and impacts of Pacific Island invasive species. 1. A worldwide review of effects of the small Indian mongoose, *Herpestes javanicus* (Carnivora, Herpestidae). *Pacific Science* 4(January): 1–16.

Hoernle, K., P. van den Bogaard, R. Weerner, B. Lissinna, F. Hauff, G. Alvarado, and D. Garbe-Schonberg. 2002. Missing history (16–71 Ma) of the Galapagos hotspot: implications for the tectonic and biological evolution of the Americas. *Geology* 30: 795–98.

Holloway, M. 2006. Democratizing taxonomy. *Conservation in Practice* 7(2): 14–21.

Hunt, J.H., R.J. Brodie, T.P. Carithers, P.Z. Goldstein, and D.H. Janzen. 1999. Dry season migration by Costa Rican lowland paper wasps to high elevation cold dormancy sites. *Biotropica* 31(1): 192–96.

Iturralde-Vincent, M.A., and R.D.E. MacPhee. 1999. Paleogeography of the Caribbean Region: implications for Cenozoic biogeography. *Bulletin of the American Museum of Natural History* 238: 1–95.

Jackson, J.B.C., and L. D'Croz. 1998. The ocean divided. In A.G. Coates, ed., *Central America: A Natural and Cultural History*, 38–71. New Haven: Yale University Press.

Jamison, J.C. 1909. *With Walker in Nicaragua, or Reminiscences of an Officer of the American Phalanx*. Columbia, MO: E.W. Stevens Publishing Company. 181 pp.

Janzen, D.H. 1967a. Synchronization of sexual reproduction of trees with the dry season in Central America. *Evolution* 21: 620–37.

Janzen, D.H. 1967b. Interaction of the bull's-horn acacia (*Acacia cornigera* L.) with an ant inhabitant (*Pseudomyrmex ferruginea* F. Smith) in eastern Mexico. *University of Kansas Science Bulletin* 47: 315–558.

Janzen, D.H. 1967c. Why mountain passes are higher in the tropics. *American Naturalist* 101: 233–49.

Janzen, D.H. 1974. The deflowering of Central America. *Natural History* 83: 48–53.

Janzen, D.H. 1975. Behavior of *Hymenaea courbaril* when its predispersal seed predator is absent. *Science* 189: 145–47.

Janzen, D.H. 1978. Seeding patterns of tropical trees. In P.B. Tomlinson and M.H. Zimmerman, eds., *Tropical Trees as Living Systems*, 83–128. New York: Cambridge University Press.

Janzen, D.H. 1980. Specificity of seed-attacking beetles in a Costa Rican deciduous forest. *Journal of Ecology* 68: 929–52.

Janzen, D.H. 1981a. *Enterolobium cyclocarpum* seed passage rate and survival in horses, Costa Rican Pleistocene seed dispersal agents. *Ecology* 62: 593–601.

Janzen, D.H. 1981b. Guanacaste tree seed-swallowing by Costa Rican range horses. *Ecology* 62: 587–92.

Janzen, D.H. 1982a. Differential seed survival and passage rates in cows and horses, surrogate Pleistocene dispersal agents. *Oikos* 38: 150–56.

Janzen, D.H. 1982b. How and why horses open *Crescentia alata* fruits. *Biotropica* 14: 149–52.

Janzen, D.H. 1982c. Attraction of *Liomys* mice to horse dung and the extinction of this response. *Animal Behaviour* 30: 483–89.

Janzen, D.H. 1982d. Cenízero tree (Leguminosae: *Pithecellobium saman*) delayed fruit development in Costa Rican deciduous forests. *American Journal of Botany* 69: 1269–76.

Janzen, D.H. 1983a. The Pleistocene hunters had help. *American Naturalist* 121: 598–99.

Janzen, D.H. 1983b. Insects. In D.H. Janzen, ed., *Costa Rican Natural History*, 619–45. Chicago: University of Chicago Press.

Janzen, D.H. 1983c. *Pseudomyrmex ferruginea* (Hormiga del Cornizuelo, Acacia-ant). In D.H. Janzen, ed., *Costa Rican Natural History*, 762–64. Chicago: University of Chicago Press.

Janzen, D.H. 1983d. No park is an island: increase in interference from outside as park size decreases. *Oikos* 41: 402–10.

Janzen, D.H. 1984a. Weather-related color polymorphism of *Rothschildia lebeau* (Saturniidae). *Bulletin of the Entomological Society of America* 30(2): 16–20.

Janzen, D.H. 1984b. Dispersal of small seeds by big herbivores: foliage is the fruit. *American Naturalist* 123: 338–53.

Janzen, D.H. 1985a. On ecological fitting. *Oikos* 45: 308–10.

Janzen, D.H. 1985b. *Spondias mombin* is culturally deprived in megafauna-free forest. *Journal of Tropical Ecology* 1: 131–55.

Janzen, D.H. 1986a. Blurry catastrophes. *Oikos* 47: 1–2.

Janzen, D.H. 1986b. Lost plants. *Oikos* 46: 129–31.

Janzen, D.H. 1986c. Mice, big mammals, and seeds: it matters who defecates what where. In A. Estrada and T.H. Fleming, eds., *Frugivores and Seed Dispersal*, 251–71. Dordrecht, Holland: Dr. W. Junk Publishers.

Janzen, D.H. 1987. How moths pass the dry season in a Costa Rican dry forest. *Insect Science and Its Application* 8: 489–500.

Janzen, D.H. 1988a. Guanacaste National Park: tropical ecological and biocultural restoration. In J.J. Cairns, ed., *Rehabilitating Damaged Ecosystems*, Vol. II, 143–92. Boca Raton, FL: CR Press.

Janzen, D.H. 1988b. Buy Costa Rican beef. *Oikos* 51: 257–58.

Janzen, D.H. 1988c. Management of habitat fragments in a tropical dry forest: growth. *Annals of the Missouri Botanical Garden* 75: 105–16.

Janzen, D.H. 1988d. Complexity is in the eye of the beholder. In F. Almeda and C.M. Pringle, eds., *Tropical Rainforests: Diversity and Conservation*, 29–51. San Francisco: California Academy of Science and AAAS.

Janzen, D.H. 1988e. Tropical dry forests: the most endangered major tropical ecosystem. In E.O. Wilson, ed., *Biodiversity*, 130–37. Washington, DC: National Academy Press.

Janzen, D.H. 1988f. Ecological characterization of a Costa Rican dry forest caterpillar fauna. *Biotropica* 20: 120–35.

Janzen, D.H. 1988g. The migrant moths of Guanacaste. *Orion Nature Quarterly* 7: 38–41.

Janzen, D.H. 1996. Prioritization of major groups of taxa for the All Taxa Biodiversity Inventory (ATBI) of the Guanacaste Conservation Area in northwestern Costa Rica, a biodiversity development project. *ASC Newsletter* 24(4): 45, 49–56.

Janzen, D.H. 1998. Conservation analysis of the Santa Elena property, Península Santa Elena, northwestern Costa Rica. Report to the Government of Costa Rica, Área de Conservación Guanacaste, ACG, Costa Rica. 129 pp. + 4 Appendices.

Janzen, D.H. 2000a. Costa Rica's Área de Conservación Guanacaste: a long march to survival through non-damaging biodevelopment. *Biodiversity* 1(2): 7–20.

Janzen, D.H. 2000b. Ingredientes esenciales de un enfoque por ecosistemas para la conservación de la biodiversidad de las áreas silvestres tropicales. Address to SBSTTA for COP 5, CBD, Montreal, Feb 1, 2000. http://www.mesoamerica.org.mx/Janzen2.htm.

Janzen, D.H. 2001. Latent extinctions—the living dead. In S.A. Levin, ed., *Encyclopedia of Biodiversity*, Vol. 3, 689–99. New York: Academic Press.

Janzen, D.H. 2002. Tropical dry forest: Área de Conservación Guanacaste, northwestern Costa Rica. In M.R. Perrow and A.J. Davy, eds., *Handbook of Ecological Restoration*, Vol. 2, *Restoration in Practice*, 559–83. Cambridge, UK: Cambridge University Press.

Janzen, D.H. 2004. Ecology of dry forest wildland insects in the Área de Conservación Guanacaste. In G.W. Frankie, A. Mata, and S.B. Vinson, eds., *Biodiversity Conservation in Costa Rica*, 80–96. Berkeley: University of California Press.

Janzen, D.H. 2005. How to conserve wild plants? Give the world the power to read them. Forward in G. Krupnick and J. Kress, eds., *Plant Conservation: A Natural History Approach*. Chicago: University of Chicago Press. 346 pp.

Janzen, D.H., M. Hajibabaei, J.M. Burns, W. Hallwachs, E. Remigio, and P.D.N. Hebert. 2005. Wedding biodiversity inventory of a large and complex Lepidoptera fauna with DNA barcoding. *Philosophical Transactions of the Royal Society B* 360(1462): 1835–46.

Janzen, D.H., and W. Hallwachs. 2011a. Philosophy, navigation and use of a dynamic database ("ACG Caterpillars SRNP") for an inventory of the macrocaterpillar fauna, and its food plants and parasitoids, of Área de Conservación (ACG), northwestern Costa Rica. http://janzen.sas.upenn.edu.

Janzen, D.H., and W. Hallwachs. 2011b. Joining inventory by parataxonomists with DNA barcoding of a large complex tropical conserved wildland in northwestern Costa Rica. *PLOS ONE* 6(8): e18123. doi: 10.1371/journal.pone.0018123.

Janzen, D.H., W. Hallwachs, P. Blandin, J.M. Burns, J. Cadiou, I. Chacon, T. Dapkey, A.R. Deans, M.E. Epstein, B. Espinoza, J.G. Franclemont, W.A. Haber, M. Hajibabaei, J.P.W. Hall, P.D.N. Hebert, I.D. Gauld, D.J. Harvey, A. Hausmann, I. Kitching, D. Lafontaine, J. Landry, C. Lemaire, J.Y. Miller, J.S. Miller, L. Miller, S.E. Miller, J. Montero, E. Munroe, S. Rab Green, S. Ratnasingham, J.E. Rawlins, R.K. Robbins, J.J. Rodriguez, R. Rougerie, M.J. Sharkey, M.A. Smith, M.A. Solis, J.B. Sullivan, P. Thiaucourt, D.B. Wahl, S.J. Weller, J.B. Whitfield, K.R. Willmott, D.M. Wood, N.E. Woodley, and J.J. Wilson. 2009. Integration of DNA barcoding into an ongoing inventory of complex tropical biodiversity. *Molecular Ecology Resources* 9(Suppl. 1): 1–26. doi: 10.1111/j.1755–0998.2009.02628.x.

Janzen, D.H., W. Hallwachs, J.M. Burns, M. Hajibabaei, C. Bertrand, and P.D.N. Hebert. 2011. Reading the complex skipper fauna of one tropical place. *PLOS ONE* 6(8): e19874. doi: 10.1371/journal .pone.0019874.

Janzen, D.H., and P.S. Martin. 1982. Neotropical anachronisms: the fruits the gomphotheres ate. *Science* 215: 19–27.

Johnson, K.P., and J.D. Weckstein. 2011. The Central American land bridge as an engine of diversification in New World doves. *Journal of Biogeography* 38: 1069–76. doi: 10.1111/j.1365–2699.2011.02501.x.

Kempter, K.A. 1997. Geologic evolution of the Rincón de la Vieja volcanic complex, northwestern Costa Rica. PhD diss., University of Texas at Austin. 189 pp.

Kerr, R.A. 2007. Mammoth killer impact gets mixed reception from earth scientists. *Science* 316: 1264–65.

Logan, W.B. 2005. Oak: the frame of civilization. New York: W.W. Norton. 320 pp.

Marshall, L.G., S.D. Webb, J.J. Sepkoski, and D.M. Raup. 1982. Mammalian evolution and the great American interchange. *Science* 215: 1351–57.

Martin, P.S. 1973. The discovery of America. *Science* 179: 969–74.

Miller, J.C., D.H. Janzen, and W. Hallwachs. 2006. 100 Caterpillars. Cambridge, MA: Harvard University Press. 264 pp.

Miller, J.C., D.H. Janzen, and W. Hallwachs. 2007. 100 Butterflies and moths. Cambridge, MA: Harvard University Press. 256 pp.

Miller, K.R. 1980. Planificación de parques nacionales para el ecodesarrollo en Latinoamérica. Madrid: Fundación para la Ecología y la Protección del Medio Ambiente. 500 pp.

Murphy, P.G., and A.E. Lugo. 1986. Ecology of tropical dry forest. *Annual Review of Ecology and Systematics* 17: 67–88.

Parsons, J.J. 1983. Beef cattle (*ganado*). In D.H. Janzen, ed., *Costa Rican Natural History*, 77–79. Chicago: University of Chicago Press.

Pennington, R.T., G.P. Lewis, and J.A. Ratter, eds. 2006. *Neotropical Savannas and Seasonally Dry Forests*. Systematics Association Special Volume Series 69. Boca Raton, FL: CRC Press. 484 pp.

Rich, P.V., and T.H. Rich. 1983. The Central American dispersal route: biotic history and paleogeography. In D.H. Janzen, ed., *Costa Rican Natural History*, 12–34. Chicago: University of Chicago Press.

Ripple, W.J., and B. Van Valkenburgh. 2010. Linking top-down forces to the Pleistocene megafaunal extinctions. *BioScience* 60: 516–26.

Sachs, J.P., and C.L. Myhrvold. 2011. A shifting band of rain. *Scientific American* 304: 60–65.

Shane, D.R. 1986. *Hoofprints on the Forest: Cattle Ranching and the Destruction of Latin America's Tropical Forests*. Philadelphia: Philadelphia Institute for the Study of Human Issues. 159 pp.

Teixeira de Moraes, M.L., and A.M. Sebbenn. 2011. Pollen dispersal between isolated trees in the Brazilian savannah: a case study of the neotropical tree *Hymenaea stigonocarpa*. *Biotropica* 43: 192–99.

Thomasa, C.D. 2011. Translocation of species, climate change, and the end of trying to recreate past ecological communities. *Trends in Ecology and Evolution* 26: 216–21.

Titiz, B., and R.L. Sanford. 2007. Soil charcoal in tropical old-growth forest from the Continental Divide to sea level. *Biotropica* 39(6): 673–82.

Tournon, J., and G. Alvarado. 1997. Carte geologique du Costa Rica/ Mapa geologico de Costa Rica. Cartago, Costa Rica: Editorial Tecnologica de Costa Rica. 79 pp. + map.

Wald, M.L. 2012. Turning wood chips into gasoline. *New York Times* blog (November 8, 2012). 1 p.

Wallace, D.R. 1992. *The Quetzal and the Macaw: The Story of Costa Rica's National Parks*. San Francisco: Sierra Club Books. 222 pp.

Wallace, D.R. 1997. *The Monkey's Bridge: Mysteries of Evolution in Central America*. San Francisco: Sierra Club Books. 277 pp.

Walter, K.S., and H.J. Gillett. 1998. *The 1997 IUCN Red List of Threatened Plants.* Cambridge: IUCN (World Conservation Union) and WCMC (World Conservation Monitoring Centre).

Weissberg, A. 2001. *Nicaragua: An Introduction to the Sandinista Revolution*. New York: Pathfinder Press. 45 pp.

Wells, W.V. 1856. *Walker's Expedition to Nicaragua: A History of the Central American War*. New York: Stringer and Townsend. 316 pp.

Wight, S.F. 1860. *Adventures in California and Nicaragua in Rhyme: A Truthful Epic*. Boston: Alfred Mudge & Son. 84 pp.

Zelenitsky, D.K., F. Therrien, G.M. Erickson, C.L. DeBuhr, Y. Kobayashi, D.A. Eberth, and F. Hadfield. 2012. Feathered non-avian dinosaurs from North America provide insight into wing origins. *Science* 338: 510–13.

Part V

The Central and Southern Pacific Seasonally Moist Lowlands and Central Valley

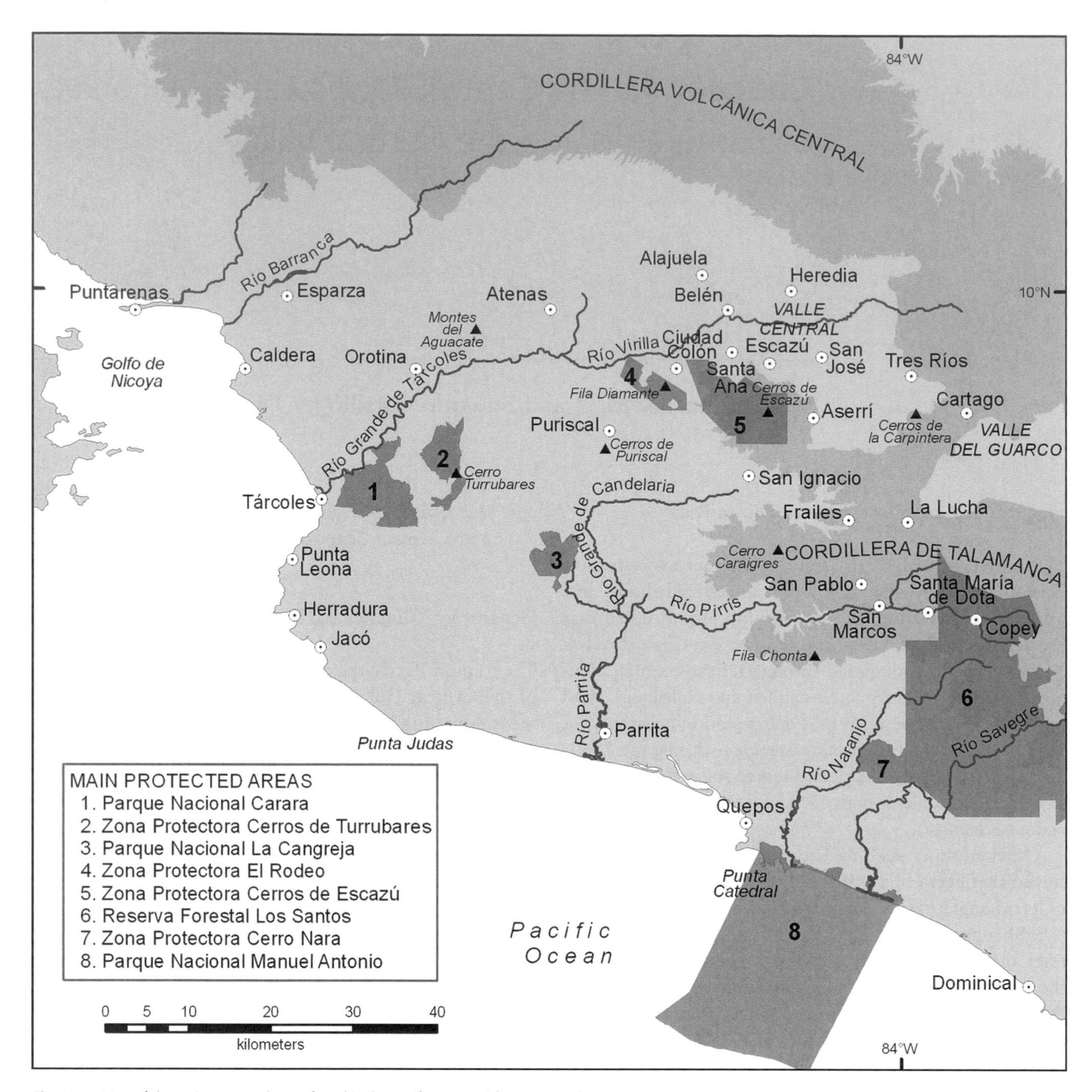

Fig. 11.1 Map of the main protected areas found in the Pacific seasonal forest zone of central Puntarenas and the Central Valley, Costa Rica. *Map prepared by Marco V. Castro.*

definition, the Golfo Dulce (from the Osa Peninsula down to Punta Burica at the border with Panama), the Central Volcanic and Talamanca cordillera highlands, and the Cocos Island regions are considered important areas of endemism. These regions occupy 20% of the national territory.

More recently researchers concluded that the Central Pacific region has an extraordinary and diverse vegetation as well, not only because it is moister due to higher levels of precipitation, but also because it forms a transition between the wet forests of the South Pacific region and the dry forests of the Northern Pacific, endowing it with a high degree of endemism. This is shown by the large number of new and endemic species that this region has recently contributed to science.

Vegetation and Flora

Unlike the Atlantic plains where the forests were removed to create extensive banana plantations and grasslands for cattle ranching (see McClearn et al., chapter 16 of this volume)—and more recently pineapple plantations—the uneven and steep terrain of the Central Pacific region of Costa Rica allowed many forests to survive. Nowadays a large part of the Central Pacific's floristic diversity is duly protected in national parks and protection zones, such as the National Parks of Carara, La Cangreja, and Manuel Antonio, and the Protection Zones of Cerros de Turrubares, El Rodeo, Nara and others. Unfortunately, they have become biological islands in seas of agricultural fields and pastureland. Other parts of the South Pacific region and the Osa Peninsula have also been heavily deforested over the past century, whereas the impressive Corcovado National Park still survives as the most important moist tropical forest along the entire Pacific coast of Central America.

Hartshorn (1993) considers that the lowland forests of Costa Rica's Central and South Pacific region are the most exuberant and diverse in Central America. He believes they are even comparable with those of the Amazon basin. In much of this Pacific moist region, vascular plants are the best-known taxonomic group. In fact, according to data collected by INBio-MNCR (2003), a total of 5,161 plant species have been recorded. They are distributed over different plant growth forms, as follows: 48.1% are herbs, 22.5% trees, 19.2% shrubs, and 10.2% vines.

Zamora et al. (2004) observe that the Área de Conservación Pacífico Central (ACOPAC) is much more heterogeneous than its counterparts along the Atlantic slope when considering types of climates and topography. Thanks to a higher sampling density, more species have been reported from the Pacific than at the wetter Atlantic slope, with 7,348 Pacific species vs. 6,454 Atlantic species. One of the reasons for the higher sampling density along the Pacific slope is the higher level of accessibility here (e.g., as a result of higher road density).

It should be emphasized, however, that ACOPAC is far more extensive than the Pacific lowland and premontane areas analyzed in this chapter. ACOPAC also includes part of the highlands of the Talamanca Mountains, which reach 3,400 m elevation in this part of the country. The highland areas of ACOPAC are protected in the Los Quetzales National Park, the Los Santos Forest Reserve and the Escazú Hills (for an in-depth analysis of the Talamancan highlands, see Kappelle, chapter 14 of this volume).

The analysis shown in Table 11.1 and Fig. 11.2 includes plant species that have been collected to date throughout this broader lowland-highland zone of ACOPAC as presented in the databases of the National Institute of Biodiversity (INBio, 2007; http//atta.inbio.ac.cr). Thus, it should be noted that the total number of species occurring in the Central Pacific region is somewhat lower than the data presented in the Table and Figure.

According to Zamora et al. (2004), the floristic pattern or composition of the Nicoya Peninsula maintains a similar and close relationship along the Pacific coast, particularly in the vicinity of Río Barranca (included in the Macacona and Esparza area) and the port of Puntarenas and Carara National Park areas, showing that this Peninsula could once have been connected to this part of the coast. In fact, the distribution of many species extends as far as the Central Valley, though at elevations above 600 m these are mixed with species typical of the Central Valley.

Ferns and fern allies have been fairly well studied and collected in the area. The most commonly reported species include *Asplenium cuspidatum* var. *triculum*, *Asplenium*

Table 11.1. Plant Genera with More than 25 Known Species in the Central Pacific Region of Costa Rica.

Genus	No. of species	Genus	No. of species
Piper	68	*Inga*	30
Peperomia	56	*Passiflora*	30
Epidendrum	55	*Paspalum*	29
Elaphoglossum	54	*Ocotea*	28
Solanum	50	*Ficus*	28
Psychotria	47	*Pleurothallis*	27
Miconia	46	*Maxillaria*	27
Anthurium	34	*Tillandsia*	27
Philodendron	32	*Polypodium*	25
Asplenium	31	*Thelypteris*	25

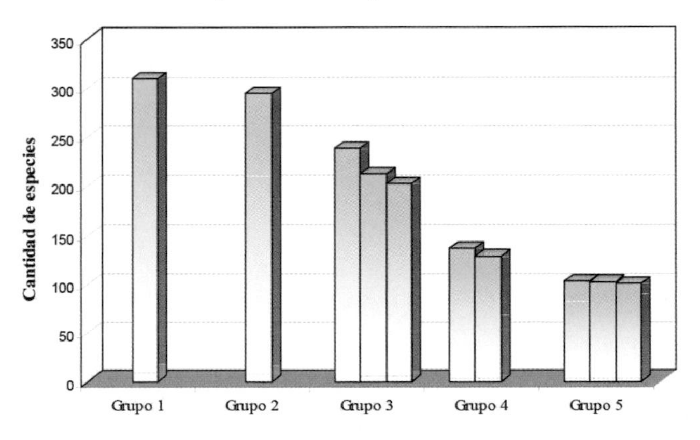

Fig. 11.2 Plant families with the largest number of species in the Central Pacific. Group 1: Orchidaceae; Group 2: Fabaceae; Group 3: Asteraceae, Rubiaceae, and Poaceae; Group 4: Melastomataceae and Piperaceae; Group 5: Araceae, Solanaceae, and Euphorbiaceae.

salicifolium var. *aequilaterale*, *Cnemidaria choricarpa*, *Bolbitis portoricensis*, *Polypodium attenuatum*, *Adiantum amplum*, *Adiantum concinnum*, *Cyclopeltis semicordata*, *Ananthacorus angustifolius*, and *Diplazium turubalense*, some of which are common in open sites and secondary forests. In Carara National Park alone, Jiménez and Grayum (2002) report 113 species while Estrada and Zamora (2004) report 243 species for the Savegre River watershed.

1. The Central Pacific Coast and Foothills

The area is diverse in trees as shown in Fig. 11.3, also based on data from the National Biodiversity Institute (INBio, 2007; http//atta.inbio.ac.cr). Moreover, Costa Rica is located in the Mesoamerican region, which is catalogued as one of the world's nine areas with a high degree of endemism

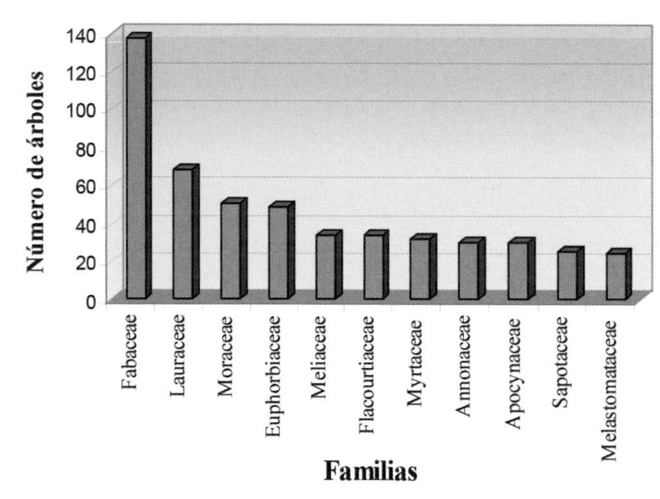

Fig. 11.3 Plant families with the largest number of trees in the Central Pacific.

and, according to Myers et al. (2000), contains 1.7% (i.e., 5,000 species) of the grand total of 300,000 vascular plant species that are believed to exist in the world. According to Zamora et al. (2004), several areas of the country, including the mountains of the Central Pacific region and the Osa Peninsula, have a high degree of endemism and Costa Rica contributes a total of 1,000 species of endemic plants (nearly 10% of its total flora).

The vegetation in some parts of this region (e.g., Carara National Park) has been extensively studied (Jiménez and Grayum 2002). Carara National Park is considered a transition zone between the dry region of the North Pacific of Guanacaste and the wetter region of the South Pacific of Puntarenas Province, with evergreen forests predominating. Lomas Entierro is the site with the largest number of deciduous species, including the same that are found in dry forest, such as *Calycophyllum candidissimum*, *Triplaris melaenodendron*, *Enterolobium cyclocarpum*, and *Bombacopsis quinata*.

In Carara, the primary forest is located in the center of the area, at Lomas Pizote and the 638 m high Montañas Jamaica (Fig. 11.4). Up to four, well-differentiated forest layers can be observed here, particularly when emergent trees such as *Caryocar costaricense* attain heights up to 40 m. In the canopy layer of 30–35 m tall, *Hieronyma oblonga*, *Brosimum utile*, and *Hura crepitans* are common trees, while *Swietenia macrophylla*, *Calophyllum brasiliense*, and *Virola koschnyi* are less common. In the middle layer, frequent species include *Simarouba amara*, *Protium panamense*, *Dendropanax arboreus*, and *Otoba novogranatensis*, among others. The understory is diverse, sometimes characterized by vines and lianas, while in other places there is

Fig. 11.4 Primary forest of Montañas Jamaica, Carara National Park. In the foreground, a *Macrocnemum roseum* (Rubiaceae) tree. *Photograph by Quírico Jiménez.*

a particular abundance of species in Rubiaceae (*Psychotria*, *Faramea*, *Hoffmannia*), Solanaceae (*Solanum*), Arecaceae (*Chamaedorea*), and Piperaceae (*Piper*). Locally, *Olmedia aspera* (Moraceae) may abound.

The secondary forests of Lomas Entierros and those along the highway that leads to the canton of Garabito have three well-differentiated forest *strata* in which *Schizolobium parahyba* is abundant, reaching heights of 30 meters; *Ficus insipida*, *Spondias mombin*, and *Luehea seemannii* are also remarkably present. In the middle *stratum* dominant species are *Bravaisia integerrima*, *Ochroma pyramidale*, *Tabebuia rosea*, and the palm *Attalea rostrata*. In the understory the endemic *Erythrochiton gymnanthus* (Rutaceae) dominates extensively.

From an ecological point of view, Carara constitutes the northernmost distribution limit for many species of trees, since it appears that the Río Grande de Tárcoles has acted as a natural barrier that did not allow the passage of some particular plant species to the north. Jiménez and Grayum (2002) mention the following species that find their distribution limit here: *Peltogyne purpurea*, *Caryocar costaricense*, *Couratari guianensis*, *Brosimum utile*, *Aspidosperma myristicifolium*, *Tachigali versicolor*, the palm *Cryosophila guagara*, and a few others, most of which have a natural distribution along the coast from the Osa Peninsula towards Carara.

The aquatic habitat in Carara, shown in Fig. 11.5, is very important for the functioning of the forest ecosystem and comprises an oxbow lake and marshes along the northeastern side of the Río Grande de Tárcoles. Here, several aquatic floating plants have been found—for example, *Eichhornia crassipes*, *Lemna* sp, *Limnobium laevigatum*, *Salvinia minima*, and *Spirodela intermedia*(Jiménez and Grayum 2002). The banks of rivers and streams also form aquatic habitats but without submerged plants (see Pringle et al, chapter 18 of this volume).

Carara is one of the very few places along the Pacific coast that has a well-developed list of inventoried plants, thanks to Jiménez and Grayum (2002). These authors recorded a total of 1,166 plant species in 633 genera and 150 families (including the ferns). Several plant genera have many species not only present in Carara, but also spreading along the entire Central Pacific coastal region. One species new to science, *Inga jimenezii* (Fabaceae), was collected during this study at a location in Carara and subsequently published by Zamora (1990). Actually, this area contains more than a dozen species new to science that have been recently recorded along the Pacific coast. Several species that resulted to be new records for both the Costa Rican and Central American floras were also collected here, including *Allosanthus trifoliatus*, *Clytostoma pterocalyx*, *Cydista lilacina*, and *Steriphoma paradoxum*.

The vegetation of Carara continues along several water courses into the lower parts of the Cerros de Turrubares Protection Zone, especially around the foothills of Cerro Bares (or Cerro Turrubares sensu stricto) and Cerro San Rafael. These foothills extend into the districts of San Rafael and San Francisco in the canton of Turrubares. However, at elevations above 800 meters the floristic com position changes owing to the presence of different Holdridge life zones that reach as far as Cerro Bares at 1,756 m elevation (Fig. 11.6). Here we find highland species that are present only in the *cordilleras*, such as *Ulmus mexicana*, *Alfaroa williamsi*, *Quercus corrugata*, *Q. seemannii*,

Fig. 11.5 View of the oxbow lake formed by the Río Grande de Tárcoles in Carara National Park. In the foreground, water hyacinth (*Eichhornia crassipes*).
Photograph by Quírico Jiménez.

Fig. 11.6 In the foreground, the secondary forest of Lomas Entierro, Carara National Park. In the background, Cerro Bares and the primary forest of the Cerros de Turrubares Protection Zone.
Photograph by Quírico Jiménez.

Toxicodendron striatum, *Brunellia standleyana*, *Panopsis costaricensis*, *Gordonia fruticosa*, *Matayba oppositifolia*, *Viburnum costaricanum*, *Styrax glabrescens*, *Freziera candicans*, among others. Several species new to science have also been published from this site, including *Psychotria turrubarensis* (Burger and Jiménez 1994), *Hyperbaena eladioana* (Jiménez 1991), and *Tetranema floribundum* (Hammel and Grayum 1995). These hills are important because of their endemism, which is the result of their isolation.

The vegetation of the area around the port of Puntarenas and the Barranca River clearly penetrates the Esparza-Orotina plateau, according to Gómez (1986), and even extends down towards the sea and the Río Grande de Tárcoles. The plateau is apparently composed of deposits of various materials established on sedimentary rocks of the lower Miocene. The entire *meseta* has a floristic composition similar to that of the Nicoya Peninsula; however, there is a noteworthy patch of secondary vegetation in the Tivives Protection Zone that occurs mainly in an extensive area called "las mesas" where *Guazuma ulmifolia*, *Tabebuia rosea*, *Tabebuia ochracea*, and *Tecoma stans* are prominent.

Remnants of other communities still occur in the area although without any protection, such as the tree savannah along both sides of the road linking Orotina and Turrubares, which stretches to the edge of the river gorge of the Río Grande de Tárcoles heading north. Savannah communities are rare and not well studied in the country, according to Elizondo and Jiménez (1988). However, the composition of this savannah, which is called Zulay (Fig. 11.7), has been studied by Jiménez (2004) and catalogued as typical of dry savannahs, with herbs of the Poaceae and Cyperaceae families predominating, and small, fire-resistant trees, growing in isolation, with *Byrsonima crassifolia*, *Curatella americana*, and *Roupala montana* dominating. The biological community that forms this savannah is found on hills with steep to undulating terrain, surrounded by remnants of secondary forest that contain important rare species such as *Swietenia macrophylla*, *Bombacopsis sessilis*, *Calophyllum brasiliense*, and an abundance of *Acosmium panamense* and *Cochlospermum vitifolium*.

The vegetation towards the east of this savannah forms part of the premontane life zone (Tosi 1969) present in several districts of the canton of Puriscal at elevations up to 1200 m a.s.l. Here, the primary forests also disappeared and the few remaining secondary patches are composed of species similar to those found in the Central Valley such as *Cinnamomum cinnamomifolium*, *Persea caerulea*, *Ficus jimenezi*, *Croton draco*, *Sapium glandulosum*, and *Zanthoxylum caribaeum*, among others. This vegetation type seems almost a continuity of the forests observed at the El Rodeo Protection Zone in Ciudad Colón, just below 900 meters elevation. The El Rodeo forest is considered the last patch of undisturbed forest near the Central Valley and has been influenced by floristic elements originating from both the dry Pacific and South Pacific regions (Zamora et al. 2004).

Studies of this forest site by Cascante and Estrada (1999, 2001) reveal that the plant families that have the largest numbers of species in this part of the country are Fabaceae, Lauraceae, and Moraceae. *Pseudolmedia oxyphyllaria*, *Clarisia racemosa*, *Nesiteria concinna*, *Brosimum alicastrum*, and *Tapirira mexicana* are the most common woody species at this site. At Fila Diamante, the forest displays three well-differentiated forest layers; *Cedrela odorata*, *Ficus obtusifolia*, and *Zinowiewia integerrima* dominate the canopy of the forest. Along riparian forest strips species like *Anacardium excelsum* and *Brosimum alicastrum* become common. Some secondary forests in this area are quite homogeneous in species make-up, which shows an abundance of *Garcinia intermedia* individuals, which may attain heights of 10–12 meters.

According to Jiménez (1999), the forest ecosystem at El Rodeo still harbors specimens of highly endangered trees such as *Cedrela salvadorensis*, *Myroxylon balsamum*, and *Sideroxylon capiri*. There is also a healthy population of the endemic *Hauya elegans*. This forest type extends along the river gorge of the Río Grande de Tárcoles towards the Central Valley, and gets even as far as the canton of Santa Ana. The floristic composition, though, certainly changes considerably there.

2. The Central Valley

The Central Valley or *Meseta Central* is a plateau flanked on the north by the Central Volcanic Mountain Range and on the south by the Cerros de Escazú (2,428 m at the highest point, Cerro Rabo de Mico). It is physiographically interconnected with the Talamanca Mountain Range and the hills of Tablazo and La Carpintera. According to Gómez (1986), the geological materials that predominate in this zone are Pleistocene volcanic deposits (in the upper layers below the soil) and sedimentary limestone (deeper layers) dating back from the middle Miocene. For a long time, the rich soils of this valley have been used by farmers (*campesinos*) for coffee and vegetables.

Unfortunately, the vegetation in this area was the first to bear the brunt of deforestation in the region, initially for the purpose of growing coffee and later for building houses, roads, highways, and ultimately cities. According to Zamora et al. (2004), the deterioration of the vegetation found above 1,000 meters was already so advanced during the first half of the past century that many early twentieth century botanists such as Wercklé and Standley did not provide a lot of descriptive data about this part of the country. These renowned scholars divided the vegetation of the Central Valley into two parts: the western Central Valley in the vicinity of San José, influenced by elements of the Central Volcanic Range to the north such as *Ficus costaricana*, *Ehretia latifolia*, *Inga punctata*, *Persea caerulea*, *Cordia eriostigma*, *Acnistus arborescens*, and other species; and the eastern Central Valley or Valle del Guarco (in the vicinity of Cartago), which has a colder, foggy climate and is influenced by species of the Talamanca Range such as *Ficus velutina*, *Toxicodendron striatum*, *Dahlia imperialis*, and *Crossopetalum tonduzii*, among others.

Some of the rivers in the Central Valley, such as the Río Virilla, still contain representatives of various tree species that existed here abundantly in the past—for example, *Ficus jimenezi*, *Trichilia havanensis*, *Erythrina costaricensis*, *Montanoa guatemalensis*, *Mauria heterophylla*, *Cassimiroa edulis*, *Albizia adinocephala*, and *Cinnamomum cinnamomifolium*. The river gorge between the cantons of Belén and Santa Ana still retains some small forest pockets with deciduous species compositions that include *Diphysa americana*, *Cedrela odorata*, *Tabebuia rosea*, and *Lysiloma dicaricata*, all being typical of more seasonal climates. Fig. 11.8 shows the Tárcoles River gorge near San Pablo de Turrubares.

Fig. 11.7 (A) Zulay tree savannahs in the canton of Turrubares; (B) In the lower parts, patches of secondary forest.
Photographs by Quírico Jiménez.

Fig. 11.8 In the center, the river gorge of the Río Grande de Tárcoles between Atenas and Turrubares, where several species of the dry Pacific region extend into the Central Valley.
Photograph by Quírico Jiménez.

Once—before intense urbanization took place—the Central Valley was an area of mere coffee plantations in which several species were used to provide shade to the coffee plants, such as the shade trees *Inga edulis* and *Inga punctata*, as well as the exotic *Albizia carbonaria* and *Erythrina poeppigiana*. Some of those individuals are still found today in the few remaining coffee fields that are like islands in a largely urbanized matrix. Along the slopes of the Central Volcanic Range and the Escazú Hills above 1,200 m elevation, but still below 1,500 m, other species of previously existing premontane forests are still found, particularly in surviving patches of riparian forests. These remnant species are *Myrcia splendens*, *Croton draco*, *Zinowiewia costaricensis*, *Hauya elegans*, *Ficus costaricana*, *Zanthoxylum caribaeum*, *Acnistus arborescens*, *Trema micrantha*, and *Myrcianthes fragrans*.

3. The Turrubares, Puriscal, and Escazú Foothills

At the western tip of the Central Valley the hills of Turrubares and La Potenciana represent the northernmost point of Fila Costeña mentioned earlier in this chapter. However, to the southwest and near the highway that connects the canton of Puriscal with Parrita, a moister kind of vegetation (with nearly 3,500 mm rainfall per year) thrives in La Cangreja National Park. According to Jiménez (1999) this is a biological paradise of special importance located in the canton of Puriscal, as it constitutes the last sizeable remnant of moist evergreen forest, which used to be widespread in this part of the country. The area contains two Holdridge life zones that form a transition between the wet tropical forest and the wet premontane forest on Cerro Cangreja (Tosi 1969). Despite its small size, the site's varied topography, high humidity associated with rainfall, and poor soils endow it with a highly diverse flora that is of special importance owing to its high levels of endemism.

The floristic studies conducted in this area (formerly a Protection Zone, then declared a National Park in 2002) have resulted in the identification of more than 350 species of trees and shrubs (Jiménez and Zamora 1989). However, more detailed data presented by Morales (1993) mention the existence of almost 1,000 species of plants. Undoubtedly this area is among those that have provided most new species to science, including *Plinia puriscalensis* and *Ayenia mastatalensis*, known to the world only from this zone. This area is the type locality for the tree known as *quira* (*Caryodaphnopsis burgeri*), a member of the Lauraceae. Moreover, after analyzing the vegetation in several sampling parcels established in this park, Acosta (1998) reported 148 different species (including 14 endemics) that are represented by individuals bigger than 10 cm diameter at breast height (DBH) per hectare, considering the site as very diverse; these values were exceeded only by sites on the Osa Peninsula studied by Castillo (1996).

Zamora et al. (2004) describe some mountain areas such as the Escazú Hills and the western part of the Los Santos region, which are geographically located between the Talamanca Range and the Pacific coast at elevations above 1,500 meters. They found that the forests thriving in these highlands are defined by a combination of species, especially dominated by oaks (*Quercus* sp). The authors noted that this pattern extends towards the southeast along the Chonta, Nara, and Costeña mountain ranges and that because of their proximity to the coast, they tend to have a greater floristic affinity with the south of the country and with vegetation found at lower elevations.

To the south of the Central Valley the landscape is characterized by the Valley of the Río Grande de Candelaria (between the cantons of Acosta and Aserrí). Here, according to Zamora et al. (2004), an important floristic peculiarity is observed owing to its microclimatic conditions, which is typical of dry forest vegetation or semi-arid sites. As these authors point out, the distinct character of its floristic composition is due to the orographic shadow effect caused by the hills of Turrubares and Caraigres and the mountains of Puriscal and San Ignacio de Acosta, which enclose this habitat along the Río Grande de Candelaria at its confluence with the Río Pirrís.

Zamora et al. (2004) also mention that this area has vicariate populations of species known only from the dry forest growing at relatively higher elevations. These species include *Astronium graveolens*, *Dalbergia retusa*, *Hyperbaena toduzii*, *Platymiscium parviflorum*, *Calycophyllum candidissimum*, *Thouinidium decandrum*, *Guettarda macrosperma*, *Plumeria rubra*, and *Eugenia hiraefolia*.

One of the region's key coastal sites is Manuel Antonio National Park. Its vegetation was studied thoroughly by Harmon (2003), who described 141 species of trees in several of the park's vegetation types, including secondary forest patches where *Apeiba tibourbou*, *Miconia argentea*, *Goethalsia meiantha*, *Cecropia peltata*, *Trema micrantha*, and *Sapium* dominated. The primary forest with up to four layers has several emergent trees (*Ceiba pentandra*, *Copaifera aromatica*, *Brosimum utile*, *Hymenaea courbaril*, *Sloanea picapica*, and *Pradosia atroviolacea*) that can attain heights up to 50 m! An outstanding feature occurs at Punta Catedral where *Myrcianthes fragrans*, a rare tree for this zone, is present. It is generally found in the higher parts of the *cordilleras*. A species new to science is observed here: *Buchenavia costaricensis*, which is also present on the Osa Peninsula.

The canopy, reaching a height of 35 m, is dominated by *Licania operculipetala*, *Cynometra hemitomophylla*, *Calophyllum longifolum*, and *Virola koschnyi*. The subcanopy, ranging from 20 to 25 m, is dominated by *Vitex cooperi*, *Tabebuia rosea*, *Virola sebifera*, *Tetrathylacium johansenii*, and *Batocarpus costaricensis*. The predominant species in the understory are *Garcinia intermedia*, *Nectandra silicifolia*, *Swartzia panamensis*, *Trichilia pallida*, *Pouteria subrotata*, and the palm *Attalea rostrata*. In the sub-xeric (drier) forest located behind Puerto Escondido on the steep cliffs of Punta Serrucho and the hillsides of Punta Catedral, most species are deciduous and some attain heights of 15–20 m, with *Bursera simaruba*, *Ochroma pyramidale*, *Plumeria rubra*, *Sapium glandulosum*, *Pseudobombax septenatum*, and *Clusia rosea* dominating. The coastal forest is a narrow strip of land with vegetation that borders the sandy beach of the park; *Hibiscus pernambucensis*, *Posoqueria latifolia*, *Bombacopsis sessilis*, *Maytenus* sp., *Tabebuia rosea*, and *Chrysobalanus icaco* grow here abundantly. Finally there are mangrove forest patches characterized by *Rhizophora mangle*, *Laguncularia racemosa*, and *Pelliciera rhizophorae*.

4. The Savegre Valley and Fila Chonta

Although the upper part of the Savegre Valley has received significant attention from botanists during the last thirty years (e.g., Kappelle 1996, 2006, and 2008, and Kappelle et al. 1989, 1991, 1994, 1995, 2000, and see chapter 14 of this volume), it was not until recently that the flora of the lower elevational zones of the Savegre Valley was also studied in detail (Acevedo et al. 2002). These latter authors assessed the floristic diversity of the 589 km^2 watershed of the Savegre River, southeast of Puerto Quepos, where the Los Santos Forest Reserve is located (Kappelle and Juárez 1994). The study reported 1,526 species of vascular plants, of which 1,110 were new records for the area and are evidence of its floristic wealth. Elevations in the watershed range from 0 to 3,491 m (Cerro de la Muerte) and the altitudinal gradient and diversity of climates give it enormous floristic richness (Kappelle et al. 1989, 2000, Acevedo et al. 2002). Details on the forests thriving in the higher parts of the Savegre Valley ($\geq$ 2,000 m elevation) are presented in a chapter by Kappelle (chapter 14 of this volume), which discusses the Talamancan montane cloud forest ecosystem.

Acevedo et al.'s (2002) study also mapped 44 ecosystems distributed over six altitudinal zones. For the scope of this chapter, however, the area of most interest is located in the lower part of the Savegre watershed, where the Cerro Nara Protection Zone is located, covering parts of the districts of Naranjillo, Quepos, and Savegre. This important protection zone receives up to 6,000 mm of precipitation per year. According to Jiménez (1999), the study reported 11 endangered tree species including *Caryocar costaricense*, *Mora oleifera*, *Parkia pendula*, *Anthodiscus chocoensis*, and *Caryodaphnopsis burgeri*. In addition, from this zone there were new reports of arboreal species that had been known previously only in the Osa Peninsula, including *Anthodiscus chocoensis*, *Newtonia suaveolens*, and *Parkia pendula*.

More recently Estrada and Zamora (2004) described and

analyzed species richness and floristic patterns along an altitudinal gradient of 0 to 3,400 meters above sea level in the Savegre River watershed, reporting a richness of 2,152 species of vascular plants, with 132 species that are endemic to our country and five that are exclusive to this watershed; the endemic species belong to families such as the Orchidaceae, Asteraceae, Melastomataceae, and Rubiaceae. It should be noted that the study by Estrada and Zamora (2004) also includes species from the high parts of the Talamanca cordillera—for example, the *páramos*, which are further discussed by Kappelle and Horn (chapter 15 of this volume).

Fungal Diversity

Fungi represent one of the least-known groups in the lower parts of the Central Pacific region; however, a recent study by Ruiz-Boyer (2006) found high taxonomic diversity. Out of a total of 310 collections of polyporoid fungi made in the zone, 131 are different species in 57 genera and 11 families, with the Coriolaceae being the best represented family with 39 genera and 72 species. The genus *Phellinus* with 10 species is the most abundant. The most common species include *Amauroderma schomburgkii*, *Coriolopsis byrsina*, *Coriolopsis floccosa*, *Coriolopsis polyzoma*, *Cyclomyces iodinus*, *Fomes fasciatus*, *Fomitopsis cupreorosea*, *Fomitopsis feei*, *Phellinus gilvus*, *Phellinus linteus*, and others.

Faunal Diversity

Although Costa Rica's Central Pacific region is considered a transition zone between tropical moist forest and tropical dry forest, with clear seasonal patterns, and despite its importance and biological richness, few mammal and bird studies have been carried out. The existing studies tend to be very specific or are limited to simple lists of species for a particular national park or geographic zone—for example, the studies by Wilson (1983) and Stiles (1983) in several parts of the country, including La Selva Biological Station in Sarapiquí and Santa Rosa National Park. With respect to the fauna, the differences between this zone and the Nicoya Peninsula are not very significant, although the fauna of the Central Pacific is obviously less well documented.

In this area, cattle ranching and deforestation for agricultural purposes (first for banana and subsequently for African oil palm plantations in the Aguirre-Parrita zone) resulted in the loss of most of the original lowland forests near the coast. Populations of species such as the jaguar (*Panthera onca*) and the tapir (*Tapirus bairdii*) were reduced and are now restricted to areas with forest remnants (such as Fila Chonta, which forms part of the Fila Costeña). These species require very large home ranges and studies in Costa Rica have shown that a jaguar needs an area of at least 20 km^2 to meet all its food and shelter needs (Carrillo 2000). One species that has been reported as endangered is the squirrel monkey (*Saimiri oerstedii*), whose populations in the Quepos and Manuel Antonio National Park region have diminished considerably (Wong and Carrillo 1996).

However, in the Central Pacific zone, the marshes and the oxbow lake that occupy an abandoned river meander formed by the Río Grande de Tárcoles to the northeast of Carara National Park—estimated to be up to 600 m long and two meters deep—are important habitats that should be rescued for diverse species of amphibians and reptiles such as American crocodiles (*Crocodylus acutus*) and aquatic birds such as roseate spoonbills (*Ajaia ajaja*), anhingas (*Anhinga anhinga*), Northern jacanas (*Jacana spinosa*), and others. The mangroves of the area also provide refuge for many animal species (see J. Cortés, chapter 5, and J. Jiménez, chapter 20 of this volume).

Mammals

In an inventory and altitudinal reconnaissance from sea level to 3,491 meters elevation conducted in the Savegre River zone, Rodríguez (2004) recorded 113 species of mammals (representing 54% of the mastofauna of the country). Of these, four were new records for the region and five were species endemic to Costa Rica, including the squirrel monkey and a pocket gopher (*Orthogeomys underwoodii*); furthermore, 19 species are ranked in some category of threat. One interesting point noted in Rodríguez' study is that with increased elevation the number of species diminishes while endemism increases, whereas altered lowland forests are the areas with highest diversity.

Of the total species reported by Rodríguez (2004) for this zone, 61 species are bats, 25 are rodents, 12 are carnivores (with four species of felines), four are primates, three are in the order Didelphimorphia (*Didelphis marsupialis*, *Marmosa mexicana*, and *Philander opossum*), three in the order Xenanthra (*Tamandua mexicana*, *Bradypus variegatus*, and *Dasypus novemcinctus*), two in the order Artiodactyla (*Pecari tajacu* and *Mazama americana*), one in the order Insectivora (*Cryptotis nigrescens*), one in the order Perissodactyla (*Tapirus bairdii*), and finally, one in the order Lagomorpha (*Sylvilagus dicei*). For more information, see the study by Rodríguez. Species such as the white-lipped peccary (*Tayassu pecari*) disappeared from the area a long time ago owing to hunting pressure.

One of the peculiarities of this area with respect to

Fig. 11.9 Crab-eating raccoon (*Procyon cancrivorus*). *Photograph by Eduardo Carrillo.*

mammals is that it has two species of raccoons (*Procyon lotor* and *Procyon cancrivorus*). The first is quite common, but the crab-eating or South American raccoon (Fig. 11.9) reaches its northernmost distribution here and has been little studied (Carrillo 1989).

Birds

Birds have also been extensively studied in the Savegre River watershed. Sánchez et al. (2004) recorded 508 species, of which 429 are resident and 79 are migratory, including 53 of Costa Rica's 75 endemic species, most of which live exclusively in high montane and subalpine-alpine forests. This study also reported 40 species that are considered endangered and some that are included in the CITES and IUCN lists, including the king vulture (*Sarcoramphus papa*), the Muscovy duck (*Cairina moschata*), the mealy amazon (*Amazona farinosa*), the great curassow (*Crax rubra*), the crested eagle (*Morphnus guianensis*), the peregrine falcon (*Falco peregrinus*), and others. Another study on the aquatic avifauna of the lower watershed of the Savegre and Naranjo rivers conducted by Alvarado (2004) reports 54 species of aquatic birds, of which 26 are resident, 21 migratory, and 7 migratory residents. For further information about the fauna of this zone, see Tiffer-Sotomayor (2003).

This area has the second largest population of scarlet macaws (*Ara macao*); the largest is found on the Osa Peninsula (Fig. 11.10), a flagship species that is common in Carara National Park where these birds nest and feed. They also have been seen resting in the mangroves near the Guacalillo zone. Other important species of the Central Pacific are the three-wattled bellbird (*Procnias tricarunculata*) that has been seen completing altitudinal migrations between Carara National Park and the Cerro Turrubares Protection Zone, as well as the chestnut-mandibled toucan (*Ramphastos swainsonii*). One rare and threatened species is the yellow-billed cotinga (*Carpodectes antoniae*) observed in La Cangreja National Park in Puriscal.

Around the Central Valley in two areas of the Escazú Hills between 1,400 and 2,100 meters elevation, Alvarado and Durán (2006) reported 104 species, of which 82% are resident, 14% are Nearctic migrants, 3% are migratory residents, and 1% concern intertropical migrants. Undoubtedly, many of the species that they identified at higher altitudes are also found at lower altitudes. Some may even reach the El Rodeo Protection Zone (see earlier in this chapter) to the west of San José. There Durán and Sánchez (2003) found that the local avifauna contains mainly species that are not really dependent on the forest and thrive well in open zones. Here, there is a strong affinity with the avifauna known from the dry Pacific area.

In this region the authors report 162 bird species belonging to 39 families, of which 67% are permanent residents, 24% are latitudinal migrants, 7% are considered nomads,

Fig. 11.10 Scarlet macaw (*Ara macao*). *Photograph by Eduardo Carrillo.*

while 2% are altitudinally migratory, mostly insectivorous and omnivorous species. Species of particular interest include the fiery-billed aracari (*Pteroglossus franzii*), which is endemic to Costa Rica's South Pacific region and western Panama, as well as the little tinamou (*Crypturellus soui*), which has disappeared from the Central Valley where it used to live in the past.

Herpetofauna

Leenders (2001) list 9,850 species of amphibians and reptiles that are found on the planet. For Costa Rica he reports a total of around 400 species whose population patterns are typical of the New World Tropics. According to Savage (2002), the Costa Rican herpetofauna consists of 396 species of amphibians and reptiles (including introduced species), which are distributed over 37 families and 140 genera, of which 174 are amphibians (44 endemics) and 222 are reptiles (17 endemics).

However, more recently Solórzano (2004) reported that 225 species of reptiles have been identified—three more than Savage recorded—of which 137 (61%) are snakes. These are grouped into 65 genera of which around 92% of the species are associated with tropical and sub-tropical regions ranging from 0 to 150 m elevation. The Pacific slope contains around 43% of all species known, including 35 species known from the northwestern Pacific region; only 18 are found with elements typically associated with xeric vegetation formations (Sasa and Solórzano 1995).

Although the reptile and amphibian groups are less well studied in this part of the country, Tiffer-Sotomayor (2003) reports that in the Savegre River watershed area and its surroundings during several field trips, a total of 39 species of reptiles and 28 species of amphibians were observed while the Harlequin toad (*Atelopus varius*) and several species of *Eleutherodactylus* had disappeared by then. The study also found that this group had a greater affinity with the herpetofauna documented for the South Pacific, and even Bolaños and Sasa (1991) estimated that 98% of the species in this zone have also been found on the Osa Peninsula.

Since the Central Pacific region has higher levels of humidity and still contains primary forests, it is actually more diverse in species than, for instance, the Nicoya Peninsula. This is consistent with the findings by Leenders (2001) who states that the herpetofauna is present in high densities in undisturbed areas of primary forests, especially in trees, palms, lianas, epiphytes, and other plants—more than in cultivated areas or secondary forests.

There is little information available about the habitat and status of herpetofauna populations of Costa Rica's Central Pacific. Savage and Villa (1986) divide the Pacific coast of Costa Rica into two herpetofaunal zones: (a) the lowlands of the northwestern Pacific, the adjacent hills, and the western central plateau; and (b) the lowlands of the southwestern Pacific and adjacent hills. Since the Central Pacific is a transition zone between the tropical dry forest and the tropical moist forest, it is possible to find amphibian and reptile species representative of both life zones in the area.

With regard to reptiles, species such as the American crocodile (*Crocodylus acutus*) (Fig. 11.11) appear to have increased their populations in the Pacific rivers. This is especially the case in the Tárcoles River, where significant crocodile concentrations can be observed near the bridge

Fig. 11.11 Group of crocodiles under the bridge at the Río Grande de Tárcoles.
Photograph by Eduardo Carrillo.

of the coastal highway that leads to the canton of Aguirre. Individuals that measure a length of more than four m can be seen basking in the sun on the sand banks of the river, allowing many tourists to stop and take their breath-taking photographs of these extraordinary animals.

In this part of the country, snakes such as the fer-de-lance (*Bothrops asper*) are abundant and can reach lengths of up to 2.2 m (Solórzano 2004). Other snakes such as the black-headed bushmaster (*Lachesis melanocephala*) may once have been observed frequently in the high areas of Puriscal. Here an individual was reportedly killed in 1967 (Solórzano 2004). This species of pit viper is endemic to Costa Rica and has possibly disappeared from the Central Pacific region due to intense deforestation and the persecution by man it has suffered because of its venomous properties.

Conservation

The Fila Chonta region harbors the last extensive patch of unprotected forest that remains in this region. It is therefore important to put it under some form of protection or strict management (e.g., as a National Park or Biological Reserve). One important characteristic of this hilly Pacific sector is that it serves as an altitudinal corridor for migratory fauna, since its forests extend from 400 m above sea level to elevations over 2,000 m.

The establishment of a biological corridor stretching between Carara National Park and Los Quetzales National Park would be very strategic because it would guarantee the protection of one of Costa Rica's most biodiverse zones, and could cover the Cerros Turrubares Protection Zone, La Cangreja National Park, the Cerro Nara Protection Zone, Fila Chonta, and Los Santos Forest Reserve.

The Executive Decree 33494-MINAE of 1997 established the key Paseo de Las Lapas Biological Corridor, which extends from the mangroves at Guacalillo up to the Candelaria River. However, this low-status corridor appears to offer insufficient protection to the many animal species that live here.

The beauty and accessibility of the beaches located in Costa Rica's Central Pacific region have led to an unregulated development of touristic infrastructure, with consequently negative effects on the natural resources that still remain in the area. Pollution, fragmentation, and high-density construction at some places are factors that must be corrected promptly if we wish to maintain functional ecosystems in this part of the country.

The flora of the Central Pacific is quite well known with the exception of the Fila Chonta region. As a common denominator, a major problem with the different fauna groups is the lack of knowledge about their natural history and the degree of threat that they face.

The pressure that is currently being exerted on the habitats of many species is increasing day by day. Chaotic tourism development is destroying forests, wetlands, and lagoons for the purpose of building tourism complexes, without taking into account the natural resources in their setting, especially in the maritime terrestrial zone and the Fila Costeña. In pursuit of "real estate development," it has been forgotten that the main attractions for the tourists who visit the country are its plants and animal species.

Unfortunately, the fauna of the Central Valley has nearly disappeared owing to the unplanned expansion of housing developments and urbanization; even the shaded coffee farms that once provided a refuge for small mammals and an infinite number of birds have disappeared. Rare riparian forests in the protection zones along some rivers could provide a basis for creating small biological corridors to protect both flora and fauna.

Today, clear-cutting, agriculture, unsustainable tourism, second-home building, and illegal hunting continue to take their toll in this part of the country, as is exemplified around the Manuel Antonio National Park. Following the analysis of multi-temporal (1985–2008) remotely sensed land cover data in and around this national park, in combination with questionnaire data from local families, Broadbent et al. (2012) found that while regeneration occurred and forest fragmentation decreased, the park is rapidly becoming isolated. Due to the rapid expansion of oil palm plantations adjacent to the park and throughout the lowland areas, ecological connectivity in the area is quickly decreasing (Broadbent et al. 2012). At the same time, these authors report, local communities perceive decreases in wildlife abundance throughout the area, possibly as a result of illegal hunting activities.

It is hoped that awareness raising and changes in attitude directed at sustainable approaches in natural resource management and conservation will significantly help reduce these threats that affect the magnificent biological diversity of these Pacific coastal forests and valleys in the central sector of Costa Rica.

References

Acevedo, H., J. Bustamante, L. Paniagua, and R. Cháves. 2002. *Ecosistemas de la Cuenca Hidrográfica del Río Savegre, Costa Rica*. Santo Domingo de Heredia: Editorial INBio. 351 pp. + maps.

Acosta, L. 1998. Análisis de la composición florística y estructura para la vegetación del piso basal de la Zona Protectora La Cangreja, Mastatal de Puriscal. *Informe de Práctica de Especialidad*. Cartago, Costa Rica: Instituto Tecnológico de Costa Rica. 69 pp.

Alvarado, G. 2004. Caracterización de la avifauna acuática en la cuenca baja de los ríos Savegre y Naranjo, Costa Rica. *Brenesia* 61: 95–103.

Alvarado, G., and F. Durán. 2006. Avifauna de los Cerros de Escazú, Costa Rica. *Brenesia* 66: 37–47.

Bolaños, F., and M. Sasa. 1991. Anfibios y reptiles de la Península de Osa, Costa Rica. In R. Soto, ed., Sondeo ecológico rápido de la Península de Osa, Informe Boscosa. 19 pp. Mimeographed.

Broadbent, E.N., A.M. Almeyda Zambrano, R. Dirzo, W.H. Durham, L. Driscoll, P. Gallagher, R. Salters, J. Schultz, A. Colmenares, and S.H. Randolph. 2012. The effect of land use change and ecotourism on biodiversity: a case study of Manuel Antonio, Costa Rica, from 1985 to 2008. *Landscape Ecology* 27(5): 731–44. DOI: 10.1007/s10980-012-9722-7.

Burger, W., and Q. Jiménez. 1994. A New Species of *Psychotria* subgenus *Psychotria* (Rubiaceae) from Costa Rica. *Novon* 4: 206–8.

Carrillo, E. 1989. Influencia del turista en los patrones de comportamiento del mapachín en el Parque Nacional Manuel Antonio. Master's thesis, Escuela de Ciencias Ambientales, Universidad Nacional. 70 pp.

Carrillo, E. 2000. Ecology and conservation of white-lipped peccaries and jaguars in Corcovado National Park, Costa Rica. PhD diss., University of Massachusetts at Amherst, Department of Wildlife and Fisheries Conservation. 128 pp.

Cascante, A., and A. Estrada. 1999. Lista con anotaciones de la flora vascular de la Zona Protectora El Rodeo, Costa Rica (un bosque húmedo premontano del Valle Central). *Brenesia* 51: 1–44.

Cascante, A., and A. Estrada. 2001. Composición florística y estructura de un bosque húmedo premontano en el Valle Central. *Revista de Biología Tropical* 49(1): 213–25.

Castillo, M. 1996. Comportamiento del bosque natural después del aprovechamiento forestal en tres sitios de la Península de Osa. Master's thesis, Instituto Tecnológico de Costa Rica. Cartago, Costa Rica. 152 pp.

Celis, R. 1998. La relación entre el desarrollo económico y la destrucción y conservación del bosque en Costa Rica. In *Memorias Conservación del Bosque en Costa Rica*, 151–60. San José, Costa Rica: Academia Nacional de Ciencias de Costa Rica.

Durán, F., and J. Sánchez. 2003. Avifauna de la Zona Protectora El Rodeo, Costa Rica: anotaciones sobre diversidad e historia natural. *Brenesia* 59–60: 35–48.

Elizondo, L.H., and Q. Jiménez. 1988. La sabana arbolada "El Escobio", Liberia, Guanacaste, Costa Rica. *Revista de Biología Tropical* 36: 175–85.

Elizondo, L.H., Q. Jiménez, R.M. Alfaro, and R. Chaves. 1989. *Contribución a la Conservación de Costa Rica*. 1. Áreas de Endemismo. 2. Vegetación Natural. San José, Costa Rica: Fundación Neotrópica. 107 pp.

Estrada, A., and N. Zamora. 2004. Riqueza, cambios y patrones florísticos en un gradiente altitudinal en la cuenca hidrográfica del río Savegre, Costa Rica. *Brenesia* 61: 1–52.

Gómez, L.D. 1986. Vegetación de Costa Rica. Vol. 1. In L.D. Gómez, ed., *Vegetación y Clima de Costa Rica*. Con 10 mapas (escala 1:200.000). San José, Costa Rica: EUNED.

Hammel, B., and M.H. Grayum. 1995. The genus *Tetranema* (Scrophulariaceae) in Costa Rica, with two new species. *Phytologia* 79(4): 269–80.

Harmon, P. 2003. *Árboles del Parque Nacional Manuel Antonio*. Santo Domingo, Costa Rica: Editorial INBio. 400 pp.

Hartshorn, G.S. 1993. Plants: Introduction. In D.H. Janzen, ed., *Costa Rican Natural History*, 118–57. Chicago: University of Chicago Press.

Herrera, W. 1986. Clima de Costa Rica. Vol. 2. In L.D. Gómez, ed., *Vegetación y Clima de Costa Rica*. Con 10 mapas (escala 1:200.000). San José, Costa Rica: EUNED.

INBio. 2003. Caracterización de la vegetación del Área de Conservación Pacífico Central (ACOPAC), Costa Rica. San José, Costa Rica: National Biodiversity Institute (INBio) and National Museum of Costa Rica (INBio-MNCR). Mimeographed.

Jiménez, Q. 1991. Una nueva especie de *Hyperbaena* Miers ex Benth. (Menispermaceae) para Costa Rica. *Brenesia* 35: 113–16.

Jiménez, Q. 1999. Importancia Biológica de la Zona Protectora La Cangreja, Puriscal. Document prepared for the Fundación Ecotrópica. 5 pp. Mimeographed.

Jiménez, Q. 2004. Importancia Biológica de la Sabana Arbolada Zulay o Cerro Rayo, Cantón de Turrubares. 14 pp. Mimeographed.

Jiménez, Q., and M.H. Grayum. 2002. Vegetación del Parque Nacional Carara, Costa Rica. *Brenesia* 57–58: 25–66.

Jiménez, Q., and N. Zamora. 1989. Lista preliminar de las especies de la Zona Protectora La Cangreja, Puriscal. 7 pp. Mimeographed.

Kappelle, M. 1991. Distribución altitudinal de la vegetación del Parque Nacional Chirripó, Costa Rica. *Brenesia* 36: 1–14.

Kappelle, M. 1996. *Los Bosques de Roble (*Quercus*) de la Cordillera de Talamanca, Costa Rica: Biodiversidad, Ecología, Conservación y Desarrollo*. Amsterdam / Santo Domingo de Heredia: Universidad de Amsterdam (UvA) / Instituto Nacional de Biodiversidad (INBio). 336 pp.

Kappelle, M., ed. 2006. *Ecology and Conservation of Neotropical Montane Oak Forests*. Ecological Studies Series, Vol. 185. Berlin: Springer Verlag. 483 pp.

Kappelle, M. 2008. *Biodiversity of the oak forests of tropical America / Biodiversidad de los bosques de roble (encino) de la América tropical*. Bilingual edition. Santo Domingo de Heredia, Costa Rica: Instituto Nacional de Biodiversidad (INBio). 336 pp.

Kappelle, M., A.M. Cleef, and A. Chaverri. 1989. Phytosociology of montane *Chusquea-Quercus* forests, Cordillera de Talamanca, Costa Rica. *Brenesia* 32: 73–105.

Kappelle, M., and M.E. Juárez. 1994. The Los Santos Forest Reserve: a bufferzone vital for the Costa Rican La Amistad Biosphere Reserve. *Environmental Conservation* 21(2): 166–69.

Kappelle, M., P.A.F. Kennis, and R.A.J. de Vries. 1995. Changes in diversity along a successional gradient in a Costa Rican upper montane *Quercus* forest. *Biodiversity and Conservation* 4: 10–34.

Kappelle, M., E. van Omme, and M.E. Juárez. 2000. Lista de la flora vascular terrestre de la cuenca superior del Río Savegre, San Gerardo de Dota, Costa Rica. *Acta Botánica Mexicana* 51: 1–38.

Kappelle, M., H.P. van Velzen, and W.H. Wijtzes. 1994. Plant communities of montane secondary vegetation in the Cordillera de Talamanca, Costa Rica. *Phytocoenologia* 22(4): 449–84.

Leenders, T. 2001. *A Guide to Amphibians and Reptiles of Costa Rica*. Miami: Distribuidores Zona Tropical. 305 pp.

Morales, J.F. 1993. Estudio preliminar de la Flórula de la Zona Protectora La Cangreja, Puriscal. Informe de Práctica de Especialidad. Cartago, Costa Rica: Instituto Tecnológico de Costa Rica. 57 p.

Myers, N., R.A. Mittermeier, C.G. Mittermeier, G.A.B. da Fonseca, and J. Kent. 2000. Biodiversity hotspots for conservation priorities. *Nature* 403: 853–58.

Rodríguez, B. 2004. Distribución altitudinal, endemismo y conservación de mamíferos en la cuenca del río Savegre, Costa Rica. *Brenesia* 61: 53–62.

Ruiz-Boyer, A. 2006. Los hongos poliporoides (Basidiomycetes) del Área de Conservación Pacífico Central (ACOPAC), Costa Rica. *Brenesia* 65: 19–41.

Sader, S., and A. Joyce. 1988. Deforestation rates and trends in Costa Rica, 1940 to 1983. *Biotropica* 20: 11–19.

Sánchez, J., G. Barrantes, and F. Durán. 2004. Distribución, ecología y conservación de la avifauna de la cuenca del río Savegre, Costa Rica. *Brenesia* 61: 63–93.

Sánchez-Azofeifa, G.A., R.C. Harris, and D.L. Skole. 2001. Deforestation in Costa Rica: a quantitative analysis using remote sensing imagery. *Biotropica* 33(3): 378–84.

Sasa, M., and A. Solórzano. 1995. The reptiles and amphibians of Santa Rosa National Park, with comments about the herpetofauna of xerophitic areas. *Herpetological Natural History* 3: 113–26.

Savage, J.M. 2002. *The Amphibians and Reptiles of Costa Rica: A Herpetofauna between Two Continents, between Two Seas*. Chicago: University of Chicago Press. 934 pp.

Savage, J.M., and J. Villa. 1986. *Introducción a la Herpetofauna de Costa Rica*. Ann Arbor, Michigan: Cushing-Malloy, Inc.

Solórzano, A. 2004. *Serpientes de Costa Rica: Distribución, Taxonomía e Historia Natural*. Santo Domingo de Heredia, Costa Rica: Editorial INBio, National Institute of Biodiversity. 781 pp.

Stiles, F.G. 1983. Aves. In D.H. Janzen, ed., *Costa Rican Natural History*, 515–55. Chicago: University of Chicago Press.

Thrupp, L.A. 1988. Environmental initiatives in Costa Rica: a political ecology perspective. *Society and Natural Resources* 3(1): 243–56.

Tiffer Sotomayor, R., ed. 2003. Inventario y análisis ecológico de fauna en la zona de influencia de la cadena de desarrollo hidroeléctrico en la cuenca del río Savegre. Final report. Fauna Savegre Project. San José, Costa Rica: Tropical Science Center (TSC) / Centro Científico Tropical (CCT). 450 pp + Anexos.

Tosi, J.A. 1969. Ecological map of Costa Rica: based on L.R. Holdridge's Classification System of Life Zones. San José, Costa Rica: Tropical Science Center.

Vaughan, C. 1988. Biodiversity. In *Memoria del Primer Congreso sobre Estrategias de Conservación para el Desarrollo Sostenible de Costa Rica*, 59–69. San José, Costa Rica: ECODES / MIRENEM.

Wilson, D.E. 1983. Checklist of mammals. In D.H. Janzen, ed., *Costa Rican Natural History*, 443–47. Chicago: University of Chicago Press.

Wong, G., and E. Carrillo. 1996. Squirrel monkey viewing and tourism in Costa Rica. In R. and C. Prescott-Allen, eds., *Assessing the Sustainability of Wild Species*, The IUCN Species Survival Commission, 12.

Zamora, N. 1990. Nuevas especies de *Inga* Miller (Mimosaceae) para Mesoamerica. *Brenesia* 33: 99–118.

Zamora, N., B.E. Hammel, and M.H. Grayum. 2004. Vegetación. In B.E. Hammel, M.H. Grayum, C. Herrera, and N. Zamora, eds., *Manual de las Plantas de Costa Rica*, Vol. 1, 91–216. St. Louis, MO: Missouri Botanical Garden, Costa Rican National Institute of Biodiversity, National Museum of Costa Rica. 299 pp.

Chapter 12 The Southern Pacific Lowland Evergreen Moist Forest of the Osa Region

Lawrence E. Gilbert[1,*], Catherine A. Christen[2], Mariana Altrichter[3], John T. Longino[4], Peter M. Sherman[3], Rob Plowes[1], Monica B. Swartz[1], Kirk O. Winemiller[5], Jennifer A. Weghorst[6], Andres Vega[7], Pamela Phillips[8], Christopher Vaughan[9], and Maarten Kappelle[10,11]

He wondered what it was about science that made men try to control and withhold a place like the Osa for the purpose of observing wildlife.
—Anderson 1989, p. 217

Dedication

We dedicate this chapter to Álvaro Ugalde (1946–2015), who passed away on February 15, 2015. Considered a father of Costa Rica's national park system, Álvaro stood at the base of the creation of Corcovado National Park—centerpiece of the extraordinarily species-rich Osa Peninsula—in October 1975. According to the *Tico Times* (February 17, 2015), Álvaro said that work remains to guarantee preservation of Costa Rica's national parks, especially Corcovado, where development pressures and gold-panners put it "in danger of extinction." A biologist by training, Álvaro worked long decades with local communities surrounding national parks to teach them benefits of preserving the land for the future. His legacy will remain for generations to come and his memory will live on in the hearts of his family and friends.

[1] Section of Integrative Biology and Brackenridge Field Laboratory, University of Texas, Austin, TX 78712, USA
[2] Smithsonian Conservation Biology Institute, Front Royal, VA 22630, USA
[3] Prescott College, Prescott, AZ 86301, USA
[4] Department of Biology, University of Utah, Salt Lake City, UT 84112, USA
[5] Department of Wildlife and Fisheries Sciences, Texas A&M University, College Station, TX 77843, USA
[6] Natural History Museum and Biodiversity Research Center, University of Kansas, Lawrence, KS 66045, USA
[7] AMBICOR, 400m E, 75m S y 75m E de la Municipalidad, Tibas, Costa Rica
[8] USDA-ARS, Livestock Insects Research Laboratory, Kerrville, TX 78028, USA
[9] ICOMVIS, Universidad Nacional, Heredia, Costa Rica, and Department of Forest and Wildlife Ecology, University of Wisconsin, Madison, WI 53706, USA
[10] World Wide Fund for Nature (WWF International), Avenue du Mont-Blanc 1196, Gland, Switzerland
[11] Department of Geography, University of Tennessee, Knoxville, TN, USA
* Corresponding author

Introduction

Philip Calvert, a contributor to the great natural history survey *Biologia Centrali Americana*, explored Costa Rica with his wife, Amelia, collecting dragonflies and making general observations between May 1909 and May 1910. Their account of the expedition (Calvert and Calvert 1917) is remarkable for what is missing. Although they made every effort to visit what at the time were remote areas for foreign visitors, they did not reach the rainforest zone on the Pacific lowlands. At the time only a few naturalists had collected in the region. Even today the flow of scientific tourism is mainly directed to field stations established in other zones long before roads led onto the Osa Peninsula.

In terms of basic ecological processes and patterns we can point to no fundamental distinctions that set this small region of rainforest apart from other Neotropical lowland areas of comparable climate. Yet these forests surrounding the Golfo Dulce exhibit a high degree of spatial heterogeneity, high species diversity, a long history of isolation with resultant endemism, unusual biogeographic affinities, and, on parts of the Osa Peninsula, an intact megafauna. Consequently, the Osa region (Fig. 12.1) has long held the interest of international conservation groups that focus on saving hotspots of biological diversity. For scientists interested in ecosystems it provides a place for comparative ecological and evolutionary studies.

Even with intense international publicity about and interest in the Osa region, the relative difficulty of access to its iconic areas, particularly Corcovado National Park, combined with the Park's modest facilities and primitive accommodations, has limited the scope of research carried out within the Park since its formation. Most such research has consisted of thesis projects of intrepid graduate students and biotic surveys by parataxonomists of Costa Rica's National Institute of Biodiversity (INBio). Large, long-term, and well-funded ecological studies are largely missing. Though several private research centers have been added to the Osa region in the past decade, the basic pattern remains. Thus point-by-point comparisons of rainforest ecosystems of the South Pacific lowlands with those of the Atlantic lowlands of Costa Rica would only highlight this lack of balance of scientific attention. In consequence, this chapter emphasizes selected organisms or systems that can be compared to parallel cases in lowland rainforests on the Atlantic side—for example, the ant communities; or that are uniquely available for study in the Osa region such as the large mammal populations or the interface of marine and terrestrial faunas. These choices point to great opportunities for the synergism of science and conservation efforts on the Osa.

People are part of ecosystems, so part of this chapter describes how a relatively intact fragment of the Southern Pacific region's once extensive forests was identified, was set aside, and continues to survive in the face of intense political, social, and economic pressures. As we elaborate below, many contingent events, cases of good fortune, and opportunism marked the creation of Corcovado Park and the subsequent growth from that seed to the larger area-wide integration of conservation and protection under the administration of ACOSA (*Área de Conservación Osa*). Some themes and lessons of this history are more widely applicable, both geographically and temporally. As many natural systems rebound from the earlier impacts of ranching, farming, and gold mining within Corcovado National Park, old problems continue and new challenges arise in the broader region integrated under ACOSA. The story is fascinating and worth telling as part of the more inclusive natural history of the region.

Human Use and Conservation

Early History

The pre-Columbian people of the Osa Peninsula inhabited an area known to archaeologists as the Diquís sub-region, part of the Greater Chiriquí cultural province, which included much of today's southern Pacific Costa Rica and western Panama. Artifacts of this period are encountered in old growth forests on the Osa (Fig. 12.2, upper left). Populations in this humid region lived in relatively dispersed settlements, practicing shifting agriculture as well as hunting, fishing, and foraging for non-cultivated food, medicinal substances, and materials used for household items and shelter construction. Crops included corn, tubers, pejibaye palm fruits, beans, gourds, cotton, chiles, and tobacco (Hall 1985, Fonseca 1986, Anchukaitis and Horn 2005). Indigenous cultivation and related practices likely affected a large proportion of the Osa's natural environment. Through interpretation of charcoal, pollen, and other microfossil records, paleoecological research in Greater Chiriquí has provided considerable evidence of such populations' ongoing, sometimes dramatic alteration of natural landscapes in favor of disturbance and successional species, through use of fire, selective promotion of favored forest flora, and other agriculture-related practices. Hunting and human settlement likely had local effects on Osa animal populations (Hall 1985, Horn 1992, Clement and Horn 2001, Anchukaitis and Horn 2005).

By the sixteenth century, Spanish colonization had introduced new landholding and land use patterns to Costa Rica, along with new social, political, and economic norms, though lasting colonial settlements were founded only in the Central Valley (*Valle Central*) and Pacific Northwest. Costa Rica's abysmal commodities' transport structure, isolation from Spanish colonial political and trade centers, lack of native labor or mineral reserves, vulnerability to privateers and raiders, and dense forests all impeded extensive development. The remote lowland moist forests of the southern Pacific region (or *Zona Sur*) did not beckon colonists already daunted by hardships in the more temperate *Valle Central*, which at least offered minimal infrastructure and rich alluvial soils. By independence (1821), most indigenous people of the *Zona Sur* were extirpated by introduced diseases, coerced labor, and flight. The Osa probably became uninhabited, and forest cover throughout

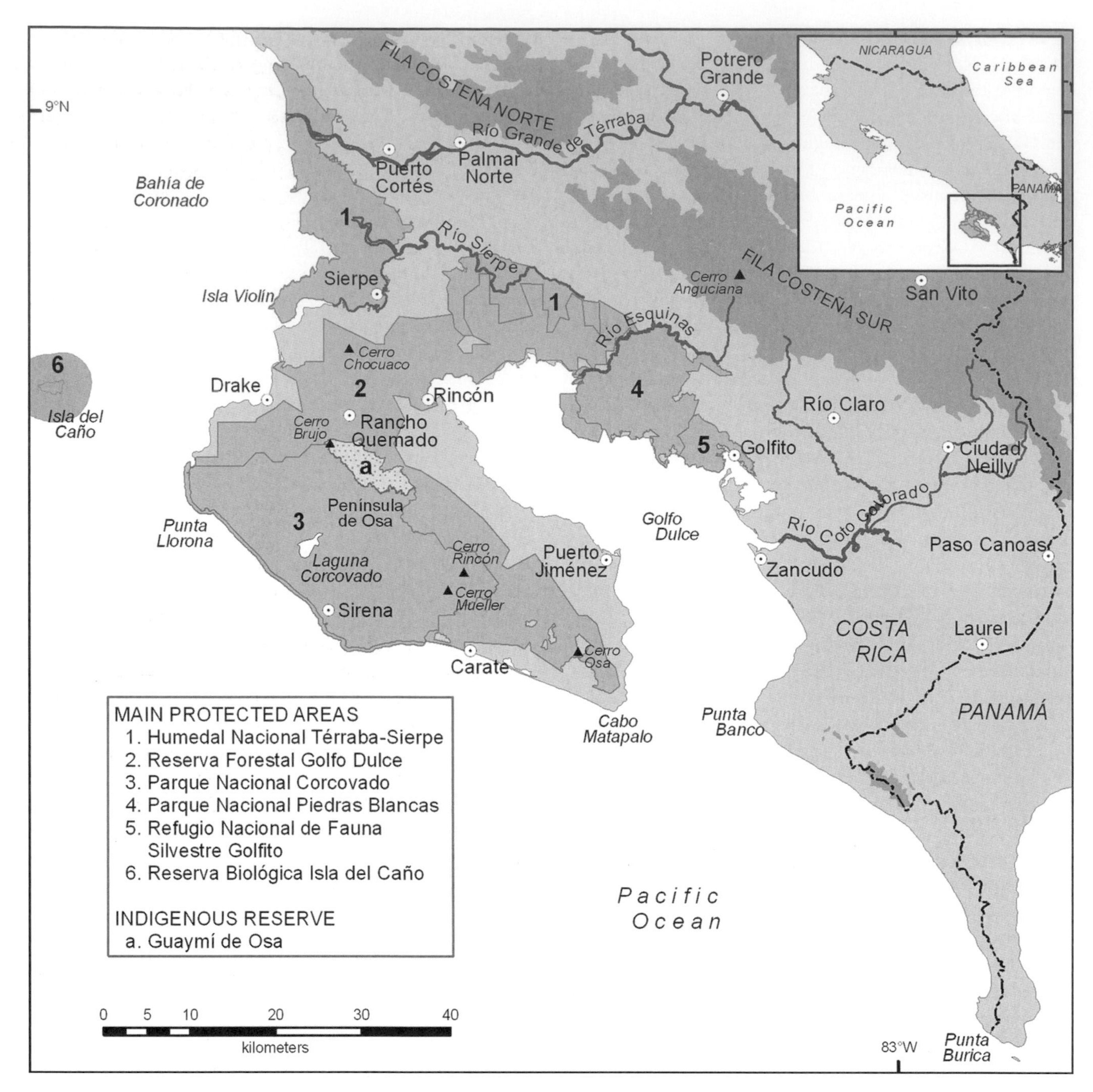

Fig. 12.1 Map of the southern Pacific moist lowland forest region on and around the Osa peninsula. Main topographic features, towns, and protected areas are shown. *Map prepared by Marco V. Castro.*

the *Zona Sur* expanded between the sixteenth and twentieth centuries (Lewis 1982–1983, Hall 1985). Nonetheless, land-use policies and practices favoring forest clearing to induce economic benefits eventually would affect even these remote lowland ecosystems (Lewis 1982–1983, Hall 1985, Anchukaitis and Horn 2005, Soluri 2005).

In colonial Costa Rica, all otherwise unassigned lands were designated as royal property. After independence, the vast areas of still-undeveloped Crown lands became state-owned *tierras baldías*, or "wastelands," where forest clearing and other development activities qualified as *mejoras*—that is, improvements (Hall 1985). During the nineteenth and twentieth centuries, to strengthen political and economic security, the national government promoted clearing, settlement, and economic development schemes on *baldíos*, some involving subsistence or smallholder agriculture, others for commercial enterprise (Christen 1994, Soluri 2005). Government-encouraged Osa agricultural settlement schemes of the mid-nineteenth century met with different results. The Frenchman Gabriel Lafond utterly failed to establish the French emigrant colony his contract with the Costa Rican government specified. The few underequipped French colonists he recruited between 1849 and 1852 evidently were stranded, penniless, somewhere in Panama en route to the Osa, and obliged to work in a Panamanian coal mine to earn passage back home (García 1988, Barrantes 2005). As with most of Costa Rica's nineteenth-century European agricultural colony projects, low recruitment, a grueling sea voyage, and colonists motivated to escape poverty and class strictures in Europe but unprepared to cope with tropical climates and non-existent infrastructure all hindered this attempt (Hall 1985).

Another category of motivated colonists, essentially "local immigrants," possessed some advantages favoring their successful resettlement in this peripheral region. Shortly before the Lafond debacle, naturalized Costa Rican Juan Mercedes Fernández led the settlement of the first modern-day Osa hamlet, Golfo Dulce. Granted citizenship and appointed the district's first *Jefe Político* (chief official), in the late 1840s Fernández oversaw the immigration of at least 88 fellow *Chiricanos* from his native Panamanian province, the closest population center. All pledged their allegiance to the Costa Rican state. Their motivations for emigration were strong; Chiriquí was characterized by a relatively dense population, large cattle ranches, and few economic opportunities for non-elites, as well as by the frequent armed civil conflicts characteristic of post-independence Colombia, to which Panama then pertained (González-Víquez 1921, Lewis 1982–1983, García 1988; Barrantes 2005 cites sources suggesting a few Lafond immigrants joined this community).

These local immigrants had strong practical advantages. They were familiar with the region's climate, flora, and fauna, and had experience with tropical crops and local fishing and hunting practices. They knew firsthand the dangers of rainy season inundations, venomous snakes,

Fig. 12.2 People in the Osa ecosystem Top left: Pre-Columbian artifacts from a ridge above Sirena Station. Bottom left: The Tropical Science Center (TSC) station near Rincón provided a site for the Organization for Tropical Studies' (OTS) courses between 1965 and 1972. Some foreign scientists thus helped promote the formation of Corcovado Park. Here Daniel Janzen and an assistant clean up after the July 1966 OTS course. Upper right: a rancho along the bulldozer path that led from Río Rincón to Sirena photographed in 1975. Lower right: people panning for gold near the mouth of Río Madrigal in 1985.
Photo credits—Top left and bottom left: L. Gilbert, upper right: C. Vaughan, lower right: S. Boinski.

and predatory cats, and were familiar with valuable local resources, including pejibaye fruits, mangrove clams, and the many wild game species. These immigrants had direct knowledge of the available local trade and transport infrastructures (waterways and paths) and conveyances (small boats and wooden carts), using these to maintain attenuated personal and economic connections with Chiriquí and, eventually, build new links with other Costa Rican populations (González-Víquez 1921, Lewis 1982–1983, García 1988, Barrantes 2005).

The isolated settlement grew slowly, with more Chiricano families arriving over the next decades. The hamlet was twice moved slightly northwest along the gulf, evidently after earthquakes caused coastal subsidence. Renamed Santo Domingo, then Puerto Jiménez, this remains the Osa's principal town (Hall 1985, Franceschi Barraza 2007). Settlers built farms and thatch-roofed homes, mainly along the Osa's gulf coast plains. They raised subsistence crops, cattle, pigs, and poultry, and hunted wild game. These homesteaders—523 by an 1892 census, 1,195 by 1927—had little reason to consider whether their clearings or hunting might deplete the game or other natural resources (González-Víquez 1921, Marin Cañas 1976, Lewis 1982–1983, García 1988, Franceschi Barraza 1997, 2007).

Early Scientific Activity and Commerce

In the nineteenth century the Osa witnessed little scientific activity. Nationwide, funding was scarce, and good natural history sites were found closer to San José. Swiss-born Henri Pittier, director of Costa Rica's new Instituto Físico-Geográfico Nacional, and his Costa Rican colleague Adolfo Tonduz, did visit Santo Domingo (Puerto Jiménez) in the 1890s and collected botanical specimens for the new national herbarium. Pittier's government-funded 1890s expeditions (on shoestring budgets) focused on accurate mapping of the southern regions, plus botanical, zoological, geological, and ethnographic studies, intended to help legislators learn "just what was theirs and how best to exploit it" (Allen 1956, Conejo 1975, Gómez and Savage 1983; quotation from Eakin 1999, p. 131).

Isolation from transport routes and markets precluded most Osa commerce. By the early twentieth century, the Osa settlers processed and sold coconut derivatives and mother-of-pearl to passing steamships. By the early 1930s, some grew bananas for commercial sale (Marin Cañas 1976, Lewis 1982–1983, García 1988, Clare 2005, Soluri 2005). In the 1930s, the US-based transnational United Fruit Company (UFCO) initiated a *Zona Sur* banana operation headquartered at Golfito, previously a fishing hamlet on the Golfo Dulce's eastern shore. United Fruit's financial and political clout and capacity for rapid regional development allowed it to negotiate highly favorable fiscal terms and territorial concessions from the government. The *Bananera* acquired enormous sections of state-owned *baldíos*. In the early twentieth century, these concessions were another means by which capital-poor Costa Rica could promote land development, raise productivity, and consolidate territorial control despite its still-small national population, then about 300,000 (Hall 1985, Bourgois 1989).

In the late 1930s, United Fruit engineers temporarily based at Puerto Jiménez directed Golfito's development. The banana industry and the 1937 discovery of placer gold in several Osa rivers brought many new migrants, among them Panamanians, Nicaraguans, and even Costa Ricans, to the *Zona Sur* and the Osa. By the 1940s great swathes of *Zona Sur* rainforest had been transformed into single-commodity banana or oil palm plantations, especially around the new regional center of Golfito, and Palmar and Esquinas (Lewis 1982–1983, Hall 1985, Bourgois 1989, Clare 2005). Puerto Jiménez had grown into a slightly larger and less demure frontier town, also gaining an airstrip, with passenger flights to San José by 1938 (Christen 1994, Franceschi Barraza 1997, Clare 2005).

During the 1920s–1930s, United Fruit subsidiary Golfo Dulce Lands Company had also acquired much of the northern Osa (Clare 2005). Wholesale plantation conversion might soon have followed, but after detailed analyses in 1943 the company determined the Osa's soils, topography, and accessibility were not apt for banana production (Vaughan 1981). Shortly after, the company deeded all 13 of its Osa *fincas* (farms, or lots) comprising 47,513 hectares (117,357 acres), about one-third of the peninsula, to a retiring company engineer (Stetson 1977).

In 1957 this Osa acreage attracted the interest of US-based timber broker Warren Stetson. After extensive on-site assessment, he estimated the property contained over 3.5 billion board feet of timber. Stetson brokered its $450,000 sale by the engineer's widow to a group of US buyers including Wilford Gonyea, of Medford, Oregon's Timber Products Company, and the Chicago-based Pritzker family, who owned half of Gonyea's company as part of their diverse and growing industrial and real-estate holdings (Christen 1994). Eventually the US owners held the Osa land through a Bahamian shell company, Transnational Trust Co., Ltd., and managed it through *Osa Productos Forestales, Sociedad Anonima* (Osa Forestal), inscribed in the Costa Rican Mercantile Registry in November 1959 (Costa Rica Regístro Mercantil, T45, F408, A18007, 11/24/59; Christen 1994).

Company headquarters were established at Rincón de Osa, with a city office in San José. Osa Forestal's holdings girdled the Peninsula from Rincón, Playa Blanca, and

Barrigones on the Golfo Dulce to the Pacific coastal plains of Salsipuedes, Sirena, Llorona, and Drake Bay, approaching the Sierpe River on the Osa's northern boundary. The southeastern portion ended just above the Río Agujas, several miles north of Puerto Jiménez. In the property's midst was a 13,462 hectare (33,251 acre) *baldío* (Vaughan 1981, Christen 1994). Both the Osa property and this inholding were then almost entirely covered with forest and other natural vegetation. Osa Forestal was authorized to conduct forestry, wood processing, mining, and "all types of agricultural, commercial and industrial activity," typical of an era when Costa Rica granted foreign interests wide powers to promote economic integration of *baldíos* (Costa Rica Registro Mercantil 1959, Christen 1994). Yale Forestry School graduate Alvin Wright, Osa Forestal's manager through the 1960s, envisioned an integrated tropical forestry operation, with a comprehensive forest inventory followed by rotational harvests and waste-free wood processing. He also explored opportunities for placer gold mining (Christen 1994).

By 1963, shortly after Wright's arrival, 83 homesteads were already established on Osa Forestal lands (Vaughan 1981, Christen 1994). Twenty-one Rincón area homesteads, eleven in the Sirena area, and forty-five around Drake Bay were recorded in Wright's 1963 settler survey. Many homesteaders had been resident for years, though none held title. Drake Bay and Sirena were settled principally by former Guanacaste residents arriving via the Sierpe River. Like Chiriquí in 1850, Guanacaste was by the 1940s a province characterized by large farms and ranches, with diminished opportunities for smallholders (Hall 1985, Christen 1994).

Osa Forestal had legal title, but these homesteaders also had longstanding legal rights. Obtaining land titles entailed costly bureaucratic processes only well-capitalized colonizers such as foreign corporations could afford. Since colonial times land improvements by untitled settlers, including forest clearing, construction, permanent cultivation, and ranching were also means for establishing ownership rights. Sale or inheritance of such land was recognized as a legitimate private property transfer, with payment for sale calculated on the basis of documented improvements (Hall 1985). In the 1960s, Costa Rica's total population was on a sharp upswing, with increasing pressure for homesteading opportunities beyond traditional population centers. The 1961 agrarian reform laws administered by the Lands and Colonization Institute (ITCO; later IDA, the Agrarian Development Institute) responded by amplifying opportunities for untitled settlers to establish ownership rights (Costa Rica Ley No. 2825 de 14 de Octubre de 1961, *Ley de Tierras y Colonización*; Christen 1994).

Wright saw these settlements as potentially detrimental to the fixed harvest rotations of integrated forestry (Christen 1994). He tried to enforce settler rental contracts, modeled on those of United Fruit. The rent was nominal but those who signed had to cede right-of-way, neither cut trees nor expand their holdings, and depart with only an indemnity payment if the company so chose (Vaughan 1981, Christen 1994). Some settlers signed, but most Drake residents immediately rejected these contracts, contested Osa Forestal's title claim, and established an agricultural cooperative. Under the new ITCO law cooperative members would be guaranteed the right to take possession of land near where they already lived. A decade of complicated negotiations between ITCO and Osa Forestal ensued, exploring potential solutions such as land swaps, but mainly resulting in heightened mistrust and antagonism between the litigants. Meantime, the impasse largely halted Osa Forestal's development activities and hindered expansion of these new Osa settlements through the 1960s (Christen 1994). This hiatus proved an historic opportunity for Osa science, and, eventually, Osa conservation.

How a Piece Was Saved

United Fruit botanists and entomologists had explored the *Zona Sur* since the 1930s, conducting company-related research and experiments, but company scientists had carried out little research on the Osa itself aside from components of United Fruit's 1943 feasibility studies. UFCO botanist Paul Allen's important dendrological key, *The Rain Forests of Golfo Dulce* (1956), was the fruit of five years' research in *Zona Sur* forests, including the Esquinas forest, a corridor linking the Osa with the mainland. Though Allen evidently did not study the Osa's forests, his 1956 volume proved a valuable resource for Alvin Wright's purposes of collecting accurate data about the Osa's forestry aptitude (Allen 1956, Skutch 1992). To fill this need, Wright soon became the first sponsor of long-term Osa botanical and ecological research. In 1962 Wright hired fellow forester Dr. Leslie Holdridge, cofounder of Tropical Science Center (TSC), a new San José-based research and consulting firm, to conduct an extensive timber cruise. While carrying out this detailed survey and marketing-oriented assessment of the Osa's timber tree species, volume, and quality, Holdridge quickly confirmed both the Osa's timbering potential and its great interest for scientific research. He likened the Osa's forest height and species diversity to that of South America's Chocó (Holdridge and Tosi 1991). Holdridge and TSC cofounder Joseph Tosi readily accepted Wright's invitation to establish a TSC field research station, a modest wooden structure next to the company's airstrip in Rincón de Osa (Fig. 12.2, lower left). For the next decade, Rincón station was an extremely active training center and research base (Holdridge and Tosi 1991, Christen 1994).

Some US universities had tried to develop tropical science field programs in Latin America during the 1950s, without much success. Their interests and frustrations prompted a May 1960 US National Academy of Sciences conference that emphasized the need for tropical field research and training facilities affording scientists and students "personal tropical experience" in wilderness areas—in politically stable countries—to facilitate research advances in taxonomy, evolution, phytogeography, ecology, morphology, and physiology (NAS-NRC 1960). Leslie Holdridge was one of the few non–US-based delegates to this Miami conference. Soon after, Rincón station became the first long-term base for wet-forest field ecology in Costa Rica, and a key location for the "Tropical Biology: An Ecological Approach," or "Fundamentals," field course run by the new Organization for Tropical Studies (OTS). Though most OTS trainees were US university students and the majority of Rincón researchers were also US-based, many others were Costa Ricans, Europeans, or from institutions elsewhere in the Americas (NAS-NRC 1960, Tosi 1975, Gómez and Savage 1983, Stone 1988, Christen 1994).

During 1964–1966, the Rincón area provided Les Holdridge with most of the wet forest study sites for his coauthored *Forest Environments in Tropical Life Zones* (1971). This volume laid out the theoretical underpinnings and practical applications of his life zones system, a method for bioclimatic classification of tropical land areas, facilitating prediction and analysis of a location's natural vegetation. Some of this Vietnam-era research was funded with US Department of Defense (DOD) grants, given its pertinence to questions regarding tropical defoliation, though chemical defoliant experiments never took place there (Holdridge et al. 1971, Holdridge and Tosi 1991). By the early 1970s, Rincón-based publications documenting environmental science, population ecology, and tropical ecosystem research from the Osa constituted a large proportion of the small corpus of literature on Western hemisphere wet tropical forests (Tosi 1975).

Also by the early 1970s, Rincón station, the Osa Forestal/settler impasse, and the status of land use and forest cover throughout Costa Rica were all undergoing dramatic changes. A nationwide surge of forest-to-agriculture conversion and of untitled settlers' "invasions" of large-holder properties, plus the extension of the Inter-American Highway to the *Zona Sur* and a subsequent regional population influx framed associated events on the Osa. Among many related developments, a politically supported organized invasion of the Sirena plains in late 1971 led Alvin Wright to bulldoze a "pioneer track" from Rincón to Sirena in March 1972 in a reactionary development plan of abandoning forestry for cattle ranching (Fig. 12.2, upper right). Rincón scientists already knew any conceivable break in the Osa Forestal-settler impasse would mean clearing Osa forests whose counterparts elsewhere in lowland Pacific and Atlantic regions were already disappearing (Christen 1994).

By the early 1970s, Rincón's diverse science constituency was debating both which Osa areas most urgently merited conservation as a natural and scientific reserve and which preservation approaches offered the best chances of success in the Osa's volatile sociopolitical circumstances (Christen 1994). OTS, an outlier in these deliberations, sought but never attained its own private Rincón research reserve. Most station veterans, government administrators, and international allies focused elsewhere, on iterations of the breathtaking, enormous ambition of securing the entire "Corcovado Basin (*Cuenca del Corcovado*)." This 29,000 hectare territory on the Osa's northwest Pacific slope encompassed several habitats and ecosystems, including the Laguna Corcovado, freshwater and palm forest wetlands, bottomland and upland rainforests, estuaries, and sandy and rocky beaches and shoreline (Figs. 12.3, 12.4, and 12.5 and accounts to follow). In the mid-1960s Joe Tosi first suggested this as the most logical Osa park or research reserve, because while it was of relatively modest size (compared with, for example, the Amazon basin) its "definable" and "defensible" natural boundaries encompassed so many habitats and animal species, especially the large mammals he knew would soon become scarce (Ewel 1991, Christen 1994).

Since then, many of Tosi's colleagues had explored this "wildlife paradise," a long day's journey over the hills from Rincón, where the roaming researcher could almost count on glimpsing a jaguar, tapir, or some other impressive mammal (Gómez 1991). The Osa's developing conservation constituency was engrossed by the near-absence of human impact and the awareness that the settler/forestry company impasse, centered here, could shift at any time, with irreversible development immediately to follow. In mid-1973, longtime Rincón researcher Jack Ewel and his University of Florida, Gainesville colleagues privately printed and widely distributed a bilingual booklet, *La Cuenca del Corcovado* (1973). Here and in accompanying correspondence they argued persuasively for protecting this entire region. Though title to alternatives such as the 13,462 ha hourglass *baldío* inside Osa Forestal's property might be easier to secure, such half-measures could not offer the ecological integrity, and hence potential conservation longevity, of the complete Corcovado basin. They contended that this endeavor could succeed only as a national park or biological reserve enhancing the national patrimony, not as a foreign/private

Fig. 12.3 Osa Peninsula, where forest and sea interact. Top left: Road from the mainland to the peninsula snakes over its narrow neck through Los Mogos. Golfo Dulce is seen at the right, and the Sierpe marshlands in the distance to the left. Top right: The Llorona plateau meets the Pacific Ocean. Bottom right: View of Sirena (airstrip visible to left of the center on the coast) from the rugged hills to the East. Las Ollas is a swampy clearing seen in the bottom left. A main park trail from Sirena passes this site on the way to Cerro Muller. Lower left: This low-energy beach near the mouth of Río Llorona was used as a landing strip for DC3s in the 1970s; it is shown at low tide.
Top left, top right, bottom right: Aerial photos by C. Foerster, 2003; lower left: photograph by L. Gilbert, July 1992.

preserve. Given the tremendous price tag, they emphasized it could succeed only with international monetary assistance for acquisition and maintenance, as befit a project offering certain benefits to humanity worldwide (Christen 1994).

Mario Boza and Álvaro Ugalde, chief administrators of Costa Rica's young and underfinanced National Park Service (SPN), were long interested in an Osa component for the growing park system. With continuing study, they became increasingly convinced of the Corcovado Basin imperative, on the basis of its scientific and conservation merits, despite its being too big and too costly, and its ownership status being too complex for strictly rational consideration. During 1974–1975, despite many setbacks, Boza, Ugalde, Tosi, Ewel, and many others in Costa Rican and US academic and government circles persistently strategized and lobbied for a Corcovado reserve (Christen 1994).

In October 1975, the 35,000 ha Corcovado National Park was created by Presidential decree, facilitated by a land exchange with Osa Forestal and the promise of emergency start-up funds from international conservation NGOs World Wildlife Fund (WWF)-US, The Nature Conservancy (TNC), and Rare Animal Relief Effort (RARE), to supplement Costa Rican government monies (Christen 1994). Further intensive lobbying led by Costa Rican scientists rallied the executive branch to direct key government agencies to commence the hard work of actual park consolidation in January 1976, at the dry season's outset and just before an anticipated wave of settler forest-clearing would have altered its landscape dramatically. Guard stations were put in place and regular patrols enacted (Christen 2006). By 1977, again with the benefit of supplemental international funding, the process of censusing, indemnifying, and relocating hundreds of settlers was completed (Vaughan 1981, Christen 1994). In a remarkably short time, Corcovado had experienced a transition from a paper park to an authentic protected conservation area, exemplifying, as Joe

Tosi had predicted, the actual if imperfect defensibility of its naturally defined borders, an attribute further tested and strained many times since 1977.

Corcovado was the first Costa Rican park justified only on the basis of its ecological and scientific merits, without reference to cultural attributes (like Santa Rosa) or recreational benefits (like Manuel Antonio). Nonetheless, circumstances discouraged the rapid growth of a robust scientific research presence there. Continuity of Osa research had been dealt a severe blow in 1973, when a new and bellicose Osa Forestal manager permanently padlocked the Rincón station, disrupting several long-term ecological studies (Christen 1994). This closing, plus the manager's ostensible intention to create foreign gated-resort communities, further galvanized conservationist efforts towards a Corcovado reserve. The political and economic vulnerability of this intense park-making effort also compelled its leaders to forgo what they judged as potentially damaging foreign-backed offers to create a research station in the Corcovado basin while the national park's own establishment was still in doubt (Christen 1994).

After the park's creation, tight budgets and administrative priorities for consolidation, patrols, and other management demands were among issues hindering research programs and facilities development, despite intentions expressed in park master plans and the 1977 biology station blueprints for Sirena, the main ranger station (Christen 1994, Christen 2006). Scientists were hesitantly welcomed from late 1976, but mostly on their own recognizance, without special access to park resources. In the late 1970s–early 1980s, a few courses came to Corcovado, and some scientists carried out work at Sirena, Llorona, and other locations (Janzen 1982). From 1979, the Park Service loaned a 6 × 9 foot room in the main Sirena building to University of Texas (UT), Austin professor Lawrence Gilbert for the use of several UT graduate students. Later, around the time of the gold-miner invasion (Fig. 12.2) that prompted the park's 1985 temporary closure (see Janzen 1985, Anderson 2003,

Fig. 12.4 Upper left: Cerro Brujo rises behind the Corcovado Basin in this view north from the coast of the park. Upper right: Swamp forest dominated by *Anacardium excelsum* with *Aechmea magdalena* understory penetrating the Laguna Corcovado marsh on ancient barrier dunes that run parallel to the coast. These streaks of trees are bounded by a dense *Raphia* palm forest that gives way abruptly to an open herbaceous marsh vegetation. Bottom: A dry season (March 1994) view of the herbaceous marsh vegetation with the *Raphia* palms to the right.
Photo credits—Upper left: L. Gilbert; Upper right: C. Foerster, 2003; bottom: C. Foerster.

Fig. 12.5 Forest type variety. Upper left: *Scheelea* palm forest in Río Sirena flood zone. Upper right: Swamp forest near Llorona. Lower left: *Brosimum*- and *Anacardium*-dominated primary forest patches near Sirena. Lower right: Primary forest on Ollas ridge trail above Sirena.
Photographs by L. Gilbert.

Ankersen et al. 2006), Gilbert, with Park Service approval, built Sirena's first lab (Fig. 12.6) with small grants from the US National Science Foundation (NSF) and WWF-US. Though funded by Gilbert and used by UT Austin students studying topics including plant-animal interactions, entomology, mammals, and uses of GIS and remote sensing techniques for evaluating forest regeneration, the lab was under National Park Service jurisdiction and has been used by many researchers and courses from numerous countries, including Costa Rica (Gilbert 1990). In the late 1990s, facilities across ACOSA including Sirena station were greatly expanded with funds from a United Nations' Global Environmental Facility (GEF) grant, mainly to handle the exponential growth in ecotourism visitors to the park, and the first Park Service–built lab was constructed (Christen 2006). Throughout these decades, though many studies have been carried out there, Corcovado has trailed far behind field research centers such as Costa Rica's La Selva and Panama's Barro Colorado Island (BCI) in terms of both scientific facilities and use. In part, this is simply a factor of Corcovado's considerably greater remoteness, which means key resources, like 24-hour electricity for generators and refrigeration, cannot be counted on (Christen 2006).

Through the 1990s, parataxonomists from INBio, Costa Rica's new National Biodiversity Institute, carried out extensive entomological/floral specimen collections and field processing in Corcovado (Fig. 12.6), and in locations throughout ACOSA, the Osa Conservation Area (e.g., see Kappelle et al. 2003), a greatly expanded multi-category reserve established in the early 1990s. ACOSA encompasses Corcovado Park, the Golfo Dulce Forest Reserve (itself created in 1978 as a Corcovado buffer, watershed conservation unit, and zone for managed forestry activities), and several other buffer zone units on and adjacent to the Osa, each with its own legal characteristics regarding natural resource conservation and use. Largely modeled on the Guanacaste Conservation Area (ACG; see Janzen and Hallwachs, chapter 10 of this volume), ACOSA's configuration and its agenda have represented a movement towards decentralization, increased emphasis on education in both

Fig. 12.6 Top left: Kit Kernan and Larry Gilbert frame the first biological field laboratory at La Sirena, August 1986. Top right: David Wescott's research seminar for a UT Austin graduate course in the field lab in 1992 included other researchers and interested ecotourists. Bottom left: INBio botanist Reynaldo Aguilar processing plant specimens in July 2000 at facilities constructed with GEF funds. Bottom right: Young ecotourists frequently flood the Sirena station, thus stressing maintenance and leaving more trash than funds. They and tour leaders are oblivious to the fact that ecotourism companies are essentially subsidized by the ACOSA office.
Photo credits—Top left: Pamela Phillips; remaining three: L. Gilbert.

conservation philosophy and environmental topics, and increased community involvement (Ankersen et al. 2006). In large measure both decentralization and emphasis on local involvement have reflected park administrators' realization, especially after the Corcovado gold miners' invasion and subsequent expulsion, that the value of national parks "was neither well-understood nor greatly respected by local communities" (Ankersen et al. 2006, p. 420). Notably, certain changes in regard to the Osa's "local communities" perhaps represent some of the most dramatic revisions of the Osa scenario in the past few decades. These changes have affected the Osa's inhabitants' range of conservation outlooks and practices as well as the demographic composition of the Osa "community."

Corcovado Park's proponents expressed sincere interest in engaging feedback from locals during the early 1970s park creation campaign. Simultaneously they also consciously avoided most such engagement. They feared that in the midst of the agrarian reform movement, and in an era when conservation concepts were unfamiliar to most members of the public, any public engagement would be too complex an undertaking at the same time as they were urgently pursuing company land swaps, enabling legislation, and other big-picture steps towards park creation (Christen 1994). Consequently, local experiences of Corcovado Park's creation and of its early years (when the park administration was said to be dominated by a "bunker" mentality) featured little inclusion of local Osa inhabitants in the park's establishment or management (Christen 1994, Ankersen et al. 2006).

Environmental education outreach in the wake of the gold-mining crisis was among the first efforts to address this gap. Shortly thereafter came the WWF-US and USAID (Agency for International Development) funding of the BOSCOSA project of the late 1980s–early 1990s, promoting community-based sustainable forestry to help slow deforestation in Corcovado's buffer areas. Ultimately suspended due to depleted funds, and widely criticized for failing to meet its self-admitted idealistic ambitions, BOSCOSA did raise local awareness of the importance of maintaining Osa forest resources and trained several Costa Rican conservationists who remained active in the region (Kurka 1994, as discussed in Ankersen et al. 2006). Another consciousness-raising experience for many Osa inhabitants was the 1994 campaign against the construction of US corporate subsidiary Ston Forestal's wood-chipping plant in an ecologically vulnerable location adjacent to the Golfo Dulce. This successful campaign, inspired and in part led by AECO (Costa Rican Ecologists' Association) became a largely Osa grassroots enterprise, one that raised the ecological and conservation consciousness of many longtime Osa inhabitants and served as a basis for interest in later local conservation agendas (van den Hombergh 1999). The watershed nature of this experience for a large segment of the region's population—replacing largely economic concerns about foreign companies (such as were expressed against Osa Forestal) with a more comprehensive combination of conservation and quality of life and livelihood and economic concerns—is somewhat reminiscent of the effects of the anti-ALCOA campaign of 1970, focused on that US company's hydroelectric dam plans in Costa Rica's El General Valley, an event that raised the environmental consciousness both of Costa Rican university students and of the region's local inhabitants (Gómez 1991, Christen 1994).

The arrival of tourism in the Osa and *Zona Sur* at various gradations of "eco," soon after the mid-1980s wholesale departure of United Fruit and its plantation jobs, has also influenced local orientations towards the conservation entities and resources that have somewhat unexpectedly brought new revenues to the Osa, and not only to large-scale foreign operations. For example, since 2004, ASEDER (the Osa's Entrepreneur Association for Responsible Development), with support from TNC and the AVINA Foundation, has helped young Osa community members launch their own ecotourism-related businesses, promoting both community development and biodiversity conservation (S. Mack, pers. comm.). Issues regarding recycling and maintaining the cleanliness of the Osa's water supply have become of central importance to local organizations and alliances based in Puerto Jiménez and other Osa towns. A consortium of local NGOs works closely with international organizations, including TNC, on park-related topics and on issues associated with conservation and well-being of the Osa's ecological and human resources. One contemporary concern to local organizations and international groups is the fate of the Golfo Dulce, one of only four tropical fjords in the world and an exemplar of marine diversity (Cortés, chapter 5 of this volume). Local conservation coalitions raised red flags about a 2007 proposal by a US-owned sport-fishing outfitter located in Puerto Jiménez for a large Gulf marina (S. Mack, pers. comm.; M. Hidalgo, pers. comm.).

These changes in interest, involvement, and outlook reflect a combination of new views by longtime Osa inhabitants and their offspring, and more than two decades of demographic transitions among the Osa's human population. In the 1980s and 1990s many people left the Osa and *Zona Sur* after the departure of the *Bananera,* United Fruit. Others with conservation interests moved to the Osa. Some members of the new "landed conservation gentry" (Ankersen et al. 2006) purchased much of the Matapalo region, maintaining it as private conservation reserves, with high-end tourism, private research stations, and similar land-extensive low-development purposes. Individuals without extensive ready capital also migrated onto the Osa, or decided to stay to participate in the Peninsula's emerging ecotourism and conservation-oriented economic activities. Extensive local community participation in implementation plans for the 2008 ACOSA/Corcovado Master Plan reflect the upswing of critical-minded community involvement in Osa conservation (M. Hidalgo, pers. comm.).

In March 1973, Jack Ewel summarized his understanding of the historical centrality of conserving the Corcovado Basin: "it covers all the major Osa ecosystems, it's still in one piece, and it's a definable, defendable piece of real estate that, for all practical purposes, would seem to be protectable for a very long time, regardless of other changes that may take place on the peninsula" (Christen 1994). For most Osa residents today, a trip to Corcovado National Park (PNC) is not an everyday practicality, nor even a conscious ambition. Yet the park, throughout its turbulent history, has consistently proven and re-proven its "protectability," essentially maintaining intact its ecological and conservation value. Corcovado National Park (and more recently, its superstructure, ACOSA) has also remained a cornerstone for ongoing conservation events that may protect the whole peninsula from the "golf-coursing" real-estate developments that have afflicted so many of Costa Rica's Pacific coastal regions.

Just as local populations have gradually, through many routes and motivations, approached and achieved their own conservation understandings and actions, so have biologists working in Corcovado and elsewhere on the Peninsula begun implementing new modes of social and conservation ecology, learning to make strong, consequential links between the questions and concerns of human constituencies and the imperative inquiries of ecological research. Poaching, for example, simultaneously presents urgent social and biological questions. Further on, this chapter will follow two examples of such Osa scientific research, in the sections on the peccary and macaw. Corcovado and ACOSA need

Fig. 12.9 Pasture and abandoned pastureland near Sirena. Upper left: The situation in 1977. Upper right: The same spot in 2003. Lower left: Santa Maria Pasture 1995, 18 years after abandonment. Lower right: Tapir browsing affects open understories in secondary and old growth forests in Corcovado Park.
Photographs by L. Gilbert.

Patches of remnant forest dominated by *Anacardium excelsum*, *Carapa guianensis*, and *Castilla tunu* can still be found in the sea of banana and oil palm plantations that today characterize the Valle de Coto Colorado (Zamora et al. 2004).

Forest Gap Dynamics, Succession, and Biological Consequences

There are few experiences more impressive in a rainforest than experiencing the fall of an old 50 m tall tree (Figs. 12.7 and 12.11). Such dramatic events mark the start of a long process of canopy gap creation and closure; the basis of forest dynamics that underlie nutrient exchange and species turnover and which are central to the many organisms that depend on patch dynamics for generation of suitable habitat and resources (Gilbert 1980). Forest gap dynamics differ in frequency, duration, and size depending on the structure, history, and composition of the forests. The diversity of habitats on the Osa reflects the outcome of a highly productive environment overlaid on resource distributions that vary with topography, aspect, soils, nutrients, and drainages. This template for high diversity has allowed many specialist communities to evolve and persist with species flow to and from similar regional biomes. Yet, the forests are not static and they undergo continuous changes in structure and composition depending on disturbances and successional processes that vary in both spatial and temporal scales.

Effects of Disturbance on Canopy Dynamics

The creation of Corcovado National Park (October 1975) and the cessation of major agricultural disturbance within its boundaries were followed (July 1979) by the arrival of researchers based at Sirena interested in the spatial ecology of animals variously dependent on plants in regenerating forest. Many current research trails around Sirena Biological Station were created for studying *Heliconius* butterfly and squirrel monkey populations (by L. Gilbert and University of Texas graduate students) and were by design placed along pasture fence lines and directed to major patches of secondary forest, banana plantations, etc., that existed at the time. The same core trail system has continued to be used for many studies through the decades and early trail markers have been updated to a georeferenced system that allows "then vs. now" comparisons of forest structure (see Figs. 12.9, 12.12, and 12.13). This historic framework makes the Sirena an important site for the study of the consequences of disturbance in lowland rainforests in the Neotropics.

Disturbances are a critical driver of community dynamics, and may include frequent small-scale events (e.g. treefalls, landslips) or infrequent large events (e.g. deforestation, hurricanes). Predictions of community response following large infrequent disturbances tend to focus on changes in species composition, but an additional consideration is the return of frequent small-scale canopy disturbances typical of older communities. In studies of canopy disturbance and recovery, old growth forests are shown to be composites of multiple regenerating phases that last many centuries (Martínez et al. 1988, Bush and Colinvaux 1990, Sheil 1999).

Large infrequent disturbances on the Osa may have natural causes (rainstorms, winds, fires, droughts, and seismic events) or anthropogenic origins such as deforestation for agriculture, timber, mining, and habitation. At Sirena, severe windstorms from the Pacific occur every few years, and one such storm in 2003 resulted in many large treefall events (Fig. 12.11). Similarly, an offshore earthquake in July 2002 was responsible for several new treefalls and landslips. Increased treefall activity is also correlated with episodes of prolonged rainfall and cloudiness driven by the El Niño Southern Oscillation (ENSO), as happened in November

Fig. 12.10 Agricultural modification of lowlands outside the protected areas of ACOSA. Small-scale banana production has returned near Palmar Sur around the rusting infrastructure of past large-scale operations abandoned in the mid-1980s. Top right: Young oil palms rise under the palm weevil–killed trunks of their predecessors. Lower left: Rice cultivation on the SW coast of Osa Peninsula. Lower right: Plantations of several tree species, typically exotic, are frequently observed along the road from Rincón to Puerto Jiménez.
Photographs by L. Gilbert.

Fig. 12.13 Sirena Station. Top left: View from the main park building to the Claro ridge in 1979. Top right: Same spot in 2001. Bottom left: View from old Comedor to main park headquarters in 1980. Bottom right: Same view in 2001.
Photographs by L. Gilbert.

Table 12.1. Treefall and Canopy Gap Densities in the Corcovado National Park, Osa Peninsula, Costa Rica

Habitat	Treefalls per ha	Gap density per ha	Fallen trees per gap	New treefall rate /ha/yr	New gap rate /ha/yr
Old pastures & clearings	3.57	1.02	3.5	0.58	0.17
Abandoned farmland	3.05	1.02	3.0	0.49	0.17
Undercut forest	1.37	1.24	1.1	0.22	0.20
Old-growth forest	1.41	1.20	1.2	0.23	0.19

NOTE. Treefall events from the 2003 field survey; gap densities from stereo aerial photographs. Gap creation rates based on mean gap duration of 6 yr.

1991 and rose to 12% in 2003 with senescence of even-aged *Ochroma*. Open canopy in farm-cleared areas peaked in 1980 at over 90% and then followed a similar pattern to regenerating forest, falling to 5% in 1991 and rising to 12% in 2003.

Overall, dynamics of frequent small-scale disturbances varied considerably during succession following this large deforestation episode. Initial regeneration was dominated by an even-aged stand of early successional species, with few treefall events until a further turnover after 30 years. This turnover consisted of gaps containing 3–40 treefalls, considerably different to small-scale events of older forest of 1–2 treefalls per gap. The study showed how recovery from disturbances at a large spatial scale resulted in large, even-aged stands of trees with partially synchronized canopy dynamics for a few generations.

Case Study 2: Consequences of Gap Dynamics for Gap-Specialist Plants

Patch dynamics (White and Pickett 1985) are especially important for gap-specialist plants, many of which have traits enabling persistence in widely dispersed, ephemeral patches. Patch connectivity is highly dependent on forest dynamics and organisms may depend on trade-offs in life history traits such as growth rates, fecundity, dispersal, dormancy, defenses, and competitive abilities. Gap-specialist plants may produce many propagules that disperse effectively, or they may have a dormant phase enabling survival until a further gap event occurs.

Unlike classic metapopulations that are primarily affected by patch size and isolation, populations living in dynamic landscapes with small transient patches may be affected by both patch quality and the rate at which patch quality changes. In a study of *Passiflora* lianas in treefall gaps in Corcovado National Park, colonization of gaps by *Passiflora vitifolia* (Fig. 12.14) was associated with high patch quality, typically exceeding 3 hours of direct sunlight at ground level (Plowes 2005). *Passiflora* seedlings occupied only 5% of new gaps owing to limitations of patch quality or dispersal. Less than 30% of new patches had suitable quality lighting for colonization. After patches were created, patch quality deteriorated rapidly resulting in a narrow temporal window for seedling establishment. Within 6 years of initial opening, seedling recruitment ceased as gap quality decreased by about 13% per year owing to canopy closure, and 8% per year owing to understory regrowth.

To survive in such a dynamic environment, these plants appear to rely on targeted seed dispersal by foraging birds and mammals, along with a dormant adult phase that plants may enter once canopies overgrow the gaps. Colonization by seed dispersal (about 80% of colonization events) was constrained by patch quality and isolation, while clonal growth from dormant plants (about 20%) was limited by a low frequency of adjacent patch events.

Case Study 3: Effects of Forest Regeneration on a Butterfly-Hostplant Relationship

Gap dynamics are expected to impact many specialized food webs when any member relies on gaps for habitat or resources (Gilbert 1980; Fig. 12.15). Since 8 resident

Fig. 12.14 Top left: *Osa pulchra*, a genus described from a small population near Rincón. Top right: *Passiflora vitafolia* occurs on both coasts but is more common in the Pacific lowlands. The all-red corona seen in the flower to the left is only seen in the Pacific forests where both it and the more widespread white form occur (right). Lower left: Two co-mimetic *Heliconius* species confined to the Pacific wet forests, *H. hewitsoni* and *H. pachinus*, replace genetic relatives and ecological counterparts *H. sapho* and *H. cydno* that are more widespread from Ecuador to Mexico. These bold yellow and black patterns are the most conspicuous indication of the Chiriquí biogeographical zone. Lower right: Research on the Passion-vine (heliconiine) butterfly community at Sirena provided leverage to fund the original biological field lab and development of a research trail system, since then used by many students, researchers, and ecotourism guides.
Photographs and graphics by L. Gilbert.

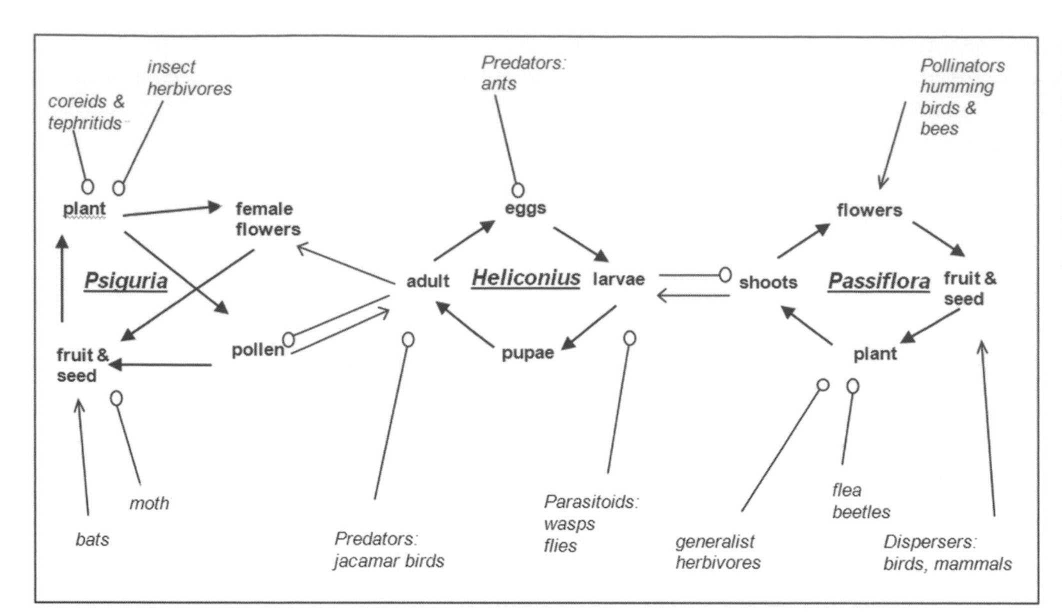

Fig. 12.15 A functional foodweb of *Heliconius–Passiflora–Psiguria* associations. Life history stages of focal species are shown and connected with dark arrows. Interactions with generalist and specialist pollinators, herbivores, predators, and dispersers are shown with light arrows. Negative interactions are shown with circles.

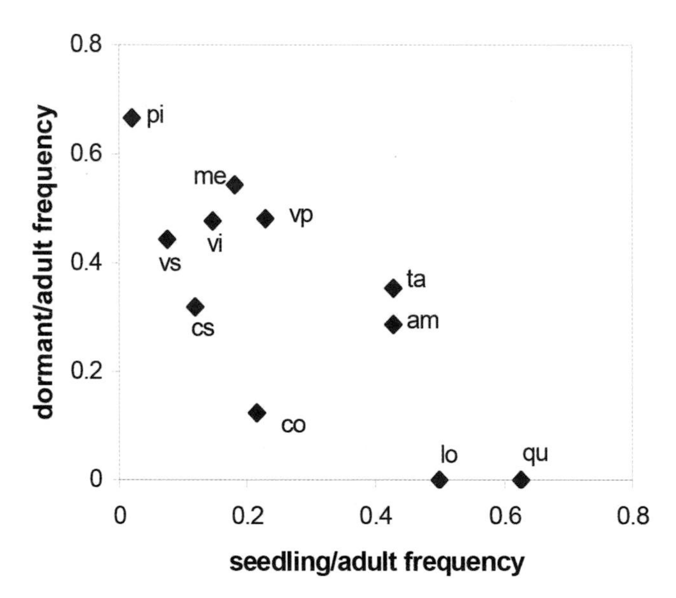

Fig. 12.16 Comparative demography of *Passiflora* at Sirena. Dispersal and dormant strategies compared using ratios of seedling/adults and dormant/adult plants. Indications: *am* ambigua, *co* coriacea, *cs* costaricensis, *lo* lobata, *me* menispermifolia, *qu* quadrangularis, *pi* pittieri, *ta* talamancensis, *vi* vitifolia (all sites), *vp* vitifolia in primary forest, and *vs* vitifolia in secondary forest.

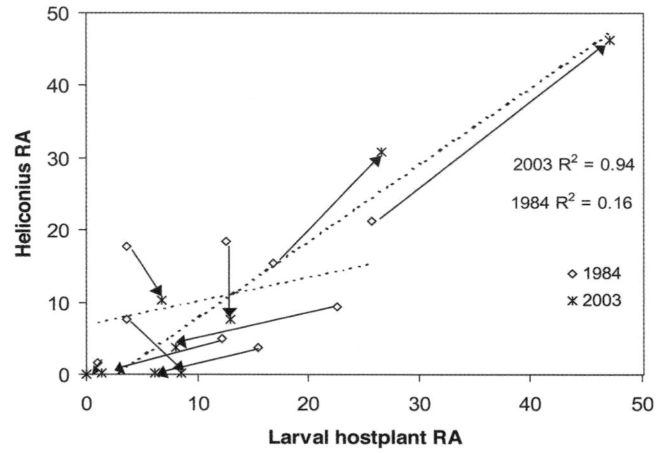

Fig. 12.17 Changes in long-term butterfly–host-plant community structure at Sirena. Comparison of relative abundance for each species during succession. Vectors show temporal change between 1984 and 2003.
Data from Gilbert and Plowes (2003).

species of *Heliconius* butterflies in the Osa (Gilbert 1991) have each specialized on one or more of the 9 *Passiflora* species (Fig. 12.14), changes in availability of suitable hostplant habitat during forest regeneration are expected to impact butterfly densities. A demographic study of the *Passiflora* group showed differences in life history traits (Fig. 12.16) that could result in differential survival under different gap dynamic regimes where species were arrayed along a dispersal-dormancy trade-off regime. Dormancy may be favored under gap regimes with slow, infrequent gap dynamics while forests with fast dynamics having high connectivity may favor species with higher investment in dispersal. As gap dynamics have changed with forest succession around the Sirena station, so too have the abundances of the *Passiflora* species and the *Heliconius* that they support. Long-term changes in the *Heliconius* community since the start of forest regeneration in 1975 show overall declines in absolute abundances, and major shifts in relative abundances of these butterflies. It appears that these trends are largely explained by changes in hostplant densities affected by the regeneration process. We have also observed a trend towards current butterfly densities being matched with their hostplant densities (Fig. 12.17). Immediately after regeneration commenced, the butterfly community was not closely correlated with hostplant densities, implying regulation by other limiting factors such as competition for adult pollen resources. As regeneration has proceeded, the butterfly and hostplant relative abundances are more closely correlated, although some important changes in rank abundances have occurred where hostplant species have been affected by gap dynamics of the regenerating forests. This case study illustrates the importance of gap dynamics and forest regeneration for a food web comprising several relatively rare species.

Fauna

In this chapter it is not possible to provide an adequate overview of the animals that inhabit the Pacific Lowland Rain Forest. In principle there are few fundamental differences in the major taxa or their ecological roles between Atlantic- and Pacific-side forests. Separate evolutionary histories caused by Costa Rica's central mountain range, the distinct seasonal patterns, and different histories of human impact combine to help focus research on different questions and on different animal species on the two lowland regions. Here we highlight a small sample from the Osa region.

Crustaceans: Land Crabs

In Costa Rica, several species of land crabs in the family Gecarcinidae inhabit both the Caribbean and Pacific coastlines. The timid and nocturnal *Gecarcinus quadratus* prefers drier sandy and sandy-loam soil substrates of beaches and adjacent forests. Conspicuous populations dot the Pacific coastline and have been studied near Sirena (Sherman 2002) and on the Nicoya Peninsula (Lindquist and Carroll 2004). Where these soils transition to clay soils frequently inundated by tidal or river effects, the larger and diurnal *Cardisoma* is the more common resident. Banks of the Río Sirena near its mouth show this transition zone (*C. crassum*) as do the salt flats of Guanacaste's Santa Rosa National Park.

Four genera of gecarcinid crabs, when dense, impact coastal forest ecosystems across the world's tropics. These crabs have varying degrees of connectedness to the sea, but most must return to the sea to breed. Planktonic larval stages explain the family's wide distribution. Three of the four genera of this family have been studied (Lindquist et al. 2009). Three years of research near Sirena Biological Station, Corcovado National Park, where land crabs emerge from their burrows at night to forage for plant materials on the forest floor, confirm general trends seen elsewhere and provide a rare perspective on land crab ecology in an intact, diverse, and essentially primary coastal forest (Sherman 2002, 2003, 2006). Several conclusions from these studies at Sirena community and ecosystem-level impacts of these organisms can be made.

In the Pacific lowlands of Costa Rica, as in other regions, gecarcinid crabs can obtain impressive population densities ranging beyond one adult or six juvenile crabs per square meter and thereby represent a dominant contribution to faunal biomass in forests along coastlines. Rough estimates at Sirena suggest population densities comparable to biomass densities seen for ungulates of the African plains or ants and termites of Neotropical forests. During the first rains of the wet season when these crabs are most active, hundreds of thousands of reproductive adult crabs migrating to the ocean can cover the ground.

Gecarcinid crabs remove much of the leaf litter layer from the soil surface down into their burrows that on the Osa range average of 0.5 to a maximum of 1.5 m deep (Sherman 2003). The cumulative effect of thousands of adult crabs per hectare consuming plant materials is the near complete removal of large patches of the leaf litter layer. The removal of leaves by individual crabs is a selective process based upon leaf-species characteristics that are poorly understood.

Gecarcinid crabs, by removing the litter layer, alter the soil chemistry and rooting density profiles for the first meter of soil depth (Sherman 2006). It is likely too that the near complete removal of litter in some areas greatly impacts litter invertebrate communities, seed establishment dynamics, and seedling competition. Excavations of crab burrow chambers revealed isolated pockets of high soil-nutrient content (organic carbon) in the subterranean soils. Presumably such outlier sites of rich soil developed through the decomposition of residual plant matter and feces in the burrow chambers. Routine findings in crab chambers of large quantities of leaf litter (in varying states of decay) further implicate such a direct mechanism. Predictably, rooting densities reflected the soil nutrient concentrations and the chambers possess significantly more root biomass than did same-depth soils unassociated with burrows. At the soil surface, the removal of the leaf litter layer by crabs was associated with reduced soil nutrient concentrations and rooting biomass relative to nearby surface soils unassociated with crab activity. Data from crab exclusion areas (one square meter or less in area) into which leaf litter was placed and allowed to decompose similarly revealed such nutrient patterns. From larger crab exclosures, after two years, the surface soils had 22% more organic carbon content than did control top soils. Although fungal hyphae, litter invertebrates, and a developing root mat were found within the exclosure treatment replicates, further research is needed to quantify impacts of litter removal from the surface.

While their most common forage is fallen leaves, gecarcinid crabs prefer to consume certain fruits, seedlings, and seeds on the forest floor, thereby affecting juvenile plant community structure and diversity (Sherman 2002). Crab diet selectivity is based both upon plant species and propagule dimensions and characteristics. While seed of some tree species are killed, others, like *Simaba cedron*, may escape rodent predators while in crab burrows then later germinate on their refuse piles (T. Lee and P. Sherman, unpublished observation). By creating a predatory gauntlet through which all plant species must pass, land crabs may ultimately reduce floristic diversity in forest zones they occupy—that is, within a half-kilometer of the coast. Two-year-long crab exclusion experiments near Sirena showed a doubling of both seedling density and litter accumulation and thus supported this hypothesis. Moreover, the crab zone at Sirena supports about half the dicotyledonous species compared with adjacent crab-free forest on the inland edge of the zone.

Corcovado's main predators of crabs include roving individuals or extended families of coati and opossum that routinely scour the crab zone as well as common black hawk (crab-hawk or *cangrejero*). These birds spend more time near the *Cardisoma* population since diurnal crabs routinely spend time at their burrow entrance or within a meter or so of the burrow during the day. It is likely that the availability of crabs in coastal forests allows higher carrying capacities for certain predator species with a cascade of indirect effects that extend further inland. Corcovado Park provides an ideal venue for researching the broader community and ecosystem importance of the crab zones in lowland rain forests that meet the sea.

Insects

Few groups of insects are sufficiently well studied to permit detailed comparisons and contrasts between the Pacific and Atlantic Lowland Rainforests. As the Pacific lowlands are more seasonal and more limited and isolated geographically, it is not unexpected that somewhat fewer species of well-known groups such as butterflies are recorded there. DeVries (1987) compared species diversity from families Papilionidae, Pieridae, and Nymphalidae between Costa Rica's Pacific and Atlantic lowland rainforests and found fewer species on the Pacific (217 vs. 261), yet Pacific and Atlantic mid-elevation forests are essentially equal (267 vs. 256). In explaining the lowland discrepancy DeVries suggested some interesting feedbacks. Lower species and pattern diversity of distasteful ithomiines (a group requiring humid conditions for adult feeding requirements) has a cascading negative effect on dismorphiines, mimics of ithomiines.

The most interesting feature of the Pacific Lowland Forest is not an exceptional richness for this group, but its relative isolation from close relatives on the Atlantic side of the central mountain range. This has allowed ecological speciation in mimetic butterflies such as *Heliconius pachinus* (Kronforst et al. 2006a,b, 2007a,b) in the Osa region. The wing pattern of this insect, its model *H. hewitsoni* (Fig. 12.14), and its mimics, *H. pachinus* and *H. sara*, may be the most easily seen indication that one has stepped into wet forests of Costa Rica's Pacific slopes. Similar cases of endemic races and species of insects are likely common but much less conspicuous.

The absence on the Pacific of other insects, like bullet ants, which are so common and dominant on the Atlantic side, present interesting ecological and evolutionary questions for investigation. Here we focus on ants as a case study since they are the most easily sampled group and have been studied in comparable detail in both lowland areas. Hogue (1993) provides a general reference for Neotropical insects. However, because he drew extensively from field studies near Rincón in assembling examples for his book, it is a very useful guide for insects of the Osa region.

Ants

The lowland forests of southwestern Costa Rica are teeming with ants. This is no surprise, since all lowland Neotropical forests are teeming with ants. But since the Osa Peninsula and vicinity is a somewhat isolated region of wet forest, with dry forest to the northwest and southeast and mountains to the north and northeast, one might expect an impoverished fauna compared to the extensive lowland rainforest on the wet Atlantic coast, and perhaps some biogeographical oddities resulting from its isolation. Here we contrast the ant fauna of Costa Rica's Pacific and Atlantic slope wet forests. We examine (1) whether overall diversity differs; (2) the degree of species overlap between the two sites; and (3) particular elements of the fauna that are missing, unique, or otherwise notable in the southern Pacific forests.

The comparison is based on a database of species occurrences in Costa Rica, where each occurrence is a separate collection event. Collection events include a direct search for foragers or nests, Winkler samples of sifted leaf litter, Malaise trap samples, and canopy fogging samples. Most of the collections were made between 1979 and the present. All the localities were below 500 m elevation. For the Pacific lowlands, most collections were from the Osa Peninsula, and a few were from Manuel Antonio National Park and Carara Biological Reserve. For the Atlantic lowlands, the great majority of occurrences were from La Selva Biological Station, and a few were from Tortuguero National Park and Hitoy Cerere Biological Reserve. The collecting effort has been far greater on the Atlantic slope (Longino and Colwell 1997, Longino et al. 2002, and see McClearn et al., chapter 16 of this volume): a total of 17,120 species occurrences were available for the Atlantic lowlands, 1,697 for the southern Pacific lowlands.

Overall species diversity does not differ between the Atlantic and Pacific lowlands. Both show a highly diverse fauna typical of lowland Neotropical forests. A total of 559 species have been recorded from the Atlantic lowlands (507 from La Selva itself), 352 species from the southern Pacific lowlands. Given the similar curves in Fig. 12.18, the lower richness observed for the Pacific lowlands is almost certainly due to undersampling.

There is a high degree of overlap in species composition. The Chao estimate of the Sørensen index, based on abundances, is 0.75 (calculated using the software application *EstimateS*, version 7.5; Colwell 2005). This index varies from zero, for no species in common between two sites, to one, for all species in common. The Chao estimate of the Sørensen index attempts to correct for undersampling effects (i.e., species shared between two sites but not sampled in one or both sites) (Chao et al. 2005).

The southern Pacific forests are more seasonal than the Atlantic slope forest, which has an impact on the ant fauna. Arboreal ants in particular show a vertical expansion of foraging zones by species adapted to drier conditions. *Camponotus sericeiventris* (Fig. 12.19), *Cephalotes minutus*, *C. umbraculatus*, *Azteca instabilis*, and *A. sericeasur* are examples of conspicuous ants that can be found foraging

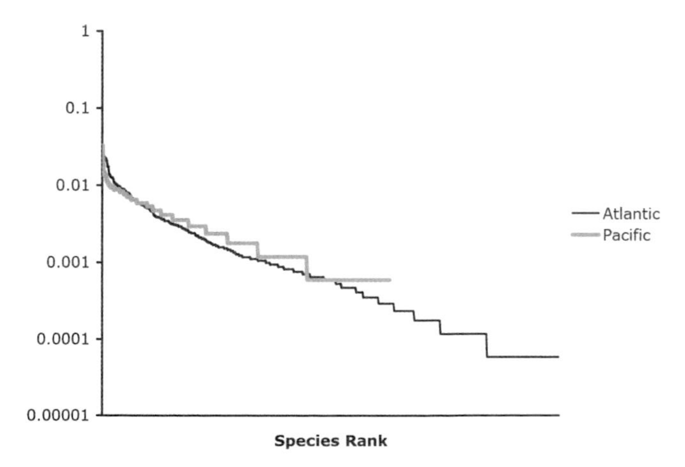

Fig. 12.18 Rank-abundance plot contrasting ant species diversity in southern Pacific versus Atlantic lowland wet forests in Costa Rica. Curve for Atlantic slope contains 559 species.

at all levels in the vegetation and even on the surface of the ground in southern Pacific forests. In Atlantic forests these species are more restricted to the high canopy, rarely occurring in the shaded understory of closed canopy forest. In Atlantic forests there are several species that inhabit the low arboreal zone in the deeply shaded and always humid understory: *Acromyrmex volcanus*, *Crematogaster longispina*, *Cyphomyrmex cornutus*, *Odontomachus erythrocephalus*, *Paratrechina caeciliae*, and *Pheidole fiorii*. These species are completely absent from the Pacific forests. In some cases taxa that are common in dry forest habitats extend into the southern Pacific forests but do not occur on the Atlantic side. These include *Dorymyrmex*, *Azteca forelii*, *Crematogaster rochai*, *Pheidole pugnax*, *Pseudomyrmex caeciliae*, and *Pyramica margaritae* (now in *Strumigenys*).

Ant plant associations in the southern Pacific forests are similar to those found elsewhere in Central America (Jiménez et al., chapter 9 in this volume). Ant acacias (*V. alleni*) occur in these areas (Fig. 12.13), inhabited by *Pseudomyrmex spinicola* and *P. particeps* (Ward 1993). The former is widespread in southern Central America, but the latter is endemic to the Osa Peninsula and vicinity. Missing species are *P. flavicornis* and *P. nigrocinctus*, species common in dry forest acacias further north. *Cecropia* trees host mainly *Azteca alfari* and *A. constructor*, which occur over large areas of the Neotropics (Longino 1991, 2007). *Azteca coeruleipennis* may also occur, which is a dry forest species whose range extends northward through the drier parts of Mesoamerica. The missing ant species are *Azteca xanthochroa* and *A. ovaticeps*, which are common inhabitants of *Cecropia* trees in the Atlantic lowlands. Specialist understory *Piper* plant species host the plant ant *Pheidole bicornis* in domatia formed from the clasping petioles. *Cordia alliodora* trees host *Azteca pittieri* and *Cephalotes setulifer* ants (Longino 1996). Understory *Ocotea* (Lauraceae) trees host *Myrmelachista lauropacifica*, an ant species endemic to the Pacific lowlands but closely related to similar *Ocotea* ants on the Atlantic slope (Longino 2006).

A common pattern is for the southern Pacific lowlands and the Atlantic lowlands to have cognate pairs of related forms. Given the arbitrariness of taxonomic rank when dealing with allopatrically differentiated forms, in some cases these have been recognized as distinct species and in other cases are considered intraspecific variation. Examples of the former include (Pacific versus Atlantic forms): *Dolichoderus curvilobus* and *D. validus*, *D. laminatus* and *D. lamellosus*, *Aphaenogaster phalangium* and *A. araneoides*, *Pheidole multispina* and *P. browni*, *Pachycondyla theresiae* (now in *Neoponera*), and *P. bugabensis*. Examples of the latter include different forms of *Pachycondyla harpax* and *Pheidole bicornis*.

Paraponera clavata (bullet ant; *bala*) is conspicuously

Fig. 12.20 Fishes that occupy very similar ecological niches as epibenthic predators in inland rivers and streams of Costa Rica. Top left above: pargo, *Lutjanus novemfasciatus*, of Corcovado National Park, a marine species within a region with very low diversity of freshwater fish species. Top left below: guapote, *Parachromis dovii*, of Tortuguero National Park, a region with high freshwater fish diversity. Sirena freshwater habitats are diverse and include, top right, the Río Claro and, below, the Quebrada Cameronal, a forest tributary of the murky Río Sirena, also shown in Fig. 12.3.
Photo credits—Top left above and below (two fish), K. Winemiller; Top right and bottom, L. Gilbert.

Winemiller and Leslie (1992) documented 8 characid, 13 cichlid, and 8 poeciliid taxa from Tortuguero National Park on the Atlantic coast (at Corcovado National Park only 2 characids, 2 cichlids, and 3 poeciliids have been documented; Winemiller 1983, Winemiller and Morales 1989). Another major contrast is the more extensive and marked longitudinal transition of fish assemblage composition that occurs in the Atlantic region. At Tortuguero, marine fish species are gradually replaced by freshwater species over a distance of several kilometers as one moves from the coastline through coastal lagoons and rivers to upland streams (Winemiller and Leslie 1992. Even among freshwater fish families, there is a gradual species replacement between brackish coastal habitats to forest streams. In contrast, marine fishes on the Osa Peninsula, such as snappers and eleotrids, are distributed fairly evenly across the longitudinal gradients of rivers and streams (Winemiller 1983, Winemiller and Morales 1989). On the Osa, these marine fishes occupy niches typical of cichlids in other regions of Central America that have greater freshwater fish diversity (e.g., Fig. 12.20). Hence, ecological opportunities for marine fishes result from the historical biogeography of the peninsula. Interestingly, the rivers in the extreme southwestern corner of Costa Rica, adjacent to the neck of the Osa Peninsula, actually contain a high diversity of freshwater fishes. These drainages mark the northern limits for a number of South American fish genera (e.g., *Aequidens*) and families (Loricariidae, Trichomycteridae) that have dispersed through Panama after the final closure of the Isthmus of Panama during the Pleistocene (Bussing 1976).

Herpetofauna

Patterns of population differentiation and speciation observed in Costa Rica's herpetofauna reflect major geographic regions of the country. The Pacific lowlands and the Osa region of interest here are termed "Southwest" (SW) by Savage and Villas (1986). This region contains about 67% of the reptile species fauna of Costa Rica and 33% of the total amphibian species fauna of Costa Rica (Savage 2002). Relative to the Atlantic side, the Osa region's high microhabitat diversity helps account for the differential richness of reptiles, but its more intense and prolonged dry season probably accounts for a relatively poor representation of amphibians. Even today across this region, areas that have not been converted to plantations or human settlements still retain fragments of natural habitat types. Forest swamps at the south end of the zone will be very similar in herpetofauna to forest swamps at the north end. The same can be said for other habitat categories. Most reptile and amphibian species are widespread in the region.

In terms of its herpetofauna, the Osa region is highlighted by species and races not found on the Atlantic side (Fig. 12.21). Its relative isolation as an "island" of lowland wet forest with strong seasonality sets it apart from similar zones across the mountain barrier. Endemic species include black-headed bushmaster (*Lachesis melanocephala*), the Osa cecilian (*Osaecilia osa*), Allen's salamander (*Oedipina alleni*), vocal rain frog (*Eleutherodactylus vocator*), granulated poison dart frog (*Dendrobates granuliferus*), Golfo Dulce poison dart frog (*Phylobates vittatus*), pugnose tree

Fig. 12.21 Herpetofauna of ACOSA, not seen in the Atlantic forests. Top left: Endemic black-headed bushmaster (*Lachesis melanocephala*), near Los Mogos. Note the red Ultisol soil below. Top right: gladiator frog (*Hyla rosenbergi*), which ranges from Peru to the Pacific lowlands of Costa Rica. Bottom right: The endemic Isthmian alligator lizard (*Coloptychon rhombifer*). Bottom left: An endemic and new, undescribed species of *Bufo* once thought to be a yellow morph of *B. melanochloris*. *Photo credits—Top left: Reynaldo Aguilar; Top right: L. Gilbert; Bottom right: Park guard A.G. Tugri; Bottom left: A. Vega.*

frog (*Smilisca sila*), bull rain frog (*Eleutherodactylus taurus*), water anole (*Norops aquaticus*), reticulated ameiva (*Ameiva leptophrys*), Isthmian alligator lizard (*Coloptychon rhombifer*), and earless lizard (*Bachia blairi*). Species ranging from South America but confined to the Pacific side forests include many-scaled anole (*Norops polylepis*), water lizard (*Neusticurus apodemus*), gladiator frog (*Hyla rosenbergi*), mangrove tree boa (*Corallus ruschenbergeri*), red-eyed tree snake (*Tripanurgus compresus*), barred cat-eyed snake (*Leptodeira rubricata*), Clark's coral snake (*Micrurus clarki*), and white-tailed hognose snake (*Porthidium porrasi*).

Some species occur widely but populations of the Pacific forests have differentiated noticeably from populations in other areas of Costa Rica. Thus, because it shows a different color pattern and size on the Pacific side, red-eyed tree frog (*Agalychnis callidyias*) was considered a different species (*A. helenae*). *Agalychnis* from the Osa Peninsula are smaller and with paler colors in the sides and limbs (Robertson and Zamudio 2009). Other amphibians that show the same trend of reduced coloration and size on the Pacific compared to Atlantic slope include dark-eyed leaf frog (*Agalychnis spurrelli*), masked tree frog (*Smilisca phaeota*), green-black poison dart frog (*Dendrobates auratus*), and Mexican tree frog (*Smilisca baudinii*). Conversely, some species previously thought to occupy both sides of Costa Rica have actually speciated while in isolation in the Pacific forest. Thus, a yellow "morph" of *Bufo melanochloris* was determined to be a different species, *Bufo aucoinae* (O'Neil 2004). There likely are other similar cases demanding further attention. Before the mid-1980s bushmasters were considered variants of a single species in Costa Rica, but later confirmed as two separate species (Fernandes et al. 2004). It is of serious concern that venom of the Pacific slope endemic black-headed bushmaster (*L. melanocephala*) is not countered by standard polyvalent antivenoms. Other recently split species include hognose pit viper (divided into *Porthidium nasutum* and *Porthidium porrasi* [Lamar and Mahmood 2003]), and the eyelash pit viper (*Bothriechis schlegelli* and *Bothriechis supraciliaris* [Solarzano et al. 1998]).

The strong dry seasons experienced in the Osa region appear to have influenced the composition of the amphibian community. Species endemic to these forests all seem adapted to an environment lacking large and predictable bodies of water. Thus, poison dart frogs lay their eggs in the leaf litter and carry them to a source of water above ground where they are tended by a parent; cecilians and rain frogs do not have a larval stage, but lay large eggs that are also tended by a parent; and tree frogs and toads can exploit ephemeral bodies of water by speeding up larval development.

Herpetologically, the Osa region is notable for producing and harboring many endemic species. Its relative isolation as an "island" of lowland wet forest and strong seasonality in rainfall help account for such diversity. Added to the mix are faunal elements moving north from South America and unable to proceed northwards beyond lowland rainforests. Presumably this flow has stopped with the deforestation of adjacent Panamanian lowlands.

Birds

The highest diversity of bird species in tropical habitats is found in mature primary rainforest, although even the most common species have population densities much lower than the most abundant temperate forest species (Terborgh et al. 1990). Low population densities of common rainforest species can be attributed to several factors. Large-bodied species, such as raptors, parrots, and woodpeckers, require large territories. Many species, such as antbirds, are associated with specialized habitats within the matrix of a mature forest, such as treefalls, vine tangles, or shaded streams (Terborgh et al. 1990). Closely related species, like woodcreepers, have significant overlap in diet but segregate by foraging height and substrate use (Chapman and Rosenberg 1991). The stable organization of understory mixed species flocks and their communal territorial defense limits the population density of core species (Jullien and Thiollay 1998).

The Osa Peninsula supports a diverse and abundant avifauna. Over 320 bird species have been recorded, including one endemic, the black-cheeked ant-tanager (*Habia atrimaxillaris*). Individual pairs and family groups of black-cheeked ant-tanagers inhabit separate drainages in hilly *terra firme* humid forest, and are now almost entirely restricted to Corcovado National Park.

Like other Neotropical humid forests, Osa's is occupied by a majority of rare bird species and a few common ones. Insectivorous species dominate the understory, while frugivorous and omnivorous species dominate the canopy (Thiollay 1994). Lekking species and regular members of mixed species flocks are among the most common in humid lowland forests. Thiollay (1994) found the number and location of lekking males to be remarkably stable in French Guiana, and core members of mixed species understory flocks maintain population stability by occupying almost all suitable habitat and maintaining a proportion of floating individuals to buffer population fluctuation (Jullien and Thiollay 1998).

As at La Selva Biological Station in the Atlantic lowlands (Loiselle 1988), the wedge-billed woodcreeper (*Glyphorynchus spirurus*) is ubiquitous in mixed species flocks and is

Fig. 12.22 Large signature birds of Neotropical lowland forests. Top left: Great currasow *Crax rubra* has recovered from hunting pressures since park creation. Bottom left: The Crested guan though abundant elsewhere is very common in Corcovado. Top right: The ACOSA, with protected reserves and the education of local populations, is helping to protect the endangered scarlet macaw from the pet trade. Bottom right: Park guard at La Sirena shows the foot of a harpy eagle confiscated from a local *campesino* who had shot it on the outside of Corcovado national park. Juvenile harpies and a nest have been sighted at or near La Sirena.
Photo credits—Upper right: Tom Jorgensen; Remaining three: L. Gilbert.

the most common forest bird. Remarkable for its abundance on the Osa, neighboring pairs of chestnut-backed antbirds (*Mymeciza exsul*) saturate the forest interior. Mealy parrot (*Amazona farinosa*) is probably the most common canopy frugivore. The Osa also supports one of Costa Rica's last healthy populations of scarlet macaw (*Ara macao*), Fig. 12.22, now increasingly restricted to the Corcovado basin (Stiles and Skutch 1989, and see below).

A stroll along Osa forest trails reveals numerous leks of displaying long-tailed hermit (*Phaethornis superciliosus*), red-capped manakin (*Pipra mentalis*), blue-crowned manakin (*P. coronata*), ochre-bellied flycatcher (*Mionectes oleaginous*), and rufous piha (*Lipaugus unirufus*). Blue-crowned manakins dominate the lower understory, while the congeneric red-capped manakin occupies the lower forest canopy (Stiles and Skutch 1989). Lek placements are determined by topography and intercept female traffic on the way to food resources (Westcott 1997).

Mixed species flocks, comprised of core members of single species pairs collectively defending large stable territories, represent a distinct foraging guild in Neotropical humid forests (Powell 1979, Munn and Terborgh 1979). With the absence from the Osa of two common nuclear species of understory mixed species flocks, white-flanked and checker-throated antwrens (*Myrmotherula quixensis* and *M. fulviventris*), the separation of distinct canopy and understory flocks seems less apparent. Understory flocks are usually dominated by pairs or family-sized groups of dotted-winged antwrens (*Microrhopias quixensis*) (Gradwohl and Greenberg 1980). Canopy flocks typically contain a pair of white-throated shrike-tanagers (*Lanio leuothorax*), which vocalize frequently and serve as sentinels within the flock (Stiles and Skutch 1989). Other common mixed species flockers include black-striped woodcreeper (*Xiphorhynchus lachrymosus*), striped foliage-gleaner (*Hyloctistes subulatus*), and tawny-crowned greenlet (*Hylophilus ochraceiceps*). Several species of tanagers accompany mixed species flocks although, with the exception of bay-headed tanager (*Tangara gyrola*) and golden-hooded tanager (*T. larvata*), tanagers of the genus *Tangara* are relatively scarce on the Osa.

Army ant followers represent another distinct foraging guild. From huge temporary nests (bivouacs) constructed of their own living bodies, the army ant *Eciton burchelli* spills out onto the forest floor in diurnal foraging raids. Moving in a fan-shaped mass, they dismember every living animal they can subdue and carry body parts back along a return trail to the bivouac as food for the colony. As the leaf litter boils with prey attempting to escape the advancing swarm, birds and other animals gather to feed at the concentrated food source. The Osa supports the highest density populations of *E. burchelli* army ant colonies recorded in the Neotropics (Swartz 1997). Virtually all bird species inhabiting the forest understory will forage opportunistically at a swarm that they encounter. However, some species have evolved the specialized behavior of memorizing the location of bivouacs in order to track multiple colonies through space and time (Swartz 2001). These species generally rely on army ant swarms for the majority of their food and are known as obligate species. The obligate ant-following bird guild of the Osa is comprised of bicolored antbird (*Gymnopithys leucaspis*), grey-headed tanager (*Eucometis penicillata*), Northern barred woodcreeper (*Dendrocolaptes sanctithomae*), and tawny-winged woodcreeper (*Dendrocincla anabatina*). Family groups of grey-headed tanagers following ant swarms adopt a foraging strategy in Central America that they do not pursue in the forests of South America. Cocoa woodcreeper (*Xiphorynchus susurrans*), black-faced ant-thrush (*Formicarius analis*) and the endemic black-cheeked ant-tanager also demonstrate the behavior of bivouac-checking (Swartz 2001), although they rely on foraging with army ants for only a portion of their food supply. A total of 45 species have been recorded foraging at army-ant swarms opportunistically, 18 of which were also observed traveling in mixed species flocks. The ocellated antbird (*Phaenostictus mcleannani*) and rufous-vented ground-cuckoo (*Neomorphus geoffroyi*), species normally found in the Central American ant-following guild, are absent from the Osa.

The Osa provides important wintering habitat for numerous Nearctic migrants. Large numbers of yellow warbler (*Dendroica petechia*), chestnut-sided warbler (*D. pensylvanica*), Tennessee warbler (*Vermivora peregrina*), and mourning warbler (*Oporonis philadelphia*), as well as Philadelphia vireo (*Vireo philadelphicus*) and red-eyed vireos (*V. olivaceus*), occupy second growth and mature forest edge. Northern waterthrush (*Seiurus noveboracensis*) and prothonotary warblers (*Protonotoria citrea*) depend on coastal mangroves. Kentucky warblers (*Oporonis formosus*) maintain individual wintering territories in the understory of the forest interior (Stiles and Skutch 1989), which also provides crucial habitat for diminishing populations of wood thrush (*Hylocichla mustelina*) (Robbins et al. 1989). Compared to the lowland forests on the Atlantic slope, the Osa supports relatively few altitudinal migrants. White-throated robins (*Turdus albicollis*) occupy the forest interior during the non-breeding season between August and December (Stiles and Skutch 1989), and three-wattled bellbirds (*Procnias tricarunculata*) can be heard practicing their songs in isolated leks of mature second growth between December and March (Powell and Bjork 2004).

If common species occur at relatively low population densities and a high proportion of species are rare, then large areas of habitat are necessary to maintain viable populations. Only on the Osa Peninsula, does Costa Rica have sufficient unfragmented lowland Pacific forest to sustain forest bird diversity, and those forests may still not be large enough for persistence of the largest species such as harpy and crested eagles (*Harpia harpia* [Fig. 12.22] and *Morphnus guianensis*, respectively).

Mammals

Most mammals are small, secretive, and nocturnal so accurate estimates of regional faunas are difficult to obtain. The total numbers of mammal species known to or expected to occur in comparable areas of the Atlantic and Pacific Lowland forests are approximately 140 and 150, respectively (Wilson 1983). Such estimates may or may not reflect a real difference in total species numbers between these forests. Certainly the sight of a troop of squirrel monkeys (*Saimiri oerstedii*) would signal entry into the Pacific lowland wet forests. In Costa Rica, only on the Osa could an eight-member graduate field course collectively see all six species of Neotropical cats within 200 meters of a field station (L. Gilbert, personal observation). Only there too can one see large troops of all four wild primates species in a one-hour canoe trip on a forest river (Río Sirena; L. Gilbert, personal observation). But the more important distinction between the Pacific and Atlantic Lowlands is not the species list per se, but the presence, on the Osa Peninsula, of an intact and abundant large mammal fauna. Corcovado Park with its relative remoteness, size, and protection allows species that are present, but rarely observed elsewhere, to exist at population densities sufficient to strongly impact the forest community. With the availability of Sirena Biological Station on the coast at a point most remote from the park's terrestrial boundaries, long-term studies of large mammals on a grand spatial scale have been and are possible. ACOSA

Fig. 12.23 Endangered mammals highly justify conservation action in ACOSA. Top left: A female jaguar with a radio collar installed by Eduardo Carrillo returns to a sea turtle it had killed the night before (covered with a *Gecarcinus quadratus* crab). This picture demonstrates that unique spotting patterns of jaguars allow individual identification and avoid problems of capture. Top right: Two of 17 Baird's tapirs that were radio-collared by Charles Foerster in a pioneering ten-year study based at the original biological station at Sirena. The cofounder of Costa Rica's park system, Alvaro Ugalde, observed his first-ever wild tapirs in 2004. Bottom left. White-lipped peccaries invariably run after bluffing with loud snapping of tusks. Bottom right: A squirrel monkey feeding on fruit is seen from a boat on Río Sirena.
Photo credits—Top left: E. McCain and L. Gilbert; Remaining three : L. Gilbert.

and the old lab at Sirena have been home for extensive population studies of tent-making bats (Chaverri and Kunz 2006a,b, Chaverri et al. 2007a,b,c, 2008). Likewise, experimental studies comparing impacts of small versus large mammals on tree seed survival according to regeneration stage of forest have been conducted there (DeMattia et al. 2004, 2006). For this chapter we focus on a few key mammals for which population studies have been carried out from the base at Sirena (Fig. 12.23).

White-Lipped Peccary

The white-lipped peccary (*Tayassu pecari*, Artiodactyla: Tayassuidae) is a Neotropical ungulate that ranges from southern Mexico to northern Argentina preferring unaltered humid and semi-deciduous forests (Mayer and Wetzel 1987, Sowls 1997). Two other species of peccary comprise the Tayassuidae. The collared peccary (*Pecari tajacu, the javelina*) is broadly distributed from the southern part of the United States of America to central Argentina and is relatively adaptable to human disturbance and encroachment. The Chacoan peccary (*Catagonus wagneri*) is the largest of the three and considered in danger of global extinction. It lives exclusively in the rapidly degrading Chaco regions of Argentina, Bolivia, and Paraguay. For an overview of peccary natural history see Sowls (1997) and for the white-lipped peccary, see Mayer and Brandt (1982).

In Costa Rica, white-lipped peccaries can be found in small isolated populations in protected areas throughout the regions of Guanacaste, Tortuguero, Caño Negro, Talamanca, and Osa Peninsula (Vaughan 2011, March 1993). All of Costa Rica's populations are in danger of local extinction. Elsewhere in the neotropics, throughout their entire range, populations of white-lipped peccaries are threatened due to the accelerating destruction of their habitat and unsustainable hunting practices (Oliver 1993, Sowls 1997).

Their propensity to form large herds that range over large areas, combined with their low reproductive rates (Gottdenker and Bodmer 1998), likely increases their susceptibility to human impact. With an average adult weight of 33 kg (Sowls 1997), the white-lipped peccary is considered omnivorous but displays a strong frugivorous preference. In addition to fruits, its diet includes stems, leaves, roots, flowers, and invertebrates (Kiltie 1981, Mayer and Wetzel 1987, Olmos 1993, Sowls 1997, Altrichter et al. 2000).

The white-lipped peccary is the only Neotropical forest mammal to form groups up to several hundred individuals (Sowls 1997) and is considered by some to be nomadic and/or migratory (Kiltie and Terborgh 1983, Bodmer 1990, Altrichter 1997, Sowls 1997). Others, however, do not categorize the herd's movements as being migratory. Researchers frequently describe disappearances of white-lipped peccary herds from their study sites, citing various causes such as local population extinctions due to hunting (Peres 1996) or disease outbreaks (Fragoso 1998, 2004). Others suggest disappearances are temporary, citing normal seasonal migratory movements (Altrichter and Almeida 2002). It is likely that all three causes are site-specifically accurate.

Over most of their distribution white-lipped peccaries are an important and favorite source of food for indigenous and rural inhabitants (Smith 1976, Kiltie 1980, Redford and Robinson 1987, Vickers 1991, Bodmer 1995, Altrichter and Jiménez 1999). Overhunting and habitat destruction threaten several populations with extinction (Cullen et al. 2000), and the subspecies *T. p. spiradens*, which inhabits southern Costa Rica's Corcovado National Park, was considered vulnerable by the IUCN/SSSC Pigs and Peccaries Specialist Group under the pre-1994 IUCN Red List categories and criteria (March 1993). A reassessment of the Red List status of the subspecies of *T. pecari* is required (Hilton-Taylor, pers. comm.). The species is listed on CITES Appendix II (2005).

Costa Rica's healthiest population of white-lipped peccary is found in the Osa Peninsula's Corcovado National Park. The Park protects 46,774 ha of mostly primary forest and is surrounded by ample buffer regions in the form of forest, indigenous and private reserves. Nevertheless, in comparison with frequent historical accounts of herds of over 500 individuals (Vaughan 1981), the Osa populations are sharply reduced. Because the white-lipped peccary is the principal prey of the Jaguar and other large cats in and around the Park, this species is considered to be key to the survival of large predators (Carrillo 2000).

Until 1995, the Osa populations of white-lipped peccaries were poorly studied. At this time, a long-term research project was started by Eduardo Carrillo later to be extended by his students (Altrichter 1997, 2000, Altrichter et al. 1999, 2000, 2001a,b, 2002, Campero 1999, Carrillo 2000, Carrillo et al. 2002, and López et al. 2006, among others). From these studies, we now more fully understand the dynamics of the Osa populations including their home ranges and movement patterns, time-budgets and activity patterns, diets and food preferences, social behaviors, genetic variability, and food-web ecology. Through years of fieldwork using methods such as radio-telemetry and continuous direct observation, an important data set has been obtained. Altrichter (2000) and Altrichter and Almeida (2002) have studied the conservation consequences of herd movements beyond the protection of Corcovado National Park in sociological research conducted with the local and indigenous inhabitants of the peninsula. In this short review, we will briefly describe the general patterns associated with these topics.

In Corcovado, the impact of seasonality (dry season from January to April and wet season from April to December) seems to strongly influence all aspects of the lives of peccaries. For that reason, the following descriptions of peccary ecology will be placed into a seasonal context.

Peccaries in Corcovado are diurnal all year round with a period of reduced activity during the middle of the day. They differentially use the forest, probably responding to food availability. They were found more often in primary forest during February to May, in secondary and coastal forest during June to September, and in herbaceous swamp during October to January (Carrillo et al. 2002).

Overall, the peccaries ranged over an area of <40 km^2, but use of the area shifted seasonally and movements were reduced when fruits were most abundant. According to Carrillo et al. (2002), peccary density may be higher where the interspersion or close proximity and mix of seasonally important habitats is high, and thus where peccaries do not have to travel as far for food. The sizes of four distinct groups followed for nearly a year varied from 21 to 70 individuals (Altrichter 1997). However, Carrillo et al. (2002), noting that the different herds' home ranges overlap and all share the same relatively small forest area, suggests that all peccaries may form one single herd.

Peccaries are mainly frugivorous in Corcovado and their behavior is highly influenced by fruit availability. They consumed around 60 plant species (37 of them fruits). Moraceae was the most preferred family, which contrasts with data collected in the Peruvian Amazon region where their diet primarily consists of palm seeds (Arecaceae). In Corcovado, they also consumed high proportions of vegetative parts of plants in the Araceae and Heliconaceae families. Earthworms were an important part of their diet during part of the year. Diets differed between months, seasons, and habitats. They consumed more fruits in coastal and primary forests and more vegetative parts in secondary forest. In the months of October and November, the consumption of vegetative parts exceeded actual fruit consumption (Altrichter et al. 2000).

Because of their high consumption of fruits and seeds, peccaries can play an important role as seed predators influencing the structure and diversity of the forest. On the basis of studies of diet and germination (Altrichter et al. 1999), it was found that peccaries destroy all seeds they consume, except those of very small size (less than 2 mm). These small-sized seeds originate mainly from fig trees (*Ficus* spp.). In other cases, they would only consume the soft part of the fruits and spit out the seed. This was the case with seeds of *Spondias* spp. This behavior can increase the dispersion of these species.

López et al. (2006) assessed the potential nutritional levels of 25 species of plants (seeds, fruits, stems, and leaves) and earthworms, which constitute an important part of the diet of white-lipped peccaries. They forage selectively for certain plant species and parts, likely reflecting the nutritional differences between them. For example, fat and carbohydrates were found more abundant in seeds and fruits while protein was more abundant in fruits and leaves. Differences of diet between white-lipped peccaries in Corcovado and in other tropical regions of Latin America could be partially explained by these results. For example, in Corcovado there are several species with higher fat and energy content than palms, which can explain the low consumption of palm seeds. Mineral content also differs among plant parts. Calcium, potassium, and magnesium contents were higher in leaves whereas copper and zinc were higher in seeds. It is possible that the regular consumption of stems and leaves of some species is related to their high mineral content (López et al. 2006).

Reproduction patterns also seem to respond to fruit availability. The highest numbers of newborns (one month old or younger) were recorded in July and August and the lowest between January and April, which correspond with the dry season. Thus, litters are born during months of high fruit availability (Altrichter et al. 2001a). Interestingly, although it was found that herds have more females than males, with an average sex ratio of 1.7:1 (Altrichter et al. 2001b), annual survival seems to be lower for females than for male peccaries. It was found that although both sexes were poached and died in accidents, only females were killed by predators or died of unknown (non-poaching) causes (Fuller et al. 2002).

The diurnal time budget also likely follows seasonal patterns. Altrichter (1997) followed four herds over 11 months and found that they spent a similar amount of their daytime (around 30%) eating, moving, and resting, and about 6% in social interactions, and other activities. The time spent resting decreased as the time spent moving and eating increased during the months of fruit scarcity. In the wet season, peccaries spent more time eating than in the dry season, probably because of the considerable time allocated to rooting. The monthly variation in time spent on social interactions and the frequency of agonistic interactions seemed to be related to breeding rather than fruit availability (Altrichter et al. 2002).

While in the study area, peccaries traveled the longest distance in October, which coincided with the lowest fruit availability. Peccaries would normally walk a few kilometers some days in search of food while other times they would spend most of the day at a single site. However, during October, the rainiest month, they would walk every day an average of 5 km (Altrichter et al. 2002). At the end of the wet season, there were very few species fruiting, and the frequent inundations kept the fruits from accumulating on the forest floor. During that period, peccaries increase the consumption of non-seasonal resources like leaves and stems (Altrichter et al. 2001b). In mid-November, the radio-marked herds left the greater study area (and protection of the park), returning in January. Analyses of the peccaries' time budget suggest that fruit scarcity at the end of the wet season affected the peccaries' behavior and probably induced them to travel long distances and increase their home-range in search of food (Altrichter et al. 2002).

With such large herds and seasonality in the availability of fruits, it is necessary for white-lipped peccaries to be constantly searching for food. Thus they are also more susceptible to human encroachment and fragmentation than collared peccaries; they need extensive, continuous tracts of land, with little or no human impact. For this reason, in order to protect the Osa population, Corcovado National Park probably is not large enough. Its protected areas may need to be expanded, while more work is done together with Osa's local communities to find ways to reduce hunting pressures (see below).

Baird's Tapir

Baird's Tapir (*Tapirus bairdii*), the largest mammal in the Neotropics, occurs from Mexico to northern South America. Loss of habitat and hunting has moved *T. bairdii* towards an endangered status. Thus, according to the IUCN, approximately 5,000 individuals survive across its original range (Myers 2005). The quality, isolation, and scale of forest habitat and relative protection from hunting provided by Corcovado National Park have provided a rare opportunity to study tapir population biology in this area. In 1994, pioneering studies on radio-collared tapir and based from the ecological field lab at Sirena Biological Station, were initiated by C. Foerster. This innovative research continued for over a decade. During the project period data were gathered on tapir home ranges, population density, diet, activity patterns, fitness, and offspring survival rates. These studies indicated that tapir live in family groups wherein single adult females are stable residents that associate with one male until another male evicts him. Telemetry demonstrated that tapir activity is 80% nocturnal and 20% diurnal (Foerster and Vaughan 2002).

The effects of these animals on forest structure can be readily observed in Corcovado Park. Understory woody plants are browsed by tapirs up to approximately one-meter height, giving a feeling of openness not seen in comparable forests lacking tapir populations (Fig. 12.9). C. Foerster (pers. comm. of unpublished data) estimated that adults eat an average of 35 kg of plant material per day or around 12,775 kg per year. To the extent that such consumption of plant biomass is differentially impacting certain plants, food webs based on those plants will be greatly impacted. Corcovado Park and its tapir population thus provide an unusual opportunity for investigating the role of tapir in rainforest ecosystem dynamics that is yet to be exploited.

Home ranges were estimated to average 124 ha per tapir and its population density at 2.26 tapirs per square mile. (Foerster and Vaughan 2002). Extrapolation of these tapir densities measured in the mix of secondary and old growth forests along the coast near Sirena to all of Corcovado Park would yield an estimate of 400–450 adults (Foerster 1998)—that is, almost 10% of the total estimated population of this species! However, it is impossible, lacking parallel studies inland from Sirena, to know whether these density estimates apply across the variety of topographic conditions and forest types that characterize other parts of Corcovado Park. Steep hilly terrain elsewhere in the park and a stand solely made up of *Raphia* palms in the center would make it difficult for tapirs to retain the same home range patterns as those close to Sirena. Increased impact from poaching near the park's boundaries would also come into play as an important determining factor.

The healthy tapir population of Corcovado National Park is a major attraction for ecotourism and an important factor in long-term external support for maintaining and protecting the park. Ironically, while on the one hand the pioneering studies at Sirena revealed the need for an opportunity for follow-up studies of tapir population biology across the park and quantitative ecological studies on the role of tapir on plant communities and food web dynamics, on the other hand research on tapir in Corcovado Park has become more difficult to conduct. As aging animals reached the end of their normal life span and as natural deaths occurred, radio collars they wore became the anecdotal cause of death. Because Sirena Biological Station is not only a center of science but of tourism, radio collars on these animals (Fig. 12.23) had become increasingly objectionable to tour guides as clients sought photos of naked tapirs. Instead of taking the opportunity to educate tourists about the role of research in providing new information about the natural history of these animals and how to protect their populations, these natural events of mortality were used in a campaign of misinformation that temporarily ended tapir research in Corcovado Park after 2007. This unfortunate episode highlights the need for research and ecotourism to be more actively engaged as mutualists rather than as competitors in recognizing common conservation goals.

Jaguar

Jaguar (*Felis onca*) populations have suffered from the dual impacts of losing natural prey and of being hunted as a predator of livestock and for pelt as farms and ranches have replaced its habitats. Stable jaguar populations persist in scattered areas throughout its range and the Osa Peninsula is one of them. Jaguars are protected in Corcovado National Park, while there is evidence that jaguars move from one national park to the next through forest reserves and other buffer zones between them. The jaguar population in Corcovado apparently persists in large part due to the availability of favored prey species: white-lipped peccary and collared peccary. Studies based in Sirena have shown that in some areas the presence of white-lipped peccary is in 88% of the scats encountered (Chincilla 1997, Carrillo et al. 2009). Studies elsewhere indicate that jaguar consumes more than 85 prey species across its geographic distribution, showing its plasticity as a hunter (Weckel et al. 2006). This attribute is commonly seen on the Osa Peninsula, where jaguars patrol the beach during the night and especially during the new moon cycles foraging for nesting sea turtles (Carrillo et al. 2009). Thus the marine ecosystem potentially buffers crashes in terrestrial food supplies for this cat in Corcovado National Park. One key question is whether only part of the population taps this resource or whether a coastal subpopulation is in control of beachfront territory.

There have been preliminary attempts to study the jaguar population in Corcovado. While capturing and collaring necessary for telemetry proved difficult and controversial, remote cameras proved capable of discriminating individuals and showed the presence of several individuals. Although little more can be said about population status, it is clear from efforts to date by E. Carrillo and his students that Corcovado National Park, with its extent of unbroken forest and populations of favored prey for jaguar, provides an unusual opportunity to conduct long-term ecological population studies on this large cat. These will be needed to understand its meta-population structure and to assess sustainability issues for its long-term population persistence on the Osa. Results from jaguar studies done in other Neotropical regions may not apply to the Osa.

Primates

The Pacific wet lowland forests of Costa Rica boast all four species of primate present in the country: mantled howler monkeys (*Alouatta palliata, mono aullador,* or *mono congo*), white-faced or white-throated capuchin monkeys (*Cebus capucinus, mono cariblanca*), Central American squirrel monkeys (*Saimiri oerstedii, mono tití*), and black-handed spider monkeys (*Ateles geoffroyi, mono araña*) (Reid 1997). Surprisingly, only the squirrel monkey and the spider monkey have been subjects of long-term behavioral studies, and there have only been four different projects. One reason for the scarcity of primate projects in wet forest sites may be that behavioral observation of arboreal primates is facilitated by little rain, low canopy height, and seasonal loss of leaves—all conditions typical of drier forest. However, studies of one taxon across a spectrum of habitats are priceless for providing a more complete picture of social and ecological variation (see Lott 1991). It is in light of within-taxon variation in social systems that studies of primates in Pacific lowland wet forests have the opportunity to shine.

Three of the four long term primate projects took place on the Osa Peninsula, with two based at Sirena Biological Station, Corcovado National Park (*S. o. oerstedii*: Boinski and Timm 1985, Boinski 1986, 1987a,b,c; *A. geoffroyi*: Weghorst 2007), and with one based just several kilometers away from the northwestern border of Corcovado at Punta Río Claro Wildlife Refuge (*A. geoffroyi*: Riba-Hernández et al. 2003, Riba-Hernández and Stoner 2005, Riba-Hernández et al. 2005). One project (*S. o. citrinellus*: Wong 1990a,b) took place in Manuel Antonio National Park, which marks the approximate northern extent of the Pacific wet lowlands (see Jiménez and Carrillo, chapter 11 in this volume). All four projects addressed basic behavioral, ecological, and conservation-related questions.

The project by Boinski (1986, 1988) on squirrel monkey socio-ecology was a crucial part of a landmark, three-forest comparison of squirrel monkey sociality that addressed variation in social systems in terms of patterns of resource distribution across the range of the species (Boinski et al. 2002, 2005). Squirrel monkeys are insectivore-frugivores, and the Costa Rican species' egalitarian, female-dispersal pattern stood in stark contrast to the female-dominant, male-dispersal and male-dominant, female-dispersal patterns of South American species (Boinski et al. 2002, 2005). The social system of *S. oerstedii* was explained as resulting primarily from the non-monopolizable nature of patches of the small, soft fruits eaten by squirrel monkeys (Boinski et al. 2002). The study of squirrel monkeys in Corcovado also delved into population viability analyses. The Central American squirrel monkey only remains in the Pacific lowland wet forests of Costa Rica (Boinski and Sirot 1997). On the basis of counts of squirrel monkeys at Sirena, surveys of forest fragments, and this species' preference for secondary forests, Boinski and Sirot (1997) estimated population sizes of squirrel monkeys in Corcovado at around 500 and the total population size in the Pacific lowlands at 4,000 (3,000 *S. o. oerstedi*, 1,000 *S. o. citrinellus*). The authors stressed the risk that exists for the endangered Central American squirrel monkey (IUCN 2006), as encroaching development, habitat loss, and fragmentation are ubiquitous pressures (Boinski and Sirot 1997).

The two socioecological projects at Sirena were separated by almost 20 years, and yesterday's early-growth, former cattle pastures have now become regenerating secondary forests that are virtual primate smorgasbords. Spider monkeys likely have experienced local population density increases since the park's creation and the subsequent forest succession, and they currently inhabit all types of forests surrounding Sirena at high densities (Weghorst 2007).

Spider monkeys are ripe fruit specialists, and the nutritional ecology study by Riba-Hernández et al. (2003, 2005) at Punta Río Claro focused on the fruit species spider monkeys chose to eat and the content of their sugar, testing hypotheses about vertebrate seed dispersal syndromes. Interestingly, spider monkeys at this site relied on nectar from trees of *Symphonia globulifera* (Clusiaceae) to such a high degree that, when this species flowered, the amount of flower destruction by spider monkeys precluded fruit development (Riba-Hernández and Stoner 2005). Nevertheless, spider monkeys are important seed dispersers in tropical forests, and they may actually aid in the regrowth of forests throughout their range (Link and Di Fiore 2006, Di Fiore and Suarez 2007). Despite having a high local population density at Sirena, the subspecies of spider monkey in the Pacific wet lowlands (*A. g. ornatus* according to Groves 2001 but referred to as *A. g. panamensis* in many studies) is endangered (IUCN 2006). Habitat loss and fragmentation are the largest threats to spider monkeys' survival;

they need relatively large tracts of forest in order to maintain viable populations (Estrada and Coates-Estrada 1996, Cowlishaw and Dunbar 2000, Zaldívar et al. 2004). A biological corridor, namely, the proposed Meso-American Biological Corridor (*Corredor Biológico Mesoamericano*, CBM; see UNDP 1999, Estrada et al. 2006), would be one way of linking forest fragments and larger, protected areas and would serve to improve gene flow between potentially isolated forest patches. Corcovado National Park clearly would play a pivotal role in any biological corridor plans, as it contains the largest tract of protected Pacific lowland wet forest and should harbor large, healthy populations of Costa Rican primates.

Conservation status assessments are made more robust by incorporating information about the behavioral and ecological variability of species. There is great potential for fruitful comparisons of primates in the dissimilar Pacific and Atlantic lowland wet forests. Multi-forest comparisons of squirrel monkey species have shown how floristic differences are integral for shaping social systems (Boinski et al. 2002). The Pacific side has seasonal forest that is not dominated by any particular tree species, while the Atlantic's forest is often dominated by just one species, such as *Pentaclethra macroloba*, and has no pronounced dry season (Herwitz 1981, Hartshorn 1983).

Long-term studies of primates in Atlantic lowland wet forests are somewhat more common than those in Pacific forests. The long history of ecological research at La Selva Biological Station and its accessibility may have aided in its hosting three long-term primate projects (capuchins and spider monkeys: Campbell and Sussman 1994; howler monkeys: Stoner 1996a,b; capuchins, spider monkeys, and howler monkeys: Bergeson 1998). One long-term study has been published on primate work conducted at La Suerte Biological Station (capuchins and howler monkeys: Bezanson 2006). Spider monkeys have been studied recently at the relatively new Atlantic lowland wet forest field station of El Zota (Lindshield 2006).

Conservation and Conflict

1. The Case of White-Lipped Peccary on the Osa

Costa Rica's national parks are in reality mixed-use areas when both legal and illegal activities are considered together. With indigenous and rural peoples living on the periphery of several parks in remote forested conditions, incursions into parks for subsistence or commercial hunting are commonplace. Moreover when more mobile animals leave the parks during periods of food shortages, they are vulnerable to harvesting. Although officially illegal, hunting around and within the national parks appears unstoppable given the currently inadequate protection, environmental education, and economic alternatives. Such human pressures on wildlife exist in Corcovado National Park as well.

As mentioned earlier, when founded in 1975, the area now known as Corcovado National Park was being settled by cattle ranchers, farmers, and squatters/speculators. Gold reserves attracted prospectors. Understandably, hunting was a way of life and a necessity among these settlers and in subsequent years for the indigenous Guaymí people who live in a reserve located on the Northeastern edge of the park. Such hunting continues today as a natural and desirable means of dietary supplement. The main animals hunted in the region are the paca, white-lipped peccary, great curassow, red brocket deer, collared peccary, agouti, coati, chachalaca, and guan (Altrichter and Almeida 2000). Many other birds, mammals, and some reptiles are also considered appropriate game by the local and indigenous peoples of the region. In other remote parts of Costa Rica, where indigenous groups such as the Bribri and Cabecar peoples live in significant numbers (e.g., the Caribbean slopes of the Talamanca Mountains), surrounding forested regions are notably empty of these highly desired species (Altrichter and Carbonell 2008).

Corcovado National Park's 40 years of existence (1975–2015), its armed patrols, however poorly funded and supplied, and the relatively low human population density in unofficial buffer zones of forest around the park, combine to maintain what is easily Costa Rica's and quite likely Central America's most intact Neotropical rain forest fauna. Within the park, although by no means untouched by hunting pressures, wildlife is thriving. The main problem for wildlife in Corcovado occurs when animals must cross the park boundary in search for food during periods of shortage in the park.

White-lipped peccary meat is the second most desired type of wild meat obtained in the forest (the paca, a nocturnal rodent that is easily hunted with trained dogs, ranks first for its superb meat). Hunting pressures can be high, both from Guaymí and rural people who live further from the park boundary in dispersed towns and homesteads. The Guaymí indigenous people do not seem to distinguish between their reserve and the park for purposes of subsistence, supplementation of domestic meat consumption, and cultural relevance. This notion implies a strong and continuous hunting pressure (sometimes five or more animals are taken out in a single day) and, given its cultural roots, one that seems difficult to control or modify (Altrichter and Almeida 2002).

During periods of low food availability during the peak and end of the wet season (when forest floors are frequently flooded), the search for food takes white-lipped peccary groups beyond the relative protection of the national park boundaries, north and east towards the small and dispersed areas where the rural and Guaymí communities live. Peccaries may travel along distances as far as 30 km (Altrichter 1997).

Expanding African oil palm plantations act as a further incentive for peccaries to leave the park. When the palm plants are relatively young, peccaries may cause significant damage to the crop. As peccaries pass through small remote towns, people alert each another about their presence and conduct opportunistic sport hunting. In these towns, people often kill as many individuals as they can and as many as one-third of a herd may be killed in a single day by these hunters. The peccary meat, although highly desired, merely supplements the domestic meat stock they have available for themselves. According to survey (or interview) data, local people weigh historical and cultural hunting practices, and hunting as counter to boredom, above hunting as a result of necessity (Altrichter and Almeida 2002). As white-lipped peccaries disperse further from the park their mortality rate due to hunting increases; hence, herd sizes reported by local observers decrease. Further away too, essential forest habitats increasingly give way to agriculture.

Consequently, the collared peccary, an abundant and widely distributed species that ranges from the southern United States to northern Argentina, becomes more frequent in these areas. Although its meat is less desirable than that of the white-lipped peccary, people also opportunistically hunt the collared peccary intensively, salting or smoking meat they cannot immediately consume. Apparently, collared peccary also supplements the diets of many who already eat relatively well—that means, it isn't a dietary necessity for them (Altrichter and Almeida 2002).

Commercial hunting pressures have also increased due to the paving of previously unimproved roads, which allowed easier transportation of meat to urban centers. On top of that, more people coming from cities are practicing illegal recreational hunting in the parks and reserves. The effects of such, more recent types of hunting are still poorly studied and remain relatively uncontrolled while they further threaten the future of these mammals. Ironically, without the license fees that would accompany legal hunting there seem to be few resources that could help law enforcement necessary to protect wildlife populations from illegal harvesting.

At a time when hunting pressures increase, the park's administration (through enhanced efforts by MINAE and with donations from NGOs) has significantly stepped up armed patrols, increasing patrol size from about 6 to roughly 40 guards in total. Dedicated guards can find the herds and actually escort them through the border areas, shepherding them away from more threatening human-inhabited zones.

With few exceptions, however, this enhanced protection system has been triggered by pulses of grant funds rather than through continued support and does not appear to form part of a larger and more holistically developed management plan. With little or no environmental education and with few alternatives for local peoples to earn their living or pass their time, such efforts, although likely essential to short-term amelioration of the population's health, are unlikely to be sustainable over the long term. In one town, park guards helped to establish a weekend soccer league, which helped to improve guard-community relations as well as take up significant portions of the weekend when local peoples would otherwise hunt out of boredom. Additionally, with known hunters playing soccer, it was a sure way of keeping an eye on things in a manner that easily could become more widely and sustainably practiced. Further creative initiatives such as this with a strongly enhanced effort towards environmental education are required for the assured security of the white-lipped peccary in Corcovado Park and the Osa Peninsula. The future of the peccary's main predator in the area, the jaguar, depends on the balance.

2. Scarlet Macaw—A Flagship Wildlife Species for the Osa Peninsula

Because of their adaptability to humans and especially altered habitats, scarlet macaws (*Ara macao, lapa roja*) represent perhaps one of the most observed and charismatic wildlife species in ACOSA (Fig. 12.22). They delight locals and tourists with their vocalizations, feeding, and aerial acrobatics. Many tourists rate observing scarlet macaws as their top nature experience on the Osa Peninsula. Scarlet macaws are often in the town of Puerto Jiménez, feeding or resting in trees around the city park and elsewhere.

This bird was abundant in the Southern Pacific region of Costa Rica until the 1940s when the banana company arrived (for a history on the banana plantations, see earlier in this chapter). This brought increased habitat destruction and overhunting of wildlife. Current estimates of population size on the Osa Peninsula range between 800–1,200 scarlet macaws, and locals believe scarlet macaw populations are increasing. Local inhabitants attribute the increase of macaws in ACOSA to a combination of factors, including (a) excellent habitat conditions in the protected areas and protection of the macaws by MINAE; (b) local citizens and businesses perceive that more money can be obtained from ecotourism than from macaw chick poaching; (c) environmental education programs, such as "worth more free than in captivity" and scarlet macaw coloring books in local schools; and (d) increasing availability of food and nests sites, particularly with exotic species such as teak, gmelina, and especially beach almond providing important food sources (Dear et al. 2010).

ACOSA scarlet macaws feed on 59 native and exotic plant species. The exotic species consumed include beach almond (*Terminalia catappa*), reported by 92% of 105 interviewed persons, teak (*Tectona grandis*), and melina (*Gmelina arborea*) (Dear et al. 2010). In Rancho Quemado's melina plantations, an estimated 85 scarlet macaws were observed feeding, but after plantations were harvested, scarlet macaws are rarely seen in the area. Planting and harvesting of teak and melina plantations should be carefully planned to provide food for scarlet macaws when food shortages exist for other species.

For nesting, interviews showed that scarlet macaws used 57 nests in 14 identified tree species. The most common nesting trees were *Caryocar costaricense* (n = 12, 24%), *Schizolobium parahyba* (n = 9, 18%), *Ceiba pentandra* (n = 7, 14%) and *Ficus sp.* (n = 5, 10%).

Perhaps the biggest threat to the Osa scarlet macaw population is poaching chicks from nests. Eleven of 57 nests were recently poached (Guittar et al. 2009). While in the past macaw nesting trees were cut down to obtain the chicks, today most trees are climbed for poaching them. It appears that about 50 chicks are poached yearly. One known poacher continues to extract about 25 chicks per year. Stopping such a devastating activity has become a priority of scarlet macaw conservation action in ACOSA. Fortunately, adult hunting, which would gravely impact the population, is practically non-existent. Reintroduction of captive birds is another potential threat to the population, although it is considered beneficial by misguided animal rescue enthusiasts. Three scarlet macaw reintroduction programs exist today. Furthermore, it has been recognized that disease brought to a native population by released macaws could in fact eliminate entire native populations. On top of that, released tame scarlet macaws could cause behavioral changes in native scarlet macaw populations (Vaughan 2006).

The scarlet macaw is endangered throughout its distribution in Latin America, but apparently has stable or increasing populations on the Osa Peninsula. This minimizes the interest in conservationists and local communities to study and conserve them, therefore most work is focused on other, more endangered species. However, scientific research in fields like population dynamics, habitat use, nesting, genetics, and diet is certainly needed to develop a successful scarlet macaw conservation program. Initially, a long-term scarlet macaw population-monitoring program will be needed, similar to the one that has been conducted in the Central Pacific region during the past 20 years (Myers and Vaughan 2004, Vaughan et al. 2005). Such a program would ideally be combined with conservation genetics research, ecological studies, and community education programs necessary to conserve this flagship species at the Osa Peninsula in the long run.

3. Ecotourism, Non-governmental Organizations, and Bureaucracy

Currently, Pacific Lowland Forests are protected and managed by two parallel government agencies or "bureaucracies" called "conservation areas": ACOPAC (Área de Conservación Pacífico Central) and ACOSA (Área de Conservación Osa). In this chapter we focus on the Osa region, while recognizing that many species range throughout both conservation areas. However, by far the bulk of large-scale and intact lowland forest sits inside ACOSA's Corcovado National Park and thus this park and the Osa Peninsula will remain a key source area for many threatened species and biotic communities present in the region (Table 12.2). What happens next in Corcovado National Park will be critical to the future of the ecosystems thriving today in Costa Rica's wet Pacific lowlands.

Since the creation of Costa Rica's National Park System (SPN), non-governmental organizations (NGOs) have provided important complementary support to governmental aid to National Parks and Wildlife Refuges. The first national environmental NGO that was established in Costa Rica is the Fundación de Parques Nacionales (FPN), which is still active today. During the first and booming decade of the park system (1970s), remote, inaccessible, and poorly visited national parks like Corcovado were essentially subsidized by fees collected inside other parks such as Volcan Poás National Park, which is easily accessible by road and heavily visited by tourists. This centralized park system was able to distribute resources to parks in need, and likewise could coordinate assistance from an NGO like the FPN.

As described earlier in this chapter, in the late 1980s, the SPN was subsumed under a broader, so-called National System of Conservation Areas (SINAC). This was done with the aim of both decentralizing national and regional conservation efforts and broadening the mission of conservation aiming at managing all natural resources in a particular region such as the Osa. In this way it was possible to create sufficient local income to maintain and protect national parks and other types of protected areas. Yet, under the following SINAC regulations, fees for entering into the parks and lodging at park stations like Sirena in Corcovado had to be collected and managed by ACOSA itself, but then

Table 12.2. Protected Areas Located in the Área de Conservación Osa (ACOSA), Costa Rica

	Size (hectares)
National Parks	
Corcovado	42,469
Piedras Blancas	14,025
Marino Ballenas	116
Forest Reserve	
Golfo Dulce	61,702
Biological Reserve	
Isla del Caño	84
Wildlife Refuges	
National	
Golfito	2,810
Pejeperro	350
Private	
Aguabuena	182
RHR Bancas	59
Hacienda Copano	260
Forestal Golfito	87
D.P. Hayes	211
Mixed	
Preciosa Platanares	226
Rancho La Merced	346
Punta Río Claro	247
Protected Wetlands	
Nacional Térraba-Sierpe	22,208
Lacustrino Pejeperrito	43
Total Area	145,425

NOTE. Outside ACOSA but still within the southern Pacific moist lowland forest bioregion (i.e., in the Central Pacific portion of it) occur two other but small protected areas that together cover less than 8,000 ha: Parque Nacional Manuel Antonio and Parque Nacional Carara (see Jiménez and Carrillo, chapter 11 of this volume).
Source: SINAC 1998, 2000.

were to be returned to the same central, protected areas bureaucracy's headquarters in San José a system that never disappeared. Rather, now it has become more a sort of absorber of funds, than distributor of resources that help the parks, unfortunately.

An additional fact that contributes to the creation of a serious conflict for large, remote parks like Corcovado is the following. National policy puts a ceiling on fees that can be charged to tourists who want to enter a national park and stay there. Thus, ACOSA administrators are on the one hand responsible for raising funds to manage, maintain, and protect the parks, reserves, and wildlife refuges in the larger Osa region, yet they cannot legally set and retain tourism fees at a level necessary to do the job. In fact, Parque Nacional Corcovado, which represents one of the world's most desirable nature experiences, is subsidizing ecotourism companies and tourists themselves are not the wiser. And, thus, because of an unforeseen consequence of bloated bureaucracy and inflexible policy, tourism at stations like Sirena is considered more as a burden and a net cost. This is because facilities must be maintained and park staff must be paid to accommodate visitors who are—involuntarily—not paying their fair share for all that is offered in Corcovado. For their part, local park staff—overburdened and underpaid—do not receive any additional compensation for dealing with hordes of tourists than when tourists do not visit the park.

Clearly when ACOSA was created the park system was decentralized only partially and not in a way that has favored park management and protection. Thus, as in the case of the white-lipped peccary, problems of poaching may arise as certain local people around the park observe a lack of adequate protection due to diminished resources to pay park guards at adequate levels.

In recent years conservation areas like ACOSA have increasingly been depending on NGOs for assistance. NGOs like the aforementioned FPN have helped by providing grants, as mentioned in an earlier section of this chapter. A program to save the squirrel monkey in Parque Nacional Manuel Antonio, for instance, was developed by ASOMOTI (Asociación Mono Tití). It did so by collecting one dollar per booked hotel room. Another example concerns AMBICOR, which is a small NGO that assisted Sirena Station to maintain its facilities. It provided bottom-up assistance to local people helping them to set up small businesses focused on tourists traveling to Corcovado Park. Another way in which NGOs have supported conservation in the Osa and other parts of the country is by participating in regional councils in the Conservation Areas themselves. Their support has varied according to the conservation area. Some NGOs like The Nature Conservancy (TNC), for instance, have implemented conservation action by purchasing and setting aside small forest blocks on the Osa Peninsula as well as in Parque Nacional Piedras Blancas, which they subsequently donated to the Costa Rican state.

Unfortunately, continued financial support from NGOs and foundations that have dealt with emergencies like the poaching crisis that occurred in 2004 (Fig. 12.24) have become less reliable since the global financial crisis happened in 2008. The future of magnificent places like Corcovado National Park will depend on developing mechanisms that allow ecotourism to sustain biodiversity and natural resources to which tourists travel to enjoy. Once they are fully aware, the ecologically minded tourists will not shy away from paying more to help sustain and protect unique places like Corcovado.

The answer to this dilemma lies perhaps in innovative roles that NGOs might help to fill and that relate to the

QUEDAN 30 INDIVIDUOS PRODUCTO DE LA CAZA Y FALTA DE COMIDA

Extinción acecha al jaguar en Corcovado

Minae no descarta declaratoria de emergencia en zona

VANESSA LOAIZA N.
vloaiza@nacion.com

CORCOVADO. Es ágil, de hábitos nocturnos, con un atractivo pelaje pardo y manchado. Se dice que ninguna presa se escapa a su andar sigiloso y a sus ataques certeros... hasta aquí, el dueño del bosque parece invencible.

Sin embargo, el carnívoro más grande de América: el jaguar, está a punto de desaparecer del Parque Nacional Corcovado, en la península de Osa, Puntarenas.

La cacería indiscriminada y la falta de alimento para este felino amenazan con extinguirlo antes de que finalice el año. Esta es la conclusión más grave del último conteo de jaguares realizado por el biólogo e investigador Eduardo Carrillo.

Este científico de la Universidad Nacional (UNA), aseguró que en la actualidad apenas 30 individuos habitan el parque de 42.400 hectáreas, hace 10 años la cifra llegaba a 150.

Las autoridades del Ministerio del Ambiente y Energía (Minae) conocen la situación, pero se declaran atadas de manos.

Los recientes recortes presupuestarios y las trabas para contratar más personal les impiden

Área de Conservación disponen este año de apenas un monto de ¢12 millones para los gastos operativos, como combustibles y recorridos. > Vea nota aparte.

Jaguares rodeados

Expuesto a estas deficiencias, el jaguar intenta sobrevivir a un enemigo que lo ataca por partida doble: al cazador y a su principal fuente de alimento, el chancho de monte.

La mayoría de los cazadores que ingresan a Corcovado van en busca de chanchos de monte, otra especie en peligro de extinción cuya carne se comercializa a un elevado precio en las cantinas de Rancho Quemado, Rincón y Cerro Brujo de Osa, o en Vanegas de Golfito.

Se cree que hoy solo quedan unos 300 chanchos, apenas un 15 por ciento de los 2.000 individuos que poblaban el parque en 1994.

Esta carencia obliga al felino a buscar ganado y perros en las poblaciones externas a Corcovado, donde los vecinos terminan disparándole por miedo.

Precisamente, de agosto a diciembre del año pasado ocho jaguares murieron en comunidades al norte del parque -Drake y Rancho Quemado-, donde los vecinos guardan la piel del animal como un trofeo, explicó Carrillo.

En una entrevista con *La Nación*, sus críticas se enfilaron a la "inacción del Minae" y al gobierno en general, que sigue promo

POCOS. Tras dos días de búsqueda *La Nación* divisó el jueves anterior a una manada de apenas 20 chanchos muy esquivos. Los cazadores aprovechan que estos animales andan en grupo para matar a varios.

México y hasta Panamá.

El conteo de los jaguares ha sido una constante de este investigador a partir de 1994. Desde entonces emplea cámaras con sensores de movimiento y los

del 4 al 7 de marzo pasado.

Junto a 10 funcionarios más, el jerarca se adentró en la montaña y detectó cuatro campamentos abandonados por cazadores ilegales y dos cabezas de chancho.

mo de Investigación Judicial tome cartas en el asunto, agregó Miguel Madrigal, gerente de áreas protegidas de Acosa.

Asimismo, el Ministro no descarta la posibilidad de declarar

IDENTIFICACIÓN
Los jaguares se identifican por las manchas de su pelaje, que no se repiten en ningún individuo de la misma especie.

MONITOREO
Existen varias formas de saber cuántos jaguares hay en Corcovado. Una de ellas es por cámaras sensoras; estas toman fotos de los animales y por la forma de las manchas se sabe cuántos jaguares pasan por diferentes áreas del parque.

Cámara sensora
Están amarradas a los árboles a 50 cm del suelo.

Hay 40 cámaras instaladas.

La otra forma de contabilizarlos es con radiocollares que emiten señales que permiten ubicar al animal geográficamente y cuáles son sus desplazamientos. Los últimos jaguares con este dispositivo fueron cazados a finales del 2002.

Zonas donde hay más desaparición de jaguares
Parque Nacional Corcovado
Golfo Dulce
Golfito
COSTA RICA

PERFIL DEL JAGUAR
Alimento: chanchos de monte, tortugas lora, perezosos y monos.
Peso: entre 50 y 100 kilos.
Hábitat: vive en las orillas de los ríos, en bosque frondosos.
Piel: Su pelo es de color pardo amarillento con machas café y negro que funcionan como las huellas digitales de los humanos, no son iguales en ningún individuo.

FREDDY SOLÍS S. / LA NACIÓN

Inacción de gobierno ahoga al parque

Hace un año espera

Cronología de atrasos

nes financieras.

"Ojalá venga más gente de campo a cuidar Corcovado"

Fig. 12.24 In 2004, poaching in Corcovado Park from neighboring towns such as La Palma and Rancho Quemado reached a crisis point. Top left: Remains of butchered white-lipped peccary found by a park patrol near La Llorona. Top right: Typical patrol of Corcovado's park guards. The numbers of patrolling guards and their resources are inadequate for the large area in need of protection. Below: Publicity in La Nación March 31, 2004, helped generate a temporary influx of grant funds to increase park patrols, but these will not continue.
Photo credits—Top left: Yamil Sáenz; Top right: R. Wilson.

operationalization of private concessions at locations like the Sirena Biological Station. Such concessions would then allow charging fees at true market value and returning more "tourist dollars" to the park itself, to cover for direct and indirect costs of conservation of this unique environment.

Conclusions

The large undisturbed lowland rain forest protected inside the Corcovado National Park and the position of the Sirena Biological Station along the coast of the Osa Peninsula account for an intact fauna of large vertebrates and unusual opportunities to study these threatened mammals at an appropriate spatial scale. The relative geographical isolation of Osa's lowland forests have provided many opportunities for plants and animals to undergo genetic differentiation from their counterparts found in the Atlantic lowlands of Costa Rica. This fact and the more seasonal rainfall occurring in this part of the country provide great opportunities for comparative biology studies.

Several authors of this chapter experienced the Osa in the 1960s and 1970s prior to road development that accelerated deforestation, and were already studying the biological richness of Corcovado Park soon after its creation. The descriptions of the Osa in the early days of Corcovado Park by Costa Rican naturalist and biologist Alvaro Wille (1983) captures many things we saw, felt, and now recall with nostalgia. While facilities and logistics at Sirena Biological Station have vastly improved since the early decades, conditions there are still primitive when compared to campus-like dorms and laboratory facilities of the Organization for Tropical Studies' (OTS) La Selva Biological Station in the Atlantic zone, or the Smithsonian Tropical Research Institute's (STRI) Barro Colorado Island (BCI) in Panama. Yet, these stations do lack access to the interface of ocean and forest and have no direct access to such a large tract of lowland rainforest where one might hope to relate pattern and process from populations to ecosystems. Thus, saving Corcovado National Park, a key example of a tropical ecosystem of importance to both science and tourism, represents an essential conservation goal in Costa Rica. The creation of Corcovado National Park truly meant saving the lowland rainforest of Costa Rica's Southern Pacific zone. In many ways, its establishment served as a cornerstone for building the country's national park system. Therefore, we decided to describe the history of its creation in full detail, and focused on the continuing challenges that affect its current and future ecological integrity. While some of these details may be as endemic to the Osa as are many of its organisms, there are lessons to learn that would apply to saving similar ecosystems at other places of our planet. The proper engagement of local communities, the ecotourism sector, and academia seems paramount to safeguard Osa's unique ecological resources for the benefit of coming generations.

References

Allen, P.H. 1956. *The Rain Forests of the Golfo Dulce*. Stanford: Stanford University Press. 417 pp.

Altrichter, M. 1997. Estrategia de alimentación y comportamiento del chancho cariblanco *Tayassu pecari* en un bosque húmedo tropical de Costa Rica. M.Sc. thesis, Universidad Nacional. Heredia, Costa Rica.

Altrichter, M. 2000. Importancia de mamíferos silvestres en la dieta de pobladores de la Península de Osa, Costa Rica. *Revista Mexicana de Mastozoología* 4: 99–107.

Altrichter, M., and R. Almeida. 2002. Exploitation of white-lipped peccaries *Tayassu pecari* (Artiodactyla: Tayassuidae) on the Osa Peninsula, Costa Rica. *Oryx* 36: 126–32.

Altrichter, M., and F. Carbonell. 2008. Plan de Acción para la Conservación del Chancho de Monte (*Tayassu pecari*) en la Reserva de la Biosfera la Amistad. Costa Rica. The Nature Conservancy (TNC), Serie Técnica 10. San José, Costa Rica. 80 pp.

Altrichter, M., C. Drews, E. Carrillo, and J. Sáenz. 2001a. Sex ratio and breeding of white-lipped peccary (*Tayassu pecari*) in a Costa Rican rain forest. *Revista de Biología Tropical* 49: 383–89.

Altrichter, M., E. Carrillo, J. Sáenz, and T. Fuller. 2001b. White-lipped peccary (*Tayassu pecari*, Artiodactyla: Tayassuidae) diet and fruit availability in a Costa Rican rain forest. *Revista de Biología Tropical* 49: 1183–92.

Altrichter, M., C. Drews, J. Sáenz, and E. Carrillo. 2002. Presupuesto de tiempo del chancho cariblanco *Tayassu pecari* en un bosque húmedo de Costa Rica. *Biotropica* 34: 136–43.

Altrichter, M., and I. Jiménez. 1999. Caza y consumo de carne de monte en la comunidad de San Juan del Norte, Reserva Biológica Indio Maíz, Nicaragua. *Boletín oficial, Sociedad Mesoamericana para la Biología y la Conservación (SMBC)* 4: 117–20.

Altrichter, M., J. Sáenz, E. Carrillo, and T. Fuller. 1999. Chanchos cariblancos (*Tayassu pecari*) como depredadores y dispersores de semillas en el Parque Nacional Corcovado, Costa Rica. *Brenesia* 52: 53–59.

Altrichter, M., J. Sáenz, E. Carrillo, and T. Fuller. 2000. Dieta estacional de *Tayassu pecari* (Artiodactyla: Tayassuidae) en el Parque Nacional Corcovado, Costa Rica. *Revista de Biología Tropical* 48: 689–702.

Anchukaitis, K., and S. Horn. 2005. A 2000-year reconstruction of forest disturbance from southern Pacific Costa Rica. *Palaeogeography, Palaeoclimatology, Palaeoecology* 221: 35–54.

Anderson, S. 1989. *Goldwalker: Tales of the Osa Peninsula from the Life of Patrick Jay O'Connell*. St. Peters, PA: Breaker Press. 375 pp.

Anderson, S. 2003. Dependent environmentalism: a case study of Oreros and the Corcovado National Park in Costa Rica. *California Geographer* 43: 22–48.

Ankersen, T., K. Regan, and S. Mack. 2006. Towards a bioregional approach to tropical forest conservation: Costa Rica's Greater Osa Bioregion. *Futures* 38: 406–31.

Barrantes, C. 2005. Historia de la región de Golfo Dulce. In J. Lobo and F. Bolaños, eds., *Historia Natural de Golfito, Costa Rica*. Santo Domingo de Heredia, Costa Rica: Editorial INBio.

Bergeson, D.J. 1998. Patterns of suspensory feeding in *Alouatta palliata, Ateles geoffroyi,* and *Cebus capucinus*. In E. Strasser, J. Fleagle, A. Rosenberger, and H. McHenry, eds., *Primate Locomotion: Recent Advances*, 45–60. New York: Plenum Press.

Bern, C.R., A.R. Townsend, and G.L. Farmer. 2005. Unexpected dominance of parent-material strontium in a tropical forest on highly weathered soils. *Ecology* 86: 626–32.

Bezanson, M.F. 2006. Leap, bridge, or ride?: ontogenetic influences on positional behavior in *Cebus* and *Alouatta*. In A. Estrada, P.A.P. Garber, M.S.M. Pavelka, and L. Luecke, eds., *New Perspectives in the Study of Mesoamerican Primates: Distribution, Ecology, Behavior, and Conservation*, 333–48. New York: Springer.

Bodmer, R. 1990. Responses of ungulates to seasonal inundations in the Amazon floodplain. *Journal of Tropical Ecology* 6: 191–201.

Bodmer, R. 1995. Managing Amazonian wildlife: biological correlates of game choice by detribalized hunters. *Ecological Applications* 5: 872–77.

Boinski, S. 1986. The ecology of squirrel monkeys in Costa Rica. PhD thesis, University of Texas, Austin.

Boinski, S. 1987a. Birth synchrony in squirrel monkeys (*Saimiri oerstedii*): a strategy to reduce neonatal predation. *Behavioral Ecology and Sociobiology* 21: 393–400.

Boinski, S. 1987b. Habitat use by squirrel monkeys (*Saimiri oerstedii*) in Costa Rica. *Folia Primatologica* 49: 151–67.

Boinski, S. 1987c. Mating patterns in squirrel monkeys (*Saimiri oerstedi*): implications for seasonal sexual dimorphism. *Behavioral Ecology and Sociobiology* 21: 13–21.

Boinski, S. 1988. Sex differences in the foraging behavior of squirrel monkeys in a seasonal habitat. *Behavioral Ecology and Sociobiology* 23: 177–86.

Boinski, S., L. Kauffman, E. Ehmke, S. Schet, and A. Vreedzaam. 2005. Dispersal patterns among three species of squirrel monkeys (*Saimiri oerstedii, S. boliviensis, and S. sciureus*). I. Divergent costs and benefits. *Behaviour* 142: 525–632.

Boinski, S., and L. Sirot. 1997. Uncertain conservation status of squirrel monkeys in Costa Rica, *Saimiri oerstedi oerstedi* and *Saimiri oerstedi citrinellus*. *Folia Primatologica* 68: 181–93.

Boinski, S., K. Sughrue, L. Selvaggi, R. Quatrone, M. Henry, and S. Cropp. 2002. An expanded test of the ecological model of primate social evolution: competitive regimes and female bonding in three species of squirrel monkeys (*Saimiri oerstedii*, *S. boliviensis*, and *S. sciureus*). *Behaviour* 139: 227–61.

Boinski, S., and R.M. Timm. 1985. Predation by squirrel monkeys and double-toothed kites on tent-making bats. *American Journal of Primatology* 9: 121–27.

Bourgois, P. 1989. *Ethnicity at Work: Divided Labor on a Central American Banana Plantation*. Baltimore: Johns Hopkins University Press.

Boza, M., and A. Bonilla. 1978. *Los Parques Nacionales de Costa Rica*. Madrid: INCAFO.

Bush, M., and P. Colinvaux. 1990. A long record of climatic and vegetation change in lowland Panama. *Journal of Vegetation Science* 1: 105–19.

Bussing, W. 1976. Geographic distribution of the San Juan ichthyofauna of Central America with remarks on its origin and ecology. In T. Thorson, ed., *Investigations of the Ichthyofauna of the Nicaraguan Lakes*, 157–75. Lincoln: University of Nebraska.

Bussing, W. 1987. *Peces de las Aguas Continentales de Costa Rica*. San José, Costa Rica: Editorial de la Universidad de Costa Rica.

Calvert, A.S., and P.P. Calvert. 1917. *A Year of Costa Rican Natural History*. New York: Macmillan.

Campbell, A.F., and R.W. Sussman. 1994. The value of radio tracking in the study of neotropical rain-forest monkeys. *American Journal of Primatology* 32: 291–301.

Campero, H. 1999. Variación y estructura genética dentro y entre grupos de chanchos de monte *Tayassu pecari* en el Parque Nacional Corcovado, Costa Rica. M.Sc. thesis, National University. Heredia, Costa Rica.

Carrillo, E. 2000. Ecology and conservation of white-lipped peccaries and jaguars in Corcovado National Park. PhD thesis, University of Massachusetts. Amherst, MA.

Carrillo, E., T. Fuller, and J. Sáenz. 2009. Jaguar (*Panthera onca*) hunting activity: effects of prey distribution and availability. Short communication. *Journal of Tropical Ecology* 25: 563–67

Carrillo, E., J. Sáenz, and T. Fuller. 2002. Movements and activities of white-lipped peccaries in Corcovado National Park, Costa Rica. *Biological Conservation* 108: 317–24.

Chao, A., R.L. Chazdon, R.K. Colwell, and T. Shen. 2005. A new statistical approach for assessing similarity of species composition with incidence and abundance data. *Ecology Letters* 8: 148–59.

Chapman, A., and K. Rosenberg. 1991. Diets of four sympatric Amazonian woodcreepers (Dendrocolaptidae). *Condor* 93: 904–15.

Chaverri, G., M. Gamba-Rios, and T.H. Kunz. 2007a. Range overlap and association patterns in the tent-making bat *Artibeus watsoni*. *Animal Behaviour* 73: 157–64.

Chaverri, G., and T.H. Kunz. 2006a. Reproductive biology and postnatal development in the tent-making bat *Artibeus watsoni* (Chiroptera: Phyllostomidae). *Journal of Zoology* (London) 270: 650–56.

Chaverri, G., and T.H. Kunz. 2006b. Roosting ecology of the tentroosting bat *Artibeus watsoni* (Chiroptera: Phyllostomidae) in southwestern Costa Rica. *Biotropica* 38: 77–84.

Chaverri, G., O.E. Quiros, M. Gamba-Rios, and T.H. Kunz. 2007b. Ecological correlates of roost fidelity in the tent-making bat *Artibeus watsoni*. *Ethology* 113: 598–605.

Chaverri, G., O.E. Quiros, and T.H. Kunz. 2007c. Ecological correlates of range size in the tent-making bat *Artibeus watsoni*. *Journal of Mammalogy* 88: 477–86.

Chaverri, G., C.J. Schneider, and T.H. Kunz. 2008. Mating system of the tent making bat *Artibeus watsoni* (Chiroptera: Phyllostomidae). *Journal of Mammalogy* 89: 1361–71.

Chazdon, R.L. 2003. Tropical forest recovery: legacies of human impact and natural disturbances. *Perspectives in Plant Ecology, Evolution and Systematics* 6: 51–71.

Chinchilla, F. 1997. La dieta del jaguar (*Panthera onca*), el puma (*Felis concolor*) y el manigordo (*Felis pardalis*) en el Parque Nacional Corcovado, Costa Rica. *Revista de Biología Tropical* 45: 1223–29.

Christen, C. 1994. Development and conservation on Costa Rica's Osa Peninsula, 1937–1977: a regional case study of historical land use policy and practice in a small Neotropical country. PhD diss., Johns Hopkins University.

Christen, C. 2006. Return of the Scientists. In *III Simposio Latinoamericano y Caribeño de Historia Ambiental*. Carmona, Spain.

CITES (Convention on the International Trade in Endangered Species of Wild Flora and Fauna). 2005. Appendix I, II, and III as adopted by the Conference of the Parties, valid from 23 June 2005. http://www.cites.org.

Clare, P. 2005. El desarrollo del banano y la palma aceitera en el Pacífico Costarricense desde la perspectiva de la ecología histórica. *Diálogos Revista Electrónica de Historia* 6: 308–46. San Pedro de Montes de Oca, Costa Rica: Universidad de Costa Rica.

Clement, R., and S. Horn. 2001. Pre-Columbian land-use history in Costa Rica: a 3000-year record of forest clearance, agriculture and fires from Laguna Zoncho. *Holocene* 11: 419–26.

Cleveland, C.C., S.C. Reed, and A.R. Townsend. 2006. Nutrient regulation of organic matter decomposition in a tropical rain forest. *Ecology* 87: 492–503.

Cleveland, C.C., and A.R. Townsend. 2006. Nutrient additions to a tropical rain forest drive substantial soil carbon dioxide losses to the atmosphere. *Proceedings of the National Academy of Sciences* (USA) 103: 10316–21.

Coen, E. 1983. Climate. In D.H. Janzen, ed., *Costa Rican Natural History*, 35–46. Chicago: University of Chicago Press.

Colwell, R.K. 2005. EstimateS: Statistical Estimation of Species Richness and Shared Species from Samples. Version 7.5. User's Guide and application. http://purl.oclc.org/estimates.

Conejo, A. 1975. *Henri Pittier*. San José, Costa Rica: Ministerio de Cultura, Juventud y Deportes.

Cowlishaw, G., and R.I.M. Dunbar. 2000. Primate Conservation Biology. Chicago: University of Chicago Press.

Cullen, L., R. Bodmer, and C. Padua. 2000. Effects of hunting in habitat fragments of the Atlantic forests, Brazil. *Biological Conservation* 95: 49–56.

Dear, F., C. Vaughan, and A. Morales. 2010. Current status and conservation of the Scarlet Macaw (*Ara macao*) in the Osa Conservation Area (ACOSA), Costa Rica. *Cuadernos de Investigacion UNED* 2(2): 7–21.

DeMattia, E.A., L.M. Curran, and B.J. Rathcke. 2004. Effects of small rodents and large mammals on neotropical seeds. *Ecology* 85: 2161–70.

DeMattia, E.A., B.J. Rathcke, L.M. Curran, R. Aguilar, and O. Vargas. 2006. Effects of small rodents and large mammal exclusion on seedling recruitment in Costa Rica. *Biotropica* 38: 196–202.

Departamento de Urbanismo. 1977. Proyecto Estación Biológica Parque Nacional Corcovado. Instituto de Vivienda y Urbanismo (INVU), Archivo Nacional de Costa Rica: SPN 476.

Devall, M., and R. Kiester. 1987. Notes on *Raphia* at Corcovado. *Brenesia* 28: 89–96.

DeVries, P. 1987. *The Butterflies of Costa Rica and Their Natural History*. Vol. I. Papilionidae, Pieridae, Nymphalidae. Princeton, NJ: Princeton University Press.

Di Fiore, A., and S.A. Suarez. 2007. Route-based travel and shared routes in sympatric spider and woolly monkeys: cognitive and evolutionary implications. *Animal Cognition*. DOI: 10.1007/s10071-006-0067-7.

Eakin, M. 1999. The origins of modern science in Costa Rica: the Instituto Fisico-Geografico Nacional, 1887–1904. *Latin American Research Review* 34: 123–50.

Estrada, A., and R. Coates-Estrada. 1996. Tropical rain forest fragmentation and wild populations of primates at Los Tuxtlas, Mexico. *International Journal of Primatology* 17: 759–83.

Estrada, A., J. Sáenz, C. Harvey, E. Naranjo, D. Muñoz, and M. Rosales-Meda. 2006. Primates in agroecosystems: conservation value of some agricultural practices in Mesoamerican landscapes. In A. Estrada, P.A.P. Garber, M.S.M. Pavelka, and L. Luecke, eds., *New Perspectives in the Study of Mesoamerican Primates: Distribution, Ecology, Behavior, and Conservation*, 437–70. New York: Springer.

Ewel, J. 1991. Interview by C. Christen, January 30, 1991. San Jose, Costa Rica, Notes.

Fernandes, D.S., F.L. Franco, and R. Fernandes. 2004. Systematic revision of the genus *Lachesis* Daudin, 1803 (Serpentes, Viperidae). *Herpetologica* 60: 245–60.

Foerster, C.R. 1998. Habitat selection and movement patterns of *Tapirus bairdii* in Costa Rican tropical forest. Master's thesis, Universidad Nacional. Heredia, Costa Rica.

Foerster, C.R., and C.Vaughan. 2002. Home range, habitat use, and activity of Baird's tapir in Costa Rica. *Biotropica* 34(3): 423–37.

Fonseca, E. 1986. Costa Rica colonial: la tierra y el hombre. San José, Costa Rica: Editorial Universitaria Centroamericana.

Fragoso, J. 1998. Home range and movement patterns of white-lipped peccary (*Tayassu pecari*) herds in the Northern Brazilian Amazon I. *Biotropica* 30:458–69.

Fragoso, J. 2004. A long term study of white-lipped peccary (*Tayassu pecari*) population fluctuations in northern Amazonia: Anthropogenic vs. "natural" causes." In K. Silvius, R. Bodmer, and J. Fragoso, eds., *People in Nature*, 286–96. New York: Columbia University Press.

Franceschi Barraza, L. 1997. Concurso de autobiografías de mujeres dirigentes campesinas e indígenas de América Latina, 10 de noviembre al 19 de diciembre, 1997. International Fund for Agricultural Development (IFAD), FIDAMERICA.

Franceschi Barraza, L. 2007. Interview with C. Christen. Puerto Jiménez, Costa Rica. Tape transcript.

Fuller, T., E. Carrillo, and J. Sáenz. 2002. Survival of protected white-lipped peccaries in Costa Rica. *Canadian Journal of Zoology* 80: 586–89.

García, M. 1988. Apuntes geohistóricos de la colonización agrícola en la Península de Osa (Costa Rica). *Geoistmo* 2: 27–40.

Gilbert, L.E. 1980. Food web organization and the conservation of neotropical diversity. In M.E. Soulé and B.A. Wilcox, eds., *Conservation Biology*. Sunderland, MA: Sinauer Associates, Inc.

Gilbert, L.E. 1990. Research in Corcovado Park and the Sirena Biological Station. Austin, TX: University of Texas at Austin. Mimeographed report.

Gilbert, L.E. 1991. Biodiversity of a Central American *Heliconius* community: pattern, process and problems. In P.W. Price, T.M. Lewinsohn, G.W. Fernandes, and W.W. Benson, eds., *Plant-animal Interactions: Evolutionary Ecology in Tropical and Temperate Regions*. New York: John Wiley and Sons.

Gilbert, L.E., and R.M. Plowes. 2003. Shifts in a butterfly community in regenerating rainforest; 25 years of change in the *Heliconius* group at Sirena. Report to ACOSA, Costa Rica.

Gómez, L.D. 1986. Vegetación de Costa Rica. Vol. 1. In L.D. Gómez, ed., *Vegetación y Clima de Costa Rica*. San José, Costa Rica: EUNED. Con 10 mapas (escala 1:200.000).

Gómez, L.D. 1991. Interview by C. Christen, July 22, 1991. San Vito de Coto Brus, Costa Rica. Tape transcript.

Gómez, L.D., and J.M. Savage. 1983. Searchers on that rich coast: Costa Rican field biology, 1400–1980. In D.H. Janzen, ed., *Costa Rican Natural History*, 1–11. Chicago: University of Chicago Press.

González-Víquez, C. 1921. Nombres geográficos de Costa Rica: Golfo Dulce. *Revista de Costa Rica* 2: 225–28.

Gottdenker, N., and R. Bodmer. 1998. Reproduction and productivity of white-lipped and collared peccaries in the Peruvian Amazon. *Journal of Zoology* (London) 245: 423–30.

Gradwohl, J., and R. Greenberg. 1980. The formation of ant wren flocks on Barro Colorado Island, Panama. *Auk* 97: 385–95.

Groves, C. 2001. *Primate Taxonomy*. Washington, DC: Smithsonian Institution Press.

Guariguata, M.R., and R. Ostertag. 2001. Neotropical secondary forest succession: changes in structural and functional characteristics. *Forest Ecology and Management* 148: 185–206.

Guittar, J., F. Dear, and C. Vaughan. 2009. Scarlet Macaw (*Ara macao*, Psittaciformes: Psittacidae) nest characteristics in the Osa Peninsula Conservation Area (ACOSA), Costa Rica. *Revista de Biologia Tropical* 57(1–2): 387–93.

Hall, C. 1985. *A Geographical Interpretation in Historical Perspective*. Boulder: Westview Press.

Hammel, B.E., and N.A. Zamora. 1993. *Ruptiliocarpon* (Lepidobotryaceae): a new arborescent genus and tropical American link to Africa, with a reconsideration of the family. *Novon* 3(4): 408–17.

Hartshorn, G.S. 1983. Plants: introduction. In D.H. Janzen, ed., *Costa Rican Natural History*, 118–57. Chicago: University of Chicago Press.

Herrera, W. 1986. Clima de Costa Rica. Vol. 2. In L.D. Gómez, ed., *Vegetación y Clima de Costa Rica*. San José, Costa Rica: EUNED. Con 10 mapas (escala 1:200.000).

Herrera, W., and L.D. Gómez. 1993. Mapa de Unidades Bióticas de Costa Rica. San José, Costa Rica: Instituto Geográfico de Costa Rica. Scale 1: 685,000.

Herrera-MacBryde, O., T.R. Maldonado, V. Jiménez, and K. Thomsen. 1997. Osa Peninsula and Corcovado National Park, Costa Rica (Central America: CPD Site MA18). In S.P. Davis, V.H. Heywood, O. Herrera-MacBryde, J. Villalobos, and A.C. Hamilton, eds., *Centres of Plant Diversity: A Guide and Strategy for their Conservation*, 215–20. Cambridge, UK: WWF—IUCN, IUCN Publications Unit.

Herwitz, S.R. 1981. *Regeneration of Selected Tropical Tree Species in Corcovado National Park, Costa Rica*. London: University of California Press.

Hogue, C.L. 1993. *Latin American Insects and Entomology*. Berkeley: University of California Press. 536 pp.

Holdridge, L.R., W.C. Grenke, W.H. Hatheway, T. Liang, and J.A. Tosi. 1971. *Forest Environments in Tropical Life Zones: A Pilot Study*. Oxford: Pergamon Press. 735 pp.

Holdridge, L.R., and J.A. Tosi. 1991. Interview by C. Christen, February 13, 1991. San José, Costa Rica. Tape transcript.

Horn, S.P. 1992. Microfossils and forest history in Costa Rica. In H.K. Steen and R.P. Tucker, eds., *Changing Tropical Forests: Historical Perspectives on Today's Challenges in Central and South America*, 16–30. Durham, NC: Forest History Society.

Huber, W. 1996. Floristische und biogeographische Untersuchungen in einem Tieflandregenwald in der pazifischen Region von Costa Rica. Diplomarbeit, Universität Wien. Vienna, Austria.

IUCN. 2006. IUCN Red List of Threatened Species. Downloaded June 14, 2007. http://www.iucnredlist.org.

Janzen, D.H. 1982. Investigaciones en biología en Parque Nacional Santa Rosa y Parque Nacional Corcovado, Costa Rica. (Draft for Memoria del Primer Simposio de Parques Nacionales y Reservas Equivalentes, UNED, San José, Costa Rica.) Turrialba, Costa Rica: INFORAT Library, CATIE.

Janzen, D.H. 1985. Corcovado National Park: A perturbed rainforest ecosystem. Mimeographed report to the World Wildlife Fund [US].

Janzen, D.H. 1990. An abandoned field is not a tree fall gap. *Vida Silvestre Neotropical* 2: 64–67.

Jiménez, J.A. 1981. The mangroves of Costa Rica: the physiognomic characterization. M.Sc. thesis, University of Miami. Miami, FL.

Jiménez, J.A., and R. Soto. 1985. Patrones regionales en la estructura y composición florística de los manglares de la costa Pacífica de Costa Rica. *Revista de Biología Tropical* 33: 25–37.

Jiménez, Q. 1999. *Árboles Maderables en Peligro de Extinción en Costa Rica*. Segunda edición revisada y ampliada. Santo Domingo de Heredia, Costa Rica: INBio. 187 pp.

Jullien, M., and J. Thiollay. 1998. Multi-species territoriality and dynamic of neotropical forest understory bird flocks. *Journal of Animal Ecology* 67: 227–52.

Kappelle, M., M. Castro, H. Acevedo, P. Cordero, L. González, E. Méndez, and H. Monge. 2002. A rapid method in ecosystem mapping and monitoring as a tool for managing Costa Rican ecosystem health. In D.J. Rapport, W.L. Lasley, D.E. Rolston, N.O. Nielsen, C.O. Qualset, and A.B. Damania, eds., *Managing for Healthy Ecosystems*. Boca Raton, FL: Lewis Publishers.

Kappelle, M., M. Castro, H. Acevedo, L. González, and H. Monge. 2003. *Ecosystems of the Osa Conservation Area, Costa Rica*. Santo Domingo de Heredia, Costa Rica: Instituto Nacional de Biodiversidad (INBio). 496 pp.

Kiltie, R. 1980. More on Amazon cultural ecology. *Current Anthropology* 21: 541–46.

Kiltie, R. 1981. Stomach contents of rain forest peccaries (*Tayassu tajacu* and *T. pecari*). *Biotropica* 13: 234–35.

Kiltie, R., and J. Terborgh. 1983. Observations on the behavior of rain forest peccaries (*Tayassu pecari*) in Perú: why do White-lipped peccaries form herds? *Zeitschrift für Tierpsychologie* 62: 214–17.

Kronforst, M.R., L.G. Young, L.M. Blume, and L.E. Gilbert. 2006a. Multilocus analyses of admixture and introgression among hybridizing *Heliconius* butterflies. *Evolution* 60: 1254–68.

Kronforst, M.R., L.G. Young, D.D. Kapan, C. McNeely, R.J. O'Neill, and L.E. Gilbert. 2006b. Linkage of butterfly mate preference and wing color preference cue at the genomic location of wingless. *Proceedings of the National Academy of Sciences of the United States of America* 103: 6575–80.

Kronforst, M.R., C. Salazar, M. Linares, and L.E. Gilbert. 2007a. No genomic mosaicism in a putative hybrid butterfly species. *Proceedings of the Royal Society B: Biological Sciences* 274: 1255–64.

Kronforst, M.R., L.G. Young, and L.E. Gilbert. 2007b. Reinforcement of mate preference among hybridizing *Heliconius* butterflies. *Journal of Evolutionary Biology* 20: 278–85.

Kurka, K. 1994. The BOSCOSA project: an experiment in rain forest conservation and rural development on the Osa Peninsula of Costa Rica. Master's thesis, Evergreen State College.

La Cuenca del Corcovado. 1973. *La Cuenca del Corcovado: Península de Osa, Costa Rica*. Gainesville, FL: privately printed [February-March, 1973], Mildred Mathias Osa File, Department of Biology, University of California, Los Angeles, CA, n.d., n.p.

Lamar, W.W., and S. Mahmood. 2003. A new species of hognose pitviper, genus *Porthidium*, from the southwestern Pacific of Costa Rica (Serpentes: Viperidae). *Revista de Biología Tropical* 51: 797–804.

Lang, S.B. 2000. Effects of logging roads on erosion in a wet tropical forest in the Río Chiquito watershed, Península de Osa, Costa Rica. Thesis M. Sc. Colorado State University. Fort Collins, CO. 87 pp.

Lew, L.R. 1983. The geology of the Osa Peninsula, Costa Rica. M.Sc. thesis, Pennsylvania State University. Philadelphia, PA. 128 pp.

Lewis, B.E. 1982–1983. Reseña histórica de la población y los recursos naturales de la Península de Osa, Pacifico Sur, 1848–1981. *Revista Geográfica de América Central* 17–18: 123–30.

Lindquist, E.S., and C.R. Carroll. 2004. Differential seed and seedling predation by crabs: impacts on tropical coastal forest composition. *Oecologia* 141: 661–71.

Lindquist, E.S., K.W. Krauss, P.T. Green, D.J. O'Dowd, P.M. Sherman, and T.J. Smith. 2009. Land crabs as key drivers in tropical coastal forest recruitment. *Biological Reviews* 84: 203–23.

Lindshield, S.M. 2006. The density and distribution of *Ateles geoffroyi* in a mosaic landscape at El Zota Biological Field Station, Costa Rica. Master's thesis, Iowa State University.

Link, A., and A. Di Fiore. 2006. Seed dispersal by spider monkeys and its importance in the maintenance of neotropical rain-forest diversity. *Journal of Tropical Ecology* 22: 235–46.

Loiselle, B. 1988. Bird abundance and seasonality in a Costa Rican lowland forest canopy. *Condor* 90: 761–72.

Longino, J.T. 1991. Taxonomy of the *Cecropia*-inhabiting *Azteca* ants. *Journal of Natural History* 25: 1571–602.

Longino, J.T. 1996. Taxonomic characterization of some live-stem inhabiting *Azteca* (Hymenoptera: Formicidae) in Costa Rica, with special reference to the ants of *Cordia* (Boraginaceae) and *Triplaris* (Polygonaceae). *Journal of Hymenoptera Research* 5: 131–56.

Longino, J.T. 2006. A taxonomic review of the genus *Myrmelachista* (Hymenoptera: Formicidae) in Costa Rica. *Zootaxa* 1141: 1–54.

Longino, J.T. 2007. A taxonomic review of the genus *Azteca* (Hymenoptera: Formicidae) in Costa Rica and a global revision of the *aurita* group. *Zootaxa* 1491: 1–63.

Longino, J.T., and R.K. Colwell. 1997. Biodiversity assessment using structured inventory: capturing the ant fauna of a lowland tropical rainforest. *Ecological Applications* 7: 1263–77.

Longino, J.T., R.K. Colwell, and J.A. Coddington. 2002. The ant fauna of a tropical rainforest: estimating species richness three different ways. *Ecology* 83: 689–702.

López, M., M. Altrichter, J. Sáenz, and E. Eduarte. 2006. Valor nutricional de los alimentos de *Tayassu pecari* (Artiodactyla: Tayassuidae) en el Parque Nacional Corcovado, Costa Rica. *Revista de Biología Tropical* 54: 687–700.

Lott, D.F. 1991. *Intraspecific Variation in the Social Systems of Wild Vertebrates*. New York: Cambridge University Press.

Madrigal, R., and E. Rojas. 1980. Manual Descriptivo del Mapa Geomorfológico de Costa Rica. San José, Costa Rica: Secretaría Ejecutiva de Planificación Sectorial Agropecuaria y de Recursos Naturales Renovables. Escala 1: 200,000.

March, I. 1993. White-lipped Peccary (*Tayassu pecari*). In W. Oliver, ed., *Pigs, Peccaries and Hippos: Status Survey and Conservation Action Plan*, 7–13. Gland, Switzerland: International Union for Conservation of Nature.

Marín Cañas, J. 1976. *Coto: La guerra del 21 con Panamá*. 2nd ed. San José, Costa Rica.

Martínez, R.M., B.E. Álvarez, J. Sarukhan, and D. Pinero. 1988. Treefall age determination and gap dynamics in a tropical forest. *Journal of Ecology* 76: 700–716.

Mayer, J., and P. Brandt. 1982. Identity, distribution and natural history of the peccaries, Tayassuidae. In M.A. Mares and H.H. Genoways, eds., *Mammalian Biology in South America*. Special Publication Series. Pittsburgh, PA: Pymatuning Laboratory of Ecology, University of Pittsburgh.

Mayer, J., and R. Wetzel. 1987. *Tayassu pecari*. *Mammalian Species* 293: 1–7.

Montero, W. 1986. Períodos de recorrencia y tipos de sequencías sísmicas de los temblores intraplaca en la regíon de Costa Rica. *Revísta Geológica de América Central* 5: 35–72.

Munn, C., and J. Terborgh. 1979. Multi-species territoriality in neotropical foraging flocks. *Condor* 81: 338–47.

Myers, M.C. 2005. Endangered wildlife conservation in the Neotropics: lessons and strategies from Costa Rica. PhD thesis, University of Minnesota.

Myers, M., and C. Vaughan. 2004. Movement and behavior of scarlet macaws (*Ara macao*) during the post-fledging dependence period: implications for *in situ* versus *ex situ* management. *Biological Conservation* 118: 411–20.

National Academy of Sciences–National Research Council (NAS-NRC). 1960. Conference on tropical botany. Washington, DC: NAS-NRC Division of Biology and Agriculture, Fairchild Tropical Garden.

O'Neil, E.M., and J.R. Mendelson III. 2004. Taxonomy of Costa Rican toads referred to *Bufo melanchlorus* Cope, with the description of new species. *Journal of Herpetology* 38: 487–94.

Oliver, W., ed. *Pigs, Peccaries and Hippos: Status Survey and Conservation Action Plan*. Gland, Switzerland: International Union for Conservation of Nature.

Olmos, F. 1993. Diet of sympatric Brazilian caatinga peccaries (*Tayassu tajacu* and *T. pecari*). *Journal of Tropical Ecology* 9: 255–58.

Peres, C. 1996. Population status of white-lipped *Tayassu pecari* and collared peccaries *T. tajacu* in hunted and unhunted amazonian forests. *Biological Conservation* 77: 115–23.

Pérez, S., A. Alvarado, and E. Ramírez. 1978. Manual Descriptivo del Mapa de Asociaciones de Subgrupos de Suelos de Costa Rica. San José, Costa Rica: Oficina de Planificación Sectorial Agropecuario, IGN / MAG / FAO. Escala 1:200,000.

Phillips, P. 1989. The relationship of successional and primary tropical rain forests to color infrared photography. MA thesis, University of Texas at Austin. Austin, TX.

Phillips, P. 1991. Key to the Vegetation Types for the Osa Peninsula, Costa Rica. Austin, TX: University of Texas.

Plowes, R.M. 2005. Tropical forest landscape dynamics: population consequences for Neotropical lianas, genus *Passiflora*. PhD diss., University of Texas. Austin, TX.

Powell, G. 1979. Structure and dynamics of interspecific flocks in a neotropical mid-elevation forest. *Auk* 96: 375–90.

Powell, G., and R. Bjork. 2004. Habitat linkages and the conservation of tropical biodiversity as indicated by seasonal migrations of three-wattled bellbirds. *Conservation Biology* 18: 500–509.

Quesada, F.J., Q. Jiménez, N. Zamora, R. Aguilar, and J. González. 1997. *Árboles de la Península de Osa*. Santo Domingo de Heredia, Costa Rica: INBio / SIDA.

Redford, K., and J. Robinson. 1987. The game of choice: patterns of Indian and colonist hunting in the Neotropics. *American Anthropology* 89: 650–66.

Reed, S.C., C.C. Cleveland, and A.R. Townsend. 2007. Controls over leaf litter and soil nitrogen fixation in two lowland tropical rain forests. *Biotropica* 39: 585–92.

Reed, S.C., C.C. Cleveland, and A.R. Townsend. 2008. Tree species control rates of free-living nitrogen fixation in a tropical rainforest. *Ecology* 89: 2924–34.

Reid, F.A. 1997. *A Field Guide to the Mammals of Central America and Southeast Mexico*. New York: Oxford University Press.

Riba-Hernández, P., and K.E. Stoner. 2005. Massive destruction of *Symphonia globulifera* (Clusiaceae) flowers by Central American spider monkeys (*Ateles geoffroyi*). *Biotropica* 37: 274–78.

Riba-Hernández, P., K.E. Stoner, and P.W. Lucas. 2003. The sugar composition of fruits in the diet of spider monkeys (*Ateles geoffroyi*) in tropical humid forest in Costa Rica. *Journal of Tropical Ecology* 19: 709–16.

Riba-Hernández, P., K.E. Stoner, and P.W. Lucas. 2005. Sugar concentration of fruits and their detection via color in the Central American spider monkey (*Ateles geoffroyi*). *American Journal of Primatology* 67: 411–23.

Robbins, C., J. Sauer, R. Greenberg, and S. Droege. 1989. Population declines in North American birds that migrate to the Neotropics. *Proceedings of the National Academy of Sciences of the United States of America* 86: 7658–62.

Robertson, J.M., and K.R. Zamudio. 2009. Genetic diversification, vicariance, and selection in a polytypic frog. *Journal of Heredity* 100: 715–31.

Savage, J.M. 2002. *The Amphibians and Reptiles of Costa Rica: A Herpetofauna between Two Continents, between Two Seas*. Chicago: University of Chicago Press.

Savage, J.M., and J. Villa. 1986. *Herpetofauna de Costa Rica*. Ohio: Cushing-Malloy Inc.

Sawyer, J.O., and A.A. Lindsay. 1971. Vegetation of the life zones in Costa Rica. Indiana Academy of Sciences Monograph no. 2. 214 pp.

Schmidt, M.R. 2001. Interactions between *Tetrathylacium macrophyllum* (Flacourtiaceae) and its live-stem inhabiting ants in the Parque Nacional Corcovado, Sección Piedras Blancas, Costa Rica Magister rerum naturarum. Austria: Faculty of Natural Sciences and Mathematics, University of Vienna.

Schneider, D.W., and T. Frost. 1986. Massive upstream migrations by a tropical freshwater neritid snail. *Hydrobiologia* 137: 153–57.

Sheil, D. 1999. Tropical forest diversity, environmental change and species augmentation: after the intermediate disturbance hypothesis. *Journal of Vegetation Science* 10: 851–60.

Sherman, P.M. 2002. Effects of land crabs on seedling densities and distributions in a mainland neotropical rain forest. *Journal of Tropical Ecology* 18(1): 67–89.

Sherman, P.M. 2003. Effects of land crabs on leaf litter distributions and accumulations in a mainland tropical rain forest. *Biotropica* 35: 365–74.

Sherman, P.M. 2006. Influence of land crabs *Gecarcinus quadratus* (Gecarcinidae) on distributions of organic carbon and roots in a Costa Rican rain forest. *Revista de Biología Tropical* 54: 149–61.

Skutch, A. 1992. Interview by C. Christen, April 29, 1992. Quizarrá, Costa Rica. Tape transcript.

Smith, N. 1976. Utilization of game along Brazil's Transamazon highway. *Acta Amazonica* 6: 455–66.

Solorzano, A., L.D. Gómez, J. Monge-Nájera, and B.I. Crother. 1998. Redescription and validation of *Bothriechis supraciliaris* (Serpentes: Viperidae). *Revista de Biología Tropical* 46: 453–62.

Soluri, J. 2005. *Banana Cultures: Agriculture, Consumption, and Environmental Change in Honduras and the United States*. Austin, TX: University of Texas Press.

Soto, R., and V. Jiménez. 1992. *Evaluación Ecológica Rápida, Península de Osa, Costa Rica*. San José, Costa Rica: Programa BOSCOSA, Fundación Neotrópica / WWF. 252 pp.

Sowls, L. 1997. *Javelinas and Other Peccaries: Their Biology, Management and Use*. 2nd ed. Tucson: University of Arizona Press.

Stetson, W. 1977. Interview by B. Lewis, June 7, 1977. Tape transcript.

Stiles, F., and A. Skutch. 1989. *A Guide to the Birds of Costa Rica*. Ithaca: Cornell University Press.

Stone, D.E. 1988. The Organization for Tropical Studies (OTS): a success story in graduate training and research. In F. Almeda and C. Pringle, eds., *Tropical Rainforests: Diversity and Conservation*. San Francisco: California Academy of Sciences.

Stoner, K.E. 1996a. Habitat selection and seasonal patterns of activity and foraging of mantled howling monkeys (*Alouatta palliata*) in northeastern Costa Rica. *International Journal of Primatology* 17(1): 1–30.

Stoner, K.E. 1996b. Prevalence and intensity of intestinal parasites in mantled howling monkeys (*Alouatta palliata*) in northeastern Costa Rica: implications for conservation biology. *Conservation Biology* 10: 539–46.

Swartz, M. 1997. Behavioral and population ecology of the army ant *Eciton burchelli* and ant-following birds. PhD diss., University of Texas. Austin, TX.

Swartz, M. 2001. Bivouac checking, a novel behavior distinguishing obligate from opportunistic species of army ant-following birds. *Condor* 103: 629–33.

Terborgh, J., S. Robinson, T. Parker, C. Munn, and N. Pierpont. 1990. Structure and organization of an Amazonian forest bird community. *Ecological Monographs* 60: 213–38.

Thiollay, J. 1994. Structure, density, and rarity in an Amazonian rainforest community. *Journal of Tropical Ecology* 10: 449–81.

Tosi, J.A. 1969. Mapa Ecológico de Costa Rica, Basado en la Clasificación Vegetal Mundial de L.R. Holdridge. San José, Costa Rica: CCT. Escala 1: 750.000.

Tosi, J.A. 1975. The Corcovado Basin on the Península de Osa. Draft manuscript chapter for *Potential National Parks, Nature Reserves, and Wildlife Sanctuary Areas in Costa Rica: A Survey of Priorities*. World Wildlife Fund Project No. 801. San José, Costa Rica: Tropical Science Center.

Tournon, J., and G. Alvarado. 1997. Mapa Geológico de Costa Rica, Folleto Explicativo con Mapa. Cartago, Costa Rica: Editorial Tecnológica de Costa Rica. 79 pp. Escala 1: 500,000.

Townsend, A.R., G.P. Asner, and C.C. Cleveland. 2008. The biogeochemical heterogeneity of tropical forests. *Trends in Ecology and Evolution* 23: 424–31.

Townsend, A. R., G.P. Asner, C.C. Cleveland, M.E. Lefer, and M.M.C. Bustamante. 2002. Unexpected changes in soil phosphorus dynamics along pasture chronosequences in the humid tropics. *Journal of Geophysical Research: Atmospheres* 107(D20): 8067.

Townsend, A.R., C.C. Cleveland, G.P. Asner, and M.M.C. Bustamante. 2007. Controls over foliar N:P ratios in tropical rain forests. *Ecology* 88: 107–18.

UNDP. 1999. Establishment of a Programme for the Consolidation of the Mesoamerican Biological Corridor Project. Document RLA/97/G31. United Nations Development Programme, Global Environment Facility. Project of the Governments of Belize, Costa Rica, El Salvador, Guatemala, Honduras, Mexico, Nicaragua, Panama.

van den Hombergh, H. 1999. *Guerreros de Golfo Dulce: Industria Forestal y Conflicto en la Península de Osa, Costa Rica*. San José, Costa Rica: Departamento Ecuménico de Investigaciones.

Vaughan, C. 1979. Plan maestro para el manejo y desarollo del Parque Nacional Corcovado, Península de Osa Costa Rica. M.Sc. thesis, Centro Agronómico Tropical de Investigación y Enseñanza. Turrialba, Costa Rica.

Vaughan, C. 1981. *Parque Nacional Corcovado: Plan de Manejo y Desarrollo*. Heredia, Costa Rica: Editorial de la Universidad Nacional. 351 pp.

Vaughan, C. 2006. Advantages and disadvantages of wildlife reintroductions. *Mesoamericana* 10(2): 88–95.

Vaughan, C. 2011. Changes in dense forest habitat for endangered wildlife species in Costa Rica from 1949 to 1977. *Cuadernos de Investigación UNED* 3(1): 15–77.

Vaughan, C., N. Nemeth, J. Cary, and S. Temple. 2005. Response of a Scarlet Macaw (*Ara macao*) population to conservation measures. *Birdlife International* 15: 119–30.

Vickers, W. 1991. Hunting yields and game composition over ten years in an Amazon Indian Territory. In J. Robinson and K. Redford, eds., *Neotropical Wildlife Use and Conservation*. Chicago: University of Chicago Press.

Ward, P.S. 1993. Systematic studies on *Pseudomyrmex* acacia-ants (Hymenoptera: Formicidae: Pseudomyrmecinae). *Journal of Hymenoptera Research* 2: 117–68.

Weber, A., W. Huber, A. Weissenhofer, N. Zamora, and G. Zimmermann, eds. 2001. *An Introductory Field Guide to the Flowering Plants of the Golfo Dulce Rain Forests, Costa Rica*. Linz, Austria: Oberösterreichisches Landes Museum. *Stapfia* 78: 1–107.

Weckel, M., W. Giuliano, and S. Silver. 2006. Jaguar (*Panthera onca*) feeding ecology: distribution of predator and prey through time and space. *Journal of Zoology* 270: 25–30.

Weghorst, J.A. 2007. High population density of black-handed spider monkeys (*Ateles geoffroyi*) in Costa Rican lowland wet forest. *Primates* 48: 108–16.

Weissenhofer, A., W. Huber, N. Zamora, A. Weber, and J. González. 2001. A brief outline of the flora and vegetation of the Golfo Dulce region. In A. Weber, W. Huber, A. Weissenhofer, N. Zamora, and G. Zimmermann, eds., *An Introductory Field Guide to the Flowering Plants of the Golfo Dulce Rain Forests, Costa Rica*. Linz, Austria: Oberösterreichisches Landes Museum. *Stapfia* 78: 15–24.

Westcott, D. 1997. Lek locations and patterns of female movement and distribution in a Neotropical frugivorous bird. *Animal Behavior* 53: 235–47.

White, P., and S. Pickett. 1985. Natural disturbance and patch dynamics: an introduction. In S. Pickett and P. White, eds., *The Ecology of Natural Disturbance and Patch Dynamics*, 3–13. New York: Academic Press.

Wieder, W.R., C.C. Cleveland, and A.R. Townsend. 2009. Controls over leaf litter decomposition in wet tropical forests. *Ecology* 90: 3333–41.

Wille, A. 1983. *Corcovado: Un Estudio Ecológico*. San José, Costa Rica: Editorial de la Universidad Estatal a Distancía. 403 pp.

Wilson, D.E. 1983. Checklist of mammals. In D.H. Janzen, ed., *Costa Rican Natural History*, 443–47. Chicago: University of Chicago Press.

Winemiller, K. 1983. An introduction to the freshwater fish communities of Corcovado National Park, Costa Rica. *Brenesia* 21: 47–66.

Winemiller, K., and M. Leslie. 1992. Fish communities across a complex freshwater-marine ecotone. *Environmental Biology of Fishes* 34: 29–50.

Winemiller, K., and N. Morales. 1989. Comunidades de peces del Parque Nacional Corcovado luego del cese de las actividades mineras. *Brenesia* 31: 75–91.

Wong, G. 1990a. Ecología del mono titi: *Saimiri oerstedi citrinellus* en el Parque Nacional Manuel Antonio, Costa Rica. Tesís de Grado, Escuela de Ciencias Ambientales, Universidad Nacional. Heredia, Costa Rica.

Wong, G. 1990b. Uso del habitát, estimación de composición y densidad poblacional del mono titi (*Saimiri oerstedi citrinellus*) en la zona de Manuel Antonio, Quepos, Costa Rica. Tesís de Grado, Escuela de Ciencias Ambientales, Universidad Nacional. Heredia, Costa Rica.

Zaldívar, M.E., O. Rocha, K.E. Glander, G. Aguilar, A.S. Huertas, R. Sánchez, and G. Wong. 2004. Distribution, ecology, life history, genetic variation, and risk of extinction of nonhuman primates from Costa Rica. *Revista de Biología Tropical* 52: 679–93.

Zamora, N., B.E. Hammel, and M.H. Grayum. 2004. Vegetation. In B.E. Hammel, M.H. Grayum, C. Herrera, and N. Zamora, eds., *Manual de Plantas de Costa Rica*, Vol. I, Introducción. *Monographs in Systematic Botany from the Missouri Botanical Garden* 97: 91–216.

Chapter 13 The Montane Cloud Forests of the Volcanic Cordilleras

Robert O. Lawton[1,*], Marcy F. Lawton[2], R. Michael Lawton[3], and James D. Daniels[4]

Introduction

The cloud forests of northern Costa Rica are now among the best studied in the world (Fig. 13.1). The Cordillera de Tilarán and the Cordillera de Guanacaste provide a landscape with dramatic relief, strong climatic contrasts, and extraordinary biodiversity. These volcanic Cordilleras are in the process, ongoing since the early Pliocene, of filling one of the three lowland gaps in the chain of Mesoamerican highlands. The Central American dry season is tempered in the Cordilleras by orographic cloud decks on the Caribbean slopes established as the northeast trade winds flow over the mountains. On the fertile volcanic soils of these Cordilleras, β-diversity in the flora and fauna, community composition, forest structure, and ecosystem functioning depend in large part on variation in dry season cloud moisture inputs.

Much of what we know of these forests is due to research in the complex of reserves in the Monteverde region of the central Cordillera de Tilarán. The system of reserves is itself a remarkable conservation success, involving the efforts of the people of the region, biologists, international and Costa Rican conservation organizations, and the Costa Rican government. Research and conservation efforts in the Monteverde area have been recently summarized in *Monteverde: Ecology and Conservation of a Tropical Cloud Forest*, edited by Nadkarni and Wheelwright (2000). We cannot hope to do justice to all the material presented there. Indeed, in most respects this chapter will be an abstract of that volume. We will, however, attempt to provide a biogeographic framework emphasizing the topographically dictated climatic constraints that underlie the biological richness of the montane forests of northern Costa Rica.

The Physical Environment

Tectonic History

The current topography of northern Costa Rica (Fig. 13.2) is the product of a complex tectonic history. During the late Cretaceous the leading edge of the Caribbean crustal plate, which is thicker than most basaltic ocean floor plates, began to wedge between the North and South American plates (Burke 1988, and see Alvarado and Cárdenas, chapter 3 of this volume). Subduction of the oceanic plate in front led to the formation of a volcanic arc, now the Greater and Lesser Antilles, on the leading edge of the Caribbean, and subduction of Pacific floor plates on the trailing edge of the Caribbean plate produced a volcanic arc there as well, with remnants now in Costa Rica and Panama. Volcanic activity on the trailing edge of the Caribbean plate was relatively quiescent from the early Tertiary until about 10 Ma (million years) when the end of the arc in what is now eastern Panama began to collide with northwestern South America and create the Atrato suture. This slowed

[1] Department of Biological Sciences, University of Alabama, PO Box 311, Huntsville, AL 35899, USA
[2] Monte Santo Learning Center, 3917 Panorama Drive, Huntsville, AL 35801, USA
[3] Department of Ecology and Evolutionary Biology, 569 Dabney Hall, University of Tennessee, Knoxville, TN 37996, USA
[4] Department of Biology and Cell Biology, Huntingdon College, 1500 Fairview, Montgomery, AL 36106, USA
* Corresponding author

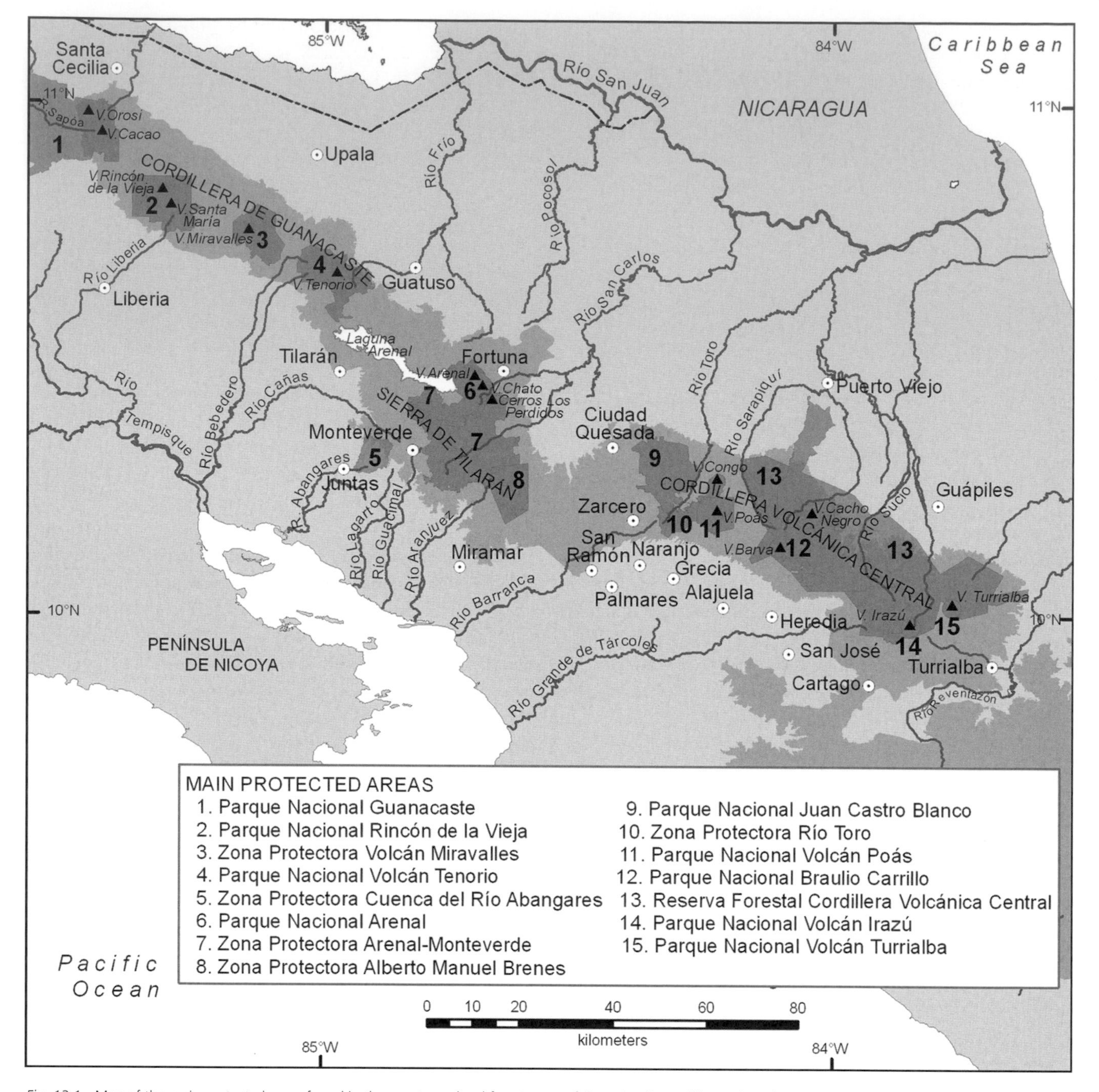

Fig. 13.1 Map of the main protected areas found in the montane cloud forest zone of the volcanic *cordilleras* of northern Costa Rica. *Map prepared by Marco V. Castro.*

the Atlantic-ward progress of the Caribbean plate, and increased the rate of subduction of the Cocos plate along the Middle America Trench. Increased subduction increased volcanic activity, faulting, and uplift along the arc. The ultimate result was the Central American land bridge, which allowed intercontinental terrestrial biotic interchange about 3.5 Ma (Marshall et al. 1982, Gómez 1986, Coates et al. 1992, Alvarado and Cárdenes, chapter 3 of this volume). At this time the Caribbean Sea was separated from the Pacific. This blocked the westward movement into the Pacific of warm surface water pushed by the trade winds (Keigwin 1982, Burke 1988).

The Cordillera de Tilarán and the Cordillera de Guanacaste have their origins in renewed volcanism and tectonic uplift along the southern Central American arc in the late Tertiary. Since the late Miocene there have been three particularly important episodes of volcanic activity. The first two produced the Aguacate Formation, a geological structure that underlies the whole Cordillera de Tilarán. First, in the late Miocene, the earliest part of the Aguacate formation, with rocks dated from 8.5 to 10.5 Ma (Cháves and Sáenz 1974), resulted from volcanic activity involving magma produced at relatively low temperatures and pressures as the subducted Cocos plate descended under the trailing edge of the Caribbean plate. The second major episode of volcanic activity produced the youngest rocks of the Aguacate formation, which date from 2.6 to 4.3 Ma (Pliocene). This involved magma produced at greater depths and pressures as the renewed subduction of the Cocos plate continued. The rocks of this uppermost member of the Aguacate formation created the dramatic cliffs at 1200–1500 m elevation throughout the Cordillera de Tilarán.

Trench parallel faulting along the trailing edge of the Caribbean plate, and compression associated with subduction, led to the creation in the late Pliocene of a *horst-graben-horst* system in which the Aguacate formation as a whole was uplifted as a horst between the Arenal fault that extends through Volcán Arenal and Cerros los Perdidos on the Caribbean side of the Cordillera de Tilarán and the Las Juntas fault that follows the Pan-American Highway along the Pacific side of the cordillera (Dengo and Levy 1969). To the southwest lies the graben that has been downthrust to create the Golfo de Nicoya, and beyond it lies the horst of the Peninsula de Nicoya. Since the elevation of the cliffs in the Cordillera de Tilarán formed by the uppermost member of the Aguacate formation declines to the northwest, this suggests that the mechanisms uplifting this horst were like those that continue to produce similar rotation and tilt of the blocks of the Nicoya and Osa peninsulas.

Finally, in the past two million years the subduction zone has continued to mature, and arc volcanic activity has shifted to sites on trench parallel faults successively farther east. Around 600,000 years ago, extensive volcanic activity in the Cordillera de Tilarán laid down the andesitic lavas and tuffs of the Monteverde Formation (Cháves and Sáenz 1974, Castillo-Muñoz 1983, Alvarado 1989, Alvarado and Cárdenas, chapter 3 of this volume). These lie discordantly upon the rocks of the Aguacate Formation, and mantle the upper part of the Cordillera to depths of up to several hundred meters. Although the volcanic foci producing the Monteverde Formation have not been precisely located (Alvarado 1989), there are several sets of trench parallel peaks that are probably relict volcanic structures. Cerros Ojo de Agua, Roble, Amigo, and Chomogo lie on a single axis, while Cerros Zapotal, la Mesa, and Frio lie on an axis 2–3 km to the northeast. Seven km farther northeast lies Pico 1790 on the divide between the Arenal and Peñas Blancas watershed as well as a series of peaks at the head of

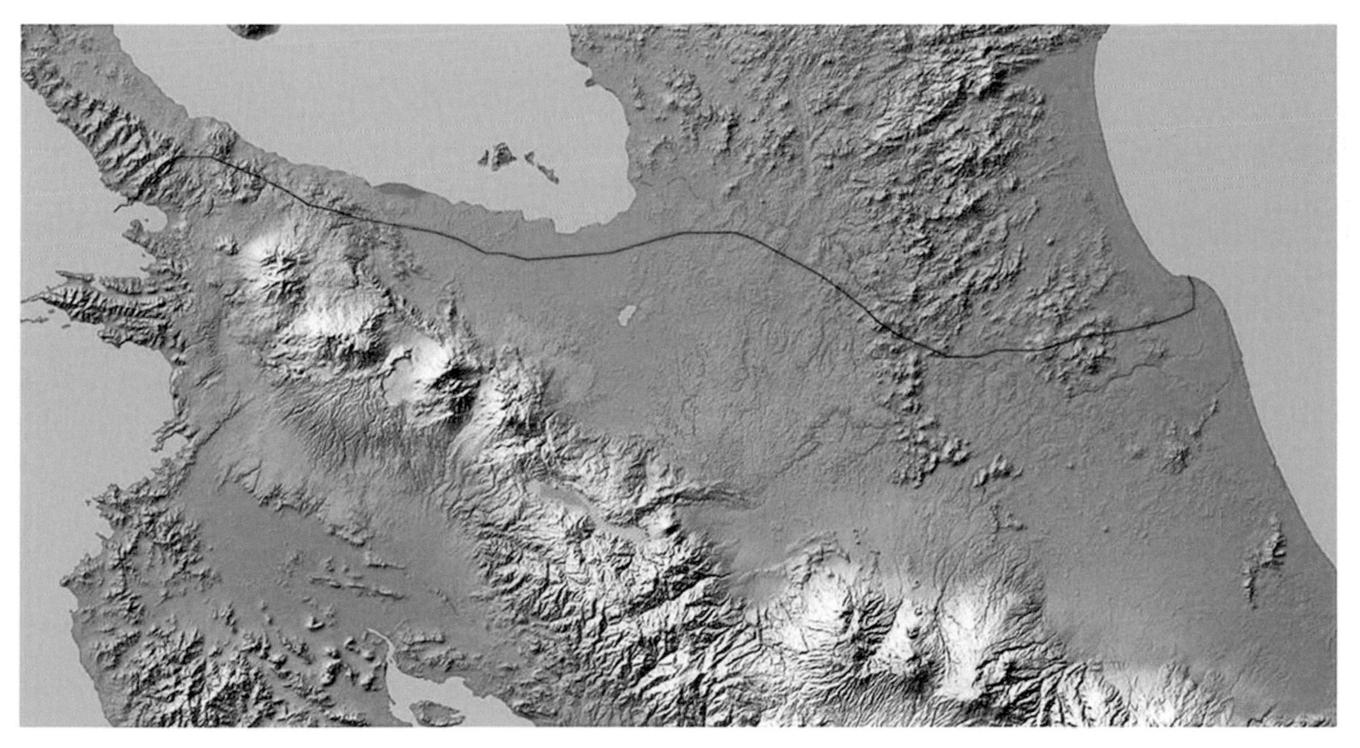

Fig. 13.2 Geography of northern Costa Rica.
From NASA's SRTM 90m resolution digital terrain model.

the Río Aranjuez valley. The currently active volcanic axis lies on the Arenal series of faults that runs through Cerro Chato and Volcán Arenal on the Caribbean margin of the Cordillera de Tilarán, and continues northwest through the volcanos of the Cordillera de Guanacaste.

The volcanic massifs of the Cordillera de Guanacaste are all Pleistocene and Holocene complex stratovolcanic structures extending in a line along a fault system from Volcán Arenal in the southeast to V. Tenorio, V. Miravalles, V. Rincón de la Vieja, and finally the Orosí complex at the northwest end (Alvarado 1989). All have been active in the past few thousand years, and Miravalles, Rincón de la Vieja, and Arenal have erupted in the past century. Each of these massifs is of similar elevation (1,670–2,028 m), and they support a similar range of life zones and habitats. These volcanic edifices are weak at the core, and breached calderas with large debris flows are conspicuous on all but the youngest massifs (Kempter et al. 1996, Alvarado et al. 2004, Cecchi et al. 2005, Siebert et al 2006), and must have major ecological consequences. For instance, the Tierras Morenas deposit southwest of the main cone of V. Tenorio is a debris flow that covers ~100 km^2 with about 2 km^3 of material (Siebert et al. 2006).

Arenal has been erupting continuously since an explosive eruption in 1968. Arenal, positioned at the northern end of the Caribbean side of the horst block of the Cordillera de Tilarán, is the youngest of the group; this volcano and the adjacent Cerro Chato apparently began erupting about 7,000 years ago (Melson 1994), blocking drainage to the Caribbean from the northern slope of the C. de Tilarán, and so creating the Laguna Arenal basin. Cerro Chato and V. Arenal are effectively contiguous with Cerro los Perdidos, and are biogeographically perhaps best considered as a part of the C. de Tilarán.

Modern Geography

The Cordillera de Tilarán today consists of a single main ridge, about 50 km long, along the Continental Divide, which winds from near San Ramon west across Cerro Zapotal (1,580 m), around the headwaters of the Río Aranjuez to Cerros La Mesa (1,820 m) and Ojo de Agua (1,800 m), then across the Brillante saddle between the Río Peñas Blancas and Río Guacimal watersheds to Cerros Roble (1,699 m), Amigo (1,849 m), and Chomogo (1,799 m) before descending toward Tilarán (Fig. 13.3). Two large ridges extend northeast from the Continental Divide flanking the Peñas Blancas valley, and dividing it from the Río Caño Negro watershed to the northwest with Lago Arenal at its base and from the Río Jamaical watershed to the southeast. The high points on these main ridges appear to be relict volcanic structures of the Monteverde Formation. These peaks of the Monteverde Formation are superimposed upon the platform provided by the uppermost member of the Aguacate Formation. In places, the roughly level surface of the Aguacate horst block is not deeply mantled by materials of the Monteverde Formation, yielding large benches or broad saddles in the mountain range. The bench occupied by the communities of Monteverde, Cerro Plano, Santa Elena, and La Cruz is an example. Among the more distinctive features of the mountain range are the "cliff edges" created by the resistant rocks of the uppermost Aguacate Formation. Below these cliffs the terrain is highly dissected into small, narrow catchments separated by narrow, steep-flanked ridges. Relief over small (10–20 ha) watersheds is often >100 m.

Valley cutting by the major rivers draining the Cordillera has had a dramatic influence on the current landscape. The whole Caribbean slope of the Cordillera is drained by tributaries of the Río San Carlos. The Río Peñas Blancas, largest of these tributaries, cuts a valley that is the dominant feature of the landscape on the Caribbean slope and lies at the core of the complex of reserves that has made the Monteverde region such an important arena for conservation. The Pacific slope is drained at its eastern end by the Río Barranca, then by a series of parallel rivers flowing, like the Barranca, into the Golfo de Nicoya. Among these are the Ríos Aranjuez, Guacimal, and Lagarto. The western Pacific slope is drained by the Ríos Cañas and Abangares, both of which flow into the mouth of the Río Tempisque at the head of the Golfo de Nicoya. The orientation of valleys in the Cordillera de Tilarán appears to be strongly directed by faulting of the Aguacate horst (Bergoeing and Brenes 1979). Many of the rivers have parallel central courses that are perpendicular to the long axis of the horst. In the upper reaches of watersheds this influence of faulting is usually less conspicuous.

Permanent streams carve deep winding ravines and gorges as they descend rapidly in stepped series of waterfalls and rapids. The gorges, or *quebradas*, in the Cordillera de Tilarán, are commonly 200 m deep as they cut through the resistant rocks of the uppermost Aguacate Formation. Following heavy rain these streams carry a heavy sediment load; boulders can be heard rumbling along the bed of the larger rivers when they are high. Little, however, is known quantitatively about rates of erosion in the area, but in volcanic ranges in Japan with similar relief and precipitation, sedimentation in small reservoirs suggests that erosion amounts to 1,000 to 10,000 m^3/km^2/year, or 1–10 mm/year (Yoshikawa et al. 1981). This is approximately the current rate of uplift on the Nicoya and Osa peninsulas (Gardner et al. 1992, Marshall and Anderson 1995).

Fig. 13.3 Physiography of the Cordillera de Tilarán in northern Costa Rica. The zones with highest elevations are colored in brown and gray.

Local terrain features, including *quebradas*, cliffs, waterfalls, landslides, hogback ridges, swamps, and permanent pools or small lakes, are a distinctive part of the montane landscape in northern Costa Rica. Little is known quantitatively about their abundance, distribution, formation, or persistence, although they contribute markedly to landscape and microclimatic diversity. The relationship of such habitats to vegetation pattern and plant distribution will be discussed below.

Soils

The soils of the montane forests of the Cordillera de Tilarán above 1300–1400 m are udic Andisols (Clark et al. 2000) or Dystrandepts (Tobón-Marín et al. 2010) with an origin in the Quaternary volcanic materials of the Monteverde Formation. Soils at lower elevations are more weathered residual Andisols developed from the Aguacate Formation, or are Inceptisols developing either on colluvium originating from both geological formations or on steep, landslide-carved slopes. Most of what is known about the montane soils of northern Costa Rica comes from a small number of local investigations on the uppermost Pacific slope and crest of the Cordillera de Tilarán (see Alvarado and Mata, chapter 4 of this volume). A systematic soil survey in the Cordillera is necessary.

Soils in the area vary greatly in depth and development, depending upon the geomorphological setting. They are shallowest and most poorly developed, or indeed nonexistent, on landslide scars scoring the cliff walls of *quebradas*. On gentle slopes of major benches, however, soils may be more than 5 m deep. On the Monteverde Formation

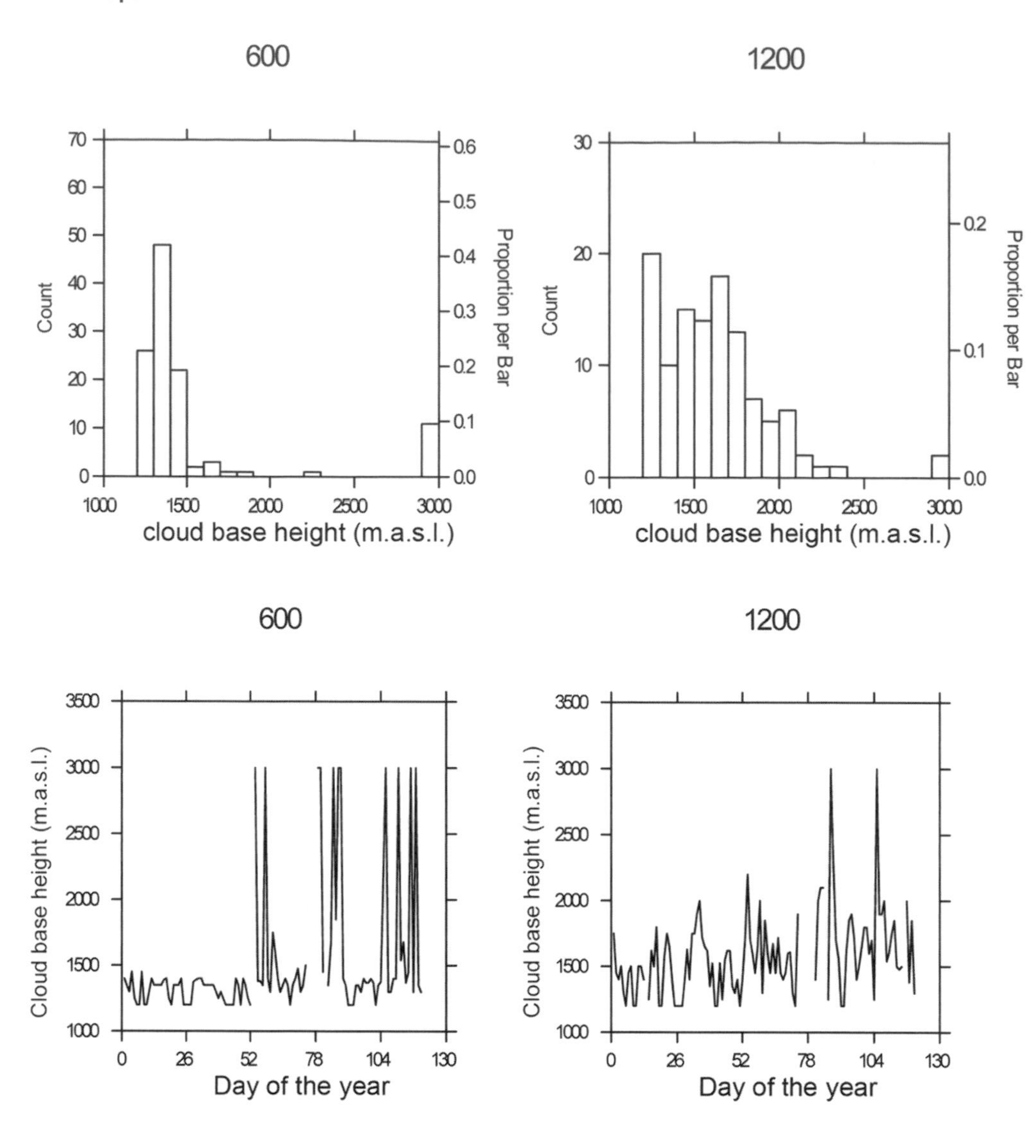

Fig. 13.7 Upper graphs: Dry season cloud base height distributions at dawn and midday on Cerro Ojo de Agua and the Brillante saddle on the continental divide in the central Cordillera de Tilarán. Lower graphs: Cloud base heights at dawn and midday from the same site on each day in the dry season of 2003. Spikes to 3,000 m indicate clear conditions or cloud bases above the trade wind inversion.
From Lawton et al. 2010.

models, and with ground observations on the distributions of forest types.

Light

Owing to the extensive cloud cover, the forests of the uppermost Pacific slope, the crest of the Cordillera, and the Caribbean slope receive less light than forests of the Pacific slope and the lowlands. In leeward cloud forest at 1,500 m elevation 1.5 km from the continental divide in the Cordillera de Tilarán, mean above-canopy incident solar radiation observed at noon is about 570 W m^{-2}, roughly 60% of that calculated for clear sky conditions (Clark et al. 2000). The diurnal courses of incident radiation are shown for days of mist and cloud, wind driven precipitation, and convectively produced rain in Fig. 13.10. Along the continental divide modal above-canopy midday photosynthetic photon flux densities (PPFD) are about 500 μmol m^2 sec^{-1}, but almost 5% of observations are >2,000 μmol m^2 sec^{-1}, owing to reflection from nearby clouds (Lawton 1990). Light penetration to the understory in the Monteverde cloud forests (Fig. 13.11) is better than reported for lowland forests, probably owing to relatively high rates of canopy damage (Lawton 1990). Canopy structure, particularly the presence of treefall gaps, strongly influences gradients of light availability (Lawton 1990; Fig. 13.12).

Hydrology

Hydrological research in the mountains of northern Costa Rica has focused on occult precipitation—the water inputs missed by ordinary rain gauges (Zadroga 1981, Clark and Nadkarni 2000, Bruijnzeel 2006, Häger 2006, Häger and Dohrenbusch 2009). Recently, however, in order to assess the impact of the conversion of cloud forest to pasture on

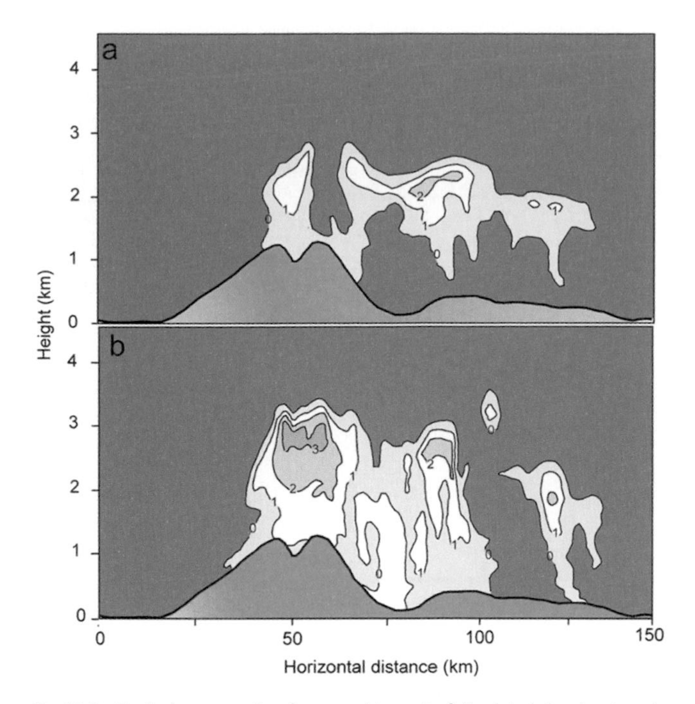

Fig. 13.8 Vertical cross-section from west to east of simulated cloud water mixing ratio (gm liquid water kg^{-1} of air) above the Cordillera de Tilarán and upwind Caribbean lowlands at midday (11:45 hr, local time) for initializing conditions of March 9, 1999, for (a) completely deforested and pastured land use scenario and (b) a completely forested landscape.
From Lawton et al. 2001.

water yield, complete hydrological budgets have been examined under paired pasture and forest local headwater catchments in lower montane rain forest of the upper Río Caño Negro basin in the northern Cordillera de Tilarán (Bruijnzeel 2006). As part of the same project, scale-up modeling to operational watersheds (100 km^2) was based on catchment-derived understanding of processes, and scattered streamflow and meteorological station data (Bruijnzeel 2006). In addition, stable-isotope analyses of cloud water, rainfall, soil water, and streamflow in the Río Guacimal basin of the upper Pacific slope of the central C. de Tilarán have been used to assess the seasonal influences of cloud/mist inputs to soil water and streamflow (Rhodes et al. 2006, Guswa et al. 2007).

Small headwater catchment hydrological budgets for lower montane rainforest and adjacent pasture are summarized in Table 13.1. The impact of terrain and wind on precipitation make estimation of areal precipitation inputs from point measurements difficult. The precipitation in Table 13.1 is calculated from rain gauge catches, adjusted for wind-caused losses around the gauge and the ~25% of the horizontal or near horizontal precipitation caught by the vegetation and transferred to the ground as canopy drip, as inputs with wind and terrain in a local geographic model. Since different models of airflow over the watershed

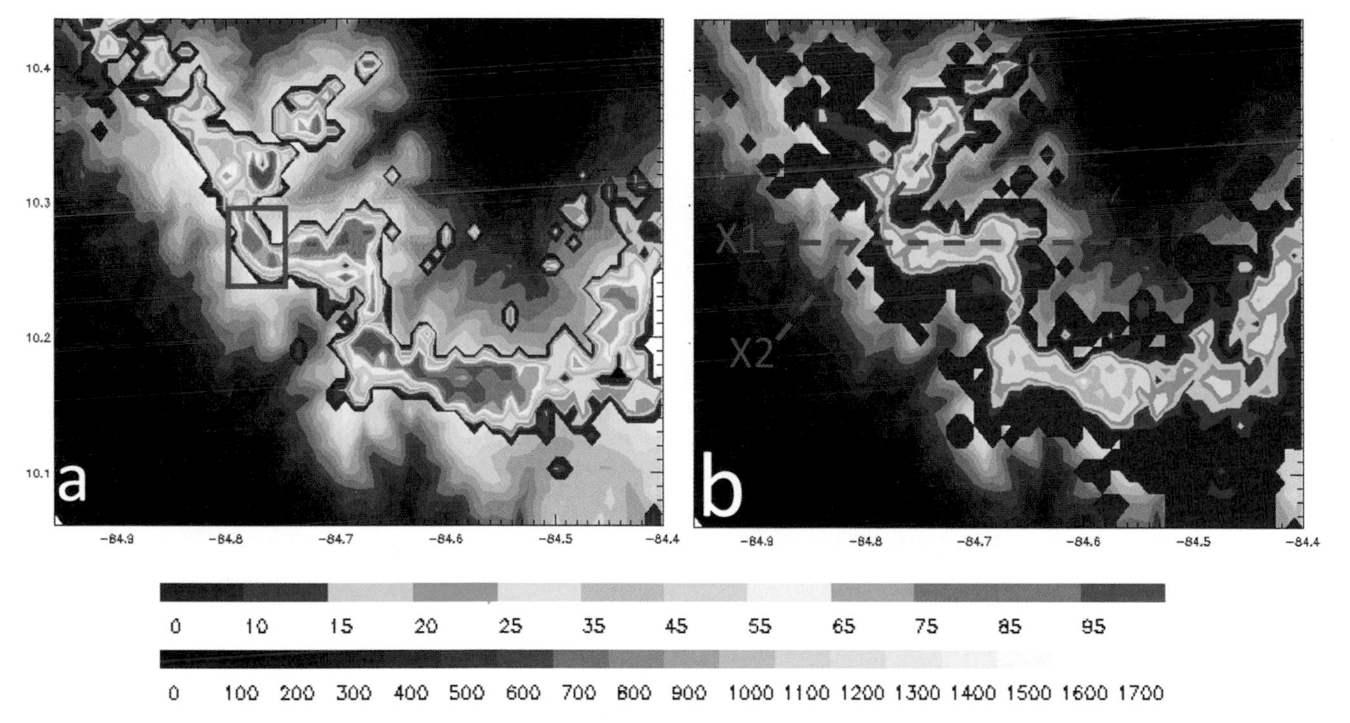

Fig. 13.9 Cloud immersion frequency over Monteverde derived from (a) MODIS data for March of 2003; and (b) MODIS data for March of 2003–2006. Note that the color shades show cloud immersion frequency in percent while gray shades depict topography (in m); the brighter the value the greater the elevation. Red rectangle on panel a shows the location of an averaging window used in validation studies. X1 and X2 in panel b indicate locations of cross-sections used to construct vertical profiles.
From Nair et al. 2008.

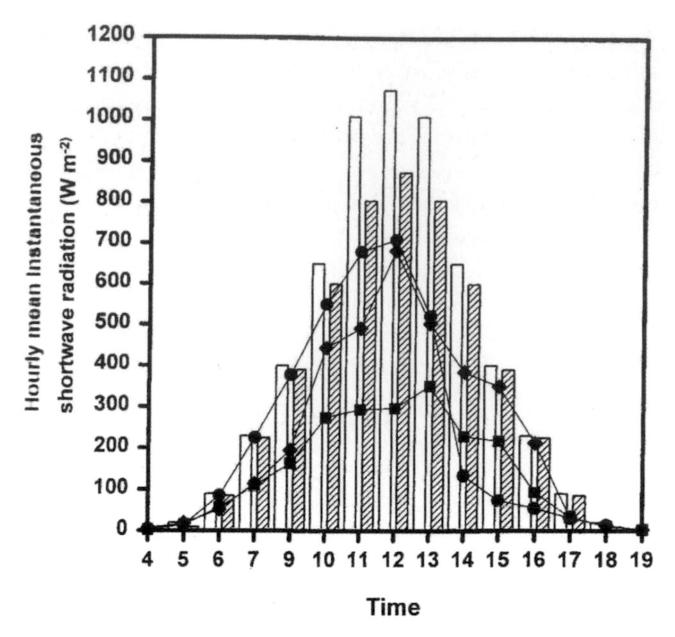

Fig. 13.10 Mean above-canopy shortwave radiation (W m²) calculated for clear sky conditions on April 21 and December 21, and measured in leeward cloud forest, 1.5 km from the continental divide, Monteverde Cloud Forest Reserve. Shown are values for days with wind-driven precipitation (■), with mist and cloud (♦), and with convectively produced rain (●).
From Clark et al. 2000.

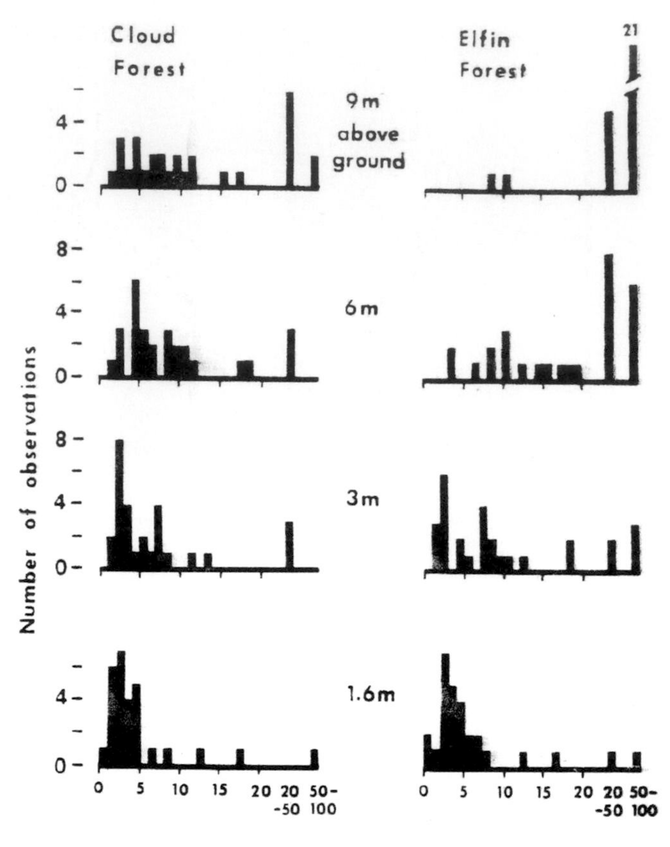

Fig. 13.11 Frequency distributions of proportions of above-canopy photosynthetic photon flux densities (PPFD) reaching four heights above the ground in elfin forest and taller cloud forest along the Continental Divide, Monteverde Cloud Forest Reserve. n = 28 at each height for each site.
From Lawton 1990.

influence the adjusted rain gauge catches and the estimates of canopy drip from captured horizontal precipitation, the areal precipitation values have considerable uncertainty (Bruijnzeel 2006).

The evapotranspiration values in Table 13.1 are quite low, since cloud cover reduces the net radiation, and the humidity is consistently high. Water yields as streamflow are high, but the apparent leakages calculated by solving the water balance equation are very large. Streamside springs in *quebradas* are common and often tapped by local farms and communities, so some of the leakage is clearly real, but some may be related to the uncertainties in estimating precipitation inputs (Bruijnzeel 2006). Baseflow in the stream draining the pasture catchment receded about twice as fast in the dry season as that from the forested catchment. The stormflow fraction, the proportion of event precipitation rapidly discharged as streamflow, from the pasture catchment is about twice that from the forested watershed (0.21 vs. 0.093), probably because soil compaction in the pasture decreases the rate of infiltration (Bruijnzeel 2006, Tobón-Marín et al. 2010).

Streamflow, in headwater creeks and major rivers, is strongly episodic; after heavy rainfalls rivers and creeks rise rapidly and *quebradas* grumble as boulders roll along river beds. Scaling up from small headwater catchments to operational watersheds involves extending the local models discussed above to a larger spatial scale. Fig. 13.13 shows measured and modeled hourly discharges of the Río Chiquito, which drains 91.4 km² on the northern end of the Cordillera de Tilarán.

Recently the stable isotopic composition of water in precipitation, the soil, and streamflow has been examined to clarify contributions of orographic and convectively produced rains to local water budgets on the upper Pacific slope of the Cordillera de Tilarán. Event to event variation in $\delta^{18}O$ and $\delta^{2}H$ is large, but monthly samples show a seasonal signal (Rhodes et al. 2006). Deuterium excess in the dry season suggests that evaporation from the Caribbean plains is important in the water supply (as wind-driven mist and cloud) to the Monteverde region. In the early rainy season the deuterium excess falls, but rises once more as the rainy season progresses. This suggests that several months of rain on the Pacific slope and lowlands are required to replenish soil waters before evapotranspiration on land produces a stable isotopic signal in rain water (Rhodes et al. 2006).

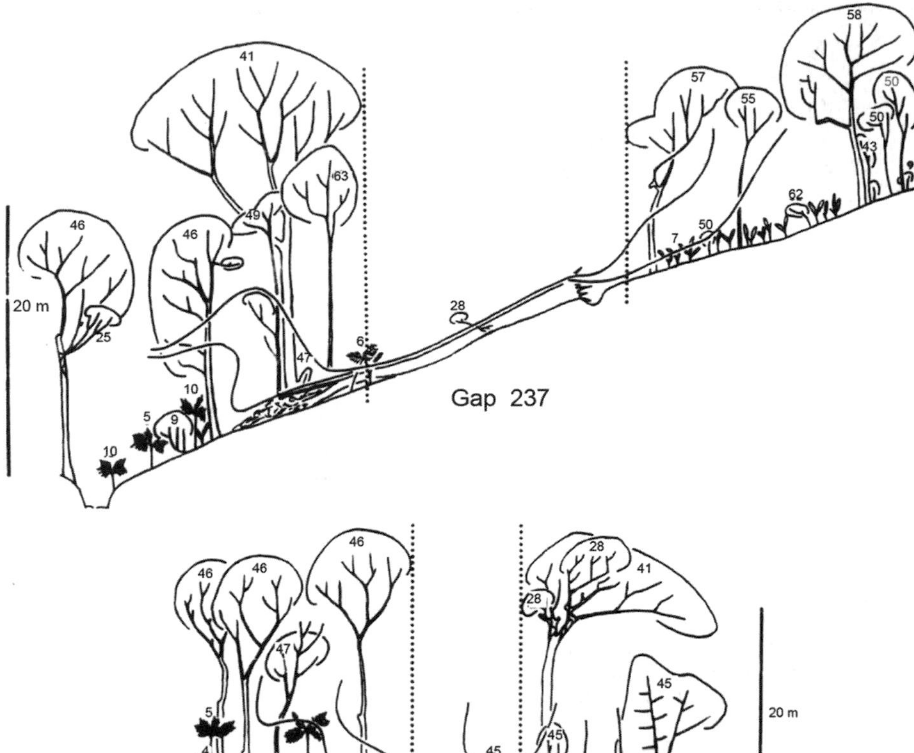

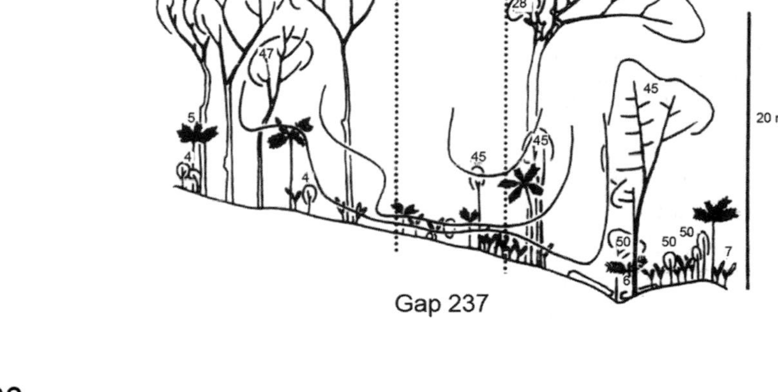

Fig. 13.12 Isoclines of the proportion of above-canopy photosynthetic photon flux densities (PPFD) and forest profile along the length and breadth of a treefall gap in cloud forest near the Continental Divide, Monteverde Cloud Forest Reserve. Isoclines are, from the bottom, 0.1, 0.25, and 0.75 of above-canopy PPFD. *From Lawton 1990.*

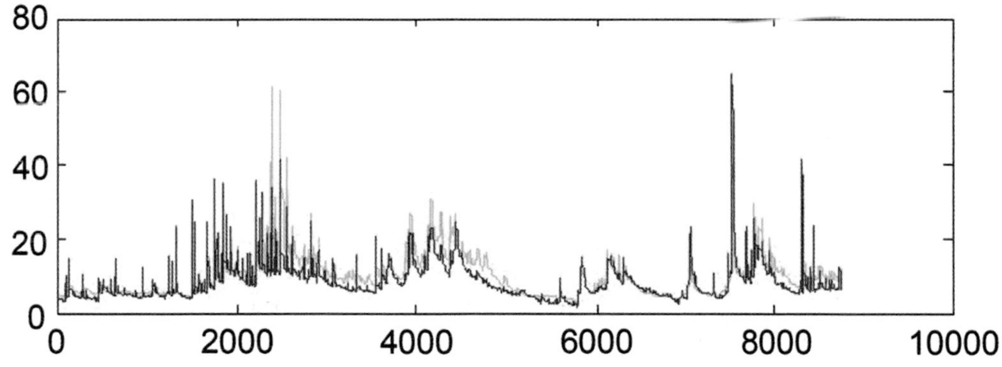

Fig. 13.13 Modeled (blue) and measured (red) stream discharge in mm hr^{-1} for the 91.4 km^2 Río Chiquito watershed, northern Cordillera de Tilarán, vs. time (in hours) between July 1, 2003, and July 1, 2004. *From Bruijnzeel 2006.*

Table 13.1. Water Budgets for Paired Lower Montane Rain Forest (3.5 ha at 1,450–1,600 m a.s.l.) and Pasture Small Headwater Catchments (8.7 ha at 1,520–1,620 m a.s.l.), Río Caño Negro Basin

Site	Precip.	Horizontal Precip.	Evapotranspiration	Streamflow	Soil Storage	Leakage
Forest	6,275[a]	685	785	2,735	+50	3,390
Pasture	6,500	815	855	2,950	+50	3,460

NOTE. Data are for the period July 1, 2003, through July 1, 2004. Values are mm yr^{-1}.

[a] Figure calculated from average net precipitation across catchment plus interception evaporation minus horizontal precipitation.

Source: Bruijnzeel 2006.

The seasonal stable isotopic signal does extend to streamflows, but is weakened (Guswa et al. 2007). Results from six catchments show that the contribution of orographic precipitation to stream baseflows, in both the dry season and the rainy season, decreases with distance from the crest of the Cordillera (Guswa et al. 2007).

Plants and Vegetation

The very diverse vascular flora of the Cordillera de Tilarán is remarkably well known, largely due to the extensive collecting coordinated by W. A. Haber (2000a,b). A checklist of the flora of the Monteverde region, probably representative of the mountain range as a whole, is provided by Haber (2000b). The vascular flora of the volcanic massifs of the C. de Guanacaste is less well known (Jiménez and Carrillo, chapter 11 of this volume); intensive floristic work there would clarify the role of arc volcanism in providing montane species stepping stones across one of the major lowland gaps in the chain of cordilleras linking northern Mesoamerica and the Andes. In particular, detailed floristic information in this setting would permit quantitative assessments of hypotheses concerning the impact of species attributes on the likelihood of colonizing isolated ranges.

Vegetation Types

The vegetation of the northern highland forests of Costa Rica is best known from extensive work in the reserve complex in the Monteverde region of the central Cordillera de Tilarán (Haber 2000). Vegetation zonation on the isolated volcanic peaks of the Cordillera de Guanacaste probably follows similar patterns, though altitudinal variation is to be expected. The vegetation of the Monteverde region spans seven Holdridge life zones, ranging from tropical moist forest on the lower Pacific slope and tropical wet forest on the lower Caribbean slope to lower montane rain forest along the continental divide (Tosi 1969). Although life zones are useful for the classification of habitat types, and numerous species are characteristic of the life zone within which they reside, boundaries are typically not well defined (see Haber 2000). Life zones in the Monteverde region are largely dependent upon the degree of exposure to the moisture-laden trade winds, which varies with topography as well as elevation.

Life Zones of the Cordillera de Tilarán

Briefly, the life zones are distributed as follows. Tropical moist forest (Premontane transition) is found on the Pacific slope below 600–800 m in elevation. This life zone experiences a dry season of 3 to 6 months. While the forest is mostly evergreen, deciduous species are common and conspicuous. Epiphytes are not a prominent component of the vegetation, although drought-hardy species such as epiphytic cacti and orchids with well-developed pseudobulbs are notable. Premontane moist forest is found on the Pacific slope between 700 and 1,100 m. This forest occupies the rain shadow in the lee of the higher peaks in the Monteverde area (Cerro Amigo, Cerro Chomogo, and Cerro Sin Nombre). The forest is generally evergreen, but with some conspicuous dry-season deciduous species. Epiphytes are not prominent, and those present are drought adapted as in the tropical moist forest below.

Premontane wet forest is found between 800 and 1,500 m on the Pacific slope and extends down as far as approximately 400 m on the Caribbean slope. The Caribbean slope forests are almost continually wet, while the Pacific slope forests experience a definite dry season. Premontane wet forest in the Monteverde area is an evergreen forest with few deciduous species and moderately abundant epiphytes.

Lower montane wet forest is restricted to the upper Pacific slopes between 1,450 and 1,600 m. This is an evergreen forest intermittently immersed in blowing cloud and mist during the dry season. Trees reach heights of 30–40 m and support many epiphytes.

Lower montane rain forest is found along the crest of the Cordillera from 1,500–1,850 m in elevation. These forests are commonly immersed in blowing mist and cloud. Epiphytes are diverse and abundant, often forming thickets on upper trunks and major limbs of canopy trees. Mats of bryophytes, filmy ferns, and the roots of vascular epiphytes trap an arboreal organic soil. Epiphyllic bryophytes are common.

Premontane rain forest, an evergreen forest with trees up to 30–40 m in height and abundant epiphytes, is found on the Caribbean slope between 700 and 1,400 m. Tropical wet forest, found below 700 m on the Caribbean slope, is evergreen, with canopy trees to 40–50 m. Epiphytes and lianas are abundant.

Distinctive Habitats

Within these life zones a series of distinctive habitats results from edaphic, microclimatic, and/or topographic factors (Lawton and Dryer 1980, Haber 2000). The most distinctive of these habitats in the Cordillera de Tilarán are (1) the steep, dry, rocky ridges descending the uppermost elements of the Aguacate Formation on the Pacific slope; (2) swampy areas of low relief on the Monteverde Formation above 1,500 m; and (3) the elfin forests on wind-exposed ridges and peaks along the crest of the Cordillera. Although these habitats are very distinctive in their extreme forms, they are not truly discrete, and grade into adjacent habitats.

Rocky ridges on the "cliff edge" formed on the Pacific slope by the uppermost member Aguacate formation are edaphically dry due to the rapid drainage and poor water storage capacity of the thin, rocky soils on steep slopes. These isolated sites are occupied by plant species, indeed a plant community, characteristic of the tropical dry forests found at much lower elevations. Dry forest trees, such as *Plumeria rubra* and *Euphorbia schlechtendalii*, are joined by xerophytes such as the large agave, *Furcraea cabuya*, the resurrection plant, *Selaginella pallescens*, and the succulent *Echeverria australis*.

Along the crest of the Cordillera de Tilarán, on level areas at the windward foot of steep slopes there are often swampy forests with seepage springs, commonly saturated soil, and standing pools in the broad but shallow pits created by uprooting trees. In areas protected from the wind there may be large trees, including *Sapium rigidifolium*, *Ocotea* cf. *viridiflora*, the oaks *Quercus corrugata* and *Q.* cf. *seemannii*, scattered in a very broken canopy often with smaller trees including the stilt-rooted *Chrysoclamys allenii*, *Viburnum venustum*, and the palm *Prestoea acuminata*. Owing to the broken canopy and extensive treefalls epiphytes are abundant and conspicuous in the understory.

Elfin forests occupy peaks, ridge crests, and the uppermost windward slopes along the crest of the Cordillera de Tilarán, and similar spots in the C. de Guanacaste, like the summit of Volcán Cacao, that haven't been recently disturbed by volcanic activity. These forests are composed of trees 3–10 m tall with crowns packed into a single, apparently wind-leveled canopy (Lawton and Dryer 1980, Lawton and Putz 1988). Many of the trees are species more widely distributed in the cloud forest as hemiepiphytes. This group includes *Schefflera rodriguesiana*, *Clusia* sp., *Cosmibuena valerii*, and *Oreopanax nubigenus*. The disturbance and regeneration dynamics of these forests are relatively well known (Lawton and Putz 1988; see below Disturbance and Regeneration).

Much of the variation in forest stature is associated with wind (Lawton and Dryer 1980, Lawton 1982, 1984, Häger 2006, Häger and Dohrenbusch 2009). Canopy height and canopy structure vary with exposure to the trade winds (Figs. 13.14–17), as do tree physiognomy (Lawton 1982), and wood densities (Lawton 1984). In general, those sites more exposed to wind support shorter forests with smoother upper canopies, composed of trees with denser wood, and thicker trunks, branches, and twigs than sites more protected from the wind.

There are also several distinctive habitats that are ephemeral, in the sense that they are created by large disturbances, and gradually lose their distinctive character in the processes of community regeneration. These include the cliffs created by landslides in steep-walled *quebradas*, riverside areas scoured in flood, and in the Cordillera de Guanacaste areas influenced by lava flows, pyroclastic flows, lahars, and debris avalanches. These are discussed below in Disturbance and Regeneration.

Plant Species Richness and Diversity

The Monteverde Flora Project, coordinated by W. A. Haber, has identified over 3,000 plant species that occur above 700 m in elevation on both slopes (Haber 2000). Although species richness at the 1 ha scale is lower in the montane forests of the Cordillera de Tilarán than in well-known neotropical lowland forests (Nadkarni et al. 1995, Haber 2000), the 350 km^2 of montane forest above 700 m elevation in the Monteverde area has a flora as rich as that of the area of 15,000 km^2 that belongs to Manú National Park, Peru (Foster 1990). The 58 km^2 above 1,200 m elevation in the Monteverde area supports a known vascular flora of 1,723 species (Haber 2000), while La Selva in

Fig. 13.14 IKONOS satellite image (1 m resolution) of forest canopy, summit of Cerro Roble (1,699 m), Monteverde Cloud Forest Reserve, central Cordillera de Tilarán. On the eastern, windward side (right) of the summit, the crown sizes are smaller and canopy roughness is less than on the western, leeward side.

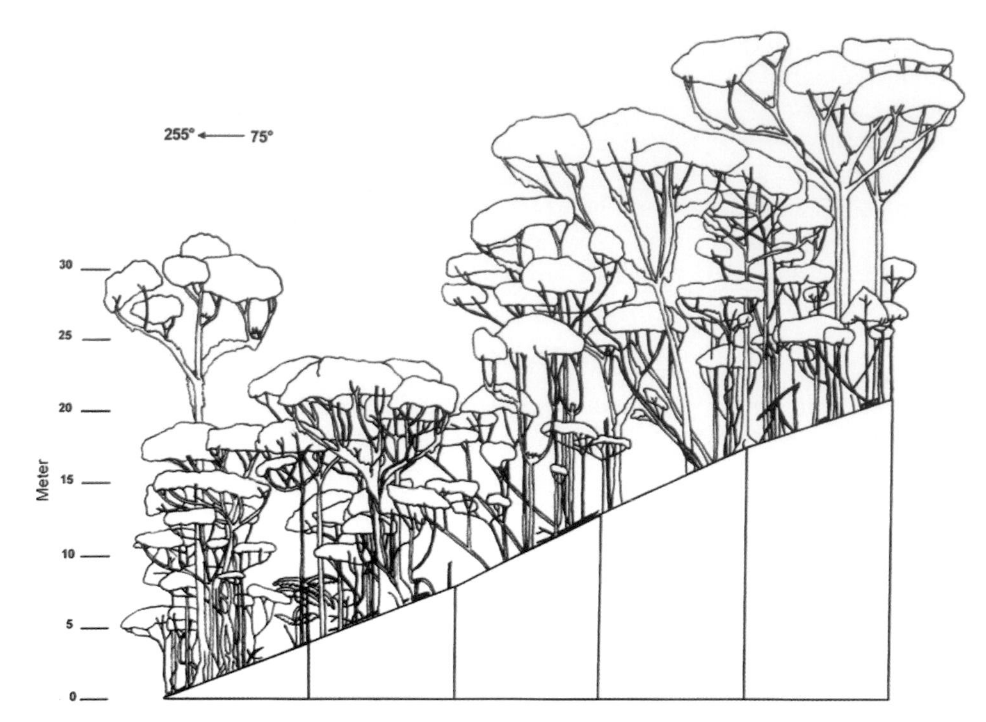

Fig. 13.15 Forest profile from 10 × 50 m plot at 1,200 m, upper San Luis valley 1.1 km to the lee of the continental divide, central Cordillera de Tilarán.
From Häger 2006.

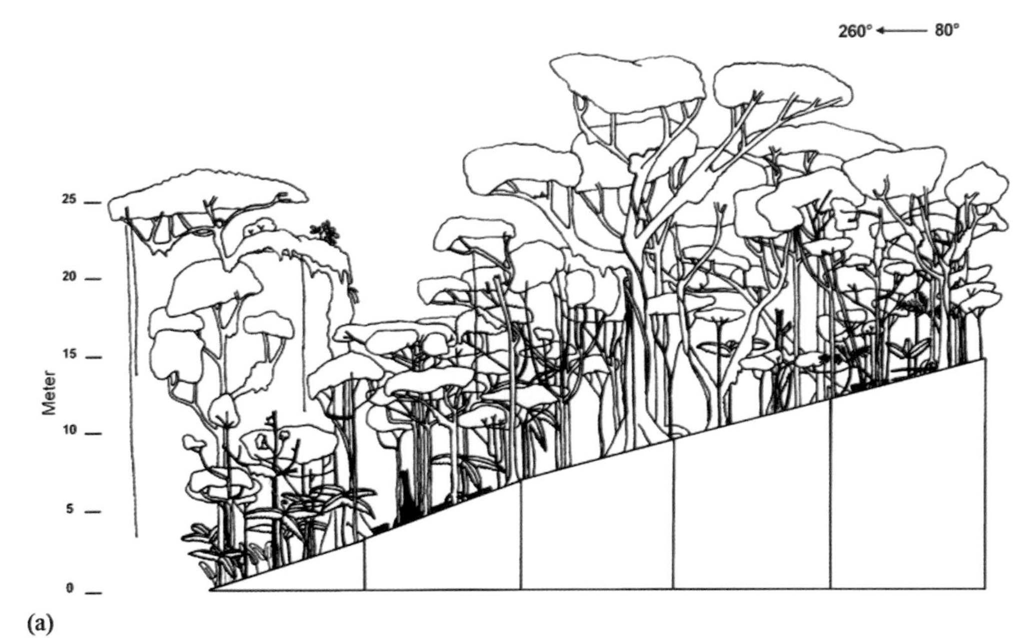

Fig. 13.16 Forest profiles from (a) 10 × 50 m plot at 1,450 m, Brillante saddle, 0.5 km to the lee of the continental divide, and (b) 10 × 50 m plot at 1,450 m, Brillante saddle, on the continental divide, central Cordillera de Tilarán.
From Häger 2006.

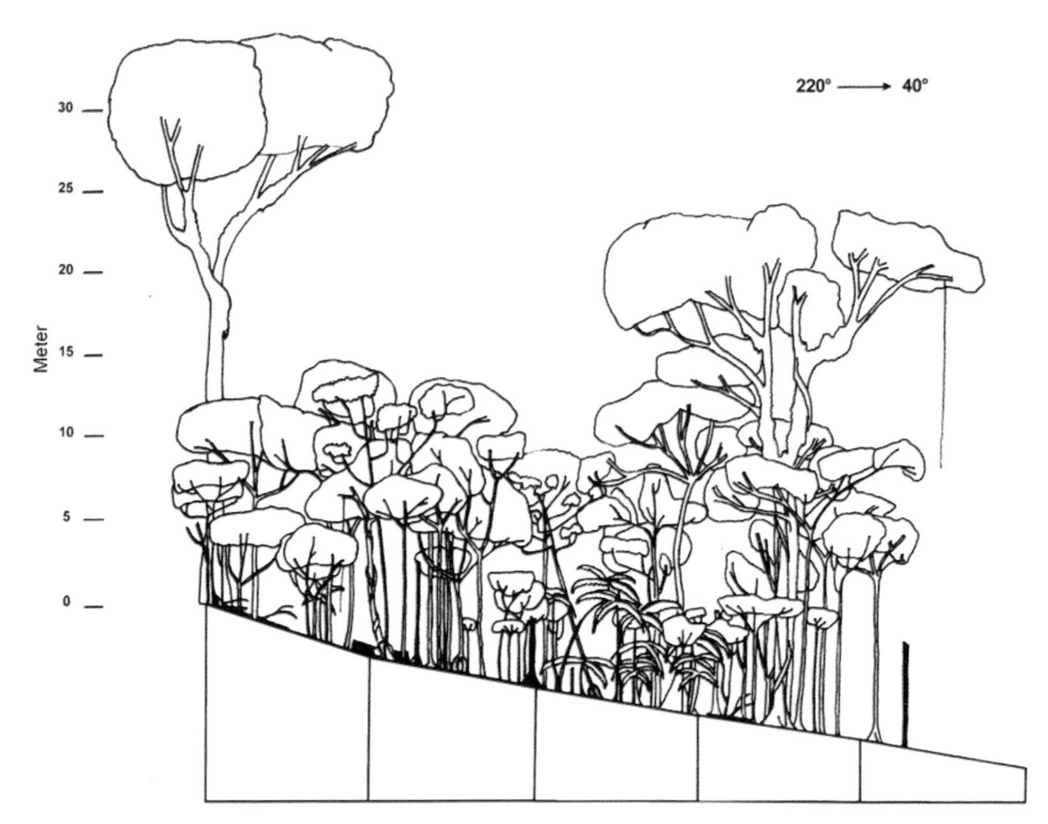

Fig. 13.17 Forest profile from 10 × 50 m plot at 1200 m, upper Peñas Blancas valley 1 km to the windward of the continental divide, Caribbean slope of the central Cordillera de Tilarán.
From Häger 2006.

Atlantic Costa Rica on 16 km^2 has a known flora of 1,678 species (Hartshorn and Hammel 1994, and see McClearn et al. chapter 16 of this volume), Barro Colorado Island, Panama, on 16 km^2 has 1,207 species (Croat 1978), and the Adolpho Ducke Reserve at Manaus, Brazil, on 100 km^2 has ~1,030 known species (Prance 1990). Since local (1 ha) diversity at Monteverde is low compared to lowland sites, but regional (10–100 km^2) diversity is comparable, there must be striking differences in the partitioning of diversity among life forms or in the way species sort themselves across the landscape, or both.

Growth Form Diversity

The diversity of plant growth forms in the cloud forests of northern Costa Rica is extraordinary, largely because regular immersion in cloud and blowing mist permits epiphytic establishment of an array of bryophytes, vascular herbs, woody shrubs, lianas, hemiepiphytes, and stranglers. In this regard cloud forests support a larger collection of life forms than do lowland forests, or even forests lower on the mountain slopes. In addition, the often broken canopy of cloud forests battered by winter storms favors development of a dense understory of giant herbs, herbaceous and woody shrubs and treelets, and palms. Finally, the biomechanical circumstances of the canopy trees result in an interesting array of trunk morphologies, in which buttressing, stilt-rooting, and the production of multiple trunks are prominent.

The bryophytes of the Monteverde cloud forests include *Anthoceros* sp., thallose liverworts of various sizes, leafy liverworts, both pendent and creeping, acrocarpous and pleurocarpous mosses, all growing in a variety of sites, including bare soil, logs on the forest floor, boulders in creeks and emergent on hillsides, tree trunks, limbs, twigs, and leaves. Cole (1983, 1984) provides guides to the Costa Rican Hepaticae, and Griffin and Morales (1983) offer a key to the mosses. Given their diversity and prominence, ecological investigations of cloud forest bryophytes are too few. At Monteverde the communities of epiphyllous bryophytes on fronds of the understory palm *Geonoma surtuba* have been investigated by Daniels (1998), and the role of bryophyte mats in the establishment of "epiphytic soils" and nutrient cycling has been examined in some detail (Nadkarni et al. 2000), but much remains to be learned about cloud forest bryophytes. Since even the cloud forest strangler fig, *Ficus crassiuscula*, exhibits host preferences, the issue of host and habitat preferences among the bryophytes begs investigation. Indeed, Sillet et al. (1995) found differences in bryophyte communities between crowns of *Ficus tuerckheimii* isolated in pastures and those that were in intact forest.

In addition to the epiphytic bryophytes there is an amazing array of epiphytic herbs and woody plants. Orchids alone account for more than 400 species in the montane forests of the C. de Tilarán, and many of the more than 200 species of ferns are epiphytes, including the filmy ferns that with bryophytes are major components of the epiphyte mats that envelope the trunks and major limbs of many of the cloud forest trees. Ferns range in size from species with pendent fronds 2 m long to *Grammitis* spp. with fronds a few cm long. The epiphyte mats with their organic soil host many microarthropods (Nadkarni and Longino 1990), and some of these are captured by the root traps of the carnivorous *Utricularia alpina*. Ericaceae provide many shrubs to the epiphytic vegetation, from 0.5 m tall *Vaccinium* spp. to *Gonocalyx costaricensis* 5 m tall.

As with the bryophytes much remains to be learned about how these communities vary with environmental gradients ranging from the large-scale geophysical gradient across the mountain range from the seasonally dry leeward slope to the ever wet forest on the windward slope, to small-scale gradients from ground to uppermost canopy, and to the qualitative differences among host species.

In cloud forests hemiepiphytes to some degree replace lianas as a life history solution to the problems of climbing upon trees (Williams-Linera and Lawton 1995). These hemiepiphytes include many herbaceous aroids, some of which are primary hemiepiphytes in that they first become established as epiphytes then extend roots to the ground, and others of which are secondary hemiepiphytes in that they first become established on the ground and subsequently grow up tree trunks. The original connection of the latter may be lost, and then replaced by hanging roots. Some of these aroids may reach large sizes; mature adult *Philodendron thalassicum* have thick creeping stems with short (2–5 cm) internodes given rise to a half-dozen leaves with 70–90 cm laminae and petioles of similar length.

The greatest diversity of growth forms is, however, among woody hemiepiphytes. Some grow erect, perched on host limbs or major crotches; of these some are shrubs and others reach tree size. Hemiepiphytic *Clusia* spp., *Schefflera rodrigueziana*, *Oreopanax nubigenus*, and *Cosmibuena valerii* may at maturity have trunks 20–30 cm in diameter and crowns >10 m broad. Other hemiepiphytes, like species of the melastome genera *Blakea* and *Topobea*, produce multiple trunked, shelf-like crowns extending 5 m or more from the trunks of their hosts. Some hemiepiphytes, such as the large herbaceous shrub *Begonia estrellensis*, are pendent from the limbs of their hosts.

The line between lianas and hemiepiphytes is often blurred in cloud forest. Plants like *Marcgravia brownei* and *Hydrangea peruviana* are scrambling climbers, with feeding roots running down the host trunk to the ground, and multiple points at which the clustered stems comprising the crowns are lashed to the host with webs of attachment roots (Williams-Linera and Lawton 1995). Such scramblers may also have specialized colonizing or traveling stems, with very long internodes and reduced leaves, and functioning much like stolons, but running as far as 10 or 15 m between adjacent tree crowns.

Stranglers are conspicuous in the montane forests of northern Costa Rica. Most are urostigmoid figs, the classical tropical strangler figs; in the Monteverde region, there are six species. The cloud forest habitat, however, seems to favor the evolution of strangling, probably because juvenile establishment is relatively easy in the well-developed and nearly permanently moist epiphytic soils. A fig in the subgenus Pharmacosycea, *F. crassiuscula*, has independently evolved the strangling habit from a free-standing ancestor (Lawton 1986, 1991). *F. crassiuscula* begins life as a sprawling vine growing on tree trunks and logs, and at this stage exhibits no host preferences. In successful individuals a single leader metamorphoses into an erect, free-standing stem, usually occurring 5–12 m up the trunk of a host canopy tree. This occurs more commonly on species of *Guarea*, particularly *G. kunthiana*, than the abundance of these host species would suggest (Daniels and Lawton 1991). Subsequent growth to maturity involves extension of the trunk above that of the host, expansion of the crown, and thickening and limited ramification of the roots into a strangling "trunk"; this too is more likely than expected on *Guarea*. In addition to the Urostigma figs and *F. crassiuscula*, a number of other tree species sometimes successfully strangle their hosts. Some are more commonly free-standing trees that facultatively grow as hemiepiphytes, but may outgrow their hosts and become self-supporting. *Weinmannia pinnata* is the most conspicuous of this group. Others are commonly hemiepiphytic, but may also outgrow their host. This is most common in the elfin forests, where this group includes *Clusia* sp., *Schefflera rodriguesiana* (= *Didymopanax pittieri*), *Cosmibuena valerii*, and *Oreopanax nubigenus*.

Quantitative comparisons of growth form diversity among habitats will require the development of clearly defined measures of growth form characteristics.

Phenology

Patterns of fruiting and flowering have wide-ranging impacts on other species. Groups as disparate as pollinators, seed dispersers, and herbivores often structure their reproductive efforts to take advantage of the predictable patterns of increased resource availability that fruits and flowers represent. In the Monteverde area at least 10% of tree species (~60 species) are fruiting or flowering in any given month

(Haber 2000). Many species (~100) bloom in the late dry season (March-May) to take advantage of the clear dry season days that make it easy for pollinators (and/or pollen) to fly, and to use the increased moisture availability of the upcoming rainy season to aid in fruit/seed set and maturation. Likewise there is a flowering minimum during the times where pollinator/pollen movement would be difficult: during the heavy rains from September to November, and during the high winds of December (Haber 2000).

Most species (excepting figs) flower and fruit only once a year, at about the same time each year (Haber 2000). Many species, however, do respond to variation in the weather. During El Niño years some species shifted flowering by 6 months, others flowered twice, and still others did not flower at all (Haber 2000). The impacts of a changing climate on patterns of flowering and fruiting remain to be explored.

Animals

Our knowledge of animals in the mountains of northern Costa Rica is the product of a long history of collection, and a diverse series of studies with particular aims rather than systematic efforts to assess the fauna as a whole. Those studies in the Cordillera de Tilarán have been well summarized in the chapters of Nadkarni and Wheelwright (2000). Our aim here is to summarize more recent research, and briefly present some of the important elements of the earlier studies, particularly as they illustrate how species activities, ranges, and interactions might be influenced by changing environmental circumstances.

Arthropods

Arthropods are staggeringly diverse. Our current knowledge, however, of the natural history, ecology, evolution, and distributions of arthropod taxa is relatively sparse. This is no less true in the mountains of northern Costa Rica than in the rest of the world. The majority of current knowledge of arthropods in this region consists of small studies of particular arthropod systems, much of which is nicely reviewed in Hanson (2000). The predominant theme in the literature seems to be lamentation about the lack of extensive sampling of arthropods. No large-scale sampling such as the ALAS project in La Selva, Costa Rica, has been conducted along cloud forest elevational gradients in northern Costa Rica. However, some taxa have received careful scrutiny. For example, Longino (2006) recently carried out a detailed examination of the neotropical arboreal ant genus *Myrmelachista*. Much of what is known about montane arthropods in northern Costa Rica has resulted from work conducted in Monteverde (750–1,850 m) and Zurquí de Moravia (1,600 m).

One of the things that sets montane cloud forest apart from lowland forest is a huge increase in epiphytic biomass. These epiphytic mats provide a very interesting microhabitat in which arthropods can specialize. Montane cloud forests in northern Costa Rica have a rich diversity of arboreal arthropods that live in epiphytic mats (Hanson 2000). Examples of recent work that has examined these arboreal arthropod communities in Monteverde are Yanoviak et al. (2004, 2007) and Schonberg et al. (2004). Yanoviak et al. (2004) examined the differences in arthropod communities in vegetative and humic portions of epiphytic mats. Schonberg et al. (2004) investigated the differences between arboreal ant assemblages in primary forest, secondary forest, and pasture trees, finding that arboreal ant species density is reduced as primary forest is converted to secondary forest and that pasture trees may serve as repositories for primary forest arboreal ant assemblages.

Though arthropod diversity generally decreases with increasing altitude, the mountains of Costa Rica are by no means depauperate. There is a large diversity of known arthropods in these mountains, and many of the more cryptic species as well as those living in habitat that is harder to access undoubtedly remain without sampling. This does not mean, however, that there is a paucity of interesting arthropods that have been collected. Among the more exotic known arthropods in the Cordillera de Tilarán is the grasshopper *Tropidacris cristata*, which has a body up to 15 cm in length, making it one of the largest grasshoppers in the world (Hanson 2000).

Altitudinal gradients have been shown to be important for arthropods in Costa Rica. For example, O'Donnell and Kumar (2006) recently showed that increasing elevation in Monteverde affects army ant community structure and behavior. Army ants (Ecitoninae, and behaviorally convergent Ponerinae) form large foraging groups that raid the forest floor, canopy, and human habitations. The rate of these raids, which appear to drive a number of interesting ecological interactions with birds and other organisms, decreases with increasing altitude. The strong altitudinal gradients in Costa Rican mountains are also very important for butterflies, over half of which are altitudinal migrants in Monteverde (360 out of 658 species) (Hanson 2000).

As most arthropods have relatively narrow altitudinal distributions they may be quite susceptible to climate change. The increasing frequency of dry spells in the dry season and lifting cloud base heights could both affect arthropod distributions and behaviors (Lawton et al. 2001). As moisture and temperature gradients change with changing regional climate, the distributions of many arthropods will shift as well. Yanoviak et al. (2007) recently found that arthropod morphospecies richness in epiphytic mats decreases in the dry season, presumably because desiccation-sensitive arthropods are going dormant or hiding very deep in the mats. Longer dry spells in the dry season could substantially alter the wetting and drying cycle of these epiphyte habitats and thereby disrupt the population and community dynamics of arboreal arthropods. Another group particularly sensitive to this sort of change may be migratory insects such as butterflies, which depend on two or more habitat types at different altitudes (Hanson 2000).

Amphibians and Reptiles

Studies in the cloud forests of northern Costa Rica have strongly influenced the ongoing discussion about the causes of the global amphibian decline (Pounds and Crump 1994, Pounds et al. 1997, 1999, 2005, Pounds 2000, 2001, Pounds and Puschendorf 2004). Such studies illustrate how important it is to know local environments and patterns of species distribution in order to interpret ecological trends at larger spatial and temporal scales.

Herpetofaunal Diversity

Knowledge of the herpetofauna rests on the work of Van Devender (1980) and Hayes et al. (1989). An updated checklist is provided by Pounds and Fogden (2000). From montane habitats in the Cordillera de Tilarán 161 species have been recorded—60 amphibians (including 2 caecilians, 5 salamanders, and 53 anurans) and 100 reptiles (29 lizards and 71 snakes). This listing is probably incomplete for several reasons (Pounds 2000). First, there has been less collecting in peripheral and hard-to-reach areas. Second, secretive taxa, and those that occur at low density, are less likely to be encountered. Finally, some diverse and numerous groups, for example, the genus *Eleutherodactylus*, may contain as yet unrecognized species.

Hayes et al. (1989) have described six zones in the Monteverde area marked by distinctive herpetological communities. These are related to elevation, and distributed as follows: Zone 1 on the lower Pacific slope from 690 to 1,300 m, Zone 2 on the mid-Pacific slope from 1,300 to 1,450 m, Zone 3 from 1,450 to 1,600 m on the upper Pacific slope, Zone 4 from 1,850 to 1,450 on the Caribbean slope (a zone commonly called "Continental Divide"), Zone 5 from 1,450 to 950 m on the upper Caribbean slope, and Zone 6 from 950 to 600 m on the lower Caribbean slope. From Zones 3 and 4, a total of 79 species of amphibians and reptiles are known, compared to 135 in a similar area at La Selva. The 161 species known above 600 m in the Cordillera compares well, however, with the lowland fauna from La Selva, which emphasizes the importance of the diversity of local environments in the maintenance of herpetological diversity (Pounds 2000). No zone has a species list containing more than 52% of the species known from the range. Sixty-four percent of the species are found on only one side of the range; 60% of the snakes are found on the Pacific slope, while only 30% of the amphibians are.

Amphibian Declines

Early studies of anuran population collapses at Monteverde focused on the golden toad, *Bufo periglenes*, and the harlequin frog, *Atelopus varius* (Crump et al. 1992, Pounds and Crump 1994). Following the initial collapse in 1987/88, a 2 x 15 km transect across the Cordillera de Tilarán was established in 1990 to monitor the herpetofauna (Pounds et al. 1997). From the pre-collapse information in Hayes et al. (1989), 50 anuran species were expected, but 25 of these were missing in 1990, although 5 reappeared in 1991–1994, and one reappeared in 1997. These "reappearances" may be the result of colonization from outlying areas, since the reappearances were on the periphery of the study area, and 5 of the 6 were species found also in the lowlands, where there is little evidence of population collapses (Pounds 2000).

Early discussions of the causes of the amphibian declines have been reviewed by Pounds (2000). Recently attention has been focused on the role of an emergent epidemic caused by the chytrid fungus, *Batrachochytrium dendrobatidis* (Pounds et al. 2005, Lips et al. 2006).

Pounds et al. (2005) suggest that, as a result of global warming, tropical montane climates are changing in ways that both stress anurans, increasing their susceptibility to infection, and favor the growth of *Batrachochytrium*. In particular, they note that, although diurnal temperatures have declined in many neotropical cloud forest settings, nocturnal temperatures have risen more, increasing the mean temperature and decreasing the diurnal temperature range. This could be caused by an increase in the base height of orographic cloud banks and an increase in cloud cover. This combination would result in drier montane habitats with cooler days and warmer nights. At this point, coupled epidemiological and cloud climatological studies are needed to investigate the coupling of chytridiomycosis outbreaks and changing local climates in tropical mountains.

Interestingly, some reptile populations declined along with the anurans in the Cordillera de Tilarán (Pounds 2000). Some of this is clearly due to trophic connections. The disappearances in 1987 of *Drymobius melanotropis* (the green frog eater), the fourth most commonly seen snake in the Peñas Blancas valley on the Caribbean slope, and in

1988 of *Chironius exoletus* (the green keelback), the third most commonly seen snake of the upper Pacific slopes, are undoubtedly due to the loss of their primary prey (Pounds 2000). More intriguing are the declines in two anoline lizards, *Norops altae* and *N. tropidolepis* (Pounds 2000). The former originally occupied zones 3 and 4, while the latter was found in zones 2–4. Both were common in primary forest at 1,540 m on the upper Pacific slope, but disappeared at that elevation, although they persist at higher sites. Their local decline and retreat to higher elevations is not directly caused by a chytrid outbreak, nor is it due to a loss of anuran prey. *N. tropidolepis* lives in shaded understory and its body temperature follows ambient air temperatures (Pounds 1988). It is active at the low temperatures of shaded cloud forest interiors, but feeds at low rates, and grows slowly. Females take nine months to reach maturity at 42 mm (snout-vent), while females of *N. intermedius*, found at warmer sites at lower elevations, reach maturity at 39 mm in only four months (Fitch 1973). The slow growth and longer generation time would hinder population recovery following collapse, but the primary factor prompting the retreat to higher elevations remains unclear. Clearly, detailed local demographic studies will be required to clarify the responses of such animals to changing climatic conditions.

Birds

Most of the research on birds in the montane areas of northern Costa Rica has been concentrated in the Monteverde area of the central Cordillera de Tilarán. The local topographic diversity and microhabitat variety found within a small area—about 20 km^2—has given ornithologists access to an area of high α- and β-diversity. Within the Monteverde area 6 Holdridge life zones are home to 425 bird species (Young and McDonald 2000, Fogden 2000). The richest life zone, Premontane Rain Forest, has 315 species, while the least rich, Lower Montane Rain Forest, has only 121, perhaps in part because of the limited area along the crest of the Cordillera in this life zone. Changes in the altitudinal ranges of bird species have contributed to the formation of the "lifting cloud base hypothesis" concerning faunal changes in the mountains of northern Costa Rica (Pounds et al. 1999), and have prompted renewed interest in the way bird community structure changes along the steep environmental gradients on the upper lee (Pacific) slope of the Cordillera de Tilarán (Jankowski 2004, Jankowski and Rabenold 2007). Auditory species counts and capture data show nearly complete turnover in avian community composition in the upper 600 m in elevation of the upper Pacific slope—a distance of ~3 km from the continental divide (Jankowski 2004). Transition in community composition is more gradual on the upper Caribbean slope, where there is a 32% Sørenson's similarity between opposite ends of a 700 m elevation transect. This contrast between the slopes appears due to a more rapid transition in dry season moisture availability, and thus a more rapid transition from cloud forest to drier forest types on the upper Pacific slope.

Much of the ornithological work in the Monteverde area has focused on birds as pollinators or seed dispersers; this work is discussed below in Pollination and in Seed Dispersal. Since many of the hummingbirds and avian frugivores are altitudinal migrants, the problems of habitat fragmentation, both on the Pacific slope of the Cordilleras and in adjacent lowlands, have been a matter of concern. There is considerable evidence that deforestation in seasonally critical habitats has put several local migrants at risk (Powell and Bjork 1994, Powell et al. 2000), and that forest fragments may play a vital role in maintaining populations of large obligate frugivores, such as the three-wattled bellbird and the resplendent quetzal (Guindon 1996, 2000).

There has, in addition, been considerable work on mixed species foraging flocks (Buskirk 1976, Powell 1979, 1980, 1985, Valburg 1992, 2000, Shopland 2000), and autecological investigations of assorted species—for example, the house wren (Young 1993, 1994a,b, 1996, Winnett-Murray 2000), the emerald toucanet (Riley 1986, Riley and Smith 1992), and the yellow-throated euphonia (Sargent 1993).

Long-term investigations have revealed much about cooperative behaviors in two systems: cooperative breeding of the brown jay (*Cyanocorax morio*), a social corvid (Lawton and Lawton 1980, 1985, 1986, Lawton and Guindon 1981, Lawton 1983, Williams et al. 1994, Williams and Lawton 2000, Hale et al. 2003, Williams 2004, Williams et al. 2004, Williams and Rabenold 2005, Williams and Hale 2006), and coordinated lek mating displays by male long-tailed manakins (*Chiroxiphia linearis*) (McDonald 1989a,b, 1993a,b, 2000, 2007, Trainer and McDonald 1993, 1995, McDonald and Potts 1994).

Study of the nesting behavior of brown jays led Skutch (1935) to describe cooperative breeding, in which nonbreeding members of a group help the breeders raise their young. Cooperative breeding and helping at the nest excited much interest, due to the evolutionary questions raised by seemingly altruistic behaviors (Lawton 1982). A series of studies at Monteverde have clarified the social structure, the activities of helpers, the genetic structure of groups, the population and group dynamics, and the nature of competition for breeding status. Brown jays live in large, territorial flocks (6–20 birds), in which all members collaborate to build the nest; feed the breeding females, nestlings, and fledglings; defend the territory; and harass potential preda-

Competitive Interactions

In the tree flora and avifauna there are a number of examples of closely related species, the ranges of which are spatially displaced, with relatively little overlap (e.g., the species of *Beilschmiedia* and *Persea* among the canopy trees, or the slate-throated and collared redstarts). Since bird ranges are changing as the regional climate changes (Pounds et al. 1999), and tree seedlings, at least, are amenable to transplant experiments, the opportunities for observational and manipulative studies of competitive interaction would seem abundant.

Predator-Prey Interactions

Studies of predation have been limited to those of the pasture pest mentioned above, and scattered observations produced in the course of other studies. Wenny (2000a), for example, found that post-dispersal rodent predation killed 98% of the seeds of *Ocotea endresiana*, 80% of those of *Guarea kunthiana*, 50% of those of *Eugenia* sp., 20% of those of *Beilschmiedia pendula*, and none of those of *Meliosma vernicosa*. And Sargent (1995) found that a captive harvest mouse readily ate mistletoe seeds collected from bird feces, perhaps explaining the 60% of dispersed mistletoe seeds that subsequently disappear. Pathogens undoubtedly influence plant and animal populations; according to settlers' accounts, the Central American yellow fever epidemic in the early 1950s caused dramatic declines in the abundance of the three monkeys in the Cordillera de Tilarán, and more recently *Impatiens sultana*, an invasive plant becoming alarmingly widespread in the range, suffered almost complete adult mortality, apparently due to an epidemic plant pathogen. But even the most basic surveys remain to be done.

Plant secondary chemistry is often related to defense against herbivores or pathogens, but chemical ecology has been explored in only a few cases (Lawton et al. 1993, Murray et al. 1994), although a series of studies exploring the phytomedicinal potential of the montane flora show a remarkable richness in secondary chemistry (Setzer et al. 1992, 1995a,b, 1998, 1999, 2000a,b, 2003, 2005, 2006, Moriarity et al. 1998, Setzer 2000). The richness of the secondary chemistry suggests that much interesting, but unexamined, chemical ecology exists.

Mutualisms

Species interactions in pollination and seed dispersal are among the best-studied elements of the biology of the montane forests in northern Costa Rica. Particular attention has been paid to the complex among the hummingbirds and the plants they pollinate, and to the large, obligately frugivorious birds, such as quetzals, and plants with specialized fruits, like the Lauraceae.

Pollination

From floral characteristics the broad outline of which plants are pollinated by what groups of pollinators is now clear (Haber 2000b, Murray et al. 2000). Approximately 9% of the flora, for example, is pollinated by hummingbirds (including 25% of the epiphytes, excluding orchids). Studies of pollination biology in the Monteverde area have been ably summarized by Murray et al. (2000). These studies have focused largely on hummingbird pollination, although several other systems have received attention. These include the pollination biology of *Ficus pertusa* and its fig wasps, pollination of the heat-generating aroid *Xanthosoma robustum* by scarab beetles (Goldwasser 2000), buzz-pollination of *Saurauia* spp. by bumblebees (Cane 1993), deceit pollination of *Begonia* (Agren and Schemske 1991), alteration of floral sex of *Centropogon solanifolius* by anther consuming larvae of *Zygothrica neolinea*, a drosophilid fly (Weiss 1996), and protection of the nectar of the hawkmoth-pollinated *Guettarda poasana* from yeasts by floral essential oils containing aromatic alcohols (Lawton et al. 1993).

The 30 species of hummingbirds and the plants they visit have been extensively investigated (Feinsinger 1976, 1978, 1987, Feinsinger and Busby 1987, Feinsinger et al. 1986, 1987, 1988a,b, 1991, 1992, Lackie et al. 1986, Linhart et al. 1987a,b, Murray et al. 1987, Feinsinger and Tiebout 1991, Tiebout 1991, Podolsky 1992). There exist two guilds of hummingbirds and flowers they pollinate on the crest and upper Pacific slope of the Cordillera de Tilarán: hummingbirds with long curved bills pollinating flowers with long, curved tubular corollae, and hummingbirds with shorter, straight bills pollinating flowers with shorter, straight tubular corollae (Feinsinger 1976, 1978). Agricultural colonization with large-scale conversion of forest to a patchy landscape of pasture, cropping systems, logged woodlots, early second-growth, and extensive edge habitats produced marked changes in the hummingbird community composition (Feinsinger 1976, 1978). Natural disturbances, however, seem not to cause large changes in hummingbird resource abundance, although a few hummingbird-pollinated plants are more abundant in gaps (Feinsinger et al. 1987, 1988b). Availability and diversity of floral resources, as well as hummingbird visitation and estimated demand on those resources, have been examined to assess diet breadth and among species overlaps in diet. Hummingbird visitation of flowers of particular species, pollen loads, and pollen deposits were similar in forest, tree fall gaps, and forest clearings cut to resemble small landslides, as, in general, were diet breadths and diet overlaps among hummingbirds (Feinsinger et al. 1988b). The differences in the response of the community of hummingbirds and the plants they pollinate to natural vs. anthropogenic disturbance points out that ecological impacts of human landscape changes may not be predicted by simply scaling up from responses to natural disturbances.

Much of this work focused on plant-plant interactions via pollinator sharing. Among the plants pollinated by each guild of hummingbirds most co-occur spatially and overlap in time of blooming, but plant species that share pollinators did not show consistent patterns of competitive interaction (Feinsinger et al. 1986). Aviary experiments did show that pollen loss can occur when hummingbirds feeding at one flower visit a different species before returning to the first (Feinsinger and Busby 1987, Feinsinger et al. 1988b), but in most cases in the field pollination success was uninfluenced by presence or density of neighboring hummingbird pollinated plants. When there were significant interactions some were competitive and others facilitative.

Nearly half of the canopy tree species, including the Lauraceae, have inflorescences with many non-descript, small, open, pale flowers, which are visited by a large array of small insects, including bees and wasps, beetles, flies, and butterflies (Haber 2000b, Murray et al. 2000). Given the diverse tree flora, many species of which occur in replacement series of closely related species, each occupying a narrow band along the gradient of moisture on the upper Pacific slope, this assemblage of generalized flowers and pollinators demands careful quantitative study.

Seed Dispersal

Most seed dispersal syndromes are present in the montane flora of northern Costa Rica, but some—for example, dispersal by water—are rare. Striking differences in the spectra of dispersal systems exist among major plant life forms: considering just the two dominant dispersal modes for each, ~65% of trees are bird dispersed and about 10% are wind dispersed, ~40% of lianas are bird dispersed and ~10% are wind dispersed, slightly more than half the shrubs are bird dispersed and ~15% are wind dispersed, ~40% of terrestrial herbs are wind dispersed and 27% are gravity dispersed, and ~65% of epiphytes are wind dispersed and a little over 25% are bird dispersed (Haber 2000b, Murray et al. 2000). Bats, arboreal mammals, terrestrial mammals, ants, and explosive dehiscence account for lesser amounts of seed dispersal in most groups.

Seed dispersal by birds has received a great deal of attention in the Monteverde area (Wheelwright 1983, 1985a,b, 1986, 1991, 1993, Wheelwright and Bruneau 1992, Wheelwright et al. 1984, Murray 1986, 1988, Murray et al. 1994, Bronstein and Hoffman 1987, Nadkarni and Matelson 1989, Sargent 2000, Wenny and Levey 1998, Wenny 2000a,b). Tanagers, finches, and thrushes are common opportunistic frugivores, which harvest mainly small, watery, carbohydrate-rich, many-seeded fruits produced by most Melastomataceae, Solanaceae, and Rubiaceae, as well as by members of many other plant families, while quetzals, toucans, and bellbirds are specialized frugivores, which consume mainly large-seeded, lipid- and protein-rich fruits like those of the Lauraceae (Wheelwright et al. 1984, Murray et al. 2000).

Fruit consumption and seed dispersal varies greatly, owing to a variety of factors including the nature of the fruit, the frugivores, the crop size, and the location of the plants (Murray et al. 2000). Some fruits have laxative agents that regulate the time the seeds spend in the digestive tract of the dispersing birds (Murray et al. 1994). The size of individual fruits and the size of a tree's crop can both influence the rate at which lauraceous fruits are removed, but crop size doesn't seem to influence the length of visits by specialized frugivores (Wheelwright 1991, 1993). Fruits of three common gap-invading pioneer shrubs are taken whole by black-faced solitaires, prong-billed barbets, and black-and-yellow silky flycatchers, which dispersed seeds widely, but fruits of the same plants were mashed by common bush tanagers, spangle-cheeked tanagers, and yellow-thighed finches, which dropped most seeds near the parent plant (Murray 1986, 1988). Similar differences have been observed for avian dispersers of *Ficus pertusa* (Bronstein and Hoffman 1987). Dispersal of *Ocotea endresiana* (Lauraceae) is remarkably well-directed in some circumstances. Fruits consumed by male three-wattled bellbirds are deposited mainly below the bird's courtship perches, which are generally on the edges of canopy gaps (Wenny and Levey 1998). Seedlings from seeds dispersed by male bellbirds are thus more likely to survive than seedlings dispersed by other birds.

A number of frugivorous birds, including the three-wattled bellbird, quetzal, and black-faced solitaire, are altitudinal migrants, but the movements of most of these are not well understood. The relationships among flowering and fruiting phenologies, pollinator and frugivore movement, and plant species ranges deserve more attention as well as broader biogeographic comparisons to determine whether the patterns seen in the Cordillera de Tilarán hold on the isolated massifs of the C. de Guanacaste.

trunk snapping accounts for 61% of the serious damage to trees in this forest, roughly a third of the trees that snap subsequently sprout. Fewer trees uproot (22% of those damaged), or are knocked down by falling neighbors (7%), and most of these die from the incident. Only 4% of trees die upright (4%). Shade-intolerant, pioneer species (*Cecropia polyphlebia*, *Hampea appendiculata*, and *Heliocarpus appendiculatus*) were roughly twice as likely to die as other species.

In the wind-exposed lower montane rain forests along the continental divide treefalls create canopy gaps covering on average 1.3% of the area each year (Nadkarni et al 2000). Most trees (80 to 90%) fall during the severe winter storms known locally as *temporales del norte*. The number and severity of these winter storms varies from year to year, and so does the number and area of treefall gaps. The percentage of the area opened each year over the decade from 1981 to 1990 varied from 0.6% to 3.8%. Although the turnover time calculated from the long-term geometric mean of the area opened per year is 77 years in this forest, turnover times calculated from 3-year periods range from 51 years to 147 years, emphasizing the importance of long-term studies. Demographic analyses of one of the dominant shade-intolerant trees (*Didymopanax pittieri*) of the windswept elfin forest on ridge crests suggest that the population is stable, and thus that forest structure at the scale of 1 km^2 is stable (Lawton 1980). This in turn suggests that the current disturbance regime has existed for some tree generations.

Gaps in the elfin forest vary considerably in many attributes (Lawton and Putz 1988). Most gaps are small; less than 5% are greater than 105 m^2. The largest 5%, however, contribute more than one-third of the canopy area opened by disturbance. Of elfin forest gaps 41% formed when trees uprooted, and 39% when trees snapped. The remainder was created by limbfall, the collapse of epiphyte masses, and lightning strikes. Gap area (log-transformed) is correlated with gap aperture (the angular opening to the sky from gap center), gap-maker trunk diameter, nurse log area (also log-transformed), and the area of disturbed mineral soil (Lawton and Putz 1988). Because many gap attributes are related to gap area it is tempting to use it as a simple measure of the overall variation among gaps, but in this forest the first principal component of variation among gap attributes, which contrasts measures of gap size and the height at which the gap maker broke, accounts for only about half of the variation among gaps, and the second principal component, a contrast of gap aperture ("openness") with position of the gap along the slope, the height of breakage, and gap area, accounts for only an additional one-sixth of the variation in gap attributes (Lawton and Putz 1988). The size of the tree that fell to create the gap, and the nurse log area in the gap, are both better correlated with the first principal component than is gap area.

Elfin forest gaps are rapidly colonized; bare ground and logs are covered within 2–3 years as a dense thicket of regrowth reaches 1–4 m in height (Lawton and Putz 1988). In small gaps (10 m^2) the leaf area index recovers in 3 years to 50% of mature forest leaf area index (LAI) of 5.1. In large gaps (120 m^2) the LAI recovers to 75% within 3 years and 90% within 6 years. Saplings of canopy tree species account for 10% of the regrowing canopy in gaps at 8 months after gap formation, and 50% at 78 months. The rate of absolute height growth is slow by lowland standards, in part because of differences in allocation. The mechanical stresses due to wind in elfin forests are very high, and even shade-intolerant, pioneer trees have dense wood, and thick stems (Lawton 1982, 1984). In relative terms, though, the rate of elfin forest regrowth is similar to that in the lowlands; regrowth of elfin forest pioneers to 50% of canopy height (3–4 m) takes about 6 years (Lawton and Putz 1988). The density of saplings of canopy tree species is greatest on disturbed mineral soil (8.3 saplings m^{-2}), owing largely to seeds, such as those of *Guettarda poasana*, germinating from the soil seed bank, but not much lower (6.4 saplings m^{-2}) on nurse logs. This is in part due to the very unusual life history of some elfin forest species, such as *Didymopanax pittieri*, *Oreopanax nubigenus*, and *Cosmibuena valerii*, which begin life as epiphytic seedlings and saplings and survive host collapse to reorient as saplings on nurse logs in gaps (Lawton and Putz 1988, Williams-Linera and Lawton 1995).

Landslides are conspicuous features of the montane landscape, particularly on the over-steepened walls of *quebradas*, or gorges, although smaller landslides occur in less precipitous terrain in wet areas in lower montane and premontane rain forest. However, the dynamics of succession on landslides have only begun to be explored (Myster 1993), although it is clear that a distinctive assemblage of species, such as *Gunnera insignis*, *Myrica phanerodonta*, and *Monochaetum* spp., are landslide specialists.

People and Nature

People have long occupied the montane forest areas of northern Costa Rica; artifacts from the area around Laguna Arenal have been dated to before 10,000 years BP (Sheets et al. 1991). The Arenal area has been particularly well investigated archeologically, in part due to the correlative dating and preservation made possible by the various eruptions of Volcán Arenal (Sheets et al. 1991, Sheets and McKee

1994). More recently montane areas have been colonized by modern farmers, and this has been followed in some areas, like Monteverde, by the growth of ecotourism. Griffith et al. (2000) discuss the modern agricultural colonization and development of the Cordillera de Tilarán, including the roles of opportunity and economic constraints.

History of Land Use

Expansion of Pre-Columbian agricultural societies after 500 BC must have created a mosaic landscape of fields and fallow lands on much of the upper Pacific slopes of the Cordillera de Tilarán and C. de Guanacaste. Pot shards are commonly found between 1,300 and 1,500 m a.s.l. in the area now occupied by Monteverde, Santa Elena, and their surrounding communities. Areas at 1,500 m, particularly the wetter cloud forest habitats, were in general not occupied (Sheets and McKee 1994). The extent to which the modern vegetation of the upper Pacific slope has been influenced by Pre-Columbian agriculture remains unknown.

Modern agricultural colonization of the upper Pacific slope of the Cordillera de Tilarán began in the 1920s and 1930s, as homesteaders established small subsistence farms between 1,200 and 1,450 m elevation (Griffith et al. 2000), although commercial beef production occurred at lower elevations. In 1953 the cheese plant at Monteverde was established, and incorporated in 1954 as the Productores de Monteverde, S. A. This led to the conversion of subsistence farms to small dairy farms, and dairy production spread along the upper Pacific slope. By the early 1970s most land on the upper Pacific slope had been occupied, and settlers had cleared patches of cloud forest along the crest of the Cordillera and in premontane rain forest on the Caribbean slope, providing considerable motivation to growing conservation efforts in the area (see below). More recently, coffee production has spread, including a number of organic coffee farms, and vegetable production has increased to serve the growing number of restaurants and hotels in the area. Following the establishment of the Monteverde Cloud Forest Preserve in 1972, ecotourism became a growing sector of the economy, and now dominates the Monteverde–Santa Elena area (Aylward et al. 1996, Chamberlain 2000).

Conservation

A detailed history of conservation activity in the Monteverde area of the Cordillera de Tilarán has been provided by Burlingame (2000), while Wheelwright (2000) discussed the problems and prospects for conservation biology in the area. The earliest conservation measures involved the protection of springs and catchments to ensure water supplies, including the headwaters of the Río Guacimal, set aside by the Quaker settlers of Monteverde, and formally protected as the Bosqueterno, S.A., in 1974. In 1970 George Powell, a graduate student studying foraging by mixed species flocks of birds, alarmed by the growing number of scattered clearings being opened in virgin cloud forest, started looking for ways to preserve critical areas. He established contacts with The Nature Conservancy (TNC), the World Wildlife Fund-U.S. (WWF), and other institutions, and with the Tropical Science Center (TSC, or CCT in Spanish) to hold and manage preserved lands as part of the Monteverde Cloud Forest Preserve, which was formally established in 1972. With land purchases funded by international conservation organizations, the Preserve rapidly grew to over 10,500 ha (Burlingame 2000), and the number of visitors grew to more than 50,000 annually by the early 1990s. By the mid-1980s the Monteverde Cloud Forest Preserve and the Tropical Science Center were focused on land and ecotourism management, and a local group at Monteverde established the *Asociación Conservacionista de Monteverde* (Monteverde Conservation League, ACM), to pursue funds for land preservation in the Peñas Blancas valley and neighboring regions. Although much of the land of interest lay within the Arenal Forest Reserve established to protect watersheds involved in hydroelectric power generation, disgruntled landowners remained in place. With funds from *Barnens Regnskog*, a Swedish organization, and The Children's Rainforest U.S., ACM bought and now administers about 18,000 ha. Currently, there exists a complex of public and private reserves, including Parque Nacional Arenal, cooperating in land preservation under the umbrella of the Área de Conservación Arenal-Tempisque (ACAT), one of Costa Rica's conservation areas administered through the National System of Conservation Areas (SINAC) of the Ministry of Environment and Energy (MINAE).

Ecotourism is not as developed in the Cordillera de Guanacaste, but much of the upper portions of the volcanic massifs are protected in a series of Costa Rican National Parks. The Orosi and Cacao volcanic complex is now included in Parque Nacional Guanacaste, and are included in the ambitious efforts to restore the series of environments from the volcanic summits to the lowland dry forests in their lee. We refer specifically to the chapter by Janzen and Hallwachs (chapter 10 of this volume) for extensive details on this extraordinary conservation and restoration effort (also see Jiménez et al., chapter 9 of this volume).

The restoration work in Guanacaste has grown from a recognition that montane systems are inextricably bound up in the fate of adjacent lowlands. Many animals, including large mammals like tapir, birds like quetzals,

wright, eds., *Monteverde: Ecology and Conservation of a Tropical Cloud Forest*, 95–147. New York: Oxford University Press.

Hartshorn, G.S., and B.E. Hammel. 1994. Vegetation types and floristic patterns. In L. McDade, K.S. Bawa, H.A. Hespenheide, and G.S. Hartshorn, eds., *La Selva: Ecology and Natural History of a Neotropical Rain Forest*, 73–89. Chicago: University of Chicago Press.

Hastenrath, S.L. 1967. Rainfall distribution and regime in Central America. *Archiv für Meteorologie, Geophysik und Bioklimatologie, Serie B* 15: 201–41.

Hayes, M.P., J.A. Pounds, and W.W. Timmerman. 1989. An annotated list and guide to the amphibians and reptiles of Monteverde, Costa Rica. *Herpetological Circulars* 17: 1–67.

Jankowski, J. 2004. Patterns of avian diversity and the interspecific abundance-distribution relationship in the Tilarán mountain range, Costa Rica. M.S. thesis, Department of Biological Sciences, Purdue University. Lafayette, IN.

Jankowski, J., and K.N. Rabenold. 2007. Endemism and local rarity in birds of neotropical montane rainforest. *Biological Conservation* 138: 453–63.

Keigwin, L. 1982. Isotopic paleoceanography of the Caribbean and East Pacific: role of Panama uplift in late Neogene time. *Science* 251: 350–53.

Kempter, K.A., S.G. Benner, and S.N. Williams. 1996. Rincón de la Vieja volcano, Guanacaste province, Costa Rica: geology of the southwestern flank and hazards implications. *Journal of Volcanology and Geothermal Research* 71: 109–27.

Koptur, S. 1985. Alternative defenses against herbivores in *Inga* (Fabaceae: Mimosoideae) over an elevational gradient. *Ecology* 66: 1639–50.

Lackie, P.M., C.D. Thomas, M.J. Brisco, and D.N. Hepper. 1986. On the pollination ecology of *Hamelia patens* (Rubiaceae) at Monteverde, Costa Rica. *Brenesia* 25–26: 203–13.

Langtimm, C.A. 1992. Specialization for vertical habitats within a cloud forest community of mice. PhD diss., University of Florida, Gainesville.

Langtimm, C.A. 2000. Arboreal mammals. In N. Nadkarni and N. Wheelwright, eds., *Monteverde: Ecology and Conservation of a Tropical Cloud Forest*, 239–40. New York: Oxford University Press.

LaVal, R.K. 1977. Notes on some Costa Rican bat communities. *Brenesia* 10/11: 77–83.

LaVal, R.K. 2004. Impact of global warming and locally changing climate on tropical cloud forest bats. *Journal of Mammalogy* 85: 237–44.

LaVal, R.K., and H. S. Fitch. 1977. Structure, movements and reproduction in three Costa Rican bat communities. *Occasional Papers of the Museum of Natural History (University of Kansas)*: 69: 1–28.

Lawton, M.F. 1982. Altruism and sociobiology: a critical look at the critical issue. PhD diss., University of Chicago. Chicago.

Lawton, M.F. 1983. *Cyanocorax morio*. In D.H. Janzen, ed., *Costa Rican Natural History*. Chicago: University of Chicago Press.

Lawton, M.F., and C. Guindon. 1981. Flock composition, breeding success, and learning in the Brown Jay. *Condor* 83: 27–33.

Lawton, M.F., and R.O. Lawton. 1980. Nestsite selection in the Brown Jay. *Auk* 97(3): 631–33.

Lawton, M.F., and R.O. Lawton. 1985. The breeding biology of the Brown Jay in Monteverde, Costa Rica. *Condor* 87: 192–204.

Lawton, M.F., and R.O. Lawton. 1986. Heterochrony and the evolution of social organization in birds. *Current Ornithology* 3: 187–224.

Lawton, R.O. 1980. Wind and the ontogeny of elfin stature in a Costa Rican lower montane rain forest. PhD diss., University of Chicago. Chicago.

Lawton, R.O. 1982. Wind stress and elfin stature in a montane rain forest tree: an adaptive explanation. *American Journal of Botany* 69: 1224–30.

Lawton, R.O. 1983. *Didymopanax pittieri*. In D.H. Janzen, ed., *Costa Rican Natural History*. Chicago: University of Chicago Press.

Lawton, R.O. 1984. Ecological constraints on wood density in a tropical montane rain forest. *American Journal of Botany* 71: 261–67.

Lawton, R.O. 1986. The evolution of strangling by *Ficus crassiuscula*. *Brenesia* 25–26: 273–78.

Lawton, R.O. 1990. Canopy gaps and light penetration into a windexposed tropical lower montane rain forest. *Canadian Journal of Forest Research* 20: 659–67.

Lawton, R.O. 1991. More on strangling by *Ficus crassiuscula*: a reply to Ramírez. *Brenesia* 32: 119–20.

Lawton, R.O., L.D. Alexander, W.N. Setzer, and K.G. Byler. 1993. Floral essential oil of *Guettarda poasana* inhibits yeast growth. *Biotropica* 25(4): 483–86.

Lawton, R.O., and V. Dryer. 1980. Vegetation of the Monteverde Cloud Forest Reserve. *Brenesia* 18: 101–16.

Lawton, R.O., U.S. Nair, R.A. Pielke Sr., and R.M. Welch. 2001. Climatic impact of tropical lowland deforestation on nearby montane cloud forests. *Science* 294: 584–87.

Lawton, R.O., U.S. Nair, D. Ray, J.A. Pounds, and R.M. Welch. 2010. Quantitative measures of immersion in cloud and the biogeography of cloud forest. In L.A. Bruijnzeel, F.N. Scatena, and L.S. Hamilton, eds., *Mountains in the Mist: Science for Conserving and Managing Tropical Montane Cloud Forests*. Cambridge, UK: Cambridge University Press.

Lawton, R.O., and F.E. Putz. 1988. Natural disturbance and gapphase regeneration in a windexposed tropical lower montane rain forest. *Ecology* 69: 764–77.

Linhart, Y.B., W.H. Busby, J.H. Beach, and P. Feinsinger. 1987b. Forager behavior, pollen dispersal, and inbreeding in two species of hummingbird-pollinated plants. *Evolution* 41: 679–82.

Linhart, Y.B., P. Feinsinger, J.H. Beach, W.H. Busby, K.G. Murray, W.Z. Pounds, S. Kinsman, C.A. Guindon, and M. Kooiman. 1987a. Disturbance and predictability of flowering plants in bird-pollinated cloud forest plants. *Ecology* 68: 1696–710.

Lips, K.R., F. Brem, R. Brenes, J.D. Reeve, R.A. Alford, J. Voyles, C. Carey, L. Livo, A.P. Pessier, and J.P. Collins. 2006. Emerging infectious disease and the loss of biodiversity in a Neotropical amphibian community. *Proceedings of the National Academy of Sciences* 103: 3165–70.

Longino, J.T. 1989. Geographic variation and community structure in an ant-plant mutualism: *Azteca* and *Cecropia* in Costa Rica. *Biotropica* 21: 126–32.

Longino, J.T. 2006. A taxonomic review of the genus *Myrmelachista* (Hymenoptera: Formicidae) in Costa Rica. *Zootaxa* 1–54.

Maffia, B., N.M. Nadkarni, and D.P. Janos. 2000. Vesicular-arbuscular mycorrhizae of epiphytic and terrestrial Piperaceae. In N. Nadkarni and N. Wheelwright, eds., *Monteverde: Ecology and Conservation of a Tropical Cloud Forest*, 338–39. New York: Oxford University Press.

Marshall, J.S., and R.S. Anderson. 1995. Quaternary uplift and seismic cycle deformation, Península de Nicoya, Costa Rica. *Geological Society of America Bulletin* 107: 463–73.

Marshall, L.G., S.D. Webb, J.J. Sepkoski, and D.M. Raup. 1982. Mammalian evolution and the Great American Interchange. *Science* 251: 1351–57.

McDonald, D.B. 1989a. Cooperation under asexual selection: age-graded changes in a lekking bird. *American Naturalist* 134: 709–30.

McDonald, D.B. 1989b. Correlates of male mating success in a lekking bird with male-male cooperation. *Animal Behaviour* 37: 1007–22.

McDonald, D.B. 1993a. Delayed plumage maturation and orderly queues for status: a manakin mannequin experiment. *Ethology* 94: 31–45.

McDonald, D.B. 1993b. Demographic consequences of sexual selection in the Long-tailed Manakin. *Behavioral Ecology* 4: 297–309.

McDonald, D.B. 2000. Cooperation between male long-tailed manakins. In N. Nadkarni and N. Wheelwright, eds., *Monteverde: Ecology and Conservation of a Tropical Cloud Forest*, 204–5. New York: Oxford University Press.

McDonald, D.B. 2007. Predicting fate from early connectivity in a social network. *Proceedings of the National Academy of Sciences* (USA) 104: 10910–14.

McDonald, D.B., and W.K. Potts. 1994. Cooperative display and relatedness among males in a lek-mating bird. *Science* 266: 1030–32.

Melson, W.G. 1994. The eruption of 1968 and tephra stratigraphy of Arenal Volcano. In P.D. Sheets and B.R. McKee, eds., *Archaeology, Volcanism, and Remote Sensing in the Arenal Region, Costa Rica*, 24–47. Austin, TX: University of Texas Press.

Moriarity, D.M., J. Huang, C.A. Yancey, P. Zhang, W.N. Setzer, R.O. Lawton, R.B. Bates, and S. Caldera. 1998. Lupeol is the cytotoxic principle in the leaf extract of *Dendropanax* cf. *querceti* from Monteverde, Costa Rica. *Planta Medica* 64: 370–72.

Murray, K.G. 1986. Consequences of seed dispersal for gap-dependent plants: relationships between seed shadows, germination requirements, and forest dynamic processes. In A. Estrada and T. H. Fleming, eds., *Frugivores and Seed Dispersal*, 187–98. Dordrecht: W. Junk Publishers.

Murray, K.G. 1988. Avian seed dispersal of three neotropical gap-dependent plants. *Ecological Monographs* 68: 271–98.

Murray, K.G., P. Feinsinger, W.H. Busby, Y.B. Linhart, J.H. Beach, and S. Kinsman. 1987. Evaluation of character displacement among plants in two tropical pollination guilds. *Ecology* 68: 1283–93.

Murray, K.G., S. Kinsman, and J.L. Bronstein. 2000. Plant-animal interactions. In N. Nadkarni and N. Wheelwright, eds., *Monteverde: Ecology and Conservation of a Tropical Cloud Forest*, 245–302. New York: Oxford University Press.

Murray, K.G., S. Russell, C.M. Piccone, K. Winnett-Murray, W. Sherwood, and M.L. Kuhlmann. 1994. Fruit laxatives and seed passage rates in frugivores: consequences for plant reproductive success. *Ecology* 75: 989–94.

Myster, R.W. 1993. Spatial heterogeneity of seedrain, seedpool, and vegetative cover on two Monteverde landslides. *Brenesia* 39–40: 137–45.

Nadkarni, N.M. 1981. Canopy roots: convergent evolution in rain forest nutrient cycles. *Science* 214: 1023–24.

Nadkarni, N.M. 1984. Epiphyte biomass and nutrient capital of a neotropical elfin forest. *Biotropica* 16: 249–56.

Nadkarni, N.M., R.O. Lawton, K.L. Clark, T.J. Matelson, and D. Schaefer. 2000. Ecosystem ecology and forest dynamics. In N. Nadkarni and N. Wheelwright, eds., *Monteverde: Ecology and Conservation of a Tropical Cloud Forest*, 303–50. New York: Oxford University Press.

Nadkarni, N.M., and J.T. Longino. 1990. Invertebrates in canopy and ground organic matter in a neotropical montane forest. *Biotropica* 22: 286–89.

Nadkarni, N.M., and T.J. Matelson. 1989. Bird use of epiphyte resources in neotropical trees. *Condor* 91: 891–907.

Nadkarni, N.M., and T.J. Matelson. 1991. Dynamics of fine litterfall within the canopy of a tropical cloud forest, Monteverde. *Ecology* 72: 2071–82.

Nadkarni, N.M., and T.J. Matelson. 1992a. Biomass and nutrient dynamics of epiphyte litterfall in a neotropical cloud forest, Costa Rica. *Biotropica* 24: 24–30.

Nadkarni, N.M., and T.J. Matelson. 1992b. Biomass and nutrient dynamics of fine litter of terrestrially rooted material in a neotropical cloud forest, Costa Rica. *Biotropica* 24: 113–20.

Nadkarni, N.M., T.J. Matelson, and W.A. Haber. 1995. Structural characteristics and floristic composition of a neotropical cloud forest, Costa Rica. *Journal of Tropical Ecology* 11: 481–94.

Nadkarni, N.M., and N. Wheelwright, eds. 2000. *Monteverde: Ecology and Conservation of a Tropical Cloud Forest.* New York: Oxford University Press.

Nair, U.S., S. Asefi, R.M. Welch, D.K. Ray, R.O. Lawton, V.S. Manoharan, M. Mulligan, T.L. Sever, D. Irwin, and J.A. Pounds. 2008. Biogeography of tropical montane cloud forests, Part II: Mapping of orographic cloud immersion. *Journal of Applied Meteorology and Climatology*, in press.

Nair, U.S., R.O. Lawton, R.M. Welch, and R.A. Pielke Sr. 2003. Impact of land use on tropical montane cloud forests: sensitivity of cumulus cloud field characteristics to lowland deforestation. *Journal of Geophysical Research* 108(D7): 4206–18.

Nair, U.S., D.K. Ray, R.O. Lawton, R.M. Welch, and R.A. Pielke Sr. 2010. The impact of deforestation on orographic cloud formation in a complex tropical environment. In L.A. Bruijnzeel, F.N. Scatena, and L.S. Hamilton, eds., *Mountains in the Mist: Science for Conserving and Managing Tropical Montane Cloud Forests*. Cambridge, UK: Cambridge University Press.

O'Donnell, S., and A. Kumar. 2006. Microclimatic factors associated with elevational changes in army ant density in tropical montane forest. *Ecological Entomology* 31: 491–98.

Peck, D.C. 1996. The association of spittlebugs with grasslands: ecology of *Prosapia* (Homoptera: Cercopidae) in upland dairy pastures of Costa Rica. PhD diss., Cornell University.

Podolsky, R.D. 1992. Strange floral attractors: pollinator attraction and the evolution of plant sexual systems. *Science* 258: 791–93.

Portig, W. 1965. Central American rainfall. *Geographical Review* 55: 68–90.

Pounds, J.A. 1988. Ecomorphology, locomotion, and microhabitat structure: patterns in a tropical mainland *Anolis* community. *Ecological Monographs* 58: 299–320.

Pounds, J.A. 2000. Amphibians and reptiles. In N. Nadkarni and N. Wheelwright, eds., *Monteverde: Ecology and Conservation of a Tropical Cloud Forest*, 149–77. New York: Oxford University Press.

Pounds, J.A. 2001. Climate and amphibian declines. *Nature* 410: 639–40.

Pounds, J.A., M.R. Bustamante, L.A. Coloma, J.A. Consuegra, M.P.L.

Fogden, P.N. Foster, E. La Marca, K.L. Masters, A. Merino-Viteri, R. Puschendorf, S.R. Ron, G.A. Sánchez-Azofeifa, C.J. Still, and B.E. Young. 2005. Widespread amphibian extinctions from epidemic disease driven by global warming. *Nature* 439: 161–67.

Pounds, J.A., and M.L. Crump. 1994. Amphibian declines and climate disturbance: the case of the Golden Toad and the Harlequin Frog. *Conservation Biology* 8: 72–85.

Pounds, J.A., and M.P.L. Fogden. 2000. Amphibians and reptiles of Monteverde. Appendix 8. In N. Nadkarni and N. Wheelwright, eds., *Monteverde: Ecology and Conservation of a Tropical Cloud Forest*, 537–40. New York: Oxford University Press.

Pounds, J.A., M.P.L. Fogden, and J.H. Campbell. 1999. Biological response to climate change on a tropical mountain. *Nature* 389: 611–14.

Pounds, J.A., M.P.L. Fogden, J.M. Savage, and G.C. Gorman. 1997. Tests of null models for amphibian declines on a tropical mountain. *Conservation Biology* 11: 1307–22.

Pounds, J.A., and R. Puschendorf. 2004. Ecology: clouded futures. *Nature* 427: 107–9.

Powell, G.V.N. 1979. Structure and dynamics of interspecific flocks in a neotropical mid-elevation forest. *Auk* 96: 375–90.

Powell, G.V.N. 1980. Migrant participation in neotropical mixed species flocks. In A. Keast and E. S. Morton, eds., *Migrant Birds in the Neotropics: Ecology, Behavior, Distribution and Conservation*, 477–83. Washington, DC: Smithsonian Institution Press.

Powell, G.V.N. 1985. Sociobiology and adaptive significance of interspecific foraging flocks in the Neotropics. *Ornithological Monographs* 36: 713–32.

Powell, G.V.N., and R.D. Bjork. 1994. Implications of altitudinal migration for conservation strategies to protect tropical biodiversity: a case study of the Resplendent Quetzal *Pharomachrus mocinno* at Monteverde, Costa Rica. *Bird Conservation International* 4: 161–74.

Powell, G.V.N., R.D. Bjork, S. Barrios, and V. Espinoza. 2000. Elevational migrations and habitat linkages: using the resplendent quetzal as an indicator for evaluating the design of the Monteverde reserve complex. In N. Nadkarni and N. Wheelwright, eds., *Monteverde: Ecology and Conservation of a Tropical Cloud Forest*, 439–42. New York: Oxford University Press.

Prance, G.T. 1990. The floristic composition of the forests of central Amazonian Brazil. In A. H. Gentry, ed., *Four Neotropical Rainforests*, 112–40. New Haven, CT: Yale University Press.

Rauscher, S.A., F. Giorgi, N.S. Diffenbaugh, and A. Seth. In press. Extension and intensification of the Meso-American mid-summer drought in the 21st Century. *Climate Dynamics.*

Ray, D.K., U.S. Nair, R.O. Lawton, R.M. Welch, and R.A. Pielke Sr. 2006. Impact of land use on Costa Rican tropical montane cloud forests: sensitivity of orographic cloud formation to deforestation in the plains. *Journal of Geophysical Research* 111. doi:10.1029/2005JD006096.

Rhodes, A.L., A.J. Guswa, and S.E. Newell. 2006. Seasonal variation in the stable isotope composition of precipitation in the tropical montane forests of Monteverde, Costa Rica. *Water Resources Research* 42: W11402. doi:10.1029/2005WR004535.

Riley, C.M. 1986. Observations of the breeding biology of Emerald Toucanets in Costa Rica. *Wilson Bulletin* 98: 585–88.

Riley, C.M., and K.G. Smith. 1992. Sexual dimorphism and foraging behavior of Emerald Toucanets *Aulacorhynchus prasinus* in Costa Rica. *Ornis Scandinavica* 23: 259–66.

Sargent, S. 1993. Nesting biology of the yellow-throated *Euphonia*: large clutch size in a neotropical frugivore. *Wilson Bulletin* 105: 285–300.

Sargent, S. 1995. Seed fate in a tropical mistletoe: the importance of host twig size. *Functional Ecology* 9: 197–204.

Sargent, S. 2000. Specialized seed dispersal: mistletoes and fruit-eating birds. In N. Nadkarni and N. Wheelwright, eds., *Monteverde: Ecology and Conservation of a Tropical Cloud Forest*, 288–89. New York: Oxford University Press.

Schonberg, L.A., J.T. Longino, N.M. Nadkarni, S.P. Yanoviak, and J.C. Gering. 2004. Arboreal ant species richness in primary forest, secondary forest, and pasture habitats of a tropical montane landscape. *Biotropica* 36: 402–9.

Setzer, M.C., D.M. Moriarity, R.O. Lawton, W.N. Setzer, G.A. Gentry, and W.A. Haber. 2003. Phytomedicinal potential of tropical cloudforest plants from Monteverde, Costa Rica. *Revista de Biología Tropical* 51(3–4): 647–74.

Setzer, W.N. 2000. The search for medicines from the plants of Monteverde. In N. Nadkarni and N. Wheelwright, eds., *Monteverde: Ecology and Conservation of a Tropical Cloud Forest*, 452–53. New York: Oxford University Press.

Setzer, W.N., M.N. Flair, K.G. Byler, J. Huang, M.A. Thompson, A.F. Setzer, D.M. Moriarity, R.O. Lawton, and D.B. Windham-Carswell. 1992. Antimicrobial and cytotoxic activity of crude extracts of Araliaceae from Monteverde, Costa Rica. *Brenesia* 38: 123–30.

Setzer, W.N., T.J. Green, R.O. Lawton, D.M. Moriarity, R.B. Bates, S. Caldera, and W.A. Haber. 1995a. Antibacterial activity of a vitamin E derivative from *Tovomitopsis psychotrifolia*. *Planta Medica* 61: 275–76.

Setzer, W.N., T.J. Green, K.W. Whitaker, D.M. Moriarity, R.O. Lawton, R.B. Bates, and S. Caldera. 1995b. A cytotoxic diacetylene from *Dendropanax arboreus* (L.) Decne. & Planchon (Araliaceae). *Planta Medica* 61: 470–71.

Setzer, W.N., J.A. Noletto, and R.O. Lawton. 2006. Chemical composition of the floral essential oil of *Randia matudae* from Monteverde, Costa Rica. *Flavor and Fragrance Journal* 21(2): 244–46.

Setzer, W.N., J.A. Noletto, R.O. Lawton, and W.A. Haber. 2005. Leaf essential oil composition of five *Zanthoxylum* species from Monteverde, Costa Rica. *Molecular Diversity* 9: 3–13.

Setzer, W.N., M.C. Setzer, A.L. Hopper, D.M. Moriarity, G.K. Lehrman, K.L. Niekamp, S.M. Moorcomb, R.B. Bates, K.J. McClure, C.C. Stessman, and W.A. Haber. 1998. The cytotoxic activity of a *Salacia* liana species from Monteverde, Costa Rica is due to a high concentration of tingenone. *Planta Medica* 64: 583.

Setzer, W.N., M.C. Setzer, D.M. Moriarity, R.B. Bates, and W.A. Haber. 1999. Biological activity of the essential oil of *Myrcianthes* sp. nov. "black fruit" from Monteverde, Costa Rica. *Planta Medica* 65: 468–69.

Setzer, W.N., M.C. Setzer, J.M. Schmidt, D.M. Moriarity, B. Vogler, S. Reeb, A.M. Holmes, and W.A. Haber. 2000b. Cytotoxic components from the bark of *Stauranthus perforatus* from Monteverde, Costa Rica. *Planta Medica* 66: 493–94.

Setzer, W.N., X. Shen, R.B. Bates, J.R. Burns, K.J. McClure, P. Zhang, D.M. Moriarity, and R.O. Lawton. 2000a. A phytochemical investigation of *Alchornea latifolia*. *Fitoterapia* 71: 195–98.

Sheets, P.D., J. Hoopes, W. Melson, B. McKee, T. Sever, M. Mueller, M. Chenault, and J. Bradley. 1991. Prehistory and volcanism in the Arenal area, Costa Rica. *Journal of Field Archaeology* 18: 445–65.

Sheets, P.D., and B.R. McKee, eds. 1994. *Archaeology, Volcanism, and Remote Sensing in the Arenal Region, Costa Rica*. Austin, TX: University of Texas Press.

Shopland, J. 2000. The cost of social foraging in mixed-species bird flocks. In N. Nadkarni and N. Wheelwright, eds., *Monteverde: Ecology and Conservation of a Tropical Cloud Forest*, 206–7. New York: Oxford University Press.

Siebert, L., G.E. Alvarado, J.W. Vallance, and B. van Wyk de Vries. 2006. Large-volume volcanic edifice failures in Central America and associated hazards. In W.I. Rose, G.J.S. Bluth, M.J. Carr, J.W. Ewert, L.C. Patino, and J.W. Vallance, eds., *Volcanic Hazards in Central America, Geological Society of America Special Paper* 412: 1–26.

Sillet, S.C., S.R. Gradstein, and D. Griffin. 1995. Bryophyte diversity of *Ficus* tree crowns from intact cloud forest and pasture in Costa Rica. *Bryologist* 98: 251–60.

Skutch, A.F. 1935. Helpers at the nest. *Auk* 52: 257–73.

Still, C.J., P.N. Foster, and S.H. Schneider. 1999. Simulating the effects of climate change on tropical montane cloud forests. *Nature* 389: 608–10.

Tiebout, H.M. 1991. Daytime energy management by tropical hummingbirds: responses to foraging constraint. *Ecology* 72: 839–51.

Timm, R.M., and R.K. LaVal. 2000a. Mammals. In N. Nadkarni and N. Wheelwright, eds., *Monteverde: Ecology and Conservation of a Tropical Cloud Forest*, 223–44. New York: Oxford University Press.

Timm, R.M., and R.K. LaVal. 2000b. Mammals of Monteverde. Appendix 10. In N. Nadkarni and N. Wheelwright, eds., *Monteverde: Ecology and Conservation of a Tropical Cloud Forest*, 553–57. New York: Oxford University Press.

Tobón-Marín, C., S.A. Bruijnzeel, A. Frumau, and J. Calvo. 2010. Changes in soil hydraulic properties and soil water status after conversion of tropical montane cloud forest to pasture in northern Costa Rica. In L.A. Bruijnzeel, F.N. Scatena, and L.S. Hamilton, eds., *Mountains in the Mist: Science for Conserving and Managing Tropical Montane Cloud Forests*. Cambridge, UK: Cambridge University Press.

Tosi, J.A. 1969. Mapa Ecológico de Costa Rica, Basado en la Clasificación Vegetal Mundial de L.R. Holdridge. Scale 1: 750,000. San Pedro de Montes de Oca, Costa Rica: Tropical Science Center (TSC).

Trainer, J.M. 2000. The roles of long-tailed manakin vocalization in cooperation and courtship. In N. Nadkarni and N. Wheelwright, eds., *Monteverde: Ecology and Conservation of a Tropical Cloud Forest*, 215–16. New York: Oxford University Press.

Trainer, J.M., and D.B. McDonald. 1993. Vocal repertoire of the long-tailed manakin and its relation to male-male cooperation. *Condor* 95: 769–81.

Trainer, J.M., and D.B. McDonald. 1995. Singing performance, frequency matching, and courtship success of long-tailed manakins (*Chiroxiphia linearis*). *Behavioral Ecology and Sociobiology* 37: 249–54.

Valburg, L.K. 1992. Flocking and frugivory: the effect of social groupings on resource use in the Common Bush-Tanager. *Condor* 94: 358–63.

Valburg, L.K. 2000. Why join mixed-species flocks?: a frugivore's perspective. In N. Nadkarni and N. Wheelwright, eds., *Monteverde: Ecology and Conservation of a Tropical Cloud Forest*, 205–6. New York: Oxford University Press.

Vance, E., and N.M. Nadkarni. 1990. Microbial biomass and activity in canopy organic matter and the forest floor of a tropical cloud forest. *Soil Biology and Biochemistry* 22: 677–84.

Vance, E., and N.M. Nadkarni. 1992. Root biomass distribution in a moist tropical montane forest. *Plant and Soil* 142: 31–39.

Van Devender, R.W. 1980. Preliminary checklist of the herpetofauna of Monteverde, Puntarenas Province, Costa Rica and vicinity. *Brenesia* 17: 319–25.

Weiss, M.E. 1996. Pollen-feeding fly alters floral phenotypic gender in *Centropogon solanifolius* (Campanulaceae). *Biotropica* 28: 770–73.

Welch, R.M., S. Asefi, U.S. Nair, Q. Han, R.O. Lawton, D.K. Ray, and V.S. Manoharan. 2008. Biogeography of tropical montane cloud forests, Part I: Remote sensing of cloud base heights. *Journal of Applied Meteorology and Climatology* 47: 960–75.

Wenny, D.G. 2000a. What happens to seeds of vertebrate-dispersed trees after dispersal? In N. Nadkarni and N. Wheelwright, eds., *Monteverde: Ecology and Conservation of a Tropical Cloud Forest*, 286–87. New York: Oxford University Press.

Wenny, D.G. 2000b. Seed dispersal, seed predation, and seedling recruitment of *Ocotea endresiana* (Lauraceae) in Costa Rica. *Ecological Monographs* 70: 331–51.

Wenny, D.G., and D.J. Levey. 1998. Directed seed dispersal by bellbirds in a tropical cloud forest. *Proceedings of the National Academy of Sciences* (USA) 95: 6204–07.

Wheelwright, N.T. 1983. Fruits and the ecology of Resplendent Quetzals. *Auk* 100: 286–301.

Wheelwright, N.T. 1985a. Competition for dispersers, and the timing of flowering and fruiting in a guild of tropical trees. *Oikos* 44: 465–77.

Wheelwright, N.T. 1985b. Fruit size, gape width, and the diets of fruit-eating birds. *Ecology* 66: 808–18.

Wheelwright, N.T. 1986. A seven-year study of individual variation in fruit production in tropical bird-dispersed tree species in the family Lauraceae. In A. Estrada and T.H. Fleming, eds., *Frugivores and Seed Dispersal*, 19–35. Dordrecht: W. Junk Publishers.

Wheelwright, N.T. 1991. How long do fruit eating birds stay in the plants where they feed? *Biotropica* 23: 29–40.

Wheelwright, N.T. 1993. Fruit size in a tropical tree species: variation, preference by birds, and heredıtability. *Vegetatio* 107/108: 163–74.

Wheelwright, N.T. 2000. Conservation biology. In N. Nadkarni and N. Wheelwright, eds., *Monteverde: Ecology and Conservation of a Tropical Cloud Forest*, 419. New York: Oxford University Press.

Wheelwright, N.T., and A. Bruneau. 1992. Population sex ratios and spatial distribution of *Ocotea tenera* (Lauraceae) trees in a tropical forest. *Journal of Ecology* 80: 425–32.

Wheelwright, N.T., W.A. Haber, K.G. Murray, and C. Guindon. 1984. Tropical fruit-eating birds and their food plants: a survey of a Costa Rican lower montane forest. *Biotropica* 16: 173–91.

Williams, D.A. 2004. Female control of reproductive skew in cooperatively breeding brown jays (*Cyanocorax morio*). *Behavioral Ecology and Sociobiology* 55: 370–80.

Williams, D.A., E.C. Berg, A.M. Hale, and C.R. Hughes. 2004. Characterization of microsatellites for parentage studies of white-throated magpie-jays (*Calocitta formosa*) and brown jays (*Cyanocorax morio*). *Molecular Ecology Notes* 4: 509–11.

Williams, D.A., and A.M. Hale. 2006. Helper effects on offspring production in cooperatively breeding brown jays (*Cyanocorax morio*). *Auk* 123: 847–57.

Williams, D.A., and M.F. Lawton. 2000. Brown Jays: complex sociality in a colonizing species. In N. Nadkarni and N. Wheelwright, eds., *Monteverde: Ecology and Conservation of a Tropical Cloud Forest*, 212–13. New York: Oxford University Press.

Williams, D.A., M.F. Lawton, and R.O. Lawton. 1994. Population growth, range expansion, and competition in the cooperatively breeding Brown Jay, *Cyanocorax morio*. *Animal Behavior* 48: 309–22.

Williams, D.A., and K.N. Rabenold. 2005. Male-biased dispersal, female philopatry, and routes to fitness in a social corvid. *Journal of Animal Ecology* 74: 150–59.

Williams-Linera, G., and R.O. Lawton. 1995. The ecology of hemiepiphytes. In M. Lowman and N. Nadkarni, eds., *Forest Canopies*, 255–82. Academic Press.

Winnett-Murray, K. 2000. Choosiness and productivity in wrens of forests, fragments and farms. In N. Nadkarni and N. Wheelwright, eds., *Monteverde: Ecology and Conservation of a Tropical Cloud Forest*, 208–10. New York: Oxford University Press.

Yanoviak, S.P., N.M. Nadkarni, and R. Solano. 2007. Arthropod assemblages in epiphyte mats of Costa Rican cloud forests. *Biotropica* 39: 202–10.

Yanoviak, S.P., H. Walker, and N.M. Nadkarni. 2004. Arthropod assemblages in vegetative vs. humic portions of epiphyte mats in a neotropical cloud forest. *Pedobiologia* 48: 51–58.

Yoshikawa, F., S. Kaizuka, and Y. Ota. 1981. *The Landforms of Japan*. Tokyo: University of Tokyo Press.

Young, B.E. 1993. Effects of the parasitic botfly *Philornis carinatus* on nestling House Wrens, *Troglodytes aedon*. *Oecologia* 93: 256–62.

Young, B.E. 1994a. The effects of food, nest predation and weather on the timing of breeding in tropical House Wrens. *Condor* 96: 341–53.

Young, B.E. 1994b. Geographic and seasonal patterns of clutch-size variation in House Wrens. *Auk* 111: 545–55.

Young, B.E. 1996. An experimental analysis of small clutch size in tropical House Wrens. *Ecology* 77: 472–88.

Young, B.E., and D.B. McDonald. 2000. Birds. In N. Nadkarni and N. Wheelwright, eds., *Monteverde: Ecology and Conservation of a Tropical Cloud Forest*, 179–222. New York: Oxford University Press.

Zadroga, F. 1981. The hydrological importance of a montane cloud forest area of Costa Rica. In R. Lal and E.W. Russell, eds., *Tropical Agricultural Hydrology*, 59–73. New York: J. Wiley and Sons.

Chapter 14 The Montane Cloud Forests of the Cordillera de Talamanca

Maarten Kappelle[1]

Dedication

I dedicate this chapter to my mother, Mary E. Mohr (b. 1936), and to the memory of my father, Dirk Kappelle (b. 1928–d. 2008), who inspired me to become a biologist. Being a dentist in his professional life, Dirk was a man of many talents and a naturalist in heart and soul. As a nature photographer he earned many prizes and participated in numerous expositions. One of the most remarkable ones was the photo exhibition in St. Andrews Hospital ("Andreas Ziekenhuis") in Amsterdam, exactly during the week that my wife Marta gave birth to our son Derk (April 1995).

During my childhood (1960s and early 1970s) every weekend Dirk and Mary took me and my two brothers out to enjoy the marvels of nature in the Netherlands. We made long hikes through the then-mushroom-laden temperate oak-beech forests and coastal sand dunes with their ever changing skies often painted by seventeenth-century Dutch masters like Vermeer. Numerous times we visited preserves in the brackish estuaries of the River Rhine's delta—a bird watchers' paradise and one of the most important European wetlands for thousands of migratory waterfowl that each boreal autumn fly to Africa with the aim to pass the cold winter at warmer latitudes southwards. It was here that my passion for nature was born, and I am still thankful to my parents for showing me all these natural wonders and wildlife spectacles.

After his retirement Dirk authored a book on the twentieth-century history of Dutch freshwater fishing in the backwaters of the Rhine delta. The work was published in Dutch in 2003 titled "Fishermen from the Inland." It narrates the personal histories of 31 artisanal fishermen, born around the year 1900, who dedicated their professional lives to traditional inland fishing in the Netherlands. Many of them specialized in capturing fish like eel and salmon that are now extinct or near-extinct in the numerous wide rivers that slowly traverse the Dutch lowlands. His nicely written *document humain* is perhaps the only account based on personal interviews that is available on this particular topic, since almost no artisanal inland fishermen remain in the country.

Being nature lovers *pur sang*, Dirk and Mary both visited me in Costa Rica on four occasions (1986, 1988–1989, 1999, and 2000–2001). During their month-long visits we made multiple trips to get a grasp of the true riches of the country: we traveled from Golfito to Gandoca, from Santa Rosa to San Gerardo de Dota, from Tortuguero to Tucurrique, from Cañas to Corcovado, and from Monteverde to Mansión. It was during these countless trips that I got a good impression of the diversity of the country's different habitats that made me undertake the endeavor to develop the current volume on Costa Rica's ecosystems, now in your hands. I am thankful to them for their continual support to my research and writings on the ecological systems of this country that I fell so much in love with back in 1985.

Introduction

The tropical evergreen cloud forest of Costa Rica's southern highlands include both true tropical montane cloud

[1] World Wide Fund for Nature (WWF International), Avenue du Mont-Blanc 1196, Gland, Switzerland, and Department of Geography, University of Tennessee, Knoxville, TN

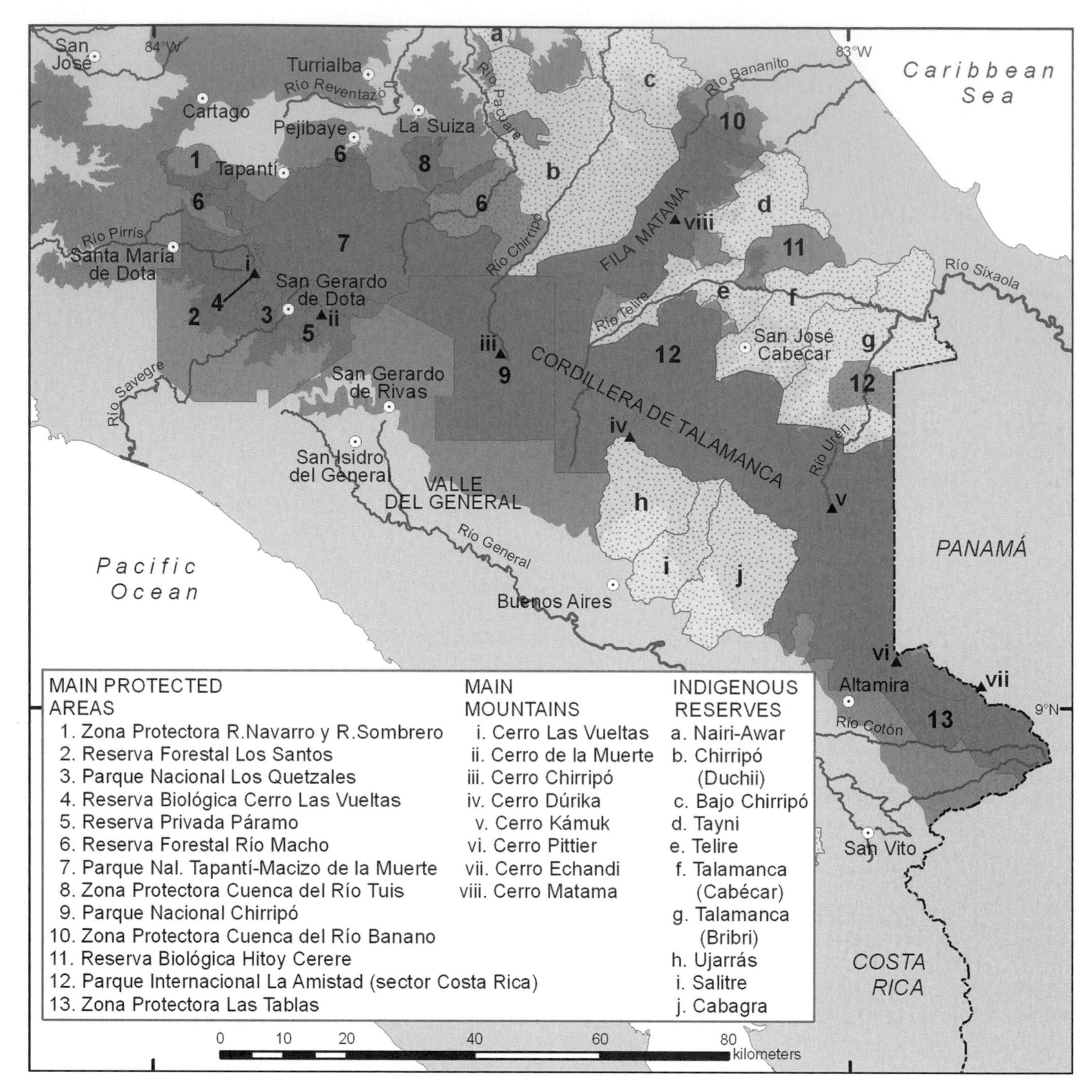

Fig. 14.1 Map of the main mountain peaks, towns, indigenous reserves and protected areas located in the Talamancan highland forest region and its adjacent areas. *Map prepared by Marco V. Castro.*

forest (TMCF) sensu Hamilton et al. (1995) and seasonal cloud forests with less cloud persistence throughout the year (Stadtmüller 1987, Kappelle et al. 1999, Kappelle and Brown 2001). These highland forests occur between 500 and 3,500 meters above sea level (m a.s.l.) and cover both the Pacific and Atlantic slopes of the Cordillera de Talamanca, the tallest and largest mountain range in southern Central America (Fig. 14.1). It runs from the Cerros de Escazú (Fig. 14.2) in central Costa Rica southwestward into western Panama (Kappelle 1996).

Costa Rica's southern evergreen cloud forest is part of the Talamanca montane forest ecoregion sensu Olson and Dinerstein (2002). It is vital to the country's society since its many rivers provide drinking and irrigation water to the populations of the main cities in the Valle Central (San José, Alajuela, Heredia, and Cartago), as well as to those in the valleys of the southern Pacific region (e.g., San Isidro del General, Buenos Aires de Puntarenas, San Vito de Java) and the southern Caribbean zone (e.g., Limón, Cahuita, Bribri, Sixaola) (Kappelle 1996, Valerio 1999). Moreover, the many hydroelectric power plants that have been built in cascade in the river systems of the Cordillera de Talamanca provide much of the country's hydropower (e.g., the hydro stations of Angostura, Cachí, and Río Macho in the middle and upper sections of the Río Reventazón basin).

Levels of species diversity and endemism are extraordinarily high in the Talamancan evergreen cloud forest owing to its relative isolation and geological past as an island archipelago, separated from the South American Andes by the Darién gap in eastern Panama, and disconnected from the Chiapas-Guatemalan mountain complex by the Nicaraguan Depression to the northwest. As a result, a large number of species is restricted to the Cordillera de Talamanca or shared only with Costa Rica's northern volcanic *cordilleras* (see Lawton et al., chapter 13 of this volume), making the Talamanca highlands and its La Amistad Biosphere Reserve of global importance (Kappelle and Brown 2001, 2003).

History of Scientific Exploration

Since the early twentieth century the evergreen montane forests of the Cordillera de Talamanca have been explored by a fair number of scientists, though not as much as the cloud forests of the volcanic *cordilleras* in the north where places like the Monteverde Cloud Forest Preserve have received significant attention from dozens of scientists over the past few decades (Nadkarni and Wheelwright 2000; Lawton et al., chapter 13 of this volume). The first naturalists to study the flora of the Talamancan montane forests were Henri Pittier (1957) and Paul Standley (1937), whose work laid the foundation for today's botanical knowledge of the

Fig. 14.2 View of a lower montane cloud forest in the disturbed northern sector of the Cordillera de Talamanca at the 2,270 m tall Pico Blanco, Cerros de Escazú. *Photograph by Marta Juárez 2006.*

country's southeastern highland forests (Hammel et al. 2004). Decades later, prominent authors such as Holdridge et al. (1971) contributed with rapid inventories of the country's forests, including the evergreen oak-dominated cloud forests near Villa Mills (2,500–3,000 m a.s.l.) just east of Cerro de la Muerte.

However, it was not until the early 1980s when more detailed studies on the biodiversity, ecology and sustainable management of the Talamancan montane forests appeared (e.g., see Hartshorn 1983, Gómez 1986, Jiménez et al. 1988, Stadtmüller and Aus der Beek 1992, Chaverri and Hernández 1995, Guariguata and Sáenz 2002, as well as Kappelle 1996, 2006a, 2008, Köhler 2002, Holl and Lulow 1997). These and other authors have contributed considerably to the current knowledge of the patterns and underlying processes that define the composition, structure, and functioning of the montane cloud forests of the Cordillera de Talamanca discussed in this chapter.

The Physical Environment

Climate

The evergreen montane cloud forest zone of southern Costa Rica is generally very moist and cool with moderate water deficits (Nuhn 1978, Coen 1983, Herrera 1986 and chapter 2 of this volume). According to the Köppen climate system, the Talamancan cloud forests enjoy a "Cf" climate with a short dry period generally running from mid-December to April (known as *verano* or "summer"), and a rainy season during the remaining seven to nine months (*invierno* or "winter").

Cloud formation and persistence is often more intense and severe along the Atlantic slopes of the *cordillera* owing to the influence of the Trade Winds that blow from the Caribbean lowlands southward (Coen 1983, and see chapter 13 of this volume by Lawton et al. on the cloud forests of the northern volcanic mountains). As a result, the Atlantic or Caribbean slopes receive much more rainfall and fog than the Pacific slopes, with differing implications for the floristic composition of the vegetation along both slopes (Kappelle 1992). Similarly, the valley of the Río Grande de Térraba near Potrero Grande is relatively dry owing to a clear rain shadow effect, which is reflected in the presence of a xeric vegetation type dominated by deciduous trees (e.g., *Bursera simarouba*).

Throughout the year but mostly in the rainy season, condensation belts with fog penetrating the forest canopy develop, often with their cloud base around 2,000 m elevation (Zadroga 1981, Mulligan 2010). As a result, the relative humidity of the air in the forest interior is normally in the range of 75 to 90%. In oak forests near San Gerardo de Dota along the Pacific slope between 2,500 and 3,000 m, for instance, relative atmospheric humidity levels oscillated between 35–95% in the 1992 dry season and between 70–90% in the 1992 rainy season (Van Dunné and Kappelle 1998).

Depending on slope orientation and cloud persistence levels the average annual temperature in Talamanca's highland forests ranges from 5–11°C at the highest forested peaks and crests over 3,000 m to 13–16°C at an altitude of 2,300 m, and 20–25°C at elevations of about 1,000 m (Herrera 1986, Chinchilla 1987, IMN 1988). Yearly mean temperatures of 25°C are common in the mid-elevation watershed of the Río Grande de Térraba, while the annual average temperature at 3,500 m elevation at the Chirripó Massif only reaches 5°C (Mora Carpio 2000). In general, average yearly temperatures decrease about 0.57°C per every 100 m increase in elevation, as has been calculated on the basis of air temperature records taken at breast height in oak forest interiors at different montane altitudes along the Pacific slope of Cerro Chirripó (Kappelle et al. 1995b).

Herrera (1986 and chapter 2 of this volume) distinguishes a total of eight climate types in the Cordillera de Talamanca: (i) a very warm, sub-humid to dry climate in, for example, the Valle del Río Grande de Térraba and Reserva Indígena Boruca-Térraba (annual averages: temperature >27°C, rainfall 1,300–1,700 mm); (ii) a very warm, sub-humid to humid climate in the Río Grande de Térraba watershed (annual averages: temperature 21–27°C, rainfall 1,550–2,050 mm); (iii) a very warm, humid climate near Paso Real and Potrero Grande in the Río Grande de Térraba watershed (annual averages: temperature 21–27°C, rainfall 1,900–3,100 mm); (iv) a very warm, humid to wet climate in the Río Coto Brus watershed (annual averages: temperature 23–27°C, rainfall 3,100–3,400 mm); (v) a warm to cool or cold, humid to wet climate in, for example, the Parque Nacional Tapantí–Macizo de la Muerte, Parque Nacional Chirripó, Parque Internacional La Amistad, and Zona Protectora Las Tablas (annual averages: temperature 6–26°C, rainfall 1,600–6,800 mm); (vi) a temperate or cold, wet climate in the northern sector of Parque Nacional Tapantí–Macizo de la Muerte and the northeastern sector of Parque Nacional Chirripó (annual averages: temperature 12–15°C, rainfall 4,500–8,000 mm).

Although January is normally the coldest month, during which frost may occur occasionally, monthly temperatures do not change a lot throughout the year. However, temperatures do vary significantly during single days—a typical characteristic of diurnal climates—as is shown by weather data collected near Villa Mills at 3,000 m. Here, the yearly mean temperature is 10.9°C, while daily temperatures can

Fig. 14.3 View of the Cordillera de Talamanca between Cerro Urán and Cerro Chirripó, as seen from the Inter-American Highway between Villa Mills and San Isidro del General. Normally, the montane forests just below the treeless *páramo* belt are bathed in clouds on a daily basis.
Photograph by Maarten Kappelle, 2004

reach minimum values of 6°C during the night and maximum values of 22°C during the day (IMN 1988).

Most precipitation in the high Cordillera de Talamanca is purely orographic and results from the formation of condensation belts (Zadroga 1981, Coen 1983, Fig. 14.3). Average rainfall in the montane forests at the Pacific slope of the *cordillera* ranges from 2,000 to 3,000 mm annually (Reserva Indígena Boruca-Térraba: ca. 2,000 mm; Ojo de Agua: 2,648 mm; Villa Mills: 2,812 mm; Tres de Junio: 3,000 mm; data from IMN 1988). Köhler et al. (2006, 2010) measured incident rainfall (gross precipitation) in montane oak forests at Talamancan Pacific slopes around 2,800 m, and recorded 2,800–2,900 mm per year. Along the Caribbean slopes in the higher parts of the Atlantic zone of the Cordillera de Talamanca, the average yearly rainfall may reach values up to 5,000 mm. This is the case in the upper watershed of the Río Grande de Orosi, in the Parque Nacional Tapantí–Macizo de la Muerte at elevations of 1,500 to 2,500 m (Mora Carpio 2000).

A large part of the total amount of rainwater available in these forests originates from fog and is scientifically known as "horizontal precipitation" (Bruijnzeel 2001). Epiphytes contribute considerably to total rainfall interception in these mist forests by intercepting cloud water from horizontal precipitation (Hölscher et al. 2004).

Geology and Geomorphology

The formation of Costa Rica's present territory began as a result of tectonic activity in the Mesozoic period (Upper Jurassic) with the appearance of the Western Archipelago—a chain of Mesoamerican islands (Lloyd 1963; Alvarado and Cárdenas, chapter 3 of this volume). Later, in the Upper Miocene, a violent uplifting of the region formed the Mesoamerican Isthmus (Lloyd 1963, Seyfried et al. 1987). Whereas the volcanic cordilleras of northern Costa Rica obtained their current form as a result of recent volcanism (Lawton et al., chapter 13 of this volume), the 320 km long Cordillera de Talamanca was formed by the accumulation of Tertiary marine sediments as well as by volcanic activity (Seyfried et al. 1987, Alvarado and Cárdenas, chapter 3 of this volume).

The Tertiary oceanic sediments that form the base of the Cordillera de Talamanca are several kilometers thick and are made up of conglomerates, sandstones, marls, limestones, and siliceous shales. Fossils, however, are rather scarce. The intercalated volcanic rocks consist of stratigraphically associated lavas and pyroclastic deposits with intrusive rocks (plutonic rocks from the Miocene) that represent gabbros, granodiorites (basic), monzonites, quartzites, and intermediate aplitic granites (Weyl 1980, Castillo 1984).

Today, Paleocene-Eocene sediments are found in the eastern part of Parque Nacional Tapantí–Macizo de la Muerte and in the northern sector of Parque Nacional Chirripó. Oligocene-Miocene sediments, in turn, are characteristic of the western sector of Parque Nacional Chirripó and the area south of San Isidro de El General. Miocene-Pliocene rocks of volcanic origin dominate the western part of part of Parque Nacional Tapantí–Macizo de la Muerte and the southern sector of Parque Nacional Chirripó, as well as the indigenous reserves of Cabagra, Salitre, and Ujarrás, and Zona Protectora Las Tablas. Finally, Tertiary and Quaternary sediments are predominant in the valleys of the Río General and Río Coto Brus (Tournon and Alvarado 1997).

In the Cordillera de Talamanca, the effects that Pleistocene glaciers had on the *páramo*-covered alpine vegetation belt over 3,200 m altitude are evident (Weyl 1955, Hastenrath 1973, Bergoeing 1977, Kappelle and Horn 2005, and chapter 15 of this volume). Below the alpine páramo zone, the montane forest belt with its (upper) Montane, Lower Montane, and Premontane zones extends down to about 500 m. It is characterized by a very rugged physiography with highly dissected fluvial land forms, a dendritic drainage pattern, narrow crests, steep slopes, and deep, V-shaped valleys (Van Uffelen 1991). Slopes are typically convex, with angles ranging from 20 to 40 degrees (sometimes up to 50 degrees) (Kappelle et al. 1989).

Today, the terrain of the Talamancan highlands is dominated by land forms of tectonic and erosive origin. Glacial forms dominate only at the highest peaks, which are covered by *páramo* vegetation. Alluvial sedimentation is the predominant geomorphology in the valley bottoms of large basins like the Río General and Río Coto Brus. Land forms of volcanic origin are restricted to the area between San Vito de Java (Coto Brus), Cañas Gordas, and Ciudad Neily (Madrigal and Rojas 1980).

The Cordillera de Talamanca is drained by an extensive network of primary and secondary rivers (Pringle et al., chapter 18 of this volume). The main rivers that originate in the Talamancan *páramos* and montane forests and that drain into the Caribbean Sea are—from north to south—the Río Reventazón, Río Grande de Orosi, Río Pacuare, Río Chirripó Atlántico, Río Telire, Río Estrella, Río Coén, and Río Urén, the latter two of which join to form the Río Sixaola at the border with Panama. In turn, the principal rivers that traverse the Talamancas towards the Pacific coast are the Río Candelaria, Río Parrita, Río Naranjo, Río Savegre, and Río Grande de Térraba or Río Grande de Diquís. The watershed basin of the 160 km long Río Grande de Térraba is the largest in the country and measures up to 507,680 ha (Mora Carpio 2000). It includes two main tributary subsystems: the Río General in the western sector (Valle del General) and the Río Coto Brus in the eastern part of the watershed area.

Soils

Talamancan montane forest soils are generally developed in ashes that originate from the volcanoes in the northern *cordilleras* (Vásquez 1983). These soils are typically dark, deep, and rich in organic material, with medium textures, low levels of fertility, and excessive drainage. Soils that have developed in mountain forest areas are often acid with pH values ranging from 3.7 to 5.5 at depths of 15 cm (Jiménez and Chaverri 1982, Kappelle 1987, Kappelle et al. 1995b).

Inceptisols predominate in the western sector of the Parque Nacional Tapantí—Macizo de la Muerte, south of Buenos Aires de Puntarenas, and around San Vito de Java. Entisols, in turn, are the main soil group in the upper part of the Parque Nacional Chirripó, and in the southern sector of the Fila Costeña. Furthermore, Ultisols are found throughout the Cordillera Talamanca and are widely spread in the eastern sector of the Parque Nacional Tapantí—Macizo de la Muerte, the Valle de El General, the indigenous reserves of Cabagra, Salitre, and Ujarrás, and Zona Protectora Las Tablas (Pérez et al. 1978).

At mid and high elevations (>2,000 m) soils may occur that contain yellow-brown to red-brown residual clays: andepts, tropepts, udults, and ustalfs (Otárola and Alvarado 1976, Vásquez 1983, Van Uffelen 1991, Kappelle et al. 1995b, Kappelle and Van Uffelen 2006, Alvarado and Mata, chapter 4 of this volume). In their pioneering soil study, Otárola and Alvarado (1976) described montane forest soils such as *Lithic Tropofolists* at higher elevations, *Tropohumods* and *Dystrandepts* at intermediate elevations, and *Dystrandepts* at lower altitudes. Similarly, Gómez (1986) relates mountain forests of the Cordillera de Talamanca with *Typic Dystrandepts*, *Typic Placandepts*, and *Andic Humitropepts* associated with *Entic Dystrandepts* and *Andic Tropohumults*.

Soils of mature old-growth oak forests generally have dark brown humus profiles composed of fine organic material, free of litter fragments, and with only little mineral material (Landaeta et al. 1978, Van Uffelen 1991, Hertel et al. 2003). In such volcanic ash–derived soils in the montane zone of the Cordillera de Talamanca (2,000–3,000 m), sand mineralogy is generally dominated by feldspars, though volcanic glass may occasionally occur in soil profiles at lower elevation; similarly, clay mineralogy is consistently amorphous and mainly allophanic (Landaeta et al. 1978). Humus layers are usually less profound along drier Pacific slopes (2,000 and 3,000 m), where they reach a thickness of only 10 to 20 cm. On the contrary, they are much better developed at similar altitudes along the water-soaked Atlantic slopes where they can reach a thickness of up to 40 cm (Kappelle et al. 1995b, Kappelle and Van Uffelen 2006). Here, soil water saturation, in combination with an anaerobic environment and low temperatures, significantly reduces belowground bioactivity and subsequently slows down decomposition rates.

In montane oak forests, the soil carbon pool size ranges from about 500 mol m^{-2} in the organic layer to 12.5 mol m^{-2} in the mineral topsoil. The molar C/N ratios in both soil layers fluctuate between 25 and 28. Similarly, N concentrations range from 100 to 150 mol m^{-3}, and P concentrations from 2.5 to 12 mol m^{-3} (Hertel et al. 2003 and 2006). Furthermore, these authors observed a very large fine root biomass (>1,300 g m^{-2}) in the soils of upper montane old-growth oak forest, which contrasts with other mature forests in the humid tropics that typically have fine root biomass levels below 1,000 g m^{-2}.

Plant Geography and Distribution

In biogeographical terms Costa Rica and its neighboring countries are part of the Central American Floristic Province as defined by Takhtajan (1986). Costa Rica's variety of environments created by different seasonal rainfall patterns, the presence of rugged, mountainous zones and gorges, together with lowlands, rich volcanic soils, the proximity of continental areas rich in species, as well as the area's geological history as an archipelago, all contribute to the country's high biodiversity, especially in the highlands (Burger 1980, Stehli and Webb 1985).

The glacial era influenced Costa Rica's present-day biodiversity considerably and generated a dynamic system of great floristic and faunal heterogeneity (Hooghiemstra et al. 1992). The immigration of plants from mountainous zones in both the north (Mexico, Guatemala) and south (the Andes) played a key role in the development of Costa Rica's montane forest flora (Gentry 1982, Graham 1989, Kappelle et al. 1992, 2000). Opportunities for species migration between North and South America, formally known as the "Great American Biotic Interchange" (GABI) (Stehli and Webb 1985, Graham 1989), increased significantly once the two subcontinents were connected by the Panamanian Isthmus, which was formed some four to five million years ago (Berry 1918, Keighwin 1982, Donnelly 1989).

With regard to modern biogeography there are a few classic studies, such as the pioneering work by Wercklé (1909) and Holdridge et al. (1971) as well as the vegetation synopsis developed by Gómez (1986). More recently, Kappelle et al. (1992) compared the floristic affinity of the highland forests of the Cordillera de Talamanca with similar forests in the Andes, while Islebe and Kappelle (1994) studied similarities between Talamancan and Guatemalan montane forests. A comparison of both studies reveals a higher level of phytogeographic affinity between Costa Rican montane forests and highland forests of the tropical Andes, than between Costa Rican highlands and their equivalents in northern Central America. This is mainly due to two reasons, one of an environmental nature and the other geographic. The first one is that the Guatemalan and Mexican forests receive less rainfall than forests with similar features in Costa Rica and, for example, Colombia; the second is that the Nicaraguan depression and lake basin that separate the mountains of northern Central America (Guatemala) from those in the south (Costa Rica, Panama) act as a barrier to highland species migration between both upland territories (Mexico/Guatemala vs. Costa Rica/Panama).

Around 75% of a total of 253 terrestrial vascular plant genera present in the montane forests of the Cordillera de Talamanca are tropical in distribution (Kappelle et al. 1992). Almost half of these tropical genera are restricted to the Neotropics (46%), with many of them—mostly shrubs and epiphytes—being centered and most diverse in the northern Andes (e.g., *Anthurium*, *Besleria*, *Burmeistera*, *Cavendishia*, *Conostegia*, *Faramea*, *Hoffmannia*, *Philodendron*, *Phoradendron*, and *Palicourea*) (Kappelle et al. 1992). A smaller number (15%) are pantropical in distribution, including treeferns (Cyatheaceae; see Rojas 1999) and genera belonging to families like Araliaceae, Euphorbiaceae, Lauraceae, Myrsinaceae, and Piperaceae. Ten percent are tropical Asian-American genera and include foremost subcanopy trees such as *Cinnamomum*, *Cleyera*, *Magnolia*, *Meliosma*, *Microtropis*, *Persea*, *Styrax*, *Symplocos*, and

Turpinia. Less common (3%) are genera shared between the American and African tropics (Meliaceae and Urticaceae) (Kappelle et al. 1992).

Temperate genera make up almost 18% of the vascular flora and include Holarctic (e.g., *Alnus*, *Quercus*, *Rhamnus*, *Vaccinium*, and *Viburnum*), Austral-Antarctic (e.g., *Drimys*, *Escallonia*, *Fuchsia*, *Gaiadendron*, *Gaultheria*, *Pernettya*, and *Weinmannia*) and wide-temperate genera (e.g., *Geranium*, *Hypericum*, *Rubus*, *Salvia*, *Senecio*, and *Valeriana*). Temperate genera are concentrated in the plant families Asteraceae, Ericaceae, and Rosaceae. Plant genera that can be found everywhere around the world (cosmopolitan genera) are mostly non-woody (e.g., ferns like *Asplenium*, *Blechnum*, *Dryopteris*, *Polypodium*, and *Pteris*, and forbs like *Gnaphalium*, *Oxalis*, *Plantago*, *Solanum*, and *Viola*) (Kappelle et al. 1992).

Holz and Gradstein (2005a) confirm these phytogeographical patterns following a review of floristic affinities of the oak forests' bryophyte flora. They noted the importance of Andean-centered moss and hepatic species, reflecting the close historical connection between the montane bryophyte floras of Costa Rica and South America. Furthermore, high percentages of Central American endemics in the bryophyte flora of these oak forests suggest the importance of climatic changes associated with Pleistocene glaciations for allopatric speciation (Holz and Gradstein 2005a).

Biodiversity at the Species Level

Fungi

The highland forests of the Cordillera de Talamanca are a true storehouse of fungi (Mueller and Mata 2001, Halling and Mueller 2005). Mycological research conducted during the past twenty years has led to a significant increase in our knowledge of mushrooms in these forests, and started with in-depth studies on agarics and boletes by Luis Diego Gómez and Rolf Singer (Halling and Mueller 2005, Mueller et al. 2006). Today, the oak forests of the high Talamancas are considered excellent laboratories for studies on fungi. It is here that mycorrhizal host trees abound (e.g., *Quercus*, *Alnus*, and *Comarostaphylis*; Kappelle 1996). Some species such as *Fistulina hepatica* are restricted to *Quercus* (oak) and *Alnus* (alder) trees (Mueller et al. 2006).

Some 22 species of polypore fungi are commonly encountered in Costa Rican oak-dominated forests. The genus *Phellinus* is richest and represented by at least six species. It is well suited to thrive on decaying oak wood. Other woody or tough macrofungal genera are *Ganoderma*, *Bjerkandera*, *Coltricia*,*Coriolopsis*, *Cyclomyces*, *Daedalea*, *Fistulina*, *Fomes*, *Fuscocerrena*, *Laetiporus*, *Perenniporia*, *Polyporus*, *Tyromyces*, and *Trametes*. Most of the polypore genera occurring in the Talamancan oak forests are cosmopolitan in distribution and seem to have adapted well to strong daily temperature fluctuations, and to high humidity levels throughout the year (Mueller et al. 2006).

The Agaricales (mushrooms and boletes) sensu Singer (1986), which include euagaric, bolete, and russuloid clades sensu Monclavo et al. (2002), is the second largest order of Basidiomycetes found in the Talamanca montane forests (Mueller et al. 2006). Greg Mueller and colleagues collected nearly 400 species of Agaricales in these forests, many of which have been identified only at genus level (e.g., *Agaricus*, *Cortinarius*, *Inocybe*, *Marasmius* sensu lato, *Mycena*, *Psathyrella*, and *Russula*). Roughly half of the 400 agarics are ectomycorrhizal, the other half being putatively saprotrophic. Mueller et al. (2006) estimate that there are perhaps up to 600 agarics that grow in the Talamancan highland forests.

Ectomycorrhizal fungal species are a common feature in Costa Rica's southern highlands. At least forty species occur frequently in the montane oak forests of the Cordillera de Talamanca (Mueller et al. 2006). The most common and diverse genera are *Amanita* (Fig. 14.4), *Boletus*, *Cantharellus*, *Hygrocybe*, *Laccaria*, *Lactarius*, *Leccinum*, *Phylloporus*, *Russula*, and *Tylopilus*. Many of the species in these genera are considered putative ectomycorrhizal macrofungal endemic to the Neotropical oak forests (Mueller et al. 2006).

Less is known about the diversity, distribution, and species composition of saprotrophic fungi. This is mainly due to identification difficulties. Saprotrophic agarics that are most common in the Talamancan highland forests belong to species in genera like *Coprinus* sensu lato, *Crepidotus*, *Galerina*, *Gymnopus*, *Hygropus*, *Hypholoma*, *Marasmiellus*, *Marasmius* sensu lato, *Mycena* sensu lato, *Phaeocollybia*, *Pleurotus*, *Psathyrella*, and *Rhodocollybia* (Mueller et al. 2006).

In another fungal study the diversity of myxomycetes (plasmodial slime molds or myxogastrids; not fungi) was studied in high-elevation oak-dominated forest at 3,100 m near the Cerro de la Muerte Biological Station. In their paper the mycological specialists listed a total of thirty-seven myxomycetic species, including eleven new records for Costa Rica, eight for Central America, and one for the neotropics (Rojas and Stephenson 2007).

Lichens

The mountain environment that prevails in the Talamancan oak forests is an ideal home for lichens. Its high precipitation, frequent fog, moderate temperatures, and excellent substrate—slow-growing hardwood oaks—are indeed very

Fig. 14.4 Fruit bodies of the mushroom *Amanita muscaria* in the oak forests of San Gerardo de Dota, Costa Rica
Photograph by Carlos Serrano, 2009.

suitable for lichen growth (Sipman 2006). As a result, oaks and other trees often show abundant lichen coverage on the bark of their trunks and branches. As Sipman (2006) states, crown twigs may carry loads of the yellowish, bushy beard lichen (*Usnea* spp.), whereas older branches are usually covered with whitish patches of leafy lichens belonging to the families Parmeliaceae and Physciaceae, in particular the genera *Hypotrachyna*, *Parmotrema*, and *Heterodermia*. In more shady situations, large individuals of the genera *Lobaria* and *Sticta* are conspicuous, and most of the bark not covered by these lichens (or by bryophytes) tends to be covered by greyish crustose lichens (Sipman 2006). In fact, macrolichens constitute a key component of the epiphytic flora of both old-growth and successional oak forests in the Cordillera de Talamanca (Holz 2003).

Kappelle and Sipman (1992) presented a first annotated checklist of the lichens that inhabit the old-growth oak forests of the Cordillera de Talamanca. They listed a total of 94 taxa distributed over 66 foliose and 28 fruticose species. The latter become gradually more abundant with increasing elevation, ranging from almost no fruticose lichens at 2,000 m to 50% of all lichen species at 3,400 m altitude. The most species-rich genera were *Hypotrachyna* (19 species), *Cladonia* (16), *Sticta* (10), *Lobaria* (9), and *Usnea* (represented by an unknown number of species). Lichen diversity peaks at both 2,500 and 3,200 m elevation. The first altitude corresponds to the transition from lower to upper montane forest, while the second one coincides with the ecotone between upper montane forest and (sub)alpine vegetation. About two-thirds of the Talamancan montane lichens are shared with the highland forests of the Colombian Andes. *Lobaria pulmonaria* was first reported for Costa Rica by Kappelle and Sipman (1992); the specimens these scholars collected in the Cordillera de Talamanca represent the southernmost distribution of this particular species.

More recent research (Holz 2003, Sipman 2006) highlights the importance of dominant genera like the foliose *Heterodermia*, *Hypotrachyna*, *Leptogium*, *Parmotrema*, and *Sticta*, and the fruticose *Cladonia* and *Ramalina*. Sipman (2006) estimates that the actual epiphytic lichen flora of the neotropical oak forests (Mexico to Colombia) is much larger than the currently known 464 species, and probably close to 1,000 species. This author reports at least 145 species of foliose and fruticose lichens for the montane oak forests of the Cordillera de Talamanca (1,500–3,500 m). If crustose and foliicolous lichens would be included, the total number of highland lichens in southern Costa Rica could rise considerably (Lücking 1992, Schubert et al. 2003).

Plants

Trees, Shrubs, Herbs, Ferns, and Vines

Costa Rica's extraordinary floristic diversity has been the focus of several studies since the beginning of the past century (Wercklé 1909; cf. Gómez and Savage 1983, Gómez 1989). The most important activities carried out by many national and international botanists include the preparation of identification keys and species checklists. With respect to the flora of the montane forests in the Cordillera de Talamanca, specific reference should be made to the works of—in alphabetical order—Alfaro and Gamboa (1999), Blaser (1987), Flores (1990), Gómez (1984, 1986), González (2005), Hartshorn and Poveda (1983), Holdridge et al. (1971), Holz et al. (2002), Kappelle (1987, 1991, 1992, 1995, 1996a, 2001, 2005, 2006b, 2006c, 2008), Kappelle and Horn (2005), Kappelle and Van Omme (1997), Kappelle and Zamora (1995), Kappelle et al. (1989, 1991, 1992, 1994, 1995a, 1995b, 1996, 2000b, 2000c), Oosterhoorn and Kappelle (2000), Orozco (1991), Van Velzen et al. (1993), and Weber (1958, 1959).

Oak forests (*Quercus* spp.)—often with dense *Chusquea* bamboo stands in the understory—dominate the Talamancan highland vegetation (Kappelle et al. 1989, 1992, Kappelle 1991, 1996). Initial studies report a total of at least 253 terrestrial plant genera in 114 families, distributed over 80 tree genera, 77 shrubs, 44 forbs and grasses, 21 vines, and 31 ferns, commonly seen in these oak forests between 2,000 and 3,200 m elevation (Kappelle et al. 1992).

Costa Rica's Instituto Nacional de Biodiversidad (INBio) reports a total of at least 1,735 plant species in 800+ genera and 200+ families for the Área de Conservación Amistad-Pacifico (ACLA-P) which includes the national parks of Tapantí–Macizo de la Muerte and Chirripó, as well as the Costa Rican sector of the international park La Amistad, which is shared with Panama (see INBio's *Atta* database at www.inbio.ac.cr). These species are spread over some 550 dicot trees, 30 palms, 320+ shrubs, 520 herbs, 170 woody lianas and herbaceous vines, 80 ferns, 10 fern-allies (clubmosses, etc.), 35 epiphytes, and some 20 parasites. If the Talamancan highland sector of the neighboring Área de Conservación Amistad-Caribe (ACLA-C) would be included (e.g., the indigenous reserves Chirripó, Tainy, and Telire, and the Reservas Hitoy Cerere and Barbilla), these numbers should perhaps be multiplied by a factor of 1.5.

Kappelle et al. (1991) listed 477 native woody species in 220 genera and 89 families for the highest parts (>2,000 m) of the Cordillera de Talamanca. At lower elevations, in the transition zone from upper montane oak forests down to mixed lower montane and premontane forests in the Amisconde area, Hooftman (1998) recorded a total of 90 genera in 49 plant families in 13,500 m^2 plots between 1,150 and 2,300 m altitude.

Recent species inventories in the Cordillera de Talamanca estimate that there are at least one thousand species of flowering plants in its montane oak forests, spread over more than 400 genera and at least 140 families (N. Zamora, INBio, pers. comm.). This would include both terrestrial and epiphytic vascular species. On average, each vascular family would be represented by an average of three genera and around seven to eight species. The total of 1,000 vascular species would represent about a ninth of the total vascular flora known from Costa Rica, on the basis of estimates made at the beginning of the twenty-first century (Hammel et al. 2004). However, the total number of vascular plants in the Cordillera de Talamanca is still expected to rise since plant species new to Costa Rica and sometimes new to science are collected, reported, and described every year. This is particularly the case in groups like epiphytic orchids and ferns that blanket the branches of tall canopy trees along the Caribbean slope (Quírico Jiménez, pers. comm., 2010), and in understory shrubs as exemplified by the recent discovery of five new *Miconia* species by Kriebel and Almeda (2012).

If the surveys in the high Talamancas would have included all plant species (that is, flowering and non-flowering vascular species, as well as non-vascular plants like bryophytes) the total number would be over 1,700 species. The great diversity of ferns in the Cordillera de Talamanca, for example, makes up one-third of the known pteridophytic flora in Costa Rica (Lellinger 1989, Alexander Rojas, pers. comm.). Furthermore, 88 out of 188 species of monocots in the mountains of south-eastern Costa Rica are orchids.

In the montane oak forests of the Cordillera de Talamanca, the most species-rich woody families are Rubiaceae, Melastomataceae, Lauraceae, Asteraceae, and Ericaceae. They represent about 30% of the total number of recorded species (477) (Kappelle et al. 1991). Several woody species are in fact treeferns (Cyatheaceae), and a few are conifers (Gymnospermae: Podocarpaceae: *Podocarpus* and *Prumnopitys*). Over 25 correspond to monocots like palms (*Geonoma*, *Chamaedorea*, and *Prestoea*), cyclanths, bamboos (*Chusquea* spp.), and vines. More than 90% of the 477 species are dicot trees or shrubs. Around 30% of the woody species belong to only five families: Rubiaceae (34 species), Melastomataceae (32), Lauraceae (28), Asteraceae (25), and Ericaceae (25). Other families of major importance are Myrsinaceae (18 species), Araliaceae (17), Solanaceae (16), Poaceae (15), Rosaceae (15), Loranthaceae (14), and Euphorbiaceae (10). The most species-rich genera are *Chusquea* (15 species), *Miconia* (14), *Ocotea* (12), *Palicourea* (10), *Oreopanax* (9), *Piper* (9), *Rubus* (9), *Solanum* (9), *Ardisia* (8), *Cavendishia* (8), *Hypericum* (8, but mainly on the upper edges of the oak forests), and *Weinmannia* (7) (Kappelle and Zamora 1995).

When evaluating the distribution of species numbers per family, it turns out there is a large number of species-poor woody families; really few families are truly species-rich. Most families (77%) are represented by less than five woody species while only a quarter is present with more than five species. However, this last category comprises two-thirds of all woody species.

While Fagaceae (*Quercus*) is the most dominant family in terms of stature, abundance, basal area, and aerial cover, the most diverse woody families that dominate the subcanopy and understory are Lauraceae (*Cinnamomum*, *Nectandra*, *Ocotea*, and *Persea*), Rubiaceae (*Hoffmannia*, *Palicourea*, *and Psychotria*), Melastomataceae (*Miconia* and *Monochaetum*), Asteraceae (*Ageratina* and *Senecio*), Ericaceae (*Cavendishia*, *Disterigma*, *Gaultheria*, *Macleania*, *Psammisia*, and *Vaccinium*), Myrsinaceae (*Ardisia*, *Cybianthus*, *Grammadenia*, *Myrsine*, and *Parathesis)*, Araliaceae (*Dendropanax*, *Oreopanax*, and *Schefflera*) and Solanaceae (*Cestrum* and *Solanum*) (Kappelle et al. 1995, 1996). Other frequently observed woody genera are *Cecropia*, *Chusquea*, *Cordia*, *Croton*, *Ficus*, *Hyptis*, *Inga*, *Machaerium*, *Peperomia*, *Piper*, *Psidium*, and *Vismia* (Luis González, pers. comm., 2002). Woody and herbaceous vines (*Blakea*, *Bomarea*, *Clematis*, *Dioscorea*, *Hydrangea*, *Mikania*, *Passiflora*, *Schlegelia*, and *Smilax*) are common at lower altitudes and at mid-elevation, but less frequent in forests above 2,800 m (Kappelle et al. 1995a, Kappelle 1996).

Talamanca's montane forests still contain wild varieties of economically important species, such as the avocado (*Persea americana*, Lauraceae). Wild individuals of *P. americana* and its close relative *P. schiedeana* may abound locally at mid-elevation in these forests. The presence of these and other wild crop relatives (e.g., *Phaseolus* beans, *Solanum* tomatoes, and *Vanilla* orchids) underscores the importance of conserving in situ the remaining wild genetic reservoirs of agrobiodiversity still found in these highland forests (Smith et al. 1991 and 1992).

Finally, according to Chaverri et al. (1997), the Cordillera de Talamanca is one of the four areas with greatest levels of endemism in Costa Rica. Perhaps 30–40% of the flora is endemic to the region (e.g., see Talamanca-Caribbean Biological Corridor Commission 1993). This is mainly due to the combination of (i) isolated patches of uncommon habitats, (ii) the presence of cloud forests, and (iii) the existence of mountainous areas that are topographically highly dissected (e.g., Gentry 1992).

Some examples of vascular plant species endemic to the Área de Conservación Amistad-Pacífico (ACLA-P) are *Bursera standleyana*, *Brunellia costaricensis*, *Calathea vinosa*, *Cavendishia talamancensis*, *Chusquea talamancensis*, *Conostegia bigibbosa*, *Dendropanax ravenii*, *Dichapetalum hammelii*, *Elaphoglossum adrianae*, *Eugenia basilaris*, *Macleania talamancensis*, *Miconia kappellei*, *Piper sagittifolium*, *Prumnopitys standleyi*, *Roldana scandens*, and *Solanum longiconicum* (see INBio's *Atta* database).

Vascular Epiphytes

Costa Rica's montane forests—those in the northern volcanic ranges as well as those in the Cordillera de Talamanca—are characterized by trunks and branches laden with vascular and non-vascular epiphytes that compete with each other for space and light (Lowman and Nadkarni 1995). The abundance of this life form in the tropical montane forest zone is mainly due to an almost continuous presence of clouds and mist (Cavalier et al. 1996, Bruijnzeel and Veneklaas 1998, Bruijnzeel et al. 2010a,b, 2011). Condensation belts, which cause persistently high relative air humidity, supply epiphytes with the water and nutrients they need for their germination, establishment, and growth. In this way, the richness of epiphytes contributes substantially to the overall diversity of tropical highland zones (Henderson et al. 1991, Wolf 1994, Bruijnzeel et al. 2010a,b).

Epiphytes occupy a fundamental position in water and nutrient cycles (Nadkarni 1984, 1986, Veneklaas 1990, Hofstede et al. 1993, Tanner et al. 1998, Bruijnzeel et al. 2010a,b). They also inhabit microsites that range from the darkest and wettest places in the understory to the sites most exposed to solar radiation and strong winds in the upper and outer forest canopy (Wolf 1993, 1994). They form mosaics of localized communities that are dominated by particular species that are typical for different microenvironments (Kappelle 2001).

During the past decades, more detailed knowledge has

been generated on the rich epiphyte flora of Costa Rica. Nadkarni (1985) cites more than 120 vascular epiphytes for the Monteverde Cloud Forest Reserve (Cordillera de Tilarán) alone. Many more are expected for the oak forests of the Cordillera de Talamanca. Undoubtedly, the orchids comprise the most diverse group among the epiphytic species present in the Talamancan forests (see Dressler 1993). The genus *Epidendrum* probably has the greatest diversity of species, accompanied by genera such as *Elleanthus*, *Lepanthes*, *Maxillaria*, *Pleurothallis*, *Scaphyglottis*, *Stelis*, and *Telipogon*.

Another notably diverse monocot family is Bromeliaceae, a strictly Neotropical family (Morales 1998). *Tillandsia* and *Vriesea* are particularly characterized by their extraordinary epiphyte diversity. Many species of the genus *Vriesea* fit the concept of "tank bromeliads," which accumulate and retain rainwater in a central tank, to sustain themselves during periods of drought (Benzing 1990, Morales 1998). These tanks often house a large fauna of insects and other animals including salamanders (e.g., see the classic study by the Costa Rican scientist Clodomiro Picado Twight [1913]). For Villa Mills (2,800–3,000 m a.s.l.), a relatively high epiphyte biomass has been calculated (715.16 g/m^2 including 49 species), due to the presence of bromeliads of medium and large size, probably belonging to a species of *Vriesea*, together with shrub species on a 2 m^2 area of stems >10 cm DBH and 1 m^2 of branches >5 cm thick (Gómez 1986). Köhler et al. (2007) calculated that epiphyte mat weight (epiphyte biomass and canopy humus) at the stand level was 16,215 kg per ha in old-growth montane oak forest near San Gerardo de Dota.

Some epiphytic monocots that frequently appear on the branches of *Quercus* spp. are *Tillandsia punctulata* and *Vriesea orosiensis*, in the Bromeliaceae, and *Epidendrum platystigma* and *Maxillaria biolleyi* in the Orchidaceae (Kappelle 1996). Araceae are also common, normally represented by (hemi)epiphytes in the genera *Anthurium*, *Monstera*, *Philodendron*, and *Syngonium*. Less diverse are the Cyclanthaceae (*Asplundia* and *Sphaeradenia*), or the Convallariaceae (*Maianthemum*). Gómez (1986) also mentions *Uncinia hamata* in the Cyperaceae as a locally important epiphytic monocot.

Numerous epiphytic dicot species occur in the Ericaceae: *Cavendishia atroviolacea*, *C. bracteata*, *Disterigma humboldtii*, *Macleania rupestris*, *Psammisia ramiflora*, *Sphyrospermum cordifolium*, and *Satyria warszewiczii*. The hemiepiphytic genus *Clusia* sometimes occurs as a terrestrial tree with stilt roots, while at other times it is found as an epiphytic shrub, occupying sites in the higher part of the oak forest (sub)canopy. For its part, the herbaceous genus *Peperomia* (Piperaceae) has more than 15 epiphytic species, each restricted to a specific vegetation layer. Other families that are represented by many epiphytic species are Araliaceae (*Oreopanax*), Asteraceae (*Liabum*, *Senecio*), Begoniaceae (*Begonia*), Campanulaceae (*Burmeistera*, *Centropogon*), Gesneriaceae (*Alloplectus*), Melastomataceae (*Blakea*, *Topobea*), Rubiaceae (*Hillia*, *Psychotria*, *Relbunium*), and probably Solanaceae (also, see Benzing 1990).

Wagner and Gómez (1983), Lellinger (1989), Kappelle and Gómez (1992), and Mehltreter (1994, 1995) have discussed the presence and abundance of epiphytic and terrestrial pteridophytes in the highlands of the Cordillera de Talamanca. The records of Wagner and Gómez (1983) refer especially to Cerro de la Muerte, which basically includes disturbed *páramo* vegetation. In his master work on the pteridophyte flora of Costa Rica, Panama, and the Chocó region of Colombia, Lellinger (1989) considered Cerro Chirripó to be a very important site for obtaining knowledge on epiphytic ferns in tropical highland areas. An inventory by Kappelle and Gómez (1992) done at this mountain concluded that the most species-rich epiphytic fern genera at Chirripó are *Asplenium*, *Grammitis*, *Hymenophyllum*, *Polypodium* sensu lato, and *Trichomanes*. The genus *Vittaria* (e.g., *V. graminifolia*) is frequently observed, but has low levels of diversity. In turn, *Elaphoglossum* is noteworthy for its incredible wealth of epiphytic ferns (e.g., *E. squamipes*) as well as terrestrial members (Kappelle et al. 1989, and Alexander Rojas, pers.comm.). The cosmopolitan genus *Huperzia*—formerly a part of *Lycopodium*—is a fernally with lots of epiphytic species, commonly found near the forest floor of the oak forests in the high Cordillera de Talamanca.

Vascular Parasites

There are also a number of heterotrophic parasites—often epiphytic at the same time—that inhabit these forests—for example, species in the Eremolepidaceae, Loranthaceae, and Viscaceae. A striking example is the shrub *Phoradendron tonduzii*, which gives a golden color to the crowns of the oaks, such as those found in the valley of San Gerardo de Dota (2,300–2,900 m). The case of the hemiparasite *Gaiadendron punctatum* is a particular one, since this species has been observed while infesting other epiphytes without attacking the phorophyte—the host tree that supports the autotrophic as well as the heterotrophic epiphytes (Benzing 1990).

Bryophytes

Bryophytes are an important component of Talamancan montane forests in terms of ecosystem functioning, biomass, and biodiversity (Holz et al. 2002). They help minimize soil erosion and occurrence of landslides through their sponge effect since they soak up and store water from rain and mist, before releasing it in regular amounts over an extended period of time (Bruijnzeel and Hamilton 2000).

Holz et al. (2002) conducted a complete bryophyte inventory of six hectares of oak forest in the Cordillera de Talamanca and identified a total of 206 species: 100 mosses, 105 hepatics, and one hornwort. Tree bases (69 species), rotten logs (70 species), and soil (70 species) are the richest habitats for bryophytes followed by trunks (61 species), branches of the inner canopy (35 species), twigs of the outer canopy (14 species), and leaves in the understory (14 species). Lejeuneaceae (31 species), Plagiochilaceae (13 species), and Lepidoziaceae (nine species) were the most important liverwort families in terms of number of species. Dicranaceae (nine species), Neckeraceae (seven species), Meteoriaceae (seven species), and Orthotrichaceae (seven species) were the most species-rich families among the mosses (Holz et al. 2002).

Here, the most speciose and abundant genera of hepatics are *Bazzania*, *Ceratolejeunea*, *Diplasiolejeunea*, *Frullania*, *Herbertus*, *Heteroscyphus*, *Lejeunea*, *Lepidozia*, *Lophocolea*, *Plagiochila*, *Porella*, and *Radula*. Similarly, the most species-rich and common moss genera are *Bryum*, *Campylopus*, *Dendropogonella*, *Fissidens*, *Holomitrium*, *Hypnum*, *Leptodontium*, *Leucobryum*, *Macromitrium*, *Meteoridium*, *Neckera*, *Pilotrichella*, *Plagiothecium*, *Polytrichadelphus*, *Porotrichum*, *Prionodon*, *Pterobryon*, *Pyrrhobryum*, *Sematophyllum*, *Squamidium*, *Syrrhopodon*, *Thuidium*, and the peat moss *Sphagnum*.

On the basis of similarities in species composition these authors report that bryophyte microhabitats in the studied forests cluster into three main groups: (1) forest floor habitats (including the tree base); (2) phyllosphere (i.e., the leaf environment); and (3) other epiphytic habitats. The distribution of species and life forms in different microhabitats reflects the vertical variation of humidity and light regimes. At the same time they show the impact of the pronounced dry season and the structural characters (tree height, stratification, number of host tree species) of these oak forests on epiphytic bryophytes compared to more humid forests and upper montane forests of lower stature (Holz et al. 2002).

The biomass of epiphytic bryophytes growing on small stems (1.8 to 2.8 cm diameter at breast height) of montane *Quercus copeyensis* trees appears mostly to be made up of mosses (54–99%), while only 14% of all recorded bryophyte species account for 90% of that "bryomass" (Van Dunné and Kappelle 1998). The most abundant moss and liverwort species that thrive on these small stems are species in the genera *Neckera*, *Pilotrichella* (Fig. 14.5), *Plagiochila*, *Porotrichodendron*, *Prionodon*, and *Rigodium*, which all seem to play a key role in stem flow of understory treelets.

Animals

Invertebrates

Very little is known about the invertebrates of the montane forests. Some studies have been conducted on soil invertebrates of cloud forests, such as those of Buskirk and Buskirk (1976), Nadkarni and Longino (1990), and Kappelle (1996, p. 26). The most abundant groups among the Arthropods (insects and spiders) are the Arachnida, Blattarida, Chilopoda, Coleoptera, Dermaptera, Diplopoda, Diplura, Diptera, Hymenoptera, Isopoda, Oligochaeta, Neuroptera, and Orthoptera (Kappelle 1996).Wesselingh et al. (2000) studied pollination by the highland bumblebee (*Bombus ephippiatus*), one of the most conspicuous insects in the oak woodlands of the Los Santos Forest Reserve. It is hoped that entomologists will increasingly focus their attention on the montane highlands of the Talamanca Mountains as many species new to science are still expected to be revealed in this part of the country.

Vertebrates

Fishes

The ichthyofauna of the Talamancan mountain rivers between 500 and 1,000 m altitude is not as diverse as in the neighboring Atlantic or Pacific lowlands (Bussing 1998). In fact, above 1,000 m freshwater fish diversity is very limited. More specifically, the Cordillera de Talamanca serves as a barrier to the Central American fish fauna, principally cichlids and poeciliids, which migrated mostly from the great lakes of Nicaragua southward along the broad lowlands of Atlantic Costa Rica (Bussing 1998).

At elevations over 2,000 m there aren't any native fish species that naturally inhabit the streams that traverse the montane oak forests or live in the small lagoons that form in the cores of peat bog areas. Exotic rainbow trout (*Oncorhynchus mykiss*), however, has been introduced at these altitudes for commercial and recreational purposes (Kappelle 2008). As in the Colombian Andes, its production in artificial ponds has led to important revenues among local small-holders making a living in towns like San Gerardo de Dota, in the western sector of the Cordillera de Talamanca (Kappelle and Juárez 1995, 2000; Fig. 14.6).

Amphibians

The mountain cloud forests of the Cordillera de Talamanca used to be very rich in amphibians like salamanders, frogs, and toads (Savage 2002). Unfortunately, a skin disease known as chytridiomycosis, first identified in 1998 and caused by a fungus (*Batrachochytrium dendrobatidis*), has triggered the decline of amphibian populations and ultimately the disappearance of a number of frogs and toads

Fig. 14.5 Hanging curtains of *Pilotrichella flexilis* mosses at 2,500 m elevation in the oak forests of San Gerardo de Dota, Costa Rica. *Photograph by Carlos Serrano, 2009.*

in the Costa Rican mountains (Pounds 2001, Pounds and Crump 1994). Apparently the fungus benefits from modern climate change that includes shifting rainfall patterns in Central America's mountains (Pounds et al. 2006). Fortunately, populations of some species seem to escape this fate and are doing rather well in the high Talamancas, including the mountain salamander (*Bolitoglossa pesrubra*, previously known as *B. subpalmata*), a leptodactilyd quark frog (*Eleutherodactylus melanostictus*), the mountain tree frog (*Hyla picadoi*), and the true mountain frog (*Rana vibicaria*) (Kappelle 2008).

The mountain salamander is a small insectivore with moderately webbed hands and feet. Populations prefer boggy sites between 1,500 and 3,500 m elevation. The species is mostly nocturnal and semi-arboreal or terrestrial. Behavioral studies showed that adults tend to hide in tank bromeliads, which often serve as oviposition sites (Vial 1968).

The less common leptodactilyd quark frog belongs to the most species-rich vertebrate genus on Earth: *Eleutherodactylus* has over 500 species (Leenders 2001). It has no toe webs but rather large truncate, emarginated disks on some of its fingers and toes. This species' range is between 1,000 and 2,500 m elevation. It is a nocturnal insectivore that serves as prey to many larger vertebrates like bats (Savage 2002). Normally it hides under rocks and logs. The similarly uncommon mountain tree frog (1,900–2,800 m alt.) is also a nocturnal insectivore, but has clear finger webs (Savage 2002). The more common true mountain frog with its webbed hands and feet, on the contrary, is both insectivorous and carnivorous and preys on arthropods and small mammals (Savage 2002). It prefers dense woods or ponds between 1,500 and 2,700 m, and breeds in shallow ponds or backwaters of very small streams (Kappelle 2008).

Reptiles

Few studies have focused on the reptiles of Talamanca's montane forests (Savage and Villa 1986, Savage 2002, Scott and Limerick 1983, Solórzano 2004). What we know is that several dozens of lizards and snakes inhabit these forests. Some common lizards are the green spiny lizard (*Sceloporus malachiticus*) and the highland alligator lizard (*Mesaspis monticola*). Snakes that can be frequently observed are

Godman's montane pit viper (*Cerrophidion godmani*) and the slender black-speckled palm pit viper (*Bothriechis nigroviridis*).

The green spiny lizard is a small, viviparous insectivore with spine-tipped scales that is preyed upon by birds and snakes (Savage 2002). It takes advantage of solar radiation during the day and is often found on perches such as fence posts, rocks, and dead logs between 600 and 3,800 m. The viviparous highland alligator lizard has yellowish-green or turquoise flecks and lines. It is also active during the day after it warms up by the sunlight. Although it is an insectivore, it may occasionally feed on juvenile *Bolitoglossa* salamanders. Adults often sit on fallen logs, decaying wood, stumps, loose bark, moss mats, or rocks between 1,800 and 3,800 m altitude (Savage 2002).

Godman's montane pit viper and the black-speckled palm pit viper are both small-sized, viviparous predators that feed on small animals including arthropods, frogs, lizards, other snakes, small birds, and rodents such as mice. While Godman's montane pit viper is active during the day and hides on the ground, in low vegetation or near logs, the palm pit viper is mostly arboreal and active during the night. Both species live in the forests between 1,400 and 3,000 m (Savage 2002, Solórzano 2004, Kappelle 2008).

Birds

The diverse avifauna of the mountain forests of the Cordillera de Talamanca has been the subject of several studies (Stiles et al. 1989, Wilms and Kappelle 2006, Gomes et al. 2008). Results show that Talamanca's premontane, lower montane, montane and sub-alpine forests are among the richest bird habitats in the country (Stiles et al. 1989). It is believed that at least 560 species of bird live in, or occasionally visit, the forests of the Parque Nacional Chirripó and the Costa Rican sector of Parque Internacional La Amistad (Boza 1984). For example, Chirripó's montane oak forests

Fig. 14.6 Local villager in San Gerardo de Dota showing a few introduced rainbow trout he has grown in his ponds for commercial purposes.

Fig. 14.7 Tall *Quercus copeyensis* (white oak) tree at 2,700 m elevation at San Gerardo de Dota, Costa Rica. The upper branches are partly covered by epiphytic bromeliads.
Photograph by Maarten Kappelle, 2007.

combining low or moderate reproductive success and a high rate of seed herbivory with factors that cause low production of viable seeds in these woody species. Average flower lifespan for these understory species is 4.4 days, in comparison with longer flower lifespan in arctic and alpine species (Wesselingh et al. 1999).

Herbivory, Frugivory, Seed Predation, and Dispersal

Holl and Lulow (1997) studied the effect of the species, type of habitat (open pasture, forest, and beneath isolated trees in pasture), and distance to the forest edge on seed depredation of 10 animal-dispersed species in the southeastern part of the Cordillera de Talamanca, near Las Alturas Biological Station. Additionally, they compared depredation caused by vertebrates and insects and noted that rabbits cut the stems of 64% of the seedlings of four native species that were planted in an abandoned pasture at 1,500 m altitude (Holl and Quirós-Nietzen 1999). Observations by Holl and Lulow (1997) showed that only 26% of the seedlings had survived two years after having been planted. Moreover, the number of seeds dispersed by animals is normally much greater below branches than in open areas such as pastures, Holl states. These results suggest that seed depredation influences regeneration of the montane forest on degraded lands, although the lack of seed dispersal is apparently the most important limiting factor in their recovery (Holl 1998, and 1999). Another study in southern Costa Rica demonstrated that tropical montane tree seeds survive through germination more often in secondary forests, with high levels of mortality occurring in abandoned pastures and forest fragments (Cole 2009). The same study highlights that the majority of seed mortality results from rodent predation in forest fragments, insects and fungal pathogens in secondary forests, and a combination of desiccation, insects, and fungal pathogens in pastures.

Plant seeds of Talamancan montane forest trees and shrubs are often dispersed by animals, a species interaction known as zoochory (Wijtzes 1990, Ten Hoopen and Kappelle 2006). Certainly, a large percentage of tropical forest bird species consumes fruits and seeds as part of their diet (Stiles 1985). Probably, frugivorous birds are the most important group of seed dispersers in Talamancan high-elevation oak forests, taking into account the low abundance of monkeys and bats at cool and cold elevations (Kappelle 1996, Wilms and Kappelle 2006, Gomes et al. 2008), although some white-faced capuchin monkeys (*Cebus capucinus*) have recently been observed at 2,800 m in the Parque Nacional Los Quetzales (Arsenio Agüero, MINAE, pers. comm., 2010). Now, regarding birds, obligate frugivores may represent around or over 10% of the whole avifauna thriving in these highland forests (Wilms and Kappelle 2006).

During field work in 2001–2002 García-Rojas (2006) evaluated the diet and habitat preference of the frugivorous resplendent quetzal in montane oak-dominated forests between 1,100 and 3,060 m in the Los Santos Forest Reserve. The diet of this subspecies of quetzal includes at least 25 species of fruit trees, thirteen of which are Lauraceae (wild avocados or "aguacatillos"). Other key diet tree species for the quetzal are *Cornus disciflora* and *Symplocos serrulata.*

García-Rojas hypothesized that a positive relationship would exist between quetzal abundance and the availability and abundance of potential food sources. His field results showed that the largest number of quetzals occurred in Lower Montane Wet Forest (33 individuals) followed by Montane Rain Forest (22). Premontane Wet and Rain Forests had the lowest levels of quetzal abundance (5 and 0, respectively). Census data indicated that quetzals span a large altitudinal gradient but concentrate between 2,000 and 3,000 m elevation—the altitude at which wild avocado trees predominate. Quetzal abundance is highest in Montane Rain Forests during the middle of the dry season (February) and highest in Premontane Forests at the end of the dry season (April). This observation led García-Rojas (2006) to the conclusion that quetzals migrate altitudinally to lower elevations as the dry season advances—a pattern that seems to correlate well with a change in food abundance during the fruiting season (e.g., the ripening of wild avocados) that occurs along the altitudinal gradient as a result of phenological differences among altitudinally restricted tree species (García-Rojas 2006).

Next to the resplendant quetzal, dozens of other frugivorous birds play a key role in the dispersal of seeds of an infinite number of ornithochorous trees and shrub genera: *Aiouea*, *Ardisia*, *Beilschmiedia*, *Billia*, *Buddleja*, *Cinnamomum* (= *Phoebe*), *Citharexylum*, *Cleyera*, *Conostegia*, *Cornus*, *Croton*, *Ficus*, *Freziera*, *Fuchsia*, *Guatteria*, *Guettarda*, *Ilex*, *Miconia*, *Monnina*, *Myrica*, *Nectandra*, *Ocotea*, *Palicourea*, *Persea*, *Sapium*, *Solanum*, *Symplocos*, *Vaccinium*, and *Viburnum* (Wilms and Kappelle 2006). Small to medium-sized bird species forage mainly on fruits of fast-growing, light-dependent trees, whereas medium- to large-sized birds prefer the fruits and seeds of slow-growing, mature forest tree species.

Tree species like *Ilex pallida*, *Freziera candicans*, *Fuchsia paniculata*, *Nectandra cufodontisii*, *Viburnum costaricanum*, and *Sapium pachystachys* are of particular importance since their fruits are consumed by multiple bird species (Wilms and Kappelle 2006; Fig. 14.8). These are the kind of trees that are of special interest to forest restoration

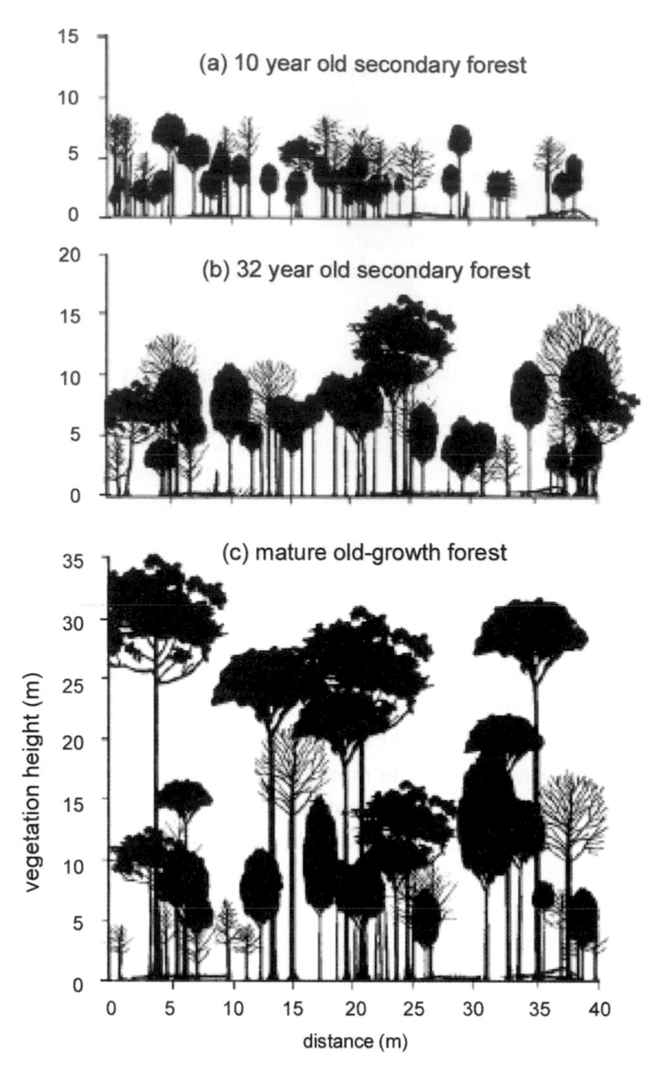

Fig. 14.10 Schematic lateral profile of three successional stages of tropical montane oak forest at around 2,800 m elevation in the Cordillera de Talamanca. (a) 10-year-old, early successional forest; (b) 32-year-old successional forest; and (c) old-growth oak forest over 250 years of age.
Reproduced from Kappelle 2004, with kind permission from Elsevier.

system for leaf blades developed by Raunkiaer and modified by Webb (1959), notophyll and microphyll leaves dominate the foliar spectrum in these oak forests: out of a total of 23 species, 4.3% were nanophylls (0.25–2.25 cm^2), 26.1% microphylls (2.25–20.25 cm^2), 47.8% notophylls (20.25–45.0 cm^2), 13.0% mesophylls (45.0–182.25 cm^2), and 8.7% macrophylls (182.25–1,640.25 cm^2). *Quercus copeyensis* as well as *Q. costaricensis* have notophyllous sizes. The same occurs with the species *Nectandra cufodontisii*, *Ocotea praetermissa*, *Prunus annularis*, *Rhamnus oreodendron*, and *Styrax argenteus*. The largest leaves (mesophylls) have been found in the species *Oreopanax capitatus* and *O. nubigenus*. Comparing successional forests along a gradient (Fig. 14.10), it turns out that average leaf size is 76.6 cm^2 in an early secondary forest (27 species), 34.2 cm^2 in a late secondary forest (31), and 57.8 cm^2 in a mature, undisturbed oak forest (23), respectively.

One notable feature is the presence of very small leaves on species represented by individuals with twisted branches. Among trees of the upper montane belt, *Weinmannia pinnata* has the smallest leaf surface area (per leaflet) and is the only nanophyll species. *Vaccinium consanguineum* (Ericaceae) may be considered a "quasi-nanophyll" species for having leaf surface areas of 3.34 to 5.52 cm^2. *Escallonia myrtilloides* has leaf blades of similar sizes to those of the two species mentioned and belongs to habitats located just below the upper limit of the forest, at 3,200–3,400 m a.s.l. (Gómez 1986, Kappelle et al. 1991, Islebe and Kappelle 1994); there it occupies sites in the sub-alpine elfin forest that borders the *páramo* vegetation (Gómez 1986). *Escallonia myrtilloides* is accompanied by many other woody species with nanophyllous leaves restricted to the sub-alpine environment, e.g., *Hypericum strictum*, *Pentacalia firmipes*, *Pernettya coriacea*, *Vaccinium floribundum, and Weinmannia trianae* (Kappelle et al. 1991). Around 10 km to the east of Jaboncillo de Dota, in the locality of Villa Mills (2,700 m a.s.l.), leaf sizes in the mesophyll class were the most prominent, especially in the sub-canopy of the forest (Dolph and Dilcher 1980). It should be noted, however, that these authors included very few species in their foliar analyses.

The average specific leaf weight of 20 forest tree species in a primary oak forest is 160.1 g m^{-2} (*specific leaf weight = leaf dry weight per leaf area*), while the specific leaf water content fluctuates around 222.1 g m^{-2}. Average levels of leaf nutrients (per unit weight) for 23 species (20 primary and three secondary) are: (i) nitrogen (N/wt): 10.8 mg g^{-1}; (ii) phosphorus (P/wt): 0.9 mg g^{-1}; and (iii) potassium (K/wt): 9.2 mg g^{-1}. The proportion of N/P is 12.0. Leaf nitrogen levels remain below 15 mg g^{-1}. The greatest concentration of phosphorus in leaves occurs in the species *Oreopanax capitatus*. The levels of potassium varied from low levels in *Miconia tonduzii*, to relatively high levels in *Oreopanax nubigenus* (Kappelle and Leal 1996). In general, nutrient concentrations in Costa Rican oak forests are similar to those observed in other tropical montane forests.

Forest Light Regime

Camacho and Bellefleur (1996) studied acclimation to two light regimes in leaves of six tree species from the oak forests in the CATIE/COSUDE Experimental Area, in Siberia de Villa Mills (2,700–2,800 m a.s.l.). These authors shaded groups of seedlings and saplings of the selected species that were growing in light (light and semi-light species sensu

Blaser [1987]: *Schefflera rodriguesiana* [light] and *Quercus copeyensis* [semi-light]), prior to expansion of the leaf blade and until their complete development. Simultaneously, selected species that were growing in shade (shade and semi-shade species sensu Blaser [1987]: *Vaccinium consanguineum* [shade] and *Drymis granadensis* [semi-shade]) were exposed to light. These authors also studied *Quercus costaricensis* and *Weinmannia pinnata*, two species that show characteristics of both groups [semi-light and semi-shade]. In their study, leaf area, blade thickness, stomatal density, specific density, specific weight, and specific water content were evaluated for individuals under the aforementioned experimental treatment as well as for those under natural conditions (the "light controls" and the "shade controls"; see Camacho and Bellefleur [1996]). The results of this analysis suggest that *Quercus copeyensis* and *Drymis granadensis* show a greater potential for acclimation to shade, while *Schefflera rodriguesiana* seems to acclimate better to brighter surroundings; the species *Vaccinium consanguineum*, *Weinmannia pinnata*, and *Quercus costaricensis* have the potential to acclimatize to both environments (Camacho and Bellefleur 1996).

Water and Nutrient Cycling

Talamancan mountain forests often experience an almost diurnal presence of clouds and mist (Kappelle 2006a,d). Although knowledge of the overall effect of clouds through fog or horizontal precipitation on the hydrological input in tropical montane forests is still scanty (Bruijnzeel and Proctor 1995), it has been widely recognized that, compared to other tropical forests, the specific atmospheric humidity regime of these forests represents one of the main factors causing the large array of differences in forest structure and functioning (Bruijnzeel 2001, Bruijnzeel et al. 2010a,b).

Köhler et al. (2006) measured incident rainfall (gross precipitation) in Talamancan montane oak forests, and recorded 2,800–2,900 mm per year, of which 70–75% corresponded to throughfall, 2–17% to stemflow, and 10–25% to canopy interception—depending on the successional age of the forest stands. Their results show that nutrient concentrations in throughfall water exceeded those measured in incident rainfall. In upper canopy trees (*Quercus*), they recorded a pH of stemflow water ranging from 4.2 to 5.7 and noted significantly higher nutrient concentrations in stemflow in these tall trees, than in lower (sub)canopy trees. Concentrations of NO_3^-, NH_4^+, Ca_2^+, and K^+ in cloud water collected in the middle and at the end of the dry season were significantly higher than at the beginning of the dry season (Köhler 2002).Total annual litter production in mature old-growth oak forest was 12,870 kg ha^{-1} per year. Leaves dominated the litter fraction, which contributed to some 56% of total litter (Köhler 2002, Köhler et al. 2006, 2008).

In a complementary study, Hölscher et al. (2003) analyzed nutrient fluxes in stemflow and throughfall in three successional stages of *Quercus copeyensis* dominated upper montane forests in the Cordillera de Talamanca: an old-growth forest stand, an early successional (10-year-old) forest stand, and a mid-successional (≥30-year-old) forest stand (Fig. 14.10). Differences in nutrient fluxes among the successional stages were related to structural characteristics of the stands (stem density, leaf area, and epiphyte abundance).

No difference in the average stand leaf area index between the old-growth forest and the early successional forest was found. However, a significantly higher leaf area was found in the mid-successional forest. Also, large differences in litterfall from non-vascular epiphytes (mosses, liverworts, and lichens) were noted. These differences reflected changes in epiphyte abundance, with highest values in the old-growth forest. Total nutrient transfer via stemflow and throughfall from the canopy to the soil showed only minor differences among the stands.

The stands studied by Hölscher et al. (2003, 2010) differed widely in the ratio of nutrient transport via stemflow to the total nutrient flux by water below the canopy. The K flux with stemflow accounted for 5% of the total in the old-growth forest but it accounted for 17% (early successional forest) and 26% (mid-successional forest) in the secondary forests. The authors concluded that differences in canopy structure and epiphyte abundance in old-growth and secondary forests resulted in large differences in the partitioning of nutrient transport into stemflow and throughfall components although total nutrient transfers via water reaching the soil were similar (Hölscher et al. 2003, 2010).

Hertel et al. (2006) report that the carbon pool (mol m^{-2}) of the organic soil layer is often highest in mature old-growth forests (533) when compared to early (80) and mid (252) successional stands. Similarly, the organic soil C/N ratio (mol mol^{-1}) increases from early (23) and mid (25) successional vegetation to old-growth (27) forests. Concentrations of N and P (mol m^{-3}) as well as the Ca, Mg, and K pools (mmol m^{-2}) also rise as forest stands develop from early into mature phases. Acidity levels (pH measured in H_2O/KCl), however, decrease from 6.0/5.6 to 4.0/3.4 over successional time (Hertel et al. 2006). Furthermore, the analysis of soil samples from the organic layer and the upper mineral soil (0–10 cm) showed that the biomass of fine roots (roots <2 mm in diameter) increased significantly with increasing stand age of the three forest types, showing a seven-times higher biomass in old-growth forest compared to early-successional forest (Hertel et al. 2006).

Recovery

Soil Seed Banks in Fragmented Forest Landscapes

Ten Hoopen and Kappelle (2006) presented the results of a study on changes in the size and composition of soil seed banks (measured as the number and diversity of germinated seedlings from collected soil), with respect to forest proximity and anthropogenic effects along a gradient from the interior of a mature montane forest into a pasture in the Cordillera de Talamanca (2,500 m altitude). For a total of 80 plant species, they recorded 4,940 germinated seeds. Average seedling density was significantly greater in pastures than in forests. Three-quarters of all seedlings corresponded to wind-dispersed seeds. Half of all seedlings recorded belonged to the genera *Hydrocotyle* or *Gnaphalium* and three of every 10 species were Asteraceae or Solanaceae. Species richness did not vary much along the gradient, from the forest to the pasture. However, there were more seedlings per species in the pastures than in the adjacent forest fragments. Most of the woody plant seeds that germinated from soil material collected in forests were dispersed by birds.

Natural Regeneration and Secondary Succession

At least twelve successional plant communities occur between 2,300 and 2,800 m in the Talamancan highlands (Kappelle et al. 1994): six types of pastures, two types of shrublands, and four types of secondary forest. These twelve communities can be grouped into two blocks of six: one representing the lower montane belt (ca. 2,300 m) and one corresponding to the upper montane belt (ca. 2,800 m).

In a complementary study, Kappelle et al. (1995a) presented a chronosequence of 8 to 20-year-old early successional, 25 to 35-year-old "late" successional, and mature forest communities, as observed in plots established around 2,900 m altitude. Along this gradient, a total of 176 vascular species was recorded, distributed over 122 genera and 75 families. Asteraceae was the most species-rich family, including the recently discovered shrub *Roldana scandens*, new to science (Poveda and Kappelle 1992). Species diversity and density was highest in the youngest forest (145 spp.), when compared to late successional (130 spp.) and mature, old-growth forest (96 spp.). This may be explained by the fact that early successional stages harbor many herb and shrub species that invade the montane clearings from subalpine shrublands and alpine *páramo* grasslands (Kappelle 1996).

Data analysis suggested a minimum time of 65 years for a cleared site to return *floristically* (not structurally!) to its original "mature" forest state, excluding the appearance of epiphytes (Kappelle et al. 1995a). However, since recovery does not necessarily follow linear processes, this hypothesis has to be tested. A complementary analysis of the possible recovery rate of the forest structure (that is, tree height and basal area) suggested a period of 84 years as the theoretically minimum time for a stand to reach maturity (Kappelle et al. 1996). In a complementary study, it was reported that species richness of cryptogamic epiphytes (mosses, liverworts, and lichens) in secondary and primary forests were nearly the same, showing that the mature (primary) oak forests are not necessarily more diverse than the successional (secondary) forests (Holz and Gradstein 2005b).

More importantly, the landscape configuration of a stand has to be taken into account when estimating recovering times of specific successional forests. Edge effects such as the proximity of seed sources and availability of seed dispersers play a key role. In fact, edge conditions in montane oak forest landscapes may clearly reduce *or* accelerate vegetation recovery of adjacent patches (Oosterhoorn and Kappelle 2000, Ten Hoopen and Kappelle 2006).

Birds as Agents in Speeding Up Forest Recovery

Gomes et al. (2008) studied the response and tolerance of a frugivorous bird community to anthropogenic habitat disturbance in a fragmented Talamancan montane oak forest landscape. As in the study done by Wilms and Kappelle (2006), bird presence and densities were compared along a disturbance gradient, ranging from open pastures to closed mature forests. During 102 hours of survey along nine transects, a total of forty frugivorous bird species were observed. Nine species responded negatively to increasing levels of disturbance while nine others reacted positively. Results indicate that large frugivores like the black guan, emerald toucanet, and resplendent quetzal are generally moderately tolerant to intermediate levels of disturbance, but intolerant to severe habitat disturbance, and that tolerance levels are often higher for medium and small-sized frugivores such as the chestnut-capped brushfinch (*Atlapetes brunneinucha*), slaty-backed nightingale-thrush (*Catharus fuscater*), Swainson's thrush (*Catharus ustulatus*), and yellow-throated brush-finch (*Atlapetes gutturalis*). In conclusion, it appears that moderately disturbed habitats in Talamancan oak-dominated cloud forests are highly suitable for restoration through natural regeneration aided by frugivorous birds (Gomes et al. 2008). However, it must be recognized that common occurrences of forest birds in human-dominated countryside landscapes like the border region of Costa Rica's highland forests does not necessarily imply that these bird species maintain sustainable populations there; in fact, many species appear to have little prospect of surviving outside the forest, as demonstrated by a study in southern Costa Rica (Daily et al. 2001). At the same time bird richness increases with restoration effort (Reid et al. 2014).

Land Use and Conservation

Human Settlements and the Forest Frontier

The history of the colonization and deforestation of the Los Santos Forest Reserve (LSFR) is exemplary for many parts of the Talamancan highlands. Colonization began in the nineteenth century with the introduction of coffee into the Central Valley, Dota region, and valley of San Isidro del General (Kappelle et al. 1999).

Initially, the migration of settlers occurred spontaneously from the central parts of the country toward the southeast. The coffee-growing village of Santa María de Dota at 1,548 m a.s.l. was the starting point for pioneers who left for the higher parts of the Cordillera de Talamanca (Ureña 1941). This urban center was originally established in 1863 by small farmers or *campesinos* coming from San Marcos de Tarrazú northwest of the town (Schubel 1980, Ureña 1990). However, Ureña (1990) reports that the valley of Santa María had already been occupied in the past by the indigenous *Huetar* people. The discovery of dispersed burial sites in the Río Pirrís valley and the presence of hieroglyphics on exposed rocks confirms this supposition (Chinchilla 1987).

From the 1920s onwards, many young coffee-growers went east in search of new lands, moving the forest frontier further into the high mountains. They followed trails along the Río Blanco and Río Brujo and decided to settle on free lands to make what they called "improvements" (Spanish: *mejoras*). Here, they established towns such as Copey and Trinidad de Dota. These and other settlements in the western highlands of the Talamanca range grew a bit after the construction of the Inter-American Highway by North American engineers in the 1940s (Merker et al. 1943, Schubel 1980). Later, in the 1950s, farmers colonized even more remote areas such as the upper part of the Río Savegre valley where today San Gerardo and Jaboncillo are located (Siles de Guerrero 1980, Geuze 1989, Ureña 1990). The town of San Vito de Java de Coto Brus in the southernmost sector of Costa Rica's Cordillera de Talamanca is another example of a recent agricultural settlement at the fringes of today's montane cloud forest.

Forest Degradation and Land Use

Studies of pollen and charcoal in sediments collected near Las Cruces Biological Station at the border of the La Amistad Biosphere Reserve reveal a nearly continuous record of human forest disturbance in the region for at least two thousand years (Anchukaitis and Horn 2005). However, particularly since the construction of the Inter-American Highway in the early 1940s, Talamancan montane cloud forests have become increasingly threatened by modern commercial exploitation and expansion of subsistence agriculture (Kappelle 1996).

In the mid-1900s, settlers first cut down the oak forest at a rate of 10–15 ha per year. They produced timber for tools, flooring, boat keels, railroad ties, buildings, and wine barrels. They also grew basic grains such as corn (*Zea mays*) and beans (*Phaseolus vulgaris*) and gathered blackberries (*Rubus* spp.) and palm heart. Later, they introduced dairy cattle and pigs, which were initially fed with corn (Schubel 1980, Geuze 1989). In addition, they produced firewood and charcoal (Schubel 1972, 1980, Siles de Guerrero 1980, Eckardt 1982, Pedroni 1991, Aus der Beek and Navas 1993). Pedroni (1991) reports a production of 4,000 tons of charcoal per year in the mid-1980s in the area between El Empalme and Villa Mills, which equaled two-thirds of the country's charcoal supply at that time.

Deforestation (clearing and burning) and landscape fragmentation in general took their toll from the 1950s until the mid-1980s (Schubel 1980, Meza and Bonilla 1990, Kappelle 1993), when changing paradigms triggered the rise of ecotourism and the development of more sustainable alternatives, including trout aquaculture. However, the loss of forest to timber, fuelwood, and land had already led to a mosaic landscape with patches of successional and remnant old-growth forest, pastures, fruit orchards (blackberry, apple, peach, plum), croplands (potato, cabbage, onion, carrot, beet, cauliflower), and settlements (Kappelle and Juárez 1995a, 1997, 2000 and 2006, and Hölscher et al. 2010; see Fig. 14.11). In fact, while until 1956, only 0.3% of the upper valley of the Savegre River around San Gerardo had been deforested, in 1984 already 13.3% of mature old-growth montane cloud forest had been lost (Van Omme et al. 1997). Overall, deforestation rates reached their peak in the 1960s and 1970s and fell subsequently by the end of the 1980s, when land abandonment led to a process of forest recovery on previously cleared lands (Kappelle et al. 1995b).

Kappelle and Juárez (1995a, 2006) studied agro-ecological zonation patterns at 30 farms in six villages located above 2,000 m elevation in the Copey district. They applied a participatory "multi-visit" survey technique, assessing cropping, livestock production, sociology, economics, health, and environmental factors. Their results discuss three agroecological zones: (i) a fruit tree zone between 2,000 and 2,400 m; (ii) a potato field zone between 2,300 and 2,600 m; and (iii) a charcoal production zone between 2,700–3,000 m (Fig. 14.12). At higher elevation (>3,000 m) the main resource use concerns moss collection for ornamental purposes (floral arrangements for Christmas rituals), while below 2,000 m coffee plantations dominate the region.

Today, ecotourism is the most important activity in the district. Tourism facilities abound, particularly in the Savegre valley. Part of the success of ecotourism can be attributed to the presence of the resplendent quetzal, for which bird watchers from around the world travel long distances to the area to see a glimpse of this magnificent and mythic bird. Of course, when exploring ecotourism opportunities in the Cordillera de Talamanca, it will be essential to assess the carrying capacity of the localities to be developed—particularly when considering the remote and fragile sectors of this important highland region (Brenes et al. 2007).

Reforestation

Until date there have been only a few reforestation activities in the high Talamancas. In general, plantations are small (several hectares) and consist of fast-growing exotic tree species like *Cupressus lusitanica* (Mexican cypress), *Pinus* (pine), and *Eucalyptus*. Sometimes, windbreaks and living fences of *Casuarina equisetifolia* (casuarina or Australian pine) are observed. Like many other tree plantations that are not being thinned, these plantations produce excessive shade and can even accelerate erosion, since few other species grow succesfully under their canopies (G. Budowski, pers. comm.). For example, Lines and Fournier (1979) demonstrated that *C. lusitanica* has an allelopathic effect on the germination of some herb seeds. Plantations of promising native species, such as the Andean alder (*Alnus acuminata*), which fixes nitrogen and helps to restore the soil, are uncommon (Kappelle 1996). Holl and Zahawi (2014)

Fig. 14.11 Landscape mosaic of the oak forest zone along the Río Savegre at about 2,300 m near San Gerardo de Dota. The mosaic is composed of old-growth *Quercus copeyensis* forest, young successional stands, pastures with isolated oak and *Buddleja* trees, living fences of cypress, and apple orchards.
Photograph by Maarten Kappelle, 1992.

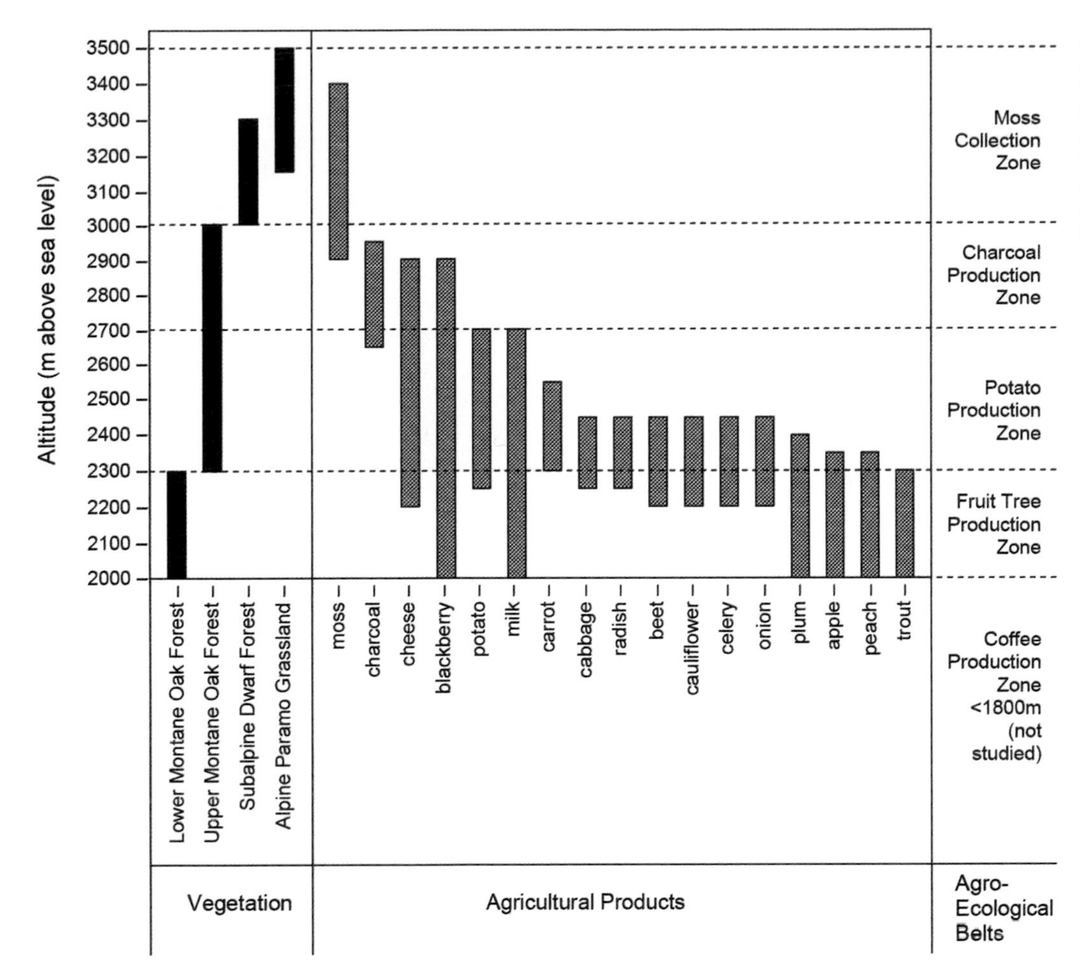

Fig. 14.12 Altitudinal zonation of mountain vegetation, agricultural products, and agroecological belts between 2,000 and 3,500 m elevation in Dota county, Costa Rica.
Reproduced from Kappelle and Juárez 2006, with kind permission from Springer.

report that planting trees at mid elevation in southern highland Costa Rica substantially increases biomass accumulation during the first several years of forest recovery in former agricultural lands.

Ethnobotany

Numerous native and introduced plant species in the montane cloud forest zone of the Cordillera de Talamanca are used by local peoples. Kappelle et al. (2000) and Kappelle and Juárez (2006) report that of a total of 189 species found in the valley of San Gerardo de Dota, 23.8% are used for medicinal purposes, 39.7% for nutrition, and 24.3% in construction (timber) or as fuel (firewood, charcoal). Other less important uses include dyes, ornamental use (ecotourism), forage, gums, oils, and poisons. Trunks (53%) and fruits (47%) are the plant parts that are most often used, followed by leaves (33%) and branches (30%). More than 27% of all the plants are used on a daily basis, while 34.9% are used occasionally and around 11.6% just rarely.

Today, however, native species are used less and less often by local people in this region. The question is whether the knowledge of medicinal and other useful plants will remain a shared good in the community—or simply will disappear while this rural community quickly modernizes (Kappelle et al. 2000).

Emerging Threats

Beyond deforestation (clear cutting, land conversion, fragmentation), many other threats are affecting—or soon will affect—the viability of the biodiversity of the montane cloud forests of southern Costa Rica. Climate change, fires, invasive species, hunting, infrastructure, exploitation, illegal *Cannabis* plantations, etc. are expected to continue to deplete plant and animal populations of this diverse region.

Today, many animal species are becoming more and more threatened by fast road traffic. An example is the killing of a female tapir in August 2009 during a road accident, when it crossed the Inter-American Highway at 2,900 m elevation near the northern limit of Parque Nacional Los

Quetzales at Ojo de Agua. The tapir was pregnant, showed milk production, and died on the spot (Arsenio Agüero, MINAE, pers. comm. 2010; Fig. 14.13).

Furthermore, along the Río Savegre near San Gerardo de Dota, palm heart extraction is still occurring in 2010 (Arsenio Agüero, MINAE, pers. comm.). However, the local population is responding swiftly and in an organized manner to this and other incidents, ensuring hunting and extraction are controlled locally by the community.

Changes in temperature and precipitation due to increased greenhouse gas (GHG) emissions into the atmosphere are already affecting tropical cloud forests and their species around the globe (e.g., Karmalkar et al. 2008, Bruijnzeel et al. 2010a,b, Laurance et al. 2011, and many sources therein). Regional climate models for Costa Rica clearly show an increase in temperature and decrease in precipitation under the A2 climate change scenario. At high elevations in the country, warming is amplified and future temperature distribution lies outside the range of present-day distribution. On the Pacific slope, temperature changes seem greater than at the Caribbean, while elevations of cloud formations are expected to increase considerably along the Pacific side (Karmalkar et al. 2008). These modern climate changes (drying of forest stands) together with human intervention may eventually cause an increase in the frequency of forest fires in cloud forests (Asbjornsen and Wickel 2009), which are historically known to occur in the Talamancan highlands (Anchukaitis and Horn 2005).

Public Protected Areas

Most areas of the montane cloud forests of the Cordillera de Talamanca have a protected status. Core areas include Parque Nacional Chirripó, Parque International La Amistad (PILA), Parque Nacional Tapantí–Macizo de la Muerte,

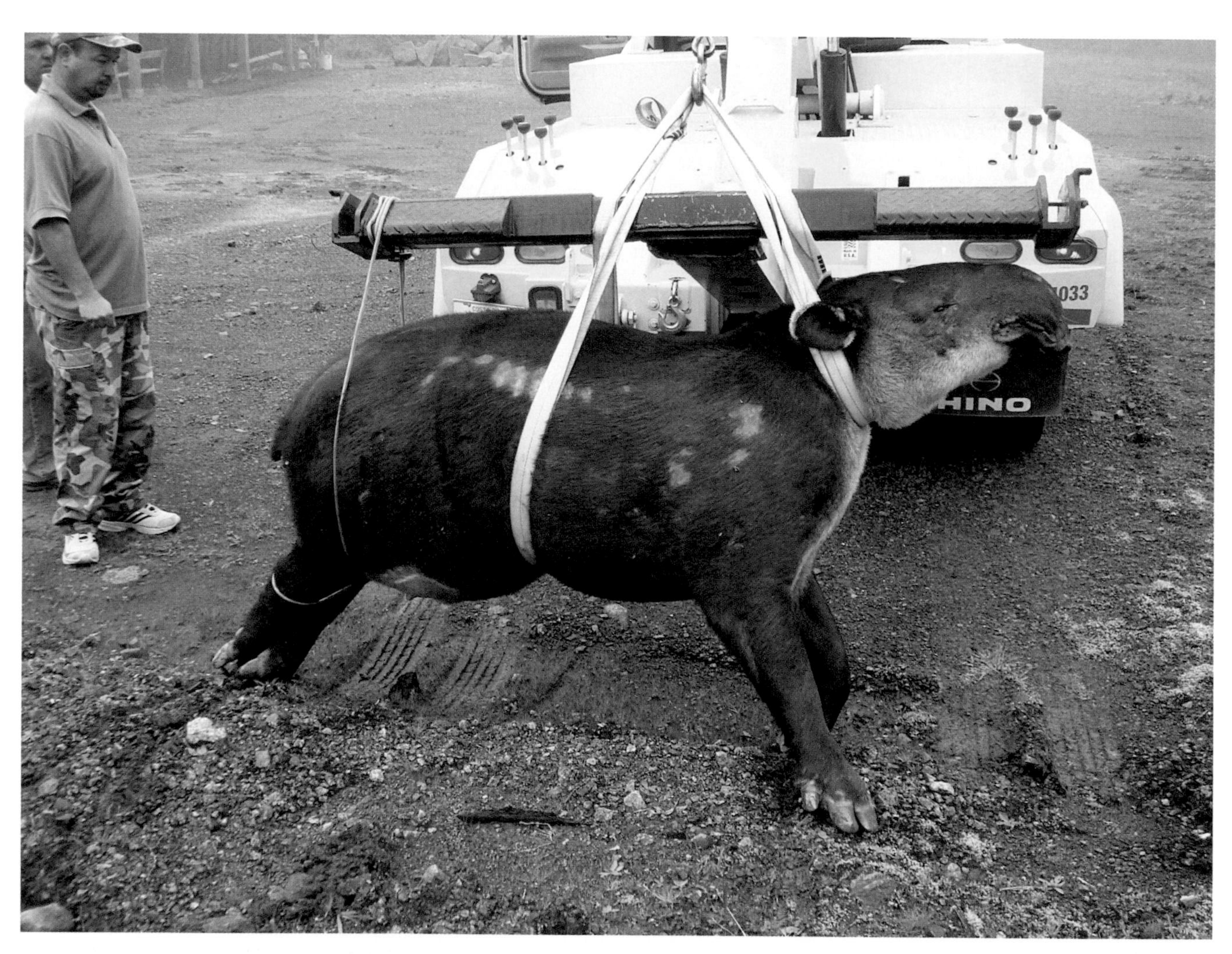

Fig. 14.13 A pregnant tapir killed in a 2009 road accident near Ojo de Agua along the Inter-American Highway is being removed by governmental officials. *Photograph by Carlos Serrano, 2009.*

Fig. 14.14 Group of protected-area managers of the ACLA-P subdivision of the Ministry of Environment, Energy and Technology (MINAET) during a biodiversity training session at Parque Nacional Los Quetzales.
Photograph by Maarten Kappelle, 2010.

Parque Nacional Los Quetzales (Fig. 14.14), Reserva Forestal Río Macho, Reserva Biológica Hitoy Cerere, Reserva Biológica Barbilla, Zona Protectora Las Tablas, Zona Protectora Río Navarro y Río Sombrero, and Reserva Forestal Los Santos (Kappelle and Juárez 1994, Kappelle 1996; Fig. 14.1). Most of these areas are part of the Reserva de la Biosfera La Amistad (RBA), a megadiverse area of 612,570 ha (Castro et al. 1995), equivalent to 12% of Costa Rica's land territory (MIRENEM 1992, Araya and De Marco 2001). It also includes seven indigenous reserves (Cabagra, Chirripó, Salitre, Talamanca, Tayní, Telire, and Ujarrás) that are populated by Bri-Bri, Cabécar, and Naso/Teribe ethnic peoples. Moreover, the RBA includes a number of protected areas in Panama across the border with Costa Rica, such as the Panamanian sector of Parque International La Amistad (PILA) (Aparicio 2006). At a national level, the protected areas belonging to the Costa Rican sector of the RBA have been administratively distributed over two Conservation Areas: Área de Conservación Amistad-Pacifico (ACLA-P) and Área de Conservation Amistad-Caribe (ACLA-C).

The RBA was designated in 1982 through UNESCO's "Man and the Biosphere Program" (MAB) (Bermúdez and Mena 1992). A year later, these forests and its adjacent *páramos* were declared a World Heritage Site (Whitmore 1990, Vernes 1992). It has also been recognized as a Center of Plant Diversity (Groombridge 1992, Harcourt et al. 1996, Chaverri et al. 1997), since it may contain even more than 10,000 species of vascular plants (L. D. Gómez, pers. comm. 1989). It is also an Endemic Bird Area (Harcourt et al. 1996) and an integral part of the species-rich Central American Biodiversity Hotspot (Myers et al. 2000).

In 2003 the highland peat bogs that are embedded within the montane cloud forest zone of the Cordillera de Talamanca were declared a Ramsar Wetland of International

Science for Conservation and Management, 268–74. Cambridge, UK: Cambridge University Press.

Köhler, L., D. Hölscher, and C. Leuschner. 2006. Above-ground water and nutrient fluxes in three successional stages of montane oak forest with contrasting abundance of epiphytes in Costa Rica. In M. Kappelle, ed., *Ecology and Conservation of Neotropical Montane Oak Forests*, 271–82. Ecological Studies Series, vol. 185. New York: Springer Verlag.

Köhler, L., D. Hölscher, and C. Leuschner. 2008. High litterfall in old-growth and secondary upper montane forest of Costa Rica. *Plant Ecology*. DOI: 10.1007/s11258-008-9421-2.

Köhler, L., C. Tobón, K.F.A. Frumau, and L.A. Bruijnzeel. 2007. Biomass and water storage dynamics of epiphytes in old-growth and secondary montane cloud forest stands in Costa Rica. *Plant Ecology* 193: 171–84.

Kriebel, R. and F. Almeda. 2012. Five new species of *Miconia* (Melastomataceae: Miconieae) from Costa Rica and Panama. *Harvard Papers in Botany* 17(1): 53–64.

Landaeta, A., C.A. López, and A. Alvarado. 1978. Caracterización de la fracción mineral de suelos derivados de cenizas volcánicas de la Cordillera de Talamanca, Costa Rica. *Agronomía Costarricense* 2(2): 117–29.

Lanzewizki, T. 1991. Populationsökologische Untersuchungen an Kleinsäugern in einem Eichen-Wolkenwald (*Quercus* spp.) der Montanstufe Costa Ricas. Thesis, Philipps-Universität. Marburg, Germany.

Laurance, W.F., D.C. Useche, L.P. Shoo, S.K. Herzog, M. Kessler, F. Escobar, G. Brehm, J.C. Axmacher, I.-C. Chen, L. Arrellano Gámez, P. Hietz, K. Fiedler, T. Pyrcz, J. Wolf, C.L. Merkord, C. Cardelus, A.R. Marshall, C. Ah-Peng, G.H. Aplet, M. del Coro Arizmendi, W.J. Baker, J. Barone, C.A. Brühl, R.W. Bussmann, D. Cicuzza, G. Eilu, M.E. Favila, A. Hemp, C. Hemp, J. Homeier, J. Hurtado, J. Jankowski, G. Kattán, J. Kluge, T. Krömer, D.C. Lees, M. Lehnert, J.T. Longino, J. Lovett, P.H. Martin, B.D. Patterson, R.G. Pearson, K.S.-H. Peh, B. Richardson, M. Richardson, M.J. Samways, F. Senbeta, T.B. Smith, T.M.A. Utteridge, J.E. Watkins, R. Wilson, S.E. Williams, and C.D. Thomas. 2011. Global warming, elevational ranges and the vulnerability of tropical biota. *Biological Conservation* 144(1): 548–57.

Lawton, R.O. 1980. Wind and the ontogeny of elfin stature in a Costa Rican lower montane rain forest. Ph. D. diss., University of Chicago. Chicago.

Lawton, R.O. 1990. Canopy gaps and light penetration into a wind-exposed tropical lower montane rain forest. *Canadian Journal of Forest Research* 20: 659–67.

Leal, M.E., and M. Kappelle. 1994. Leaf anatomy of a secondary montane *Quercus* forest in Costa Rica. *Revista de Biología Tropical* 42(3): 473–78.

Leenders, T. 2001. *A Guide to Amphibians and Reptiles of Costa Rica*. Miami, FL: Zona Tropical. 305 pp.

Lellinger, D.B. 1989. *The Ferns and Fern-Allies of Costa Rica, Panama, and the Choco*, Part I: Psilotaceae through Dicksoniaceae. *Pteridologia* 2A. The American Fern Society. 364 pp.

Lines, N., and L.A. Fournier. 1979. Efecto alelopático de *Cupressus lusitanica* Mill. sobre la germinación de semillas de algunas hierbas. *Revista de Biología Tropical* 27: 223–29.

Lloyd, J.J. 1963. Tectonic history of the south Central American orogen. *Memoir—American Association of Petroleum Geologists* 2: 88–100.

López-Barrera, F., and R.H. Manson. 2006. Ecology of acorn dispersal by small mammals in montane forests of Chiapas, Mexico. In M. Kappelle, ed., *Ecology and Conservation of Neotropical Montane Oak Forests*, 165–76. Ecological Studies Series, vol. 185. New York: Springer Verlag.

Lowman, M.L., and N.M. Nadkarni. 1995. *Forest Canopies*. San Diego, CA: Academic Press.

Lücking, R. 1992. Foliicolous lichens—a contribution to the knowledge of the lichen flora of Costa Rica, Central America. *Beihefte zur Nova Hedwigia* 104: 1–179.

Madrigal, R., and E. Rojas. 1980. Manual Descriptivo del Mapa Geomorfológico de Costa Rica. Escala: 1:200,000. Oficina de Planificación Sectorial Agropecuario, Secretaría Ejecutiva de Planificación Sectorial Agropecuaria y de Recursos Naturales Renovables. San José, Costa Rica: IGN / MAG / FAO.

Mehltreter, K.V. 1994. Biogeographie und Ökologie der Pteridophyten der Hochgebirge Costa Ricas. Dr. thesis, Universität Ulm. Ulm, Germany. 95 pp.

Mehltreter, K.V. 1995. Species richness and geographical distribution of montane pteridophytes of Costa Rica, Central America. *Feddes Repertorium* 106(5–8): 563–84.

Merker, C.A., W.R. Barbour, J.A. Scholten, and W.A. Dayton. 1943. *The Forests of Costa Rica: A General Report on the Forest Resources of Costa Rica*. Washington, DC: USDA Forest Service and Office of the Coordinator of Inter-American Affairs. 84 + 49 pp.

Meza, T., and A. Bonilla. 1990. *Áreas Naturales Protegidas de Costa Rica*. Cartago, Costa Rica: Editorial Tecnológica de Costa Rica (ETCR). 318 pp.

Miller, K., E. Chang, and N. Johnson. 2001. *Defining Common Ground for the Mesoamerican Biological Corridor*. World Resources Institute (WRI). 45 pp.

Monro, A.K., O. Chacón, N. Zamora, N. Brown, F. González, D. Santamaría, and M. Correa. 2009. Zonas de Biodiversidad del Parque Internacional La Amistad (PILA) Costa Rica-Panama. London, UK: The Natural History Museum, Instituto Nacional de Biodiversidad, and Universidad de Panamá.

Mooring, M., J. Yee, C. Bryce, W. Taylor, and B. Perry. 2010. *Savegre Valley Mammal Study: A Report of Preliminary Results*. San Diego, CA: Point Loma Nazarene University.

Mora Carpio, J. 2000. *Atlas Geográfico del Área de Conservación La Amistad Pacífico (ACLAP)*. Proyecto Proarca/Capas. San Isidro de Pérez Zeledón, Costa Rica: Ministerio de Ambiente y Energía (MINAE) / CCAD / USAID / SINAC / MINAE. 66 pp.

Morales, J.F. 1998. *Bromelias de Costa Rica*. Santo Domingo de Heredia, Costa Rica: Editorial INBio. 180 pp.

Morales, J.F., N. Zamora, and B. Herrera. 2007. Análisis de la vegetación en la franja altitudinal de 800–1500 m.s.n.m. en la vertiente pacífica del Parque Internacional La Amistad (PILA), Costa Rica. *Brenesia* 68: 1–15.

Mueller, G.M., R.E. Halling, J. Carranza, M. Mata, and J.P. Schmit. 2006. Saprotrophic and ectomycorrizal macrofungi of Costa Rican oak forest. In M. Kappelle, ed., *Ecology and Conservation of Neotropical Montane Oak Forests*, 55–68. Ecological Studies Series, vol. 185. New York: Springer Verlag.

Mueller, G., and M. Mata. 2001. The Costa Rican National Fungal Inventory: A Large Scale Collaborative Project. Inoculum. Supplement to *Mycologia* 52 (5): 1–4.

Mulligan, M. 2010. Modelling the tropics-wide extent and distribution of cloud forest and cloud forest loss, with implications for conservation priority. In L.A. Bruijnzeel, F.N. Scatena, and L. S. Hamilton, eds., *Tropical Montane Cloud Forests: Science for Conservation and Management*. Cambridge, UK: Cambridge University Press.

Myers, N., R.A Mittermeier, C.G. Mittermeier, G.A.B. da Fonseca, and J. Kent. 2000. Biodiversity hotspots for conservation priorities. *Nature* 403: 853–58.

Nadkarni, N.M. 1984. Epiphyte biomass and nutrient capital of a neotropical elfin forest. *Biotropica* 16: 249–56.

Nadkarni, N.M. 1985. *Canopy Plants of the Monteverde Cloud Forest Reserve*. San José, Costa Rica: Tropical Science Center (TSC).

Nadkarni, N.M. 1986. The effects of epiphytes on precipitation chemistry in a neotropical cloud forest. *Selbyana* 9: 47–52.

Nadkarni, N.M., and J.T. Longino. 1990. Invertebrates in canopy and ground organic matter in a neotropical montane forest. *Biotropica* 22: 286–89.

Nadkarni, N.M., and N. Wheelwright, eds. 2000. *Monteverde: Ecology and Conservation of a Tropical Cloud Forest*. New York: Oxford University Press.

Naranjo, E.J., and C. Vaughan. 2000. Ampliación del ámbito altitudinal del tapir centroamericano (*Tapirus bairdii*). *Revista de Biología Tropical* 48: 724.

Nuhn, H. 1978. Atlas Preliminar de Costa Rica. Información Geográfica Regional. San José, Costa Rica: Instituto Geográfico Nacional (IGN) and Ofiplan.

Obando, V. 2002. *Biodiversidad en Costa Rica: Estado del Conocimiento y Gestión*. Santo Domingo de Heredia, Costa Rica: INBio / SINAC. 81 pp.

Oldeman, R.A.A. 1983. Tropical rain forest, architecture, silvigenesis and diversity. In S.L. Sutton, T.C. Whitmore, and A.C. Chadwick, eds., *Tropical Rain Forest Ecology and Management*, 139–50. Oxford, UK: Blackwell.

Olson, D.M., and E. Dinerstein. 2002. The Global 200: priority ecoregions for global conservation. *Annals of the Missouri Botanical Garden* 89: 199–224.

Oosterhoorn, M., and M. Kappelle. 2000. Vegetation structure and composition along an interior-edge-exterior gradient in a Costa Rican montane cloud forest. *Forest Ecology and Management* 126: 291–307.

Orozco, L. 1991. Estudio ecológico y de estructura horizontal de seis comunidades boscosas de la Cordillera de Talamanca, Costa Rica. *Colección Silvicultura y Manejo de Bosques Naturales* 2: 1–34. Turrialba, Costa Rica: CATIE.

Otárola, C.E., and A. Alvarado. 1976. Caracterización y clasificación de algunos suelos del Cerro de la Muerte, Talamanca, Costa Rica. Int. Rep. Fac. Agronomía. San Pedro, Costa Rica: Universidad de Costa Rica. 57 pp.

Pedroni, L. 1991. Sobre la producción de carbón en los robledales de altura de Costa Rica. *Colección Silvicultura y Manejo de Bosques Naturales* 3: 1–27. Turrialba, Costa Rica: CATIE.

Pérez, S., A. Alvarado, and E. Ramírez. 1978. Manual Descriptivo del Mapa de Asociaciones de Subgrupos de Suelos de Costa Rica. Escala: 1:200,000. Oficina de Planificación Sectorial Agropecuario. San José, Costa Rica: IGN / MAG / FAO.

Picado Twight, C. 1913. Les bromeliacees epiphytes. *Bulletin des Sciences de France et Belgique* 47: 215–360.

Pittier, H. 1957. *Ensayo Sobre las Plantas Usuales de Costa Rica*. Segunda edición revisada. San Pedro de Montes de Oca, Costa Rica: Universidad de Costa Rica. 264 pp.

Pohl, R.W. 1991. Blooming history of the Costa Rican bamboos. *Revista de Biología Tropical* 39(1): 111–24.

Pounds, J.A. 2001. Climate and amphibian declines. *Nature* 410: 639–40.

Pounds, J.A., M.R. Bustamante, L.A. Coloma, J.A. Consuegra, M.P.L. Fogden, P.N. Foster, E. La Marca, K.L. Masters, A. Merino-Viteri, R. Puschendorf, S.R. Ron, G.A. Sánchez-Azofeifa, C.J. Still, and B.E. Young. 2006. Widespread amphibian extinctions from epidemic disease driven by global warming. *Nature* 439: 161–67. DOI: 10.1038/nature04246.

Pounds, J.A., and M.L. Crump. 1994. Amphibian declines and climate disturbance: the case of the Golden Toad and the Harlequin Frog. *Conservation Biology* 8: 72–85.

Poveda, L.J., and M. Kappelle. 1992. *Roldana scandens* (Asteraceae), una especie nueva de arbusto escandente para Costa Rica. *Brenesia* 37: 157–60.

Powell, V.N.G., and R. Bjork. 1994. Implications of altitudinal migration for conservation strategies to protect tropical biodiversity: a case study of the resplendant quetzal *Pharomachrus mocinno* at Monteverde, Costa Rica. *Conservation International* 4: 161–74.

Reid, J.L., C.D. Mendenhall, J.A. Rosales, R.A. Zahawi, and K.D. Holl. 2014. Landscape context mediates avian habitat choice in tropical forest restoration. *PLOS ONE* 9(3): e90573.doi:10.1371/journal.pone.0090573.

Rojas, A.F. 1999. *Helechos Arborescentes de Costa Rica*. Santo Domingo, Heredia, Costa Rica: Editorial INBio. 173 pp.

Rojas, C., and S.L. Stephenson. 2007. Distribution and ecology of myxomycetes in the high-elevation oak forests of Cerro Bellavista, Costa Rica. *Mycologia* 99(4): 534–43.

Romero, C. 1999. Reduced-impact logging effects on commercial non-vascular pendant epiphyte biomass in a tropical montane forest in Costa Rica. *Forest Ecology and Management* 118: 117–25.

Sánchez, J. 2002. *Aves del Parque Nacional Tapantí, Costa Rica*. Santo Domingo de Heredia, Costa Rica: Instituto Nacional de Biodiversidad (INBio). 235 pp.

Savage, J.M. 2002. *The Amphibians and Reptiles of Costa Rica*. Chicago: University of Chicago Press. 934 pp.

Savage, J.M., and J. Villa. 1986. *Introduction to the Herpetofauna of Costa Rica. Contributions to Herpetology*, No. 3. Athens, OH: Society for the Study of Amphibians and Reptiles.

Schubel, R.J. 1972. A Preliminary Study of Charcoal Making in the Talamanca Mountains of Costa Rica. Technical Report. San Pedro de Montes de Oca, Costa Rica: Organization for Tropical Studies (OTS). 7 pp.

Schubel, R.J. 1980. The Human Impact on a Montane Oakforest, Costa Rica. PhD thesis, University of California at Los Angeles (UCLA). Los Angeles, CA.

Schubert, R., R. Lücking, and H.T. Lumbsch. 2003. New species of foliicolous lichens from "La Amistad" Biosphere Reserve, Costa Rica. *Willdenowia* 33: 459–65.

Scott, N.J., and S. Limerick. 1983. Reptiles and amphibians. In D.H. Janzen, ed., *Costa Rican Natural History*, 351–74. Chicago: University of Chicago Press.

Seyfried, H., A. Astorga, and C. Calvo. 1987. Sequence stratigraphy of deep and shallow water deposits from an evolving island arc: the upper cretaceous and tertiary of Central America. *Facies* 17: 203–14.

Siles de Guerrero, G. 1980. Estudio Socioeconómico y Técnico de Productores de Carbón y Recolectores de Mora y Lana en las Reservas de Río Macho y Los Santos. Informe Técnico 10. San José, Costa Rica: Ministerio de Agricultura y Ganadería (MAG). 32 pp.

Sillett, T.S. 1994. Foraging ecology of epiphyte-searching insectivorous birds in Costa Rica. *Condor* 96: 863–77.

Sipman, H.J.M. 2006. Diversity and biogeography of lichens in Neotropical montane oak forests. In M. Kappelle, ed., *Ecology and Conservation of Neotropical Montane Oak Forests*, 69–81. Ecological Studies Series, vol. 185. New York: Springer Verlag.

Smith, N.J.H., J.T. Williams, and D.L. Plucknett. 1991. Conserving the tropical Cornucopia. *Environment* 33(6): 7–9 + 30–32.

Smith, N.J.H., J.T. Williams, D.L. Plucknett, and J.P. Talbot. 1992. *Tropical Forests and Their Crops*. Ithaca, NY: Comstock Publishing Associates (Cornell University Press). 568 pp.

Solórzano, A. 2004. *Serpientes de Costa Rica*. Santo Domingo de Heredia, Costa Rica: Editorial INBio. 791 pp.

Solórzano, S., A.J. Baker, and K. Oyama. 2004. Conservation Priorities for Resplendant Quetzals based on analysis of mitochondrial DNA control-region sequences. *Condor* 106: 449–56.

Stadtmüller, T. 1987. *Los Bosques Nublados en el Trópico Húmedo*. Turrialba: University of the United Nations (Tokyo) and Centro Agronómico Tropical de Investigación y Enseñanza (CATIE). 85 pp.

Stadtmüller, T., and R. aus der Beek. 1992. Development of forest management techniques for tropical high mountain primary oak-bamboo forest. In F.R. Miller and K.L. Adam, eds., *Wise Management of Tropical Forests: Proceedings of the Oxford Conference of Tropical Forests*, 245–59. Oxford, UK: Oxford Forestry Institute, University of Oxford.

Standley, P.C. 1937. Flora of Costa Rica. *Field Museum of Natural History Botanical Series* 18: 1–1571.

Stehli, F.G., and S.D. Webb, eds. 1985. *The Great American Biotic Interchange*. New York: Plenum Press.

Stiles, F.G. 1985. Conservation of forest birds in Costa Rica: problems and perspectives. In A.W. Diamond and T.E. Lovejoy, eds., *Conservation of Tropical Forest Birds*, 141–68. Cambridge, UK: International Council for Bird Preservation.

Stiles, F.G., A.F. Skutch, and D. Gardner. 1989. *A Guide to the Birds of Costa Rica*. Ithaca, NY: Cornell University Press.

Takhtajan, A.L. 1986. *Floristic Regions of the World*. Translated by T.J. Crovello and A. Cronquist. Los Angeles, CA: University of California Press.

Talamancan-Caribbean Biological Corridor Commission and The Nature Conservancy (TNC). 1993. Work Report Phase 1 of the Talamancan-Caribbean Biological Corridor Project. With five annexes. Limón, Costa Rica: Talamancan-Caribbean Biological Corridor Commission / The Nature Conservancy (TNC).

Tanner, E.V.J., P.M. Vitousek, and E. Cuevas. 1998. Experimental investigation of nutrient limitation of forest growth on wet tropical mountains. *Ecology* 79: 10–22.

Tavares de Almeida, R. 2000. *Evaluando y Valorizando las Áreas Silvestres Protegidas para la Conservación del Jaguar en Mesoamérica: Costa Rica, Área de Conservación La Amistad Pacífico*. Ministerio de Ambiente y Energía (MINAE) and Instituto Nacional de Biodiversidad (INBio). San José, Costa Rica: Universidad Latina de Costa Rica. 63 pp.

Ten Hoopen, G.M., and M. Kappelle. 2006. Soil seed bank size and composition in disturbed and old growth montane oak forests in Costa Rica. In M. Kappelle, ed., *Ecology and Conservation of Neotropical Montane Oak Forests*, 299–308. Ecological Studies Series, vol. 185. New York: Springer Verlag.

Tobler, M.W. 2002. Habitat use and diet of Baird's tapirs (*Tapirus bairdii*) in a montane cloud forest of the Cordillera de Talamanca, Costa Rica. *Biotropica* 34(3): 468–74.

Tobler, M.W., E.J. Naranjo, and I. Lira-Torres. 2006. Habitat preference, feeding habits and conservation of Baird's Tapir in Neotropical montane oak forests. In M. Kappelle, ed., *Ecology and Conservation of Neotropical Montane Oak Forests*, 347–59. Ecological Studies Series, vol. 185. New York: Springer Verlag.

Tosi, J.A., Jr. 1969. Mapa Ecológico de Costa Rica, Basado en la Clasificación Vegetal Mundial de L.R. Holdridge. Escala: 1:750,000. San José, Costa Rica: Centro Científico Tropical (CCT).

Tournon, J., and G. Alvarado. 1997. Mapa Geológico de Costa Rica. Escala: 1:500,000. Cartago, Costa Rica: Editorial Tecnológico de Costa Rica.

Ureña, A. 1990. Reseña Histórica del Cantón de Dota. San José, Costa Rica: Editorial Serrano Elizondo. 379 pp.

Ureña, E. 1941. Monografía de Santa María de Dota. *Revista del Archivo Nacional de Costa Rica* 5(1–2): 69–85.

Valerio, C.E. 1999. *Costa Rica: Ambiente y Biodiversidad*. Santo Domingo de Heredia, Costa Rica: Editorial INBio. 139 pp.

Van den Bergh, M.B., and M. Kappelle. 1998. Diversity and distribution of small terrestrial rodents along a disturbance gradient in montane Costa Rica. *Revista de Biología Tropical* 46(2): 331–38.

Van den Bergh, M., and M. Kappelle. 2006. Diversity and distribution of small terrestrial rodents in disturbed and old growth montane oak forests in Costa Rica. In M. Kappelle, ed., *Ecology and Conservation of Neotropical Montane Oak Forests*, 337–45. Ecological Studies Series, vol. 185. New York: Springer Verlag.

Van Dunné, H.J.F., and M. Kappelle. 1998. Biomass-diversity relations of epiphytic bryophytes on small *Quercus copeyensis* stems in a Costa Rican montane cloud forest. *Revista de Biología Tropical* 46(2): 35–42.

Van Leeuwen, E.M.M. 1988. Estudios Preliminares de Estructura y Dinámica de un Bosque Robledal (*Quercus costaricensis* Liebmann) en las Montanas Altas de la Cordillera de Talamanca, Costa Rica. M.Sc. thesis, Wageningen Agricultural University. Wageningen, Netherlands. 73 pp.

Van Omme, E., M. Kappelle, and M.E. Juárez. 1997. Land cover/use changes and deforestation trends over 55 years (1941–1996) in a Costa Rican montane cloud forest watershed area. In ISSS, AISB, IBG, and ITC, eds., *Abstracts of the 1997 Conference on Geo-Information for Sustainable Land Management*. Enschede, The Netherlands: ITC.

Van Uffelen, J.G. 1991. A geological, geomorphological and soil transect study of the Chirripó Massif and adjacent areas, Cordillera de Talamanca, Costa Rica. M.Sc. thesis, Wageningen Agricultural University. Wageningen, The Netherlands.

Van Velzen, H.P., W.H. Wijtzes, and M. Kappelle. 1993. Lista de especies de la vegetación secundaria del piso montano pacífico, Cordillera de Talamanca, Costa Rica. *Brenesia* 39–40: 147–61.

Vargas, A. 2010. Roble de Dota es el Árbol Excepcional del 2010. *La Nación*, Saturday, June 12, 2010, 18.

Vásquez, A. 1983. Soils. In D.H. Janzen, ed., *Costa Rican Natural History*, 63–65. Chicago: University of Chicago Press.

Veneklaas, E.J. 1990. Rainfall Interception and Above-ground Nutrient Fluxes in Colombian Montane Tropical Rain Forest. PhD thesis, Utrecht University. Utrecht, Netherlands. 109 pp.

Vernes, J.R. 1992. Biosphere Reserves: Relations with Natural Heritage Sites. *Parks* (IUCN) 3(3): 29–34.

Vial, J.L. 1968. The ecology of the tropical salamander *Bolitoglossa subpalmata* in Costa Rica. *Revista de Biología Tropical* 15: 13–115.

Wagner, W.H., and L.D. Gómez. 1983. Pteridophytes. In D.H. Janzen, ed., *Costa Rican Natural History*, 311–18. Chicago: University of Chicago Press.

Webb, L.J. 1959. A physiognomic classification of Australian rain forests. *Journal of Ecology* 47: 551–70.

Weber, H. 1958. Die *Páramos* von Costa Rica und Ihre pflanzengeographische Verkettung mit den Hochanden Südamerikas. Akademie der Wissenschaften, *Abhandlungen der Mathematisch-Naturwissenschaftlichen Klasse* 1956: 120–94.

Weber, H. 1959. Los páramos de Costa Rica y su concatenación fitogeográfica con los Andes Sudamericanos. San José, Costa Rica: Instituto Geográfico Nacional (IGN).

Wercklé, C. 1909. La subregión fitogeográfica costarricense. San José, Costa Rica: Sociedad Nacional de Agricultura, Tipografía Nacional. 55 pp.

Wesselingh, R.A. 1998. Plant reproduction and pollination in a tropical montane forest in Costa Rica. *Acta Botanica Neerlandica* 47(1): 155.

Wesselingh, R.A., H.C.M. Burgers, and J.C.M. den Nijs. 2000. Bumblebee pollination of understory shrub species in a tropical montane forest in Costa Rica. *Journal of Tropical Ecology* 16(5): 657–72.

Wesselingh, R.A., M. Witteveldt, J. Morissette, and H.C.M. den Nijs 1999. Reproductive ecology of understory species in a tropical montane forest in Costa Rica. *Biotropica* 31(4): 637–45.

Weyl, R. 1955. *Contribución a la geología de la Cordillera de Talamanca*. San José, Costa Rica: Instituto Geográfico Nacional (IGN). 77 pp.

Weyl, R. 1980. *Geology of Central America*. Stuttgart, Germany: Gebr. Borntraeger.

Whitmore, T.C. 1990. *An Introduction to Tropical Rain Forests*. Oxford, UK: Clarendon. 226 pp.

Widmer, Y. 1993. Bamboo and gaps in the oak forests of the Cordillera de Talamanca, Costa Rica. *Verhandlungen der Gesellschaft für Ökologie* (Freising) 22: 329–32.

Widmer, Y. 1994. Distribution and flowering of six *Chusquea* bamboos in the Cordillera de Talamanca, Costa Rica. *Brenesia* 41–42: 45–57.

Widmer, Y. 1998. Pattern and performance of understory bamboos (*Chusquea* spp.) under different canopy closures in old-growth oak forests in Costa Rica. *Biotropica* 30(3): 400–415.

Widmer, Y., and L.G. Clark. 1991. New species of *Chusquea* (Poaceae: Bambusoideae) from Costa Rica. *Annals of the Missouri Botanical Garden* 78: 164–71.

Wijtzes, W.H. 1990. Dispersal strategies of upper montane primary forest and secondary vegetation on the Pacific side of the Cordillera de Talamanca, Costa Rica. Technical Report. Amsterdam: Hugo de Vries Laboratory, University of Amsterdam.

Wilms, J.J.A.M., and M. Kappelle. 2006. Frugivorous birds and seed dispersal in disturbed and old growth montane oak forests in Costa Rica. In M. Kappelle, ed., *Ecology and Conservation of Neotropical Montane Oak Forests*, 309–24. Ecological Studies Series, vol. 185. New York: Springer Verlag.

Wolf, J.H.D. 1993. Diversity patterns and biomass of epiphytic bryophytes and lichens along an altitudinal gradient in the Northern Andes. *Annals of the Missouri Botanical Garden* 80(4): 928–60.

Wolf, J.H.D. 1994. Factors controlling the distribution of vascular and non-vascular epiphytes in the northern Andes. *Vegetatio* 112: 15–28.

Zadroga, F. 1981. The hydrological importance of a montane cloud forest area of Costa Rica. In R. Lal and E.W. Russell, eds., *Tropical Agricultural Hydrology*, 59–73. New York: J. Wiley.

Chapter 15 The *Páramo* Ecosystem of Costa Rica's Highlands

Maarten Kappelle[1,2,*] and Sally P. Horn[2]

Definition and Global Distribution of Páramos

Páramo is a grass- or shrub-dominated ecosystem that occupies the cool and wet upper slopes of tropical mountains, at alpine elevations above the timber or tree line and below the snow limit, if present (Cuatrecasas 1968, Cleef 1981, Monasterio and Vuilleumier 1986, Luteyn 1999, Hofstede et al. 2003). At the global level, páramos belong to the biome or major habitat type known as *Tropical Montane Grasslands and Shrublands* (Olson et al. 2001, Hoekstra et al. 2005), which, in structure and physiognomy, is similar to the arctic tundra biome, although páramos are characterized by large diurnal, rather than annual, shifts in temperature.

Páramos in the broadest sense of the biome are found in Latin America (Cuatrecasas 1958, Cleef 1981, Luteyn 1999), Africa (Hedberg 1964, 1992), and Southeast Asia (Smith and Cleef 1988, Hope et al. 2003). In Africa, the term "afroalpine" is used to refer to the páramo zone ("moorlands") along the slopes of Mt. Kilimanjaro (5,895 m elevation), Mt. Kenya (5,199 m), Mt. Elgon (4,321 m), Mt. Rwenzori (5,109 m), and the high mountains of Ethiopia (to 4,550 m) (Hedberg 1964, 1992, Beck et al. 1981), while in Asia the term "tropical alpine" is most common, particularly on 4,884 m high Mt. Jaya (Mt. Carstenz), Irian Jaya, Indonesia, and Mt. Wilhelm (4,509 m), Papua New Guinea (Van Royen 1980, Smith and Cleef 1988, Hnatiuk 1994, Rundel et al. 1994a,b, Johns et al. 2006). Though species groups differ by continent, a number of plant families (e.g., Asteraceae and Ericaceae) and plant growth forms (e.g., stem rosettes, tussock grasses, and cushion plants) occur in tropical alpine ecosystems around the globe (Smith and Cleef 1988, Luteyn 1992, Hofstede et al. 2003).

In Latin America, páramos are mainly restricted to the latitudinal zone between the parallels of 11°N and 8°S (Cuatrecasas 1979, Luteyn 1999), where they cover some 35,000 km^2 (Hofstede et al. 2003) (Fig. 15.1). In Central America, they occur in Costa Rica (Fig. 15.2) and Panama (Weber 1958, Luteyn 1999, Kappelle 2003, Kappelle and Horn 2005, Samudio and Pino 2006), where they have been termed "Isthmian Páramos" (Vargas and Sánchez 2005) to distinguish them from their Andean equivalents (Luteyn 1999). In the Andean cordilleras of South America páramos occur on high peaks and plateaus in Venezuela (Salgado-Laboriau 1979, 1980, Monasterio 1980, Monasterio and Molinillo 2003, Vareschi 1970, 1992), Colombia (Cuatrecasas 1958, 1968, Cleef 1981, Sturm and Rangel 1985, Rangel 2000), and Ecuador (Mena et al. 2001). Continuing southward, the biome extends to the Huancabamba depression in northern Peru (Brack and Mendiola 2000, Recharte et al. 2002, Mostacero et al. 2007), where páramo takes the form of a drier tussock grass community, the *jalca* (Sarmiento 2007). Some of the highest Andean peaks with páramo vegetation on their slopes are Mt. Chimborazo (6,310 m), Mt. Cotopaxi (5,896 m), and Mt. Cayambe (5,840 m) in Ecuador, the "twin peaks" of Cristóbal Colón (5,776 m) and Simón Bolívar (5,776 m) in the isolated Santa Marta Mountains in northern Colombia, Mt. Nevado El Ruiz in central Colombia (5,335 m), and Pico Bolivar (5,007 m) in the Venezuelan Cordillera de Mérida.

An interesting example of non-Andean páramo or páramo-like vegetation occurs in coastal Brazil, on the mountain summits in the Atlantic Forest region where

[1] World Wide Fund for Nature (WWF International), Avenue du Mont-Blanc 1196, Gland, Switzerland
[2] Department of Geography, University of Tennessee, Knoxville, TN, USA
* Corresponding author

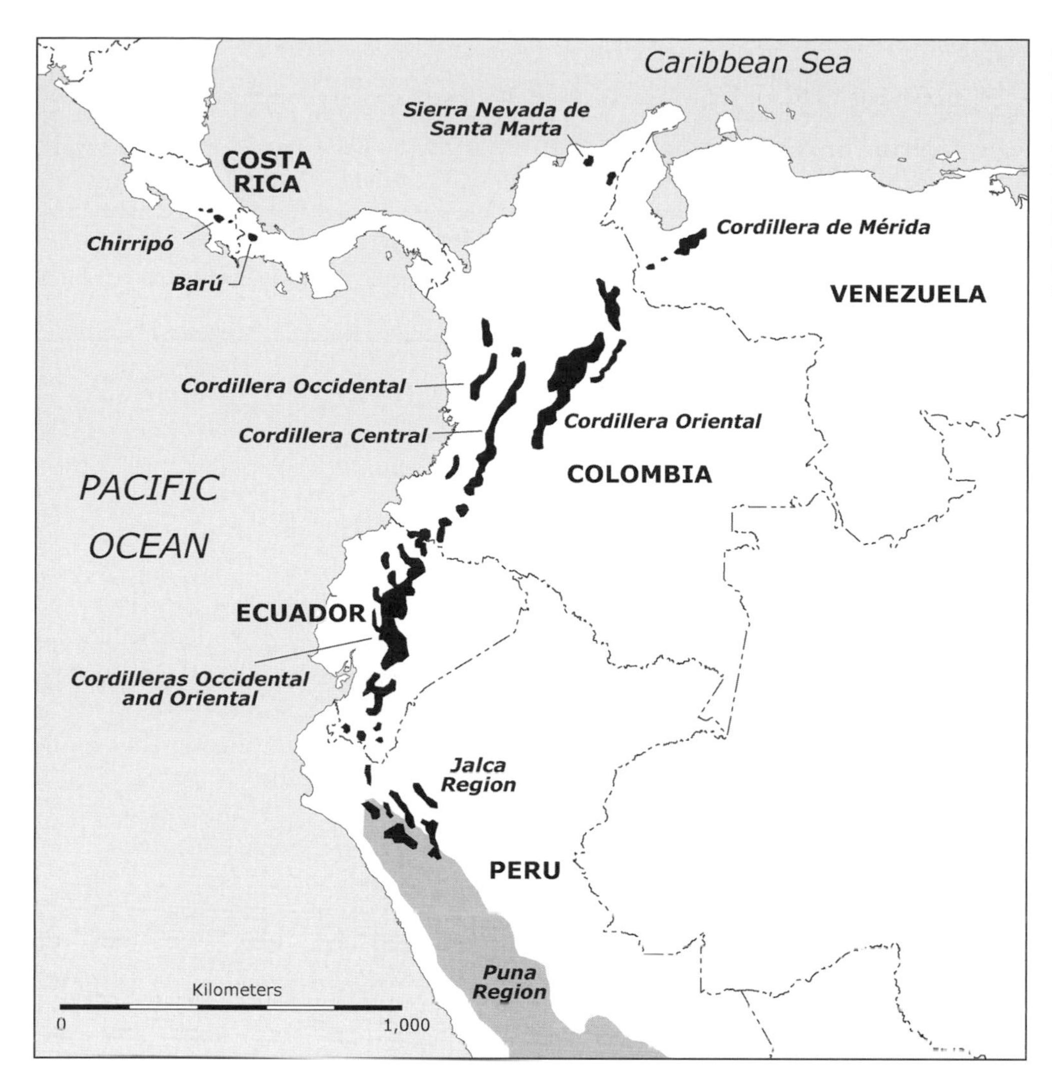

Fig. 15.1 The distribution of *páramo* vegetation in Central America and the northern Andes. The black shading indicates elevations above 3,000 m that are potentially *páramo*, as originally mapped by Jim Luteyn.
Modified from Luteyn 1999 with the permission of the author. Originally published in Horn and Kappelle (2009) and reprinted with the permission of Springer.

Safford (1999a,b) has carried out detailed studies of vegetation dynamics following fires. Full understanding of the affinity of these highland ecosystems with Andean-centered páramos awaits more detailed botanical study (A.M. Cleef, pers. comm.). These Brazilian coastal mountain páramos may potentially show more floristic similarity with non-páramo vegetation on elevated sandstone plateaus in the Amazon basin of Colombia and Venezuela (Cleef and Duivenvoorden 1994).

Neotropical alpine ecosystems that are similar to Andean and Isthmian páramo, though much drier and less species-rich, are found at northern latitudes in Guatemala on the Tajamulco Volcano (at 4,220 m the highest point in Central America), the Tacaná Volcano (4,110 m), and the high peaks of the Altos de Chiantla in the Sierra de los Cuchumatanes (Hastenrath 1968, Islebe and Kappelle 1994, Islebe and Cleef 1995, Steinberg and Taylor 2008), as well as in Mexico along the slopes of the volcanoes Citlaltepetl (Pico de Orizaba, 5,754 m), Popocatépetl (5,452 m), and Iztaccíhuatl (5,286 m) (Beaman 1959, Hastenrath 1968, Rzedowski 1978, Almeida et al. 1994). These alpine meadows are often dominated by bunch grasses and are locally known as *zacatonales* ("zacate" means grass in the Aztec language Nahuatl). Steinberg and Taylor (2008) consider the zacatonales in the Sierra de los Cuchumatanes to be páramo grasslands, but most researchers place the northern limit of true páramo in Costa Rica (Kappelle and Horn 2005).

Beginning at approximately 8°S and extending to 30°S, neotropical wet páramo is replaced by arid *puna* ecosystems characterized by low precipitation, cold temperatures, and sparse, species-poor vegetation of low stature (Cabrera 1968, Quintanilla 1983). The puna stretches over a high plateau (*altiplano*) and covers the Andean slopes of central and southern Peru, Bolivia, northern Chile, and northwestern Argentina.

Isthmian and Andean páramos have their lowest altitudinal occurrences close to 3,000 m on smaller mountains and around 3,500 m on higher peaks, where—under natural conditions—they border montane (Andean) or subalpine (elfin) closed-canopy forests (Luteyn et al. 1992, Luteyn 1999, Kappelle and Brown 2001, Kappelle 2005d). In Costa Rica the transition from closed-canopy montane forest to alpine, treeless páramo vegetation occurs at around 3,100 m on the Buenavista massif of the Cordillera de Talamanca (informally known as the "Cerro de la Muerte"; Fig. 15.3), and near ca. 3,300 m on the higher Chirripó massif, with variations between windward and leeward slopes (Weber 1959, Kappelle 1991, 1992, Chaverri 2008), and at some sites between burned and unburned mountainsides (Chaverri et al. 1976).

On high mountains in the Andes, neotropical páramos

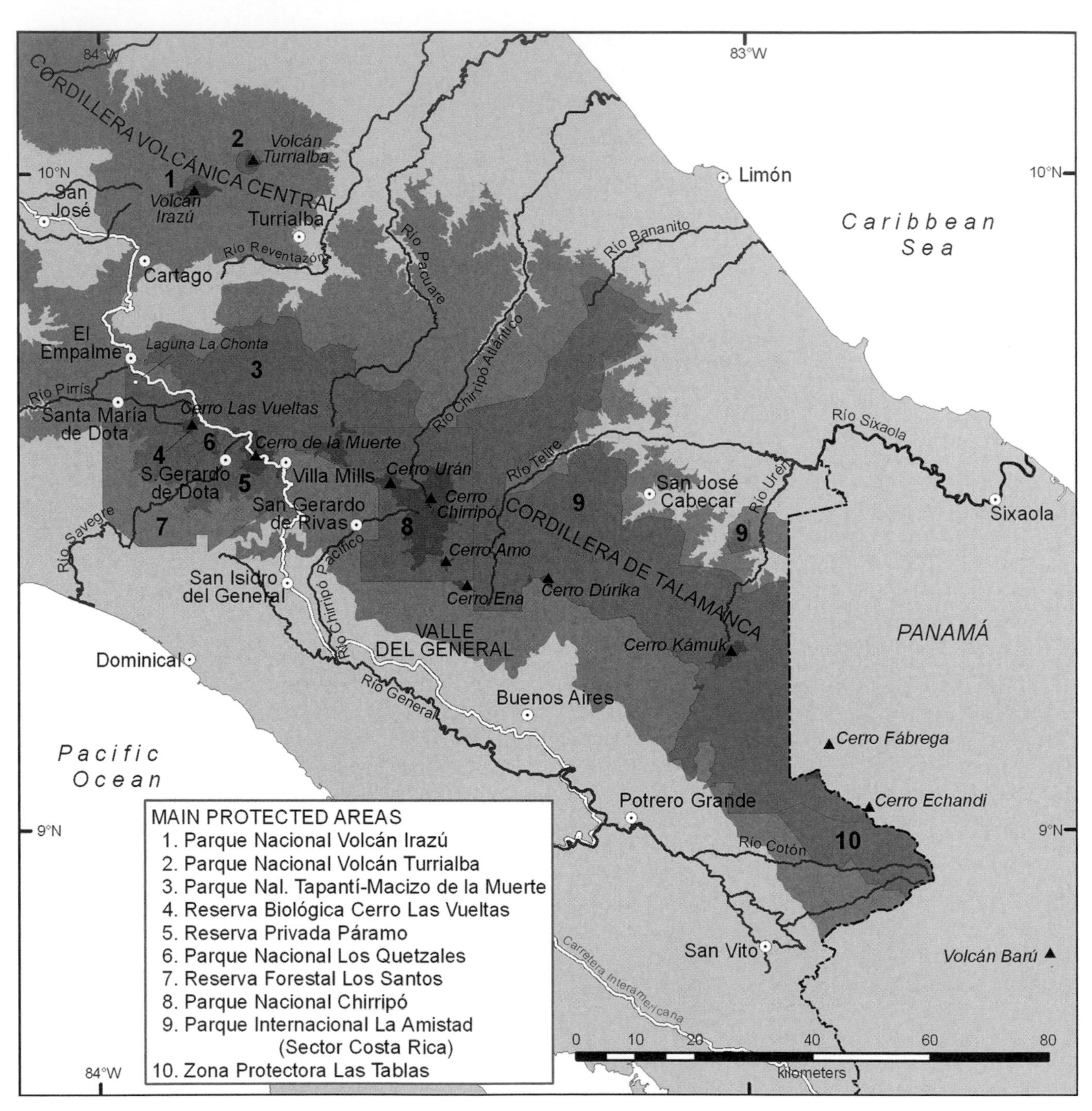

Fig. 15.2 Map showing the main peaks covered by *páramo* and the location of protected areas that conserve *páramo* ecosystems in the Costa Rican highlands. *Map prepared by Marco V. Castro.*

Fig. 15.3 View of the Buenavista massif (Cerro de la Muerte), looking towards the southeast with the town of Quepos visible in the distance. *Photograph by Carlos Serrano.*

have their upper limits between 4,500 and 5,000 m elevation, where they border the nival belt characterized by snow-capped peaks (Cuatrecasas 1958, Weber 1958, Troll 1959, 1968, Cleef 1981, Sturm and Rangel 1985, Luteyn 1999). On lower neotropical mountains with peaks between 3,000 to 4,500 m, as in Costa Rica and Panama, páramos may lack the fully developed upper páramo belt known as *superpáramo* (Chaverri and Cleef 1996, 2005, Kappelle et al. 2005b).

The páramo landscape in Costa Rica underwent repeated glaciation during the late Quaternary (Weyl 1955a, Hastenrath 1973, Horn 1990a, Orvis and Horn 2000, Lachniet and Seltzer 2002), as did Andean páramos (e.g., Smith et al. 2005). The buildup, advance, and subsequent decay of valley glaciers and small ice caps throughout the highland neotropics shaped páramo landscapes during repeated cycles of cooling and warming. In Costa Rica the effects of past glaciation are particularly evident on the Chirripó massif, where glacial erosion and deposition left behind a picturesque alpine landscape dotted by over 30 glacial lakes (Horn et al. 1999, 2005; Fig. 15.4; see also Horn and Haberyan in this volume). Páramo lakes in Costa Rica, as throughout the Andes, form the headwaters of major rivers upon which lowland populations depend (Luteyn 1999).

History of Scientific Exploration in Costa Rican Páramos

Scientific exploration of Costa Rican páramos by European and North American scholars started at the end of the nineteenth century when a first geological profile of the páramo-dotted Cordillera de Talamanca was prepared by geologist Robert T. Hill (1898). He based his ideas on observations previously made by paleontologist William M. Gabb (Gabb 1874a,b, 1895) who had visited Costa Rica after being hired in 1873 by entrepreneur Minor C. Keith, who led the construction of the railroad between Puerto Limón and San José (Denyer and Soto 1999). According to Swiss botanist Henri F. Pittier (1891), Costa Rica's highest peak, Cerro Chirripó, was still completely unexplored at that time. In Guatemala, however, German naturalist Moritz F. Wagner

(1866) had already noted the presence of tropical alpine vegetation on that country's highest mountains.

Costa Rica's páramo-covered peaks were believed to be sacred by the indigenous peoples of the area, the Cabécar and Bri-Bri tribes (Stone 1961). For that reason, the peaks were still mostly unexplored upon the arrival of the Spanish colonists in the early sixteenth century, although indigenous trails existed across the Talamancan mountain range, and were used by early explorers crossing from the Atlantic to the Pacific slope and vice versa (Gomez 2005a). The first non-indigenous explorer to ascend the summit of Cerro Chirripó was likely the German priest and missionary Agustín Blessing Presinger, who climbed the peak in 1904 (Kohkemper 1968, Gómez 2005a). Five years later, though, the botanist Karl Wercklé (1909) stated that alpine páramo vegetation was absent in Costa Rica. However, Wercklé did report the presence of some alpine plant life forms—samples of which are still stored in Costa Rica's National Herbarium—in the more accessible northwestern sector of the Cordillera de Talamanca (Gómez 1978). The area Karl Wercklé visited was today's Dota County, which includes Cerro de las Vueltas, previously known as *Páramo del Abejonal*, and other peaks of the Buenavista massif (Gómez 2005a). Wercklé stressed the need for further exploration to confirm this assumption (Gómez 1978). Apparently, Wercklé had not met with Blessing or heard about the impressions from his visit to Chirripó's peak (Gómez 2005a).

In 1920, the American mining geologist Wickland and German colleague Ruin made the first geological journey to the Chirripó páramo, for the purpose of collecting rock samples. They were followed by the Swiss botanist Walter Kupper, who ascended the high peaks in 1932 and collected the first plant specimens (Kohkemper 1968, Grayum et al. 2004). These specimens were later identified and described as species new to science by German taxonomist Karl Süssenguth (1942). Only five years later, the US botanist Paul Standley (1937) of Chicago's Field Museum was the first to note the significant similarity between the flora of Costa Rica's páramo and the plant species composition of highland Andean South America (Gómez 2005a). Standley was able to reject Wercklé's hypothesis that no páramo was present in Costa Rica.

Fig. 15.4 Photo mosaic of the Chirripo *páramo*, showing evidence of past glaciation. Clockwise from upper left: Lago Chirripó, a glacial tarn occupying a bedrock basin in the cirque floor at the head of the Valle de los Lagos, and two downstream tarns dammed by rock thresholds; Cerro Chirripó, a glacial horn produced by the headward erosion of glaciers in three cirques; glacial striations in the Valle Talari; Lago de las Morrenas 1, the largest lake in the Valle de las Morrenas, also a tarn quarried into bedrock (see also Figs. 19.7 and 19.8 in chapter 19 of this volume, by Horn and Haberyan); the U-shaped valley carved by the Pirámide glacier, on the eastern flank of Cerro Chirripó; glacially smoothed rock exposure in the upper Valle Talari (Valle de los Conejos).
Photo credits—Lake photographs by Chad Lane, others by Sally Horn. (Photo of U-shaped valley carved by the Pirámide glacier from Lachniet and Seltzer 2002.)

German geologist Richard H. Weyl visited Cerro Chirripó in 1955, accompanied by Hannes Ihrig, and prepared a first map of the Chirripó massif (Gómez 2005a). Together with Costa Rican engineer Féderico Gutiérrez Braun he made a reconnaissance flight and gave names to several high peaks (Gutiérrez-Braun 1955). Weyl (1955a,b, 1956a,b, 1957) was the first to report evidence of past glaciation on the massif; his initial observations of glacial features have been confirmed and extended through field studies by other earth scientists (e.g., Hastenrath 1973, Horn 1990a, Orvis and Horn 2000, 2005, Lachniet and Seltzer 2002, Lachniet et al. 2005).

In the same period that Weyl conducted his geological research, German botanist Hans Weber visited Chirripó with Alfonso Jiménez Muñoz, at that time curator at Costa Rica's National Herbarium (Gómez 2005a). Weber studied Chirripó's páramo flora in detail (Grayum et al. 2004) and produced an extensive report on his botanical observations (Weber 1958, 1959). In his publications he stressed the striking floristic affinity between Costa Rica's (isthmian) páramo and its Andean equivalents, reconfirming Paul Standley's earlier conclusion. Weber's botanical account marked a change in the study of Costa Rican páramos, catalyzing field trips aimed at the collection and description of flowering plants, ferns, lichens, and fungi (Gómez 2005a).

Between 1963 and 1968, Costa Rican botanist Luis Diego Gómez Pignataro—to whom this book is dedicated—organized a series of botanical collection trips to the Buenavista (or Cerro de la Muerte) massif (including the 3,491 m high Cerro Buenavista), Cerro Cuericí (also known as Cerro Chirripocillo), Cerro Urán, and Cerro Chirripó (Kohkemper 1968, Gómez 1986). Luis Gómez was inspired by Warren Herbert Wagner Jr., who in 1967 taught him pteridology on field trips to Cerro de la Muerte, during which Gómez collected, identified, and listed numerous fern species (Gómez 1978, Gómez 1994a,b).

On several of his later trips Gómez was accompanied by Australian-based, US-born botanist Arthur S. Weston, who collected several thousand plant specimens in the páramos over repeated visits in the 1970s. Among his collections was a new genus, *Westoniella*, in the Asteraceae (Cuatrecasas 1977), which has 5 species in Costa Rica (four of which are endemic) and 2 in Panama. Weston is among only a handful of botanists to have visited and collected on the more remote páramo-covered peaks in Southeast Costa Rica, including Mts. (or *cerros*) Amo, Dúrika, Dudu, and Kámuk, as well as on nearby Panamanian Cerro Fábrega. Weston described his observations of the páramo flora of these previously unexplored mountains in two highly valu-

environment (Holdridge 1967). The map of Costa Rican Life Zones shows a presence of subalpine rain páramo on the Chirripó massif and the summit of Cerro Kámuk (Tosi 1969), covering only 0.2% of Costa Rica's total land area (Holdridge et al. 1971). However, both Holdridge et al.'s analysis and Tosi's life zone map did not cover all páramo patches that exist in Costa Rica—which actually add up to around 0.3% of the country's territory (Kappelle and Horn 2005).

Isthmian páramos are normally surrounded by tropical montane cloud forests (Kappelle 2001, chapter 14 of this volume), generally dominated by tall oaks in the genus *Quercus* (Blaser 1987, Kappelle 1995, 1996, 2006). These oaks are of Holarctic origin and migrated southward through North America hundreds of thousands of years ago (Hooghiemstra et al. 1992, Kappelle et al. 1992; Kappelle, chapter 14 of this volume). Today, the isthmian páramos share many plant species with adjacent montane oak forests (Cleef and Chaverri 1992, 2005, Kappelle et al. 1992, Vargas and Sánchez 2005).

Some authors consider the isthmian mountains of Costa Rica and Panama (e.g., the Cordillera de Talamanca) to be a disconnected, northwestern extension of the Andean *cordilleras* of South America (F. Sarmiento, pers. comm.). These isthmian páramos on the summits of Central American mountains can be considered biogeographic "islands in the sky" located in the northwestern part of the Neotropical páramo "archipelago" (Luteyn 1999, Cleef and Chaverri 2005, Vargas and Sánchez 2005).

Costa Rica's northwesternmost páramos occur on the peaks of the Irazú (3,432 m) and Turrialba (3,340 m) volcanoes in the Cordillera Volcánica Central, while the core páramo area is situated further southeast, in the Cordillera de Talamanca, and distributed over three protected areas: Tapantí-Macizo de la Muerte National Park (e.g., Cerro Vueltas, 3,156 m; Cerro Buenavista, 3,491 m), Chirripó National Park (Cerro Cuericí, 3,345 m; Cerro Urán, 3,280 m; Cerro Chirripó, 3,819 m; and another 25 peaks over 3,000 m; for details see map in Kappelle and Horn, 2005, p. 765), and the western sector of the binational La Amistad International Park (e.g., Cerro Kamuk, 3,554 m; Cerro Dúrika, 3,280 m) (Fig. 15.2). The páramos in neighboring Panama are found in the eastern sector of La Amistad International Park (Cerro Fábrega, 3,340 m; Cerro Itamut, 3,279 m; and Cerro Echandi, 3,162) and on the summit of Barú Volcano (3,475 m) (Kappelle and Horn 2005, Samudio and Pino 2006).

In Costa Rica, the páramo ecosystem covers a mere 15,205 ha, of which almost 10,000 ha is found on Cerro Chirripó and neighboring peaks (Kappelle 2005a). That is less than 0.3% of Costa Rica's total land surface (51,100 km^2). These areas of páramo in Costa Rica account for only 0.4% of the total area of 35,000 km^2 of páramo estimated for the Neotropics as a whole (Hofstede et al. 2003). About half (52.7%) of the páramo area in Costa Rica occurs in the subalpine belt (3,100–3,300 m), while the other half (47.3%) is located in the alpine belt (3,300–3,819 m).

In Costa Rica, a clear altitudinal zonation of páramo vegetion types can be observed, especially along the slopes of Cerro Chirripó (Kappelle 1991, Chaverri and Cleef 1996). The lowest páramo zone (ca. 3,100–3,300 m), known as *subpáramo* (Chaverri and Cleef 1996, Luteyn 1999), borders the upper montane oak-dominated cloud forests (Kappelle 1996). The subpáramo is a transition zone composed of species from the forest below and grass páramo above, in which shrubs and small trees form mosaics that alternate with shrubby grasslands dominated by bunch grasses or dwarf bamboos (Horn and Kappelle 2009). The subpáramo shrublands (*arbustales, matorrales*) are 2 to 8 m tall (Fig. 15.6). Here, dwarfish, sparsely-distributed, epiphyte-loaded trees (e.g., *Buddleja*, *Comarostaphylis*, *Escallonia*, *Myrsine*) mix with small-leaved, gnarled shrubs (e.g., *Diplostephium*, *Hypericum*, *Miconia*, *Solanum*) (Weber 1958, Kappelle 1991, Chaverri and Cleef 1996).

Above the subpáramo, the truly alpine belt (>3,300 m) in its strict sense is known as *páramo proper*—a term coined by Cuatrecasas (1958)—and consists of 0.5 to 2.5 m tall, bamboo-dominated (*Chusquea subtessellata*) grasslands (*chuscales;* Fig. 15.7), at drier spots replaced by bunch or tussock grasses (e.g., *Calamagrostis*, *Festuca*, *Muhlenbergia*.) The broomlike *Chusquea subtesellata* is the most common species in Costa Rican páramos and perhaps covers up to 60% of the total páramo area in the country (Kappelle 1991).

Above 3,600 m elevation on the Chirripó massif, scattered occurrences of a type of transitional *superpáramo* may be observed on rocky and sandy soils. In this vegetation zone on the highest and coldest slopes, patches of small plants capable of enduring low temperatures, high diurnal temperature variation, and strong radiation (e.g., *Acaena*, *Draba*, *Lupinus*, *Senecio*, *Westoniella*) grow amidst exposed rocks covered by dense mats of mosses and lichens (Weber 1958, Kappelle 1991, Chaverri and Cleef 1996).

The Physical Environment

Climate

As in other tropical mountainous countries, such as Venezuela and Papua New Guinea (Troll 1959, Lauer 1981, Rundel et al. 1994a,b), Costa Rica's páramo-blanketed high

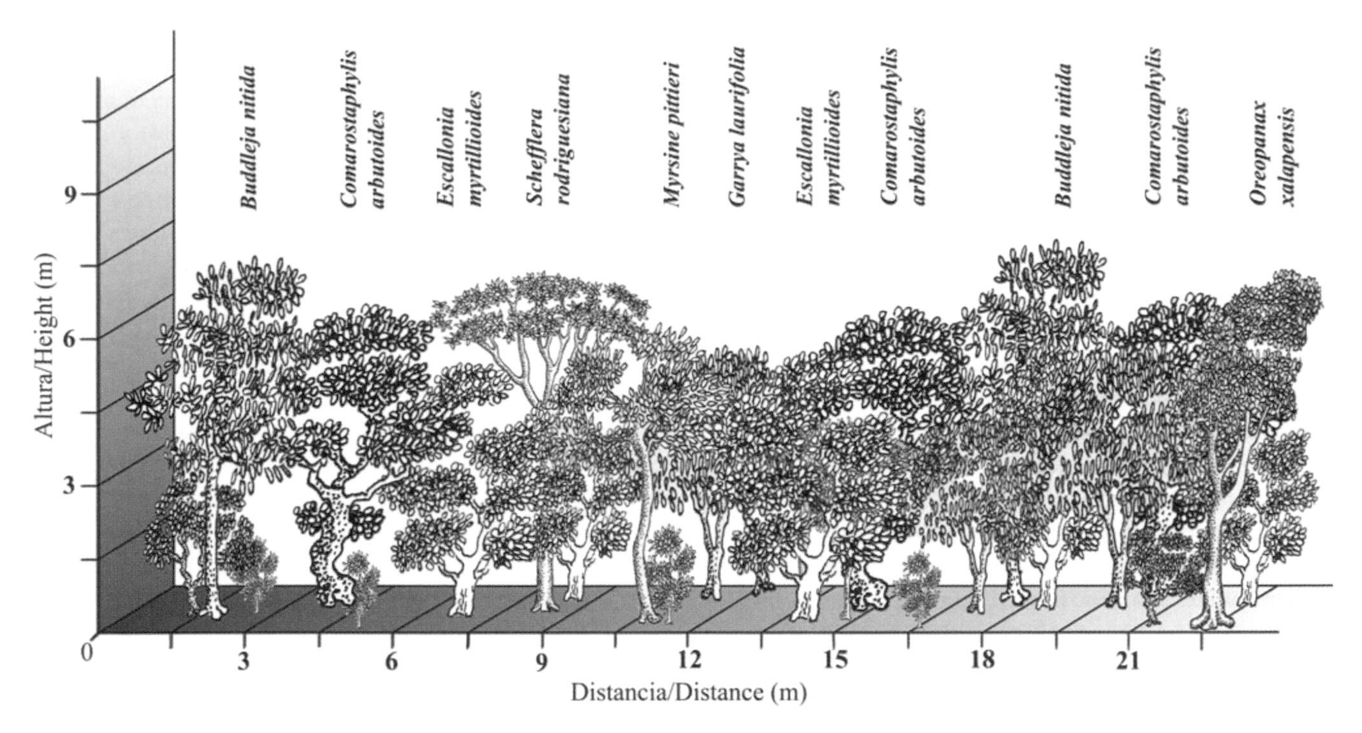

Fig. 15.6 Schematic lateral profile of a 6 m tall woody *subpáramo* shrubland in the subalpine belt around 3,100 m elevation at Buenavista massif. The small trees and shrubs depicted include the genera *Buddleja*, *Comarostaphylis*, *Escallonia*, *Garrya*, *Myrsine*, *Oreopanax*, and *Schefflera*.
Drawing by Francisco Quesada and Luis González Arce.

Fig. 15.7 Clumps of *Chusquea subtessellata* bamboos on the Buenavista massif (Cerro de la Muerte).
Photograph by Carlos Serrano.

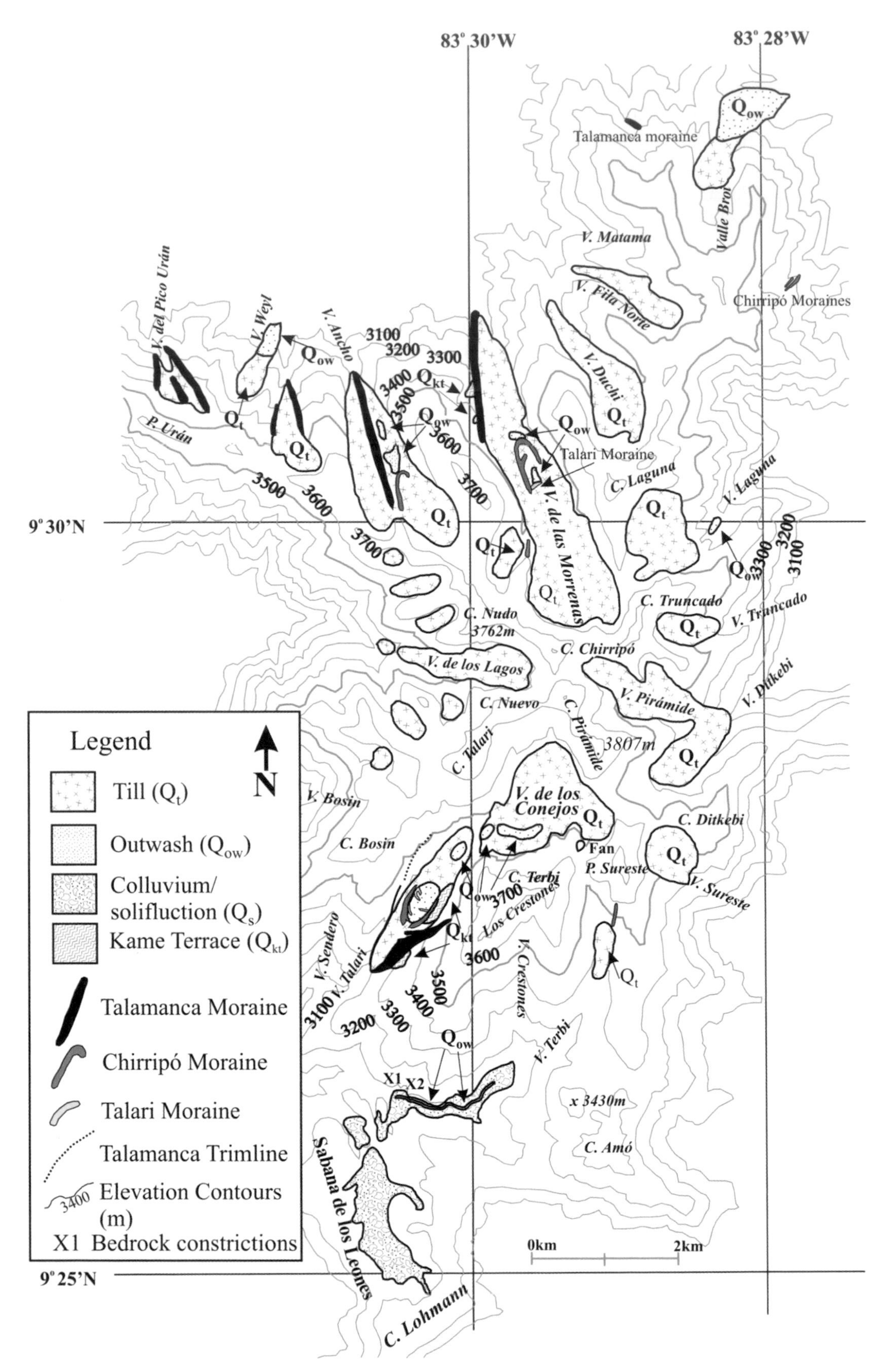

Fig. 15.9 Map of Quaternary glacial geology in the Chirripó *páramo*, showing distribution of glacial till (unsorted, unstratified sediment) and selected moraines (ridges of till) deposited by paleoglaciers, together with other glacial features. This map from Lachniet and Seltzer (2002) is based on extensive field reconnaissance and aerial photo interpretation. The abbreviations P, C, and V stand for Pico (Peak), Cerro (Mount), and Valle (Valley), respectively.
Reproduced from the Geological Society of America Bulletin, *vol. 114, p. 922 (July 2002), with the permission of the Geological Society of America (GSA).*

(Weber 1958, Kappelle 2005a). Irazú volcano is a complex stratovolcano (500 km^2) with an irregular subconic shape and multiple craters, including a mustard-colored crater lake (Alvarado 2000; Alvarado and Cárdenes, chapter 3 of this volume). Lavas at Irazú may be classified as basalts, including augitic basalts, and basaltic andesites (Alvarado 1993). Turrialba is also a stratovolcano (500 km^2), mainly constructed of pyrogenic and basaltic andesites, dacites, and rare basalts (Alvarado 2000). Its upper structure is conic and it possesses three well-defined craters (Alvarado 1993). Turrialba and Irazú volcanoes share their bases and can be considered twin volcanoes. Both have been active in historic time. Irazu's last eruptions were between 1963 and 1965, including the day of US President John F. Kennedy's visit to Costa Rica in 1963. Fumaroles were observed at Irazú from the mid-1960s through the late 1970s and appeared again in 1994 (Alvarado 2000). Turrialba showed signs of increased activity in 1996; eruptions of ash began in 2010 (Avard et al. 2012) and continue in 2015.

The soils of Costa Rican páramos have received limited scientific attention (e.g., see Alvarado and Mata, chapter 4 of this volume). Otárola (1976) and van Uffelen (1991) investigated páramo soils for their thesis research, but existing publications deal mainly with páramo soil taxonomy (Kappelle and van Uffelen 2005). A comparison of soil classification studies demonstrates that Costa Rican páramo soils are similar to soils of the Andean páramos, particularly those of the Colombian and Venezuelan highlands (Walter and Medina 1969, Thouret 1983, Malagón and Pulido 2000).

Soil taxonomic studies from the Chirripó and Buenavista páramos report the presence of Histosols (Lithic Tropofolist, Lithic Troposaprist, Cryic Sphagnofibrist), Entisols (Lithic Troporthent), Inceptisols (Lithic Humitropept, Lithic Placandept), and Andisols (Acric Hapludand-Typic Hapludand, Lithic Placaquand, Lithic Placudand) (Otárola 1976, Otárola and Alvarado 1976, Pérez et al. 1978, van Uffelen 1991, Kappelle and van Uffelen 2005). Most of the Histosols are ill-drained, organic, shallow soils with a thin iron pan, associated with soils derived from volcanic ashes. The soil type known as Sphagnofibrist is typical of páramo peat bogs known as *paramillos*, which contain turf layers predominantly built up of mosses such as *Sphagnum* (peat moss). In contrast to the Histosols, the páramo Entisols are young, stony, mineral soils with little development and a lack of distinct horizons. They occur on steep slopes on mountain tops. Inceptisols in the páramos are moderately developed, display horizons with altered mineral material, and may contain iron pans as in Humitropepts. Ultimately, the Andisols of the páramos are thick, dark brown to black soils developed from mainly andesitic and rhyolitic volcanic material, rich in bases (Otárola and Alvarado 1976, van Uffelen 1991, Kappelle and van Uffelen 2005). A mineralogical study of Dystrandepts and Placandepts near the Buenavista massif revealed that the sand fraction of these Andisols is dominated by volcanic glass, feldspar, and quartz, with a rise in feldspar content with increasing elevation (Landaeta et al. 1978). The clay fraction is dominated by amorphous and allophane clay-size materials (Landaeta et al. 1978).

Bioclimatic conditions of generally low temperatures, low precipitation, and low evapotranspiration influence the formation of páramo soils (e.g., Walter and Medina 1969, Lauer 1981, Malagón and Pulido 2000). Other fundamental soil-forming factors are andolization (presence of volcanic ashes) and hydromorphism (soil water saturation, thick humus layer), responsible for the pedogenesis of Andisols and Histosols, respectively (Malagón and Pulido 2000; Alvarado and Mata, chapter 4 of this volume). Soil temperatures measured between 3,100 and 3,300 m elevation at 0.5 m soil depth range from 8 to 10°C and coincide with the average annual atmospheric temperature at this elevation (van Uffelen 1991; and see Thouret 1983, for Colombia, and Walter and Medina 1969, for Venezuela). These soil temperatures mark the upper forest limit, perhaps because they constrain the growth and activity of tree roots (Körner 1998).

Limnology

The name "Chirripó" comes from the indigenous Cabécar language and means *Land of Eternal Waters*. These "eternal waters" refer to the numerous lakes in the highest parts of the Chirripó massif and the several major rivers that originate here, which include the Chirripó Atlántico, Telire, Chirripó Pacífico, and Ceibo rivers.

Today, 30-odd lakes of glacial origin exist within the páramo area of the Chirripó massif, constituting one of Costa Rica's few lake districts (Horn and Haberyan, chapter 19 of this volume). In general, these lakes have cool temperatures, and are dilute, polymictic, and very clear due to the low productivity that accompanies their low nutrient levels. Horn et al. (2005) collected water chemistry data for 19 glacial lakes from 3,450 to 3,570 m elevation in 1998, 2000, and 2001. They found values to be similar from lake to lake and year to year, and to be consistent with sparse prior measurements between 1966 and 1991. Phytoplankton in the lakes at Chirripó is composed of cyanobacteria, chlorophytes, and chrysophytes (Haberyan et al. 1995, Zeeb et al. 1996, Wujek et al. 1998, Horn et al. 1999, 2005), while benthic fauna consists of copepods, amphipods, and larvae of chironomids and other insects (Gocke et al. 1981).

The two largest glacial lakes are the 22 m deep Lago

Fig. 15.10 The summit of Cerro Asunción on the Buenavista massif (Cerro de la Muerte). *Photograph by Carlos Serrano.*

Chirripó in the Pacific-facing Valle de los Lagos, previously studied by Gocke et al. (1981), and the 8.3 m deep Lago de las Morrenas 1 in the Valles de las Morrenas, which drains to the Caribbean. The diatom *Aulacoseira* is the most common diatom in the sediments of Lago Chirripó and Lago de las Morrenas 1 (Haberyan and Horn 1999, 2005, Haberyan et al. 1997). The limnology of both lakes has been described in detail by Haberyan, Umaña, and Horn in different publications (e.g., Umaña et al. 1999, Haberyan et al. 2003; Horn and Haberyan, chapter 19 of this volume). Temperature measurements in 1998 and 2000 showed the lakes to be warmer than reported by earlier researchers (Löffler 1972, Gocke et al. 1981, Jones et al. 1993), possibly indicating a change in one or more aspects of the páramo climate (see Horn and Haberyan, chapter 19 of this volume).

West of Chirripó National Park, at elevations around 2,300 to 2,700 m along the Inter-American Highway, páramo-like ponds with peat bogs known as *paramillos* are found (e.g., Brak et al. 2005). An example is the 1 m deep, acidic and tannin-stained Tres de Junio pond with abundant *Sphagnum* moss, described by Horn and Haberyan (chapter 19 of this volume). Also, some artificial ponds are located within or near the Buenavista páramo, including the Asunción Pond on the south slope of Cerro Asunción (Fig. 15.10), which has particularly low pH levels (Horn and Haberyan, chapter 19 of this volume).

Quaternary History

Glacial Climate and Vegetation History

Reconstructions of past ice extent, as well as pollen assemblages in a high-elevation bog, provide evidence of glacial climate and vegetation history in the Costa Rican páramos. Field studies on the Chirripó massif have documented sets of moraines that correspond to different glacial stages. In the Valle de las Morrenas, a glacial trough on the north face of Cerro Chirripó, Orvis and Horn (2000, 2005) identified four such stages. Chirripó I was mainly constrained within the cirque, and Chirripó II–IV corresponded to three moraine groups found at different elevations within the valley. Lachniet and Seltzer (2002) examined moraines and other glacial features throughout the Chirripó massif and identified three moraine groups (Talari, Chirripó, and Talamanca, Fig. 15.9) that appear to correspond to the Chirripó II–IV moraines of Orvis and Horn (2000). Radiocarbon dating of basal lake sediments revealed that the most recent glacier in the Valle de las Morrenas receded rapidly near the end of the Younger Dryas, a global cooling event at the end of the Pleistocene, ca. 12,900–11,600 cal yr BP (calibrated years before present; Orvis and Horn 2000), but efforts to date earlier stages are ongoing. From reconstructions of equilibrium line altitudes (ELAs) of paleoglaciers, Orvis and Horn (2000) estimated minimum temperature depressions of 7.4–8.0°C for stages I–IV in the Valle de las Morrenas; and Lachniet and Seltzer (2002) estimated temperature depressions of 7.4–8.7°C for paleoglaciers in this valley and the Valle Talari. Orvis and Horn (2000) also used modern tropical-glacier hypsometries to model ice fluxes in glaciers in the Valle de las Morrenas and estimate relative climatic moisture conditions. Their results indicated that the Chirripó II stage was colder, but drier, than the Chirripó III stage (producing a smaller glacier), but that the Chirripó IV stage was both colder and wetter than Chirripó stages II and III.

Pollen analysis of sediments from the Turbera La Chonta (Hooghiemstra et al. 1992, Islebe and Hooghiemstra 1997, 2005) provided a record of highland vegetation and climate, covering more than 50,000 and perhaps ca. 80,000 years, that appears consistent with some inferences from glacier reconstructions. Turbera La Chonta is a peat bog at 2,310 m altitude in the Cordillera de Talamanca, located close to the Inter-American Highway near El Empalme, and presently surrounded by montane forest. Pollen assemblages revealed successive fluctuations of the upper forest line between ca. 2,300 m and 2,800 m altitude during the Early Glacial, prior to 50,000 years ago. During this interval the La Chonta peat bog was alternately situated close to the subalpine forest belt or within the upper montane forest belt. During the Pleniglacial (ca. 50,000–13,000 yr BP) the upper forest line fell to ca. 2,100 m altitude, and páramo vegetation surrounded La Chonta. A temperature depression of 7–8°C is inferred for this period, in close accord with reconstructions from glacial ELAs. During the Late Glacial, the forest line rose rapidly (in just a few hundred years) to 2,700 to 2,800 m elevation, and upper montane forest dominated by oak trees (*Quercus* spp.) surrounded the site for ca. 1,700 yr (Hooghiemstra et al. 1992). Subsequent climate cooling (estimated at 1.5 to 2.5°C) occurred during the Younger Dryas, when the upper forest line dropped to ca. 2,400 m, and successional stands of alder (*Alnus*) developed at the site. Renewed warming during the Holocene led to a rise in the upper forest line to 3,300–3,500 m, shrinking the area of páramo (Hooghiemstra et al. 1992). Extrapolating their pollen-based changes in forest limit to the entire Talamancan cordillera, Islebe and Hooghiemstra (2005) prepared distribution maps of páramo and montane vegetation in Costa Rica for 10,000 and 18,000 years ago that show that during the Last Glacial Maximum, a continuous corridor of páramo vegetation may have existed between the northern highlands of the Cordillera de Talamanca in Costa Rica and western Panama.

Postglacial History of Páramo Vegetation and Aquatic Communities

Horn (1993) examined pollen evidence of postglacial vegetation history in the Chirripó páramo preserved in a 5.6 m long lake sediment core from Lago de las Morrenas 1 (3,477 m elevation), the largest lake in a chain of glacial lakes in the Valle de las Morrenas. The pollen spectra indicate that treeless páramo vegetation has surrounded the site since the last retreat of ice 10,000–12,000 cal yr BP (Horn 1993, Orvis and Horn 2000). Percentages of grass pollen and other páramo taxa decline upward in the profile, while certain lowland, premontane, and montane taxa increase slightly. These changes may in part reflect upslope migrations of forest taxa owing to postglacial climate warming, as more strongly evidenced at the La Chonta site; they may also signal possible human impacts on forests during the late Holocene (Horn 1993, Horn and League 2005).

Compound-specific carbon isotope analyses of terrestrially derived *n*-alkanes isolated from a parallel core from Lago de las Morrenas 1 reveal shifts over the Holocene in proportions of C_3 and C_4 plants in the páramo (Lane et al. 2011). The general pattern is one of higher C_4 plant abundance early in the Holocene, followed by increased C_3 dominance during the middle Holocene, and then a return to greater C_4 plant abundance in the late Holocene. These changes are hypothesized to have resulted from environmental factors including habitat availability and atmospheric CO_2 concentrations in the early postglacial period, and from changes in precipitation delivery to the high elevation ecosystem as a result of millennial-scale ITCZ dynamics (Lane et al. 2011). Compound-specific hydrogen isotope analyses of the Lago de las Morrenas 1 sediments support the inferred shifts in precipitation on the massif (Lane and Horn 2013).

Haberyan and Horn (1999, 2005) examined diatoms in the Lago de las Morrenas 1 sediments to reconstruct possible changes in the aquatic community of a páramo lake. They found that at least ninety-five percent of all diatoms belong to a single species of *Aulacoseira* that appeared to be a new species in the *A. alpigena* / *A. lirata* complex. Its dominance throughout the Lago de las Morrenas 1 record suggested that the lake has always been cold, polymictic, and clear. Two peaks in the diatom accumulation rate seemed to correlate with peaks in charcoal in this particular core and one from nearby Lago Chirripó, and may be related to fire effects in the watershed or possibly to lower lake levels resulting from local drought. However, neither fire nor local droughts appear to have affected the composition of the diatom flora, which may be insensitive to any limnological changes arising from such occurrences.

Fire History

Charcoal fragments preserved in sediments and soils provide information on long-term fire history in the Costa Rican páramos. Evidence comes both from the modern páramos and from sites within the present montane forest zone that supported páramo vegetation during past intervals of cooler climate. The earliest charcoal from a páramo interval at La Chonta dates to ca. 37,500 cal yr BP (Horn and Kappelle 2009). As this age greatly predates human settlement of Costa Rica, the fire that produced the charcoal must have been set by lightning or volcanism.

The glacial lakes of the Chirripó páramo provide evidence of recurrent fires during the Holocene. Horn and League (2005) examined charcoal in the 5.6 m profile from Lago de las Morrenas 1 and in a 1.1 m long sediment core from Lago Chirripó. Both lake cores contained abundant microscopic charcoal (examined on microscope slides) as well as macroscopic charcoal (quantified by sieving). Results indicate that fires set by people or lightning have repeatedly burned the Chirripó páramo. The microscopic charcoal record from Lago Chirripó extends to ca. 4,600 cal yr BP and shows peaks in fire activity that generally match peaks in the corresponding section of the Lago de las Morrenas 1 microscopic charcoal record (Horn 1993). The uppermost sections of both sediment cores show lower charcoal influx rates than some deeper sections, suggesting that recent fire recurrence intervals in the Chirripó páramo are not unprecedented. A high-resolution analysis of macroscopic charcoal in contiguous 1-cm intervals of the Lago de las Morrenas 1 core confirmed that fires burned within the lake watershed throughout the Holocene, and revealed variations in charcoal influx likely driven in part by precipitation variability (League and Horn 2000; Lane and Horn 2013).

The long history of fire in the Neotropical páramos—whether natural or anthropogenic—and the responses of modern páramo species and communities to fire leads to the characterization of páramos as *fire-dependent ecosystems*, which are those in which fire is an essential process (Myers 2006, Horn and Kappelle 2009).

Plant Geography and Endemism

Cleef and Chaverri (1992, 2005) studied the phytogeographic origin of 150 vascular plant genera present in the páramos of the Cordillera de Talamanca. They recognized seven geographic floral elements on the basis of current distributions of these genera: (a) true páramo elements (4% of all genera); (b) Neotropical montane genera, which include Mexican-Guatemalan and tropical Andean subelements (25%); (c) wide-tropical genera present on at least two tropical continents (7%); (d) Holarctic genera (15%); (e) Austral-Antarctic genera (14%); (f) wide temperate genera present on at least two temperate continents (24%); and (g) cosmopolitan genera distributed worldwide (11%).

The Cordillera de Talamanca constitutes the southern limit of a number of Holarctic plant genera such as *Cirsium*, *Comarostaphylis*, *Garrya*, *Helianthemum*, and *Mahonia*. Wide temperate genera such as *Gentiana* and *Hieracium* arrived from the north, while *Hypericum* and *Plantago* apparently moved into the Talamancas from the Colombian Andes. A comparison with the vascular flora of Colombia shows that about 95% of the vascular genera found in Costa Rican páramos are shared with the northern tropical Andes (Cleef and Chaverri 1992).

In sub-*páramo* shrublands—sometimes with dwarfish trees—growing at the lower limit of Costa Rica's páramos around 3,100 m elevation, Islebe and Kappelle (1994, 2005) documented a total of 127 vascular plant genera. Fifty percent are woody. Most genera are Neotropical shrubs and wide tropical trees. A small number correspond to wide temperate herbs. A comparison with similar subalpine dwarf forests in Guatemala revealed that 30% of the 178 vascular plant genera found in both Guatemala and Costa Rica are shared between the subalpine forests of these countries. Most of these genera are temperate in distribution, with the Holarctic element twice as important as the Austral-Antarctic element (Islebe and Kappelle 1994, 2005).

The affinities between Costa Rican high elevation floras and those in Guatemala and Colombia, respectively, are the result of a long geological history that includes the formation of the Central American isthmus millions of years ago, subsequent intervals of uplift of the Cordillera de Talamanca, and, ultimately, the Plio-Pleistocene sequence of glacial intervals, which caused páramo belts to move up and down along elevational gradients (Hooghiemstra et al. 1992, Islebe and Hooghiemstra 1997, 2005). Other factors that have influenced the development of the Central American páramo floras are the large floristic diversity of the numerous montane and lowland vegetation types, the great environmental heterogeneity of the highlands with different local climate regimes, and the large migration potential of individual species (Cleef and Chaverri 1992, 2005, Hooghiemstra et al. 1992).

Five flowering plant genera are endemic to the páramos of the Cordillera de Talamanca: *Iltisia*, *Jessea*, *Laestadia*, *Talamancalia*, and *Westoniella*—all in the Asteraceae (Compositae). At the species level, a total of 146 flowering plants are endemic to the isthmian páramos of Costa Rica and Panama (Vargas and Sánchez 2005). This value represents 35% of all flowering plants known from the páramos. One hundred seven are dicots while 39 are monocots. Asteraceae is the family with the most endemic species, followed by Poaceae and Orchidaceae. The genera with most endemic species are *Ageratina* and *Westoniella* in the Asteraceae, and *Telipogon* in the Orchidaceae (Vargas and Sánchez 2005).

Among the ferns and fern-allies almost half of the species have a broad distribution, but by contrast 25% are endemic to the páramos of Costa Rica and Panama (Barrington 2005). The Andean element is prominent among the true páramo ferns. Comparison of the Andean páramo fern flora with that of Costa Rica reveals that most genera are common in both regions. However, in the Costa Rican páramos there are also some species derived from the alpine zone of southern Mexico and Chiapas (Barrington 2005).

Endemism at the species level among páramo bryophytes (mosses and liverworts) is, in contrast, very low. Of 230 species, only *Cryptothallus hirsutus* is endemic to the Costa Rican páramos. Four bryophyte species are restricted to Central America, 19 to Neotropical páramos, and 25 to the northern Andes and Talamancan highlands (Gradstein and Holz 2005). The páramo mosses and liverworts are made up of a mix of tropical and temperate taxa; almost all species also occur in Colombia.

Biodiversity at the Species Level

Protists and Fungi

Algae

A recent inventory of fossil and modern eukariotic algae in the lakes and ponds of Costa Rica's páramo revealed the presence of 41 genera (Kappelle et al. 2005a). A forty-second genus, the green alga *Botryococcus*, was missed in the compilation but also occurs (Haberyan et al. 1995). These algal genera are distributed over four divisions: 1) Chlorophyta or green algae; 2) Dinophyta or dinoflagellates; 3) Ochrophyta or heterokont algae, which include the diatoms (Bacillariophyta) and golden algae (Chrysophyta); and 4) Rhodophyta or red algae.

To date, the Ochrophyta or Heterokontophyta, with a total of 32 genera, represents the most diverse group. Twenty-six ochrophytes belong to the diatoms (2 centric genera and 24 pennate genera) while 6 are golden algae. All heterokont algae but one (a golden alga) are identified from lake sediment samples. Furthermore, about six recent genera of green algae, two dinoflagellates, and one red alga were discovered. All together these results imply a potentially high diversity of modern phytoplankton in the glacial lakes and ponds located in the Costa Rican páramo region.

Fungi and Lichens

Luis D. Gómez (2005b) presented a preliminary list of the Eumycota (true fungi) known from the páramos of Costa Rica and Panama. His inventory listed 272 species in 162 genera and 66 families. Thirty-seven families are Basidiomycetes and 29 are Ascomycota. Forty species are considered plant parasites.

At least 204 lichen species have been reported from Costa Rica's páramos (Sipman 2005). Another 15 taxa remain unidentified and may eventually result in new species for the isthmus. Dominant lichen growth forms are foliose and fruticose. Sipman (2005) noted that these growth forms facilitate water uptake and drying, and thus allow maintenance of poikilohydric growth conditions in the mostly humid páramo climate. The most species-rich lichen genera in Costa Rican páramos are *Cladonia*, *Erioderma*, *Everniastrum*, *Heterodermia*, *Hypotrachyna*, *Leptogium*, *Oropogon*, *Peltigera*, *Stereocaulon*, *Sticta*, and *Usnea*.

Most lichen species are widespread in the neotropics, and only a few are restricted to Costa Rica. Some reach their northern or southern limit in this part of the continent. The páramo lichen flora differs strongly from that of the surrounding forests, as is evidenced by the scarcity of orders like Arthoniales, Graphidales, and Pyrenulales, and the abundance of species of the order Lecanorales (Sipman 2005).

Plants

Bryophytes: Mosses and Liverworts

The páramos of Costa Rica have a rich bryophyte flora. By 2005, 230 species (117 mosses, 113 liverworts) had been identified (Gradstein and Holz 2005), which was about 50% more than revealed by a previous census only six years earlier (Luteyn 1999). The 230 species of Costa Rican páramo bryophytes represent about 28% of the entire bryoflora known from all Neotropical páramos together. Gradstein and Holz (2005) estimated that future inventories will reveal many more species, potentially leading to a total of 300 species.

Bryophyte genera that are most species-rich in Costa Rican páramos are *Campylopus* and *Plagiochila*, followed by *Adelanthus*, *Bartramia*, *Bryum*, *Daltonia*, *Frullania*, *Leptodontium*, *Leptoscyphus*, *Macromitrium*, *Metzgeria*, *Radula*, and *Sphagnum*.

Ferns and Fern-Allies

Costa Rican páramos harbor 80 species of pteridophytes in 25 genera, among which *Elaphoglossum* (tongue ferns) with 12 species and *Huperzia* (formerly in the genus *Lycopodium*, the clubmosses) with nine species are the most diverse (Barrington 2005). Many of these páramo ferns display morphological features that protect the plants from damage due to cold temperatures and frequent fires. For example, some ferns have deep-buried rhizomes, thick, recurved leaf segments, and a heavy covering of scales. Similarly, the only genus (*Blechnum*) of treeferns found in the isthmian páramos has a thick trunk resistant to low temperatures and recurrent fires (Horn 1988, Barrington 2005).

Flowering Plants

Luteyn (1999) described the páramo flora as relatively young on a geological time scale. Many taxa evolved during the Late Pliocene or Early Pleistocene, some 2 to 4 million years ago (Van der Hammen and Cleef 1986). According to the detailed inventory that Luteyn carried out over several decades, there are at least 4,697 species of plants in the Latin American páramos, spread over 3,399 vascular plants and 1,298 non-vascular species. Flowering plants make up the bulk of the vascular species: at least 3,045 species (634 monocots and 2,411 dicots) thrive in the páramos of Central America and the northern Andes.

Several lists of páramo plant species are available for Costa Rica, some more exhaustive than others (Weber 1959, Weston 1981a, Kappelle et al. 1991, Gómez 1994a, b, Luteyn 1999, Zamora et al. 2004, Vargas and Sánchez 2005). The most recent inventory and review reported up to 416 flowering plant species (in 216 genera and 72 families) from the isthmian páramos of Costa Rica and Panama (Vargas and Sánchez 2005). This amount corresponds to 14% of the 3,045 species of flowering plants, around 71% of the families, and 50% of the genera listed by Luteyn (1999). A total of 121 Costa Rican/Panamanian páramo species are monocots (53 genera in 11 families), while 295 are dicots (163 genera in 61 families). The most species-rich plant family is Asteraceae (73 species), followed by Poaceae (52), Cyperaceae (19), Orchidaceae (18), Scrophulariaceae (18), Rosaceae (18), Apiaceae (16), and Ericaceae (12) (Chaverri and Cleef 1996; Vargas and Sánchez 2005). The most diverse genera are *Ageratina* (Asteraceae), *Festuca* (Poaceae), and *Lachemilla* (Rosaceae), each with eight species. Most flowering plants (230 to 250 species) have been observed with flowers or fruits in either March or August (Fig. 15.11).

One hundred twenty-seven species of flowering plants are restricted to elevations over 3,000 m altitude and can be considered true páramo elements. Fifty of these are endemic to the páramos, while twenty are restricted to the uppermost elevations within the páramos (3,500–3,819 m). These include *Azorella biloba*, *Castilleja quirosii*, *Draba jorullensis*, *Lewisia megarhiza*, *Lysipomia acaulis*, *Poa chirripoensis*, *Ranunculus crassirostratus*, *Senecio kuhbieri*, *Stevia westonii*, *Uncinia koyamae*, *Westoniella chirripoensis*, and *W. eriocephala* (Vargas and Sánchez 2005).

In general, the Chirripó and Buenavista massifs are the richest in flowering plants compared to the more remote páramos such as at Cerro Kámuk, Cerro Echandi, and Cerro Fábrega. These differences in flowering plant diversity may in part reflect the larger size and greater habitat heterogeneity of the Chirripó and Buenavista massifs. But the greater intensity of collecting on the Chirripó and Buenavista peaks must be taken into account when making such a comparison.

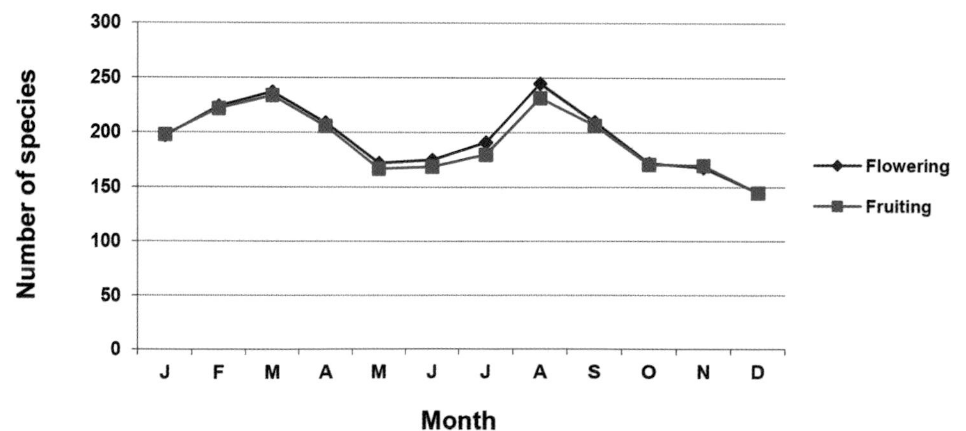

Fig. 15.11 Number of species of flowering plants that have been observed while flowering or fruiting in the *páramos* of Costa Rica and Panama.
From Vargas and Sánchez 2005. Reproduced with permission from Editorial INBio, Costa Rica.

Animals

Invertebrates

Nematodes

An assessment of nematode diversity, abundance, and community structure in the Costa Rican páramos revealed seven orders, 26 families, 54 genera, and 4 species (Esquivel 2005). Nematode abundance was highest in the upper 10 cm of soils under *Comarostaphylis*-dominated dwarf forests. The observed nematode communities are characterized by a high percentage of plant, hyphal, and bacterial feeders.

Tardigrades

Six out of twelve species of Tardigrada reported for Costa Rica are known from the páramos of the Cerro de la Muerte, where three occur on bryophytes (*Echiniscus arctomys*, *Hypsibius scoticus*, and *Macrobiotus occidentalis*), and three on lichens (*Echiniscus bigranulatus*, *Hypsibius convergens*, and *Milnesium tardigradum*) (Kappelle 2005b).

Insects

Collections stored at Costa Rica's Instituto Nacional de Biodiversidad (INBio) document 7 orders, 19 families, 55 genera, and 71 insect species in the páramos of Costa Rica. The specimens originated from the higher parts (≥3100 m) of the Cordillera de Talamanca and the Cordillera Volcánica Central. Lepidoptera is the most diverse order. The most abundant species are *Gonodonta pyrgo*, *Hortensia similis*, and the bumblebee *Bombus ephippiatus* (Kappelle 2005c).

Barrientos and Monge (2005) compared insect communities of Cerro Chirripó and other Neotropical páramos on the basis of field data and literature reports. Field sampling using 8,000 net sweeps in the Chirripó páramo yielded a total of 144 morphospecies within 16 orders. Most morphospecies (70) belonged to the Diptera, followed by Hymenoptera (23), Lepidoptera (18), and Coleoptera (15). Samples from humid microhabitats were more diverse. Adult nectarivores, and immature saprophages, herbivores, and parasites, were most abundant.

Mollusks

Barrientos (2005) collected terrestrial mollusks in an opportunistic way during a twelve-day period at different páramo sites (Chirripó massif, Cerro Cuerici, Cerro Buenavista massif, Irazú Volcano). In total, she found 27 morphospecies in eight families. The best represented malacofaunal groups were the Helicarionidae with 14 morphospecies and Orthalicidae with four. It was striking that over half of the recorded species belonged to the Helicarionidae (Barrientos 2005).

Vertebrates

Amphibians and Reptiles

The herpetofauna of the páramos and adjacent dwarf forests of Costa Rica includes three orders, eight families, nine genera, and nineteen species (Kappelle and Savage 2005). Only three are true páramo species: the mushroom-tongue salamander (*Bolitoglossa pesrubra*), the montane alligator lizard (*Mesaspis monticola*), and the green spiny lizard (*Sceloporus malachiticus*).

Birds

Seventy bird species have been observed in the páramos of Costa Rica (Barrantes 2005). Twelve species are considered true páramo species as they abound throughout the year: the carnivorous red-tailed hawk (*Buteo jamaicensis*); the insectivorous black-cheeked warbler (*Basileuterus melanogenys*), wrenthrush (*Zeledonia coronata*), flame-throated warbler (*Parula gutturalis*), and timberline wren (*Thryorchilus browni*); the nectarivorous volcano hummingbird (*Selasphorus flammula*); the frugivorous black-billed nightingale-thrush (*Catharus gracilirostris*) and sooty robin (*Turdus nigrescens*); and the frugivorous-insectivorous sooty-capped bush-tanager (*Chlorospingus pileatus*), volcano junco (*Junco vulcani*), large-footed finch (*Pezopetes capitalis*), and rufous-collared sparrow (*Zonotrichia capensis*). Another 34 species are commonly observed at the lower páramo border while the remaining are occasional visitors. Stiles et al. (1989) also mentioned the slaty flowerpiercer (*Diglossa plumbea*) as a characteristic species of Costa Rican páramo.

Overall, 8% of the plant species in the páramo rely on birds for pollination, and 20% for seed dispersal. These figures contrast with other tropical ecosystems in which much higher percentages of plant species depend on birds for seed dispersal (Barrantes 2005).

Mammals

Seven orders of mammals—excluding humans—inhabit the páramos of Costa Rica: the Insectivora, Chiroptera, Rodentia, Lagomorpha, Carnivora, Perissodactyla, and Artiodactyla (Carrillo et al. 2005). They include up to 14 families, 25 genera, and 32 species, many of which remain abundant in the páramos, particularly in the more remote areas. The genera *Cryptotis* (shrews), *Sylvilagus* (rabbits), and *Leopardus* (ocelot and margay) are represented by two species each. Three large mammals are considered threatened and have been listed in Appendix I of the CITES Convention (http://www.cites.org/eng/app/index.shtml): *Puma concolor* ssp. *costaricensis* (puma or mountain lion), *Leopardus pardalis* (ocelot), and *Tapirus bairdii* (Baird's tapir) (Carrillo et al. 2005).

that is found in narrow, ecotonal fringes along the upper forest line around 3,100 m elevation. These *C. arbutoides*-dominated dwarf forests may harbor up to 200 different species of vascular plants. Often, *C. arbutoides* is accompanied by other gnarled trees and shrubs, in the genera *Buddleja*, *Clethra*, *Escallonia*, *Garrya*, *Hesperomeles*, *Miconia*, *Monnina*, *Myrica*, *Myrsine*, *Oreopanax*, *Rhamnus*, *Schefflera*, *Vaccinium*, *Viburnum*, and *Weinmannia*. Normally, all these gnarled trees have pubescent, nanophyllous, coriaceous, and/or sclerophyllous leaves bundled at the top of twisted branches that are often densely covered by nonvascular epiphytes (mosses, liverworts, and lichens).

Brak et al. (2005) characterized the *paramillo* communities of Turbera La Chonta (2,300 m elevation), the bog from which Hooghiemstra et al. (1992) collected cores for pollen analysis (see previous sections in this chapter). Brak et al. (2005) described and mapped nine plant communities along a gradient from the center of the bog to its outer, forested border. Four of the plant communities are true peat bog associations with the presence of aquatic, submerged, or floating plants. These communities are dominated by *Eleocharis acicularis*, *Hieracium irasuense*, *Hypericum strictum*, *Paepalanthus costaricensis*, *Rhynchospora schaffneri*, *Utricularia* aff. *subulata*, and *Xyris nigrescens*, in association with peat moss (*Sphagnum magellanicum* and *S. recurvum*). Three other communities are typical for the transition from the peatbog to the well-drained forest. They include the previously mentioned *Puya-Blechnum* community, and associations dominated by the grass *Cortaderia nitida* (forming tussocks or bunches), shrubs including *Disterigma humboldtii*, *Hesperomeles obtusifolia*, *Pernettya prostrata*, *Vaccinium consanguineum*, and *Ugni myricoides*, numerous mosses, and trees such as *Escallonia myrtilioides* and *Drimys granadensis* (Brak et al. 2005). The adjoining forest is dominated by tall oak trees (*Quercus seemannii*) (Kappelle 2006).

Conservation and Sustainable Use

Land Use History and Recent Fire Dynamics

Humans have not historically occupied the Costa Rican páramos. Today a handful of farmers live just below the lower limit of páramo along the Inter-American highway crossing of the Buenavista massif, and a few workers occupy living quarters next to the communication towers on Cerro Buenavista, within the páramo. The very low population density within and adjacent to the Costa Rican páramos differs from the situation in many Andean páramos, which support human settlements, agricultural fields, and extensive herds of livestock that graze páramo plants (Balslev and Luteyn 1992; Luteyn 1999).

On the Chirripó massif, horses that are used to transport gear to the national park facilities graze páramo vegetation, though impacts are constrained by park policies on transport animals. A few horses and cows are sometimes seen grazing páramo plants along the Inter-American Highway, but impacts are very low owing to low numbers of animals. Grazing impacts were locally higher in the late nineteenth and early twentieth centuries, prior to the construction of the Inter-American Highway in the 1940s, when cattle and pigs from San Isidro were driven over the summit for sale in the capital, along a rough track that crossed the páramo close to the present highway (Horn 1989c).

Fires periodically affect the Costa Rican páramos, most often ignited inadvertently by careless visitors, but perhaps sometimes set intentionally. Aircraft accidents have ignited fires, and over long time spans, lightning has played a role. Field studies following twentieth-century fires in the Buenavista and Chirripó páramos demonstrated differential species responses to fire and rates of regrowth (Janzen 1973, Chaverri et al. 1976, Williamson et al. 1986, Horn 1989a, 1997, 1998a). The bamboo *Chusquea subtessellata* and the ericaceous shrubs *Vaccinium consanguineum* and *Pernettya prostrata* typically resprout vigorously after fire, but rarely, if ever, recolonize burn sites by seeding (Horn 1989a, Horn 2005). The shrub *Hypericum irasuense*, in contrast, suffers high mortality in fires, but reestablishes successfully by seed following all but the largest fires.

Preexisting vegetation, fire characteristics, and site differences both before and after burning seem to affect rates of shrub and herb survival, colonization, and growth in páramo burn sites (Horn 1989a, 2005). Among woody species, *Chusquea subtessellata* shows the fastest postfire recovery rates; clumps of bamboo that were 1–2 m high prior to burning will have regained their prefire heights within 10 years. Associated shrub species may require a decade or more to recover comparable postfire statures; some will have regained prefire height within a decade but most will not have regained their prefire stem diameter. Bare patches of ground between regenerating shrubs and bamboo clumps may persist for a decade or more following burning (Horn 2005).

Protected Area Management and Conservation

All Costa Rican páramos are legally protected within national protected wildlife areas. The 50,150 ha Chirripó National Park, established in 1975, protects Costa Rica's largest area of páramo, on the slopes of the country's highest summit, Cerro Chirripó (3,819 m), and surrounding peaks. The La Amistad International Park (PILA) (199,147 ha), a binational park and World Heritage site four times larger, was created in 1982 and protects patches of páramo and páramo-like peatbogs on peaks that stretch southeastward from Cerro Chirripó, such as Eli, Dúrika, and Kamuk in Costa Rica, and Fábrega, Itamut, and Echandi in Panama. La Amistad park also protects a large area of surrounding montane forest. The páramos of the Buenavista massif are protected within the 58,500 ha Tapantí–Macizo de la Muerte National Park, decreed in 1999. Páramo on the volcanic summits of the Cordillera Central is protected in the Irazú and Turrialba national parks. The national parks of Tapantí-Macizo de la Muerte, Chirripó, and Amistad have been jointly declared by UNESCO as a Man And Biosphere (MAB) Reserve. Additionally, the Talamancan peatlands (*Turberas de Talamanca*) that occur in a matrix of oak-bamboo forests (Kappelle 1996) at elevations ranging from 700 to 3,819 m were declared as a Ramsar Wetland Site of International Importance in 2003 (Ramsar site #1286).

Chaverri and Esquivel (2005) discussed the vital importance of Chirripó National Park for páramo conservation in Costa Rica. They presented a brief history of the establishment of this magnificent national park, which followed a concerted effort of a group of scientists, local community leaders, and park rangers (Chaverri 2008).

Visitation in Chirripó National Park ranged from 100 persons per year immediately after its creation to 600 per year in the late 1970s, fluctuated between 300 and 2,000 people per year throughout the 1980s, and reached over 8,000 visitors in 1995. Chaverri and Esquivel (2005) documented problems related to over-visitation of the park, including the trampling of vegetation along trails, inappropriate waste management, and increased risk of fire, and suggested the need to control the annual number of park visitors to respect the carrying capacity of the páramo ecosystem (see also Horn 1998b, Fürst et al. 2004). Because of its remoteness, La Amistad International Park is much less visited by tourists than Chirripó National Park.

A number of threats are of continued concern in Costa Rica's páramos. Hunting and poaching, extraction of palm heart and orchids, moss gathering for ornamental arrangements, and changes in fire frequency and extent are current or potential stressors that can negatively impact the páramos of Costa Rica. Future climate change is an additional concern, as warming could have a major impact on páramo distribution and aquatic communities (Nogués-Bravo et al. 2007). If the elevational ranges of suitable habitats for montane tree and shrub species migrate upslope with changes in climate, and if the plants are able to follow, montane woodlands might in some areas replace the typical bamboo and grass páramos of the Cordillera de Talamanca. Just a few degrees of climate warming could potentially induce a regional rise in treelines that would result in the contraction and fragmention of the páramo habitat. Concomitant shifts in moisture availability and disturbance regimes might lead also, or instead, to the colonization of today's páramo zone by weedy herbaceous species characteristic of disturbed sites at lower elevations in Costa Rica, or by non-native, invasive herbs or shrubs. Climate change over the coming centuries could radically alter Costa Rican páramo vegetation and ecosystems.

Chaverri and colleagues (Chaverri and Esquivel 2005, Chaverri 2008) recommended a series of conservation actions that should be undertaken to preserve the Costa Rican páramos in the long run, apart from the climate change issue. They stressed the importance of education about the páramos, even in a world in which modern, human-induced climate change is brought under control. Raising awareness among páramo visitors and people who live in nearby areas about the importance and potential fragility of the páramo environment is considered a key strategy necessary to ensure that this unique tropical alpine ecosystem can survive into the distant future.

Acknowledgments

We are much indebted to Antoine M. Cleef, Oscar Esquivel G., and the late Adelaida Chaverri P., Luis Diego Gómez P., and Alfonso Mata J. for their continuous support of our páramo research since the early 1980s. We also would like to thank the many colleagues and students who worked with us in the field and lab, in particular Henry Hooghiemstra, Gerald Islebe, Roger Horn, Ken Orvis, Maureen Sánchez, Kurt Haberyan, Brandon League, Lisa Kennedy, Chad Lane, Luis Poveda, Nelson Zamora, Gerrit van Uffelen, and Marta E. Juárez Ruiz. The Sistema Nacional de Áreas de Conservación (SINAC) of the Costa Rican Ministerio de Ambiente y Energía (MINAE), and its precursor, the Servicio de Parques Nacionales (SPN), are gratefully acknowledged for permitting access to the protected páramos of Costa Rica. We are pleased to also acknowledge the institutions and organizations that have funded our explorations of the Costa Rican páramos. Kappelle's research has been funded by the Netherlands government (DGIS, NWO) and several Dutch universities (Amsterdam, Utrecht) as well as Costa Rican institutions (INBio, Museo Nacional de Costa Rica, Universidad de Costa Rica, Universidad Nacional) and The Nature Conservancy (TNC), and Horn has been supported by the A.W. Mellon Foundation, the National Geographic Society, the National Science Foundation (SBR-9512484), the University of Tennessee, the Association of American Geographers, the US Information Agency (Fulbright program), and the University of California, Berkeley. Finally, we would like to thank the many park guards and *baqueanos* of Chirripó National Park, and the people of the communities of San Gerardo de Rivas, and San Gerardo de Dota, for their hospitality and unconditional support during the past three decades.

References

Alfaro, E., and B. Gamboa. 1999. *Plantas Comunes del Parque Nacional Chirripó*. Santo Domingo de Heredia, Costa Rica: Instituto Nacional de Biodiversidad (INBio). 283 pp.

Almeida, L., A.M. Cleef, A. Herrera, A. Velázquez, and I. Luna. 1994. El zacatonal alpino en el volcán Popocatépetl, México, y su posición en las montañas tropicales de América. *Phytocoenología* 22(3): 391–436.

Alvarado, G.E. 1993. *Costa Rica: Land of Volcanoes*. Cartago, Costa Rica: Editorial Tecnológica de Costa Rica, Gallo Pinto Press. 181 pp.

Alvarado, G.E. 2000. *Los Volcanes de Costa Rica: Geología, Historia y Riqueza Natural*. 2nd ed. San José, Costa Rica: EUNED.

Avard, G., J. Pacheco, E. Fernández, M. Martínez, E. Menjívar, J. Brenes, R. van der Laat, E. Duarte, and W. Sáenz. 2012. Turrialba volcano: Opening of a new fumarolic vent on the southeast flank of the West Crater on January 12th, 2012, as a consequence of a shallow decompression. Heredia, Costa Rica: Observatorio Vulcanológico y Sismológico de Costa Rica (OVSICORI), Universidad Nacional (UNA). 14 pp.

Badilla, M.B., and M. Kappelle. 1992. La composición florística del Chirripó. *Talamanca, Boletín Informativo* (San José) 3: 11.

Ballman, P. 1976. Eine geologische Traverse des Ostteils der Cordillera de Talamanca. Costa Rica (Mittelamerika). *Neues Jahrbuch Geologische und Paläontologische Mithandlungen*: 502–12.

Balslev, H., and J.L. Luteyn. 1992. *Páramo: An Andean Ecosystem under Human Influence*. London: Academic Press.

Barquero, J., and L. Ellenberg. 1982–83. Geomorfología del piso alpino del Chrirripó en la Cordillera de Talamanca, Costa Rica. *Revista Geográfica de América Central* 17/18: 293–99.

Barquero, J., and L. Ellenberg. 1986. Geomorphologie der alpinen Stufe des Chirripó in Costa Rica. *Eiszeitalter und Gegenwart* 36: 1–9.

Barrantes, G. 2005. Aves de los páramos de Costa Rica. In M. Kappelle and S.P. Horn, eds., *Páramos de Costa Rica*, 521–32. Santo Domingo de Heredia, Costa Rica: Instituto Nacional de Biodiversidad (INBio).

Barrientos, Z. 2005. Moluscos terrestres de los páramos de Costa Rica. In M. Kappelle and S.P. Horn, eds., *Páramos de Costa Rica*, 501–12. Santo Domingo de Heredia, Costa Rica: Instituto Nacional de Biodiversidad (INBio).

Barrientos, Z., and J. Monge. 1995. Geographic homogeneity among insect communities in neotropical páramos: a hypothesis test. *Caldasia* 18(86): 49–56.

Barrientos, Z., and J. Monge. 2005. Homogeneidad geográfica en comunidades de insectos en los páramos neotropicales: prueba de una hipótesis. In M. Kappelle and S.P. Horn, eds., *Páramos de Costa Rica*, 657–66. Santo Domingo de Heredia, Costa Rica: Instituto Nacional de Biodiversidad (INBio).

Barrington, D.S. 2005. Helechos de los páramos de Costa Rica. In M. Kappelle and S.P. Horn, eds., *Páramos de Costa Rica*, 375–95. Santo Domingo de Heredia, Costa Rica: Instituto Nacional de Biodiversidad (INBio).

Beaman, J.H. 1959. The alpine flora of Mexico and Central America. *Yearbook of the American Philosophical Society* 1959: 266–68.

Beck, E., H. Rehder, P. Pongratz, R. Scheibe, and M. Senser. 1981. Ecological analysis of the boundary between the afroalpine vegetation types '*Dendrosenecio* Woodlands' and '*Senecio-Lobelia keniensis* Community' on Mt. Kenya. *Journal of the East African Natural History Society and National Museum* 172: 1–11.

Bergoeing, J.P. 1977. Modelado glaciar en la Cordillera de Talamanca, Costa Rica. Instituto Geográfico Nacional (IGN). *Informe Semestral* (julio–diciembre): 33–44.

Blaser, J. 1987. Standörtliche und waldkundliche Analyse eines Eichen-Wolkenwaldes (*Quercus* spp.) der Montanstufe in Costa Rica. PhD thesis, Georg-August Universität. Göttingen, Germany. 235 pp.

Boza, M.A. 1988. *Parques Nacionales de Costa Rica*. Fundación Neotrópica. San José, Costa Rica: Editorial Heliconia. 271 pp.

Brack, A., and C. Mendiola. 2000. *Ecología del Perú*. United Nations Development Program (UNDP). Lima, Perú: Editorial Bruño.

Brak, B., M. Vroklage, M. Kappelle, and A.M. Cleef. 2005. Comunidades vegetales de la Turbera de La Chonta en Costa Rica. In M. Kappelle and S.P. Horn, eds., *Páramos de Costa Rica*, 607–29. Santo Domingo de Heredia, Costa Rica: Instituto Nacional de Biodiversidad (INBio).

Braun-Blanquet, J. 1979. *Fitosociología. Bases para el estudio de las comunidades vegetales*. 3rd ed. Madrid: Blume Ediciones. 820 pp.

Bravo, J., A. Chaverri, and G. Solano. 1991. *Plan de Manejo del Parque Nacional Chirripó*. San José, Costa Rica: Instituto Geográfico Nacional (IGN) / Universidad Nacional (UNA) / Servicio de Parques Nacionales (SPN). 83 pp.

Cabrera, A.L. 1968. Ecología vegetal de la puna. In C. Troll, ed., *Geo-ecology of the Mountainous Regions of the Tropical Americas. Colloquium Geographicum* 9: 91–116. Bonn: Ferdinand Dümmers Verlag.

Calvo, G. 1987. Geología del Macizo del Chirripó, Cordillera de Talamanca, Costa Rica. Tesis. Campaña Geológica G 5216. Universidad de Costa Rica (UCR). San Pedro, Costa Rica. 37 pp.

Carrillo, E., G. Wong, and J.C. Sáenz. 2005. Mamíferos de los páramos de Costa Rica. In M. Kappelle and S.P. Horn, eds., *Páramos de Costa Rica*, 533–45. Santo Domingo de Heredia, Costa Rica: Instituto Nacional de Biodiversidad (INBio).

Castillo, R. 1984. Geología de Costa Rica. San José, Costa Rica: Universidad Estatal a Distancia. 182 pp.

Chaverri, A. 2008. *Historia Natural del Parque Nacional Chirripo, Costa Rica*. Santo Domingo de Heredia, Costa Rica: Instituto Nacional de Biodiversidad (INBio). 141 pp.

Chaverri, A., and A.M. Cleef. 1996. Las comunidades vegetales en los páramos de los macizos del Chirripó y Buenavista. *Revista Forestal Centroamericana* 5(17): 44–49.

Chaverri, A., and A.M. Cleef. 2005. Comunidades vegetales de los páramos de los macizos de Chirripó y Buenavista, Costa Rica. In M. Kappelle and S.P. Horn, eds., *Páramos de Costa Rica*, 577–92. Santo Domingo de Heredia, Costa Rica: Instituto Nacional de Biodiversidad (INBio).

Chaverri, A., and O. Esquivel. 2005. Conservación, visitación y manejo del Parque Nacional Chirripó, Costa Rica. In M. Kappelle and S.P. Horn, eds., *Páramos de Costa Rica*, 669–99. Santo Domingo de Heredia, Costa Rica: Instituto Nacional de Biodiversidad (INBio).

Chaverri, A., B. Herrera, and O. Herrera-MacBryde. 1997. La Amistad Biosphere Reserve: Costa Rica and Panama. In S.D. Davis, V. Heywood, O. Herrera-MacBryde, J. Villalobos, and A.C. Hamilton, eds., *Centres of Plant Diversity: A Guide and Strategy for their Conservation*, vol. 3: *The Americas*, 209–14. Cambridge, UK: World Conservation Union (IUCN) / Smithsonian Institute (SI).

Chaverri, A., C. Vaughan, and L.J. Poveda. 1976. Informe de la gira efectuada al Macizo de Chirripó a raíz del fuego ocurrido en marzo de 1976. *Revista Costa Rica* 11: 243–79.

Clark, L.G. 1997. Diversity, biogeography and evolution of *Chusquea* (Poaceae: Bambusoideae). In G.P. Chapman, ed., *The Bamboos*, 33–44. London: Academic Press.

Cleef, A.M. 1981. The Vegetation of the Páramos of the Colombian Cordillera Oriental. *Dissertationes Botanicae* 61: 1–320.

Cleef, A.M., and A. Chaverri. 1992. Phytogeography of the páramo flora of the Cordillera de Talamanca, Costa Rica. In H. Balslev and J.L. Luteyn, eds., *Páramo: An Andean Ecosystem Under Human Influence*, 45–60. London: Academic Press.

Cleef, A.M., and A. Chaverri. 2005. Fitogeografía de la flora del páramo de la Cordillera de Talamanca, Costa Rica. In M. Kappelle and S.P. Horn, eds., *Páramos de Costa Rica*, 287–304. Santo Domingo de Heredia, Costa Rica: Instituto Nacional de Biodiversidad (INBio).

Cleef, A.M., and J.F. Duivenvoorden. 1994. Phytogeographic analysis of a vascular species sample from the Araracuara sandstone plateau, Colombian Amazonia. *Mém. Soc. Biogéogr.* (3 série) 4: 65–81. R. Schnell Festschrift.

Coen, E. 1983. Climate. In D.H. Janzen, ed., *Costa Rican Natural History*, 35–46. Chicago: University of Chicago Press.

Cuatrecasas, J. 1934. Observaciones geobotánicas en Colombia. *Trabajos del Museo Nacional de Ciencias Naturals, Serie Botánica* 27: 1–144. Madrid.

Cuatrecasas, J. 1958. Aspectos de la vegetación natural de Colombia. *Revista de la Acadamia Colombiana de Ciencias Exactas* 10(40): 221–68.

Cuatrecasas, J. 1968. Páramo vegetation and its life forms. In C. Troll, ed., *Geoecology of the Mountainous Regions of the Tropical Americas. Colloquium Geographicum* 9: 163–86. *Colloquium Geographicum* 9: 91–116. Bonn: Ferdinand Dümmers Verlag.

Cuatrecasas, J. 1977. *Westoniella*, a new genus of the Asteraceae from the Costa Rican páramos. *Phytologia* 35: 471–87.

Cuatrecasas, J. 1979. Comparación fitogeográfica de los páramos entre las varias cordilleras desde Costa Rica al Perú. Proceedings of the Seminario Internacional sobre Medio Ambiente Páramo. Mérida, Venezuela.

Denyer, P., and G.J. Soto 1999. Contribución pionera de William M. Gabb a la geología y cartografía de Costa Rica. *Anuario de Estudios Centroamericanos* (Universidad de Costa Rica) 25(2): 103–38.

Dinerstein, E., D.M. Olsen, D.J. Graham, A.L. Webster, S.A. Primm, M.P. Bookbinder, and G. Ledec. 1995. *A Conservation Assessment of the Terrestrial Ecoregions of Latin America and the Caribbean*. Washington, DC: WWF / The World Bank.

Drummond, M.S., M. Bordelon, J.Z. de Boer, M.J. Defant, H. Bellon, and M.D. Feigenson. 1995. Igneus petrogenesis and tectonic setting of plutonic and volcanic rocks of the Cordillera de Talamanca, Costa Rica–Panamá, Central American Arc. *American Journal of Science* 295: 875–919.

Esquivel, A. 2005. Nemátodos de los páramos de Costa Rica. In M. Kappelle and S.P. Horn, eds., *Páramos de Costa Rica*, 477–88. Santo Domingo de Heredia, Costa Rica: Instituto Nacional de Biodiversidad (INBio).

Fürst, E., M.L. Moreno, D. García, and E. Zamora. 2004. *Sistematización y Análisis del Aporte de los Parques Nacionales y Reservas Biológicas al Desarrollo Económico y Social en Costa Rica: Los Casos del Parque Nacional Chirripó, Parque Nacional Cahuita y Parque Nacional Volcán Poás*. Heredia, Costa Rica: Centro Internacional de Política Económica para el Desarrollo Sostenible (CINPE), Universidad Nacional (UNA). 218 pp.

Gabb, W.M. 1874a. Note on the geology of Costa Rica. *American Journal of Science* 7: 438–39.

Gabb, W.M. 1874b. Notes on the geology of Costa Rica. *American Journal of Science* 8: 388–390.

Gabb, W.M. 1895. Informe sobre la exploración de Talamanca 1873–1874. *Anales del Instituto Fisico-Geográfico Nacional de Costa Rica* 5: 71–90.

Gocke, K., E. Lahmann, G. Rojas, and J. Romero. 1981. Morphometric and basic limnological data of Laguna Grande de Chirripó, Costa Rica. *Revista de Biología Tropical* 27: 165–74.

Gómez P., L.D.1978. Contribuciones a la pteridología costarricense. XII. Carlos Wercklé. *Brenesia* 14/15: 361–93.

Gómez P., L.D. 1986. *Vegetación de Costa Rica*. San José, Costa Rica: Editorial de la Universidad Estatal a Distancia (EUNED). 327 pp.

Gómez P., L.D. 1994a. Checklist of Costa Rican Páramo Plants and the Cuericí Forest. Typed. Moravia, Costa Rica: Organization for Tropical Studies (OTS). 21 pp.

Gómez P., L.D. 1994b. Checklist of Plants of Cerro de la Muerte and other Costa Rican Páramos and Adjacent Forests (Cryptogams excluded). Typed. San Vito, Coto Brus, Costa Rica: Las Cruces Biological Station, Organization for Tropical Studies (OTS). 21 pp.

Gómez P., L.D. 2005a. La exploración científica de los páramos costarricenses. In M. Kappelle and S.P. Horn, eds., *Páramos de Costa Rica*, 101–10. Santo Domingo de Heredia, Costa Rica: Instituto Nacional de Biodiversidad (INBio).

Gómez P., L.D. 2005b. Hongos verdaderos (Eumycota) de los páramos de Costa Rica. In M. Kappelle and S.P. Horn, eds., *Páramos de Costa Rica*, 323–30. Santo Domingo de Heredia, Costa Rica: Instituto Nacional de Biodiversidad (INBio).

Gradstein, S.R., and I. Holz. 2005. Briófitas de los páramos de Costa Rica. In M. Kappelle and S.P. Horn, eds., *Páramos de Costa Rica*, 361–74. Santo Domingo de Heredia, Costa Rica: Instituto Nacional de Biodiversidad (INBio).

Grayum, M.H., B.E. Hammel, S. Troyo, and N. Zamora. 2004. History: botanical exploration and floristics in Costa Rica. In B.E. Hammel, M.H. Grayum, C. Herrera, and N. Zamora, eds., *Manual de Plantas de Costa Rica*, Vol. I. *Monographs in Systematic Botany from the Missouri Botanical Garden* 97: 1–45.

Gutiérrez-Braun, F. 1955. Expedición del Doctor Richard Weyl al macizo del Chirripó. San José, Costa Rica: Instituto Geográfico Nacional (IGN).

Haberyan, K.A., and S.P. Horn. 1999. A 10,000-year diatom record from a glacial lake in Costa Rica. *Mountain Research and Development* 19: 63–68.

Haberyan, K.A., and S.P. Horn. 2005. Un registro de diatomeas que cubre 10.000 años del Lago de las Morrenas 1, Parque Nacional Chirripó, Costa Rica. In M. Kappelle and S.P. Horn, eds., *Páramos de Costa Rica*, 275–85. Santo Domingo de Heredia, Costa Rica: Instituto Nacional de Biodiversidad (INBio).

Haberyan, K.A., S.P. Horn, and B.F. Cumming. 1997. Diatom assemblages from Costa Rican lakes: an initial ecological assessment. *Journal of Paleolimnology* 17: 263–74.

Haberyan, K.A., S.P. Horn, and G. Umaña. 2003. Basic limnology of fifty-one lakes in Costa Rica. *Revista de Biología Tropical* 51(1): 107–22.

Haberyan, K.A., G. Umaña, C. Collado, and S.P. Horn. 1995. Observations on the plankton of some Costa Rican lakes. *Hydrobiologia* 312: 75–85.

Hastenrath, S. 1968. Certain aspects of the three-dimensional distribution of climate and vegetational belts in the mountains of Central America and Mexico. In C. Troll, ed., *Geoecology of the Mountainous Regions of the Tropical Americas. Colloquium Geographicum* (Ferd. Dümmers Verlag, Bonn) 9: 122–30.

Hastenrath, S. 1973. On the Pleistocene glaciation of the Cordillera de Talamanca, Costa Rica. *Zeitschrift für Gletschertkunde und Glazialgeologie* 9: 105–21.

Hedberg, O. 1964. Features of afroalpine plant ecology. *Acta Phytogeographica Suecica* 49: 1–144.

Hedberg, O. 1992. Afroalpine vegetation compared to páramo: convergent adaptations and divergent differentiation. In H. Balslev and J.L. Luteyn, eds., *Páramo: An Andean Ecosystem Under Human Influence*, 15–29. London: Academic Press.

Herrera, W. 1986. Clima de Costa Rica. Vol. 2. In L.D. Gómez, ed., *Vegetación y Clima de Costa Rica*. With 10 maps (scale: 1:200.000). San José, Costa Rica: EUNED.

Herrera, W. 2005. El clima de los páramos de Costa Rica. In M. Kappelle and S.P. Horn, eds., *Páramos de Costa Rica*, 113–28. Santo Domingo de Heredia, Costa Rica: Instituto Nacional de Biodiversidad (INBio).

Hill, R.T. 1898. The geological history of the Isthmus of Panama and portions of Costa Rica. *Bulletin of the Museum of Comparative Zoölogy at Harvard College* 28(5): 149–285.

Hnatiuk, R.J. 1994. Plant form and function in alpine New Guinea. In P.W. Rundel, A.P. Smith, and F.C. Meinzer, eds., *Tropical Alpine Environments: Plant Form and Function*, 307–18. Cambridge, UK.

Hoekstra, J.M., T.M. Boucher, T.H. Ricketts, and C. Roberts. 2005. Confronting a biome crisis: global disparities of habitat loss and protection. *Ecology Letters* 8: 23–29.

Hofstede, R., P. Segarra, and P. Mena, eds. 2003. *Los Páramos del Mundo: Proyecto Atlas Mundial de los Páramos*. Quito, Ecuador: Global Peatland Initiative / NC / IUCN / EcoCiencia. 299 pp.

Holdridge, L.R. 1967. *Life Zone Ecology*. San José, Costa Rica: Centro Científico Tropical (CCT). 206 pp.

Holdridge, L.R., W.C. Grenke, W.H. Hatheway, T. Liang, and J.A. Tosi. 1971. *Forest Environments in Tropical Life Zones: A Pilot Study*. Oxford, UK: Pergamon Press. 747 pp.

Hooghiemstra, H., A.M. Cleef, G. Noldus, and M. Kappelle. 1992. Upper Quaternary vegetation dynamics and palaeoclimatology of the La Chonta bog area (Cordillera de Talamanca, Costa Rica). *Journal of Quaternary Science* 7(3): 205–25.

Hope, G., R. Hnatiuk, and J. Smith. 2003. Asia y Oceanía. In R. Hofstede, P. Segarra, and P. Mena, eds., *Los Páramos del Mundo: Proyecto Atlas Mundial de los Páramos*, 245–53. Quito, Ecuador: Global Peatland Initiative / NC / IUCN / EcoCiencia.

Horn, S.P. 1988. Effect of burning on a montane mire in the Cordillera de Talamanca, Costa Rica. *Brenesia* 30: 81–92.

Horn, S.P. 1989a. Postfire vegetation development in the Costa Rican páramos. *Madroño* 36(2): 93–114.

Horn, S.P. 1989b. Prehistoric fires in the Chirripó highlands of Costa Rica: sedimentary charcoal evidence. *Revista de Biología Tropical* 37: 139–48.

Horn, S.P. 1989c. The Inter-American Highway and human disturbance of páramo vegetation in Costa Rica. *Yearbook, Conference of Latin American Geographers* 15: 13–22.

Horn, S.P. 1990a. Timing of deglaciation in the Cordillera de Talamanca, Costa Rica. *Climate Research* 1: 81–83.

Horn, S.P. 1990b. Vegetation recovery after the 1976 páramo fire in Chirripó National Park, Costa Rica. *Revista de Biología Tropical* 38(2): 267–75.

Horn, S.P. 1991. Fire history and fire ecology in the Costa Rican páramos. In S.C. Nodvin and T.A. Waldrop, eds., *Fire and the Environment: Ecological and Cultural Perspectives*, 289–96. Proceedings of an International Symposium. March 20–24, 1990, Knoxville, TN.

Horn, S.P. 1992. Microfossils and forest history in Costa Rica. In H.K. Steen and R.P. Tucker, eds., *Changing Tropical Forests: Historical Perspectives on Today's Challenges in Central and South America*, 16–30. Durham, NC: Forest History Society.

Horn, S.P. 1993. Postglacial vegetation and fire history in the Chirripó Páramo of Costa Rica. *Journal of Quaternary Research* 40(1): 107–16.

Horn, S.P. 1997. Postfire resprouting of *Hypericum irazuense* in the Costa Rican páramos: Cerro Asunción revisited. *Biotropica* 29(4): 529–31.

Horn, S.P. 1998a. Postfire regrowth of *Vaccinium consanguineum* Klotzch (Ericaceae) in the Costa Rican Páramos. *Revista de Biología Tropical* 46(4): 1117–20.

Horn, S.P. 1998b. Fire management and natural landscapes in the Chirripó Páramo, Chirripó National Park, Costa Rica. In K. Zimmerer and K.R. Young, eds., *From Nature's Geography: Biogeographical Landscapes and Conservation in Developing Countries*, 125–46. Madison, WI: University of Wisconsin Press.

Horn, S.P. 2005. Dinámica de la vegetación después de fuegos recientes en los páramos de Buenavista y Chirripó, Costa Rica. In M. Kappelle and S.P. Horn, eds., *Páramos de Costa Rica*, 631–55. Santo Domingo de Heredia, Costa Rica: Instituto Nacional de Biodiversidad (INBio).

Horn, S.P., and L.G. Clark. 1992. Pollen viability in *Chusquea subtessellata* (Poaceae: Bambusoideae). *Biotropica* 24: 577–79.

Horn, S.P., and K.A. Haberyan. 1993. Physical and chemical properties of Costa Rican lakes. *National Geographic Society Research and Exploration* 9(1): 86–103.

Horn, S.P., and M. Kappelle. 2009. Fire in páramo ecosystems of Central and South America. In M.A. Cochrane, ed., *Tropical Fire Ecology: Climate Change, Land Use and Ecosystem Dynamics*, 505–39. Berlin, Germany: Springer-Praxis.

Horn, S.P., and B.L. League. 2005. Registros de sedimentos lacustres de la vegetación del holoceno e historia del fuego en el páramo de Costa Rica. In M. Kappelle and S.P. Horn, eds., *Páramos de Costa Rica*, 253–73. Santo Domingo de Heredia, Costa Rica: Instituto Nacional de Biodiversidad (INBio).

Horn, S.P., K.H. Orvis, and K.A. Haberyan. 1999. Investigación limnológica y geomorfológica de lagos y glaciares del Parque Nacional Chirripó, Costa Rica. Instituto Geográfico Nacional (IGN). *Revista Informe Semestral* 35: 95–106.

Horn, S.P., K.H. Orvis, and K.A. Haberyan. 2005. Limnología de las lagunas glaciales en el páramo del Chirripó, Costa Rica. In M. Kappelle and S.P. Horn, eds., *Páramos de Costa Rica*, 161–81. Santo Domingo de Heredia, Costa Rica: Instituto Nacional de Biodiversidad (INBio).

Horn, S.P., and R.L. Sanford Jr. 1992. Holocene fires in Costa Rica. *Biotropica* 24: 354–61.

Islebe, G.A. 1996. Vegetation, phytogeography and paleo-ecology of the last 20,000 years of montane Central America. PhD thesis, Universidad de Amsterdam. 179 pp.

Islebe, G.A., and A.M. Cleef. 1995. Alpine plant communities of Guatemala. *Flora* 190: 79–87.

Islebe, G.A., and H. Hooghiemstra. 1997. Vegetation and climate history of montane Costa Rica since the last glacial. *Quaternary Science Reviews* 16: 589–604.

Islebe, G.A., and H. Hooghiemstra. 2005. Historia del clima y de la vegetación montañosa de Costa Rica desde el último glaciar. In M. Kappelle and S.P. Horn, eds., *Páramos de Costa Rica*, 215–35. Santo Domingo de Heredia, Costa Rica: Instituto Nacional de Biodiversidad (INBio).

Islebe, G.A., H. Hooghiemstra, and K. van der Borg. 1995a. A cooling event during the Younger Dryas Chron in Costa Rica. *Paleogeography, Paleoclimatology, Paleoecology* 117: 73–80.

Islebe, G.A., H. Hooghiemstra, and R. van't Veer. 1996. Holocene vegetation and water level history in two bogs of the Cordillera de Talamanca, Costa Rica. *Vegetatio* 124: 155–71.

Islebe, G.A., H. Hooghiemstra, and R. van't Veer. 2005. Historia holocénica de la vegetación y del nivel de agua en dos turberas de la Cordillera de Talamanca, Costa Rica. In M. Kappelle and S.P. Horn, eds., *Páramos de Costa Rica*, 237–52. Santo Domingo de Heredia, Costa Rica: Instituto Nacional de Biodiversidad (INBio).

Islebe, G.A., and M. Kappelle. 1994. A phytogeographical comparison between the subalpine forests of Guatemala and Costa Rica. *Feddes Repertorium* 105: 73–87.

Islebe, G.A., and M. Kappelle. 2005. Comparación fitogeográfica entre los bosques subalpinos de Guatemala y Costa Rica. In M. Kappelle and S.P. Horn, eds., *Páramos de Costa Rica*, 305–19. Santo Domingo de Heredia, Costa Rica: Instituto Nacional de Biodiversidad (INBio).

Islebe, G.A., K. van den Borg, and H. Hooghiemstra. 1995b. The Younger Dryas climatic event in the Cordillera de Talamanca, Costa Rica. *Biologie en Mijnbouw* 74: 281–83.

Janzen, D.H. 1973. Rate of regeneration after a tropical high-elevation fire. *Biotropica* 5(2): 117–22.

Janzen, D.H. 1976. Why bamboos wait so long to flower. *Annual Review of Ecology and Systematics* 7: 347–91.

Janzen, D.H. 1983. *Swallenochloa subtessellata* (Chusquea, Batamba, Matamba). In D.H. Janzen, ed., *Costa Rican Natural History*, 330–31. Chicago: University of Chicago Press.

Johns, R., T. Utteridge, H. Hopkins, and P. Edwards. 2006. *A Guide to the Alpine and Sub-Alpine Flora of Mt Jaya*. Kew, UK: Royal Botanic Gardens. 687 pp.

Jones, J.R., K. Lohman, and G. Umaña. 1993. Water chemistry and trophic state of eight lakes in Costa Rica. *Verhandlungen, Internationale Vereinigung fur Theoretische und Angewandte Limnologie* 25: 899–905.

Kappelle, M. 1991. Distribución altitudinal de la vegetación del Parque Nacional Chirripó, Costa Rica. *Brenesia* 36: 1–14.

Kappelle, M. 1992. Structural and floristic differences between wet Atlantic and moist Pacific montane *Myrsine-Quercus* forests in Costa Rica. In H. Balslev and J.L. Luteyn, eds., *Páramo: An Andean Ecosystem under Human Influence*, 61–70. London: Academic Press.

Kappelle, M. 1995. Ecology of Mature and Recovering Talamancan Montane *Quercus* Forests, Costa Rica. PhD thesis, Universidad de Amsterdam. 270 pp.

Kappelle, M. 1996. *Los Bosques de Roble (*Quercus*) de la Cordillera de Talamanca, Costa Rica: Biodiversidad, Ecología, Conservación y Desarrollo*. Santo Domingo de Heredia, Costa Rica: Instituto Nacional de Biodiversidad (INBio). 336 pp.

Kappelle, M. 2001. Bosques nublados de Costa Rica. In M. Kappelle and A.D. Brown, eds., *Bosques Nublados del Neotrópico* [*Neotropical Cloud Forests*], 301–70. Santo Domingo de Heredia, Costa Rica: Instituto Nacional de Biodiversidad (INBio). 698 pp.

Kappelle, M. 2003. Costa Rica. In R. Hofstede, P. Segarra, and P. Mena, eds., *Los Páramos del Mundo: Proyecto Atlas Mundial de los Páramos*, 87–90. Quito, Ecuador: Global Peatland Initiative / NC / IUCN / EcoCiencia.

Kappelle, M. 2005a. Hacia una breve definición del concepto "páramo." In M. Kappelle and S.P. Horn, eds., *Páramos de Costa Rica*, 29–36. Santo Domingo de Heredia, Costa Rica: Instituto Nacional de Biodiversidad (INBio).

Kappelle, M. 2005b. Tardígrados de los páramos de Costa Rica. In M. Kappelle and S.P. Horn, eds., *Páramos de Costa Rica*, 489–91. Santo Domingo de Heredia, Costa Rica: Instituto Nacional de Biodiversidad (INBio).

Kappelle, M. 2005c. Insectos de los páramos de Costa Rica. In M. Kappelle and S.P. Horn, eds., *Páramos de Costa Rica*, 493–99. Santo Domingo de Heredia, Costa Rica: Instituto Nacional de Biodiversidad (INBio).

Kappelle, M. 2005d. Bosques enanos subalpinos de Costa Rica. In M. Kappelle and S.P. Horn, eds., *Páramos de Costa Rica*, 593–605. Santo Domingo de Heredia, Costa Rica: Instituto Nacional de Biodiversidad (INBio).

Kappelle, M. 2006. Structure and composition of Costa Rican montane oak forests. In M. Kappelle, ed., *Ecology and Conservation of Neotropical Montane Oak Forests*, Ecological Studies Series, vol. 185, 127–39. New York: Springer Verlag.

Kappelle, M., and A.D. Brown, eds. 2001. *Bosques Nublados del Neotrópico* [*Neotropical Cloud Forests*]. Santo Domingo de Heredia, Costa Rica: Instituto Nacional de Biodiversidad (INBio). 698 pp.

Kappelle, M., M. Castro, A. Garita, L. González, and H. Monge. 2005b. Ecosistemas de los páramos del Área de Conservación La Amistad Pacífico en Costa Rica. In M. Kappelle and S.P. Horn, eds., *Páramos de Costa Rica*, 549–75. Santo Domingo de Heredia, Costa Rica: Instituto Nacional de Biodiversidad (INBio).

Kappelle, M., and A.M. Cleef. 2003. Memorias acerca de una científica en el páramo costarricense: Adelaida Chaverri-Polini (21 de mayo del 1947–20 de setiembre del 2003). *Brenesia* 59–60: 3–5.

Kappelle, M., and A.M. Cleef. 2004a. In memoriam: Adelaida Chaverri–Polini, May 21, 1947–September 20, 2003. *Revista de Biología Tropical* 52(1): XIII–XVI.

Kappelle, M., and A.M. Cleef. 2004b. Adelaida Chaverri: ecóloga de tierras altas, conservacionista genuina. *Manejo Integrado de Plagas y Agroecología* (Costa Rica) 73: 1–7.

Kappelle, M., A.M. Cleef, and A. Chaverri. 1989. Phytosociology of montane *Chusquea-Quercus* forests, Cordillera de Talamanca, Costa Rica. *Brenesia* 32: 73–105.

Kappelle, M., A.M. Cleef, and A. Chaverri. 1992. Phytogeography of Talamancan montane *Quercus* forest, Costa Rica. *Journal of Biogeography* 19: 299–315.

Kappelle, M., K.A. Haberyan, and S.P. Horn. 2005a. Algas fósiles y recientes de los páramos de Costa Rica. In M. Kappelle and S.P. Horn, eds., *Páramos de Costa Rica*, 331–41. Santo Domingo de Heredia, Costa Rica: Instituto Nacional de Biodiversidad (INBio).

Kappelle, M., and S.P. Horn, eds. 2005. *Páramos de Costa Rica*. Santo Domingo de Heredia, Costa Rica: Instituto Nacional de Biodiversidad (INBio). 767 pp.

Kappelle, M., and J. Savage. 2005. Anfíbios y reptiles de los páramos de Costa Rica. In M. Kappelle and S.P. Horn, eds., *Páramos de Costa Rica*, 513–19. Santo Domingo de Heredia, Costa Rica: Instituto Nacional de Biodiversidad (INBio).

Kappelle, M., and E. van Omme. 1997. Lista de las plantas de los bosques nubosos subalpinos de la Cordillera de Talamanca en Costa Rica. *Brenesia* 47–48: 55–71.

Kappelle, M., E. van Omme, and M.E. Juárez. 2000. Lista de la flora vascular terrestre de la cuenca superior del Río Savegre, San Gerardo de Dota, Costa Rica. *Acta Botánica Mexicana* 51: 1–38.

Kappelle, M., and J.G. van Uffelen. 2005. Los suelos de los páramos de Costa Rica. In M. Kappelle and S.P. Horn, eds., *Páramos de Costa Rica*, 147–59. Santo Domingo de Heredia, Costa Rica: Instituto Nacional de Biodiversidad (INBio).

Kappelle, M., and J.G. van Uffelen. 2006. Montane oak forest distribution along temperature, humidity and soil gradients in Costa Rica. In M. Kappelle, ed., *Ecology and Conservation of Neotropical Montane Oak Forests*, Ecological Studies Series, vol. 185, 39–54. New York: Springer Verlag.

Kappelle, M., J.G. van Uffelen, and A.M. Cleef. 1995. Altitudinal zonation of montane *Quercus* forests along two transects in the Chirripó National Park, Costa Rica. *Vegetatio* 119: 119–53.

Kappelle, M., N. Zamora, and T. Flores. 1991. Flora leñosa de la zona alta (2000–3819 m) de la Cordillera de Talamanca, Costa Rica. *Brenesia* 34: 121–44.

Kohkemper, M. 1968. *Historia de las Ascensiones al Macizo del Chirripó*. San José: Instituto Geográfico Nacional (IGN). 120 pp. + 1 mapa.

Kohkemper, M. 1983. Nueve Expediciones a la Cordillera de Talamanca. Instituto Geográfico Nacional (IGN). *Informe Semestral* (enero–junio). 33 pp.

Körner, C. 1998. A re-assessment of high elevation treeline positions and their explanation. *Oecologia* 115: 445–59.

Lachniet, M.S., and G.O. Seltzer. 2002. Late Quaternary glaciation of Costa Rica. *Geological Society of America Bulletin* 114(5): 547–58. Correction, vol. 114, 921–22.

Lachniet, M.S., G.O. Seltzer, and L. Solís. 2005. Geología, geomorfología y depósitos glaciares en los páramos de Costa Rica. In M. Kappelle and S.P. Horn, eds., *Páramos de Costa Rica*, 129–46. Santo Domingo de Heredia, Costa Rica: Instituto Nacional de Biodiversidad (INBio).

Landaeta, A., C.A. López, and A. Alvarado. 1978. Caracterización de la fracción mineral de suelos derivados de cenizas volcánicas de la Cordillera de Talamanca, Costa Rica. *Agronomía Costarricense* 2(2): 117–29.

Lane, C.S., S.P. Horn, C.I. Mora, K.H. Orvis, and D.B. Finkelstein. 2011. Sedimentary stable carbon isotope evidence of late Quaternary vegetation and climate change in highland Costa Rica. *Journal of Paleolimnology* 45: 323–38.

Lane, C.S., and S.P. Horn. 2013. Terrestrially derived n-alkane δD evidence of shifting Holocene paleohydrology in highland Costa Rica. *Arctic, Antarctic, and Alpine Research* 45(3): 342–49.

Lauer, W. 1981. Ecoclimatological conditions of the páramo belt in tropical high mountains. *Mountain Research and Development* 1: 209–21.

League, B.L., and S.P. Horn. 2000. A 10,000 year record of páramo fires in Costa Rica. *Journal of Tropical Ecology* 16: 747–52.

Löffler, H. 1972. Contribution to the limnology of high mountain lakes in Central America. *Internationale Revue der Gesamten Hydrobiologie* (Viena) 57(3): 397–408.

Luteyn, J.L. 1992. Páramos: why study them? In H. Balslev and J.L. Luteyn, eds., *Páramo: An Andean Ecosystem Under Human Influence*, 1–14. London: Academic Press.

Luteyn, J.L. 1999. Páramos: a checklist of plant diversity, geographic distribution and botanical literature. *Memoirs of the New York Botanical Garden* 84: 1–278.

Luteyn, J.L., A.M. Cleef, and O. Rangel. 1992. Plant diversity in páramo: towards a checklist of páramo plants and a generic flora. In H. Balslev and J.L. Luteyn, eds., *Páramo: An Andean Ecosystem Under Human Influence*, 71–84. London: Academic Press.

Malagón, D., and R. Pulido. 2000. Suelos del páramo colombiano. In J.O. Rangel, ed., *Colombia: Diversidad Biótica*, III: La Región de Vida Paramuna, 37–84. Instituto de Ciencias Naturales (ICN). Bogotá, Colombia: Facultad de Ciencias, Universidad Nacional de Colombia.

Mena, P., G. Medina, and R. Hofstede. 2001. *Los Páramos del Ecuador: Particularidades, Problemas y Perspectivas*. Quito, Ecuador: Editorial Abya Yala. 311 pp.

Monasterio, M., ed. 1980. *Estudios Ecológicos de los Páramos Andinos*. Mérida, Venezuela: Universidad de los Andes.

Monasterio, M., and M. Molinillo. 2003. Venezuela. In R. Hofstede, P. Segarra, and P. Mena, eds., *Los Páramos del Mundo: Proyecto Atlas Mundial de los Páramos*, 205–36. Quito, Ecuador: Global Peatland Initiative / NC / IUCN / EcoCiencia.

Monasterio, M., and F. Vuilleumier. 1986. Introduction: high tropical mountain biota of the world. In F. Vuilleumier and M. Monasterio, eds., *High Altitude Tropical Biogeography*, 3–7. Oxford, UK: Oxford University Press.

Mostacero, J., F. Mejía, W. Zelada, and C. Medina. 2007. *Biogeografía del Perú*. Lima, Peru: Editorial del Pacífico.

Myers, N. 1988. Threatened biotas: 'Hotspots' in tropical forests. *Environmentalist* 8: 1–20.

Myers, N. 1990. The biodiversity challenge: Expanded hot-spots analysis. *Environmentalist* 10: 243–56.

Myers, N. 2003. Biodiversity hotspots revisited. *BioScience* 53: 916–17.

Myers, N., R.A. Mittermeier, C.G. Mittermeier, G.A.B. da Fonseca, and

J. Kent. 2000. Biodiversity hotspots for conservation priorities. *Nature* 403: 853–58.

Myers, R.L. 2006. *Living with Fire: Sustaining Ecosystems and Livelihoods through Integrated Fire Management.* Tallahassee, FL: Global Fire Initiative, The Nature Conservancy.

Nogués-Bravo, D., M.B. Araújo, M.P. Errea, and J.P. Martínez-Rica. 2007. Exposure of global mountain systems to climate warming during the 21st century. *Global Environmental Change* 17: 420–28.

Obando, L.G. 2004. Geología y petrografía del Cerro Buenavista (Cerro de la Muerte) y Alrededores, Costa Rica. *Revista Geológica de América Central* 30: 31–39.

Olson, D.M., and E. Dinerstein. 1998. The Global 200: a representation approach to conserving the Earth's most biologically valuable ecoregions. *Conservation Biology* 12: 502–15.

Olson, D.M., and E. Dinerstein. 2002. The Global 200: priority ecoregions for global conservation. *Annals of the Missouri Botanical Garden* 89: 199–224.

Olson, D.M., E. Dinerstein, E.D. Wikramanayake, N.D. Burgess, G.V.N. Powell, E.C. Underwood, J.A. D'Amico, I. Itoua, H.E. Strand, J.C. Morrison, C.J. Loucks, T.F. Allnutt, T.H. Ricketts, Y. Kura, J.F. Lamoreux, W.W. Wettengel, P. Hedao, and K.R. Kassem. 2001. Terrestrial ecoregions of the world: a new map of life on Earth. *Bioscience* 51(11): 933–38.

Orvis, K.H., and S.P. Horn. 2000. Quaternary glaciers and climate on Cerro Chirripó, Costa Rica. *Quaternary Research* 54: 24–37.

Orvis, K.H., and S.P. Horn. 2005. Los glaciares cuaternarios y el clima del cerro Chirripó, Costa Rica. In M. Kappelle and S.P. Horn, eds., *Páramos de Costa Rica*, 185–213. Santo Domingo de Heredia, Costa Rica: Instituto Nacional de Biodiversidad (INBio).

Otárola, C.E. 1976. Caracterización y clasificación de algunos suelos de la Cordillera de Talamanca. Lic. thesis, Universidad de Costa Rica.

Otárola, C.E., and A. Alvarado. 1976. Caracterización y clasificación de algunos suelos del Cerro de la Muerte, Talamanca, Costa Rica. Int. Rep. Universidad de Costa Rica, Facultad de Agronomía. San Pedro, Costa Rica. 57 pp.

Pérez, S., A. Alvarado, and E. Ramírez. 1978. Manual Descriptivo del Mapa de Asociaciones de Subgrupos de Suelos de Costa Rica. Escala: 1:200,000. Oficina de Planificación Sectorial Agropecuario. San José, Costa Rica: IGN / MAG / FAO.

Pittier, H. 1891. Viaje de Exploración al Valle Del Río Grande de Térraba. San José, Costa Rica: Tip. Nacional.

Pittier, H. 1938. Apuntaciones Etnológicas sobre los Indios Bribri. Serie Etnológica, vol. I, part I. San José, Costa Rica: Museo Nacional and Imprenta Nacional.

Quintanilla, V.G. 1983. Comparación entre dos ecosistemas tropoandinos: la puna chilena y el páramo ecuatoriano. *Informaciones Geográficas* (Chile) 30: 25–45.

Rangel, J.O. 2000. *Colombia: Diversidad Biótica III: La Región de Vida Paramuna.* Universidad Nacional de Colombia, Facultad de Ciencias. Bogotá, Colombia: Instituto de Ciencias Naturales (ICN). 902 pp.

Recharte, J., L. Albán, R. Arévalo, E.R. Flores, L. Huerta, M. Orellana, L. Oscanoa, and P. Sánchez. 2002. El Grupo Páramos / Jalcas y Punas del Perú: Instituciones y Acciones en Beneficio de Comunidades y Ecosistemas Alto Andinos. In *Anales de la Reunión del Grupo Internacional de Páramos*, 785–811. Bogotá, Colombia.

Rich, P.U., and T.H. Rich. 1983. The Central American dispersal route: biotic history and paleogeography. In D.H. Janzen, ed., *Costa Rican Natural History*, 12–34. Chicago: University of Chicago Press.

Rundel, P.W. 1994. Tropical alpine climates. In P.W. Rundel, A.P. Smith, and F.C. Meinzer, eds., *Tropical Alpine Environments: Plant Form and Function*, 21–43. Cambridge, UK: Cambridge University Press.

Rundel, P.W., F.C. Meinzer, and A.P. Smith. 1994b. Tropical alpine ecology: progress and priorities. In P.W. Rundel, A.P. Smith, and F.C. Meinzer, eds., *Tropical Alpine Environments: Plant Form and Function*, 355–63. Cambridge, UK: Cambridge University Press.

Rundel, P.W., A.P. Smith, and F.C. Meinzer, eds. 1994a. Tropical Alpine Environments: Plant Form and Function. Cambridge, UK: Cambridge University Press.

Rzedowski, J. 1978. *Vegetación de México*. México DF: Editorial Limusa.

Safford, H.D. 1999a. Brazilian páramos I. Introduction to the physical environment and vegetation of the campos de altitude. *Journal of Biogeography* 26: 713–38.

Safford, H.D. 1999b. Brazilian páramos II. Macro- and meso-climate of the campos de altitude and affinities with high mountain climates of the tropical Andes and Costa Rica. *Journal of Biogeography* 26: 739–60.

Salgado-Labouriau, M.L., ed. 1979. *El Medio Ambiente Páramo*. Caracas, Venezuela: Centro de Estudios Avanzados.

Salgado-Labouriau, M.L. 1980. Paleoecología de los páramos venezolanos. In M. Monasterio, ed., *Estudios Ecológicos de los Páramos Andinos*. Mérida, Venezuela: Universidad de los Andes.

Samudio, R., and J.L. Pino, eds. 2006. *Evaluación Biológica del Ecosistema de Páramo de Panamá: Cerros Fábrega—Itamut*. Panama City: ANAM / SOMASPA / TNC. 54 pp.

Sarmiento, F. 2007. Book Review: Páramos de Costa Rica. *Mountain Research and Development* 27(1): 95–97.

Schneidt, J., U. Stein, B. Furchheim-Weberling, S. Wiedmann, and F. Weberling. 1996. Estudios sobre formas de crecimiento de algunas especies típicas del páramo de Costa Rica. *Brenesia* 45–46: 51–112.

Schneidt, J., and F. Weberling. 1993. Wuchsformuntersuchungen im Páramo Costa Ricas. III. Untersuchungen an Ericaceen-Arten. *Flora* (Germany) 187: 403–27.

Seyfried, H., A. Astorga, and C. Calvo. 1987. Sequence stratigraphy of deep and shallow water deposits from an evolving island arc: the Upper Cretaceous and Tertiary of Central America. *Facies* 17: 203–14.

Shimizu, C. 1992. Glacial landforms around Cerro Chirripó in Cordillera de Talamanca, Costa Rica. *Journal of Geography* (Japan) 101(7): 615–21.

Sipman, H.J.M. 1999. Lichens. In J.L. Luteyn, ed., *Páramos: A Checklist of Plant Diversity, Geographic Distribution and Botanical Literature. Memoirs of the New York Botanical Garden* 84: 41–53.

Sipman, H.J.M. 2005. Líquenes de los páramos de Costa Rica. In M. Kappelle and S.P. Horn, eds., *Páramos de Costa Rica*, 343–60. Santo Domingo de Heredia, Costa Rica: Instituto Nacional de Biodiversidad (INBio).

Smith, J.A., G.O. Seltzer, D.T. Rodbell, D.L. Farber, and R.C. Finkel. 2005. Early local last glacial maximum in the tropical Andes. *Science* 308: 678–81.

Smith, J.M.B., and A.M. Cleef. 1988. Composition and origins of the world's tropicalpine floras. *Journal of Biogeography* 15: 631–45.

Soderstrom, T.R., and C.E. Calderón. 1978. The species of *Chusquea* (Poaceae: Bambusoideae) with verticillate buds. *Brittonia* 30: 154–64.

Standley, P.C. 1937. Flora of Costa Rica. *Field Museum of Natural History, Botanical Series* 18: 1–1571.

Stehli, F.G., and S.D. Webb., eds. 1985. *The Great American Biotic Interchange*. New York City: Plenum Press.

Stein, U., and F. Weberling. 1993. Wuchsformuntersuchungen im Páramo Costa Ricas. I. Einführung. II. *Acaena cylindristachya* Ruiz and Pavón, *Acaena elongata* L. (Rosaceae–Sanguisorbeae) und *Chusquea subtessellata* Hitchcock (Poaceae). *Flora* (Germany) 187: 369–402.

Steinberg, M., and M. Taylor. 2008. Guatemala's Altos de Chiantla: changes on the high frontier. *Mountain Research and Development* 28(3/4): 255–62.

Stiles, G., A. Skutch, and D. Gardner. 1989. *A Guide to the Birds of Costa Rica*. Ithaca, NY: Cornell University Press. 511 pp.

Stone, D. 1961. *Las Tribus Talamanqueñas de Costa Rica*. San Jose, Costa Rica: Editorial Lehmann.

Sturm, H., and J.O. Rangel, eds. 1985. *Ecología de los Páramos Andinos: Una Visión Preliminar Integrada*. Biblioteca José Jerónimo Triana. Instituto de Ciencias Naturales (ICN). Bogotá, Colombia: Universidad Nacional de Colombia.

Süssenguth, K. 1942. Neue Pflanzen aus Costa Rica, insbesondere vom Chirripó Grande, 3837 m. *Botanische Jahrbücher fur Systematik* 72: 270–302.

Thouret, J.C. 1983. La temperatura de los suelos: temperatura estabilizada en profundidad y correlaciones térmicas y pluviométricas. In T. Van der Hammen, A. Pérez-P., and P. Pinto, eds., *La Cordillera Central Colombiana: Transecto Parque Los Nevados. Estudios de Ecosistemas Tropandinos* 1: 142–49. Vaduz: J. Cramer.

Tosi, J.A., Jr. 1969. Mapa Ecológico de Costa Rica, Basado en la Clasificación Vegetal Mundial de L.R. Holdridge. Scale: 1:750,000. San José, Costa Rica: Centro Científico Tropical (CCT).

Troll, C. 1959. Die tropische Gebirge: Ihre dreidimensionale klimatische und pflanzengeographische Zonierung. *Bonner Geographische Abhandlungen* 25: 1–93.

Troll, C. 1968. The cordilleras of the tropical Americas: aspects of climatic, phytogeographical and agrarian ecology. In C. Troll, ed., *Geoecology of the Mountainous Regions of the Tropical Americas. Colloquium Geographicum* 9: 15–56. Bonn: Ferdinand Dümmers Verlag.

Troll, C., and W. Lauer, eds. 1978. Geoecological relations between the southern temperate zone and the tropical mountains. *Erdwissenschaftliche Forschung* 11: xxx, 1–563.

Umaña, G., K.A. Haberyan, and S.P. Horn. 1999. Limnology in Costa Rica. In R.G. Wetzel and B. Gopal, eds., *Limnology in Developing Countries*, vol. 2, 33–62. International Association for Limnology (SIL).

Van der Hammen, T., and A.M. Cleef. 1986. Development of the high Andean páramo flora and vegetation. In F. Vuilleumier and M. Monasterio, eds., *High Altitude Tropical Biogeography*, 153–201. Oxford, UK: Oxford University Press.

Van Royen, P. 1980. *The Alpine Flora of New Guinea*. Vaduz, Liechtenstein: Cramer.

van Uffelen, J.G. 1991. A Geological, Geomorphological and Soil Transect Study of the Chirripó Massif and Adjacent Areas, Cordillera de Talamanca, Costa Rica. M.Sc. thesis, Wageningen Agricultural University. Wageningen, Holanda. 2 vol. 72 pp.

Vareschi, V. 1970. *Flora de los Páramos*. Mérida, Venezuela: Universidad de los Andes.

Vareschi, V. 1992. *Ecología de la Vegetación Tropical*. Caracas, Venezuela: Sociedad Venezolana de Ciencias Naturales. 306 pp.

Vargas, G. 1987. Estudio de la Vegetación del Páramo Costarricense. Thesis, Universidad de Costa Rica (UCR). San Pedro, Costa Rica.

Vargas, G., and J.J. Sánchez. 2005. Plantas con flores de los páramos de Costa Rica y Panamá: el páramo ístmico. In M. Kappelle and S.P. Horn, eds., *Páramos de Costa Rica*, 397–435. Santo Domingo de Heredia, Costa Rica: Instituto Nacional de Biodiversidad (INBio).

Wagner, M. 1866. Über den Character und Höhenverhältinisse der Vegetation in der Cordillera von Veraguas und Guatemala. München: Sitzungsberichte der (Ber. Kgl.) Bayerischen Akademie der Wissenschaften.

Wallace, D.R. 1992. *The Quetzal and the Macaw: The Story of Costa Rica's National Parks*. San Francisco, CA: Sierra Club Books.

Walter, H. 1985. *Vegetation of the Earth and Ecological Systems of the Geobiosphere*. 3rd ed. New York: Springer. 318 pp.

Walter, H., and E. Medina. 1969. La temperatura del suelo como factor determinante para la caracterización de los pisos subalpino y alpino en los Andes de Venezuela. *Sociedad Venezolana de Ciencias Naturales* 28(115–116): 201–10.

Weber, H. 1958. Die páramos von Costa Rica und ihre Pflanzengeographische Verkettung mit den Hochanden Südamerikas. *Akademie der Wissenschaften und der Literatur, Abhandlungen der Mathematisch-Naturwissenschaftlichen Klasse* 1956(3): 120–95.

Weber, H. 1959. *Los Páramos de Costa Rica y su Concatenación Fitogeográfica con los Andes Sudamericanos*. San José, Costa Rica: Instituto Geográfico Nacional (IGN). Traducción al español de Weber, 1958.

Weberling, F., and B. Weberling. 1993. Wuchsformuntersuchungen im Páramo Costa Ricas. IV: Untersuchungen an *Senecio*–und *Hypericum*–Arten. *Flora* (Jena) 188: 291–320.

Weberling, F., and B. Furchheim-Weberling. 2005. El mosaico de formas de crecimiento en los páramos de Costa Rica. In M. Kappelle and S.P. Horn, eds., *Páramos de Costa Rica*, 438–73. Santo Domingo de Heredia, Costa Rica: Instituto Nacional de Biodiversidad (INBio).

Wercklé, C. 1909. La subregión fitogeográfica costarricense. San José, Costa Rica: Sociedad Nacional de Agricultura, Tipografía Nacional. 55 pp.

Weston, A. 1981a. The Vegetation and Flora of the Chirripó Páramo. Unpublished report. San José, Costa Rica: Tropical Science Center (TSC). 10 pp.

Weston, A. 1981b. Páramos, Ciénagas and Subpáramo Forest in the Eastern Part of the Cordillera de Talamanca. Unpublished report. San José, Costa Rica: Tropical Science Center (TSC). 32 pp.

Weyl, R. 1955a. Vestigios de una glaciación del Pleistoceno en la Cordillera de Talamanca, Costa Rica, América Central. Instituto Geográfico Nacional (IGN). *Informe Trimestral* (julio a setiembre) III: 9–32. San José, Costa Rica.

Weyl, R. 1955b. *Contribución a la Geología de la Cordillera de Talamanca*. San José, Costa Rica: Instituto Geográfico Nacional (IGN). 77 pp.

Weyl, R. 1956a. Eiszeitliche Gletscherspuren in Costa Rica (Mittelamerikas). *Zeitschrift für Gletscherkunde und Glazialgeologie* 3: 317–25.

Weyl, R. 1956b. Spuren eismitlicher Vergletscherung in der Cordillera de Talamanca Costa Ricas (Mittelamerika). *Neues Jahrbuch Geologische und Paläontologische Mithandlungen* 102: 283–94.

Weyl, R. 1957. Beiträge zur Geologie der Cordillera de Talamanca Costa

Ricas (Mittelamerika). *Neues Jahrbuch für Geologie und Palaeontologie* 105: 123–204.

Weyl, R. 1980. *Geology of Central America*. Berlin, Germany: Gebrüder Borntraeger. 371 pp.

Wiedmann, S., and F. Weberling. 1993. Wuchsformuntersuchungen im Páramo Costa Ricas. V: Untersuchungen an *Escallonia poasana* Donnell-Smith, *Coriaria thymifolia* Humboldt & Bonpland und *Myrrhidendron donnell-smithii* Coulter & Rose. *Flora* (Germany) 188: 321–42.

Williamson, G.B., G.A. Schatz, A. Alvarado, C.S. Redhead, A.C. Stam, and R.W. Sterner. 1986. Effects of repeated fire on tropical páramo vegetation. *Tropical Ecology* 27: 62–69.

Wujek, D.E., R.E. Clancy, and S.P. Horn. 1998. Silica-scaled Chrysophyceae and Synurophyceae from Costa Rica. *Brenesia* 49/50: 11–19.

Wunsch, O., G. Calvo, B. Willscher, and H. Seyfried. 1999. Geologie der Alpinen Zone des Chirripó-Massives (Cordillera de Talamanca, Costa Rica, Mittelamerika). *Profil* 16: 193–210.

Zamora, N., B.E. Hammel, and M.H. Grayum. 2004. Vegetation. In B.E. Hammel, M.H. Grayum, C. Herrera, and N. Zamora, eds., *Manual de Plantas de Costa Rica*, vol. I. Introducción. *Monographs in Systematic Botany from the Missouri Botanical Garden* 97: 91–216.

Zeeb, B.A., J.P. Smol, and S.P. Horn. 1996. Chrysophycean stomatocysts from Costa Rican tropical lake sediments. *Nova Hedwigia* 63(3–4): 279–99.

Part VII

The Wet Caribbean Lowlands

Chapter 16 The Caribbean Lowland Evergreen Moist and Wet Forests

Deedra McClearn[1,*], J. Pablo Arroyo-Mora[2], Enrique Castro[3], Ronald C. Coleman[4], Javier F. Espeleta[5], Carlos García-Robledo[6], Alex Gilman[3], José González[3], Armond T. Joyce[7], Erin Kuprewicz[8], John T. Longino[9], Nicole L. Michel[10], Carlos Manuel Rodríguez[11], Andrea Romero[12], Carlomagno Soto[3], Orlando Vargas[3], Amanda Wendt[13], Steven Whitfield[14], Robert M. Timm[15]

Introduction

The saying "geography is destiny" has been used by historians and economists to explain large-scale phenomena such as human trading routes, migration patterns, technological innovation, spread of disease, and the motivation for and outcomes of wars. In the context of Costa Rican ecosystems, one might also invoke the expression to frame more than five million years of the history of Costa Rica's Caribbean lowlands. We do not imply that events have been or will be predictable, but rather that geography has played a critical role in Costa Rica's Caribbean lowlands with respect to initial origin of the landmass, climate, patterns of vegetation, sites of early human settlement, colonial expansion, the construction of access roads, colonial and postcolonial exploitation of natural resources, conservation efforts, and current political and environmental issues. In this chapter we offer a brief overview of these topics, encapsulate major biological research efforts in the Caribbean lowlands over the past 50 years, and look into the future—always within the framework of geography at different scales.

Today the Caribbean lowland portion of Costa Rica, broadly defined, includes the entire Province of Limón (with its Caribbean Sea margin), plus portions of the Provinces of Heredia, Alajuela, and Guanacaste (with their northern edges along the Río San Juan and other stretches of Costa Rica's border with Nicaragua) (Fig. 16.1). For this map, we define the upper altitudinal limit of the area as 300 m, although some sections of this chapter will use slightly different elevations, depending on the standards of a particular

[1] Organization for Tropical Studies, Box 90630, Durham, NC 27708, USA
[2] McGill University, Department of Geography, 805 Sherbrooke Street West, Montreal, Quebec, H3A 2K6, Canada
[3] La Selva Biological Station, Organization for Tropical Studies, Apartado 676-2050, San Pedro de Montes de Oca, Costa Rica
[4] Department of Biological Sciences, California State University Sacramento, 6000 J Street, Sacramento, CA 95819, USA
[5] Tropical Science Center, Apartado 8-3870-1000, San José, Costa Rica
[6] Laboratory of Interactions and Global Change, Department of Multitrophic Interactions, Institute of Ecology (INECOL), Mexico and National Museum of Natural History, Departments of Botany and Entomology, Smithsonian Institution, PO Box 37012, Washington, DC 20013, USA
[7] 1408 Eastwood Drive, Slidell, LA 70458, USA
[8] National Museum of Natural History, Smithsonian Institution, Department of Botany, PO Box 37012, Washington, DC 20013, USA
[9] Department of Biology, University of Utah, 257 S 1400 E, Salt Lake City, UT 84112, USA
[10] University of Saskatchewan, Department of Animal and Poultry Science, 51 Science Place, Saskatoon, SK S7N 5A6, Canada
[11] Conservation International, Apartado 8-3870, San José, Costa Rica
[12] University of Wisconsin-Whitewater, Whitewater, WI 53190, USA
[13] Department of Ecology and Evolutionary Biology, University of Connecticut, 75 N. Eagleville Road, Unit 3043, Storrs, CT 06269, USA
[14] Zoo Miami, Conservation and Research Department, Miami, FL 33177, USA
[15] Department of Ecology and Evolutionary Biology & Biodiversity Institute, University of Kansas, 1345 Jayhawk Drive, Lawrence, KS 66045, USA
* Corresponding author. Current address: DKU Program Office, Duke University, Box 90036, Durham, NC 27708, USA

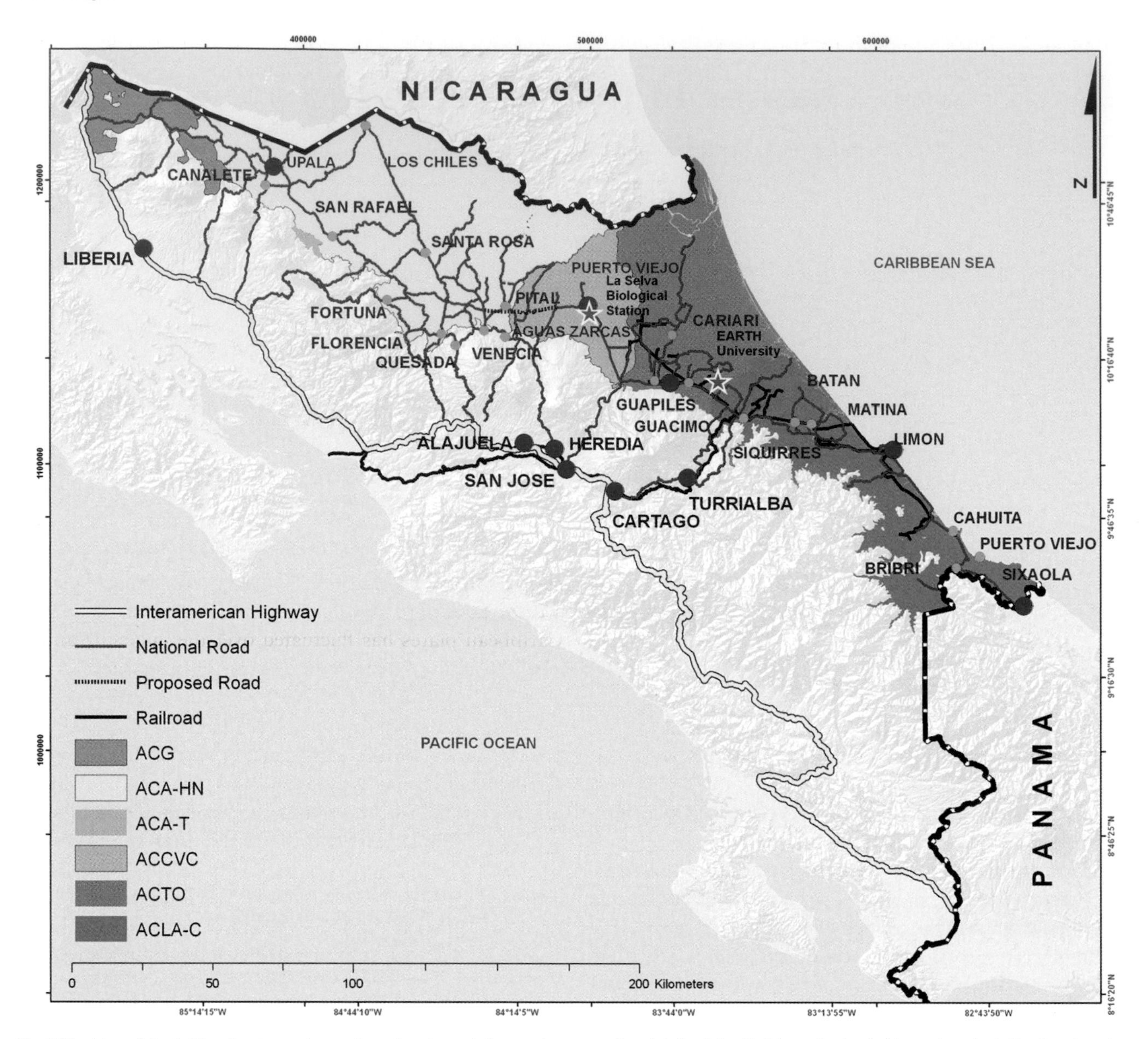

Fig. 16.1 Map of Costa Rica showing major roads, national population centers, as well as details of the Caribbean lowlands (shown in colors). The Pan-American Highway is shown as a thick yellow line. The northern edge of the zone is the border with Nicaragua (including the Río San Juan) and the eastern edge is the Caribbean Sea. Although the railway is no longer in use, its extent (solid black lines) is depicted here to illustrate the route taken by highland coffee from the Central Plateau to the port city of Limón during the late nineteenth and early twentieth centuries. Note that there are two towns labeled Puerto Viejo. One is in the canton of Sarapiquí (near La Selva Biological Station) and the other is near Limón. Unwary researchers heading to La Selva have been known to catch the wrong bus in San José and to end up on the Caribbean coast. The Caribbean lowlands have contributions from six of the country's 11 conservation areas. ACG (Guanacaste), ACA-HN (Arenal-Huertar Norte), ACA-T (Arenal-Tilarán), ACCVC (Cordillera Volcánica Central), ACTo (Tortuguero), and ACLA-C (La Amistad-Caribe). The La Selva Biological Station and EARTH University are indicated by stars.

field of study. At the northern end of the zone, we chose the watershed of the Río Sapoa as the boundary. This river drains into Lake Nicaragua, whereas all the rivers to the north and west drain into the Pacific Ocean. A more readily identifiable border is the northern stretch of the Panamerican Highway, which is virtually congruent with the Sapoa watershed boundary.

The Caribbean lowlands cover approximately 13,760 km^2 or about 27% of the terrestrial landmass of Costa Rica. The major population center of this area is the port city of Limón, Costa Rica's second largest city, with a population of about 63,000 people. Other locations that receive frequent mention throughout this chapter include the OTS-administered La Selva Biological Station and EARTH University (Escuela de la Agricultura de la Región Tropical Húmeda). Six of the country's 11 conservation areas are represented in the

Caribbean lowlands, although three of them have only a small area represented in the lowlands. Fig. 16.2 shows the distribution of protected areas and indigenous reserves in the Caribbean evergreen, forested lowlands.

Historical Overview

We treat the history of the Costa Rican Caribbean lowlands in five sections:

1. Early tectonic activity (70 million years ago [mya]) through the Miocene shoaling of the Central American Seaway (CAS) and incipient closure of the Isthmus (from approximately 24 mya to 5.3 mya) but see Bacon et al. (2015)
2. Post-Miocene development and closure of the Isthmus until human arrival (from approximately 5.3 mya to 16,000 years before present [ybp])
3. Early human (indigenous) settlement until the arrival of the first Europeans (from 16,000 ybp to 1502 AD)
4. Colonial and postcolonial trends (from 1502 to the 1960s)
5. Recent history (from the 1960s on)

The major topics considered for all of these time segments have been radically rethought in the past 25 years. New sites (geological, paleontological, archaeological, and ecological), new techniques (high-resolution remote sensing, climate modeling, dated molecular phylogenies, in addition to analysis of isotopes, phytoliths, and starch grains), and new paradigms (origins of tropical agriculture, demographic collapse, the pristine myth) have all influenced current thinking. We cannot offer definitive answers to many of the most engaging questions or even cover the complexities of the debates. Additionally, we are only too aware that our treatment now will seem woefully out of date in a few years. We do hope that our presentation here will capture some of the excitement and fullness of ongoing work, guide the reader to unexpected sources of information, and perhaps even prompt new lines of investigation.

Early Tectonic Activity through the Miocene–Pliocene Closure of the Isthmus

Today the Caribbean Sea waters of Costa Rica and Panama are warmer (Lessios 2008), saltier (Benway and Mix 2004, Haug et al. 2001, Keigwin 1982), and less productive (Cannariato and Ravelo 1997, Jain and Collins 2007) than are the Pacific coastal waters. These differences have accumulated (but fluctuated) for the past many millions of years, as the wide abyssal Central American Seaway (CAS) between the oceans underwent a series of shoaling events throughout the Miocene and into the Pliocene. The primary engine for these changes has been tectonic activity for more than the past 70 my in one of the world's most complex intersections of oceanic and continental plates (Alvarado et al. 2009, Van Avendonk et al. 2011, Buchs et al. 2010, Hoernle et al. 2002, Hoernle et al. 2008, Sadofsky et al. 2009, Wegner et al. 2011, Werner et al. 1999).

The Caribbean seafloor (Caribbean Large Igneous Province or CLIP) is generally (but not universally) acknowledged to be of Pacific origin and to have shoved its way through the gap between North and South America about 65 mya. The Farallón Plate fractured 23 mya (Lonsdale 2005), at which time the resultant Cocos Plate portion began subducting beneath the Caribbean Plate along the Middle American Trench (MAT), which is a stretch of Pacific coast from Mexico to Panama. This subduction caused the lifting of the seafloor and the volcanic activity that created the Costa Rican and Panamanian portions of the isthmian land bridge between North and South America (Sak et al. 2009). The convergence rate of the Cocos and Caribbean plates has fluctuated over the millennia and is currently about 90 mm/year offshore of Costa Rica (Harris et al. 2010, Hoernle et al. 2008). The thickened but buoyant section of the Cocos Plate identified as the Cocos Ridge is positioned off the Osa Peninsula in southern Costa Rica (Hoernle et al. 2002).

The Talamanca Mountains represent the arc highlands and the Caribbean coast area around Limón is the back-arc basin of this ridge (MacMillan et al. 2004). The failure of smooth subduction of this ridge has been implicated by some geologists as the cause of the highly destructive 1991 Limón/Cahuita earthquake on the Caribbean side of Costa Rica (Fernández-Arce 2009; Cortés, chapter 17 of this volume), but other authors consider that the slab subducting beneath the Talamanca Mountains is behaving "normally" and that the buoyant Cocos Ridge piece is only beginning to enter the MAT 50 km offshore from the Osa Peninsula (Dzierma et al. 2011).

The periodic upheavals in and final cessation of the deep water CAS, along with the formation of the Caribbean shores of Costa Rica and Panama, deflected the Atlantic waters northward along the east coast of North America. By mechanisms still not definitively elucidated, the establishment of the Isthmus is thought by many scientists to have created the Gulf Stream and to have initiated the series of Pleistocene Ice Ages (Bartoli et al. 2005, Haug et al. 2004, Nof and Van Gorder 2003, Steph et al. 2006—but see Molnar 2008 for other perspectives), from the last of which the world emerged about 10,000 years ago. From this deep historical perspective, therefore, the Caribbean

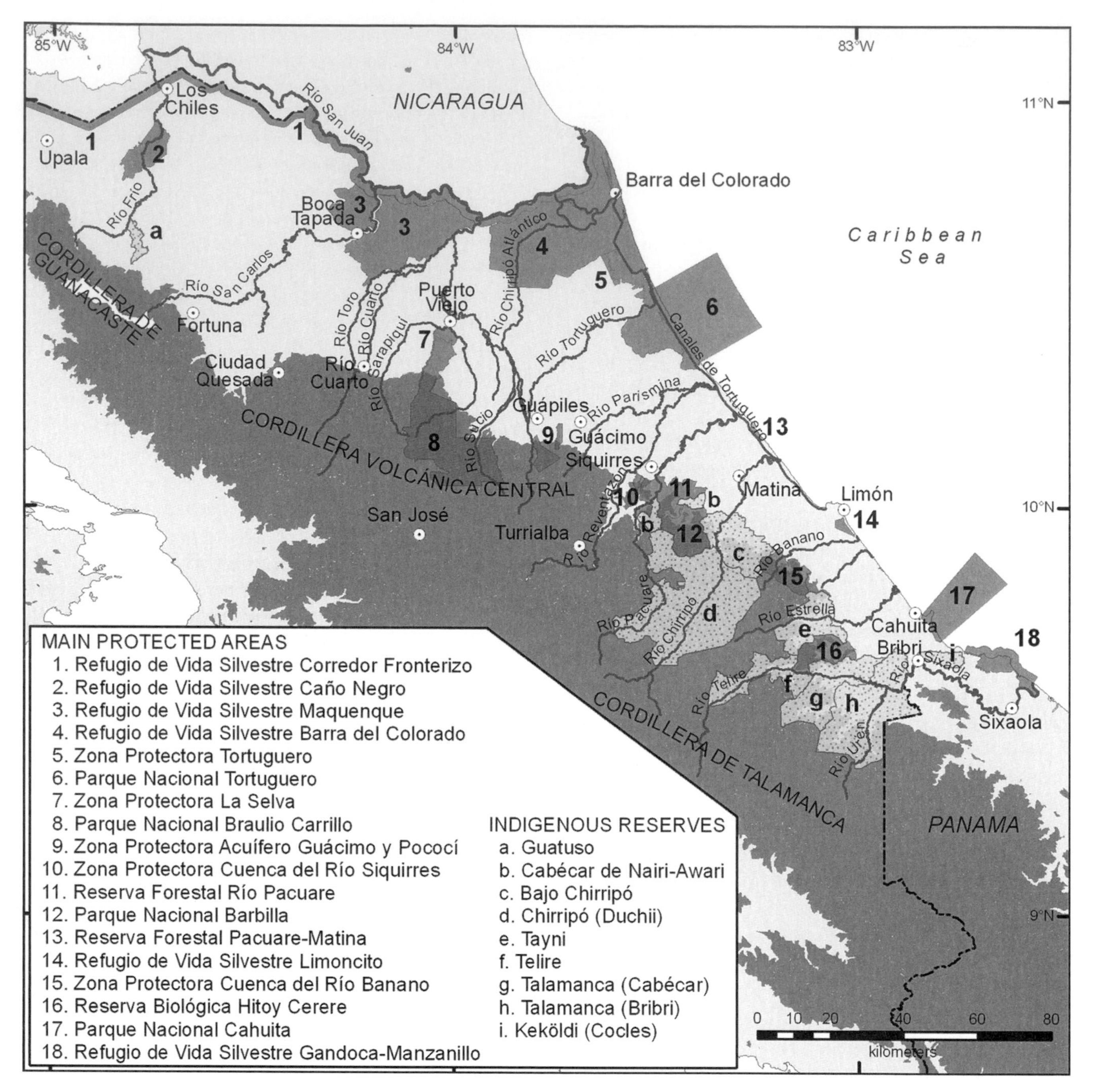

Fig. 16.2 Map of the main protected areas and indigenous reserves found in Costa Rica's Caribbean evergreen moist and wet forest lowlands. *Map prepared by Marco V. Castro.*

lowlands of Costa Rica can be assigned a major role in "the most important natural event to affect the surface of the earth in the past 60 million years" (Coates 1997).

The Miocene Epoch was a 20-million-year stretch of time (between about 25 and 5 mya) during which the space now occupied by Panama and Costa Rica was transformed from an open seaway to an island arc (or peninsula) and finally to a continuous isthmus. The climate fluctuated, from cool Early Miocene to the Middle Miocene Climatic Optimum 12–15 mya and through a few more warming and cooling cycles after that (Kürschner et al. 2008, You et al. 1996). Maximum temperatures at 12 mya are usually described as about 5 degrees C warmer than they are now in the lowland tropics. The bits of basalt and limestone that were being pushed above sea level in the area that is now Caribbean-side Costa Rica immediately became potential colonization sites for all manner of terrestrial organisms waiting in the wings (North and South America) and in the winds.

Our knowledge of plant communities in the New World lowland tropics of the Miocene has been and continues to be fragmentary. Good fossil sites are rare and widely scattered in space and time. The current lowland tropical vegetation of Mexico and Central America was previously considered to be primarily of South American origin (Gentry 1982, Hammel and Zamora 1993, Wendt 1993). Ancient transatlantic dispersal (in both directions) had been postulated for fern spores, based on the closest living relatives of some fern genera being found in Africa and South America (Moran and Smith 2001), but this mechanism was invoked for few angiosperms. North and South American floras shared only 2.6% of their palynomorphs (fossil pollen morphotypes) in the Middle Eocene (45 mya), but shared 30% after establishment of the Isthmus (Burnham and Graham 1999).

These data were previously taken as evidence of isolation of North and South American floras before the land bridge connection, followed by plant dispersal across the land bridge, with a south to north flow predominating. The fossil palynomorphs from a late Miocene site near Limón, Costa Rica, led to a reconstruction of the site as a moist, warm, closed-canopy forest (with an abundance of ferns and lycophytes in the understory) bordered by a *Rhizophora* mangrove swamp—no palms, no grasses, and no evidence of savanna-like habitat (Graham 1987, 1992, Graham and Dilcher 1998). The heights of the exposed land formations near Limón were estimated to be about 1,400 m in the early Miocene and 1,700 m in the late Miocene—this low relief would also imply that there was not yet a distinction between vegetation assemblages on the two sides of the Isthmus (Graham and Dilcher 1998). During the Miocene there was probably greater divergence in marine organisms than in the terrestrial vegetation on the Pacific and Caribbean sides.

Two new techniques are expanding and changing our image of the evolution of New World tropical vegetation. The proliferation of dated molecular phylogenies of plant groups and the study of phytoliths—small slivers of silica ("plant opal") in plant tissues that are diagnostic to the family, genus, or even species level—are providing higher resolution and recalibrated dates for the evolution of tropical plant communities. Recent molecular biogeographic studies of several plant groups suggest multiple instances of long-distance dispersal between Africa and South America (Lavin et al. 2004, Pennington et al. 2006, Renner 2004a, b, 2005) and across the Central American Seaway before isthmian closure (Cody et al. 2010). These techniques could greatly enhance and modify our understanding of the early colonization and subsequent evolution of the Caribbean lowland vegetation (Pennington et al. 2006).

An intriguing but little explored potential source of inorganic material and life forms for the emerging isthmus would have been transoceanic winds. Much has being made in recent years of the enormous African dust clouds (from the Sahara and Sahel deserts) that are affecting the Amazon Basin (Ansmann et al. 2009, Bristow et al. 2010, Kallos et al. 2006), contributing mineral elements to Jamaica and Barbados (Caquineau et al. 2002, Muhs and Budahn 2009), spreading crop diseases (Brown and Hovmøller 2002), possibly contributing to Caribbean coral diseases (Garrison et al. 2003), threatening human health (Griffin 2007, Mundt et al. 2009), and probably modifying global climate (Perkins 2001, Washington et al. 2009). Dust fluxes across the tropical Pacific have been less well documented (McGee et al. 2007) and the deep history of all of these winds remains largely unknown. It is reasonable to imagine dust, pollen, spores, and pathogens travelling on these winds, but apparently larger organisms can be borne along as well. A well-documented, wind-assisted locust plague in 1988 brought a swarm of these insects 5,000 km across the Atlantic from Western Africa to the Caribbean in four or five days (Lorenz 2009).

In summary, the scenario for the Miocene island arc phase of the Caribbean lowlands involves the following factors: overall increase in land area caused by seafloor lifting and volcanism (modified by fluctuations in sea level), warm temperatures peaking at 12 mya, and colonization of land by plants (probably dominated by southern forms), mammals (dominated by northern forms), and birds (both northern and southern contributions). For other taxa, a dated molecular phylogeny strongly suggests that the turtle *Rhinoclemmys* (the only New World genus of the family Geoemydidae) crossed to the New World via Beringia during the Eocene, and then into South America during the Miocene before land bridge closure (Le and McCord 2008). A similar hypothesis has been promoted for the New World *Polyommatus* butterflies (the "blues" of the family Lycaenidae) having moved across Beringia from Asia and then into South America about 10.7 mya (Vila et al. 2011). South American taxa thought to have arrived in Central America before the completion of the isthmus include túngara frogs (Weigt et al. 2005), certain vipers (Zamudio and Greene 1997), and ctenosaur lizards (Hasbún et al. 2005). The pygmy rain frog (*Pristimantis*) appeared to have radiated from the Pacific side of Costa Rica and Panama over to the Caribbean side during the Miocene (Wang et al. 2008).

The Reserva Talamanca, a 54,000 ha lowland parcel between Puerto Limón and the Cordillera de Talamanca, has been the territory of the Bribri and Cabécar indigenous groups for several hundred years (Borge and Castillo 1997) and is currently the home to approximately 8,000 of the remaining 21,000 members of these groups. In 83 homegardens, 46 cultivated plant species were growing, including native species (*Inga*), non-native species with a long history in Central America (plantain, cacao, peach palm), and recent arrivals (orange, mango) (Zaldivar et al. 2002). The local diet includes 84 plant species—60 grown in "near" space (homegardens and fields that can be manipulated) and 24 grown in "far" space (natural forest from which products may be harvested sustainably but which cannot—according to traditional practices—be otherwise manipulated) (García-Serrano and Del Monte 2004). The genetic diversity of cassava (*Manihot esculenta*) found in these homegardens is higher than that found in commercial crops, and includes cultivars not found in South American indigenous homegardens (Zaldivar et al. 2004).

In summary, changes in neotropical vegetation over the past 10,000 years may be tied to at least five significant and possibly overlapping influences: fluctuations in climate (Farrera et al. 1999, Hughen et al. 2004, Islebe et al. 1996, Islebe and Sánchez 2002, Lozano-García et al. 2007), natural colonization of plants across the Central American Isthmus from both North and South America, fire (Avnery et al. 2011), loss of browsing pressure from mammalian megafauna (Johnson 2009), and direct human modification of the landscape by forest clearing, domestication of crop plants, and other agricultural practices (Iriarte 2011, Lane et al. 2008).

Colonial and Postcolonial Development

The fourth voyage of Columbus in 1502 included an oft-cited stop at the site of the current city of Limón, or rather at a small island (Isla Uvita) just offshore. The Europeans did not initially find this coastal area a particularly profitable region for exploitation, in addition to which they encountered strong resistance against their incursions from the indigenous people. The reader is guided to several treatments of the early history of Spanish settlement of the Caribbean area of Costa Rica as well as to discussions of more recent history (Molina and Palmer 2007, Palmer and Molina 2004, Ross and Capelli 2003). For the past 20 years, there has been much talk about the extent of the demographic collapse of indigenous populations due to diseases brought by the Europeans. Denevan (1992, 2003) and Newson (1993) estimated that as many as 90% of the indigenous people died upon contact with Europeans in areas as geographically widespread as the Caribbean Islands, Nicaragua, and the Amazon Basin. We consider the issue of human populations of the Caribbean lowlands below.

Recent History

Although many of the fine points of early human impact on the Caribbean lowland landscape have been lost (or hidden) in the mists of time, the record of land use change since the 1960s is remarkably complete. This knowledge is largely due to data from aerial photographs and remote sensing maps and also to concerted efforts to document the rapid changes that occurred during these years. This time period marks the beginning of at least three landscape-altering trends in Costa Rica that had a particularly strong impact on the Caribbean side of the country—rapid population growth, rampant deforestation, and incipient conservation efforts. One important point has already been made—that there are probably virtually no pristine or virgin forests in the Caribbean lowlands or anywhere in Costa Rica.

However, as recently as the early 1950s, most of the Caribbean lowlands were still covered in primary forest and older secondary forest, regrown from previous centuries of episodic and localized cutting, including felling trees for railroad ties in the nineteenth century.

Increased access to an area has invariably led to an increase in human settlement (Rosero-Bixby and Palloni 1998), from the construction of the Alajuela–Limón railroad at the end of the nineteenth century (Augelli 1987) to the completion of the San José–Guápiles highway through Parque Nacional Braulio Carrillo in the late 1970s (Sader and Joyce 1988) and to the completion of other roads over the past 30 years. Many areas in the Caribbean lowlands are still considered to be relatively inaccessible, and new roads continue to be planned and constructed even as the pressures from population growth and agricultural intensification increase. We treat the topics of deforestation, agriculture, and land conservation in separate sections below.

Current Conditions

The Physical Environment

Climate and Weather

The climate in the Caribbean lowland area is greatly influenced by the trade winds (*vientos alisios*) that blow onshore in a northeast to southwest direction. Local annual rainfall varies over a 2-fold range from about 2,500 mm to about 5,000 mm per year. Rainfall and other aspects of climate are also affected by the passage of the Intertropical Convergence Zone (ITCZ) over the area twice a year. Rainfall is generally lower during the months of January through March, but the dry season is not as pronounced on the Caribbean side of the central mountain ranges as it is on the Pacific side (see also Herrera, chapter 2 of this volume).

The Costa Rican Instituto Meteorológico Nacional (IMN) has divided the Caribbean lowlands into six different zones, three of which border Nicaragua to the north (N1, N2, N3) and three of which run along the Caribbean coast (A1, A2, A3) (Fig. 16.3). The national data records indicate very similar temperatures in these zones (maximum, minimum, and median), but the rainfall patterns are distinctly different (inset, Fig. 16.3). The zone with highest annual rainfall is in the north coastal zone N3 that includes the town of Guápiles (4,860 mm per year), with the driest zones at either end of the lowland stretch (2,722 mm around Upala and 2,470 mm in the zone that includes Cahuita and Sixaola). La Selva Biological Station is in a zone that receives an average of 3,710 mm per year, although the station itself receives 4,000 mm of rain per year (50-year mean annual rainfall).

Temperature and rainfall are the most commonly kept records for government and corporate weather stations, but there are many other relevant measures of weather and

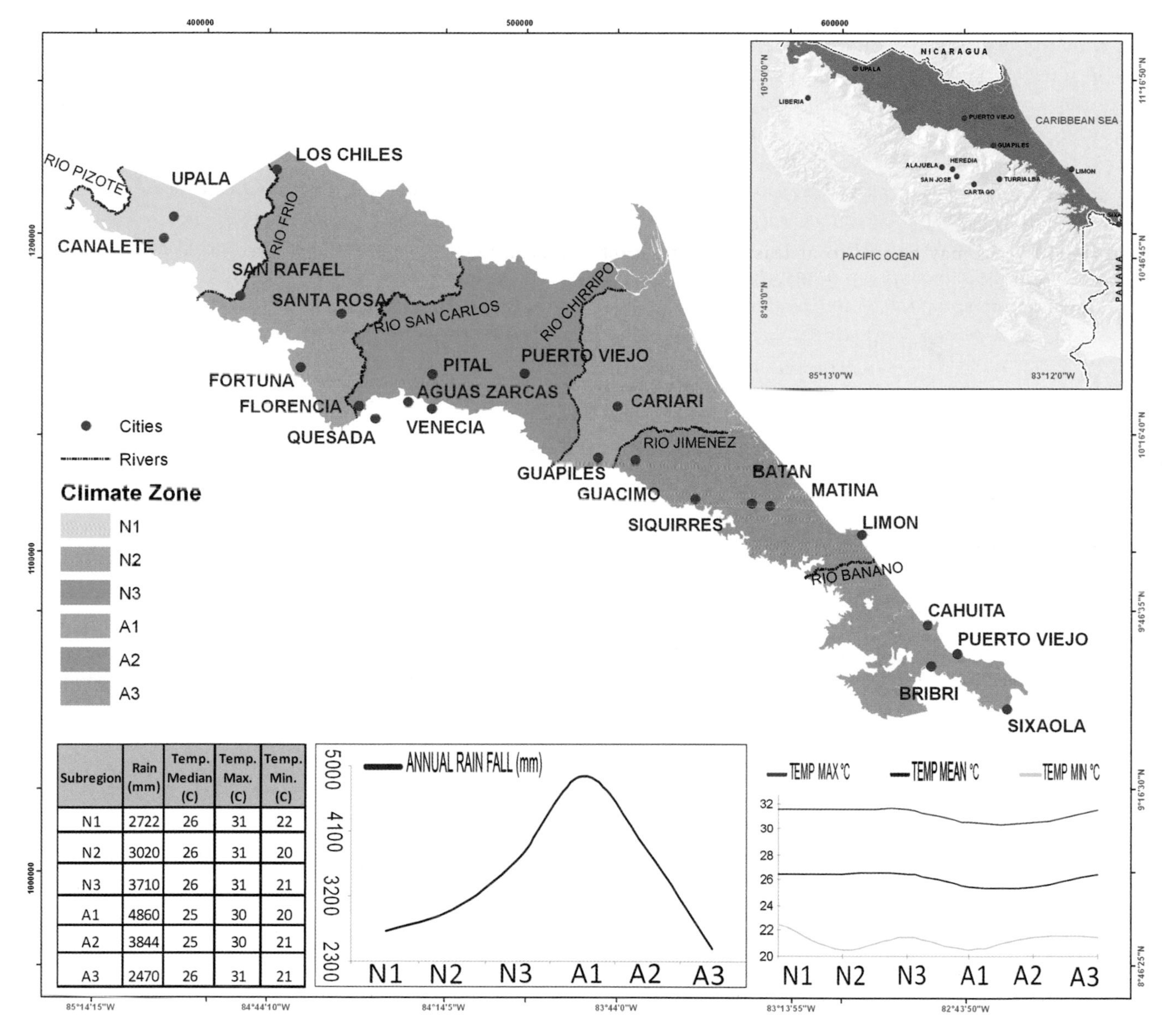

Subregion	Rain (mm)	Temp. Median (C)	Temp. Max. (C)	Temp. Min. (C)
N1	2722	26	31	22
N2	3020	26	31	20
N3	3710	26	31	21
A1	4860	25	30	20
A2	3844	25	30	21
A3	2470	26	31	21

Fig. 16.3 Map of the climate zones of Costa Rica's Caribbean lowlands as defined by the National Meteorological Institute (IMN [Instituto Meteorológico Nacional]). The zones labeled N1, N2, and N3 are along Costa Rica's northern border with Nicaragua and the zones marked A1, A2, and A3 are the Atlantic sea margin areas. Temperature regimes are broadly similar across the entire area but annual rainfall has a distinct peak in the center of the region.

climate, including relative humidity, wind speed, wind direction, soil moisture, and PAR (photosynthetically active radiation). PAR is an important component of most climate models. Usually only the best-equipped weather stations maintain the full range of data. These additional records seem to be particularly prone to missing data, faulty or uncalibrated sensors, and human error in quality control (Clark and Clark 2011, Senna et al. 2005).

Patterns of tropical vegetation have been tied to climate at the medium and long-term temporal scales (from hundreds to millions of years) and also to a spatial scale of kilometers to hundreds of kilometers. On a geological time scale, the climate has fluctuated in ways that usually (but not always) pair relatively warm and moist conditions during one phase and cool and dry conditions during alternate phases (see Historical Overview, this chapter). In the tropical forests of today, the climate variables most frequently associated with specific vegetation types are temperature and rainfall. A mesoscale temperature gradient is generally associated with an altitudinal gradient (Cavelier 1996, Cavelier et al. 1999, Lieberman et al. 1996), whereas a rainfall gradient may be linked to or independent of elevation. Across lowland neotropical forests (eliminating the altitudinal variable), the floristic composition at a regional scale is nearly always linked to some aspect of water availability, although other factors such as soil fertility and historical accident may also play a role in creating and maintaining plant beta-diversity (Condit et al. 2002, Marques et al. 2011, Pyke et al. 2001, Toledo et al. 2011).

Primary productivity, biomass, global terrestrial carbon, and biodiversity are all concentrated in the tropics (Dirzo and Raven 2003, Dixon et al. 1994, Hartshorn 2006, Malhi et al. 2002, Melillo et al. 2002, Saugier et al. 2001). The impact of ongoing and future global climate change on tropical forests is therefore of great concern (Malhi et al. 2009). The likely effects of changes in temperature, increased atmospheric CO_2, and rainfall have attracted considerable attention (Beer et al. 2010, Clark 2004, Lewis et al. 2009, Körner 2009). Empirical data indicate that temperatures in the tropics have increased by about 0.26 degrees C per decade since the 1960s, largely driven by increases in nighttime minimum temperatures (Malhi and Wright 2004), and climate models predict continued trends in this direction (Cramer et al. 2004, Diffenbaugh and Scherer 2011, IPCC 2007, Malhi and Phillips 2004). If tropical forests are already near a high temperature threshold (Reed et al. 2012), a small increase in temperature could dramatically change forest-wide carbon stocks by mechanisms such as altering the balance between photosynthesis and respiration (Clark et al. 2003, Doughty and Goulden 2008), stimulating release of carbon from soils (Raich et al. 2006), or modifying floristics in favor of liana abundance (Schnitzer and Bongers 2002, Wright et al. 2004). The likelihood of an atmospheric CO_2 fertilization effect in the tropics is still a matter of intense debate (Cernusak et al. 2011, Holtum and Winter 2010, Lewis et al. 2009).

Rainfall is an interesting case. Many climate models predict drought in the tropics (Cox et al. 2008, Neelin et al. 2006), but other models predict increased rainfall in areas that already receive high rainfall (the-rich-get-richer hypothesis) (Chou et al. 2009). Empirical data show declining rainfall across the tropics over the past 40 years, but this decrease is mostly concentrated in Africa and Asia (Malhi and Wright 2004). Still other models, and some recent empirical data, point to an increase in climate extremes that would intensify the severity of storms, floods, and droughts without necessarily changing annual rainfall (Aguilar et al. 2005, Jentsch et al. 2007, Marengo et al. 2011, Sun et al. 2007). Decrease in water availability on an annual or seasonal basis is of concern to tropical ecologists and climate scientists because many species of tropical trees and understory plants in lowland moist and wet forests of the neotropics are particularly drought-sensitive (Bunker and Carson 2005, Engelbrecht et al. 2007, Engelbrecht and Kursar 2003, Enquist and Enquist 2011, Feeley et al. 2011, Tyree et al. 2003).

In order to integrate information from a single well-documented site in the Caribbean lowlands of Costa Rica, we present the case of the La Selva Biological Station, situated at the base of Volcán Barva (Fig. 16.4). The weather records for the past 50 years from the meteorological stations on the property indicate an increase in temperature of 0.25 degrees C per decade, largely driven by increase in nighttime low temperatures (Clark and Clark 2011), consistent with global tropical patterns. The rainfall has been variable but has not changed over the 50-year period, maintaining a mean of 4,000 mm annually (Fig. 16.5). Interestingly, a shorter record (of the past 29 years) appears to show an increase in rainfall—but that increase disappears when the entire 50 year record is analyzed. A similar cycle is seen with the long-term weather records for Barro Colorado Island (BCI) in Panama (S. Paton, pers. comm.). Annual wood production in the La Selva old-growth forest is positively correlated with dry season rainfall and negatively associated with mean annual nighttime temperature for both stand-level (ten years of data) and focal tree species (24 years of data) analyses (Clark et al. 2010). In other words, the big trees grow less when the dry season is a bit drier than usual and also when the nights are a bit warmer than usual—and these two effects are independent. No evidence emerged for a carbon fertilization effect (Clark et al. 2010). Growth of La Selva trees is negatively correlated with global atmospheric CO_2 and with temperature (Clark

Fig. 16.4 View from the top of the CARBONO research tower on the SSO trail at the La Selva Biological Station, July 2010, in the middle of the rainy season. The view is looking southward towards San José (which is 55 km away in a straight-line distance on the other side of the peaks on the horizon). The base of the tower is at 97 m a.s.l. The peak on the left is Cacho Negro, a dormant volcano 2,100 m high. The peak on the far right is Volcán Barva, which provides the run-off for most of the streams and rivers of La Selva. Much of the forest in the middle ground is part of the San Juan–La Selva Biological Corridor.
Photo by Orlando Vargas.

et al. 2003), implying a causal linkage among tree growth, temperature, and atmospheric CO_2.

Some authors have predicted that high-elevation forests and low-elevation dry forest will be the most vulnerable of Costa Rica's ecosystems under conditions of climate change (Enquist 2002). Other authors predict "lowland biotic attrition" at La Selva and other lowland sites because, although lowland species can move upslope to track cooler temperatures, there is no source for species adapted to climates warmer than those of the existing lowlands (Colwell et al. 2008, Feeley and Silman 2010).

Soils

The parent material for most of the Caribbean lowland soils is derived from the episodic volcanic lava flows from the mountains of the Central Cordillera, in addition to contributions made by volcanic mud flows (lahars) and volcanic ash (Sollins et al. 1994; Alvarado and Mata, chapter 4 of this volume). Even the sands of the Caribbean beaches are composed of volcanic material due to a process whereby: (1) the rivers have carried gravel, fine sediments, and ash from the volcanoes to the coast; and (2) coastal wave action has redistributed the material along the shore in a series of beach ridges (Nieuwenhuyse and Kroonenberg 1994). Small sections of the coast are composed of ancient reefs, but these are the exception rather than the rule (Battistini and Bergoeing 1984). Material from the offshore continental shelves apparently did not contribute significantly to the coastal sands. Coastal deposition has more than kept pace with coastal erosion in recent centuries, so that even with

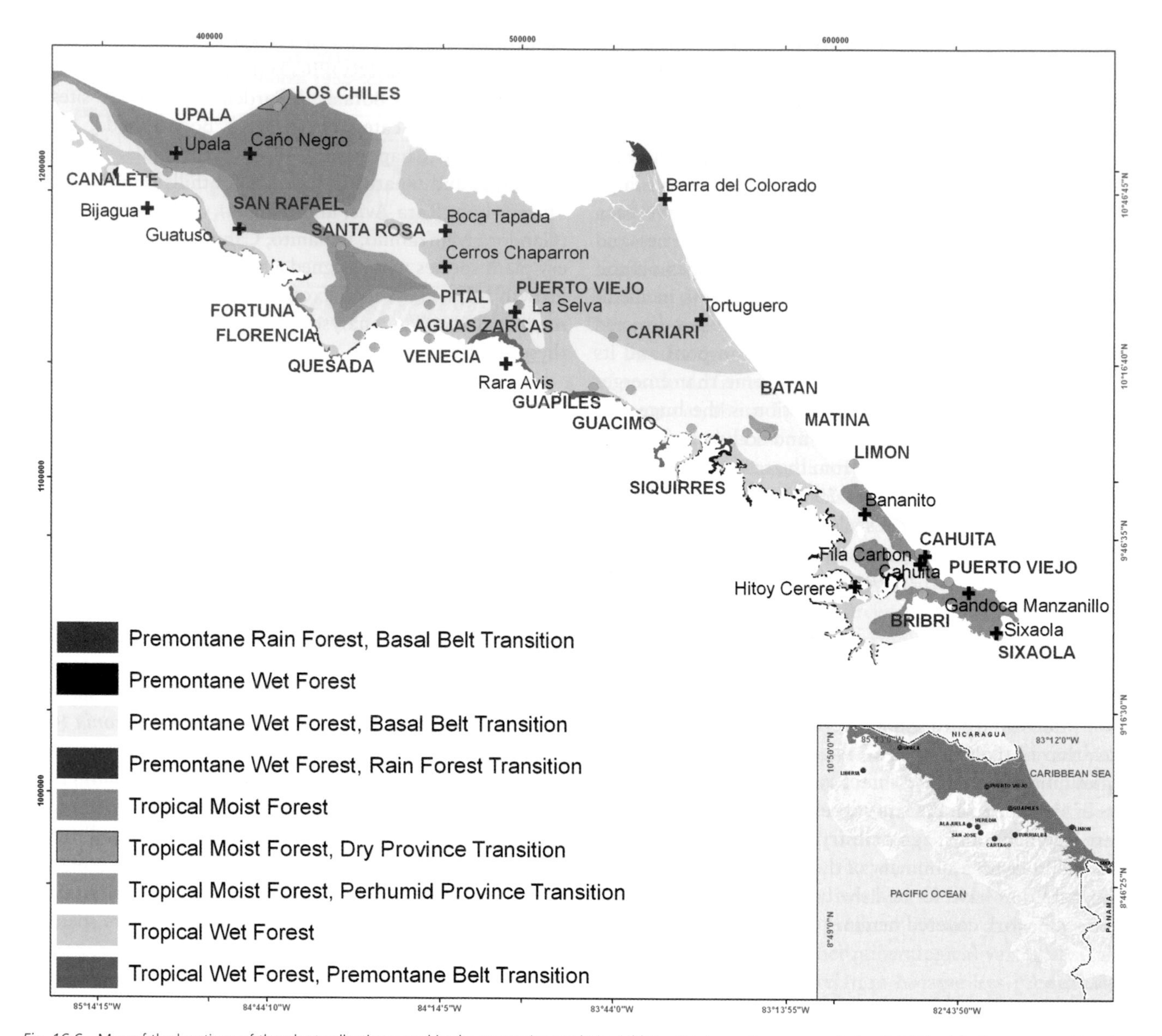

Fig. 16.6 Map of the locations of the plant collections used in the vegetation analysis. Additional color coding represents the Holdridge Life Zones covered herein.

Half of the species (1,508) are known from only a single one of the 16 sites. All other species are known from two to seven sites. This high degree of beta-diversity is striking but also in keeping with similar patterns seen across other Neotropical areas (Hartshorn and Hammel 1994, Marques et al. 2011, Pyke et al. 2001, Toledo et al. 2011).

The growth forms of the plant species considered by geomorphological unit follow the same general pattern that is seen in the complete database—herbs are the dominant type of plant, followed by trees, shrubs, vines, and palms.

If La Selva is taken as a nearly complete flora, it is instructive to consider patterns of vegetation specific to that site. Of the trees, the most abundant is *Pentaclethra macroloba*, followed by the palms *Welfia regia* and *Socratea exorrhiza* and then by *Protium pittieri* and *Warszewiczia coccinea*. The dominant canopy emergent trees are *Dipteryx panamensis*, *Hymenolobium mesoamericanum*, *Lecythis ampla*, *Vochysia ferruginea*, *Hieronyma alchorneoides*, and *Ceiba pentandra*. The three dominant families in terms of species numbers are Orchidaceae, Araceae, and Bromeliaceae. Note that the dominant families at La Selva are not the same as those reported for the Caribbean lowlands as a whole. This may be a real phenomenon or it may be a function of the incomplete plant lists from the other sites.

At La Selva, the number of non-native vascular plants currently stands at 158. Many of these were introduced intentionally (including crops planted by Leslie R. Holdridge in the 1950s and 1960s, and trees planted by Gary Hartshorn in the Arboretum in subsequent years). A few are recently escaped ornamental plants, including the highly invasive red banana (*Musa velutina*) that arrived at La Selva in 2003 or 2004.

La Selva has a nearly complete plant list, it represents a sizeable chunk (1,500 ha) of forest, and it is often used as a "control" or "intact" forest site in studies of nearby fragments. It is important to bear in mind, however, that La Selva cannot be considered pristine (see previous sections in this chapter). Indigenous people modified the landscape for at least 3,000 years and more recently the forest has felt the negative impact of the arrival of invasive species, modern climate change, pesticide residues, and a host of other influences.

Changes in Forest Cover

The production and harvesting of bananas, sugar cane, cattle, timber, non-timber forest products, heart of palm, cacao, and pineapple have all left their mark on the Caribbean lowlands in terms of deforestation and the resultant vegetation of modified landscapes (Fig. 16.7). Countrywide deforestation rates in Costa Rica reduced the coverage of primary forest from 67% (1940), to 56% (1950), to 45% (1961), to 32% (1977), to 17% (1983) (Sader and Joyce 1988). Several other estimates of Costa Rican deforestation rates are available based on slightly different methods (FAO 2000, Lutz 1993, Rosero-Bixby and Palloni 1998) but all agree on the rapid conversion of primary forest to other uses during the 1960s, 1970s, and 1980s.

Deforestation occurred both on a small scale and on a large scale. Farmers took over a few hectares of land on an informal basis or were given land by the government and encouraged to "improve" their holdings (*mejoras*) by clearing them (Augelli 1987, Brockett and Gottfried 2002). These were years of a rapidly burgeoning human population—Costa Rica's population quadrupled between 1950 and 2000 (Rosero-Bixby and Palloni 1998)—and the government actively promoted the settling at Costa Rica's frontiers. At the same time that small holdings were expanding, large-scale deforestation was carried out for the purpose of cattle ranching and planting of agricultural crops. The amount of pastureland in Costa Rica doubled between 1950 and 1973 (Augelli 1987). Over time cattle ranching proved to be unsustainable or economically unfeasible in most areas of Costa Rica and many cattle pastures were abandoned or converted to other land uses by the 1990s.

These countrywide trends have been clearly evident in the Caribbean lowlands. Much of the banana plantation expansion occurred in the area northwest of Limón to the Río Sarapiquí basin, although banana plantations were also established to the south of Limón in the Estrella and Sixaola river valleys. In the northern area of the Caribbean lowlands 2,300 km^2 were deforested from 1961 to 1977 through clearing for a combination of agriculture, human colonization, roads, railroads, and logging (Pérez and Protti 1978). Detailed descriptions of changes in land use starting

Fig. 16.7 Aerial view in February 2011 (dry season) showing the border between pineapple fields and secondary forest on the property of the Costa Rican—Swiss Tropical Fruit Marketing Company in Horquetas, Sarapiquí. This property is known as Finca Roswita. *Photograph taken on a joint photography mission of OTS, Panthera Costa Rica, and Lighthawk.*

in the 1960s around La Selva in Puerto Viejo de Sarapiquí make illuminating albeit sobering reading (Butterfield 1994, Montagnini 1994, Read 1999). The population of the Sarapiquí canton (with La Selva near its center) skyrocketed from 2,169 people in 1950 to 45,218 people in 2000 (Schelhas and Sánchez-Azofeifa 2006).

To determine the loss of forest by ecological zone, Sader and Joyce (1988) digitized the black and white aerial photographs from various dates collected by the Costa Rican Ministry of Agriculture (MAG) (at a scale of 1:1,000,000) as well as the 1977 (Pérez and Protti 1978) and 1983 (Flores Rodas 1984) images created from aerial photography and Landsat MSS satellite data. They overlaid these maps with a digitized 1:750,000 scale map of ecological zones (Tosi 1969) to show that most of the remaining Caribbean lowland forest in 1983 pertained to the Tropical Wet, Tropical Moist, and Premontane Wet life zones, in that order. The highest rates of forest disturbance (8.8% to 16.4%) occurred in these life zones during the period from 1977 to 1983. Between 1940 and 1983, Costa Rica lost 78% of its Tropical Wet forest, most of which was in the Caribbean lowlands. Especially high rates of clearing were recorded on the 0–5% slope category in the northeastern Caribbean lowlands. Another land-use evaluation based on agricultural censuses and satellite images concluded that in the year 2000 forest in the Caribbean lowlands was restricted to lands with poor biophysical characteristics (mostly in national parks) and that "banana plantations have claimed the best soils in flat, non-flooded areas" (Veldkamp et al. 2001).

More recent publications have revealed additional information on a localized geographic scale. Sánchez-Azofeifa et al. (2003) used data acquired by the Landsat TM sensor during 1996 and 1997 to analyze forest changes in buffer strips of one and 10 kilometers in width around all protected areas. These data indicated that deforestation was "under control" in the one kilometer buffer in protected areas within most of the Caribbean lowlands, but that widespread deforestation within the 10 km buffer around the Tortuguero National Park in the Caribbean coastal area north of the city of Limón was turning the park into an isolated island of forest within an agricultural landscape. In the area near EARTH University in Guápiles, as in many other regions of the Caribbean lowlands, abandoned pasture lands (many of which had become young secondary forests) are undergoing a wave of conversion to pineapple plantations (Joyce 2006).

An additional, and surely unintended, source of deforestation in the Caribbean lowlands during the 1980s and early 1990s is associated with projects on forest inventory and sustainable forestry. The original hope was that primary and older secondary forests could be logged selectively to provide marketable timber and enough income for the landholders in order to prevent the conversion of the forests to agricultural and grazing lands (Finegan 1992, Howard 1995, Howard and Valerio 1996). In many cases, however, the plans for selective logging did not work out as intended and the forest designated for several of these forestry projects has now almost all been cleared (Rodríguez unpublished data) because the logging practices proved not to be sustainable, because the access roads built in the zone greatly increased the vulnerability of these trees (these were the years before the passage of the Forestry Law), or because of other factors. Analyses of two different forestry projects are presented in the Forestry section below.

Costa Rica's deforestation rates have declined in recent years, due partly to protection derived from the Forestry Law of 1996 (De Camino et al. 2000), which both prohibits the transformation of land that was forest in 1996 to any other sort of land cover *and* provides Payment for Environmental Services (using the acronyms PES in English and PSA [*pago por servicios ambientales*] in Spanish) for intact forest. On lands adjacent to Parque Nacional Braulio Carrillo and La Selva, the rates of deforestation went from 19.4% per year (8,319 hectares lost) between 1986 and 1996 to 6.7% per year (2,332 hectares lost) between 1996 and 2000 (Schelhas and Sánchez-Azofeifa 2006). Any young forest regrown since 1996 is not protected by the Forestry Law, and it is these areas that are under pressure from expanding pineapple agriculture in the lowlands.

Forestry and Natural Regeneration

Two main forestry activities have occurred during the past decades in areas outside the protected area network in the Caribbean lowlands—selective logging and forestry plantations—and both have economic and conservation importance.

Natural Forest Management

Natural forest management (selective logging) in Costa Rica started as a regulated activity in 1969 (Law 4465). It was not until 1996 however, with Forestry Law 7575, that this activity became highly regulated by the Forestry Authority of the State (De Camino et al. 2000) with a more standardized methodology following principles, criteria, and indicators of forest sustainability. In order to assess the potential value of natural forest management plans (NFMPs) in biodiversity studies, Arroyo-Mora collected 130 NFMPs submitted to the Ministerio de Ambiente y Energía (MINAE) between 1995 and 2005. The parcels were scattered over an area of 65,234 ha of Caribbean lowlands of the Cordillera Volcánica Central Conservation Area (CVCCA). Sixty-two of the 130 NFPMs were approved by MINAE and the total area under management with these forest parcels was 1,840 ha, with most of the forest management carried out in the largest patches within the study area. More NFMPs were submitted and approved in the first half of the 1995–2005 period in comparison to the second half, with 1997 and 1998 as the peak years for NFMPs submitted and approved (Arroyo-Mora et al. 2009). A potential explanation for the lower number of plans submitted in the second half of the period is that incentives shifted from natural forest management towards PSA after the passage of the Forestry Law in 1996. As is the case with plantations (see below), the future of natural forest management is uncertain.

An additional example highlights the fate of the internationally funded forestry projects from this era. Aerial photographs acquired in January 1983 were interpreted to stratify a 14,489 ha forest area between the Río Sucio on the north and the Braulio Carrillo National Park on the south, the Río Puerto Viejo on the east, and the Río Guacima on the west into six categories on the basis of the amount of forest disturbance and canopy height. This forest inventory project was funded by the USAID (Programa CORENA GCR-AID-515-T-032) and conducted in conjunction with the Ministry of Agriculture and, at the time, was known as the 032 Project. Tree measurements were made on 74 one-hectare plots that were randomly located in the six forest disturbance strata. Inventory results were summarized by species and showed that *Pentaclethra macroloba* (the dominant tree at La Selva) represented the greatest biomass of any tree across all of these 74 plots. A follow-up study was mounted three years later, funded by NASA and conducted with the ITCR School of Forestry, to verify the status of these parcels and to use the 032 Project tree measurement data to estimate the aboveground biomass in the plots that were not altered (Sader et al. 1989). By the end of 1986, however, 29 of the plots had been altered through logging and/or clearing.

Plantations

Forestry plantations, as a productive activity, started in Costa Rica at the end of the 1960s, and by 1979 there were 171,000 hectares of forestry plantations in the country (Arias 2004). From the Costa Rican Forest Cover Assessment for the year 2005, Calvo et al. (2006) report a total of 100,547 hectares of plantations for the country, based on data from NGOs, the Forestry Authority of the State, and private companies. From this total, 20,435 ha are located in the Caribbean lowlands, corresponding to approximately 20% of the area of the country's timber plantations. The primary stock of tree species that provide the current and future timber products from plantations encompasses 19 species, of which the most important (in terms of area that the plantations cover) are *Tectona grandis*, *Gmelina arborea*, *Cordia alliodora*, *Vochysia guatemalensis*, and *Hieronyma alchorneoides*. *T. grandis* and *G. arborea* plantations cover 78.3% of the total area (Corella 1999). Seventy-three percent of these plantations were established between the years 2000 and 2006, whereas the other 17% were created between the years 1992 and 2000. Currently, at the national level and probably also at the local level in the Caribbean lowlands, older plantations have been harvested and replaced by cash crops such as pineapple, sugar cane, and bananas (Calvo 2008), resulting in an uncertain future for forestry plantations in general.

In the Caribbean lowlands, reforestation with exotic species (gmelina and teak) and research with native species (Butterfield and Espinoza 1995, Butterfield and González 1996, Montagnini et al. 1995) developed almost simultaneously. Plantations with gmelina and teak were favored for their fast growing characteristics, as well as the low maintenance of gmelina and the good market prices of teak. Studies with native tree species, which began in the mid-1980s in and around the La Selva Biological Station with the TRIALS project, showed the adaptability and growth potential of several native species in degraded soils and abandoned areas, providing a base for the development of reforestation programs with small and medium farmers in the region. Recent studies on native plantations have focused on their potential for carbon sequestration (Redondo-Brenes and Montagnini 2006, Russell et al. 2010).

Secondary Forests

During the years that deforestation rates slowed, secondary forest regeneration from abandoned cattle pastures and agricultural plots increased. Between 2000 and 2005, secondary forest regrowth in Central America, as defined by the United Nations Food and Agriculture Organization (FAO; 2007), was estimated to be 2.4% (Asner et al. 2009). In Costa Rica, over 2,000 km^2 of forest regrowth occurred between 1960 and 2000 (Arroyo-Mora et al. 2005), mostly in hilly regions, resulting in net afforestation at the country scale. At the turn of the century in Costa Rica, secondary forests covered an estimated 400,000 ha (Berti 2000), although this number may be low because of imprecise techniques to detect secondary forest in tropical dry and moist forests both on the ground and from remote sensing data (Calvo et al. 2006).

In Costa Rica's Caribbean lowlands, the regrowth of secondary forest after deforestation, some of it haphazard and some of it managed, has resulted in a mosaic of forest patches of different sizes, ages, distance from nearest seed source, and history of use before regrowth. At the turn of the century, in the Huetar Norte region (in the area between the La Selva Biological Station and the Río San Juan),

largely from the research and personal observations of one of the coauthors (R.C.C.).

Invertebrates

The most comprehensive invertebrate inventory project in the Caribbean lowlands was the 14-year ALAS (Arthropods of La Selva) enterprise led by Rob Colwell and Jack Longino. During this time period (starting in 1991), hundreds of thousands of specimens were collected, processed, and deposited in museums around the world, including at Costa Rica's INBio (National Biodiversity Institute) (http://purl.oclc.org/alas). The ALAS project also developed the software programs EstimateS and Biota, used for calculating biodiversity and for managing biodiversity databases, respectively (Colwell 2005, Colwell et al. 2004, Gotelli and Colwell 2001). A long-term tropical butterfly project run by Phil DeVries and Isidro Chacón moved its Costa Rican center of operations from La Selva to the nearby Tirimbina Biological Reserve and continues to provide country-wide information on lepidopteran diversity (Chacón and Montero 2007, DeVries 1979, 1997, DeVries et al. 2012, Walla et al. 2004). Angel Solís from INBio and his colleagues have been monitoring dung beetle populations throughout the country for several decades, with a 35-year record for La Selva (Escobar et al. 2008). Nematode diversity in different soil types at La Selva (including canopy soils) has recently become a noteworthy topic (Powers et al. 2009). Treatments of the more ecological invertebrate projects are to be found in the Plant-Animal Interactions sections below.

Fish

Although the species comprising the fish fauna of the Caribbean lowlands of Costa Rica have been identified, most aspects of their biology in the wild are poorly studied (e.g., see Pringle et al., chapter 18 of this volume). The total biodiversity of fishes in the Caribbean lowlands is minute (Bussing 1998; Bussing 1994, Appendix 4 lists 43 species at the La Selva Biological Station) compared to the thousands of species in the Amazon (1300++; Goulding 1980), and the fish fauna is dominated by just a few families—the cichlids (Cichlidae), the characins (Characidae), and the livebearers (Poeciliidae). Even these families are represented by only a handful of species each. The various catfish families that achieve incredible diversity further south (e.g., the pimelodid catfishes, the callichthyids, and the suckermouth catfishes of the family Loricariidae) are represented by just three species of pimelodids. Other families have token representation, including single species of the Carcharhinidae, Elopidae, Gobiesocidae, Rivulidae, Atherinidae, Syngnathidae, Synbranchidae, Haemulidae, Centropomidae, Eleotridae, and a couple of members of the Mugilidae and Gobiidae (Table 16.1).

Despite low biodiversity, the ichthyofauna has many fascinating characteristics. For example, the pipefish *Pseudophallus mindii* is found near the coast but has also been recorded at least twice at La Selva (Bussing 1998; Coleman, personal observation), a full 75 km from the coast, and certainly a long swim for a small pipefish. The freshwater clingfish (*Gobiesox nudus*) is from a family, Gobiesocidae, often associated with the intertidal zone of the eastern Pacific Ocean including Washington, British Columbia, and Alaska. Fishes such as the American eel (*Anguilla rostrata*) and the pike killifish (*Belonesox belizanus*) are considered distinctly coastal and have wide latitudinal ranges on the Atlantic coast north and south of Costa Rica. The blackbelt cichlid, *Theraps maculicauda*, always appears to be within sight of the surf line. Other "coastal" fish, such as the tarpon (*Megalops atlanticus*), penetrate far upstream into the lowlands. The bigmouth sleeper (*Gobiomorus dormitor*) is usually seen near the coast but can be found far inland at La Selva. It is not known whether individual sleepers move this distance—given their sedentary lifestyle, it seems unlikely.

The distribution of species in the region is no doubt a product of the overall river patterns in the region. There appear to be three major assemblages. The large river bounding the north side of the region is the Río San Juan, which flows from Lake Nicaragua into the Caribbean. Because this large river drains Lake Nicaragua, it brings fishes that are not typically found in the rest of the zone into the northern edge of the region. For instance, several cichlids such as the midas cichlid *Amphilophus citrinellum* and the jaguar cichlid *Parachromis managuense* are common in the Lake and are also found at the river mouth at the far northeastern corner of the lowlands at Refugio Nacional de Vida Silvestre Barra del Colorado. In fact, it is interesting that these species have not penetrated further south. The bull shark *Carcharhinus leucas* and the sawfish *Pristis pristis* are in the Lake and in the Río San Juan, and extremely rarely a bull shark will swim up the Río Sarapiquí to be spotted at La Selva. Gars (*Astractosteus tropicus*) are found in the Río San Juan and are common in the sluggish waters of Caño Negro. We cannot compare the fishes of the Costa Rican lowlands to the fishes north of the Río San Juan because that portion of Nicaragua is largely unstudied with regard to fishes.

The bulk of the northern portion of the lowlands is drained by several large river systems that originate in the volcanic highlands to the west and flow roughly north or northeast to the Río San Juan. These include, from west

Table 16.1. List of Fishes of the Caribbean Lowlands of Costa Rica Ordered per Sector

Family	Species	English Name	Spanish Name
Fishes of the Río San Juan sector			
Carcharinidae	*Carcharhinus leucas*	bull shark	tiburón
Pristidae	*Pristis pristis*	sawfish	pez sierra
Lepisosteidae	*Astractosteus tropicus*	gar	gaspar
Clupeidae	*Dorosoma chavesi*	shad	sabalete
Characidae	*Astyanax nasutus*	Colcibolca tetra	sardina lagunera
Gymnotidae	*Gymnotus maculosus*	knifefish	madre de barbudo
Pimelodidae	*Rhamdia nicaraguensis*	catfish	barbudo
Rivulidae	*Rivulus isthmensis*	Isthmian rivulus	olomina
Poeciliidae	*Alfaro cultratus*	alfaro	olomina
	Brachyraphis parismina	olomina	olomina
	Neoheterandria umbratilis	olomina	olomina
	Phallichthys amates	merry widow	olomina
	Phallichthys tico	olomina	olomina
	Poecilia gillii	molly	olomina
	Poecilia mexicana	shortfin molly	olomina
Atherinidae	*Atherinella hubbsi*	silverside	sardina
Synbranchidae	*Synbranchus marmoratus*	marbled swamp eel	anguila de pantano
Cichlidae	*Amphilophus citrinellus*	midas cichlid	mojarra
	Archocentrus centrarchus	flier cichlid	mojarra
	Archocentrus nigrofasciatus	convict cichlid	congo
	?Astatheros alfari	pastel cichlid	mojarra
	Astatheros longimanus	redbreast cichlid	cholesca
	Herotilapia multispinosa	rainbow cichlid	mojarrita
	Parachromis dovii	wolf cichlid	guapote
	Parachromis managuensis	jaguar cichlid	guapote tigre
Gerreidae	*Eugerres plumieri*	mojarra	mojarra prieta
Haemulidae	*Pomadasys croco*	Atlantic grunt	roncador
Mugilidae	*Agonostomus monticola*	mountain mullet	tepemechín
Gobiidae	*Awaous banana*	lamearena	lamearena
Fishes of the coastal sector			
Elopidae	*Megalops atlanticus*	tarpon	sábalo real
Anguillidae	*Anguilla rostrata*	American eel	anguila
Poecillidae	*Belonesox belizanus*	pike killifish	pepesca gaspar
	Poecilia mexicana	shortfin molly	olomina
Atherinidae	*Atherinella hubbsi*	silverside	sardina
Cichlidae	*?Amphilophus citrinellus*	midas cichlid	mojarra
	?Archocentrus centrarchus	flier cichlid	mojarra
	Theraps maculicauda	blackbelt cichlid	pis pis
Centropomidae	*Centropomus parallelus*	fat snook	calva
	Centropomus pectinatus	blackfin snook	gualaje Atlántico
	Centropomus unidecimalis	common snook	róbalo
Carangidae	*Caranx latus*	jack	jurel
Gerreidae	*Eugerres plumieri*	mojarra	mojarra prieta
Haemulidae	*Pomadasys croco*	Atlantic grunt	roncador
Eleotridae	*Dormitator maculatus*	fat sleeper	guarasapa
	Eleotris amblyopsis	sleeper	pez perro
	Eleotris pisonis	spiny-cheek sleeper	pez perro
	Gobiomorus dormitor	big-mouth sleeper	guavina
Paralichthyidae	*Citharichthys spilopterus*	flounder	lenguado
	Citharichthys uhleri	flounder	lenguado
Achiridae	*Trinectes paulistanus*	sole	lenguado redondo
Fishes of the central sector			
Characidae	*Astyanax aenus (= fasciatus)*	banded tetra	sardina
	Bramocharax bransfordii	longjaw tetra	sardina picuda
	Brycon guatemalensis	machaca	machaca
	Bryconamericus scleroparius	creek tetra	sardina de quebrada
	Carlana eigenmanni	carlana tetra	sardinita
	Hyphessobrycon tortuguerae	Tortuguero tetra	sardinita
	Roeboides bouchellei	glass headstander	sardinita
Gymnotidae	*Gymnotus cylindricus*	knifefish	madre de barbudo
Pimelodidae	*Rhamdia nicaraguensis*	catfish	barbudo

continued

Table 16.1. Continued

Family	Species	English Name	Spanish Name
	Rhamdia guatemalensis	catfish	bardudo
	Rhamdia rogersi	catfish	barbudo
Rivulidae	*Rivulus isthmensis*	Isthmian rivulus	olomina
Poeciliidae	*Alfaro cultratus*	alfaro	olomina
	Brachyraphis holdridgei	olomina	olomina
	Brachyraphis olomina	olomina	olomina
	Brachyraphis parismina	olomina	olomina
	Neoheterandria umbratilis	olomina	olomina
	Gambusia nicaraguensis	mosquito fish	olomina
	Phallichthys amates	merry widow	olomina
	Phallichthys quadripunctatus	olomina	olomina
	Phallichthys tico	olomina	olomina
	Poecilia gillii	molly	olomina
	Priapichthys annectens	olomina	olomina
Atherinidae	*Atherinella chagresi*	silverside	sardina
	Atherinella hubbsi	silverside	sardina
Gobiesocidae	*Gobiesox nudus*	clingfish	chupapiedra
Syngnathidae	*Pseudophallus mindii*	pipefish	pez pipa
Synbranchidae	*Synbranchus marmoratus*	marbled swamp eel	anguila de pantano
Cichlidae	*Archocentrus nigrofasciatus*	convict cichlid	congo
	Archocentrus septemfasciatus		mojarra
	Astatheros alfari	pastel cichlid	mojarra
	Astatheros rostratus		
	Herotilapia multispinosa	rainbow cichlid	mojarrita
	Hypsophrys nicaraguense		moga
	Neetoplus nematopus		moga
	Parachromis dovii	wolf cichlid	guapote
	Parachromis loisellei		guapotillo
	Tomocichla underwoodi	tuba	vieja
Mugilidae	*Agonostomus monticola*	mountain mullet	tepemechin
	Joturus pichardi	hog mullet	bobo
Gobiidae	*Awaous banana*	lamearena	lamearena
	Sicydium adelum	goby	chupapiedra
	Sicydium altum	goby	chupapiedra
Eleotridae	*Gobiomorus dormitor*	big-mouth sleeper	guavina
Fishes of the Río Sixaola sector			
Characidae	*Astyanax aenus* (= *A. fasciatus*)	banded tetra	sardina
	Astyanax orthodus	largespot	sardina blanca
Gymnotidae	*Gymnotus cylindricus*	knifefish	madre de barbudo
Rhamphichthyidae	*Hypopomus occidentalis*	knifefish	madre de barbudo
Pimelodidae	*Rhamdia guatemalensis*	catfish	barbudo
	Rhamdia rogersi	catfish	barbudo
Poeciliidae	*Alfaro cultratus*	alfaro	olomina
	Brachyraphis parismina	olomina	olomina
	Phallichthys amates	merry widow	olomina
	Poecilia gillii	molly	olomina
	Priapichthys annectens	olomina	olomina
Atherinidae	*Atherinella chagresi*	silverside	sardina
Syngnthidae	*Pseudophallus mindii*	pipefish	pez pipa
Synbranchidae	*Synbranchus marmoratus*	marbled swamp eel	anguila de pantano
Cichlidae	*Archocentrus myrnae*	topaz cichlid	mojarra
	Archocentrus nigrofasciatus	convict cichlid	congo
	Astatheros bussingi		mojarra
	Astatheros rhytisma		
	Parachromis loisellei		guapote amarillo
Haemulidae	*Pomadasys croco*	Atlantic grunt	roncador
Mugilidae	*Agonostomus monticola*	mountain mullet	tepemechin
	Joturus pichardi	hog mullet	bobo
Gobiidae	*Awaous banana*	lamearena	lamearena
	Sicydium adelum	goby	chupapiedra
	Sicydium altum	goby	chupapiedra
Eleotridae	*Gobiomorus dormitor*	big-mouth sleeper	guavina

to east, the Río San Carlos system and the Río Sarapiquí-Chirripó system (Bussing 1998). East of the Chirripó the rivers flow into the Caribbean, possibly an important difference in terms of the movement of fishes because fishes in the latter rivers must pass through intertidal, brackish, or marine waters to connect to other river systems. Interestingly, there do not appear to be significant differences between the fish fauna of these two types of rivers, although there are some. This may be because in the area around the city of Guápiles, numerous tributaries of the Río Chirripó (draining north to the Río San Juan) and the Tortuguero and Parismina systems (draining east into the Caribbean) are found within a few kilometers of each other. It is easy to imagine that one of the regular flooding events common to this area could move fishes between the north-draining systems and those that drain to the Caribbean.

To the east of Siquirres, from Matina to the port of Limón, as the topography flattens out into the coastal plain, there is a distinct lack of rivers. The only water now flowing through the seemingly endless banana plantations are tiny silt-filled streams. As one heads south, near Cahuita and down to the Panamanian border, there are smaller rivers that flow into the Caribbean and others that drain into the Sixaola system. This break is significant because there is a notable change in the fish communities between the Sixaola system and the northern part of the lowlands. The Río Sixaola and its tributaries have similar, related, but distinctly different species, than the rest of the region. For example, with regard to cichlids, although *Archocentrus septemfasciatus* is common throughout the entire northern quadrant, it is replaced in the Sixaola region by *Archocentrus myrnae* (Tobler 2007). Similarly, *Amphilophus alfari* is replaced by *Amphilophus bussingi*. This is not to say that no northern sector fish are found in the Sixaola. For example, although there is a different tetra, namely the largespot tetra, *Astyanax orthodus*, the banded tetra *A. aenus* is also there, as is the convict cichlid, which is found over much of northern Costa Rica (Bussing 1998).

Other species will be found once someone looks for them. For example, again with regard to cichlids, Bussing does not indicate *Neetroplus nematopus* in the Río San Juan, yet it is common both in the main part of the Caribbean lowlands and in Nicaragua. Other fishes, such as *Tomocichla underwoodi* (= *T. tuba*) are almost certainly not found in the Sixaola region because we have looked extensively for them and not found them there. Finally, there is the possibility that some fishes are located now where they have been introduced by humans. This is a common occurrence around the world both currently and historically and there is little reason to imagine that the Caribbean lowlands have been exempt from this influence.

The second key contrast between the Costa Rican Caribbean lowlands and the Amazon drainage concerns the changes in water volume and depth. The Amazonian flood cycle is present in the Costa Rican lowlands in almost similar magnitude, but is completely different in periodicity and tempo. The rivers of the main part of the Caribbean lowlands (there are no natural lakes as such, except for Lake Nicaragua on the northern boundary) are the product of constantly changing hydrodynamics. If anything, it is the variability that characterizes these habitats. The variation originates in the large-scale patterns of rainfall—heavy rains in late September through early January, the dry season from mid-January through April, the "little wet" season from May through August and then the "little dry" season through August and September. The timing of each of these periods changes from year to year. The important point is that even during the dry season, large storms can drop a substantial amount of water both on the lowlands and on the mountains to the west, which then rapidly moves down the streams and into the rivers. Rivers can swell in a matter of hours to twice their volume or more. For example, the height of the Río Puerto Viejo rose almost ten meters within 24 hrs at the La Selva Biological Station in January of 2006 and then dropped back to its original level over the next couple of days.

This variation in water volume can radically alter the physical environment, moving large woody debris tremendous distances and reshaping river substrates in minutes. The water itself also changes in temperature, dropping to near 20°C or lower, or rising up to 27°C or higher. With changes in water depth, areas that were slow-moving become rapids and vice versa. The rivers can flood huge areas of land, then recede, moving fish into places far from the original river. This is important because many of the fishes associate themselves with particular substrates, depths, and temperatures. While the physical conditions change, sometimes daily, so too does the distribution and activity of many of the fishes. In summary, although various species can be nominally placed as being native to the region, the exact location at which a species is found can vary daily. Almost two decades of underwater snorkel studies have revealed that just because a fish is found at a particular location today means little as to whether that species will be found there tomorrow or next week. An open question remains whether the fish move into and out of desirable portions of a river or whether they simply "hunker down" and wait out the bad weather. Clearly at least some fishes are moving large distances because stretches of some rivers will become disconnected or entirely dewatered during certain times of the year, yet have fishes at other times. Although *Rivulus* (a killifish) can occasionally be spotted "flipping" along the ground from one body of water to the next, most of the fishes in the region are confined to living and dispersing in water.

A final key contrast with other aquatic environments is the almost complete lack of aquatic macrophytes in most of these rivers and streams. There simply are no plants in most places (see also Pringle et al., chapter 18 of this volume). The substrate is rock, gravel, sand, or compacted clay. The exception is during flooding events when terrestrial vegetation becomes inundated. Nor are there large-leaved aquatic or emergent plants on which the fish can lay eggs. Small fishes, such as many livebearers, persist by living along the margins of the rivers. Woody debris is critical for hiding, as well as for creating microclimates of different velocity, substrate, and temperature.

The taxonomy of some groups is a major challenge for understanding the fish communities of the Caribbean lowlands. For example, the convict cichlid, *Archocentrus nigrofasciatus*, has been placed in at least five genera within the last two decades (*Herichthys*, *Heros*, *Cichlasoma*, *Cryptoheros*, *Amatitlania*). These name changes reflect ongoing attempts to understand phylogenetic relationships that are not clearly resolved, largely because the biogeographic origins of these fishes are complicated (Chakrabarty and Albert 2011). Certainly some of the fishes are the result of radiations north from South America with the rise of the Isthmus of Panama, but the timing and numbers of these radiations are unclear. Other fishes have likely come from the north. The wolf cichlid (*Parachromis dovii*), or *guapote* as it is called locally, comes from a group with origins in the West Indies, yet the West Indian fishes are likely derived from older South American ancestors. No doubt further research will resolve some of these difficulties, particularly concerning closely related cichlid taxa such as *Poecilia gillii* and *P. mexicana*.

Research on fishes in the Caribbean lowlands has been hampered by several obstacles. The large amount of moving water and the large sediment load also mean that visibility varies from excellent to non-existent. Certain rivers, such as the Río Chirripó, have never had any appreciable visibility, and therefore our knowledge is limited to the results of sampling (fishing, electroshocking). Others, such as the Río Puerto Viejo or the Río San José, can range from almost clear to no visibility from one day to the next. Names of rivers in the region also pose a challenge in that names on maps do not always agree with names on road signs, with GIS systems, or with local names for a river. Also, certain names are used repeatedly for different rivers, even two rivers located relatively close to each other. For instance, Río Toro appears as the name of several rivers in the region and there are two different rivers named Río Sardinal within 15 kilometers of each other, yet they are entirely unconnected. Similarly, different segments of a river often bear a different name. Finally, river connectedness is complex and may involve stream capture, reticulation, and intermittent flows. This complexity could be important for understanding the evolution and relatedness of populations and subpopulations of fishes.

The region and the fishes in it face various threats. Intense agriculture (particularly banana and pineapple) close to the rivers means that pesticides and sedimentation are always potential problems. Fishing by humans is present and likely responsible for local reduction in populations of some species—for example, *guapote* are increasingly rare and also wary of humans. *Tomocichla underwoodi* numbers near the town of Río Frio have decreased markedly over the last decade. Spear fishing, though illegal, is increasingly common and brazen and potentially a major threat because it specifically targets the largest individuals.

On January 8, 2009, a previously unrecognized threat demolished the Río Sarapiquí when a magnitude 6.1 earthquake on the Caribbean slope of Volcán Poas caused massive landslides of material into the headwaters and tributaries of the Río Sarapiquí. The main river channel became saturated with mud and debris such that it virtually solidified all the way down to the confluence with the Río Puerto Viejo (about 25 km). It appeared that most, possibly all, of the fish in the main river were killed at that time, apparently from oxygen deprivation. How frequently such events occur, and whether their magnitude is amplified by human effects on the terrain, is unknown.

On a positive note, the past decade has seen a notable increase in Costa Ricans using donning masks and snorkeling, simply to appreciate the beauty and environment of the lowland rivers. This is a positive development.

Amphibians

The lowland forests (below 300 m a.s.l.) of the Caribbean slope of Costa Rica host a diverse amphibian assemblage, with 91 species of frogs in 38 genera, 10 species of salamanders in two genera, and three species of caecilians in two genera (Table 16.2). Frogs are distributed among 14 families, yet salamanders and caecilians are each represented in the region by a single family. Of Costa Rica's 187 amphibian species, 104 occur in the Caribbean lowlands. There are no amphibian species that are endemic to the Caribbean lowlands of Costa Rica, largely because amphibian species in the lowland tropics generally have large geographic ranges (Savage 2002, Whitfield et al. in press). The amphibian fauna of the lowlands is quite distinct from

Table 16.2. (a) Diversity and Habitat Affiliations of Amphibians of the Caribbean Lowlands of Costa Rica

Taxon	Family	Genera	Species	Larval habitat	Adult habitat
Frogs	Rhinophrynidae	1	1	P	F
	Bufonidae	4	6	P,S	T
	Aromobatidae	1	1	S	T
	Dendrobatidae	5	5	S, Ph	T
	Eleutherodactylidae	1	2	D	A, T
	Brachycephalidae	1	17	D	A, Aq, T
	Strabomantidae	1	6	D	A, T
	Leiuperidae	1	1	P	T
	Leptodactylidae	1	4	P	F, T
	Amphignathodontidae	1	1	D	A
	Centrolenidae	3	11	S	A
	Hylidae	15	30	P, Ph, S	A
	Microhylidae	2	2	P	F
	Ranidae	1	4	P, S	Aq, T
Salamanders	Plethodontidae	2	10	D	A, F, T
Caecilians	Caeciliidae	2	3	D	F
Total		42	104		

NOTE. Larval habitat indicates pond or pool breeding (P), stream-breeding (S), phytotelm-breeding (Ph), and direct-developing (D). Adult habitat indicates arboreal (A), aquatic (Aq), fossorial (F), or terrestrial (T).
Source: Data derived from Savage 2002 and IUCN 2006.

montane forests, but is rather similar to those amphibian faunas found from moist forests from southern Mexico through central Panama (Savage 2002). The amphibian fauna from the most intensively studied site in the region, La Selva Biological Station, hosts 55 species (52.9% of the Caribbean lowlands assemblage).

Three major ecological assemblages of amphibians found in Caribbean lowlands may be distinguished based primarily upon their use of reproductive resources: (1) lotic-breeding species with tadpoles in streams or rivers, (2) lentic-breeding species with larvae in ponds or pools, and (3) species with specialized forms of terrestrial reproduction. The stream-breeding assemblage is much less abundant and diverse in the lowlands than upslope, and in the lowlands glass frogs (Family Centrolenidae) are the most diverse representatives. The assemblage of lentic-breeding species may utilize habitats ranging from small forest pools to large permanent water bodies, and is particularly diverse in temporary wetlands (Donnelly and Guyer 1994). A few lentic-breeding species—including the dendrobatid *Oophaga pumilio* (previously *Dendrobates pumilio*) and the hylid *Cruziohyla calcarifer*—reproduce in phytotelmata. Terrestrial species with direct development (including the species-rich genera *Craugastor* and *Pritimatnis*) are particularly diverse and abundant in the Caribbean lowlands, likely because the high rainfall and lack of a severe dry season prevent terrestrial eggs from desiccating (Scott 1976, Whitfield et al. 2007).

Fortunately, for the most part, amphibian assemblages in the lowland tropics have not been as severely affected by the rapid and devastating species loss occurring in montane sites (Lips et al. 2005, 2006, 2008, Whitfield et al. in press). Thirty-three percent of Costa Rican amphibians are listed by the IUCN as threatened with extinction, but only 22.1% of species in the Caribbean lowlands region are threatened, and these are mostly higher-elevation species whose ranges only marginally extend into the lowlands.

There are nonetheless many significant threats to lowland amphibian faunas, including widespread land-use change, emerging infectious diseases, and climate change (Butterfield 1994, Bell and Donnelly 2006, Whitfield et al. in press). The apparently non-native amphibian chytrid fungus, *Batrachochytrium dendrobatidis*, appears to be broadly distributed in this region and the neighboring volcanic cordilleras (Puschendorf et al. 2006, Puschendorf et al. 2009, and see Lawton et al., chapter 13 of this volume), but has not been reported to cause widespread declines here. However, at least one frog species (*Craugastor ranoides*) seems to have suffered extirpation throughout the region even close to sea level—possibly due to chytridiomycosis (Puschendorf et al. 2005, Puschendorf et al. 2006). Invasive competitors currently appear to represent little threat. Although at least one non-native frog has established small but persistent populations in this region (*Osteopilus septentrionalis* surrounding Puerto Limón), it is probably restricted to highly disturbed residential areas (Savage 2002). Whitfield et al. (2007) reported assemblage-wide declines for terrestrial frogs at La Selva, but the geographic extent and proximate causes for these declines remain unclear. The most serious future threats to amphibian populations in addition to continued habitat loss will likely result from anthropogenic climate change, in particular if increases in temperature are accompanied by directional shifts in precipitation regimes (Aguilar et al. 2005, IPCC 2007, Lawler et al. 2009, Whitfield et al. in press).

Reptiles

The reptiles of the Caribbean lowlands of Costa Rica encompass a diverse assemblage including 36 species of lizards in 22 genera, 83 species of snakes in 47 genera, 6 species of turtles in 4 genera, and 2 species of crocodilians each in their own genus. In all, 127 of Costa Rica's 222 reptile species occur in the lowlands of the Caribbean versant. Over half of the diversity in this region belongs to colubrid snakes, but the anoles (Family Polychrotidae) constitute another conspicuously diverse group. Savage (2002) suggested that the reptile fauna of this region is composed of species forming a lowland moist forest assemblage that ranges from eastern Mexico through Panama and is differentiated from other Mesoamerican reptile assemblages occurring in upland or more xeric habitats. The reptile fauna of the most intensively studied site in the Caribbean lowlands, La Selva Biological Station, includes 89 species (70.0% of the regional pool; Guyer and Donnelly 2005).

The ecology of the reptiles is extremely variable, both within and among the major groups. The lizards in the region range from the large-bodied and primarily herbivorous *Iguana* to very small-bodied forest floor arthropod-eating lizards (*Lepidoblepharis*, *Norops*). A diverse assemblage of leaf-litter lizards includes anoles, geckos, teiids, gymnophthalmids, anguids, and skinks (Lieberman 1986). The anoles sort ecologically along a vertical gradient from leaf-litter to trunk-dwelling and canopy species (Irschick et al. 1997). The boid and viperid snakes are semi-arboreal (*Boa*), arboreal (*Corallus*, *Bothriechis*), or terrestrial (*Epicrates*, *Bothrops*, *Lachesis*, *Porthidium*) and feed primarily on vertebrates. The elapids feed mostly on other snakes. The highly diverse assemblage of colubrids ranges dramatically in body form and habitat preferences, with many arboreal, terrestrial, and fossorial species but relatively few aquatic species (Guyer and Donnelly 1990, 2005, Savage 2002). The colubrids also vary greatly in diet, preying upon arthropods (*Tantilla*), mollusks (*Dipsas*, *Sibon*), amphibians (*Chironius*, *Leptodeira*, *Leptophis*, *Urotheca*), lizards (*Scaphiodontophis*, *Imantodes*, *Oxybelis*), snakes (*Clelia*, *Drymarchon*, *Erythrolamprus*), birds (*Pseustes*), or mammals (*Lampropeltis*). The non-marine turtle fauna includes stream-dwelling species (*Rhinoclemmys funerea*, *Chelydra*, and *Trachemys*), pond-inhabiting species (two *Kinosternon* and *Chelydra*), and a single terrestrial species (*Rhinoclemmys annulata*). The crocodilians include the spectacled caiman, *Caiman crocodilus*, which inhabits swamps, ponds, streams, and rivers, and the American crocodile, *Crocodylus acutus*, which occurs primarily in larger rivers.

The greatest threat to reptiles in the Caribbean lowlands is rampant habitat modification. Although some species of reptiles fare well in lands cleared of forests (especially heliothermic lizards such as *Ameiva* sp. and *Basilicus vittatus*), most reptiles prefer intact forests. Turtles (particularly *Chelydra* and *Trachemys*) and green iguanas are hunted and consumed by humans. Crocodiles and caiman have been actively hunted by humans for meat and hides; and while crocodiles were extirpated through much of their range, conservation efforts have been extremely effective and crocodile populations have been recovering in the past two decades (Savage 2002). Three non-native species of lizards have established populations in the area (the house geckos *Hemidactylus frenatus* and *Hemidactylus garnotti*, and the Caribbean anole *Ctenotus cristatellus*) but appear to be confined to highly disturbed habitats and currently do not appear to pose a threat to native faunas. Huey et al. (2009) suggested that many forest-dwelling lizards are thermoconformers whose optimal body temperatures are low relative to ambient temperatures, and that human-induced increases in temperature are likely to have particularly adverse effects on these ectotherms in the near future.

Birds

The Atlantic lowland evergreen forests (below 300 m a.s.l.) along the Caribbean slope of Costa Rica are home to a highly diverse avifauna. These forests, as discussed earlier, are characterized by high rainfall, limited seasonality, a variety of habitat types, and complex vegetation structure within forests, all of which combine to support high bird species richness (Orians 1969, Slud 1976, Stiles 1983). These biotic and climatic factors ensure the availability of a wide variety of food resources such as fruits, flowers, nectar, and insects throughout the year, and also provide many unique resources for birds, such as army ant swarms (Blake and Loiselle 2000). The close proximity of the Caribbean slope, North America, and the once-continuous Caribbean forest belt connecting the region to the species-rich South American forests has further contributed to the origin of high species diversity in the Caribbean lowlands of Costa Rica, with species diversifying along elevational (e.g., golden-crowned and white-throated spadebills, *Platyrinchus coronatus* and *P. mystaceus*, respectively) and latitudinal (e.g., the northern and southern nightingale-wrens, *Microcerculus philometa* and *M. marginatus*, respectively) gradients (Levey and Stiles 1994). Finally, as with the birds

Table 16.3. (b) Diversity and Habitat Affiliations of Reptiles of the Caribbean Lowlands of Costa Rica

Taxon	Family	Genera	Species	Habitat
Lizards	Corytophanidae	2	3	A, T
	Iguanidae	2	2	A, T
	Polychrotidae	4	13	A, T
	Gekkonidae	5	7	A, T
	Teiidae	1	2	T
	Gymnopthalmidae	3	3	T
	Anguidae	2	3	T
	Scincidae	2	2	T
	Xantusiidae	1	1	T
Snakes	Boidae	3	3	A, T
	Ungaliophidae	1	1	A, T
	Colubridae	37	69	A, Aq, F, T
	Viperidae	5	7	A, T
	Elapidae	1	3	F, T
Turtles	Chelydridae	1	1	Aq
	Kinosternidae	1	2	Aq, F
	Emydidae	2	3	Aq, T
Crocodilians	Alligatoridae	1	1	Aq
	Crocodylidae	1	1	Aq
Total		**75**	**127**	

NOTE. Habitat indicates arboreal (A), aquatic (Aq), fossorial (F), or terrestrial (T).
Source: Data derived from Savage 2002.

of the Osa Peninsula (see Gilbert et al., chapter 12 of this volume), the birds of the Caribbean lowlands tend to have low population densities but high species packing, with co-occurring similar species segregating by prey type, foraging height, and substrate use, among other variables (Terborgh et al. 1990, Chapman and Rosenberg 1991).

As a result of these factors, the lowland Caribbean forests have the highest avian diversity in Costa Rica. A total of 484 species in 309 genera and 61 families has been recorded to date (Table 16.4). Of these, 289 are known or strongly suspected to breed in the Caribbean lowlands (Stiles and Skutch 1989, Garrigues and Dean 2007, Sigel et al. 2010). Yet, although the region hosts the highest species diversity, endemism rates are lower than on either the Osa Peninsula or the higher-elevation forests, likely due to the once-continuous extent of lowland forest from Mexico south into South America (Levey and Stiles 1994). The Costa Rican lowland Caribbean region hosts, during part to all of the year, 20 species that are endemic to southern Central America and northern Colombia. Of these, three species are also found in the Pacific lowlands and foothills of both coasts and 14 range into the Caribbean foothills. Four species—snowy cotinga (*Carpodectus nitidus*), plain-colored tanager (*Tangara inornata*), sulphur-rumped tanager (*Heterospingus rubrifrons*), and Nicaraguan seed-finch (*Oryzobus nuttingi*)—are restricted to the Caribbean lowlands, rarely ranging up to elevations above 700 m.

An additional 33 species of elevational migrants traverse the Caribbean slope, presumably to track the availability of food resources. For example, the three-wattled bellbird (*Procnias tricarunculata*) and the resplendent quetzal (*Pharomachrus mocinno*) move downslope in the non-breeding season, following the availability of their preferred Lauraceae (avocado) fruits (Hamilton et al. 2003, Powell and Bjork 1994). However, one species of hummingbird—the violet-crowned woodnymph (*Thalurania colombica*)—reverses the trend by breeding in the lowlands and migrating upslope (Stiles and Skutch 1989). Elevational migrants are highly dependent upon forested habitat in both their breeding and non-breeding grounds, as well as along their migratory pathway, and perhaps as a consequence of forest loss and fragmentation, 27% of these species (9 of 33) are considered threatened (IUCN 2010).

Lowland Caribbean forests and disturbed areas (including secondary forests, open areas, and gardens) provide important overwintering habitat to many species of latitudinal migrants. Most latitudinal migrants breed in North America and either pass through Costa Rica on migration (76 species) or spend the non-breeding (winter) season in the lowland Caribbean area (49 species; Stiles and Skutch 1989, Parker et al. 1996, Garrigues and Dean 2007). These species have broader geographic ranges and are not as forest-dependent (and accordingly are not as threatened) as elevational migrants, with 7% of the species (9 of 125) considered threatened. The avifauna also includes three species of Austral migrants, which move annually between Costa Rica and South America. The streaked flycatcher (*Myiodynastes maculatus*) breeds in South America and spends the non-breeding season in the Costa Rican lowlands. An additional two species, the piratic flycatcher (*Legatus leucophaius*) and the yellow-green vireo (*Vireo flavoviridis*), breed in Costa Rica and migrate to South America during the late wet season (October–April).

The greatest avian diversity in the lowland Caribbean region is found in primary forests and older, tall secondary forests, with a total of 166 species. An additional 66 species are generalists, found in forests as well as in other habitats. Open areas, including fields, pastures, young second growth, gardens, and developed areas, have the next-highest diversity, with a total of 118 species. Edge habitats, between open and forested areas, are used by an additional 52 species. Finally, the region is home to 59 species using aquatic habitats (including lakes, rivers, and marshes), and 23 aerial species (mostly swallows and nightjars that spend the majority of their time on the wing; Sigel et al. 2010).

The avifauna takes advantage of the wide variety of food resources available within the lowland Caribbean forests throughout the year. Insectivores numerically dominate the lowland Caribbean region, with 167 species. Omnivores are the second most-diverse dietary guild with 155 species, followed by carnivores with 97 species. The avifauna also includes 33 species of frugivores, 25 species of nectarivores, and 6 species of granivores.

One of the most notable features of tropical lowland avifauna, including that of the lowland Caribbean region of Costa Rica, is the single- and multispecies flocks. Five types of flocks are commonly found in lowland Caribbean forests: single species flocks, understory antwren/antvireo mixed flocks, understory-midstory tanager mixed flocks, canopy mixed flocks, and ant-followers. Thirty-two species of landbirds form single-species flocks, including the great green (*Ara ambiguus*) and scarlet (*A. macao*) macaws, chestnut-mandibled (*Ramphastos swainsonii*) and keel-billed (*R. sulfuratus*) toucans, and dusky-faced (*Mitrospingus cassinii*) and plain-colored (*Tangara inornata*) tanagers. Forest understory antwren-antvireo mixed flocks form around several nuclear species, including white-flanked (*Myrmotherula axillaris*), checker-throated (*Epinecrophylla fulviventris*), and dot-winged (*Microrhopias quixensis*) antwrens and streak-crowned antvireos (*Dysithamnus striaticeps*), which were among the most abundant understory species at La Selva in the 1970s, but are now scarce (Sigel et al. 2006). These flocks are attended by numerous species of wrens, gnatwrens, flycatchers, and woodcreepers, including the sulphur-rumped flycatcher (*Myiobius sulphureipygius*), buff-throated foliage-gleaner (*Automolus ochrolaemus*), and plain xenops (*Xenops minutus*). Tanager flocks form around two nuclear species, tawny-crested (*Tachyphonus delattrii*) and olive (*Chlorothraupis carmioli*) tanagers, and are attended by many species of tanagers, finches, flycatchers, and woodcreepers, including the yellow-margined flycatcher (*Tolmomyias assimilis*), the white-throated shrike-tanager (*Lanio leucothorax*), and the orange-billed sparrow (*Arremon aurantiirostris*). Canopy mixed flocks include cacique flocks formed around the scarlet-rumped cacique (*Cacicus uropygialis*), greenlet-honeycreeper flocks formed around the lesser greenlet (*Hylophilus decurtatus*) and the shining honeycreeper (*Cyanerpes lucidus*), and grosbeak flocks centered around the nuclear species black-faced grosbeak (*Caryothraustes poliogaster*); the latter two canopy flocks are frequently attended by Nearctic migrant warblers as well as resident species. Finally, ant-following flocks form around three nuclear species: bicolored (*Gymnopithys leucaspis*), ocellated (*Phaenostictus mcleannani*), and spotted antbirds (*Hylophylax naevioides*). These flocks are joined by several obligate and facultative attendant species, including northern barred-woodcreeper (*Dendrocolaptes sanctithomae*), plain-brown woodcreeper (*Dendrocincla fuliginosa*), bare-crowned (*Gymnocichla nudiceps*) and immaculate (*Myrmeciza immaculata*) antbirds, red-throated ant-tanagers (*Habia fuscicauda*), and rarely the rufous-vented ground-cuckoo (*Neomorphus geoffroyi*).

The lowland Caribbean avifauna faces a wide variety of threats, including global climate change (Clark et al. 2003, Colwell et al. 2008), pesticides (Matlock et al. 2002), and cascading effects of alterations to other components of the lowland rainforest community (Feeley and Terborgh 2008, Young et al. 2008, Michel 2012). However, the greatest threat is the loss and fragmentation of lowland evergreen forests. Forested land cover in the Caribbean lowlands has decreased from nearly 100% to 70% in 1963 to 35% by 1983, and forest loss has continued since this time (Read et al. 2001). As a result, a total of 55 species are considered near-threatened, threatened, or vulnerable according to the IUCN (2010), and/or of medium, high, or urgent conservation priority (Parker et al. 1996). The majority of these species (36, or 65%) are associated with forest habitats, and another 5 species of generalists spend some time in forested habitats.

Fragmentation-associated threats to lowland Caribbean forest birds also disproportionately affect species of various dietary guilds, specifically insectivores. Although the distribution of threatened species amongst dietary guilds of all species in all habitats in the lowland Caribbean is relatively even (8 carnivores, 14 frugivores, 3 granivores, 14 insectivores, 3 nectarivores, and 13 omnivores), understory insectivores have been particularly hard hit at the La Selva Biological Station in the Sarapiquí region. At La Selva, 51 species have experienced moderate or severe declines since 1960, including 8 species that are believed extirpated. Of these, 65% (33 of 51) are insectivores, 41% (21 of 51) are ant-followers or associated with mixed flocks, and 47% (24 of 51) are residents of the forest understory. The nuclear species of both the antwren–antvireo understory mixed flocks (*M. axillaris*, *E. fulviventris*, and *M. quixensis*) and the tanager mixed flocks (*T. delattrii* and *C. carmioli*) are among the species that have experienced declines at La Selva, which in turn has apparently led to declines of attendant species, including the sulphur-rumped flycatcher (*M. sulphureipygius*) and white-throated shrike-tanager (*L. leucothorax*; Sigel et al. 2006). It has not yet been established with certainty whether the trends seen at La Selva are widespread throughout the Caribbean lowlands. In any case, insectivore declines have important implications for arthropod and plant communities, as insectivorous birds (along with bats) at La Selva protect plants by consuming herbivorous arthropods (Michel et al. 2014).

Mammals

Costa Rica is one of the few countries in the Western Hemisphere in which the entire mammalian fauna that was present

Table 16.4. Diversity, Migratory Status, Habitat Guilds, Dietary Guilds, and Number of Threatened Species among Birds of the Caribbean Lowlands of Costa Rica

Family	Genera / species	Permanent residents[a]	Latitudinal migrants[b]	Elevational migrants[c]	Visitants[d]	Habitat guilds[e]	Dietary guilds[f]	Threatened species[g]
Tinamidae	2/3	3	0	0	0	F, O	O	1
Anatidae	5/6	2	3	0	1	Aq	O	0
Cracidae	3/3	2	0	0	1	F	O	2
Odontophoridae	2/3	2	0	0	1	F	O	2
Podicipedidae	2/2	2	0	0	0	Aq	C	0
Phalacrocoracidae	1/1	1	0	0	0	Aq	C	0
Anhingidae	1/1	1	0	0	0	Aq	C	0
Ardeidae	11/17	11	6	0	0	Aq, F, O	C, I	1
Threskiornithidae	4/4	2	0	0	2	Aq, F	C	0
Ciconiidae	1/1	0	0	0	1	Aq	C	0
Cathartidae	3/3	3	0	0	0	F, O	C	0
Accipitridae	19/33	19	11	1	2	Ae, Aq, E, F, G, O	C, I	6
Falconidae	5/10	6	3	0	1	E, F, O	C	1
Rallidae	8/9	6	2	0	1	Aq, F	O	1
Heliornithidae	1/1	1	0	0	0	Aq	O	0
Eurypygidae	1/1	1	0	0	0	F	C	0
Aramidae	1/1	0	0	0	1	Aq	C	0
Charadriidae	3/3	0	2	0	1	Aq	I	0
Recurvirostridae	1/1	0	1	0	0	Aq	C	0
Jacanidae	1/1	0	1	0	0	Aq	O	0
Scolopacidae	6/13	0	13	0	0	Aq, O	C, I	1
Columbidae	7/14	11	0	0	3	F, O	E, F, O	2
Psittacidae	8/12	9	0	3	0	E, F, G, O	F	6
Cuculidae	5/8	4	3	0	1	F, G, O	C, I, O	1
Tytonidae	1/1	1	0	0	0	O	C	0
Strigidae	6/7	7	0	0	0	F, G, O	C, I	0
Caprimulgidae	4/7	4	3	0	0	Ae, F, O	I	0
Nyctibiidae	1/2	2	0	0	0	F, O	C, I	0
Apodidae	4/8	3	1	0	4	Ae	I	1
Trochilidae	21/26	12	1	8	5	E, F, G, O	N	3
Trogonidae	1/5	3	0	2	0	F	O	0
Momotidae	3/3	2	0	0	1	F, G	O	0
Alcedinidae	2/6	5	1	0	0	Aq, F	C	0
Bucconidae	3/4	4	0	0	0	E, F, G	I	0
Galbulidae	2/2	2	0	0	0	E, F	I	0
Ramphastidae	4/5	3	0	2	0	F, G	O	2
Picidae	7/9	8	1	0	0	E, F, G, O	I, O	1
Furnariidae	11/14	12	0	0	2	F, G, O	I	1
Thamnophilidae	13/18	16	0	2	0	E, F, O	I	3
Formicariidae	1/1	1	0	0	0	F	I	0

continued

at the time of European settlement is still largely extant, at least in well-protected parks. The Caribbean lowland area of Costa Rica has a diverse mammalian fauna that is characteristic of Neotropical lowland rain forests and consists of approximately 125 species representing 10 orders and 30 families. The majority of species found in the lowlands are broadly distributed in the northern Neotropics—lowland tropical mammals tend to be distributed both in latitude and elevation in a manner more widespread than is the case for amphibians, reptiles, and birds.

The Costa Rican Caribbean lowland mammal fauna includes 5 marsupials, 71 (possibly more) bats, 3 (possibly 4) primates, 2 armadillos, 3 anteaters, 1 rabbit, 3 squirrels, 10 (possibly more) long-tailed rats and mice, 1 pocket gopher, 1 porcupine, 1 paca, 1 agouti, 4 mustelids, 1 skunk, 4 procyonids, 5 cats, 2 peccaries, 2 deer, and 1 tapir. The West Indian manatee (*Trichechus manatus*) historically occurred in the rivers and canals along and well inland from the coast and it is still found there now, although in reduced numbers (Jiménez, chapter 20 of this volume). This species list is likely to be complete with the exception of the orders Chiroptera and Rodentia, where cryptic or difficult-to-capture species may not yet have been observed. All of the species in this historical species list, with the exception of the giant anteater (*Myrmecophaga tridactyla*), still occur in the lowlands and represent more than half (55%) of Costa Rica's mammalian fauna.

Although Costa Rica's Caribbean lowlands occupy an

Table 16.4. Continued

Family	Genera / species	Permanent residents[a]	Latitudinal migrants[b]	Elevational migrants[c]	Visitants[d]	Habitat guilds[e]	Dietary guilds[f]	Threatened species[g]
Grallariidae	1/2	2	0	0	0	E, F	I	0
Tyrannidae	36/54	35	13	3	3	E, F, G, O	F, I, O	3
Cotingidae	5/5	2	0	2	1	F	F, O	3
Pipridae	3/4	1	0	3	0	E, F	F	1
Vireonidae	3/10	3	7	0	0	E, F, G, O	I, O	0
Corvidae	1/2	2	0	0	0	G, O	O	0
Hirundinidae	6/9	3	6	0	0	Ae, O	I	0
Troglodytidae	6/10	10	0	0	0	E, F, O	I	1
Sylviidae	3/3	3	0	0	0	E, F, G	I	0
Turdidae	4/8	1	4	2	1	F, G, O	O	1
Mimidae	2/2	1	1	0	0	E, O	O	0
Bombycillidae	1/1	0	1	0	0	G	O	0
Parulidae	13/34	3	31	0	0	E, F, G, O	I, O	3
Genus Incertae Sedis (Coerebidae)	1/1	1	0	0	0	O	O	0
Thraupidae	14/27	17	3	3	4	E, F, G, O	I, O	3
Genus Incertae Sedis (Saltator)	1/5	5	0	0	0	E, F, O	O	1
Emberizidae	8/11	8	2	0	1	F, O	G, O	2
Cardinalidae	5/7	2	4	1	0	F, G, O	G, O	1
Icteridae	9/15	12	3	0	0	E, F, G, O	I, O	0
Fringilidae	1/4	3	0	1	0	G, O	F	1
Passeridae	1/1	1	0	0	0	O	O	0
Total	**309/484**	**286**	**128**	**33**	**37**			**55**

[a] Permanent residents live in the lowland Caribbean year-round.
[b] Latitudinal migrants include passage migrants and breeding and non-breeding part-year residents that spend the remainder of the year in North or South America.
[c] Elevational migrants include breeding and non-breeding part-year residents that spend the remainder of the year at higher elevations.
[d] Visitants occur accidentally or visit occasionally, though not on a seasonal or annual basis.
[e] Habitat guild codes: Ae: Aerial (spend most time on the wing); Aq: Aquatic (associated with lakes, rivers, and marshes); E: edge (associated with forest edge and canopy); F: Forest (associated with primary or tall secondary forest); G: Generalist (associated with multiple habitat types); O: Open (associated with fields, pastures, young secondary forest, suburban, and urban areas). Codes from Sigel et al. 2010.
[f] Dietary guild codes: C: Carnivores (diet primarily or entirely vertebrates, carrion, snails, and large arthropods); F: Frugivores (diet primarily or entirely fruit); G: Granivores (diet primarily or entirely seeds); I: Insectivores (diet primarily or entirely insects and other arthropods); N: nectarivores (diet primarily or entirely floral nectar); O: Omnivore (diet includes food from multiple categories). Codes from Sigel et al. 2010.
[g] Threatened species include species assigned Conservation Priority of Urgent (1), High (2), or Medium (3) by Parker et al. 1996 (per Blake and Loiselle 2000), and/or species assigned an IUCN (2010) Red List Category of Near Threatened or higher.
Source: Data derived from Slud 1960, Stiles and Skutch 1989, Ridgely and Gwynne 1992, Parker et al. 1996, Blake and Loiselle 2000, Sigel et al. 2006, 2010, Garrigues and Dean 2007, and IUCN 2010.

extensive area from the Nicaraguan to Panamanian borders, until recently little research has been undertaken on the region's mammals other than in the La Selva–Parque Nacional Braulio Carrillo region and to a lesser extent in the Maquenque and Tortuguero protected areas. More than 100 scientific papers on various aspects of mammals at La Selva have been published since Slud (1960) first mentioned white-lipped peccaries, monkeys, and tapirs there. As part of the elevational transect that became the La Selva–Parque Nacional Braulio Carrillo biological corridor, Timm et al. (1989) conducted a faunal survey of the elevational transect from 35 m to 2,600 m on Volcán Barva, documenting that at least 141 species of mammals occurred in the protected area. They reviewed the historical and present-day distributions, systematics, and ecology of each species. This research was the first such elevational transect undertaken in the Neotropics.

Three species of primates inhabit the Caribbean lowlands of Costa Rica, the mantled howler monkey, Central American spider monkey, and white-faced capuchin monkey. All three species were widely distributed throughout the Caribbean lowlands historically. During the early 1950s, primate populations throughout Central America were ravaged by an epidemic of mosquito-borne yellow fever (Fishkind and Sussman 1987, Timm 1994). Primates were observed only infrequently at La Selva between the late 1960s and early 1980s. Fortunately, all three species rebounded and they became more abundant than they were in the 1960s and late 1970s. Today capuchin, howler, and spider monkeys can be seen almost daily in protected areas and nearby forest fragments, often in large groups. Capuchin and howler monkeys can occupy quite small fragments, moving along fencerows. Capuchins living in close proximity to banana plantations feed on ripe bananas from the plantations. Historically, howler monkeys were the most abundant primates in the Caribbean lowlands, and this still may be the case today. A sighting of the night monkey (*Aotus*) at La Selva has never been confirmed or repeated.

The order Carnivora deserves special mention, not because there have been many studies on these animals, but because of their critical role in ecosystems (Terborgh 1988) and because they represent some of the animals most sought after by researchers and visitors alike. The most commonly seen species of the Carnivora are actually omnivorous or frugivorous—the white-nosed coati (*Nasua narica*), kinkajou (*Potos flavus*), and tayra (*Eira barbara*). The northern raccoon (*Procyon lotor*) occurs in low density in this area, which is interesting given its extremely wide distribution from southern Canada to central Panama and its high abundance in other areas of its range. The Neotropical otter (*Lontra longicaudis*) can be seen along streams and its fecal remains of fish and crustacean debris can be found on rocks. The ocelot (*Leopardus pardalis*) is the most frequently seen of the cats; it adapts well to disturbed habitats and can be found in small fragments, although little is known about how these populations are affected by fragmentation and the development of roads. Jaguars (*Panthera onca*) and pumas (*Puma concolor*) are the two large cats found in the area. The jaguar is now extremely rare, especially in the lowlands, although wandering individuals are occasionally seen. The puma, on the other hand, although not abundant, is more common than the jaguar and is seen regularly at La Selva. Given the number of observations in recent years in many areas of the country, it appears as if puma numbers are increasing. The majority of sightings at La Selva are in areas where collared peccaries are frequently observed. Most of the data collected on felines in the Caribbean lowlands come from camera-trapping efforts by TEAM (the Tropical Ecology Assessment and Monitoring project of Conservation International, CI), a long-term biodiversity monitoring program that has been successful in capturing pictures of elusive animals from the lowlands (Ahumada et al. 2011). The coyote (*Canis latrans*), a generalist predator, recently expanded its range in Costa Rica to include the Caribbean lowlands.

Costa Rica has a rich and diverse bat fauna with all of the families and feeding niches found in the New World represented. There are currently 117 species of bats known from the country, which means that more than 50% of Costa Rica's terrestrial mammals are bats. The middle and higher elevation slopes have received far less study than have the lowlands. The Caribbean lowland zone has a fauna of at least 71 bat species that includes all of the Neotropical families and most of the genera, as well as all of the feeding niches. More research has been undertaken on bats in the Caribbean lowlands than has been undertaken on all other Costa Rican mammals combined throughout all of the country. La Selva, with its rich bat fauna and ready access to both mature forest and varying stages of second growth, has been the center of bat research in Costa Rica for nearly five decades. The Tirimbina Biological Reserve has also become an important center of bat research in the past decade.

Early studies on bats at La Selva focused primarily on distributions, basic ecology, and systematics, providing a background for in-depth studies of behavior and ecology, as well as for assessing recent changes in distribution and abundance. LaVal (1977) reported on the distribution of several then poorly known species that he found at La Selva, including the first records of the thumbless bat (*Furipterus horrens*). Inexplicably, this species has never again been detected in Costa Rica despite the intensive netting and acoustical monitoring efforts in the lowlands. Critical to later studies on ecology and conservation of bats has been the presence of voucher specimens that the early researchers deposited in scientific collections and detailed keys for the identification of species, which were based heavily upon studies undertaken at La Selva (Timm and LaVal 1998, Timm et al. 1999).

When the white tent-making bat (*Ectophylla alba*) was discovered at La Selva, it was considered one of the rarest of New World bats and little was known of its biology (Timm 1982). Research has been undertaken on this species over several decades there and at Tirimbina, documenting that it alters the shape of *Heliconia* leaves by cutting the midrib along the length of the leaf, causing the sides of the leave to collapse down around the bats and forming a roost site that protects the bats from predators as well as acting as a rain shield. Research on the tent-making bat species (including *Dermanura phaeotis*, *D. watsoni*, *Uroderma bilobatum*, *Vampyressa thyone*) has provided several exciting new insights into bat biology (LaVal and Rodríguez-Herrera 2002, Rodríguez-Herrera et al. 2007, 2008, Rodríguez-Herrera and Tschapka 1999, Timm 1982, 1984, 1985, 1987).

Numerous other studies with nectar-feeding bats (Greiner et al. 2013, Sperr et al. 2009, Tschapka 2003, Voigt 2004, 2013), sac-winged bats (Voigt 2005, Voigt et al. 2008), vampire bats (Voigt et al. 2012), and others make this group of lowland Caribbean vertebrates one of the best studied in all of the New World tropics.

There have been few ecological studies on rodents in the Caribbean lowlands almost certainly because most species

both large and small seeds over long distances via endozoochory. Though usually found in the forest canopy, *Pteroglossus torquatus*, *Ramphastos sulfuratus*, and *R. swainsonii* have been observed consuming fruits and seeds of the ginger *Renealmia alpinia* in the understory (García-Robledo and Kuprewicz 2009). Toucans may also play an integral role in the dispersal and establishment of some large-seeded palms commonly found in the Costa Rican lowlands (e.g., *Iriartea deltoidea*). Mixed parentage aggregations of *I. deltoidea* seedlings located beneath adult trees are governed by the seed handling behaviors and movement patterns of these large ramphastids (toucans) (Sezen et al. 2009).

Seed dispersal by bats has received much attention in the Neotropics with many studies conducted in the Caribbean lowland forests of Costa Rica. Bats can disperse small seeds (e.g., *Ficus* spp., *Cecropia* spp.) via endozoochory, whereas they disperse large seeds (e.g., *Dipteryx panamensis*, palm seeds) by dropping them during flight or under feeding sites and roosts (Fleming and Heithaus 1981, Kelm et al. 2008, and references therein). Bats may be highly effective seed dispersers owing to their capability for long distance seed dispersal among various habitats (Heithaus and Fleming 1978).

Most phyllostomid bats found in Caribbean lowland forests are highly frugivorous. *Artibeus*, *Carollia,* and *Dermanura* consume fruits and seeds from many understory and canopy plants in the region (Levey et al. 1994). *Artibeus jamaicensis* and *Dermanura watsoni* have been evaluated as highly effective dispersers of seeds from the fig *Ficus insipida* (Banack et al. 2002). Within the Sarapiquí Basin of the Caribbean slope, Melo et al. (2009) found 46 species of large (>8 mm) seeds beneath leaf tents constructed by *D. watsoni*. Their findings had implications on seed and seedling distributions near bat tents within forested habitats—seed densities and seedling abundances were higher under bat tents than in areas away from tents (Melo et al. 2009). Bats, owing to their abilities to forage, feed, and roost in diverse habitats, may serve as effective natural reforestation agents. Artificial roosts constructed throughout a forest–pasture mosaic in the Caribbean lowlands successfully recruited 10 species of bats, including five frugivores/nectarivores (Kelm et al. 2008). These bats transported many seeds from early-successional plants into degraded lands, potentially leading to forest succession within these pasturelands (Kelm et al. 2008). However, further evaluations of seed germination success and seedling growth are needed to evaluate the effectiveness of bats as reforestation agents (Holl 2008).

Seed dispersal and frugivory by terrestrial mammals is a common phenomenon in the forests of the Costa Rican Caribbean slope. Numerous terrestrial species are known fruit-eaters, including agoutis (*Dasyprocta punctata*), collared peccaries (*Pecari tajacu*), armadillos (*Dasypus novemcinctus*), coatis (*Nasua narica*), tayras (*Eira barbara*), kinkajous (*Potos flavus*), pacas (*Cuniculus paca*), and numerous species of opossums. Most Carnivora likely also consume fruits and seeds to supplement their diets, though their effectiveness as seed dispersers is virtually unknown in this region. Small rodents, particularly the heteromyid rodent *Heteromys desmarestianus*, consume and cache small and large seeds, potentially serving as effective seed dispersers for some plants (especially palms) (Fleming 1974). Many terrestrial rodents consume and destroy seeds while foraging, acting as significant seed predators and only incidental seed dispersers (Smythe 1986).

The seeds of the palm *Socratea exorrhiza* that fall to the forest floor are rapidly encountered by small (*H. desmarestianus*, *Proechimys semispinosus*) and mid-sized (*D. punctata*) terrestrial mammals (Kuprewicz 2010, 2013). Predation of seeds by terrestrial mammals tracks fruiting patterns, with lower predation levels during peak fruit set compared to higher levels at the end of the fruiting season (Notman and Villegas 2005). Seed caching, however, does not appear to follow these fruiting patterns (Notman and Villegas 2005), and hoarding events by agoutis at La Selva are relatively rare (Kuprewicz 2010, 2013).

In forests protected from human hunting in the Caribbean lowlands (such as La Selva), local populations of collared peccaries have recently increased, likely due to direct effects (fewer peccaries killed by hunters than previously) as well as indirect effects (release from predators [large felids] that remain uncommon in the area despite reduced human hunting). Collared peccaries forage singly or in small groups (Romero et al. 2013) and are seed predators that consume and kill many large seeds (e.g., *Dipteryx panamensis*, *Iriartea deltoidea*, *Mucuna holtonii*, *Socratea exorrhiza*) (Kuprewicz and García-Robledo 2010, Kuprewicz 2010, 2013). Peccaries may disperse some seeds via endozoochory or expectoration (Beck 2005), but in the Caribbean forest of La Selva, they act primarily as seed predators; this behavior negatively affects seedling recruitment and may have dramatic implications for future tree distributions and plant propagation (Kuprewicz 2010, 2013). A comparison of seed removal and seed fates in two Caribbean lowland forests (Tirimbina Rain Forest Center, a hunted forest, and La Selva, a forest protected from hunting) found that some seed species (*Carapa nicaraguensis*, *Lecythis ampla*, *Pentaclethra macroloba*) had higher seed removal rates in La Selva when compared to Tirimbina. Overall seed dispersal was also higher at La Selva than at Tirimbina (Guariguata et al. 2000). Removal of large terrestrial frugivorous mammals through hunting can have complex effects on the seed dispersal, seed survival, and resultant seedling demography in defaunated regions (Wright et al. 2000).

People and Nature

Human Populations and Demography

As indicated earlier, precontact indigenous populations were substantial but not nearly the size of Maya, Aztec, and Inca populations to the north and south. The Caribbean lowlands constitute part of the Costa Rican archaeological subregion identified as Atlantic Highlands and Watershed. The best-known archaeological site in this area is the Las Mercedes site, uncovered by the railroad construction of Minor C. Keith in the 1870s, and currently located on the campus of EARTH University in Guápiles. Keith removed more than 15,000 items and distributed them among several museums in the United States. In the 1890s the Swedish archaeologist Carl V. Hartman also worked at Las Mercedes and sent many valuable objects to Sweden. This site is dated at about 1,000–2,000 years old and therefore does not provide insight into the earliest indigenous populations (Snarskis 1976). Many additional archaeologically valuable locations are scattered around the Caribbean lowlands, some of them currently under excavation and others undiscovered or untouched.

The human population of Costa Rica's Caribbean lowlands grew slowly throughout the colonial period after the demographic collapse of the indigenous populations (Augelli 1987). Spanish settlers tended to establish themselves in the Central Valley (Palmer and Molina 2004). The Caribbean coast is known during these years for the exploits of a few colorful English and Dutch pirates—the Nicaraguan Caribbean town of Bluefields, for instance, is named after the Dutch pirate Abraham Blauvelt—and not for any substantial settlements of colonists or indigenous people (Ross and Capelli 2003). After Costa Rica's independence from Spain in 1821, the Central Valley population centers (San José, Cartago, Alajuela, Heredia) continued to expand—a trend that was enhanced as coffee production became the dominant economic force in the country. The entire human population of Costa Rica is estimated to have been about 19,000 people in 1770, then 52,000 in 1801, then 200,000 in 1900, then 875,000 in 1950 (Augelli 1987). Sixty years later Costa Rica's population size had been multiplied by a factor of five, reaching a total of 4,564,000 inhabitants (Centro Centroamericano de Población 2011).

Significant populations of Europeans and North Americans did not arrive in the Caribbean lowlands until the construction of the railroad between Alajuela and the Caribbean coast in the late nineteenth century. The railroad was built to transport coffee, Costa Rica's major agricultural export crop, from the central highlands (where it was grown) to the Caribbean port of Limón for export to Europe. The builder and financier of this railroad, US investor Minor C. Keith, started the Caribbean lowland banana industry to help finance the railroad project. Excellent, extensive treatments of both the coffee and the banana industries of Costa Rica are recommended (Chapman 2007, Chomsky 1996, Koeppel 2008, Paige 1997, Palmer and Molina 2004, Ross and Capelli 2003). West Indian Afro-Caribbean workers were brought to the Costa Rican lowlands to work on the railroad and on the banana plantations. By 1883 there were 902 Jamaicans in Limón Province, and by 1927 there were 18,003. These workers were not granted Costa Rican citizenship until 1948, and they were not allowed to leave Limón Province, even to work on the Pacific banana plantations. Today about one-third of the population of Limón Province is made up of West Indian immigrants and their descendants (Biesanz et al. 1987). The distinctive language and culture of the Limón area are due to the influence of these Afro-Caribbean workers and their descendents.

As mentioned previously, Puerto Limón is the largest city in the Caribbean lowlands. Although it is the second largest city in Costa Rica, its population of 63,000 puts it at a distant second to the capital of San José with its population of about 335,000. Other Caribbean-side population centers include Guápiles (~19,000), Siquirres (~18,000), Guácimo (~7,000), and Puerto Viejo de Sarapiquí (~6,000). Human settlement in the Caribbean lowlands increased dramatically during the 1960s, 1970s, 1980s, and 1990s owing to at least three factors: Costa Rica's extremely high birth rate before 1960 (Coale 1983), a deliberate governmental policy to move people from the Central Valley to the "hinterlands," and farm labor immigration from other Central American countries (particularly from Nicaragua). The birthrate has fallen in recent years (Coale 1983) but the Caribbean lowlands continue to see increasing populations.

Agriculture and Pesticides

The Caribbean lowland area of Costa Rica is threatened by the same dangers that affect other tropical systems—deforestation, loss of biodiversity, genetic fragmentation of plant and animal populations, spread of alien invasive species, climate change, human population increase, and the pressure to value short-term economic gains over long-term sustainability. Some of these subjects are treated in other sections of this chapter. Here we focus on one particular theme, that of agricultural development and pesticide use, because it is a topic that is critically important in its own right and also because it brings many of these other areas into clear focus. A consideration of agriculture and pesticide use is also intimately related to emerging issues such as valuation of ecosystem services, global food security, poverty alleviation, the objectives of the Millennium Ecosystem Assessment, and the new Sustainable Development Goals (SDGs).

Globally, the use of pesticides (insecticides, fungicides, rodenticides, herbicides, and germicides) has increased dramatically in the past seventy years (Carvalho 2006) and every corner of the planet, including Arctic ice, bears traces of these chemicals (Chernyak et al. 1996, Cone 2006, Pelley 2006). Much early use of insecticides was related to mosquito/malaria control (Jaga and Dharmani 2003, Lubick 2007), but a significant proportion of pesticide application has now been turned toward agricultural food production (Monge et al. 2005, Wesseling et al. 1999). Although the use of pesticides has certainly led to increased food production—the so-called Green Revolution—(Cooper and Dobson 2007), some of these positive effects are being lost through evolving resistance to the chemicals by pests and non-target organisms (Brausch and Smith 2009, Jansen et al. 2011, Nolte 2011, Raymond et al. 2001) or offset by various direct and indirect calculated costs of pesticide use. Pesticides have been shown to have serious human health consequences, both acute (Soares and Porto 2009) and chronic (Abhilash and Singh 2009, Charboneau and Koger 2008, Nag and Raikwar 2011, Yearout et al. 2008), as well as a range of environmental impacts including soil and water contamination (Liess and von der Ohe 2005) and non-target organism toxicity (Berny 2007, Kendall and Smith 2003, Pisani et al. 2008). The role of pesticides in the endocrine and developmental disruption of wildlife and humans (see Buchanan et al. 2009, Casals-Casas and Desvegne 2011, Mnif et al. 2011, Soin and Smagghe 2007 for reviews) is also becoming an increasingly powerful concern.

Both the negative and positive aspects of global agricultural development and pesticide use are seen in microcosm with the specific case of the Caribbean lowlands of Costa Rica over the past several decades. DDT was used widely and heavily in Costa Rica from 1957 to 1985 (to the tune of 1,387 total tons during that period) to control mosquito-borne malaria (Pérez-Maldonado et al. 2010, Duszeln 1991). Today DDT and other banned organochloride pesticides (OCPs)—widely known as persistent organic pollutants (POPs)—have a relatively low and uniform distribution in soils across the country (Daly et al. 2007a) but, as recently as 2009, levels of DDT in soil and children's blood serum were considered above acceptable levels in two rural communities in the Limón area (Pérez-Maldonado et al. 2010). Pesticide use in Central America doubled between 1980 and 2000 (Wesseling et al. 2005). During the 1990s, Costa Rica's annual use of 4 kg of pesticides per inhabitant and 38 kg per agricultural worker was the highest in all of Central America (Wesseling et al. 2001). In recent years, Costa Rica has been at the top of the world's countries in pesticide use (Polidoro et al. 2008, World Resources Institute 2007), on the basis of the number of kg of active ingredients applied annually to each hectare of agricultural land (52 kg a.i. [active ingradients]/ha/yr in 2000). Raw pesticide import data from 1977 to 2009 have been analyzed for Costa Rica (de la Cruz et al. 2014) and indicate increases in the variety of chemicals used, as well as their environmental hazards. Pesticides are used on large volume agricultural crops such as banana, coffee, pineapple, rice, and heart-of-palm (also called palm heart), as well as on non-traditional agricultural export crops (NTAEs) such as plantain, carrot, cassava, squash, and cut flowers. Pesticides are applied to crops for export as well as for local (within Costa Rica) consumption.

A third to a half of Costa Rica's imported pesticides are used in banana production (Castillo et al. 2006) and most of Costa Rica's banana production today is in the Caribbean lowlands. Bananas are the world's number one fruit consumed and in the past few years Costa Rica has been among the world's top three banana-exporting countries, along with Ecuador and the Philippines (Barraza et al. 2011). In 2009 Costa Rica's banana plantations covered more than 50,000 hectares and employed 35,000 workers (CORBANA 2009). The effects of chemicals used by the United Fruit Company in Costa Rica during the sigatoka (*Mycosphaerella musicola*) fungus outbreak of the 1930s led to an extended workers' strike (Marquardt 2002). Decades later, the claims of worker sterility from the use of the nematicide DBCP (marketed under the name Nemagon) led to several class-action lawsuits settled out of court or in favor of the plaintiffs (Ling and Jarocki 2003). DBCP has been linked to several types of cancer among Costa Rica's banana workers (Wesseling et al. 1996).

Chlorothalonil is a pesticide currently in use on Costa Rican banana plantations, with application rates of up to 45 times per year (Chaves et al. 2007). This fungicide is considered to be highly toxic to humans (Caux et al. 1996, Margni et al. 2002, Sherrard et al. 2003) and in Costa Rica has also been shown to be toxic to fish, birds, and aquatic invertebrates (Castillo et al. 2000). Fortunately, unlike DDT, it breaks down rapidly in Costa Rica's tropical climate (45% degradation after 24 hrs), but some residues are still present after 85 days and toxic metabolites are also persistent (Chaves et al. 2007). In Costa Rica, as has been the case elsewhere, the pesticide industry has changed from using a few broadly acting chemicals to using a wide range of active ingredients with shorter active lives (Galt 2008a).

Chemicals from agricultural production find their way into the soils and rivers of the Caribbean lowlands. Sediments from clearing land for plantations also drain into the area's rivers and modify habitat for local freshwater organisms and for more distant marine organisms. The streams receiving water from Caribbean lowland banana-processing plants are home to altered macroinvertebrate communities in the presence of pesticide levels that are well below those responsible for acute toxicity (Castillo et al. 2006). A more distant effect is produced by the Río Estrella, which drains a large area of banana plantations and discharges into the Caribbean Sea 10 km north of Parque Nacional Cahuita and the offshore coral reef (Roder et al. 2009). Aerosolized pesticides are carried from their lowland sites of application to higher elevations on the slopes of Volcán Barva, where they precipitate out over "pristine" mid-elevation forests (Daly et al. 2007b).

Cassava growers in an indigenous Bribri community near Limón indicated that they had had no training in use of agricultural chemicals and that less than one-third of them used protective clothing during spraying (Polidoro et al. 2008). In this same community, Bribri women who worked with chemically sprayed plantains reported more respiratory problems than women who did not work in agriculture (Fieten et al. 2009).

In the face of worrisome data regarding pesticide effects, there is potentially some good news. The most toxic chemicals used in Costa Rican agriculture do have less toxic alternatives (Humbert et al. 2007). An organic treatment for the post-harvest control of banana crown rot has been developed by scientists at EARTH University (Demerutis et al. 2008). Some international banana companies have sought Rainforest Alliance certification, which lends the required stamp of "sustainability" to products sold to segments of the European market. In Costa Rica, education and extension services also effectively reduce the intensity of pesticide use (Galt 2007). For example, on a potato farm near Cartago pesticide use was reduced if there were minors (children) in the household or if the farmer had taken an agricultural course (Galt 2008b).

Recent studies indicate that the current standards for pesticide residues in the destination countries are creating pressure for less pesticide use in the producing countries, including Costa Rica (Galt 2008c). Consumer groups around the world indicate a "willingness to pay" more for bread and produce in exchange for less pesticide use (Batte et al. 2007, Boccaletti and Nardella 2000, Chalak et al. 2008, Florax et al. 2005, Foster and Maurato 2000, Kahn 2009). Adherence to "developed country pesticide standards" in developing countries (Okello and Swinton 2010) is one powerful way to break the ominous "circle of poison" of the recent past whereby chemicals banned in developed nations were sold and used in the Third World, only to return to the developed world as residue on imports (Weir and Shapiro 1981). In a study of pesticide use on five vegetables (carrot, chayote, corn, green bean, and squash) grown in Costa Rica both for national use and for export, pesticide use was less on the produce bound for export to the United States (Galt 2008), reportedly because of EPA guidelines on pesticide residues.

Although large-scale completely pesticide-free agriculture is probably not possible, it is entirely possible to imagine a scenario that combines the following: (1) economic incentives for the use of chemicals with lower toxicity and with lower application rates, (2) development of more organically grown and free-trade products, (3) industrial, academic, and extension training to maximize worker knowledge and safety, (4) continued and intensified consumer pressure for healthy products, (5) enforced guidelines on pesticide residues in consumer nations, (6) environmental monitoring of air, soil, and water by government and the scientific community, (7) epidemiological studies on the relation between disease and exposure to chemicals, and (8) creative high technology solutions to food security such as removing pesticides from fruit juices in the final stages of processing. Many of these efforts are currently underway, albeit not universally, in the Caribbean lowlands of Costa Rica. Costa Rica is in a position to attempt a unified approach to intensified but non-destructive agriculture because of its long history of agricultural activity, its world-recognized conservation efforts, and the well-established multi-institutional and international scientific research being conducted within the country.

Conservation

The conservation history of Costa Rica since the 1950s has been well documented and analyzed. Excellent treatments are available regarding the founding of the national park and reserve system (Boza 1993, Evans 1999, Wallace 1992), the proliferation of conservation efforts during the past 60+ years (Calvo-Alvarado 1990, Campbell 2002a, Johnson and Clisby 2009, Powell et al. 2000), and the establishment of specific projects and private reserves (Butterfield 1994, Chornook and Guindon 2008, Nadkarni and Wheelwright 2000, Wheelwright and Nadkarni 2014). Case studies of Costa Rican conservation initiatives such as payment for environmental services (Dick et al. 2010, Morse et al. 2009, Pagiola 2008, Sánchez-Azofeifa et al. 2007, Sierra and Russman 2006, Snider et al. 2003), the Mesoamerican Biological Corridor (Dettman 2006, Sader et al. 2004), the Biodiversity Law of 1998 (Miller 2011), participatory resource management (Sims and Sinclair 2008), and carbon sequestration (Lansing 2011) are also available.

For many decades Costa Rica has been held up as an example to the world for the amount of its land under protection, its sustainable ecotourism, and the strong environmental ethic among its citizens. Critics and scholars have pointed out some of the discrepancies between this image and stark reality, including illegal logging and harvesting of forest products, underfunding of national parks, negative tourism impacts on protected areas, duplicated or disorganized conservation efforts, and unequal distribution of the economic benefits of conservation initiatives (Campbell 2002a,b, Hoffman 2011, Silva 2003, Sylvester and Avalos 2009, Vivanco 2006). Several unquestionable conservation success stories do exist, however, as does a continuing willingness on the part of interested parties to invent new strategies to cope with problems that arise. The title of an article by one of the founders of Costa Rica's National Park system, Mario A. Boza, remains as apt today as it was upon its publication—with regard to environmental solutions, "Costa Rica is a laboratory, not Ecotopia" (Boza et al. 1995).

More than a dozen national parks, wildlife refuges, and biological corridors have been created in the Caribbean lowlands since 1970, when the Tortuguero area was first protected and then made into a National Park in 1975 (for a map of the region's protected areas, see Fig. 16.2). As a heuristic exercise, we have chosen to compare the Tortuguero area to the San Juan–La Selva (SJLS) Biological Corridor because of several fascinating similarities and instructive differences in the establishment and current operations of these two areas. Among the similarities in the scenario for the creation of these two entities are (1) the realization that an iconic vertebrate species was about to be lost, (2) rapid legal action to protect the nesting areas of these animals, (3) cooperation among scientists, government officials, and representatives of non-governmental organizations (NGOs) to create and consolidate the protected areas, (4) the large size and ecological diversity of the area under protection, (5) the welcome outcome—a conservation success story—at least in terms of halting the precipitous decline of the iconic vertebrate species, and (6) increasing pressure on the protected areas from other sources in recent years. Notable differences in the two stories are (1) the dominant conservation paradigms at the time of establishment of the protected areas, (2) the funding and management of the areas, (3) the ownership of the land contained within the protected areas, (4) the extent of community income from ecotourism, and (5) the strength of international scientific research conducted in the protected areas.

The 35 km stretch of the Caribbean coast between the Tortuguero and Parismina rivers has probably been a nesting ground for the Atlantic population of the green turtle *Chelonia mydas* and other sea turtles for many hundreds if not thousands of years (Spotila 2004). Early Spanish explorers noted the vast turtle populations in this area and named the zone "Tortuguero" (place of turtles) (Ross and Capelli 2003). Indigenous people and colonial settlers harvested eggs and adult turtles for local consumption and for export (Jackson 1997, Lefever 1992), but large-scale exportation of eggs, meat, and calipee in the mid-twentieth century dramatically increased the pressure on the Costa Rican green turtle population almost to the point of extinction. Dr. Archie Carr, a turtle specialist from the University of Florida, had been monitoring these animals since 1955 and became alarmed at their rapid decline in the early 1970s (Carr 1967, Carr et al. 1978). He spoke urgently with people in the international conservation community and the Costa Rican government and prompted a series of actions over the next several years. First, an executive decree was issued in 1970, prohibiting the harvesting of turtles and their eggs along the beach and protecting a portion of the nesting area. The 19,000 ha Parque Nacional Tortuguero was established as one of the first of Costa Rica's national parks by Law 5680 in 1975, during the presidency of Daniel Oduber. A contiguous area of 92,000 ha was established in 1985 as the Refugio Nacional de Vida Silvestre Barra del Colorado. In 1994, a 50 ha parcel of land wedged between the two protected areas became the Dr. Archie Carr Wildlife Refuge and the site of the John H. Phipps Biological Station. Two marine turtle laws, 7906 (1999) and 8325 (2002), banned catching the turtles at sea.

Owing to these conservation efforts, Tortuguero now has the largest group of nesting green turtles in the entire Atlantic population (see also Cortés, chapter 17 of this volume). From 1971 to 2003, there was a 417% increase in the number of nests, with estimates of 104,000 nests per year and 17,000–37,000 nesting females per year during the 1999–2003 period (Troëng and Rankin 2005). A very active international research effort, including a graduate program based at the University of Florida (supported by the Phipps Biological Station in Costa Rica), has contributed important information on the Tortuguero green turtles, including data on genetic diversity of nesting females (Bjorndal et al. 2005), spatial distribution of nests (Tiwari et al. 2005), adult survival (Troëng and Chaloupka 2007), effect of sea surface temperatures on nesting (Solow et al. 2002), hatchling success (Tiwari et al. 2006), levels of bacteria in nests (Santoro et al. 2006), and the genetic composition of juvenile foraging groups (Monzón-Argüello et al. 2010).

The green turtle is still considered to be endangered over its entire range (Rieser 2012, Seminoff and Wallace 2012), although some populations are in better shape than others (Broderick et al. 2006, Seminoff and Shanker 2008). The Tortuguero population, although rebounded, is still nowhere near the levels estimated for Precolombian populations (Jackson 1997, McClenachan et al. 2006, Troëng and Rankin 2005). These turtles take nearly 30 years to reach sexual maturity (Frazer and Ladner 1986) and the overharvesting of juveniles in the open ocean (both intentionally and as bycatch of commercial fishing and shrimping operations) could lead to reduced numbers of mature adults available for breeding in the future (Lahanas et al. 1998, Lagueux 1998, Mortimer 1995). Despite these caveats, the population of green turtles at Tortuguero in recent years appears healthy (Chaloupka et al. 2008). The hope persists that the Tortuguero/Barra del Colorado protected area will also benefit other threatened and endangered vertebrates such as manatees (Smethurst and Nietschmann 1999), hawksbill turtles (Bjorndal et al. 1993, Troëng et al. 2005), and leatherback turtles (Troëng et al. 2007), but the recovery of these species seems less optimistic.

Since its inception, the scientific research in the Tortuguero protected area has been combined with ecotourism and revenue generation for the local community (Meletis 2007, Lee and Snepenger 1992). Turtle watching tourism has increased greatly over the years, as have the revenues generated by ecotourism (Ballantyne et al. 2009, Jacobson and Figueroa Lopez 1994). A training program for local tourist guides was developed (Jacobson and Robles 1992) and turtle watching guidelines for sustainable ecotourism have been suggested (Landry and Taggart 2010). Every year scores of volunteers arrive to patrol the beach during the laying season (Campbell and Smith 2006), in order to tag adults, count nests, and watch over hatchlings. The basic tourism package has changed over the years from a "hard" (rugged) to a "soft" (more comfortable) experience (Harrison and Meletis 2010, Place 1991). With tourism has come a certain amount of disturbance for the human community as well as for the turtles (Harrison and Meletis 2010, Jacobson and Figueroa López 1994), such as waste management problems, trampling of beaches, and light from houses distracting hatchlings during their migration to the sea. Adjustments to the tourism offerings are made periodically in an adaptive management paradigm (Meletis and Campbell 2009, Harrison and Meletis 2010). Many segments of society are involved in the turtle conservation enterprise, including scientists, local guides, tourists, volunteers, citizen scientists, donors, funding agencies, government officials, and tourism-dependent business people. Tortuguero turtle conservation has drawn the attention of social scientists nearly to the extent that it has attracted the attention of biologists (Campbell 2002b).

Twenty years after the establishment of Parque Nacional Tortuguero, a parallel scenario arose in the Caribbean lowland forest along the Río San Juan with the great green macaws (*Ara ambiguus*). Where once there had existed large flocks of these birds in Costa Rica's Caribbean lowland feeding and nesting areas, their numbers were reduced to about 50 breeding pairs by the early 1990s (Monge et al. 2003). Other populations still existed in reproductively isolated groups from Honduras to Ecuador but the total species population was down to around 5,000 individuals when the species was designated as CITES I, critically endangered in 1985 (Müller 2000) and was then reduced further to about 3,700 individuals over the next ten years (Monge et al. 2009). The decline of this macaw in Costa Rica appeared to be due more to habitat destruction than to direct harvesting of the animals, although illegal collecting of eggs and hatchlings for the pet trade was documented. Land-clearing for cattle pastures and banana plantations resulted in a loss of 80% of suitable habitat for these birds by the 1990s (see Deforestation section, this chapter). Of particular concern was the logging of *Dipteryx panamensis* trees, a critical food and nesting tree for the great green macaw in Costa Rica (Stiles and Skutch 1989). Because of its extremely dense wood, this tree had been virtually immune to logging pressures until the introduction of a particularly strong carbon steel chainsaw blade. George V.N. Powell, a noted ornithologist who had been working in Monteverde with highland birds, was also working in the Caribbean lowlands with the great green macaw population. He sounded the alarm about the macaws (Powell et al. 1999) in much the same manner that Archie Carr had done with the green turtles at Tortuguero.

Starting in the early 1990s, there was much talk of biological corridors as a conservation strategy (Beier and Noss 1998, Newcomer 2002, Sánchez-Azofeifa et al. 2003). The concept of a Paseo Pantera (Path of the Panther), connecting protected areas from Mexico to Colombia, had been proposed by Archie Carr III (son of the Archie Carr who promoted the creation of Parque Nacional Tortuguero) and other scientists from the Wildlife Conservation Society (WCS) and the Caribbean Conservation Corporation (CCC), along with USAID and other funding agencies (Kaiser 2001). The GEF (Global Environmental Facility) was established in 1991 with World Bank funds to support conservation initiatives (Clémençon 2006, Ervine 2007, 2010, Griffiths 2004), including the Mesoamerican Biological Corridor (MBC; a reformulated version of the Paseo Pantera) (Miller et al. 2001, Minc et al. 2001). This concept was formalized in 1992 at the UN Conference on Environment and Development (UNCED) in Río de Janeiro (MBC 2002). In 1997, an official treaty to create the MBC was signed by Central American countries, with Mexico joining later. Each country was tasked with developing its own plans—Costa Rica chose to create a series of smaller corridors, each administered by a user group with the involvement of at

Fig. 16.8 Entrance to the Boca Tapada community in the Refugio Nacional Vida Silvestre Maquenque at the northern end of the San Juan–La Selva Biological Corridor. The community lies along the Río San Carlos near the confluence with the Río San Juan and the border with Nicaragua. This area is the breeding and nesting area of Costa Rica's remaining population of Great Green Macaws.
Photo by Deedra McClearn.

least one representative from MINAE (Ministerio de Ambiente y Energía). The treaty version of the MBC now had two principal aims: physical connectivity of protected areas and sustainable economic development in the communities around the protected areas (Finley-Brook 2007).

To return to the plight of the great green macaw, the San Juan–La Selva Biological Corridor (246,608 ha) was defined in 1997, as part of the larger national corridor initiative and also to protect the nesting area of the great green macaw. This is a mixed-use corridor, with private lands and agricultural fields. The northern portion of the corridor was set aside as the Refugio Nacional de Vida Silvestre Maquenque in 2005 after it became clear that the political will and the financing did not exist to make this a fully protected national park. The refuge is contiguous with the Nicaraguan Indio Maíz national park across the Río San Juan, and is the only remaining nesting area of the great green macaw in Costa Rica. The *Dipteryx panamensis* trees were protected from logging throughout the entire country in 2008 by executive decree. The management of the SJLS Corridor is handled through the Tropical Science Center (TSC) by Olivier Chassot (who had been a research assistant working for George Powell in 1994) and Guiselle Monge. The establishment of Maquenque and the SJLS Corridor can be considered a conservation success story in the sense that the decline of this population of macaws seems to have been arrested and their numbers may even be increasing after 15 years of protection. There is a modest biological field station at the Lagarto Lodge in Boca Tapada (on the edge of the Maquenque wildlife refuge) (Fig. 16.8) which hosts groups of Costa Rican and international students. Small-scale efforts to combine conservation and ecotourism have met with some success, but revenues from ecotourism have not radically transformed the local economy, which remains based in agriculture, particularly pineapples and yucca (Jackiewicz 2006). Ironically, the sort of development that the SJLS corridor communities are expecting and demanding is exactly the sort of development that has caused social and environmental problems in Tortuguero.

The publications pertaining to the SJLS Corridor, the Refugio Nacional de Vida Silvestre Maquenque, and the Costa Rican portions of the MBC include technical reports from governments and from NGOs (Lampman 2000, Sader et al. 2004), as well as popular articles about the great green macaw (Chassot and Monge 2002, Chassot et al. 2005, 2009, Monge et al 2005). The scientific literature contains a number of analyses of land use, notably those employing remote sensing (satellite and airplane overflight imagery) to assess forest coverage, forest biomass, and the distribution of particular tree species (Chun 2008, Kalacska et al. 2008, Wang et al. 2008).

Every park, wildlife refuge, and private reserve has its unique story, just as every conservation program has its champions and its critics (Fig. 16.9). Parque Nacional Tortuguero was created during the time of the early park system, just after the "closing of the frontier" in the 1960s (Augelli 1987). The SJLS Corridor came into being twenty years later, during an era with a much larger Costa Rican human population, no more freely available, government-sponsored land for small-holder farming, and the expectation of economic benefits from any and all biological conservation projects. A transfrontier mixed-use biological corridor is a very different beast from a discrete national park. After 20 years of existence and billions of dollars spent on the MBC, the GEF has drawn its share of criticism regarding the practical implementation of conservation programs and the neoliberal agenda that it represents (Clémençon 2006, Ervine 2007, 2010). Nonetheless, at a smaller spatial scale the SJLS Corridor and the Refugio Nacional de Vida Silvestre Maquenque can be credited with protecting critical habitat. The maintenance of forest cover in these areas will have a lasting impact on the conservation of Costa Rica's biodiversity. The first fifteen years represent a promising start towards the optimal scenario that biodiversity conservationists envision, but agricultural pressure and a local sense of unfulfilled economic expectations pose formidable risks to this vision.

Fig. 16.9 These images of Costa Rican wildlife stamps from the 1980s depict the green turtle and the great green macaw, and represent an increasing environmental awareness by both the Costa Rican government and the public in general.

Perspectives on the Future

We began this chapter with the theme of geography but quickly added the theme of time—the deep time of the geological perspective as well as the short time frame of such human enterprises as road building and forest clearing. The coauthors are scientists and conservationists and all are aware of the importance of spatial and temporal scale in our research activities. We also recognize that our own efforts are often limited in both space and time, but that the maintenance of intact ecosystems in today's world requires planning and implementation activities that are temporally prolonged and spatially extensive. The rapidity of the transformation of the Costa Rican Caribbean lowlands has been astonishing. We wonder whether there is common ground for an ecologist, a hotel owner in Maquenque, a turtle-tagging volunteer at Tortuguero, a local subsistence farmer, a NASA remote sensing expert, a MINAE park ranger, a Bribri landowner with a homegarden and a field of cassava, and a corporate responsibility executive at a banana company all to work together to ensure that ecosystem services and biodiversity are maintained in the long run. We certainly believe there are common ground and a common cause that will trigger our joint attention to address these issues successfully in an integrated manner in the near future.

References

Abhilash, P.C., and N. Singh. 2009. Pesticide use and application: an Indian scenario. *Journal of Hazardous Materials* 165: 1–12.

Acuña-Mesén, R.A. 1998. *Las Tortugas Continentales de Costa Rica*. San José, Costa Rica: Universidad de Costa Rica.

Aguilar, E., T.C. Peterson, P.R. Obando, R. Frutos, J.A. Retana, M. Solera, J. Soley, I.G. Garcia, R.M. Araujo, A.R. Santos, V.E. Valle, M. Brunet, L. Aguilar, L. Álvarez, M. Bautista, C. Castañon, L. Herrera, E. Ruano, J.J. Sinay, E. Sánchez, G.I.H. Oviedo, F. Obed, J.E. Salgado, J.L. Vázquez, M. Baca, M. Gutiérrez, C. Centella, J. Espinosa, D. Martínez, B. Olmedo, C.E.O. Espinoza, R. Nuñez, M. Haylock,

imoquinto informe Estado de la Nación en desarrollo humano sostenible, 1–26.

Calvo, J., V. Jiménez, and J.C. Solano. 2006. Estudio de monitoreo de cobertura forestal de Costa Rica 2005. II. Parte Coberturas de áreas reforestadas, plantadas con café y frutales en Costa Rica para el estudio de cobertura forestal. Proyecto Ecomercados, 114 Fondos GEF—Ecomercados convenio de donación TF FONAFIFO, ITCR, EOSL. San José, Costa Rica.

Calvo-Alvarado, J.C. 1990. The Costa Rican national conservation strategy for sustainable development: exploring the possibilities. *Environmental Conservation* 17: 355–58.

Campbell, L.M. 2002a. Conservation narratives in Costa Rica: conflict and co-existence. *Development and Change* 33: 29–56.

Campbell, L.M. 2002b. Science and sustainable use: views of marine turtle conservation experts. *Ecological Applications* 12: 1229–46.

Campbell, L.M., and C. Smith. 2006. What makes them pay? Understanding values of conservation volunteers in Tortuguero, Costa Rica. *Environmental Management* 38: 84–98.

Cannariato, K.G., and A.C. Ravelo. 1997. Pliocene-Pleistocene evolution of eastern tropical Pacific surface water circulation and thermocline depth. *Paleoceanography* 12: 805–20.

Capers, R.S., R.L. Chazdon, A. Redondo-Brenes, and A.B. Vílchez. 2005. Successional dynamics of woody seedling communities in wet tropical secondary forests. *Journal of Ecology* 93: 1071–84.

Caquineau, S., A. Gaudichet, L. Gómes, and M. Legrand. 2002. Mineralogy of Saharan dust transported over northwestern tropical Atlantic Ocean in relation to source regions. *Journal of Geophysical Research* 107: 1–14.

Cardelús, C.L., M.C. Mack, C. Woods, J. DeMarco, and K.K. Treseder. 2009. The influence of tree species on canopy soil nutrient status in a tropical lowland wet forest in Costa Rica. *Plant and Soil* 318: 47–61.

Carr, A.F. 1967. *So Excellente a Fishe: A Natural History of Sea Turtles*. Garden City, New York: Natural History Press.

Carr, A., M.H. Carr, and A.B. Meylan. 1978. The ecology and migrations of sea turtles, 7. The west Caribbean green turtle colony. *Bulletin of the American Museum of Natural History* 162: 1–46.

Carvalho, F. 2006. Agriculture, pesticides, food security and food safety. *Environmental Science and Policy* 9: 685–92.

Casals-Casas, C., and B. Desvergne. 2011. Endocrine disruptors: from endocrine to metabolic disruption. *Annual Review of Physiology* 73: 135–62.

Castillo, L.E., E. Martínez, C. Ruepert, C. Savage, M. Gilek, M. Pinnock, and E. Solís. 2006. Water quality and macroinvertebrate community response following pesticide applications in a banana plantation, Limón, Costa Rica. *Science of the Total Environment* 367: 418–32.

Castillo, L.E., C. Ruepert, and E. Solís. 2000. Pesticide residues in the aquatic environment of banana plantation areas in the North Atlantic zone of Costa Rica. *Environmental Toxicology and Chemistry* 19: 1942–50.

Caux, P.Y., R.A. Kent, G.T. Fan, and G.L. Stephenson. 1996. Environmental fate and effects of chlorothalonil: a Canadian perspective. *Critical Reviews in Environmental Science and Technology* 26: 45–93.

Cavelier, J. 1996. Environmental factors and ecophysiological processes along altitudinal gradients in wet tropical mountains. In S.S. Mulkey, R.L. Chazdon, and A.P. Smith, eds., *Tropical Forest Plant Ecophysiology*, 399–439. New York: Chapman and Hall.

Cavelier, J., S.J. Wright, and J. Santamaria. 1999. Effects of irrigation on fine root biomass and production, litterfall and trunk growth in a semideciduous lowland forest in Panama. *Plant and Soil* 211: 207–13.

Centro Centroamericano de Población. 2011. Estimación de la población de Costa Rica. Centro Centroamericano de Población, Instituto Nacional de Estadística y Censos, Universidad de Costa Rica. San José, Costa Rica. Accessed June 21, 2011, http://ccp.ucr.ac.cr.

Cernusak, L., K. Winter, C. Martínez, E. Correa, J. Aranda, M. García, C. Jaramillo, and B.L. Turner. 2011. Responses of legume versus nonlegume tropical tree seedlings to elevated CO_2 concentration. *Plant Physiology* 157: 372–85.

Chacón, I.A., and J. Montero. 2007. *Mariposas de Costa Rica*. Costa Rica: INBio.

Chakrabarty, P., and J.S. Albert. 2011. Not so fast: a new take on the Great American Biotic Interchange. In J.S. Albert and R.E. Reis, eds., *Historical Biogeography of Neotropical Freshwater Fishes*, 293–305. Berkeley: University of California Press.

Chalak, A., K. Balcombe, A. Bailey, and I. Fraser. 2008. Pesticides, preference heterogeneity and environmental taxes. *Journal of Agricultural Economics* 59: 537–54.

Chaloupka, M., K.A. Bjorndal, G.H. Balazs, and A.B. Bolten. 2008. Encouraging outlook for recovery of a once severely exploited marine megaherbivore. *Global Ecology and Biogeography* 17: 297–304.

Chapman, A., and K. Rosenberg. 1991. Diets of four sympatric Amazonian woodcreepers (Dendrocolaptidae). *Condor* 93: 904–15.

Chapman, P. 2007. *Bananas: How the United Fruit Company Shaped the World*. Edinburgh: Canongate Publishing.

Charboneau, J.P., and S.M. Koger. 2008. Plastics, pesticides and PBDEs: endocrine disruption and developmental disabilities. *Journal of Developmental and Physical Disabilities* 20: 115–28.

Chassot, O., and G. Monge. 2002. La biodiversidad amenazada en el Corredor Biológico San Juan–La Selva. *Ambientico* 107: 14–15.

Chassot, O., G. Monge, G. Powell, P. Wright, and S. Palminteri. 2005. San Juan–La Selva Biological Corridor, Costa Rica: a project of the Mesoamerican Biological Corridor for the protection of the great green macaw and its environment. San José, Costa Rica: Tropical Science Center.

Chassot, O., J.G. Monge, and G. Powell. 2009. Biología de la conservación de la lapa verde (1994–2009), 15 años de experiencia. San Pedro, Costa Rica: Centro Cientifico Tropical.

Chaves, A., D. Shea, and W.G. Cope. 2007. Environmental fate of chlorothalonil in a Costa Rican banana plantation. *Chemosphere* 69: 1166–74.

Chazdon, R.L., S. Careaga, C. Webb, and O. Vargas. 2003. Community and phylogenetic structure of reproductive traits of woody species in wet tropical forests. *Ecological Monographs* 73: 331–48.

Chazdon, R.L., and F.G. Coe. 1999. Ethnobotany of woody species in second-growth, old-growth, and selectively logged forests of northeastern Costa Rica. *Conservation Biology* 13: 1312–22.

Chazdon, R.L., R.K. Colwell, J.S. Denslow, and M.R. Guariguata. 1998. Statistical methods for estimating species richness of woody regeneration in primary and secondary rain forests of NE Costa Rica. In F. Dallmeier and J. Comiskey, eds., *Forest Biodiversity Research, Monitoring and Modeling: Conceptual Background and Old World Case Studies*, 285–309. Paris: Parthenon Publishing.

Chazdon, R.L., S.G. Letcher, M. van Breughel, M. Martínez-Ramos,

F. Bongers, and B. Finegan. 2007. Rates of change in tree communities of secondary tropical forests following major disturbances. *Philosophical Transactions of the Royal Society B: Biological Sciences* 362: 273–89.

Chazdon, R.L., A. Redondo-Brenes, and B.A. Vílchez. 2005. Effects of climate and stand age on annual tree dynamics in tropical second-growth rain forests. *Ecology* 86: 1808–15.

Chernyak, S.M., C.P. Rice, and L.L. McConnel. 1996. Evidence of currently-used pesticides in air, ice, fog, seawater and surface microlayer in the Bering and Chuckchi Seas. *Marine Pollution Bulletin* 32: 410–19.

Cherrett, J.M. 1986. History of the leaf-cutting ant problem. In C.S. Lofgren and R.K. van der Meer, eds., *Fire Ants and Leaf-cutting Ants, Biology and Management*, 10–17. Boulder: Westview Press.

Chomsky, A. 1996. *West Indian Workers and the United Fruit Company in Costa Rica 1870–1940*. Baton Rouge: Louisiana State University Press.

Chornook, K., and W. Guindon. 2008. *Walking with Wolf*. Hamilton, Ontario: Wandering Words Press.

Chou, C., J.D. Neelin, C.-A. Chen, and J.-Y. Tu. 2009. Evaluating the "rich-get-richer" mechanism in tropical precipitation change under global warming. *Journal of Climate* 22: 1982–2005.

Chun, S.L.M. 2008. The utility of digital aerial surveys in censusing *Dipteryx panamensis*, the key food and nesting tree of the endangered great green macaw (*Ara ambigua*) in Costa Rica. PhD diss., Duke University.

Clark, D.A. 2004. Sources or sinks? The response of tropical forests to current and future climate and atmospheric composition. *Philosophical Transactions of the Royal Society of London, Series B: Biological Sciences* 359: 477–91.

Clark, D.A., S. Brown, D.W. Kicklighter, J.Q. Chambers, J.R. Thomlinson, and J. Ni. 2001. Measuring net primary production in forests: concepts and field methods. *Ecological Applications* 11: 356–70.

Clark, D.A., and D.B. Clark. 2011. Assessing tropical forests' climatic sensitivities with long-term data. *Biotropica* 43: 31–40.

Clark, D.A., D.B. Clark, R. Sandoval, and M.V. Castro. 1995. Edaphic and human effects on landscape-scale distributions of tropical rain forest palms. *Ecology* 76: 2581–94.

Clark, D.A., S.C. Piper, C.D. Keeling, and D.B. Clark. 2003. Tropical rain forest tree growth and atmospheric carbon dynamics linked to interannual temperature variation during 1984–2000. *Proceedings of the National Academy of Sciences* 100: 5852–57.

Clark, D.B. 1996. Abolishing virginity. *Journal of Tropical Ecology* 12: 735–39.

Clark, D.B., and D.A. Clark. 2000. Landscape scale variation in forest structure and biomass in a tropical rain forest. *Forest Ecology and Management* 137: 185–98.

Clark, D.B., D.A. Clark, and S.F. Oberbauer. 2010. Annual wood production in a tropical rain forest in NE Costa Rica linked to climatic variation but not to increasing CO_2. *Global Change Biology* 16: 747–59.

Clark, D.B., M.W. Palmer, and D.A. Clark. 1999. Edaphic factors and the landscape-scale distributions of tropical rain forest trees. *Ecology* 80: 2662–75.

Clark, M.L., D.A. Roberts, and D.B. Clark. 2005. Hyperspectral discrimination of tropical rain forest tree species at leaf to crown scales. *Remote Sensing of Environment* 96: 375–98.

Clémençon, R. 2006. What future for the Global Environment Facility? *Journal of Environment Development* 15: 50–74.

Clement, C.R., and A.B. Junqueira. 2010. Between a pristine myth and an impoverished future. *Biotropica* 42: 534–36.

Coale, A.J. 1983. Recent trends in fertility in less developed countries. *Science* 221: 828–32.

Coates, A.G. 1997. The forging of Central America. In A.G. Coates, ed., *Central America: A Natural and Cultural History*, 1–37. New Haven, London: Yale University Press.

Cody, S., J.E. Richardson, V. Rull, C. Ellis, and R.T. Pennington. 2010. The Great American Biotic Interchange revisited. *Ecography* 33: 326–32.

Coley, P.D., and J.A. Barone. 1996. Herbivory and plant defenses in tropical forests. *Annual Review of Ecology and Systematics* 27: 305–35.

Colwell, R.K. 2005. EstimateS 7.5 User's Guide. http://viceroy.eeb.uconn.edu/estimates.

Colwell, R.K., G. Brehm, C.L. Cardelús, A.C. Gilman, and J.T. Longino. 2008. Global warming, elevational range shifts, and lowland biotic attrition in the wet tropics. *Science* 322: 258–61.

Colwell, R.K., C.X. Mao, and J. Chang. 2004. Interpolating, extrapolating, and comparing incidence-based species accumulation curves. *Ecology* 85: 2717–27.

Condit, R., N. Pitman, E.G. Leigh Jr., J. Chave, J. Terborgh, R.B. Foster, V.P. Núñez, S. Aguilar, R. Valencia, G. Villa, H.C. Muller-Landau, E. Losos, and S.P. Hubbell. 2002. Beta-diversity in tropical forest trees. *Science* 295: 666–69.

Cone, M. 2006. Silent snow: the slow poisoning of the Arctic. Grove Weidenfeld Publishers.

Cooper, J., and H. Dobson. 2007. The benefits of pesticides to mankind and the environment. *Crop Protection* 26: 1337–48.

CORBANA (Corporación Bananera Nacional CR). 2009. Estadísticas bananeras (online). http://www.corbana.co.cr.

Corella, O. 1999. Valoración de la base forestal de las plantaciones forestales y su contribución al abastecimiento de madera en la zona del Atlántico Norte de Costa Rica. M.Sc. thesis, CATIE. Turrialba, Costa Rica.

Cox, P.M., P.P. Harris, C. Huntingford, R.A. Betts, M. Collins, C.D. Jones, T.E. Jupp, J.A. Marengo, and C.A. Nobre. 2008. Increasing risk of Amazonian drought due to decreasing aerosol pollution. *Nature* 453: 212–15.

Cramer, W., A. Bondeau, S. Schaphoff, W. Lucht, B. Smith, and S. Sitch. 2004. Tropical forests and the global carbon cycle: impacts of atmospheric carbon dioxide, climate change and rate of deforestation. *Philosophical Transactions of the Royal Society of London, Series B* 359: 331–43.

Daly, G.L., Y.D. Lei, C. Teixeira, D.C.G. Muir, L.E. Castillo, L.M.M. Jantunen, and F. Wania. 2007a. Organochlorine pesticides in the soils and atmosphere of Costa Rica. *Environmental Science and Technology* 41: 1124–30.

Daly, G.L., Y.D. Lei, C. Teixeira, D.C.G. Muir, L.E. Castillo, and F. Wania. 2007b. Accumulation of current-use pesticides in neotropical montane forests. *Environmental Science and Technology* 41: 1118–23.

De Camino, R., O. Segura, L.G. Arias, and I. Pérez. 2000. Costa Rica: forest strategy and the evolution of land use. Washington, DC: World Bank.

De la Cruz, E., V Bravo-Durán, F. Ramírez, and L.E. Castillo. 2014. Environmental hazards associated with pesticide import into Costa Rica, 1977–2009. *Journal of Environmental Biology* 35: 44–55.

García-Robledo, C., and C.C. Horvitz. 2009. Host plant scents attract rolled-leaf beetles to Neotropical gingers in a Central American tropical rain forest. *Entomologia Experimentalis et Applicata* 131: 115–20.

García-Robledo, C., and C.C. Horvitz. 2011. Experimental demography and the vital rates of generalist and specialist insect herbivores on novel and native host plants. *Journal of Animal Ecology* 80: 976–89.

García-Robledo, C., and C.C. Horvitz. 2012a. Jack of all trades masters novel host plants: positive genetic correlations in specialist and generalist insect herbivores expanding their diets to novel hosts. *Journal of Evolutionary Biology* 25: 38–53.

García-Robledo, C., and C.C. Horvitz. 2012b. Parent-offspring conflicts, "optimal bad motherhood" and the "mother knows best" principles in insect herbivores colonizing novel host plants. *Ecology and Evolution* 25: 38–53.

García-Robledo, C., and E.K. Kuprewicz. 2009. Vertebrate fruit removal and ant seed dispersal in the Neotropical ginger *Renealmia alpinia* (Zingiberaceae). *Biotropica* 41: 209–14.

García-Robledo, C., and C.L. Staines. 2008. Herbivory in gingers from latest Cretaceous to present: is the ichnogenus *Cephaloleichnites* (Hispinae, Coleoptera) a rolled-leaf beetle? *Journal of Paleontology* 82: 1035–37.

García-Robledo, C., D.L. Erickson, C.L. Staines, T.L. Erwin, and W. J. Kress. 2013. Tropical plant–herbivore networks: reconstructing species interactions using DNA barcodes. *PLOS ONE* 8: e52967. doi:52910.51371/journal.pone.0052967.

García-Serrano, C.R., and J.P. Del Monte. 2004. The use of tropical forest (agroecosystems and wild plant harvesting) as a source of food in the Bribri and Cabecar cultures in the Caribbean Coast of Costa Rica. *Economic Botany* 58: 58–71.

Garrigues, R., and R. Dean. 2007. *The Birds of Costa Rica: A Field Guide*. Ithaca, NY: Cornell University Press.

Garrison, V.H., E.A. Shinn, W.T. Foreman, D.W. Griffin, C.W. Holmes, C.A. Kellogg, M.S. Majewski, L.L. Richardson, K.B. Ritchie, and G.W. Smith. 2003. African and Asian dust: from desert soils to coral reefs. *BioScience* 53: 469–80.

Genereux, D., and C. Pringle. 1997. Chemical mixing model of streamflow generation at La Selva Biological Station, Costa Rica. *Journal of Hydrology* 199: 319–30.

Gentry, A.H. 1982. Neotropical floristic diversity: phytogeographical connections between Central and South America, Pleistocene climatic fluctuations, or an accident of the Andean Orogeny? *Annals of the Missouri Botanical Garden* 69: 557–93.

Gentry, G.L., and L.A Dyer. 2002. On the conditional nature of Neotropical caterpillar defenses against their natural enemies. *Ecology* 83: 3108–19.

Goebel, T., M.R. Waters, and D.H. O'Rourke. 2008. The late Pleistocene dispersal of modern humans in the Americas. *Science* 319: 1497.

Gotelli, N.J., and R.K. Colwell. 2001. Quantifying biodiversity: procedures and pitfalls in the measurement and comparison of species richness. *Ecology Letters* 4: 379–91.

Goulding, M. 1980. *The Fishes and the Forest: Explorations in Amazonian Natural History*. Berkeley: University of California Press.

Graham, A. 1987. Miocene communities and paleoenvironments of Southern Costa Rica. *American Journal of Botany* 74: 1501–18.

Graham, A. 1992. Utilization of the isthmian land bridge during the Cenozoic: paleobotanical evidence for timing, and the selective influence of altitudes and climate. *Review of Palaeobotany and Palynology* 72: 119–28.

Graham, A., and D.L. Dilcher. 1998. Studies in Neotropical paleobotany. XII. A palynoflora from the Pliocene Río Banano formation of Costa Rica and the Neogene vegetation of Mesoamerica. *American Journal of Botany* 85: 1426–38.

Greiner, S., M. Nagy, M. Knörnschild, H. Hofer, and C.C. Voigt. 2013. Sex-biased senescence in a polygynous bat species. *Ethology* 119: 1–9.

Griffin, D.W. 2007. Atmospheric movement of microorganisms in clouds of desert dust and implications for human health. *Clinical Microbiology Reviews* 20: 459–77.

Griffiths, T. 2004. Help or hindrance?: the Global Environmental Facility, biodiversity conservation, and indigenous peoples. *Cultural Survival Quarterly* 28: 1–11.

Guariguata, M.R., R.L. Chazdon, J.S. Denslow, J.M. Dupuy, and L. Anderson. 1997. Structure and floristics of secondary and old-growth forest stands in lowland Costa Rica. *Plant Ecology* 132: 107–20.

Guariguata, M.R., and J.M. Dupuy. 1997. Forest regeneration in abandoned logging roads in lowland Costa Rica. *Biotropica* 29: 15–28.

Guariguata, M.R., J. Rosales, and B. Finegan. 2000. Seed removal and fate in two selectively logged lowland forests with contrasting protection levels. *Conservation Biology* 14: 1046–54.

Guillén, A.L. 1993. Inventario comercial y análisis silvicultural de bosques húmedos secundarios en la region Huétar Norte de Costa Rica. Thesis, Instituto Tecnológico de Costa Rica. Cartago, Costa Rica.

Guyer, C., and M.A. Donnelly. 1990. Length-mass relationships among an assemblage of tropical snakes in Costa Rica. *Journal of Tropical Ecology* 6: 65–76.

Guyer, C., and M.A. Donnelly. 2005. *Amphibians and Reptiles of La Selva, Costa Rica, and the Caribbean Slope*. Berkeley: University of California Press.

Haines, B.L. 1978. Element and energy flows through colonies of the leaf-cutting ant, *Atta colombica*, in Panama. *Biotropica* 10: 270–77.

Hamilton, D., V. Molina, P. Bosques, and G.V.N. Powell. 2003. El estatus del pájaro campana (*Procnias tricarunculata*): un ave en peligro de extinción. *Zeledonia* 7: 15–24.

Hammel, B.E., and N.A. Zamora. 1993. *Ruptiliocarpon* (Lepidobotryaceae): a new arborescent genus and tropical American link to Africa, with a reconsideration of the family. *Novon* 3: 408–17.

Harris, R.N., G. Spinelli, C.R. Ranero, I. Grevemeyer, H. Villinger, and U. Barckhausen. 2010. Thermal regime of the Costa Rican convergent margin: 2. Thermal models of the shallow Middle America subduction zone offshore Costa Rica. *Geochemistry, Geophysics, Geosystems* 11: 1–22.

Harrison, E., and Z. Meletis. 2010. Tourists and turtles: searching for a balance in Tortuguero, Costa Rica. *Conservation and Society* 8: 26–43.

Hartshorn, G.S. 1983. Plants: introduction. In D.H. Janzen, ed., *Costa Rican Natural History*, 118–57. Chicago: University of Chicago Press.

Hartshorn, G.S. 2006. Understanding tropical forests. *BioScience* 56: 264–65.

Hartshorn, G.S., and B.E. Hammel. 1994. Vegetation types and floristic patterns. In L.A. McDade, K.S. Bawa, H. Hespenheide, and G.S. Hartshorn, eds., *La Selva: Ecology and Natural History of a Neotropical Rain Forest*, 73–89. Chicago: University of Chicago Press.

Hasbún, C.R., A. Gómez, G. Köhler, and D.H. Lunt. 2005. Mitochondrial DNA phylogeography of the Mesoamerican spiny-tailed lizards (*Ctenosaura quinquecarinata* complex): historical biogeography, species status and conservation. *Molecular Ecology* 14: 3095–3107.

Haug, G.H., R. Tiedemann, and L.D. Keigwin. 2004. How the Isthmus of Panama put ice in the Arctic. *Oceanus* 42: 95–98.

Haug, G.H., R. Tiedemann, R. Zahn, and A.C. Ravelo. 2001. Role of Panama uplift on oceanic freshwater balance. *Geology* 29: 207–10.

Heckenberger, M.J., A. Kuikuro, U.T. Kuikuro, J.C. Russell, M. Schmidt, C. Fausto, and B. Franchetto. 2003. Amazonia 1492: pristine forest or cultural parkland? *Science* 301: 1710–14.

Heithaus, E.R., and T.H. Fleming. 1978. Foraging movements of a frugivorous bat, *Carollia perspicillata* (Phyllostomatidae). *Ecological Monographs* 48: 127–43.

Herrera, B., and B. Finegan. 1997. Substrate conditions, foliar nutrients and the distributions of two canopy tree species in a Costa Rican secondary rain forest. *Plant and Soil* 191: 259–67.

Hespenheide, H.A. 1985. The visitor fauna of extrafloral nectaries of *Byttneria aculeata* (Sterculiaceae): relative importance and roles. *Ecological Entomology* 10: 191–204.

Hespenheide, H.A. 1991. Bionomics of leaf-mining insects. *Annual Review of Entomology* 36: 535–60.

Hiremat, A.J., J.J. Ewel, and T.G. Cole. 2002. Nutrient use efficiency in three fast-growing tropical trees. *Forest Science* 48: 662–71.

Hirth, H.F. 1963. Some aspects of the natural history of *Iguana iguana* on a tropical strand. *Ecology* 44: 613–15.

Hoernle, K., D.L. Abt, K.M. Fischer, H. Nichols, F. Hauff, G.A. Abers, P. Bogaard, K. van den Heydolph, G. Alvarado, M. Protti, and W. Strauch. 2008. Arc-parallel flow in the mantle wedge beneath Costa Rica and Nicaragua. *Nature* 451: 1094–97.

Hoernle, K., P. Bogaard, R. van den Werner, B. Lissinna, F. Hauff, G. Alvarado, and D. Garbe-Schöngberg. 2002. Missing history (16–71 Ma) of the Galapagos hotspot: implications for the tectonic and biological evolution of the Americas. *Geology* 30: 795–98.

Hoffman, D.M. 2011. Do global statistics represent local reality and should they guide conservation policy?: examples from Costa Rica. *Conservation Society* 9: 16–24.

Holl, K.D. 2008. Are there benefits of bat roosts for tropical forest restoration? *Conservation Biology* 22: 1090–91.

Holtum, J.A.M., and K. Winter. 2010. Elevated CO_2 and forest vegetation: more a water issue than a carbon issue? *Functional Plant Biology* 37: 694–702.

Horn, M.H. 1997. Evidence for dispersal of fig seeds by the fruit-eating characid fish *Brycon guatemalensis* Regan in a Costa Rican tropical rain forest. *Oecologia* 109: 259–64.

Horn, S.P. 2006. Pre-Columbian maize agriculture in Costa Rica: pollen and other evidence from lake and swamp sediments. In J. Staller, R. Tykot, and B. Benz, eds., *Histories of Maize: Multidisciplinary Approaches to the Prehistory, Biogeography, Domestication, and Evolution of Maize*, 367–80. San Diego, CA: Elsevier Press.

Horn, S.P., and L.M. Kennedy. 2001. Pollen evidence of maize cultivation 2700 years ago at La Selva Biological Station, Costa Rica. *Biotropica* 33: 191–96.

Horn, S.P., and R.L. Sanford. 1992. Holocene fires in Costa Rica. *Biotropica* 24: 354–61.

Horn, S.P., R.L.J. Sanford, D. Dilcher, T. Lott, P.R. Renne, M.C. Wiemann, D. Cozadd, and O. Vargas. 2003. Pleistocene plant fossils in and near La Selva Biological Station, Costa Rica. *Biotropica* 35: 434–41.

Horvitz, C.C. 1981. Analysis of how ant behaviors affect germination in a tropical myrmecochore *Calathea microcephala* (P and E) Koernicke (Marantaceae). Microsite selection and aril removal by neotropical ants, *Odontomachus*, *Pachycondyla*, and *Solenopsis* (Formicidae). *Oecologia* 51: 47–52.

Howard, A.F. 1995. Price trends for stumpage and selected agricultural products in Costa Rica. *Forest Ecology and Management* 75: 101–10.

Howard, A.F., and J. Valerio. 1996. Financial returns from sustainable forest management and selected agricultural land-use options in Costa Rica. *Forest Ecology and Management* 81: 35–49.

Howe, H.F. 1977. Bird activity and seed dispersal of a tropical wet forest tree. *Ecology* 58: 539–50.

Howe, H.F., and J. Smallwood. 1982. Ecology of seed dispersal. *Annual Review of Ecology and Systematics* 13: 201–28.

Huey, R.B., C.A. Deutsch, J.J. Tewksbury, L.J. Vitt, P.E. Hertz, H.J.A. Perez, and T. Garland. 2009. Why tropical forest lizards are vulnerable to climate warming. *Proceedings of the Royal Society B: Biological Sciences* 276: 1939–48.

Hughen, K.A., T.I. Eglinton, L. Xu, and M. Makou. 2004. Abrupt tropical vegetation response to rapid climate changes. *Science* 304: 1955–59.

Humbert, S., M. Margni, R. Charles, O.M. Torres Salazar, A.L. Quirós, O. Jolliet. 2007. Toxicity assessment of the main pesticides used in Costa Rica. *Agriculture, Ecosystems and Environment* 118: 183–90.

IPCC. 2007. Summary for policymakers. In S. Solomon, D. Qin, M. Manning, Z. Chen, M. Marquis, K.B. Averyt, M. Tignor, and H.L. Miller, eds., *Climate Change 2007: The Physical Sciences*. Contribution of Working Group 1 to the Fourth Assessment Report of the Intergovernmental Panel on Climate Change. Cambridge, UK: Cambridge University Press.

Iriarte, J. 2011. Narrowing the gap: exploring the diversity of early food-production economies in the Americas. *Current Anthropology* 50. 677–80.

Iriarte, J., and E. Alonso. 2009. Phytolith analysis of selected native plants and modern soils from southeastern Uruguay and its implications for paleoenvironmental and archeological reconstruction. *Quaternary International* 193: 99–123.

Iriarte, J., I. Holst, O. Marozzi, C. Listopad, E. Alonso, A. Rinderknecht, and J. Montaña. 2004. Evidence for cultivar adoption and emerging complexity during the mid-Holocene in the La Plata basin. *Nature* 432: 614–17.

Irschick, D.J., L.J. Vitt, P.A. Zani, and J.B. Losos. 1997. A comparison of evolutionary radiations in mainland and Caribbean *Anolis* lizards. *Ecology* 78: 2191–203.

Islebe, G., H. Hooghiemstra, M. Brenner, J.H. Curtis, and D.A. Hodell. 1996. A Holocene vegetation history from lowland Guatemala. *Holocene* 6: 265–71.

Islebe, G., and O. Sánchez. 2002. History of Late Holocene vegetation at Quintana Roo, Caribbean coast of Mexico. *Plant Ecology* 160: 187–92.

IUCN. 2006. The 2006 IUCN Red List of Threatened Species. www.iucnredlist.org.

IUCN. 2010. The 2010 IUCN Red List of Threatened Species. www.iucnredlist.org.

Jackiewicz, E.L. 2006. Community-centered globalization: modernization under control in rural Costa Rica. *Latin American Perspectives* 33: 136–46.

Jackson, J.B.C. 1997. Reefs since Columbus. *Coral Reefs* 16(suppl.): S23–S32.

Jacobson, S.K., and A. Figueroa López. 1994. Biological impacts of ecotourism: tourists and nesting turtles in Tortuguero National Park, Costa Rica. *Wildlife Society Bulletin* 22: 414–19.

Jacobson, S.K., and R. Robles. 1992. Ecotourism, sustainable development, and conservation education: development of a tour guide training program in Tortuguero, Costa Rica. *Environmental Management* 16: 701–13.

Jaga, K., and C. Dharmani. 2003. Global surveillance of DDT and DDE levels in human tissues. *International Journal of Occupational Medicine and Environmental Health* 16: 7–20.

Jain, S., and L.S. Collins. 2007. Trends in Caribbean paleoproductivity related to the Neogene closure of the Central American Seaway. *Marine Micropaleontology* 63: 57–74.

Jansen, M., A. Coors, R. Stoks, and L. De Meester. 2011. Evolutionary ecotoxicology of pesticide resistance: a case study in *Daphnia*. *Ecotoxicology* 20: 543–51.

Jentsch, A., J. Kreyling, and C. Beierkuhnlein. 2007. A new generation of climate-change experiments: events, not trends. *Frontiers in Ecology and the Environment* 5: 365–74.

Johnson, C.N. 2002. Determinants of loss of mammal species during the Late Quaternary 'megafauna' extinctions: life history and ecology, but not body size. *Proceedings of the Royal Society of London, Series B* 269: 2221–27.

Johnson, C.N. 2009. Ecological consequences of Late Quaternary extinctions of megafauna. *Proceedings of the Royal Society of London, Series B* 276: 2509–19.

Johnson, D.M. 2004a. Life history and demography of *Cephaloleia fenestrata* (Hispinae : Chrysomelidae : Coleoptera). *Biotropica* 36: 352–61.

Johnson, D.M. 2004b. Source-sink dynamics in a temporally heterogeneous environment. *Ecology* 85: 2037–45.

Johnson, D.M. 2005. Metapopulation models: an empirical test of model assumptions and evaluation methods. *Ecology* 86: 3088–98.

Johnson, D.M., and C.C. Horvitz. 2005. Estimating postnatal dispersal: tracking the unseen dispersers. *Ecology* 86: 1185–90.

Johnson, M., and S. Clisby. 2009. Naturalising distinctions: the contested field of environmental relations in Costa Rica. *Landscape Research* 34: 171–87.

Joyce, A.T. 2006. *Land use change in Costa Rica: 1966–2006, as influenced by social, economic, political, and environmental factors*. San José, Costa Rica: Lil SA.

Kahn, M. 2009. Economic evaluation of health cost of pesticide use: willingness to pay method. *Pakistan Development Review* 48: 459–70.

Kaiser, J. 2001. Bold corridor project confronts political reality. *Science* 293: 2196–99.

Kalacska, M., G.A. Sánchez-Azofeifa, B. Rivard, J.C. Calvo-Alvarado, and M. Quesada. 2008. Baseline assessment for environmental services payments from satellite imagery: a case study from Costa Rica and Mexico. *Journal of Environmental Management* 88: 348–59.

Kallos, G., A. Papadopoulos, P. Katsafados, and S. Nickovic. 2006. Transatlantic Saharan dust transport: model simulation and results. *Journal of Geophysical Research* 111: 1–11.

Kaspari, M. 1993. Removal of seeds from Neotropical frugivore droppings: ant responses to seed number. *Oecologia* 95: 81–88.

Keigwin, L. 1982. Isotopic paleoceanography of the Caribbean and East Pacific: role of Panama uplift in Late Neogene time. *Science* 217: 350–52.

Kelm, D.H., K.R. Wiesner, and O. von Helversen. 2008. Effects of artificial roosts for frugivorous bats on seed dispersal in a Neotropical forest pasture mosaic. *Conservation Biology* 22: 733–41.

Kendall, R.I., and P.N. Smith. 2003. Wildlife toxicology revisited. *Environmental Science and Technology* 37: 178A–183A.

Kennedy, L.M., and S.P. Horn. 1997. Prehistoric maize cultivation at the La Selva Biological Station Costa Rica. *Biotropica* 29: 368–70.

Kennedy, L.M., and S.P. Horn. 2008. A Late Holocene pollen and charcoal record from La Selva Biological Station, Costa Rica. *Biotropica* 40: 11–19.

Keyeux, G., C. Rodas, N. Gelvez, and D. Carter. 2002. Possible migration routes into South America deduced from mitochondrial DNA studies in Colombian Amerindian populations. *Human Biology* 74: 211–33.

Kleber, M., L. Schwendenmann, E. Veldkamp, and R. Rößner Jahn. 2007. Halloysite versus gibbsite: silicon cycling as a pedogenetic process in two lowland neotropical rain forest soils of La Selva, Costa Rica. *Geoderma* 138: 1–11.

Koeppel, D. 2008. *Banana: The Fate of the Fruit That Changed the World*. New York: Penguin.

Körner, C. 2009. Responses of humid tropical trees to rising CO_2. *Annual Review of Ecology, Evolution, and Systematics* 40: 61–79.

Kraenzel, M., A. Castillo, T. Moore, and C. Potvin. 2003. Carbon storge of harvest-age teak (*Tectona grandis*) plantations in Panama. *Forest Ecology and Management* 173: 213–25.

Krech, S., III. 1999. *The Ecological Indian*. New York: W.W. Norton.

Kubitzki, K., and A. Ziburski. 1994. Seed dispersal in flood plain forests of Amazonia. *Biotropica* 26: 30–43.

Kumar, B.M., and P.K.R. Nair. 2010. *Tropical Homegardens*. The Netherlands: Springer.

Kuprewicz, E.K. 2010. The effects of large terrestrial mammals on seed fates, hoarding, and seedling survival in a Costa Rican rain forest. PhD diss., University of Miami.

Kuprewicz, E.K. 2013. Mammal abundances and seed traits control the seed dispersal and predation roles of terrestrial mammals in a Costa Rican forest. *Biotropica* 45: 333–42.

Kuprewicz, E.K., and C. García-Robledo. 2010. Mammal and insect predation of chemically and structurally defended *Mucuna holtonii* (Fabaceae) seeds in a Costa Rican rain forest. *Journal of Tropical Ecology* 26: 263–69.

Kürschner, W.M., Z. Kvacek, and D.L. Dilcher. 2008. The impact of Miocene atmospheric fluctuations on climate and the evolution of terrestrial ecosystems. *Proceedings of the National Academy of Sciences* 105: 449–53.

Lagueux, C. 1998. Marine turtle fishery of Caribbean Nicaragua: human use patterns and harvest trends. Ph.D. diss., University of Florida.

Lahanas, P.N., K.A. Bjorndal, A.B. Bolten, S.E. Encalada, M.M. Miyamoto, R.A. Valverde, and B.W. Bowen. 1998. Genetic composition of a green turtle (*Chelonia mydas*) feeding ground population: evidence for multiple origins. *Marine Biology* 130: 345–52.

Lampman, S.C. 2000. Environmental change in the Mesoamerican Biological Corridor. *Journal of Forestry* 98: S3.

Landry, M.S., and C.T. Taggart. 2010. "Turtle watching" conservation guidelines: green turtle (*Chelonia mydas*) tourism in nearshore coastal environments. *Biodiversity and Conservation* 19: 305–12.

Lane, C.S., C.I. Mora, S.P. Horn, K.H. Orvis. 2008. Sensitivity of bulk sedimentary stable carbon isotopes to prehistoric forest clearance and maize agriculture. *Journal of Archaeological Science* 35: 2119–32.

Lansing, D.M. 2011. Realizing carbon's value: discourse and calculation in the production of carbon forestry offsets in Costa Rica. *Antipode* 43: 731–53.

LaVal, R.K. 1977. Notes on some Costa Rican bats. *Brenesia* 10/11: 77–83.

LaVal, R.K., and B. Rodríguez-Herrera. 2002. *Murciélagos de Costa Rica: Bats*. Santo Domingo de Heredia, Costa Rica: Instituto Nacional de Biodiversidad (INBio).

Lavin, M., B.P. Schrire, G. Lewis, R.T. Pennington, A. Delgado-Salinas, M. Thulin, C.E. Hughes, A. Beyra Matos, and M.F. Wojciechowski. 2004. Metacommunity process rather than continental tectonic history better explains geographically structured phylogenies in legumes. *Philosophical Transactions of the Royal Society B: Biological Sciences* 359: 1509–22.

Lawler, J.J., S.L. Shafer, and A.R. Blaustein. 2009. Projected climate impacts for the amphibians of the Western Hemisphere. *Conservation Biology* 24: 38–50.

Le, M., and W.P. McCord. 2008. Phylogenetic relationships and biogeographical history of the genus *Rhinoclemmys* Fitzinger, 1835 and the monophyly of the turtle family Geoemydidae (Testudines: Testudinoidea). *Zoological Journal of the Linnean Society* 153: 751–67.

Le Corff, J., and C.C. Horvitz. 1995. Dispersal of seeds from chasmogamous and cleistogamous flowers in an ant-dispersed Neotropical herb. *Oikos* 73: 59–64.

Lee, D.N.B., and D.J. Snepenger. 1992. An ecotourism assessment of Tortuguero, Costa Rica. *Annals of Tourism Research* 19: 367–70.

Lefever, H.G. 1992. Turtle bogue: Afro-Caribbean life and culture in a Costa Rican village. Selingsgrove, London.

Leigh, E.G. 1999. *Tropical Forest Ecology: A View from Barro Colorado Island*. Oxford: Oxford University Press.

Lessios, H.A. 2008. The Great American Schism: divergence of marine organisms after the rise of the Central American Isthmus. *Annual Review of Ecology, Evolution, and Systematics*.

Letcher, S.G. 2010. Phylogenetic structure of angiosperm communities during tropical forest succession. *Proceedings of the Royal Society of London, Series B* 277: 97–104.

Letcher, S.G., and R.I. Chazdon. 2009a. Rapid recovery of biomass, species richness, and species composition in a forest chronosequence in northeastern Costa Rica. *Biotropica* 41: 608–17.

Letcher, S.G., and R.I. Chazdon. 2009b. Lianas and self-supporting plants during tropical forest succession. *Forest Ecology and Management* 257: 2150–56.

Levey, D.J. 1987. Seed size and fruit-handling techniques of avian frugivores. *American Naturalist* 129: 471–85.

Levey, D.J., and M.M. Byrne. 1993. Complex ant plant interactions: rainforest ants as secondary dispersers and postdispersal seed predators. *Ecology* 74: 1802–12.

Levey, D.J., T.C. Moermond, and J.S. Denslow. 1994. Frugivory: an overview. In L.A. McDade, K.S. Bawa, H.A. Hespenheide, and G.S. Hartshorn, eds., *La Selva: Ecology and Natural History of a Neotropical Rain Forest*, 282–94. Chicago: University of Chicago Press.

Levey, D.J., and F.G. Stiles. 1994. Birds: ecology, behavior, and taxonomic affinities. In L.A. McDade, K.S. Bawa, H.A. Hespenheide, and G.S. Hartshorn, eds., *La Selva: Ecology and Natural History of a Neotropical Rainforest*, 217–28. Chicago: University of Chicago Press.

Lewis, S.L., J. Lloyd, S. Sitch, E.T.A. Mitchard, and W.F. Laurance. 2009. Changing ecology of tropical forests: evidence and drivers. *Annual Review of Ecology, Evolution, and Systematics* 40: 529–49.

Lieberman, D., M. Lieberman, R. Peralta, and G. Hartshorn. 1996. Tropical forest structure and composition on a large-scale altitudinal gradient in Costa Rica. *Journal of Ecology* 84: 137–52.

Lieberman, S.S. 1986. Ecology of the leaf litter herpetofauna of a neotropical rain forest: La Selva, Costa Rica. *Acta Zoológica Mexicana* 15: 1–72.

Liess, M., and P. von der Ohe. 2005. Analyzing effects of pesticides on invertebrate communities in streams. *Environmental Toxicology and Chemistry* 24: 954–65.

Ling, A., and M.O. Jarocki. 2003. Pesticide justice. *Multinational Monitor* 24: 6.

Lips, K.R., F. Brem, R. Brenes, J.D. Reeve, R.A. Alford, J. Voyles, C. Carey, L. Livo, A.P. Pessier, and J.P. Collins. 2006. Emerging infectious disease and the loss of biodiversity in a Neotropical amphibian community. *Proceedings of the National Academy of Sciences* 103: 3165–70.

Lips, K.R., P.A. Burrowes, J.R. Mendelson, and G. Parra-Olea. 2005. Amphibian declines in Latin America: widespread population declines, extinctions, and impacts. *Biotropica* 37: 163–65.

Lips, K.R., J. Diffendorfer, J.R. Mendelson, and M.W. Sears. 2008. Riding the wave: reconciling the roles of disease and climate change in amphibian declines. *PLOS Biology* 6: 441–54.

Loiselle, B.A., and J.G. Blake. 1999. Dispersal of melastome seeds by fruit-eating birds of tropical forest understory. *Ecology* 80: 330–36.

Longino, J.T., J. Coddington, and R.K. Colwell. 2002. The ant fauna of a tropical rain forest: estimating species richness three different ways. *Ecology* 83: 689–702.

Lonsdale, P. 2005. Creation of the Cocos and Nazca plates by fission of the Farallon Plate. *Tectonophysics* 404: 237–64.

Lorenz, M.W. 2009. Migration and trans-Atlantic flight of locusts. *Quaternary International* 196: 4–12.

Lovelock, C.E., and J.J. Ewel. 2005. Links between tree species, symbiotic fungal diversity and ecosystem functioning in simplified tropical ecosystems. *New Phytologist* 167: 219–28.

Lowman, M.D. 1995. Herbivory as a canopy process in rain forest trees. In M.D. Lowman and N.M. Nadkarni, eds., *Forest Canopies*. New York: Academic Press.

Lozano-García, M.d.S., M. Caballero, B. Ortega, A. Rodriguez, and S. Sosa. 2007. Tracing the effects of the Little Ice Age in the tropical lowlands of eastern Mesoamerica. *Proceedings of the National Academy of Sciences* 104: 16200–16203.

Lubick, N. 2007. DDT's resurrection. *Environmental Science and Technology* (September 15): 6323–25.

Lumpkin, H.A., and W.A. Boyle. 2009. Effects of forest age on fruit composition and removal in tropical bird-dispersed understorey trees. *Journal of Tropical Ecology* 25: 515–22.

Lutz, E. 1993. Interdisciplinary fact-finding on current deforestation in Costa Rica. The World Bank Environmental Department, Environmental Working Paper No. 61.

Lyons, S.K., F.A. Smith, and J.H. Brown. 2004. Of mice, mastodons and men: human-mediated extinctions on four continents. *Evolutionary Ecology Research* 6: 339–58.

MacFadden, B.J. 2006. Extinct mammalian biodiversity of the ancient New World tropics. *Trends in Ecology and Evolution* 21: 157–65.

MacMillan, I., P.B. Gans, and G. Alvarado. 2004. Middle Miocene to present plate tectonic history of the southern Central American Volcanic Arc. *Tectonophysics* 392: 325–48.

Malhi, Y., L.E.O.C. Aragão, D. Galbraith, C. Huntingford, R. Fisher, P. Zelazowski, S. Sitch, C. McSweeney, and P. Meir. 2009. Exploring the likelihood and mechanism of a climate-change-induced dieback of the Amazon rainforest. *Proceedings of the National Academy of Sciences* 106: 20610–15.

Malhi, Y., and O.L. Phillips. 2004. Tropical forests and global atmospheric change: a synthesis. *Philosophical Transactions of the Royal Society of London, Series B* 359: 549–55.

Malhi, Y., O.L. Phillips, J. Lloyd, et al. 2002. An international network to monitor the structure, composition and dynamics of Amazonian forests (RAINFOR). *Journal of Vegetation Science* 13: 439–50.

Malhi, Y., and J. Wright. 2004. Spatial patterns and recent trends in the climate of tropical rainforest regions. *Philosophical Transactions of the Royal Society of London, Series B* 359: 311–29.

Mann, C.C. 2005. *1491: New Revelations of the Americas before Columbus*. New York: Vintage Books.

Marengo, J.A., J. Tomasella, L.M. Alves, W.R. Soares, and D.A. Rodríguez. 2011. The drought of 2010 in the context of historical droughts in the Amazon region. *Geophysical Research Letters* 38: 12703–7.

Margni, M., D. Rossier, P. Crettaz, and O. Jolliet. 2002. Life cycle impact assessment of pesticides on human health and ecosystems. *Agriculture, Ecosystems and Environment* 93: 379–92.

Marquardt, S. 2002. Pesticides, parakeets, and unions in the Costa Rican banana industry, 1938–1962. *Latin American Research Review* 37: 3–36.

Marques, M.C.M., M.D. Swaine, and D. Liebsch. 2011. Diversity distribution and floristic differentiation of the coastal lowland vegetation: implications for the conservation of the Brazilian Atlantic forest. *Biodiversity and Conservation* 20: 153–68.

Marquis, R. 1990. Genotypic variation in leaf damage in *Piper arieianum* (Piperaceae) by a multispecies assemblage of herbivores. *Evolution* 44: 104–20.

Marquis, R. 1992. A bite is a bite is a bite?: Constraints on response to folivory in *Piper arieianum* (Piperaceae). *Ecology* 73: 143–52.

Marquis, R.J. 1984. Leaf herbivores decrease fitness of a tropical plant. *Science* 226: 537–39.

Marquis, R.J., and E. Braker. 1994. Plant–herbivore interactions: diversity, specificity and impact. In L.A. McDade, K.S. Bawa, H.A. Hespenheide, and G.S. Hartshorn, eds., *La Selva: Ecology and Natural History of a Neotropical Rain Forest*, 261–81. Chicago: University of Chicago Press.

Martin, P.S. 1967. Prehistoric overkill. In P.S. Martin and H.E. Wright Jr., eds., *Pleistocene Extinctions: The Search for a Cause*, 75–120. New Haven: Yale University Press.

Martin, P.S. 1984. Prehistoric overkill: the global model. In P.S. Martin and R.G. Klein, eds., *Quaternary Extinctions: A Prehistoric Revolution*, 354–403. Tucson, AZ: University of Arizona Press.

Matlock, R.B., Jr., D. Rogers, P.J. Edwards, and S.G. Martin. 2002. Avian communities in forest fragments and reforestation areas associated with banana plantations in Costa Rica. *Agriculture, Ecosystems and Environment* 91: 199–215.

MBC. 2002. The Mesoamerican Biological Corridor: a platform for sustainable development. Managua, Comisión Centroamericana de Ambiente y Desarrollo (CCAD). United Nations Development Program/Global Environmental Facility.

McClenachan, L., J.B.C. Jackson, and M.J.H. Newman. 2006. Conservation implications of historic sea turtle nesting beach loss. *Frontiers in Ecology and the Environment* 4: 290–96.

McDade, L.A., K.S. Bawa, H.A. Hespenheide, and G.S. Hartshorn, eds. 1994. *La Selva: Ecology and Natural History of a Neotropical Rainforest*. Chicago: University of Chicago Press.

McGee, D., F. Marcantonio, and J. Lynch-Stieglitz. 2007. Deglacial changes in dust flux in the eastern equatorial Pacific. *Earth and Planetary Science Letters* 257: 215–30.

McKenna, D.D., and B.D. Farrell. 2005. Molecular phylogenetics and evolution of host plant use in the Neotropical rolled leaf 'hispine' beetle genus *Cephaloleia* (Chevrolat) (Chrysomelidae: Cassidinae). *Molecular Phylogenetics and Evolution* 37: 117–31.

McKenna, D.D., and B.D. Farrell. 2006. Tropical forests are both evolutionary cradles and museums of leaf beetle diversity. *Proceedings of the National Academy of Sciences* 103: 10947–51.

Meletis, Z.A. 2007. Wasted visits?: ecotourism in theory vs. practice at Tortuguero, Costa Rica. Ph.D diss., Duke University.

Meletis, Z.A., and L.M. Campbell. 2009. Call it consumption!: (Re)conceptualizing ecotourism as consumption and consumptive. *Geography Compass* 1: 850–70.

Melillo, J.M., P.A. Steudler, J.D. Aber, K. Newkirk, H. Lux, F.P. Bowles, C. Catricala, A. Magill, T. Ahrens, and S. Morrisseau. 2002. Soil warming and carbon-cycle feedbacks to the climate system. *Science* 298: 2173–76.

Melo, F.P.L., B. Rodríguez-Herrera, R.L. Chazdon, R.A. Medellin, and G.G. Ceballos. 2009. Small tent-roosting bats promote dispersal of large-seeded plants in a Neotropical forest. *Biotropica* 41: 737–43.

Michel, N.L. 2012. Mechanisms and consequences of avian understory insectivore population decline in fragmented Neotropical rainforest. PhD diss., Tulane University.

Michel, N.L., T.W. Sherry, and W.P. Carson. 2014. The omnivorous collared peccary negates an insectivore-generated trophic cascade in Costa Rican wet tropical forest understorey. *Journal of Tropical Ecology* 1: 1–11.

Miller, K., E. Chang, and N. Johnson. 2001. Defining common ground for the Mesoamerican Biological Corridor. Washington, DC: World Resources Institute.

Miller, M.J. 2011. Persistent illegal logging in Costa Rica: the role of corruption among forestry regulators. *Journal of Environment Development* 20: 50–68.

Minc, G., D. Rodríguez, S. Sakai, R. Quiroga, L. Taber, and A. Rodríguez. 2001. The Mesoamerican Biological Corridor as a vector for sustainable development in the region: the role of international financing, preliminary considerations. Washington, DC: Inter-American Development Bank and World Bank.

Mnif, W., A.I.H. Hassine, A. Bouaziz, A. Bartegi, O. Thomas, and B. Roig. 2011. Effect of endocrine disruptor pesticides: a review. *International Journal of Environmental Research and Public Health* 8: 2265–303.

Molina, I., and S. Palmer. 2007. *The History of Costa Rica*. San José, Costa Rica: Editorial UCR.

Molnar, P. 2008. Closing of the Central American Seaway and the Ice Age: a critical review. *Paleoceanography* 23: 1–15.

Monge, G., O. Chassot, G.V.N. Powell, S. Palminteri, U.A. Zelaya, and P. Wright. 2003. Ecología de la lapa verde (*Ara ambigua*) en Costa Rica. San José, Costa Rica: Centro Cientifico Tropical.

Monge, P., T. Partanen, C. Wesseling, V. Bravo, C. Ruepert, and I. Burstyn. 2005. Assessment of pesticide exposure in the agricultural population of Costa Rica. *Annals of Occupational Hygiene* 49: 375–84.

Montagnini, F. 1994. Agricultural systems in the La Selva region. In L.A. McDade, K.S. Bawa, H.A. Hespenheide, and G.S. Hartshorn, eds., *La Selva: Ecology and Natural History of a Neotropical Rain Forest*, 307–16. Chicago: University of Chicago Press.

Montagnini, F., E. González, C. Porras, and R. Rheingans. 1995. Mixed and pure forest plantations in the humid tropics. *Commonwealth Forestry Review* 74: 306–14.

Montagnini, F., K. Ramstad, and F. Sancho. 1993. Litterfall, litter decomposition and the use of mulch of four indigenous tree species in the Atlantic lowlands of CR. *Agroforestry Systems* 23: 39–61.

Monzón-Argüello, C., L.F. López-Jurado, C. Rico, A. Marco, P. López, G.C. Hays, and P.L.M. Lee. 2010. Evidence from genetic and Lagrangian drifter data for transatlantic transport of small juvenile green turtles. *Journal of Biogeography* 37: 1752–66.

Moran, R., and A.R. Smith. 2001. Phytogeographic relationships between neotropical and African-Madagascan pteridophytes. *Brittonia* 53: 304–51.

Morrison, G., and D.R. Strong. 1981. Spatial variations in egg density and the intensity of parasitism in a Neotropical chrysomelid (*Cephaloleia consanguinea*). *Ecological Entomology* 6: 55–61.

Morse, W.C., J.L. Schedlbauer, S.E. Sesnie, B. Finegan, C.A. Harvey, S.J. Hollenhorst, K.L. Kavanagh, D. Stoian, and J.D. Wulfhorst. 2009. Consequences of environmental service payments for forest retention and recruitment in a Costa Rican biological corridor. *Ecology and Society* 14: 23–43.

Mortimer, J.A. 1995. Teaching critical concepts for the conservation of sea turtles. *Marine Turtle Newsletter* 71: 1–4.

Muhs, D.R., and J.R. Budahn. 2009. Geochemical evidence for African dust and volcanic ash inputs to terra rossa soils on carbonate reef terraces, northern Jamaica, West Indies. *Quaternary International* 196: 13–35.

Müller, M. 2000. Review of the *in situ* status of the great green or Buffon's macaw *Ara ambigua* and the European Endangered Species Programme (EEP). *International Zoo Yearbook* 37: 183–90.

Mundt, C.C., K.E. Sackett, L.D. Wallace, C. Cowger, and J.P. Dudley. 2009. Aerial dispersal and multiple-scale spread of epidemic disease. *EcoHealth* 6: 546–52.

Nadkarni, N.M., and N.T. Wheelwright. 2000. *Monteverde: Ecology and Conservation of a Tropical Cloud Forest*. New York: Oxford University Press.

Nag, S.K., and M.K. Raikwar. 2011. Persistent organochlorine pesticide residues in animal feed. *Environmental Monitoring and Assessment* 174: 327–35.

Neelin, J.D., M. Munnich, H. Su, J.E. Meyerson, and C.E. Holloway. 2006. Tropical drying trends in global warming models and observations. *Proceedings of the National Academy of Sciences* 103: 6110–15.

Newcomer, Q. 2002. Path of the tapir: integrating biological corridors, ecosystem management, and socioeconomic development in Costa Rica. *Endangered Species Update* 19: 186–93.

Newson, L. 1993. The demographic collapse of native peoples of the Americas, 1492–1650. In W. Bray, ed., *The Meeting of Two Worlds*, 247–88. British Academy.

Nichols-Orians, C.M. 1991a. Condensed tannins, attine ants, and the performance of a symbiotic fungus. *Journal of Chemical Ecology* 17: 1177–95.

Nichols-Orians, C.M. 1991b. Differential effects of condensed and hydrolyzable tannin on polyphenol oxidase activity of attine symbiotic fungus. *Journal of Chemical Ecology* 17: 1811–19.

Nichols-Orians, C.M. 1991c. The effects of light on foliar chemistry, growth and susceptibility of seedlings of a canopy tree to an attine ant. *Oecologia* 86: 552–60.

Nichols-Orians, C.M. 1991d. Environmentally induced differences in plant traits: consequences for susceptibility to a leaf-cutter ant. *Ecology* 72: 1609–23.

Nicotra, A.B., R.L. Chazdon, and S. Iriarte. 1999. Spatial heterogeneity of light and woody seedling regeneration in tropical wet forests. *Ecology* 80: 1908–26.

Nieuwenhuyse, A., A.G. Jongmans, and N. van Breemen. 1993. Andisol formation in a Holocene beach ridge plain under the humid tropical climate of the Atlantic coast of Costa Rica. *Geoderma* 57: 423–42.

Nieuwenhuyse, A., and S.B. Kroonenberg. 1994. Volcanic origin of Holocene beach ridges along the Caribbean coast of Costa Rica. *Marine Geology* 120: 13–26.

Nof, D., and S. Van Gorder. 2003. Did an open Panama Isthmus correspond to an invasion of Pacific water into the Atlantic? *Journal of Physical Oceanography* 33: 1324–36.

Nolte, P. 2011. Pesticide resistance: what action will you take? *American Vegetable Grower* 59: 28.

Norden, N., R.L. Chazdon, A. Chao, Y.-H. Jiang, and B.A. Vílchez. 2009. Resilience of tropical rain forests: tree community reassembly in secondary forests. *Ecology Letters* 12: 385–94.

Notman, E.M., and A.C. Villegas. 2005. Patterns of seed predation by vertebrate versus invertebrate seed predators among different plant species, seasons and spatial distributions. In P. M. Forget, J.E. Lambert, P.E. Hulme, and S.B. Van der Wall, eds., *Seed Fate, Predation, Dispersal and Seedling Establishment*, 55–75. Wallingford: CABI Publishing.

Novotný, V., and Y. Basset. 2000. Rare species in communities of tropical insect herbivores: pondering the mystery of singletons. *Oikos* 89: 564–72.

Novotný, V., S.E. Miller, Y. Basset, L. Cizek, P. Drozd, K. Darrow, and J. Leps. 2002. Predictably simple: assemblages of caterpillars (Lepidoptera) feeding on rainforest trees in Papua New Guinea. *Proceedings of the Royal Society B* 269: 2337–44.

Okello, J.J., and S.M. Swinton. 2010. From circle of poison to circle of virtue: pesticides, export standards and Kenya's green bean farmers. *Journal of Agricultural Economics* 61: 209–24.

Orians, G.H. 1969. The number of birds species in some tropical forests. *Ecology* 50: 783–801.

Pagiola, S. 2008. Payments for environmental services in Costa Rica. *Ecological Economics* 65: 712–24.

Paige, J.M. 1997. *Coffee and Power: Revolution and the Rise of Democracy in Central America*. Cambridge, MA: Harvard University Press.

Palmer, S., and I. Molina, eds. 2004. *The Costa Rica Reader*. Durham: Duke University Press.

Parker, G.G. 1994. Soil fertility, nutrient availability, and nutrient cycling. In L.A. McDade, K.S. Bawa, H.A. Hespenheide, and G.S. Hartshorn, eds., *La Selva: Ecology and Natural History of a Neotropical Rain Forest*, 54–64. Chicago: University of Chicago Press.

Parker, T.A., III, D.F. Stotz, and J.W. Fitzpatrick. 1996. Ecological and distributional databases. In D.F. Stotz, J.W. Fitzpatrick, T.A. Parker III, and D.K. Moskovits, eds., *Neotropical Birds: Ecology and Conservation*, 115–436. Chicago: University of Chicago Press.

Passos, L., and P.S. Oliveira. 2002. Ants affect the distribution and performance of seedlings of *Clusia criuva*, a primarily bird-dispersed rain forest tree. *Journal of Ecology* 90: 517–28.

Pelley, J. 2006. DDT's legacy lasts for many decades. *Environmental Science and Technology* 40: 4533–34.

Peñaloza, C., and A.G. Farji-Brener. 2003. The importance of treefall gaps as foraging sites for leaf-cutting ants depends on forest age. *Journal of Tropical Ecology* 19: 603–5.

Pennington, R.T., J.E. Richardson, and M. Lavin. 2006. Insights into the historical construction of species-rich biomes from dated plant phylogenies, neutral ecological theory and phylogenetic community structure. *New Phytologist* 172: 605–16.

Pérez, S., and F. Protti. 1978. Comportamiento del sector forestal durante el periodo 1950–1977. DOC-OPSA-15zs Oficina de Planificación Sectorial Agropecuaria, San José, Costa Rica.

Pérez-Maldonado, I.N., A. Trejo, C. Ruepert, R.D.C. Jovel, M.P. Méndez, M. Ferrari, E. Saballos-Sobalvarro, C. Alexander, L. Yáñez-Estrada, D. Lopez, S. Henao, E.R. Pinto, and F. Díaz-Barriga. 2010. Assessment of DDT levels in selected environmental media and biological samples from Mexico and Central America. *Chemosphere* 78: 1244–49.

Perfecto, I., and J. Vandermeer. 1993. Distribution and turnover rate of a population of *Atta cephalotes* in a tropical rain forest in Costa Rica. *Biotropica* 25: 316–21.

Perkins, S. 2001. Dust, the thermostat. *Science News* 160: 200–202.

Pinto-Tomas, A.A., M.A. Anderson, G. Suen, D.M. Stevenson, F.S.T. Chu, W.W. Cleland, P.J. Weimer, and C.R. Currie. 2009. Symbiotic nitrogen fixation in the fungus gardens of leaf-cutter ants. *Science* 326: 1120–23.

Piperno, D.R., and T.D. Dillehay. 2008. Starch grains on human teeth reveal early broad crop diet in northern Peru. *Proceedings of the National Academy of Sciences* 105: 19622–27.

Piperno, D.R., and I. Holst. 1998. The presence of starch grains on prehistoric stone tools from the humid Neotropics: indications of early tuber use and agriculture in Panama. *Journal of Archaeological Science* 25: 765–76.

Piperno, D.R., and D.M. Pearsall. 1998. *The Origins of Agriculture in the Lowland Neotropics*. San Diego: Academic Press.

Piperno, D.R., A.J. Ranere, I. Holst, and P. Hansell. 2000. Starch grains reveal early root crop horticulture in the Panamanian tropical forest. *Nature* 407: 894–97.

Pisani, J.M., W.E. Grant, and M. Mora. 2008. Simulating the impact of cholinesterase-inhibiting pesticides on non-target wildlife in irrigated crops. *Ecological Modelling* 210: 179–92.

Place, S.E. 1991. Nature tourism and rural development in Tortuguero. *Annals of Tourism Research* 18: 186–201.

Polidoro, B.A., R.M. Dahlquist, L.E. Castillo, M.J. Morra, E. Somarriba, and N.A. Bosque-Pérez. 2008. Pesticide application practices, pest knowledge, and cost-benefits of plantain production in the Bribri–Cabécar Indigenous Territories, Costa Rica. *Environmental Research* 108: 98–106.

Porder, S., D. Clark, and P.M. Vitousek. 2006. Persistence of rock-derived nutrients in the wet tropical forests of La Selva. Costa Rica. *Ecology* 87: 594–602.

Poux, C., P. Chevret, D. Huchon, W.W.D.J.E. Jong, P. Douzery, and S. Url. 2006. Arrival and diversification of caviomorph rodents and platyrrhine primates in South America. *Systematic Biology* 55: 228–44.

Powell, G., and R. Bjork. 1994. Implications of altitudinal migration for conservation strategies to protect tropical biodiversity: a case study of the resplendent quetzal (*Pharomachrus mocinno*) at Monteverde, Costa Rica. *Bird Conservation International* 4: 161–74.

Powell, G., P. Wright, C. Guindon, and R. Alemán Bjork. 1999. Resultados y recomendaciones para la conservación de la lapa verde (*Ara ambigua*) en Costa Rica. San José, Costa Rica: Centro Cientifico Tropical.

Powell, G.V.N., J. Barborak, and S.M. Rodriguez. 2000. Assessing representativeness of protected natural areas in Costa Rica for conserving biodiversity: a preliminary gap analysis. *Biological Conservation* 93: 35–41.

Powers, J.S., M.D. Corre, T.E. Twine, and E. Veldkamp. 2011. Geographic bias of field observations of soil carbon stocks with tropical land-use changes precludes spatial extrapolation. *Proceedings of the National Academy of Sciences* 108: 6318–22.

Powers, J.S., and W.H. Schlesinger. 2002a. Geographic and vertical patterns of stable carbon isotopes in tropical rain forest soils in Costa Rica. *Geoderma* 109: 141–60.

Powers, J.S., and W.H. Schlesinger. 2002b. Relationships among soil carbon distributions and biophysical factors at nested spatial scales in rain forests of northeastern Costa Rica. *Geoderma* 109: 165–90.

Powers, T.O., D.A. Neher, P. Mullin, A. Esquivel, R.M. Giblin-Davis, N. Kanzaki, S.P. Stock, M.M. Mora, and L. Uribe-Lorio. 2009. Tropical nematode diversity: vertical stratification of nematode communities in a Costa Rican humid lowland rainforest. *Molecular Ecology* 18: 985–96.

Puschendorf, R., F. Bolaños, and G. Chaves. 2006. The amphibian chytrid fungus along an altitudinal transect before the first reported declines in Costa Rica. *Biological Conservation* 132: 136–42.

Puschendorf, R., A.C. Carnaval, J. van der Wal, H. Zumbado-Ulate, G. Chaves, F. Bolaños, and R.A. Alford. 2009. Distribution models for the amphibian chytrid *Batrachochytrium dendrobatidis* in Costa Rica: proposing climatic refuges as a conservation tool. *Diversity and Distributions* 15: 401–8.

Puschendorf, R., G. Chaves, A.J. Crawford, and D.R. Brooks. 2005. *Eleutherodactylus ranoides* (NCN): dry forest population, refuge from decline? *Herpetological Review* 36: 53.

Pyke, C.R., R. Condit, A. Salomón, and S. Lao. 2001. Floristic composition across a climatic gradient in a neotropical lowland forest. *Journal of Vegetation Science* 12: 553–66.

Raich, J.W., A.E. Russell, K. Kitayama, W.J. Parton, and P.M. Vitousek. 2006. Temperature influences carbon accumulation in moist tropical forests. *Ecology* 87: 76–87.

Ramírez, O.A., C.E. Carpio, R. Ortíz, and B. Finegan. 2002. Economic value of carbon sink services of tropical secondary forests and its management implications. *Environmental and Resource Economics* 21: 23–46.

Raymond, M., C. Berticat, M. Weill, N. Pasteur, and C. Chevillon. 2001.

Insecticide resistance in the mosquito *Culex pipiens*: what have we learned about adaptation? *Genetica* 112: 287–96.

Read, J.M. 1999. Land-cover change detection for the tropics using remote sensing and geographic information systems. PhD diss., Louisiana State University.

Read, J.M., J.S. Denslow, and S.M. Guzman. 2001. Documenting land cover history of a humid tropical environment in northeastern Costa Rica using time-series remotely sensed data. In A.C. Millington, S.J. Walsh, and P.E. Osborne, eds., *GIS and Remote Sensing Applications in Biogeography and Ecology*. Boston, MA: Kluwer Academic Publishers.

Redondo-Brenes, A., and F. Montagnini. 2006. Growth, productivity, aboveground biomass, and carbon sequestration of pure and mixed native tree plantations in the Caribbean lowlands of Costa Rica. *Forest Ecology and Management* 232: 168–78.

Redondo-Brenes, A., A. Vílchez, and R.L. Chazdon. 2001. Estudio de la dinámina y composición de cuatro bosques secundarios en la región Huetar Norte, Sarapiquí, Costa Rica. *Revista Forestal Centroamericana* 36 (Oct–Dec): 21–26.

Reed, S.C., T.E. Wood, M.A. Cavaleri. 2012. Tropical forests in a warming world. *New Phytologist* 193: 27–29.

Reguero, M., A.M. Candela, and R.N. Alonso. 2007. Biochronology and biostratigraphy of the Uquía Formation (Pliocene–early Pleistocene, NW Argentina) and its significance in the Great American Biotic Interchange. *Journal of South American Earth Sciences* 23: 1–16.

Reich, A., J.J. Ewel, N.M. Nadkarni, T. Dawson, and R.D. Evans. 2003. Nitrogen isotope ratios shift with plant size in tropical bromeliads. *Oecologia* 137: 587–90.

Rieser, A. 2012. *The Case of the Green Turtle: An Uncensored History of a Conservation Icon*. Baltimore: Johns Hopkins University Press.

Renner, S.S. 2004a. Bayesian analyses of combined data partitions, using multiple calibrations, supports recent arrivals of Melastomataceae in Africa and Madagascar. *American Journal of Botany* 88: 1290–300.

Renner, S.S. 2004b. Plant dispersal across the tropical Atlantic by wind and sea currents. *International Journal of Plant Sciences* 165(Suppl.): S23–S33.

Renner, S.S. 2005. Relaxed molecular clocks for dating historical plant dispersal events. *Trends in Plant Science* 10: 550–58.

Ridgely, R.S., and J.A. Gwynne. 1992. *A Guide to the Birds of Panama: With Costa Rica, Nicaragua, and Honduras*. Princeton, NJ: Princeton University Press. 656 pp.

Rinker, B.H., and M.D. Lowman. 2001. Canopy herbivory and soil ecology, the top–down impact of forest processes. *Selbyana* 22: 225–31.

Roder, C., J. Cortés, C. Jiménez, and R. Lara. 2009. Riverine input of particulate material and inorganic nutrients to a coastal reef ecosystem at the Caribbean coast of Costa Rica. *Marine Pollution Bulletin* 58: 1937–43.

Rodríguez-Herrera, B., R.A. Medellín, and M. Gambia-Ríos. 2008. Roosting requirements of white tent-making bat *Ectophylla alba* (Chiroptera: Phyllostomidae). *Acta Chiropterologica* 10: 89–95.

Rodríguez-Herrera, B., R.A. Medellín, and R.M. Timm. 2007. Murciélagos neotropicales que acampan en hojas: neotropical tent-roosting bats. Santo Domingo de Heredia, Costa Rica: Instituto Nacional de Biodiversidad (INBio).

Rodríguez-Herrera, B., and M. Tschapka. 1999. Tent use by *Vampyressa nymphaea* (Chiroptera: Phyllostomidae) in *Cecropia insignis* (Moraceae) in Costa Rica. *Acta Chiropterologica* 6: 171–74.

Romero, A., B.J. O'Neill, R.M. Timm, K.G. Gerow, and D. McClearn. 2013. Group dynamics, behavior, and current and historical abundance of peccaries in Costa Rica's Caribbean lowlands. *Journal of Mammalogy* 94: 771–91.

Romero, A., and R.M. Timm. 2013. Reproductive strategies and natural history of the arboreal, Neotropical vesper mouse, *Nyctomys sumichrasti*. *Mammalia* 77: 363–70.

Rosero-Bixby, L., and A. Palloni. 1998. Population and deforestation in Costa Rica. *Population and Environment* 20: 149–85.

Ross, Y., and L. Capelli. 2003. *Passion for the Caribbean*. San José, Costa Rica: Grupo Santillana SA.

Rowell, C.H.F. 1985a. The feeding biology of a species-rich genus of rainforest grasshoppers (*Rhachicreagra*: Orthoptera, Acrididae), I. Foodplant use and foodplant acceptance. *Oecologia* 68: 87–98.

Rowell, C.H.F. 1985b. The feeding biology of a species-rich genus of rainforest grasshoppers (*Rhachicreagra*: Orthoptera, Acrididae), II. Foodplant preference and its relation to speciation. *Oecologia* 68: 99–104.

Rowell, H.F. 1978. Food plant specificity in neotropical rain-forest acridids. *Entomologia Experimentalis et Applicata* 24: 651–62.

Rowell, H.F. 1983a. Checklist of acridoid grasshoppers. In D.H. Janzen, ed., *Costa Rican Natural History*, 651–53. Chicago: University of Chicago Press.

Rowell, H.F. 1983b. *Drymophilacris bimaculata*. In D.H. Janzen, ed., *Costa Rican Natural History*, 714–16. Chicago: University of Chicago Press.

Rowell, H.F. 1983c. *Osmilia flavolineata*. In D.H. Janzen, ed., *Costa Rican Natural History*, 750–51. Chicago: University of Chicago Press.

Rowell, H.F. 1983d. *Tropidacris cristata*. In D.H. Janzen, ed., *Costa Rican Natural History*, 772–73. Chicago: University of Chicago Press.

Ruiz-Narváez, E.A., F.R. Santos, D. de Carvalho-Silva, and J. Azofeifa. 2005. Genetic variation of the Y chromosome in Chibcha speaking Amerindians of Costa Rica and Panama. *Human Biology* 77: 71–91.

Russell, A.E., C.A. Cambardella, J.J. Ewel, and T.B. Parkin. 2004. Species, rotation, and life-form diversity effects on soil carbon in experimental tropical ecosystems. *Ecological Applications* 14: 47–60.

Russell, A.E., J.W. Raich, A.R. Bedoya, O. Valverde-Barrantes, and E. González. 2010. Impacts of individual tree species on carbon dynamics in a moist tropical forest environment. *Ecological Applications* 20: 1087–100.

Russell, A.E., J.W. Raich, O.J. Valverde-Barrantes, and R.F. Fisher. 2007. Tree species effects on soil properties in experimental plantations in tropical moist forest. *Soil Science Society of America Journal* 71: 1389–97.

Sader, S.A., and A.T. Joyce. 1988. Deforestation rates and trends in Costa Rica—1940 to 1983. *Biotropica* 20: 11–19.

Sader, S.A., T. Sever, D. Irwin, and S. Saatchi. 2004. Monitoring the Mesoamerican Biological Corridor: a NASA/CCAD cooperative research project. Final Report submitted to NASA-ESE (NAG5-8712).

Sader, S.A., R.T. Waide, W.T. Lawrence, and A.T. Joyce. 1989. Tropical forest biomass and successional age class relationships to a vegetation index derived from Landsat TM data. *Remote Sensing of Environment* 28: 143–56.

Sadofsky, S., K. Hoernle, S. Duggen, F. Hauff, R. Werner, and D. Garbe-Schönberg. 2009. Geochemical variations in the Cocos Plate subducting beneath Central America: implications for the composition of arc volcanism and the extent of the Galápagos Hotspot influence on the Cocos oceanic crust. *International Journal of Earth Sciences* (Geologische Rundschau) 98: 901–13.

Sak, P.B., D.M. Fisher, T.W. Gardner, J.S. Marshall, and P.C. Lafemina. 2009. Rough crust subduction, forearc kinematics, and Quaternary uplift rates, Costa Rican segment of the Middle American Trench. *Geological Society of America Bulletin* 121: 992–1012.

Sánchez-Azofeifa, G.A., G.C. Daily, A.S.P. Pfaff, and C. Busch. 2003. Integrity and isolation of Costa Rica's national parks and biological reserves: examining the dynamics of land-cover change. *Biological Conservation* 109: 123–35.

Sánchez-Azofeifa, G.A., A. Pfaff, J.A. Robalino, and J.P. Boomhower. 2007. Costa Rica's payment for environmental services program: intention, implementation, and impact. *Conservation Biology* 21: 1165–73.

Santoro, M., G. Hernández, M. Caballero, and F. García. 2006. Aerobic bacterial flora of nesting green turtles (*Chelonia mydas*) from Tortuguero National Park, Costa Rica. *Journal of Zoo and Wildlife Medicine* 37: 549–52.

Saugier, B., J. Roy, and H.A. Mooney. 2001. Estimations of global terrestrial productivity: converging toward a single number? In J. Roy, B. Saugier, and H.A. Mooney, eds., *Terrestrial Global Productivity*, 543–57. San Diego, CA: Academic Press.

Savage, J.M. 2002. *The Amphibians and Reptiles of Costa Rica*. Chicago: University of Chicago Press.

Schelhas, J., and G.A. Sánchez-Azofeifa. 2006. Post-frontier forest change adjacent to Braulio Carrillo National Park, Costa Rica. *Human Ecology* 34: 407–31.

Schnitzer, S.A., and F. Bongers. 2002. The ecology of lianas and their role in forests. *Trends in Ecology and Evolution* 17: 223–30.

Schrago, C.G. 2007. On the time scale of New World primate diversification. *American Journal of Physical Anthropology* 132: 344–54.

Schwendenmann, L., E. Veldkamp, T. Brenes, J.J. O'Brien, and J. Mackensen. 2003. Spatial and temporal variation in soil CO_2 efflux in an old-growth neotropical rain forest, La Selva, Costa Rica. *Biogeochemistry* 64: 111–28.

Scott, N.J. 1976. The abundance and diversity of the herpetofaunas of tropical forest litter. *Biotropica* 8: 41–58.

Seminoff, J.A., and K. Shanker. 2008. Marine turtles and IUCN Red Listing: a review of the process, the pitfalls, and novel assessment approaches. *Journal of Experimental Marine Biology and Ecology* 356: 52–68.

Seminoff, J.A., and B.P. Wallace. 2012. Sea turtles of the eastern Pacific: advances in research and conservation. Tucson: University of Arizona Press.

Senna, M.C.A., M. Costa, and Y.E. Shimabukuro. 2005. Fraction of photosynthetically active radiation absorbed by Amazon tropical forest: a comparison of field measurements, modeling, and remote sensing. *Journal of Geophysical Research* 110: 1–8.

Sezen, U.U., R.L. Chazdon, and K.E. Holsinger. 2009. Proximity is not a proxy for parentage in an animal-dispersed Neotropical canopy palm. *Proceedings of the Royal Society B* 276: 2037–44.

Sherrard, R., C. Murray-Gulde, J.J. Rodgers, and Y. Shah. 2003. Comparative toxicity of chlorothalonil: *Ceriodaphnia dubia* and *Pimephales promelas*. *Ecotoxicology and Environmental Safety* 56: 327–33.

Sierra, R., and E. Russman. 2006. On the efficiency of environmental service payments: a forest conservation assessment in the Osa Peninsula, Costa Rica. *Ecological Economics* 59: 131–41.

Sigel, B.J., D.W. Robinson, and T.W. Sherry. 2010. Comparing bird community responses to forest fragmentation in two lowland Central American reserves. *Biological Conservation* 143: 340–50.

Sigel, B.J., T.W. Sherry, and B.E. Young. 2006. Avian community response to lowland tropical rainforest isolation: 40 years of change at La Selva Biological Station, Costa Rica. *Conservation Biology* 20: 111–21.

Silva, E. 2003. Selling sustainable development and shortchanging social ecology in Costa Rican forest policy. *Latin American Politics and Society* 45: 93–127.

Sims, L., and A.J. Sinclair. 2008. Learning through participatory resource management programs: case studies from Costa Rica. Adult Education Quarterly 58: 151–69.

Slocum, M.G., and C.C. Horvitz. 2000. Seed arrival under different genera of trees in a neotropical pasture. *Plant Ecology* 149: 51–62.

Slud, P. 1960. The birds of finca "La Selva," Costa Rica: a tropical wet forest locality. *Bulletin of the American Museum of Natural History* 121: 55–148.

Slud, P. 1976. Geographical and climatic relationships of avifaunas, with special reference to avian distribution in the Neotropics. *Smithsonian Contributions to Zoology* 212: 1–149.

Smethurst, D., and B. Nietschmann. 1999. The distribution of manatees (*Trichechus manatus*) in the coastal waterways of Tortuguero, Costa Rica. *Biological Conservation* 89: 267–74.

Smythe, N. 1986. Competition and resource partitioning in the guild of Neotropical terrestrial frugivorous mammals. *Annual Review of Ecology and Systematics* 17: 169–88.

Snarskis, M.J. 1976. Stratigraphic excavations in the eastern lowlands of Costa Rica. *American Antiquity* 41: 342–53.

Snider, A.G., S.K. Pattanayak, E.O. Sills, and J.L. Schuler. 2003. Policy innovations for private forest management and conservation in Costa Rica. *Journal of Forestry* (July/August): 17–23.

Soares, E.L., and M. Firpo de Souza Porto. 2009. Estimating the social cost of pesticide use: an assessment from acute poisoning in Brazil. *Ecological Economics* 68: 2721–28.

Soin, T., and G. Smagghe. 2007. Endocrine disruption in aquatic insects: a review. *Ecotoxicology* 16: 83–93.

Sol Castillo, R.F. 2000. Asentamientos prehispánicos en la reserva biológica La Selva, Sarapiquí, Costa Rica. Tésis de licenciado no. 339 en antropología, University of Costa Rica.

Sollins, P., M.F. Sancho, C.R. Mata, and R.L. Sanford. 1994. Soils and soil process research. In L.A. McDade, K.S. Bawa, H.A. Hespenheide, and G.S. Hartshorn, eds., *La Selva: Ecology and Natural History of a Neotropical Rain Forest*, 34–53. Chicago: University of Chicago Press.

Solow, A.R., K.A. Bjorndal, and A.B. Bolten. 2002. Annual variation in nesting numbers of marine turtles: the effect of sea surface temperature on re-migration intervals. *Ecology Letters* 5: 742–46.

Sperr, E.B., E.A. Fronhofer, and M. Tschapka. 2009. The Mexican mouse opossum (*Marmosa mexicana*) as a flower visitor at a neotropical palm. *Mammalian Biology* 74: 76–80.

Spotila, J.R. 2004. *Sea Turtles: A Complete Guide to Their Biology, Behavior, and Conservation*. Baltimore, MD: Johns Hopkins University Press.

Stahl, P.W. 2004. Greater expectations. *Nature* 432: 561–63.

Staines, C.L. 1996. The genus *Cephaloleia* (Coleoptera: Chrysomelidae) in Central America and the West Indies. *Revista de Biología Tropical, Special Publication* 3: 3–87.

Steph, S., R. Tiedemann, M. Prange, and J. Groeneveld. 2006. Changes in Caribbean surface hydrography during the Pliocene shoaling of the Central American Seaway. *Paleoceanography* 21: 1–25.

Stiles, F., and A. Skutch. 1989. *A Guide to the Birds of Costa Rica*. Ithaca, NY: Cornell University Press.

Stiles, F.G. 1983. Birds: introduction. In D.H. Janzen, ed., *Costa Rican Natural History*, 502–29. Chicago: University of Chicago Press.

Strong, D.R. 1981. The possibility of insect communities without competition: Hispine beetles on *Heliconia*. In R.F. Denno, ed., *Insect Life History Patterns: Habitat and Geographic Variation*, 183–94. New York: Springer.

Strong, D.R. 1982a. Harmonious coexistence of hispine beetles on *Heliconia* in experimental and natural communities. *Ecology* 63: 1039–49.

Strong, D.R. 1982b. Potential interspecific competition and host specificity: hispine beetles on *Heliconia*. *Ecological Entomology* 7: 217–20.

Strong, D.R., and M.D. Wang. 1977. Evolution of insect life histories and host plant chemistry: hispine beetles on *Heliconia*. *Evolution* 31: 854–62.

Stuart, A.J., P.A. Kosintsev, T.F.G. Higham, and A.M. Lister. 2004. Pleistocene to Holocene extinction dynamics in giant deer and woolly mammoth. *Nature* 431: 684–89.

Sun, Y., S. Solomon, A. Dai, and R.W. Portmann. 2007. How often will it rain? *Journal of Climate* 20: 4801–18.

Surovell, T., N. Waguespack, P.J. Brantingham, and G.C. Frison. 2005. Global archaeological evidence for proboscidean overkill. *Proceedings of the National Academy of Sciences* 102: 6231–36.

Sylvester, O., and G. Avalos. 2009. Illegal palm heart (*Geonoma edulis*) harvest in Costa Rican national parks: patterns of consumption and extraction. *Economic Botany* 63: 179–89.

Tamm, E., T. Kivisild, M. Reidla, M. Metspalu, D.G. Smith, C.J. Mulligan, C.M. Bravi, O. Rickards, C. Martínez-Labarga, E.K. Khusnutdinova, S.A. Fedorova, M.V. Golubenko, V.A. Stepanov, M.A. Gubina, S.I. Zhadanov, L.P. Ossipova, L. Damba, M.I. Voevoda, J.E. Kipierri, R. Villems, and R.S. Malhi. 2007. Beringian standstill and spread of Native American founders. *PLOS ONE* 9: 1–6.

Terborgh, J., S. Robinson, T. Parker, C. Munn, and N. Pierpont. 1990. Structure and organization of an Amazonian forest bird community. *Ecological Monographs* 60: 213–38.

Timm, R.M. 1982. *Ectophylla alba. Mammalian Species* 166: 1–4.

Timm, R.M. 1984. Tent construction by *Vampyressa* in Costa Rica. *Journal of Mammalogy* 65: 166–67.

Timm, R.M. 1985. *Artibeus phaeotis. Mammalian Species* 235: 1–6.

Timm, R.M. 1987. Tent construction by bats of the genera *Artibeus* and *Uroderma*. In B.D. Patterson and R.M. Timm, eds., Studies in neotropical mammalogy: essays in honor of Philip Hershkovitz. *Fieldiana: Zoology* (New Series) 39: viii + 1–506.

Timm, R.M. 1994. The mammal fauna. In L.A. McDade, K.S. Bawa, H.A. Hespenheide, and G.S. Hartshorn, eds., *La Selva: Ecology and Natural History of a Neotropical Rain Forest*, 229–37 + 394–98. Chicago: University of Chicago Press.

Timm, R.M., and R.K. LaVal. 1998. A field key to the bats of Costa Rica. Occasional Publication Series, *University of Kansas Center of Latin American Studies* 22: 1–30.

Timm, R.M., R.K. LaVal, and B. Rodríguez-Herrera. 1999. Clave de campo para los murciélagos de Costa Rica. *Brenesia* (*Museo Nacional de Costa Rica*) 52: 1–32.

Timm, R.M., D.E. Wilson, B.L. Clauson, R.K. LaVal, and C.S. Vaughan. 1989. Mammals of the La Selva–Braulio Carrillo Complex, Costa Rica. *North American Fauna* 75: 1–162.

Tiwari, M., K. Bjorndal, A.B. Bolten, and B.M. Bolker. 2005. Intraspecific application of the mid-domain effect model: spatial and temporal nest distributions of green turtles, *Chelonia mydas*, at Tortuguero, Costa Rica. *Ecology Letters* 8: 918–24.

Tiwari, M., K. Bjorndal, A. Bolten, and B. Bolker. 2006. Evaluation of density-dependent processes and green turtle *Chelonia mydas* hatchling production at Tortuguero, Costa Rica. *Marine Ecology Progress Series* 326: 283–93.

Tobler, M. 2007. Reversed sexual dimorphism and courtship by females in the topaz cichlid *Archocentrus myrnae* (Chichlidae, Teleostei) from Costa Rica. *Southwestern Naturalist* 52: 371–77.

Toledo, M., L. Poorter, M. Peña-Claros, A. Alarcón, J. Balcázar, J. Chuviña, C. Leaño, J.C. Licona, H.ter Steege, and F. Bongers. 2011. Patterns and determinants of floristic variation across lowland forests of Bolivia. *Biotropica* 43: 405–13.

Torroni, A., J.V. Neel, R. Barrantes, T.G. Schurr, and D.C. Wallace. 1994. Mitochondrial DNA "clock" for the Amerinds and its implications for timing their entry into North America. *Proceedings of the National Academy of Sciences* 91: 1158–62.

Tosi, J.A. 1969. Mapa ecológica de Costa Rica. San José, Costa Rica: Tropical Science Center.

Troëng, S., and M. Chaloupka. 2007. Variation in adult annual survival probability and remigration intervals of sea turtles. *Marine Biology* 151: 1721–30.

Troëng, S., P.H. Dutton, and D. Evans. 2005. Migration of hawksbill turtles *Eretmochelys imbricata* from Tortuguero, Costa Rica. *Ecography* 28: 394–402.

Troëng, S., E. Harrison, D. Evans, A. de Haro, and E. Vargas. 2007. Leatherback turtle nesting trends and threats at Tortuguero, Costa Rica. *Chelonian Conservation and Biology* 6: 117–22.

Troëng, S., and E. Rankin. 2005. Long-term conservation efforts contribute to positive green turtle *Chelonia mydas* nesting trend at Tortuguero, Costa Rica. *Biological Conservation* 121: 111–16.

Tschapka, M. 2003. Pollination of the understorey palm *Calyptrogyne ghiesbreghtiana* by hovering and perching bats. *Biological Journal of the Linnean Society* 80: 281–88.

Tyree, M.T., B.M.J. Engelbrecht, G. Vargas, and T.A. Kursar. 2003. Desiccation tolerance of five tropical seedlings in Panama: relationship to a field assessment of drought performance. *Plant Physiology* 132: 1439–47.

USDA–NRCS, Soil Survey Staff. 1999. Soil Taxonomy. Washington, DC: US Government Printing Office.

Van Avendonk, H.J.A., W.S. Holbrook, D. Lizarralde, and P. Denyer. 2011. Structure and serpentinization of the subducting Cocos plate offshore Nicaragua and Costa Rica. *Geochemistry, Geophysics, Geosystems* 12: 1–23.

Van der Wall, S.B., and W.S. Longland. 2004. Diplochory: are two seed dispersers better than one? *Trends in Ecology and Evolution* 19: 155–61.

Veldkamp, E., A. Becker, L. Schwendenmann, D.H. Clark, and H. Schulte-Bisping. 2003. Substantial labile carbon stocks and microbial activity in deeply weathered soils below a tropical wet forest. *Global Change Biology* 9: 1171–84.

Veldkamp, A., P.H. Verburg, K. Kok, G.H.J. de Koning, J. Priess, and A.R. Bergsma. 2001. The need for scale sensitive approaches in spatially explicit land use change modeling. *Environmental Modeling and Assessment* 6: 111–21.

Vila, R., C.D. Bell, R. Macniven, B. Goldman-Huertas, R.H. Ree, C.R. Marshall, Z. Bálint, K. Johnson, D. Benyamini, and N.E. Pierce. 2011. Phylogeny and palaeoecology of *Polyommatus* blue butterflies show Beringia was a climate-regulated gateway to the New World. *Proceedings of the Royal Society B: Biological Sciences* 278: 2737–44.

Vílchez, A., R.L. Chazdon, and A. Redondo. 2004. Fenología reproductiva de cinco especies forestales del bosque secundario tropical. *Kurú: Revista Forestal* (Costa Rica) 1: 1–10.

Vivanco, L.A. 2006. *Green Encounters: Shaping and Contesting Environmental Relations in Costa Rica*. Oxford: Berghahn Books.

Voigt, C.C. 2004. The power requirements (Glossophaginae: Phyllostomidae) in nectar-feeding bats for clinging to flowers. *Journal of Comparative Physiology B* 174: 541–48.

Voigt, C.C. 2005. The evolution of wing sacs and perfume-blending in the emballonurid bats. In R. Mason, M. LeMaster, and D. Müller-Schwarze, eds., *Chemical Signals in Vertebrates*, vol. X, 93–100. Springer Verlag.

Voigt, C.C. 2013. Sexual selection in neotropical bats. In R.H. Macedo and G. Machado, eds., *Sexual Selection–Perspectives and Models from the Neotropics*, 409–32. Amsterdam: Elsevier.

Voigt, C.C., O. Behr, B. Caspers, O. von Helversen, M. Knörnschild, F. Mayer, and M. Nagy. 2008. Songs, scents, and senses: sexual selection in the greater sac-winged bat, *Saccopteryx bilineata*. *Journal of Mammalogy* 89: 1401–10.

Voigt, C.C., S.L. Voigt-Heucke, and K. Schneeberger. 2012. Isotopic data do not support food sharing within large networks of female vampire bats (*Desmodus rotundus*). *Ethology* 118: 260–68.

Walla, T.R., S. Engen, P.J. DeVries, and R. Lande. 2004. Modeling vertical beta-diversity in tropical butterfly communities. *Oikos* 107: 610–18.

Wallace, D.R. 1992. *The Quetzal and the Macaw*. San Francisco: Sierra Club Books.

Wang, I.J., A.J. Crawford, and E. Bermingham. 2008. Phylogeography of the pygmy rain frog (*Pristimantis ridens*) across the lowland wet forests of isthmian Central America. *Molecular Phylogenetics and Evolution* 47: 992–1004.

Washington, R., C. Bouet, G. Cautenet, E. Mackenzie, I. Ashpole, S. Engelstaedter, G. Lizcano, G.M. Henderson, K. Schepanski, and I. Tegen. 2009. Dust as a tipping element: the Bodélé Depression, Chad. *Proceedings of the National Academy of Sciences* 106: 20564–71.

Webb, S.D. 1991. Ecogeography and the Great American Interchange. *Paleobiology* 17: 266–80.

Wegner, W., G. Wörner, R.S. Harmon, and B.R. Jicha. 2011. Magmatic history and evolution of the Central American Land Bridge in Panama since Cretaceous times. *Geological Society of America Bulletin* 123: 724.

Weigt, L.A., A.J. Crawford, A.S. Rand, and M.J. Ryan. 2005. Biogeography of the túngara frog, *Physalaemus pustulosus*: a molecular perspective. *Molecular Ecology* 14: 3857–76.

Weir, D., and M. Schapiro. 1981. *Circle of Poison*. Oakland, CA: Institute for Food and Development Policy.

Wendt, T. 1993. Composition, floristic affinities and origins of the Mexican Atlantic slope rain forests. In T.P. Ramamoorthy, R. Bye, A. Lot, and J. Fa, eds., *Biological Diversity of Mexico: Origins and Distribution*, 595–680. Oxford University Press.

Werner, P. 1985. La reconstitution de la forêt tropicale humide au Costa Rica: analyse de croissance et dynamique de la végetation. Diss., University of Lausanne. Switzerland.

Werner, R., K. Hoernle, P. van den Bogaard, C. Ranero, and R. von Huene. 1999. Drowned 14-m-y old Galápagos archipelago off the coast of Costa Rica: implications for tectonic and evolutionary models. *Geology* 27: 499–502.

Wesseling, C., A. Ahlbom, D. Antich, A.C. Rodríguez, and R. Castro. 1996. Cancer in banana plantation workers in Costa Rica. *International Journal of Epidemiology* 25: 1125–31.

Wesseling, C., D. Antich, C. Hogstedt, A.C. Rodríguez, and A. Ahlbom. 1999. Geographical differences of cancer incidence in Costa Rica in relation to environmental and occupational pesticide exposure. *International Journal of Epidemiology* 28: 365–74.

Wesseling, C., A. Aragon, L. Castillo, M. Corriols, F. Chaverri, E. de la Cruz, M. Keifer, P. Monge, T.J. Partanen, C. Ruepert, and V. van Wendel de Joode. 2001. Hazardous pesticides in Central America. *International Journal of Occupational and Environmental Health* 7: 287–94.

Wesseling, C., M. Corriols, and V. Bravo. 2005. Acute pesticide poisoning and pesticide registration in Central America. *Toxicology and Applied Pharmacology* 207: 697–705.

Wheelwright, N.T., and N.M. Nadkarni, eds. 2014. *Monteverde: ecología y conservación de un bosque nuboso tropical.* Bowdoin Scholars' Bookshelf. Book 3. http://digitalcommons.bowdoin.edu /scholars-bookshelf/3

Whitfield, S.M., K.E. Bell, T. Philippi, M. Sasa, F. Bolaños, G. Chaves, J.M. Savage, and M.A. Donnelly. 2007. Amphibian and reptile declines over 35 years at La Selva, Costa Rica. *Proceedings of the National Academy of Sciences* 104: 8352–56.

Whitfield, S.M., K.R. Lips, and M.A. Donnelly. In press. Decline and conservation of amphibians in Central America. In H.H. Heatwole, C. Barrio-Amoros, and J.W. Wilkenson, eds., *Status of Conservation and Declines of Amphibians: Western Hemisphere*. Sydney: Surrey Beatty and Sonos, Pty. Ltd.

Wilf, P., C.C. Labandeira, W.J. Kress, C.L. Staines, D.M. Windsor, A.L. Allen, and K.R. Johnson. 2000. Timing the radiations of leaf beetles: hispines on gingers from latest Cretaceous to recent. *Science* 289: 291–94.

Willis, K.J., L. Gillson, and T.M. Brncic. 2004. How "virgin" is virgin rainforest? *Science* 304: 402–3.

Wirth, R., W. Beyschlag, R.J. Ryel, and B. Holldöbler. 1997. Annual foraging of the leaf-cutting ant *Atta colombica* in a semideciduous rain forest in Panama. *Journal of Tropical Ecology* 13: 741–57.

Wood, T.E., D. Lawrence, and D.A. Clark. 2005. Variation in leaf litter nutrients of a Costa Rican rain forest is related to precipitation. *Biogeochemistry* 73: 417–37.

Wood, T.K. 1993. Diversity in the New World Membracidae. *Annual Review of Entomology* 38: 409–35.

Woodburne, M.O. 2010. The Great American Biotic Interchange: dispersals, tectonics, climate, sea level and holding pens. *Journal of Mammalian Evolution* 17: 245–64.

Woodburne, M., A.L. Cione, and E.P. Tonni. 2006. Central American provincialism and the Great American Biotic Interchange. In

O. Carranza-Castañeda and E.H. Lindsay, eds., Advances in late Tertiary vertebrate paleontology in Mexico and the Great American Biotic Interchange. Publicación Especial del Instituto de Geología y Centro de Geociencias de la Universidad Nacional Autónoma de México, vol. 4, 73–101.

Wooton, J.T., and M.P. Oemke. 1992. Latitudinal differences in fish community trophic structure, and the role of fish herbivory in a Costa Rican stream. *Environmental Biology of Fishes* 35: 311–19.

World Resources Institute. 2007. http://www.wri.org.

Wright, S.J., O. Calderón, A. Hernandéz, and S. Paton. 2004. Are lianas increasing in importance in tropical forests?: a 17-year record from Panama. *Ecology* 85: 31–45.

Wright, S.J., H. Zeballos, I. Domínguez, M.M. Gallardo, M.C. Moreno, and R. Ibañez. 2000. Poachers alter mammal abundance, seed dispersal, and seed predation in a neotropical forest. *Conservation Biology* 14: 227–39.

Wu, S.T., and A.T. Joyce. 1988. The application of an integrated knowledge-based Geographic Information System for the study of mountainous tropical forest. In *Technical Papers of the 1988 ACSM-ASPRS Annual Convention* 6: 12–20.

Yearout, R., X. Game, K. Krumpe, and C. Mckenzie. 2008. Impacts of DBCP on participants in the agricultural industry in a third world nation (an industrial health and safety case study of a village at risk). *International Journal of Industrial Ergonomics* 38: 127–34.

You, C.F., P.R. Castillo, J.M. Gieskes, L.H. Chan, and A.J. Spivack. 1996. Trace element behavior in hydrothermal experiments: implications for fluid processes at shallow depths in subduction zones. *Earth and Planetary Science Letters* 140: 41–52.

Young, A.M. 1980. Environmental partitioning in lowland tropical rain forest cicadas. *Journal of the New York Entomological Society* 88: 86–101.

Young, A.M. 1984. On the evolution of cicada × host-tree associations in Central America. *Acta Biotheoretica* 33: 163–98.

Young, B.E., T.W. Sherry, B.J. Sigel, and S. Woltmann. 2008. Edge and isolation effects nesting success of Costa Rican forest birds. *Biotropica* 40: 615–22.

Zaldivar, M.E., O.J. Rocha, G. Aguilar, L. Castro, E. Castro, and R. Barrantes. 2004. Genetic variation of cassava (*Manihot esculenta* Crantz) cultivated by Chibchan Amerindians of Costa Rica. *Economic Botany* 58: 204–13.

Zaldivar, M.E., O.J. Rocha, E. Castro, and R. Barrantes. 2002. Species diversity of edible plants grown in homegardens of Chibchan Amerindians from Costa Rica. *Human Ecology* 30: 301–16.

Zamora, C.O., and F. Montagnini. 2007. Seed rain and seed dispersal agents in pure and mixed plantations of native trees and abandoned pastures at La Selva Biological Station, Costa Rica. *Restoration Ecology* 15: 453–61.

Zamudio, K.R., and H.W. Greene. 1997. Phylogeography of the bushmaster (*Lachesis muta*: Viperidae) implications for Neotropical biogeography, systematics, and conservation. *Biological Journal of the Linnean Society* 62: 421–42.

Zhang, J., B. Rivard, A. Sánchez-Azofeifa, and K. Castro-Esau. 2006. Intra- and inter-class spectral variability of tropical tree species at La Selva, Costa Rica: implications for species identification using HYDICE imagery. *Remote Sensing of Environment* 105: 129–41.

Part VIII

The Caribbean Sea and Shore

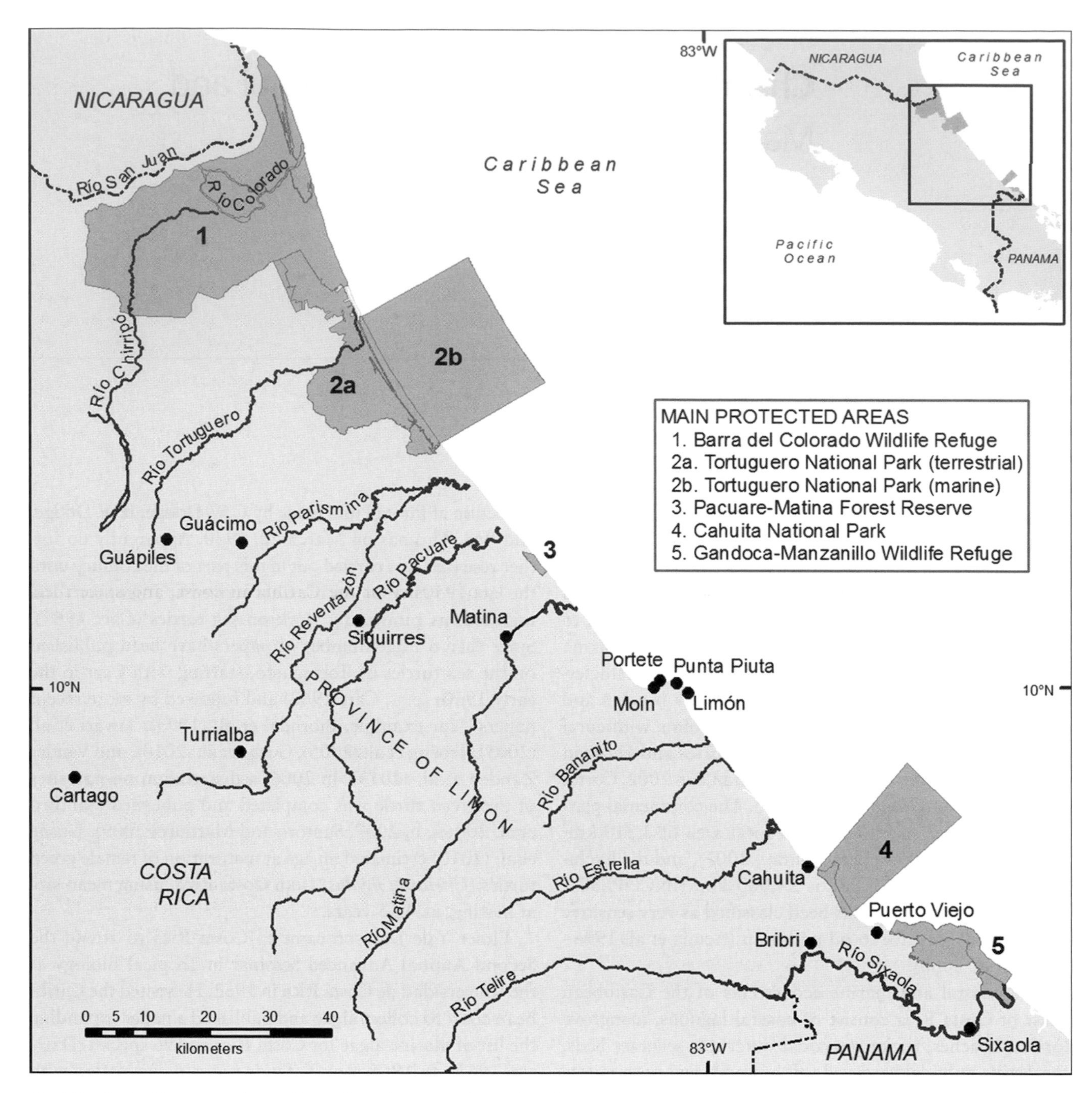

Fig. 17.1 The Caribbean coast of Costa Rica and its Marine Protected Areas (MPAs).
Map prepared by Marco V. Castro.

same year, also at Portete, the sea urchins *Echinometra lucunter* (Linnaeus, 1758), *Tripneustes ventricosus* (Lamarck, 1816), and *Lytechinus variegatus* (Lamarck, 1816), to determine lipid reserves in their guts (Lawrence 1967). Joseph Richard Houbrick reported on 229 species of mollusks that he collected in 1966 in Portete and Barra del Colorado. The collection consisted of 181 species of gastropods, 35 pelecypods, 3 scaphopods, 8 chitons, and 2 cephalopods. Fifty-one species were new records for the western Caribbean (Houbrick 1968). In 1968, a plankton tow was taken offshore as part of a US Navy project, and Michel and Foyo (1976) found species of the following groups: Siphonopora, Heteropoda, Copepoda, Euphausicea, Chaetognatha, and Salpidae. In 1971, Deborah M. Dexter, sponsored by the Organization for Tropical Studies (OTS), collected samples from five sandy beaches between Limón and Cahuita with the aim to study the macro-infauna, and reported 52 species, mainly polychaetes and crustaceans (Dexter 1974); the polychaetes were described by Fauchald (1973). As part of the RV *Pillsbury* Cruise to Central America in 1971, bottom trawls and plankton tows were done at 11 sites along the coast and specimens of sponges, black corals, mollusks, crabs, crinoids, sea stars, holothurians, polychaetes, pennatulids, and nemerteans were collected (Voss 1971). Other than the cruise report by Gilbert L. Voss, only one publication has been found in which specimens collected during that expedition are mentioned: Williams' (1993) paper on mud shrimps collected east of Limón. The RV *Alpha Helix*, Scripps Institution of Oceanography, during the Belém to Belize City Expedition, collected one kilometer south of the Port of Limón on July 9, 1977, and a paper was published on four species of pelagic copepods (Ferrari and Bowman 1980).

The most important marine ecosystems of the Caribbean, in terms of area, species richness, and economic value, are the coral reefs. The first study was by Gerard M. (Jerry) Wellington while with the Peace Corps in 1970. He described the coral reef at Cahuita and its associated flora and fauna (Wellington 1974a); and he published on the algae of the area (Wellington 1973, 1974b). The next investigation was also carried out at Cahuita by Risk et al. (1980) in 1978, when Michael J. (Mike) Risk came to Costa Rica to teach the first course on coral reefs offered in the country. Both of these studies noted that excessive terrigenous sedimentation was negatively affecting the coral reef at Cahuita. This motivated research on the effect of sediments on corals, and a thesis was done (Cortés 1981), and papers published (Cortés and Risk 1984, 1985). Work on the impact of human activity on the Caribbean coral reefs has continued (Cortés 1994, Fonseca and Cortés 2002, Roder 2005, Roder et al. 2009). After the initial studies at Cahuita, other reef areas of the Caribbean of Costa Rica were described: all the reefs along the coast (Cortés and Guzmán 1985a), at the Refugio Nacional de Vida Silvestre Gandoca-Manzanillo (Cortés 1992a), and at Punta Cocles (Fernández and Alvarado 2004). The largest coral reef is at Parque Nacional Cahuita, followed by the reefs at the Refugio Nacional de Vida Silvestre Gandoca-Manzanillo (Fig. 17.1).

In the late 1970s the first thesis on the Caribbean coast related to marine organisms was done by Claudia Charpentier. She studied the seasonal variation in the chemical composition of five species of algae from Cahuita (Charpentier 1980).

The early 1980s are marked by several events that significantly transformed the reefs of the Caribbean. The first was the 1982–83 El Niño Southern Oscillation (ENSO) event, during which the southern Caribbean experienced doldrums conditions with clear skies resulting in abnormal warming of the waters that led to extensive coral bleaching and death (Cortés et al. 1984). The second event was the mass mortality of the black sea urchin, *Diadema antillarum* (Philippi, 1845), a very important herbivore (Murillo and Cortés 1984). Finally, around that time, the first observation of massive death of the sea fan, *Gorgonia flabellum* Linnaeus, 1758, was recorded by Guzmán and Cortés (1984). These disturbances combined with the extreme terrigenous sedimentation resulted in a significant coral reef decline (Cortés 1994). Another bleaching event impacted the reefs in 1995 (Jiménez 2001).

The 1991 Limón Earthquake (7.5 Richter Scale), which uplifted the coast as much as 1.9 m in some areas, resulted in the death of intertidal and shallow subtidal organisms (Cortés et al. 1992, 1994). A significant impact was observed on the burrowing sea urchin, *E. lucunter*. Skeletons of hundreds of this urchin were cast on the rocky areas, and many were collected and measured to determine population structure (R. Soto et al., unpublished data).

The 1980s were also marked by an interest in the marine biodiversity of the Caribbean coast, and since then several theses have been written and papers published. A series of papers on algae were published during that decade: Kemperman and Stegenga (1983, 1986), Soto (1983), Stegenga and Kemperman (1983), and Soto and Ballantine (1986); and more recently, a paper by Thomas and Freshwater (2001), an inventory of macroalgal epiphytes on the seagrass *Thalassia testudinum* in Parque Nacional Cahuita (Samper-Villareal et al. 2008), a species list by Bernecker (2009), and new reports by Bernecker and Wehrtmann (2009).

Sponges from Cahuita and Isla Uvita were identified as part of a thesis (Loaiza-Coronado 1989) and later published (Loaiza-Coronado 1991), while Van der Hal (2006) identified and studied the distribution of sponges in seagrass beds. Cortés (1996) published a compilation of reported sponges from the Caribbean coast of Costa Rica, and an update (Cortés et al. 2009).

Some groups of cnidarians have been studied and papers published since the 1980s include hydroids (Cortés 1992b, Cortés 1996–1997, Kelmo and Vargas 2002, Cortés 2009a), octocorals (Guzmán and Cortés 1985, Guzmán and Jiménez 1989, Sánchez 2001, Breedy 2009), stony corals (Cortés and Guzmán 1985b, Cortés 1992c, 2009a), other cnidarians (Cortés 1996–1997, 2009b), and sea anemones (Acuña et al. 2013).

There has been an interest in mollusks for a long time, starting with the pioneering work by Houbrick (1968), followed by the collections done by David G. Robinson while working on his dissertation research: the systematics and paleoecology of Pleistocene gastropods (Robinson 1991). Robinson published with Michel Montoya a list of marine mollusks collected between 1982 and 1986, at ten localities of the Caribbean coast of Costa Rica: Barra del Colorado, Río Matina, Moín, Portete, Puerto Limón, Río Banano, Río Estrella, Cahuita, Puerto Viejo, and Manzanillo. Their inventory included 395 species: 288 gastropods, 100 bivalves, five polyplacophorans, and two cephalopods (Robinson and Montoya 1987). José Espinosa from Cuba and Jesús Ortea from Spain, in conjunction with the Instituto Nacional de Biodiversidad (INBio), did extensive collections between Cahuita and Gandoca from 1999 to 2001, and published many papers (e.g., Espinosa and Ortea 1999, 2000, 2001, Ortea et al. 1999, 2001, Ortea 2001, Espinosa et al. 2006). Finally, Rodríguez-Sevilla and coworkers (2003) published a compilation of all known mollusks of the Caribbean coast of Costa Rica. García-Ríos and Álvarez-Ruiz (2011) added five new records of chitons (Polyplacophora) to the list of 8 previously known species. Ornelas-Gatdula et al. (2012) studied the sea slug, *Navanax aenigmaticus*, and found that it consisted of three species, *N. gemmatus* being the species found on the Caribbean coast. Yolanda Camacho-García et al. (2014) reported on the diversity and distribution of sea slugs in this area.

The crustaceans have also received some attention and lists of species have been published, and new species described. Dora P. Henry and Patsy A. McLaughlin reported four species of barnacles collected from boat hulls in Portete (Henry and McLaughlin 1975). The first thesis on crustaceans from the Caribbean was by Odalisca Breedy on benthic microcrustaceans (Breedy 1986); she later published on the isopods (Breedy and Murillo 1995). The other paper on isopods of the Caribbean was by Regina Wetzer and Niel L. Bruce, in which they described a new genus and species that inhabit the shallow waters of Parque Nacional Cahuita, with specimens collected by Richard C. Brusca and P.M. Delaney (Wetzer and Bruce 1999). Paul S. Young described a new species of coral-inhabiting barnacle found in Brazil and Cahuita (Young 1989). Several papers have been published on copepods, starting with the report of four species from the Caribbean of Costa Rica by Ferrari and Bowman (1980), and another on a calanoid species (Walter 1989). Wolfgang Mielke collected benthic copepods in Costa Rica between August and September 1990, and published a series of papers that report and describe new species and subspecies from the Caribbean (Mielke 1992, 1993, 1994). Álvaro Morales Ramírez in his compilation of copepods of Costa Rica included 13 species from the Caribbean (Morales-Ramírez 2001, Morales-Ramírez et al. 2014). Dennis A. Moran and Ana I. Dittel in their publication on anomuran and brachyuran crabs of Costa Rica included the known Caribbean species (Moran and Dittel 1993). Austin B. Williams (1993) in his paper on mud shrimps from the western Atlantic included species from Costa Rica. Rita Vargas and Jorge Cortés published a series of compilations on several groups of crustaceans (Vargas and Cortés 1997, 1999, 2006), while Ingo S. Wehrtmann and Rita Vargas reported new records and range extensions of shrimps (Wehrtmann and Vargas 2003). Iorgu Petrescu and Richard W. Heard described a new species of cumacean collected in Parque Nacional Cahuita (Petrescu and Heard 2004). Eduardo Suárez-Morales et al. (2013) reported on a group of planktonic copepods from Cahuita. Other papers published on crustaceans include a study of the coral biocroder shrimp *Alphaeus simus* Guérin-Meneville, 1856, at Cahuita (Cortés 1985). Ronald Umaña and Didiher Chacón studied post-larvae of the lobster *Panulirus argus* (Latreille, 1804) (Umaña and Chacón 1994). Wehrtmann and Albornoz (2002) found differences in the reproductive traits of two species of *Alpheus*, one from the Caribbean and another from the Pacific. Terossi et al. (2010a), studying populations of the shrimp *Hippolyte obliquimanus* in Brazil and Costa Rica, found that it has a high plasticity of reproductive features, not only as adults, but also in the larval development (Terossi et al. 2010b). Azofeifa-Solano et al. (2014) reported on the reproductive biology of an anemone shrimp.

Three papers have been published on fisheries-related topics; the first was on crustaceans: an assessment of pink shrimp populations, *Farfantepenaeus brasiliensis* (Latreille, 1817) (reported as *Penaeus brasiliensis*) (Tabash-Blanco 1995). The second was on mollusks: a comparison of populations of the West Indian topshell, *Cittarium pica*, in exploited and protected sites (Schmidt et al. 2002), and the third, on post-larval settlement of the spiny lobster, *Panulirus argus* (Latreille, 1804) in Cahuita National Park (González and Wehrtmann 2011).

There have been a few taxonomical studies of echinoderms. Juan José Alvarado and Jorge Cortés published a list of known echinoderms of Costa Rica including Caribbean species (Alvarado and Cortés 2004). Natalie Bolaños and coworkers, with the help of J.J. Alvarado, published on the diversity and abundance of echinoderms in the reef lagoon at Cahuita (Bolaños et al. 2005). Gordon Hendler described a species of ophiuroid that he collected, and dedicated it to the Centro de Investigación en Ciencias del Mar y Limnología (CIMAR) of the Universidad de Costa Rica (Hendler 2005). Another study on echinoderms was carried out by Marta F. Valdez and Carlos Villalobos, who determined the spatial distribution, correlation with the substrate, and aggregations of the black sea urchin *D. antillarum* at Cahuita (Valdez and Villalobos 1978). Papers were published on the mass mortality of *D. antillarum* in 1983 and 1992 (Murillo and Cortés 1984, Cortés 1994), and on its subsequent recovery (Alvarado et al. 2004, Fonseca et al. 2006, Myhre and Acevedo-Gutiérrez 2007). The diversity of echinoderms of the Caribbean coast must be higher than reported, when compared to the species lists of neighboring countries (Alvarado 2011).

The first paper published on fishes of the Caribbean coast of Costa Rica was by Gilbert and Kelso (1971), covering the fishes of the Tortuguero area, mainly freshwater species. Another paper from that area was prepared by Winemiller and Leslie (1992), who studied the fish communities in the freshwater-marine ecotone at the river mouths and coastal lagoons. Phillips and Perez-Cruet (1984) compared reef fishes from Cahuita and the Pacific. Didiher Chacón of the non-govermental organization ANAI (National Association for Indigenous Matters) published on the tarpon (Chacón 1993, Chacón and McLarney 1992). This NGO has also published on the leatherback turtle population that nests in Gandoca (Chacón et al. 1996, Chacón 1999). Bussing and López (2009) in their chapter on marine fishes of Costa Rica included a list of species of the Caribbean coast. Salas et al. (2010) found that there is weak genetic connectivity between populations of a damselfish, *Stegastes partitus*, from Costa Rica-Panamá (CR-PAN) and the Mesoamerican Barrier Reef System. They also noticed strong self-recruitment in CR-PAN. Benavides-Morera and Brenes (2010) reported 13 species of fishes captured at Laguna Gandoca, with 77% of the catch consisting of one species, *Centropomus pectinatus*. They also found that salinity was important in the temporal and spatial distribution of the icthyofauna of the lagoon. An important contribution to knowledge of the biodiversity of the Caribbean coast of Central America is the Bussing and López (2010) compendium on the coastal and marine fishes of the Caribbean Coast of Lower Central America, which includes all the species known from the Caribbean coast of Costa Rica.

Ignacio Jiménez did a thesis on the ecology and conservation of the West Indian manatee *Trichechus manatus* Linnaeus, 1778, in the northeast of Costa Rica (Jiménez 1998). He later published two papers, one on the state of conservation, ecology, and popular knowledge of the manatee (Jiménez 1999), and the other on a predictive model of the distribution of the manatee based on studies along the northern Caribbean coast of Costa Rica (Jiménez 2005). He found that manatees prefer lagoons to other watercourses and indicated the threat of forest clearing to manatee conservation. Previously, Reynolds et al. (1995) found that the population of the manatee has been declining and attributed this to illegal hunting, increases in motorboat traffic, high levels of pollutants, and the ingestion of plastic bags. Ignacio Jiménez has established the Fundación Salvemos al Manatí de Costa Rica (Save the Costa Rican Manatee Foundation).

Laura May-Collado and colleagues studied dolphin whistles along the southern Caribbean coast of Costa Rica. May-Collado and Wartzok (2008) determined that the common bottlenose dolphins (*Tursiops truncatus*) can change their whistle structure depending on local conditions. The Guyana dolphin's (*Sotalia guianensis*) whistles have a higher frequency in Costa Rica than in well-studied populations in Brazil (May-Collado and Wartzok 2009). Finally, May-Collado (2010) found changes in the whistle structure of populations in Guyana and bottlenose dolphins when interacting along the Caribbean coast of Costa Rica.

Other studies on marine biodiversity include one on sipunculids (Cutler et al. 1992), and a thesis on polychaetes of Parque Nacional Cahuita by Victoria Bogantes Aguilar, increasing the number of species to over 50 (Bogantes-Aguilar 2014). The zooplankton of Parque Nacional Cahuita with emphasis on echinoderm larvae was studied by Álvaro Morales-Ramírez (1987) in his Master thesis and later published (Morales and Murillo 1996). Allan Carrillo-Baltodano (2012) re-surveyed 25 years later, the same stations as Morales-Ramírez and found significant differences regarding some but not all groups. Dominici-Arosemena et al. (2000) published a paper on the ichthyoplankton of the Limón area. A compilation on marine biodiversity of the Caribbean coast of Costa Rica was published by Cortés and Wehrtmann (2005). The book "Marine Biodiversity of Costa Rica, Central America" includes numerous chapters with lists of species from the Caribbean (Wehrtmann and Cortés 2009).

Other topics studied include primary productivity and biomass of phytoplankton in the coral reef at Parque Nacional Cahuita (Silva-Benavides 1986), mangrove ecology and pollution assessment (Coll et al. 2001, 2004), mangrove monitoring (Fonseca et al. 2007a, Cortés et al. 2010), seagrass bed ecology (Paynter et al. 2001, Krupp 2006, Nielsen-Muñoz 2007, Nielsen-Muñoz and Cortés 2008, Krupp et al. 2009), and seagrass bed monitoring (Fonseca et al. 2007b, Cortés et al. 2010, Van Tussenbroek et al. 2014). Schmidt et al. (2002) published on the population ecology and fishery of the gastropod *Cittarium pica* (Linnaeus, 1758). Jiménez and Cortés (1993) published on the compressive strength and density of the coral *Siderastrea siderea* (Ellis and Solander, 1786), while Muller-Parker and Cortés (2001) focused on the distribution of light and nutrients in Cahuita, and Acevedo-Gutiérrez et al. (2005) addressed the social interactions between bottlenose and tucuxis dolphins in the Gandoca-Manzanillo area.

Most of the research on marine pollution has been car-

Fig. 17.2 Notches indicating previous sea levels in Punta Portete. The lower one was the level before the Limón Earthquake of 1991.
Photograph by Jorge Cortés.

following sequence is the Suretka Formation (Pliocene-Pleistocene) and is made up of volcanic and plutonic blocks and coarse grains. The Late Pleistocene-Holocene deposits known as the Limón Formation are from reef origin, and form the present-day rocky points on the southern Caribbean coast of Costa Rica, where extant coral reefs grow (Taylor 1975, Coates et al. 1992, Fernández et al. 1994, Denyer 1998, Linkimer and Aguilar 2000, McNeill et al. 2000, Denyer et al. 2003, Brandes et al. 2007). For additional details on the geology of this region reference is made to Alvarado and Cárdenes (chapter 3 of this volume).

Waves, Tides, and Circulation

Waves are normally from the northeast, but when there are storms or hurricanes in the Caribbean Sea waves may come from other directions as well (Lizano 2007). If they are strong enough, they can break corals and lift sediments from the reef bottom (Cortés 1981). The tides on the Caribbean coast of Costa Rica are mixed or semidiurnal depending on the moon phase, and have a range of less than 40 cm. The tides are affected by wind direction and force, atmospheric pressure, currents, and waves (Lizano 2006).

The main currents along the coast are from the northwest to the southeast, as part of the persistent cyclonic Panama-Colombia Gyre, the coastal branch of which is sometimes referred to as the Panama-Colombia Countercurrent (Mooers and Maul 1998, Andrade et al. 2003, Centurioni and Niiler 2003). Strong eddies in the opposite direction—that is, SE to NW, along the coast—have been observed. These currents distribute sediments, nutrients, pollutants, and organisms along the coast. The transport of sediments and pollutants is impacting the coral reefs in a negative manner (Cortés and Risk 1985, Cortés et al. 1998).

Ecosystems

Coastal Lagoons

Characteristic of the northern section of the Caribbean coast of Costa Rica are the elongated coastal lagoons parallel to the coastline (Parkinson et al. 1998, Denyer and Cárdenas 2000) (Fig. 17.3). The coastal lagoons, including a few dredged sections, extend from Moín to Barra del Colorado, connecting all the northern section of the Caribbean coast. The flooded lands along the channels are covered by the tree *Pterocarpus officinalis* Jacq. and the palm *Raphia taedigera* Mart., 1828.

Studies of the fauna and flora of the coastal lagoons have been scarce, although a few papers have been published on the lagoons' fish (Gilbert and Kelso 1971, Winemiller and Leslie 1992), as well as on the manatees. Two studies have shown that manatee populations are declining mainly owing to human action (Reynolds et al. 1995, Jiménez 1999). Jiménez (2005) found that forest clearings are a threat to manatee conservation because they expose manatees more to human populations.

A paper was published on hydrocarbon pollution levels that were higher near the refinery and low at the northern end of the Canales de Tortuguero (Acuña et al. 1986).

On the basis of geomorphic, sedimentologic, and stratigraphic data obtained in the Barra de Colorado area, Parkinson et al. (1998) concluded that the north Caribbean region is a wave-dominated delta that has formed along a passive continental margin. This tectonic setting contrasts with the southern section of the Caribbean coast and the Pacific coast that are seismically active continental margins. Much more study is needed in these important coastal lagoons.

Fig. 17.3 Canales de Tortuguero, a coastal lagoon parallel to the shoreline. *Photograph by Jorge Cortés.*

Mangrove Forests

The reduced tidal range and the geomorphology of the Caribbean coast of Costa Rica are not propitious for the development of mangrove forests (Coll et al. 2001, Cortés et al. 2001). Only two well-developed mangrove forests exist on the Caribbean coast, while there are isolated trees in several areas (Cortés 1991, Cortés et al. 2001, Sánchez-Vindas 2001, personal observation). One of the mangrove forests is found near the main port of the Caribbean, Moín, and it has been greatly impacted by construction and pollution (Cortés et al. 2001). This mangrove forest was studied in the mid-1970s and was found to be structurally more complex than Pacific stands with the tallest trees belonging to *Pterocarpus officinalis*, which in fact is not considered a nuclear species of the mangrove forest (Pool et al. 1977).

The other mangrove forest of the Caribbean coast of Costa Rica is in much better condition. It is located at Laguna Gandoca (Figs. 17.1, 17.4) within the Refugio Nacional de Vida Silvestre Gandoca-Manzanillo (Coll et al. 2001) where pollution has been very low (Coll et al. 2004). The mangrove forest in Laguna Gandoca is the largest on the Caribbean coast of Costa Rica (12.5 ha), and this area is three times the area it had in 1976 (Coll et al. 2001). The forest consisted of *Rhizophora mangle* Linnaeus, 1753, *Avicennia germinans* (Linnaeus) Stearn, 1958, *Laguncularia racemosa* (Linnaeus) C.F. Gaertn., 1805, and *Conocarpus erectus* Linnaeus, 1753.

The mangrove forest at Gandoca has been monitored since 1999 following the Caribbean Coastal Marine Productivity or CARICOMP protocol (CARICOMP 1997). The peak in productivity and flowering was in July, the warmest month. Productivity apparently declined during the study period while the mean temperature went up. Biomass (14 kg/m^2) and density (9 trees/10 m^2) were low compared to other CARICOMP sites, but productivity in

July (4 g/m²/day) was similar to most CARICOMP sites (Fonseca et al. 2007a).

Beaches

The entire northern section of the Caribbean coast comprises a series of high-energy sandy beaches (Fig. 17.5), which are very important for turtle nesting (Troëng and Rankin 2005), while in the southern section beaches are present between rocky points (Cortés and Guzmán 1985a, Cortés and Jiménez 2003). The color of the beaches ranges from white—consisting almost exclusively of carbonates—to pitch-black beaches made up of magnetite. The presence of pink and golden beach colors depends on the proportions of marine organism fragments, terrigeneous sediments, and weathering (Cortés et al. 1998).

Dexter (1974) has published the only paper that addresses sandy-beach fauna. She identified and quantified invertebrates from five sandy beaches along the southern section of the Caribbean coast (Playa Bonita, Airport Beach, Cahuita North, Cahuita South, Puerto Viejo). The most abundant species was the isopod *Excirolana braziliensis* Richardson, 1912 (reported as *Cirolana salvadorensis*) followed by the polychaete *Scolelepis* (*Scolelepis*) *squamata* (Müller, 1806) (reported as *Scolelepis agilis*) and the bivalve *Donax denticulatus* Linnaeus, 1758. Other groups found were nematodes, nemerteans, oligochaetes, cumaceans, amphipods, decapods, and gastropods.

Four species of marine turtles nest on the beaches of the Caribbean coast: *Chelonia mydas* (Linné, 1758) (green turtle), *Dermochelys coriacea* (Vandelli, 1761) (leatherback), *Eretmochelys imbricata* (Linné, 1766) (hawksbill), and sporadically *Caretta caretta* (Linné, 1758) (loggerhead) (Savage 2002, Troëng 2005). Some beaches at the Caribbean coast of Costa Rica are considered turtle nesting sites of global importance—for example, Tortuguero for the green (Troëng

Fig. 17.4 Rhizophora mangle, the predominant mangrove tree in the Laguna Gandoca, Refugio Nacional de Vida Silvestre Gandoca-Manzanillo. *Photograph by Jorge Cortés.*

Fig. 17.5 A beach within the Parque Nacional Tortuguero, an important nesting site for the green turtle, *Chelonia mydas*. *Photograph by Peter O. Baumgartner.*

and Rankin 2005), and Gandoca for the leatherback turtle (Chacón et al. 1996, Chacón 1999, Troëng et al. 2004).

Rocky Intertidal Zone

Rocky points are present only in the southern section of the Caribbean coast of Costa Rica and are made up of carbonate rocks, fossil coral reefs, and beachrocks (Fig. 17.6). The assemblages of the rocky intertidal zones have not been studied extensively. There are some reports of mollusks collected here (Houbrick 1968, Robinson and Montoya 1987, Espinosa and Ortea 2001) and there is one paper on the endolithic fauna of carbonate substrates observed in Cahuita (Pepe 1985). Nothing has been published on the geographic distribution or vertical zonation of organisms that inhabit the rocky intertidal zone.

Schmidt et al. (2002) studied the West Indian topshell, *C. pica*, at three locations along the Caribbean coast, two where these snails are collected artisanally for food, and one where it is protected. They found that the average population density at the unexploited site, being 14 ind/m^2, was three times higher than at sites where it was extracted. In the exploited sites the animals were smaller, but growth rates were not significantly different from the non-exploited site. They recommended that the minimum size for extraction should be 40 mm and that the fishery should be closed from July to November, which coincides with the reproductive season.

The entire intertidal and shallow subtidal zones were seismically uplifted during the 1991 Limón Earthquake, in some areas as much as 1.9 m (Denyer et al. 1994a,b, Denyer 1998). The subaerial exposure of marine organisms resulted in their massive death (Cortés et al. 1992, 1994). This process of coastal uplift has been going on for a long time as evidenced by notches in the rocks of the Limón area and fossil coral reefs found inland (Denyer 1998, Cortés and Jiménez 2003) (Figs. 17.2, 17.6).

Seagrass Beds

Four species of seagrasses have been reported from the Caribbean coast of Costa Rica (Cortés and Salas 2009):

Thalassia testudinum Banks ex König, 1805 (turtle grass), *Syringodium filiformis* Kützing, 1860 (manatee grass), *Halophila decipiens* Ostenfeld, 1902, and *Halodule wrightii* Ascherson, 1868. The first two species form extensive beds mainly in the reef lagoons (Cortés and Guzmán 1985a, Cortés et al. 1992, Nielsen-Muñoz and Cortés 2008, Krupp et al. 2009). Seagrass beds are shallow, very productive environments (Fig. 17.7). In Cahuita the peak of production was in July (Fonseca et al. 2007b, Cortés et al. 2010), and higher at sites with intermediate environmental conditions and sediment grain size (Paynter et al. 2001).

A seagrass bed in Parque Nacional Cahuita has been monitored using the above-mentioned CARICOMP protocol. Average productivity of *T. testudinum* was 2.7 ± 1.15 $g/m^2/day$, which is intermediate when compared to other CARICOMP sites. Total biomass of *T. testudinum* (822.8 ± 391.84 g/m^2) was intermediate to high, and shoot density (1,188 ± 335.5 $shoots/m^2$) was higher than other CARICOMP sites (Fonseca et al. 2007b). Flowering of *T. testudinum* was observed starting in April and continuing until June with a peak in May, while fruit production continued until August. The sex ratio was two males for every female flower in 2004, and four males/female in 2005 and 2006 (Nielsen-Muñoz 2007, Nielsen-Muñoz and Cortés 2008). When compared to other CARICOMP sites the productivity in Costa Rica results to be intermediate although it has been declining (Van Tussenbroek et al. 2014).

Seagrass beds, almost exclusively made up of *T. testudinum*, in the Refugio Nacional de Vida Silvestre Gandoca-Manzanillo covered an area of about 16 ha (Krupp 2006, Krupp et al. 2009). Environmental conditions and especially depth controlled the average canopy cover and aboveground biomass of *T. testudinum*, as well as levels of shoot densities, productivity, and leaf sizes (Krupp et al. 2009).

Coral Reefs

As indicated above, the southern section of the Caribbean coast of Costa Rica has carbonate promontories, mainly comprised of fossil coral reefs (Pleistocene, Holocene) (Fig. 17.8), and beachrock in some sections (Cahuita, Punta Uva). Present day coral reefs are growing over those rocky outcrops. The coral reefs are distributed in three discrete sections (Cortés and Guzmán 1985a, Cortés and Jiménez 2003) (Fig. 17.1): (1) fringing reefs between Moín and Limón, including Isla Uvita; (2) the largest fringing reef of the coast plus patch reefs and carbonate banks at Cahuita National Park (Fig. 17.9); and (3) fringing and patch reefs, carbonate banks, and an algal ridge between Puerto Viejo and Punta Mona. The two most important economic activities along the southern Caribbean coast of Costa Rica, tourism and fisheries, both depend on these coral reefs.

Natural and anthropogenic disturbances have negatively impacted the coral reefs at the Caribbean coast. The main recent threats to the reefs are coral bleaching (Cortés et al. 1984, Jiménez 2001), the massive death of the black sea urchin, *D. antillarum* (Murillo and Cortés 1984, Alvarado et al. 2004), and the 1991 Limón Earthquake, which uplifted the coast significantly (Cortés et al. 1992), all of which have taken their toll on the reefs. But the most important impacts along the Caribbean coral reefs today are related to human activities: increased terrigenous sediment loads

Fig. 17.6 Rocky intertidal zone (fossil reefs) at Isla Uvita. Note the erosional notches that mark the sea level before the 1991 Limón earthquake. *Photograph by Jorge Cortés.*

Fig. 17.7 Seagrass beds in the reef lagoon at Parque Nacional Cahuita. Monospecific bed of *Thalassia testudinum* on the left, and *Syringodium filiformis* on the right. *Photo credits—Left, Ingo S. Wehrtmann; right: Vanessa Nielsen.*

Fig. 17.8 Fossil corals at Isla Uvita: *Diploria strigosa* on the left and piles of *Acropora palmata* on the right. *Photographs by Jorge Cortés.*

resulting from deforestation and bad agricultural practices (Cortés and Risk 1985, Cortés 1994), overexploitation of resources (J. Cortés, personal observation), tourism, and damage by anchoring (unpublished data) (Fig. 17.10).

One coral reef in Parque Nacional Cahuita has been monitored using the CARICOMP protocol continously since 1999. Live coral cover increased from 15 to 17% while coralline algal cover decreased from 17 to 5%. Unfortunately, macroalgal cover has increased from 63 to 74% (Fonseca et al. 2006). Other coral reefs along the coast have also shown signs of recovery (unpublished data) but live coral coverage is still far below what it was or should be for a similar region with lower levels of anthropogenic impact.

Isla Uvita

Isla Uvita is the only relatively large island on the Caribbean coast of Costa Rica (Figs. 17.1, 17.11). It is historically important since it was Christopher Columbus' only landing site in Costa Rica when he visited the coast during his

Fig. 17.11 Isla Uvita, a natural laboratory just off the Port of Limón. *Photograph by Jorge Cortés.*

Even today, more than 22 years after the earthquake, there are sections of the mountains that have not recovered completely, and consequently generate sediments during strong storms that are deposited downslope affecting the coast (E. Junier, pers. comm., 2007).

Human activity inland, including pollution and sedimentation, does have a significant impact on the coastal area. Acuña et al. (1986) found high levels of pollutants where the Río Parismina—which receives water from towns located inland—meets the Canales de Tortuguero. Other sources of pollution are close to the shoreline, at the ports and refinery (Mata et al. 1987, Acuña-González et al. 2004). High levels of terrigenous sedimentation from deforested mountains inland are found along the Caribbean coast of Costa Rica and are negatively impacting coral reefs (Cortés and Risk 1985, Fonseca and Cortés 2002, Cortés and Jiménez 2003, Mora-Cordero 2005). The impact of those sediments on the reefs is the main anthropogenic disturbance to the coast. In recent years coastal alteration (urban growth, development of tourist facilities) and possibly overexploitation of marine resources have impacted coastal and marine ecosystems along the Caribbean coast, although there are no studies on this issue available yet. Schmidt et al. (2002) have published the only paper that exists on overexploitation of a marine organism at the Caribbean coast. However, the author of this chapter has observed a significant decline in lobster populations and those of commercially important fish.

Declines in fish populations have been reported Caribbean-wide by Paddock et al. (2009). Central America, including the Caribbean coast of Costa Rica, turns out to be the Caribbean subregion that has the highest levels of decline (about 6% per year between 1996 and 2007).

Climate Change

The impact of climate change on Costa Rica's coastal areas has not yet been studied. However, several changes in coastal climates are expected. First, sea level rise will occur and result in greater levels of coastal erosion, which actually is already taking place along the Caribbean coast (Cortés and León 2002). This will result in the generation of more sediment. Also, there will be intrusions of saline water towards the inland. Furthermore, seawater is expected to get warmer while acidity (pH) levels are expected to go down (Kleypas et al. 2006, IPCC 2007). Both processes will have a severe impact on Costa Rica's Caribbean coral reefs over the coming period.

Conservation

Marine Protected Areas

There are three Marine Protected Areas (MPAs) at the Caribbean coast that include both terrestrial and marine areas [note: only the size of the marine portion is indicated in parentheses here] (Fig. 17.1): Parque Nacional Tortuguero (52,681 ha), Parque Nacional Cahuita (23,290 ha), and Refugio Nacional de Vida Silvestre Gandoca-Manzanillo (4,983 ha) (Mora et al. 2006). The first protected area is intended to safeguard turtle nesting beaches and fishing

Fig. 17.12 Main natural impacts on the coral reefs of the Caribbean coast of Costa Rica. Top left: Bleached elkhorn coral, *Acropora palmata*, in 1983; the coral has lost its symbiotic algae (Zooxanthellae) and the white skeleton is visible. Top right: Dead black sea urchin, *Diadema antillarum*, in Cahuita National Park in 1983. Bottom: Uplifted coral, *Siderastrea siderea*, five days after the 1991 Limón Earthquake.
Photographs by Jorge Cortés.

ticular to inform decision-making on policies for resource use and management.

Other Needs

The afore-mentioned presence of private reserves and the engagement of NGOs have proven to be important factors that contribute to successful protection, conservation, and sustainable use of the coastal and marine systems and their organisms. A more extensive protected area coverage and stronger stakeholder involvement should be encouraged, though, especially in relation to fisheries. There is also a need for all kinds of educational materials concerning marine ecosystems found at the Caribbean coast. And there should be much more active participation of local communities in fund-raising, managing, and protecting the coastal and marine ecosystems in the area. Also, there is a strong need for field stations that could facilitate the study of Caribbean marine environments in Costa Rica in situ. Isla Uvita and Cahuita seem to be ideal sites for the establishment of such research stations. And finally, better international cooperation is needed in order to advance research on topics for which there is no in-country expertise or equipment available.

Conclusions

The Costa Rican Caribbean coastal and marine ecosystems are species rich (Cortés and Wehrtmann 2005, Wehrtmann et al. 2009), economically important for local communities, and key to sustainable development in the region. However, they are being threatened by natural and anthropogenic factors. Hence, human disturbances must be eliminated or at least minimized, to ensure a healthy future for all coastal and marine ecosystems that thrive along Costa Rica's Caribbean shores.

Acknowledgments

I thank Maarten Kappelle and the late Luis Diego Gómez for the invitation to contribute to this book. Economic support to write this chapter was provided by the Universidad de Costa Rica (UCR) through the Escuela de Biología. Fieldwork was supported by the Vicerrectoría de Investigación and the Centro de Investigación en Ciencias del Mar y Limnología (CIMAR), both at UCR. Juan José Alvarado, Laura May-Collado, and Sebastian Tröeng provided valuable information. Eric Alfaro, Omar Lizano, Percy Denyer, and Teresita Aguilar helped with literature searches and commented on different sections of this chapter. Xochilt Lezama Cáceres helped with the maps. My thanks to Richard Petersen, Harlan Dean, Earl Junier Wade, and José A. Vargas for doing a critical review of the chapter. I appreciate the support over many years of teachers, students, colleagues, assistants, protected areas personnel, and local inhabitants who have helped me get to know the Caribbean region of Costa Rica.

References

Acevedo-Gutiérrez, A., A. DiBerardinis, S. Larkin, L. Larkin, and P. Forestell. 2005. Social interactions between tucuxis and bottlenose dolphins in Gandoca-Manzanillo, Costa Rica. *Latin American Journal of Aquatic Mammals* 4: 49–54.

Acuña, F.H., A. Garese, A.C. Excoffon, and J. Cortés. 2013. New records of sea anemones (Anthozoa: Actiniaria) from Costa Rica. *Revista de Biología Marina y Oceanografía* 48: 177–84.

Acuña, J., M.M. Murillo, and F. Araya. 1986. Estudio preliminar sobre la presencia de hidrocarburos de petróleo en la zona fluvial Río Moín-Canales Tortuguero. *Ingeniería y Ciencia Química* 10: 59–60.

Acuña, J.A., J. Cortés, and M.M. Murillo. 1996–1997. Mapa de sensibilidad ambiental para derrames de petróleo en las costas de Costa Rica. *Revista de Biología Tropical* 44(3)/45(1): 463–70.

Acuña-González, J.A., J.A. Vargas-Zamora, E. Gómez-Ramírez, and J. García-Céspedes. 2004. Hidrocarburos de petróleo, disueltos y dispersos, en cuatro ambientes costeros de Costa Rica. *Revista de Biología Tropical* 52(Suppl. 2): 43–50.

Aguilar, T. 2009. Marine fossils. In I.S. Wehrtmann and J. Cortés, eds., *Marine Biodiversity of Costa Rica, Central America*, 81–94. Berlin: Springer + Business Media BV.

Aguilar, T., and P. Denyer. 1994. Bioestratigrafía del parche arrecifal de la Quebrada Brazo Seco, Plio-Pleistoceno, Limón, Costa Rica. *Revista Geológica de América Central* 17: 55–66.

Alfaro, E.J. 2002. Some characteristics of the annual precipitation cycle in Central America and their relationships with its surrounding tropical oceans. *Tópicos Meteorológicos y Oceanográficos* 9: 88–103.

Alfaro, E.J. 2007. Escenarios climáticos para temporadas con alto y bajo número de huracanes en el Atlántico. *Revista de Climatología* 7: 1–13.

Alfaro, E.J., A. Quesada, and F. Solano. 2010. Análisis del impacto en Costa Rica de los ciclones tropicales ocurridos en el Mar Caribe desde 1968 al 2007. *Diálogos: Revista Electrónica de Historia* 11: 22–38.

Alvarado, J.J. 2011. Echinoderm diversity in the Caribbean Sea. *Marine Biodiversity* 41: 261–85.

Alvarado, J.J., and J. Cortés. 2004. The state of knowledge on echinoderms of Costa Rica and Central America. In T. Heinzeller and J.H. Nebelsick, eds., *Echinoderms: Munchen*, 149–55. Leiden: A.A. Balkema Publ.

Alvarado, J.J., J. Cortés, and E. Salas. 2004. Status of the sea urchin

Diadema antillarum Philippi (Echinodermata: Echinoidea) at Cahuita National Park (1977–2003), Costa Rica. *Caribbean Journal of Science* 40: 257–59.

Alvarado, J.J., B. Herrera, L. Corrales, J. Asch, and P. Paaby. 2011. Identificación de las prioridades de conservación de la biodiversidad marina y costera en Costa Rica. *Revista de Biología Tropical* 59: 829–42.

Alvarado, L., and E.J. Alfaro. 2003. Frecuencia de los ciclones tropicales que afectaron a Costa Rica durante el siglo XX. *Tópicos Meteorológicos y Oceanográficos* 10: 1–11.

Andrade, C.A., E.D. Barton, and C.N.K. Mooers. 2003. Evidence for an eastward flow along the Central and South American Caribbean Coast. *Journal of Geophysical Research* 108: 3185–96.

Azofeifa-Solano, J.C., M. Elizondo-Coto, and I.S. Wehrtmann. 2014. Reproductive biology of the sea anemone shrimp *Periclimenes rathbunae* (Caridea, Palaemonidae, Pontoniinae), from the Caribbean coast of Costa Rica. *ZooKeys* 457: 211–25.

Banta, W.C., and R.J.M. Carson. 1977. Bryozoa from Costa Rica. *Pacific Science* 31: 381–424.

Barrantes-Rojas, N. 2010. Programa interpretativo del ecosistema marino de la Isla Quiribrí (Uvita), Limón. Licentiate thesis, Universidad de Costa Rica. San Pedro, Costa Rica.

Benavides-Morera, R., and C.L. Brenes. 2010. Análisis hidrográfico e ictiológico de las capturas realizadas con una red de trampa fija en la laguna de Gandoca, Limón, Costa Rica. *Revista de Ciencias Marinas y Costeras* 2: 9–26.

Bergoeing, J.P. 1998. *Geomorfología de Costa Rica*. San José, Costa Rica: Instituto Geográfico Nacional.

Bernecker, A. 2009. Marine benthic algae. In I.S. Wehrtmann and J. Cortés, eds., *Marine Biodiversity of Costa Rica, Central America*. Berlin: Springer + Business Media BV. Text: pp. 109–117, Species List: CD pp. 17–70.

Bernecker, A., and I.S. Wehrtmann. 2009. New records of benthic marine algae and Cyanobacteria for Costa Rica, and a comparison with other Central American countries. *Helgoland Marine Research* 63: 219–29.

Bjorndal, K.A., A.B. Bolten, and C.J. Laguex. 1993. Decline of the nesting population of hawkbill turtles at Tortuguero, Costa Rica. *Conservation Biology* 7: 925–27.

Blair, N., C. Geraghty, G. Gund, and B. Jones. 1996. An economic evaluation of Cahuita National Park: establishing the economic value of an environmental asset. Unpublished report. Boston: Kellogg Graduate School of Management.

Bogantes-Aguilar, V. 2014. Poliquetos (Annelida: Polychaeta) del Parque Nacional Cahuita, Limón, Costa Rica. Licentiate thesis, Universidad de Costa Rica. San Pedro, Costa Rica.

Bolaños, N., A. Bourg, J. Gómez, and J.J. Alvarado. 2005. Diversidad y abundancia de equinodermos en la laguna arrecifal del Parque Nacional Cahuita, Caribe de Costa Rica. *Revista de Biología Tropical* 53(suppl. 3): 285–90.

Brandes, C., A. Astorga, S. Bak, R. Littke, and J. Winsemann. 2007. Fault controls on sediment distribution patterns, Limón Basin, Costa Rica. *Journal of Petroleum Geology* 30: 25–40.

Breedy, O. 1986. Contribución al estudio de los microcrustáceos bentónicos (Isopoda y Tanaidacea) en el arrecife coralino del Parque Nacional Cahuita, Limón, Costa Rica. Licentiate thesis, Universidad de Costa Rica. San Pedro, Costa Rica.

Breedy, O. 2009. Octocorals. In I.S. Wehrtmann and J. Cortés, eds., *Marine Biodiversity of Costa Rica, Central America*. Berlin: Springer + Business Media BV. Text: pp. 161–67, Species List: CD pp. 108–11.

Breedy, O., and M.M. Murillo. 1995. Isópodos (Crustacea: Peracarida) de un arrecife del Caribe de Costa Rica. *Revista de Biología Tropical* 43: 219–29.

Budd, A.F., K.G. Johnson, T.A. Stemann, and B.H. Tompkins. 1999. Pliocene to Pleistocene reef coral assemblages in the Limon Group of Costa Rica. *Bulletins of American Paleontology* 357: 119–58.

Bussing, W.A., and M. López. 2009. Marine fish. In I.S. Wehrtmann and J. Cortés, eds., *Marine Biodiversity of Costa Rica, Central America*. Berlin: Springer + Business Media BV. Text: pp. 453–58, Species List: CD pp. 412–73.

Bussing, W.A., and M.I. López. 2010. Peces costeros del Caribe de Centroamérica Meridional/Marine Fishes of the Caribbean Coast of Lower Central America. *Revista de Biología Tropical* 58(Suppl. 2): 1–234.

Camacho-García, Y.E., M. Pola, L. Carmona, V. Padula, G. Villani, and J.L. Cervera. 2014. Diversity and distribution of the heterobranch sea slug fauna on the Caribbean of Costa Rica. *Cahiers de Biologie Marine* 55: 109–27.

CARICOMP. 1997. Structure and productivity of mangrove forests in the Greater Caribbean Region. *Proceedings of the 8th International Coral Reef Symposium, Panamá* 1: 669–72.

Carr, A.F. 1962. Orientation problems in the high seas travel and terrestrial movement of marine turtles. *American Scientist* 50: 359–74.

Carr, A.F. 1967. *So Excellent a Fishe: A Natural History of Sea Turtles*. Graden City, NY: Natural History Press.

Carrillo-Baltodano, A.M. 2012. Diversidad, abundancia, composición y biomasa del zooplancton de la zona arrecifal del Parque Nacional Cahuita, Limón ¿Cuál es la disponibilidad de larvas de invertebrados bentónicos 25 años después? Licentiate thesis, Universidad de Costa Rica. San Pedro, Costa Rica.

Centurioni, L.R., and P.P. Niiler. 2003. On the surface currents of the Caribbean Sea. *Geophysical Research Letters* 30: 1279–82.

Chacón, D. 1993. Aspectos biométricos de una población de sábalo *Megalops atlanticus* (Pisces, Megalopidae). *Revista de Biología Tropical* 41(Suppl. 1): 13–18.

Chacón, D. 1999. Anidación de la tortuga *Dermochelys coriacea* (Testudines: Dermochelyidae) en Playa Gandoca, Costa Rica (1990 a 1997). *Revista de Biología Tropical* 47: 225–36.

Chacón, D., and W.O. McLarney. 1992. Desarrollo temprano del sábalo *Megalops atlanticus* (Pisces, Megalopidae). *Revista de Biología Tropical* 40: 171–77.

Chacón, D., W. McLarney, C. Ampie, and B. Venegas. 1996. Reproduction and conservation of the leatherback turtle *Dermochelys coriacea* (Testudines: Dermochelyidae) in Gandoca, Costa Rica. *Revista de Biología Tropical* 44: 853–60.

Charpentier, C. 1980. Variaciones estacionales en la composición química de cinco algas de la costa Caribe de Costa Rica y su posible utilización. Licentiate Thesis, Universidad de Costa Rica. San Pedro, Costa Rica.

Coates, A.G., J.B.C. Jackson, L.S. Collins, T.M. Cronin, H.J. Dowsett, L.M. Bybell, P. Jung, and J.A. Obando. 1992. Closure of the Isthmus of Panama: the near-shore marine records of Costa Rica and western Panama. *Geological Society of America Bulletin* 104: 14–28.

Coll, M., J. Cortés, and D. Sauma. 2004. Características físico-químicas y determinación de plaguicidas en el agua de la laguna de Gandoca, Limón, Costa Rica. *Revista de Biología Tropical* 52(Suppl. 2): 33–42.

Coll, M., A.C. Fonseca, and J. Cortés. 2001. Caracterización del manglar de Laguna Gandoca, Refugio Nacional de Vida Silvestre Gandoca-Manzanillo, Limón, Costa Rica. *Revista de Biología Tropical* 49(Suppl. 2): 321–29.

Collins, L.S., and A.G. Coates, eds. 1999. A Paleobiotic Survey of Caribbean Faunas from the Neogene of the Isthmus of Panama. *Bulletins of American Paleontology* 357: 1–351.

Collins, L.S., A.G. Coates, J.B.C. Jackson, and J.A. Obando. 1995. Timing and rates of emergence of the Limón and Bocas del Toro basins: Caribbean effects of Cocos Ridge subduction? In P. Mann, ed., *Geologic and Tectonic Development of the Caribbean Plate Boundary in Southern Central America. Geological Society of America Special Papers* 295: 291–307.

Cortés, J. 1981. The Coral Reef at Cahuita, Costa Rica, A Reef Under Stress. M.Sc. thesis, McMaster University. Hamilton, Ontario, Canada.

Cortés, J. 1985. Preliminary observations of *Alphaeus simus* Guérin-Meneville, 1856 (Crustacea: Alphaeidae) a little known Caribbean bioeroder. *Proceedings of the 5th International Coral Reef Congress, Tahiti* 5: 351–53.

Cortés, J. 1991. Ambientes y organismos marinos del Refugio Nacional de Vida Silvestre Gandoca-Manzanillo, Limón, Costa Rica. *Geoistmo* 5: 61–68.

Cortés, J. 1992a. Los arrecifes coralinos del Refugio Nacional de Vida Silvestre Gandoca-Manzanillo, Limón, Costa Rica. *Revista de Biología Tropical* 40: 325–33.

Cortés, J. 1992b. Organismos de los arrecifes coralinos de Costa Rica: V. Descripción y distribución geográfica de hydrocorales (Cnidaria; Hydrozoa: Milleporina and Stylasterina) de la costa Caribe. *Brenesia* 38: 45–50.

Cortés, J. 1992c. Nuevos registros de corales (Anthozoa: Scleractinia) para el Caribe de Costa Rica: *Rhizosmilia maculata* y *Meandrina meandrites*. *Revista de Biología Tropical* 40: 243–44.

Cortés, J. 1994. A reef under siltation stress: a decade of degradation. In R.N. Ginsburg, compiler, *Proceedings of the Colloquium on Global Aspects of Coral Reefs: Health, Hazards and History, 1993*, 240–46. Miami, FL: RSMAS, University of Miami.

Cortés, J. 1996. Biodiversidad marina de Costa Rica: Filo Porifera. *Revista de Biología Tropical* 44: 911–14.

Cortés, J. 1996–1997. Biodiversidad marina de Costa Rica: Filo Cnidaria. *Revista de Biología Tropical* 44(3)/45(1): 323–34.

Cortés, J. 2007. Coastal morphology and coral reefs. In J. Bundschuh and G.E. Alvarado, eds., *Central America: Geology, Resources and Hazards*, vol. 1, 185–200. London: Taylor and Francis.

Cortés, J. 2009a. Stony corals. In I.S. Wehrtmann and J. Cortés, eds., *Marine Biodiversity of Costa Rica, Central America*. Berlin: Springer + Business Media BV. Text: pp. 169–73, Species List: CD pp. 112–18.

Cortés, J. 2009b. Zoanthids, sea anemones, and corallimorpharians. In I.S. Wehrtmann and J. Cortés, eds., *Marine Biodiversity of Costa Rica, Central America*. Berlin: Springer + Business Media BV. Text: pp. 157–59, Species List: CD pp. 105–7.

Cortés, J., A.C. Fonseca, M. Barrantes, and P. Denyer. 1998. Type, distribution and origin of sediments of the Gandoca-Manzanillo National Wildlife Refuge, Limón, Costa Rica. *Revista de Biología Tropical* 46(Suppl. 6): 251–56.

Cortés, J., A.C. Fonseca, and M. Coll. 2001. Descripción del manglar del Refugio de Vida Silvestre Gandoca-Manzanillo, Limón, Costa Rica. In F. Pizarro, C. Gómez-Fuentes, and R. Córdoba- Muñoz, eds., *Humedales de Centroamérica*, 90–91. San José, Costa Rica: ORMA-UICN.

Cortés, J., A.C. Fonseca, J. Nivia, V. Nielsen-Muñoz, J. Samper-Villareal, E. Salas, S. Martínez, and P. Zamora-Trejos. 2010. Monitoring coral reefs, seagrasses and mangroves in Costa Rica (CARICOMP). *Revista de Biología Tropical* 58(Suppl. 3): 1–22.

Cortés, J., and H.M. Guzmán. 1985a. Arrecifes coralinos de la costa Atlántica de Costa Rica. *Brenesia* 23: 275–92.

Cortés, J., and H.M. Guzmán. 1985b. Organismos de los arrecifes coralinos de Costa Rica. III: Descripción y distribución geográfica de corales escleractinios (Cnidaria: Anthozoa: Scleractinia) de la costa Caribe. *Brenesia* 24: 63–124.

Cortés, J., and C.E. Jiménez. 2003. Corals and coral reefs of the Caribbean of Costa Rica: past, present and future. In J. Cortés, ed., *Latin American Coral Reefs*, 223–39. Amsterdam: Elsevier Science.

Cortés, J., and A. León. 2002. *The Coral Reefs of Costa Rica's Caribbean Coast*. Santo Domingo, Heredia, Costa Rica: Editorial INBio.

Cortés, J., M.M. Murillo, H.M. Guzmán, and J. Acuña. 1984. Pérdida de zooxantelas y muerte de corales y otros organismos arrecifales en el Caribe y Pacífico de Costa Rica. *Revista de Biología Tropical* 32: 227–32.

Cortés, J., and M.J. Risk. 1984. El arrecife coralino del Parque Nacional Cahuita, Costa Rica. *Revista de Biología Tropical* 32: 109–21.

Cortés, J., and M.J. Risk. 1985. A reef under siltation stress: Cahuita, Costa Rica. *Bulletin of Marine Science* 36: 339–56.

Cortés, J., and E. Salas. 2009. Seagrasses. In I.S. Wehrtmann and J. Cortés, eds., *Marine Biodiversity of Costa Rica, Central America*. Berlin: Springer + Business Media BV. Text: pp. 119–22, Species List: CD pp. 71–72.

Cortés, J., R. Soto, and C. Jiménez. 1994. Efectos ecológicos del Terremoto de Limón. *Revista Geológica de América Central, Volumen Especial: Terremoto de Limón*: 187–92.

Cortés, J., R. Soto, C. Jiménez, and A. Astorga. 1992. Death of intertidal and coral reef organisms as a result of a 7.5 earthquake. *Proceedings of the 7th International Coral Reef Symposium, Guam* 1: 235–40.

Cortés, J., N. Van der Hal, and R.W.M. Van Soest. 2009. Sponges. In I.S. Wehrtmann and J. Cortés, eds., *Marine Biodiversity of Costa Rica, Central America*. Berlin: Springer + Business Media BV. Text: pp. 137–42, Species List: CD pp. 83–93.

Cortés, J., and I.S. Wehrtmann. 2005. Costa Rica. In P. Miloslavich and E. Klein, eds., *Caribbean Marine Biodiversity: The Known and the Unknown*, 169–79. Lancaster, PA: DEStech Publications.

Cortés, J., and I.S. Wehrtmann. 2009. Diversity of marine habitats of the Caribbean and Pacific of Costa Rica. In I.S. Wehrtmann and J. Cortés, eds., *Marine Biodiversity of Costa Rica, Central America*, 1–45. Berlin: Springer + Business Media BV.

Cutler, N., E. Cutler, and J.A. Vargas. 1992. Peanut worms (Phylum Sipuncula) from Costa Rica. *Revista de Biología Tropical* 40: 153–58.

Dawson, E.Y. 1962. Additions to the marine flora of Costa Rica and Nicaragua. *Pacific Naturalist* 3: 375–95.

Denyer, P. 1998. Historic-prehistoric earthquakes, seismic hazards, and Tertiary and Quaternary geology of the Gandoca-Manzanillo National Wildlife Refuge, Limón, Costa Rica. *Revista de Biología Tropical* 46(Suppl. 6): 237–50.

Denyer, P., O. Arias, and S. Personius. 1994a. Efectos tectónicos del terremoto de Limón. *Revista Geológica de América Central, Volumen Especial: Terremoto de Limón*: 39–52.

Denyer, P., and G. Cárdenas. 2000. Costas marinas. In P. Denyer and S. Kussmaul, eds., *Geología de Costa Rica*, 185–218. Cartago: Editorial Tecnológica de Costa Rica.

Denyer, P., W. Montero, and G.E. Alvarado. 2003. Atlas tectónico de Costa Rica. San Pedro, Costa Rica: Editorial Universidad de Costa Rica.

Denyer, P., S. Personius, and O. Arias. 1994b. Generalidades sobre los efectos geológicos del terremoto de Limón. *Revista Geológica de América Central, Volumen Especial: Terremoto de Limón*: 29–38.

Dexter, D.M. 1974. Sandy-beach fauna of the Pacific and Atlantic coasts of Costa Rica and Colombia. *Revista de Biología Tropical* 22: 51–66.

Dick, B.M. 2004. Prioridades de conservación en la gestión integrada de los recursos naturales en la zona costera de la Reserva Pacuare, Limón, Costa Rica. M.Sc. thesis, Universidad de Costa Rica. San Pedro, Costa Rica.

Dominici-Arosemena, A., E. Brugnoli-Olivera, S. Solano-Ulate, H. Molina-Ureña, and A.R. Ramírez-Coghi. 2000. Ictioplancton en la zona portuaria de Limón, Costa Rica. *Revista de Biología Tropical* 48: 439–42.

Espinosa, J., and J. Ortea. 1999. Descripción de nuevas Marginelas (Mollusca: Neogastropoda: Marginellidae) de Cuba y del Caribe de Costa Rica y Panamá. *Avicennia* 10/11: 165–76.

Espinosa, J., and J. Ortea. 2000. Descripción de un género y once especies nuevas de Cystiscidae y Marginellidae (Mollusca: Neogastropoda) del Caribe de Costa Rica. *Avicennia* 12/13: 95–114.

Espinosa, J., and J. Ortea. 2001. Moluscos del mar Caribe de Costa Rica: desde Cahuita hasta Gandoca. *Avicennia* (Suppl. 4): 1–76.

Espinosa, J., J. Ortea, and J. Magaña. 2006. Nuevas especies de la Familia Eulimidae Philippi, 1853 (Mollusca: Prosobranchia) con caracteres singulares, recolectadas en Costa Rica, Cuba y Bahamas. *Revista de la Academia Canaria de Ciencias* XVII: 137–41.

Fauchald, K. 1973. Polychaetes from Central American sandy beaches. *Bulletin of the Southern California Academy of Sciences* 72: 19–31.

Fernández, C., and J.J. Alvarado. 2004. El arrecife coralino de Punta Cocles, costa Caribe de Costa Rica. *Revista de Biología Tropical* 52(Suppl. 2): 121–29.

Fernández, J.A., G. Bottazi, G. Barboza, and A. Astorga. 1994. Tectónica y estratigrafía de la Cuenca de Limón Sur. *Revista Geológica de América Central, Volumen Especial: Terremoto de Limón*: 15–28.

Ferrari, F.D., and T.E. Bowman. 1980. Pelagic copepods of the Family Oithonidae (Cyclopoida) from the east coast of Central and South America. *Smithsonian Contributions to Zoology* 312: 1–27.

Fischer, R., and T. Aguilar. 2007. Invertebrate paleontology. In J. Bundschuh and G.E. Alvarado, eds., *Central America: Geology, Resources and Hazards*, vol. 1, 453–66. London: Taylor and Francis.

Fonseca, A.C., and J. Cortés. 2002. Land use in the La Estrella River basin and soil erosion effects on the Cahuita reef system, Costa Rica. In B. Kjerfve, W.J. Wiebe, H.H. Kremer, W. Salomons, and J.I. Marshall, eds., *Caribbean Basins: LOICZ Global Change Assessment and Synthesis of River Catchment/Island-Coastal Sea Interaction and Human Dimension. LOICZ Reports & Studies* 27: 68–82 + 114–18. Texel, Netherlands: LOICZ.

Fonseca, A.C., J. Cortés, and P. Zamora. 2007a. Monitoreo del manglar de Gandoca, Costa Rica (Sitio CARICOMP). *Revista de Biología Tropical* 55: 23–31.

Fonseca, A.C., V. Nielsen, and J. Cortés. 2007b. Monitoreo de pastos marinos en Perezoso, sitio CARICOMP en Cahuita, Costa Rica. *Revista de Biología Tropical* 55: 55–66.

Fonseca, A.C., E. Salas, and J. Cortés. 2006. Monitoreo del arrecife coralino Meager Shoal, Parque Nacional Cahuita (sitio CARICOMP). *Revista de Biología Tropical* 54: 755–63.

García, V., J.A. Acuña-González, J.A. Vargas-Zamora, and J. García-Céspedes. 2006. Calidad bacteriológica y desechos sólidos en cinco ambientes costeros de Costa Rica. *Revista de Biología Tropical* 54(Suppl. 1): 35–48.

García-Céspedes, J., J.A. Acuña-González, and J.A. Vargas-Zamora. 2004. Metales traza en sedimentos de cuatro ambientes costeros de Costa Rica. *Revista de Biología Tropical* 52(Suppl. 2): 51–60.

García-Ríos, C.I., and M. Álvarez-Ruiz. 2011. Diversidad y microestructura de quitones (Mollusca: Polyplacophora) del Caribe de Costa Rica. *Revista de Biología Tropical* 59: 129–36.

Gilbert, C.R., and D.P. Kelso. 1971. Fishes of the Tortuguero area, Caribbean Costa Rica. *Bulletin of the Florida State Museum, Biological Sciences* 16: 1–54.

González, O., and I.S. Wehrtmann. 2011. Postlarval settlement of spiny lobster, *Panulirus argus* (Latreille, 1804) (Decapoda: Palinuridae), at the Caribbean coast of Costa Rica. *Latin American Journal of Aquatic Research* 39: 575–83.

Goshe, L.R., L. Avens, F.S. Scharf, and A.L. Southwood. 2010. Estimation of age at maturation and growth of Atlantic green turtles (*Chelonia mydas*) using skeletochronology. *Marine Biology* 157: 1725–40.

Guzmán, H.M., and J. Cortés. 1984. Mortandad de *Gorgonia flabellum* en la costa Caribe de Costa Rica. *Revista de Biología Tropical* 32: 305–8.

Guzmán, H.M., and J. Cortés. 1985. Organismos de los arrecifes coralinos de Costa Rica. IV. Descripción y distribución geográfica de octocoralarios de la costa Caribe. *Brenesia* 24: 125–74.

Guzmán, H.M., and E.M. García. 2002. Mercury levels in coral reefs along the Caribbean coast of Central America. *Marine Pollution Bulletin* 44: 1415–20.

Guzmán, H.M., and C.E. Jiménez. 1989. *Pterogorgia anceps* (Pallas) (Octocorallia: Gorgoniidae): nuevo informe para la costa caribeña de Costa Rica. *Revista de Biología Tropical* 37: 231–32.

Guzmán, H.M., and C.E. Jiménez. 1992. Contamination of coral reefs by heavy metals along the Caribbean coast of Central America (Costa Rica and Panama). *Marine Pollution Bulletin* 24: 554–61.

Hendler, G. 2005. Two new brittle star species of the genus *Ophiothrix* (Echinodermata: Ophiuroidea: Ophiotrichidae) from coral reefs in the southern Caribbean Sea, with notes on their biology. *Caribbean Journal of Science* 41: 583–99.

Henry, D.P., and P.A. McLaughlin. 1975. The barnacles of the *Balanus amphitrite* complex (Cirripedia, Thoracica). *Zoologische Verhandelingen* 141: 1–254.

Herrera, W. 1985. *Clima de Costa Rica*. San José, Costa Rica: Editorial UNED.

Houbrick, J.R. 1968. A survey of the litoral marine molluscs of the Caribbean coast of Costa Rica. *Veliger* 11: 4–23.

INCOPESCA. 2006. *Memoria Institucional 2002–2006: Instituto Costarricense de Pesca y Acuicultura*. San José, Costa Rica: Imprenta Nacional.

IPCC (Intergovernmental Panel on Climate Change). 2007. *Climate Change 2007: The Physical Science Basis*. Paris: WMO, UNEP.

Jiménez, C. 2001. Bleaching and mortality of reef organisms during a warming event in 1995 on the Caribbean coast of Costa Rica. *Revista de Biología Tropical* 49(Suppl. 2): 233–38.

Jiménez, C., and J. Cortés. 1993. Density and compressive strength of the coral *Siderastrea siderea* (Ellis and Solander, 1786) (Scleractinia: Siderastreidae): intraspecific variability. *Revista de Biología Tropical* 41(Suppl. 1): 39–43.

Jiménez, I. 1998. Ecología y conservación del manatí (*Trichechus manatus*, L.) en el noreste de Costa Rica. Base de datos de los humedales del noreste de Costa Rica asociada a un Sistema de Información Geográfica. M.Sc. thesis, Universidad Nacional. Heredia, Costa Rica.

Jiménez, I. 1999. Estado de conservación, ecología y conocimiento popular del manatí (*Trichechus manatus*) en Costa Rica. *Vida Silvestre Neotropical* 8: 18–30.

Jiménez, I. 2005. Development of predictive models to explain the distribution of the West Indian manatee *Trichechus manatus* in tropical watercourses. *Biological Conservation* 125: 491–503.

Kelmo, F., and R. Vargas. 2002. Anthoathecatae and Leptothecatae hydroids from Costa Rica (Cnidaria: Hydrozoa). *Revista de Biología Tropical* 50: 599–627.

Kemperman, T.C.M., and H. Stegenga. 1983. A new *Caulerpa* species (Caulerpaceae, Chlorophyta) from the Caribbean side of Costa Rica, Central America. *Acta Botanica Neerlandica* 32: 271–75.

Kemperman, T.C.M., and H. Stegenga. 1986. The marine bentic algae of the Atlantic side of Costa Rica. *Brenesia* 25–26: 99–122.

Kleypas, J.A., R.A. Feely, V.J. Fabry, C. Langdon, C.L. Sabine, and L.L. Robbins. 2006. Impacts of Ocean Acidification on Coral Reefs and Other Marine Calcifiers: A Guide for Future Research. Report of a Workshop. NSF / NOAA / USGS.

Krupp, L.S. 2006. Distribution, Ecology and State of the Seagrass Beds in the Gandoca-Manzanillo National Wildlife Refuge, Caribbean Costa Rica. M.Sc. thesis, Universität Bremen. Bremen, Germany.

Krupp, L.S., J. Cortés, and M. Wolff. 2009. Growth dynamics and state of the seagrass *Thalassia testudinum* in the Gandoca-Manzanillo National Wildlife Refuge, Caribbean Costa Rica. *Revista de Biología Tropical* 57(Suppl. 1): 187–201.

Lawrence, J.M. 1967. Lipid reserves in the gut of three species of tropical sea urchins. *Caribbean Journal of Science* 7: 65–68.

Lessios, H.A., D.R. Robertson, and J.D. Cubit. 1984. Spread of *Diadema* mass mortality through the Caribbean. *Science* 226: 335–37.

Linkimer, L., and T. Aguilar. 2000. Estratigrafía sedimentaria. In P. Denyer and S. Kussmaul, eds., *Geología de Costa Rica*, 43–62. Cartago: Editorial Tecnológica de Costa Rica.

Lizano, O.G. 1996. Un método gráfico para el pronóstico de oleaje durante huracanes en el Caribe adyacente a Costa Rica. *Tópicos Meteorológicos y Oceanográficos* 3: 11–17.

Lizano, O.G. 2006. Algunas características de las mareas en las costa Pacífica y Caribe de Centroamérica. *Ciencia y Tecnología* 24: 51–64.

Lizano, O.G. 2007. Climatología del viento y oleaje frente a las costas de Costa Rica. *Ciencia y Tecnología* 25: 43–56.

Lizano, O.G., and W. Fernández. 1996. Algunas características de las tormentas tropicales y de los huracanes que atravesaron o se formaron en el Caribe adyacente a Costa Rica durante el período 1886–1988. *Tópicos Meteorológicos y Oceanográficos* 3: 3–10.

Lizano, O.G., and R.J. Moya. 1990. Simulación de oleaje durante el huracán Joan (1988) a su paso por el Mar Caribe de Costa Rica. *Revista de Geofísica* 33: 105–26.

Loaiza-Coranado, B. 1989. Generalidades del Phylum Porifera y bases para su identificación con sinopsis de algunas de ellas, en Limón, Costa Rica. Licentiate thesis, Universidad Nacional. Heredia, Costa Rica.

Loaiza-Coronado, B. 1991. Estudio taxonómico de las esponjas del Parque Nacional Cahuita, sector Puerto Vargas e Isla Uvita, Limón, Costa Rica. *Brenesia* 36: 21–62.

Loría, L.G., A. Banichevich, and J. Cortés. 1998. Radionucleidos en corales de Costa Rica. *Revista de Biología Tropical* 46(Suppl. 5): 81–90.

Marshall, J.S. 2007. Geomorphology and physiographic provinces. In J. Bundschuh and G.E. Alvarado, eds., *Central America: Geology, Resources, Hazards*, 75–122. London: Taylor and Francis.

Mata, A., J.A. Acuña, M.M. Murillo, and J. Cortés. 1987. La contaminación por petróleo en el Caribe de Costa Rica: 1981–1985. *Caribbean Journal of Science* 23: 41–49.

May-Collado, L.J. 2010. Changes in whistle structure of two dolphin species during interspecific associations. *Ethology* 116: 1–10.

May-Collado, L.J., and D. Wartzok. 2008. A comparison of bottlenose dolphin whistles in the Atlantic Ocean: factors promoting whistle variation. *Journal of Mammalogy* 89: 1229–40.

May-Collado, L.J., and D. Wartzok. 2009. A characterization of Guyana dolphin (*Sotalia guianensis*) whistles from Costa Rica: the importance of broadband recording systems. *Journal of the Acoustical Society of America* 125: 1202–13.

McNeill, D.F., A.G. Coates, A.F. Budd, and P.F. Borne. 2000. Integrated paleontologic and paleomagnetic stratigraphy of the upper Neogene deposits around Limon, Costa Rica: a coastal emergence record of the Central American Isthmus. *Geological Society of America Bulletin* 112: 963–81.

Michel, H.B., and M. Foyo. 1976. Caribbean Zooplankton. Part I. Siphonophora, Heteropoda, Copepoda, Euphausiacea, Chaetognatha and Salpidae. Washington, DC: Office of Naval Research Department, Navy.

Mielke, W. 1992. Six representatives of the Tetragonicipitidae (Copepoda) from Costa Rica. *Microfauna Marina* 7: 101–46.

Mielke, W. 1993. Species of the taxa *Orthopsyllus* and *Nitroca* (Copepoda) from Costa Rica. *Microfauna Marina* 8: 247–66.

Mielke, W. 1994. Two co-occurring new *Karllangia* species (Copepoda: Ameriidae) from the Caribbean coast of Costa Rica. *Revista de Biología Tropical* 42: 141–53.

Miloslavich, P., J.M. Díaz, P.E. Klein, J.J. Alvarado, C. Díaz, J. Gobin, E. Escobar-Briones, J.J. Cruz-Motta, E. Weil, J. Cortés, A.C. Bastidas, R. Robertson, F. Zapata, A. Martín, J. Castillo, A. Kazandjan, and M. Ortiz. 2010. Marine biodiversity in the Caribbean: regional estimates and distribution patterns. *PLOS ONE* 5(8): e11916. DOI:10.1371/journal.pone. 0011916.

Mooers, C.N.K., and G.A. Maul. 1998. Chapter 7. Intra-Americas Sea circulation. In A.R. Robinson and K.H. Brink, eds., *The Sea*, vol. 11, 183–208. New York: John Wiley.

Mora, A., C. Fernández, and A.G. Guzmán. 2006. *Áreas Marinas Protegidas y Áreas Marinas de Uso Múltiple de Costa Rica: Notas para una Discusión*. San José, Costa Rica: MarViva.

Mora, D.A., J.C. Rojas, A.V. Mata, and M.A. Sequeira. 1987. Calidad sanitaria de las aguas de la playa de Limón en el período 1981–1984. *Tecnología en Marcha* 8: 15–22.

Mora-Cordero, C. 2005. Factores que afectan la cuenca del Río La Estrella y recomendaciones para mejorar la gestión en la zona costera. Limón, Costa Rica. M.Sc. thesis, Universidad de Costa Rica. San Pedro, Costa Rica.

Morales, A., and M.M. Murillo. 1996. Distribution, abundance and composition of coral reef zooplankton, Cahuita National Park, Limon, Costa Rica. *Revista de Biología Tropical* 44: 619–30.

Morales-Ramírez, A. 1987. Caracterización del zooplancton del arrecife en el Parque Nacional Cahuita, Limón, Costa Rica. M.Sc. thesis, Universidad de Costa Rica. San Pedro, Costa Rica.

Morales-Ramírez, A. 2001. Biodiversidad marina de Costa Rica: Los microcrustáceos: Subclase Copepoda (Crustacea: Maxillopoda). *Revista de Biología Tropical* 49(Suppl. 2): 115–33.

Morales-Ramírez, A., E. Suárez-Morales, M. Corrales, and O. Esquivel-Garrote. 2014. Diversity of the free-living marine and freshwater Copepoda (Crustacea) in Costa Rica: a review. *ZooKeys* 457: 15–33.

Moran, D.A., and A.I. Dittel. 1993. Anomura and brachyuran crabs of Costa Rica: annotated list of species. *Revista de Biología Tropical* 41: 599–617.

Muller-Parker, G., and J. Cortés. 2001. Spatial distribution of light and nutrients in coral reefs of the Pacific and Caribbean coasts of Costa Rica during January 1997. *Revista de Biología Tropical* 49(Suppl. 2): 251–63.

Murillo, M.M., and J. Cortés. 1984. Alta mortalidad en la población del erizo de mar *Diadema antillarum* Philippi (Echinodermata: Echinoidea), en el Parque Nacional Cahuita, Limón, Costa Rica. *Revista de Biología Tropical* 32: 167–69.

Myhre, S., and A. Acevedo-Gutiérrez. 2007. Recovery of sea urchin *Diadema antillarum* populations is correlated to increased coral cover and reduced macroalgal cover. *Marine Ecology Progress Series* 329: 205–10.

Nagelkerken, I., K. Buchan, G.W. Smith, et al. 1997. Widespread disease in Caribbean sea fans: I. Spreading and general characteristics. *Proceedings of the 8th International Coral Reef Symposium, Panamá* 1: 679–82.

Nielsen-Muñoz, V. 2007. Abundancia, biomasa y floración de *Thalassia testudinum* (Hydrocharitaceae) en el Parque Nacional Cahuita, Caribe de Costa Rica. Thesis, Universidad de Costa Rica. San Pedro, Costa Rica.

Nielsen-Muñoz, V., and J. Cortés. 2008. Abundancia, biomasa y floración de *Thalassia testudinum* (Hydrocharitaceae) en el Caribe de Costa Rica. *Revista de Biología Tropical* 56(Suppl. 4): 175–89.

Ornelas-Gatdula, E., Y. Camacho-García, M. Schrödl, V. Padula, Y. Hooker, T.M. Gosliner, and A. Valdés. 2012. Molecular systematics of the "*Navanax aenigmaticus*" species complex (Mollusca, Opisthobranchia): coming full circle. *Zoologica Scripta* 41: 374–85.

Ortea, J. 2001. El género *Doto* Oken, 1815 (Mollusca: Nudibranchia) en el mar Caribe: Historia natural y descripción de nuevas especies. *Avicennia* (Suppl. 3): 1–46.

Ortea, J., J. Espinosa, and Y. Camacho. 1999. Especies del género *Polycera* Cuvier, 1816 (Mollusca: Nudibranchia) recolectados en la epifauna de algas rojas del Caribe de Costa Rica y Cuba. *Avicennia* 10/11: 157–64.

Ortea, J., J. Espinosa, and L. Moro. 2001. Descripción de una nueva especie de *Philine* Ascanius, 1772. *Avicennia* (Suppl. 4): 38–40.

Paddack, M.J., J.D. Reynolds, C. Aguilar, et al. 2009. Recent region-wide declines in Caribbean reef fish abundance. *Current Biology* 19: 1–6.

Parkinson, R.W., J. Cortés, and P. Denyer. 1998. Passive margin sedimentation on Costa Rica's north Caribbean coastal plain, Río Colorado. *Revista de Biología Tropical* 46(Suppl. 6): 221–36.

Paynter, C.K., J. Cortés, and M. Engels. 2001. Biomass, productivity and density of the seagrass *Thalassia testudinum* at three sites in Cahuita National Park, Costa Rica. *Revista de Biología Tropical* 49(Suppl. 2): 265–72.

Pepe, P.J. 1985. Littoral endolithic fauna of the Central American Isthmus. *Revista de Biología Tropical* 33: 191–94.

Pereira-Chaves, J., and L. Sierra-Sierra. 2009. Estrategia de manejo de los recursos marinos y costeros en Isla Uvita, Limón, Costa Rica. *Revista de Ciencias Marinas y Costeras* 1: 127–43.

Pérez-Reyes, C.R. 2003. Interpretación ambiental de un sendero autoguiado en Isla Uvita de Puerto Limón, como un aporte al desarrollo turístico de la Región Caribe de Costa Rica. Licentiate thesis, Universidad de Costa Rica. San Pedro, Costa Rica.

Petrescu, I., and R.W. Heard. 2004. Three new Cumacea (Crustacea: Peracarida) from Costa Rica. *Zootaxa* 721: 1–12.

Phillips, P.C., and M.J. Perez-Cruet. 1984. A comparative survey of reef fishes in Caribbean and Pacific Costa Rica. *Revista de Biología Tropical* 32: 95–102.

Pool, D.J., S.C. Snedaker, and A.E. Lugo. 1977. Structure of mangrove forests in Florida, Puerto Rico, Mexico, and Costa Rica. *Biotropica* 9: 195–212.

Reynolds, J.E., W.A. Szelistowski, and M.A. León. 1995. Status and conservation of manatees *Trichechus manatus manatus* in Costa Rica. *Biological Conservation* 71: 193–96.

Risk, M.J., M.M. Murillo, and J. Cortés. 1980. Observaciones biológicas preliminares sobre el arrecife coralino en el Parque Nacional Cahuita, Costa Rica. *Revista de Biología Tropical* 28: 361–82.

Robinson, D.G. 1991. The systematics and paleoecology of the prosobranch gastropods of the Pleistocene Moín formation of Costa Rica. PhD diss., Tulane University. New Orleans, LA.

Robinson, D.G., and M. Montoya. 1987. Los moluscos marinos de la costa Atlántica de Costa Rica. *Revista de Biología Tropical* 35: 375–400.

Roder, C. 2005. Land-based pollution on the Caribbean coast of Costa Rica: nutritional and skeletal characteristics of the reef-building coral *Siderastrea siderea* as a response. M.Sc. thesis, Universität Bremen. Bremen, Germany.

Roder, C., J. Cortés, C. Jiménez, and R. Lara. 2009. Riverine input of particulate material and inorganic nutrients to a coastal reef ecosystem at the Caribbean coast of Costa Rica. *Marine Pollution Bulletin* 58: 1937–43.

Rodríguez-Sevilla, L., R. Vargas, and J. Cortés. 2003. Biodiversidad marina de Costa Rica: Gastrópodos (Mollusca: Gastropoda) de la costa Caribe. *Revista de Biología Tropical* 51(Suppl. 3): 302–99.

Rojas, M.T. 1990. Determinación del cadmio, cromo, cobre, hierro, manganeso, plomo y zinc en el pepino de mar, *Holothuria* sp. (Echinodermata) del arrecife coralino del Parque Nacional Cahuita, Costa Caribe, Costa Rica. Licentiate thesis, Universidad de Costa Rica. San Pedro, Costa Rica.

Rojas, M.T., J.A. Acuña, and O.M. Rodríguez. 1998. Metales traza en el pepino de mar *Holothuria (Halodeima) mexicana* del Caribe de Costa Rica. *Revista de Biología Tropical* 46(Suppl. 6): 215–20.

Salas, E., H. Molina-Ureña, R.P. Walter, and D.D. Heath. 2010. Local and regional genetic connectivity in a Caribbean coral reef fish. *Marine Biology* 157: 437–45.

Salazar, A., O.G. Lizano, and E.J. Alfaro. 2004. Composición de sedimentos en las zonas costeras de Costa Rica utilizando Fluorescencia de Rayos-X (FRX). *Revista de Biología Tropical* 52(Suppl. 2): 61–75.

Samper-Villarreal, J., A. Bernecker, and I.S. Wehrtmann. 2008. Inventory of macroalgal epiphytes on the seagrass *Thalassia testudinum* (Hydrocharitaceae) in Parque Nacional Cahuita, Caribbean coast of Costa Rica. *Revista de Biología Tropical* 56(Suppl. 4): 163–74.

Sánchez, J.A. 2001. Systematics of the southwestern Caribbean *Muriceopsis* Aurivillius (Cnidaria: Octocrallia), with the description of a new species. *Bulletin of the Biological Society of Washington* 10: 160–80.

Sánchez-Vindas, P.E. 2001. *Flórula Arborescente del Parque Nacional Cahuita*. San José, Costa Rica: Editorial EUNED.

Sandí, G. 1990. Determinación de Zn, Cd, Pb, Cu, Fe, Mn, y Cr en aguas del arrecife coralino del Parque Nacional Cahuita, Costa Rica. Licentiate thesis, Universidad de Costa Rica. San Pedro, Costa Rica.

Santoro, M., E.C. Greiner, J.A. Morales, and B. Rodríguez-Ortíz. 2006a. Digenetic trematode community in nesting green sea turtles (*Chelonia mydas*) from Tortuguero National Park, Costa Rica. *Journal of Parasitology* 92: 1202–6.

Santoro, M., G. Hernández, M. Caballero, and F. García. 2006b. Aerobic bacterial flora of nesting green turtles (*Chelonia mydas*) from Tortuguero National Park, Costa Rica. *Journal of Zoo and Wildlife Medicine* 37: 549–52.

Santoro, M., and S. Mattiucci. 2009. Sea turtle parasites. In I.S. Wehrtmann and J. Cortés, eds., *Marine Biodiversity of Costa Rica, Central America*. Berlin: Springer + Business Media B.V. Text: pp. 507–19, Species List: CD pp. 497–500.

Santoro, M., J.A. Morales, B. Stacy, and E.C. Greiner. 2007. *Rameshwarotrema uterocrescens* trematode parasitism of the oesophageal glands in green sea turtles (*Chelonia mydas*). *Veterinary Record* 159: 59–60.

Savage, J.M. 2002. *The Amphibians and Reptiles of Costa Rica*. Chicago: University of Chicago Press.

Schmidt, E., M. Wolff, and J.A. Vargas. 2002. Population ecology and fishery of *Cittarium pica* (Gastropoda: Trochidae) on the Caribbean coast of Costa Rica. *Revista de Biología Tropical* 50: 1079–90.

Silva-Benavides, A.M. 1986. Productividad primaria, biomasa del fitoplancton y la relación con parámetros físico-químicos en el arrecife coralino del Parque Nacional Cahuita. M.Sc. thesis, Universidad de Costa Rica. San Pedro, Costa Rica.

Smith, G.W., L.D. Ives, I.A. Nagelkerken, and K.B. Ritchie. 1996. Caribbean sea-fan mortalities. *Nature* 383: 487.

Soto, R. 1983. Nuevos informes para la flora bentónica marina de Costa Rica. *Brenesia* 21: 365–70.

Soto, R., and D.L. Ballentine. 1986. La flora bentónica del Caribe de Costa Rica (Notas preliminares). *Brenesia* 25/26: 123–62.

Spongberg, A.L. 2004. PCB contamination in surface sediments in the coastal waters of Costa Rica. *Revista de Biología Tropical* 52(Suppl. 2): 1–10.

Stegenga, H., and T.C.M. Kemperman. 1983. Acrochaetiaceae (Rhodophyta) new to the Costa Rican Atlantic flora. *Brenesia* 21: 67–91.

Suárez-Morales, E., A. Carrillo, and A. Morales-Ramírez. 2013. Report on some monstrilloids (Crustacea: Copepoda) from a reef area off the Caribbean coast of Costa Rica, Central America with description of two new species. *Journal of Natural History* 47: 619–38.

Tabash-Blanco, F.A. 1995. An assessment of pink shrimp, *Penaeus brasiliensis*, populations, in three areas of the Caribbean coast of Costa Rica. *Revista de Biología Tropical* 43: 239–50.

Taylor, G.D. 1975. The Geology of the Limón Area of Costa Rica. PhD diss., Louisiana State University. Baton Rouge, LA.

Taylor, M.A., and E.J. Alfaro. 2005. Climate of Central America and the Caribbean. In J.E. Oliver, ed., *Encyclopedia of World Climatology*, 183–89. Netherlands: Springer.

Taylor, W.R. 1933. Notes on algae from the tropical Atlantic Ocean, II. Papers Mich Acad Sci Arts Lett 17:395–407.

Terossi, M., J.A. Cuesta, I.S. Wehrtmann, and F.L. Mantelatto. 2010b. Revision of the larval morphology (Zoea I) of the family Hippolytidae (Decapoda, Caridea), with a description of the first stage of the shrimp *Hippolyte obliquimanus* Dana, 1852. *Zootaxa* 2624: 49–66.

Terossi, M., I.S. Wehrtmann, and F.L. Mantelatto. 2010a. Interpopulation comparison of reproduction of the Atlantic shrimp *Hippolyte obliquimanus* (Caridea: Hippolytidae). *Journal of Crustacean Biology* 30: 571–79.

Thomas, D.T., and D.W. Freshwater. 2001. Studies of Costa Rican Gelidiales (Rhodophyta): four Caribbean taxa including *Pterocladiella beachii* sp. nov. *Phycologia* 40: 340–50.

Tiwari, M., K.A. Bjorndal, A.B. Bolten, and B.M. Bolker. 2005. Intraspecific application of the mid-domain effect model: spatial and temporal nest distributions of green turtles, *Chelonia mydas*, at Tortuguero, Costa Rica. *Ecology Letters* 8: 918–24.

TNC (The Nature Conservancy). 2008. Evaluación de ecorregiones marinas en Mesoamérica. Sitios prioritarios para la conservación en las ecorregiones Bahía de Panamá, Isla del Coco y Nicoya del Pacífico Tropical Oriental, y en el Caribe Suroccidental de Costa Rica y Panamá. San José, Costa Rica: The Nature Conservancy.

Troëng, S. 2005. Migration of sea turtles from Caribbean Costa Rica: implications for management. PhD diss., Lund University. Lund, Sweden.

Troëng, S., D. Chacón, and B. Dick. 2004. Possible decline in leatherback turtle *Dermochelys coriacea* nesting along the coast of Caribbean Central America. *Oryx* 38: 395–403.

Troëng, S., D. Evans, E. Harrison, and C. Lagueux. 2005. Migration of green turtles *Chelonia mydas* nesting at Tortuguero, Costa Rica. *Marine Biology* 148: 435–47.

Troëng, S., and E. Rankin. 2005. Long-term conservation of the green turtle *Chelonia mydas* nesting population at Tortuguero, Costa Rica. *Biological Conservation* 121: 111–16.

Tsuda, R.T. 1968. Additional records of marine benthic algae from Costa Rica. *Caribbean Journal of Science* 8: 103–4.

Umaña, R., and D. Chacón. 1994. Asentamiento en estadios de postlarvales de la langosta *Panulirus argus* (Decapoda: Palinuridae), en Limón, Costa Rica. *Revista de Biología Tropical* 42: 585–94.

Valdez, M.F., and C.R. Villalobos. 1978. Distribución espacial, correlación con el substrato y grado de agregación en *Diadema antillarum* Phillipi (Echinodermata: Echinoidea). *Revista de Biología Tropical* 26: 237–45.

Valverde, J. 2000. Descentralización y comanejo de recursos en el Caribe Tico. *Ciencias Ambientales* 19: 45–59.

Vander Zanden, H.B., K.E. Arthur, A.B. Bolten, B.N. Popp, C.J. Lagueux, E. Harrison, C.L. Campbell, and K.A. Bjorndal. 2013. Trophic ecology of a green turtle breeding population. *Marine Ecology Progress Series* 476: 237–49.

Van der Hal, N. 2006. Presence and Diversity of Sponge Species at the

Caribbean coast of Costa Rica. M.Sc. thesis, University of Amsterdam. Netherlands.

Van Tussenbroek, B.I., J. Cortés, R. Collins, A.C. Fonseca, P.M.H. Gayle, H.M. Guzmán, G.E. Jácome, R. Juman, K.H. Koltes, H.A. Oxenford, A. Rodríguez-Ramírez, J. Samper-Villareal, S.R. Smith, J.J. Tschirky, and E. Weil. 2014. Caribbean-wide, long-term study of seagrass beds reveals local variations, shifts in community structure and occasional collapse. *PLOS ONE* 9(3): e90600. doi:10.1371/journal.pone.0090600

Vargas, R., and J. Cortés. 1997. Biodiversidad marina de Costa Rica: Orden Stomatopoda (Crustacea: Malacostraca: Hoplocarida). *Revista de Biología Tropical* 45: 1531–39.

Vargas, R., and J. Cortés. 1999. Biodiversidad marina de Costa Rica: Crustacea: Decapoda (Penaeoidea, Sergestoidea, Stenopodidae, Caridea, Thalassinidea, Palinura) del Caribe. *Revista de Biología Tropical* 47: 877–85.

Vargas, R., and J. Cortés. 2006. Biodiversidad marina de Costa Rica: Crustacea: Infraorden Anomura. *Revista de Biología Tropical* 54: 461–88.

Vaughan, T.W. 1919. Fossil corals from Central America, Cuba, and Porto Rico, with an account of the American Tertiary, Pleistocene, and Recent coral reefs. *United States National Museum Bulletin* 103: 189–524.

Voss, G.L. 1971. Narrative of RV John Elliot Pillsbury Cruise P-7101: Central America, January 20–February 5, 1971. Miami, FL: School of Marine and Atmospheric Science, University of Miami.

Walter, T.C. 1989. Review of the new world species of *Pseudodiaptomus* (Copepoda: Calanoida) with a key to the species. *Bulletin of Marine Science* 45: 590–628.

Wehrtmann, I.S., and L. Albornoz. 2002. Evidence of different reproductive traits in the transisthmian sister species, *Alpheus saxidomus* and *A. simus* (Decapoda, Caridea, Alpheidae): description of the first postembryonic stage. *Marine Biology* 140: 605–12.

Wehrtmann, I.S., and J. Cortés, eds. 2009. *Marine Biodiversity of Costa Rica, Central America*. Berlin: Springer + Business Media BV. Text: 538 pp., Species Lists CD: 500 pp.

Wehrtmann, I.S., J. Cortés, and S. Echeverría-Sáenz. 2009. Marine biodiversity of Costa Rica: perspectives and conclusions. In I.S. Wehrtmann and J. Cortés, eds., *Marine Biodiversity of Costa Rica, Central America*, 521–33. Berlin: Springer + Business Media BV.

Wehrtmann, I.S., and R. Vargas. 2003. New records and range extensions of shrimps (Decapoda: Penaeoidea, Caridea) from the Pacific and Caribbean coasts of Costa Rica, Central America. *Revista de Biología Tropical* 51: 268–74.

Weitzer, V.A. 2000. From conflict to collaboration: the case of Cahuita National Park, Costa Rica. M.Sc. thesis, University of Manitoba. Winnipeg, Manitoba, Canada.

Wellington, G.M. 1973. Additions to the Atlantic benthic flora of Costa Rica. *Brenesia* 2: 17–20.

Wellington, G.M. 1974a. An ecological description of the marine and associated environments at Monumento Nacional Cahuita. San José, Costa Rica: Sub-dirección de Parques Nacionales, MAG.

Wellington, G.M. 1974b. The benthic flora of Punta Cahuita: annotated list of species with additions to the Costa Rican Atlantic flora. *Brenesia* 3: 19–30.

Wetzer, R., and N.L. Bruce. 1999. A new genus and species of sphaeromatid isopod (Crustacea) from Atlantic Costa Rica. *Proceedings of the Biological Society of Washington* 112: 368–80.

Williams, A.B. 1993. Mud shrimps, Upogebiidae, from the western Atlantic (Crustacea: Decapoda: Thalassinidea). *Smithsonian Contributions to Zoology* 544: 1–77.

Winemiller, K.O., and M.A. Leslie. 1992. Fish communities across a complex freshwater-marine ecotone. *Environmental Biology of Fishes* 34: 29–50.

Young, P.S. 1989. *Ceratoconcha paucicostata*, a new species of coral-inhabiting barnacle (Cirripedia, Pyrgomatidae) from the western Atlantic. *Crustaceana* 56: 193–99.

Part IX

The Rivers, Lakes, and Wetlands

Chapter 18 Rivers of Costa Rica

Catherine M. Pringle[1,*], Elizabeth P. Anderson[2], Marcelo Ardón[3], Rebecca J. Bixby[4], Scott Connelly[1], John H. Duff[5], Alan P. Jackman[6], Pia Paaby[7], Alonso Ramírez[8], Gaston E. Small[9], Marcia N. Snyder[1], Carissa N. Ganong[1], and Frank J. Triska[5]

Introduction

Rivers of Costa Rica drain the narrow isthmus between the continents of North and South America, discharging into either the Pacific Ocean or the Caribbean Sea. The country is thus characterized by a high ocean-to-coast ratio and has 34 major watersheds (Fig. 18.1). The orientation of mountain ranges along the longitudinal axis of the country contributes to the high number of smaller watersheds (>100) that are characterized by steep gradients in their headwaters. Most rivers have their headwaters located in volcanic mountains and are characterized by cold clear waters which evolve into turbid and warm waters in the lowlands. The narrow mountainous region of Costa Rica results in multiple parallel river systems with relatively small drainage areas. Most of the rivers that discharge into the Caribbean and Pacific are relatively short, with the exception of the Río San Juan (37 km). The Río San Juan is the largest (41,600 km^2) watershed in Central America, located on the border between Costa Rica (which contains ~30% of the watershed) and Nicaragua (which contains ~70% of the watershed) and discharging into the Caribbean.

Much of Costa Rica has relatively young soils due to recent volcanic activity (Alvarado and Mata, chapter 4 of this volume). Moreover (as discussed in detail later in this chapter), ongoing geothermal activity (Alvarado and Cárdenes, chapter 3 of this volume) occurs along the volcanic spine of Costa Rica, creating spatial patterns in stream solute levels that influence the ecology of streams draining volcanic mountainous areas (Fig. 18.2; Pringle and Triska 2000).

Rivers of Costa Rica provide a legacy of historical changes in land use since they integrate the watersheds that they drain, reflecting changes in terrestrial ecosystems. Before the arrival of the Spanish, Costa Rica was inhabited by scattered Indian populations that practiced shifting agriculture, and effects on river ecosystems can be assumed to have been minimal relative to subsequent effects associated with expanding human populations. Agricultural settlement began in the Central Valley and in Nicoya (primarily to produce sugar cane, tobacco, and coffee). Coffee exports began in 1825 and permanent crops replaced the forests of the Central Valley by the end of the century. The end of the nineteenth century was also characterized by a wave of colonization towards the coasts, particularly the Pacific Coast. Deforestation increased exponentially in the early 1920s (WRI 1991). Expansion of the cattle-ranching frontier in

[1] Odum School of Ecology, University of Georgia, Athens, GA 30602, USA
[2] Department on Earth & the Environment, Florida International University, Mlami, FL, USA
[3] Department of Biology, East Carolina University, Greenville, NC, USA
[4] Department of Biology, University of New Mexico, Albuquerque, NM, USA
[5] US Geological Survey, Menlo Park, CA, USA
[6] Department of Chemical Engineering, University of California, Davis, CA, USA
[7] Organization for Tropical Studies (OTS), Apartado 676-2050, San Pedro de Montes de Oca, Costa Rica
[8] Institute for Tropical Ecosystem Studies, University of Puerto Rico, Río Piedras, Puerto Rico
[9] University of St. Thomas, Saint Paul, MN, USA
* Corresponding author

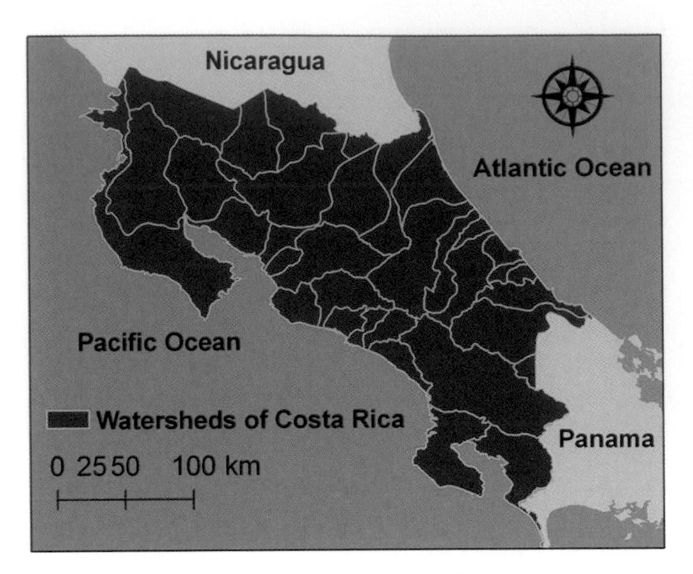

Fig. 18.1 Map displaying the 34 major watersheds of Costa Rica. *Figure created by M. Snyder using Digital Atlas of Costa Rica 2004 data.*

the 1950s has had many deleterious effects on forests and water resources. Costa Rica lost 50% of its forest cover between 1940 and 1986, resulting in significant problems in soil erosion, river siltation, and water quality. Costa Rican rivers are vulnerable because of the intensity and duration of rainfall (1.5–6 m yr^{-1}). Severe erosion is often a consequence of forest removal in highland and mountainous regions: from 20–200 tons of soil ha^{-1} yr^{-1} are eroded from deforested slopes (Hartshorn et al. 1982). Sediment loads in rivers have negative effects on fish and other organisms and are ultimately discharged into coastal regions. In the early 1980s Costa Rica had one of the highest annual rates of deforestation in the world, peaking at 100,000 acres per year (e.g., see Sader and Joyce 1988). Current rates of deforestation are relatively low, since much of the remaining primary forest lies within protected areas; however, pressure to exploit timber and water resources in national parks and forest reserves is increasing with human population growth

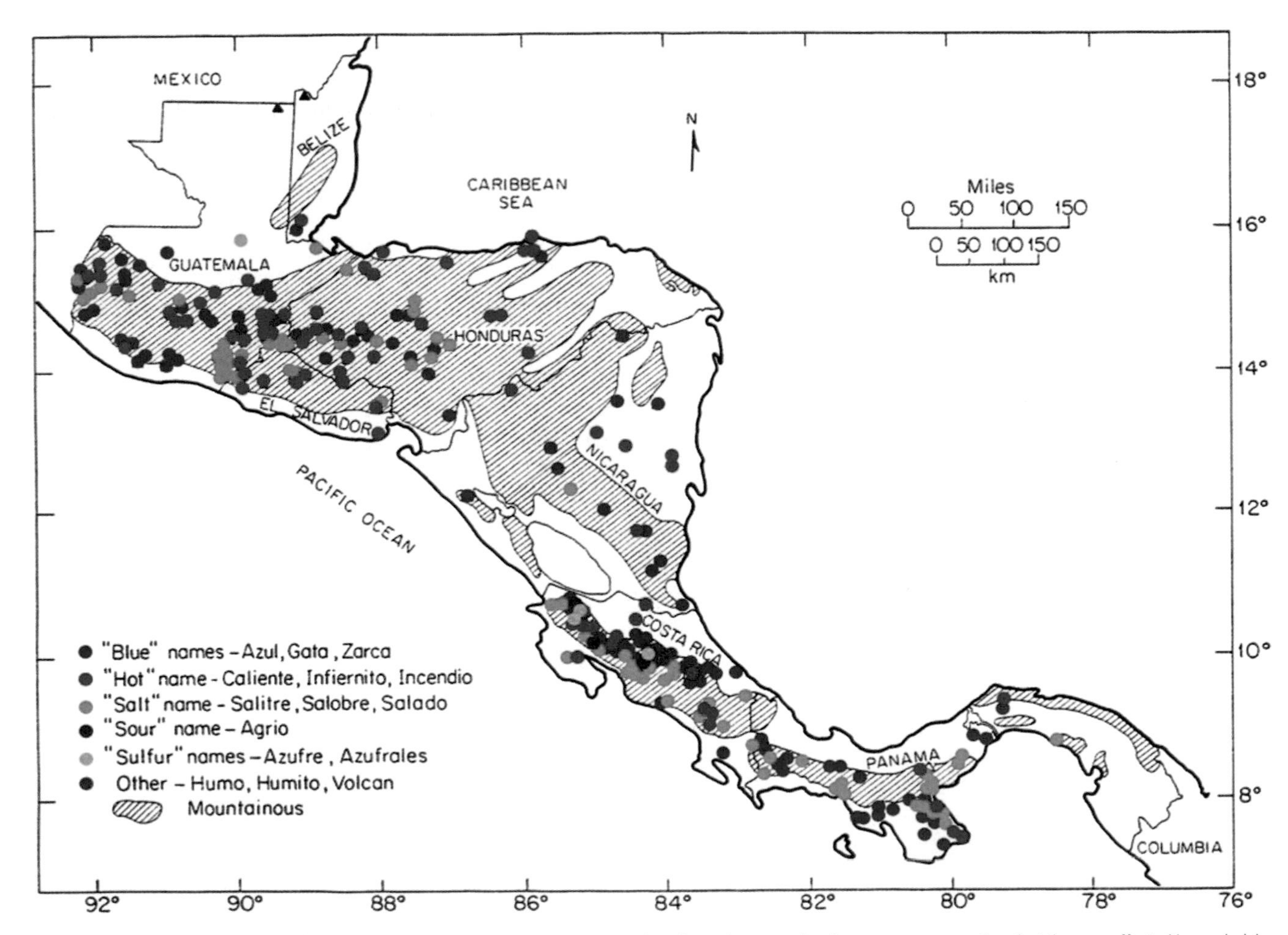

Fig. 18.2 Map of Central America showing mountainous areas and approximate locations of streams that have names suggesting that they are affected by underlying geothermal activity. Volcanic activity is common throughout Central America as a result of the northeastward subduction of the Cocos ridge beneath the Caribbean plate. Thus, geothermally modified ground waters occur frequently in the landscape. The distribution of these streams is clustered along the central axis of the mountains. *Modified from Pringle and Triska 2000.*

Fig. 18.3 Río Sarapiquí–San Juan drainage on the Caribbean Slope of the Cordillera Central. This is a 2001 Landsat true color satellite image draped over a digital elevation model with the relief exaggerated. The forest in this image is dark green and is mostly contained in the protected areas of Braulio Carrillo National Park, and the Maquenque Wildlife Refuge. *Image created by M. Snyder.*

(Fig. 18.3). Shortages of potable water for expanding human populations are becoming increasingly problematic, even in wet lowland areas such as Puerto Viejo de Sarapiquí (Vargas 1995), which receive as much rainfall as 4 m yr^{-1}. For a more detailed history of river development in Costa Rica, see Pringle and Scatena (1999b).

Today, the rivers of Costa Rica appear and function very differently than they did just a decade ago and their structural and functional properties are likely to continue to change in response to future anthropogenic influences.

Numerous contributions to riverine research in Costa Rica (reviewed in this chapter) have been made by investigators at the Universidad de Costa Rica (UCR) and its Centro de Investigación en Ciencias del Mar y Limnología (CIMAR), the Universidad Nacional (UNA), EARTH College, and the Instituto Costarricense de Electricidad (ICE). The Organization for Tropical Studies (OTS) has also facilitated riverine research at its biological stations throughout the country—with a significant number of riverine studies conducted at La Selva Biological Station on Costa Rica's Caribbean slope (e.g., La Selva STREAMS Project: http://www.streamslaselva.net) and an increasing number of studies occurring at OTS's Palo Verde Biological Station on the Pacific coast. Additionally, studies conducted at Maritza Biological Station (located in Guanacaste on the Pacific slope and developed with the Stroud Water Research Center) have significantly added to our knowledge of Costa Rican rivers.

In the sections that follow, we discuss (1) riverine *plants* with a focus on algal periphyton (given the conspicuous lack of information on aquatic macrophytes); (2) riverine *animals* including both vertebrates and invertebrates; (3) *species interactions* in rivers ranging from mutualism, competition, and predation to frugivory and seed dispersal by fishes; (4) *ecosystem functioning and dynamics* with emphasis on biogeochemistry, nutrient cycling, primary productivity, and decomposition; (5) *people and nature* with a focus on environmental effects of hydropower, land use changes, water withdrawals, and invasive species; and finally (6) *future perspectives*, a section which discusses research needs and the importance of maintaining the biological integrity of rivers for human survival.

Plants

In this section we primarily focus on algal periphyton; however, we will first briefly discuss aquatic and semi-aquatic vascular plants. In the Caribbean lowlands, emergent aquatic vegetation is composed of aquatic grasses like *Panicum maximum*, *Oryza latifolia* (small rice, *arrocillo*), *Hymenochne amplexicaulis*, *Brachiaria* sp., as well as floating vegetation such as water hyacinth (*Eichhornia crassipes*; Jiménez et al. 2005, and see Jiménez, chapter 20 of this volume). Three species of *Eichhornia* occur in Costa Rica with the native species (*E. azurea* and *E. heterosperma*) commonly occurring in swamps and ponds. *E. crassipes* is abundant (especially in Limón and Guanacaste [Janzen

1983]) and is a widespread invasive weed in reservoirs such as Lago Arenal.

In general, tropical algae are poorly understood (Mann and Droop 1996). This is likely a function of unexplored areas and the prevalence of a cosmopolitan paradigm long held in the phycological community. Algae are often reported as having cosmopolitan distributions, with worldwide occurrences (Finlay et al. 2002); however, research now indicates that diatom species demonstrate biogeographical patterns that reflect endemism and regional distributions based on environmental factors and historical constraints (Kociolek and Spaulding 2000, Kilroy et al. 2007). These biogeographic patterns hold especially true for many tropical regions where very high levels of diatom endemism have been documented (e.g., Metzeltin and Lange-Bertalot 1998, Moser et al. 1998, Metzeltin and Lange-Bertalot 2002).

Although algal distributions are poorly documented in Central America including Costa Rica, regional patterns have been reported that are associated with distinct stream solute chemistries resulting from geothermal activity (Pringle et al. 1993). In a broad landscape-scale survey, Pringle et al. (1993) collected algal samples from multiple streams draining the Barva, Poás, and Arenal Volcanoes along the Cordillera Central (Figs. 18.2, 18.3; and see Lawton et al., chapter 13 of this volume, for a detailed account on the forested ecosystem of this volcanic mountain range). These streams receive different types of geothermally modified groundwater inputs. Diatoms were dominant in most samples, except for thermal (hot) springs, very low pH environments (acidic), and streams with mineral precipitation. In hot springs at Tabacón near Arenal and Tucarón near Platanar, cyanobacteria were prevalent, but with low diversity including the genera *Phormidium*, *Lyngbya*, *Dermocarpa*, *Oedogonium*, and *Oscillatoria*, and the diatom genus *Pinnularia*. Green algae (Chlorophyta) were diverse and occurred in many samples, including *Microspora* in low pH, acid sulfate waters of Río Azufre (pH = 4.6). A more recent taxonomic survey recognized nineteen new diatom taxa described from the acidic Río Agrio that drains Volcán Poás, including taxa from the diatom genera *Chamaepinnularia*, *Eunotia*, *Frustulia*, and *Stauroneis* (Wydrzycka and Lange-Bertalot 2001).

Other habitats and regions of Costa Rica are less studied but research shows diatoms (Bacillariophyceae) to be the dominant group, with red algae (Rhodophyta), green algae (Chlorophyta), and blue greens (Cyanobacteria) as commonly represented algal divisions in rivers. It also should be noted that aerophilic taxa, which grow in moist subaerial habitats (including bromeliads and wet mosses), form a significant algal component in the wet tropical forests of Costa Rica, and can wash into streams secondarily.

Many diatom taxa that occur in rivers in the Caribbean lowland streams (R. Bixby, unpublished data) and Pacific Coast (Silva-Benavides 1996a) are regionally endemic (*Navicula incarum* Lange-Bert. & Rumrich, *N. ingapirca* Rumrich & Lange-Bert., *N. kohlenbachii* Lange-Bert. & Rumrich, and *Stenopterobia pumila* Lange-Bert. & Rumrich, all described from Ecuador) or pantropical (e.g., *Cocconeis feuerbornii* Hustedt described from Java and *Rhopalodia gibberula* var. *argentina* Brun., collected from South America, the Caribbean, and Japan [Bourrelly and Manguin 1952]). Overall, we expect the number of endemic, Neotropical, and Pantropical distribution records for all algae to increase—as more floristic and taxonomic works are completed in Costa Rica.

Very little systematic and phylogenetic research has been completed on Costa Rican algae. One notable exception is the inclusion of freshwater red algae (Rhodophyta) into North/Central American systematics analyses (Sheath et al. 1993a,b, Sherwood and Sheath 1999). Among freshwater red algae, interesting patterns have been documented in tropical areas of Central America. Sheath et al. (1993b) initiated a study to survey the North American specimens in the Order Ceramiales. Parts of this Order belong to a group of freshwater red algae thought to be secondary invaders from marine environments. This survey reported the first definitive Central American records in Costa Rican streams of the freshwater red algae *Ballia prieurii* Kütz. and *Caloglossa ogasawaerensis* Okam based on morphological and molecular analyses.

Algae are light-limited in lowland tropical streams that drain primary forest (Fig. 18.4). In the Caribbean lowlands (specifically at La Selva Biological Station; see McClearn et al., chapter 16 of this volume), standing crop accrual of algal periphyton has been shown to be significantly higher at sites bordered by pasture, with high light levels (14–45 mg m^{-2} chl *a*), versus shaded forested watersheds (4 mg m^{-2} chl *a*, $P < .001$) (Paaby and Goldman 1992). This degree of light limitation and resulting low algal biomass response is similar to that found in streams of temperate regions where light can be a limiting factor even with increased nutrient levels.

Periphyton biomass and species assemblages are often affected by stream solute (nutrients and ions) levels. Solutes can occur in naturally elevated conditions or can be in higher concentrations owing to anthropogenic inputs. As previously mentioned in this section and discussed in more detail later (see Geochemistry section), many streams draining volcanic regions of Costa Rica receive groundwater with elevated levels of solutes resulting from underlying geothermal activity (Pringle and Triska 1991, 2000). Detailed taxonomic studies of algae that are ongoing at

Fig. 18.4 Quebrada Salto (60 m a.s.l.), a lowland interior forest stream that is heavily shaded by a multi-strata canopy. The Salto drains La Selva Biological Station, which is located at the gradient break between the mountains and foothills on the Caribbean slope. This river is one of the two main drainages of the station and it receives natural inter-basin transfers of geothermally modified groundwaters (of the sodium chloride bicarbonate type) in its lower reaches.
Photo by C. Pringle.

La Selva Biological Station (R. Bixby, University of New Mexico) indicate that euryhaline algal taxa are dominant in high-solute streams at La Selva. High-solute streams receive interbasin transfers of geothermally modified groundwater and are classified as a dilute *sodium-chloride-bicarbonate* geothermal water type. Euryhaline taxa (Fig. 18.5) reflect the presence of salts and chloride in these streams. In contrast, low-solute streams that do not receive groundwater inputs are poorly buffered and dominated by acidophilic taxa (Bixby et al. 2005; Fig. 18.5).

Biotic parameters such as competition, predation, and grazing can also play a role in influencing algal growth and species composition in Costa Rican streams and rivers. These studies are discussed later in this chapter (see Species Interactions section).

Because algae function well as indicators of the status of

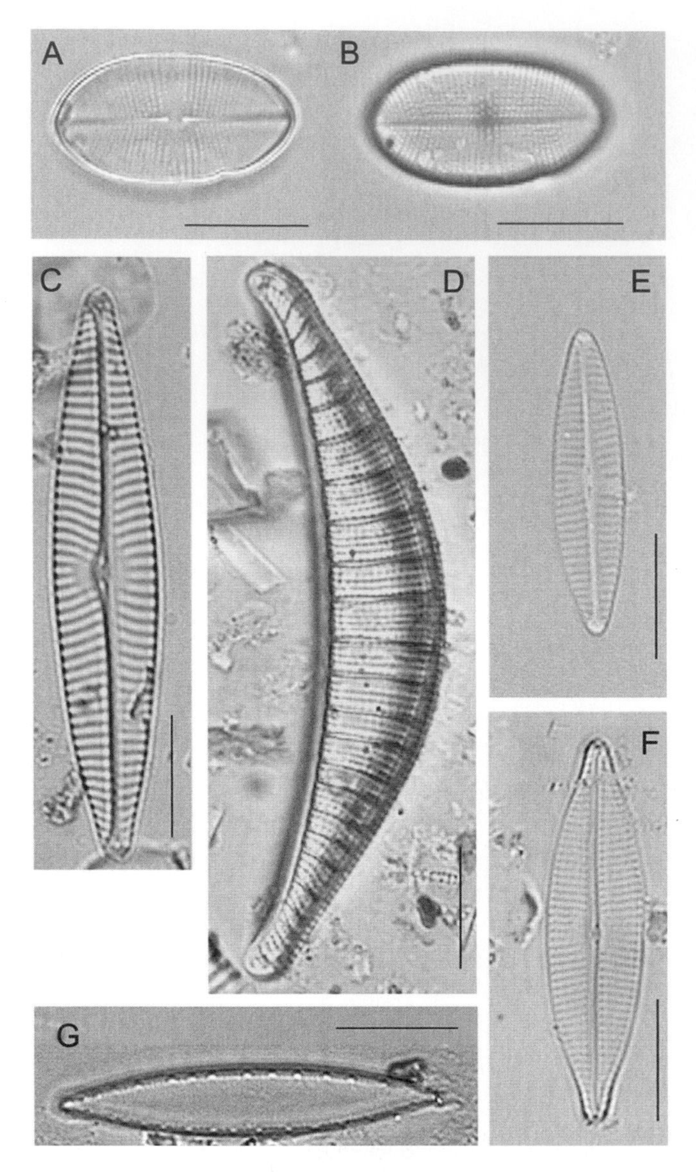

Fig. 18.5 Light micrographs (scale bar = 10 µm) of Neotropical diatoms found in streams draining La Selva Biological Station, Costa Rica. Some of the taxa are good indicators of stream chemistry, as indicated below: (A) *Cocconeis feuerbornii* Hustedt, raphe valve; (B) pseudoraphe valve; (C) *Navicula kohlenbachii* Lange-Bert. & Rumrich; (D) *Rhopalodia gibberula* var. *argentina* Brun.; (E) *Navicula incarum* Lange-Bert. & Rumrich; (F) *Navicula ingapirca* Rumrich & Lange-Bert.; (G) *Stenopterobia pumila* Lange-Bert. & Rumrich. This taxon is found in solute-poor unbuffered streams at La Selva with low bicarbonates and is an indicator of acidic conditions.
Photos by R. Bixby.

their environment and the quality of its waters, they have been used to assess anthropogenic impacts on Costa Rican streams and rivers. Anthropogenic impacts on algal communities include decreased dissolved oxygen levels, organic inputs (including elevated nitrogen compound levels), and increased sediment load (Silva-Benavides 1996b, Michels et al. 2006). Michels (1998a,b) studied the effects of pig farm sewage on periphyton communities in two streams in south-central Costa Rica. The diatom species composition broadly reflected increases in agricultural pollution (i.e., increases in soluble reactive phosphorus [SRP], ammonium, and nitrate, and low dissolved oxygen), although diatom species groups did not correlate well with common diatom indices of pollution developed in Europe. This study demonstrates the need for understanding the ecological requirements for Neotropical diatoms and the development of more regionally based biomonitoring indices.

From a species-level perspective, studies looking at organic pollution in Costa Rican streams note that cosmopolitan diatom taxa (*Gomphonema parvulum* [Kütz.] Kütz., *Navicula subminuscula* Manguin, *Navicula seminulum* Grun., and *Nitzschia palea* [Kütz.] W. Sm.) were indicative of organic pollution in streams (Silva-Benavides 1996b, Michels 1998a). In contrast, pristine stream sites were found to reflect more variability in environmental parameters and were dominated by different and more regionally distributed diatoms (Quebrada La Pita, Michels, 1998a; Río Savegre, Silva-Benavides, 1996b). Michels et al. (2006) examined diatom communities from 23 stream sites with differing riparian land use in Golfo Dulce on the Osa Peninsula. Diatom species distributions were highly correlated with riparian shading, which separated the first-order forested streams from the larger rivers. This shading gradient was associated with decreased canopy cover as a result of deforestation and agriculture. Turbidity and biological oxygen demand also explained much of the variation among species assemblages.

Animals

The relatively recent geologic history of Costa Rica contributes to the lower species diversity of these rivers compared to those of South America. The close proximity of the ocean also creates a distinct signature and marine-derived fauna is abundant in lowland rivers. The presence of marine fauna far inland is a good example, such as the sightings of sharks and flounder in the San Juan–Sarapiquí river drainages.

If Costa Rica's river fauna is compared to that of the Caribbean islands, which also have short and steep river drainages, the pattern is different. Rivers in Costa Rica are much more diverse and species-rich than those of the islands, which are more difficult to colonize by freshwater species. Therefore, Costa Rica is somewhat of an intermediate place, having a low diversity relative to continents and a rich fauna when compared to insular systems.

The diversity of river types (e.g., steep mountain streams,

meandering lowland rivers) and sizes provides for a wide array of faunal community structures and compositions. Many lowland river systems are fish- and shrimp-dominated, while insects and amphibians often predominate in mountainous headwater streams. Animals inhabit a wide variety of habitats within a river: cobble-covered riffles; sandy and muddy pools; and the deep stretches of large rivers. Even the vertical walls of waterfalls are the habitat of some insect species and decapods (Calvert 1915).

The amount of information available about the diversity and ecology of the different animal groups varies significantly among taxa. Similar to other parts of the tropics, groups with large body sizes are better known than small organisms. As expected, the diversity of vertebrates is well known, with fishes being the most species-rich group and perhaps the most important for the functioning of riverine ecosystems. In contrast, invertebrates are poorly studied: we lack complete species inventories and have little information on invertebrates at the levels of population, community, and ecosystem (Ramírez and Pringle 1998a,b).

River fauna differs considerably among and within drainages. Geomorphological characteristics of rivers provide the setting for the establishment of distinct faunal assemblages. The large percentage of mountainous terrain in Costa Rica results in a significant fraction of river drainages characterized by steep slopes, the presence of waterfalls, and many torrentially flowing waters. Areas of coastal lowlands are relatively limited, in particular on the Pacific slope, and few rivers have developed the typical complex series of meanders and floodplains that characterizes coastal rivers. Some animal groups have distributions that clearly reflect changes in river geomorphology. Fish assemblages, for example, are diverse and abundant in lowland areas, but decrease toward headwaters, and completely disappear above ~800 m a.s.l. In addition, waterfalls play a major role defining fish distribution, with a lack of fish upstream of large waterfalls. Marine-related fauna is also abundant in lowland rivers and decrease or disappear toward the mountains. Marine fishes, such as tarpon (*Megalops atlanticus*) and even some shark species, are known to travel upriver for several kilometers in large lowland rivers (Bussing 1993) and shrimp are more abundant near the coast (Ramírez et al. 1998).

Here we focus on the animal fauna that inhabit rivers or that rely on river resources for food. Thus, while birds are included, we touch upon only those that consume aquatic resources. We discuss (1) general species richness of different faunal groups (our purpose is not to provide extensive species checklists); (2) riverine habitats, where habitat refers to the type of river system inhabited by a particular faunal group and the specific locations used within a river (e.g., riffles, pools, riparian areas); and (3) available information about animal function within the ecosystem, which is rather limited for most taxa and quite variable among species within faunal groups.

Fishes

Fishes are perhaps the most important vertebrate group inhabiting river ecosystems. There are 174 fish species reported for Costa Rica's freshwater ecosystems, including introduced species, distributed in 38 families (see www.fishbase.org). The most diverse families are the Cichlidae, Poeciliidae, and Characidae. Twenty fish species, mainly in the families Poeciliidae and Characidae, are considered endemic to Costa Rica. River fishes of Costa Rica are well documented by Bussing (1998) and information is also available on the web at www.fishbase.org. Ecological aspects of fish communities at La Selva Biological Station are discussed in detail by Burcham (1980) and Bussing (1994). Winemiller (1983) provides an overview of the fish assemblages of Corcovado National Park located in the Osa Peninsula on Costa Rica's Pacific Coast. For a detailed description of the ecology of this lowland rainforest peninsula, see Gilbert al. (chapter 12 of this volume).

Fishes are more diverse in lowland rivers, decline in abundance and diversity with elevation, and completely disappear from most rivers above 800 m a.s.l. Within a river, they use a wide variety of habitats (e.g., pools, undercut banks; Lyons and Schneider 1990) but tend to be more abundant in slow-water areas such as pools. Deep pools with woody debris provide shelter and food resources for many fish species and they tend to be more abundant and diverse in this type of habitat (Fig. 18.6).

Fishes play a key role in ecosystem processes such as nutrient cycling, primary production, and decomposition. Predatory fishes can be important controlling populations of invertebrates and other fish species and can have direct or indirect effects on other resources (e.g., detritus, algae; Pringle and Hamazaki 1998). Species that consume algae and fine organic matter from the river bottom are also important influences on primary production and the movement of organic matter within river systems. Wootton and Oemke (1992) discuss the importance of fish herbivory in a lowland tropical stream (draining La Selva Biological Station and adjacent environments) in the context of latitudinal differences in fish community structure. Inputs of terrestrial insects (e.g., ants) support high densities of fishes (3–14 individuals m^{-2}) in forested headwater streams (Small et al. 2013a). In a larger, fourth-order stream, the contribution of terrestrial insects to the diet of the omnivorous fish *Astyanax*

Fig. 18.6 Theraps underwoodi, a common cichlid found in the Río Sarapiquí drainage that is vulnerable to displacement by invasive tilapia. *Photo by R. Coleman.*

aeneus (Characidae) resulted in this species playing a disproportionally important role in stream nutrient cycling. *Astyanax* represented 12% of the fish population and 18% of the fish biomass in this stream, but accounted for 90% of phosphorus supplied by fish excretion (Small et al. 2011a).

Twelve species of non-native fishes are now established in many rivers throughout Costa Rica. The best known examples are trout and tilapia, both introduced for aquaculture. Trout inhabit cool-water mountain streams and are now well established in many streams that used to lack fishes (e.g., see Kappelle, chapter 14 of this volume). Trout have been responsible for the extirpation of stream-dwelling frogs in most of the high-elevation streams where they have become established. The negative environmental consequences of the spread of tilapias in lowland streams of Costa Rica are discussed in detail in the final section of this chapter (People and Nature: Non-native Fish Introductions; and see Jiménez, chapter 20 of this volume). Increased abundance and distribution of tilapia could mean the endangerment and ultimately extinction of native cichlids (Fig. 18.6) in Costa Rica: the Cichlidae family is the most diverse fish family found in inland waters in Costa Rica, with 9 genera and 24 species (Bussing 1998). Displacement of native cichlids by tilapias will undoubtedly have major effects on the structure and function of river ecosystems. Other invasive species include armored catfish (*Pterygoplichthys* spp.: Loricariidae), or "peces diablo" ("devilfish"), which have recently been identified in the Reventazón River basin (Solano and Arias 2011). *Pterygoplichthys*, common aquarium fish native to South America, are invasive in other Central American countries and are phosphorus-rich fish that form large benthic aggregations and alter stream nutrient dynamics (Capps and Flecker 2013).

Amphibians

Of the 179 amphibian species reported for Costa Rica, ~15–20 are aquatic, with larval tadpole stages that dwell in rivers (Young et al. 2004). Stream-dwelling frog taxa are primarily in the genera *Atelopus* (e.g., *Atelopus varius*; Fig. 18.7) and *Rana* (e.g., *Rana warszewitschii*), and in the families Centrolenidae (glass frogs) and Hylidae (tree frogs), reaching highest levels of speciation and abundance in rivers at higher elevations where predatory fishes are absent.

Species-specific body and mouthpart morphology allows larval frogs to occupy virtually all stream microhabitats, including riffles, runs, pools, bank-side marginal pools, and anoxic detrital accumulations. Although tadpoles are a ubiquitous component of many upland Costa Rican streams, the larval stages of stream-breeding anurans are relatively under-studied, and many of the species are yet to be formally described. Savage (2002) provides an overview of most amphibians in Costa Rica. Additional information can be found at the website of Costa Rica's National Biodiversity Institute, INBio (www.inbio.ac.cr).

Catastrophic amphibian declines in Costa Rica, like throughout much of the Neotropics, are well documented (Stuart et al. 2004), with perhaps the earliest and best-known example being the extinction of the golden toad of Monteverde (*Bufo periglenes*) in 1989 (see Lawton et al., chapter 13 of this volume). While some amphibian declines are considered enigmatic, such as the dramatic decrease in amphibians inhabiting protected old-growth lowland habitat at La Selva Biological Station (Whitfield et al. 2007), the fungal pathogen *Batrachochytrium dendrobatidis* has been implicated with the extirpations or extinctions of most highland stream-dwelling species (i.e., frogs with larval tadpole stages) in Costa Rica (Lips et al. 2003). Particularly decimated are species of the once common stream-breeding genus *Atelopus*, many species of which are now believed to be extinct in the wild (La Marca et al. 2005). Although generally sub-lethal to tadpoles, the chytrid fungus kills adult frogs by inhibiting electrolyte transport (Voyles et al. 2009), and is considered to be an emerging infectious disease of amphibians. Additional information on amphibian declines can be found in Lips (1998, 1999), Young et al. (2004), and Whitfield et al. (2007, 2012, 2013).

Until the early to mid-1980s, tadpoles were a conspicuous component of highland river ecosystems above 800 m a.s.l in Costa Rica. They undoubtedly played a key role in stream and riparian foodwebs. Stream-dwelling amphibians create a linkage of energy and nutrients between aquatic and terrestrial habitats, initially due to the eggs that adults deposit into the streams, and again when the young froglets metamorphose and move to the riparian area. However, little is known about the trophic status of most larvae and adults, and how tadpoles affect stream ecosystem structure and function is poorly understood. Unfortunately, there is presently little chance to study their ecological roles within the country of Costa Rica. Realizing that the ongoing spread of chytrid fungus had not yet decimated highland frogs throughout parts of Panama, investigators have consequently pursued studies in Panama on the role of stream-dwelling frogs on ecosystem function as part of the Tropical Amphibian Declines in Streams (TADS) Project (Whiles et al. 2006). Tadpoles were found to reduce algae, alter algal community composition, increase biomass-specific primary production, and alter macroinvertebrate functional structure. It has become clear, from TADS' studies in Panama, that stream-dwelling frogs were undoubtedly once key players in structuring highland streams in Costa Rica and that their loss has significantly affected ecosystem function (see Ranvestel et al. 2004, Whiles et al. 2006, Verburg et al. 2007, Connelly et al. 2008, Colon-Gaud et al. 2009, Whiles et al. 2013).

Reptiles

Caimans, crocodiles, and turtles are the main reptilian fauna inhabiting rivers, and their biology is poorly known.

Fig. 18.7 Stream-dwelling frogs (i.e., with larval tadpole stages) have largely disappeared from high elevation streams throughout Costa Rica. (A) Sexually dimorphic harlequin frogs, *Atelopus varius*, in amphiplexus on the banks of the Río Sardinalito in Braulio Carrillo National Park (~650 m a.s.l.) in 1983 before the taxon was extirpated. This taxon is vulnerable to a fungal pathogen that has since resulted in its extirpation from most high elevation streams throughout Costa Rica. (B) Dying harlequin frog, *Atelopus zeteki*, at high elevation stream site in El Cope, Panama, 2004.
(A) photograph by C. Pringle; (B) photograph by S. Connelly.

Caimans (*Caiman crocodilus*) and crocodiles (*Crocodylus acutus*) can be found in rivers up to 350 m a.s.l. Caimans are small (up to 2.5 m) and inhabit many interior lowland swamps and slow rivers. The crocodile is a coastal species that was formerly common in mangrove swamps and coastal rivers but it can also be found in rivers up to 50 km inland. The reader is directed to other references for more detailed life history information about caimans and crocodiles in Costa Rica (Dixon and Staton 1983, Scott and Limerick 1983, Allsteadt and Vaughan 1988, 1992, Savage 2002; website of Costa Rica's National Biodiversity Institute, INBio [www.inbio.ac.cr]; see also Jiménez, chapter 20 of this volume).

There are several species of turtles that use rivers as their main habitats, all tending to increase in diversity in lowland rivers. They fall into three families: (1) the snapping turtle *Chelydra serpentina* (Chelydridae); (2) the aquatic and terrestrial Embydidae; and (3) the semi-aquatic mud turtles (Kinosternidae).

Most reptiles favor slow-flowing rivers and streams in lowland areas as their habitats (e.g., see McClearn et al., chapter 16 of this volume, for reptiles thriving in the Caribbean lowlands). Within these they can be observed on wood debris and along the river margins. At least some turtles prefer to inhabit fast-flowing waters. Among them, the white-lipped mud turtle (*Kinosternon leucostomum*) can be found in fast-flowing streams with clear waters where it feeds on aquatic invertebrates. Most riverine reptiles are predatory; however, some turtles are omnivorous and consume significant quantities of plant material as well as animals. More detailed information about riverine turtles can be found in Savage (2002) and at the website of Costa Rica's National Biodiversity Institute, INBio (www.inbio.ac.cr).

Birds

Birds are a major component of the faunal diversity associated with rivers of Costa Rica. There are 910 species reported for Costa Rica (as of 2014) and ~150 species are associated with aquatic environments—many with rivers. Detailed information about the life histories of specific bird species common in riverine ecosystems can be found in Stiles (1983) and Stiles and Skutch (1989).

Birds play important roles in rivers as predators and some have a role in disturbing the bottom in their search for food. Common wading birds include six species of herons, green ibis (*Mesembrinibus cayennensis*), and tiger herons (*Tigrisoma* spp.). Costa Rica's six species of kingfisher (*Ceryle* and *Chloroceryle* spp.) dive for fish from riverside trees, and anhingas (*Anhinga anhinga*) spear fish with their bills while swimming underwater and are often seen drying their wings in the sun on riverside snags.

Many bird species associated with rivers have particular habitat requirements. For example, jacanas (*Jacana spinosa*) are common in rivers where aquatic and riparian vegetation is abundant, several species of sand pipers (e.g., *Actitis macularis*) can be found in rivers with sandy or muddy shores, and ducks (e.g., *Cairina moschata*) can be found even in small streams with low flow. The strikingly patterned sunbittern (*Eurypyga helis*) prefers forested streams, as does the sungrebe, or finfoot (*Heliornis fulica*).

Mammals

A few species of mammals are closely associated with rivers. The Neotropical river otter (*Lontra longicaudus*) inhabits riparian habitats from small streams to large rivers in lowland areas of Costa Rica as well as other parts of Central and South America. Otters feed almost exclusively on crustaceans and fish and represent an important predator of larger bodied shrimp (e.g., *Macrobrachium;* Figs. 18.8, 18.9) and fish (e.g., cichlids, Fig. 18.6). Otters prefer deeper streams with high arboreal and shrub cover and large rocks and logs (Spinola-Parallada and Vaughan-Dickhaut 1995). The otter is considered a threatened species due to habitat loss, but is classified as Data Deficient on the IUCN Red List of Threatened Species.

Another important mammal in river ecosystems is the fishing bat or greater bulldog bat, *Noctilio leporinus*, in the family Noctilionidae. *N. leporinus* is a relatively large bat with a mass of 60–90 g and a wingspan of up to 60 cm, with a bulldog-like face, narrow pointed ears, and very small eyes (Brandon 1983). Fishing bats are common in lowland areas of Costa Rica and can be found searching for food over slow-moving rivers and ponds. Their diet consists primarily of fish but also includes insects and crustaceans picked from the water surface and insects caught in the air. They fly close to the water surface, with their greatly enlarged hind feet and claws occasionally dipping into the waters to grab small fish and other organisms that break the water surface. *Noctilio albiventris,* the lesser bulldog bat, is smaller than *N. leporinus*, with a mass of 30–40 g and a wingspan of 40–45 cm. It primarily feeds on insects caught on the wing or from the river surface (Brandon 1983).

Although now at the verge of extinction, manatees (*Trichechus manatus*) used to inhabit riverine ecosystems near coastal areas in Costa Rica. Reports from the 1800s include the presence of manatees in the Sarapiquí and San Juan River drainages (Ligon 1983). Recent work with predictive

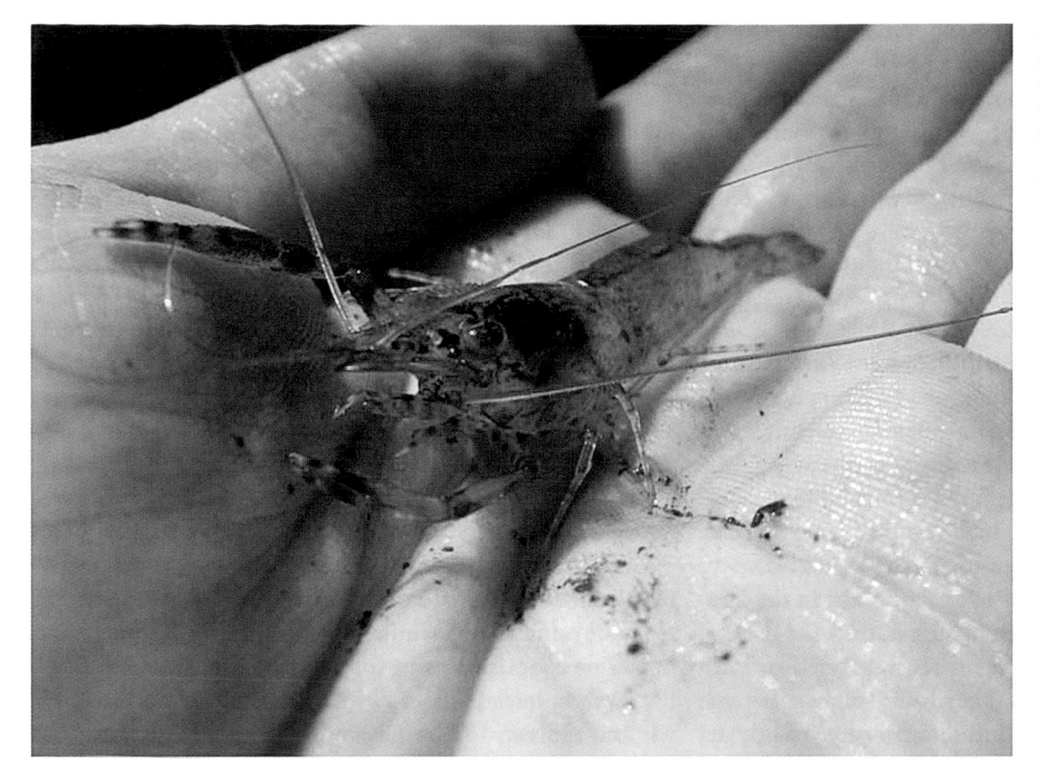

Fig. 18.8 A freshwater migratory shrimp in the genus *Macrobrachium*, collected from the Sábalo River in the Caribbean lowlands. *Macrobrachium* spp. can reach sizes of greater than 15 cm.
Photo by M. Snyder.

Fig. 18.9 The freshwater shrimp, *Macrobrachium carcinus*, about to be consumed at a restaurant in La Virgen, Costa Rica.
Photo by M. Snyder.

Fig. 18.11 Marcelo Ardón measuring microbial respiration in Quebrada Carapa at La Selva Biological Station.

shredders in leaf breakdown. Low abundance of invertebrate shredders in leaf packs has been reported in various streams in La Selva (Ramírez and Pringle 1998b, Rosemond et al. 1998, 2001, Ardón et al. 2006, Stallcup et al. 2006, Ardón and Pringle 2008). Also providing support for the importance of microbes, studies have reported high fungal biomass (Rosemond et al. 2002, Ardón et al. 2006, Stallcup et al. 2006), bacterial biomass (Ardón and Pringle 2008), and microbial respiration (Ramírez et al. 2003, Ardón et al. 2006, Stallcup et al. 2006, Ardón and Pringle 2007) on decomposing leaves in streams at La Selva (Fig. 18.11). Benstead (1996), on the other hand, reported a higher abundance of insect shredders (range, 2–8 individuals per leaf bag) in a higher elevation (675 m) stream than the abundance reported for streams at La Selva (range, 0–2 individuals per leaf bag; Rosemond et al. 1998, Ardón et al.

2006, Stallcup et al. 2006, Ardón and Pringle 2008). Future studies at different geographical locations are required to determine the relative importance of microbial versus insect detritivores active in Costa Rican streams.

Using the natural stream phosphorus gradient at La Selva and an experimentally P-enriched stream, Ramírez et al. (2003) found that microbial respiration on leaves was positively related to SRP concentration, saturating around 15 μg SRP L^{-1}. Leaf decay rate, fungal biomass, and invertebrate biomass showed similar saturating responses to dissolved P concentrations (Rosemond et al. 2002). Even in streams with elevated P-levels, nitrogen does not appear to secondarily limit organic matter breakdown or microbial activity (Stallcup et al. 2006).

Leaf breakdown rates are characteristically high in warm tropical streams (Stout 1980, Benstead 1996). It has been generally thought that defensive secondary compounds slow down decomposition rates of tropical leaves (Janzen 1975, Stout 1980, Wantzen et al. 2002). However, in studies conducted in lowland Costa Rica at La Selva, Ardón and Pringle (2008) found that defensive secondary compounds rapidly leach from leaves in streams, and the structural composition and nutrient content of leaves determine their breakdown rates (Ardón et al. 2006, Stallcup et al. 2006). A recent comparative study between streams at La Selva, Costa Rica, and Coweeta, North Carolina, found that temperate leaves from Coweeta had higher concentrations of secondary compounds than tropical species at La Selva, and that structural compounds were important in determining breakdown rates in both sites (Ardón et al. 2009).

Large macroconsumers such as fishes and shrimps can also affect the rate of organic matter processing. Experimental exclusion of fishes and shrimps from leaf packs led to increased biomass and density of colonizing insects, but leaf decomposition rate was slower in the absence of these macroconsumers (Rosemond et al. 2001).

Production of Higher Trophic Levels

Ramírez and Pringle (1998b) measured the abundance, biomass, and secondary production of insects in pool and riffle habitats in a lowland stream at La Selva. While all of these parameters were low compared to subtropical and temperate streams, the biomass turnover rate was high, indicating rapid population turnover. This finding is consistent with studies of stream insect life histories in northwestern Costa Rica (Jackson and Sweeney 1995) that have documented rapid insect development.

At La Selva, insect biomass standing stocks have been related to the number of days since the last large rainstorm, indicating that abiotic disturbance is a major structuring force of the stream insect community (Ramírez and Pringle 1998b). The dominant consumer taxon, in terms of biomass and production, was larval Chironomidae (Diptera) (Ramírez and Pringle 2006). Although Chironomid biomass and secondary production rates were similar among streams that varied in P concentrations (Ramírez and Pringle 2006), chironomids have a higher dietary P requirement, suggesting evolutionary adaptation to local environmental conditions (Small et al. 2011b).

Studies in lowland streams at La Selva Biological Station have demonstrated that shrimps represent the majority of the standing invertebrate biomass, in both the wet season and dry seasons, when compared to insects, as well as being a significant component to overall secondary invertebrate production (Snyder 2012).

People and Nature

Human activities have dramatically modified stream and river ecosystems of Costa Rica. Dams, mainly for hydropower, regulate the flow of many large rivers and have become a common feature in mountainous regions of the country's northern Caribbean slope. An expanding tourism industry has led to a greater demand for surface water concessions to support resort-like developments, especially in areas along the northern and central Pacific coast. The area of irrigated cropland is also increasing, as are domestic and industrial demands for water. The following section discusses the primary human uses of rivers in Costa Rica as well as some of the major development trends affecting water resources and their consequences for river ecosystems. Creative solutions to protect water resources and efforts to encourage more integrated water resources management are also briefly summarized here.

Human Uses of Rivers

Human uses of rivers in Costa Rica can be divided into extractive and non-extractive uses. In terms of extractive uses, rivers provide numerous ecosystem-produced goods to human populations, including water for domestic and industrial uses, irrigation, channel materials, shrimp, and fish. In many parts of the country, potable water comes from rivers, especially in areas where homes and businesses are connected to a water service network administered by either the national-level utility company Acueductos y Alcantarillados (AyA), municipal governments, or local water administration offices. Mining operations extract river bed materials (boulders and cobbles) from many rivers (e.g., the Sucio River on the northern Caribbean slope) for use in the construction of roads. Fishing takes many forms in Costa Rica. For example, some subsistence fishing is practiced among native communities in southern Costa Rica and along the eastern slopes of the Talamanca Mountains. River fish also form a supplemental part of the diet of many Costa Ricans in rural, lowland areas where recreational fishing of species like *Joturus pichardi*, *Agonostomus monticola*, *Brycon guatemalensis*, *Theraps underwoodi*, and *Parachromis dovii* is common. In addition, a growing sport-fishing industry in the region near Barro Colorado and the lower San Juan River along the Nicaraguan border attracts thousands of international tourists annually. Species commonly sought after include *Megalops atlanticus* and *Centropomus undecimalis*.

Hydropower recently has emerged as a primary use of rivers in Costa Rica; here it is considered an extractive use on the basis of the way that most hydropower plants operate—that is, by diverting water or restricting flows from river channels for extended periods of time. Among Central American countries, Costa Rica is a regional leader in hydropower and more than 80% of the country's electricity is currently generated by hydropower dams. Since the early 1990s, the country has experienced explosive hydropower development, with more than 30 plants constructed in the decade of the 1990s alone (Anderson et al. 2006b). Costa Rica has plans to double its installed hydropower capacity over the period 2004–2014 through a combination of several new large storage dams and smaller water diversion projects (ICE 2004).

Rivers provide a number of non-extractive uses to humans in Costa Rica. Non-extractive uses are loosely classified here as ecosystem services and whose value to human populations is difficult to quantify. Waste assimilation is among the most important of these uses. Wastewater is usually managed at a household or individual industry level and, if a septic tank is not in place, is discharged directly into rivers generally with little to no treatment. Rivers like the Virilla and Tárcoles assimilate and carry waste away from the densely populated Central Valley.

Navigation is another important non-extractive use of rivers in Costa Rica, and in the past, large lowland rivers were primary transportation routes between coastal areas and the interior of the country. For example, the San Juan–Sarapiquí River corridor existed until the middle of the twentieth century as a major thoroughfare for moving goods and people between the Caribbean coast and the Central Valley; the town of Puerto Viejo de Sarapiquí marks the point to which the river was navigable year-round and from there goods were moved by mule track and dirt road over the mountains to the San José region (Butterfield 1994). While today roads have replaced the need for river-based travel in most areas, there are still human communities that are primarily accessible by water (e.g., Tortuguero on the northern Caribbean coast). In addition, rivers also act as a sort of "water park" for local human communities throughout the country. The importance of rivers for recreational uses can typically be observed on any sunny weekend or holiday when families gather to swim and picnic along river banks. Also, ecotourism now comprises a substantial portion of Costa Rica's economy and rivers play a crucial role in attracting thousands of international tourists. For example, one major tour company, *Ríos Tropicales*, offers river rafting trips of varying difficulty on nine rivers across the country (Fig. 18.12).

Threats to Costa Rican Rivers and Trends in Development

Much of what is currently known about the natural history and ecology of Costa Rican streams and rivers is based on studies conducted in forested catchments upstream from major human settlements. However, intact forest cover and the absence of human influences no longer characterize the majority of the country's river basins. Here we discuss four major trends that are reshaping Costa Rica's landscape and leaving an indelible imprint on river ecosystems: (1) hydropower development; (2) land use conversion; (3) increasing concessions of water for human uses; and (4) the introduction of non-native species such as tilapia.

Hydropower Development

Topography and climate have created a considerable hydropower potential for Costa Rica. The longitudinal orientation of mountains and large amounts of annual precipitation (>5 meters in some areas) has resulted in hundreds of high-gradient streams with sufficient discharge for electricity generation. Costa Rica's theoretical hydroelectric potential is estimated at 25,500 megawatts (CFIA 2005); practical hydropower potential is estimated at 10,000 megawatts (FAO 2005). Numerous hydropower plants of different types (e.g., storage, water diversion) and both large and small dams currently harness the flow of Costa Rican rivers (Fig. 18.13), with a collective installed generation capacity of approximately 1,300 megawatts (Anderson et al. 2006b). However, the majority of Costa Rica's hydropower potential remains untapped and more projects are either under construction or proposed to take advantage of this natural potential as a means for meeting the country's future needs for electricity.

The list of large dams currently in operation (or nearing completion) includes the Cachí, Angostura, and Reventazón Dams in the Reventazón River Basin; the Toro I and II,

Fig. 18.12 Uses of the Sarapiquí River include transportation, hydropower, recreation, and ecotourism (i.e., sightseeing tours and whitewater rafting). *Photos by E. Anderson.*

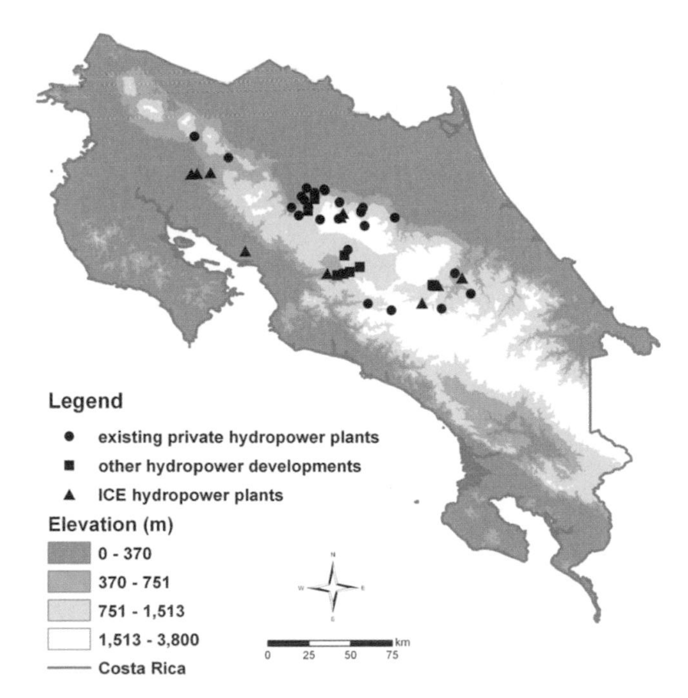

Fig. 18.13 Despite restrictions on private companies, the results of electricity sector reform are visible throughout Costa Rica. Hydropower has been a beneficiary of reforms and 28 private dams have been constructed since 1990. In addition, the Costa Rican Instituto Costarricense de Electricidad (ICE) operates 11 dams and nine other hydropower developments as subsidiaries of ICE have also built many dams. This map shows that private dams are concentrated on gradient breaks and many are located on the wet, northern Caribbean slope.

Cariblanco, and General Dams in the Sarapiquí River Basin; the Peñas Blancas Dam in the San Carlos River Basin; and the Arenal Dam complex in northwestern Costa Rica. Perhaps the best known of these is the Arenal Dam Complex, a multipurpose project that generates a large portion of Costa Rica's electricity, provides irrigation water, and has created new recreational areas. This Complex has changed the face of the landscape in northwestern Costa Rica: an interbasin transfer across the continental divide (from the Caribbean to the Pacific slope) provided additional water to fill a large reservoir (Fig. 18.14) that flooded several towns and required the resettlement of hundreds of people. After being used to generate electricity, water from the dam is diverted

Fig. 18.14 Arenal reservoir, Costa Rica, is one of the largest hydropower projects in Central America and supplies a substantial portion of Costa Rica's energy supply.
Photo by E. Anderson.

to a canal system and used for irrigation by rice and sugarcane farmers in the area near Bagaces, Costa Rica. The Arenal Dam complex also influences the physical, chemical, and biological characteristics of streams and wetlands in the Tempisque River Basin around the area of Palo Verde National Park; these water bodies are the eventual recipients of water from the Arenal reservoir downstream from agricultural operations.

The trend of proliferation of small dams for hydropower, which began in the early 1990s, has disrupted riverine connectivity and altered river ecosystems throughout the country. Of the >30 hydropower plants constructed during the 1990s, the majority have dams <15 meters high (accepted international standard for small dams; World Commission on Dams 2000) and operate as water diversion projects—where water is removed from the channel, piped down a natural gradient, run through turbines to generate electricity, and then returned to the river several kilometers downstream from the original diversion (Anderson et al 2006b). Most of these plants were constructed by private companies following the passage of legislation in 1990 and 1995 that partially privatized electricity generation in Costa Rica. On a local level, three of the principal ecological consequences of these small dams are (1) the break in hydrologic connectivity created by the dam at the water diversion site; (2) substantial flow reduction or de-watering of the river reach between the dam / diversion site and the turbines; and (3) unnatural fluctuations in flow and water temperature downstream from the turbines.

With respect to connectivity, water-mediated transport of matter and organisms is possible across the dam in a downstream direction during high flow events where discharge exceeds the capacity of the water diversion canal or pipeline. Nevertheless, movement of organisms upstream is restricted by these dams (Figs. 18.15, 18.16). Evidence suggests that migratory species in Costa Rica have been adversely affected: few or no individuals of migratory shrimp and fish species (e.g., *Macrobrachium* spp., *Joturus pichardi*, *Sicydium altum*) have been reported in areas upstream from dams (Anderson et al. 2006a; Snyder et al. 2011, W. McLarney, pers. comm.). De-watered reaches between water diversion sites and turbines are often several kilometers long and usually carry from 5 to 10% of average annual flow, or sometimes even less river water, depending on how the hydropower plant is managed. Flow reductions of this magnitude virtually transform a large river into a small stream (Fig. 18.17). Subsequent changes in aquatic habitat along the de-watered reach may affect the distribution of fish species (Figs. 18.18, 18.19) with life history characteristics linked to specific flow or habitat conditions (Anderson et al. 2006a). Downstream from the de-watered reach, hydropower plant operations continue to influence physical river conditions over several kilometers. Here, release of water from turbines following electricity generation causes increases in discharge and water temperature, with larger than natural changes over shorter time scales, which are predictable according to generation schedules and non-natural patterns. Discharge fluctuations in river reaches receiving production waters result in habitat instability for aquatic biota, especially benthic organisms, and can also compromise recreational uses in downstream areas, mainly for safety reasons.

Roughly half of the small hydropower plants constructed during the 1990s in Costa Rica are located on rivers in the San Carlos and Sarapiquí River Basins, both of which drain parts of the northern Caribbean slope (Fig. 18.13). Cumulatively, these hydropower plants have fragmented the drainage networks of the two river basins. The dams of most hydropower plants are located in middle elevation areas of high topographic relief (400–1,000 m a.s.l.; Fig. 18.13); thus the barrier they present has disrupted connectivity between headwater streams and downstream areas at multiple points in the network. For example, in the Sarapiquí River Basin (Fig. 18.20), >300 river kilometers, mostly headwater streams, were located upstream from the dams of eight hydropower plants in operation in 2006 (Anderson et al.

Fig. 18.15 The El Toro dam on the Sarapiquí River.
Photo by E. Anderson.

Fig. 18.16 Stream de-watering by hydropower projects changes the hydrological and thermal regimes directly below the dam in a de-watered reach of the Puerto Viejo River.
Photos by E. Anderson.

Fig. 18.17 De-watered reach of the Puerto Viejo River downstream from the Doña Julia dam. Note the small amount of watered area on the stream bottom and the limited habitat for aquatic fauna such as fishes, shrimps, and otters.
Photo by E. Anderson.

2007). De-watered reaches also represent breaks in riverine connectivity on the basis of differences in physical conditions between these reaches and upstream and downstream river segments. Again, the Sarapiquí River Basin provides an example, with approximately 31 kilometers of de-watered reaches resulting from the operation of eight hydropower plants in 2006 (Anderson et al. 2007).

Although there are more small, water diversion–type hydropower plants proposed for Costa Rican rivers, the government's development focus since the turn of the twenty-first century has started to favor larger dam projects. The social and ecological implications of these projects, as with other large dams, are more complex and far-reaching than those of small, water diversion hydropower plants (World Commission on Dams 2000; McCully 2001). The most noteworthy of the large dam projects is the Diquís Hydropower Project currently proposed by the ICE in southern Costa Rica. The original scheme for this project was first proposed in the 1970s (Reisner and McDonald 1986); since then the project has been presented and retracted from national electricity sector plans in multiple forms. Past schemes of the project proposed the location of the dam to be downstream of the confluence of the Coto Brus and

Fig. 18.18 Mountain mullet, *Agonostomus monticola*, a migratory fish species whose distribution is disrupted by dams.
Photo by E. Anderson.

Fig. 18.19 Theraps underwoodi (Cichlidae) has been characterized as an equilibrium species whose presence may be a good indication of whether hydropower plants leave sufficient water for Neotropical stream fishes.
From Anderson et al. 2006a; photo by E. Anderson.

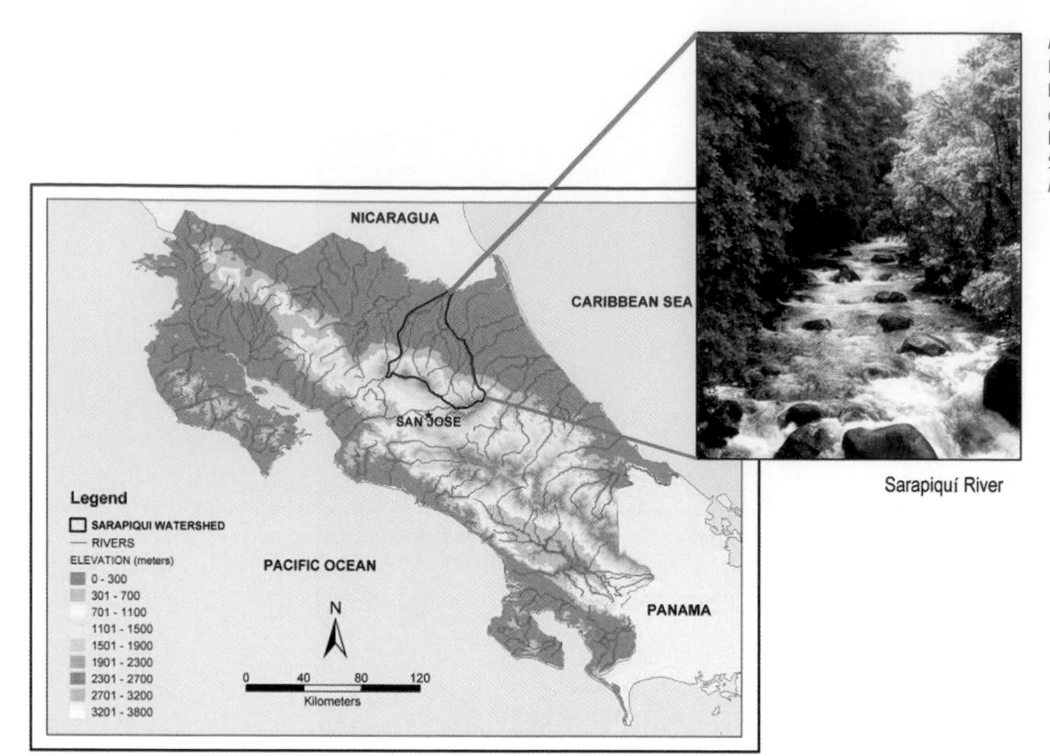

Fig. 18.20 Location of the Sarapiquí River Basin on Costa Rica's Caribbean Slope. This is one of the most developed watersheds in terms of hydropower plants. Inset photo of Sarapaquí taken near La Virgen. *Inset photo by E. Anderson.*

General Rivers. These plans were criticized on the basis of projected environmental and social impacts: the dam would have flooded a large area of tropical forest and agricultural lands in an area inhabited by native communities and sedimentation behind the proposed dam was likely to affect the long-term utility of the project. The current plan for the Diquís Hydropower Project has taken these considerations into account and has relocated the project site to the General River, near the town of Buenos Aires. If completed as currently envisioned, the project will create a deep reservoir behind a 150 meter-high dam on the General River and will generate approximately 632 megawatts of electricity (J. Picado, Instituto Costarricense de Electricidad [ICE], pers. comm.). In terms of installed generation capacity, this project will be the largest in Central America.

The strong possibility that the Diquís Hydropower Project will be constructed, as well as other current trends, suggests that hydropower will continue to be the primary source of electricity for Costa Rica well into the future. Construction of dams in Costa Rica and throughout Central America is expected to escalate over the next two decades in response to expanding human populations, increased rural electrification, and growing demands for electricity (Scatena 2004). In fact, a report by the Conservation Strategy Fund (2005) documented approximately 400 potential new hydropower projects, in various stages of study from feasibility to investment opportunity (Burgues-Arrea 2005). In this economic climate, the Diquís Hydropower Project would help reinforce Costa Rica's role as a regional leader in hydropower development.

Land Use Conversion

With much of the country's original forest having been cleared in the latter half of the twentieth century (e.g., see Janzen and Hallwachs, chapter 10 of this volume), rivers in Costa Rica today drain a mosaic of pastures, croplands, and urban areas, interspersed with patches of remnant natural forest and secondary growth (Fig. 18.21). Based on their position in the landscape, rivers and streams are on the receiving end of the changes in land cover and land use that have occurred (Fig. 18.3).

Increasingly, conversion from one agricultural land use to another, or from one crop to another, is a trend that can be observed throughout Costa Rica. For example, on the country's Caribbean slope, large areas originally covered by tropical forest were cleared for cattle pastures during the period 1960–1980; these pasturelands were then converted to banana plantations in the 1990s (Vargas 1995) or, more recently, to pineapple plantations (Fagan et al. 2013). The Tempisque River Basin also presents an interesting example of changing land use patterns that have occurred over the past half-century: in the 1950s, natural forest and pasture each covered approximately half of the land area of the basin. By the year 2000 croplands, which were almost non-existent in 1950, comprised a significant area of the basin (roughly 25%), with rice being one of the dominant crops in terms of land area and economic importance to the region. Since then, sugarcane has emerged as a more lucrative crop and many rice fields in the Tempisque River Basin are now being converted to sugarcane (Jiménez et al. 2005; Jiménez, chapter 20 of this volume).

What do all of these changes in land cover and land use mean for the structure and function of river ecosystems in Costa Rica? In general, agricultural land uses in Costa Rica have been linked to water pollution from sediment, agricultural chemicals, and animal waste. However, specific effects depend on the type, intensity, and location of the agricultural land use, and the ecological consequences of some land uses pose more of a threat to river ecosystems than others. For example, forest clearing for pasture has been linked not only to increased stream sediment loads, but also to the increased presence of fecal coliforms in stream waters from cattle. The ecological effects of banana agriculture on river ecosystems may present more cause for concern than pastures (Castillo et al. 2006, Grant et al. 2013). Banana plantations usually involve major modifications to the way in which water moves through the landscape through the construction of canals that channelize runoff; agricultural chemicals, mainly fungicides, are also applied in large quantities to bananas through aerial spraying and ground application (Castillo et al. 1997, 2000). Both of these actions influence the quantity and quality of regional water resources (Vargas 1995, Castillo et al. 2006). Like bananas, rice plantations also usually contain a network of canals that supply the fields with irrigation water. The cocktail of chemicals applied to rice negatively affects the quality of water in rivers adjacent to or downstream from rice plantations (Rizo-Patron 2005). Pineapple and sugarcane are also highly chemically-intensive crops.

Fig. 18.21 Examples of land cover and land use change in Costa Rica: (a) forest; (b) forest clearing; (c) pineapple plantation; (d) cattle grazing; and (e) banana plantation.

Concessions for Water

Expanding human populations, agriculture, and a growing tourism industry exert increasing pressure on the quantity of surface water resources in Costa Rica. Concessions for surface water are processed by the Departamento de Aguas of the Ministry of Environment and Energy (MINAE). However, demands for water for domestic and commercial uses are not always in line with real needs, nor is the real value of water incorporated in the price typically paid for the concession, though some exceptions exist (Barrantes Moreno 2006). Further, concessions for water often exceed available water supply in some areas or during certain parts of the year. The cases of the Tempisque River Basin and the Monteverde region illustrate how concessions for water are affecting the quantity of surface water resources in Costa Rica and the potential for conflicts in areas where water scarcity is anticipated in the future.

The Tempisque River Basin, the largest in Costa Rica in terms of land area, covers more than half of the province of Guanacaste and approximately 10% of Costa Rica's national territory. The basin drains a mosaic of agricultural, urban, and forested areas with a highly seasonal climate that includes marked differences in precipitation and river discharge between wet and dry periods. Diverse economic activities, such as the cultivation of rice, sugarcane, and tilapia, depend on the basin's water resources, as does the tourism industry of the northern Pacific coast. These water users draw from both surface and groundwater resources, and water demands for cropland irrigation, aquaculture, and commercial uses (tourism) often exceed natural supply. For example, to date more than 20 m^3/s of water from the Tempisque River has been legally concessioned; this amount is nearly triple the mean discharge during most of the dry season (February–April), estimated at 7 m^3/s (Jiménez et al. 2005). In addition, plans to construct a water supply dam on the upper Tempisque River have been presented as a means for meeting projected water demands of tourism-related developments on the northern Pacific coast; this project would only exacerbate existing pressures on water resources in the basin (J. Calvo, Instituto Tecnológico de Costa Rica [ITCR], pers. comm.). Scientists concerned about the ecological implications that current and future concessions of water may have on aquatic ecosystems of the basin initiated a process in 2003 to develop environmental flow recommendations for the Tempisque River, based on the needs of two common riverine species: *Crocodilus acutus* and *Parachromis dovii* (Jiménez et al. 2005).

In the area of Monteverde, local protests erupted in 2004 when a group of local businessmen were granted a concession to draw water from the La Cuecha stream. Water rights were formally solicited for irrigation of crops and grasslands; however, some Monteverde residents claimed that the businessmen were actually planning to use the water for other purposes that might include creation of tilapia farms or development of a hotel or other tourism-related infrastructure. In January 2005, opponents blocked roads and hindered construction of a pipeline that would draw water from the La Cuecha stream, arguing that water withdrawals would threaten the integrity of natural resources in the Monteverde area and that an appropriate environmental impact assessment had not been completed (*Tico Times*, January 28, 2005).

Non-native Fish Introductions

After habitat destruction, non-native species introductions are a principal cause of imperilment and extinction of freshwater fishes worldwide. Exotic fish introductions, in particular tilapiine fishes (generally known as tilapias), have the potential to influence community and population dynamics of freshwater biota in Costa Rican streams. Native to Africa and the southwestern Middle East, tilapias (Family: Cichlidae) are now pan-tropically distributed following intentional and unintentional introductions most often associated with aquaculture projects (Canonico et al. 2005). Two species, *Oreochromis mossambicus* and *Oreochromis aereus*, were introduced to Costa Rica in the 1960s for small-scale rural aquaculture; an additional tilapia species, *O. niloticus*, was introduced in 1979 (Fitzsimmons 2000). Tilapia production surged in Costa Rica (and worldwide) during the 1990s, as aquaculture practices intensified. By 1998, tilapia farms in Costa Rica produced >6,000 metric tons of tilapia annually, and the country was home to one of the single largest tilapia aquaculture operations in the Americas near the city of Cañas (Fitzsimmons 2000). Alongside Honduras and Ecuador, Costa Rica is now among the top exporters in the Americas of tilapia to international markets. In 2002, Costa Rica produced an estimated 17,000 metric tons of farmed tilapia and an increase in production to 21,000 metric tons is projected by 2010 (Fitzsimmons 2000, 2004).

Tilapia introductions have been linked to the decline of native species in freshwater systems throughout the tropics. These unintended consequences relate to the fact that the same characteristics that make tilapia good for aquaculture virtually predispose them to success as an invasive species (Canonico et al. 2005). Tilapia are fast growing, and widely tolerant of a range of environmental conditions such as salinity, dissolved oxygen, and temperature. They also have the ability to feed at different trophic levels. The reproductive biology of tilapia also gives them an advantage over native cichlid fishes in many places (van Breukelen 2015). Tilapia have short reproductive cycles and spawn year round, and *Oreochromis* practice parental care through mouthbrooding, which allows them to protect their young and easily colonize new habitats. Genetic effects of introduced tilapia on native cichlids are also a concern, as is habitat alteration and eutrophication caused by waste produced in cage farming.

Little information documenting the current range of tilapias in Costa Rica is available. Since 1990 tilapias have been commonly found along the San Juan River drainage from Lake Nicaragua to the Caribbean Sea. These tilapias were presumably escapees following stocking of Lake Nicaragua during the 1980s (McKaye et al. 1995). It is assumed tilapia inhabit major tributaries of the San Juan River (like the Sarapiquí and San Carlos Rivers) and it is highly likely that other drainages harbor tilapia as well, particularly those of the Guanacaste region where many intensive tilapia farms are located. While long-term effects of the presence and spread of tilapia on river ecosystems in Costa Rica remains to be seen, predictions are not good based on experiences from other countries, which have shown that native cichlid populations often decline in the presence of introduced tilapias (McKaye et al. 1995; Canonico et al. 2005).

Future Perspective

Importance of Conserving the Biointegrity of Costa Rican Rivers

There is clearly a strong need for integrated management of water resources and concerted efforts to balance human demands for water with the needs of aquatic ecosystems. Very little infrastructure for monitoring basic water quality parameters exists in Costa Rica. It is hoped that the information provided in this chapter will provide a foundation to inspire efforts to ensure riverine biointegrity throughout the country. Monitoring programs, adaptive management, and creative solutions are needed to balance the many demands that humanity is placing on these running water ecosystems.

The previous section (People and Nature) exemplifies the types of development issues and conflicts likely to dominate the future agenda of water resources management in many parts of Costa Rica. Issues range from river fragmentation associated with dam construction to land use changes, water withdrawals, and displacement of native biota by exotic species (e.g., see TNC 2009). Tradeoffs between development and environmental effects must be carefully considered. What are the cumulative effects of dam construction within a river's drainage on migratory stream biota, fisheries, potable water supply, and ecotourism? How will future changes in land use within specific watersheds affect nonpoint source nutrient and pesticide inputs into rivers—and how will this affect stream biota and human communities? To what extent will excessive water withdrawals from rivers, which are being facilitated by concessions for water (and often granted on the basis of demand rather than supply), reduce river discharge and compromise the ability of rivers to provide key ecosystem goods and services (e.g., waste assimilation, transportation, fish) upon which residents of Costa Rica rely?

Conclusions: Research Needs

As this chapter illustrates, while our understanding of Costa Rican river ecosystems is growing there are many gaps in our knowledge, ranging from lack of information at the species level (regarding basic natural history) to population-, community-, and ecosystem-level information. Nonetheless, Costa Rica stands out, given the amount of information on rivers that is available relative to many other tropical countries.

It is widely acknowledged among freshwater ecologists that we have a limited understanding of how tropical rivers function relative to our knowledge of temperate streams and rivers (Pringle 2000, Ramírez et al. 2008). Ecological and hydrological characteristics of tropical rivers are wide-ranging and often very different from temperate streams. It is unwise to extend paradigms generated from temperate stream research to the tropics without ample data—particularly with respect to conservation and management decisions. For example, the applicability of hydropower technology developed for rivers in the temperate zone must be evaluated with respect to differing hydrological and biological features in tropical Costa Rican rivers (Pringle et al.2000). There is clearly a critical need for studies on the ecological consequences of dam construction in Costa Rica although some inroads have been made (see Anderson et al. 2006a,b, 2007). Developing new methods for assessing and monitoring water quality for tropical rivers is also important (but see Springer et al. 2007, Umaña-Villalobos and Springer 2006), since biological and chemical indices of water quality developed for temperate zone systems are often not appropriate.

Further discussion on priority research and management needs for tropical rivers in general can be found in Pringle (2000), which provides a comprehensive review of river conservation in Latin America and the Caribbean.

Acknowledgments

The authors gratefully acknowledge the support of the National Science Foundation (NSF), which has almost continually supported the STREAMS project in lowland Costa Rica since 1987 through the following awards: BSR-87-17746, BSR-91-07772, DEB9528434, DEB0075339, and DEB0545463. The Organization for Tropical Studies (OTS) has provided continuous strong logistical support, which has made our research in Costa Rica possible.

References

Allsteadt, J., and C. Vaughan. 1988. Ecological studies of the Central American caiman (*Caiman crocodilus fuscus*) in Caño Negro Wildlife Refuge, Costa Rica. *Bulletin of the Chicago Herpetological Society* 23: 123–26.

Allsteadt, J., and C. Vaughan. 1992. Population status of *Caiman crocodilus* (Crocodylia: Alligatoridae) in Caño Negro, Costa Rica. *Brenesia* 38: 57–64.

Anderson, E.P., M.C. Freeman, and C.M. Pringle. 2006a. Ecological consequences of hydropower development in Central America: impacts of small dams and water diversion on neotropical stream fish assemblages. *River Research and Applications* 22: 397–411.

Anderson, E.P., C.M. Pringle, and M. Rojas. 2006b. Transforming tropical rivers: an environmental perspective on hydropower development in Costa Rica. *Aquatic Conservation: Marine and Freshwater Systems* 16: 679–93.

Anderson, E.P., C.M. Pringle, and M.C. Freeman. 2008. Quantifying the cumulative extent of river fragmentation by dams in the Sarapiquí River Basin, Costa Rica: a first step towards evaluating environmental tradeoffs. *Aquatic Conservation: Marine and Freshwater Systems* 18: 408–17.

Ardón, M., and C.M. Pringle. 2007. The quality of organic matter mediates the response of heterotrophic biofilms to phosphorus enrichment of the water column and substratum. *Freshwater Biology* 52: 1762–72.

Ardón, M., and C.M. Pringle. 2008. Do secondary compounds inhibit microbial- and insect-mediate leaf breakdown in a tropical rainforest stream, Costa Rica? *Oecologia* 155: 311–23.

Ardón, M., C.M. Pringle, and S.L. Eggert. 2009. Does leaf chemistry differentially affect breakdown in tropical versus temperate streams?: importance of using standardized analytical techniques to measure leaf chemistry. *Journal of the North American Benthological Society* 28: 440–53.

Ardón, M., L.A. Stallcup, and C.M. Pringle. 2006. Does leaf quality mediate the stimulation of leaf breakdown by phosphorus in Neotropical streams? *Freshwater Biology* 51: 618–33.

Ardón, M., J.H. Duff, A. Ramírez, G.E. Small, A.P. Jackman, F.J. Triska, and C.M. Pringle. 2013. Experimental acidification of two biogeochemically-distinct neotropical streams: buffering mechanisms and macroinvertebrate drift. *Science of the Total Environment* 443: 267–77.

Banack, S.A., M.H. Horn, and A. Gawlicka. 2002. Disperser- vs. establishment-limited distribution of a riparian fig tree (*Ficus insipida*) in a Costa Rican tropical rain forest. *Biotropica* 34: 232–43.

Barbee, N.C. 2005. Grazing insects reduce algal biomass in a neotropical stream. *Hydrobiologia* 532: 153–65.

Barrantes Moreno, G. 2006. Economic valuation of water supply as a key environmental service provided by montane oak forest watershed areas in Costa Rica. In M. Kappelle, ed., *Ecology and Conservation of Neotropical Montane Oak Forests*, 436–46. *Ecological Studies Series*, vol. 185. New York: Springer Verlag.

Benstead, J.P. 1996. Macroinvertebrates and the processing of leaf litter in a tropical stream. *Biotropica* 28: 367–75.

Benstead, J.P., J.G. March, C.M. Pringle, and F.N. Scatena. 1999. Effects of a low-head dam and water abstraction on migratory tropical stream biota. *Ecological Applications* 9: 656–68.

Bixby, R.J., U. Wydrzycka, and C.M. Pringle. 2005. Diatom assemblages as indicators of solute levels in lowland Neotropical streams. *Bulletin of the North American Benthological Society* 22: 425–26.

Bourrelly, P., and E. Manguin. 1952. Algues d'Eau Douce de la Guadeloupe et Dépendances. Centre National de la Recherche Scientifique. Paris: Société d'Edition d'Enseignement Supérieur. 281 pp.

Boyero, L., and J. Bosch. 2002. Spatial and temporal variation of macroinvertebrate drift in two neotropical streams. *Biotropica* 34: 567–74.

Brandon, C. 1983. *Noctilio leporinus*. In D.H. Janzen, *Costa Rican Natural History*, 480–81. Chicago: University of Chicago Press.

Brookshire, E.N.J., L.O. Hedin, J.D. Newbold, D.M. Sigman, and J.K. Jackson. 2012. Sustained losses of bioavailable nitrogen from montane tropical forests. *Nature Geoscience* 5: 123–126.

Brookshire, E.N.J., S. Gerber, D.N.L. Menge, and L. Hedin. 2012b. Large losses of inorganic nitrogen from tropical rainforests suggest a lack of nitrogen limitation. *Ecology Letters* 15:9–16.

Burcham, J. 1980. Fish communities and environmental characteristics of two lowland streams in Costa Rica. *Revista de Biología Tropical* 36: 273–85.

Burgues-Arrea, I. 2005. Inventario de Proyectos de Infraestructura en Mesoamérica. Conservation Strategy Fund, Proyecto: Integración de la Infraestructura y la Conservación de la Biodiversidad en Mesoamérica. Fondo de Alianzas para los Ecosistemas Críticos, The Nature Conservancy (TNC).

Bussing, W.A. 1993. Fish communities and environmental characteristics of a tropical rainforest river in Costa Rica. *Revista de Biología Tropical* 41: 791–809.

Bussing, W.A. 1994. Ecological aspects of the fish community. In L.A. McDade, K.S. Bawa, H.A. Hespenheide, and G.S. Hartshorn, eds., *La Selva: Ecology and Natural History of a Neotropical Rain Forest*, 195–98. Chicago: University of Chicago Press.

Bussing, W.A. 1998a. Peces de las aguas continentales de Costa Rica. Editorial de la Universidad de Costa Rica.

Bussing, W.A. 1998b. Freshwater fishes of Costa Rica. *Revista de Biología Tropical* 46(Suppl. 2): 1–458.

Butterfield, R.P. 1994. The regional context: land colonization and conservation in Sarapiquí. In L.A. McDade, K.S. Bawa, H.A. Hespenheide, and G.S. Hartshorn, eds., *La Selva: Ecology and Natural*

History of a Neotropical Rain Forest, 299–306. Chicago: University of Chicago Press.

Calvert, P.P. 1915. Studies on Costa Rican Odonata, VI—The waterfall-dwellers: the transformation, external features, and attached diatoms of *Thaumatoneura* larva. *Entomological News* 26: 295–305.

Canonico, G.C., A. Arthington, J.K. McCrary, and M.L. Thieme. 2005. The effects of introduced tilapias on native biodiversity. *Aquatic Conservation: Marine and Freshwater Systems* 15: 463–83.

Capps, K. A., and A. S. Flecker. 2013. Invasive aquarium fish transform ecosystem nutrient dynamics. *Proceedings of the Royal Society B* 280: 20131520.

Castillo, L.E., E. de la Cruz, and C. Ruepert. 1997. Ecotoxicology and pesticides in tropical aquatic ecosystems of Central America. *Environmental Toxicology and Chemistry* 16(1): 41–51.

Castillo, L.E., E. Martínez, C. Ruepert, C. Savage, M. Gilek, M. Pinnock, and E. Solís. 2006. Water quality and macroinvertebrate community response following pesticide applications in a banana plantation, Limón, Costa Rica. *Science of the Total Environment* 367: 418–32.

Castillo, L.E., C. Ruepert, and E. Solis. 2000. Pesticide residues in the aquatic environment of banana plantation areas in the North Atlantic zone of Costa Rica. *Environmental Toxicology and Chemistry* 19: 1942–50.

Cedeño-Obregón, F. 1986. Contribución al conocimiento de los camarones de agua dulce de Costa Rica. Colección perteneciente al Museo de Zoología de la Escuela de Biología de la Universidad de Costa Rica. Lic. San Pedro de Montes de Oca, Costa Rica: Universidad de Costa Rica.

CFIA. 2005. Panorama Nacional y Regional de la Industria Eléctrica. Colegio Federado de Ingenieros y Arquitectos de Costa Rica (CFIA). Accessed December 2005. http://www.cfia.or.cr/informes.htm.

Chace, F.A., and H.H. Hobbs. 1966. The freshwater and terrestrial decapod crustaceans of the West Indies with special reference to Dominica. *United States National Museum Bulletin* 292: 1–258.

Chandler, M., L.J. Chapman, and C.A. Chapman. 1995. Patchiness in the abundance of metacercariae parasitizing *Poecilia gilli* (Poecilidae) isolated in pools of an intermittent tropical stream. *Environmental Biology of Fishes* 42: 313–21.

Colon-Gaud, C., M. Whiles, S.S. Kilham, K.R. Lips, C.M. Pringle, S. Connelly, and S. Peterson. 2009. Assessing ecological responses to catastrophic amphibian declines: patterns of macroinvertebrate production and food web structure in upland Panamanian streams. *Limnology and Oceanography* 54: 331–43.

Connelly, S., C.M. Pringle, R.J. Bixby, R. Brenes, M. Whiles, K.R. Lips, S.S. Kilham, and A.D. Huryn. 2008. Changes in stream primary producer communities resulting from large-scale catastrophic amphibian declines: Can small-scale experiments predict effects of tadpole loss? *Ecosystems* 8: 1262–76.

Crook, K.E., C.M. Pringle, and M. C. Freeman. 2009. A method to assess longitudinal riverine connectivity in tropical streams dominated by migratory biota. *Aquatic Conservation* 19: 714–23.

Dixon, J.R., and M.A. Staton. 1983. *Caiman crocodilus*. In DH. Janzen, ed., *Costa Rican Natural History*, 387–88. Chicago: University of Chicago Press.

Drewe, K.E., M.H. Horn, K.A. Dickson, and A. Gawlicka. 2004. Insectivore to frugivore: ontogenetic changes in gut morphology and digestive enzyme activity in the characid fish *Brycon guatemalensis*. *Journal of Fish Biology* 64: 890–902.

Duff, J.H., C.M. Pringle, and F.J. Triska. 1996. Nitrate reduction in sediments of lowland tropical streams draining swamp forest in Costa Rica: an ecosystem perspective. *Biogeochemistry* 33: 179–96.

Duft, M., K. Fittkau, and W. Traunspurger. 2002. Colonization of exclosures in a Costa Rican stream: effects of macrobenthos on meiobenthos and the nematode community. *Journal of Freshwater Ecology* 17: 531–41.

Fagan, M.E., R.S. DeFries, S.E. Sesnie, J.P. Arroyo, W. Walker, C. Soto, R.L. Chazdon, and A. Sanchun. 2013. Land cover dynamics following a deforestation ban in northern Costa Rica. *Environmental Research Letters* 8(3): 034017.

Finlay, B.J., E.B. Monaghan, and S.C. Maberly. 2002. Hypothesis: the rate and scale of dispersal of freshwater diatom species is a function of their global abundance. *Protist* 153: 261–73.

Fitzsimmons, K. 2000. Future trends of tilapia aquaculture in the Americas. In B.A. Costa-Pierce and J.E. Rakocy, eds., *Tilapia Aquaculture in the Americas*, vol. 2, 252–64. Baton Rouge, LA: The World Aquaculture Society.

Fitzsimmons, K. 2004. Development of new products and markets for the global tilapia trade. In R. Bolivar, G. Mair, and K. Fitzsimmons, eds., *Proceeding of the Sixth International Symposium on Tilapia in Aquaculture*, 624–33. Manila, Philippines: BFAR.

Flowers, R.W. 1992. Review of the genera of mayflies of Panama with a checklist of Panamanian and Costa Rican species (Ephemeroptera). In D. Quintero and A. Aiello, eds., *Insects of Panama and Mesoamerica: Selected Studies*, 37–51. Oxford, England: Oxford University Press.

Flowers, R.W., and C. De la Rosa. 2010. Ephemeroptera. *Revista de Biologia Tropical* 58(Suppl. 4): 63–93.

Food and Agriculture Organization (FAO). 2005. Dirección de Fomento de Tierras y Aguas. AQUASTAT, Costa Rica. Accessed December 10, 2005. http://www.fao.org/ag/agl/aglw/aquastat/countries/costa_rica/indexesp.stm.

Fureder, L. 1994. Drift patterns in Costa Rica streams. PhD thesis, Innsbruck University. Austria.

Ganong, C.N., G.E. Small, M. Ardón, W.H. McDowell, D.P. Genereux, J.H. Duff, and C.M. Pringle. 2015. Interbasin flow of geothermally modified ground water stabilizes stream exports of biologically important solutes against variation in precipitation. *Freshwater Science* 34 (1): 276–86.

García-Guerrero, M. U., F. Becerril-Morales, F.Vega-Villasante, and L.D. Espinosa-Chaurand, 2013. Los langostinos del género *Macrobrachium* con importancia económica y pesquera en América Latina: conocimiento actual, rol ecológico y conservación. *Latin American Journal of Aquatic Research* 41: 651–75.

Genereux, D.P., S.J. Wood, and C.M. Pringle. 2002. Chemical tracing of interbasin groundwater transfer in the lowland rainforest of Costa Rica. *Journal of Hydrology* 258: 163–78.

Gómez, L.D. 1984. *Las Plantas Acuáticas y Anfíbios de Costa Rica y Centroamérica*. I. Liliopsida. San José, Costa Rica: EUNED. 430 pp.

Goulding, M. 1980. *The Fishes and the Forest: Explorations in Amazonian Natural History*. University of California, Berkeley.

Grant, P. B., Woudneh, M. B., & Ross, P. S. 2013. Pesticides in blood from spectacled caiman (*Caiman crocodilus*) downstream of banana plantations in Costa Rica. *Environmental Toxicology and Chemistry* 32: 2576–83.

Greathouse, E.A., C.M. Pringle, and J.G. Holmquist. 2006a. Conservation and management of migratory fauna and dams in tropical streams of Puerto Rico. *Aquatic Conservation* 16: 695–712.

Greathouse, E.A., C.M. Pringle, W.H. McDowell, and J.G. Holmquist. 2006b. Indirect upstream effects of dams: consequences of migratory consumer extirpation in Puerto Rico. *Ecological Applications* 16: 339–52.

Gutiérrez-Fonseca, P.E., and M. Springer. 2011. Description of the final instar nymphs of seven species from *Anacroneuria* Klapálek (Plecoptera: Perlidae) in Costa Rica, and first record for an additional genus in Central America. *Zootaxa* 2965: 16–38.

Hartshorn, G., L. Hartshorn, A. Atmella, L.D. Gómez, A. Mata, R. Morales, R. Ocampo, D. Pool, C. Quesada, C. Solera, R. Solorzano, G. Stiles, J. Tosi, A. Umaña, C. Villalobos, and R. Wells. 1982. *Costa Rica Country Environmental Profile: A Filed Study*. San José, Costa Rica: Tropical Science Center.

Henley, R.W., and J. Ellis. 1983. Geothermal systems, ancient and modern. *Earth Science Reviews* 19: 1–50.

Holmquist, J.G., J.M. Schmidt-Gengenbach, and B.B. Yoshioka. 1998. High dams and marine-freshwater linkages: effects on native and introduced fauna in the Caribbean. *Conservation Biology* 12: 621–30.

Holzenthal, R.W. 1988. Catálogo sistemático de los Tricópteros de Costa Rica. *Brenesia* 29: 51–82.

Horn, B.W., and R.W. Lichtwardt. 1981. Studies on the nutritional relationship of larval *Aedis aegypti* (Diptera: Culicidae) with *Smittium culisetae* (Trichmycetes). *Mycologia* 73: 724–940.

Horn, M.H. 1997. Evidence for dispersal of fig seeds by the fruit-eating characid fish *Brycon guatemalensis* Regan in a Costa Rican tropical rain forest. *Oecologia* 109: 259–64.

Instituto Costarricense de Electricidad (ICE). 2004. Plan de la Expansión de la Generación Eléctrica. San Jose, Costa Rica: ICE.

Instituto Costarricense de Electricidad (ICE). 2009. Déjenos contarle, revista informativa del Proyecto Hidroeléctrico El Diquís. Costa Rica: ICE. 15 pp.

Irons, J.G., M.W. Oswood, R.J. Stout, and C.M. Pringle. 1994. Latitudinal patterns in leaf litter breakdown: is temperature really important? *Freshwater Biology* 32: 401–11.

Jackson, J.K., and B.W. Sweeney. 1995. Egg and larval development times for 35 species of tropical stream insects from Costa Rica. *Journal of the North American Benthological Society* 14: 115–30.

Janzen, D.H. 1975. Tropical blackwater rivers, animals, and mast fruiting by the Dipterocarpaceae. *Biotropica* 6: 69–103.

Janzen, D.H. 1983. *Costa Rican Natural History*. Chicago: University of Chicago Press. 816 pp.

Jiménez, J.A., J. Calvo, F. Pizarro, and E. González. 2005. Conceptualización de caudal ambiental en Costa Rica: Determinación inicial para el Río Tempisque. San José, Costa Rica: UICN-Mesoamerica. 42 pp.

Kilroy, C., B.J.F. Biggs, and W. Vyverman. 2007. Rules for macroorganisms applied to micoorganisms: patterns of endemism in benthic freshwater diatoms. *Oikos* 116: 550–64.

Kociolek, J.P., and S.A. Spaulding. 2000. Freshwater diatom biogeography. *Nova Hedwigia* 71: 223–41.

Krammer, K., and H. Lange-Bertalot. 1986. Bacillariophyceae 1. Teil: Naviculaceae. In H. Ettl, J. Gerloff, H. Heynig, and D. Mollenhauer, eds., *Süsswasserflora von Mitteleuropa* 2/1: 1–876. Jena, Germany: Gustav Fischer Verlag.

La Marca, E., K.R. Lips, S. Lotters, R. Puschendorf, R. Ibañez, J.V. Rueda-Almonacid, R. Schulte, C. Marty, F. Castro, J. Manzanilla-Puppo, J.E. García-Perez, F. Bolaños, G. Cháves, J.A. Pounds, E. Toral, and B.E. Young. 2005. Catastrophic population declines and extinctions in neotropical harlequin frogs (Bufonidae: *Atelopus*). *Biotropica* 37: 190–201.

Lara, L.R., I.S. Wehrtmann, C. Magalhães, and F.L. Mantelatto. 2013. Species diversity and distribution of freshwater crabs (Decapoda: Pseudothephusidae) inhabiting the basin of the Río Grande de Térraba, Pacific slope of Costa Rica. *Latin American Journal of Aquatic Research* 41:685–95.

Lichtwardt, R.W. 1997. Costa Rican gut fungi (Trichomycetes) infecting lotic insect larvae. *Revista de Biología Tropical* 45: 1349–83.

Ligon, S. 1983. *Trichechus manatus*. In D.H. Janzen, ed., *Costa Rican Natural History*. Chicago: University of Chicago Press.

Lips, K.R. 1998. Decline of a tropical montane amphibian fauna. *Conservation Biology* 12: 106–17.

Lips, K.R. 1999. Mass mortality and population declines of anurans at an upland site in western Panama. *Conservation Biology* 13: 117–25.

Lips, K.R. 2001. Reproductive trade-offs and bet-hedging in *Hyla calypso*, a Neotropical treefrog. *Oecologia* 128:509–18.

Lips, K.R., D.E. Green, and R. Papendick. 2003. Chytridiomycosis in wild frogs from southern Costa Rica. *Journal of Herpetology* 37: 215–18.

Lyons, J., and D.W. Schneider. 1990. Factors influencing fish distribution and community structure in a small coastal river in southwestern Costa Rica. *Hydrobiologia* 203: 1–14.

Magalhães, C., L.R. Lara, and I.S. Wehrtmann. 2010. A new species of freshwater crab of the genus *Allacanthos* (Crustacea, Decapoda, Pseudothelphusidae) from southern Costa Rica, Central America. *Zootaxa* 2604: 52–60.

Mann, D.G., and S.J.M. Droop. 1996. Biodiversity, biogeography and conservation of diatoms. *Hydrobiologia* 336: 19–32.

March, J.G., J.P. Benstead, C.M. Pringle, and F.N. Scatena. 2003. Damming tropical island streams: problems, solutions, and alternatives. *BioScience* 53: 1069–78.

March, J.G., C.M. Pringle, M.J. Townsend, and A.I. Wilson. 2002. Effects of freshwater shrimp assemblages on benthic communities along an altitude gradient of a tropical island stream. *Freshwater Biology* 47: 1–14.

McCully, P. 2001. *Silenced Rivers: The Ecology and Politics of Large Dams*. London: Zed Books. 359 pp.

McKaye, K.R., J.D. Ryan, J.R. Stauffer, L.J.L. Pérez, G.I. Vega, and E.P. van den Berghe. 1995. African tilapia in Lake Nicaragua: ecosystem in transition. *Bioscience* 45: 406–11.

Metzeltin, D., and H. Lange-Bertalot. 1998. Tropical diatoms of South America I. *Bibliotheca Diatomologica* 5: 1–695.

Metzeltin, D., and H. Lange-Bertalot. 2002. Diatoms from the "Island Continent" Madagascar. *Iconographia Diatomologica* 11: 1–286.

Michels, A. 1998a. Effects of sewage water on diatoms (Bacillariophyceae) and water quality in two tropical streams in Costa Rica. *Revista de Biología Tropical* 46(Suppl. 6): 153–75.

Michels, A. 1998b. Use of diatoms (Bacillariophyceae) for water quality assessment in two tropical streams in Costa Rica. *Revista de Biología Tropical* 46(Suppl. 6): 143–52.

Michels, A., G. Umaña, and U. Raeder. 2006. Epilithic diatom assemblages in rivers draining into Golfo Dulce (Costa Rica) and their relationship to water chemistry, habitat characteristics and land use. *Archiv für Hydrobiologie* 165: 167–90.

Moll, D., and K.P. Jansen. 1995. Evidence for a role in seed dispersal by two tropical herbivorous turtles. *Biotropica* 27: 121–27.

Moser, G., H. Lange-Bertalot, and D. Metzeltin. 1998. Insel der Endemiten-Geobotanisches Phänomen Neukaledonien. *Bibliotheca Diatomologica* 38: 1–464.

Newbold, J.D., B.W. Sweeney, J.K. Jackson, and L.A. Kaplan. 1995. Concentrations and export of solutes from six mountain streams in northwestern Costa Rica. *Journal of the North American Benthological Society* 14: 21–37.

Paaby, P., and C.R. Goldman. 1992. Chlorophyll, primary productivity, and respiration in a lowland Costa Rican stream. *Revista de Biología Tropical* 40: 185–88.

Power, M.E. 1990. Effects of fish in river food webs. *Science* 250: 811–14.

Pringle, C.M. 1997. Exploring how disturbance is transmitted upstream: going against the flow. *Journal of the North American Benthological Society* 16: 425–38.

Pringle, C.M. 2000. Riverine conservation in tropical versus temperate regions: ecological and socioeconomic considerations. In P. J. Boon, B. Davies, and G. Petts, eds., *Global Perspectives on River Conservation: Science Policy and Practice*, chap. 15, 367–78. New York: John Wiley and Sons.

Pringle, C.M., and T. Hamazaki. 1997. Effects of fishes on algal response to storms in a tropical stream. *Ecology* 78: 2432–42.

Pringle, C.M., and T. Hamazaki. 1998. The role of omnivory in a neotropical stream: separating diurnal and nocturnal effects. *Ecology* 79: 269–80.

Pringle, C.M., P. Paaby-Hansen, P.D. Vaux, and C.R. Goldman. 1986. *In situ* nutrient assays of periphyton growth in a lowland Costa Rican stream. *Hydrobiologia* 134: 207–13.

Pringle, C.M., and A. Ramírez. 1998. Use of both benthic and drift sampling techniques to assess tropical stream invertebrate communities along an altitudinal gradient, Costa Rica. *Freshwater Biology* 39: 359–73.

Pringle, C.M., G.L. Rowe, F.J. Triska, J. Fernández, and J. West. 1993. Landscape linkages between geothermal activity and solute composition and ecological response in surface waters draining the Atlantic slope of Costa Rica. *Limnology and Oceanography* 38: 753–74.

Pringle, C.M., and F.N. Scatena. 1999a. Factors affecting aquatic ecosystem deterioration in Latin America and the Caribbean. In U. Hatch and M.E. Swisher, eds., *Tropical Managed Ecosystems: New Perspectives on Sustainability*, 104–13. Oxford: Oxford University Press.

Pringle, C.M., and F.N. Scatena. 1999b. Freshwater resource development: case studies from Puerto Rico and Costa Rica. In U. Hatch and M.E. Swisher, eds., *Tropical Managed Ecosystems: New Perspectives on Sustainability*, 114–21. Oxford: Oxford University Press.

Pringle, C.M., F. Scatena, P. Paaby, and M. Nuñez. 2000. River conservation in Latin America and the Caribbean. In P.J. Boon, B. Davies, and G. Petts, eds., *Global Perspectives on River Conservation: Science, Policy and Practice*, chap. 2, 39–75. New York: John Wiley and Sons.

Pringle, C.M., and F.J. Triska. 1991. Effects of geothermal groundwater on nutrient dynamics of a lowland Costa Rican stream. *Ecology* 72: 951–65.

Pringle, C.M., and F.J. Triska. 2000. Emergent biological patterns and surface-subsurface interactions at landscape scales. In J.B. Jones and P.J. Mulholland, eds., *Stream and Groundwaters*, 167–93. Academic Press.

Pringle, C.M., F.J. Triska, and G. Browder. 1990. Spatial variation in basic chemistry of streams draining a volcanic landscape on Costa Rica's Caribbean slope. *Hydrobiologia* 206: 73–85.

Pringle, C.M., M. Freeman, and B. Freeman. 2000. Regional effects of hydrologic alterations on riverine macrobiota in the New World: Tropical-temperate comparisons. *BioScience* 50: 807–23.

Ramírez, A. 2010. Odonata. *Revista de Biologia Tropical* 58(Suppl. 4): 97–136.

Ramírez, A., P. Paaby, C.M. Pringle, and G. Aguero. 1998. Effect of habitat type on benthic macroinvertebrates in two lowland tropical streams, Costa Rica. *Revista de Biología Tropical* 46: 201–13.

Ramírez, A., D.R. Paulson, and C. Esquivel. 2000. Odonata of Costa Rica: diversity and checklist of species. *Revista de Biología Tropical* 48: 247–54.

Ramírez, A., and C.M. Pringle. 1998a. Invertebrate drift and benthic community dynamics in a lowland neotropical stream, Costa Rica. *Hydrobiologia* 386: 19–26.

Ramírez, A., and C.M. Pringle. 1998b. Structure and production of a benthic insect assemblage in a neotropical stream. *Journal of the North American Benthological Society* 17: 443–63.

Ramírez, A., and C.M. Pringle. 2001. Spatial and temporal patterns of invertebrate drift in streams draining a Neotropical landscape. *Freshwater Biology* 46: 47–62.

Ramírez, A., and C.M. Pringle. 2004. Do macroconsumers affect insect responses to a natural stream phosphorus gradient? *Hydrobiologia* 515: 235–46.

Ramírez, A., and C.M. Pringle. 2006. Fast growth and turnover of chironomid assemblages in response to stream phosphorus levels in a tropical lowland landscape. *Limnology and Oceanography* 51: 189–96.

Ramírez, A., C.M. Pringle, and M. Douglas. 2006. Temporal and spatial patterns in stream physicochemistry and insect assemblages in tropical lowland streams. *Journal of the North American Benthological Society* 25: 108–25.

Ramírez, A., C.M. Pringle, and L. Molina. 2003. Effects of stream phosphorus levels on microbial respiration. *Freshwater Biology* 48: 88–97.

Ramirez, A., C.M. Pringle, and K.M. Wantzen 2008. Tropical river conservation. In D. Dudgeon, ed., *Tropical Stream Ecology*, 285–304. London: Academic Press, 316 pp.

Ranvestel, A.W., K.R. Lips, C.M. Pringle, M.R. Whiles, and R.J. Bixby. 2004. Neotropical tadpoles influence stream benthos: evidence for the ecological consequences of decline in amphibian populations. *Freshwater Biology* 49: 274–85.

Reisner, M., and R.H. McDonald. 1986. The high costs of high dams. In A. Maguire and J.W. Brown, eds., *Bordering on Trouble*, 270–307. MD: Adler & Adler for World Resources Institute.

Rizo-Patron, F.L.S. 2005. Estudio de los arrozales del Proyecto Tamarindo: agroquimicos y macroinvertebrados bentónicos en relación al Parque Nacional Palo Verde, Guanacaste, Costa Rica. Tesis, Universidad Nacional. Heredia, Costa Rica.

Rosemond, A.D., C.M. Pringle, and A. Ramírez. 1998. Macroconsumer effects on insect detritivores and detritus processing in a tropical stream. *Freshwater Biology* 39: 515–24.

Rosemond, A.D., C.M. Pringle, A. Ramírez, and M.J. Paul. 2001. A test of top-down and bottom-up control in streams. *Ecology* 82: 2279–93.

Rosemond, A.D., C.M. Pringle, A. Ramírez, M.J. Paul, and J.L. Meyer. 2002. Landscape variation in phosphorus concentration and effects on detritus-based tropical streams. *Limnology and Oceanography* 47: 278–89.

Sader, S., and A. Joyce. 1988. Deforestation rates and trends in Costa Rica, 1940–1983. *Biotropica* 20: 11–19.

Savage, J.M. 2002. *Amphibians and Reptiles of Costa Rica: A Herpetofauna between Two Continents, between Two Seas*. Chicago: University of Chicago Press. 954 pp.

Scatena, F.N. 2004. A survey of methods for setting the minimum instream flow standards in the Caribbean basin. *River Research and Applications* 20: 127–35.

Schneider, D.W., and J. Lyons. 1993. Dynamics of upstream migration in 2 species of tropical freshwater snails. *Journal of the North American Benthological Society* 12: 3–16.

Scott, N.J., and S. Limerick. 1983. Reptiles and amphibians. In D.H. Janzen, ed., *Costa Rican Natural History*, 351–73. Chicago: University of Chicago Press.

Sheath, R.G., D. Kaczmarczyk, and K.M. Cole. 1993a. Distribution and systematics of freshwater *Hildenbrandia* (Rhodophyta, Hildenbrandiales) in North America. *European Journal of Phycology* 28: 115–21.

Sheath, R.G., M.L. Vis, and K.M. Cole. 1993b. Distribution and systematics of freshwater Ceramiales (Rhodophyta) in North America. *Journal of Phycology* 29: 108–17.

Sherwood, A.R., and R.G. Sheath. 1999. Biogeography and systematics of *Hildenbrandia* (Rhodophyta, Hildenbrandiales) in North America: inferences from morphometrics, and rbcL and 18S gene sequence analyses. *European Journal of Phycology* 34: 523–32.

Silva-Benavides, A.M. 1996a. The epilithic diatom flora of a pristine and a polluted river in Costa Rica, Central America. *Diatom Research* 11: 105–42.

Silva-Benavides, A.M. 1996b. The use of water chemistry and benthic diatom communities for qualification of a polluted tropical river. *Revista de Biología Tropical* 44: 395–416.

Small, G.E., C.M. Pringle, M. Pyron, and J.H. Duff. 2011a. Role of the fish *Astyanax aeneus* (Characidae) as a keystone nutrient recycler in low-nutrient Neotropical streams. *Ecology* 92: 386–97.

Small, G.E., J.P. Wares, and C.M. Pringle. 2011b. Differences in phosphorus demand among detritivorous chironomid larvae reflect intraspecific adaptations to differences in food resource stoichiometry across lowland tropical streams. *Limnology & Oceanography* 56: 268–78.

Small, G.E., R.J. Bixby, C. Kazanci, and C. M. Pringle. 2011c. Partitioning stoichiometric components of epilithic biofilms using mixing models. *Limnology and Oceanography: Methods* 9: 185–93.

Small, G.E., P.J. Torres, L.M. Schweizer, J.H. Duff, and C.M. Pringle. 2013a. Importance of terrestrial arthropods as subsidies in lowland Neotropical rain forest stream ecosystems. *Biotropica* 45: 80–87.

Small, G.E., J.H. Duff, P.J. Torres, and C.M. Pringle. 2013b. Insect emergence as a nitrogen flux in Neotropical streams: comparisons with microbial denitrification across a stream phosphorus gradient. *Freshwater Science* 32:1178–87.

Snyder, M.N. 2012. Abundance, distribution, nutrient cycling and energy flow of freshwater Palaemonid shrimps in lowland Costa Rica. PhD diss., University of Georgia, Athens, GA.

Snyder, M.N., E.A. Anderson, and C.M. Pringle. 2011. A migratory shrimp's perspective on habitat fragmentation in the neotropics: extending our knowledge from Puerto Rico. In A. Asakura, ed., *New Frontiers in Crustacean Biology: Proceedings of the TCS Summer Meeting, Tokyo, 20–24 September 2009*, pp. 109–62. Crustaceana Monographs, vol. 15. Boston: Brill.

Snyder, M.N., C.M. Pringle, and R.T. Soto-Mayer. 2013. Landscape-scale disturbance and protected areas: long-term dynamics of populations of the shrimp, *Macrobrachium olfersii* in lowland neotropical streams, Costa Rica. *Journal of Tropical Ecology* 29:81–85.

Snyder, M.N., G.E. Small, and C.M. Pringle. 2015. Diet-switching by omnivorous freshwater shrimps diminishes differences in nutrient recycling rates and body stoichiometry across a food quality gradient. Freshwater Biology 60: 526–36.

Solano, D.H., and A.M. Arias. 2011. Peces diablo (Teleósteo: Siluriformes: Loricariidae) en la cuenca del río Reventazón, Costa Rica. Biocenosis 25(1–2): 79–86.

Spínola-Parallada, R.M., and C. Vaughan-Dickhaut. 1995. Abundancia relativa y actividad de marcaje de la nutria neotropical (*Lutra longicaudis*) en Costa Rica / Relative abundance and spraint-marking activity of the neotropical river otter (*Lutra longicaudis*) in Costa Rica. *Vida Silvestre Neotropical* 4(1): 38–45.

Springer, M. 2006. A taxonomic key to the families of caddisfly larvae (Insecta: Trichoptera) of Costa Rica. *Revista de Biología Tropical* 54: 273–86.

Springer, M. 2008. Aquatic insect diversity of Costa Rica: state of knowledge. *Revista de Biologia Tropical* 56(Suppl. 4): 273–95.

Springer, M. 2010. Trichoptera. *Revista de Biologia Tropical* 58(Suppl. 4): 151–98.

Springer, M., D. Vásquez, A. Castro, and B. Kohlmann. 2007. Bioindicadores de la calidad de agua. Special publication of EARTH University, Costa Rica.

Springer, M., A. Ramírez, and P. Hanson. 2010. Macroinvertebrados de Agua Dulce de Costa Rica I. *Revista de Biologia Tropical* 58(Suppl. 4): 1–200.

Springer, M., S. Echeverría-Sáenz, and P.E. Gutiérrez-Fonseca. 2014. Costa Rica. In: P. Alonso-EguíaLis, J.M. Mora, B. Campbell, and M. Springer, eds. *Diversidad, conservación y uso de los macroinvertebrados dulceacuícolas de México, Centroamérica, Colombia, Cuba y Puerto Rico*. Instituto Mexicano de Tecnología del Agua, Jiutepec, Morelos, México. 442 pp.

Stallcup, L.A., M. Ardón, and C.M. Pringle. 2006. Does nitrogen become limiting under high-P conditions in detritus-based tropical streams? *Freshwater Biology* 51: 1515–26.

Stark, B.P. 1998. The *Anacroneuria* of Costa Rica and Panama (Insecta: Plecoptera: Perlidae). *Proceedings of the Biological Society of Washington* 111: 551–603.

Stiles, F.G. 1983. Birds. In D.H. Janzen, ed., *Costa Rican Natural History*, 502–618. Chicago: University of Chicago Press.

Stiles, F.G., and A.F. Skutch. 1989. *A Guide to the Birds of Costa Rica*. Comstock Publishing. 656 pp.

Stout, J. 1980. Leaf decomposition rates in Costa Rican lowland tropical rainforest streams. *Biotropica* 12: 264–72.

Stout, R.J. 1989. Effects of condensed tannins on leaf processing in mid-latitude and tropical streams: a theoretical approach. *Canadian Journal of Fisheries and Aquatic Sciences* 46: 1097–106.

Stuart, S.N., J.S. Chanson, N.A. Cox, B.E. Young, A.S.L. Rodrigues, D.L. Frishman, and R.W. Waller. 2004. Status and trends of amphibian declines and extinctions worldwide. *Science* 306: 1783–86.

TNC (The Nature Conservancy). 2009. *Evaluación de Ecorregiones de Agua Dulce de Mesoamérica: Sitios Prioritarios para la Conservación en las Ecorregiones de Chiápas a Darién*. San José, Costa Rica: The Nature Conservancy. 515 pp.

Triska, F.J., C.M. Pringle, J.H. Duff, R.J. Avanzino, A. Ramírez, M. Ardón, and A.P. Jackman. 2006a. Soluble reactive phosphorus transport and retention in tropical rainforest streams draining a volcanic and geothermally active landscape in Costa Rica: long term concentration patterns, pore water environment and response to ENSO events. *Biogeochemistry* 81: 131–43.

Triska, F.J., C.M. Pringle, J.H. Duff, R.J. Avanzino, and G. Zellweger. 2006b. Soluble reactive phosphorus (SRP) transport and retention in tropical rain forest streams draining a volcanic and geothermally active landscape in Costa Rica: *in situ* amendment and laboratory studies. *Biogeochemistry* 81: 145–57.

Triska, F.J., C.M. Pringle, G.W. Zellweger, J.H. Duff, and R.J. Avinzino. 1993. Dissolved inorganic nitrogen composition, transformation, retention, and transport in naturally phosphate-rich and phosphate-poor tropical streams. *Canadian Journal of Fisheries and Aquatic Sciences* 50: 665–75.

Umaña-Villalobos, G., and M. Springer. 2006. Environmental variation in the Grande de Térraba River and some of its tributaries, south Pacific of Costa Rica. *Revista de Biología Tropical* 54: 265–72.

van Breukelen, N. A. 2015. Interactions between native and non-native cichlid species in a Costa Rican river. *Environmental Biology of Fishes* 98:885–89.

Vargas, R.J. 1995. History of municipal water resources in Puerto Viejo de Sarapiquí, Costa Rica: a socio-political perspective. Master's thesis, University of Georgia. Athens, GA.

Verburg, P., S.S. Kilham, C.M. Pringle, K.R. Lips, and D.L. Drake. 2007. A stable isotope study of a neotropical stream food web prior to the extirpation of its large amphibian community. *Journal of Tropical Biology* 23: 643–51.

Villalobos, C.R., and E. Burgos. 1975. *Potamocarcinus* (*Potamocarcinus*) *nicaraquensis* (Pseudothelphusidae: Crustacea) en Costa Rica. *Revista de Biología Tropical* 22(2): 223–37.

Voyles, J., S. Young, L. Berger, C. Campbell, W.F. Voyles, A. Dinudom, D. Cook, R. Webb, R.A. Alford, L.F. Skerratt, and L. Speare. 2009. Pathogenesis of chytriomycosis, a cause of catastrophic amphibian declines. *Science* 326: 582–85.

Wantzen, K.M., R. Wagner, R. Suetfeld, and W.J. Junk. 2002. How do plant-herbivore interactions of tress influence coarse detritus processing by shredders in aquatic ecosystems of different latitudes? *Verhandlungen der Internationalen Vereinigung für Theoretische und Angewandte Limnologie* 28: 815–21.

Webster, J.R., and E.F. Benfield. 1986. Vascular plant breakdown in freshwater ecosystems. *Annual Review of Ecology and Systematics* 17: 567–94.

Whiles, M., R.O. Hall Jr., W.K. Dodds, P. Verburg, A.D. Huryn, C.M. Pringle, K.R. Lips, S.S. Kilham, C. Colon-Gaud, A.T. Rugenski, S. Peterson, and S. Connelly. 2013. Disease-driven amphibian declines alter ecosystem processes in a tropical stream. *Ecosystems* 16: 146–57.

Whiles, M., K. Lips, C.M. Pringle, S.S. Kilham, R. Bixby, R. Brenes, S. Connelly, J.C. Colon-Gaud, M. Hunte-Brown, A.D. Huryn, C. Montgomery, and S. Peterson. 2006. The effects of amphibian population declines to the structure and function of Neotropical stream ecosystems. *Frontiers* 4(1): 27–34.

Whitfield, S.M., K.E. Bell, T. Philippi, M. Sasa, F. Bolaños, G. Chaves, J.M. Savage, and M.A. Donnelly. 2007. Amphibian and reptile declines over 35 years at La Selva, Costa Rica. *Proceedings of the National Academy of Sciences* 104: 8352–56.

Whitfield, S.M., E. Geerdes, I. Chacon, E.B. Rodriguez, R.R. Jiménez, M.A. Donnelly, and J.L. Kerby. 2013. Infection and co-infection by the amphibian chytrid fungus and ranavirus in wild Costa Rican frogs. *Diseases of Aquatic Organisms* 104: 173–78.

Whitfield, S.M., J. Kerby, L.R.Gentry, and M.A. Donnelly. 2012. Temporal variation in infection prevalence by the amphibian chytrid fungus in three species of frogs at La Selva, Costa Rica. *Biotropica* 44: 779–84.

Winemiller, K.O. 1983. An introduction to the freshwater fish communities of Corcovado National Park, Costa Rica. *Brenesia* 21: 47–66.

Winemiller, K.O., and M.A. Leslie. 1992. Fish assemblages across a complex, tropical fresh-water marine ecotone. *Environmental Biology of Fishes* 34: 29–50.

Winemiller, K.O., and B.J. Ponwith. 1998. Comparative ecology of eleotrid fishes in Central American coastal streams. *Environmental Biology of Fishes* 53: 373–84.

Wootton, T., and M.P. Oemke. 1992. Latitudinal differences in fish community trophic structure, and the role of fish herbivory in a Costa Rican stream. *Environmental Biology of Fishes* 35: 311–19.

World Commission on Dams (WCD). 2000. Dams and development: a framework for decision-making. Accessed February 26, 2007. http://www.damsreport.org

WRI (World Resources Institute). 1991. Accounts overdue: natural resource depreciation in Costa Rica. Washington, DC: WRI.

Wydrzycka, U., and H. Lange-Bertalot. 2001. Las diatomeas (Bacillariophyceae) acidófilas del Río Agrio y sitios vinculados con su cuenca, Volcán Poás, Costa Rica. *Brenesia* 55–56: 1–68.

Young, B.E., S.N. Stuart, J.S. Chanson, N.A. Cox, and T.M. Boucher. 2004. *Disappearing Jewels: The Status of New World Amphibians*. Washington, DC: NatureServe.

Chapter 19 Lakes of Costa Rica

Sally P. Horn[1,*] and Kurt A. Haberyan[2]

Introduction

Costa Rica has an abundance of lakes, distributed from sea level to the nation's highest peaks and within each of the major terrestrial ecosystems. We focus here on permanent, predominantly fresh, water bodies of all sizes, including water bodies that would be "ponds" in the classification of Horne and Goldman (1994). Costa Rica has lakes that exemplify nearly every natural process of lake formation, including volcanic activity, fluvial dynamics, glaciation, and landslides and other forms of mass movement. Humans have also created many lakes, for a variety of reasons: hydroelectric power, water storage, recreation, aquaculture, livestock, and as consequences of road construction and other activities that blocked original drainages. Natural lakes in Costa Rica are popular hiking and tourist destinations today, and had practical and possibly symbolic importance for prehistoric cultures. They contribute to Costa Rica's high habitat and biological diversity, and their sediments provide key evidence of ecosystem history. High-elevation lakes in Costa Rica, as throughout the world (Messerli 2001), may be harbingers of global climate change (Horn et al. 2005).

We begin this chapter with a short history of research on Costa Rican lakes and a consideration of lake distribution in the country as a whole. We then present a regional survey of Costa Rican lakes, following the classification of terrestrial ecosystems used throughout this book (Kappelle, chapters 1 and 21 of this volume). For each of the seven principal terrestrial ecosystem regions we describe the common mechanisms of lake formation and highlight two to five lakes that are particularly well known. While much of this information is drawn from our published work, we include additional observations by ourselves and others on aquatic biology and other lake characteristics. We also summarize the contributions that paleolimnological studies have made to understanding ecosystem and environmental history in each region.

Research on Costa Rican Lakes

Scientific investigations of Costa Rican lakes began in the latter half of the twentieth century, carried out by Costa Rican scientists and students as well as by visiting foreigners (Umaña et al. 1999). The first detailed study was a thesis on the basic limnology and biology of Laguna de Río Cuarto, submitted to the University of Costa Rica by Kohkemper (1954). The Austrian limnologist Löffler (1972) described the limnology and planktonic communities of high-elevation lakes visited in 1966, while Bumby (1982), a student from the United States of America, focused on the chemical characteristics and macrophytes of mainly low- to mid-elevation lakes during fieldwork in 1973. In the late 1970s, Bergoeing (1978) and Bergoeing and Brenes (1978) interpreted the geomorphology and history of several Costa Rican lake basins from aerial photography, and Bolaños (1979) published an early paper on *Tilapia* aquaculture.

Limnological studies within Costa Rica increased dramatically in the 1980s, propelled in part by the establishment of CIMAR (Centro de Investigación en Ciencias de Mar y Limnología), a research center at the University of

[1] Department of Geography, University of Tennessee, Knoxville, TN 37996, USA
[2] Department of Natural Sciences, Northwest Missouri State University, Maryville, MO 64468, USA
* Corresponding author

Costa Rica that began activities in 1979 (Umaña et al. 1999, Vargas 2004). Students and faculty at the Universidad Nacional (UNA) in Costa Rica also initiated limnological studies at this time, as did we ourselves and other foreign visitors. Papers on lakes from this decade focused on physical and chemical properties (Gocke et al. 1981, 1987, Baker 1987, Charpentier et al. 1988), sediments (Horn 1989a), phytoplankton (Hargraves and Víquez 1981, Wujek 1984, Camacho 1985, Umaña 1985, 1988, Dickman and Nanne 1987), and zooplankton (Collado 1983, Collado et al. 1984a,b, Ramírez 1985, Dussart and Fernando 1986). A large number of papers on freshwater fish also appeared (see review by Umaña et al. 1999, and Pringle et al., chapter 18 of this volume); most studies concerned fish in streams but Ulloa et al. (1988, 1989) described fish communities in the Arenal Reservoir.

Publications on lakes have grown in both number and scope from the 1990s onward. While some investigations have continued to explore the limnology and biology of particular lakes (for example, Gocke et al. 1990, Gocke 1996–97, Ramírez et al. 1990, Ramírez and Camacho 1991, Umaña 1990, 1993, 1997a,b, 2001, Umaña and Collado 1990, Umaña and Jiménez 1995, Umaña et al. 1997, Jiménez and Springer 1994, 1996, Petersen and Umaña 2003), new efforts have focused on the diversity of lakes and their geographic distribution. One of the first broad surveys was presented in 1993 (Horn and Haberyan 1993), and geographic expansion continues to the present, with the publication of basic limnological data from additional lakes in many regions (e.g., Jones et al. 1993, Horn et al. 1999, 2005, Haberyan et al. 2003, Tassi et al. 2009). The 1990s also saw the publication of regional surveys on plankton in Costa Rican lakes (Haberyan et al. 1995, Wujek et al. 1998) and on diatoms (Haberyan et al. 1997), chrysophyte cysts (Zeeb et al. 1996), and pollen grains (Rodgers and Horn 1996) in surface lake sediments.

The documentation of limnological parameters and biological communities continues to be a strong focus, with recent publications exploring microbial diversity in lake waters (Cabassi et al. 2014) and chironomid sub-fossils in surface sediments (Wu et al. 2015). More studies today include repeat measurements to elucidate seasonal and interannual variations (e.g., Umaña 1997b, 2001, 2010a,b, 2014a,b). With basic limnological and ecological information now in hand, a secondary focus of lake research in Costa Rica is developing an understanding of the history of lake basins, biota, and surroundings through paleolimnological studies. Such studies are based on evidence preserved in lake sediments, including diatoms, pollen grains, charcoal fragments, and other biological and geochemical indicators (Cohen 2003). The study of sediment profiles from lakes and swamps in Costa Rica is a key tool for understanding Quaternary climate change (Horn 2007), especially during the Holocene (past 11,700 years) and late Pleistocene (ca. 126,000 to 11,700 years ago). Sediment records from Costa Rican lakes complement and extend what can be learned from the study of ancient soils (Driese et al. 2007), cave speleothems (Lachniet et al. 2004), tree rings (Anchukaitis et al. 2008), and glacial geomorphology (Orvis and Horn 2000, Lachniet and Seltzer 2002). Lake-sediment records also contain abundant evidence of the activities and impacts of pre-Columbian people (Horn 2006, 2007).

The following sections provide a brief introduction to selected lakes of Costa Rica. While we refer to many publications, a detailed review of the literature is beyond the scope of this chapter. For additional information and references see reviews by Umaña et al. (1999) and Haberyan et al. (2003), and the bibliography of CIMAR publications developed by Fuentes et al. (2006).

Distribution and Origins of Costa Rican Lakes

Lakes exist where there are topographic depressions and water to fill them (Cohen 2003). Costa Rica's high rainfall, and wide array of basin-forming geomorphic and human processes, have produced a very large number of natural and artificial lakes. Most of these lakes are too small to appear on maps that show the country as a whole but they are readily apparent on the 1:50,000 scale topographic maps produced by the Instituto Geográfico Nacional (IGN) from aerial photographs. In the 1980s, IGN staff members meticulously compiled information on the location and size of all water bodies over 0.25 ha in area on each of the country's 139 1:50,000-scale topographic map sheets. The total number of lakes tallied, including seasonal ponds, brackish lagoons, and reservoirs, was 652. Although many of these lakes have since been drained for agriculture, our limnological investigations have revealed some two dozen lakes that were not mapped, either because they are of a more recent origin or because they were obscured by shadows, glare, clouds, or forest cover on the original photographs. Thus, even accounting for the draining of many lakes and the inclusion of seasonal and brackish water bodies in the IGN database, it is likely that several hundred permanent freshwater lakes still exist in Costa Rica.

The 652 water bodies included in the IGN database are distributed throughout the country, from both coasts to the crests of the mountain ranges (Fig. 19.1). Most of the lakes are located in the lowlands; 51% are below 40 m in elevation, 86% are below 500 m, and 92% are below 1,000 m. In terms of lake area, 13% of the IGN water bodies are less

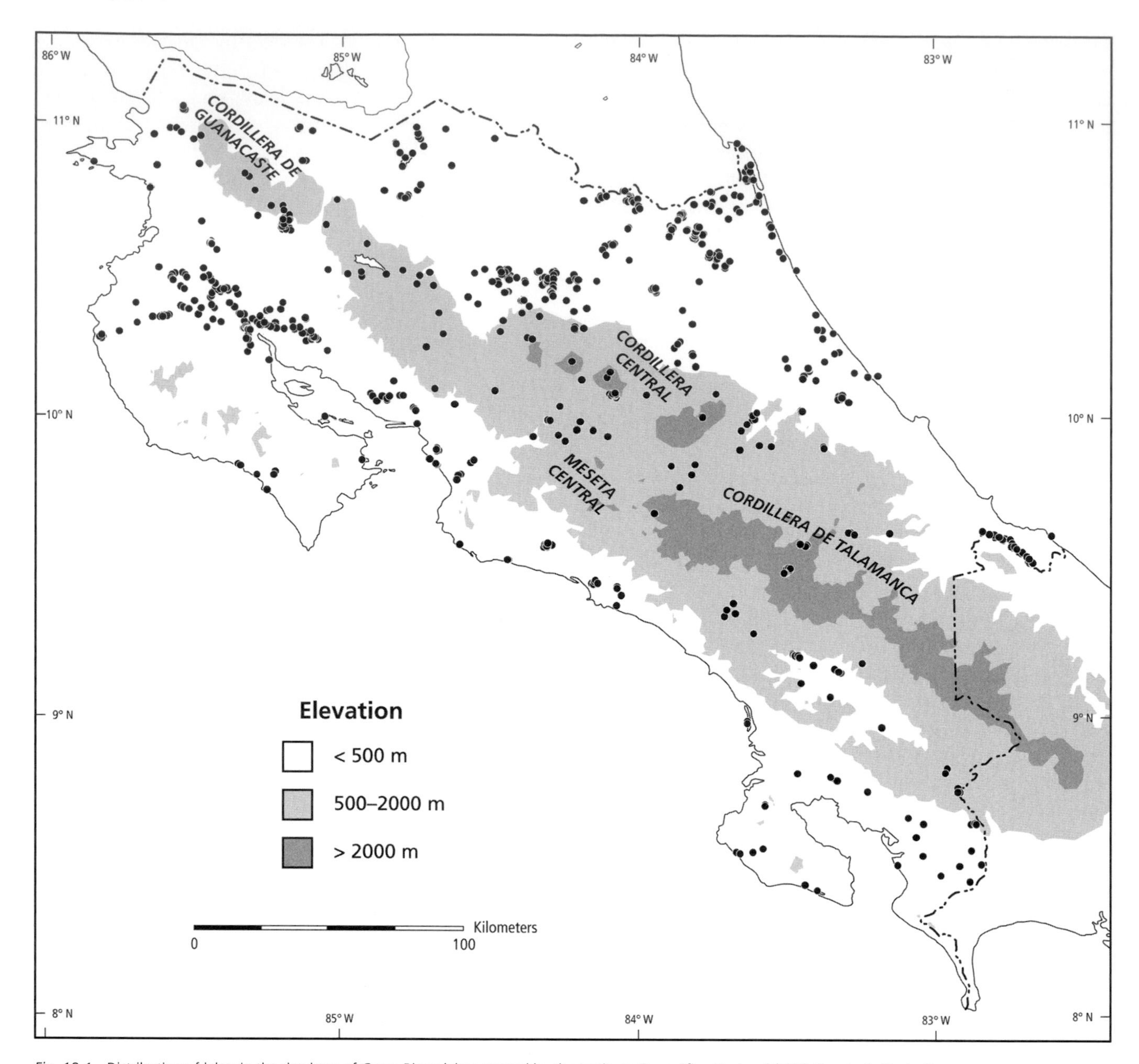

Fig. 19.1 Distribution of lakes in the database of Costa Rican lakes created by the Instituto Geográfico Nacional (IGN), San José, Costa Rica. *Map produced by the Cartographic Services Laboratory of the University of Tennessee, Knoxville, TN, USA.*

than 1 ha; 66% are between 1 and 10 ha, 17% are between 10 and 100 ha, 4% are between 100 and 1,000 ha, and less than one percent are larger than 1,000 ha.

The dominant mechanisms of lake formation vary with elevation, geology, and terrain, with some lake-forming processes limited to small areas of the country, and others relatively widespread. Along both coasts sandbars have created coastal lagoons that are freshened by inflowing streams (Umaña et al. 1999, Haberyan et al. 2003). In the floodplains of large rivers, stream courses have shifted over time, leaving behind old channels and other depressions that become lakes. Where slopes are steeper at low to mid-elevations, landslides of various types have blocked drainage of former stream valleys, creating lakes. Volcanism has produced lava flows and lahars that have created impoundments at these elevations, in addition to craters of various types found along a wide elevational gradient. Several volcanic craters at high elevations along the crests of the Cordillera de Guanacaste and Cordillera Central hold lakes (Alvarado 2000, and see Alvarado and Cárdenes, chapter 3 of this volume). Even the highest peaks in the country, reaching over 3,800 m in the Cordillera de Talamanca, are surrounded by lakes, here formed by glaciers (Horn et al. 1999, 2005). For some lakes, the origins of the basins they occupy remain uncertain (Alvarado 2000).

In more modern times, human activity has created abundant lakes. Farm ponds have been built for aquaculture, agriculture, and livestock. A multitude of small water bodies has been formed as a secondary impact of road or railroad construction, either by grading across a drainage or by excavating fill for roadbeds. Large reservoirs store water for human consumption (e.g., the Orosi Reservoir, which supplies much of San José: Sánchez-Azofeifa et al. 2002), while others also provide hydroelectric power, recreation, and other services, including the Arenal (8000 ha), Cachí (283 ha), and Angostura (220 ha) reservoirs. The environmental impacts of reservoir construction on Costa Rican streams are described by Pringle et al., chapter 18 of this volume; see also Anderson et al. (2006).

Regional Survey of Costa Rican Lakes

Lakes of the Pacific Dry Forest

We consider together here lakes in the two sectors of the Pacific dry forest ecosystem: the Northern Pacific lowland deciduous dry forest, and the Nicoya-Tempisque Pacific dry forest (Fig. 19.2). For extensive descriptions of the Pacific Dry Forest ecosystem and its conservation we refer to Jiménez et al. and Janzen and Hallwachs, chapters 9 and 10 of this volume, respectively. Although seasonally arid, the northern Pacific lowlands of Costa Rica support several permanent, natural water bodies in addition to a number of seasonal lakes and artificial ponds maintained by irrigation. Natural lakes in this region occur in two clusters: one on the southern slope of Miravalles volcano, and one near the mouth of the Río Tempisque. These two areas are among a small number of "lake districts" (Horne and Goldman 1994) in Costa Rica in which multiple lakes exist that share a common mode of formation. Such areas provide important settings for limnological research. Although formed by the same processes, often at the same time, and occupying the same general climate setting, individual lakes within particular lake districts are likely to differ somewhat in basin form, size, microclimate, and watershed conditions including land use and disturbance history, making lake districts valuable sites for comparative research in both modern lake biology and paleolimnology.

Lakes in the Miravalles lake district occupy depressions between 330 and 570 m in elevation, in an area of undulating topography created by a volcanic debris avalanche from Miravalles Volcano about 8,300 years ago (Alvarado et al. 2004, Siebert et al. 2006). The eight we have investigated range from less than one hectare to 4.4 ha in size, and are surrounded by cattle pastures and remnant areas of lowland deciduous dry forest. Despite the seasonal rainfall in this part of Costa Rica, these lakes hold water year-round, although water levels may drop significantly during years of exceptionally severe drought, as we observed in March 1998 (Haberyan et al. 2003). The three we highlight here, **Laguna San Pablo**, **Estero Blanco**, and **Laguna Martínez** (Fig. 19.2, Table 19.1), are fairly broad, shallow, and turbid. These lakes vary in many characteristics, but tend to have somewhat elevated levels of calcium (about 18–28 mg/L), sodium (~12–25 mg/L), and silica (~15–33 mg/L). Aquatic macrophytes are present in all three lakes. Laguna San Pablo (Fig. 19.3) and Estero Blanco (Fig. 19.4) are largely open water, but San Pablo has two large (>50 m^2) floating islands of grass and other plants that are pushed across the lake by strong dry season winds, and Estero Blanco has abundant submerged macrophytes. Laguna Martínez, a smaller and more sheltered lake, had open water in 2001 but was covered by water hyacinth at the time of our last visit in June 2003.

Lakes in the Tempisque lake district are found near sea level at the head of the Gulf of Nicoya. Depressions here were formed by the natural migration of the Río Tempisque and its tributaries, which leaves behind abandoned channels, and by the build-up of natural levees along the main river that dam the mouths of tributary streams. Although many of these lakes are shown on topographic maps as permanent water bodies, most probably dry down substantially during the dry season. Land managers, biologists, and conservationists generally refer to these sites as seasonal wetlands rather than lakes (Bravo and Ocampo 1993; Jiménez, chapter 20 of this volume). Examples include the "lagunas" of Palo Verde National Park and of the Laguna Mata Redonda Wildlife Refuge, located 10 km to the southwest (Boza and Cevo 1998). It is unknown whether the seasonal desiccation of these floodplain lakes is the natural long-term condition or the result of stream diversions and agricultural development. In both Palo Verde National Park and the Laguna Mata Redonda Wildlife Refuge, the question of the natural hydrological state is further complicated by the recent spread of aquatic macrophytes, particularly cattail (*Typha*), through areas of formerly open water (Somarribas and Bravo 1999, Horn and Kennedy 2006). Outside these protected reserves, the IGN maps show lakes that no longer exist, having been drained to make way for the cultivation of rice or other crops. Although these former lakes may be impossible to recognize in the field, their faint outlines are evident in recent satellite imagery.

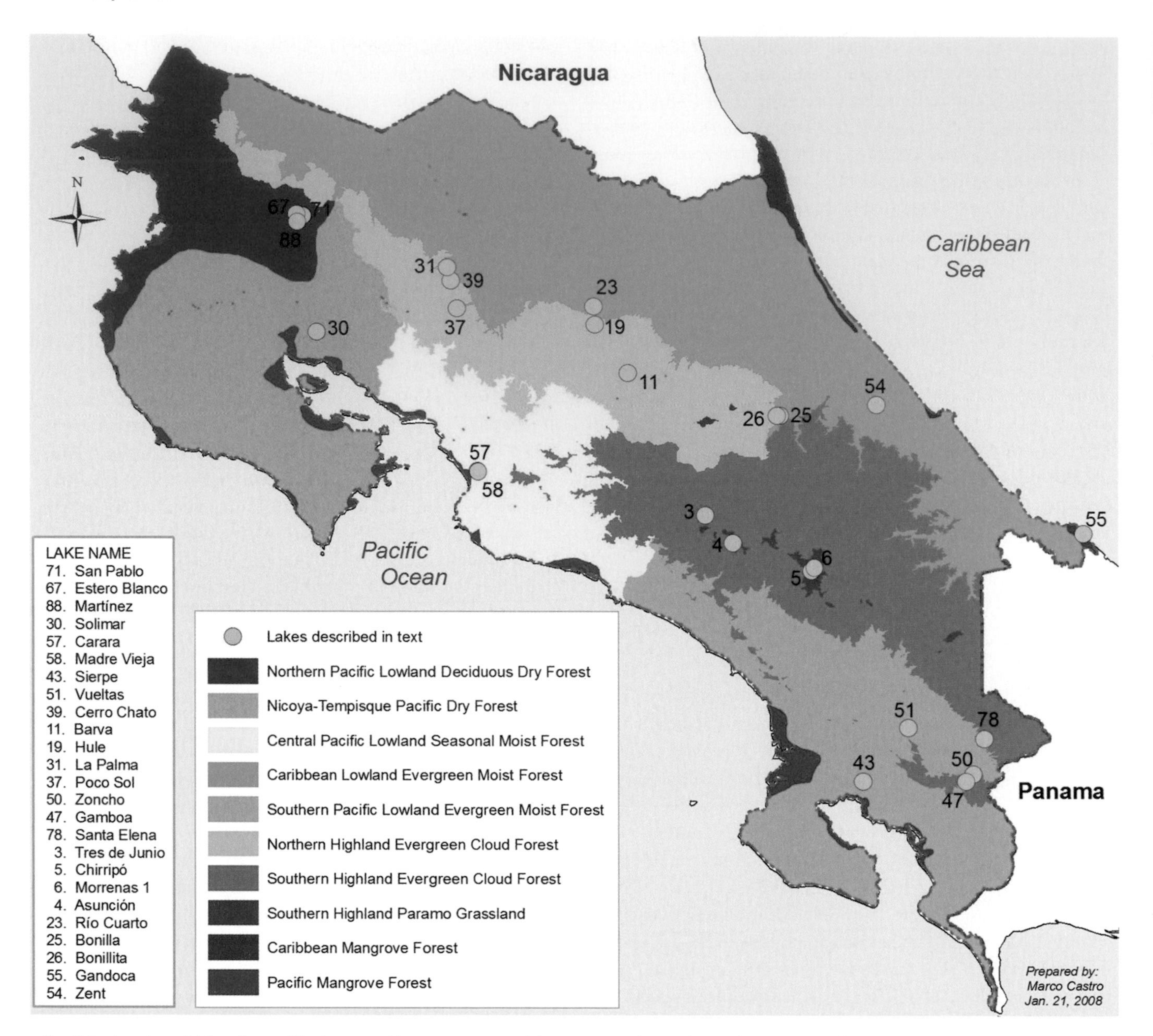

Fig. 19.2 Locations of lakes discussed in the text, with respect to terrestrial ecosystem regions. Lake numbers correspond to numbers in the authors' database of limnological observations from approximately one hundred Costa Rican lakes.
Map produced by Marco V. Castro.

Aside from the Miravalles lakes and the Tempisque lakes, farm ponds are scattered across the Pacific dry forest region. We have sampled two in the Nicoya-Tempisque Pacific dry forest sector: Laguna Palmita, along the Inter-American highway about 28 km SSE of the town of Cañas (Horn and Haberyan 1993), and **Laguna Solimar,** 16 km SSW of Cañas, in the lowlands at the head of the Gulf of Nicoya. Laguna Solimar (Table 19.1) is a turbid, eutrophic lake characterized by warm temperatures (32°C), and relatively high alkalinity and conductivity, in keeping with the lake's location at low elevation in the driest part of Costa Rica. Haberyan et al. (1995) classified Solimar as a cyanophyte-dominated lake, with *Cylindrospermum* the most common phytoplankton. The zooplankton community was composed largely of the copepod *Mesocylcops thermocyclopoides*, with some ostracods.

Sediment cores for paleolimnological study have been recovered from the Miravalles lakes and from the Bocana wetland in Palo Verde National Park. Pollen, microscopic charcoal, and sediment characteristics in profiles from Lagunas San Pablo, Estero Blanco, Martínez, and three other lakes in the Miravalles district document initial forest development following the debris avalanche and associated eruptive phenomena that formed the lake basins, and subsequent shifts in moisture availability, vegetation composition, and fire incidence (Arford and Horn 2004; Horn et al., unpublished data). Past fires in these lowland dry forests resulted from natural ignition (from volcanism and lightning) as well as human activity; analyses of macroscopic (>250 μm) charcoal fragments too large to be readily dispersed by wind indicate that many of the fires occurred within lake watersheds. The upper sections of all six lake sediment profiles contain maize pollen and charcoal from agricultural fires. The earliest maize pollen, in the sediment core from Laguna Martínez, is associated with charcoal dated to 5,500 cal yr BP (calibrated years before present) and constitutes the earliest evidence for maize cultivation in all of Costa Rica (Arford and Horn 2004). [Note: cal yr BP ages reported throughout the text are the weighted means of the calibration probability distributions (Telford et al. 2004a), derived using the Calib 5.0.1 program (Stuiver and Reimer 1993) and the dataset of Reimer et al. 2004.] Maize is a fully domesticated plant that can persist on the landscape only with human assistance; the distribution of its pollen in the Miravalles sediments thus demonstrates the presence of settled humans within the northern Pacific lowland deciduous dry forest for over five millennia (Horn 2006). Diatom assemblages in the San Pablo core suggest changes in lake level that generally match climate interpretations from pollen and sediment characteristics (Haberyan et al. 2005), but seem not to show any effects of prehistoric agriculture.

The dense and sticky sediments of the Bocana wetland have proven difficult to core, and we have recovered and examined only a partial profile from this site 7 km east of the Organization for Tropical Studies field station in Palo Verde National Park. The most important finding is the presence of pollen of cattail (*Typha*) in two samples adjacent to and below a charcoal fragment dated to 4,500 cal yr BP (Horn and Kennedy 2006). Beginning in the 1980s, areas of formerly open water in the seasonal wetlands of Palo Verde

Table 19.1. Selected Data from Lakes of the Pacific Dry Forest of Costa Rica

Lake Name	San Pablo	Estero Blanco	Martínez	Solimar
Lake Number	71	67	88	30
Parameter				
Latitude (°N)	10.6594	10.6657	10.6405	10.2724
Longitude (°W)	85.1782	85.2012	85.1961	85.1301
Elevation (m a.s.l.)	450	430	340	8
Area (ha)	4.38	1.50	1.5	4.33
Depth (m)	2.5	2.8	3.6	2.5
Secchi (m)	0.5	1.6	1.3	0.4
Temperature (°C)	29.1	28.8	25.5	32.0
Stratified?	yes	yes	nd	yes
Macrophyte cover	0	4	9	2
pH	9.50	7.90	6.36	7.73
Conductivity (uS/cm)	176	214	nd	120
O_2 (mg/L)	12.0	7.6	nd	13
CO_2 (mg/L)	nd	6	nd	0
Alkalinity (mg/L $CaCO_3$)	95	75	nd	119
Ca^{+2} (mg/L)	17.72	18.10	28.02	11.9
Mg^{+2} (mg/L)	2.37	6.39	8.87	3.54
K^+ (mg/L)	6.46	4.63	5.70	7.12
Na^+ (mg/L)	15.71	11.69	24.54	6.68
Si (mg/L)	28.33	15.07	33.32	12.7
Cl^- (mg/L)	7.24	4.47	4.81	4.5
S total (mg/L)	nd	7.88	nd	0.6

NOTE. Lagunas San Pablo, Estero Blanco, and Martínez of the Miravalles lake district are located in the northern Pacific lowland deciduous dry forest sector, and Laguna Solimar is located in the Nicoya-Tempisque Pacific dry forest sector of the Pacific dry forest ecosystem. Our previous publications reported coordinates based on the Ocotepeque 1935 datum, as used on Costa Rican topographic maps (Orvis 2002), but here we provide coordinates determined using Google Earth, which uses a WGS84 datum. Some limnological data represent a composite of multiple visits, combined so as to represent dry and wet seasons equally. Macrophyte cover is estimated using a linear scale ranging from 0 to 9, reflecting the proportion of lake surface that is covered by or underlain by macrophytes (coverage 0 reflects 0 to 10%, coverage 1 reflects 10 to 19%, etc.). nd = no data are available. See Horn and Haberyan (1993) and Haberyan et al. (2003) for additional data.

Fig. 19.3 Dry-season view of Laguna San Pablo in the northern Pacific lowland deciduous dry forest ecosystem. This lake occupies a depression in undulating topography formed by volcanic debris flows and lahars from Volcán Miravalles, visible in the background. Watershed vegetation consists of cattle pasture with scattered trees. One of the lake's two floating islands can be seen in the lower left of the image; a portion of the other island is visible on the far right.
Photo by Sally Horn.

Fig. 19.4 Laguna Estero Blanco in the northern Pacific lowland deciduous dry forest ecosystem, photographed during the wet season.
Photo by Christine Lafrenz.

National Park began to be choked with dense stands of cattail, a process that has greatly reduced migratory bird habitat (see also Jiménez, chapter 20 of this volume). The rapid spread of cattail led to its description as an "invasive species" and one report (Cochard and Jackes 2005) that the plant was not native to the area but had been introduced from Eurasia. However, the sediment profile from the Bocana marsh shows that *Typha* is native and of some antiquity in the park: if regarded as an invasive, it belongs in a special category as a *native* invasive species (Horn and Kennedy 2006).

A third paleolimnological site in the Pacific dry forest ecosystem consists of an ancient lake bed, now dry and exploited for its diatom-rich sediment (diatomite). Chávez and Haberyan (1996) described diatom assemblages in 23 samples taken at various depths within the Camastro Diatomite, located on the lower western slope of the Rincón de la Vieja volcanic complex. The deposit as a whole is likely Pleistocene in age and may span more than 30,000 years. Seven assemblages were recognized, all suggesting that the lake was similar to many modern lakes of the region: shallow, eutrophic, and slightly basic (pH, 7.8 to 8.5), with low to medium concentrations of silica, and low conductivity.

Table 19.2. Selected Data from Lakes of the Central Pacific Lowland Seasonal Moist Forest of Costa Rica

Lake Name	Carara	Madre Vieja
Lake Number	57	58
Parameter		
Latitude (°N)	9.7984	9.8103
Longitude (°W)	84.5932	84.5949
Elevation (m a.s.l.)	16	16
Area (ha)	2.13	1.60
Depth (m)	3.5	3.0
Secchi (m)	0.6	0.2
Temperature (°C)	30.8	32.9
Stratified?	no	no
Macrophyte cover	1	9
pH	7.99	7.60
Conductivity (uS/cm)	362	412
O_2 (mg/L)	4.4	0.6
CO_2 (mg/L)	12	27
Alkalinity (mg/L $CaCO_3$)	169	202
Ca^{+2} (mg/L)	35.33	44.01
Mg^{+2} (mg/L)	9.24	15.05
K^+ (mg/L)	5.37	3.17
Na^+ (mg/L)	18.85	12.20
Si (mg/L)	15.05	18.38
Cl^- (mg/L)	4.70	2.00
S total (mg/L)	3.39	0.42

NOTE. Our previous publications reported coordinates based on the Ocotepeque 1935 datum, as used on Costa Rican topographic maps (Orvis 2002), but here we provide coordinates determined using Google Earth, which uses a WGS84 datum. Macrophyte cover is estimated using a linear scale ranging from 0 to 9, reflecting the proportion of lake surface that is covered by or underlain by macrophytes (coverage 0 reflects 0 to 10%, coverage 1 reflects 10 to 19%, etc.). See Haberyan et al. (2003) for additional data.

Lakes of the Central Pacific Lowland Seasonal Moist Forest

The Central Pacific lowland seasonal moist forest region (Jiménez et al., chapter 11 of this volume) has fewer natural lakes than the dry and moist forest regions of the northern and southern Pacific lowlands, respectively. Both of the lakes we have sampled were formed by fluvial action (Table 19.2). **Laguna Carara** occupies a meander scar of the Río Tárcoles in the Carara Biological Reserve; it shows the classic oxbow shape of a cut-off meander, but reserve staff indicated that it receives overflow from the Tárcoles during the wet season, functioning as an alternate channel. The muddy appearance of the lake confirms this connection, at least when we visited in July 1997. The abundance of CO_2 and paucity of O_2 suggest that primary production in the lake itself is low; rather, much carbon probably enters from the surrounding plants. In 1997 we also visited **Laguna Madre Vieja,** located across the river from L. Carara. This lake also occupied an old meander scar, but other characteristics differed: its waters were not muddy, and water lilies covered the surface. These observations, together with the elevated concentrations of ions, indicated that Madre Vieja was not flushed annually by river overflow in the 1990s. However, satellite images document shifts in the course of the Río Tárcoles between 2002 and 2013 that led to the stream reoccupying the abandoned channel in which Laguna Madre Vieja had formed. The lake no longer exists in 2015, its creation and destruction both a consequence of natural fluvial processes in this dynamic floodplain environment.

We have not collected sediment cores from Laguna Carara or Laguna Madre Vieja. Abandoned channel lakes generally have relatively short life spans (Cohen 2003), but their sediments can provide valuable records of ecosystem history. The likelihood that sediments are flushed from Laguna Carara during flood events suggests that this site may be a poor prospect for coring, but records might be preserved in wetlands and other small depressions within agricultural fields on the floodplain, identifiable from satellite imagery as locations of former stream channels. The presence of large crocodiles here (Herrera 1992) will require special vigilance, as these reptiles have endangered sediment-coring teams in other parts of the world (Richardson and Livingstone 1962, Tyson 2000).

Lakes of the Southern Pacific Lowland Evergreen Moist Forest

Lakes formed by mass movement and fluvial dynamics are numerous in the southern Pacific lowland evergreen moist forest region described by Gilbert et al. (chapter 12 of this volume), as are small artificial lakes. Here we describe two contrasting natural lakes of very different origin (Table 19.3).

Laguna Sierpe is located at the southern edge of the Térraba-Sierpe delta, where it abuts the hills that link the Osa Peninsula to the mainland. It comprises a large, shallow (2.2 m depth) expanse of water surrounded by floating mats of grasses and other plants. Despite its flat bottom and the great exposure of the lake, we found evidence of weak temperature stratification in July 1997. The lake may owe its formation to the build-up of sediments and vegetation along the Río Sierpe or along the coast. The composition of its waters reflect inputs from the surrounding vegetation. The modern diatom assemblage in its surface sediments is largely composed of *Nitzschia amphibia* and *Pseudostaurosira brevistriata*.

Table 19.3. Selected Data from Lakes of the Southern Pacific Lowland Evergreen Moist Forest of Costa Rica

Lake Name	Sierpe	Vueltas
Lake Number	43	51
Parameter		
Latitude (°N)	8.7867	8.9662
Longitude (°W)	83.3265	83.1757
Elevation (m a.s.l.)	16	270
Area (ha)	102.7	0.30
Depth (m)	2.2	3.0
Secchi (m)	1.2	0.6
Temperature (°C)	33.0	29.9
Stratified?	yes	yes
Macrophyte cover	1	1
pH	7.09	8.72
Conductivity (uS/cm)	102	233
O_2 (mg/L)	6.6	10.1
CO_2 (mg/L)	6	0
Alkalinity (mg/L $CaCO_3$)	48	125
Ca^{+2} (mg/L)	8.88	21.72
Mg^{+2} (mg/L)	3.50	8.43
K^+ (mg/L)	0.00	3.86
Na^+ (mg/L)	4.35	7.02
Si (mg/L)	9.90	20.95
Cl^- (mg/L)	2.60	1.50
S total (mg/L)	0.37	1.48

NOTE. Our previous publications reported coordinates based on the Ocotepeque 1935 datum, as used on Costa Rican topographic maps (Orvis 2002), but here we provide coordinates determined using Google Earth, which uses a WGS84 datum. Some limnological data represent a composite of multiple visits, combined so as to represent dry and wet seasons equally. Macrophyte cover is estimated using a linear scale ranging from 0 to 9, reflecting the proportion of lake surface that is covered by or underlain by macrophytes (coverage 0 reflects 0 to 10%, coverage 1 reflects 10 to 19%, etc.). See Umaña et al. (1999) and Haberyan et al. (2003) for additional data.

Laguna Vueltas overlooks the Río Limón, which joins the Río Térraba below the confluence of the General and Coto Brus rivers. It apparently formed by ponding behind a block slide, and is turbid and warm. *Tilapia* were added to the lake for sport fishing, and these fish are prey to boat-billed herons whose Spanish name (chocuaco) gives the lake its local name (we use Laguna Vueltas for consistency with the topographic map). The lake is rather shallow; this, plus inputs from the herons and other sources, contributes to the elevated oxygen levels we observed. The most common diatom in its surface sediments is *Nitzschia frustulum*.

A sediment core from Laguna Vueltas documents late Holocene shifts in vegetation and fire incidence resulting from human activities (Horn et al., unpublished data). Sedimentary pollen assemblages and stable carbon isotope signatures document extensive forest clearance and maize cultivation around 1,100 cal yr BP. Charcoal and maize disappear from the record about 450 cal yr BP, indicating site abandonment following the Conquest, which was marked by population decline throughout Costa Rica. At that point in the record, tree pollen sharply rebounds, indicating recovery of the southern Pacific lowland evergreen moist forest.

Lakes of the Northern Highland Evergreen Cloud Forest

The northern highlands are dominated by two chains of Quaternary volcanoes, the Cordillera de Guanacaste to the north and the Cordillera Central to the south, with an intervening area of older, Tertiary volcanic rocks in the Cordillera de Tilarán (Lawton et al., chapter 13 of this book). Not surprisingly, crater lakes of various types are best developed in this part of the country. However, the northern highland evergreen cloud forest region is also home to several lakes formed by lava or lahar flows and by landslides (Table 19.4).

Permanent, freshwater lakes in craters atop volcanoes in this ecosystem include Lagunas Santa María, Cerro Chato, and Barva, on the volcanoes for which they are named, and Laguna Botos on Volcán Poás. We highlight Laguna Cerro Chato and Laguna Barva, which have been sampled repeatedly. **Laguna Cerro Chato** at the southern end of the Cordillera de Guanacaste occupies an inactive crater associated with Volcán Arenal (2.5 km to the northwest). The lake is relatively deep (17.9 m) and surrounded by crater walls that rise 60–120 m above the lake surface (Umaña 2010a). During four visits during the wet season Umaña

Table 19.4. Selected Data from Lakes of the Northern Highland Evergreen Cloud Forest

Lake Name	Cerro Chato	Barva	Hule	La Palma	Poco Sol
Lake Number	39	11	19	31	37
Parameter					
Latitude (°N)	10.4428	10.1337	10.2948	10.4857	10.3501
Longitude (°W)	84.6883	84.1053	84.2100	84.7137	84.6699
Elevation (m a.s.l.)	1050	2840	740	570	776
Area (ha)	2.75	0.77	54.71	5.00	2.88
Depth (m)	17.9	7.9	22.5	10.8	11.3
Secchi (m)	2.3	1.7	3.0	1.4	2.1
Temperature (°C)	21.2	11.7	21.1	25.7	23.7
Stratified?	yes	no	yes	yes	yes
Macrophyte cover	0	0	1	1	0
pH	7.24	7.45	6.55	8.16	7.26
Conductivity (uS/cm)	28	60	78	293	107
O_2 (mg/L)	7.0	8.0	9.2	11.3	5.9
CO_2 (mg/L)	3	4	5	6	3
Alkalinity (mg/L $CaCO_3$)	8	9	60	130	48
Ca^{+2} (mg/L)	0.32	0.63	7.88	21.64	11.03
Mg^{+2} (mg/L)	0.12	0.32	2.48	17.21	1.67
K^+ (mg/L)	0.00	1.39	1.47	2.82	0.43
Na^+ (mg/L)	1.15	7.41	3.97	15.02	4.46
Si (mg/L)	0.27	0.84	13.47	24.53	10.19
Cl^- (mg/L)	2.00	9.49	2.82	10.68	2.30
S total (mg/L)	0.35	2.00	0.63	5.36	1.50

NOTE. Our previous publications reported coordinates based on the Ocotepeque 1935 datum, as used on Costa Rican topographic maps (Orvis 2002), but here we provide coordinates determined using Google Earth, which uses a WGS84 datum. Some limnological data represent a composite of multiple visits, combined so as to represent dry and wet seasons equally. Macrophyte cover is estimated using a linear scale ranging from 0 to 9, reflecting the proportion of lake surface that is covered by or underlain by macrophytes (coverage 0 reflects 0 to 10%, coverage 1 reflects 10 to 19%, etc.). See Horn and Haberyan (1993), Umaña et al. (1999), and Haberyan et al. (2003) for additional data.

et al. (1997) found the lake to be stratified, but hypothesized that it may overturn during the windier dry season, when whitecaps are visible on the lake despite the sheltering provided by the high crater walls. The bottom of the lake is strewn with rocks and boulders, and we were able to retrieve only a bottom sample of leaves and small amounts of mud. Diatoms in this sample consisted largely of *Brachysira brachysira*.

Laguna Cerro Chato's waters are dilute. Repeated measurements of lake pH by Umaña and Jiménez (1995) revealed strong variation over time and space, from 3.3 to 6.0, a condition influenced by the very low alkalinity of the lake (8 mg L^{-1}).

The Laguna Cerro Chato watershed supports unbroken northern highland cloud forest. The shade cast by the forest, together with the steep slopes, precludes the development of littoral vegetation. Umaña and Jiménez (1995) found the lake's phytoplankton to be dominated by chlorophytes, especially *Arthrodesmus bifidus*, *Monorhaphidium griffithii*, and small coccoid cells. Zooplankton was dominated by the copepod *Tropocyclops prasinus prasinus* (85% of all zooplankters observed), and also included four species of cladocerans. The mesh size used for sampling zooplankton probably prevented the collection of rotifers, but even accounting for that, zooplankton diversity was very low. The lake lacks fish, but tadpoles were observed to be abundant in the open water of the lake. Umaña and Jiménez (1995) examined the stomach content of one tadpole and suggested that zooplankton was not a large part of the diet.

Jiménez and Springer (1994, 1996) examined benthic macrofauna in Laguna Cerro Chato. Chironomids were the most abundant organisms, followed by oligochaetes and nematodes. Samples from deeper locations in the lake had lower species diversity and evenness. Overall patterns of abundance and diversity correlated with the oxygen content of the water and the organic content of surface and near-surface sediments.

Laguna Barva lies atop the higher Volcán Barva in the Cordillera Central, also in an old crater with well-developed cloud forest on both the inner and outer crater walls. The lake is much smaller and shallower than Laguna Cerro Chato, and given its high elevation (2,860 m) might be expected to follow the pattern of frequent mixing characteristic of other tropical lakes in cool, cloudy highlands (Löffler 1964). However, Umaña (1990, 1997b, 2010a,b) observed temperature stratification during some visits, especially between May and August. Abundant allochthonous organic matter contributed by the surrounding forest results in low hypolimnetic oxygen levels and releases humic compounds that stain the water brown. In general, Barva's waters are poor in dissolved minerals, but high ammonium values have been detected. As in Laguna Cerro Chato, pH values show considerable variability, from 5.5 to 7.5 (Umaña et al. 1999).

Phytoplankton dominance has been found to vary over time between small cryptophytes (*Cryptochrisis minor*, *Chroomonas* sp.), desmids (*Cosmarium sphaerosporum*, *Staurastrum paradoxum*), cyanobacteria (*Gloeocapsa* sp.), and chlorophytes (*Eutetramorus tetrasporus*). Over 70 species have been documented in repeated sampling of Barva's phytoplankton (Umaña et al. 1999), including three species of silica-scaled Chrysophyceae and Synurophyceae (Wujek et al. 1998). Interestingly, the phytoplankton community does not seem to follow an annual pattern, perhaps due to the varying influence of Caribbean and Pacific weather systems (Umaña 2010b). Surface sediment contains abundant diatoms of *Brachysira brachysira*, accompanied by *Gomphonema gracile*, along with a diversity of chryosphyte cyst morphotypes (Zeeb et al. 1996).

The zooplankton of Laguna Barva consists mainly of the copepods *Tropocyclops prasinus* and *Thermocyclops tunuis*, and the cladoceran *Ceriodaphnia cornuta*. Water boatmen (*Notonecta* sp.) are also common in open water. The biota of the benthos and littoral zones include representatives of at least six groups: the amphipod *Hyalella azteca*, the copepod *Paracyclops chiltonii*, the hydrocarinid orbatid *Trimalaconothus novus*, the odonates *Libellula* sp. and *Anax* sp., the trichoptera *Oxyethira* sp. and *Limnephilus* sp., and chironomids (Umaña et al. 1999). Fish are absent.

Fig. 19.5 Laguna Hule, one of three lakes in a volcanic explosion crater in the northern highland evergreen cloud forest ecosystem. The forest on the interior crater walls is little modified by human activity.
Photo by Kurt Haberyan.

Laguna Hule (Fig. 19.5) is the largest of three natural lakes in a large *maar*, or volcanic explosion crater, located 11 km north of Volcán Poás. The *maar* that contains Laguna Hule is over 1.5 km in diameter, so qualifies as a *caldera* (Alvarado 2000); it is known as the Hule or Bosque Alegre Caldera or Crater (Gocke 1996–97), and has protected status as the Bosque Alegre National Wildlife Refuge. Unlike the several crater lakes atop volcanoes along the crests of the NW-SE trending Cordillera de Guanacaste and Cordillera Central, the floor of the Hule Crater is lower in elevation than much of the terrain beyond the crater rim. This is the typical configuration of maars, which result from the violent degassing of magma or the interaction of groundwater and magma. Recent investigations indicate that the explosion that formed the Hule maar occurred about 6,200 cal yr BP (Salani and Alvarado 2010, Alvarado et al. 2011).

A secondary pyroclastic cone and lava flow in the middle of the Hule caldera separates Laguna Hule (c. 55 ha) from Laguna Congo (15 ha), and the much smaller Laguna Bosque Alegre (<1 ha). Laguna Hule is fed by two permanent streams that originate to the south, above the caldera walls on the north slope of Volcán Congo. During the wet season, it may also receive flow from Laguna Bosque Alegre, into which Laguna Congo may flow. Laguna Hule drains to the north via the Río Hule, but during the dry season there is often no outflow. The steep walls of the Hule caldera support relatively intact forest, but the slopes beyond the crater have been deforested for agriculture and cattle grazing.

Laguna Hule is deep (22.5 m) and relatively clear (Secchi depth, 3.0 m). The lake is especially noteworthy in the variations of CO_2 content of its deep waters (27 to 65 mg/L; also see Tassi et al. 2009). Local residents report periodic fish kills and a rotten-egg smell in the air; these suggest occasional lake turnover on a decadal time scale. One such turnover apparently occurred a few months before our July

1997 visit, at which time the lake had weakly restratified (Cabassi et al. 2014). Gocke and collaborators (1996–1997) reported a partial turnover that had occurred in January–February 1989, while in 1979 its deep water may have had enhanced oxygenation. Buildup of CO_2 in deep waters is suggestive of a volcanic source, and rapid degassing of large, deep lakes can be a threat to humans (e.g., Lakes Nyos and Monoun, Cameroon: Kling et al. 1987). The smaller size and shallower depth of Laguna Hule suggests that its deep waters could accumulate only enough CO_2 to blanket the bottom of the crater to about 2 m. This is cause for concern, but the periodic turnover of the lake (often in the cooler, windier months when stratification is weakest), prevents the waters from becoming fully saturated. Interestingly, a similar turnover event occurred at the same time in nearby Laguna de Río Cuarto (see below).

Laguna Hule supports well-developed littoral vegetation along most of its shore, especially in shallower water near the outlet. Chlorophytes and cyanobacteria dominate the plankton. On the basis of work by Umaña (1993), common taxa include *Ankistrodesmus braunii*, *Arthrodesmus phinus*, *Dictyosphaerium ehrenbergianum*, *Mougeotia* sp., *Oocystis* sp., *Scenedesmus ecornis*, *Staurodesmus* sp., and *Staurastrum* spp. (all chlorophytes), *Trachelomonas volvocina* (a euglenophyte), and the cyanobacteria *Coelosphaerium* sp., *Merismopedia tenuissima* and *M.* cf. *chondroidea;* Wydrzycka (1996) listed additional *Trachelomonas* spp. and suggested that Umaña's *T. volvocina* is instead *T. volvocinopsis*. The zooplankton is dominated by cladocera (especially *Daphnia laevis*, with *Ceriodaphnia cornuta*, *Bosmina hagmanni*, and *B. longirostris*), and also includes cyclopoid copepods (*Microcyclops ceibaensis*), rotifers (*Keratella* sp., *Polyarthra* sp., and *Ptygura* sp.), and *Choroborus* larvae (Haberyan et al. 1995, Umaña et al. 1999). Fish species in Laguna Hule include the cichlid *Parachromis dovii* ("guapote"), *Astyanax aeneus*, and *Poecilia gillii*. Tabash and Guadamuz (2000) investigated feeding patterns and growth and recruitment of guapote, as a contribution to maintaining the sport fishery of this species. They found that the fish feed primarily on other fishes (including juvenile guapote), insects, and microcrustaceans, and have a long life cycle (8–10 years).

Laguna La Palma also owes its origin to a volcano, but by a different mechanism: Laguna La Palma and nearby Laguna Cedeño were formed when lava and lahar flows from Volcán Arenal dammed stream courses. Laguna La Palma, dammed by a lava flow erupted between 1535 and 1650 AD (Alvarado 2000), is fairly deep for its size (10.8 m), and develops a well-defined thermocline around 2.5 m deep. In the deep water, CO_2 and Si are noticeably enriched, and O_2 is scarce; these data indicate some degree of persistence of stratification. The diatom assemblage in surface sediments of La Palma is dominated overwhelmingly by *Thalassiosira pseudonana*.

Laguna Poco Sol is a beautiful lake in the Bosque Eterno de los Niños, a large private forest reserve in the Cordillera de Tilarán contiguous with the Monteverde Cloud Forest Reserve (Nadkarni and Wheelwright 2000a; Lawton et al., chapter 13 of this volume). Its formation is unclear. The circular shape of the lake and nearby sulfur deposits and hot springs suggest that it could occupy a volcanic crater (Alvarado 2000). An early claim, now known to be incorrect, that the lake was over 64 m deep, made volcanism the only likely mode of origin; Alvarado (1989) consequently classified it as an explosion crater, similar to the very deep Laguna Río Cuarto, described below. A report by students in an OTS course (Armoudlian and De Moraes 1994) that the lake was only 11.5 m deep, confirmed by us in 1997 (Haberyan et al. 2003), opened the possibility that the lake was formed in another way. A depth of 11 m is still significant for a lake of less than 3 ha, but given the abrupt topography of the area it would be possible to create such a lake by damming a stream valley (Alvarado 2000, Haberyan et al. 2003). The lake has a relatively flat bottom, but is slightly deeper adjacent to the slope that would have been the source of the landslide or slump. In 1998, we cored the lake to resolve the formation issue (and to obtain a pollen record of forest history), but our field results were inconclusive (see below). We favor the landslide hypothesis, but the question of lake formation remains open. The remote location of the lake, at the end of an unpaved road that turns to deep mud in the wet season, and includes a swinging suspension bridge, has deterred frequent limnological sampling, but interestingly the lake always seems to be stratified. The thermocline seems to vary from 3 to 8 m, and is certainly enhanced by the input of cool stream water. The concentration of CO_2 is an order of magnitude greater in deep water (31 mg/L), suggesting long-term stratification. *Acnanthidium catenata* accounts for the vast majority of diatoms in surface sediments.

Several sediment cores have been raised from lakes in the northern highland evergreen cloud forest. Cores from Lagunas Hule, María Aguilar (Horn and Haberyan 1993), and Barva in the Cordillera Central, and Laguna Cote in the Cordillera de Guanacaste (Horn and Haberyan 1993), contain layers of volcanic tephra that reveal the history of volcanic events that disturbed or destroyed forests of the region (Horn 2001, Arford 2001), and also may have affected aquatic communities (Haberyan 1998, Telford et al. 2004b). The Laguna Hule core spans the last ca. 2,500 years and is noteworthy for the high percentages of *Cecropia* pollen in samples. In most of Costa Rica, pollen of this secondary forest taxon is found together with other microfossil and sedimentary evidence of prehistoric human disturbance and agriculture. However, the high *Cecropia* percentages in the Laguna Hule sediments are not associated with other evidence of human disturbance; they instead signal secondary succession following natural disturbance from periodic slumping of the steep crater walls (Horn 2007).

In their review of the physical geography of the Monteverde Cloud Forest area, Clark et al. (2000) mentioned the desirability of obtaining pollen records from a transect of lakes in the region. Unfortunately, some of the lakes they cite as possible coring targets are inappropriate or problematical. Laguna Cerro Chato is strewn with boulders and contains very little fine sediment, while Laguna Cocoritos appears to be an artificial lake (Haberyan et al. 2003). We visited Laguna Arancibia and a smaller nearby lake ("Arancibita") in 1992, but in 2000, before we could return to core them, a major debris avalanche occurred at the site. Informal accounts shared with us at the time indicated that both lakes were buried in the debris. However, a later report by Alvarado et al. (2003) showed the debris avalanche to be about 300 m northwest of Laguna Arancibia. This lake thus remains a potential future target for paleolimnological study. From Laguna Poco Sol we recovered a 1.1 m core in 1998, but could go no further because of heavy rainfall and an abundance of downed trees in the lake, which snagged the coring cable. Pollen is well preserved in the sediments, with a high diversity of cloud forest types, but samples of wood from 63 cm depth and sediment from 97 cm depth both returned modern (post-1950) radiocarbon dates, showing that the profile we recovered represents very little time. We are exploring options to make a return coring visit with geophysical imaging equipment that would reveal the structure of the basin and sediments. This could help us pinpoint the best coring spot in the lake and establish the mode of formation of the lake. We strongly agree with Clark et al. (2000) on the desirability of obtaining paleoecological records from this region of Costa Rica, in which modern biota and ecological processes are increasingly well known (Nadkarni and Wheelwright 2000b; Lawton et al., chapter 13 of this book).

Lakes of the Southern Highland Evergreen Cloud Forest

Evergreen cloud forests of the southern highlands described by Kappelle (chapter 14 of this volume) occupy areas of Tertiary sedimentary and volcanic rocks intruded by granitic stocks in the Cordillera de Talamanca, and Tertiary volcanic rocks in the Cordillera Costera (Alvarado and Cárdenes, chapter 3 of this volume). Landslides and related earth movements are the major lake forming processes. A number of artificial lakes also exist in the region. We focus here on three sites near the Organization for Tropical Studies' Las Cruces Biological Station near the Panamanian border, and one site along the Inter-American Highway at the northwestern end of the Cordillera de Talamanca (Table 19.5).

Laguna Zoncho is a small natural lake located 3 km north of the Las Cruces station, on the grounds of Finca Cántaros, a private nature reserve open to the public. The lake is bordered by an ornamental garden, and by slopes reforested in 1994 following prior use for coffee production and cattle grazing. Laguna Zoncho formed approximately 3,200 years ago by large scale slumping or faulting, or both in combination. Despite an abundance of submerged vegetation, oxygen and carbon dioxide values in the lake are moderate (7.1 and 3 mg/L, respectively). Normally the lake's depth is around 2.6 m, but it was some 2 m deeper in June 2007 than in previous years. Although surfaces for epiphytic diatoms abound in the lake, modern diatom assemblages are dominated by the planktonic diatom *Aulacoseira italica*.

On the ridge above the Las Cruces station, about 1 km west of the upper boundary of the forest reserve, is **Laguna Gamboa**, also likely formed by earth movements, and bordered by an extensive herbaceous marsh. Cloud forest covers a hill that abuts the east side of the lake, and other uplands to the west, but a considerable portion of the watershed has been cleared for cattle grazing. Relative to Zoncho, Gamboa is higher in elevation, slightly shallower, clearer, richer in tannins, and more acidic, but most other ions are comparably abundant. In Gamboa, low oxygen and high carbon dioxide levels suggest significant allochthonous input of organic matter from the adjacent marsh. Despite its shallow depth, Gamboa was stratified (at 1.3 m) in July 1997. Bart (1999) documented a drop in water level of over 1 m during the dry season of 1999. At least five species of sedges occur in the marsh, along with several grasses, *Polygonum segetum*, and other herbaceous dicots (Bart 1999). Areas of the marsh with standing water were found to have *Utricularia* sp. (Lentibulariaceae) and *Elodea canadensis* (Hydrocharitaceae) (Bart 1999); these aquatic plants are probably also present in the open-water portion of the marsh that we designate as Laguna Gamboa. The recent diatom assemblage in Laguna Gamboa is composed of comparable proportions of *Eunotia minor*, *Eunotia intermedia*, *Pinnularia braunii*, and a species of *Aulacoseira*.

Laguna Santa Elena is a small landslide-dammed lake located about 15 km north of Las Cruces, in the foothills of the Cordillera de Talamanca. When we visited in 2000 the lake was only 0.25 ha in size, but local residents indicated that the lake was larger (perhaps 1 ha in size) as recently as the mid-twentieth century. In 2000 water entered the lake

Table 19.5. Selected Data from Lakes of the Southern Highland Evergreen Cloud Forest

Lake Name	Zoncho	Gamboa	Santa Elena	Tres de Junio
Lake Number	50	47	78	3
Parameter				
Latitude (°N)	8.8121	8.7893	8.9290	9.6647
Longitude (°W)	82.9607	82.9877	82.9257	83.8498
Elevation (m a.s.l.)	1190	1460	1055	2670
Area (ha)	0.75	0.25	0.13	0.05
Depth (m)	2.6	2.0	3.8	0.7
Secchi (m)	1.1	1.3	1.4	1.5
Temperature (°C)	24.6	21.8	28.7	15.0
Stratified?	no	yes	nd	no
Macrophyte cover	6	2	3	5
pH	7.37	6.91	nd	5.23
Conductivity (uS/cm)	16	17	nd	30
O_2 (mg/L)	7.1	2.8	nd	5.8
CO_2 (mg/L)	3	14	nd	12
Alkalinity (mg/L $CaCO_3$)	12	16	nd	3
Ca^{+2} (mg/L)	0.66	0.64	0.30	1.68
Mg^{+2} (mg/L)	0.20	0.29	0.45	0.33
K^{+} (mg/L)	0.00	0.00	0.00	5.20
Na^{+} (mg/L)	0.84	0.57	0.19	7.28
Si (mg/L)	0.69	0.86	2.44	0.36
Cl^{-} (mg/L)	0.10	0.40	nd	9.38
S total (mg/L)	0.00	0.00	0.00	0.94

NOTE. Our previous publications reported coordinates based on the Ocotepeque 1935 datum, as used on Costa Rican topographic maps (Orvis 2002), but here we provide coordinates determined using Google Earth, which uses a WGS84 datum. Some limnological data represent a composite of multiple visits, combined so as to represent dry and wet seasons equally. Macrophyte cover is estimated using a linear scale ranging from 0 to 9, reflecting the proportion of lake surface that is covered by or underlain by macrophytes (coverage 0 reflects 0 to 10%, coverage 1 reflects 10 to 19%, etc.). nd = no data are available. See Haberyan et al. (2003) for additional data.

from an adjacent marsh fed by a small stream, and exited by a modified channel at the opposite end. Santa Elena is moderately deep (3.8 m) for its small size—almost twice as deep as Laguna Gamboa, while roughly the same in area. Of the three lakes we highlight in the southern evergreen highland cloud forest, Santa Elena is notably richer in Si. Among the recent diatoms, *Pinnularia braunii*, *Frustulia saxonica*, and *Luticola mutica* are common.

Although the pond we call **Tres de Junio** (Fig. 19.6) on the northwestern crest of the Cordillera de Talamanca may have been created by the construction of the Inter-American Highway, we include it here because similar natural ponds within high-elevation bogs of the region have attracted the attention of biologists since Standley (1936–37) made what may be the first plant collections in the area (see also Weber 1959). At 2,670 m elevation, it lies roughly 500 m below the lower modern limit of páramo vegetation, but the pond's edges and hummocks within the open water support unique assemblages of vascular and non-vascular plants that overlap with those of the higher páramos (Kappelle and Horn, chapter 15 of this volume), and are sometimes denoted as "paramillo" vegetation (Weber 1959, Chaverri and Cleef 2005). The pond is less than 1 m deep, with abundant *Sphagnum* on hummocks and edges, and consequently acidic and tannin-stained. Conductivity is low, owing to frequent cloud cover and reduced evaporation. The modern diatom assemblage is unique in its abundance of *Actinella* (*A. brasiliensis* and *A. punctata*) and several species of *Eunotia*.

Sediment cores from Laguna Zoncho and Laguna Santa Elena document the clearing of the southern highland evergreen cloud forest by prehistoric inhabitants for maize agriculture, followed by forest recovery after the Conquest (Clement and Horn 2001, Lane et al. 2004, Anchukaitis and Horn 2005, Taylor et al. 2013). Maize pollen at the base of the ca. 3,200-year sediment record from Laguna Zoncho constitutes the oldest botanical evidence of maize in southern Pacific Costa Rica and adjacent Panama, predating by a millennium maize macrofossils from archaeological sites (Horn 2006). Stable carbon isotope ratios (Lane et al. 2004, Taylor et al. 2013), diatom assemblages (Haberyan and Horn 2005), and phosphorous fractions (Filippelli et al. 2010) in the Zoncho sediment core show close correspondence with pollen and charcoal evidence of human

Fig. 19.6 Tres de Junio pond, at the edge of the Inter-American highway in the southern highland evergreen cloud forest ecosystem.
Photo by Sally Horn.

impacts in the watershed. Diatoms in Zoncho also indicate variations in lake levels during the past few thousand years that appear to reflect climate shifts.

Bogs near Tres de Junio have been drilled by several researchers, including our research team. Sediments and pollen found in cores from the La Chonta and Parque Vicente Lachner bogs show that these sites were lakes during portions of the last glacial interval, and were at that time surrounded by treeless páramo vegetation, rather than montane cloud forest, owing to glacial depression of treeline elevations (Martin 1964, Hooghiemstra et al. 1992, Islebe and Hooghiemstra 1997). Preliminary analyses of diatoms in glacial-age sediments of La Chonta bog deposited between 16,600 and 34,300 cal yr BP revealed dominance (≥98%) by a species of *Aulacoseira*. Late-Holocene bog sediments deposited ca. 3,000 cal yr BP in contrast include an assemblage of *Frustulia saxonica*, *Tabellaria binalis*, *Brachysira brachysira*, and *Eunotia guyanense*.

Lakes of the Southern Highland Páramo Grassland

The southern highland páramo grassland ecosystem in Costa Rica (see Kappelle and Horn, chapter 15 of this volume) is home to a third lake district of great scenic beauty as well as scientific interest—the suite of glacial lakes that surround the nation's highest mountain peak, Cerro Chirripó (3819 m) in Chirripó National Park (Fig. 19.7). Some 30-odd lakes occupy basins scoured by glacial ice, dammed by glacial debris, or both—remnants of the repeated glaciation of the high peaks that ended its last cycle some 11,000 years ago (Orvis and Horn 2000, Horn et al. 2005). Here we describe the two largest lakes (Table 19.6) in the Chirripó lake district and the records of ecosystem history their sediments have yielded; for information on other lakes of the district, see Horn et al. (2005). We also describe an artificial pond located 28 km WNW of Cerro Chirripó in the more accessible Buenavista páramo along the Inter-American highway.

Lago Chirripó, the largest and deepest of the glacial lakes, is a classic tarn occupying one of three glaciated valleys that surround Cerro Chirripó, itself a glacial horn (Fig. 19.7). Lago Chirripó is the uppermost lake in a chain of three glacial lakes in the Valle de los Lagos, which drains to the Pacific via the Río Chirripó del Pacífico. The lake is 7.8 ha in area and has a maximum depth of 22 m. Temperature measurements by Gocke et al. (1981) to a depth of 14 m revealed complete mixing to this depth; weak stratification (1°C difference) may occur for brief periods but the lake is polymictic, as all glacial lakes in the district appear to be (Horn et al. 2005). Cool temperatures, coupled with low nutrient levels (which are, in turn, related to the poorly weathered rocks here), cause very low levels of productivity in the lake. Consequently, the lake is very clear. Secchi depths have not been measured, but likely often exceed 5 m,

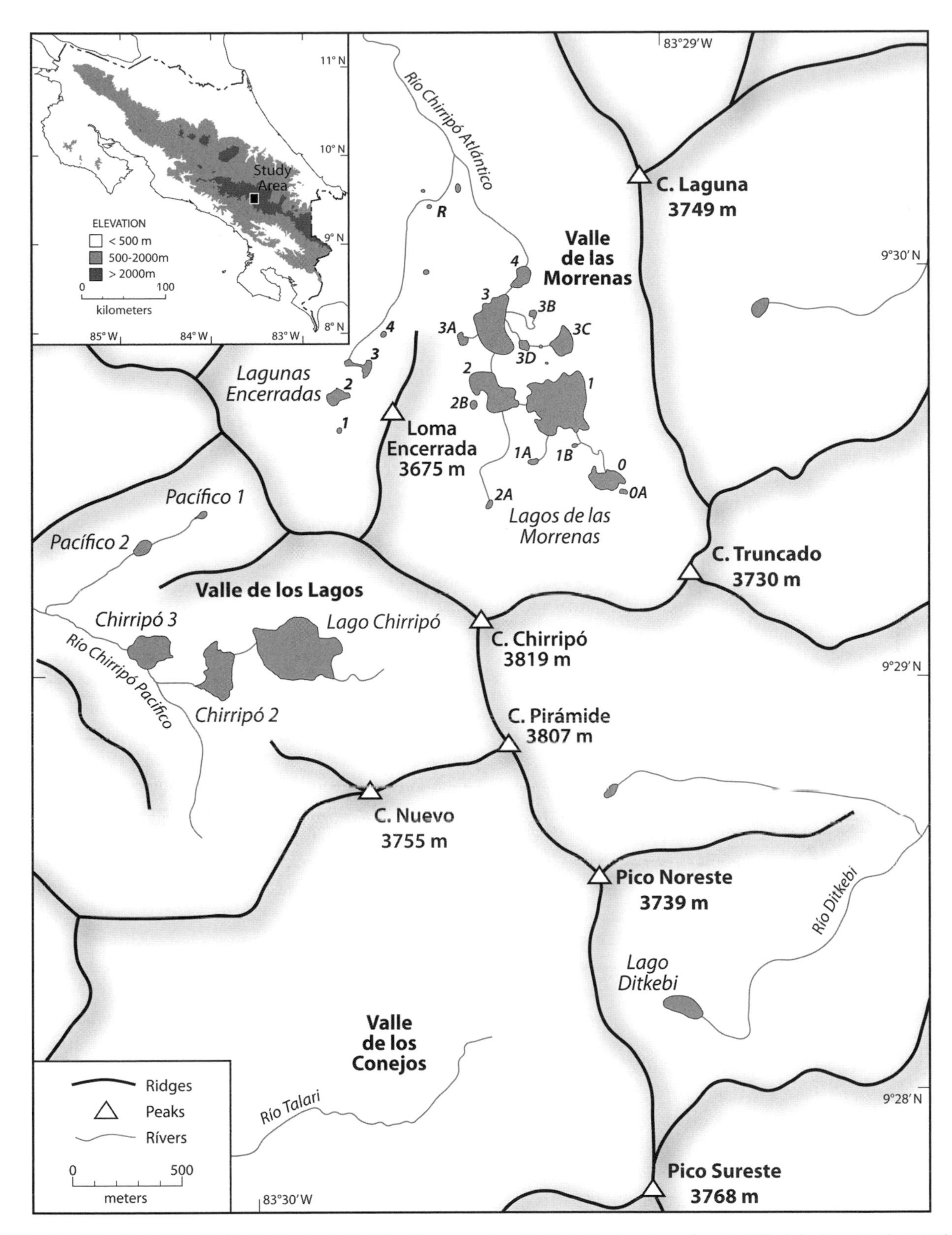

Fig. 19.7 Map of the Chirripó lake district. The Centro Ambientalista Páramo mentioned in the text is downstream from the Valle de los Conejos, about 1.3 km SW of the point where the Río Talari exits the main map.
Map produced by the Cartographic Services Laboratory of the University of Tennessee, Knoxville.

Table 19.6. Selected Data from Lakes of the Southern Highland Páramo Grassland

Lake Name	Chirripó	Morrenas 1	Asunción
Lake Number	5	6	4
Parameter			
Latitude (°N)	9.4825	9.4925	9.5731
Longitude (°W)	83.4958	83.4848	83.7572
Elevation (m a.s.l.)	3520	3480	3340
Area (ha)	7.76	5.20	0.05
Depth (m)	22.0	8.3	1.0
Secchi (m)	9.0	7.5	2.0
Temperature (°C)	12.8	14.6	12.3
Stratified?	no	no	no
Macrophyte cover	0	0	5
pH	5.40	7.29	4.93
Conductivity (uS/cm)	0	0	0
O_2 (mg/L)	5.2	6.6	6.9
CO_2 (mg/L)	3	3	3
Alkalinity (mg/L $CaCO_3$)	8	10	0
Ca^{+2} (mg/L)	1.23	1.86	0.76
Mg^{+2} (mg/L)	0.17	0.20	0.17
K^+ (mg/L)	0.00	0.57	1.13
Na^+ (mg/L)	0.81	1.37	3.00
Si (mg/L)	0.93	1.84	0.72
Cl^- (mg/L)	0.42	1.27	9.43
S total (mg/L)	0.24	0.59	1.57

NOTE. Our previous publications reported coordinates based on the Ocotepeque 1935 datum, as used on Costa Rican topographic maps (Orvis 2002), but here we provide coordinates determined using Google Earth, which uses a WGS84 datum. Some limnological data represent a composite of multiple visits, combined so as to represent dry and wet seasons equally. Macrophyte cover is estimated using a linear scale ranging from 0 to 9, reflecting the proportion of lake surface that is covered by or underlain by macrophytes (coverage 0 reflects 0 to 10%, coverage 1 reflects 10 to 19%, etc.). See Horn and Haberyan (1993), Horn et al. (1999, 2005), and Haberyan et al. (2003) for additional data.

based on the ease with which individual bamboo leaves on portions of the lake bottom can be discerned.

Quillworts (*Isoetes storkii*) are abundant along the edge of Lago Chirripó and extend to depths of several m. The phytoplankton is composed primarily of cyanobacteria (*Microcystis* sp.) along with chlorophytes (*Selenastrum* sp., *Oocystis* sp., *Kirchneriella* sp., *Botrycoccus braunii*, *Quadrigula* sp.), and the pyrrhophyte *Peridinium willei* (Gocke et al. 1981, Jones et al. 1993, Haberyan et al. 1995). Wujek et al. (1998) reported three species of silica-scaled chrysophytes (*Mallomonas acaroides*, *M. crassisquama*, and *M. cyathellata* var. *chilensis*). Gocke et al. (1981) noted fairly high numbers of a pink-colored copepod, but zooplankton is otherwise unstudied. These researchers also described a benthic fauna consisting of amphipods and larvae of chironomids and other insects. The pink color of the copepods mentioned by Gocke et al. (1981) is consistent with the absence of fish in Lago Chirripó and other glacial lakes of the area. When fish are present, copepods are typically clear, which makes them harder to see. In lakes without fish, copepods develop red or pink coloration, which provides some protection from ultraviolet light (Hairston 1979).

The bottom of Lago Chirripó is also home to an extensive population of *Aulacoseira*, which accounts for over 99% of the diatoms in modern surface sediment. *Aulacoseira* is usually planktonic, dependent on water turbulence to remain suspended in the water, but in Lago Chirripó, it survives quite successfully on the bottom, likely due to the clarity of the water and to nutrient regeneration from sediments. In addition, *Aulacoseira* usually requires elevated levels of silica, but we measured less than 3 mg/L Si in surface waters here. Perhaps Si is rapidly regenerated from dead diatoms on the bottom and forms a local Si-rich layer, or perhaps this population has become adapted to lower silica levels.

Lago Chirripó and other lakes in the Valle de los Lagos are separated from the adjacent Valle de las Morrenas by the continental divide; the latter valley drains via the Río Chirripó Atlántico to the Caribbean. The largest lake in this valley is **Lago de las Morrenas 1** (5.2 ha, Fig. 19.8), with a depth of 8.3 m. This lake, like L. Chirripó and all others in the district, is dilute and very clear; when we cored the lake in 1989 it was possible to see the core site at a depth of 7.5 m. Lago de las Morrenas 1 also supports a benthic population of *Aulacoseira*, but has a higher diversity of diatoms, largely owing to *Gomphonema gracile* and *Sellaphora pupula*. The phytoplankton is composed mainly of the chlorophytes *Eutetramorus tetrasporus* and *Quadrigula* sp. (Haberyan et al. 1995). Wujek et al. (1998) did not find silica-scaled chrysophytes in the plankton sample, but Zeeb et al. (1996) found 11 morphotypes of chrysophycean cysts in the surface sediment.

No fish exist at present in lakes of the Chirripó district, but trout were introduced by a hiker or a park employee to the Río Talari near the shelter and office complex known as the Centro Ambientalista, downstream from the bowl-shaped Valle de los Conejos (Fig. 19.7). In 2003, our colleague Ken Orvis was told that the fish had been eliminated, but he observed at least two at the site. We hope that no further fish introductions take place and that introduced fish can be removed before they spread naturally or are spread by humans to lakes of the district, as fish could greatly alter the native populations at all levels of the food web.

Aquatic communities in the Chirripó lake district may be vulnerable to other potential human actions, both local and global. During February–March 1998, we found lake water temperatures to be significantly warmer than measured by Löffler (1972) in August 1966, Gocke et al. (1981) in March 1979, and Jones et al. (1993) in August 1986. We initially attributed these elevated water temperatures to the effects of a strong El Niño (positive ENSO [El

Fig. 19.8 Lago de las Morrenas 1 (center) and other lakes of the Valle de las Morrenas in the southern highland *páramo* grassland ecosystem of Chirripó National Park. *Photo by Chad Lane.*

Niño–Southern Oscillation] conditions) (Horn et al. 1999). However, in 2000 we again measured high lake water temperatures under strongly *negative* ENSO conditions (Horn et al. 2005). Records of past ENSO events (Woodruff et al. 1998, Woodruff 2001) reveal that the earlier, much colder lake temperature observations were made under neutral to positive ENSO conditions, indicating that some factor other than El Niño must have been responsible for the warm temperatures in the year 2000 (Horn et al. 2005). In January 2003, under positive ENSO conditions, temperatures in glacial lakes were lower than our 1998 and 2000 values, but were still generally higher than temperatures measured two to four decades earlier (Löffler 1972, Gocke et al. 1981, Jones et al. 1993). In these clear, high-altitude lakes water temperatures are controlled partly by air temperature, but also by daytime cloudiness (which blocks insolation) and nighttime cloudiness (which blocks cooling by re-radiation). Our repeat temperature measurements may indicate that one or more of these aspects of the Chirripó climate are changing, perhaps triggered by global or more-local changes (Pounds et al. 1999, Lawton et al. 2001). Admittedly, we cannot with the data at hand rule out the influence of accidents of the timing of our temperature measurements (and those of previous researchers)—for example, local weather conditions during the months to days before sampling took place. What we do know from existing measurements is that temperatures in lakes of the Chirripó district *can* change dramatically over short time scales. A multi-year program of continuous monitoring of lake temperatures using data loggers would reveal whether past and future changes are indications of secular changes in climate.

No natural lakes exist in the southern highland páramo grasslands outside of Chirripó National Park, but two artificial ponds are located within or near the Buenavista páramo (Cerro de la Muerte) along the Inter-American highway: Quebrador Pond, a site quarried for road building materials (Horn and Haberyan 1993), and **Asunción Pond.** Located at the crest of the highway, on the south side of Cerro Asunción, this pond is artificial but at least 66 years old, based on a photograph taken by botanist Hugh Iltis in 1949 (Horn 1989b). Asunción Pond is similar in size and depth to Tres de Junio Pond (15 km distant), but is 670 m higher in elevation and more acidic (pH, 4.93 vs. 5.23). Its pH was lower than has been measured in any lake in Costa Rica other than the two crater lakes Laguna Cerro Chato and Laguna Botos, the latter of which is on Volcán Poás and is likely acidified by high concentrations of sulfur dioxide emitted from the adjacent active crater (Nicholson et al. 1996, Umaña 2001). Diatoms in Asunción Pond are distinct from those of both Tres de Junio and of other páramo lakes we have sampled, dominated by *Pinnularia braunii* and *Encyonema lunatum*. The presence of *Spaghnum* may contribute to the low pH of Asunción pond; nitrogen oxides in vehicle exhaust could also be a factor at this site on the very crest of the Inter-American highway.

Sediment cores from the glacial lakes of Cerro Chirripó provide information on the timing of deglaciation in the highland and on postglacial climate, vegetation, and fire regimes. Pollen assemblages in core samples from Lago de las Morrenas 1 show that the lake has been surrounded by treeless páramo since the last retreat of ice 10,000–12,000 cal yr BP (Horn 1993, Orvis and Horn 2000). Compound-specific carbon isotope analyses of terrestrially derived *n*-alkanes in the sediments indicate higher abundance of C_4 plants early in the Holocene, increased C_3 plant dominance during the middle Holocene, and then a return to greater C_4 plant abundance in the late Holocene (Lane et al. 2011). Cores from both Lago de las Morrenas 1 and Lago Chirripó contain abundant charcoal particles ranging in size from <50 μm to over 1 mm, indicating that fires set by people or lightning have repeatedly burned the páramo (Horn 1989a, 1993, League and Horn 2000, Horn and League 2005). Charcoal influx is highest in the last 4,800 cal yr BP, suggesting drier climate as well as increased human activity. Two distinct layers of macroscopic charcoal in the sediments of Lago Chirripó appear to indicate intervals of lower lake level about 1,100 and 2,500 cal yr BP that may be associated with drought intervals noted at those times elsewhere in Costa Rica (Horn and Sanford 1992; Horn et al., unpublished data) and the circum-Caribbean region (Hodell et al. 2000, 2001, Lane et al. 2014). The period from about 7,700–4,800 cal yr BP is in contrast marked by very low charcoal influx and appears to represent a wetter period in the Chirripó páramo and regionally (League and Horn 2000, Horn 2007). These inferred hydrological shifts in the Chirripó páramo during the Holocene are supported by recent compound-specific hydrogen isotope analyses (Lane and Horn 2013).

Diatom assemblages in the Morrenas 1 core show remarkable stability over time (Haberyan and Horn 1999). At each of 52 levels examined, at least ninety-five percent of all diatoms belong to a single species of *Aulacoseira* that appears to be a new species in the *A. alpigena / A. lirata* complex. This taxon is common in other lakes on the Chirripó massif. It may be related to the *Aulacoseira* species that occurs in glacial-age sediments of the La Chonta bog, and to those that occur in high altitude lakes in the Andes. On the basis of what we know of the ecology of this diatom, its dominance throughout the Lago de las Morrenas 1 record suggests that the lake has always been cold, polymictic, and clear. The great Secchi depth of the lake (7.5 m) may allow the *Aulacoseira* filaments' adequate illumination while they lie on the bottom. Two peaks in the diatom accumulation rate seem to correlate with peaks in charcoal, and may be related to fire effects in the watershed or to lower lake levels. However, neither fire nor local droughts appear to have affected the composition of the diatom flora, which may be insensitive to any limnological changes arising from such events.

Lakes of the Caribbean Lowland Evergreen Moist Forest

The evergreen moist forests of the Caribbean lowlands described by McClearn et al. (chapter 16 of this volume) occupy a dynamic area with many lake-forming processes. We review one lake formed by volcanism, two lakes formed by fluvial dynamics (one in combination with coastal dynamics), and two lakes whose histories appear to involve both fluvial dynamics and volcanism (Table 19.7).

Laguna de Río Cuarto is the deepest natural lake in Costa Rica, reaching 66 m. Its basin is an approximately circular, steep-sided volcanic maar (Gocke et al. 1987, Alvarado 2000). The lake is situated in a fracture zone that extends northward from the active crater of Volcán Poás through Volcán Congo and the Laguna Hule Caldera, also a maar (Horn 2001). The lake is warm (surface temperature >25°C) and is stratified year-round for several years at a time, mixing only sporadically (e.g., oligomictic). The near-perpetual stratification of the lake is due to its great depth and relatively small surface area, and the relative protection from wind provided by the crater walls, which extend 20 m above the lake along the west rim. The stable stratification causes surface waters to be low in nutrients, while sedimentation has increased most nutrient levels in deep water.

Like Laguna Hule, Laguna de Río Cuarto occasionally experiences turnover in the cooler, windier months (Gocke

Table 19.7. Selected Data from Lakes of the Caribbean Lowland Evergreen Moist Forest

Lake Name	Río Cuarto	Bonilla	Bonillita	Gandoca	Zent
Lake Number	23	25	26	55	54
Parameter					
Latitude (°N)	10.3560	9.9926	9.9921	9.5898	10.0293
Longitude (°W)	84.2167	83.6019	83.6114	82.5986	83.2769
Elevation (m a.s.l.)	380	380	450	0	17
Area (ha)	31.52	30.79	5.98	7.25	1.25
Depth (m)	66.0	27.0	20.0	6.4	2.0
Secchi (m)	3.9	1.7	2.7	1.1	>2.0
Temperature (°C)	27.0	27.4	23.9	29.4	29.7
Stratified?	yes	yes	yes	yes	no
Macrophyte cover	0	0	0	1	0
pH	7.88	6.98	6.98	7.19	7.47
Conductivity (uS/cm)	133	139	19	266	383
O_2 (mg/L)	5.7	6.7	5.8	4.8	1.0
CO_2 (mg/L)	5	4	5	10	18
Alkalinity (mg/L $CaCO_3$)	104	97	23	26	188
Ca^{+2} (mg/L)	13.76	8.60	2.36	7.40	56.24
Mg^{+2} (mg/L)	4.78	5.16	1.16	4.90	6.04
K^+ (mg/L)	2.84	2.84	1.13	2.70	1.34
Na^+ (mg/L)	6.40	4.90	2.10	29.40	7.28
Si (mg/L)	20.07	16.45	3.78	5.70	10.65
Cl^- (mg/L)	3.44	1.70	1.60	48.00	5.60
S total (mg/L)	0.33	0.11	0.19	nd	2.93

NOTE. Our previous publications reported coordinates based on the Ocotepeque 1935 datum, as used on Costa Rican topographic maps (Orvis 2002), but here we provide coordinates determined using Google Earth, which uses a WGS84 datum. Some limnological data represent a composite of multiple visits, combined so as to represent dry and wet seasons equally. Macrophyte cover is estimated using a linear scale ranging from 0 to 9, reflecting the proportion of lake surface that is covered by or underlain by macrophytes (coverage 0 reflects 0 to 10%, coverage 1 reflects 10 to 19%, etc.). nd = no data are available. See Horn and Haberyan (1993), Umaña et al. (1999), and Haberyan et al. (2003) for additional observations.

et al. 1987, Charpentier et al. 1988). Such a turnover, which may have been nearly complete, occurred prior to our visit in July 1997. At this time, the water was turbid (Secchi depth, 0.8 m, compared to 6.1 m in July 1991) and flecked with plate-like mineral crystals of unknown composition. Also in 1997, the thermocline was slightly weaker (1.7 vs. 2.8°C) and shallower (4.0 vs. 13.3 m), and most ions were more evenly distributed than in 1991. Finally, local residents reported a fish kill and rotten-egg smell in January of 1997. Although we did notice a build-up of deep water CO_2 in 1991 (29 mg/L), it did not reach the levels we recorded at that time in Laguna Hule (65 mg/L). Further comparisons are in Umaña et al. (1999) and Cabassi et al. (2014).

Gocke et al. (1990) studied the annual cycle of primary productivity in Laguna de Río Cuarto. Based on light/dark bottle studies, they reported that net primary productivity was 163 gC/m²/y, with maxima in March–April and September–October and minima in July and December–February. Because net annual photosynthesis was insufficient to account for total lake respiration, it appears that much respiration (some 20%) in the lake is of terrestrial carbon.

Fish are present in Laguna de Río Cuarto (Haberyan et al. 1995), but communities have not been described in detail. Men fishing on the shore of the lake in March 2005 reported the presence of guapote (*Parachromis dovii*) and machaca (possibly *Brycon guatemalensis*), some very large (scientific names are from www.fishbase.org). Ramírez et al. (1990) carried out repeated sampling of zooplankton between February 1984 and March 1985, and found rotifers to be the most abundant group. They reported the presence of *Keratella americana*, *Polyarthra vulgaris*, *Pompholix complanata*, *Hexarthra intermedia*, *Euchlanis dilatata*, and *Lecane* sp., along with the cladocerans *Diaphanosoma spinulosum* and *Bosmina longirostris* and the copepod *Microcyclops varicans*. In July 1991 we sampled zooplankton with a 160 µm net, too coarse to catch rotifers, and found the cladocerans *Ceriodaphnia cornuta*, *D. spinulosum*, and *Bosmina hagmanni* to dominate the zooplankton (Haberyan et al. 1995). The phytoplankton in July 1991 was dominated by the cyanobacteria *Merismopedia* along with the chlorophytes *Dactylococcopsis* and *Scenedesmus* (Haberyan et al. 1995); a separate analysis of silica-scaled

Fig. 19.9 Laguna Bonillita in the Caribbean lowland evergreen moist forest ecosystem.
Photo by Sally Horn.

chrysophytes in the sample (Wujek et al. 1998) documented the presence of *Mallomonas acaroides*, *M. crassisquama*, and *Paraphysomonas vestita*. Analyses of surface sediments collected from near the center of the lake in 1991 revealed a modern diatom assemblage dominated by *Fragilaria tenera* with *Nitzschia* cf. *vitrea* and *Navicula* cf. *laevissima* (Haberyan et al. 1997), and a diversity of chrysophycean cysts (Zeeb et al. 1996).

Lagunas Bonilla and Bonillita (Fig. 19.9) are located on a terrace of the Río Reventazón, where it cuts across the lower southeast slope of Volcán Turrialba. Their formation may have involved landslides and lahars (volcanic mudflows) as well as stream dynamics. On the basis of radiocarbon dating of plant fragments at the base of a lake-sediment core, Laguna Bonillita, the smaller and higher of the two lakes, formed about 2,500 cal yr BP; the formation of Laguna Bonilla has not been reliably dated (Northrop and Horn 1996). Both lakes are relatively deep (Bonilla, 27 m, and Bonillita, 20 m), and are stratified, at least seasonally (Horn and Haberyan 1993). Neither of the lakes has permanent surface inlets, and only Bonilla has a stream outlet. Of the two, Laguna Bonilla has been studied in more detail. Umaña (1997a) determined that the lake is oligomictic, mixing sometime between November and February, during the coolest and windiest time of the year. The lake is mildly eutrophic, with abundant cyanobacteria, low Secchi disk transparency (1.5–1.9 m), and almost permanent anoxia in the deepest portion. Umaña (1997a) measured N to P ratios and on the basis of observed patterns suggested that the lake shifted between phosphorous limitation in the wet season and nitrogen limitation in the dry season. Further study (Petersen and Umaña 2003) revealed that nitrogen limitation becomes more important months after the onset of stratification in the late dry season.

The phytoplankton of Laguna Bonilla is co-dominated by chlorophytes and cyanobacteria, while that of Bonillita is dominated by pyrrophytes and chlorophytes (Haberyan et al. 1995, Umaña 1997a). Silica-scaled chrysophytes in the Synurophyceae and Chrysophyceae have also been documented in Bonillita (Wujek et al. 1998). The zooplankton community in both lakes includes the cladocerans *Bosmina hagmanni* and *Ceriodaphnia cornuta*, but *B. hagmanni* dominates in Bonilla and *C. cornuta* in Bonillita, and copepods were found only in Bonilla (Haberyan et al. 1995, Umaña 1997a). Modern diatom assemblages in surface sediments are also distinct in the two lakes (*Fragilaria tenera* and *F. exigua* in Bonilla, versus *Encyonema silesiacum* and *Rhapalodia gibba* in Bonillita), despite their proximity and similar watershed conditions (Haberyan et al. 1997). Fish are present in both lakes, but have not been studied scientifically.

Laguna Gandoca is a sinuous lake located at sea level along the Caribbean coast near the border with Panama, in the Gandoca-Manzanillo National Wildlife Refuge. The lake occupies a former channel of the Río Sixaola (Denyer 1998), now occupied by the much smaller Río Gandoca. Laguna Gandoca is separated from the Caribbean by a narrow sandbar that impounds the lake, but allows seawater to percolate through. River water, being less dense, forms a stable layer on top. Consequently, stratification is very strong and deep waters are quite distinct from surface waters, being especially saline, oxygen-free, and sulfide-rich (Haberyan et al. 2003; additional data in Umaña et al. 1999). Surprisingly, the several grab samples we took of surface sediment were nearly devoid of fine organic matter. We found no diatoms in the sediments, likely due to the corrosive nature of sea water.

At its seaward end, Laguna Gandoca is surrounded by Caribbean mangrove forest, but inland of the mangrove area where we sampled the lake, it is surrounded by agricultural fields and remnant areas of Caribbean lowland evergreen moist forest. The intensive banana cultivation near the lake involves significant ground and aerial spraying of pesticides, a situation that prompted Coll et al. (2004) to analyze water samples from the lake for the presence of these chemicals. They tested for the presence of 20 organochlorated and organophosphorated pesticides in six samples from each of three sampling sites in the lake and found none whatsoever. They interpreted this result in a positive light as potentially indicating that these chemicals were not reaching the lake; however, they cautioned that sampling at other times is necessary to confirm this and that the high rainfall and flooding during their sampling period could have flushed contaminants from the lake.

Although preliminary analyses of Laguna Gandoca detected no pesticides, our observations at another lake in the region suggest that pesticides are influencing lake biota. **Laguna Zent** is an oxbow of the Río Chirripó Atlántico, positioned just downstream from the river's emergence onto the coastal plain; like Laguna Gandoca, it is in an area of banana cultivation—actually within a plantation. Trees along the banks provide shade, but at least at our sampling site there were few aquatic macrophytes, and even considering the shade, the lake was surprisingly clear (Secchi depth, greater than the lake depth of 2.0 m). Leaf litter in the lake displayed a fine white coating, and leaves were hardly decomposed. We attribute this to the effects of pesticide runoff on aquatic decomposers, though we did not test our water sample for pesticide residues. Lake waters were very low in oxygen but enriched in carbon dioxide and sulfur. Recent diatoms in the lake are also unusual, being mainly represented by species of *Navicula* (*N. radiosa* and *N. cryptocephala*) and the greatest abundance (13%) of *Gyrosigma spenceri* we have found in Costa Rica.

Sediment cores from Lagunas Bonilla and Bonillita and from swamps at the La Selva Biological Station provide evidence of the late Holocene history of the Caribbean lowland evergreen moist forest (Northrop and Horn 1996, Horn and Kennedy 2001, Kennedy and Horn 2008). As in the Pacific lowlands, pollen and charcoal in cores reveal forest clearance, maize cultivation, and agricultural burning. Stable carbon isotope signatures provide additional evidence of maize agriculture, and a means for gauging the extent of prehistoric cultivation within the watersheds (Lane et al. 2004, 2009). The continuous presence of *Pentaclethra* pollen in the 3,200-year record from the Cantarrana swamp at La Selva Biological Station puts a constraint on the extent of possible changes in late Holocene moisture in the region. Past drought intervals evident in paleolimnological and other records from the circum-Caribbean could not have been so intense or so long-lasting in the Caribbean lowland evergreen moist forest as to have eliminated mature *Pentaclethra* trees from the Cantarrana watershed.

Conclusions

Modern lakes in Costa Rica range in age from a few decades or less, for most formed by human activity, to over 10,000 years, in the case of the glacial lakes atop Cerro Chirripó. Lakes are ephemeral features over geological time scales, and lakes that predate Costa Rica's modern glacial lakes have been changed or buried by time and by geomorphic processes. A few old lakes have been filled in to become bogs; the existence of others is marked by the presence of ancient lake sediments in areas now too dry to support lakes or wetlands of any kind. We can expect that nearly every lake in Costa Rica has been affected, to a greater or lesser degree, by changes in its watershed, whether due to climate change or to humans. From the study of lake sediments we are developing a nationwide chronology of such changes, and from this it is clear that prehistoric humans had major influences, beginning by at least 5,500 years ago in northwestern Costa Rica, where lake sediments document the first cultivation of maize and associated forest clearance and agricultural burning, and extending to the Panamanian border and across the Central Highlands to the Caribbean lowlands.

It is equally clear that human impacts will continue, and will become increasingly widespread and of great potential threat to aquatic environments and biota, whether due to sewage, pesticides, eutrophication, aquaculture, erosion, the introduction of alien species, or resource depletion (e.g., see TNC 2009). Such localized threats can, we hope, be managed in part through community-based efforts to protect resources, and through improved land use policies and enforcement. In contrast, the larger, more subtle threats of global warming will be more difficult to mitigate; even now, the high-altitude lakes on Cerro Chirripó may be already responding.

Both sets of threats require much more detailed, and more frequent, scientific observations, because sound responses must be based on ecological knowledge. The challenge for Costa Rica's freshwater resources is, as in many nations, the balancing of water quality and aquatic biodiversity in the long term with human needs in the short term.

Acknowledgments

Our research on Costa Rican lakes and their sediments has been supported by grants from the National Geographic Society, the National Science Foundation (SES-9111588, BCS-0242286, DGE-0538420), The A.W. Mellon Foundation, the Association of American Geographers, the University of Tennessee, Northwest Missouri State University, and Troy State University. We thank the many people who co-authored or otherwise contributed to the work we reviewed here, most especially Gerardo Umaña, Maureen Sánchez, and Ken Orvis, without whose collaboration our work would have been vastly more difficult, and much less fun. We also thank the government of Costa Rica and many private landowners for granting permission for our studies.

References

Alvarado, G.E. 1989. *Los Volcanes de Costa Rica: Geología, Historia y Riqueza Natural*. 1st ed. San José, Costa Rica: Editorial Universidad Estatal y Distancia.

Alvarado, G.E. 2000. *Los Volcanes de Costa Rica: Geología, Historia y Riqueza Natural*. 2nd ed. San José, Costa Rica: Editorial Universidad Estatal y Distancia.

Alvarado, G.E., R. Mora, and G. Peraldo. 2003. The June 2000 Arancibia debris avalanche and block slide. *Landslide News* 14/15: 29–32.

Alvarado, G.E., G.J. Soto, F.M. Salani, P. Ruiz, and L. Hurtado de Mendoza. 2011. The formation and evolution of Hule and Río Cuarto maars, Costa Rica. *Journal of Volcanology and Geothermal Research* 201: 342–56.

Alvarado, G.E., E. Vega, J. Chaves, and M. Vásquez. 2004. Los grandes deslizamientos (volcánicos y novolcánicos) de tipo *debris avalanche* en Costa Rica. *Revista Geológica de América Central* 30: 83–99.

Anchukaitis, K.J., M.N. Evans, N.T. Wheelwright, and D.P. Schrag. 2008. Stable isotope chronology and climate signal calibration in neotropical montane cloud forest trees. *Journal of Geophysical Research* 113: G03030. DOI: 10.1029/2007JG000613.

Anchukaitis, K.J., and S.P. Horn. 2005. A 2000-year reconstruction of forest disturbance from southern Pacific Costa Rica. *Palaeogeography Palaeoclimatology Palaeoecology* 221(1–2): 35–54.

Anderson, E.P., C.M. Pringle, and M. Roja. 2006. Transforming tropical rivers: an environmental perspective on hydropower development in Costa Rica. *Aquatic Conservation: Marine and Freshwater Ecosystems* 16: 679–73.

Arford, M.R. 2001. Late Holocene environmental history and tephrostratigraphy in northwestern Costa Rica: a 4000 year record from Lago Cote. M.S. thesis, University of Tennessee. Knoxville, TN.

Arford, M.R., and S.P. Horn. 2004. Pollen evidence of the earliest maize agriculture in Costa Rica. *Journal of Latin American Geography* 3(1): 108–15.

Armoudlian, A., and C. De Moraes. 1994. Preliminary limnological study of Lake Pocosol. In B. Young and S. Sargent, eds., *Coursebook for Tropical Biology: An Ecological Approach*, 99–103. San José, Costa Rica: Organization for Tropical Studies.

Baker, R.G.E. 1987. Notes on the high altitude freshwater lakes of Volcán Rincón de la Vieja National Park, northern Costa Rica. *Brenesia* 27: 47–54.

Bart, D. 1999. Report on the reference conditions for the restoration of the marshes near the Las Cruces Biological Field Station, Costa Rica. Unpublished report submitted to the Las Cruces Biological Station, Organization for Tropical Studies, Costa Rica.

Bergoeing, J.P. 1978. *La Fotografía Aérea y su Aplicación a la Geomorfología de Costa Rica*. San José, Costa Rica: Instituto Geográfico Nacional.

Bergoeing, J.P., and L.G. Brenes. 1978. Laguna de Hule, una caldera volcánica. *Informe Semestral*, Instituto Geográfico Nacional de Costa Rica (Jul.–Dec.): 59–63.

Bolaños, B.J. 1979. Estudio preliminar sobre el cultivo de híbridos de Tilapia (*Tilapia hornorum x T. mossambica*) con gallinaza y superfosfato triple, en Costa Rica. *Revista Lationamericana de Acuicultura* 2: 22–28.

Boza, M.A., and J.H. Cevo. 1998. *Parques Nacionales y Otras Áreas Protegidas de Costa Rica / Costa Rican National Parks and Other Protected Areas*. San José, Costa Rica: INCAFO Costa Rica.

Bravo, J., and L. Ocampo. 1993. Humedales de Costa Rica. Heredia, Costa Rica: Universidad Nacional de Costa Rica.

Bumby, M.J. 1982. A survey of aquatic macrophytes and chemical qualities of nineteen localities in Costa Rica. *Brenesia* 19/20: 487–535.

Cabassi, J., F. Tassi, F. Mapelli, S. Borin, S. Calabrese, D. Rouwet, G. Chiodini, R. Marasco, B. Chouaia, R. Avino, O. Vaselli, G. Pecoraino, F. Capecchiacci, G. Bicocchi, S. Caliro, C. Ramírez, and R. Mora-Amador. 2014. Geosphere-biosphere interactions in bio-activity volcanic lakes: evidences from Hule and Río Cuarto (Costa Rica). *PLOS ONE* 9(7): e102456.

Camacho, L. 1985. Variación estacional de las algas planctónicas del Lago de Río Cuarto, Alajuela, Costa Rica. Lic. thesis, University of Costa Rica. San José, Costa Rica.

Charpentier, C., F.A. Tabash, I.A. Fallas, J.C. Zumbado, L. Camacho, and E. Ramírez. 1988. Variación estacional en el lago de Río Cuarto, provincia de Alajuela, Costa Rica. I. Limnología físico-química. *Uniciencia* 5(1/2): 77–85.

Chaverri, A., and A.M. Cleef. 2005. Comunidades vegetales de los páramos de los macizos de Chirripó y Buenavista, Costa Rica. In M. Kappelle and S.P. Horn, eds., *Páramos de Costa Rica*, 577–92. Santo Domingo de Heredia, Costa Rica: Instituto Nacional de Biodiversidad.

Chávez, L., and K.A. Haberyan. 1996. Diatom assemblages from the Camastro diatomite, Costa Rica. *Revista de Biología Tropical* 44(2): 899–902.

Clark, K.L., R.O. Lawton, and P.R. Butler. 2000. The physical environment. In N.M. Nadkarni and N.T. Wheelwright, eds., *Monteverde: Ecology and Conservation of a Tropical Cloud Forest*, 15–38. New York: Oxford University Press.

Clement, R.M., and S.P. Horn. 2001. Pre-Columbian land use history in Costa Rica: a 3000-year record of forest clearance, agriculture, and fires from Laguna Zoncho. *Holocene* 11: 419–26.

Cochard, R., and B.R. Jackes. 2005. Seed ecology of the invasive tropical tree *Parkinsonia aculeata*. *Plant Ecology* 180: 13–31.

Cohen, A.S. 2003. *Paleolimnology: The History and Evolution of Lake Systems*. New York: Oxford University Press.

Coll, M., J. Cortés, and D. Sauma. 2004. Características físicos-químicas y determinación de plaguicidas en el agua de la laguna de Gandoca, Limón, Costa Rica. *Revista de Biología Tropical* 52(suppl. 2): 33–42.

Collado, C.M. 1983. Costa Rican freshwater zooplankton and zooplankton distribution in Central America and the Caribbean. M.S. thesis, University of Waterloo. Ontario, Canada.

Collado, C.M., D. Defaye, B.H. Dussart, and C.H. Fernando. 1984a. The freshwater copepoda (Crustacea) of Costa Rica with notes on some species. *Hydrobiologia* 119: 89–99.

Collado, C.M., C.H. Fernando, and D. Sephton. 1984b. The freshwater zooplankton of Central America and the Caribbean. *Hydrobiologia* 113: 105–19.

Denyer, P. 1998. Historic-prehistoric earthquakes, seismic hazards, and Tertiary and Quaternary geology of the Gandoca-Manzanilla National Wildlife Refuge, Limón, Costa Rica. *Revista de Biología Tropical* 46(suppl. 6): 237–50.

Dickman, M., and H. Nanne. 1987. Impacts of tilapia grazing on plankton composition in artificial ponds in Guanacaste province, Costa Rica. *Journal of Freshwater Ecology* 4(1): 93.

Driese, S.G., K.H. Orvis, S.P. Horn, Z.H. Li, and D.S. Jennings. 2007. Paleosol evidence for Quaternary uplift and for climate and ecosystem changes in the Cordillera de Talamanca, Costa Rica. *Palaeogeography, Palaeoclimatology, Palaeoecology* 248(1–2): 1–23.

Dussart, B.H., and C.H. Fernando. 1986. Remarks on two species of copepods in Costa Rica, including a description of a new species of *Tropocyclops*. *Crustaceana* 50(1): 39–44.

Filippelli, G., C. Souch, S.P. Horn, and D. Newkirk. 2010. The pre-Columbian footprint on terrestrial nutrient cycling in Costa Rica: insights from phosphorous in a lake sediment record. *Journal of Paleolimnology* 43: 843–56.

Fuentes, G., A.B. Azofeifa, and S. Aguilar. 2006. Bibliografía sobre la producción científica del Centro de Investigación en Ciencias del Mar y Limnología—CIMAR de la Universidad de Costa Rica. Costa Rica: Organization for Tropical Studies, Ciudad de la Investigación.

Gocke, K. 1996–97. Basic morphometric and limnological properties of Laguna Hule, a caldera lake in Costa Rica. *Revista de Biología Tropical* 44(3)/45(1): 537–48.

Gocke, K., W. Bussing, and J. Cortés. 1987. Morphometric and basic limnological properties of the Laguna de Río Cuarto, Costa Rica. *Revista de Biología Tropical* 35(2): 277–85.

Gocke, K., W. Bussing, and J. Cortés. 1990. The annual cycle of primary productivity in Laguna de Río Cuarto, Costa Rica. *Revista de Biología Tropical* 38(2B): 387–94.

Gocke, K., E. Lahman, G. Rojas, and J. Romero. 1981. Morphometric and basic limnological data of Laguna Grande de Chirripó, Costa Rica. *Revista de Biología Tropical* 29(1): 165–74.

Haberyan, K.A. 1998. The effect of volcanic ash influx on the diatom community of Lake Tanganyika, East Africa. *Transactions of the Missouri Academy of Science* 32: 102–5.

Haberyan, K.A., and S.P. Horn. 1999. A 10,000-year diatom record from a glacial lake in Costa Rica. *Mountain Research and Development* 19(1): 63–70.

Haberyan, K.A., and S.P. Horn. 2005. Diatom paleoecology of Laguna Zoncho, Costa Rica. *Journal of Paleolimnology* 33(3): 361–69.

Haberyan, K.A., S.P. Horn, and M.R. Arford. 2005. Diatom shifts at Laguna San Pablo, Costa Rica, over the last 8000 years. Poster presented at the 18th North American Diatom Symposium, Mobile, Alabama.

Haberyan, K.A., S.P. Horn, and B.F. Cumming. 1997. Diatom assemblages from Costa Rican lakes: an initial ecological assessment. *Journal of Paleolimnology* 17: 263–74.

Haberyan, K.A., S.P. Horn, and G. Umaña. 2003. Basic limnology of fifty-one lakes in Costa Rica. *Revista de Biología Tropical* 51(1): 107–22.

Haberyan, K.A., G. Umaña, C. Collado, and S.P. Horn. 1995. Observations on the plankton of some Costa Rican lakes. *Hydrobiologia* 312: 75–85.

Hairston, N.G., Jr. 1979. The adaptive significance of color polymorphism in two species of *Diaptomus* (Copepoda). *Limnology and Oceanography* 24: 15–37.

Hargraves, P.E., and R. Víquez. 1981. Dinoflagellate abundance in the Laguna Botos, Poás Volcano, Costa Rica. *Revista de Biología Tropical* 29(2): 257–64.

Herrera, W. 1992. *Mapa-Guía de la Naturaleza Costa Rica / Costa Rican Nature Atlas-Guidebook*. San José, Costa Rica: INCAFO Costa Rica.

Hodell, D.A., M. Brenner, and J.H. Curtis. 2000. Climate change in the northern American tropics and subtropics since the last ice age. In D.L. Lentz, ed., *Landscape Transformations in the Pre-Columbian Americas*, 13–38. New York: Columbia University Press.

Hodell, D.A., M. Brenner, J.H. Curtis, and T.P. Guilderson. 2001. Solar forcing of drought frequency in the Maya lowlands. *Science* 292: 1367–70.

Hooghiemstra, H., A.M. Cleef, G.W. Noldus, and M. Kappelle. 1992. Upper Quaternary vegetation dynamics and palaeoclimatology of the La Chonta bog area (Cordillera de Talamanca, Costa Rica). *Journal of Quaternary Science* 7(3): 205–25.

Horn, S.P. 1989a. Prehistoric fires in the Chirripó highlands of Costa Rica: sedimentary charcoal evidence. *Revista de Biología Tropical* 37(2): 139–48.

Horn, S.P. 1989b. The Inter-American Highway and human disturbance of páramo vegetation in Costa Rica. *Yearbook of the Conference of Latin Americanist Geographers* 15: 13–22.

Horn, S.P. 1993. Postglacial vegetation and fire history in the Chirripó páramo of Costa Rica. *Quaternary Research* 20: 107–16.

Horn, S.P. 2001. The age of the Hule explosion crater, Costa Rica, and the timing of subsequent tephra eruptions: evidence from lake sediments. *Revista Geológica de América Central* 24: 57–66.

Horn, S.P. 2006. Pre-Columbian maize agriculture in Costa Rica: pollen and other evidence from lake and swamp sediments. In J. Staller, R. Tykot, and B. Benz, eds., *Histories of Maize: Multidisciplinary Approaches to the Prehistory, Biogeography, Domestication, and Evolution of Maize*, 367–80. San Diego: Elsevier Press.

Horn, S.P. 2007. Late Quaternary lake and swamp sediments: recorders of climate and environment. In J. Bundschuh and G.E. Alvarado, eds., *Central America: Geology, Resources, Hazards*, vol. 1. Leiden, Netherlands: Taylor & Francis/Balkema.

Horn, S.P., and K.A. Haberyan. 1993. Physical and chemical properties of Costa Rican Lakes. *National Geographic Research and Exploration* 9(1): 86–103.

Horn, S.P., and L.M. Kennedy. 2001. Pollen evidence of maize cultivation 2700 B.P. at La Selva Biological Station, Costa Rica. *Biotropica* 33(1): 191–96.

Horn, S.P., and L.M. Kennedy. 2006. Pollen evidence of the prehistoric presence of cattail (Typhaceeae: *Typha*) in Palo Verde National Park, Costa Rica. *Brenesia* 66: 85–87.

Horn, S.P., and B.L. League. 2005. Registros de sedimentos lacustres de la vegetación del Holoceno e historia del fuego en el páramo de Costa Rica. In M. Kappelle and S.P. Horn, eds., *Páramos de Costa Rica*, 253–73. Santo Domingo de Heredia, Costa Rica: Instituto Nacional de Biodiversidad.

Horn, S.P., K.H. Orvis, and K.A. Haberyan. 1999. Investigación limnológica y geomorfológica de lagos glaciares del Parque Nacional Chirripó, Costa Rica. *Revista Informe Semestral* (Instituto Geográfico Nacional de Costa Rica) 35: 95–106.

Horn, S.P., K.H. Orvis, and K.A. Haberyan. 2005. Limnología de las lagunas glaciales en el páramo del Chirripó, Costa Rica. In M. Kappelle and S.P. Horn, eds., *Páramos de Costa Rica*, 161–81. Santo Domingo de Heredia, Costa Rica: Instituto Nacional de Biodiversidad.

Horn, S.P., and R.L. Sanford. 1992. Holocene fires in Costa Rica. *Biotropica* 24: 354–61.

Horne, A.J., and C.R. Goldman. 1994. *Limnology*. 2nd ed. New York: McGraw Hill.

Islebe, G.A., and H. Hooghiemstra. 1997. Vegetation and climate history of montane Costa Rica since the last glacial. *Quaternary Science Reviews* 16(6): 589–604.

Jiménez, C.E., and M. Springer. 1994. Vertical distribution of benthic macrofauna in a Costa Rican crater lake. *Revista de Biología Tropical* 42(1/2): 175–79.

Jiménez, C.E., and M. Springer. 1996. Depth related distribution of benthic macrofauna in a Costa Rican crater lake. *Revista de Biología Tropical* 44(2): 673–78.

Jones, J.R., K. Lohman, and G. Umaña. 1993. Water chemistry and trophic state of eight lakes in Costa Rica. *Verhandlungen Internationale Vereinigung für Theoretische und Angewandte Limnologie* 25(2): 899–905.

Kennedy, L.M., and S.P. Horn. 2008. A late Holocene pollen and charcoal record from La Selva Biological Station, Costa Rica. *Biotropica* 40(1): 11–19.

Kling, G.W., M.A. Clark, H.R. Compton, J.D. Devine, W.C. Evans, A.M. Humphrey, E.J. Koenigsberg, J.P. Lockwood, M.L. Tuttle, and G.N. Wagner. 1987. The 1986 Lake Nyos gas disasters in Cameroon, West Africa. *Science* 236: 169–75.

Kohkemper, J. 1954. Bosquejo limnológico de la laguna "Del Misterio." Ing. thesis, University of Costa Rica. San José, Costa Rica.

Lachniet, M.S., Y. Asmerom, S.J. Burns, W.P. Patterson, V.J. Polyak, and J.O. Seltzer. 2004. Tropical response to the 8200 yr B.P. cold event? Speleothem isotopes indicate a weakened early Holocene monsoon in Costa Rica. *Geology* 32(11): 957–60.

Lachniet, M.S., and G.O. Seltzer. 2002. Late Quaternary glaciation of Costa Rica. *Geological Society of America Bulletin* 114(5): 547–58.

Lane, C.S., and S.P. Horn. 2013. Terrestrially derived n-alkane δD evidence of shifting Holocene paleohydrology in highland Costa Rica. *Arctic, Antarctic, and Alpine Research* 45(3): 342–49.

Lane, C.S., S.P. Horn, and M.T. Kerr. 2014. Beyond the Mayan lowlands: impacts of the Terminal Classic Drought in the Caribbean Antilles. *Quaternary Science Reviews* 86: 89–98.

Lane, C.S., S.P. Horn, and C.I. Mora. 2004. Stable carbon isotope ratios in lake and swamp sediments as a proxy for prehistoric forest clearance and crop cultivation in the neotropics. *Journal of Paleolimnology* 32(4): 375–81.

Lane, C.S., S.P. Horn, C.I. Mora, K.H. Orvis, and D.B. Finkelstein. 2011. Sedimentary stable carbon isotope evidence of late Quaternary vegetation and climate change in highland Costa Rica. *Journal of Paleolimnology* 45(3): 323–38.

Lane, C.S., S.P. Horn, Z.P. Taylor, and C.I. Mora. 2009. Assessing the scale of prehistoric human impact in the neotropics using stable carbon isotope analyses of lake sediments: a test case from Costa Rica. *Latin American Antiquity* 20(1): 120–33.

Lawton, R.O., U.S. Nair, R.A. Pielke Sr., and R.M. Welch. 2001. Climatic impact of tropical lowland deforestation on nearby montane cloud forests. *Science* 294: 584–87.

League, B.L., and S.P. Horn. 2000. A 10,000 year record of páramo fires in Costa Rica. *Journal of Tropical Ecology* 16: 747–52.

Löffler, H. 1964. The limnology of tropical high-mountain lakes. *Verhandlungen Internationale Vereinigung für Theoretische und Angewandte Limnologie* 15: 176–93.

Löffler, H. 1972. Contribution to the limnology of high mountain lakes in Central America. *Internationale Revue der Gesamten Hydrobiologie* 57(3): 397–408.

Martin, P.S. 1964. Paleoclimatology and a tropical pollen profile. In *Report on the VI International Congress on the Quaternary* (Warsaw 1961) 2: 319–23.

Messerli, B. 2001. Editorial: The International Year of Mountains (IYM), the Mountain Research Initiative (MRI) and PAGES. *PAGES (Past Global Changes) News* 9(3): 2.

Nadkarni, N.M., and N.T. Wheelwright. 2000a. Introduction. In N.M.

Nadkarni and N.T. Wheelwright, eds., *Monteverde: Ecology and Conservation of a Tropical Cloud Forest*, 3–13. New York: Oxford University Press.

Nadkarni, N.M., and N.T. Wheelwright, eds. 2000b. *Monteverde: Ecology and Conservation of a Tropical Cloud Forest*. New York: Oxford University Press.

Nicholson, R.A., P.D. Roberts, and P.J. Baxter. 1996. Preliminary studies of acid and gas contamination at Poás volcano, Costa Rica. In A.J. Appleton, R. Fuge, and G.J.H. McCall, eds., *Environmental Geochemistry and Health*, 239–40. Geological Society of America Special Publication 113.

Northrop, L.A., and S.P. Horn. 1996. PreColumbian agriculture and forest disturbance in Costa Rica: palaeoecological evidence from two lowland rainforest lakes. *Holocene* 6(3): 289–99.

Orvis, K.H. 2002. GPS Locations and Costa Rican Topo Maps. Unpublished report. Knoxville, Tennessee: University of Tennessee. 23 pp. http://trace.tennessee.edu/utk_geogpubs/5/

Orvis, K.H., and S.P. Horn. 2000. Quaternary glaciers and climate on Cerro Chirripó, Costa Rica. *Quaternary Research* 54: 24–37.

Petersen, R.R., and G. Umaña. 2003. Nitrogen and phosphorous limitation in lakes Bonilla and Bonillita, Costa Rica. *Verhandlungen Internationale Vereinigung für Theoretische und Angewandte Limnologie* 28(3): 1520–25.

Pounds, J.A., M.P.L. Fogden, and J.H. Campbell. 1999. Biological response to climate change on a tropical mountain. *Nature* 398: 611–15.

Ramírez, E. 1985. Variaciones estacionales de la comunidad zooplanctónica del Lago de Río Cuarto, Alajuela, Costa Rica. Lic., informe de trabajo dirigida, University of Costa Rica. San José, Costa Rica.

Ramírez, E., and E. Camacho. 1991. Estudio limnológico preliminar de la Laguna Hule, Costa Rica. *Uniciencia* 8(1–2): 17–25.

Ramírez, E., F. Tabash, and C. Charpentier. 1990. Variación estacional en el lago de Río Cuarto, provincia de Alajuela, Costa Rica. *Uniciencia* 7(1/2): 19–25.

Reimer, P.J., M.G.L. Baillie, E. Bard, A. Bayliss, J.W. Beck, C.J.H. Bertrand, P.G. Blackwell, C.E. Buck, G.S. Burr, K.B. Culter, P.E. Damon, R.L. Edwards, R.G. Fairbanks, M. Friedrich, T.P. Guilderson, A.G. Hogg, K.A. Hughen, B. Kromer, F.G. McCormac, S.W. Manning, C.B. Ramsey, R.W. Reimer, S. Remmele, J.R. Southon, M. Stuiver, S. Talamo, F.W. Taylor, J.J. van der Plicht, and C.E. Weyhenmeyer. 2004. IntCal04 terrestrial radiocarbon age calibration, 26–0 ka BP. *Radiocarbon* 46: 1029–58.

Richardson, J.L., and D. Livingstone. 1962. An attack by a Nile crocodile on a small boat. *Copeia* 1962(1): 203–4.

Rodgers, J.C., III, and S.P. Horn. 1996. Modern pollen spectra from Costa Rica. *Palaeogeography, Palaeoclimatology, Palaeoecology* 124(1–2): 53–71.

Salani, F.M., and G.E. Alvarado. 2010. El maar poligenético de Hule (Costa Rica). Revisión de su estratigrafía y edades. *Revista Geológica de América Central* 43: 97–118.

Sánchez-Azofeifa, G.A., R.C. Harriss, A.L. Storrier, and T. Camino-Beck. 2002. Water resources and regional land cover change in Costa Rica: impacts and economics. *Water Resources Development* 18(3): 409–24.

Siebert, L., G.E. Alvarado, J.W. Vallance, and B. van Wyk de Vries. 2006. Large-volume volcanic edifice failures in Central America and associated hazards. In W.I. Rose, G.J.S. Bluth, M.J. Carr, J.W. Ewert, L.C. Patino, and J.W. Vallance, eds., *Volcanic Hazards in Central America*, 1–26. Geological Society of America Special Paper 412.

Somarribas, G., and J. Bravo. 1999. Una propuesta metodológica participativa para la protección y conservación de humedales. Avance del Proyecto: "Refugio de Vida Silvestre 'Laguna' Mata Redonda, Guanacaste, Costa Rica."

Standley, P.C. 1936–37. Flora of Costa Rica. *Field Museum of Natural History Botany Series* 18(1–4): 1–1571.

Stuiver, M., and P.J. Reimer. 1993. Extended C-14 database and revised Calib 3.0 C-14 age calibration program. *Radiocarbon* 35: 215–30.

Tabash, F.A., and E. Guadamuz. 2000. A management plan for the sport fishery of *Parachromis dovii* (Pisces: Cichlidae) in Hule lake, Costa Rica. *Revista de Biología Tropical* 48(2/3): 473–85.

Tassi, F., O. Vaselli, E. Fernández, E. Duarte, M. Martínez, A. Delgado Huertas, and F. Bergamaschi. 2009. Morphological and geochemical features of crater lakes in Costa Rica: an overview. *Journal of Limnology* 68(2): 193–205.

Taylor, Z.P., S.P. Horn, and D.B. Finkelstein. 2013. Pre-Hispanic agricultural decline prior to the Spanish Conquest in southern Central America. *Quaternary Science Reviews* 73: 196–200.

Telford, R.J., P. Barker, S. Metcalfe, and A. Newton. 2004a. Lacustrine responses to tephra deposition: examples from Mexico. *Quaternary Science Reviews* 23: 2337–53.

Telford, R.J., E. Heegaard, and H.J.B. Birks. 2004b. The intercept is a poor estimate of a calibrated radiocarbon age. *Holocene* 14(2): 296–98.

TNC (The Nature Conservancy). 2009. *Evaluación de Ecorregiones de Agua Dulce de Mesoamérica: Sitios Prioritarios para la Conservación en las Ecorregiones de Chiápas a Darién*. San José, Costa Rica: The Nature Conservancy. 515 pp.

Tyson, P. 2000. *The Eighth Continent: Life, Death, and Discovery in the Lost World of Madagascar*. New York: William Morrow.

Ulloa, J., O. Alpírez, and J. Cabrera. 1988. Presencia de *Bryconamericus scleroparius*, *Poecilopsis turrubarensis* y *Cichlasoma nicaraguense* en el Embalse Arenal, Costa Rica. *Revista de Biología Tropical* 36: 171–72.

Ulloa, J., P. Cabrera, and M. Mora. 1989. Composición, diversidad y abundancia de peces en el Embalse Arenal, Guanacaste, Costa Rica. *Revista de Biología Tropical* 37: 127–32.

Umaña, G. 1985. Phytoplankton species diversity of 27 lakes and ponds of Costa Rica (Central America). M.S. thesis, Brock University. Ontario, Canada.

Umaña, G. 1988. Fitoplancton de las lagunas Barba, Fraijanes y San Joaquín, Costa Rica. *Revista de Biología Tropical* 36(2B): 471–77.

Umaña, G. 1990. Limnología básica de la Laguna del Barva. *Revista de Biología Tropical* 38(2B): 431–35.

Umaña, G. 1993. The planktonic community of Laguna Hule, Costa Rica. *Revista de Biología Tropical* 41(3A): 499–507.

Umaña, G. 1997a. Basic limnology of Lago Bonilla, a tropical lowland lake. *Revista de Biología Tropical* 45(4): 1429–37.

Umaña, G. 1997b. Variabilidad temporal en tres lagos volcánicos. In *Memoria, Jornadas de Investigación 1997*. San José, Costa Rica: Universidad de Costa Rica. 25 pp.

Umaña, G. 2001. Limnology of Botos lake, a tropical crater lake in Costa Rica. *Revista de Biología Tropical* 49(suppl. 2): 1–10.

Umaña, G. 2010a. Comparison of basic limnological aspects of some crater lakes in the Cordillera Volcánica Central, Costa Rica. *Revista Geológica de America Central* 43: 137–46.

Umaña, G. 2010b. Temporal variation of phytoplankton in a small tropical crater lake. *Revista de Biología Tropical* 58(4): 1405–19.

Umaña, G. 2014a. Phytoplankton variability in Lake Fraijanes, Costa Rica, in response to local weather variation. *Revista de Biología Tropical* 62(2): 483–94.

Umaña, G. 2014b. Ten years of limnological monitoring of a modified natural lake in the tropics: Cote Lake, Costa Rica. *Revista de Biología Tropical* 62(2): 567–78.

Umaña, G., and C. Collado. 1990. Asociación planctónica en el Embalse Arenal, Costa Rica. *Revista de Biología Tropical* 38(2A): 311–21.

Umaña, G., K.A. Haberyan, and S.P. Horn. 1999. Limnology in Costa Rica. In B. Gopal and R.W. Wetzel, eds., *Limnology in Developing Countries*, vol. 2, 33–62. New Delhi: New Delhi International Scientific Publications.

Umaña, G., and C.E. Jiménez. 1995. The basic limnology of a low altitude tropical crater lake: Cerro Chato, Costa Rica. *Revista de Biología Tropical* 43(1–3): 131–38.

Umaña, G., F. Villalobos, and B. Bofill. 1997. Distribución vertical de zooplancton en el Embalse Arenal, Costa Rica. *Revista de Biología Tropical* 45(2): 923–26.

Vargas, J.A. 2004. Preface: aquatic ecosystems of Costa Rica III. *Revista de Biología Tropical* 52(suppl. 2): xi–xiii.

Weber, H. 1959. *Los Páramos de Costa Rica y su Concatenación Fitogeográfico con los Andes Suramericanos*. San José, Costa Rica: Instituto Geográfico Nacional.

Woodruff, S.D. 2001. COADS updates including newly digitized data and the blend with the UK Meteorological Office marine data bank and quality control in recent COADS updates. In Proceedings of Workshop on Preparation, Processing and Use of Historical Marine Meteorological Data, Tokyo (November 28–29, 2000), 9–13 & 49–53. Japan Meteorological Agency and the Ship & Ocean Foundation, Tokyo.

Woodruff, S.D., H.F. Díaz, J.D. Elms, and S.J. Worley. 1998. COADS release 2 data and metadata enhancements for improvements of marine surface flux fields. *Physics and Chemistry of the Earth* 23: 517–27.

Wu, J., D.F. Porinchu, S.P. Horn, and K.A. Haberyan. 2015. The modern distribution of chironomid sub-fossils (Insecta: Diptera) in Costa Rica and the development of a regional chironomid-based temperature inference model. *Hydrobiologia* 742: 107–27.

Wujek, D.E. 1984. Scale bearing Chrysophyceae (Mallomonadaceae) from north-central Costa Rica. *Brenesia* 22: 309–13.

Wujek, D.E., R.E. Clancy Jr., and S.P. Horn. 1998. Silica-scaled Chrysophyceae and Synurophyceae from Costa Rica. *Brenesia* 49–50: 1–19.

Wydrzycka, U.M. 1996. Las especies de *Trachelomonas* (Algas: Euglenophyta) en tres lagunas volcánicas de Costa Rica. *Revista de Biología Tropical* 44(2): 477–84.

Zeeb, B.A., J.P. Smol, and S.P. Horn. 1996. Chrysophycean stomatocysts from Costa Rican tropical lake sediments. *Nova Hedwigia* 63(3–4): 279–99.

Chapter 20 Bogs, Marshes, and Swamps of Costa Rica

Jorge A. Jiménez[1]

Introduction

Costa Rica has a high diversity of freshwater wetlands, ranging from the highland peat bogs in the mountainous Talamancas and the marshes of the Pacific lowlands to the often-seasonal swamps dominated by palms and other forest trees thriving along the coasts (Fig. 20.1).

Most of the available information regarding bogs, marshes, and swamps in Costa Rica is restricted to descriptive studies where lists of plant species and studies on forest structure abound. Close to 400 plant species have been listed for these ecosystems (Crow 2002, Córdoba et al. 1998, Gómez 1984, Soto and Arias 1994). Studies on their ecological or functional aspects are, however, mostly absent. Fauna is listed in a few studies but most fauna cited is not restricted to marsh or swamp ecosystems.

Even knowledge on the total extent of these ecosystems in the country is highly fragmentary and controversial. Bogs, restricted mostly to the Talamanca mountains, cover around 235 ha according to estimations based on satellite imagery and information provided by organizations such as INBio, SINAC/MINAE, OTS, and IUCN.

Estimates for other types of freshwater wetlands coverage are highly variable (Rodríguez 2004, Bravo et al. 1996, Córdoba et al. 1998, Sylvander 1978, Ellison 2004). Back in the mid-seventies, close to 300,000 ha were reported: about 139,000 ha of these as forest swamps, close to 96,200 ha as marshes and nearly 60,000 ha as palm swamps (Sylvander 1978). More recent estimates using satellite image analysis indicate a total of around 222,000 ha, where forest swamps comprised 79,650 ha, palm swamps about 92,800 ha, and marshes around 49,650 ha. Serrano-Sandí et al. (2013) estimated that palm swamps in Costa Rica totaled 53,931 ha, which they reported to be 16.2% of the total wetland area in the country (332,000 ha). Most of these data are, however, lacking field verification and they are likely to be overestimations.

Present total coverage of freshwater wetlands is most likely in the order of 140,000 ha for the whole country (Programa Estado de la Nación 2003), with forested wetlands comprising about half of this area and marshes covering close to 30,000 ha.

Estimations of vegetation cover have always been a challenge: while different wetland types are distinguished in such estimations, in reality they do not always form clear and distinct units in the field. Many marshes as well as palm and forest swamps are found growing together in a single area with considerable overlapping areas. Even very small changes in topographic elevation and sedimentation may result in structural and functional changes within such wetlands. Geo-morphological processes trigger those changes and fluctuations in water flow quantity and quality and may also interact in determining the dominant type of wetland vegetation.

Excluding bogs that are found between 1,200–3,100 m a.s.l, most freshwater wetlands are found below 200 m a.s.l., and in the coastal and northern plains. Here, wetlands have been giving way to agriculture and pasturelands at very rapid rates. Particularly marshes under seasonal climates have shown a significant decrease in area, during the past three decades. In the Tempisque watershed, for instance, seasonal marshes in 1975 amounted to around 33,000 ha. By the year 2000 only 7,500 ha (22%) remained. Drainage for sugar cane and rice plantations, as well as for pasture-

[1] Fundación MarViva, Apartado 020-6151, Santa Ana, Costa Rica

lands, has been the main cause for the loss of the Tempisque wetlands.

It is, therefore, not surprising that existing cover estimations are highly controversial depending on the methods used, year of analysis, and extent of field verification done.

Highland Bogs

Highland bogs, also known as "raised ombrothermic bogs" (Gómez 1986) or "*turberas de altura*," are found between 1,200–3,100 m a.s.l., along the continental divide of the Cordillera de Talamanca (Brak et al. 2005). These peat-forming bogs are restricted to small and poorly drained depressions subject to seasonal flooding. With an area just over 235 ha, this ecosystem is very restricted in the country. The largest patches are found in the Dúrika sabanas (Fig. 20.2) and the Cerro Utyum area (Weston 1981, cited by Brak et al. 2005, and information available at INBio's website, www.inbio.ac.cr, and maps prepared by the IUCN).

Flooding in these bogs is closely related to the seasonal precipitation patterns but is also dependent on vegetation coverage in the surrounding areas (Islebe et al. 2005). Water table levels determine the intensity and duration of floodings, which cause an accumulation of peat under anoxic conditions. Acid waters typify these bogs (Horn and Haberyan, chapter 19 of this volume). The low acidity (pH) levels of the bogs' waters usually affect embryonic and larval development of anuran amphibians (Saber and Dunson 1978) as well as the development of tall vegetation (Brak et al. 2005).

Most bog areas in the country are found at the headwaters of the Lari, Volcán, Telire, Ceibo, Orosi, and Savegre rivers. However, the limited extension of thesc bogs result in a relatively low impact on the overall hydrology of these watersheds.

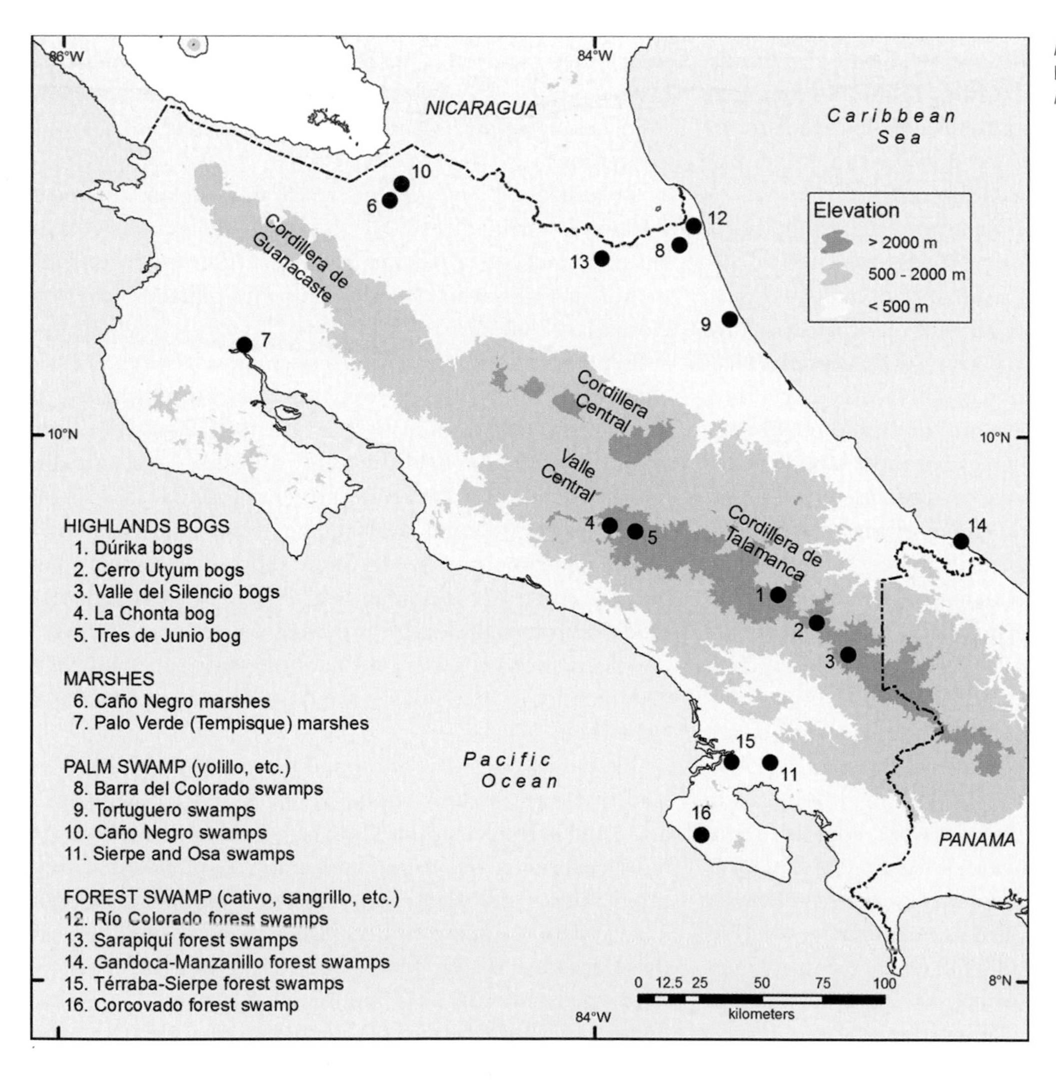

Fig. 20.1 Map of the main wetlands in Costa Rica.
Map prepared by Marco V. Castro.

Fig. 20.2 Highland peat bog in the Sabana de Dúrika, La Amistad International Park (PILA), Costa Rica. Two arborescent *Blechnum buchtieni* ferns with erect living and pendant decaying fronds dominate in the front.
Photo by Luis González Arce.

Palynological and paleoecological studies in these ecosystems are available from bogs adjacent to the Inter-American Highway (Hooghiemstra et al. 1992, Islebe et al. 1996, Rodgers and Horn 1996). These studies indicate that changes in hydrological conditions have resulted in vegetational changes. Fire events that took place during the Holocene have been connected to climatic variability in that period (Horn 1989, 1993). Repeated fire in the surrounding *páramos* can affect these bogs by affecting their water retention properties in the soil. Although Horn and Haberyan (chapter 19 of this volume) present limnological data for some pond-like bogs (e.g., the Tres de Junio pond, along the Inter-American Highway south of Cartago; see Fig. 20.3), no detailed hydrological studies are available for these ecosystems yet.

While the highland bogs share a large amount of structural and floristic elements with the *páramo* vegetation surrounding them, they can be differentiated, however (Kappelle and Horn, chapter 15 of this volume). Over 100 species of forbs, grasses, and sedges have been reported from the high-elevation bogs of Costa Rica (Gómez 1984). These species are typically distributed in a zonal pattern that relates to the bog's water depth. Brak et al. (2005), Kappelle (1996), Hooghiemstra et al. (1992), and Gómez (1984) have described in detail this zonal pattern: at the bog edges, there is a zone containing species such as the bamboo *Chusquea subtessellata*, the grass *Cortaderia nitida*, the shrubs *Myrsine coriacea*, *Escallonia myrtilloides*, and *Hesperomeles heterophylla*, and ericads such as *Macleania rupestris* and *Vaccinium consanguineum*, which occur together with the hepatic *Frullania* and the lichen *Usnea*.

Fig. 20.3 Highland peat bog at Tres de Junio in the Tapantí-Macizo de la Muerte National Park, Costa Rica. The terrestrial bromeliad *Puya dasylirioides* dominates the aspect of this vegetation type.
Photo by Anouk Paulissen.

In the neighboring zone, closer to the center of the bog, large, cycad-like ferns in the Blechnaceae dominate (*Blechnum buchtieni*, *B. loxense*) as well as the endemic bromelia *Puya dasylirioides* together with the small (sub)shrubs *Pernettya prostrata*, *Hypericum strictum*, and the fern *Elaphoglossum*, which is found together with the yellow-eyed grasses *Xyris nigrescens and X. subulata.*

At the center of the bog one can observe *Hypericum strictum*, *Paepalanthus costaricensis*, *Hieracium irasuense*, *Utricularia* aff. *subulata*, *Isoetes storkii*, and *Pernettya prostrata*, which occur in association with mosses such as *Campylopus* sp., *Sphagnum magellanicum*, *S. recurvum*, and *Breutelia subarcuata*. Sedges such as *Carex donnell-smithii*, *C. jamesonii*, *C. bonplandii*, and *Rhynchospora schaffneri*, together with Juncaceae (*Juncus* sp.), thrive together with the lichen *Cladina confusa* (Brak et al. 2005, Kappelle 1996, Gómez 1986).

Endemic species reported from this ecosystem are largely shared with the *páramo* and other nearby ecosystems. The salamander *Bolitoglossa pesrubra* (formerly known as *P. subpalmata*) has been recorded from the bogs, although thriving mostly in adjacent ecosystems (Kappelle and Savage 2005). This amphibian is usually found among the leaves of another endemic, the bromeliad *Puya dasylirioides*, also found in adjacent *páramo* (Vial 1966, 1968). The demography of these bromeliad populations has been studied by Augspurger (1985). Other endemics shared with the *páramo* and highlands in general are the orchid *Gomphichis adnata*, the lichens *Lobariella pallida*, *Menegazzia neotropica*, *Nephroma helveticum*, and the thrush *Turdus nigrescens*. Very recently, even a new species of dink frog (Anura: Eleutherodactylidae: *Diasporus ventrimaculatus*) was discovered in the bog area of the Valle del Silencio (La Amistad International Park, PILA) along the Caribbean slope of the Cordillera de Talamanca (Chaves et al. 2009).

While bogs have suffered human disturbance as early as ca. 4,900 years ago (Islebe et al. 2005), the total area of these ecoystems has remained relatively constant in recent times. Their conservation has benefited from the inclusion of most sites within the National System of Conservation Areas (SINAC). The 2003 Ramsar declaration of the Costa Rican Talamanca Peatlands (*Turberas de Talamanca*) at elevations of 700–3,819 m and intermingled with oak-bamboo forests (Kappelle 1996) as a Wetland Site of International Importance (Ramsar Site no. 1286, Costa Rica's eleventh Ramsar Site), further contributes to the conservation of this fragile ecosystem (see: www.ramsar.org/wn/w.n.costarica_talamanca.htm). Fortunately, this Ramsar site covers areas where the Central American tapir *Tapirus bairdii*, the ocelot *Felis pardalis*, and the red brocket deer *Mazama americana* are still found today.

According to studies on past changes, modern anthropogenic fire events (Horn 1988 and 2005) and climate change (Islebe et al. 2005) might, however, threat the long-term health of these fragile highland bog ecosystems. Some important sites, such as La Chonta, are not part of the protected area system and should be incorporated as soon as possible, since current land-use changes are affecting its hydrological conditions that maintain the integrity of this ecosystem and may ultimately cause eutrophication of the bog waters and loss of its fragile species (M. Kappelle, personal observation).

Marshes

The Flora of the Marshes

Marshes are confined to basins and depressions where rainfall and run-off water accumulate. Herbs, grasses, and sedges dominate these ecosystems. Their non-woody vegetation composition reflects climatic, hydrological, and geomorphological variations as they change from site to site.

Under rainy climates affecting the Caribbean coastal plains (Herrera 1986, and chapter 2 of this volume), marshes are dominated by herbs such as *Calathea lagunae*, *Calathea lutea*, *Thalia geniculata*, and *Montrichardia arborescens*, sedges like *Cyperus giganteus* and *Lasciacis procerruna*, grass genera such as *Panicum* and *Scleria*, and vines like *Ipomoea* and *Solanum lancefolia*. In the Sarapiquí region *Calathea lutea* and *Spathiphyllum friedrichstallii* and shrubs such as *Acalypha diversifolia* dominate the more open, marshy areas. The Corcovado marsh in the core of the Osa peninsula (Gilbert et al., chapter 12 of this volume), is dominated by the grass *Pennisetum* and herbs like *Jussiaea*, *Polygonum*, and *Aeschynomene*. Other species occurring in marshes under rainy climates are *Hymenachne amplexicaulis* and *Oryza latifolia* (Hartshorn 1983, Hartshorn and Hammel 1994, Myers 1990).

The headwaters of the Sierpe River along the southwestern Pacific coast harbor about 2,500 ha of marshes (Fig. 20.4), dominated by *Thalia geniculata*, *Cyperus papyrus*, *C. gigantus*, *Fimbristylis spadicea*, *Paspalidium germinatum*, and *Gynerium sagittatum* that occur together with floating herbs such as *Nymphaea blanda* and the leather fern *Acrostichum aureum* (Figs. 20.5 and 20.6) (Álvarez et al. 1999, Kappelle et al. 2003).

The largest extensions of marshes under seasonal climates (Herrera 1986, and chapter 2 of this volume) are mainly concentrated in just two areas: the Lower Tempisque region that harbors over 7,000 ha, and the Caño Negro–Río Frío–Guatuso Plains that house some 16,000 ha of seasonal marshes. Functional aspects in these marshes are highly seasonal.

Most of the information available on seasonal marshes comes from the two best studied marsh regions of the country: Caño Negro and Palo Verde. In the core of the

Fig. 20.4 Freshwater marsh with floating vegetation at Laguna Sierpe. *Photo by Luis González Arce.*

Fig. 20.5 Freshwater marsh dominated by the leather fern *Acrostichum aureum* at Estero Caballo in the Sierpe-Térraba wetland.
Photo by Luis González Arce.

Fig. 20.6 Detail of the vegetation dominated by the leather fern *Acrostichum aureum* at Estero Caballo in the Sierpe-Térraba wetland.
Photo by Luis González Arce.

Fig. 20.7 View of the open waters of Lago Caño Negro, Costa Rica. *Photo by Garret Crow.*

9,969 ha Refugio Nacional de Vida Silvestre Caño Negro, the 800 ha-sized Lago Caño Negro is located (Fig. 20.7). Its open waters have a depth of up to 3 m in the rainy season and may disappear almost completely at the end of the dry season. While there is some similarity in species composition between the marshes of Caño Negro and Palo Verde (Fig. 20.8), there are also important differences regarding the species that (co)dominate each area.

Preliminary analyses of the Caño Negro wetlands reported close to 80 species of aquatic plants (Zamora and Bravo 1992, Castillo and March 1993). Vegetation in the Caño Negro ponds is highly seasonal with *Nymphaea ampla*, *Nymphaea blanda*, *Neptunia plena*, *Pistia stratiotes*, *Polygonum hispidium*, and *Salvinia sprucei* dominating during the rainy season. *Salvinia auriculata* may locally dominate in the more open waters (Fig. 20.9). *Ludwigia sedioides*, *L. peploides*, and *Pontederia rotundifolia* grow in the pond's shallower areas. At the edges of the ponds clumps of *Ambrosia cumanensis* coexist with a number of grasses: *Hymenachne amplexicaulis*, *Echinochloa polystachya*, *Paspalum repens*, *Eragrostis hypnoides*, and *Panicum parvifolium*, as well as other species like *Luziola subintegra* and the rice plant *Oryza latifolia* (Castillo and March 1993, Zamora and Bravo 1992). Mixed with these grasses occurs a series of sedges: *Cyperus papyrus*, *C. imbricatus*, *C. holoschoenoides*, *Oxycarium cubense*, *Rhynchospora corymbosa*, *Fuirena umbellata*, and *Scleria macrophylla*. The edges dominated by tall grasses and sedges appear to be highly dependent on geomorphological processes such as erosion and accretion. Changes in their total coverage happen from year to year.

During the dry season, when water levels are lower, conditions allow the growth of a large variety of herbs. Among them, *Polygonum segetum*, *P. punctatum*, *Aeschynomene virginica*, and *Justicia comata* grow on more exposed soil. In deeper areas, *Polygonum hispidum* and *P. acuminatum* form rather dense mats. *Hydrolea spinosa*, *Solanum campechiense*, *Ceratopteris richardii*, *Echinodorus subalatus* subsp. *andrieuxii*, and *Tonina fluviatilis* have also been found here. Noteworthy, Zamora and Bravo (1992)

Fig. 20.8 Overview of the marshes in Parque Nacional Palo Verde, Costa Rica. In the back at the left, the Río Tempisque. *Photo by Garret Crow.*

reported that the Caño Negro wetlands serve as the southernmost habitat of the palm *Acoelorraphe wrightii*.

Phytoplankton in the marshes' ponds is composed of around 240 species. Chlorophyta are the most diverse group with 98 species, followed by Euglenophyta, which is represented by 44 species. Plant diversity decreases considerably during the dry season. Apparently the Cyanophyta group, while less diverse in terms of species, is the most abundant and dominating group (Umaña 1991).

The Lower Tempisque area harbors the second largest concentration of marshes in Costa Rica. An extensive mix of interconnected flatlands and ponds dominates the floodplains of the Tempisque River, from which over 130 species of aquatic plants have been reported (Crow 2002).

Topographical variations in these marshes are minimal. In the Palo Verde Lagoon, for example, ground elevation varies within a 60 cm range along the 8 km length of the lagoon. In spite of these small variations, changes in vegetation types become evident over a depth gradient. The Palo Verde treelet (*Parkinsonia aculeata*)—which gives its name to the lagoon and the national park—, *Echinodorus paniculatus*, and the sedges *Eleocharis mutata*, *E. interstincta*, *Cyperus digitatus*, and *C. giganteus* are found at higher grounds where the water is usually less than 30 to 40 cm deep (Fig. 20.10). In the deeper parts of the lagoons (40–80 cm) emergents like *Thalia geniculata*, *Canna glauca*, *Paspalum repens*, *Paspalidium germinatum*, *Ludwigia inclinata*, and *Typha dominguensis* are found. In areas with more open water one observes species such as *Nymphaea pulchella*, *N. amazonum*, *Eichhornia crassipes*, *E. heterosperma*, *Neptunia natans*, *Sagittaria guyanensis*, *Pistia stratiotes*, *Limnobium laevigatum*, *Salvinia auriculata*, *Azolla microphylla*, *Nymphoides humboldtianum*, *Rynchospora corymbosa*, *Oxicrym cubense*, *Ludwigia pepliodes*, and *Pontederia rotundifolia* (Crow 2002, Guzmán 2007).

Most of the vegetation cover has a seasonal appearance. Reduced soil cover (60–63%) is mostly observed during the peak of the dry season (April–May), while highest cover percentages are found in November, at the end of the rainy season (Hernández 1990).

The marshes' water depth, geomorphological processes, human disturbance history, and seasonal hydrology jointly define locally which specific emergent species will dominate which marsh. At some sites the herb *Mimosa pigra* may dominate while in other, more elevated places species such as *Thalia geniculata* and *Typha dominguensis* often compose a mixture with isolated *Parkinsonia aculeata* trees. In other areas, *Thalia geniculata* and some species of *Canna* can be found, which co-dominate with *Typha*.

Variations in species assemblages may result from anthropogenic changes in the area. *Typha dominguensis*, for example, is a native weed that has been observed in the palynological record available from the Palo Verde marsh (1,200 ha). It appeared already at least some 4,500 years ago (Horn and Kennedy 2006). During the past few decades, it has responded very aggressively to human-induced changes in hydrological conditions and fire regimes. Over the past century local people periodically used fire to modify the marsh, as part of their cattle-ranching practices. These fires favored the occurrence of areas with open water in which floating species started to dominate during the rainy season.

These open waters have served as habitats of great importance to at least 60 species of waterfowl (Trama 2005). For instance, they have been heavily used by anatidae ducks (Hernández 1993) with *Dendrocygna autumnalis* being the most abundant species in the area. Also, the endangered bird *Jabiru mycteria* is considered highly dependent on these marshes (Villarreal-Orias 1997). This large stork often visits several marshes in a single day (Gamboa 2003).

The Fauna of the Marshes

Marshes harbor an interesting fauna that is worthwhile mentioning though often not endemic to wetland areas. For instance, the common caiman (*Caiman crocodylus*) constitutes an important component of the dynamics of the Caño Negro marshes. In the early 1990s its total population was estimated at about 3,000 individuals (Allsteadt and Vaughan 1992). Caiman density is high during the dry season when 83 to 166 individuals are observed per ha, concentrated in ponds and watercourses that maintain high water levels.

Fig. 20.9 The eared watermoss (*Salvinia auriculata*), a small, free-floating, aquatic fern-ally that forms extensive carpets on still waters at Caño Negro, Costa Rica. Reproduction is often vegetative and rapid.
Photo by Garret Crow.

During the rainy season, as flooding expands, they expand their range into the adjacent marshes and channels.

In the Tempisque marshes similar patterns are observed among American crocodile (*Cocodrylus acutus*) populations. These display one of the highest densities ever reported for Costa Rica (between 15.8 and 21.9 individuals per km of shoreline). This high density represents a fourfold increase in the crocodile population of this area that happened during the past two decades (Sánchez 2001).

In the Caño Negro marshes, fish communities in seasonal ponds are composed of some 21 species dominated by *Poecilia gillii*, *Astyanax aeneus*, and *Rhamdia nicaraguensis*. Less abundant is *Ophisternon aenigmaticum*. During the rainy season when flooding occurs and ponds become deeper and more extended, levels of fish diversity become higher and fish species abundance gets lower (Sáenz et al. 2006). One of the distinct and endangered fish species in the marshes of Caño Negro is *Atractosteus tropicus* (Lepisosteidae, *gaspar*), a large, 50–60 cm long carnivorous fish that feeds on other fishes (Poeciliidae, Characidae, Pimelodidae, Gymnotidae) as well as on crustaceans (Mora et al. 1997). Exotic, invasive fish species such as *Oreochromis niloticus* (Nile tilapia), a cichlid fish of African origin, has already been found in this marsh. It has been introduced for aquaculture purposes in the early 1960s and is now reported as abundant in these marshes (Cabrera et al. 1992). Exotic tilapias (*Oreochromis sp.*) have been reported from other marshes, including those in the Lower Tempisque basin (Pizarro and Rojas 1993). Its impact on the foodweb and other fish populations in those habitats is still unknown.

Close to 307 species of birds (aquatic and terrestrial), including the endangered jabiru stork (*Jabiru mycteria)* and the great white egret (*Egretta alba*, Fig. 20.11), are associated with the marshes of Caño Negro. A total of 101 of them are migratory (Hidalgo 1993a). Over 60 species of waterfowl (including some 23 migrants) have been reported for the marshes of the Lower Tempisque region, including the green heron (*Butorides virescens*, Fig. 20.12). Anatidae ducks heavily use these marshes. *Dendrocygna autumnalis* is the most abundant species here (Hernández 1993).

Fig. 20.10 Marshes in Parque Nacional Palo Verde with aquatic emergent *Echinodorus paniculatus* plants and a few *palo verde* trees (*Parkinsonia aculeata*), Costa Rica. *Photo by Garret Crow.*

Fig. 20.11 A great white egret (*Egretta alba*) at Caño Negro, Costa Rica.
Photo by Garret Crow.

In the Caño Negro marshes, a population of around 47 individuals of the endangered jabiru has been reported (Villarreal-Orias 2000). The jabiru stork feeds on local fish species such as *Synbranchus marmoratus*, *Cichlasoma* sp., and *Ariopsis seemanni* and snails (*Pomacea costaricana*). Jabirus foraging in groups achieve higher fish capture frequencies and appear to be more succesful (Villarreal-Orias 1997). Rare avian components such as *Anas clypeata*, *Rynchops niger* and *Bartramia longicauda* are also observed in these marshes (Villarreal-Orias 1997).

One of the most important Costa Rican centers of mollusk diversity and endemism concerns the Tempisque river basin and its marshes (Barrientos 2003). Here, the endemic bivalve *Nephronias tempisquensis* has been observed in the Mata Redonda marsh adjacent to the Tempisque River (Scott and Carbonell 1986). These marshes also harbor a high diversity of macro-invertebrates. Over a hundred species of benthic invertebrates have been reported from the Palo Verde Marshes so far (Trama et al. 2009).

Fig. 20.12 Close-up of the green heron (*Butorides virescens*) along the Río Tempisque at Parque Nacional Palo Verde, Costa Rica.
Photo by Pablo Elizondo.

Swamps

Palm Swamps

Palm swamps are a dominant element of lowland wetlands, representing mature steady-state wetlands with at least 2,800 years of presence in the region (Urquhart 1997, 1999). Studies on this particular ecosystem are still rather scarce.

Most of the palm swamps reported from Costa Rica are located in and around the Barro Colorado Wildlife Refuge and the Tortuguero National Park. Other palm swamps are known from smaller areas around the Caño Negro Wildlife Refuge, and in the Sierpe-Osa region. Over half (55%) of Costa Rica's existing palm swamps are conserved within protected areas today (Serrano-Sandi et al. 2013). Because palm and forest swamps grow together, significant overlap between these two ecosystems is to be expected when estimations of their coverage are done.

Which species dominate and how the overall structure of a palm swamp appears will depend on the seasonality and the hydrological regime of a particular site under study. In general, the better drained a swamp is, the greater its plant diversity is. Also, non-constant flooding conditions favor higher plant species diversity than permanent flooding does. There is a tendency for palm species to lose dominance over hardwood species as flooding depth and flood prolongation decrease (Myers 1990). Dominant palm species in such swamps are *Raphia taedigera* (locally known as *yolillo*, Figs. 20.13 and 20.14) and *Manicaria saccifera* (royal palm, *napa* or *cabecinegro*).

Fig. 20.13 Palm swamp dominated by raffia or *yolillo* (*Raphia taedigera*) at Quebrada Taboga in the Sierpe-Térraba wetland. *Photo by Luis González Arce.*

Fig. 20.14 Close-up of the *Raphia taedigera*–dominated palm swamp at Quebrada Taboga in the Sierpe-Térraba wetland. *Photo by Luis González Arce.*

In *M. saccifera* and *R. taedigera* palm swamps, the maximum level of flooding can be close to 1 m above ground level, with short dry spells throughout the year. Soils are flooded during 53 to 92% of the year but water tables are very close to soil level during most of the year (Myers 1990).

Plant diversity tends to be low, with 3–27 species with stems over 10 cm diameter at breast height (dbh) per ha (Anderson and Mori 1967, Devall and Kiester 1987, Myers 1990, Chavarría and Valverde 2000). Structural development varies significantly: *Raphia*-dominated swamps exhibit stem densities from 527 to 640 stems/ha and basal areas from 224 to 338 m²/ha (Myers 1990); *Manicaria*-dominated swamps exhibit higher stem densities (715–910 stems/ha) but lower basal areas (47–94 m²/ha (Myers 1990, Holdridge et al. 1971). In *Manicaria* palm swamps a large part of the stems belong to hardwood species (stem densities around 230 stems per ha) while in *Raphia* swamps stem abundance is mostly dominated by palms (Myers 1990).

In *Raphia*-dominated swamps, *Raphia* is accompanied by disperse trees belonging to species such as *Amanoa potamophila*, *Andira inermis*, *Annona glabra*, *Ardisia compressa*, *Calophyllum brasiliense*, *Campnosperma panamensis*, *Carapa guianensis*, *Cassipourea guianensis*, *Crataeva tapia*, *Crudia acuminata*, *Grias fendleri*, *Guatteria amphifolia*, *Homalium racemosum*, *Inga* sp., *Ixora nicaraguensis*, *Luehea seemannii*, *Ocotea* sp., *Pentaclethra macroloba*, *Pithecellobium latifolium*, *Posoqueria latifolia*, *Prioria copaifera*, *Pterocarpus officinalis*, *Rauvolfia* sp., *Sickingia maxonii*, *Sloanea picapica*, *Spondias mombim*,

Symphonia globulifera, and *Trichilia tuberculata* (e.g., see Kappelle et al. 2003, for details on the flora of the *Raphia* swamps found in the Osa Conservation Area).

The understory is relatively open and dominated by sedges together with *Becquerelia cymosa*, *Calyptrocarya glomerulata*, *Scleria microcarpa*, *Thalia geniculata*, the shrub *Psychotria chagrensis*, small trees such as *Alchornea costaricense*, *Casearia* sp., *Mabuetia guatemalensis*, *Posoqueria grandiflora*, *Senna reticulata*, *Urera caracasana*, the ginger *Renealmia aromatica*, the herbs *Palicourea fastigiata*, *Jussiaea latifolia*, *Calathea lagunae*, *Calathea lutea*, *Calathea foliosa*, *Cyclanthus bipartitus*, the ferns *Adiantum latifolium* and *Acrostichum aureum*, as well as *Tabernaemontana chrysocarpa*, the palms *Calyptrogyne glauca*, *Asterogyne martiana*, the aroids *Montrichardia arborescens*, *Dieffenbachia davidsei*, and *Spathiphyllum friedrichsthalii* and the floating *Pistia stratiotes* (Myers 1990, Chavarria and Valverde 2000, Hartshorn 1983, Kappelle et al. 2003).

In *Manicaria*-dominated swamps along the Caribbean coastal plains, this palm is found associated with other palms such as *Astrocaryum alatum* and *Euterpe* sp., and mixed with tree species such as *Calophyllum brasiliense*, *Symphonia globulifera*, *Carapa guianensis*, and *Dialium guianensis* (Gómez 1985). In the Sierpe and Golfo Dulce areas, at the Pacific coast, *Raphia* grows in small patches together with *Symphonia globulifera* (*cerillo*) and other palm species such as *Elaeis oleifera* and *Scheelea rostrata* together with the non-palm tree *Pterocarpus officinalis* (Allen 1977, Kappelle et al. 2003).

Both *Elaeis oleifera* and *Scheelea rostrata* are found dominating swamps of much smaller extension. *E. oleifera* is found in small patches behind mangroves in the wet areas of both coastal plains while *Scheelea rostrata* stands are found in the meanders of the Tempisque and Palma rivers in the Lower Tempisque Basin. The small stands dominated by these two species are rapidly disappearing owing to drainage and fire events.

While fauna restricted to these environments is scarce and little studied, the occurrence of manatees in the channels associated with *Raphia* and *Manicaria* swamps along the Caribbean coast is worthwhile mentioning. In these channels small populations of the manatee (*Trichechus manatus*) are reported. They feed on many types of plants including the benthic *Ludwigia* and *Hydrilla* herbs (Jiménez 2000).

Tapirs and peccaries feed on *Raphia* fruits in the Osa Peninsula (Janzen 1983) and seasonal migrations of these mammals between *Raphia* swamps and adjacent rain forests have been reported.

In *Raphia*-dominated swamps total numbers of amphibian and reptile species are around 14 and 17, respectively (Bonilla-Murillo et al. 2013). All these species are also found in adjacent non-wetland habitats but seem to be well adapted to the harsh conditions of the palm-dominated wetlands.

Forest Swamps

Forest swamps not dominated by palms are characterized by low levels of plant diversity, compared with adjacent dryland forests. Species dominance varies from site to site in response to differences in hydrological regimes and geomorphological settings.

Among the most common forest swamps in the country are those dominated by *Pterocarpus officinalis* (Fabaceae). In such *sangrillo* swamps, *P. officinalis* exhibits stem densities of about 57 trees/ha and basal areas of 10.6 m^2/ha. This tree species shares the canopy layer with other tall-growing species such as *Carapa guianensis* (Meliaceae) and *Astrocaryum alatum*, which display similar densities (57 and 143 stems/ha) and basal areas (9.7 and 1.6 m^2/ha, respectively). Other minor elements in those stands are *Pentaclethra macroloba* and *Virola multiflora*. Close to the coastline *P. officinalis*–dominated swamps grow, even containing mangrove forest elements. At Moín, at the Caribbean coast, *P. officinalis* density was 212 stems/ha with basal areas of 7 m^2/ha. The two other main woody constituents (*Avicennia germinans* and *Rhizophora mangle*) showed densities of 84 and 118 stems/ha and basal areas of 13.4 and 3.3 m^2/ha, respectively (Jiménez 1981). The understory at the site is dominated by *Crinum erubescens*, with herbs such as *Rustia occidentalis*, *Pavonia spicata*, *Hymenocalis litoralis*, and the fern *Acrostichum aureum* (Jiménez 1981).

Along the Pacific coast, within the Corcovado National Park, *Pterocarpus* swamps exhibit total densities of 660 saplings and adult trees per ha. In transitional zones where *Mora oleifera* and *P. officinalis* grow together, tree densities ranged from 400 stems per ha for *Pterocarpus* to 155 for *Mora*. At the same site seedlings of *P. officinalis* exhibit densities of 72/m^2 (Janzen 1978). At those sites where *Mora oleifera* dominates over *P. officinalis* the tidal influence seems to be more prominent. Here, *Mora oleifera* clearly dominates over other species, including *Amphitecna latifolia*, *Avicennia germinans*, *Tabebuia rosea*, and *Luehea seemannii* (Hartshorn 1983, 1988). The structural development of these forests exhibits stem densities of 235/ha and basal areas of 35 m^2/ha (Holdridge et al. 1971). At locations deeper into the inland, forest swamps are composed of a mixture of *Carapa guianensis*, *Cryosophila guagara*, *Erythrina lanceolata*, *Grias fendleri*, *Mouriri* sp., *Prestoea decurrens*, *Pterocarpus officinalis*, and *Virola koschnyi* (Hartshorn 1983, Kappelle et al. 2003)

Fig. 20.15 Forest swamp dominated by *Symphonia globulifera* along a tributary of the Sierpe River.
Photo by Luis González Arce.

In the Sierpe-Térraba wetland area, some forest swamps are dominated by *Symphonia globulifera* (Figs. 20.15 and 20.16) though they thrive at places mixed with other important woody species including *Anacardium excelsum*, *Caryocar costarricense*, *Terminalia oblonga*, *Calophyllum brasiliense*, *Pterocarpus officinalis*, *Erythrina lanceolata*, *Inga vera*, *Carapa guianensis*, *Hernandia didymantha*, and herbs such as *Montrichardia arborescens* and the floating *Pistia stratiotes* (Bravo et al. 2000, Kappelle et al. 2003). *Cecropia* sp. occurs in gaps and is quite abundant in these forest swamps, while *Pachira aquatica* is found growing along the margins of the river channels and lagoons. At the Osa Peninsula itself, some small isolated patches of forest swamps are dominated by *Prioria copaifera* trees (Holdridge et al. 1971). In these *Prioria* swamps, the endemic herb *Calathea longiflora* has been observed (Gómez 1984).

Prioria copaifera (*cativo*) is also flourishing along the Caribbean coast, where it forms extensive swamps called "cativales." These are usually growing behind the narrow mangrove belts that occur on the coastal plains. Here, on loamy soils that accumulate large amounts of detritus, or on seasonally flooded riverbanks, *Prioria copaifera* individuals can reach significant sizes and basal area values. Trees with a dbh of 160 cm have been reported here. While *P. copaifera* dominates, stand composition may vary from site to site. Along the Río Colorado, *Prioria copaifera* exhibits significant dominance over other arboreal components such as *Stemmadenia donnell-smithii*, *Pithecellobium latifolium*, *Grias fendleri*, and *Ixora finlaysoniana* (Hartshorn 1983, 1988). At other sites, *P. copaifera* is found growing in association with *Pentaclethra macroloba*, *Iriartea deltoidea*, and *Eschweilera calyculata* (Chavarria and Valverde 2000).

Fig. 20.16 Detail of the *Symphonia*-dominated forest swamp in the Sierpe River watershed.
Photo by Luis González Arce.

Structural attributes also fluctuate among sites. At the Río Colorado *P. copaifera* densities were high (290/ha) with basal areas close to 55 m^2/ha (Holdridge et al. 1971). Southward sites exhibit *P. copaifera* densities of 79 trees/ha (21% of the total) and 29 m^2/ha (63% of the basal area of the forest) (Chavarria and Valverde 2000). *Carapa guianensis* represents 16% of the stems thicker than 10 cm dbh and about 10% (4.9 m^2/ha) of the total basal area, while *Pentaclethra macroloba* has some 44 stems/ha and is responsible for a basal area of 4.2 m^2/ha (Chavarria and Valverde 2000). Whereas it is usually a sub-dominant tree of the forest swamps in northeastern Costa Rica, *C. guianensis* shows a highly synchronous phenology, with flowering happening from September onwards and fruiting occurring during the following May (McHargue and Hartshorn 1983). From 750 to 3,950 seeds are produced by each individual tree, many of which (54–98%) are removed by rodents such as agoutis (*Agouti paca*) (McHargue and Hartshorn 1983). As a light-demanding species, growth of *C. guianensis* is significantly faster in light gaps (Webb 1999). *Campnosperma panamensis* (*orey*) is a swamp tree usually restricted to better-drained swamps usually in the most elevated portions of the swamps, indicating intolerance to strong inundation (Phillips 1995).

Orey swamps occur in small patches at the Gandoca-Manzanillo Wildlife Refuge and at the mouth of the Sixaola River. Gómez (1985) reports orey growing in association with *Dialyanthera otoba*, *Symphonia globulifera*, *Calophyllum brasiliense*, *Carapa guianensis*, *Grias fendleri*, and *Saccoglotis thrichogyna*.

In those sections of the forest swamps where *C. panamensis* dominates, this species shows densities of

203 stems/ha (37% of trees above 10 cm dbh) and basal areas of 13.8 m^2 ha (45% of the total basal area). Other components like the palm *Euterpe precatoria* and *Cassipourea* aff. *killipii* have stem densities of 120 and 140 per ha, and basal areas of 3.9 and 4.4 m^2/ha, respectively. Locally less important elements, such as *Symphonia globulifera* and *Pentaclethra macroloba* account for 26 and 20 trees per ha, with basal areas of 3.5 and 2.9 m^2/ha, respectively (Chavarria and Valverde 2000). In this forest type, the undergrowth is dominated by *Psychotria poeppigiana*, *Voyria* sp., *Tococa guianensis*, and the shrub *Ourartea* sp. (Chavarria and Valverde 2000).

Pentaclethra macroloba is a tree that may become very important in some forest swamps. While not exclusive to forest swamps this species becomes the dominant element in many swamps in the Caribbean lowlands. In the Sarapiquí region in northern Costa Rica, for instance, many swamps are dominated by *P. macroloba*, which grows together with *Carapa nicaraguensis*, *Luehea seemannii*, *Otoba novogranatensis*, *Pachira aquatica*, and *Pterocarpus officinalis* (Hartshorn and Hammel 1994). The subcanopy is composed of trees such as *Grias cauliflora*, *Pithecellobium valerioi*, and palms like *Iriartea deltoidea*, and *Welfia georgii*. At the same time, the understory is dominated by *Adelia triloba*, *Astrocaryum alatum*, *Bactris longiseta*, *Chione costaricensis*, and *Psychotria chagrensis* (Hartshorn 1983, 1988). In such *Pentaclethra* swamps a total of 115 species have been observed (Hartshorn 1983) though just 10 of them account for more than 75% of the basal area. Here, *P. macroloba* accounts for 30% of the total basal area (31 m^2/ha) and densities of 265 stems per ha. Furthermore, *Carapa nicaraguensis* exhibits basal area values of 13 m^2/ha and stem densities of 69/ha, while the palm species *Iriartea deltoidea* is present with high stem densities (133/ha), but its basal area is only 3.38 m^2/ha (Hartshorn and Hammel 1994). It appears that these *P. macroloba* swamps are highly dynamic and demonstrate fast growth and regeneration rates, with a half-life of about 34 years (Liebermann et al. 1985).

The economic value of some of these forest swamps—especially those dominated by *Prioria copaifera* and *Carapa guianensis*—has attracted interest of foresters and catalyzed research, particularly on the wood extraction potential of these swamps (Webb 1998, 1999).

Conservation

Wetland Protection

During the past few decades the seasonally flooded wetlands in Costa Rica have been significantly degraded owing to factors such as seasonal fires, artificial drainage, and water diversion. Moreover, illegal fishing and crocodile hunting have been reported periodically (e.g., in March–April 2010; M. Kappelle, pers. comm.); seasonal fires have considerably affected the distribution and coverage of wetland vegetation types; heavy sedimentation, resulting from the degradation of associated watersheds, has caused significant habitat change; and other threats including pollution with pesticides, eutrophication with fertilizers, and the introduction of alien species such as *Tilapia* have significantly degraded Costa Rica's freshwater ecosystems, including its bogs, marshes, and swamps (e.g., see TNC 2009).

Changes in the Río Frío watershed, associated with the Caño Negro marshes, have promoted the reduction of open water areas that are now being occupied by grasses that colonize recently created mud flats (Castillo and March 1993, Brenes et al. 2001). With channel siltation and losses in wetland coverage, the interconnectivity previously found among Caño Negro, Río Frío, and the Guatuso floodplains is gradually being lost. The exchange of fauna, water, and nutrients among these systems is therefore disappearing. Similarly, in the Tempisque River the severe extraction of water for irrigation purposes has dramatically reduced the connectivity between this river and the adjacent marshes, particularly during the dry season. Drainage and widening of the riverbed, for flood control purposes, have impacted riparian communities, crocodile nesting, and riverine vegetation in most lowland areas.

The development of agriculture at the expense of wetland areas is generating significant agrochemical contamination. Pesticides have been reported in eggs of the woodstork (*Mycteria americana*) common in the seasonal marshes of the lower Tempisque River. High correlations of p,p'-DDE (dichlorodiphenyldichloroethylene), a persistent organic pollutant (POP), have been related to reductions in eggshell thickness (Hidalgo 1993b). Similarly, accumulation of lead, cadmium, selenium, and manganese in woodstork adults was reported for the same locality (Burger et al. 1993).

About 60% of freshwater wetlands in the country are located within protected areas, including 11 Ramsar Sites: Palo Verde (1991), Caño Negro (1991), Tamarindo (1993), Térraba-Sierpe (1995), Gandoca-Manzanillo (1995), Humedal Caribe Noreste (1996), Isla del Coco (1998), Manglar de Potrero Grande (1999), Laguna Respringue (1999), Cuenca Embalse Arenal (2000), and Turberas de Talamanca (2003). However, even these protected wetlands are affected by changes in water flow and quality, originated outside the protected areas. Many marshes in the lower Tempisque Basin, for example, are closely interconnected hydrologically. Within the Palo Verde National Park, water flows downwards, starting at Poza Verde, through Varillall, then via Piedra Blanca and Palo Verde, in order to finally reach the Nicaragua marshes. This cascade-like connection makes "downstream" marshes highly dependent on alterations in "upstream" marshes. Unfortunately, the marsh that is most upstream (Poza Verde) is located partially outside the Park and receives large amounts of waters flowing in from upstream sugar cane and rice fields. Water quality in all "downstream" marshes, even within a National Park like Palo Verde, is likely to be affected by agriculture-derived contaminants, including pesticides and fertilizers, as well as sediments originating from soil erosion.

Interconnectivity among the various Tempisque marshes is also observed in the movements of animals. It is known that birds and fishes migrate seasonally and annually from one marsh to another. When open areas in some of the marshes disappear, duck populations leave for other, more open marshes in the region (McCoy and Rodríguez 1993), underlining the vital connection among marshes and their species populations at regional scales.

With the onset of the rainy season *gaspar* fish (*Atractosteus tropicus*) migrate upstream to reproduce in the marshes and ponds of the Caño Negro refuge. Similarly, at the onset of the dry season large populations of sardine (*Astyanax fasciatus*) migrate between the large wetland areas of the lower Tempisque River and the headwaters of its tributaries (López, 1978). Most of these connections have now disappeared or are significantly degraded.

Waterfowl that used to visit different marshes in the Lower Tempisque region now fly between wetlands and adjacent rice fields, so as to deal with the loss of some of the marshes. In a recent case study, over 50 species of waterfowl were recorded in a rice field, while only 31 species were observed in a natural marsh (Hurtado-Astaiza 2003). Not only was the diversity higher in the rice field but also over 70% of the common species turned out to be more abundant in rice fields than in adjacent marshes (Hurtado-Astaiza 2003). Clearly, farming-related activities enhance foraging opportunities for waterfowl while providing the birds with shallow open-water sites.

To some extent, the rice fields have mitigated marsh habitat loss in the Lower Tempisque River. They are, however, being replaced by sugarcane fields, a habitat not very suitable for waterfowl. By the year 2000, around 45% of the cropland in the Tempisque-Bebedero watershed areas was covered by sugarcane, while by 2007 sugarcane fields occupied a total of 70% of the available cropland in that area. In 2008, price increments in imported rice triggered the establishment of new rice fields in the area but in 2009 restrictions in credit access reduced again the amount of cropland covered by rice.

The dramatic losses in wetland areas observed in Costa Rica and the related loss of interconnectivity among the remaining sites (Fig. 20.17) underline the need to strengthen the management of protected wetland areas. A watershed approach is urgently needed to safeguard the remaining wetlands and maintain healthy ecological flows between those wetlands (Jiménez et al. 2005).

Fig. 20.17 Dredging at rivers and construction of levies are disconnecting wetlands from the riverine systems in most watersheds of the country. *Photo by Jorge Jiménez.*

Habitat losses are negatively affecting many of the ecological services normally provided by marshes. Water quality enhancement, a key regional function of marshes, is rapidly being lost. Large tracts of marshes, historically associated with the main watersheds in Costa Rica, have disappeared to give room to croplands. Dikes built along riverbanks have isolated remaining patches of marsh from the main river channels. All these transformations have reduced the capacity of marshes to trap sediments, pesticides, and persistent organic pollutants (POPs). Likely, economically important activities such as the artisanal fisheries in the Térraba-Sierpe Delta and the artisanal green clam (*Polymesoda radiata*) fisheries in the Tempisque River have already been impacted negatively by those changes.

The legal framework for protection of freshwater wetland resources in Costa Rica is very recent. It includes the adherence to the Ramsar Convention in 1991 and several articles in the national Wildlife Law (law nr. 7317) and Environmental Law (law nr. 7554), both established in 1995, as well as in the Forest Law (law nr. 7575) approved in 1996, and the Biodiversity Law (law nr. 7788) created in 1999.

Since wetland ecosystems have been declared as being of Public Interest (Environmental Law, article 41), one would expect freshwater wetlands to be better conserved. However, for decades legal voids have allowed freshwater wetland reclamation in the country. The definition of wetlands, for example, created over decades substantial vagueness around wetland conservation. In Costa Rica wetlands are legally considered to be "*areas of marsh, fen, peatland or water, whether natural or artificial, permanent or temporary, with water that is static or flowing, fresh, brackish or salt, including areas of marine water, the depth of which at low tide does not exceed six metres*" (Ramsar Convention 2009). This definition is identical to the one provided by the Convention on Wetlands (Ramsar, Iran 1971). Such an ample concept resulted in a ministerial requirement to delimit and declare any specific wetland, before it will be considered as a legally recognized wetland (Wildlife Law, article 7). Thus, under this interpretation, wetlands were not receiving any protected area status until a delimination and declaration process had been completed. Therefore, wetland areas within private lands—that is, not state-owned—were part of the national system of protected areas only after they were legally expropriated and the private owner had received the corresponding compensation (Res 2005–0461, Tribunal de Casación Penal). It was only recently (2011) that the Constitutional Court granted full protection to wetlands by declaring all of them part of the State's Natural Heritage, requiring those occuring in private properties to be protected.[1]

A study conducted by The Nature Conservancy in close collaboration with its partners (TNC 2009), clearly identifies key freshwater sites of important conservation value, including the Tempisque delta, the Caño Negro wetlands, the Tárcoles river mouth, the Osa swamps, and the Sixaola and San Juan River deltas, among others. It is in these areas that we will need to focus our attention in order to be able to protect the remaining marshes and swamps that still contain so much biodiversity.

Degradation and Restoration: The Case of the Palo Verde Marshes

Complementary to the protection of still-existing though threatened wetlands in Costa Rica, the restoration of degraded bogs, marshes, and swamps will become more and more important in the coming decades. Here, a case study on the ecological restoration of the main marshes of the Palo Verde National Park is presented. It discusses a series of steps that were undertaken to recover the original extent of open water, vegetation communities, and faunal assemblages that once characterized this important lowland wetland along the Pacific coast.

During the mid-1970s open water areas were the dominating habitat in the marshes of the Palo Verde National Park. After decades of periodic fires about 85% of its surface was dominated by open water, floating vegetation, and grasses. The elimination of periodic fires and the diversion of surface run-off water from the lagoon (meant to build roads inside the park) coincided with the elimination of the marsh's grazing and an eight-year dry-weather spell. These factors fostered the subsequent colonization of the marsh by *Typha dominguensis*. By 1986 most open-water areas had disappeared and *Typha* occupied around 57% of the marsh while the shrub *Parkinsonia aculeata* covered about 6%. Six years later *Typha* occupied 53% of the marsh and *P. aculeata* 18%. By the early 1990s, most of the open-water areas had disappeared (Castillo and Guzmán 2004).

Marsh degradation was further accelerated by the high evapotranspiration rates of *Typha* that further altered the hydrological regimes and ultimately affected the marsh considerably. During the rainy season evapotranspiration rates in *Typha*-dominated areas were 57–80% higher than in open waters, while these numbers were even higher during the dry season (90 to 128%) (Calvo and Arias 2006). Average actual evapotranspiration for this species has been calculated at 3.2 to 4.8 mm per day, one of the highest in vegetated areas in the region (Guzmán 2007). Reduced run-off and rainfall in combination with higher evapotranspiration rates have reduced water depth and the number of months that the marsh is flooded. These conditions favored the expansion of *Parkinsonia aculeata*, which covered less than 2% of the marsh in the 1970s, to almost 20% by 2002 (Solano 2004). In *Parkinsonia*-covered areas around 15% of the annual rainfall was intercepted by the shrubs' crowns (Calvo and Arias 2006). This amount of water represents about 8% of the total volume required to filll the marsh (Calvo and Arias 2004). Rainfall interception further reduced the lagoon's water levels and enhanced invasion by the bulrush or cattail *Typha*.

The rapid reduction in areas with open water resulted in a decrease of waterfowl species and floating aquatic plant diversity. By 1988 only 3,000 ducks and 500 teals were found where previously over 35,000 ducks and 25,000 teals were observed (McCoy and Rodríguez 1994). Highest waterfowl density occurred during December–February when migratory populations arrived from higher latitudes. Ultimately, the reduction of flooded areas in the wider wa-

[1] Constitutional Court Vote # 2011-016938.

tershed area promotes congregations of both migratory *and* local waterfowl populations, resulting in extraordinary concentrations of waterfowl in marshes with remaining open waters.

These dramatic reductions in open-water areas and the consequent decrease in waterfowl concentrations have raised important concerns in the conservation sector. Rehabilitation attempts of the Palo Verde marsh started in 1986, when creeks were dragged to increase drainage towards the lagoon. Cattle were partially reintroduced to the lagoon's area in the assumption that grazing pressure would open up the marsh (McCoy 1994, Burnidge 2000). Removal of *Typha* stands was also attempted in small areas of the marsh using mechanical crushing, a method that proved to be cheap and effective (McCoy and Rodríguez 1994). These efforts brought partial recovery of the waterfowl populations but lost ground to the re-expansion of *Typha* and *Parkinsonia* in the following years.

Additional efforts were started to eradicate *Typha* and *Parkinsonia* from the Palo Verde marsh at the start of this century. Restoration of the hydrological flows that feed into the marsh was the first step in this effort (Jiménez et al. 2003). Restoring the flow discharge (about 1,350,000 m^3 per year in rainy years) of the Huertón Creek back into the marsh resulted in a significant hydrological improvement. This creek had been previously diverted as a result of road construction back in the early eighties. Rebuilding the creek bed and constructing culverts at the road allowed the seasonal flows of this creek to discharge back into the Palo Verde marsh. Furthermore, the Chamorro Creek, which also had discharged water from the Piedra Blanca Marsh into the Palo Verde Marsh, had beeen diverted by the same road. The Chamorro Creek introduced significant amounts of water (between 2.3 and 9.4 million m^3 per year in rainy years) into the Palo Verde Marsh. Improving the existent culverts and cleaning the creek bed allowed the flows to reach the Palo Verde marsh again.

As a consequence of flow restoration, high water levels in the lagoon were reached earlier in the year and maintained for a period at least one month longer—a pattern most likely similar to that observed before the divertions. Once the hydrological flows were restored *Typha* and *P. aculeata* stands that had invaded the marsh and eliminated other types of vegetation were removed. By applying mechanical crushing and fire it was possible to remove dense mats of *Typha* in about 370 ha of marsh. Large open-water areas occurred after these interventions. The response of waterfowl was immediate. The next dry season over 25,000 ducks, 6,000 teals, and more than 3,500 woodstork (*Mycteria americana*) were observed in the marsh with hundreds of individuals belonging to some 60 other species of waterfowl (Trama 2004). Aquatic plant diversity also increased dramatically. Both floating and emergent species colonized the open water areas. Close to 80 species of floating and emergent aquatic plants colonized areas previously occupied exclusively by *Typha*.

While *Typha* dominance was eliminated, the long-term maintenance of wide open-water areas is still an open question. It is necessary to define whether those open-water areas depend in the long term on habitat disturbance. The existence of large seasonal concentrations of waterfowl in this marsh was the main reason for the establishment of this protected area. We need to ask if these seasonal concentrations were the result of periodic disturbances such as fire and grazing generated by previous cattle ranching operations. Or are they the result of crushing and fire as generated by the rehabilitation project? During the three years following the removal of the emergent vegetation significant open-water areas could naturally persist. It might, however, still be too soon to answer this question fully.

This particular case of wetland restoration highlights the need to restore the seasonal, hydrological, and fire regimes and at the same time recover the system's flora and fauna, in order to ensure bogs, marshes, and swamps are ecologically healthy and continue to provide much-needed environmental services for human well-being.

References

Allen, P.H. 1977. *The Rainforest of Golfo Dulce*. Redwood City, CA: Stanford University Press. 417 pp.

Allsteadt, J., and C. Vaughan. 1992. Population status of *Caiman crocodylus* (Crocodylia: Alligatoridae) in Caño Negro, Costa Rica. *Brenesia* 38: 57–64.

Álvarez, J., C. Asch, G. Oconitrillo, and S. Vargas. 1999. Plan de ordenamiento territorial para la gestión ambiental del humedal Sierpe de Osa, Costa Rica. San José, Costa Rica: Instituto Geográfico Nacional (IGN). 155 pp.

Anderson, R., and S.A. Mori. 1967. A preliminary investigation of *Raphia* palm swamps, Puerto Viejo, Costa Rica. *Turrialba* (2): 221–24.

Augspurger, C.K. 1985. Demography and life history variation of *Puya dasylirioides*, a long-lived rosette in tropical subalpine bogs. *Oikos* 45: 341–52.

Barrientos, Z. 2003. Estado actual del conocimiento y la conservación de los moluscos continentales de Costa Rica. *Revista de Biología Tropical* 51(suppl. 3): 285–92.

Bonilla-Murillo, F., D. Beneyto, M. Sasa. 2013. Amphibians and reptiles in the swamps dominated by the palm *Raphia taedigera* (Arecaceae) in northeastern Costa Rica. *Revista de Biología Tropical* 61(suppl. 1): 143–61.

Brak, B., M. Vroklage, M. Kappelle, and A.M. Cleef. 2005. Comunidades vegetales de la turbera de altura "La Chonta" en Costa Rica. In M. Kappelle and S.P. Horn, eds., *Páramos de Costa Rica*, 607–30. Costa Rica: Editorial INBio.

Bravo, J., J. González, G. Quiros, M. Alvarado, J. Sandí, and L. Piedra. 2000. Inventario de los bosques inundables de cerillo (*Symphonia globulifera*) en el humedal Térraba Sierpe, Costa Rica. *Boletín de humedales y zonas costeras* 2(5). San José. Costa Rica: IUCN, Oficina para Mesoaméica.

Bravo, J., M. Romero, A.J. Sánchez, and J. Reynolds. 1996. Inventario y evaluación de los humedales de la cuenca baja del río Tempisque, Guanacaste, Costa Rica. In *Workshop on the Utilization and Sustainable Management of the Hidric Resources* (November 28–December 1, 1996), 237. Heredia, Costa Rica: EUNA.

Brenes, L.G., F.J. Solano, and D.M. Salas. 2001. Degradación del sistema lagunar Caño Negro (norte costarricense) por sedimentación. *Ciencias Ambientales* 21: 36–41.

Burger, J., J.A. Rodgers, and M. Gochfeld. 1993. Heavy metal and selenium levels in endangered wood storks, *Mycteria americana*, from nesting colonies in Florida and Costa Rica. *Archives of Environmental Contamination and Toxicology* 24(4): 417–20.

Burnidge, W.S. 2000. Cattle and management of freshwater neotropical wetlands in the Palo Verde National Park, Guanacaste, Costa Rica. M.Sc. thesis, The University of Michigan, School of Natural Resources and Environment. Ann Arbor, MI. 89 pp.

Cabrera, J., C.L. Ampie, and G. Galeano. 1992. Presencia de *Oreochromis niloticus* (Pisces: Cichlidae) en lagunas estacionales del Refugio Nacional de Vida Silvestre Caño Negro, Costa Rica. *Brenesia* 38: 169–70.

Calvo, J.C., and O. Arias. 2004. Restauración hidrológica del humedal Palo Verde. *Ambientico* 129: 7–8.

Calvo, J.C., and O. Arias. 2006. Estudio de evapotranspiración de la tifa (*Typha dominguensis*) en el Parque Nacional Palo Verde, Guanacaste, Costa Rica. Research Report. San José, Costa Rica: Organization for Tropical Studies (OTS). 14 pp.

Castillo, M., and J.A. Guzmán. 2004. Cambios en la cobertura vegetal en el Humedal Palo Verde según SIG. *Ambientico* 129: 4–6.

Castillo, R., and J. March. 1993. Cambios en los habitats ecológicos del Refugio Nacional de Vida Silvestre de Caño Negro 1961–1992. *Revista de Ciencias Sociales de la Universidad de Costa Rica* 62: 51–67.

Chavarría, C.R., and O. Valverde. 2000. Delimitación y muestreo florístico del humedal de Punta Mona, Gandoca-Manzanillo, Costa Rica. *Boletin de Humedales y Zonas Costeras* 2(5). San José, Costa Rica: IUCN.

Chaves, G., A. García-Rodríguez, A. Mora, and A. Leal. 2009. A new species of dink frog (Anura: Eleutherodactylidae: *Diasporus*) from Cordillera de Talamanca, Costa Rica. *Zootaxa* 2088: 1–14.

Córdoba, R., J.C. Romero, and N.J. Windevoxhel. 1998. *Inventario de los Humedales de Costa Rica*. San José, Costa Rica: IUCN. 380 pp.

Crow, G.E. 2002. *Plantas Acuáticas del Parque Nacional Palo Verde y el Valle del Río Tempisque*. Santo Domingo de Heredia, Costa Rica: INBio. 300 pp.

Devall, M., and R. Kiester. 1987. Notes on *Raphia* at Corcovado. *Brenesia* 28: 89–96.

Ellison, A.M. 2004. Wetlands of Central America. *Wetlands Ecology and Management* 12: 3–55.

Gamboa, M.E. 2003. El comportamiento de *Jabiru mycteria* durante la época reproductiva en humedales de la zona norte de Costa Rica. *Zeledonia* 7(1): 25–32.

Gómez, L.D. 1984. Las plantas acuáticas y anfibias de Costa Rica y Centroamérica. 1. Liliopsida. San José, Costa Rica: Editorial Universidad Estatal a Distancia (EUNED). 430 pp.

Gómez, L.D. 1986. Vegetación de Costa Rica. San José, Costa Rica: Editorial Universidad Estatal a Distancia (EUNED).

Guzmán, J.A. 2007. Effects of land cover changes on the water balance of the Palo Verde Wetland, Costa Rica. Enschede, the Netherlands: International Institute for Geo-Information Science and Earth Observation. 85 pp.

Hartshorn, G. 1983. Introduction: Plants. In D.H. Janzen, ed., *Costa Rican Natural History*, 118–83. Chicago: University of Chicago Press.

Hartshorn, G. 1988. Tropical and subtropical vegetation of Mesoamerica. In M.G. Barbour and W.D. Billing, eds., *North American Terrestrial Vegetation*, 365–90. Cambridge, UK: Cambridge University Press.

Hartshorn, G.S., and B.E. Hammel. 1994. Vegetation types and floristic patterns. In L.A. McDade, K.S. Bawa, H.A. Hespenheide, and G.S. Hartshorn, eds., *La Selva: Ecology and Natural History of a Neotropical Rain Forest*, p. 73–89. Chicago: University of Chicago Press.

Hernández, D. 1990. Cambios anuales en la composición y distribución de la vegetación acuática en el humedal estacional de Palo Verde, Costa Rica. Master's thesis, Programa Regional en Manejo de Vida Silvestre, Universidad Nacional. Heredia, Costa Rica. 25 pp.

Hernández, D. 1993. Uso de habitat por patos Anatidae en un humedal estactional de Costa Rica. In *Proceedings of the Ornithological Congress of Costa Rica*, 10. San José, Costa Rica: CIPA / MNCR / PRMVS / UNA.

Herrera, W. 1986. Clima de Costa Rica. San José, Costa Rica: Editorial Universidad Estatal a Distancia (EUNED).

Hidalgo, C. 1993a. Avifauna del Refugio Nacional de Vida Silvestre Caño Negro. In *Proceedings of the Ornithological Congress of Costa Rica*, 29. San José, Costa Rica: CIPA / MNCR / PRMVS / UNA / UCR.

Hidalgo, C. 1993b. Residuos de plaguicidas organoclorados en huevos de *Mycteria americana*. In *Proceedings of the Ornithological Congress of Costa Rica*, 22. San José, Costa Rica: CIPA / MNCR / PRMVS / UNA.

Holdridge, L.R., W.C. Grenke, W.H. Hatheway, T. Liang, and J.A. Tosi. 1971. *Forest Environments in Tropical Life Zones: A Pilot Study*. Oxford: Pergamon Press. 746 pp.

Hooghiemstra, H., A.M. Cleef, G.W. Noldus, and M. Kappelle. 1992. Upper Quaternary vegetation dynamics and paleoclimatology of the La Chonta bog area. *Journal of Quaternary Science* 7(3): 205–25.

Horn, S.P. 1988. Effect of burning on a montane mire in the Cordillera de Talamanca, Costa Rica. *Brenesia* 30: 81–92.

Horn, S.P. 1989. Prehistoric fires in the Chirripó páramo of Costa Rica: sedimentary charcoal evidence. *Revista de Biología Tropical* 37: 139–48.

Horn, S.P. 1993. Postglacial vegetation and fire history of the Chirripó páramo of Costa Rica. *Quaternary Research* 40: 107–16.

Horn, S.P. 2005. Dinámica de la vegetación después de fuegos recientes en los páramos de Buenavista y Chirripó, Costa Rica. In M. Kappelle and S.P. Horn, eds., *Páramos de Costa Rica*, 631–55. Santo Domingo de Heredia, Costa Rica: Instituto Nacional de Biodiversidad (INBio).

Horn, S.P., and L.M. Kennedy. 2006. Pollen evidence of the prehistoric presence of cattail (*Typha* sp.) in Palo Verde National Park, Costa Rica. *Brenesia* 66: 85–87.

Horn, S.P., K.H. Orvis, and K.A. Haberyan. 2005. Limnología de las lagunas glaciales en el páramo del Chirripó, Costa Rica. In M. Kappelle and S.P. Horn, eds., *Páramos de Costa Rica*, 161–81. Santo Domingo de Heredia, Costa Rica: Instituto Nacional de Biodiversidad (INBio).

Hurtado-Astaiza, J. 2003. Abundancia, diversidad, riqueza, uso de hábitat y comportamiento de aves acuáticas: una comparación entre un humedal seminatural y un arrozal con riego en Costa Rica. M.Sc. thesis, Wildlife Management Regional Program (PRMVS), Universidad Nacional (UNA). Heredia, Costa Rica. 108 pp.

Islebe, G., H. Hooghiemstra, and R. van't Veer. 1996. Holocene vegetation and water level history in two bogs of the Cordillera de Talamanca, Costa Rica. *Vegetatio* 124:155–71.

Islebe, G., H. Hooghiemstra, and R. van't Veer. 2005. Historia holocénica de la vegetación y el nivel de agua en dos turberas de la Cordillera de Talamanca, Costa Rica. In M. Kappelle and S.P. Horn, eds., *Páramos de Costa Rica*, 237–52. Santo Domingo de Heredia, Costa Rica: Instituto Nacional de Biodiversidad (INBio).

Janzen, D.H. 1978. Description of a *Pterocarpus officinalis* monoculture in Corcovado National Park. *Brenesia* 14–15: 305–9.

Janzen, D.H., ed. 1983. *Costa Rican Natural History*. Chicago: University of Chicago Press. 816 pp.

Jiménez, I. 2000. Los manatíes del río San Juan y los canales de Tortuguero: ecología y conservación. San José, Costa Rica: Amigos de la Tierra. 120 pp.

Jiménez, J.A. 1981. The mangroves of Costa Rica: a physiognomic characterization. M.Sc. thesis, University of Miami. Miami, FL. 130 pp.

Jiménez, J.A., J.C. Calvo, E. González, and F. Pizarro. 2005. Conceptualización de caudal ambiental en Costa Rica: determinación inicial para el río Tempisque. San José, Costa Rica: IUCN.

Jiménez, J.A., E. González, and J. Calvo. 2003. Recomendaciones técnicas para la restauración hidrológica del Parque Nacional Palo Verde. San Pedro, Costa Rica: Organización para Estudios Tropicales (OTS). 11 pp.

Kappelle, M. 1996. *Los Bosques de Roble (*Quercus*) de la Cordillera de Talamanca, Costa Rica: Biodiversidad, Ecología, Conservación y Desarrollo*. Santo Domingo de Heredia, Costa Rica: Instituto Nacional de Biodiversidad (INBio). 336 pp.

Kappelle, M., M. Castro, H. Acevedo, L. González, and H. Monge. 2003. *Ecosystems of the Osa Conservation Area, Costa Rica*. Bilingual edition (English-Spanish). Santo Domingo de Heredia, Costa Rica: Instituto Nacional de Biodiversidad (INBio). 496 pp.

Kappelle, M., and J. Savage. 2005. Anfíbios y reptiles de los páramos de Costa Rica. In M. Kappelle and S.P. Horn, eds., *Páramos de Costa Rica*, 513–19. Santo Domingo de Heredia, Costa Rica: Instituto Nacional de Biodiversidad (INBio).

Lieberman, D., M. Lieberman, R. Peralta, and G.S. Hartshorn. 1985. Mortality patterns and stand turnover rates in a wet tropical forest in Costa Rica. *Journal of Ecology* 73: 915–24.

López, M.I. 1978. Migración de la sardina *Astyanax fasciatus* (Characidae) en el río Tempisque, Guanacaste, Costa Rica. *Revista de Biología Tropical* 26(1): 261–75.

McCoy, M. 1994. Seasonal, freshwater marshes in the tropics: a case where cattle grazing is not bad. In G. Meffe, C.R. Carroll, and M.A. Sunderland, eds., *Principles of Conservation Biology*, 352–53. Sunderland, MA: Sinauer.

McCoy, M.B., and J.M. Rodríguez. 1993. Números mensuales de patos silvestres en los pantanos de la cuenca baja del Río Tempisque, Guanacaste, durante las estaciones secas de 1986–1990. In *Proceedings of the Orthnitological Congress of Costa Rica*, 9. San José, Costa Rica: CIPA / MNCR / PRMVS / UNA.

McCoy, M.B., and J.M. Rodríguez. 1994. Cattail (*Typha dominguensis*) eradication methods in the restoration of a tropical seasonal, freshwater marsh. In W.J. Mitsch, ed., *Global Wetlands: Old World and New*, 469–82. Dordrecht, Netherlands: Elsevier Science.

McHargue, L.A., and G.S. Hartshorn. 1983. Seed and seedling ecology of *Carapa guianensis*. *Turrialba* 33(4): 399–404.

Mora, M., J. Cabrera, and G. Galeano. 1997. Reproducción y alimentación del gaspar *Atractosteus tropicus* (Pisces: Lepisosteidae) en el Refugio Nacional de Vida Silvestre Caño Negro, Costa Rica. *Revista de Biología Tropical* 45(2): 861–66.

Myers, R.L. 1990. Palm Swamps. In A.E. Lugo, M. Brinson, and S. Brown, eds., *Forested Wetlands*, 267–86. Amsterdam: Elsevier Science Publishers. 527 pp.

Phillips, S. 1995. Holocene evolution of the Changuinola Peat Deposit, Panama: Sedimentology of a marine-influenced tropical peat deposit on a tectonically active coast. PhD diss., University of British Columbia. British Columbia, Canada.

Pizarro, J.F., and J.R. Rojas. 1993. Presencia de tilapia, *Oreochromis* (Pisces: Cichlidae) en la desembocadura del río Bebedero, Golfo de Nicoya, Costa Rica. *Revista de Biología Tropical* 41(3b): 921–24.

Programa Estado de la Nación. 2003. Noveno Informe del Estado de la Nación en Desarrollo Humano Sostenible. San José, Costa Rica: Programa Estado de la Nación.

Ramsar Convention. 2009. The Ramsar Convention definition of "wetland" and classification system for wetland type. http://www.ramsar.org.

Rodgers, J.C., and S.P. Horn. 1996. Modern pollen spectra from Costa Rica. *Palaeogeography, Palaeoclimatology, Palaeoecology* 124: 53–71.

Rodríguez, G. 2004. Importancia de los Humedales en Costa Rica: Investigadores relacionados con el tema de Humedales. San José, Costa Rica: Asamblea Legislativa. 14 pp.

Saber, P.A., and W.A. Dunson. 1978. Toxicity of bog water to embryonic and larval anuran amphibians. *Journal of Experimental Zoology* 204(1): 33–42.

Sáenz, I., M. Protti, and J. Cabrera. 2006. Composición de especies y diversidad de peces en un cuerpo de agua temporal en el Refugio Nacional de Vida Silvestre Caño Negro, Costa Rica. *Revista de Biología Tropical* 54(2): 639–45.

Sánchez, J. 2001. Estado de la población de Cocodrilos (*Cocrodylus acutus*) en el río Tempisque, Guanacaste. Santo Domingo de Heredia, Costa Rica: Instituto Nacional de Biodiversidad (INBio). 49 pp.

Scott, D.A., and M. Carbonell, eds. 1986. *A Directory of Neotropical Wetlands*. Cambridge, UK: IUCN. 684 pp.

Serrano-Sandí, J., F. Bonilla-Murillo, and M. Sasa. 2013. Distribution, surface and protected area of palm swamps in Costa Rica and Nicaragua. *Revista de Biología Tropical* 61(suppl. 1): 25–33.

Solano, J. 2004. Estudio de la distribucion y abundacia de *Parkinsonia aculata* y *Typha dominguensis* en el Humedal Palo Verde. Research

Report for the B.Sc. Degree, Instituto Tecnológico de Costa Rica (ITCR). Cartago, Costa Rica. 97 pp.

Soto, R., and E. Arias. 1994. Informe de humedales de la Península de Osa. Agua Buena de Rincón de Osa, Costa Rica: Fundación Neotrópica, Proyecto Boscosa. 29 pp.

Sylvander, R.B. 1978. Los bosques del país y su distribución por provincia. Desarrollo integral de los recursos forestales. PNUD/FAO-COS/72/013. Dirección General Forestal, San José, Costa Rica: Working Paper No. 15. 64 pp.

TNC (The Nature Conservancy). 2009. *Evaluación de Ecorregiones de Agua Dulce de Mesoamérica: Sitios Prioritarios para la Conservación en las Ecorregiones de Chiápas a Darién*. San José, Costa Rica: The Nature Conservancy. 515 pp.

Trama, F. 2004. Restauración del Humedal Palo Verde para aves. *Ambientico* 129: 11–12.

Trama, F. 2005. Manejo Activo y Restauración del Humedal Palo Verde: Cambios en las coberturas de vegetación y respuesta de las aves acuáticas. M.Sc. thesis, Universidad Nacional, Heredia, Costa Rica: International Institute on Wildlife Management and Conservation (ICOMVIS). 154 pp.

Trama, F., F.L. Rizo-Patrón, and M. Springer. 2009. Macroinvertebrados bentónicos del Humedal de Palo Verde, Costa Rica. *Revista de Biología Tropical* 57(suppl. 1): 275–84.

Umaña, G. 1991. Fitoplancton de Caño Negro: un llano de inundación tropical, Costa Rica, América Central. *Proceedings of the III National Congress on Biology*. San Pedro, Costa Rica: University of Costa Rica. 31 pp.

Urquhart, G.R. 1997. Paleoecological evidence of *Raphia* in the precolumbian Neotropics. *Journal of Tropical Ecology* 13(6): 783–92.

Urquhart, G.R. 1999. Long-term persistence of *Raphia taedigera* Mart. swamps in Nicaragua. *Biotropica* 31(4): 565–69.

Vial, J.L. 1966. Variation in altitudinal populations of the salamander, *Bolitoglossa subpalmata*, on the Cerro de la Muerte, Costa Rica. *Revista de Biología Tropical* 14: 111–21.

Vial, J.L. 1968. The ecology of the tropical salamander *Bolitoglossa subpalmata* in Costa Rica. *Revista de Biología Tropical* 15: 13–115.

Villarreal-Orias, J.A. 1997. Estado actual, presas y uso del hábitat del jabirú (*Jabirú mycteria*) en la cuenca baja del río Tempisque, Costa Rica. M.Sc. thesis, Regional Wildlife Management Program (PRMVS), Universidad Nacional. Heredia, Costa Rica.

Villarreal-Orias, J.A. 2000. Tamaño poblacional, reproducción y habitat del jabirú (*Jabiru mycteria*) en el Área de Conservación Tempisque, Costa Rica. Regional Wildlife Management Program (PRMVS), Universidad Nacional. Heredia, Costa Rica. 24 pp.

Webb, E.L. 1998. Gap-phase regeneration in selectively-logged lowland swamp forest, northeastern Costa Rica. *Journal of Tropical Ecology* 14(2): 247–60.

Webb, E.L. 1999. Growth ecology of *Carapa nicaraguensis* Aublet (Meliaceae): implications for natural forest management. *Biotropica* 31(1): 102–10.

Weston, A.S. 1981. Páramos, Ciénagas and Subpáramo Forest in the Eastern Part of the Cordillera de Talamanca. San José, Costa Rica: Tropical Science Center (TSC). 15 pp.

Zamora, N., and J. Bravo. 1992. Caracterización de la Vegetación del Refugio Nacional de Vida Silvestre Caño Negro, Alajuela, Costa Rica. *Revista de Ciencias Ambientales* 9: 4–21.

Part X

Conclusion

Chapter 21 Costa Rican Ecosystems: A Brief Summary

Maarten Kappelle[1,2]

Introduction

The present volume, *Costa Rican Ecosystems*, offers an extensive panorama of the main terrestrial, freshwater, coastal, and marine ecosystems that inhabit the small tropical country of Costa Rica. Actually, all key ecosystems that occur in the tropics—with exception of arid deserts and snowy mountain caps—are found in this tiny piece of land and sea along the Central American Isthmus—probably the most species-dense country on Earth (Kappelle, chapter 1 of this volume).

In this closing chapter, a summary of Costa Rica's main ecosystems is given, using the preceding chapters as its foundation. It tries to offer the reader a synopsis of Costa Rica's overwhelming diversity of life at the highest organizational level: the ecosystem. Hence, this chapter touches briefly upon alpha, beta, and gamma levels of biodiversity, and points out some of the main ecological processes that act at the interface of species and their environment. However, the chapter is not meant to give a full overview of each individual ecosystem; to better understand the ins and outs of each ecological system, the reader is referred to the corresponding chapter for further reading.

Oceanic and Coastal-Marine Ecosystems

Costa Rica's marine and coastal ecosystems include a wealth of plant and animal associations, from the deep sea bottom up to the Pacific and Caribbean shores. These ecosystems contain almost 7,000 species of vertebrates, invertebrates, plants, and microorganisms; almost 75% of all species occur in the Pacific salty waters. Invertebrates are most abundant with nearly 4,000 species at the Pacific coast and Ocean (Cortés, chapter 5 of this volume). At sea level the mean annual temperature is approximately 27°C (Herrera, chapter 2 of this volume).

Rocky intertidal shores make up for more than half of the Pacific coast and offer a home to both carnivorous and herbivorous gastropods. At Cahuita on the Caribbean coast a kind of endolithic fauna on top of carbonate substrates is observed (Cortés, chapter 17 of this volume). At Isla del Coco the rocky intertidal environment colors pink and black owing to incrusting algae that grow over barnacles on rock outcrops; here, caves abound along the island's steep rocky cliffs (Cortés, chapter 7 of this volume). Intertidal barnacles like *Tetraclita rubescens* and *T. stalactifera* as well as the porcellanid crab, *Petrolisthes armatus*, have been observed at rocky outcrops along the Golfo de Nicoya (Vargas, chapter 6 of this volume).

Sandy beaches occur widely along both coasts and include some of the last, globally significant nesting sites of marine turtles. Four species nest along the Caribbean shores: *Chelonia mydas* (green turtle), *Dermochelys coriacea* (leatherback), *Eretmochelys imbricata* (hawksbill), and sporadically *Caretta caretta* (loggerhead) (Cortés, chapter 17 of this volume). At Ostional along the Pacific coast (Península de Nicoya) a significant portion of the world's population of olive ridley turtles (*Lepidochelys olivacea*) nests.

In the northern sector of the Caribbean shores, between Barra del Colorado and Moín, a number of elongated

[1] World Wide Fund for Nature (WWF International), Avenue du Mont-Blanc 1196, Gland, Switzerland

[2] Department of Geography, University of Tennessee, Knoxville, TN 37996, USA

coastal lagoons occur parallel to the coastline. They form part of a wave-dominated delta that has formed along a passive continental margin. The northern lagoons are home to manatees and fish species alike. Sadly, their manatee populations are declining because of human action. The lagoons include a few dredged sections and are all interconnected. On the flooded lands along the channels *Pterocarpus officinalis* trees and *Raphia taedigera* palms dominate the vegetation structure. The southern section of the Caribbean coast lacks these typical lagoons and has—just like the Pacific coast—seismically active continental margins (Alvarado and Cárdenes, chapter 3, and Cortés, chapter 17, of this volume).

At the ecotone of the river mouths of the Río Tempisque and Río Grande de Térraba occur large swaths of mangrove forests with a variety of salt-tolerant trees such as *Avicennia germinans* and *Rhizophora mangle*. They represent some of the best developed, most diverse, and largest mangroves in Central America. These cradles of biodiversity act as filters between river watersheds and marine systems and serve as nurseries for numerous prawn, bivalves, and fishes (Cortés, chapter 5 of this volume). At the same time, seasonality of the mangrove-rich Golfo de Nicoya seems to be determined by salinity changes as a result of rainfall differences, rather than by changes in water temperature (Vargas, chapter 6 of this volume). Here, primary production is among the highest reported for tropical estuaries.

On the other hand, mangrove forests along the Caribbean coast are less well developed and less productive. The small patch of mangroves found close to Moín, the main port at the Caribbean close to Puerto Limón, has been negatively impacted by construction and pollution. Fortunately, the largest mangrove forest on the Caribbean coast—the one at Laguna Gandoca—is rather well preserved. Next to the previously mentioned mangrove species, two other mangrove species are found at Gandoca, *Laguncularia racemosa* and *Conocarpus erectus* (Cortés, chapter 17 of this volume).

At other places, like in the upper Golfo de Nicoya, intertidal sand and mud flats exist, which form essential habitat to tropical tidal flat benthos and a variety of migratory shorebirds (e.g, waders). Subtidal sediments—those that are permanently submerged—abound in the Golfo de Nicoya and provide a home to infaunal communities of polychaete worms, crustaceans, and mollusks. The flats near the port of Punta Morales are dominated by deposit-feeding polychaete worms. Nematoda, Foraminifera, Copepoda, and Ostracoda have been reported from this site (Vargas, chapter 6 of this volume).

Highly productive seagrass beds dominated by marine flowering plants are relatively rare in Costa Rica but do occur along the Pacific coast—for instance, along the mouth of the Sierpe river (Cortés, chapter 5 of this volume). At the Caribbean coast seagrass species *Thalassia testudinum* and *Syringodium filiformis* form extensive beds, mainly in reef lagoons. In Cahuita, average productivity of *T. testudinum* was 2.7 ± 1.15 g/m^2/day (Fonseca et al. 2007, cited by Cortés, chapter 17 of this volume).

Coral reefs occur along both the Pacific and Caribbean shores. Coral community patches and isolated coral colonies cover only a few hectares, have a low coral diversity, and are discontinuous (Cortés, chapter 5 of this volume). Pacific coastal reefs may be constructed and dominated by coral species such as *Pavona gigantea*, *P. clavus*, and *Pocillopora eydouxi*. Unfortunately, several reefs such as those found at places between Cabo Blanco and Golfo de Nicoya are mostly dead, while recovery is slow. Isla del Caño off the west coast of Península de Osa has some of the most extensive coral reefs of the country. Here, pocilloporids dominate reef associations, although twenty species of octocorals, two black corals, seventeen reef-building corals, and four ahermatypic coral species have been identified from this island (Cortés, chapter 5 of this volume).

Along the Caribbean coast, fringing reefs are found between Moín and Limón, including Isla Uvita. However, the largest Caribbean fringing reef plus patch reefs and carbonate banks are observed inside Parque Nacional Cahuita. Additionally, smaller patches are located between Puerto Viejo and Punta Mona. Unfortunately, the 1991 Limón Earthquake, which uplifted the coast, together with coral bleaching and the massive dying of sea urchins have taken their toll on the Caribbean reefs, as Cortés explains (Cortés, chapter 17 of this volume).

In turn, Isla del Coco has extensive coral reefs and coral communities made up of eighteen species of zooxanthellate corals and fifteen azooxanthelate species. The main reef builder is *Porites lobata*. Isla del Coco's reefs are partially eroded biologically by corallivores that feed on them, like the crown-of-thorns starfish *Acanthaster planci* and the pufferfish *Arothron meleagris*. Unfortunately, thirty years ago the island's coral reefs were severely impacted by the 1982–1983 El Niño Southern Oscillation (ENSO) event, from which the reefs still recover (Cortés, chapter 7 of this volume).

Rhodolith beds composed of coralline red macroalgae are found at Isla del Coco. Costa Rica's pelagic ecosystem—which includes the neritic or coastal zone and the oceanic zone—is rich in plankton, fish, whales, and dolphins (e.g., the spotted coastal dolphin, *Stenella attenuata*; Cortés, chapter 7 of this volume). In the Golfo de Nicoya some dinoflagellates that are part of the local phytoplankton community, like *Cochlodinium catenatum*, may occasionally cause harmful red tides (*marea roja*, a kind of toxic algal bloom) that lead to increased fish mortality (Vargas, chapter 6 of this volume).

Little is known about the deep benthos of Costa Rica. However, bacterial mats and specific bivalves have been observed in some deep, low-oxygen waters in the Pacific Ocean (Cortés, chapter 5 of this volume). The deep Golfo Dulce serves as a low-oxygen tropical fjord. The Golfo de Nicoya is rich in commercial fish species such as *corvina* (Sciaenidae), red snapper (Lutjanidae), sea catfishes (Ariidae), and flatfishes (several families). Other common groups include sharks (e.g., *Mustelus dorsalis*, *M. lunulatus*), rays (e.g., *Urotrygon cimar*), flounders, gobies, morays, and congers (Vargas, chapter 6 of this volume).

At Isla del Coco there is no gradual change from a shallow (<50 m) to a deep sea fauna, but an abrupt change occurs around 50 m. As Cortés (chapter 5 of this volume) points out, below that depth many species, including entire groups of organisms never seen in Isla del Coco's shallow waters, have been observed. Small islands and islets like Isla Bolaños at the Pacific coast form rocky outcrops and serve as important nesting areas for seabirds. Furthermore, in the deep waters of the Caribbean, the presence of copepods, siphonophorans, heteropods, copepods, euphausids, chaetognaths, and salps has been reported (Cortés, chapter 17 of this volume).

The Lowland Seasonally Dry Forests

Northwestern Costa Rica harbors the southernmost sector of Meso-America's dry deciduous lowland forests: Guanacaste's *bosque seco*. Its seasonal climate is characterized by a very noticeable and regularly occurring relatively rain-free dry season (the *época seca*) of four to eight months duration, while in the remaining period a total of 500 to 3,000 mm rain falls (Janzen and Hallwachs, chapter 10, and Herrera, chapter 2, of this volume). During the dry season in February and March, at least sixty deciduous and semi-deciduous trees lose their foliage and start to blossom (e.g., *Bursera simaruba* and *Tabebuia rosea*). *Bursera* trees frequently show signs of exfoliation (i.e., the peeling off of their bark). The average annual temperature in the dry deciduous forests along the North Pacific coast is 27.6°C, while occasional maximum extremes of 41°C have been recorded in the hot, dry lowlands of the Río Tempisque watershed area (Herrera, chapter 2 of this volume).

Adjacent to the dry deciduous forests of Guanacaste—including central Nicoya—are the seasonal forests of northern and central Puntarenas and the neighboring Central Valley or *Meseta Central*. These forests are partially dry and partially moist, and are characterized by a more or less clear dry season. Many of its deciduous species are also found in the northern Pacific region, but intermingle with trees and shrubs that flourish in somewhat wetter environments (Jiménez and Carrillo, chapter 11 of this volume). This Central Pacific seasonal forest region includes the La Cangreja, Carara, and Manuel Antonio national parks, as well as the *zonas protectoras* of El Rodeo, Nara, and Turrubares. Parque Nacional Carara—with semi-evergreen and evergreen forest patches—is probably the best example of the transition zone that gradually links the Northern dry deciduous forests in Guanacaste with the wet and rain forests of the Pacific coastal zone in the south (e.g., the Osa Peninsula).

Researchers cited by Jiménez et al. (chapter 9 of this volume) estimate that prior to 1940 the total dry forest area in Costa Rica covered almost 8% of the national territory. By 1950 this area was reduced to about 40,000 hectares, mainly owing to logging and clearing for pastures to raise cattle. During the following decades, practically all of Costa Rica's dry forests disappeared, with the exception of a few remaining, protected fragments in some parts of Guanacaste (e.g., Parque Nacional Santa Rosa). Janzen and Hallwachs (chapter 10 of this volume) highlight the subsequent ecological restoration efforts that were undertaken to reestablish the dry deciduous forest in the Área de Conservación Guanacaste (ACG) since the mid-1980s.

The seasonally dry deciduous forests of the Northern Pacific lowlands of Guanacaste and the northern sector of the Nicoya Peninsula range from sea level up to average elevations of 400 m, and up to 700 m at some local peaks. They are less complex than Costa Rica's wet lowland forests in terms of floristic composition and structure (Jiménez et al., chapter 9 of this volume, and references therein). According to D.H. Janzen (pers. comm.), apparently some 300,000 species (65% of the estimated number of species in Costa Rica) call the Area de Conservación Guanacaste (ACG) their home. Intermingled with the truly seasonal, dry deciduous forest, another forest type can be observed in the Guanacaste landscape: the riparian forest that grows in strips along streams and often houses larger trees that are evergreen and keep their leaves throughout the year. Such trees (e.g., *Anacardium*, *Brosimum*, *Guarea*, *Hymenaea*, and *Ocotea*) benefit year-round from the waters of the rivers at the edges of which they thrive well.

Guanacaste's dry deciduous forest has two vegetation layers, the canopy and the understory. Its canopy stratum is 20 to 30 m tall and is composed of short, stout trunks and large, often spreading, flat-topped crowns often with small leaves or compound leaves with leaflets (e.g., in the case of legume trees). The crowns are usually not in lateral contact with each other (Jiménez et al., chapter 9 of this volume). Many woody species are equipped with thorns and spines. This forest type harbors at least 250 species of ground-rooted vascular plants including lots of woody lianas. Epiphytes are few, though epiphytic cacti like *Hylocereus costaricensis* may locally abound. Fabaceae represent the dominant tree and shrub family in most stands. Important tree species that contribute significantly to dry forest basal area are *Quercus oleoides*, *Hymenaea courbaril*, *Manilkara zapota*, *Sloanea terniflora*, *Luehea* spp., *Ficus* spp., *Dilodendron costaricense*, and *Calycophyllum candidissimum* (Jiménez et al., chapter 9 of this volume).

Near Bagaces and Palo Verde close to the Río Tempisque, current secondary successional forests are co-dominated by tree species like *Calycophyllum candidissimum*, *Licania arborea*, *Brosimum alicastrum*, *Spondias mombin*, *Guazuma ulmifolia*, *Thouinidium decandrum*, *Caesalpinia eriostachys*, *Luehea candida*, *Tabebuia ochracea*, and *Pachira quinata*. Further to the south, at the Lomas Entierro site in the seasonal forests of Parque Nacional Carara, also a fair number of deciduous species are found—for example, *Calycophyllum candidissimum*, *Triplaris melaenodendron*, *Enterolobium cyclocarpum*, and the thorny *Bombacopsis quinata* (Jiménez and Carrillo, chapter 11 of this volume). Here, locally emergent trees such as *Caryocar costaricense* can grow as tall as 40 m. Eastward of Carara, in the Escazú mountains at the southern slopes of the Central Valley, seasonal premontane forests at 1,200 m elevation may harbor trees like *Croton draco*, *Ficus costaricana*, *Myrcianthes fragrans*, *Trema micrantha*, *Zanthoxylum caribaeum*, among others. Here, in the intramontane valleys of the Central Pacific Conservation Region (ACOPAC), fungi may locally abound; at least a hundred thirty different species are known to reside in the seasonal forests of this part of the country.

While Guanacaste's lowlands are drier than the central and southern sector of the Península de Nicoya, the latter has a greater variety of ecosystems and flora owing to its combination of dry plant communities and more moist vegetation types. In the Nandayure area of central Nicoya at least 665 species of plants occur. At the same time, numerous dry forest species known from the Guanacaste plains do not reach southern Nicoya where a moister climate with annual rainfall up to 2,000 mm serves as a natural barrier to these species (Jiménez et al., chapter 9 of this volume, and references therein).

The vertebrate diversity of Costa Rica's northern dry deciduous forests and adjacent seasonal forests is extraordinary. Around 110 species of mammals are present in Guanacaste's dry forests alone, including bats, rodents, marsupials, weasels, cats, raccoons, primates, artiodactyls, canids, edentates, a rabbit, and a tapir (Jiménez et al., chapter 9 of this volume). Squirrel monkeys (*Saimiri oerstedi*) used to abound in the seasonal forests of Parque Nacional Manuel Antonio in the Central Pacific region (Jiménez and Carrillo, chapter 11 of this volume). Dry forest bird diversity includes some 345 species, including the impressive jabiru wood stork. The three-wattled bellbird (*Procnias tricarunculata*) is known to migrate altitudinally between Carara and Cerro Turrubares in the Central Pacific region.

The herpetofauna is also quite remarkable with some 18 amphibian and 59 reptile species known from the Parque Nacional Santa Rosa alone, including the dry forest dweller *Agkistrodon bilineatus*, a pit viper (*serpiente moccasin*, cantil snake), and the herbivorous green iguana (*Iguana iguana*). Fer-de-lances (*Bothrops asper*) and black-headed bushmasters (*Lachesis melanocephala*) are locally common. Crocodiles (*Crocodylus acutus*) are frequently observed in the larger Pacific river mouths (e.g., Río Grande de Tárcoles). The freshwater fish fauna of the northern dry deciduous forest streams is relatively poor compared to wet forest river ichthyofauna. There are at least 53 species of freshwater fish in the ecological drainage units of Santa Elena, Nosara, and Nicoya, but perhaps up to sixty, with only four being endemic to the region. The locally most diverse freshwater fish families here are Cichlidae, Characidae, and Poeciliidae (Jiménez et al., chapter 9 of this volume).

Over the past centuries, habitat conversion, hunting, and poaching have significantly reduced the original populations of jaguar, puma, and Baird's tapir (Jiménez et al., chapter 9 of this volume). Fortunately, forest protection and ecological restoration, including natural regeneration and man-induced reforestation, have led to a revival of some local populations of species such as the spider monkey (*Ateles geoffroyi*) and two species of peccary (Janzen and Hallwachs, chapter 10 of this volume).

Next to the considerable vertebrate diversity, the variety of insects and spiders in the dry deciduous lowlands of northern Costa Rica is extraordinary. There are at least 13,000 insect species in the Parque Nacional Santa Rosa alone, among which over three thousand species are moths and butterflies (Janzen and Hallwachs, chapter 10 of this volume). Other important groups are dragonflies, true bugs, beetles, bees, wasps, and ants. Many serve as prey; feed on leaves, fruits, and seeds; and pollinate flowers (Jiménez et al., chapter 9 of this volume). Among the crustaceans, the freshwater prawn *Macrobrachium americanum* has been studied in detail.

Animal-pollinated trees and shrubs of the dry deciduous forests, often with large, brightly colored flowers, are mostly visited by bees (e.g., *Centris*, *Epicharis*, *Mesoplia*, and *Mesocheira*), moths, and nectarivorous bats. The large Guanacaste tree, *Enterolobium cyclocarpium*, for instance, is pollinated by a large array of moths, beetles, and bees. Moreover, tens of thousands of bees have been estimated to visit a single large flowering crown of *Andira inermis* at peak foraging periods. Bats, on the other hand, may pollinate ten to twenty different tree and shrub species at a single location in places like Santa Rosa and Palo Verde.

As Janzen and Hallwachs (chapter 10 of this volume) point out, in Guanacaste's dry deciduous forests herbivory by large vertebrates is probably trivial when compared to that of insects. Dry forest leaf-cutter ants (e.g., *Atta* spp.) consume far bigger volumes of leaves than any other herbivorous insect group (e.g., beetles). However, leaf-eating beetle diversity can be considerable with over a hundred different species known from these deciduous forests. Certain dry forest insects maintain mutualistic relations with specific plants, suggesting potential coevolution between the involved species. The example of *Pseudomyrmex* ants, which live within the swollen thorns of *Vachellia* shrubs (formerly in the genus *Acacia*, locally known as bull horn acacia), feed on sugar from extrafloral nectaries, and defend the plants against other invaders and herbivores, is illustrative (Jiménez et al., chapter 9 of this volume).

Over the past 25 years efforts have been undertaken to restore the reduced dry deciduous forest area of Guanacaste, using the Parque Nacional Santa Rosa established in 1972 as a starting point. The focus was on stopping manmade fires, planting trees in pastures, and stimulating natural regeneration. As soon as the fire frequency was significantly reduced natural woody succession took place and shaded out introduced African grasses like jaragua (*Hyparrhenia rufa*). Janzen and Hallwachs (chapter 10 of this volume)—who stood at the basis of this extraordinary restoration effort—narrate how common, often wind-dispersed dry forest trees, shrubs, and vines returned, how frugivorous vertebrates came back in the recovering dry forests bringing in new seeds to the successional patches, and how woody plant seedlings encountered spores of mycorrhizal fungi for their roots to grow successfully in the treeless pastures. As a result, secondary dry forest emerged at many places in northern Guanacaste, allowing the dry forest biodiversity with its specific ecological patterns and processes to restore itself surprisingly well. However, still considerable time is needed to achieve the maturity levels of structure and diversity that are known from the full-grown dry deciduous forest still thriving in the core of Parque Nacional Santa Rosa.

The Lowland Rain Forests

Costa Rica's 30–50 m tall tropical lowland evergreen rain forests are found on both the Pacific and Caribbean sides of the country. While the Caribbean or Atlantic rain forests run from the Río San Juan at the border with Nicaragua all the way down to the Río Sixaola at the Panamanian frontier, the Pacific rainforests are found only in the southern sector of the country. The latter have their northern limit near Dominical and run along the shores of the Península de Osa and Golfo Dulce (Fig. 21.1). Both the Pacific and Atlantic lowland rain forests run into the foothills of the central mountain ranges of Costa Rica, where they intergrade with premontane and montane forests around 300 to 500 m above sea level. The climate of Costa Rican rain forests is generally very humid and very warm, often without a dry season or with a rather short, rain-free period. Average annual temperatures range from 27°C near the coast to 20°C in the foothills (Herrera, chapter 2 of this volume).

The southern Pacific lowland evergreen rain or moist forests are concentrated in the iconic 1975-established Parque Nacional Corcovado on the Península de Osa and in protected fragments around the Golfo Dulce, being the core areas of the Área de Conservación Osa (ACOSA). They are characterized by a high degree of spatial heterogeneity, high species diversity, a long history of isolation with resultant endemism, unusual biogeographical affinities, and, in parts of the Osa Peninsula, an intact megafauna (Gilbert et al., chapter 12 of this volume). Here, mean annual rainfall values range from 3,000 mm on the plains to 6,000 mm on the peninsula's peaks, while soil types belong mostly to Entisols, Inceptisols, Mollisols, and Ultisols (Alvarado and Mata, chapter 4 of this volume).

The remaining forest on Isla del Coco in the Pacific Ocean—Costa Rica's single oceanic island—is classified as a species-poor tropical lowland rain forest dominated by a very small set of tree species: *Sacoglottis holdridgei*, *Ocotea insularis*, *Clusia rosea*, *Henriettella fascicularis*, and *Miconia dodecandra*. Some other tree species like *Ficus pertusa*, *Eugenia cocosensis*, and *Brosimum* sp. have only been found sporadically in the northern sector of the island. The common rainforest treeferns *Cyathea alfonsiana* and *C. notabilis* are endemic to the island and not found elsewhere. *Euterpe precatoria* var. *longevaginata* is the single palm species thriving on Isla del Coco. Overall the tiny forest has a green-red appearance caused by a reddish bromeliad that

Fig. 21.1 Costa Rican lowland rain forest at the mouth of a small river on the Península de Osa. At the left side grow some coconut palm trees (*Cocos nucifera*) on a sandy soil close to the beach.
Photo by Yamil Sáenz.

covers the branches of almost every green tree (Montoya, chapter 8 of this volume).

The Caribbean lowland rain forest zone, on the other hand, covers about a quarter of Costa Rica's terrestrial surface and includes portions of the provinces of Limón, Heredia, Alajuela, and Guanacaste. It is distributed over six of the country's 11 conservation areas and harbors several major tropical research and education facilities including the renowned OTS-administred La Selva Biological Station at Puerto Viejo de Sarapiquí. It is the region with the largest number of indigenous reserves and highest protected area coverage. Here, the climate is largely determined by seasonal trade winds. Mean annual rainfall ranges from about 2,500 to 5,000 mm per year, while soils are mostly derived from volcanic lava, mud flows, and ashes from the neighboring Cordillera Volcánica Central (McClearn et al., chapter 16 of this volume).

The rain forests of Costa Rica's southern Pacific region are among the most complex and diverse in the world and can easily be compared to the Colombian Chocó rain forests or Malaysian dipterocarp forests (Gilbert et al., chapter 12 of this volume, and references therein). They are well stratified with a number of forest layers, going down from emergent, canopy, and subcanopy layers to understory and ground strata. In these dense, closed-canopy rain forests not more than 5% of the sunlight that falls on the canopy actually pours down to ground level, giving way to forest floor plants that are adapted to thrive well in the dark shade. With at least 2,600 species of vascular plant species of which 80 are endemic to the peninsula and 700 are different tree species, the flora of the Osa region is among the richest of the country—and perhaps the world. The extensive Caribbean rain forests are similarly rich and contain over 3,000 species of vascular plants—a quarter of which are trees—in almost 200 families (McClearn et al., chapter 16 of this volume).

Tall trees with spectacular buttresses and large woody lianas dominate the diverse lowland rain forest of the Península de Osa. Some individuals of *Minquartia guianensis* and *Ceiba pentandra* trees reach heights over 70 m. Other common trees in Corcovado are *Ardisia cutteri*, *Aspidosperma megalocarpon*, *Brosimum utile*, *Carapa guianensis*, *Caryocar costaricense*, *Heisteria longipes*, *Manilkara* sp., *Peltogyne purpurea*, *Protium* spp., *Qualea paraensis*, *Symphonia globulifera*, *Trichilia* sp., *Virola* sp., and *Vochysia hondurensis*, as well as the palms *Iriartea gigantea*, *Poulsenia armata*, *Socratea durissima*, *Sorocea cufodontisii*, and *Welfia georgii*.

Small-scale disturbances in the rain forests of the Osa and Sarapiquí regions are frequent, dynamic, and varied, and may change the forest structure and its diversity considerably. Normally, post-disturbance regeneration in treefall gap spaces in these rain forests is dominated in the beginning by fast-growing pioneer plants (e.g., *Cecropia* and *Ochroma* trees) and colonizing lianas like *Passiflora vitifolia*, and may end with large, even-aged stands of trees after 30 to 50 years. The nine *Passiflora* liana species known from the Península de Osa are visited and pollinated by a similar number of host-specific species of resident, often mimetic *Heliconius* butterflies (Gilbert et al., chapter 12 of this volume).

At La Selva in the Sarapiquí area in the northern Caribbean plains the most abundant rain forest trees are *Pentaclethra macroloba*, *Protium pittieri*, *Warszewiczia coccinea*, and the palms *Welfia regia* and *Socratea exorrhiza*. They are accompanied by emergent canopy trees like *Dipteryx panamensis*, *Hymenolobium mesoamericanum*, *Lecythis ampla*, *Vochysia ferruginea*, *Hieronyma alchorneoides*, and *Ceiba pentandra*, which in some cases may reach stem diameters of up to 150 cm. Here, the most speciose plant genera (with 20 to 70 species per genus) are, ordered from species-richest to species-poorest: *Piper*, *Psychotria*, *Miconia*, *Philodendron*, *Anthurium*, *Inga*, *Peperomia*, *Solanum*, *Calathea*, and *Pleurothallis* (McClearn et al., chapter 16 of this volume). Probably over 1,500 plant species thrive at the grounds of the La Selva Biological Station.

Animal assemblages in Costa Rica's rain forests are extremely varied. Along the southern Pacific coast rain forest species range from abundant plant-consuming land crabs (Gecarcinidae) on the sandy beaches and tent-making bats in abandoned shelters to solitary jaguars (*Felis onca*) that still prey on white-lipped peccaries (*Tayassu peccary*) in the core zone of Corcovado's tall and dense rain forests. In the Sarapiquí lowlands, on the other hand, centrolenid glass frogs and small-bodied arthropod-eating lizards share the rain forest habitat with a large range of mammals including marsupials, bats, primates, armadillos, anteaters, squirrels, rats and mice, cats, peccaries, and deer, among others (McClearn et al., chapter 16 of this volume).

The ant diversity of Costa Rica's lowland rain forests is exceptional—559 species are known from the Caribbean lowlands and 352 from the southern Pacific rain forests. Between both lowland regions there is a high degree of overlap in terms of species composition (Gilbert et al., chapter 12 of this volume). In the southern Pacific lowland region one can observe ant-plant associations between *Pseudomyrmex* ants and acacia-like *Vachellia* shrubs, similar to those found in Guanacaste's dry deciduous forests (Jiménez et al., chapter 9 of this volume).

The fauna of Osa's pristine freshwater habitats such as those observed in the Río Sirena basin is dominated by abundant mollusks, prawns (*Macrobrachium* spp.), fishes, crocodiles, wading birds, and tapirs (*Tapirus bairdii*). Large southern Pacific coastal rivers (e.g., Río Térraba and Río Coto) are richer in fish diversity than the smaller streams of the Osa itself. The most common freshwater fish families on the Península de Osa as well as at the La Selva Biological Station are Characidae, Cichlidae, and Poeciliidae. In the freshwater streams at La Selva a total of 43 fish species have been recorded.

The herpetofauna of the southern Pacific lowlands, on the other hand, contains about 33% of all amphibian species known to occur in Costa Rica and 67% of all reptile species in the country (Gilbert et al., chapter 12 of this volume). A fair number of reptiles and amphibians are endemic to the peninsula, including the black-headed bushmaster, a salamander, several frogs, an anole, and some lizards. Out of the 187 amphibian species known to Costa Rica, at least a hundred thrive in the Caribbean lowland forests. They include over ninety frog species, around ten salamanders, and a few caecilians. Caribbean lowland reptiles are similarly diverse. Over half of Costa Rica's 127 reptiles are found in the Atlantic rain forest zone. They include over thirty lizards, at least eighty snakes—mostly colubrids—, six turtles, and two crocodilians (McClearn et al., chapter 16 of this volume).

Costa Rica's lowland rain forest bird diversity is huge as well. Raptors, parrots, woodpeckers, antbirds, and woodcreepers are just a few of the avifaunal groups that one can often observe in these forests. In general, rain forest bird diversity tends to be locally high, though relatively low population densities are observed per single species. Small population sizes are often the result of specialized species requirements in terms of microhabitat or diet. At the Península de Osa some 320 bird species have been recorded, including the endemic black-cheeked ant-tanager (*Habia at-*

rimaxillaris) and the charismatic but now locally threatened scarlet macaw (*Ara macao*). Numerous Nearctic migratory birds find important wintering habitat on the peninsula (Gilbert et al., chapter 12 of this volume). The harpy eagle (*Harpia harpia*) used to find a home on the Osa but has locally disappeared decades ago, most probably owing to habitat loss and hunting. On the other side of the country, in the Atlantic lowlands, 484 rain forest bird species have been observed, 289 of which appear to breed here (McClearn et al., chapter 16 of this volume). At La Selva alone at least 400 species of bird have been observed over the past five decades.

Some 140 to 150 mammals occur in the lowland rain forests of Costa Rica's Atlantic and Pacific slopes. In the magnificent rain forests of Corcovado at La Sirena one may run into a troop of squirrel monkeys (*Saimiri oerstedii*), just like one might do in the seasonal lowland forest of Parque Nacional Manuel Antonio (Jiménez and Carrillo, chapter 11 of this volume). In the rain forest channels along the Atlantic coast, the legendary West Indian manatee (*Trichechus manatus*) is still found, although in reduced numbers.

Omnivorous though often mostly frugivorous white-lipped peccaries may occasionally form large herds with 20 to 70 individuals in the southern Pacific rain forests. They have become vulnerable and are frequently hunted by man if not preyed upon by large cats like the jaguar, as is the case in Parque Nacional Corcovado. Peccaries may significantly influence the structure and plant diversity of these forests because they behave as voracious herbivores, consuming large amounts of viable fruits and seeds. While peccaries are principally diurnal, Corcovado's tapirs are mostly active during the night. They appear to live in family groups with single adult females serving as its stable residents that would allow one single male in their herd (Gilbert et al., chapter 12 of this volume, and references therein).

Although jaguars are severely threatened owing to hunting, habitat loss, and disappearance of prey species, stable populations with 30 and 50 individuals still persist on the Osa Peninsula. Apparently individuals move from Parque Nacional Corcovado to other protected areas on the peninsula. They prey on dozens of species, including nesting sea turtles at Corcovado's beaches, but have a preference for white-lipped peccaries, which are found in more than two-thirds of scats recorded by researchers (E. Carrillo, pers. comm.). Other conspicuous mammals observed in these southern Pacific wet forests are the primates; in fact, all four Costa Rican monkeys are found here: the mantled howler monkey, the white-headed, white-throated, or white-faced capuchin monkey, the Central American squirrel monkey, and the black-handed spider monkey, a ripe fruit specialist that locally relies on nectar from *Symphonia globulifera* trees (Gilbert et al., chapter 12 of this volume). On the Atlantic side, however, the rain forests provide habitat to capuchin, howler, and spider monkeys only—the squirrel monkey is not found here. Howler and capuchin monkeys have been observed occupying Caribbean forest fragments.

Occasionally capuchins invade banana plantations to feed on ripe bananas. Howlers, however, appear to have been historically the most abundant species in Caribbean rain forests (McClearn et al., chapter 16 of this volume).

The Highland Cloud Forests and Treeless Páramos

Costa Rica's mid- and high-elevation zones above 500 m altitude support different kinds of evergreen tropical forests and natural grasslands, ranging from moist/humid/wet/rain mountain forests to true tropical montane cloud forest (TMCF) characterized by diurnal mist, and foggy *páramo* grasslands dominated by thickets of bamboo and rocky outcrops. The cloud forests belong to the richest mountain forests globally and are part of the Meso-American/Andean biodiversity hotspot (Fig. 21.2). Its species density—that is, the number of species per area—is among the highest in the world.

The climate is cool and windy, and often determined by a persistent, frequent, or seasonal occurrence of clouds (Herrera, chapter 2 of this volume). Cloud formation and persistence is often more intense and severe along the Atlantic slopes of the mountains owing to the influence of trade winds (*vientos alisios*) that blow from the Caribbean lowlands southward. As a result, tree crowns are regularly bathed in mist and fog ("canopy wetting"). This frequent, orographic cloud and mist presence causes a considerable increase in atmospheric humidity within the forest interior—a phenomenon known as "horizontal precipitation" (Lawton et al., chapter 13 of this volume). In fact, these forests filter global air masses in such a way that they seize and incorporate water and nutrients from the mist itself into their cycles. The intense water cycle of the Costa Rican cloud forests feeds the country's many rivers that drain into the Caribbean Sea or Atlantic Ocean. Average annual temperatures range from 8°C at 3,400 m at the upper forest limit at Cerro Chirripó to about 20°C at 1,000 m and 24°C at around 500 m elevation. The yearly mean rainfall is correlated with slope orientation and fluctuates between approximately 1,000 and 3,000 mm, although at some place at mid-elevation along the Atlantic slopes of the *cordilleras* annually up to 6,000 mm may fall (Herrera, chapter 2 of this volume).

Costa Rican mountain soils are often yellowish, acid,

Fig. 21.2 Costa Rican highland cloud forest at 1,250 m elevation along the Río Grande de Orosí in the Tapantí sector of Parque Nacional Tapantí–Macizo de la Muerte. Bromeliads and other epiphytes festoon the branches of an overhanging riparian tree.
Photo by Maarten Kappelle.

and peaty with important organic (humus) upper horizons on steep slopes in rugged terrain (Histisols). They are frequently water-saturated and generally poor since they suffer from nutrient leaching (podsols). Young volcanic soils, however, may be locally rich in nutrients (Andisols). The water-logged soil environment is frequently anaerobic, which impedes root respiration and causes a reduction in belowground bioactivity, with consequently low decomposition levels. In the Cordillera de Talamanca, Inceptisols predominate in the western sector of the Parque Nacional Tapantí–Macizo de la Muerte, Entisols in the Parque Nacional Chirripó, and Ultisols in different sectors of the Talamanca highlands (Alvarado and Mata, chapter 4, and Kappelle, chapter 14, of this volume).

The mountain forests of Costa Rica serve as the country's highland backbone (Alvarado and Cárdenes, chapter 3 of this volume), and stretch from the northern tip of the volcanic Cordillera de Guanacaste at the border with Nicaragua, through the volcano-rich Cordillera de Tilarán and Cordillera Volcánica Central, southward via the Central Valley, into the sediment-rich Cordillera de Talamanca, southeastward down to the border with Panama. These high mountains are carved by V-shaped valleys with permanent streams that often descend in stepped series of waterfalls and rapids along steep, winding ravines (Alvarado and Cárdenes, chapter 3, and Lawton et al., chapter 13, of this volume).

Unfortunately, montane forest coverage in many places along the Pacific and Atlantic slopes of these mountain chains has been reduced drastically since the Second World War. Luckily, at a number of spots, highland rain and cloud forest acreage is recovering owing to natural regeneration (secondary succession) and planting with native trees (ecological restoration). However, research in the Cordillera de

Talamanca indicates that disturbed montane forests will need at least sixty to ninety years following a period of clearing, burning, and grazing in order to recover in terms of structure, and probably an additional fifty years to fully restore its plant diversity including the epiphytes it originally housed (Kappelle, chapter 14 of this volume). At landslides in the Cordillera de Tilarán, natural succession often starts with specialist genera such as *Gunnera*, *Myrica*, and *Monochaetum* (Lawton et al., chapter 13 of this volume).

Highland deforestation and habitat loss is already leading to species extinction (e.g., plants, amphibians) and loss of environmental services such as crystal clear drinking water that people in mountain towns near Monteverde, along the slopes of Volcán Irazú, or in the valley of Santa María de Dota depend on.

The structure of Costa Rica's wet cloud forests—and particularly those found at Monteverde or at the summit of Volcán Barva—is characterized by its elfin nature expressed in a reduced or dwarfish tree stature, an increased tree stem density, stunted trees with gnarled trunks and branches holding dense crowns with thick (pachyphyllous), hard (sclerophyllous), leathery (coriaceous), and tiny (microphyllous to nanophyllous) leaves (Lawton et al., chapter 13 of this volume). Tree branches are generally draped with dozens of pendant mosses and liverworts covered with water droplets and filmy ferns (*Hymenophyllum*), while sustaining vascular epiphytes such as orchids, bromeliads (*Tillandsia*, *Vriesea*), aroids, ericads, and ferns. Occasionally, loranthaceous parasitic shrubs abound on branches of, for instance, oak trees.

The maximum cloud forest canopy tree height in Costa Rica ranges from approximately 20 m high at 1,000 m elevation near Monteverde to 40 or even 50 m at about 2,500 m in the montane oak forests near San Gerardo de Dota and Villa Mills in the Cordillera de Talamanca. Here, the mature forest density of stems with a diameter at breast height (dbh) >5 cm is around 500 stems per hectare at 2,700 m, while 2,500 stems per hectare are recorded in young (20 to 30 yr old) successional cloud oak forests (Kappelle, chapter 14 of this volume). At an altitude of about 1,500 m in the 18–25 m tall cloud forests of Monteverde there are approximately 555 stems >10 cm dbh per hectare, responsible for a basal area of almost 75 m^2 and an aboveground, terrestrially rooted biomass of almost 500 Mg (Lawton et al., chapter 13 of this volume, and references therein). As a general rule, however, cloud forest structure may differ significantly between slopes depending on slope orientation (windward vs. leeward), rainfall (rainy vs. rain shadow), elevation (temperature variations), and soil (nutrient rich vs. leached horizons).

Species diversity in Costa Rican cloud forests is tremendous. Together with the Andean montane forests, Costa Rican (and Panamanian) cloud forests are among the richest in the world and serve as a center of speciation from which many species originated and radiated into northern Meso-America, the Caribbean, and southern South America. Hence, endemism is locally low at the genus level but high at species level. Plant genera such as *Epidendrum* (an orchid), *Elaphoglossum* (a fern), *Peperomia*, *Senecio* sensu lato, and *Anthurium* (herbs), *Psychotria*, *Miconia*, and *Solanum* (shrubs), and *Chusquea* (a bamboo) are particularly rich in endemic species (Kappelle, chapter 14 of this volume). In the Monteverde region (area: 350 km^2) over 3,000 plant species have been recorded, which is about a quarter of all plant species known from Costa Rica. Here, above 700 m elevation more than 750 plant species in some 100 genera are trees, of which at least 10% are fruiting or flowering in any given month, while around a hundred bloom especially during the drier months of March, April, and May. Orchid diversity (e.g., *Elleanthus*, *Epidendrum*, *Lepanthes*, *Maxillaria*, *Pleurothallis*, *Scaphyglottis*, *Stelis*, and *Telipogon*) is extraordinary with over 30 species known from Monteverde alone (Lawton et al., chapter 13 of this volume). Further south, in the Cordillera de Talamanca, plant species diversity is even higher and possibly up to 5,000 species have their home in Talamanca's rain and cloud forests at elevations over 500 m above sea level. Around 250 ground-rooted vascular plant genera dominate the Talamancan montane oak forests between 2,000 and 3,200 m elevation. It is here that fungal and lichen diversity is greatest, with some 400 species of Agaricales and over a hundred known foliose and fruticose lichens (Kappelle, chapter 14 of this volume).

The average Costa Rican cloud forest flora is composed of many canopy and subcanopy tree genera including *Alchornea*, *Alnus*, *Brunellia*, *Buddleja*, *Clethra*, *Cleyera*, *Clusia*, *Drymis*, *Escallonia*, *Ficus* (some figs like the hemi-epiphytic *Ficus crassiuscula* show a strangling behavior), *Guarea*, *Inga*, *Magnolia*, *Meliosma*, *Myrcianthes*, *Myrsine*, *Ocotea*, *Oreopanax*, *Persea*, *Prunus*, *Quercus*, *Rhamnus*, *Schefflera*, *Symplocos*, *Trichilia*, *Weinmannia*, *Zanthoxylum*, and the gymnosperm *Podocarpus*. In the Cordillera de Talamanca, mountain forests between 2,000 and 3,200 m elevation are mostly dominated by species of *Quercus* (oak), often with understories characterized by *Chusquea* bamboos (Kappelle, chapter 14 of this volume). One particular endemic cloud forest tree species is the 20–30 m tall *Ticodendron incognitum*, a remnant of the Tertiary Laurasian flora. Being the only species in the new family Ticodendraceae it was discovered in 1989 and resembles alder (*Alnus*). In general, seeds of a third of all tree species are bird-dispersed while seeds of around a tenth are dispersed by wind. Cloud forest trees are accompanied in the understory by shrubs (*Ardisia*, *Ilex*, *Macleania*, *Miconia*, *Palicourea*, *Psychotria*, *Rubus*, *Senecio*, *Solanum*, and *Vaccinium*) and dwarf palms (*Chamaedorea*, *Geonoma*, *Prestoea*) as well as by numerous pteridophytes including tree ferns (*Culcita*, *Cyathea*, *Dicksonia*, *Lophosoria*). The most species-rich woody families are Rubiaceae, Melastomataceae, Lauraceae, Asteraceae, and Ericaceae, which together represent about 30% of the total number of known woody species in Talamancan montane oak forests. Seeds of Melastomataceae, Solanaceae, and Rubiaceae found at Monteverde are mostly dispersed by opportunistic frugivorous tanagers, finches, and thrushes (Lawton et al., chapter 13 of this volume). Lianas are less common (*Hydrangea*, *Passiflora*). Hundreds of mosses and liverworts thrive here and may form impressive curtains in the forest interior (*Dendropogon*, *Pilotrichella*) or cover large sections of the bark of trees (e.g., *Frullania*, *Plagiochila*) (Kappelle, chapter 14 of this volume). Rodents predate on seeds of many of these cloud forest trees and shrubs including species in the genera *Ocotea*, *Guarea*, *Eugenia*, and *Beilschmiedia* (Lawton et al., chapter 13 of this volume).

Arthropod diversity in Costa Rica's mountain forests is immense, with thousands of insects known from Monteverde, for instance. However, there is no clear insight in species numbers owing to the lack of extensive sampling at mid- and high elevation. Arboreal arthropod communities are particularly rich in species, some of which have been studied in detail (e.g., species in the ant genus *Myrmelachista*). Another ant genus (*Azteca*) forms mutualistic relations with *Cecropia* trees, similar to the protective cooperation that happens between ants and acacia-like species in Costa Rica's dry lowlands (Jiménez et al., chapter 9 of this volume). A large amount of insects including numerous bees, beetles, butterflies, flies, and wasps pollinate the flowers of nearly half of the cloud forest canopy tree species, including the Lauraceae, at Monteverde (Lawton et al., chapter 13 of this volume). The highland bumblebee (*Bombus ephippiatus*) is a key pollinator in the Talamancan montane oak woodlands.

Fish diversity is extremely low in Costa Rican highland forest streams over 2,000 m elevation. The only common species is the exotic rainbow trout (*Oncorhynchus mykiss*) that has been introduced for commercial and recreational purposes. On the contrary, native amphibians and reptiles form an essential component of Costa Rica's montane cloud forests. Lawton et al. (chapter 13 of this volume) mention the presence of sixty amphibians in the Cordillera de Tilarán, of which over fifty are anurans (frogs, toads) and a hundred are reptiles (70% of which are snakes [e.g., vipers] and the remainder lizards). The green spiny lizard (*Sceloporus malachiticus*) and the highland alligator lizard (*Mesaspis monticola*) are commonly observed in the Talamancan oak forests. *Eleutherodactylus* is one of the most species-rich genera of frogs on Costa Rica's mountains. Populations of other anurans such as the golden toad, *Bufo periglenes*, and the harlequin frog, *Atelopus varius*, seem to have collapsed over the past decades, probably owing to an emergent, climate-change–related epidemic caused by a chytrid fungus. The golden toad, which used to be endemic to the Monteverde Cloud Forest Reserve, is now considered extinct and consequently has become the symbol of cloud forest destruction globally.

In Monteverde alone 425 species of bird have been spotted including the spectacular resplendent quetzal (*Pharomachrus mocinno*), the three-wattled bellbird (*Procnias tricarunculata*), the well-studied brown jay (*Cyanocorax morio*), and some 30 species of hummingbirds (Trochilidae). A large number of these birds play roles as flower pollinators or seed dispersers. At least 560 species of bird have been recorded in Parque Nacional Chirripó and Parque Internacional La Amistad. Common bird families are Emberizidae, Parulidae, Thraupidae, Trochilidae, Turdidae, and Tyrannidae. Over the past three decades the quetzal, an altitudinal migrant, has become the flagship bird species in conserving remnant cloud forest fragments in Costa Rica.

Cloud forest mammal diversity is equally spectacular with over 120 species living in the central Cordillera de Tilarán, including 58 bats, 15 murid rodents, and two endemic mammals (a harvest mouse and a shrew). Eight of the bat species (e.g., *Artibeus toltecus*) are frugivorous and feed on fruits of some forty plant species, while about 10% of the Monteverde's plant species are pollinated by bats. Large cats such as pumas, ocelots, and margays thrive here, as well as brocket deer, sloths, pacas, coyotes, squirrels, collared peccaries, and Baird's tapirs. Monkeys are a common feature in these mountains (e.g., capuchin, howler, and spider monkeys). Rodents like Muridae and Heteromyidae may abound during years following mast seeding of oak trees in the upper valley of the Río Savegre (Kappelle, chapter 14 of this volume).

At (sub)alpine elevations above the timber or treeline at 3,000 to 3,400 m elevation cloud forest is replaced by treeless grasslands and shrublands, often dominated by clump-forming *Chusquea subtessellata* bamboos. This is the so-called *páramo*, a tropical alpine, tundra-like, cool and wet ecosystem (in fact, it is a biome in its own!), composed of bamboos, shrubs, herbs, grasses, ferns, mosses, liverworts, and lichens (Kappelle and Horn, chapter 15 of this volume). The presence of the ericad shrub *Comarostaphylis arbutoides* marks the border between cloud forest and *páramo*. During the late Quaternary the Costa Rican *páramo* landscape underwent repeated glacial cycles of cooling and warming that left their traces on the rocky outcrops and formed U-shaped valleys and glacial lakes (Alvarado and Cárdenes, chapter 3 of this volume). Today, *páramo* is found at the summits of the Irazú and Turrialba volcanoes and at the higher mountain peaks along the Cordillera de Talamanca, including the *cerros* of Vueltas, Muerte, Cuericí, Urán, Chirripó, Amo, Dúrika, Dudu, and Kámuk. Two-thirds of Costa Rica's 15,000 hectares of *páramo* vegetation is found in the heart of Parque Nacional Chirripó, which is dominated by the 3,819 m high Cerro Chirripó. The climate is characterized by strong daily fluctuations in temperature and cloud cover (Herrera, chapter 2 of this volume). The average annual *páramo* temperature ranges from 11.0°C at 3,000 m to about 6°C at 3,800 m. Rainfall may vary significantly among mountain slopes. Depending on slope orientation the yearly mean rainfall is somewhere between 1,000 and 4,000 mm (Kappelle and Horn, chapter 15 of this volume). *Páramo* soils belong mostly to Histisols, Entisols, Inceptisols, and Andisols (Alvarado and Mata, chapter 4 of this volume).

Páramo species diversity is extraordinary if we take into account the small surface that this ecosystem occupies in Costa Rica. Eukaryotic algae (41 genera) abound in the *páramo* lakes and ponds, while true fungi (272 species) are extremely diverse as well. The same applies to lichens (204 *páramo* species), mosses (117 species), liverworts (113 species), and ferns (80 species including fern-allies). To date 416 flowering plants have been recorded, out of which 146 are endemic to Costa Rica and Panama. At the genus level several plant taxa are endemic to these *páramos*: *Iltisia*, *Jessea*, *Laestadia*, *Talamancalia*, and *Westoniella*. The endemic *Westoniella* is represented by six species. Other vascular plant genera that are true *páramo* inhabitants are *Alchemilla*, *Azorella*, *Castilleja*, *Draba*, *Gnaphalium*, *Lewisia*, *Lysipomia*, *Poa*, *Ranunculus*, *Senecio*, *Stevia*, and *Uncinia*. Most of the flowering *páramo* plants have special adaptations (e.g., short stems, hairy leaves) to withstand the harsh climate. A number of plant species also display adaptations to cope successfully with *páramo* fires that occur naturally at regular intervals. Unfortunately, *páramo* fire frequency has increased over the past decades owing to human activity. This has negatively impacted *páramo* recovery in a number of places. Fortunately, the bamboo *Chusquea subtessellata* seems to recover quickly after a fire has taken place (Kappelle and Horn, chapter 15 of this volume).

In terms of animal diversity a total of 71 insects have been reported, as well as 27 mollusks. Nineteen species of amphibians and reptiles are known, including the mushroom-tongue salamander, the montane alligator lizard, and the green spiny lizard. Birds are omnipresent and include seventy species, twelve of which are considered true *páramo* species. Most *páramo* birds are carnivorous (e.g., the red-tailed hawk), nectarivorous (for instance, the volcano hummingbird), or frugivorous (such as the sooty robin). Birds are responsible for the pollination of 8% of all *páramo* plant species and for seed dispersal of 20% of all plants. Mammals, in turn, are represented by 32 species, many of which are abundant in Costa Rican *páramo*. Shrews, rabbits, tapirs, ocelots, margays, and even pumas can be observed in remote areas (Kappelle and Horn, chapter 15 of this volume).

The Wetlands

Wetlands ("*humedales*") are areas in which water covers the soil, or is present either at or near the surface of the soil, year-round or for varying periods of time. Costa Rica's wide variety of natural wetlands includes rivulets, rivers, lagoons, ponds, lakes, coastal channels, seasonal swamps dominated by forest trees or palms, marshes, water-logged peat bogs, and hot springs. A separate category of Costa Rican wetlands concerns the artificial reservoirs that result from river damming for the generation of hydroelectric power, as is the case for Lago Arenal (Costa Rica's biggest water body), Lago de Cachí, and Lago Angostura. Some of these wetlands are determined by freshwater (the inland or non-tidal wetlands) while others are characterized by brackish or more saline waters (coastal or tidal wetlands). The latter include mangrove forests which have been dealt with earlier in this chapter (see under "Oceanic and Coastal-Marine Ecosystems"). They are spread over 34 major watersheds (Pringle et al., chapter 18 of this volume) and drain into the Caribbean Sea or Pacific Ocean.

Costa Rica's limnology is characterized by multiple parallel river systems with relatively small drainage areas. They run from the highest peaks to the coastal zones. An exception to this rule is the Río San Juan watershed, which is the largest watershed in Central America and shared with Nicaragua. Hundreds of permanent water bodies are found in these watershed basins. Horn and Haberyan (chapter 19 of this volume) report a total of 652 lakes in Costa Rica, including seasonal ponds and brackish lagoons. They originate from volcanic activity, fluvial dynamics, past glaciations, or landslides. Many streams and lakes have a geochemical solute composition which is determined by the input from acid-sulfate and sodium-chloride-bicarbonate springs that originate along volcano slopes.

Algal groups like Cyanobacteria (blue greens), which include the genera *Phormidium*, *Lyngbya*, *Dermocarpa*, *Oe-*

dogonium, and *Oscillatoria*, inhabit hot springs at Tabacón at Lago Arenal. Laguna Solimar, a turbid eutrophic lake near Cañas north of the Golfo de Nicoya, is cyanophyte-dominated, with *Cylindrospermum* as the most common phytoplankton and the copepod *Mesocyclops thermocyclopoides* as the principal zooplankton species (Horn and Haberyan, chapter 19 of this volume). At the southern tip of the Cordillera de Guanacaste, phytoplankton of Laguna Cerro Chato is dominated by chlorophytes, especially *Arthrodesmus bifidus* and *Monorhaphidium griffithii*. The lake's zooplankton is determined by the copepod *Tropocyclops prasinus prasinus* (Horn and Haberyan, chapter 19 of this volume, and references therein). Diatom genera, on the other hand, such as *Chamaepinnularia*, *Eunotia*, *Frustulia*, and *Stauroneis* thrive particularly in acidic rivers like the Río Agrio, which drains Volcán Poás (Pringle et al., chapter 18 of this volume). Laguna Barva at the summit of the volcano of the same name is dominated by small cryptophytes, desmids, cyanobacteria, and chlorophytes. Glacial lakes in the *páramo* of Parque Nacional Chirripó harbor phytoplankton that includes cyanobacteria, chlorophytes, pyrrhophytes, and diatoms such as *Aulacoseira* sp. Some diatom species in the genus *Navicula* are indicative of organic pollution in Costa Rican streams.

Costa Rican highland bogs cover only 235 ha and are distributed from 1,200 to 3,100 m elevation. A good example is the series of bogs at the Sabanas de Dúrika in the Cordillera de Talamanca. Here, repeated fire in the neighboring *páramo* vegetation may negatively affect water retention properties at ground level and consequently impact the aquatic bog vegetation.

High-elevation bogs in the Talamanca mountains harbor up to a hundred species of aquatic and semi-aquatic forbs, grasses, and sedges. Normally, these peaty bogs show a clear zonation of plants from the semi-aquatic edge to the wetter core of the bog. At the bog edge grows a plant community that includes the *páramo* bamboo *Chusquea subtessellata*, the bunch grass *Cortaderia nitida*, and shrubs like *Myrsine coriacea*, *Escallonia myrtillioides*, and *Hesperomeles heterophylla*. Halfway, closer to the bog's center, one observes the large, cycad-like fern *Blechnum buchtieni*, the endemic bromeliad *Puya dasylirioides*, shrubs like *Pernettya prostrata* and *Hypericum strictum*, and yellow-eyed grasses in the genus *Xyris*. In the core of the bog one may observe *Hypericum* together with *Paepalanthus*, *Hieracium*, *Utricularia*, *Isoetes*, and *Pernettya*. Here, *Carex* sedges, *Juncus* rushes, mosses like *Campylopus*, *Sphagnum*, and *Breutelia*, and *Cladina* lichens abound.

The acidic and tannin-stained Tres de Junio pond and its bog vegetation at 2,670 m elevation in the Cordillera de Talamanca (close to Cerro de la Muerte) is less than 1 m deep and inhabited by abundant *Sphagnum* mosses on hummocks and edges. Further south is the largest glacial lake of Costa Rica, the 22 m deep Lago Chirripó. Sitting at the base of Cerro Chirripó's *páramo* vegetation it harbors abundant quillworts (*Isoetes storkii*) that grow along the edge of the lake (Horn and Haberyan, chapter 19 of this volume). The bog salamander *Bolitoglossa pesrubra* thrives among the wet leaves of *Puya dasylirioides* bromeliads. This bog environment is shared with a dink frog (*Diasporus ventrimaculatus*) as well as with the sooty thrush or robin *(Turdus nigrescens)* (Jiménez, chapter 20 of this volume, and references therein).

Marshes are more widespread in the country than bogs and reach a total surface of 30,000 ha. They occur in basins and depressions where rainfall and run-off water accumulate. Palm-dominated swamps and seasonally flooded forest swamps together cover about 140,000 ha of land in the country. The marshes of the Caribbean coastal lowlands are dominated by large *Calathea*, *Montrichardia*, and *Thalia* forbs in combination with *Cyperus* and *Lasciacis* sedges, *Panicum* and *Scleria* grasses, and *Ipomoea* and *Solanum* vines. More-open marshes in the Sarapiquí region provide a niche for *Spathiphyllum friedrichstallii* and *Acalypha diversifolia* plants. Caribbean wetlands also provide a home to aquatic grasses like *Brachiaria* and *Panicum* and to floating water hyacinths (three species of *Eichhornia*).

Similar species assemblages are found in the marshes of Caño Negro and Palo Verde in the north, and at the headwaters of Río Sierpe in the southern part of the country. Open waters at Caño Negro are seasonally covered by *Nymphaea* water lilies, the water lettuce *Pistia stratiotes*, and species in the genera *Neptunia*, *Polygonum*, and *Salvinia*. In turn, at the borders of the 1,200 ha Palo Verde marsh one may observe the *palo verde* treelet (*Parkinsonia aculeata*), the Amazon sword plant *Echinodorus paniculatus*, and *Cyperus* or *Eleocharis* sedges. Here, at deeper sites, *Nymphaea* and *Eichhornia* species co-dominate the floating vegetation (Jiménez, chapter 20 of this volume). Aggressive cattails (*Typha dominguensis*) show invasive behavior in the freshwater marshes of Parque Nacional Palo Verde (Horn and Haberyan, chapter 19 of this volume).

Costa Rican palm swamps are mostly found in the wildlife refuges of Barro Colorado and Caño Negro, as well as in Parque Nacional Tortuguero and the Sierpe-Osa region. They are seasonally flooded and dominated by the palm species *Raphia taedigera* ("yolillo") and *Manicaria saccifera*. In palm swamps plant species diversity is normally low, with less than thirty species found at a single spot. True forest tree–dominated swamps are often characterized by

Pterocarpus officinalis, *Carapa guianensis*, *Astrocaryum alatum*, *Pentaclethra macroloba*, or *Symphonia globulifera* trees. All in all, these bogs, marshes, and swamps house up to 400 different species of plant (Jiménez, chapter 20 of this volume).

Costa Rica's river fauna is locally diverse and may differ considerably among and within drainages. Pringle et al. (chapter 18 of this volume) state that freshwater fish assemblages in Costa Rica are diverse and abundant in lowland areas, but decrease toward headwaters, and completely disappear above ~800 m above sea level. So far, a total of 174 fish species have been reported for the country's freshwater ecosystems, with Cichlidae, Poeciliidae, and Characidae being the most speciose families. They play essential roles in nutrient cycling, primary production, and decomposition in Costa Rican wetlands. *Poecilia gillii*, for instance, is a diatom-feeding specialist, affecting the structure and composition of algal communities of Laguna Hule. In the latter lake the cichlid *Parachromis dovii* ("guapote") and the banded tetra *Astyanax aeneus* are also observed.

An important number of fish species feed on fruits that fall in the water; these fish serve as seed dispersers of many riparian shrubs and trees such as figs. The alien fish *Oreochromis niloticus* (Nile tilapia) has been introduced in a number of lakes, including Laguna Chocuaco—which is also known as Laguna Vueltas—close to the Río Grande de Térraba (Horn and Haberyan, chapter 19 of this volume). Furthermore, the seasonal ponds of Caño Negro house ichtyofaunal communities that may be composed of at least 21 different fish species, including the endangered, 50–60 cm long, carnivorous *Atractosteus tropicus*, which is locally known as *gaspar* (Jiménez, chapter 20 of this volume).

Big, marine-derived freshwater shrimps in the genus *Macrobrachium*, a macroconsumer, are abundant at a number of places and serve as prey for other species. Algal assemblages in lowland streams are highly influenced by such omnivorous shrimp assemblages. Other freshwater invertebrates include all major orders of aquatic insects (e.g., dragonflies, caddisflies, stoneflies, and mayflies), which may be highly diverse at many places (Pringle et al., chapter 18 of this volume).

Of all known amphibians in Costa Rica, 15–20 are truly aquatic. Examples are the stream-dwelling frogs *Atelopus varius* and *Rana warszewitschii* (Pringle et al., chapter 18 of this volume). The most conspicuous freshwater reptiles are the caimans (*Caiman crocodilus*), American crocodiles (*Crocodylus acutus*), and aquatic turtles (Chelydridae, Embydidae, and Kinosternidae), which can be found in rivers up to 350 m elevation. Caimans are still abundant in the Caño Negro marshes where about 3,000 individuals were recorded in the early 1990s. Crocodiles inhabit the Tempisque marshes and mouth of the Río Grande de Tárcoles. River turtles in the genus *Rhinoclemmys* inhabiting Parque Nacional Tortuguero have been observed feeding on fruits from *Dieffenbachia* and *Ficus*, for which they appear to serve as seed dispersers (Pringle et al., chapter 18, and Jiménez, chapter 20, of this volume).

A total of 150 bird species are associated with rivers and other freshwater environments. Good examples are the kingfishers, jacanas, ducks (*Cairina moschata*), and sand pipers (*Actitis macularis*). The Palo Verde marsh provides a home to sixty species of waterfowl, including *Dendrocygna autumnalis* ducks and the endangered stork *Jabiru mycteria*. The marshes of Caño Negro are even richer in terms of avifauna and provide habitat to over 300 species including the jabiru stork and the great white egret (*Egretta alba*). The green heron (*Butorides virescens*) is a common feature in the marshes of the Lower Tempisque region (Jiménez, chapter 20 of this volume).

Freshwater mammals in Costa Rican lowland areas are few. Some of the most conspicuous ones are the river otter (*Lutra longicaudus*), the large fishing bat (*Noctilio leporinus*), and the manatee (*Trichechus manatus*) that used to live in rivers and channels near coastal areas, such as found in the Tortuguero region and near Gandoca-Manzanillo (Pringle et al., chapter 18 of this volume). Unfortunately, manatees have been decimated in the country over the past fifty years and are rarely observed today.

Closing Remarks

A lot remains to be done to fully understand the diversity and complexity of the structure and functioning of Costa Rican ecosystems. The present book just highlights the main features of the patterns and processes that characterize the ecosystems of this incredible tropical country—and the current chapter only touches quickly upon the most important aspects of those features. It is hoped that future generations will further investigate the ecological systems of this *Rich Coast*, this immensely rich country, which is arguably the most diverse spot on Earth. It will be in-depth ecological knowledge that is ultimately needed to successfully raise awareness amongst peoples to conserve and where necessary restore these ecological systems on which humans, and particularly Costa Rica's society, depend so greatly for future survival.

Acronyms

AAAS American Association for the Advancement of Science
AC Área de Conservación
ACAHN Área de Conservación Arenal Huetar Norte
ACAT Área de Conservación Arenal-Tempisque
ACCVC Área de Conservación Cordillera Volcánica Central
ACE abundance-based cover estimation
ACG Área de Conservación Guanacaste
ACLAC Área de Conservación La Amistad-Caribe
ACLAP Área de Conservación La Amistad-Pacífico
ACM Asociación Conservacionista de Monteverde
ACMIC Área de Conservación Marina Isla del Coco
ACOPAC Área de Conservación Pacífico Central
ACOSA Área de Conservación Osa
ACT Área de Conservación Tempisque
ACTo Área de Conservación Tortuguero
AD Anno Domini
AECO Asociación Ecologista Costarricense
AICA Área Importante para la Conservación de Aves
ANAI Asociación Nacional de Asuntos Indígenas
ALAS Arthropods of La Selva
ASCONA Asociación Costarricense para la Conservación de la Naturaleza
ASEDER Asociación de Emprendedores para el Desarrollo Responsable, Osa
ASOMOTI Asociación Mono Tití
ATBI All-Taxa Biodiversity Inventory
AyA Instituto Costarricense de Acueductos y Alcantarillados
BCI Barro Colorado Island, Panama
BINGO Big International Non-Governmental Organization
BP Before Present
CARICOMP Caribbean Coastal Marine Productivity Program
CARIPOL Marine Pollution Monitoring Program in the Caribbean
CAS Central American Seaway
CATIE Centro Agronómico Tropical de Investigación y Enseñanza
CBD Convention on Biological Diversity
CBM Corredor Biológico Mesoamericano
CCAD Comisión Centroamericana de Ambiente y Desarrollo
CCC Caribbean Conservation Corporation
CCT Centro Científico Tropical
CCW counterclockwise
CEC cation exchange capacity
CENAT Centro Nacional de Alta Tecnología
CI Conservation International
CIMAR Centro de Investigación en Ciencias del Mar y Limnología
CITES Convention on International Trade of Endangered Species
CLIP Caribbean Large Igneous Province
CONARE Consejo Nacional de Rectores
CONICIT Consejo Nacional para Investigaciones Científicas y Tecnológicas
CORBANA Corporación Bananera Nacional
CORENA Conservación de Recursos Naturales Renovables
COSUDE Agencia Suiza para el Desarrollo y la Cooperación
CRCC Costa Rican Coastal Current
CRD Costa Rican Dome
CSO civil society organization
CSR corporate social responsibility
DBCP dibromochloropropane
DBH diameter at breast height
DCR Domo de Costa Rica
DDE dichlorodiphenyldichloroethylene
DDT dichlorodiphenyltrichloroethane
DGF Dirección General Forestal
DGIS Directorate-General for International Cooperation, Netherlands
DGVS Dirección General de Vida Silvestre
DIVERSITAS International Programme of Biodiversity Science
DNA deoxyribonucleic acid
DOC dissolved organic carbon
DOM dissolved organic matter
EARTH Escuela de la Agricultura de la Región Tropical Húmeda
EBA Endemic Bird Area
ECEC effective cation exchange capacity
ECOMA Programa Ecología y Manejo de Tierras Altas, UNA
ECOMAPAS Proyecto de Mapeo de Ecosistemas, INBio-SINAC
EDU ecological drainage unit
EEZ Exclusive Economic Zone
ELA equilibrium line altitude
ENSO El Niño Southern Oscillation
EPA Environmental Protection Agency
ESPH Empresa de Servicios Públicos de Heredia
ETP Eastern Tropical Pacific
EUC Equatorial Undercurrent
FAICO Fundación Amigos de la Isla del Coco
FAO United Nations Food and Agriculture Organization
FFEM Fonds Français pour l'Environnement Mondial
FONAFIFO Fondo de Financiamiento Forestal de Costa Rica
FPN Fundación de Parques Nacionales
FUNDECOR Fundación para el Desarrollo de la Cordillera Volcánica Central
GABI Great American Biotic Interchange
GDFCF Guanacaste Dry Forest Conservation Fund
GEF Global Environmental Facility
GHG greenhouse gases
GIS geographic information system
GOES Geostationary Operational Environmental Satellite
GSA Geological Society of America
HAB harmful algal bloom
HLDG Herbario Luis Diego Gómez
HMS His Majesty's Ship

ICE Instituto Costarricence de Electricidad
ICOMVIS Instituto Internacional de Conservación y Manejo de Vida Silvestre
ICRAF World Agroforestry Centre
ICT Instituto Costarricense de Turismo
IDA Instituto de Desarrollo Agrario
IFGN Instituto Físico-Geográfico Nacional
IGN Instituto Geográfico Nacional
IICA Instituto Interamericano de Cooperación para la Agricultura
IMN Instituto Meteorológico Nacional
INBio Instituto Nacional de Biodiversidad
INCOP Instituto Costarricense de Puertos del Pacífico
INCOPESCA Instituto Costarricense de Pesca y Acuicultura
INEC Instituto Nacional de Estadísticas y Censos
IPCC Intergovernmental Panel on Climate Change
ITCO Instituto de Tierras y Colonización
ITCR Instituto Tecnológico de Costa Rica
ITCZ Intertropical Convergence Zone
IUBS International Union of Biological Sciences
IUCN International Union for Conservation of Nature
IUFRO International Union of Forest Research Organizations
LAI leaf area index
LOICZ Land-Ocean Interactions in the Coastal Zone
LSFR Los Santos Forest Reserve
MA Millennium Ecosystem Assessment
MAB Man and the Biosphere Program
MAG Ministerio de Agricultura y Ganadería
MAT Middle American Trench
MBC Mesoamerican Biological Corridor
MBRS Mesoamerican Barrier Reef System
MCFP Monteverde Cloud Forest Preserve
MCL Monteverde Conservation League
MINAE Ministerio del Ambiente y Energía
MINAET Ministerio del Ambiente, Energía y Telecomunicaciones
MIRENEM Ministerio de Recursos Naturales, Energía y Minas
MNCR Museo Nacional de Costa Rica
MODIS Moderate Resolution Imaging Spectroradiometer
MPA Marine Protected Area
MSS Multi-Spectral Scanner
MSY Maximum Sustainable Yield
MY Motor Yacht
NASA National Aeronautics and Space Administration
NEC North Equatorial Current
NECC North Equatorial Counter Current
NFMP natural forest management plan
NGO non-governmental organization
NGS National Geographic Society
NOAA National Oceanic and Atmospheric Administration
NPA Natural Protected Area
NPK nitrogen, phosphorus, and potassium
NPP net primary production
NRC National Research Council
NRCS Natural Resources Conservation Service
NSF National Science Foundation
NTAE non-traditional agricultural export crop
NWO Nederlandse Organisatie voor Wetenschappelijk Onderzoek
OAS Organization of American States
OCPs organochloride pesticides
OET Organización para Estudios Tropicales
OMZ Oxygen-Minimum Zone
OSPAR Convention for the Protection of the Marine Environment of the North-East Atlantic
OTS Organization for Tropical Studies
PAR photosynthetically active radiation
PBDE polybrominated diphenyl ether
PCB polychlorinated biphenyl
PdO Península de Osa
PES payment for environmental services
PET potential evapotranspiration ratio
PILA Parque Internacional La Amistad
PNC Parque Nacional Corcovado
PNG Parque Nacional Guanacaste
PNSR Parque Nacional Santa Rosa
POPs persistent organic pollutants
PPCPs pharmaceutical and personal care products
PPFD photosynthetic photon flux density
PPNG Proyecto Parque Nacional Guanacaste
PPP public-private partnership
PSA pago por servicios ambientales
RAMSAR Convention on Wetlands of International Importance, especially as Waterfowl Habitat
RARE Rare Animal Relief Effort
RBA Reserva de la Biósfera La Amistad
RBT Revista de Biología Tropical
REA Rapid Ecological Assessment
RECOPE Refinadora Costarricense de Petróleo
REDD Reduced Emissions from Deforestation and forest Degradation
RV Research Vessel
SAR synthetic aperture radar
SCP Site Conservation Plan
SDGs Sustainable Development Goals
SEC South Equatorial Current
SEPSA Servicios Electricos Potosi S.A.
SGS spatial genetic structure
SI Smithsonian Institution
SICA Sistema de la Integración Centroamericana
SINAC Sistema Nacional de Áreas de Conservación
SJLS San Juan - La Selva biological corridor
SOC soil organic content
SOM soil organic matter
SPOT Satellite Pour l'Observation de la Terre
SPN Servicio de Parques Nacionales
SRP soluble reactive phosphorus
SRTM Shuttle Radar Topography Mission
SSC Species Survival Commission, IUCN
STRI Smithsonian Tropical Research Institute
TADS Tropical Amphibian Declines in Streams
TEAM Tropical Ecology Assessment and Monitoring
TM Thematic Mapper, Landsat
TMCF Tropical Montane Cloud Forest
TNC The Nature Conservancy
TRF Tropical Rain Forest
TSC Tropical Science Center
TSOC total soil organic carbon
UCP University of Chicago Press
UCR Universidad de Costa Rica
UFCO United Fruit Company
UICN Unión Internacional para la Conservación de la Naturaleza
UN United Nations
UNA Universidad Nacional

UNAM Universidad Nacional Autónoma de México
UNCED United Nations Conference on Environment and Development
UNCLOS United Nations Convention on the Law of the Sea
UNCSD United Nations Conference on Sustainable Development
UNDP United Nations Development Programme
UNED Universidad Estatal a Distancia
UNEP United Nations Environment Programme
UNESCO United Nations Educational, Scientific and Cultural Organization
UNFCCC United Nations Framework Convention on Climate Change
USAID United States Agency for International Development
USDA United States Department of Agriculture
USGS United States Geological Survey
UT University of Texas
UTK University of Tennessee at Knoxville
UvA Universiteit van Amsterdam
WCD World Commission on Dams
WCS Wildlife Conservation Society
WMO World Meteorological Organization
WRI World Resources Institute
WUR Wageningen University and Research Centre
WWF World Wide Fund for Nature *or* World Wildlife Fund

Subject Index

Page numbers in italics refer to figures and tables.

ACAT. *See* Área de Conservación Arenal-Tempisque (ACAT)
Acevedo, H., 353
Acevedo-Gutiérrez, A., 169–70, 223, *595*
ACG. *See* Área de Conservación Guanacaste (ACG)
ACLA-C. *See* Área de Conservation Amistad-Caribe (ACLA-C)
ACLA-P. *See* Área de Conservación Amistad-Pacífico (ACLA-P)
ACMIC. *See* Área de Conservación Marina Isla del Coco (ACMIC)
ACOPAC. *See* Área de Conservación Pacífico Central (ACOPAC)
ACOSA. *See* Área de Conservación Osa (ACOSA)
Acosta, L., 352
Acueductos y Alcantarillados, 639
Acuña, F. H., 606
Acuña-González, Jenaro, 596
Adams, P. A., 231
Adelson, G. S., 266
Agassiz, Alexander, 165, 177
Agrarian Development Institute (IDA), 365
agriculture
 agrarian reform and, 365, 370
 alley cropping and, 85–86
 in cloud forests of volcanic Cordilleras, 443
 crop insurance and, 68
 deforestation and, 541, *541*
 in dry forest areas, 276, 277
 fertilizer and, 68, 70–71, 85
 Green Revolution in, 564
 land use conversion and, 646–47, *647*
 new practices in, 87
 nitrogen fixing and, 70–71
 organic inputs and, 85–86
 pesticides and, 563–65
 pollution and, 563–65, 647, 677
 sediment loads and, 602–3
 slash-and-burn, 68, 73, 84, 276
 soils' organic content and, 85
 sustainability labeling and, 565
 wetlands and, 699, 700
 See also banana industry; soils of Costa Rica
Aguado, M. T., 118
Agüero, J. M., 81, 86
Aguilar, G., 273
Aguilar, Reynaldo, *370*
Albatross expeditions, 99, 116, 165, 167, 177, 231
Albornoz, L., 594
ALCOA (company), 371
Alfaro, Anastasio, 166, 195
Alfaro, E., 142, 596
Alfaro, J. P., 78
Alfaro, R., 249
Allan Hancock Foundation, 100, 123, 168
Allen, Paul, 7, 365, 373–74
Almeda, F., 460
Almeida, R., 396
Altrichter, M., 197, 396
Alvarado, A., 68–71, 78, 81, 86
Alvarado, G., 32, 201, 355, 456, 667
Alvarado, J. J., 102, 122, 123, 594, 608
Alvarado García, Joaquín, 180, *184*
Álvarez, M. D., 271
Álvarez-Ruiz, M., 594
AMBICOR (NGO), 403
Anderson, James J., 100
Anderson, R. C., 223
Anderson, R. S., 231
Angulo, A., 171
Apsit, V. J., 272–73
Arauz, Randall, 170
Araya, L. M., 78
Ardón, Marcelo, *638*, 639
Área de Conservación Amistad-Pacífico (ACLA-P), 9, 460, 461
Área de Conservación Arenal-Tempisque (ACAT), 443
Área de Conservación Guanacaste (ACG), *297–98*, *299*, *301*, *306*, *308*, *315–16*, *323–24*
 academic study of, 339
 agriculture in, 303, 304
 All Taxa Biodiversity Inventory (ATBI) for, 250
 biodiversity in, 250, 292–93, 297–98, 304, 319–22, 331, 335–36, 711
 birds of, 265
 climate change and, 293–94, 320–22, *320–21*
 climate in, 311, 313–15, *317*, 319–322, *329*
 conservation fund for, 281
 creation of national park in, 305–6
 deforestation in, *298*, 300, 303, 305
 disappearance of megafauna in, 293–94
 distorted population dynamics in, 337–38
 ecological succession in, 337–39
 environmental education and, 336
 Europeans' arrival in, 294, 296–97
 even-aged forest in, 296
 expropriation of land for, 335
 fire control as conservation strategy in, 332, 334
 first humans and their impact in, 293–95
 floristic composition in the Península de Nicoya and, 277, 279
 forest and grassland fires in, 296, *299*, 300, 303, 307, *308*, *312–14*
 four primary ecosystems of, *292*
 geology of, 251, 290–93, 320
 growth of, 281
 human impact on, 331, 335, 337–38
 hurricanes and, 314–15
 indigenous people's use of fire in, 300
 insect species composition in, 315
 invasive species in, 335–36
 jaguars in, 262
 landscape-level restoration in, 297–98
 life zones of, 252
 livestock raising in, 300, 302, 304
 mammals in, 277
 maps of, *291*, *292*, *294–95*, *326*
 meteor impact in, 293
 as model for Área de Conservación Osa (ACOSA), 369
 moving threatened species to new locations and, 322
 as national monument and national park, 311
 national parks and protected areas within, 279, 280
 old-growth forest in, 297
 Oliver North's airstrip in, 297
 parasitism in, 275
 pasture in, *309–10*
 petroglyphs in, 298
 plant-animal interactions and, 328–31
 plant community distribution and, 254
 plant reproduction seasonality in, 328–31
 Pleistocene anachronisms in, 293
 poor soils in, 331, 334, 335
 prehistoric megafauna in, *305*, 333, 337
 proposals for expansion of, 322
 protection and restoration of marine portion of, 335
 purchase of land for inclusion in, 336–37
 reasons for creation of predecessor park and, 331
 reforestation in, 303, 305
 research sites and, xix

Área de Conservación Guanacaste (*continued*)
social proximity to centers of power and, 304, 331
species surviving meteor hit and, 290
thermal equator and, 311
tourism and, 304
trade winds and, 293
travel routes through, 304, 305
tree species composition in, 315
wood harvesting in, 296
as World Heritage site, 280
See also dry forests; Proyecto Parque Nacional Guanacaste (PPNG)
Área de Conservación Marina Isla del Coco (ACMIC), 162, 182, 193, 237
Área de Conservación Osa (ACOSA)
biodiversity in, 374
climate types in, 372
community involvement in master plan for, 371
configuration and agenda of, 369–70
conservation work in, *395*, 402
endemism in, *388*, *391*
funding for research in, 369
geology of, 372–73
herpetofauna of, *391*
human impact on, 361
landforms in, 373
mapping of, 9
mapping of ecosystems in, *376*
Osa community and, 370
park fees in, 402–3
protected areas in, 403, 713
scarlet macaws in, *393*, 401–2
soils of, 373
study of tent-making bats in, 394–95
tourism in, 403
Área de Conservación Pacífico Central (ACOPAC), 9, 347, 402, 712
Área de Conservación Tempisque (ACT), 264, 277, 280, 281
Área de Conservation Amistad-Caribe (ACLA-C), 481
Área de Conservation Amistad-Pacífico (ACLA-P), 481, 482
Arias, C., 148
Arias Sánchez, Oscar, 4
Arroyo Mora, J. P., 279, 542–43
Arroyo Mora, P., 277
ASEDER. *See* Entrepreneur Association for Responsible Development (ASEDER)
Asociación MarViva, 180
Asociación Mono Tití (ASOMOTI), 403
Atkin, L., 77
Atlantic lowlands
biodiversity in, *387*
fauna of, *387*, 389–90, *390*, 392, 394–400
vegetation of, 347, *390*
AVINA Foundation, 371
Azofeifa-Solano, J. C., 594

Baert, L., 231
Bakus, G. J., 107, 169, 173–74
Ball, B., 153
banana industry
in Caribbean lowlands, 563
habitat destruction and, 567
in Osa region, 364–65, 371
pesticide use and, 564–65, 677
Banks, N., 231
Banta, William C., 100, 591
Barbee, N. C., 635
Barclay, George W., 209
Barnens Regnskog, 443
Barrantes, G., 224, 363
Barrantes-Rojas, N., 604
Barrientos, Z., 511
Barry, R., 20
Bartels, C., 148, 149
Bartsch, P., 168
Bawa, K. S., 271, 272
Beard, J. S., 8
Beaudette Foundation, 100
Beebe, William, 100, 123, 168
Belcher, Edward, 194, 209
Bellamy, C. L., 231
Bellefleur, P., 474–75
Benavides-Morera, R., 595
Bennett, H. H., 64
Bentham, G., 209
Benumof, B. J., 206
Bergoeing, J. P., 202, 656
Bernecker, A., 208, 211
Berry, William, 52
Bertsch, F., 86
Bianchi, G., 118
Bickel, D. J., 231
Bigelow, H. B., 168
biodiversity
All Taxa Biodiversity Inventory (ATBI) and, 250
alpha, beta, and gamma levels of, 6–7, 709
in Área de Conservación Guanacaste (ACG), 292–93, 304, 319–22, 331, 335–36, 711
in Área de Conservación Osa (ACOSA), 374
in Área de Conservación Pacífico Central (ACOPAC), 712
in Atlantic versus Pacific lowlands, 387, *387*
biased collection efforts and, 539
biological corridors and, 567
Caribbean coastal and marine ecosystems and, 595
Central American land bridge and, 5
in Central Pacific region, 346, 347, 352–56
climate change and, 28
in cloud forests, 428–29, 433–35, 453, 457–69, 718–19
Convention on Biological Diversity (CBD) and, 482
Costa Rica's innovations in preservation of, 4
definitions of, 7
deforestation and, 3
degree of in Costa Rica, 5–6
in dry forests, 250, 276–77, 309, 711, 712–13
at ecosystem level, 6–7
endemism and, 6
of fish in Caribbean lowlands, 546
Great American Biotic Interchange (GABI) and, 5, 457, 531
in growth forms, 431–32
habitat loss and, 467
hours of sunlight and solar radiation and, 21
human impact on coral reefs and, 178–79
illegal fishing and, 236
improving inventories of, 5, 8–9
information about large versus small organisms and, 627
invasive species and, 336
Isla del Coco and, 169, 171–72, 194–95, 200, 236–38
livestock raising and, 302, 303
mangrove ecosystems and, 108
in marshes, 692–93
megadiverse countries and, 6
Mesoamerican biodiversity hotspot and, 499, 716
1999 Biodiversity Law and, 701
of oceanic and costal-marine ecosystems, 709, 710
of oceanic islands, 192–93
in Osa region, 361, 373–75, 387, 392, 394
in Oxygen-Minimum Zones, 118
on Pacific coast, 119
in palm swamps, 694–95
in the *páramos*, 509–14, *510*, 720
in Parque Nacional Santa Rosa, 254
political and legal frameworks supporting, xvii
as priceless, 3
progress on in Costa Rica, xvii
in rain forests, 714–16
of rivers of Costa Rica, 626–27, 630, 633
software programs for study of, 546
Sørenson Index and, 387
in South Pacific region, 347
Stockholm Environment Institute on, xv
Sustainable Development Goals (SDGs) and, xv–xvi
Taxonomic Impediment and, 5
threats to, 6
in wet forests of the Caribbean lowlands, 539–40, 550–53, 555–56, 561
in wetlands, 692–93, 694–95, 721–22
Biodiversity Support Program, 193
biofuels, 336–37
Biolley, Paul, 167
Birkeland, Charles, 100
Bishop Museum, 169, 227
Bjarte, H. J., 229
Blair, N., 608
Blake, J. G., 561
Blanco, M. A., 211
Blaser, H., 81

Blaser, J., 475
Blauvelt, Abraham, 563
Blessing Presinger, Augustín, 497
Bogantes-Aguilar, Victoria, 595
Bogarín, D., 211
bogs, marshes, and swamps
  agriculture and, 699
  biodiversity in, 692–93, 694–95, 721–22
  conservation efforts and, 699–701
  definition of wetlands and, 701
  degradation and restoration of, 701–2
  disconnection of from watersheds and, *700*
  ecological services provided by, 700
  endangered and threatened species in, 692–93
  extent of, 683
  fauna of the marshes and, 691–93, *693*
  fire and, 721
  fish communities in, 692
  flora of the marshes and, 687, *687–692*, 689–691
  forest swamps and, 696–99, *697–98*
  highland bogs and, 684–87, *685–86*
  human disturbance of, 686, 690–91, *700*
  interconnectivity of, 699–700
  invasive species in, 692
  loss of wetlands and, 683–84
  map of, *684*
  in Osa region, 376, 377
  palm swamps and, 694–96, *694–95*
  Palo Verde marshes and, 701–2
  Ramsar Convention on Wetlands of International Importance and, 686
  research on, 683
  rice cultivation and, 700
  sugarcane cultivation and, 700
  threats to, 686–87, 699–701
  topographical variations in, 690
  water flow and quality and, 699–700
  wetlands ecosystems and, 720–22
Boinski, S., 399
Bolaños, F., 356
Bolaños, J., 79
Bolaños, Natalie, 594
Bolaños, Rafael Angel, 8
Bonito, Benito "Bloody Sword," 162
Boone, Lee, 168
Bosque Nacional Diriá, 267, 280
Bosque Puerto Carrillo, 279
Bourillon, Roger, 247
Bowman, T. E., 604
Boza, Mario, 305, 367, 566
Brak, B., 513, 685
Brandes, C., 597
Braun-Blanquet, J., 8, 9, 512
Breedy, O., 102, 118
Breedy, Odalisca, 594
Brenes, C., 207–8, 595
Brenes, L. G., 656
Bright, D. E., 231
Brinkman, M., 52
Broadbent, E. N., 358
Broenkow, W. W., 100
Brölemann, E. W., 235
Brooks, D. R., 269
Browman, H. L., 156
Brown, J. W., 230, 231
Brown, S., 473
Bruce, Niel L., 594
Brugnoli-Oliveira, E., 148
Bruijnzeel, L. A., 422–23
Brusca, Richard C., 594
Bumby, M. J., 656
Bundschuh, J., 32
Burcham, J., 627
Burkenroad, M. D., 168
Burlingame, L. J., 443
Burnham, R. J., 254
Buskirk, R., 463
Buskirk, W., 463
Bussing, W. A.
  on fish communities at La Selva Biological Station, 627
  fish of Caribbean coast and, 595
  fish of Caribbean lowlands and, 549
  Isla del Coco and, 169, 171, 219–20
  specimen collection and, 116
Buurman, P., 83
Byers, G. W., 231

Cabalceta, G., 83
Cabeças de Grado, Joan, 162
Cahoon, L. B., 143
Cajiao-Jiménez, M. V., 122
Calderón, R., 10
California Academy of Sciences, 100, 167–68
California Current, 200
California Gold Rush (1848), 298
Calvert, Amelia, 360
Calvert, Philip, 360
Calvo, J., 543
Camacho, M., 81, 474–75
Camacho-García, Yolanda, 594
Campos, J., 149, 151–52, 153, 154
Carazo Odio, Rodrigo, 164
carbon cycle. *See* climate change
carbon sequestration, 81, 85, 109
Caribbean coastal and marine ecosystems
  beaches and, 600–601, *601*
  biodiversity in, 595
  bottom profile of, *103*
  climate and weather and, 596, 597
  climate change and, 606
  coastal lagoons and, 598, *599*
  conservation efforts in, 606, 608–9, 610
  coral reefs of, 593, 602–3
  crustaceans and, 594
  current state of knowledge on, 596–97
  declining fish populations and, 606
  dolphin studies and, 595
  echinoderms of, 594
  Economic Exclusive Zone (EEZ) of, 591
  El Niño–Southern Oscillation (ENSO) warming events and, 604–5
  fish studies and, 595
  future of, 609–10
  geography of, 591
  geology and geomorphology of, *596*, 597–98, *598*, 601
  history of, 591–96
  human population and demography and, 604
  Isla Uvita and, 603–4, *606*, 608
  Limón earthquake (1991) and, 593
  manatee studies and, 595
  mangrove forests and, 599–600
  maps of, *592*
  mass mortality of organisms in, 593, 595, 601, 602, 604–5
  open waters and, 604
  versus Pacific coast, 97
  pollution and, 595–97, 606
  protected areas and, 593, 606, 608
  research expeditions and, 592–93
  research needs in, 609–10
  rocky intertidal zone and, 601, *602*
  seagrass beds in, 111, 601–2
  summary of oceanic and coastal-marine ecosystems and, 709–10
  threats to, 602–3, 604–6, *605*, *607*
  waves, tides, and water circulation in, 598
  See also wet forests of Caribbean lowlands
Caribbean Coastal Marine Productivity (CARICOMP) protocol, 599–600, 602, 603, 609
Caribbean Conservation Corporation (CCC), 567, 608
Caribbean lowlands. *See* wet forests of Caribbean lowlands
Caribbean region, as Biodiversity Superpower, xvi
CARICOMP protocol. *See* Caribbean Coastal Marine Productivity (CARICOMP) protocol
CARIPOL. *See* Marine Pollution Monitoring Program in the Caribbean (CARIPOL)
Carr, Archie, 566, 591
Carr, Archie, III, 567
Carrillo, Eduardo, *395*, 396, 398–99
Carrillo-Baltodano, Allan, 595
Carson, Renate J. M., 591
CARTA Project, 257
Cascante, A., 350
Castillo, M., 352
Castillo, P., 172, 201, 203, 205
Castro, R., 28
Castro Campos, Marco V., 8
CATIE (Centro Agronómico Tropical de Investigación y Enseñanza). *See* Tropical Agricultural Research and Higher Education Center (CATIE)
Celis, R., 345
CENAT, 257
Central America
  biogeographic history of, 5
  Central American Floristic Province and, 457
  closing of the isthmus and, 5, 55, 97, 290–91, 391, 416, 529–32

Central America (*continued*)
  Great American Biotic Interchange (GABI) and, 5, 457
  isthmian and Andean mountain ranges and, 500
  routes across the isthmus and, 298, 300
  transisthmian sister species and, 97
Central Pacific region
  aquatic habitats in, 349, *349*, 354
  biodiversity in, 346, 347, 352, 353, 354–56
  birds in, 355–56
  climate of, 345
  closing of the isthmus and, 503
  conservation efforts in, 357–58
  deforestation in, 351, 357
  endangered and newly discovered tree species in, 353
  endangered species in, 355
  endemism in, 345–46, 348, 350, 355, 356
  extent of Central Valley in, 351
  fungal diversity in, 354
  geography of, 345–46
  geology of, 351
  herpetofauna in, 356–57
  human impact on, 357–58
  mammals in, 354–55
  northern distribution limit for some trees in, 349
  property development in, 357–58
  protected areas in, 345, *346*
  river gorge in, *352*
  savannahs in, 350, *351*
  species new to science in, 349, 350
  tourism in, 357–58
  vegetation of, 347–353
Centro de Investigación en Ciencias del Mar y Limnología (CIMAR)
  contributions to riverine research and, 623
  development of marine science in Costa Rica and, 100–101
  echinoderm studies and, 594
  ecoregional assessment and, 181
  Isla del Coco studies and, 171, 172, 175–76
  limnological studies and, 656–57
  marine pollution studies and, 595–96
Chacón, Didiher, 594, 595
Chacón, Isidro, 546
Chandler, M., 634
Charpentier, Claudia, 593
Chassot, Olivier, 568
Chavarría, U., 257
Chaverri, A., 247, 461, 498–99, 508, 513, 515
Chávez, L., 663
Cheek, A. O., 155
Chemsak, J. A., 231
Chevalier-Skolnikoff, S., 263–64
Child, Allan, 100
Choe, J. C., 231
Chorley, R., 20
Choudhury, A., 269
Chubb, L. J., *206*
CIMAR. *See* Centro de Investigación en Ciencias del Mar y Limnología (CIMAR)
CITES. *See* Convention on International Trade of Endangered Species (CITES)
Clark, K. L., 422, 668
Clark, L. G., 512
classification systems
  biotic units and, 8
  climate and, 25, 26
  Costa Rican ecoregions and, 10
  DNA barcoding and, 270
  for Isla del Coco vegetation, 211, 213
  Luis Diego Gómez's work and, 8, 9
  taxonomy of fish communities and, 550
  UNESCO's vegetation classification system and, 9, 377
  volume editor's choice of, 10
  World Wildlife Fund (WWF) and, 250
Cleef, A. M., 499, 508, 512, 513
Cleveland, C. C., 84
climate and weather
  altitudinal variable and, 536
  in Área de Conservación Guanacaste (ACG), *317*, 319–322, *320–21*, *329*
  on Caribbean coast, 596, 597
  climate regions and, 23–27
  in cloud forests, 420–24, *421–22*, 454–55, 716
  in dry forests, 311, 313–15, 711
  in dry lowlands, 267
  El Niño effect and, 102, 104
  Föhn effect and, 27
  Gulf Stream and, 529
  Hadley Cell and, 197
  hours of sunlight and solar radiation and, 21, *22*
  hydric balance and, 23, *23–27*
  Intertropical Convergence Zone (ITCZ) and, 173, 197–99, *198*, 420, 503, 535
  Isla del Coco and, 197–99
  lake temperatures and, 672–73, 674–75
  lake turnover and, 667, 674–75
  latitudinal position and, 19–20
  meteorological monitoring and, 422–23, 425–26, 535–36
  orography and, 19
  of Osa region, 372
  of Pacific coast, 104
  in *páramos*, 500, 502–3
  in Pleistocene and Holocene Eras, 529–30, 532, 533
  precipitation and, 21–22, *22*
  production of higher trophic levels and, 639
  relative humidity and, 22–23, *23*
  rivers of Costa Rica and, 622
  temperature and, 21, *21*
  temporal and spatial variations in, 21–23
  temporales del norte and, 442
  tourism and, 182
  wet forests of Caribbean lowlands and, 534–37, *535*
  See also climate change; El Niño events
climate change
  altitudinal distributions of arthropods and, 433–34
  Área de Conservación Guanacaste (ACG) and, 293, 320–22, 336
  biodiversity and, 28
  carbon dioxide neutrality and, 4
  Caribbean coastal and marine ecosystems and, 606
  changing equilibrium states and, 334
  cloud forests of the Talamanca and, 480
  Costa Rican ecosystems and, 27–29
  dormancy of moths and butterflies and, 314
  effects of on dry forests, 310
  fossil evidence of past change and, 27–28
  on geological time scale, 536
  highland bogs and, 686
  interruption and calendar-shifting of seasonal events and, 330–31
  Isla del Coco and, 178, 199
  lake ecosystems and, 678
  mangrove ecosystems and, 109
  mitigation of, 28–29
  Pacific coastal waters and, 121
  on Pacific versus Atlantic slope, 480
  páramos and, 506, 515
  plant reproductive seasonality and, 328
  during Pleistocene and Holocene, 53
  rain forest restoration and, 335
  seagrass beds ecosystems and, 111
  stabilization of greenhouse gas concentrations and, 8
  Stockholm Environment Institute on, xv
  Sustainable Development Goals (SDGs) and, xv–xvi
  thermoconforming animals and, 552
  wet forests of Caribbean lowlands and, 536–37
  See also climate and weather
Cline, Joel D., 100
Cloern, J. E., 142
cloud forests, *717*
  biodiversity in, 718–19
  comparative study of, 457
  deforestation and restoration in, 717–18
  forest invasion and restoration and, 334
  lakes in, 664–70
  in Osa region, *376*, 377
  seed dispersal in, 718
  summary of ecosystem, 716–19
  See also cloud forests of the Talamanca; cloud forests of the volcanic Cordilleras
cloud forests of the Talamanca, *453*, *455*
  agro-ecological zonation in, 477–78, *479*
  altitudinal zonation in, *479*
  amphibians in, 463–64
  biodiversity in, 453, 457–467
  biological corridors in, 482–83
  birds in, 465–66
  bryophytes in oak forests and, 462–63
  climate of, 454–55

ecosystem functioning and dynamics in, 473–77
emerging environmental threats in, 479–80
endangered and threatened species in, 482
endemism in, 453, 461
environmental recovery in, 475–77
epiphytes in, 461–62
ethnobotany in, 479
fishes in, 463
forest structure in, 473
forest succession in, *474*, 475, 476–77
fungi in, 458
future of, 482–83
geology and geomorphology of, 453, 455–56
history of scientific exploration in, 453–54
human settlement in, 477–78
Inter-American Highway and, 477, 479–80, *480*
leaf characteristics in, 473–74
lichens in, 458–60
mammals in, 466–67
map of, *452*
plants of, 457–58, 460–63, 468–69
public protected areas in, 480–82
rainfall, mist, and clouds in, 475
Ramsar Wetland of International Importance in, 481–82
reforestation in, 478–79
reptiles in, 464–65
rivers of, 456
seed predation in, 471, 472–73
societal uses of water from, 452–53
soils of, 456–57
species interactions and, 469, 471–73
successional stages of, *474*
tourism in, 477, 478
vascular parasites in, 462
vegetation zones in, 467–68
water and nutrient cycles and, 475–76
See also cloud forests; cloud forests of the volcanic Cordilleras

cloud forests of the volcanic Cordilleras
above-canopy photosynthetic photon flux densities in, *426*
above-canopy shortwave radiation in, *426*
amphibians and reptiles in, 434–35
arthropods in, 433
biodiversity in, 428, 429, 431–32, 433, 434, 435
birds in, 435–37
canopy structure and treefall gaps in, 424, 429, 442
climate of, 420–24, *421*
conservation in, 443–44
distinctive habitats in, 428–29
endemism in, 437
forest light regime in, 474–75
forest profiles of, *430–31*
growth form diversity in, 431–32
history of land use in, 443
Holdridge Life Zones in, 428
hunting and land use changes in, 437
hydrology of, 424–26, *425–27*, 428
landslides in, 442
light in, 424
versus lowland forests, 433
mammals in, 437
map of, *416*
modern geography of, 418–19
Monteverde Flora Project in, 429, 431
nutrient cycling in, 440–41, *441*
people and nature in, 442–44
phenology of vegetation in, 432–33
physiography of, *419*
plants and vegetation of, 428–433
protected areas in, *416*
rainfall, mist, and clouds in, 422–24, *424*, 425, *425–26*, 440–41, *441*
research and conservation in, 415
satellite image of canopy in, *429*
soils in, 419–20
species interactions in, 437–40
tectonic history of, 415–18
temperature changes and, 434
trophic structure and energy flow in, 441
water cycle in, 420–21
wind in, 421–23, *421–22*, 425
See also cloud forests; cloud forests of the Talamanca

Coan, E. V., 168
Coates, A. G., 5
Cocos Anticyclonic Eddy, 200
coffee industry, 621
Cole, M. C., 431
Colnett, James, 194, 225, 226
Colombia Current, 200
Columbus, Christopher, 3, 534, 603–4
Colwell, Rob, 546
CONARE. *See* Consejo Nacional de Rectores (CONARE)
Conquest, L., 156
Consejo Nacional de Rectores (CONARE), 171, 172, 182
conservation
biofuels as threat to, 336
biological corridors and, 357, 400, 482–83, *537*, 545, *556*, 566–69
in bogs, marshes, and swamps, 699–701
Caribbean coastal and marine ecosystems and, 606, 608–9
in cloud forests of volcanic Cordilleros, 443–44
conflict over, 400–404
conservation organizations and, 122, 180
Costa Rica's innovations in, 4
crocodile populations and, 552
debt for nature swaps and, 279
definition of wetlands and, 701
versus development, 281
dry forests and, 279, 331–33, 334
Eastern Pacific Marine Corridor Initiative and, 181
ecological restoration and, 122
ecosystem approach and, 322
elevational transects and, 556
endangered species and support for, *395*, 398
environmental laws and, 700–701
factors motivating conservation efforts and, *566*
funding for, 402–4
gap analysis and, 8, 9
green turtles and, 566–67
highland bogs and, 686
image versus reality of in Costa Rica, 565–66
importance of nonthreatened plants and animals in, 332
Important Areas for Bird Conservation and, 218
Isla del Coco and, 180–82, 236–38
laboratory cultivation and, 152–53
legislation and, 122
lists of endangered and threatened species and, 482
local interventions to strengthen, 281
local people's involvement in, 608–9, 610
management of, 608–9, 610, 649
mangrove ecosystems and, 109
Marine Protected Areas (MPAs) and, 122, 123, 156
NGOs and, 402–4
Noah's Ark efforts and, 322, 336
no-take versus protected areas and, 179
paper parks and, 122
protected areas and, 4, 8, 9, 121–22, 180
protests against water concessions and, 648
radio collars and, 398
Ramsar Convention on Wetlands of International Importance and, 686
scarlet macaws and, 401–2
status assessment and, 400
structural drivers of, 280
sustainable forestry and, 370–71
threat abatement and, 122
tourism and, 402–4, 568–69
UN conventions on, 122
valuation of and payment for ecological services and, 122–23, 182, 277, 280, 542–43, 608
in wet forests of the Caribbean lowlands, 565–69
wetland protection and, 699–701
white-lipped peccaries and, 400–401

Conservation International (CI), 181, 608
Conservation Strategy Fund, 646
Convention for the Protection of the Marine Environment of the North-East Atlantic (OSPAR Convention), 116
Convention on Biological Diversity (CBD), 10, 482
Convention on International Trade of Endangered Species (CITES), 276, 482, 567, 608
Coolidge, K. R., 231
Córdoba-Muñoz, R., 142
corporate social responsibility (CSR), xvii

Corrales, A., 149
Corredor Biológico Mesoamericano (CBM), 482
Cortés, J.
coral reef ecosystems and, 111–12
Costa Rica's coastal-marine ecosystems and, 10, 98
crustacean studies and, 594
Isla del Coco ecosystems and, 174
Isla Uvita and, 604, 605
lists of mollusks and, 153
on mangrove systems, 143
marine biodiversity studies and, 595
rocky intertidal ecosystems and, 107
synthesis of marine research and, 102
Costa Rica
beaches of, *41*
biogeographic history of, 5
biological richness of, 3
biotic units of, 8, 19
cattle boom and end of cattle industry in, 304–5
climate and latitudinal position of, 19–20
climate and orography of, 19, 30
climate regions of, 23–27, 30
deforestation rates throughout, 541, 542
degree of biodiversity in, 5–6
Economic Exclusive Zones (EEZ) of, *98*, 121, 170, 178
ecosystems of, 6–11, 27–29
endemicity of species in, 6
environmental publications on, xix–xx
as first carbon dioxide–neutral country, 4
five terrestrial ecoregions of, 10
geographical coordinates of, 19
government departments responsible for environmental management in, 4
as Green Republic, xv
growing environmental awareness in, 569
image versus reality of conservation in, 565–66
increasing impact on ecosystems in, xvii
independence from Spain, 162
indigenous people of, 532–33, 534
integrated environmental management in, xvii–xviii
as laboratory, 30
land area of, 19
life zones of, 7–8
marine versus land territory of, 6
as most species-dense country on Earth, 709
name of, 3
National Herbarium of, 497
Pacific versus Caribbean coast of, 6
paleographic reconstruction of, *54*
political history of, 300
population of, 563
progress in conservation in, xvii
as regional leader in hydropower, 640
tectonics of, 31–32
temporal and spatial climate variations in, 21–23
threats to environment of, 3–4, 6
three terrestrial biomes of, 10
wetland systems in, 9–10
Costa Rican Coastal Current, 199–200
Costa Rican Dome, 199
Costa Rican Electricity Institute (ICE), 21
Costa Rican Fisheries Institute (INCOPESCA), 608
Costa Rican Institute of Tourism (ICT), 156
Costa Rican port authority, 156
Cousteau, Jean Jacques, 164
Cousteau, Philippe, 164
Cowardin, L. M., 9
Crisp, D. J., 107
Crocker, Templeton, 100
Cromwell Current, 199
Crossland, Cyril, 168
Crow, G. E., 257
Cuming, Hugh, 98
Cutler, N., 146
Cwikla, P. S., 231

Dall, W. H., 177
Dampier, William, 162
Daniels, J. D., 431
Darwin, Charles, 112, 222, 224, *226*, 230
Dauphin, G., 211, 256
Davidse, Gerrit, 498
Davies, Edward, 162
Davis, D. R., 231
Dawson, C. E., 100
Dawson, Elmer Yale, 100, 101, 102, 591
Dean, H. K., 110, 144, 146
deforestation
for agriculture, 303–4, 354, 541, *541*
in Área de Conservación Guanacaste (ACG), *298*, 300, 303–4, 305
biodiversity loss and, 3
birds and, 265
for cattle ranching, 354
in Central Pacific region, 357
in Central Valley, 351
cloud forests and, 717–18
Costa Rica's Grand Contradiction and, 4
degree of in Costa Rica, 4
dry forests and, 248, 276–77, 300, 711
end of northern migration of herpetofauna and, 392
extinctions and, 718
Gulf of Nicoya ecosystem and, 139
historical land use and, 622
howler monkeys and, 264
human settlement in the cloud forests of the Talamancas and, 477
Isla del Coco and, 196, 218
for livestock raising, 300
loss of ecological services and, 718
measurement of forest loss and, 541–42
migrating birds and, 435
multiple causes of, 308, 309
in Parque Nacional Santa Rosa, *307*
rate of in Costa Rica, 345, 541, 542, 622–23
reversal of, 4–5
sediment loads and, 602–3
selective logging and, 542
slash-and-burn practices and, 276
small- versus large-scale, 541
soil erosion and, 622
soil organic content and, 84
timber industry and, 305
in wet forests of the Caribbean lowlands, 541–42
de Goeris, Luis, 226
Delaney, P. M., 594
Denevan, W. M., 534
Denyer, P., 39–40, 597
Desliens, Nicolas, 162
DeVries, P., 270, 386, 546
Dexter, D. M., 100, 105, 145, 592, *601*
Díaz-Fergusson, E., 105, 146
Dinerstein, E., 452–53
Dittel, A. I., 146, 147, 594
Dittmann, S., 110, 145
Dodge, B. W., 591
Dodge, C. W., 591
Dohrenbusch, A., 422
Dominici-Arosemena, A., 595
Drewe, K. E., 635–36
dry forests, *251*, *253*, *255*
abandonment of agricultural lands and, 267
apparently evergreen trees in, 333
in Área de Conservación Guanacaste (ACG), 300, 303, 308–11, *315–16*, 322
bees, wasps, and ants in, 271
biodiversity in, 250, 276–77, 309, 711, 712–13
biological corridors and, 280–81
birds of, 265–67
burning of pastures in, 255
climate and weather and, 250, 267, 271, 311, 313–15
climate change and, 310
conservation, management, and sustainability in, 279
defining, 308–11
defoliation by insects in, 273, 330–31
deforestation and, 248, 276–77, 711
disappearance of, 247–48, 249–250
dry season in, 267
as ecoregion, 10
edge effect with rain forests and, 319–20
environmental stresses and, 276
epiphytism in, 275
floristic composition of, 254–56, 260, 277, 279
flowering, pollination, and breeding patterns in, 271–73
forest cover in, 277, *278*, 279
forest fires and, 248–49, 273, 276
forest islands and, 309, 310
forest structure and, 252–54
fragmentation and restoration of, 331–37
freshwater fishes in, 268–69
freshwater shrimps in, 271

geology, geomorphology, and soils of, 250–52
herbivory and frugivory in, 273–74
human impact on, 337–38
hunting in, 276–77
insects in, 269–71
invasion of pastures by, 332–33, 334
lakes of, 659–61, 663
land use history of, 276, 277
life zones of, 252
mammals in, 262
versus moist forests, 250
old-growth ecosystems and, 337
parasitism in, 275–76
in Parque Nacional Palo Verde, *252*
pasture and, *248*
plant-animal mutualism in, 274
poor scientific understanding of, 309–10
property development and, 276, 279
protected areas and, *249*
protected areas of dry lowlands and, *249*
versus rain forests, 311
reforestation in, 248, 277, 279–80
reptiles and amphibians in, 267–68
restoration of, 309–10, 712–13
seasonality in, 267, 310–11, 322–31, *323–24*
secondary forest and, *261*
seed dispersal in, 273–74, 471
soils and, 252, 254
spatial genetic structure of trees in, 273
species interactions in, 271–76
summary of ecosystem, 711–13
threats to, 712
types of trees and, 252–53
vegetation and flora and, 252
vigor of seedlings in, 273
vulnerability of versus cloud and rain forests, 309
of the Western Hemisphere, 247
Dudzik, K. J., 223
Duff, J. M., 637
Duran, F., 355
Durham, J. W., 168
Dutch Agency for International Cooperation (DGIS), 513

Earle, Sylvia, 123
EARTH College, 623
earthquakes
coastal subsidence and, 364
frequency of in Osa region, 372
1991 Limón Earthquake and, 49, 593, *597*, *598*, 601, *602*, 605–6, 710
recent treefalls and landslips in Osa region and, 379
routes across Central America and, 298
sea levels and, *598*
2009 earthquake in Caribbean lowlands and, 550
Earth Summit (United Nations Conference on Environment and Development, Rio de Janeiro, 1992), xv
EARTH University, 563, 565
Eastern Pacific Marine Corridor Initiative, 181
Eberhard, G. W., 231
Echeverría-Sáenz, S., 153–54
ecological services
bogs, marshes, and swamps and, 700
deforestation and, 718
ecosystem management and, 156
mangrove forests and, 108–9
marshes and, 700
valuation of and payment for, 122–23, 182, 277, 280, 542–43, 608
ecology, fundamentals of, 10
ECOMAPAS Project, 9
Economic Exclusive Zones, *98*, 121, 170, 178, 591
ecosystems
anthropic biomes and, 13
benthic, 145
biodiversity at ecosystem level and, 6–7
classifications of in Costa Rica, 7–10
climate change and, 27–29
communities within, 143
definitions of, 10, 143
ecosystem concept in this volume and, 10, 13
ecosystem goods and services and, 156
life zones and, 7–8
mapping of, 9
need for ecosystem-based management and, 156
oceanic and coastal-marine, 709–11
in organization of this volume, 13
photographs of, *12*
structure of this volume and, 11, 13
See also specific ecosystems
Edgar, G. J., 179
Edwards-Widmer, Y., 469, 473
Elizondo, L. H., 345–46, 350
Ellenberg, H., 9
El Niño events
climate change and, 121
coral reefs and, 113, 114, 120, 171, 593, 604–5, 710
crustaceans and, 111
drought and, 23, 26, 27
Gulf of Nicoya ecosystem and, 148
Isla del Coco and, 199–200
lake water temperatures and, 672–73
ocean temperatures and, 2, 104, 219
species composition and, 102
stream phosphorus levels and, 637
threats to Pacific coast environment and, 120–21
treefall activity and, 379–80
tree flowering in 1998 and, 271
Emery, C., 228
endangered and threatened species
coral reefs and, 109
deforestation and, 718
lists of, 396, 482, 608
mangrove forests and, 109
in marshes, 692–93
moving up and out of sight, 333
support for conservation work and, *395*, 398
threats to cichlids and, 628
Enquist, C. A. F., 27–28
Enrique II, 162
Entrepreneur Association for Responsible Development (ASEDER), 371
environmental education, 336
environmental laws, 4, 9, 700, 701
Epifanio, C. E., 142, 147
Epler, B. C., 194
equator, thermal, 311, 313, 314, 319
Equatorial Current of the South, 199–200
Espinosa, José, 594
Espinoza Mendiola, M., 149
Esquivel, A., 78
Esquivel, O., 515
Estero Real–Tempisque Freshwater Ecoregion, 268
Estrada, A., 260, 348, 350, 353–54, 468
estuaries, 150–51, *151*
Ewel, J., 77, 366–67, 371

FAICO. *See* Foundation of Friends of Isla del Coco (FAICO)
Faxon, W., 177
Fernández, C., 116
Fernández, Juan Mercedes, 363
Fernández, W., 596
Fernández-Alamo, M. A., 100
Fernández de Oviedo, Gonzalo, 101, 151
Ferrari, F. D., 604
Fertilizantes de Centroamérica (Costa Rica) S.A. (FERTICA), 64
Feutry, O., 102, 109
FFEM. *See* French Fund for the World Environment (FFEM)
fire. *See* forest and grassland fires
Fischer, S., 146
Fisher, A. K., 168
Fisheries and Aquaculture Institute, 156
fishing and fisheries, *182*
along rivers, 451, 640
artisanal, 152
ban on catching turtles and, 566
commons and, 139
fishery potential and, 153
fish farms and, 648–49
fishing lines collected by national park personnel and, *183*
human impact on Isla del Coco and, 178
illegal, 182, 236, 699
management strategies and, 152–53, 156
Maximum Sustainable Yield and, 153
modified circle hooks and, 116
no-take versus protected areas and, 179
overfishing and, 148–49, 151–54, *151*, 171
Pacific coast's human population and, 120
restriction of access to fishing boats and, 180–81, 182

fishing and fisheries (*continued*)
restriction of in Área de Conservación Guanacaste (ACG), *335*
war's displacement of, *335*
in wet forests of the Caribbean lowlands, 550
wetland habitat loss and, 700
Foerster, C., *395*, 398
Fogden, M. P. L., 434
Föhn effect, 27
Fonseca, A. C., 103
Forel, A., 228
forest and grassland fires, *312*
in Área de Conservación Guanacaste (ACG), 296, *299*, 300, 303, 307, *308*, *312–14*
clearing for agriculture and, 303
in consecutive years, 334
in dry forest areas, 276
fire control and, 280, 306, 307, 332–34, 713
natural versus anthropogenic, 334
in *páramos*, 498–99, *498*, 508, 514, 721
pastures and, 300, 302, 303, 307, *309–10*
payment for ecological services and, 542, 543
recovery from, 514
wetlands and, 686, 691, 699, 701, 702, 721
forestry and forest management, 542–45, 569
Forestry Law of 1996, 542, 543
Forsythe, W., 86
fossils, 50–52, *51*
Foundation of Friends of Isla del Coco (FAICO), 180, 237
Fournier, L. A., 77, 211, 478
Fournier, M. L., 151–52
Foyo, M., 592, 604
FPN. *See* Fundación de Parques Nacionales (FPN)
Fraile, J., 77, 79
Frankie, G. W., 247, 254, 271–72
Freer, E., 154
French Fund for the World Environment (FFEM), 171–72, 175–76, 182, 211, 237
Freytag, P. H., 231
Fuller, C. C., 155
Fundación Delfín Talamanca, 608
Fundación de Parques Nacionales (FPN), 402, 403
Fundación KETO, 608
Fundación Salvemos al Manatí de Costa Rica, 595, 608
Furchheim-Weberling, B., 512

Gabb, William More, 3, 495
Gage, Thomas, 152
Gámez Lobo, Rodrigo, *xviii*
García, V., 596
García-Méndez, K., 102, 108
García-Rios, C. I., 594
García-Robledo, C., 561
García-Rojas, M., 471
Garita, J., 154
Garrison, G., 169, 219–20
Garvin, T., 281
Gasca, R., 100
GEF. *See* Global Environmental Facility (GEF)
Genereux, D. P., 636
Gentry, Al, 468
geological regions of Costa Rica, *31*
Abyssal Plain and, *32*, 33, *34*
Arenal depression and, 48
back-arc region and, 49–50
Baja Talamanca and, 49
Barranca estuarine system and, 39–40, *39*
coastal geomorphology and, *42–43*
Cordillera Central and, 46
Cordillera Costeña range and, 49
Cordillera de Guanacaste and, 45
Cordillera del Coco and, 34
Cordillera de Talamanca and, 46–48
Cordillera de Tilarán and, 45–46
Coto Brus valley and, 48
Fila Bustamente range and, 49
forearc ophiolitic promontories and, 35–37
General Valley and, 48
geological glossary and, 61–63
Golfo Dulce Basin and, 45
intra-magmatic axis basins and, 48
literature on, 32
Magmatic Arc and, 45–48
Middle America Subduction Trench and, *32*, 33, *34*, 36–37
Montes del Aguacate, 45–46
Nicoya Basin and, 37–40
Nicoya Peninsula and, 35
North Limón basin and, 50
Orotina Basin and, 37–40
Osa Peninsula and, 36–37, *37*
Parrita Basin and, 40, 43, 45
Parrita-Quepos area and, 40, *40*
Punta Burica and, 36–37, *37*
Punta Judas erosive platform and, *41*
Puntarenas sand bar and, *38*, 39–40
Quepos Promontory and, 36
San Carlos–Caño Negro–Tortuguero plain and low hills and, 49–50
Santa Elena Peninsula and, 35
Santa Rosa Ignimbrite Plateau and, 45
South Limón basin and, 50
summary of geological history and, 53–56
Tempisque Basin and, 37–40
Térraba Basin and, 40, 43, 45
thrust-fold deformation belts and, 48–49
Turrubares Block and, 36
Valle Central and, 48
Gichuru, M. P., 84
Gilbert, C. R., 595
Gilbert, Lawrence, 368–69, *370*
Gillespie, T. W., 254
Gissler, August, 162, 164, 196
Gladstone, D. E., 275
Glasstetter, M., 175
Global Environmental Facility (GEF), 9, 211, 237, 369, *567*, *569*
Gocke, K.
exploration of *páramos* and, 498
Gulf of Nicoya and, 142, 143
lake studies of, 666–67, 672, 675
on red tides, 154
Goffredi, S. K., 118
Goldman, C. R., 656
Golfo Dulce ecosystem, *143*, 154
Gomes, L. G., 476
Gómez, J. R., 225
Gómez, L. D.
biogeography of cloud forests and, 457
classification of vegetation macrotypes by, 8, 9, 10
classification schemes for Isla del Coco vegetation and, 211
on epiphytes in cloud forests of the Talamanca, 462
exploration of *páramos* and, 497–98
on fungi in cloud forests of the Talamanca, 458
on fungi in the *páramos*, 509
geomorphology of Costa Rica's dry forests and, 251
illness and death of, xix–xx
influences on, 497
Isla del Coco crabs and, 232
paleobotanical research and, 52
publications on Costa Rica and, xix
on vegetation in dry forests, 252, 253, 254, 256
on vegetation of Central Pacific region, 350
vegetation studies in the cloud forests of the Talamanca and, 467
on zonal pattern in highland bogs, 685
Gonyea, Wilford, 364
Gonzales, E., 273
González, C., 207–8
González, E., 156
González, M. A., 83
González, W., 107
González D'Ávila, Gil, 3
Goodnight, J. C., 231
Goodnight, M. L., 231
Goshe, L. R., 591
Gradstein, S. R., 458, 509–10
Grant, P. R., 223
Grant, R., 223
Grayum, M., 256, 348, 349
Great American Biotic Interchange (GABI), 5, 457, 532
greenhouse gases. *See* climate change
Griffin, D., 431
GRUAS I and II projects, 8, 9
Grubb, P. J., 8
Guadamuz, E., 667
Guanacaste Dry Forest Conservation Fund, 281
Guariguata, M. R., 544
Guayamí de Osa, *362*
Gueydon, Henri Louis de, 194

Guillén, C., 77
Gulf of Nicoya ecosystem
  boundaries of, 139–41
  as commons, 139
  comparable environments in Australia and, 145
  concentration of sediments in, 141–42
  development of coastline and, 139
  eutrophy in, 142
  expeditions and, 147, 148, 155
  fish communities in, 148–49
  fishing in, 153
  history of, 150–52
  hypertrophy in, 142
  intertidal sediments in, 144–45
  loss of suspension feeders in, 151–52
  mangrove community in, 143
  maps of, *140*, *141*, 151
  need for ecosystem-based management in, 156
  overfishing and, 154
  photographs of, 139–41, *140*
  physical characteristics of, 141–42
  phytoplankton in, 146–47
  pollution in, 155–56
  red tides in, 146, 154–55
  research needs and, 149
  rocky intertidal substrates and, 146
  salinity of, 141–42, *141*
  sardine migration and, 140–41
  seasonality and, 142, 148, 149
  shipping in, 139–40
  shrimp trawling and bottom damage in, 153–54
  as single ecosystem, 143
  steady state in, 149
  subtidal sediments in, 145–46
  tidal levels in, *147*
  toxic algal blooms and, 155
  trophic modeling and, 149–50, *150*, 154
  tropho-dynamics of, 148
  vertical stratification in, 141
  water chemistry and primary productivity of, 142–43, *143*, 149–50, 151
  water currents in, 142
  worldwide estuarine trends and, 150–56
  zooplankton in, 147–48
Gutiérrez Braun, Féderico, 497
Guzman, H. M., 102, 604, 605
Guzmán, Héctor, 596

Haber, W. A., 428, 429
Haberyan, K. A., 663, *685*
Haberyan, Kurt, 498–99
habitat fragmentation, 264
Häger, A., 422
Haggar, J. P., 77
Halffter-Salas, G., 231
Hallwachs, W., 252, 270, 274, 275, 279, *315*
Halstead, B. W., 168
Hamazaki, T., 634–35
Hamilton, L. S., *452*
Hammel, B. E., 8
Hamrick, J. L., 272
Hancock, Allan, 100
Hansa-Luftbild (German company), 9
Hansen, K. L., 153
Hanson, P., 433
Harbor Branch Oceanographic Institution, 169
Hargraves, P., 146, 154
Harmon, P., 353
Hartman, Carl V., 563
Hartman, O., 168
Hartshorn, G., 65, 254, 345, 347, 374–75, 541
Hass, Hans, 168
Hayes, M. P., 434
Heal, O. W., 83
Heard, Richard W., 594
Hedgpeth, J. W., 156
Heithaus, E. R., 273
Helmer, E. H., 473
Hendler, Gordon, 594
Henry, Dora P., 594
Heppner, J. B., 231
Hernáez-Bové, Patricio P., 118
Herrera, Gerardo, 498
Herrera, M. E., 77
Herrera, W., 8, 198, 345, 372
Herrick, J. E., 81
Hertel, D., 475–76
Hertlein, L. G., 168, 171, 235
Hill, Robert T., 495
Hine, Walter, 305
Hoffman, R. L., 235
Hoffmann, Karl, 3, 7
Hogue, C. L., 227, 231, 386
Holdridge, Leslie R., *8*
  biogeography of cloud forests and, 457
  classification system of, 10
  crops planted at La Selva Biological station by, 541
  on forests of Osa region, 365, 374
  foundational wet forest studies by, 366
  land use recommendations of, 65
  1960 National Academy of Sciences (US) conference and, 366
  predictive power of Holdridge model and, 8
  publications on Costa Rica and, xix
  Tropical Science Center (TSC) and, xv
  See also Holdridge Life Zones
Holdridge Life Zones
  in Área de Conservación Guanacaste (ACG), 252, 256, 262, 322, *326*
  in Central Pacific region, 349, 352
  classic field study introducing, 7–8
  in cloud forests of volcanic Cordilleros, 428, 435
  in Nicoya Peninsula, 260
  páramo ecosystem and, 499–500
  soil organic matter and, 82
Holl, K. D., 471, 478–79
Hölscher, D., 475
Holz, I., 458, 463, 509–10
Hooftman, D. A. P., 460
Hooghiemstra, H., 27, 515, 685
Hopkins, Timothy, 166
Hopkins Stanford Galapagos Expedition (1898–1899), 99, 166
Horn, M. H., 635, *685*
Horn, S. P., 502, 506–7, 512
Horn, Sally, 498–99
Horne, A. J., 656
Hossfeld, B., 148
Houbrick, Joseph, 592, 593–94
Huber, B., 231
Huertas, J. A., 224
Huey, R. B., 552
Humbert, S., 155
Humboldt Current, 199–200
Humedal Nacional Térraba-Sierpe, *99*, *362*
hunting
  antihunting efforts and, 401, *404*
  in cloud forests of volcanic Cordilleras, 437
  cultural and historical reasons for, 401
  in dry forest areas, 276–77
  extinction of New World megafauna and, 293–94
  illegal, 358, 699
  in and near national parks, 400–401
  Pleistocene Era extinctions and, 532
  of reptiles, 552
  seed dispersal and, 562
  threats to *páramos* and, 515
  threat to tapir populations and, 467
hurricanes, 314–15, *321*, *324*
hydroelectric power, 632, 639–646, *641–46*, 649

Ibarra, E., 122
IDA. *See* Agrarian Development Institute (IDA)
Iltis, Hugh, 674
INBio. *See* Instituto Nacional de Biodiversidad (INBio)
INCOPESCA. *See* Costa Rican Fisheries Institute (INCOPESCA)
Instituto Costarricense de Electricidad (ICE), 623, *641*
Instituto Físico-Geográfico Nacional, 166–67
Instituto Meteorológico Nacional (IMN), 535, *535*
Instituto Nacional de Biodiversidad (INBio)
  ALAS (Arthropods of La Selva) research and, 546
  on cloud forest biodiversity, 460
  ECOMAPAS Project and, 9, 513
  growth of ecosystems knowledge and, xv, 8–9
  INBioparque and, xv, 13
  inventories in Parque Nacional Palo Verde and, 257
  Osa region research and, 361, 369, 377
  páramo research and, 499, 511
  plant species lists from, 539

Instituto Nacional de Biodiversidad (*continued*)
specimen collection on Caribbean coast and, 594
Inter-American Highway, *321*
Área de Conservación Osa (ACOSA) and, 366
commercial exploitation of cloud forests and, 477
construction of human settlements and, 477, 514
Cordillera de Talamanca and, *455*, 507
extension of, 366
lakes and ponds along, 660, 668–69, *670*, 674, 685
Mesa Santa Rosa and, 291, *301*
páramos along, 514, 670, 673
Parque Nacional Santa Rosa and, 305
peat deposits along, 82, 506
protected forest along, 4
ranching and, 331
tapir killed in road accident and, 479, *480*
through Área de Conservación Guanacaste (ACG), 304
Inter American Institute of Agricultural Sciences (IICA), 64
International Union of Biological Sciences (IUBS), 5
Intertropical Convergence Zone (ITCZ), 20–21, 104, 173, 197, *198*, 372, 420, 503
invasive species
in Área de Conservación Guanacaste (ACG), 335–36
biodiversity and, 336
cattail as, 659, 663
Isla del Coco ecosystems and, 211
tilapia as, 628, *628*, 640, 648–49, 692
Irons, J. G., 637–38
Isla del Coco and its ecosystems, *164*, *166–67*, *204*, *206*, *207*
age of the island and, 201
ants and, 228
arachnids and, 231, *232*
arthropods and, 235
bay communities of, 214–15, 218
beaches and, 173, *174*
beetles and, 228
biodiversity and, 171–72, 193–95, 200–211, 236–38
birds of, 221–24, *224*, *225*
butterflies and, *230*
caves, tunnels, and arches and, 175, *176*, *177*, *178*, 206
climate and weather and, 162, 172–73, 197–99
coastal cliff communities of, 215–16, 218, *218*
coconut groves on, 214–15
conservation and, 180–82
coral reefs and, 175, 178–79, *179*, *181*, 710
Costa Rican government presence on, 236
current state of knowledge about, 171–72, 196–97
deforestation and, 196, 218
disharmonious floras and, 209
El Niño Southern Oscillation (ENSO) and, 178, 199, 200, 219
endemicity of species and, 172, 193, 208–11, 219
expeditions and, 165–68, 171, 177, 194–96, 197, 209, 226, 228, 231
fishes of, 219–20
fishing near, 182, *182*
freshwater invertebrates and, 233–35
fungi, Myxomycetes, and lichens on, 208–9
future sustainability and, 236
geographical coordinates of, 162, 172, 197
geology and geomorphology of, 172, 192, 200–201, *201*, 205–6
history of, 162, 164, 165–71, 193–96
human impact on, 178–79, 196, 214–15, 219, 235–36
hydroelectric power plant and, 196
importance of for research, 193, 211
as Important Area for Bird Conservation, 218
Indo-Pacific species and, 165, 184, 193
insects and other terrestrial invertebrates on, 227–32, *229*
as internationally important wetland, 165, 193, 236, 238
Intertropical Convergence Zone (ITCZ) and, 173, 197–98, 199
introduced vertebrates as threat to, 224–25, 226–27
invasive species and, 211
Isla del Coco Bioregion and, 193
land cover percentages and, 218
landslide vegetation and, 217, 218, *218*
mammals and, 224–27, *228*, 236
maps of, 162, *163*, *164*, *195*, 203
as Marine Conservation Area, 236
national exploration and tourism and, 196
national park and, 164–65, 236
need for environmental management on, 236–38
North Equatorial Countercurrent and, 165, 173, *173*
as oceanic island, 192–93, 199–200, 209, 211
open ocean and deep waters and, 177–78
origins of plants on, 210–11
penal and agricultural colonies and, 196, 226
rain forests and, 713–14
reptiles of, 220–21, *221*, *222*
research needs and, 182–83, 237–38
riparian communities and, 216–17, 218
rocky intertidal zones and, 173, *175–76*
sedimentary deposits from the Holocene in, 201–3, *202*
shallow and deep sea fauna and, 711
soils of, 207–8
sources of species on, 229
species introduced to, 196
sphingids or hawk moths and, 229–31
terrestrial crustaceans and, 231–33, *233*, *234*, *235*
threats to, 178–79
topography and hydrology of, 205–7
tourism and, 236
treasure hunting and, 195–96
Treasure Island and, 162
tropical cloud forest on, 213–14, *214*, *215*, 218
tropical rain forest on, 213, *213*, *216*, 218
vegetation and plant communities of, 209–11, 217–18, *217*, *219*
vertebrates on, 218–19
volcanic rocks and, 203–5
waste from human residents and visitors and, 236
water circulation around, 173, *173*
as World Heritage Site, 165, 184, 193, 236, 238
See also Pacific coast of Costa Rica
Isla del Coco Marine Conservation Area (ACMIC). *See* Área de Conservación Marina Isla del Coco (ACMIC)
Islebe, G. A., 27, 457, 508
ITCO. *See* Lands and Colonization Institute (ITCO)
IUCN Red List, 482, 608

Jackson, J. B., 150
Janss Foundation, 169
Janzen, D. H.
All Taxa Biodiversity Inventory (ATBI) and, 250
beetle studies and, 269, 273
conservation advocacy of, 280
dry forest studies and, 248, 252, 270, 274
growth of ecosystems knowledge and, xv
Guanacaste studies and, 252, 260, 711
natural history of Costa Rica and, xix, xx, 270
páramos studies and, 498, 512
plant-animal mutualism and, 274
reforestation projects and, 279
J. Craig Venter Institute, 169
Jennings, S., 156
Jiménez, C., 113, 123, 175–76, 595–96, 665
Jiménez, Ignacio, 595
Jiménez, J., 9, 143, 156, 349
Jiménez, Q., 260, 348, 350–53
Jiménez, W., 473
John A. Phipps Biological Station, 566
Johnson, N. C., 85, 231
Jordal, B. H., 231
Joyce, A., 345, 542
Juárez, M. E., 477–78, 479

Kalácska, M., 277, 279
Kappelle, Dirk, 451
Kappelle, M., *xx*, *8*
on birds in forest recovery, 476
cloud forest studies and, 457, 460, 462–63, 467–69, 473, 476–79
on lichens in oak forests, 459

mapping of marine ecosystems and, 9
páramo studies and, 499, 508
on tree plantations in Osa region, 377
on zonal pattern in highland bogs, 685
Kass, D. L., 78
Kasteleijn, H. W., 194
Keith, Minor C., 495, 563
Kelso, D. P., 595
Kennedy, John F., 505
Kernan, Kit, *370*
Kimble, J. M., 83
Kirkendall, L. R., 229
Kohkemper, J., 656
Kohkemper, M., 502
Köhler, L., 462, 475
Kramp, P. L., 168
Kriebel, R., 460
Kroodsma, D. E., 223–24
Kumar, A., 433
Kupper, Walter, 497
Kuprewicz, E. K., 561
Kury, A. B., 231
Kyoto Protocol, 8

Lachniet, M. S., 503, *504*, 506–7
Lafond, Gabriel, 363
Lahmann, E. J., 107
lakes of Costa Rica
Asunción Pond and, 674
Caribbean lowland moist forest and, 674–77
Central Pacific moist forest and, 663
Chirripó lake district and, *671*
climate change and, 678
diatom assemblages in, 663, 668, 669–70, 676
distribution and origins of, 657–59, *658*
El Niño–Southern Oscillation (ENSO) warming events and, 672–73
Estero Blanco and, 659, 660–61, *662*
fish kills and, 666, 675
formation of, 656, 663–64, 666–68, *666*, 674–77
geological history of, 657, 663–64, 670, 674, 677–78
human impact on, 678
Lago Asunción and, 672
Lago Chirripó and, 670, 672, 674
Lago de las Morrenas and, 672, *673*, 674
Laguna Arancibia and, 668
Laguna Barva and, 664, 665–66
Laguna Bonilla and, 676–77
Laguna Bonillita and, 675, 676–77, *676*
Laguna Bosque Alegre and, 666
Laguna Botos and, 674
Laguna Carara and, 663
Laguna Cerro Chato and, 664–65, 668, 674
Laguna Congo and, 666
Laguna Cote and, 667
Laguna de Río Cuarto and, 667, 674–76
Laguna Gamboa and, 668–69
Laguna Gandoca and, 675, 677
Laguna Hule and, 665, 666–68, *666*, 674
Laguna La Palma and, 665, 667
Laguna Madre Vieja and, 663
Laguna María Aguilar and, 667
Laguna Martínez and, 659, 660–61
Laguna Palmita and, 660
Laguna Poco Sol and, 665, 667–68
Laguna San Pablo and, 659, 660–61, *662*
Laguna Santa Elena and, 668–69
Laguna Sierpe and, 664
Laguna Solimar and, 660, 661
Laguna Vueltas and, 664
Laguna Zent and, 675, 677
Laguna Zoncho and, 668, 669–70
lake temperatures and, 674–75
lake turnover and, 674–75
maps of, *658*, *660*
northern highland evergreen cloud forest and, 664–68
of the Pacific dry forest, 659–61, 663
páramo grasslands and, 670, 672–74
pollen records in study of, 668, 669–70, 674
presence versus absence of fish in, 672
Quebrador Pond and, 673–74
research on, 656–57
sediment cores and, 661, 663–64, 667–70, 674, 677
Southern Highland cloud forest and, 668–70
Southern Pacific moist forest, 664
Tres de Junio Pond and, 669–70, 674
turnover in, 666–67
water pollution and, 677
water temperatures and, 672–73
wetlands ecosystems and, 720–22
wet versus dry season and, 662
Lal, R., 81, 83
Lamont Doherty Earth Observatory, 168
Lands and Colonization Institute (ITCO), 365
Lang, S. B., 86
Lanteri, A. A., 231
Lapied, E., 79
Las Cruces Biological Station, 477, 668
La Selva Biological Station, *568*
ALAS (Arthropods of La Selva) research at, 433, 546
archaeological artifacts at, 532
Arthropods of La Selva (ALAS) Project in, 560
bats at, 437, 557
biodiversity at, 429, 431, 434
butterfly inventories at, 560
contributions to riverine research and, 623, 624–25
as control forest site, 541
decline in amphibians of, 629
decline of terrestrial frogs in, 551
fish communities at, 627
food web in streams of, *634*
forest studies in and around, 543–45
fossils at, 532
frugivory and seed dispersal at, 561, 635–36
herbivory at, 558, 634–35
hydrodynamics at, 549
insect community at, 558
invertebrate studies at, 546, 635
lakes of, 677
microbial respiration studies at, *638*
organic matter processing at, 637–39
patterns of vegetation at, 540–41
plant species lists from, 539
predator-prey interactions at, 635
primate studies at, 400, 557
production of higher trophic levels at, 639
as renowned research site, 369, 372, 404, 714
reptiles of, 552
research needs and, 556
rivers and streams of, *625*
San Juan–La Selva Biological Corridor and, 566, 568–69, *568*
seed dispersal at, 562
soils at, 77, 80, 86, 538–39
species collection at, 386–87, 539
stream chemistry at, 636
stream nutrient dynamics at, 637
terminus of Pleistocene lava flows at, 636
travel to, xv, *528*
view from research tower at, *537*
weather at, 535–37, *538*
Laurencio, D., 268
Laurito, C., 52
LaVal, R. K., 272, 437, 557
Lavelle, P., 79
Lavenberg, R., 219–20
Law of the Sea of the United Nations (UNCLOS), 164–65
Lawrence, John M., 591–92
Lawton, M. F., 423, 473
Layton, W. E., 224
Leenders, T., 356
Leiva, J. A., 85
Lellinger, D. B., 462
León, P., 148, 149, 305
León, S., 79
Leslie, M., 390, 595, 634
Lessios, H. A., 165
Levi, H. W., 231
Levin, L. A., 118
Liesner, R., 252, 260, 273
Lièvre, D., 165, 193
Lines, N., 478
Link, J. S., 156
Linsley, E. G., 231
Lips, K. R., 634
livestock raising
abandonment of ranching lands and, 279
in Área de Conservación Guanacaste (ACG), 302
cattle slaughter industry and, 300
clearing land for, 300
cross-isthmus transport and, 331
decline in ranching and, 277

livestock raising (*continued*)
deforestation and, 541
economic viability of, 302–3
environmental damage caused by, *315*
fire and, 302, *315*, 691
grasses chosen for, 302
habitat destruction and, 567
historical land use and, 276, 621–22
land use conversion and, 646–47, *647*
reduction of biodiversity and, 302, 303
Lizano, O., 142, 155, 171, 596
Lobo, L. A., 272
Lockwood, J. P., 206
Löffler, H., 498, 656
Lohmann, W., 52
Loiselle, B. A., 561
Longino, J., 433, 463, 546
López, F. L., 78
López, M., 116, 148, 171, 219, 397, 595
López Pozuelo, F., 218
Loría-Naranjo, M., 109
Los Angeles County Museum of Natural History, 169
Lourenço, W. R., 231
Lovejoy, Thomas Eugene, *xv*, 4
Lücking, A., 208, 211, 224
Lücking, R., 208, 211, 224
Lugo, A. E., 248
Luke, Spencer R., 100
Lulow, M. E., 471
Luteyn, J. L., *510*
Lyons, J., 635

Macintyre, I. G., 175
Madrigal, R., 32
Madrigal-Castro, E., 102
MAG. *See* Ministerio de Agricultura y Ganadería (MAG)
Malavassi, E., 205
Maldonado, T., 256
Manning, R. B., 100, 102, 165, 168
maps and mapping
Área de Conservación Guanacaste (ACG) and, *291*, *292*, *294–95*, *326*
Área de Conservación Osa (ACOSA) and, 377
bogs, marshes, and swamps and, *684*
Caribbean coast and, *592*
Caribbean lowlands and, *528*
Chirripó lake district and, *671*
climate zones of Caribbean lowlands and, *535*
cloud forests of the Talamanca and, *452*
cloud forests of the volcanic Cordilleras and, *416*
of Costa Rica, *11*, *31*
of currents in Eastern Tropical Pacific, 173
distribution of lakes and, *658*, *660*
distribution of *páramo* vegetation and, *493*
ECOMAPAS Project and, 513
of Economic Exclusive Zones (EEZ), *98*
forest cover in Chorotega region and, *278*
funding for, 9
geological, *31–32*, *504*
Gulf of Nicoya and, *140*, *141*, 151
hydropower plants and, *641*, *646*
Isla del Coco and, 162, *163*, *164*, 171, *195*, 203
locations of plant collections in Caribbean lowlands and, *540*
of marine ecosystems, 9, 103
of marine protected areas, *592*
mega-mammal localities and, *50*
methodology for, 9
northern Costa Rica and, *416*, *417*
Osa region vegetation and, 375, *376*, *377*
of Pacific coast, *99*
paleographic reconstruction of Costa Rica and, *54*
páramo ecosystem of Costa Rican highlands and, *494*, 497
of *páramo* plant communities, 513
phytogeographic units and, 9
of potential land use, 64
of protected areas of Caribbean moist forest, *530*
of protected areas of montane cloud forest zone, *416*
of protected areas of Northwestern dry lowlands, *249*
of protected areas of Pacific moist lowland forests, *362*
of protected areas of Pacific seasonal forest zone, *346*
of protected areas of Talamancan high forest region, *452*
of protected areas on Pacific coast, *99*
of protected areas that conserve *páramo* ecosystem, *494*
soil maps and, 64, *65*, *66–67*, *69*, *72*, *74–75*
Southern Pacific region and Osa peninsula and, *362*
Tosi's life zone map and, *500*
variable place names and, *550*
watersheds of Costa Rica and, *622*
world maps and atlases and, 162
Marine Pollution Monitoring Program in the Caribbean (CARIPOL), 596, 609
Maritza Biological Station, 623
Marr, Wilhelm, 154
marshes. *See* bogs, marshes, and swamps
Martin, J. W., 171–72
Martin, P. S., 532
Martínez, A., 20, 116
MarViva Foundation, 237
Mata, A., 101, 156
Mata Jiménez, Alfonso, 247
Mather, J. R., 23, 25, 26
Mathis, W. N., 231
Mattiucci, S., 106
May-Collado, L., 116, 595
McGlynn, T. P., 77
McLaughlin, Patsy A., 594
McLennan, B., 281
McNeill, D. F., 597
Mediterranean Red Book of threatened habitats, 116
Mehltreter, K. V., 462
Mesoamerican Biological Corridor, *545*, *565*, *567–68*
Michel, H. B.
marine research and, *592*
on open waters off the Caribbean coast, 604
Mielke, W.
crustacean studies and, 594
Gulf of Nicoya ecosystem and, 144
migration of species
altitudinal, 355, 507
Great American Biotic Interchange (GABI), 457
of invertebrates, 635
latitudinal, 553
migration potential and, 508–9
oceanic islands and, 192, 197, 208
sardines and, 141
sea-level rise and, 109
seasonal, 323–28, 696
sharks and, 183
of shrimp, 632
of turtles, 567
weather events as cues for, 322
Mikheyev, A. S., 228
Millennium Development Goals (MDGs), 563–64
Millennium Ecosystem Assessment (MA, 2006), xvii
Miller, Kenton, 305–6
Miller, S. E., 227, 231
MINAE. *See* Ministerio de Ambiente y Energía (MINAE)
mining, 639–40
Ministerio de Agricultura y Ganadería (MAG), 65
Ministerio de Ambiente y Energía (MINAE)
administration of protected areas by, 608, 609
anti-hunting efforts of, 401
ECOMAPAS Project and, 9, 513
ecosystem management and, 156
freshwater ecosystem monitoring and, 633
mapping of ecosystems in Área de Conservación Osa (ACOSA) and, 377
marine issues and, 123
natural forest management and, 542–43
water concessions and, 648
Missouri Botanical Garden, plant species lists from, 539
Mohr, Mary E., 451
moist forests. *See* cloud forests; dry forests; rain forests
Molina-Ureña, H., 148
Monge, Guiselle, 568
Monge, J., 78, 79–80, 511
Montes, C., 97
Monteverde Cloud Forest Preserve, 443, 462
Montoya, M., 165, 169–70, 211, 218–19, 223–24, 230, 594

Monumento Nacional Santa Rosa, 298
Mora, A., 114, 178–79
Mora, Dagner, 596
Mora, J. M., 224
Mora, José Joaquín, 300
Mora, S., 32
Morales, A., 147
Morales, J. F., 275, 352, 468
Morales, M. I., 431
Morales-Ramírez, A., 102, 116, 147, 148, 594, 595
Moran, Dennis A., 594
Moreno-Díaz, Mary Luz, 182
Motta, P. J., 170
Moya, Jorge, 247
Mueller, Greg, 458
Mueller-Dombois, D., 9
Mug, M., 6, 9–10
Muller-Parker, G., 595
Murphy, P. G., 248
Murray, K. G., 438
Museo Nacional of Costa Rica, 166
Myers, N., 348

Nadkarni, N., 422, 433, 441, 462, 463
NASA, 257
Nason, J. D., 272
National Academy of Sciences (US), 366
National Aqueduct and Sewerage System (AyA), 21
National Association for Indigenous Matters (ANAI), 595, 608
National Geographic Society, 171
National Institute of Biodiversity (INBio). *See* Instituto Nacional de Biodiversidad (INBio)
National Meteorological Institute (IMN), 21
national park system
  administration of protected areas by, 608
  ECOMAPAS Project and, 9
  founding of, 360
  funding and, 402
  Isla del Coco and, 236, 237
  NGOs and, 402
  Osa region research facilities and, 369
  overvisiting in, 515
  removal of vestiges of human residence in, 306
  See also protected areas; *and specific parks*
National System of Conservation Areas (SINAC). *See* national park system; protected areas
Nature Conservancy, The
  acquisition and donation of land by, 403
  Caribbean coast conservation efforts and, 608
  cloud forest conservation and, 443
  Costa Rican–Panamanian site conservation plan and, 483
  Costa Rican wetland systems and, 9–10
  ecoregional assessment and, 181
  Isla del Coco Bioregion and, 193
  Parque Nacional Corcovado and, 367
  sustainable economic development and, 371
  wetland conservation and, 701
NECC. *See* North Equatorial Countercurrent (NECC)
Newbold, J. D., 636
Newsom, L., 534
New York Zoological Society, 168
New Zealand Threat Classification System, 116
Nickle, D. A., 231
nitrogen cycle, xv–xvi
Nixon, S. W., 142
NOAA Southwest Fisheries Science Center, 170
Norrbom, A. L., 231
North, Oliver, 297, *299*
North Equatorial Countercurrent (NECC), 165, 173, 199–200
Nova-Bustos, N., 107

Obando, J. A., 5, 6, 9
Obando-Calderón, G., 224
oceanic and coastal-marine ecosystems, summary of, 709–11
O'Donnell, S., 433
Oduber, Daniel, 305, 566
Odum, Eugene P., 10
Oelbermann, M., 85
Oemke, M. P., 627
Oldeman, R. A. A., 473
Olsgard, F., 155
Olson, D. M., 10, 452–53
Opresko, D. M., 118
Organization for Tropical Studies (OTS)
  contributions to riverine research and, 623
  dry forest conservation and, 280
  lake studies and, 661, 667
  land use assessments and, 64–65
  marine algae research and, 591
  North American students and, xv
  origins of researchers at, 366
  research facilities of, 404
  research reserves and, 366
  specimen collection and, 592–93
  See also La Selva Biological Station
Ornelas-Gatdula, E., 103, 594
Ørsted, Anders Sandøe, 3, 7
Ortea, Jesús, 594
Ortega, S., 107
Orvis, K., 498–99, 502, 506–7, 672
Osa Forestal (Osa Productos Forestales), 364–68, 371
Osa region
  ants in, 386–88
  biodiversity in, 361, 373–75, 387, 392, 394
  birds of, 392–94
  butterflies in, 386
  climate of, 372
  commerce in, 364–65, 371
  conservation efforts in, 365–72
  early scientific activity in, 364–65
  El Niño Southern Oscillation (ENSO) and, 379–80
  endemism in, 388, 391–92
  fate of Golfo Dulce and, 371
  fishes of, 388–91
  forest gap dynamics and, 378–85
  forest swamps in, 697
  geomorphology of Costa Rica and, 373
  gold mining in, *363*, 364, 365
  herpetofauna in, 391–92
  human impact on, 361–64, 371, 389
  importance of for science, 361, 404–5
  insects in, 386–88
  land crabs in, 385–86
  land use policies in, 362–63, 365, 366
  land use types in forests of, 382–83
  limited exploration of, 360–61
  mammals in, 394–400
  maps of, *362*, 375, *376*, 377
  Osa Peninsula and, *367*
  padlocking of research station in, 368
  palm swamps in, 694
  pastureland in, *378*
  pre-Columbian artifacts in, *363*
  primates in, 399
  rain forests in, *714*
  rancho in, *363*
  small- and large-scale disturbances in, 379–80, 382
  soils of, 373
  stature of trees in, *375*
  streams and rivers of, 389
  threatened tree species in, 374
  timber industry in, 364–66
  tourism and local economies in, 371
  treefalls in, 379–81, *380*, 382
  Tropical Science Center (TSC) station in, *363*
  vegetation of, *368–69*, *373–75*, 377–78
  See also Área de Conservación Osa (ACOSA)
OSPAR Convention. *See* Convention for the Protection of the Marine Environment of the North-East Atlantic (OSPAR Convention)
Otárola, C. E., 456, 505
OTS. *See* Organization for Tropical Studies (OTS)

Pacific coast of Costa Rica
  area of coastal waters and, 97, 165
  Bahía Culebra and, 102, 103, 105, 111, 113, 120, 123
  beach ecosystems and, 105–6, *105*, *106*
  bottom profile of, *103*
  versus Caribbean coast, 97
  circulation, tides, and waves and, 104–5
  climate and, 97, 104
  climate change and, 121
  cold seeps and, 117, 118
  conservation efforts and, 122–23
  coral reef ecosystems and, 111–13

Pacific coast of Costa Rica (*continued*)
Costa Rican Thermal Dome and, 100, 104, 116, 119
Costa Rica's inability to patrol waters and, 121, 122
current state of knowledge about, 101–3
deep benthos zone and, 116–19
deep water fisheries and, 118
early research on, 101
ecosystem health and human survival and, 123
environmental threats along, 120–21
expeditions along, 99–101, 102, 116–17, 118, 123
geology and geomorphology of, 103–4
Golfo de Nicoya and, 101, 103, 104, 109, 110, *110*, 112
Golfo Dulce and, 102, 104, 111, 112, 118, 123
historical overview of marine research and, 98–101
human population and demography of, 120
hydrography of, 104
intertidal flats ecosystems and, 109–11, *110*
Intertropical Convergence Zone (ITCZ) and, 104
Isla del Caño and, 102–3
Isla del Coco expeditions and, 169
islands and islets of, *118*, 119
mangrove forests and, 108–9
marine biodiversity of, 119
marine protected areas and, 121–22
offshore islands and, 112
Oxygen-Minimum Zones and, 104, 117–18
pelagic ecosystem and, 116
protected areas of, *99*
real-estate development along, 371
research needs and, 123
rhodolith beds in, 114, *115*, 116
rocky intertidal ecosystems and, 106–7, *107*
seagrass beds ecosystems and, 111
species diversity and, 119
summary of oceanic and coastal-marine ecosystems and, 709
sustainability and, 123
upwelling and, 105, 107, 111, 113
weather and, 104, 111
See also Gulf of Nicoya ecosystem
Pacific Equatorial Undercurrent, 199
Paddack, M. J., 606
Paine, R. T., 146
Palo Verde Biological Station, 623
Palter, J., 142
páramo ecosystem of Costa Rican highlands, *495*, *496*, *506*
algae in, 509
altitudinal zonation and, *455*, 467, 479, 494–95, 499–500
animals in, 511
biodiversity in, 509–14, *510*, 720
bogs of, *685*, 686
bryophytes in, 509–10
climate change and, 506
climate in, 500, 502–3, *502*
conservation and sustainable use in, 514–15
definition of *páramos* and, 492
effects of fire in, 498–99, *498*
endemism in, 508–9
ferns and fern-allies in, 509, 510
as fire-dependent ecosystem, 508
fire in, 508, 514
flowering plants in, 510, *510*
forest succession and, 474, 476
fungi and lichens in, 509
geographical distribution of in Central America, 500
geology and geomorphology of, 495, 497, 503, *504*, 505–8
global distribution of *páramos* and, 492–95, *493*
history of scientific exploration in, 495, 497–99
Holdridge Life Zones in, 499–500
inland waters of, 505–6
as islands in the sky, 500
lakes of, 670, 672–74
land use history in, 514
lateral profile of, *501*
limnology of, 505–6
mapping of, 497
Mesoamerican biodiversity hotspot and, 499
montane cloud forests surrounding, 500
paramillo communities and, 513
plant specimen collection and, 497–98
protected areas in, *494*
quarternary history of, 506–9
Ramsar Wetland of International Importance in, 515
rivers originating in, 456
sacred sites of indigenous people and, 497
soils of Costa Rica, *505*
summary of, 719–20
Talamancan Montane Forest ecoregion and, 499
threats to, 515
vegetation of, 508–9, 512–14, *513*
World Heritage Site and, 481, 499
Parker, N., 170
Parkinson, R. W., 598
Parque Internacional La Amistad
Amistosa corridor initiative and, 9
biodiversity in, 460
biological corridors and, 482
birds in, 465
climate in, 454
conservation and, 514–15
as core protected area, 480
forest zonation of, 468
highland peat bog in, *685*
location of, *452*, *494*
mammals in, 466, 467
Man and Biosphere Reserve and, 515
The Nature Conservancy's site conservation plan for, 483
Panama and, 481
páramos in, 500
plant communities in, *513*
visitation levels in, 515
World Heritage Sites and, 515
Parque Nacional Arenal, *416*, 443
Parque Nacional Barbilla, *530*
Parque Nacional Barra Honda, *249*, 251, 280
Parque Nacional Bosque Diriá, *249*
Parque Nacional Braulio Carrillo, *416*, *530*, *556*, *560*, *623*
Parque Nacional Cahuita
conservation efforts in, 606, 608–9
coral reefs in, 593, 603, *604*, 608
crustacean studies in, 594
location of, *530*, *592*
marine biodiversity studies and, 595
marine pollution studies in, 596
protected areas and, 4
seagrass beds in, 602, *603*
size of, 606
Parque Nacional Carara
biological corridors and, 357
birds in, *355*
dry and moist forest in, 711
floristic composition of, 347
forest layers in, 348–49
location of, *346*, 403
oxbow lake in, *349*
primary forest in, *348*
scarlet macaws and, *355*
as transition zone, 348
tree species in, 260
Parque Nacional Chirripó
biodiversity in, 460
birds in, 465–66
conservation and, 514–15
as core protected area, 480
elevational changes in woody species richness in, 468
establishment of, 247
fires in, 498, *498*
glacier morphology in, *47*
lakes of, 670
location of, *452*, *494*
mammals in, 466, 467
management planning for, 499
páramos in, 500
protected areas and, 4
Talamancan-Caribbean Biological Corridor and, 482
visitation levels in, 515
Parque Nacional Corcovado
Baird's tapir in, 397–98
biological corridors and, 400, 482
birds in, 392, *393*
BOSCOSA project in, 370
butterfly-hostplant relationship and, 383–85, *383–84*
canopy gap densities in, 382

cessation of agricultural disturbance in, 378
conditions before creation of, 389
conservation work in, 402
creation of, 361, 367–68, 400
difficulty of accessing, 361
ecosystems in, 375
ecotourism in, 369, *370*
environmental activism in, 371
environmental education and, 370–71
expansion of needed, 397
fishes of, 390, *390*, 627
forest swamps in, 696
funding and, 367, 370–71, 403–4
gold mining in, 368–69, 370, 389, 400
human impact in, 361
human pressures on wildlife in, 400–401
importance of for science, 404–5
jaguars in, 398–99
land crabs in, 385–86
location of, *99*, *362*
mammals in, 394–95, 716
Man and Biosphere Reserve and, 515
mapping of vegetation types and, *376*
moist tropical forest in, 347
park fees and, 402–3
patch dynamics in, 383
poaching crisis in, 403, *404*
primate studies in, 399
protectability of, 371
protected areas and, 4
public engagement and, 370
relocation of settlers and, 367, 380–81
research hindered at, 368
research projects in, 368–70
research stations in, *380*, *381*, *382*
research trails in, 378–79
size of, 403
sustainable forestry in, 371
treefall events in, 382
white-lipped peccaries in, 396
Parque Nacional Guanacaste
biological corridors and, 281
forest fires and, 249
insects in, 269
location of, *249*, *416*
rattlesnakes in, 268
reforestation and, 279
turtles in, 268
volcanic complexes in, 443
See also Área de Conservación Guanacaste (ACG)
Parque Nacional Isla del Coco
administration station in, *307*
before and after aerial photos of, *381*
biodiversity in, 171–72
biomass of fish in, 171
buffer zone around, 180
conservation and, 180–82
Costa Rica's claim of sea around, 164–65
creation of, 164, 180, 193
deforestation in, *307*
discovery of new species and, 171
fish diversity studies at, 169–70
fishing lines retrieved at, *183*
importance of for research, 183–84
maps and, 171
new species found at, 169
personnel of, *184*
removal of cattle and horses from, 306–7
secondary forest in, *350*
shark tagging and, 170
threats to, 360
valuation of environmental goods and services and, 182
as World Heritage site, 165, 184, 193
See also Isla del Coco and its ecosystems
Parque Nacional Juan Castro Blanco, *416*
Parque Nacional La Cangreja, *346*, 352, 355, 357, 711
Parque Nacional Los Quetzales, 357, *452*, 471, 481, *494*
Parque Nacional Manuel Antonio
ants in, 387
dry and moist forest in, 711
Eastern Tropical Pacific Marine Corridor and, 165
gastropods at, 107
justification for creation of, 368
location of, *99*, *346*, 403
as paper park, 122
primate studies in, 399
property development around, 358
squirrel monkeys in, 354, 403, 716
vegetation of, 353
Parque Nacional Maquenque, 8
Parque Nacional Marino Ballena, *99*, 113, 403
Parque Nacional Marino Las Baulas, *99*, 106, *106*
Parque Nacional Palo Verde
Arenal Dam complex and, 642
biological corridors and, 281
birds of, 265
cactus in, *252*
climate of, 250
dry forest conservation and, 280
fires in, 249, 255, 701, 702
freshwater avifauna in, 266
geomorphology and, 251–52
location of, *249*
marshes in, *690*
natural hydrological state in, 659
plant genera with largest number of species in, 257
pollination systems in, 273
restoration of marshes in, 701–2
species inventories in, 257
water flow and quality in, 699–700
wetlands in, *258*, *259*, 660, 661, 663
Parque Nacional Piedras Blancas, *99*, *362*, 403, 482
Parque Nacional Rincón de la Vieja, 249, *249*, *416*, *635*
Parque Nacional Santa Rosa
Área de Conservación Guanacaste (ACG) and, 281, 331
beetles in, 269
biodiversity in, 254
butterflies in, 270
creation and growth of, 305–6
dry forest conservation and, 280
fire control in, 307
focus of guards' efforts in, 306
insects of, 269
justification for creation of, 368
landscape-level restoration and, 298
lianas in, 256
location of, *99*, *249*
monkeys and capuchins in, 264
moths in, 270
protected areas and, 4
reptile and amphibian species in, 267
restoration efforts in, 713
as right-now act of conservation, 331–37
seed dispersal and, 274
sphingids in, 270
turtles in, 268
white-faced capuchins in, 263–64
Parque Nacional Tapantí–Macizo de la Muerte
biodiversity in, 460
birds in, 466
climate in, 454
as core protected area, 480
forest communities of, 469
highland peat bog in, *686*
location of, *452*, *494*
mammals in, 466, 467
Man and Biosphere Reserve, 515
páramos in, 500
soils in, 455, 456
Parque Nacional Tortuguero, *601*
ants in, 387
conservation efforts in, 606, 608
creation of, 566, 567, 569
fishes of, 390, *390*
frugivory and seed dispersal in, 636
as isolated island of forest, 542
as key protected area, 4
location of, *530*, *592*
palm swamps in, 694
size of, 606
Parque Nacional Volcán Irazú, 4, *416*, *494*
Parque Nacional Volcán Poás, 402, *416*
Parque Nacional Volcán Tenorio, *416*
Parque Nacional Volcán Turrialba, 4, *416*, *494*
Pascal, M., 218–19
Paseo Pantera, 567
Pearse, John S., 169
Pedroni, L., 477
Pereira, A. I., 145
Pereira-Chaves, J., 604
Pérez, E. A., 52
Pérez-Cruet, M. J., 595
Pérez-Reyes, C. R., 604
Peruvian Current, 199–200
pesticides, 564–65
Petren, K., 224

Petrescu, Iorgu, 594
pharmaceuticals, 169
Phillips, P., 148–49, 375, *376*, 377, *595*
Pilsbry, Henry A., 168
Pinchot, Gifford, 168
Piperno, D. R., 533
Pittier, Henri
  cloud forests of the Talamanca and, 453–54
  expeditions of, 3, 166, 167, 195, *195*, 364, 495, 503
  on Isla del Coco climate, 172
Pizarro, Francisco, 296
plant reproduction
  in cloud forests of the Talamanca, 469, 471–73
  cloud forest species interactions and, 437–39
  crown-to-crown vegetative competition and, 330
  dry forest seed dispersal and, 332, 333
  endozoochory versus ectozoochory and, 560
  frugivory and, 560–62
  nonreproductive juveniles and, 333–34
  pollination by birds and, 435
  pollination systems and, 439, 469, 471, 719
  prehistoric dispersal agents and, 329
  seasonality of migration of, 328–31
  seed dispersal and, 397, 435, 439, 471–73, 560–62, 635–36, 718
  seed predation and, 471, 472–73
Platnick, N. I., 231
pollution, 150–51, 155, 608, 647
Ponwith, B. J., 634
Porsch, Otto, 7
Pounds, J. A., 434
Powell, George, 443, 567
Powers, J. S., *82*, 84, 85, 279
PPNG. *See* Proyecto Parque Nacional Guanacaste (PPNG)
Prescott, S. C., 64
PRETOMA, 116, 170
Pringle, C. M., 634–35, 636, 639
Pritzker family, 364
Proctor, J., 77
Programa Cooperativo Oficina de Café-MAG, 64
Programa Estado de la Nación en Desarrollo Humano Sostenible, xvii
protected areas
  administration of, 608
  as biogeographical islands, 347
  bogs, marshes, and swamps and, 686–87, 694
  of Caribbean coast, 592
  of Caribbean moist and wet forests, *530*
  in Central Pacific region, *346*
  in cloud forests of the Talamanca, 481
  indigenous reserves and, 481
  of northwestern dry lowlands, *249*
  of the Pacific coast, *99*
  páramo ecosystem of Costa Rican highlands and, *494*
  rain forests and, 714
  in Southern Pacific region, *362*
  wetland areas within private lands and, 701
Protti, E., 172
Proyecto Parque Nacional Guanacaste (PPNG), 331
Puente de Piedra, *47*
Punta Río Claro Wildlife Refuge, 399

Quesada, A. J., 107–8
Quesada-Alpízar, M. A., 102
Quiros Herrera, R., 279

Radulovich, P., 86
Raich, J. W., 85
railroads, 563
Rainforest Alliance, 565
rain forests
  biodiversity of, 714–16
  challenges of primate studies in, 399
  climate change and, 335
  disturbances and restoration in, 715
  versus dry forests, 311
  edge effects and, 319–20
  forest gap dynamics and, 378–85
  forest invasion and restoration and, 334, 335
  land-use types in, 382–83
  large mammals in, 716
  low population densities of birds in, 392–94
  in Osa region, *376*, *377*, *714*
  patch dynamics in, 382–83
  protected areas and, 714
  similarities in Osa region and elsewhere, 374
  small- and large-scale disturbances in, 379–80, 382
  summary of lowland rain forest ecosystems and, 713–16
  treefalls in, 378, 379–81, *380*
Ramírez, A., 635, 639
Ramírez, A. L., 77
Ramírez, A. R., 148
Ramírez, C., 77
Ramsar Convention on Wetlands of International Importance, 165, 193, 236, 686, 699–701
ranching. *See* livestock raising
Rare Animal Relief Effort (RARE), 367
Raunkiær, C., 473–74
Reaka, Marjorie, 100, 102
RECOPE. *See* Refinadora Costarricense de Petróleo (RECOPE)
Refinadora Costarricense de Petróleo (RECOPE), 596
reforestation
  in Área de Conservación Guanacaste (ACG), 303, 305
  butterfly-hostplant relationship and, 383–85, *383–84*
  canopy dynamics and, 380–82
  cattle and, 81
  cloud forests and, 478–79, 717–18
  versus creating a tree plantation, 309
  dry forests and, 248, 277, 279–80, 309–10
  research on regenerating tropical forests and, 380
  soil organic content and, 84–85
  in wet forests of the Caribbean lowlands, 543–45
  young forest ecosystem and, 13
Refugio de Fauna Silvestre Rafael Lucas Rodríguez Caballero, 280
Refugio de Vida Silvestre Bosque Alegre, 666
Refugio de Vida Silvestre Corredor Fronterizo, *530*
Refugio de Vida Silvestre Curú, *249*, 262, 281
Refugio de Vida Silvestre Junquillal, *249*
Refugio de Vida Silvestre Laguna Mata, 659
Refugio de Vida Silvestre Limoncito, *530*
Refugio Nacional de Fauna Silvestre Golfito, *362*
Refugio Nacional de Vida Silvestre Barra del Colorado, *530*, *592*, 694
Refugio Nacional de Vida Silvestre Bosque Nacional Diriá, 265
Refugio Nacional de Vida Silvestre Caño Negro, *530*, 689, 694, 700
Refugio Nacional de Vida Silvestre Gandoca-Manzanillo, *530*, *592*
  conservation efforts in, 606, 608–9
  coral reefs and, 593
  forest swamps in, 698
  lakes in, 677
  mangrove forests and, 599, *600*
  seagrass beds in, 602
  size of, 606
Refugio Nacional de Vida Silvestre Maquenque, *530*, 568–69, *568*, *623*
Refugio Nacional de Vida Silvestre Ostional, *106*, *249*, 280
Rehder, H. A., 168
Reid, F. A., 79
Reserva Biológica Barbilla, 481
Reserva Biológica Carara, 387, 663
Reserva Biológica Cerro Las Vueltas, *452*
Reserva Biológica Hitoy Cerere, 387, *452*, 481, *494*, *530*
Reserva Biológica Isla del Caño, *99*, *362*, 403
Reserva Biológica Lomas de Barbudal, *249*, 255, 280
Reserva Biológica Tirimbina, 546, *557*, 560
Reserva de la Biosfera La Amistad, 481
Reserva Forestal Arenal, 443
Reserva Forestal Cordillera Volcánica Central, *416*
Reserva Forestal Golfo Dulce, *362*
Reserva Forestal Los Santos
  as core protected area, 481
  floristic diversity of, 353
  forest communities of, 469

human settlements in, 477
insects in, 463
location of, *346*, *452*, *494*
Reserva Forestal Pacuare-Matina, *530*, *592*
Reserva Forestal Río Macho, *452*, 481
Reserva Forestal Río Pacuare, *530*, 608
Reserva Natural Absoluta Cabo Blanco
biodiversity loss in, 153, 277
biological corridors and, 281
dry forest conservation and, 280
evergreen trees in, 262
location of, *99*, *249*
mammals in, 265
as protected area, 4
survey of amphibians and reptiles in, 268
Reserva Privada Páramo, *452*, *494*
Reyes-Castillo, P., 231
Reynolds, J. E., 595
Riba-Hernández, P., 399
Richards, Francis A., 100
Richardson, R., 206
Riehl, C., 266
Rio+20. *See* United Nations Conference on Sustainable Development (UNCSD, Rio de Janeiro, 2012)
Ríos Tropicales (tourism company), 640
Rio Summit. *See* Earth Summit (United Nations Conference on Environment and Development, Rio de Janeiro, 1992)
Risk, Michael J. (Mike), 593
Rivera, D. I., 257
rivers of Costa Rica
algae and, 624–26
amphibians of, 628–29
animals of, 626–33
aquatic insects of, 633
biodiversity of, 626–27, 630, 633
biointegrity of, 649
birds of, 630
changing appearance and function of, 623
climate and weather and, 622
conservation efforts and, 648, 649
contributions to riverine research and, 623
crabs of, 632
cyanobacteria and, 624
dams on, 632, 640–45, *641*, *643–45*, 648
descriptive river names and, 636
diatoms and, 624, 626, *626*
diversity of river types and, 626–27
ecosystem processes in, 627–28, 633, 634–35, 636–39
endemism and, 624
extinctions of river species and, 629
extractive and nonextractive human uses of, 639–40
fish of, 627–28, *628*
food web in, *634*
frugivory and seed dispersal in, 635–36
geochemistry of, 636
geothermally modified groundwaters and, *622*
herbivory in, 634–35
historical land use and, 621–22
hot springs and, 624
hydroelectric power and, 632, 639, 640–44, *641–46*, 646, 649
invasive species in, 628, 648–49
invertebrate drift in, 635
land use conversion around, 646–47, *647*
mammals of, 630
marine fauna far inland and, 626
organic matter processing in, 637–39
people and nature around, 639–44, 646–49
plants and, 623–26
pollution in, 626
predator-prey interactions in, 635
primary production in, 637
production of higher trophic levels in, 639
reptiles of, 629–30
research needs and, 649
riverine connectivity and, 632, 642–44
shortages of potable water and, 623
shrimps of, 632–33
species interactions in, 633–36
springs in, 636
stream nutrient cycling and, 628
stream nutrient dynamics in, 637
stream solute and, 624–25
threatened species in, 629, *629*, 630, 632
tourism and recreation and, 639, 640, *641*
transportation on, 640, *641*
upstream versus downstream fauna and, 627
waste assimilation and, 640
water concessions and, 648
water pollution and, 647
watersheds of Costa Rica and, 621, *622*
wetlands ecosystems and, 720–22
Robinson, David C., 593–94
Rocha, O. J., 273
Rodríguez, A., 260
Rodríguez, B., 354
Rodríguez, J. M., 80
Rodríguez-Fonseca, J., 154
Rodríguez-Sevilla, L., 594
Rojas, C., 208
Rojas, E., 32, 154
Rojas, J., 149
Rojas, M. T., 596
Rojas-Acuña, O. W., 172
Rojas-Figueroa, 110
Rolim, G., 23
Romero, A., 558
Roosevelt, Franklin D., 168
Rostad, T., 153
Roth, B., 168
Rouse, G. W., 118
Ruiz-Boyer, A., 354
Russell, A. E., 545
Ryther, J. M., 142

Sader, S., 345, 542
Safford, H. D., 493
Salas, E., 595
Salazar, A., 155
Samper-Villareal, J., 102
Sánchez, J., 355
Sánchez, P. A., 81
Sánchez-Azofeifa, G. A., 345
Sánchez-Navas, 100
Sandoval, L., 224
Santoro, M., 106
Sapper, Karl, 3, 64
Sargent, S., 438
Sasa, M., 267, 356
Sato, A., 224
Sauerbeck, D. R., 83
Savage, J. M., 356, 391, 552
savannahs, 350, *351*
Savitsky, B., 8
Sayre, R., 9
Schall, E. W., 168
Scheer, Georg, 168
Schlesinger, W. H., *82*, 83, 84
Schmidt, E., 595, 601, 606
Schmitt, Waldo L., 168
Schneider, D. W., 635
Schneidt, J., 512
Schonberg, L. A., 433
Scripps Institution of Oceanography, 100
Segura-Puertas, L., 100
Seltzer, G. O., 503, *504*, 506–7
Senn, D. G., 175
Serafino, A., 77
Servicio de Parques Nacionales (SPN), 306–7
Servicios Eléctricos Potosí S.A. (SEPSA), 65
Shannon-Wiener diversity function *H*, 144, 145
Sherry, T. W., 224
Sibaja-Cordero, J. A.
beach ecosystems and, 106
Gulf of Nicoya ecosystem and, 144, 146
intertidal flats ecosystems and, 110
on intertidal gastropods, 107–8
Isla del Coco ecosystems and, 173–74, 206–7
Pacific coast studies and, 102
red tides and, 155
Sierra, C., 225
Sierra-Sierra, L., 604
Sillet, S. C., 431
Sillett, T. S., 466
SINAC (Sistema Nacional de Áreas de Conservación). *See* national park system
Sinclair, B. J., 231
Singer, Rolf, 458
Sipman, H. J. M., 459, 509
Sirot, L., 399
Sistema Nacional de Áreas de Conservación (SINAC). *See* national park system
Sixth International Botanical Congress (1935), 8
Skutch, A. F., 224, 265
Slater, J. A., 231
Slud, P., 223, 265, 556
Small, G. E., 637
Smith, J. N. M., 224
Smith, M. A., 275

Smithsonian Institution, 168
Smithsonian Tropical Research Institute (STRI), 169, 181, 404, 596
Soil Science Society of Costa Rica, 65
soils of Costa Rica
  Alfisols and, 73–74, *75–76*, 207
  Andisols and, 68–71, *69*, *76*, 78, 83, 419, 420, 505, 717
  ants and termites and, 77–78
  Aquepts and, 67–68
  arthropods and, 77
  bacteria and fungi and, 75–76
  carbon sequestration and, 85, 544–45
  cattle and, 80–81, *81*
  in Central Valley, 351
  cloud forests and, 419–20, 426, 456–57, 716–17
  crabs and, 79, 373
  dating of, 538
  decomposition and, 83–84
  deforestation and, 345, 622
  density of litter fauna and, 77
  dry forests and, 251, 252, 254
  Dystrandepts and, 419, 505
  Dystrudands and, 81
  ecosystem management and, 84–85
  ecotourism and, 87
  Entisols and, 65–66, *66*, 83, 85, 207, 373, 505, 713
  erosion and, 86–87
  fertilizer and, 68, 70–71, 80–81
  Histisols and, 82–83, 505, 717
  history of soil science in Costa Rica and, 64–66
  Humitropepts, 505
  Inceptisols and, 66–68, *67*, 79, 83, 85, 207, 373, 419, 456–57, 505, 713, 717
  international soil science meetings and, 65
  Isla del Coco and, 207–8
  Life Zone approach and, 83
  Mollisols and, 71, 373, 713
  nematodes and, 76–77
  new agricultural practices and, 87
  nutrient availability and, 86
  Orthents and, 65, 66
  in the Osa region, 373
  Oxisols and, 538
  Placandepts and, 505
  in *páramos*, 505
  parent material of, 67
  Placudands and, 81
  poor soils of Área de Conservación Guanacaste (ACG) and, 331, 334, 335
  in rain forests, 713
  rodents and, 79–80
  soil as a habitat and, 74–75
  soil compaction and, 80–81, 86–87
  soil orders and, 83
  soil organic matter (SOM) and, 82–85, *82*, *83*
  Sphagnofibrist and, 505
  springtails and, 77
  types of soil orders and, 65
  Udepts and, 67
  Ultisols and, 73–74, *74*, 83, 373, *391*, 456–57, 538, 713, 717
  Ustepts and, 67
  variability of, 87
  vegetation and, 81
  Vertisols and, 71–73, *72*
  volcanic activity and, 621
  in wet forests of the Caribbean lowlands, 537–39
  worms and, 78
Solano, S., 110, 144
Solís, Angel, *546*
Sollins, P., 86
Solomon, S. E., 228
Solórzano, A., 267, 356
Somoza, Anastasio, 305
Soto, F., 77
Soto, R., 143–44
South Asian tsunami (2004), 109
South Equatorial Current, 199–200
Southern Pacific region, 347, *362*, 664. *See also* Osa region
Sowerby, George, 99
Special Areas of Conservation, 114
Spight, Tom M., 100, 107, 123
Spongberg, A. L., 155, 156, 596
Springer, M., 77, 633, 665
Standley, P. C., 3, 351, 453–54, 497
Stanford Oceanographic Expeditions, 100, 168–69
Steinberg, M., 493
Stergiou, K. I., 156
Stern-Pirlot, A., 153
Stetson, Warren, 364
Stevens, G. C., 256
Stevenson, Robert Louis, 162
Stiles, F. G., 224, 265, 354
Still, C. J., 423
Stockholm Environment Institute, xv
Stoner, K. E., 272, 273
Ston Forestal, 371
Stroud Water Research Center, 623
Suárez, E., 100
Suárez-Morales, Eduardo, 594
Süssenguth, Karl, 497
Sustainable Development Goals (SDGs), xv–xvi
swamps. *See* bogs, marshes, and swamps
Sweatman, H. P. A., 224
Swimmer, Y., 116
Szelistowski, W. A., 154

Tabash, F. A., 667
Tabash-Blanco, F., 142, 150
Tafur, N., 86
Takhtajan, A. L., 457
Talamancan-Caribbean Biological Corridor, 482
Talamancan Montane Forests Ecoregion, 10, 482. *See also* cloud forests of the Talamanca
Taylor, M., 493
Taylor, William Randolf, 591
tectonics, 31–32, *32*, 43, *44*. *See also* earthquakes; geological regions of Costa Rica
Ten Hoopen, G. M., 476
Térraba-Sierpe National Wetland, 120
Terrosi, M., 594
Thiollay, J., 392
Thomas, D. B., 231
Thomas, W. S., 591
Thompson, William, 162
Thornthwaite, C. W., 23, 25, 26
Thrupp, L. A., 345
Thurber, A. R., 118
Tiffer-Sotomayor, R., 355, 356
timber industry, 364–65, 370–71, 374, 542, 568
Timber Products Company, 364
Timm, R. M., 272, 273, 437, 558
Tivives Protection Zone, 350
Tokioka, T., 100
Tomlin, J. R. le B., 168
Tonduz, Adolfo, 364
Tosi, Joseph, xv, 8, 65, 365–68, 375, 500
tourism
  benefits of, xvii
  biodiversity loss and, 277
  in Central Pacific region, 357–58
  challenges for conservationists and, 403
  climate and, 182
  in cloud forests of the Talamanca, 477, 478, 483
  in cloud forests of the volcanic Cordilleras, 443
  conservation and, 568–69
  deforestation and, 308
  environmental impact of, *567*
  fish and, 122
  importance of rivers for, 640
  Isla del Coco and, 236
  local economies in Osa region and, 371
  mangroves and, 108
  Pacific coast's human population and, 120
  in Parque Nacional Corcovado, 369, *370*
  pollution and, 608
  property development and, 280, 357–58
  sediment loads and, 602–3
  soil health and, 87
  turtle watching and, *567*
  water demands for, 639, 648
Transnational Trust Co., Ltd., 364–65
Trevithick, Richard, 151
Troll, C., 9
trophic modeling, 149–50, *150*
Tropical Agricultural Research and Higher Education Center (CATIE), xv, 64, 474–75
Tropical Amphibian Declines in Streams Project (TADS), 629
Tropical Ecology Assessment and Monitoring Network (TEAM), 560
Tropical Science Center (TSC)
  Caribbean coast conservation efforts and, 608

cloud forest conservation and, 443
erosion and, 87
growth of ecosystems knowledge and, xv
land use assessments and, 64–65
in Osa region, *363*, *365–66*
San Juan–La Selva Biological Corridor and, *568*
Trusty, J., 211, 213–17
Tsuchiya, M., 197
Tsuda, Roy T., 591

Ugalde, Álvaro, 305, 360, 367, *395*
UK Biodiversity Action Plan, 114, 116
Ulate, Vargas, 254
Umaña, G., 665, 667
Umaña, Ronald, 594, 676
UNA. *See* Universidad Nacional (UNA)
UNCLOS. *See* Law of the Sea of the United Nations (UNCLOS)
UNDP. *See* United Nations Development Programme (UNDP)
UNEP. *See* United Nations Environment Programme (UNEP)
UNESCO. *See* United Nations Educational, Scientific, and Cultural Organization (UNESCO)
United Fruit Company, 364, 365, 371, 564
United Nations Conference on Environment and Development (UNCED), 567
United Nations Conference on Sustainable Development (UNCSD, Rio de Janeiro, 2012), xv, 13
United Nations Development Programme (UNDP), xvi, 182, 211
United Nations Educational, Scientific, and Cultural Organization (UNESCO)
Amistad Biosphere Reserve and, 499
Man and Biosphere Reserves and, 515
Reserva de la Biosfera La Amistad and, 481
vegetation classification system and, 9, 377
World Heritage Programme of, 165, 181, 184, 193, 236–38, 280
World Heritage Sites and, 481, 499, 515
United Nations Environment Programme (UNEP), 4–5
United States, military action in Central America by, 300
Universidad de Costa Rica, 64, 499, 594, 596, 623, 633. *See also* Centro de Investigación en Ciencias del Mar y Limnología (CIMAR)
Universidad Nacional (UNA), xix, 499, 623, 657
Universidad Nacional Autónoma de México (UNAM), 100
University of California, Santa Barbara, 169
University of Delaware, 101
University of Southern California, 168
University of Texas at Austin, 368–69, *370*
USAID (Agency for International Development), 370
US Civil War, 300
US Department of Defense, 366
US Environmental Protection Agency (EPA), *565*
US Fish Commission, 231
US Navy Galápagos Expedition, 168

Valdez, Marta F., 594
Vancouver, George, 194
Van Dam, D., 84
Vanderbilt, William K., 99–100, 168
Van Devender, R. W., 434
Van Leewen, E. M. M., 473
Van Uffelen, J. G., 505
Vargas, J. A., 101, 110, 142, 144–47, 156
Vargas, R., 169, 594
Vargas-Castillo, R., 102, 169
Vargas-Montero, M., 154
Vargas-Zamora, J. A., 105, 110–11, 144, 146, 155
Vaughan, C., 80, 247, 345, 375
Vaughan, Thomas Wayland, 596
Veldkamp, E., 84
Verrill, Addison, 99
Vicencio-Aguilar, M. E., 100
Villa, J., 356, 391
Villalobos, A. F., 235
Villalobos, C., 106, 146, 267, 594
Villarreal Orias, J., 265
Vinson, S. B., 247, 271
Víquez, C., 231
Víquez, R., 146, 154
volcanoes, *44*
archaeological preservation and, 442
Caribbean lowland soils and, 538
geochemistry of streams and, 636
geological history of Área de Conservación Guanacaste (ACG) and, 291–93, *294–95*, *320*
geological history of Isla del Coco and, 172, 192, 200–201, 203–5
geology of Caribbean coast and, 597–98
lake formation and, 658, 664, 666, *666*, 667, 674, 676
national parks and, 443
recent eruptions of, 46, 56, 296, 505
soils of Costa Rica and, 621
techtonic activity and, *622*
volcanic edifices and, *37*
Volcán Orosí and, *297–98*
See also cloud forests of the volcanic Cordilleras
Von Frantzius, Alexander, 3
Von Humboldt, Alexander, 7
Von Wangelin, M., 147
Voss, Gilbert L., 593

Wagner, Moritz F., 495, 497
Wagner, W. H., 462, 497
Walker, William, 300
Wallace, Alfred Russell, 5
Wartzok, D., 595
weather. *See* climate and weather
Webb, L. J., 473–74
Weber, H., 497, 502
Weber, W. A., 209
Weberling, F., 499, 512
Wedin, D. A., 85
Wehrtmann, I., 10, 106, 118, 149, 153, 594
Wellington, Gerard M. (Jerry), 593
Wenny, D. G., 438
Wercklé, C., 351
Werklé, Karl, 3, 7, 457, 497
Werner, T. K., 224
Wescott, David, *370*
Wesselingh, R. A., 463
Weston, Arthur S., 497–98
Weston, J. C., 165
wet forests of Caribbean lowlands
African dust clouds, 531
agriculture and pesticide use in, 565
altitudinal and latitudinal migration and, 553, 555, 556
amphibians and, 550–52
animals of, 545–58
archaeological sites in, 563
biodiversity in, 539–40, 550–53, 555–56, 561
birds of, 552–54
changes in forest cover in, 541–42
climate and weather in, 534–37, 549
climate change and, 536–37
climate zones of, *535*
coastal beaches and, 537–38
colonial and postcolonial development in, 534
conservation efforts in, 552, 565–69
decline of insectivores in, 554
deforestation in, 541–42
DNA barcoding in research in, 545
fishing in, 550
fish of, 546–50
frugivory and seed dispersal in, 560–62
future of, 569
geographic extent of, 527–29
geography as destiny in, 527
geological history of, 529–32
geomorphological units of, 539
habitat modification in, 552
herbivory in, 558–60
historical overview of, 529–34
Holdridge Life Zones in, *540*
human populations and demography of, 532–34, 563
hunting in, 552
hydrodynamics of, 549–50
invertebrates of, 546
lack of plants in rivers and streams of, 550
lakes of, 663, 674–77
Lepidoptera studies in, 560
locations of plant collections in, *540*
locust swarm in, 531
mammals of, 554–58
precipitation in, *538*
primates of, 556–57
protected areas and indigenous reserves in, *530*

wet forests of Caribbean lowlands (*continued*)
recent land use history of, *534*
remote sensing for forest studies in, *545*, *569*
reptiles of, *552*
secondary forests and, *543–45*
snorkeling in, *550*
soils in, *537–39*
sources of biomass in, *558*
transoceanic winds and, *531*
variable place names and, *550*
vegetation of, *539–41*
water pollution in, *550*
See also cloud forests; rain forests
wetlands. *See* bogs, marshes, and swamps; lakes of Costa Rica; rivers of Costa Rica
Wetzer, Regina, 594
Weyl, R., 32, 497
Wheeler, W. M., 228
Wheelwright, N., 433, 443
Whitfield, S. M., 551
Whitney, N. M., 170
Whoriskey, S., 116
Wicksten, M. K., 169
Widmer, Y., 469
Wilder, D. D., 231
Wildlife Conservation Society (WCS), 567
Wille, Alvaro, 404
Williams, A. B., 593, 594
Williamson, G. B., 498
Wilms, J., 466, 476
Wilson, D. E., 354
Wilson, Edward O., 3, 7
Wilson, H. V., 177
Winemiller, K., 390, 595, 627, 634
Wirth, W. W., 231
Witter, J., 155
Wolff, M., 146, 147, 148, *150*, 153, 154
Woodley, N. E., 231
Woodrow G. Krieger Expedition, 168
Woods Hole Oceanographic Institutions, 100
Wootton, T., 627
World Bank, 567
World Resources Institute (WRI), 193
World War II, 100
World Wide Fund for Nature. *See* World Wildlife Fund (WWF)
World Wildlife Fund (WWF)
Biodiversity Support Program and, 193
classification of ecosystems by, 250
cloud forest conservation and, 443
ecosystems recognized by, 499
Isla del Coco and, 193
Parque Nacional Corcovado and, 367
Wright, Alvin, 365, 366
Writki, Klaus, 100
Wujek, D. E., 672
Würsig, B., 170
WWF. *See* World Wildlife Fund (WWF)

Yanoviak, S. P., 433, 434
Yeaton, R. I., 275
York, T. R., 231
Young, B. E., 170
Young, Paul S., 594

Zahawi, R. A., 478–79
Zamora, Jesús Jiménez, 162
Zamora, N., 8–10, 254, 260, 347–49, 351–54, 468
Zamparo, D., 275
Zimmerman, T. L., 171–72
Zona Protectora Acuífero Guácimo y Pococí, *530*
Zona Protectora Alberto Manuel Brenes, *416*
Zona Protectora Arenal-Monteverde, *416*
Zona Protectora Cerro Nara, *346*, *353*, *357*
Zona Protectora Cerros de Escazú, *346*
Zona Protectora Cerros de Turrubares, *346*, *349*, *355*, *357*
Zona Protectora Cuenca del Río Abangares, *416*
Zona Protectora Cuenca del Río Banano, *452*, *530*
Zona Protectora Cuenca del Río Siquirres, *530*
Zona Protectora Cuenca del Río Tuis, *452*
Zona Protectora El Rodeo, *346*, 350, *355*, 711
Zona Protectora La Selva, *530*
Zona Protectora Las Tablas, *452*, 454, 456, 467, 481, *494*
Zona Protectora Nara, 711
Zona Protectora Península de Nicoya, 281
Zona Protectora Río Toro, *416*
Zona Protectora R. Navarro y R. Sombrero, *452*, 466, 481
Zona Protectora Tortuguero, *530*
Zona Protectora Turrubares, 711
Zona Protectora Volcán Miravalles, *416*
Zonneveld, I. S., 9
Zoological Society of New York, 100. *See also* Wildlife Conservation Society (WCS)

Systematic Index of Common Names

Page numbers in italics refer to figures and tables.

English Common Names

acacia, 274. *See also specific species of acacia*
acacia ant, 338
acorn woodpecker, 466
acridid, 559
African grass, 300, 302, 303, 307, 332, 334, 713
Africanized honeybee, 335–36
African oil palm, 354, 401
agaric, 458
agariciid, 175
agave, 429
agouti. *See* Central American agouti
ahermatypic coral, 710
alder, 75, 458, 506, 718. *See also specific species of alder*
alga, 100, 101, 102, 105, 107, 111, 121, *147*, 150, 154, 155, 168, 173, *175*, 218, 509, 591, 593, *605*, 624–26, 627, 629, 632, 634–35, 636, 709, 711, 720–21, 722. *See also specific species of alga*
Allen's salamander, 391
almond, 196. *See also specific species of almond*
Amazon sword plant, 721
amblyopinine beetle, 440
ambrosia beetle, 228–229
ameiva, 392. *See also specific species of ameiva*
American crocodile, 354, 356, 552, 722
American eel, 546, 547
American oil palm, 377
amphibian, 434–35, 463, 550–52, 627, 628–29, 684, 686, 696, 712, 715, 719, 720, 722
amphioxus, 153
amphipod, 105, 167, 168, 170, 600, 672. *See also specific species of amphipod*
anchovy, 148
Andean alder, 478
angiosperm, 111, 531
anglerfish, 148, 149. *See also specific species of anglerfish*
anguid, 552
anhinga, 267, 354, 630
annelid, 119
anole, 552, 715. *See also specific species of anole*
anole lizard, 267, 435
anoline lizard. *See* anole lizard
anomuran crab, 594
ant, 75, 77–78, 228, 271, 321, 361, 386–88, *387*, *388*, 389, 433, 439, 440, 545, 561, 712, 713, 715, 719. *See also specific species of ant*
ant acacia, 338, 387
antbird, 392, 715. *See also specific species of antbird*
anteater, *555*, *715*
ant-follower, 554
ant-tanager, 392, 394, 554, 715. *See also specific species of ant-tanager*
ant-thrush, 394. *See also specific species of ant-thrush*
antvireo, 554
antwren, 393, 554. *See also specific species of antwren*
anuran, 719
apatelodid, 560
aphid, 269
apple, 477, *478*, *479*
aquatic bird, 266, 354, 355
aquatic grass, 623, 721
aracari, 561–62
arachnid, 232
arbuscular mycorrhiza, 76, 440, 545
Arenal rivulus, 268
ark clam, 151, *151*, 152–53
armadillo, 277, 532, 555, 562, 715. *See also specific species of armadillo*
army ant, 321, 327, 328, 394, 433
aroid, 216, 696, 718
arrow worm, 147, 148
arthropod, 153, 227, 235, 266, 433–34, 465, 552, 558, 719
artiodactyl, 262, 712
ascidian, 100, 107
Atlantic grunt, 547, 548
atyid, 632. *See also specific species of atyid*
Australian pine, 478
avocado, 68, 73, 461, 471–72, 472, 482, 553
azooxanthelate, 175, 710

bacteria, 596, 711. *See also specific species of bacteria*
bactroid, 559
Baird's tapir, 262, 263, 277, 279, 337, *395*, 397–98, 467, 473, 482, 511, 559, 712, 719
balsa, *380*
bamboo, 461, 468, 469, 500, *500*, 511–12, 514, 515, 672, 685, 686, 718, 719, 720, 721
banana, 65, 66, 68, 77, 79, 83, 85, 196, 303, 347, 354, 364, 377–78, 379, *379*, 401, 541, 542, 543, 549, 557, 563, 564–65, 567, 646, 647, *647*, 677, 716. *See also specific species of banana*
banded peacock, 270
banded tetra, 268, 269, 276, 547, 548, 549, 722
band-tailed pigeon, 466
barbet, 439. *See also specific species of barbet*
bare-crowned antbird, 554
bark beetle, 228–29
barnacle, 106, 107, 143, 146, *147*, 173, 594, 709
barred cat-eyed snake, 392
basidiomycete, 559
basidiomycotic fungus, *209*
bass, 269. *See also specific species of bass*
bat, 262, 264–65, 271, 272, 273, 329, 330, *332*, *338*, *354*, *384*, 437, 439, 464, 466, 467, 471, 545, 555, 557, 562, 712, 713, 715, 719, 722. *See also specific species of bat*
bay-headed tanager, 394
bean, 66, 75, 78, 85–86, 196, 303, 361, 461, 477, 533. *See also specific species of bean*
bear, 5
beard lichen, 459
bee, 271–72, 311, *384*, 439, 712, 713, 719. *See also specific species of bee*
beet, 477, *479*
beetle, 228, 269–70, 272, 273, 323, *384*, 439, 440, 545, 560, 712, 713, 719. *See also specific species of beetle*
bellbird, 439, 443–44. *See also specific species of bellbird*
benthic alga, 143, 637
benthic fauna, 100, 153, 505, 672
benthic macroalga, 172
benthos, 110, 116–19, 142, 143, 145, 153, 177, 183, 666, 710, 711
bicolored antbird, 394, 554
bigeye thresher shark, 170
bigmouth sleeper, 389, 546, 547, 548

bird, 231, 332, 355–56, 392–94, 435–37, 439, 443–44, 465–66, 471–73, 476–77, 483, 531, 545, 552–54, 630, 632, 663, 691, 692, 700, 712, 715–16, 718, 719, 720, 722. *See also specific species of bird*
bivalve, 108, 117, 145, 147, 151, 153, 594, 600, 710, 711. *See also specific species of bivalve*
black-and-yellow silky flycatcher, 439
black-bellied whistling duck, 266–67
blackbelt cichlid, 546, 547
blackberry, 477, *479*
black-billed nightingale-thrush, 511
black cambute, 153
black-cheeked ant-tanager, 392, 394, 715–16
black-cheeked warbler, 511
black coral, 118, 593, 608, 710
black-faced ant-thrush, 394
black-faced grosbeak, 554
black-faced solitaire, 439, 466
blackfin snook, 547
black guan, 466, 476
black-handed spider monkey, 399, 716
black hawk, 386
black-headed bushmaster, 357, 391, 392, 712, 715
black-headed trogon, 266
black iguana, 268
black mangrove, 108, 143–44
black noddy, 222
black rat, 193–94, 224
black river turtle, 636
black sea turtle, 106, 220
black sea urchin, 593, 594, *607*
black-speckled palm pit viper, 465
black-striped woodcreeper, 393
black-thighed grosbeak, 466
black vulture, 266
black wood turtle, 558
blue crabs, 147, 153, 232, *235*
blue-crowned manakin, 393
blue green (alga), 624, 634, 720–21
blue-winged teal, 266
boa, 392. *See also specific species of boa*
boat-billed heron, 664
boatman. *See specific species of boatman*
boat-tailed grackle, 303
boid, 552
bolete, 458
booby, 170. *See also specific species of booby*
bottlenose dolphin, 169, 595
bovid, 51
brachiopod, 145
brachyuran crab, 147, 169, 594
bregmacerotid, 148
brittle star, 110
brocket deer, 719
bromeliad, 213, *216*, 257, 275, 466, *470*, 633, 686, *686*, 713–14, *717*, 718, 721. *See also specific species of bromeliad*
brown booby, 222
brown-capped vireo, 466
brown cattle, *303*
brown jay, 435–36, 719
brown noddy, 222, *225*
brown rat, 224, 225
brown seabird, 170
brown shrimp, 153
brown-throated three-toed sloth, 277, 558
brown wood turtle, 558, 636
bruchid, 269
brush-footed butterfly, 270
bryophyte, 211, *212*, 256, 431, 458, 462–63, 466, 509–10, 511, 512–13
bryozoan, 100, 168
buff-throated foliage-gleaner, 554
buffy tufted-cheek, 466
bug, 269, 313, 559, 712. *See also specific species of bug*
bull, 302
bullet ant, 387–88, *388*
bull horn acacia, 713
bull rain frog, 392
bull shark, 389, 546, 547
bulrush, 257, 701
bumblebee, 463, 469, 511, 719
bunch grass, 721
burrowing sea urchin, 593
bushmaster, 392. *See also specific species of bushmaster*
bush tanager, 439
butterfly, 227–28, 229–30, 269, 270, 272, 314, 323, 326–27, 379, 383–85, *383–84*, 386, 433–34, 439, 443–44, 531, 545, 546, 560, 712, 719

cabbage, 477, *479*
cacao, 78, 534, 541
cacique, 554. *See also specific species of cacique*
cactus, *252*, 256, 259, 275, 712
caddisfly, 633, 722
caecilian. *See* cecilian
caesalpinoid, 273
caiman, 552, 629–30, *634*, 691–92, 722. *See also specific species of caiman*
calabash, 274, 275
calanoid, 148, 594
callichthyid, 546
camel, 532
camelid, 51
canid, 51, 262, 712
cantaloupe, 68
cantil snake, 267, *267*
capuchin. *See* white-faced capuchin monkey
carangid, 148, 149
Caribbean anole, 552
carnivore, 106, 107, 147, 149, 171, 265, 269, 293, 354, 466, 532, 554, 556, 636
carrot, 477, *479*, 564, *565*
cashew, 73, 196, 377
cassava, 303, 533, 534, 564
cat, 196, 220, 226, 262, 364, 394, 555, 712, 715. *See also specific species of cat*
caterpillar, 275, 559
catfish, 148, 149, 268, 389, 546, 547, 548, 628. *See also specific species of catfish*
cattail, 257, 280, 659, 661, 701, 721
cattle, 73, 80–81, *81*, 86, 255, 257, 262, 274, 276, 279, 280, 300, 302, 303, *303*, 304–5, 306–7, 311, 331, 332–33, 338, 347, 354, 364, 443, 477, 514, 541, 567, 621–22, 646, 647, *647*, 659, 662, 663, 666, 702. *See also specific species of cattle*
cattle egret, 220, 267
cauliflower, 477, *479*
Cauque River prawn, 233–34
cebus monkey, 264
cecilian, 392, 434, 550, 715
celery, *479*
Central American agouti, 79–80, *80*, 260, 265, 274, 329–30, 400, 467, 555, 562, 698
Central American spider monkey, 559
Central American squirrel monkey, 265, 354, 379, 394, *395*, 399, 400, 403, 712, 716
centrolenid, 715
cephalochordate, 101, 110
cephalopod, 154, 168, 592, 594
cervid, 51
cetacean, 102, 116, 154
Chacoan peccary, 395
chaetognath, 102, 148, 604, 711
characid, 268, 390
characin, 268, 546
charadriiform bird, 145
chayote, *565*
checker-throated antwren, 393, 554
chestnut-backed antbird, 393
chestnut-capped brushfinch, 476
chestnut-mandibled toucan, 355, 554
chestnut-sided warbler, 394
chicken, 196
chili pepper, 71, 361
chilopod, 235
chimaeroid, 103
chironomid, 657, 665, 666, 672
chiton, 107, 153, 592
chlorophyte, 665, 667, 672, 675, 676, 690, 721
chrysophyte, 672, 676
chytrid, 434, 551, 629, 719
cicada, 269. *See also specific species of cicada*
cichlid, 268, 389, 390, 463, 546, 547, 548, 549, 550, 628, *628*, 630, 649, 667, 692, 722. *See also specific species of cichlid*
citrus, 73
cladoceran, 665, 666, 667, 675, 676–77
clam, 33, *117*, 118, 151, 153, 364. *See also specific species of clam*
Clark's coral snake, 392
clingfish, 546, 548. *See also specific species of clingfish*
clinostomatid, 634
clubmoss, 211, 213, 509
cnidarian, 100, 149, 593
coati, 264, 332, 386, 400, 562. *See also specific species of coati*

cocoa, 66, 68, 196
cocoa woodcreeper, 394
coconut, 194, 196, 214, 215, 273, 364, 377, *714*
coconut palm, 194, 196, 214–15, 364, *714*
Cocos atyid, 235
Cocos clingfish, 220
Cocos cuckoo, 222, *223*
Cocos finch, 222–23, 224, *226*
Cocos flycatcher, 222, 224
Cocos gecko, 220, *222*, 226
Cocos goby, 220
Cocos lizard, 220, *221*, 226
Cocos orange land crab, *233*
Cocos prawn, 234–35
Cocos sleeper, 220
cod, 148
coelenterate, 168
coffee, 64, 68, 71, 77, 78, 150, 196, 303–4, 351, 352, 358, 377, 443, 477, 478, *479*, 483, 563, 564, 621
coliform bacteria, 155
collared peccary, 264, *264*, 277, 332, 395, 400, 401, 437, 467, 558, 559, 562, 719
collared redstart, 438, 466
collared trogon, 466
colubrid, 715
common noddy, 222, *225*
common northern raccoon, 466
conger, 148, 149, 711
conifer, 461, 469
convict cichlid, 268, 547, 548, 549, 550
copepod, 110, 144, 147, 593, 594, 604, 660, 666, 667, 672, 675, 677, 711, 721
coral, 100, 102–3, 104, 109, 111–14, *112*, 120–21, *120*, 123, 165, 168, 169, 171, 172, 175–76, 178–79, *179*, *180*, *181*, 591, 593, 595, 596, 597, 598, 601, 602–3, *603*, 604, *604*, 605, *605*, 606, *607*, 608–9, 710. *See also specific species of coral*
coralline alga, 100, 113, 603, *605*
corallivore, 102, 114, 176, 710
coreid, *384*
corn, 86, 196, 303, 361, 477, 533, 565
corvid, 435
cotinga, 355, 553
cotton, 71, 276, 303, 361
cotton boll weevil, 329
cotton rat, 277
cotton-tail rabbit, 264, 277
cougar, 467, 482
cow, 467
coyote, 264, 277, 332, 333, 437, 466, 557, 719
crab, 75, 79, *80*, 105, 144, 146, 149, 174, 215, 232, *395*, 593, 632, *634*. *See also specific species of crab*
crab-eating raccoon, 355, *355*
crab hawk, 386
creek tetra, 269, 276, 547
crested eagle, 355, 394
crested guan, 265, *393*, 466
crinoid, *593*
croaker, 154
crocodile, 268, 356–57, *357*, 388, 552, 629–30, *634*, 663, 692, 699, 712, 715, 722. *See also specific species of crocodile*
crown-of-thorns starfish, 176, 710
crustacean, 100, 101, 102, 103, 110–11, 118, 144–45, 146, 147, 149, 153, 154, 155, 168, 171, 177, 233, 388, 593, 594, 630, 632, 710, 712–13
crustose lichen, 459
cryptogamic epiphyte, 476
cryptophyte, 666, 721
ctenosaur, 268, 303, 332, 531
cuckoo, 222, *223*, 394, *554*. *See also specific species of cuckoo*
cumacean, 110, 594, 600
curassow, 466. *See also specific species of curassow*
cutlass fish, 148
Cuvier's beaked whale, 169
cyanobacterium, 119, 154, 440, 624, 666, 667, 672, 675, 676, 721
cyanophyte, 660, 721
cyclanth, 461
cyclopoid, 667
cypress, *478*. *See also specific species of cypress*

damselfish, 114, 595
damselfly, 633
dark-eyed leaf frog, 392
Darwin's finch. *See* Cocos finch
dasypodid, 51
decapod, 116–17, 146, 153, 154, 600
decapod crustacean, 144, 145
deer, 75, 260, 555, 715. *See also specific species of deer*
dendrobatid, 551
desmid, 666, 721
detritivore, 144, 639
devilfish, 168, 628
diatom, 142–43, 146, 147, 154, 506, 509, 624, 626, *626*, 637, 657, 663, 664, 666, 668, 669, 670, 672, 674, 676, 677, 721, 722
Dice's cottontail rabbit, 466
dicot, 460, 461, 462, 468, 509, 510, 668
dink frog, 686, 721
dinoflagellate, 146, 150, 154, 509, 711
dinosaur, 290
diplopod, 235
dipterocarp, 374, 714
dismorphiine butterfly, 386
diurnal butterfly, 227, *230*
dog, 196, 437
dolphin, 170, 710. *See also specific species of dolphin*
domestic cat, 224
domestic pigeon, 221
donkey, 151
dotted-winged antwren, 393, 554
dragonfly, 269, 633, 712, 722
drosophilid, 438
drum, 154
duck, 265, 691, 692, 700, 701, 722. *See also specific species of duck*
dung beetle, 546
dusk-faced tanager, 554
dwarf palm, 467, 719

eagle, 355, *393*, 394, 716. *See also specific species of eagle*
eared seal, 219
eared watermoss, *691*
earless lizard, 392
earthworm, 75, 78–79, *79*
echinoderm, 100, 101, 116–17, 145, 146, 149, 165, 168, 169, 171, 594, 595
echnoid, 177
ectomycorrhizal fungus, 257
ectosymbiont bacteria, 118
edentate, 262, 712
eel, 451. *See also specific species of eel*
egret, 265, 266. *See also specific species of egret*
eleotrid, 390, 634
elephant ear tree, 274
elephantid, 51
elkhorn coral, *607*
emerald toucanet, 435, 466, 476
engraulid, 148
entomofauna, 227, 231
epibenthos, 149
epiphyllic lichen, 208–9
epiphyllous bryophyte, 211, 428, 431
epiphyte, 275, 377, 429, 431–32, 433, 438, 439, 440–41, *441*, 442, 455, 457, 459, 460, 461–62, 463, 466, 468, 469, *470*, 475, 476, 500, 539, 593, 712, *717*, 718. *See also specific species of epiphyte*
equid, 51
ericad, 513, 685, 718
euagaric, 458
euglenophyte, 667
euglossine bee, 271
eukariotic, 509
euphausid, 604, 711
euryhaline alga, 625
eyelash pit viper, 392

falcon, 355. *See also specific species of falcon*
false killer whale, 169, 170
false vampire, 264–65
fat sleeper, 269, 389, 547
felid, 51
feline, 354
feral pig, 224, *227*
fer-de-lance, 357, 712
fern, 9, 68, 210, 211, 213, 215, 216, 275, 347–48, 377, 431, 460, 462, 468, 469, 509, 510, *513*, 531, 539, *685*, 686, 718, 719, 720, 721. *See also specific species of fern*
fern-ally, 509, 510
fiddler crab, 145

fiery-billed aracari, 355
fig, 722. *See also specific species of fig*
fig wasps, 272, 438
filmy fern, 428, 432, 718
finch, 222, 224, 226, 230, 439, 554, 719. *See also specific species of finch*
finescale sleeper, 268
finfoot, 630
firefly, 269
fish, 149, 183, 388–91, *390*, 463, 546–50, 558, 561, 595, 606, 627–28, 630, 632, 634–35, *634*, 637, 639, 642–43, *644–45*, 657, 664, 665, 666, 667, 672, 692, 693, 700, 710, 711, 712, 715, 719, 722. *See also specific species of fish*
fishing bat, 630
flagellate, 147, 155
flamboyant, 273
flame-throated warbler, 466, 511
flame tree, 273
flatfish, 148, 149, 389
flatworm, 119, 268, 275, 596, 609
flea, *384*, 440
flea beetle, 560
flounder, 148, 149, 547, 626, 711
fly, 232, 275, 323, *384*, 438, 439, 719. *See also specific species of fly*
flycatcher, 554. *See also specific species of flycatcher*
foliicolous lichen, 460
foliose lichen, 466
foraminifer, 110, 147
forb, 721
forest rabbit, 437, 559
four-footed butterfly, 270
fowl, 196, 451, 691, 692, 700, 701, 702, 722. *See also specific species of fowl*
fox, 265, 277, 466. *See also specific species of fox*
freshwater prawn. *See* freshwater shrimp
freshwater shrimp, 271
frigatebird, 170. *See also specific species of frigatebird*
frog, 463, 465, 473, 545, 550, 551, 628, 629, *629*, 715, 719, 722. *See also specific species of frog*
frog lung fluke, 275
fruticose lichen, 459, 460, 718
fungus, 208–9, *209*, 509, 634, 638–39, 713, 718, 720. *See also specific species of fungus*
fur seal, 101, 219

Galápagos sea lion, 169
gar, 547
gastropod, 101, 106, 107, 123, 168, 173, 233, 592, 594, 595, 600
gecarcinid, 385–86
gecko, 552. *See also specific species of gecko*
Geoffroy's spider monkey, 262, 277, 399–400, 437, 482, 556–57, 712, 716, 719
geometrid, 560
geonomid, 559
ghost crab, 232
giant anteater, 5, 262, 265, 277, 331, 437, 555
ginger, 696
gladiator frog, *391*, 392
glass frog, 551, 628, 715
glass headstander, 268, 547
glass sponge, 169, 172
glyptodont, 50, *51*, 293
glyptodontid, 51
gmelina, 334–35, 402
gnatwren, 554
goat, 194, 196, 224, 225–26, 467
gobiid, 148
goby, 148, 149, 389, 548, 711. *See also specific species of goby*
Godman's montane pit viper, 465
golden alga, 509
golden-crowned spadebill, 552
golden-hooded tanager, 394
golden toad, 6, 434, 629, 719
Golfo Dulce poison dart frog, 391
gomphoteriid, 51
gomphothere, 274, 293, 532
gopher, 555. *See also specific species of gopher*
gourd, 361
grackle, 303. *See also specific species of grackle*
granulated poison dart frog, 391
grass, 71, 196, 211, 218, 249, 255, 269, 276, 296, 299, 300, *301–2*, 302–3, 306, *306*, 307, *308*, *311*, *314*, 332–35, 460, 492–93, 500, 507, 512–15, 531, 591, 602, 623, 659, 664, 668, 685–87, 689, 699, 701, 713, 719, 721. *See also specific species of grass*
grasshopper, 433, 559
gray fox, 265, 277
great ape, 263–64
great curassow, 265, 355, *393*, 400
great egret, 267. *See also* great white egret
greater bulldog bat, 630
great frigatebird, 222
great green macaw, 8, 554, 567, 568–69, *568*, *569*
great tinamou, 466
great white egret, 692, *693*, 722. *See also* great egret
green alga, 113, 173–74, 509, 624
green bean, 565
green-black poison dart frog, 392
green clam, 700
green frog eaters, 434–35
green heron, 692, *693*, 722
green ibis, 630
green iguana, 268, 552, 558, 712
green keelback, 435
greenlet, 393, 554. *See also specific species of greenlet*
greenlet-honeycreeper, 554
green spiny lizard, 464, 465, 511, 719, 720
green tree, 257
green turtle, 122, 566–67, 591, 600–601, *601*, 609, 709
grey-headed tanager, 394
grey teak, 377
grosbeak, 554. *See also specific species of grosbeak*
ground sloth, 50, 274, 293, 532, 533
grunt, 389
guan, 400. *See also specific species of guan*
guapote, *390*, 667, 675
guinea fowl, 196
gull, 170. *See also specific species of gull*
Guyana dolphin, 595
gymnophthalmid, 552
gymnosperm, 468

hammerhead shark, 170
harlequin frog, *629*, 719
harlequin toad, 356
harpy eagle, *393*, 394, 716
harvestman, 231
harvest mouse, 262, 437, 438, 440, 719
hawk, 386, 466, 511, 720. *See also specific species of hawk*
hawk moth, 229–31, 270, 272
hawksbill sea turtle, 106, 220, 600, 709
headstander, 268, 547. *See also specific species of headstander*
helminth, 275
hemichordate, 145
hemiepiphyte, 429, 431, 432, 462, 718
hemiparasite, 462
hepatic, 211, 256, 458, 463, 685
herb, 539, 687, 691, 696, 718, 719
herbaceous vine, 374, 460
hermatypic coral, 114
hermit, 232–33, 393. *See also specific species of hermit*
hermit crab, 232–33
heron, 265, 266. *See also specific species of heron*
herpetofauna, 391–92, 434
hesperiid, 270
heterokont alga, 509
heteromyid, 562
heteropod, 604, 711
heterotrophic parasite, 462
highland alligator lizard, 464, 465, 719
highland yellow-shouldered bat, 467
hispid cotton rat, 300
Hoffmann's two-toed sloth, 277, 558–59
hog mullet, 548
hognose pit viper, 392
hogplum, 377
holothurian, 169, 593
honeybee, 271
honeycreeper, 554. *See also specific species of honeycreeper*
hornworm, 270
hornwort, 463
horse, *51*, 274, 293, 300, 303, *304*, *305*, 306, 333, 338, 467, 514, 532
house gecko, 552
howler monkey. *See* mantled howler monkey
hummingbird, 230, 266, 272, 275, *384*, 438–39, 443–44, 469, 553, 719. *See also specific species of hummingbird*

humpback whale, 169, 170
hyacinth, *249*, 257, *258*, 623, 659, 721. *See also specific species of hyacinth*
hydrochoerid, 51
hydroid, 593
hylid, 551
hyperiid amphipod, 170
hyperparasitoid wasp, 275

ibis, 265. *See also specific species of ibis*
ichthyoplankton, 220, 595
ichtyofauna, 722
iguana, 268, 552, 558, 712. *See also specific species of iguana*
iguanian lizard, 276
immaculate antbird, 554
infauna, 105, 145, 593, 710
insect, 269–71, 315, 323–28, *384*, 386–88, 439, 462, 463, 469, 471, 473, 511, 627, 632, 633, 634, *634*, 635, 637–39, 667, 712, 713, 719, 720, 722. *See also specific species of insect*
insect grazer, 635
insectivorous bat, 265
invertebrate, 77, 100, 102, 106–7, 110–11, 114, 117, 119, 143, 145, 147–48, 153–55, 169, 227–35, 269–71, 385–86, 389, 396, 463, 511, 545–46, 559–60, 565, 596, 600, 623, 627, 629–30, 632–33, 635, 636, 638–39, 693, 709, 722
iron stick tree, 194
isopod, 143, 145, 594, 600
Isthmian alligator lizard, *391*, 392
Isthmian rivulus, 547, 548
ithomiine butterfly, 386

jabiru stork, 692, 693, 712, 722
jacana, 265, 630, 722
jaguar, 262, 277, 305, 354, 366, *395*, 396, 398–99, 401, *404*, 437, 467, 482, 557, 712, 715, 716
jaguar cichlid, 546, 547
jaguarundi, 262
jay, 473. *See also specific species of jay*
junco, 511. *See also specific species of junco*

kapok, 374
keelback, 435. *See also specific species of keelback*
keel-billed toucan, 554
Kentucky warbler, 394
kikuyu grass, 71, 86
killifish, 549–50. *See also specific species of killifish*
kingfisher, 266, 630, 722
king vulture, 355
kinkajou, 557, 562
kinorhynch, 117, 144
kite, 466. *See also specific species of kite*
knifefish, 268, 547, 548

lancelet, 144
land crab, 231, 232, 373, 385–86, 715
large-footed finch, 466, 511
large-spot tetra, 548, 549
leaf beetle, 269
leaf-cutter ant, 78, *78*, 271, 273, 559, 713
leafhopper, 231
leaf-miner, 559
leaf-mining beetle, 559
leatherback sea turtle, 106, *106*, 220, 268, 567, 595, 600, 608, 709
leather fern, *688*
legume, 272
lemon, 196
leporid, 51
leptodactilyd quark frog, 464, 482
lesser bulldog bat, 630
lesser greenlet, 554
lettuce, 721. *See also specific species of lettuce*
liana, 213, 253, 256, 257, 272, 275, 383, 431, 432, 439, 469, 715, 719
lichen, 208–9, 218, 440, 458–60, 466, 476, 500, 509, 511, 512, *513*, 685, 686, 718, 719, 720, 721. *See also specific species of lichen*
lightning beetle, 269
lightning bug, 313
lily, 663, 721. *See also specific species of lily*
limacodid, 560
limpet, 107, *634*
lithodid crab, 118
little tinamou, 355
littorinid gastropod, *147*
littorinid snail, 144
livebearer, 268, 546, 550
liverwort, 209, 211, 463, 476, 509–10, 512, 513, 718, 719, 720. *See also specific species of liverwort*
lizard, 264, 465, 473, 552, 715, 719. *See also specific species of lizard*
lizardfish, 148
lobster, 146, 594. *See also specific species of lobster*
locust, 531
loggerhead turtle, 600, 709
longhorn beetle, 269
long-tailed hermit, 393
long-tailed manakin, 265, 435, 436–37
long-tailed weasel, 467
louse, 174. *See also specific species of louse*
lycaenid, 270
lycophyte, 531

macaw, 8, 266–67, 355, *356*, 371, *393*, 401–2, 554, 567–69, *568*, *569*, 716. *See also specific species of macaw*
machaca, 547, 635, 675
macroalga, 114, *115*, 119, 172, 593, 603, 710
macrobacteria, *117*, 118
macrofauna, 102, 110, 144–45, 635, 665
macroinvertebrate, *117*, 271, 565, 629, 633, 636
macrolichen, 459
macrophyte, 623, 659, 661, 663, 665, 669, 672, 675
maërl, 114, 116
magpie jay, 303, 338
mahi-mahi, 116
mahogany, 259, 276, 305, 482
maize, 85–86, 276, 533, 661, 664, 669, 677, 678
mammal, 394–400, 437, 439, 443–44, 466–67, 473, 531, 630, 632, 696, 712, 715, 719, 720
mammutid, 51
manakin, 266, 393, 435, 436–37. *See also specific species of manakin*
manatee, 265, 567, 608, 696, 710, 716, 722. *See also specific species of manatee*
manatee grass, 602
manefish, 171
mango, 68, 73, 534
mangrove, 9, 10, *12*, 66, 67, 97, 101, 102, 103, 108–9, *108*, 112, 113, 120, 139, 142, 143–44, 147, 148, 149, *151*, 153, 215, 252, 256, 265, *299*, *308*, 354, 355, 373, 375, *376*, 377, 394, 531, 595, 596, 599–600, *600*, 608, 609, 696, 697, 710. *See also specific species of mangrove*
manioc, 196, 533
manta ray, 178
mantis shrimp, 146
mantled howler monkey, 260, 264, 265, 279, 399, 556–57, 558, 716, 719
many-scaled anole, 392
marbled swamp eel, 268, *547*, 548
margay, 262, 467, 482, 511, 719, 720
marine mammal, 6, 169, 219
marine Pacific snake, 221
marine turtle, 268, 608, 709
marlin, 178
marsupial, 262, *555*, 712, 715
masked booby, 222
masked tree frog, 392
mastodon, *51*, *55*, 533
mayfly, 633, 722
mealy amazon, 355
mealy parrot, 393
medusa, 100, 167, 168
megafauna, 274, 293–95, 297, *304*, *305*, 329, 337–38, 361, 532–34, 713
megalonychid, 51
megatheriid, 50, 51, *51*
meiofauna, 101, 102, 110, 144–45, 635
melastome, 216
melon, 71, 276, 336
merry widow, 547, 548
metalmark, 270
Mexican cypress, 478
Mexican hairy porcupine, 466, 558–59
Mexican tree frog, 392
mice, 79–80, 265, 466, 467, 472–73, 555, 715. *See also specific species of mice*
microalga, 142, 146
microarthropod, 77, 432
microcrustacean, 667

microdinosaur, 290
midas cichlid, *546*, *547*
mimosoid, 253, 272
mistletoe, 438
mite, 440
mold, 458. *See also specific species of mold*
mollusk, 100, 101, 102, 103, 107, 110, 116–17, 120, 145–46, 149, 151, 152, 153, 155, 165, 167, 168, 169, 170, 275, 305, 388, 511, 552, 592, 593, 594, 609, 632, 693, 710, 715, 720
molly, 268, 547, 548
monarch butterfly, 327
mongoose, 336
monkey, 438, 471, 556, 719. *See also specific species of monkey*
monocot, 460–62, 468, 509, 510, 544
montane alligator lizard, 511, 720
moray, 148, 149, 711
mosquito, 231, 556
mosquito fish, 548
moss, 209, 211, 218, 431, 463, *464*, 469, 476, *479*, 500, 506, 509–10, 512, 513, 515, 686, 718, 719, 720, 721. *See also specific species of moss*
moth, 231, 265, 269, 270, 271, 272, 314, 323–26, *330*, *384*, 545, 712, 713. *See also specific species of moth*
motmot. *See specific species of motmot*
mountain lion, 437, 467, 511
mountain mullet, 220, 269, 389, 547, 548, *645*
mountain salamander, 464, 482
mountain thrush, 466
mountain tree frog, 464, 482
mourning warbler, 394
mouse. *See* mice; *and specific species of mouse*
mud shrimp, 594, 604
mule, 331
mullet, 220, 269, 389, 547, 548, *645*. *See also specific species of mullet*
murex, 152, *152*, 153
muricid snail, *152*
murid, 437
Muscovy duck, 267, 276, 355
mushroom, 257, 458, *459*
mushroom-tongue salamander, 511, 720
mussel, *117*
mustelid, 555
mycorrhizal fungus, 75, 76, 334, 713
myctophid, 148
mylodontid, 50, 51, *51*
myriapod, 235
mysid, 147
mytilid bivalve, 33, 117
myxogastrid, 458

Nearctic migrant warbler, 554
nematode, 110, 223, 276, 511, 546, 600, 665
nemertean, 119, 593, 600
Neotropical otter, 557
neritid, 389
neuston, 169
Nicaraguan seed-finch, 553
nightingale-wren, 552
night monkey, 557
Nile tilapia, 692
nine-banded armadillo, 467
noddy. *See specific species of noddy*
Northern barred woodcreeper, 394, 554
northern jacana, 276, 354
Northern raccoon, 557
Northern waterthrush, 394
nudibranch, 103
nymphalid, 270

oak, 81, 257, 293, 377, 429, 458–61, 462–64, *464*, 465–67, 469, 471–75, *474*, *478*, 482, 500, 506, 515, 686, 718, 719. *See also specific species of oak*
ocellated antbird, 394, 554
ocelot, 262, 263, 467, 482, 511, 557, 686, 719, 720
ochraceous wren, 466
ochre-bellied flycatcher, 393
ochrophyte, 509
octocoral, 102, 103, 171, 172, 593, 710
odonate, 666
oil palm, 66, 68, 358, 364, 377–78, *379*
oligochaete, 155, 600, 665
olive ridley sea turtle, 106, 220, 268, 280, 709
olive tanager, 554
olomina, 268, 547, 548
omnivore, 149, 389, 553–54, 556, 635
onion, 477, *479*
onuphid polychaete, 156
onychophoran, 78
ophiidid, 148
ophiuroid, 116, 149, 153
opossum, 262, 265, 277, 332, 386, 532, 562. *See also specific species of opossum*
orange, 196, 336, 534
orange-billed sparrow, 554
orbatid, 666
orchid, 211, 213, 275, 335, 438, 460, 461, 462, 515, 686, 718
oriole, 222, 265. *See also specific species of oriole*
orthopteran, 560
Osa cecilian, 391
ostracod, 110, 147, 660
otariid, 170, 219
otariid seal, 194
otter, 630, *644*. *See also specific species of otter*
owl, 311
oyster, 151–52

paca, 196, 265, 400, 437, 467, 555, 562, 719
Pacific fat sleeper, 269
palaemonid, 632
palm, 9, *12*, 68, 73, 211, 213, 260, 315, 336, *368–69*, 377, 398, 429, 461, 477, 479–480, 515, 533, 534, 539, 559, 562, 564, 598, 690, 694–696, *694–95*, 699, 710, 713, 715, 719, 721. *See also specific species of palm*
palm weevil, *379*
palo verde, 690, *692*
pampathere, 50
Panamanian fiddler crab, 232
pandalid shrimp, 153
panther, 567
papaya, 76, 303
paper wasp, 271
papilionid, 270
parasite, 269, 462. *See also specific species of parasite*
pargo, *390*
parrot, 392, 715. *See also specific species of parrot*
passerine, 276
passion-vine butterfly, *383*
pastel cichlid, 268, 547, 548
peach, 534
peach palm, 83, 377, 477, *479*
peanut, 303, 533
peanut-head bug, 559
peanut worms, 146
peat moss, 463, 505, 514
peccary, 262, 263, 555, 558, 562, 696, 712, 715, 716. *See also specific species of peccary*
pectinidid bivalve, 168
pelagic thresher shark, 170
pelecypod, 592
penaid shrimp, 144, 146, 150, 153, 154
pennatulid, 593
pepper, 68. *See also specific species of pepper*
peregrine falcon, 355
phorophyte, 462
phyllostomid, 273
phytobenthos, 143, 149
phytoplankton, 103, 104, 113, 114, *120*, 121, 142, 143, 146–47, 148, 149, 509, 595, 660, 665, 666, 672, 675, 676, 690, 711, 721
pierid, 270, 327
pig, 194, 196, 207–8, 221, 225, 226, 364, 477, 514. *See also specific species of pig*
pigeon, 221, 466. *See also specific species of pigeon*
pike killifish, 546, 547
pimelodid, 546
pine, 478. *See also specific species of pine*
pineapple, 73, 303, 347, 541, 542, 543, 564, 568, 646, 647, *647*
pink shrimp, 153, 594
pinniped, 219
pipefish, 268, 389, 546, 548
piper, 630
piratic flycatcher, 553
pit viper, 357, 712
plain-brown woodcreeper, 554
plain-colored tanager, 553, 554
plain xenops, 554
plankton, 100, 101, 102, 104, 116, 149, 169, 199, 592–93, 656, 657, 667, 672, 710

planktonic microalga, 146
plantain, 534, 564
plum, 477, *479*
pocilloporid, 710
pocket gopher, 79–80, 354
podocarp, 468
poeciliid, 268, 269, 390, 463
pogonophoran, *117*
poison dart frog, 392
polychaete, 101, 102, 103, 105, 110, *110*, 116, 118, 144–45, 146, *147*, 149, 153, 155, 156, 168, 593, 595, 600, 710. *See also specific species of polychaete*
polyplacophoran, 594
porcellanid crab, 106, 146, 709
porcupine, 79, 264, 555. *See also specific species of porcupine*
portunid crab, 146
potato, 71, 477, 478, *479*
poultry, 364
prawn, 108, 233–34, 235, 389, 710, 713, 715. *See also specific species of prawn*
primate, 262, 354, 394, 399, 400, 532, 555, 556, 712, 715
proboscidean, 50
procyonid, 555
prong-billed barbet, 439
prothonotary warbler, 394
pteridophyte, 462
pufferfish, 144, 148, 176, 389, 635
pugnose tree frog, 391–92
puma, 262, 263, *263*, 482, 511, 557, 712, 719, 720
purple (murex), 152
purple crab, *234*
pygmy rain frog, 531
pyrrhophyte, 672, 676, 721

quetzal. *See* resplendent quetzal
quillwort, 672, 721

rabbit, 196, 262, 471, 511, 532, 555, 712, 720. *See also specific species of rabbit*
raccoon, 262, 264, 265, 355, 712. *See also specific species of raccoon*
radish, *479*
raffia, *694*
rainbow bass, 269
rainbow cichlid, 547, 548
rainbow trout, 463, *465*, 719
rain frog, 392
ramphastid, 562
raptor, 392, 715
rat, 79–80, 219, 221, 224, 225, 231, 236, 265, 466, 467, 555, 715. *See also specific species of rat*
rattlesnake, 268
ray, 145, 149, 154, 711
razor clam, 153
red alga, 174, 509, 624
red banana, 541
red-breast cichlid, 268, 547
red brocket deer, 277, 467, 559, 686
red-capped manakin, 393
red ceiba, 377
red-eyed tree frog, 392
red-eyed tree snake, 392
red-eyed vireo, 394
red-footed booby, 222
red-footed seabird, 170
red-furred cattle, 300
red macroalga, 710
red mangrove, 108, 143
red snapper, 148, 711
redstart, 438, 466. *See also specific species of redstart*
red-tailed hawk, 466, 511, 720
red-tailed squirrel, 466
red-throated ant-tanager, 554
reef-building coral, 171, 710
reptile, 434–35, 552, 629–30, 696, 712, 715, 719, 720, 722
resplendent quetzal, 435, 438, 439, 443–44, 466, 471–72, *472*, 476, 478, 482, 553, 719
resurrection plant, 429
reticulated ameiva, 392
rhino, 532
rhodolith, 114, *115*, 116, 121
rice, 68, 71, 76, 196, 276, 303, 336, 377, *379*, 642, 647, 648, 659, 683, 689, 700
riodinid, 270
river otter, *634*, 722
river sardine, 268
robin, 148, 394, 511, 720, 721. *See also specific species of robin*
rock louse, 174
rodent, 79–80, 262, 265, 273, 354, 386, 465, 466, 471, 472–73, 532, 557–58, 698, 712, 719
roseate spoonbill, 266, *266*, 267, 354
rotifer, 665, 667, 675
royal palm, 694
ruddy-capped nightingale-thrush, 466
ruddy treerunner, 466
rufous-collared sparrow, 511
rufous piha, 393
rufous-vented ground-cuckoo, 394, 554
rush, 216, 721. *See also specific species of rush*
russuloid, 458

sabellariid polychaete, *147*
saber-tooth cat, 55
salamander, 462, 463, 464, 465, 473, 550, 551, 686, 715, 721. *See also specific species of salamander*
sally lightfoot crab, 233
salp, 604, 711
Salvin's spiny pocket mouse, 262, 265
sand crab, 145
sand dollar, 110
sand piper, 722
sardine, 140–41. *See also specific species of sardine*
saturniid, 270, 560
sawfish, 268, 546, 547
scaphopod, 592
scarab beetle, 438
scarlet macaw, 266, 267, 355, *356*, 393, *393*, 401–2, 554, 716
scarlet-rumped cacique, 554
sciaenid, 148, 149, 155
scleractinian coral, 608
scorpion, 227, *229*, 231
scorpion fish, 148
scorpion mud turtle, 268
sea anemone, 107–8
sea bird, 711
sea catfish, 148, 711
sea cucumber, 596
sea fan, 593, 605
seagrass, 102, 103, 111, 120, 593, 595, 596, 601–2, *603*, 604, 608, 609, 710
seal, 101, 194, 219. *See also specific species of seal*
sea lion, 170, 219. *See also specific species of sea lion*
sea robin, 148
sea slug, 103, 594
sea spider, 100
sea star, 593. *See also specific species of sea star*
sea turtle, 105, 106, *106*, 123, 305, *395*, 398, 566, 591, 596, 608, 609, 716. *See also specific species of sea turtle*
sea urchin, 105, 113, 165, 176, 592, 710
sea weed, 114
sedge, 686, 687, 689, 696, 721
seed beetle, 269
serranid, 148
shark, 116, 148, 149, 154, 168, 170, 171, 178, 183, 626, 627, 711. *See also specific species of shark*
shellfish, 121, 151, *151*, 156
shield bug, 269
shining honeycreeper, 554
ship rat, 224
short-tailed hawk, 265
shrew, 437, 511, 719, 720
shrimp, 68, 118, 148, 149, 153–54, 156, 169, 233, 234, 594, 626, 630, *631*, 632–33, 634–35, *634*, 639, 642, 644, 722. *See also specific species of shrimp*
siboglinid tubeworm, 117
silky anteater, 265
silky shark, 170
silverside, 547, 548
silvery-throated jay, 466
siphonophore, 100, 167, 168, 171, 604, 711
sipunculid, 146, 153, 155, 595
skink, 552
skipper, 560
skipper butterfly, 270, 275
skunk, 262, 555. *See also specific species of skunk*
slate-throated redstart, 438
slaty-backed nightingale thrush, 476
slaty flowerpiercer, 511

sleeper, 269, 389, 547. *See also specific species of sleeper*
slender black-speckled palm pit viper, 465
slender gray fox, 466
slender harvest mouse, 437
slime mold, 458
sloth, 719. *See also specific species of sloth*
slug, 103, 594. *See also specific species of slug*
small rice, 623
snail, 75, 76, 145, 153, 155, 165, 232, 389, 605, 634, 635. *See also specific species of snail*
snake, 75, 76, 363, 465, 552, 715, 719. *See also specific species of snake*
snapper, 148, 154, 389, 390. *See also specific species of snapper*
snapping turtle, 630
snook, 389, 547
snout beetle, 269
snowy cotinga, 553
social wasp, 327–28, *328*
sole, 547
solemyid, 33
solifugid, 231
solitaire, 439, 466. *See also specific species of solitaire*
sooty-capped bush-tanager, 511
sooty robin, 511, 720, 721
sooty tern, 222
sooty thrush, 721
Southern river otter, 277
soybean, 71
spadebill, 552. *See also specific species of spadebill*
spangle-cheeked tanager, 439
spermatophyte, 210
sperm whale, 194
sphingid, 229–31, 270
spider, 231, 712. *See also specific species of spider*
spider monkey. *See* Geoffroy's spider monkey
spiny-cheek sleeper, 389, 547
spiny lobster, 594
spiny pocket mouse, 274
spiny-tailed iguana, 268
spirochaete, *117*
sponge, 166, 177, 593. *See also specific species of sponge*
spoonbill, 266–67, *266*, 354. *See also specific species of spoonbill*
spot-breasted oriole, 222, 265
spot-crowned woodcreeper, 466
spotted antbird, 554
spotted dolphin, 102, 116, 710
spotted skunk, 467
spotted sleeper, 220
springtail, 77
squash, 533, 564, 565
squat lobster, 118
squid, 149
squirrel, 79, 264, 273, 472–73, 555, 715, 719
squirrel monkey. *See* Central American squirrel monkey
starfish, 176, 710. *See also specific species of starfish*
star grass, 255
stem borer, 560
stingray, 148, 149
stomatopod, 100, 153, 154, 165, 168
stonefly, 633, 722
stony coral, 103, 172, 593
stork, 265, 691, 722. *See also specific species of stork*
strangler fig, *375*, 431, 432
strawberry, 68
streak-crowned antvireo, 554
streaked flycatcher, 553
striped foliage-gleaner, 393
striped hog-nosed skunk, 467
stylaster, 169
succulent, 429
suckermouth catfish, 546
sugar cane, 64, 68, 71, 73, 196, 276, 303, 336, 541, 543, 642, 647, 648, 683, 700
sulphur (butterfly), 270
sulphur-rumped flycatcher, 554
sulphur-rumped tanager, 553
sunbittern, 630
sundown cicada, 559
sungrebe, 630
Swainson's thrush, 476
swallowtail butterfly, 270
swallow-tailed gull, 170
swallow-tailed kite, 466
sweet potato, 86
swollen-thorn acacia, 274

tachinid, 275
tamarind, 73
tanager, 393–94, 439, 554, 719. *See also specific species of tanager*
tank bromeliad, 462, 464
tapir, 262, *263*, *318*, 354, 366, *378*, 388, 443–44, 467, 479–80, *480*, 555, 556, 696, 712, 715, 716, 720. *See also specific species of tapir*
tapirid, 51
tardigrade, 511
tarpon, 546, 547, 595, 627
tawny-crested tanager, 554
tawny-crowned greenlet, 393
tawny-winged woodcreeper, 394
tayassuid, 51
tayra, 265, 467, 557, 562
teak, *76*, 279, 543. *See also specific species of teak*
teal, 701. *See also specific species of teal*
teiid, 552
teleost, 149
Tennessee warbler, 394
tent-making bat, 395, 557, 715
termite, 75, 77–78, *78*
tern, 222. *See also specific species of tern*
tetra, 547. *See also specific species of tetra*
thallose liverwort, 431
thicket tinamou, 276
threadfin anglerfish, 118
three-toed sloth, 467
three-wattled bellbird, 355, 394, 435, 439, 553, 712, 719
thrush, 439, 686, 719. *See also specific species of thrush*
thumbless bat, 557
tick, 440
tiger heron, 630
tilapia, 336, 628, *628*, 640, 648–49, 692. *See also specific species of tilapia*
timberline wren, 511
tinamou, 266, 466. *See also specific species of tinamou*
toad, 463, 719. *See also specific species of toad*
tobacco, 196, 361
tomato, 71, 461
Tome's rice rat, 440
tongue fern, 509
tortoise beetle, 269
tortoise shell. *See specific species of shell*
toucan, 439, 561–62. *See also specific species of toucan*
toxodont, 51
toxodontid, 51
tree boa, 392
tree frog, 628, 634
treefern, 211, 214, *215*, 377, 457, 461, 468, 469, 509, 513, *513*, 719
trematode, 275–76
trichiurid, 148
trogon, 266, 466. *See also specific species of trogon*
tropical almond, 273
trout, 336, 477, *479*, 628. *See also specific species of trout*
true bug, 712
true mountain frog, 464, 482
true spider, 231
tuba, 548
tuber, 361
tubeworm, 117. *See also specific species of tubeworm*
tucuxis dolphin, 595
tuna, 100, 102, 178
tungara frog, 531
tunicate, 609
turkey, 267. *See also specific species of turkey*
turkey vulture, 266
turquoise-browed motmot, 266
turtle, 103, 116, 153, 168, 178, 531, 552, 600, 606, 629, 630, 715, 722. *See also specific species of turtle*
tussock grass, 492, 500, 512
two-toed sloth, 467

ungulate, 5, 385, 395, 532
urchin, 105, 113, 165, 176, 592, 593, 594, 602, 605, 607, 710. *See also specific species of urchin*
urostigmoid fig, 432

Vaillant's frog, 268
vampire bat, 264, 557
variegated squirrel, 273
vascular epiphyte, 428, 461–62, 469, 718
vascular parasite, 462
velvet worm, 78
vertebrate, xix, 55, 119, 143, 166, 168, 218–19, 224, 226, 262–65, 269, 273, 275, *315*, 332, 333, 336, 399, 404, 463–67, 471, 511, 545, 552, *556*, 557, 558, 566–67, 623, 627, 709, 712, 713
vesicomyid, 33
vesicular-arbuscular mycorrhiza, 440
vesper mouse, 558
vesper rat, 437
vine, 211, 253, 254, 257, 269, 332, 333, 338, 347, 348, 374, *383*, 392, 432, 460–61, 469, 713, 721
violet-crowned woodnymph, 553
viper, 531. *See also specific species of viper*
viperid, 552
vireo, 394. *See also specific species of vireo*
Virginia opossum, 262, 265
vocal rain frog, 391
volcano hummingbird, 511, 720
volcano junco, 511
vulture, 306–7. *See also specific species of vulture*

wading bird, 715
warbler, 222, *224*, 394, 466, 511, 554. *See also specific species of warbler*
wasp, 269, 271, 272, 275, 323, 327–28, *384*, 439, 712, 719. *See also specific species of wasp*
water anole, 392
water boatman, 666
waterfowl, 700, 701–2, 722
water hyacinth, 257, *258*, *349*, 623, 659, 721
water lettuce, 721
water lily, 663, 721
water lizard, 392
water opossum, 277
water turkey, 267
weasel, 262, 712. *See also specific species of weasel*
wedge-billed woodcreeper, 392–93
weed, 81, 114, 196, 211, 335, 515, 624, 691. *See also specific species of weed*
weevil, 269–70, 273–74, 329, 330. *See also specific species of weevil*
West Indian manatee, 555, 595
West Indian topshell, 594
whale, 154, 162, 168, 194, 710. *See also specific species of whale*
whip spider, 231
white-capped noddy, 222
white clover, 71
white-faced capuchin monkey, 260, 263–64, 328, 399, 466, 471, 473, 556–57, 716, 719
white-flanked antvireo, 554
white-flanked antwren, 393
white-headed capuchin, 559, 716. *See also* white-faced capuchin monkey
white ibis, 267
white-lipped mud turtle, 630
white-lipped peccary, 277, 354, 395–97, *395*, 400–401, *404*, 437, 556, 558, 559, 715, 716
white-nosed coati, 466, 557
white oak, 81, *470*
white-tailed deer, 196, 224, 226, *228*, 265, 277, 332, 559
white-tailed hognose snake, 392
white tent making bat, 557
white tern, 222
white-throated capuchin monkey. *See* white-faced capuchin monkey
white-throated magpie-jay, 266
white-throated robin, 394
white-throated shrike-tanager, 393, 554
white-throated spadebill, 552
white-tip reef shark, 170, 171
wide-head sea catfish, 268
wild sweet potato, 230
wolf cichlid, 547, 548, 550
wood borer, 227
woodcreeper, 266, 392, 554, 715. *See also specific species of woodcreeper*
woodnymph, 553. *See also specific species of woodnymph*
woodpecker, 392, 715. *See also specific species of woodpecker*
woodstork, 266, 267, 276, 699
woodthrush, 394
worm, 78, 153. *See also specific species of worm*
wren, 435, 554. *See also specific species of wren*
wrenthrush, 511

yam, 533
yellow-billed cacique, 466
yellow-billed cotinga, 355
yellow-eyed grass, 686, 721
yellow-margined flycatcher, 554
yellow-naped parrot, 266
yellow-olive flycatcher, 275
yellow-shouldered bat, 467, 553
yellow-thighed finch, 439, 466
yellow-throated brush-finch, 476
yellow-throated euphonia, 435
yellow toad, 267
yellow warbler, 222, *224*, 394
Yeti crab, 118
yucca, 303, 568

Zaca's fiddler crab, 232
zooplankton, 100, 102, 103, 104, 147–48, 169, 595, 604, 660, 665, 666, 667, 672, *675*, *676*, 721
zooxanthellate coral, 165, 175, 710

Spanish Common Names

agave, 429
agouti, 79–80, *80*, 260, 265, 274, 329–30, 400, 467, 555, 562, 698
aguacatillo, 471
alfaro, 547, 548
alga, 100, 101, 102, 105, 107, 111, 121, *147*, 150, 154, 155, 168, 173, *175*, 218, 509, 591, 593, *605*, 624–26, 627, 629, 632, 634–35, 636, 709, 711, 720–21, 722
almeja blanca, *151*
almendro, 273
anguila, 547
anguila de pantano, 268, 547, 548
anhinga, 267, 354, 630
aracari, 561–62
árbol de la llama, 273
armadillo, 277, 532, 555, 562, 715
arrayán, 513
arrocillo, 623

bagre cuatete, 268
bala, 387–88
bala de cañon, 260
balsa, *380*
banana, 65, 66, 68, 77, 79, 83, 85, 196, 303, 347, 354, 364, 377–78, 379, *379*, 401, 541, 542, 543, 549, 557, 563, 564–65, 567, 646, 647, *647*, 677, 716
barbudo, 268, 547, 548
batamba, 512
bobo, 548
bromelia, 213, *216*, 257, 275, 466, *470*, 633, 686, *686*, 713–14, *717*, 718, 721
burro, 300
caimito, 73
calva, 547
cambute, 153
cangrejero, 386
caoba, 259, 276
caro caro, 274
cascabel, 268
casco de burro, 107
castellana, 267, *267*
cativo, 66
cebu, 302, *303*
cedro, 276
cerillo, 696
chachalaca, 400
chile, 361
chocuaco, 361
choreja, 257
chucheca, 151, *151*, 152

chupapiedra, 548
cicada, 269
cigueñón, 266
coati, 264, 332, 386, 400, 562
coco, 273
cocobolo, 276
cocotero, 273, 377
congo, 268, 547, 548
copey, 215
coral, 100, 102–3, 104, 111–14, 112, 120–21, *120*, 123, 165, 168, 169, 171, 172, 175–76, 178–79, *179*, *180*, *181*, 591, 595, 596, 597, 601, 602–3, *603*, 604, *604*, 605, *605*, 606, 607, *608–9*, 710
corvina, 148, 154, 711
coyote, 264, 277, 332, 333, 437, 466, 557, 719
cristóbal, 259, 262, 276
curculio, 269

danta, 319
dormilón, 269

encino, 313, *315*, *325*
encino negro, 272
enea, 257
espatula rosada, 266
espavel, 315

frijol, 66

ganado, 300
garceta grande, 267
garcilla bueyera, 267
garrobo, 268
gaspar, 547, 700, 722
gavilán colicorto, 266
gmelina. *See* melina
grisón, 277
guacamayo rojo, 266
guacimo, 334, 338
gualaje Atlántico, 547
guanábana, 230
guanacaste, 259, 265, 274, 279, 333, 334
guapinol, 273–74, 328–29
guapote, 269, 547, 548, 550
guapotillo, 548
guarasapa, 547
guarumo, 215, 217, 228, *230*
guatusa, 79
guavina, 269, 389, 547, 548
guayacán real, 259, 276
gumbo limbo, 256

hormiga bala. *See* bala
hormiga del cornizuelo, 274, 338

ibis blanco, 267
indigofera, 300
indio desnudo, 256

jabirú, 266–67
jacana, 265, 630, 722
jaguar, 262, 277, 305, 354, 366, 395, 396, 398–99, 401, 404, 437, 467, 482, 557, 712, 715, 716
jaguarundi, 262
jaragua, 255, 276, 300, *301*, 302, *302*, 306, *306*, *308*, 311, *312–14*, 335, 713
javelina, 395
jícaro, 274, *304*, *315*
jurel, 547

lagartija, 267
lamearena, 547, 548
lapa roja, 401
laurel negro, 276
lechuga, 257
lengua de vaca, 710
lenguado, 547
lenguado redondo, 547
liana, 213, 253, 256, 257, 272, 275, 383, 431, 432, 439, 469, 715, 719
lirio de agua, 257
llama del bosque, 273
luciernaga, 313

machaca, 547
madre de barbudo, 268, 547
majagua, 214, *217*
majagual, 225
mapache, 264
marañon, 377
madre de barbudo, 547, 548
medusa, 100, 167, 168
melina, 334–35, 402
mocasín, 267, 712
moga, 548
mojarra, 268, 269, 547, 548
mojarrita, 547, 548
mono araña, 399
mono aullador, 264, 399
mono cariblanca, 263–64, 399
mono congo, 264, 399
mono tití, 399
mucuna, 78
murex, 152, *152*, 153

napa, 694

ocelote, 262
ojoche, 292
olomina, 268, 547, 548
orey, 698

paca, 196, 265, 400, 437, 467, 555, 562, 719
palmito, 73
palo de hierro, 194, 213
palo verde, 257, *258*, 690, *692*, 721
papaya, 76, 303
pargo, *390*
pato aguja, 267
pato real, 267
pava moñuda, 266
pecho rojo, 268
pejibaye, 377
pez diablo, 628
pez perro, 547
pez pipa, 268, 548
pez sierra, 268, 547
piangua, *151*, 153
picaculo, 389
pie de burro, 151
pijije común, 266
pis pis, 547
pitahaya, 275
pochote, 254, 279, 377
pocoyo, 389
poró, 75
puma, 262, 263, *263*, 482, 511, 557, 712, 719, 720
púrpura, *152*

quetzal, 435, 438, 439, 443–44, 466, 471–72, *472*, 476, 478, 482, 553, 719
quira, 352

róbalo, 389, 547
roble negro, 272
roncador, 547, 548
ron-ron, 276

sabalete, 547
sábalo real, 547
saltarín colilargo, 266
sangrillo, 696
sapo amarillo, 267
sardina, 268, 389, 547, 548
sardina blanca, 548
sardina de quebrada, 547
sardina lagunera, 547
sardina picuda, 547
sardinita, 268, 547
serpiente mocasín, 267

tachinid, 275
taltuza, 79
tayra, 265, 467, 557, 562
tempisque, 276
tepemechín, 269, 389, 547, 548
tepezcuintle, 79, 265, 277
tiburón, 547
tigrillo, 262
tolomuco, 467
tortuga baula, 106, 268
tortuga lora, 106, 268
tucuxi, 595, 608
túngara, 531

vampiro falso, 264–65
vieja, 548

yayo, *315*
yolillo, 66, 377, 694, *694*
yucca, 303, 568

zacate, 568

Systematic Index of Scientific Names

Page numbers in italics refer to figures.

Acacia, 257, 271, 274, 713
  Acacia allenii, 374
  Acacia collinsii, 338
  Acacia farnesiana, 256
  Acacia villosa, 255
  See also *Vachellia*
Acaena, 500, 512
  Acaena cylindristachya, 512
Acalypha
  Acalypha diversifolia, 687, 721
  Acalypha pittieri, 211
Acanthaceae, 254, 469
Acanthaster planci, 176, 710
Acanthobotrium nicoyaense, 149
Acanthocereus tetragonus, 256
Acarina, 77
Acartia lilljenborgii, 147
Acaulospora, 440
Accipitridae, 555
Achiridae, 547
Acnanthidium catenata, 667
Acnistus arborescens, 351, 352
Acoelorraphe wrightii, 690
Acosmium panamense, 256, 350
Acrididae, 559
Acrocomia aculeata, 335
Acromyrmex volcanus, 387
Acropora palmata, 603, 607
Acrostichum aureum, 9, 215, 377, 687, 688, 696
Actinella, 669
  Actinella brasiliensis, 669
  Actinella punctata, 669
Actitis macularis, 630, 722
Acuariidae, 276
Adelanthus, 510
Adelia triloba, 699
Adiantum
  Adiantum amplum, 348
  Adiantum concinnum, 275, 348
  Adiantum latifolium, 696
  Adiantum trapeziforme, 275
Aechmea, 275
  Aechmea magdalena, *368*
Aellopos fadus, 325
Aequidens, 391
Aeschynomene, 255, 377, 687
  Aeschynomene virginica, 689
Agalychnis
  Agalychnis callidyias, 392
  Agalychnis helenae, 392
  Agalychnis spurrelli, 392
Agaricales, 458, 718
Agaricus, 458
Agave seemanniana, 255
Ageratina, 461, 509, 510
Agkistrodon bilineatus, 267, *267*, 712
Agonostomus monticola, 220, 269, 389, 547, 548, 640, *645*
Agoutidae, 79
Agouti paca, 79, *80*, 196, 265, 467, 698
Agrius cingulatus, 230
Agrostis
  Agrostis bacillata, 512
  Agrostis tolucensis, 513
Aguna asander, 270
Aiouea, 471
Ajaia ajaja, 266, *266*, 354
Albizia
  Albizia adinocephala, 351
  Albizia carbonaria, 352
  Albizia caribaea, 272
Alcedinidae, 555
Alchemilla, 720
Alchornea, 52, 718
  Alchornea costaricense, 696
Alcyonacea, *181*
Alcyoniidae, *181*
Alfaroa, 473
  Alfaroa costaricensis, 469
  Alfaroa guanacastensis, 377
  Alfaroa williamsi, 349
Alfaro cultratus, 547, 548
Alligatoridae, 553
Alloplectus, 462
Allosanthus trifoliatus, 349
Alnus, 52, 458, 507, 718
  Alnus acuminata, 75, 76, *76*, 478
Alophia silvestris, 255
Alopias
  Alopias pelagicus, 170
  Alopias superciliosus, 170
Alouatta palliata, 264, 399, 558
Alpheus, 594
  Alpheus simus, 594
Alsophila notabilis, 211
Amanita, 458
  Amanita muscaria, 458
Amanoa potamophila, 695
Amatitlania, 550
Amauroderma schomburgkii, 354
Amazilia saucerrottei, 275–76
Amazona
  Amazona auropalliata, 266
  Amazona farinosa, 355, 393
Ambates, 560
Amblycercus holosericeus, 466
Amblycerus spondiae, 273
Amblyopinus
  Amblyopinus emarginatus, 440
  Amblyopinus tiptoni, 440
Amblypygi, 231
Ambrosia cumanensis, 689
Ameiva, 552
  Ameiva leptophrys, 392
Americonuphis reesei, 110, 156
Ammotrecha solitaria, 231
Amphignathodontidae, 551
Amphilophus
  Amphilophus alfari, 549
  Amphilophus bussingi, 549
  Amphilophus citrinellus, 546, 547
Amphipholis geminata, 110
Amphitecna latifolia, 696
Anacardium, 369, 711
  Anacardium excelsum, 260, 262, 315, 350, 368, 378, 697
  Anacardium occidentalis, 377
Anacroneuria, 633
Anadara, 153
  Anadara tuberculosa, *151*, 153
Anamura, 232
Ananthacorus angustifolius, 348
Anartia fatima, 270
Anas
  Anas clypeata, 693
  Anas discors, 266
Anatidae, 555
Anax, 666
Anchoa, 154
Andira inermis, 260, 377, 695, 713
Anguidae, 553
Anguilla rostrata, 546, 547
Anguillidae, 547
Anhinga anhinga, 267, 354, 630
Anhingidae, 555
Ankistrodesmus braunii, 667
Annelida, 119
Annona
  Annona costaricana, 52
  Annona glabra, 214, 215, *217*, 230, 695
  Annona muricata, 230
  Annona reticulata, 254
Annonaceae, 348, 468
Anolis biporcaus, 339
Anomocora carinata, 179
Anous
  Anous minutus, 222
  Anous stolidus, 222, 225

Anteos
Anteos clorinde, 327
Anteos maerula, 327
Anthoceros, 431
Anthodiscus, 374
Anthodiscus chococense, 353
Anthonomus, 329, 330
Anthophoridae, 271
Anthophorinae, 271
Anthopleura nigrescens, 107–8
Anthurium, 347, 457, 462, 469, 539, 715, 718
Anthurium bakeri, 539
Antillesoma antillarum, 146
Anura, 686
Anyphaenoides cocos, 231
Aotus, 556
Apeiba tibourbou, 254, 257, 353
Aphaenogaster, 561
Aphaenogaster araneoides, 387
Aphaenogaster phalangium, 387
Aphelandra scabra, 257, 262
Aphrissa statira, 327
Apiaceae, 52, 499, 510
Apidae, 271
Apis mellifera, 335–36
Apocynaceae, 230, 254, 348
Apodidae, 555
Ara
Ara ambigua, 8, 554, 567
Ara macao, 266, 355, *356*, 393, 400, 554, 716
Araceae, 347, 396, 462, 539, 540
Arachis pintoi, 85
Arachnida, 463
Araliaceae, 52, 457, 461, 462
Aramidae, 555
Araneae, 231
Archaeatya, 233
Archaeatya chacei, 235
Archaeolithophyllum, 52
Archaeoprepona, 270
Archaeoprepona demophon, 325
Archocentrus
Archocentrus centrarchus, 547
Archocentrus myrnae, 548, 549
Archocentrus nigrofasciatus, 268, 547, 548, 550
Archocentrus septemfasciatus, 548, 549
Arctocephalus galapagoensis, 219
Arcytophyllum, 512
Arcytophyllum lavarum, 512, 513
Ardeidae, 555
Ardisia, 269, 461, 469, 471, 719
Ardisia compressa, 695
Ardisia cuspidata, 217
Ardisia cutteri, 374, 715
Ardisia glandulosa-marginata, 81
Arecaceae, 349, 396, 468, 539
Arenaria, 512
Ariidae, 148, 268, 711
Ariopsis seemanni, 693
Arius guatemalensis, 268
Aromobatidae, 551
Arothron meleagris, 176, 710
Arrabidaea, 257
Arremon aurantiirostris, 554
Arsenura arianae, 323
Arthoniales, 509
Arthrodesmus
Arthrodesmus bifidus, 665, 721
Arthrodesmus phinus, 667
Arthropoda, 119
Artibeus
Artibeus jamaicensis, 265
Artibeus toltecus, 437, 719
Artiodactyla, 354, 511
Asclepias, 255
Asclepias woodsoniana, 255
Ascomycetes, 208
Ascomycota, 6, 509
Aspidosperma
Aspidosperma megalocarpon, 260, 374, 715
Aspidosperma myristicifolium, 349
Asplenium, 347, 458, 462
Asplenium cuspidatum var. *triculum*, 347
Asplenium salicifolium var. *aequilaterale*, 347–48
Asplundia, 462
Astatheros
Astatheros alfari, 268, 547, 548
Astatheros bussingi, 548
Astatheros diquis, 389
Astatheros longimanus, 268, 547
Astatheros rhytisma, 548
Astatheros rostratus, 548
Asteraceae, 52, 62, 254, 347, 354, 458, 460, 461, 462, 467, 468, 469, 476, 492, 497, 499, 509, 510, 539, 719
See also Compositae
Asterogyne martiana, 696
Astractosteus tropicus, 546, 547, 722
Astrangia dentata, 179
Astrocaryum alatum, 696, 699, 722
Astronium graveolens, 257, 259, 260, 276, 353
Astyanax, 628
Astyanax aeneus, 269, 276, 389, 547, 548, 549, 627–28, 692, 722
Astyanax aeneus costaricensis, 268
Astyanax fasciatus, 140–41, 268, 547, 548, 558, 700
Astyanax nasutus, 547
Astyanax orthodus, 548, 549
Ateles geoffroyi, 262, 399, 482, 559, 712
Atelopus, 628, 629
Atelopus varius, *356*, 434, 628, *629*, 719, 722
Atelopus zeteki, *629*
Atherinella
Atherinella chagresi, 548
Atherinella hubbsi, 547, 548
Atherinidae, 546, 547, 548
Atlapetes
Atlapetes brunneinucha, 476
Atlapetes gutturalis, 476
Atractosteus tropicus, 692, 700
Atta, 273, 559, 713
Atta cephalotes, 78, 559
Atta colombica, 271
Attalea rostrata, 315, 349, 353
Attini, 561
Atya, 632, 634
Atya crassa, 632
Atya innocous, 632
Atya margaritacea, 632
Atya scabra, 632
Atyia, 235
Atyidae, 233, 235
Aulacorhynchus prasinus, 466
Aulacoseira, 507, 668, 670, 672, 674, 721
Aulacoseira alpigena, 507, 674
Aulacoseira italica, 668
Aulacoseira lirata, 507, 674
Auricularia
Auricularia auricula, 257
Auricularia delicata, 257
Auricularia fuscosuccinea, 257
Automeris
Automeris tridens, 323
Automeris zozimanaguana, 323
Automeris zugana, 323
Automolus ochrolaemus, 554
Avicennia
Avicennia bicolor, 109, 143
Avicennia germinans, 108, 109, 143–44, 377, 599, 696, 710
Awaous
Awaous banana, 547, 548
Awaous transandeanus, 389
Axonopus aureus, 255
Ayenia mastatalensis, 352
Azolla microphylla, 690
Azorella, 512, 720
Azorella biloba, 510, 513
Azteca, 440, 719
Azteca alfari, 387
Azteca chartifex, 388, *388*
Azteca coeruleipennis, 387
Azteca constructor, 387
Azteca forelii, 387
Azteca instabilis, 387
Azteca ovaticeps, 387
Azteca pittieri, 387
Azteca sericeasur, 387
Azteca xanthochroa, 387

Bachia blairi, 392
Bacillariophyceae, 624
Bacillariophyta, 509
Bactris
Bactris gasipaes, 377
Bactris hondurensis, 539
Bactris longiseta, 699
Bactris major, 377
Baetidae, 635
Balaenoptera
Balaenoptera edeni, 154
Balaenoptera musculus, 154

Balaenopteridae, 116
Balanus, 143
Ballia prieurii, 624
Banara guianensis, 339
Bartramia, 510
Bartramia longicauda, 693
Basidiomycetes, 208, 458, 509
Basileuterus melanogenys, 511
Basilicus vittatus, 552
Bassaricyon gabbii, 482
Bassariscus sumichrasti, 482
Bathymodiolus, 117
Batocarpus, 374
Batocarpus costaricensis, 353, 374
Batrachochytrium dendrobatidis, 434, 463, 551, 629
Bauhinia
Bauhinia glabra, 259
Bauhinia paulettia, 260
Bauhinia ungulata, 270, 338
Bazzania, 463
Becquerelia cymosa, 696
Begonia, 438, 462
Begonia estrellensis, 432
Begoniaceae, 462
Beilschmiedia, 438, 471, 719
Beilschmiedia pendula, 438
Belonesox belizanus, 546, 547
Bernoullia flammea, 260, 261
Bertiera angustifolia, 230
Besleria, 457
Biblis hyperia, 339
Bignoniaceae, 254
Billia, 471, 473
Billia hippocastanum, 472
Bjerkandera, 458
Blakea, 461, 462
Blattarida, 463
Blechnaceae, 482, 686
Blechnum, 12, 458, 510, 512, 514
Blechnum buchtieni, 513, 685, 686, 721
Blechnum loxense, 686
Blechnum occidentale, 215
Boa, 552
Bocconia, 473
Boidae, 553
Bolbitis portoricensis, 348
Boletus, 458
Bolitoglossa, 465
Bolitoglossa pesrubra, 464, 511, 686, 721
Bolitoglossa subpalmata, 464, 686
Bomarea, 461, 469
Bombacopsis
Bombacopsis quinata, 73, 254, 259, 272, 348, 377, 712
Bombacopsis sessilis, 350, 353
See also *Pachira*; *Pochota quinata*
Bombus ephippiatus, 463, 469, 511
Bombycillidae, 556
Boraginaceae, 254
Borreria
Borreria ocymoides, 230
Borreria prostrata, 230
Bos
Bos indicus, 300, *303*
Bos taurus, 300, *303*
Bosmina
Bosmina hagmanni, 667, 675, 676–77
Bosmina longirostris, 667, 675
Bothriechis, 552
Bothriechis nigroviridis, 465
Bothriechis schlegelii, 392
Bothriechis supraciliaris, 392
Bothrops, 552
Bothrops asper, 357, 712
Botryococcus, 509
Botryococcus braunii, 672
Brachiaria, 623, 721
Brachycephalidae, 551
Brachymenium spirifolium, 256
Brachyrhaphis, 389
Brachyrhaphis holdridgei, 548
Brachyrhaphis olomina, 268, 548
Brachyrhaphis parismina, 547, 548
Brachyrhaphis rhabdophora, 389
Brachysira brachysira, 665, 666, 670
Brachyura, 232, 632
Braconidae, 275
Bradypus variegates, 177, 354, 467, 558
Bramocharax bransfordii, 547
Branchiostoma, 144
Branchiostoma californiense, 110, 144
Brassavola nodosa, 275
Bravaisia integerrima, 349
Breutelia, 721
Breutelia subarcuata, 686
Bromelia, 275
Bromeliaceae, 52, 260, 462, 469, 540
Bromus, 512
Brosimum, 213, 292, 294, 369, 374, 711, 713
Brosimum alicastrum, 254, 260, 350, 712
Brosimum costaricanum, 260, 374
Brosimum utile, 348, 349, 353, 374, 715
Bruchidae, 269, 273
Brunellia, 473, 718
Brunellia costaricensis, 461
Brunellia hygrotermica, 377
Brunellia standleyana, 350
Bryconamericus, 634
Bryconamericus scleroparius, 269, 276, 547, 558
Brycon guatemalensis, 547, 558, 561, 635–36, 640, 675
Bryum, 463, 510
Bubulcus ibis, 220, 267
Bucconidae, 555
Buchenavia costaricensis, 353
Buchnera pusilla, 255
Buddleja, 467, 471, 478, 500, 512, 514, 718
Buddleja cordata, 472
Bufo, 391
Bufo aucoinae, 392
Bufo luetkenii, 267
Bufo melanochloris, *391*, 392
Bufo periglenes, 6, 434, 629, 719
Bufonidae, 551
Bulbostylis paradoxa, 255
Buprestidae, 231, 559
Burmeistera, 457, 462
Bursera, 257
Bursera graveolens, 255
Bursera schlechtendalii, 260
Bursera simarouba, 256, 257, 259, 260, 261, 262, 353, 454, 711
Bursera standleyana, 461
Buteo
Buteo brachyurus, 266
Buteo jamaicensis, 466, 511
Butorides virescens, 692, *693*, 722
Byrsonima, 52
Byrsonima crassifolia, 254, 255, 338, 350, 377

Cacicus uropygialis, 554
Caeciliidae, 551
Caenobita compressus, 215
Caesalpinia, 294
Caesalpinia eriostachys, 254, 261, 712
Caiman, 634
Caiman crocodilus, 552, 630, 722, 691
Cairina moschata, 267, 276, 355, 630, 722
Calamagrostis, 500, 512
Calandrina, 512
Calathea, 539, 561, 715, 721
Calathea foliosa, 696
Calathea inocephala, 261
Calathea lagunae, 687, 696
Calathea longiflora, 697
Calathea lutea, 687, 696
Calathea vinosa, 461
Calatola costaricensis, 292
Caligo memnon, 270
Calliandra tergemina, 255
Callichlamys latifolia, 327
Callicore pitheas, 325
Callinectes, 146
Callinectes arcuatus, 146, 147
Callinectes sapidus, 146, 153
Calocitta formosa, 266, 338
Caloglossa ogasawaerensis, 624
Calophyllum
Calophyllum brasiliense, 348, 350, 377, 695, 696, 697, 698
Calophyllum longifolium, 353
Calvatia rugosa, 257
Calycophyllum candidissimum, 254, 257, 259, 348, 353, 712
Calyptogena, 33, 117
Calyptogena costaricana, 117
Calyptrocarya glomerulata, 216, 696
Calyptrogyne glauca, 696
Camaridium micracanthum, 211
Campanulaceae, 462, 469
Campnosperma panamensis, 695, 698–99
Camponotus
Camponotus biolleyi, 228
Camponotus sericeiventris, 387–88
Campylopus, 463, 510, 512, 686, 721

Canacidae, 231
Canavalia maritima, 214
Canis
Canis familiaris, 196
Canis latrans, 51, 264, 333, 466, 557
Canna, 691
Canna glauca, 690
Cannabis, 479
Cantharellus, 458
Capitellidae, 110
Capparaceae, 254
Capparis, 257
Capparis incana, 255
Capra aegagrus hircus, 194, 224
Caprimulgidae, 555
Carangidae, 547
Caranx, 154
Caranx latus, 547
Carapa
Carapa guianensis, 80, 374, 377, 378, 695, 696, 697, 698, 699, 715, 722
Carapa nicaraguensis, 699
Carcharhinidae, 546, 547
Carcharhinus
Carcharhinus falciformis, 170
Carcharhinus leucas, 389, 546, 547
Carcharhinus melanopterus, 170
Cardinalidae, 556
Cardisoma, 385, 386
Cardisoma crassum, 215, 232, 235, 385
Caretta caretta, 600, 709
Carex, 512, 721
Carex bonplandii, 686
Carex donnellsmithii, 686
Carex jamesonii, 686
Carex lehmanniana, 513
Cariniana, 374
Carlana eigenmanni, 547
Carludovica drudei, 261
Carnivora, 511, 557, 562
Carollia
Carollia perspicillata, 265
Carollia subrufa, 265
Carpodectes
Carpodectes antoniae, 355
Carpodectes nitidus, 553
Caryocar, 374
Caryocar costaricense, 348, 349, 353, 374, 402, 482, 697, 712, 715
Caryodaphnopsis burgeri, 352, 353
Caryophyllia
Caryophyllia diomedeae, 179
Caryophyllia perculta, 179
Caryothruastes poliogaster, 554
Casearia, 696
Casearia corymbosa, 338
Casearia nitida, 338
Casmerodius albus, 267
Cassia
Cassia biflora, 338
Cassia grandis, 335, 338
Cassidinae, 269, 560
Cassimiroa edulis, 351
Cassipourea
Cassipourea aff. *killipii*, 699
Cassipourea guianensis, 695
Castilla tunu, 378
Castilleja, 720
Castilleja irazuensis, 512
Castilleja quirosii, 510
Casuarina equisetifolia, 478
Catagonus wagneri, 395
Cathartes aura, 266
Cathartidae, 555
Catharus
Catharus frantzii, 466
Catharus fuscater, 476
Catharus gracilirostris, 511
Catharus ustulatus, 476
Catopsis, 275
Cattleya, 275
Caulerpa sertularioides, 113
Cavendishia, 457, 461, 469
Cavendishia atroviolacea, 462
Cavendishia bracteata, 462
Cavendishia talamancensis, 461
Cayaponia racemosa, 338
Ceanothus caeruleus, 513
Cebus
Cebus capucinus, 264, 399, 466, 471, 473, 559
Cebus capucinus ssp. *imitator*, 263
Cecropia, 381, 387, 440, 461, 473, 667–68, 697, 715, 719
Cecropia obtusifolia, 339
Cecropia peltata, 353
Cecropia pittieri, 215, 217, 228, 230
Cecropia polyphlebia, 440, 442
Cedrela
Cedrela odorata, 86, 273, 276, 350, 351, 482
Cedrela salvadorensis, 276, 351
Cedrela tonduzii, 374
Ceiba pentandra, 262, 265, 272, 339, 353, 374, 402, 540, 715
Celastraceae, 468
Centris, 271, 713
Centris aethyctera, 271
Centris bicornuta, 271
Centris flavifrons, 271
Centris flavofasciata, 271
Centrolenidae, 551, 628
Centropogon, 462, 469
Centropogon solanifolius, 438
Centropomidae, 268, 546, 547
Centropomus, 389
Centropomus nigrescens, 389
Centropomus parallelus, 547
Centropomus pectinatus, 547, 595
Centropomus undecimalis, 547, 640
Cephalochordates, 153
Cephaloleia, 560
Cephaloleia belti, 560
Cephaloleia fenestrata, 560
Cephalotes
Cephalotes minutus, 387
Cephalotes setulifer, 387
Cephalotes umbraculatus, 387
Cephaloziella subtilis, 256
Cerambycidae, 231, 269, 273
Cerapachys neotropicus, 388
Ceratium
Ceratium dens, 154
Ceratium furca, 146
Ceratium fusca, 154
Ceratium fusus, 154
Ceratolejeunea, 463
Ceratopteris richardii, 689
Ceriodaphnia cornuta, 666, 667, 675, 676–77
Cerrophidion godmani, 465
Cervidae, 5
Ceryle, 630
Cestrum, 461
Chaetocalyx, 256
Chaetoceros, 146, 147
Chaetognatha, 119, 147, 148, 592
Chamaedorea, 349, 461, 469, 719
Chamaepetes unicolor, 466
Chamaepinnularia, 624, 721
Characidae, 268, 269, 389, 546, 547, 548, 558, 627, 628, 692, 712, 715, 722
Charadriidae, 555
Charidotella tuberculata, 269
Chaunochiton, 374
Cheilanthes brachypus, 275
Chelonia
Chelonia agassizii, 106, 220
Chelonia mydas, 566, 591, 600, *601*, 709
Chelonia mydas ssp. *agassizii*, 268
Chelydra, 552
Chelydra serpentina, 630
Chelydridae, 553, 630, 722
Chichlidae, *645*
Chilopoda, 463
Chiococca semipilosa, 256
Chione costaricensis, 699
Chironectes minimus, 277
Chironius, 552
Chironius exoletus, 435
Chironomidae, 634, 635, 639
Chiroptera, 265, 272, 511, 555
Chiroxiphia linearis, 266, 435
Chloranthaceae, 214, 468
Chloris
Chloris paniculata, 218, *219*
Chloroceryle, 630
Chlorophyta, 119, *147*, 509, 624, 690
Chlorospingus pileatus, 511
Chlorothraupis carmioli, 554, 561
Choloepus hoffmanni, 277, 467, 482, 558
Chordata, 119
Choreutidae, 231
Chroborus, 667
Chromacris colorata, 339
Chroomonas, 666
Chrysobalanus icaco, 353
Chrysochromulina polylepis, 155
Chrysoclamys allenii, 429

Chrysomelidae, 269, 560
Chrysophyceae, 666, 676
Chrysophyta, 199, 509
Chthamalus, 106
  Chthamalus fissus, 107
Chusquea, *12*, 460, 461, 467, 468, 469, 473, 512, 718
  Chusquea subtessellata, 500, *501*, 512, 513, 514, 685, 719, 720, 721
  Chusquea talamancensis, 81, 461, 469
  Chusquea tomentosa, 81, 469
  Chusquea tonduzii, 512
Cicadellidae, 231
Cicadidae, 269, 559
Cichlasoma, 550, 693
  Cichlasoma alfari, 558
  Cichlasoma nicaraguense, 558
  Cichlasoma nigrofasciatum, 558
  Cichlasoma septemfasciatum, 558
  Cichlasoma tuba, 558
Cichlidae, 268, 269, 389, 546, 547, 548, 558, 627, 628, 648, 712, 715, 722
Ciconiidae, 555
Cimaria vargasi, 153
Cinnamomum, 457, 461, 469, 471
  Cinnamomum cinnamomifolium, 350, 351
Cipura campanulata, 255
Cirolana salvadorensis, 145, 600
Cirripedia, *147*
Cirsium, 508
Cissus, 469
Citharexylum, 471
Citharichthys
  Citharichthys gilberti, 389
  Citharichthys spilopterus, 547
  Citharichthys uhleri, 547
Citrus, 335
Cittarium pica, 594, 595, *601*
Cladina, 513, 721
  Cladina confusa, 686
Cladocora
  Cladocora debilis, 179
  Cladocora pacifica, 179
Cladonia, 459, 460, 509, 513
  Cladonia corymbosula, 512
Clarisia racemosa, 350
Clelia, 552
Clematis, 461
Clethra, 514, 718
  Clethra gelida, 469
Cleyera, 457, 471, 718
  Cleyera theaeoides, 472
Clupeidae, 547
Clusia, 429, 432, 462, 469, 718
  Clusia rosea, 213, 215, 216, 217, *219*, 353, 713
Clusiaceae, 399, 468, 499
Clytostoma pterocalyx, 349
Cnemidaria choricarpa, 348
Cnidaria, 119
Coccoidea, 77
Coccoloba caracasana, 260
Cocconeis feuerbornii, 624, 626
Coccyzus ferrugineus, 222, *223*
Cochlodinium
  Cochlodinium catenatum, 146, 154, 711
  Cochlodinium catenatum cf. *polykrikoides*, 154
Cochlopinia tryoniana, 635
Cochlospermum vitifolium, 254, 255, 338, 350
Cocodrylus acutus, 692
Cocos nucifera, 214, 273, 377, *714*
Cocytius anteus, 230
Coelosphaerium, 667
Coendou mexicanus, 79, 558
Coenobita compressus, 232–33
Coenobitidae, 232
Coenocyathus bowersi, 179
Coerebidae, 556
Coffea, 377
Coleoptera, 86, 231, 269–70, 273, 463, 511, 559
Colibri thalassinus, 469
Collema, 440
Collembola, 77
Coloptychon rhombifer, *391*, 392
Coltricia, 458
Colubridae, 553
Colubrina arborescens, 256
Columba
  Columba fasciata, 466
  Columba livia, 221
Columbidae, 555
Comarostaphylis, 458, 467, 500, 508, 511, 512
  Comarostaphylis arbutoides, 468, *501*, 512, 513–14, 719
Combretum decandrum, 259
Commelinaceae, 469
Compositae, 509
  See also Asteraceae
Conepatus semistriatus, 467
Conocarpus erectus, 109, 215, 377, 599, 710
Conostegia, 457, 471
  Conostegia bigibbosa, 461
Convallariaceae, 462
Convolvulaceae, 230, 254, 257, 273
Cookeina
  Cookeina speciosa, 257
  Cookeina tricholoma, 257
Copaifera
  Copaifera aromatica, 262, 353
  Copaifera camibar, 374, 482
Copepoda, 144, 592, 710
Copiocerinae, 559
Coprinus, 458
Coptocycla
  Coptocycla dorsoplagiata, 269
  Coptocycla rufonotata, 269
Coragyps atratus, 266
Corallus, 552
  Corallus ruschenbergeri, 392
Corapipo
  Corapipo altera, 561
  Corapipo leucorrhoa, 561
Cordia, 257, 461
  Cordia alliodora, 254, 387, 543
  Cordia eriostigma, 351
  Cordia gerascanthus, 255, 276
Coriaria, 512
Coricuma nicoyensis, 110
Coriolaceae, 354
Coriolopsis, 458
  Coriolopsis byrsina, 354
  Coriolopsis floccosa, 354
  Coriolopsis polyzoma, 354
Cornus, 471
  Cornus disciflora, 469, 471, 472, 482
Corollia perspicillata, 338
Cortaderia, 512
  Cortaderia bifida, 513
  Cortaderia nitida, 514, 685, 721
Cortinarius, 458
Corvidae, 556
Corytophanidae, 553
Cosmarium sphaerosporum, 666
Cosmibuena valerii, 429, 432, 442
Cotingidae, 556
Couratari, 374
  Couratari guianensis, 349, 374, 482
Couropita nicaraguensis, 260
Coursetia elliptica, 255
Cracidae, 555
Crassostrea prismatica, 151
Crataeva tapia, 377, 695
Craugastor, 551
  Craugastor ranoides, 551
Crax rubra, 266, 355, 393, 466
Crematogaster
  Crematogaster cruces, 388
  Crematogaster erecta, 388
  Crematogaster evallans, 388
  Crematogaster longispina, 387
  Crematogaster rochai, 387
  Crematogaster tenuicula, 388
Crematosperma, 374
Crepidotus, 458
Crescentia alata, 256, 274, 275, *304*, *315*, 335, 338
Cricetidae, 558
Criconemella palustris, 76
Crinum erubescens, 696
Crocodylidae, 553
Crocodylus, 634
  Crocodylus acutus, 268, 354, *356*, 552, 630, 648, 712, 722
Crossopetalum tonduzii, 351
Crotalus durissus, 268
Croton, 461, 471
  Croton axillaris, 257
  Croton draco, 260, 350, 352, 712
  Croton xalapensis, 472
Crotophaga ani, 222
Crudia acuminata, 695
Crustacea, 100, 144, 166, 271, 389
Cruziohyla calcarifer, 551
Cryptoheros, 550
  Cryptoheros sajica, 389

Cryptonemiales, 52
Cryptothallus hirsutus, 509
Cryptotis, 511
  Cryptotis gracilis, 482
  Cryptotis nigrescens, 354
Crypturellus
  Crypturellus cinnamomeus, 276
  Crypturellus soui, 356
Crysophila, 260
Crysophila guagara, 349, 377, 696
Crysopidae, 231
Crytochrisis minor, 666
Ctenophora, 119
Ctenosaura similis, 268
Ctenotus cristatellus, 552
Cuculidae, 555
Culcita, 719
Culicia stellata, 179
Cuniculus paca, 562
Cupressus lusitanica, 478
Curatella americana, 254, 255, 350, 377
Curculionidae, 86, 228–29, 269, 273, 329, 559
Cuvieronius, 53
  Cuvieronius hyodon, 50
Cyanerpes lucidus, 554
Cyanobacteria, 119, 624, 720–21
Cyanocompsa pallerina, 222, 223
Cyanocorax morio, 435, 719
Cyanolyca argentigula, 466, 472
Cyathea, 13, 210, 215, 719
  Cyathea alfonsiana, 210, 213, 214, 713
  Cyathea nesiotica, 210, 216
  Cyathea notabilis, 210, 211, 213, 713
Cyatheaceae, 52, 377, 457, 461, 468, 459
Cyathus striatus, 257
Cybianthus, 461
Cyclanthaceae, 462, 469
Cyclanthus bipartitus, 696
Cyclomyces, 458
  Cyclomyces iodinus, 354
Cyclopeltis semicordata, 275, 348
Cyclopes didactylus, 265
Cydista
  Cydista diversifolia, 257
  Cydista lilacina, 349
Cylindrospermum, 660, 721
Cylindrotheca closterium, 146
Cynometra hemitomophylla, 353
Cynoscion, 154
  Cynoscion squamipinnis, 155
Cyperaceae, 254, 462, 482, 510
Cyperus, 257, 721
  Cyperus digitatus, 690
  Cyperus giganteus, 687, 690
  Cyperus holoschoenoides, 689
  Cyperus imbricatus, 689
  Cyperus papyrus, 687, 689
Cyphomyrmex cornutus, 387
Cyprideis pacifica, 110

Dactylococcopsis, 675
Daedalea, 458
Dahlia imperialis, 351
Dalbergia retusa, 256, 257, 276, 332, 353, 482
Daltania, 510
Danaea nodosa, 216
Danaus plexippus, 327
Daphnia laevis, 667
Dasyprocta punctata, 79–80, 265, 274, 329, 467
Dasyproctidae, 79
Dasypus novemcinctus, 277, 339, 354, 467
Decapoda, 118, 389
Declieuxia fruticosa, 255
Delonix regia, 273
Delphinidae, 116
Dendrobates
  Dendrobates auratus, 392
  Dendrobates granuliferus, 391
  Dendrobates pumilio, 551
Dendrobatidae, 551
Dendrocincla
  Dendrocincla anabatina, 394
  Dendrocincla fuliginosa, 554
Dendrocolaptes sanctithomae, 394, 554
Dendrocygna autumnalis, 266, 691, 692, 722
Dendroica
  Dendroica pensylvanica, 394
  Dendroica petechia, 394
  Dendroica petechia ssp. *aureola*, 222, 224
Dendropanax, 461, 473
  Dendropanax arboreus, 348
  Dendropanax querceti, 469
  Dendropanax ravenii, 374, 461
Dendrophyllia oldroydae, 179
Dendropogon, 719
Dendropogonella, 463
  Dendropogonella rufescens, 469
Dermanura
  Dermanura phaeotis, 557
  Dermanura watsoni, 557
Dermaptera, 463
Dermocarpa, 624, 720
Dermochelys coriacea, 106, 220, 268, 600, 709
Desmodium, 257
Desmodus rotundus, 264
Desmophyllum dianthus, 179
Diadema
  Diadema antillarum, 593, 594–95, 602, 605, 607, 609
  Diadema mexicanum, 113, 169, 171, 176
Dialium guianensis, 696
Dialyanthera otoba, 698
Diaphanosoma spinulosum, 675
Diasporus ventrimaculatus, 686, 721
Dichapetalum hammelii, 461
Dichogaster, 79
Dicksonia, 719
Dicranaceae, 463
Dicranopteris
  Dicranopteris flexuosa, 217
  Dicranopteris pectinata, 217
Dictyosphaerium ehrenbergianum, 667
Didelphimorphia, 265, 354
Didelphis
  Didelphis marsupialis, 354
  Didelphis virginiana, 262, 339
Didymopanax pittieri, 432, 442
Dieffenbachia, 722
  Dieffenbachia davidsei, 696
  Dieffenbachia longispatha, 636
Digenea, 275, 634
Diglossa plumbea, 511
Dilleniaceae, 269
Dilodendron costaricense, 254, 712
Dinophyta, 119, 509
Dioscorea, 461, 469
Diospyros salicifolia, 256
Diphysa
  Diphysa americana, 351
  Diphysa humilis, 255
Diplasiolejeunea, 463
Diplazium turubalense, 348
Diplopoda, 77, 463
Diploria strigosa, 603
Diplostephium, 500, 512
Diplura, 463
Dipsas, 552
Diptera, 231, 232, 275, 463, 511, 559, 634, 635, 639
Dipteryx panamensis, 540, 567, 568, 715
Disocactus amazonicus, 275
Disterigma, 461, 512
  Disterigma humboldtii, 462, 514
Dolichoderus
  Dolichoderus curvilobus, 387
  Dolichoderus lamellosus, 387
  Dolichoderus laminatus, 387
  Dolichoderus validus, 387
Donax
  Donax denticulatus, 600
  Donax panamensis, 145
Dormitator
  Dormitator latifrons, 269, 389
  Dormitator maculatus, 547
Dorosoma chavesi, 547
Doryfera ludoviciae, 469
Dorymyrmex, 387
Draba, 500, 512, 720
  Draba jorullensis, 510
Drimys, 52, 458, 718
  Drimys granadensis, 469, 475, 514
Drymarchon, 552
Drymobius melanotropis, 434
Dryopteris, 458
Dysithammus striaticeps, 554
Dysodia speculifera, 323

Echeverria australis, 429
Echimyidae, 79
Echiniscus
  Echiniscus arctomys, 511
  Echiniscus bigranulatus, 511
Echinochloa polystachya, 689

Echinodorus
Echinodorus paniculatus, 690, 692, 721
Echinodorus subalatus ssp. *andrieuxii*, 256, 689
Echinometra lucunter, 592, 593
Echinopepon paniculatus, 257
Echinothrix diadema, 165
Eciton, 327, 328
Eciton burchelli, 321, 394
Ecitoninae, 433
Ectatomma
Ectatomma edentatum, 388
Ectatomma ruidum, 561
Ectatommini, 561
Ectophylla alba, 557
Egretta alba, 692, 692, 722
Ehretia latifolia, 351
Eichhornia, 721
Eichhornia azurea, 623
Eichhornia crassipes, 257, *258*, 349, *349*, 377, 623, 690
Eichhornia heterosperma, 623, 690
Eira barbara, 265, 467, 557
Elaeis
Elaeis guianensis, 377
Elaeis oleifera, 377, 696
Elanoides forficatus, 466
Elaphoglossum, 347, 462, 468, 510, 686, 718
Elaphoglossum adrianae, 461
Elaphoglossum reptans, 214
Elaphoglossum squamipes, 462
Elapidae, 553
Eleocharis, 721
Eleocharis acicularis, 514
Eleocharis interstincta, 690
Eleocharis mutata, 257, 690
Eleotridae, 269, 389, 546, 547, 548, 634
Eleotris, 220
Eleotris amblyopsis, 547
Eleotris picta, 220, 389
Eleotris pisonis, 547
Eleotris tubularis, 220
Eleutherodactylidae, 551, 686
Eleutherodactylus, 356, 434, 464, 719
Eleutherodactylus melanostictus, 464
Eleutherodactylus taurus, 392
Eleutherodactylus vocator, 391
Elleanthus, 275, 462, 718
Elodea canadensis, 668
Elopidae, 546, 547
Emberizidae, 196, 466, 556, 719
Embydidae, 630, 722
Emerita rathbunae, 145
Emydidae, 553
Encyclia, 275
Encyclia cordigera, 275
Encyonema
Encyonema lunatum, 674
Encyonema silesiacum, 677
Endopachys grayi, 179
Engraulis, 154
Entada gigas, 213
Enterolobium
Enterolobium cyclocarpium, 81, 259, 261, 265, 272, 274, 319, 333, 335, 338, 348, 712, 713
Enterolobium schomburkgii, 319
Enyo ocypete, 325
Eois, 560
Ephemeroptera, 633, 634, 635
Epicharis, 271, 713
Epicrates, 552
Epidendrum, 275, 347, 462, 718
Epidendrum cocoense, 211, 213
Epidendrum insularum, 211, 213
Epidendrum jimenezii, 211
Epidendrum platystigma, 462
Epimecis, 560
Epinannolene pittieri, 235
Epinecrophyla fulviventris, 554
Epiperipatus biolleyi, 78
Equisetum aff. *giganteum*, 52
Equus, 53
Equus conversidens, 51
Eragrostis hypnoides, 689
Erebidae, 324
Eremolepidaceae, 462
Eremotherium, 50
Erethizontidae, 79
Eretmochelys imbricata, 106, 152, 220, 600, 709
Ericaceae, 52, 432, 458, 460, 461, 462, 467, 468, 469, 473, 474, 482, 492, 499, 510, 719
Erinnyis obscura, 230
Erioderma, 509
Eryngium scaposum, 512
Erythrina, 75–76, 78
Erythrina costaricensis, 351
Erythrina fusca, 214
Erythrina lanceolata, 696, 697
Erythrina poeppigiana, 85, 352
Erythrochiton gymnanthus, 349
Erythrolamprus, 552
Erythroxylum havanense, 257, 259, 260
Escallonia, 458, 467, 500, 512, 514, 718
Escallonia myrtilioides, 474, *501*, 685, 721
Escalloniaceae, 499
Eschweilera calyculata, 697
Esenbeckia berlandieri, 256
Eucalanus
Eucalanus attenuatus, 147
Eucalanus elongatus, 147
Eucalyptus, 478
Euchlanis dilatata, 675
Eucometis penicillata, 394
Eucynorta insularis, 231
Eudocimus albus, 267
Eugenia, 213, 438, 469, 719
Eugenia basilaris, 461
Eugenia cocosensis, 213, 217, 713
Eugenia hiraefolia, 353
Eugenia salamensis, 482
Eugerres plumieri, 547
Euglenophyta, 690
Euglossinae, 271
Eulepidotis, 232
Eumomota superciliosa, 266
Eumycota, 509
Eunica monima, 270, 325
Eunotia, 624, 669, 721
Eunotia guyanense, 670
Eunotia intermedia, 668
Eunotia minor, 668
Euphausicea, 592
Euphonia gouldi, 561
Euphorbiaceae, 254, 347, 348, 457, 461, 468, 539
Euphorbia schlechtendalii, 255, 260, 429
Eurema daira, 270, 325
Eurypyga helis, 630
Eurypygidae, 555
Eurytides, 270
Eurytides epidaus, 270
Eurytides philolaus, 270
Euscirrhopterus poeyi, 323
Eutelia furcata, 323
Euterpe, 696
Euterpe precatoria, 699
Euterpe precatoria var. *longevaginata*, 211, 213, 713
Eutetramorus tetrasporus, 666, 672
Everniastrum, 509
Evolvulus alsinoides, 255
Excirolana braziliensis, 600

Fabaceae, 254, 257, 259, 270, 276, 300, 328, 347, 348, 349, 350, 374, 473, 539, 559, 696, 712
See also Leguminosae
Fagaceae, 461
Falconidae, 555
Falco peregrinus, 355
Faramea, 349, 457
Farfantepenaeus, 154
Farfantepenaeus brasiliensis, 594
Farfantepenaeus brevirostris, 153
Farfantepenaeus californiense, 153
Felis
Felis onca, 398, 715
Felis pardalis, 686
Felis silvestris, 224, 226
Festuca, 500, 510, 512
Festuca dolichophylla, 512
Ficus, 230, 254, 347, 374, 397, 402, 461, 471, 561, 712, 718, 722
Ficus costaricana, 351, 352, 712
Ficus crassiuscula, 431, 432, 718
Ficus glabrata, 561
Ficus insipida, 349, 561, 635–36
Ficus jimenezi, 350, 351
Ficus obtusifolia, 350
Ficus padifolia, 52
Ficus pertusa, 213, 230, 438, 439, 713
Ficus talamanca, 52
Ficus tuerckheimii, 431
Ficus velutina, 351
Fidicina mannifera, 559

Fimbristylis spadicea, 687
Fissidens, 463
  Fissidens juruensis var. *juruensis*, 256
  Fissidens radicans, 256
  Fissidens yucatanensis, 256
Fistulina, 458
  Fistulina hepatica, 458
Flacourtiaceae, 348
Fomes, 458
  Fomes fasciatus, 354
Fomitopsis
  Fomitopsis cupreorosea, 354
  Fomitopsis feei, 354
Foraminifera, 61, 119, 144, 710
Formicariidae, 555
Formicarius analis, 394
Formicidae, 228, 271, 274
Fragilaria
  Fragilaria exigua, 677
  Fragilaria tenera, 676, 677
Frankia, 76
  Frankia acuminata, 75, 76, *76*
Fregata minor, 170, 222
Freziera, 471
  Freziera calophylla, 214
  Freziera candicans, 350, 471, 472
  Freziera grisebachi, 377
Fringilidae, 556
Frullania, 463, 510, 685, 719
Frustulia, 624, 721
  Frustulia saxonica, 669, 670
Fuchsia, 458, 469, 471
  Fuchsia paniculata, 471, 472
Fuirena umbellata, 689
Fulgora laternaria, 559
Fulgoridae, 559
Fungi, 119
Fungia
  Fungia curvata, 113, *179*
  Fungia distorta, *179*
Furcraea cabuya, 429
Furipterus horrens, 557
Furnariidae, 555
Fuscocerrena, 458

Gaiadendron, 458
  Gaiadendron punctatum, 462
Galbulidae, 555
Galerina, 458
Galictis vittata, 277
Gallus gallus, 196
Gambusia nicaraguensis, 548
Ganoderma, 458
Garcinia intermedia, 256, 351, 353
Gardineroseris planulata, 113, 175, *179*
Garrya, 508, 514
  Garrya laurifolia, *501*
Gaultheria, 458, 461, 512
Geastrum
  Geastrum javanicum, 257
  Geastrum saccatum, 257
Gecarcinidae, 79, 232, 385, 715
Gecarcinus quadratus, 79, *234*, 373, 385, 394
Gekkonidae, 553
Gentiana, 508
Gentianaceae, 52, 216
Geometridae, 231
Geomyidae, 79
Geonoma, 260, 461, 469, 719
  Geonoma hoffmanniana, 469
  Geonoma surtuba, 431
Geranium, 458
Gerreidae, 547
Gesneriaceae, 462, 469
Gigaspora, 440
Gliricidia sepium, 255
Gloeocapsa, 666
Glomus, 440
Glossophaga, 329
  Glossophaga leachii, 265
  Glossophaga soricina, 265, 272, 338
Glossotherium aff. *tropicorum*, 50
Glottidia audebarti, 145
Glycine max, 75
Glyphorynchus spirurus, 392–93
Glyptotherium
  Glyptotherium aff. *texanum*, 50
  Glyptotherium cf. *arizonae*, 50
Glyricidia sepium, 85
Gmelina arborea, 73, 334–35, 377, 402, 543
Gnaphalium, 476, 720
  Gnaphalium americanum, 512
Gobiesocidae, 546, 548
Gobiesox
  Gobiesox fulvus, 220
  Gobiesox nudus, 546, 548
Gobiidae, 389, 546, 547, 548
Gobiomorus
  Gobiomorus dormitor, 546, 547, 548
  Gobiomorus maculatus, 269, 389
  Gobiomorus polylepis, 268
Goethalsia meiantha, 353
Gomphichis adnata, 686
Gomphonema
  Gomphonema gracile, 666, 672
  Gomphonema parvulum, 626
Gomphrena, 256
Goniolithon, 52
Gonocalyx costaricensis, 432
Gonodonta pyrgo, 511
Gonyaulax digitale, 146
Gordonia frutiscosa, 350, 377
Gorgonia, 605
  Gorgonia flabellum, 593
Grallariidae, 556
Grammadenia, 461
  Grammadenia myricoides, 81
Grammitis, 52, 432, 462
Grandiarca grandis, 151, *151*, 152, 153
Graphidales, 509
Grapsus grapsus, 174, 233
Grias
  Grias cauliflora, 699
  Grias fendleri, 377, 695, 696, 697, 698
Guadua paniculata, 256
Guaiacum sanctum, 259, 260, 276
Guarea, 432, 473, 711, 718, 719
  Guarea glabra, 272
  Guarea kunthiana, 432, 438
  Guarea tonduzii, 469
Guatteria, 471
  Guatteria amphifolia, 695
  Guatteria oliviformis, 469
Guazuma, 294
  Guazuma ulmifolia, 254, 261, 303, 334, 338, 350, 712
Guettarda, 471
  Guettarda macrosperma, 353
  Guettarda poasana, 438, 442
Gunnera, 440, 718
  Gunnera insignis, 442
Guzmania sanguinea, 213, *216*
Gygis alba, 222, 226
Gymnocichla nudiceps, 554
Gymnophthalmidae, 553
Gymnopithys leucaspis, 394, 554
Gymnopus, 458
Gymnospermae, 461
Gymnostomiella vernicosa, 256
Gymnotidae, 268, 547, 548, 692
Gymnotus
  Gymnotus cylindricus, 547, 548
  Gymnotus maculosus, 268, 547
Gynerium sagittatum, 687
Gyrosigma spenceri, 677

Habia
  Habia atrimaxillaris, 392, 715–16
  Habia fuscicauda, 554
Habronematoidea, 276
Haematoloechidae, 275
Haematoloechus meridionalis, 275
Haematoxylum brasiletto, 255, 296
Haemulidae, 546, 547, 548
Halictinae, 271
Halipegus eschi, 275
Halodule wrightii, 602
Halophila
  Halophila baillonii, 111
  Halophila decipiens, 602
Hampea appendiculata, 442
Harpella tica, 634
Harpia harpyia, 394, 716
Hauya elegans, 351, 352
Hebeloma, 257
Hedyosmum, 52
  Hedyosmum brenesii, 377
  Hedyosmum racemosum, 214
Heisteria longipes, 374, 715
Helianthemum, 508
Helicarionidae, 511
Heliconaceae, 396
Heliconia, 52, 557
Heliconiinae, *383*
Heliconius, 379, 384, *384*, 715
  Heliconius cydno, *383*
  Heliconius hewitsoni, *383*, 386
  Heliconius pachinus, *383*, 386

Heliconius sapho, 383
Heliconius sara, 386
Heliocarpus appendiculatus, 442
Heliornis fulica, 630
Heliornithidae, 555
Hemicyclops thalassius, 147
Hemidactylus
Hemidactylus frenatus, 552
Hemidactylus garnotti, 552
Hemileucinae, 270
Hemiptera, 231, 269
Hemiuridae, 275
Henriettella fascicularis, 213, 713
Hepaticae, 211, 431
Herbertus, 463
Herichthys, 550
Hernandia didymantha, 697
Heros, 550
Herotilapia multispinosa, 269, 547, 548
Herpetotheres cachinnans, 339
Hesperiidae, 270, 275
Hesperomeles, 514
Hesperomeles heterophylla, 512, 685, 721
Hesperomeles obtusifolia, 514
Heterocarpus vicarius, 149, 153–54
Heterodermia, 459, 460, 509
Heterokontophyta, 509
Heteromyidae, 79, 80, 467, 719
Heteromys, 467
Heteromys desmarestianus, 466
Heteromys oresterus, 482
Heteropoda, 592
Heteropogon contortus, 256
Heteroscyphus, 463
Heterospingus rubifrons, 553
Hexagonia glaber, 257
Hexaplex radix, 153
Hexarthra intermedia, 675
Hibiscus pernambucensis, 353
Hidrolea elatior, 257
Hieracium, 508, 721
Hieracium irasuense, 514, 686
Hieronyma
Hieronyma alchorneoides, 543, 715
Hieronyma oblonga, 348
Hillia, 462
Hillia loranthoides, 377
Hippolyte obliquimanus, 594
Hirundinidae, 556
Hispinae, 559
Historis
Historis acheronta, 325
Historis odius, 227–28, 230
Hoffmannia, 349, 457, 461
Hoffmannia piratarum, 216
Holochroa ochra, 323
Holomitrium, 463
Homalium racemosum, 695
Homoptera, 231, 269, 559
Homo sapiens, 335
Hortensia similis, 511
Huberodendron, 374
Huperzia, 462, 510
Huperzia branchiata, 211, 213
Huperzia pittieri, 213
Hura, 294
Hura crepitans, 348
Hyalella azteca, 666
Hydrangea, 461, 469, 719
Hydrangea peruviana, 432
Hydrilla, 696
Hydrocharitaceae, 668
Hydrocotyle, 476
Hydrocotyle umbellata, 214
Hydrolea spinosa, 256, 689
Hydrozoa, 172
Hygrocybe, 458
Hygropus, 458
Hyla
Hyla calypsa, 634
Hyla picadoi, 464
Hyla rosenbergi, 391, 392
Hylesia lineata, 270
Hylidae, 551, 628
Hylocereus costaricensis, 275, 712
Hylocichla mustelina, 394, 561
Hyloctistes subulatus, 393
Hylophilus
Hylophilus decurtatus, 554
Hylophilus ochraceiceps, 393
Hylophylax nevioides, 554
Hymenachne, 377
Hymenachne amplexicaulis, 687, 689
Hymenaea, 711
Hymenaea courbaril, 254, 273–74, 328–29, 330, 333, 353, 712
Hymenocalis litoralis, 696
Hymenochne amplexicaulis, 623
Hymenolobium mesoamericanum, 540, 715
Hymenophyllum, 52, 462, 718
Hymenoptera, 228, 271, 275, 463, 511, 559
Hyparrhenia rufa, 255, 276, 300, 306, 306, 335, 713
Hyperbaena
Hyperbaena eladioana, 350
Hyperbaena tonduzii, 353
Hypericum, 52, 458, 461, 500, 508, 512
Hypericum cardonae, 513
Hypericum costaricense, 512
Hypericum irazuense, 512, 514
Hypericum stenopetalum, 512
Hypericum strictum, 474, 512, 514, 686, 721
Hyphessobrycon
Hyphessobrycon savagei, 389
Hyphessobrycon tortuguerae, 547
Hypholoma, 458
Hypnum, 463
Hypochoeris, 512
Hypolytrum amplissimum, 213
Hypopomus occidentalis, 548
Hypotrachyna, 458, 459, 460, 509
Hypsibius
Hypsibius convergens, 511
Hypsibius scoticus, 511
Hypsophrys nicaraguense, 548
Hyptis, 461

Icacinaceae, 292
Icteridae, 196, 556
Icterus
Icterus mesomelas, 196
Icterus pectoralis, 196, 221, 222, 265
Iguana, 552
Iguana iguana, 268, 339, 558, 712
Iguanidae, 553
Ilex, 52, 81, 471, 719
Ilex lamprophylla, 468, 469
Ilex pallida, 469, 471, 472, 482
Ilex vulcanicola, 482
Iltisia, 509, 720
Imantodes, 552
Impatiens sultana, 438
Indigofera, 300
Inga, 347, 440, 461, 534, 539, 695, 715, 718
Inga bella, 374
Inga edulis, 352
Inga jimenezii, 349
Inga punctata, 351, 352
Inga sheroliensis, 52
Inga vera, 327, 697
Inocybe, 458
Insectivora, 354, 511
Ipomoea, 215, 217, 257, 687, 721
Ipomoea alba, 230
Ipomoea batatas, 86
Ipomoea digitata, 338
Ipomoea indica, 230
Ipomoea pes-caprae, 215, 230
Ipomoea philomega, 230
Iriartea
Iriartea deltoidea, 562, 697, 699
Iriartea gigantea, 374, 715
Iryanthera, 374
Ischnochiton dispar, 107
Ischnocodia annulus, 269
Isoetes, 513, 721
Isoetes panamensis, 256
Isoetes storkii, 672, 686, 721
Isopoda, 463
Itaballia demophile, 270
Ixora
Ixora finlaysoniana, 697
Ixora nicaraguensis, 695

Jabiru mycteria, 266, 691, 692, 722
Jacana spinosa, 276, 354, 630
Jacanidae, 555
Jacquemontia mexicana, 255
Jacquinia nervosa, 256, 257
Jamesonia, 52
Jatropha costaricensis, 255
Javania cailleti, 179
Jessea, 509, 720
Johngarthia cocoensis, 232
Joturus pichardi, 548, 640, 642
Juglans olanchana, 52
Juncaceae, 482, 686

Junco vulcani, 511
Juncus, 686, 721
Jussiaea, 687
Jussiaea latifolia, 696
Justicia comata, 689

Karatophyllum bromeliodes, 52
Karwinskia calderoni, 255
Keratella, 667
Keratella americana, 675
Killinga nudiceps, 211
Kinorhyncha, 119, 144
Kinosternidae, 553, 630, 722
Kinosternon, 552
Kinosternon leucostomum, 630
Kinosternon scorpioides, 268
Kirchneriella, 672
Kiwa puravida, 118
Kogiidae, 116
Kohleria spicata, 216
Krameria ixine, 255
Kricogonia lyside, 270
Kronitta, 148

Laccaria, 458
Lachemilla, 510, 513
Lachesis, 552
Lachesis melanocephala, 357, 391, *391*, 392, 712
Lacistema aggregatum, 260, 261
Lactarius, 458
Ladenbergia
Ladenbergia brenesii, 469
Ladenbergia sericophylla, 377
Laelia, 275
Laelia rubescens, 275
Laestadia, 509, 720
Laetiporus, 458
Lagomorpha, 354, 511
Laguncularia racemosa, 109, 143, 353, 377, 599, 710
Lamellibrachia, 117
Lampropeltis, 552
Lampyridae, 269, 313
Lanio leucothorax, 393, 554
Laportea aestuans, 216
Lasciacis, 721
Lasciacis procerruna, 687
Laterallus ambigularis, 222
Laubierus alivini, 118
Lauraceae, 52, 348, 350, 352, 387, 438, 439, 457, 460, 461, 467, 468, 471, 553, 719
Lecane, 675
Lecanorales, 509
Leccinum, 458
Lecythis ampla, 715
Legatus leucophaius, 553
Leguminosae, 273, 274, 473
See also Fabaceae
Leiuperidae, 551
Lejeunea, 463
Lejeuneaceae, 211, 463
Lemna, 349
Lennoa madreporoides, 256
Lentibulariaceae, 668
Leopardus, 511
Leopardus pardalis, 262, 467, 511, 557
Leopardus tigrinus, 482
Leopardus wiedii, 262, 467
Lepanthes, 462, 718
Lepidoblepharis, 552
Lepidobotryaceae, 374
Lepidochelys olivacea, 106, 220, 268, 709
Lepidocolaptes affinis, 466
Lepidoptera, 6, 229–30, 231, 270, 275, 511, 557, 559–60
Lepidozia, 463
Lepidoziaceae, 463
Lepisosteidae, 547, 692
Leptobasis guanacaste, 633
Leptodactylidae, 551
Leptodeira, 552
Leptodeira rubricata, 392
Leptodesmus folium, 235
Leptodontium, 463, 510
Leptogium, 440, 460, 509
Leptophis, 552
Leptoscyphus, 510
Leptoseris
Leptoseris papyracea, 113, *179*
Leptoseris scabra, 165, 175, *179*
Letis, 325
Leucauge argyra, 232
Leucobryum, 463
Lewisia, 512, 720
Lewisia megarhiza, 510
Liabum, 462
Libellula, 666
Licania
Licania arborea, 254, 260, 712
Licania operculipetala, 353
Ligia, 174
Limacodidae, 323
Limnephilus, 666
Limnobium laevigatum, 349, 690
Lingula, 145
Liomys salvini, 262, 274
Lipaugus unirufus, 393
Lithophorella, 52
Lithothamnion, 52
Lithothamnion corallioides, 114
Littorina modesta, 174
Littorinidae, 106
Lobaria, 458, 459
Lobaria pulmonaria, 459
Lobariella pallida, 686
Loganiaceae, 260
Lonchocarpus, 257
Lonchocarpus hughesii, 259
Lonchocarpus minimiflorus, 257, 482
Longidorus, 76
Lontra
Lontra longicaudis, 482, 557, 630
Lontra provocax, 277
See also *Lutra*
Lophiodes spilurus, 149
Lophocolea, 463
Lophosoria, 719
Loranthaceae, 461, 462, 469
Loricariidae, 391, 546, 628
Lozania mutisiana, 469
Ludwigia, 377, 696
Ludwigia inclinata, 690
Ludwigia pepliodes, 689, 690
Ludwigia sedioides, 689
Luehea, 254, 712
Luehea candida, 254, 712
Luehea seemannii, 349, 377, 695, 696, 699
Luehea speciosa, 254
Lupinus, 500
Luticola mutica, 669
Lutjanidae, 148, 711
Lutjanus, 154, 389
Lutjanus argentiventris, 389
Lutjanus colorado, 389
Lutjanus novemfasciatus, 389, *390*
Lutra, 634
Lutra longicaudis, 277, 722
See also *Lontra*
Luziola subintegra, 689
Lycopodium, 462, 510
Lycopodium brachiatum, 211
Lygaidae, 231
Lygodium venustum, 275
Lyngbya, 624, 634, 720
Lysiloma
Lysiloma auritum, 272
Lysiloma discaritu, 351
Lysiloma divaricatum, 257
Lysipomia, 513, 720
Lysipomia acaulis, 510
Lytechinus variegatus, 592

Mabuetia guatemalensis, 696
Machaerium, 461
Machaerium kegelii, 256
Machaerium robiniifolium, 257
Macleania, 461, 469, 719
Macleania rupestris, 462, 685
Macleania talamancensis, 461
Macrobiotus occidentalis, 511
Macrobrachium, 233, 389, 630, *631*, 632–33, *634*, 642, 715, 722
Macrobrachium acanthurus, 632
Macrobrachium amazonicum, 632
Macrobrachium americanum, 233–34, 271, 632, 713
Macrobrachium carcinus, *631*, 632
Macrobrachium cocoensis, 234–35
Macrobrachium digueti, 632
Macrobrachium hancocki, 234, 632
Macrobrachium heterochirus, 632
Macrobrachium occidentale, 632
Macrobrachium olfersii, 632, 633
Macrobrachium panamense, 632
Macrobrachium tenellum, 632
Macrocnemum roseum, *348*
Macrolepidoptera, 229–30
Macromitrium, 463, 510

Magnolia, 457, 718
Magnolia poasana, 469
Magnolia sororum, 469
Mahonia, 508
Maianthemum, 462, 469
Mallomonas
Mallomonas acaroides, 672, 676
Mallomonas crassisquama, 672, 676
Mallomonas cyathellata var. *chilensis*, 672
Malpighiaceae, 254
Malvaceae, 254, 257, 539
Malvaviscus arborea, 338
Mammuthus columbi, 50, 51–52
Manacus candei, 561
Manduca
Manduca dilucida, 314, 323
Manduca lanuginosa, 314, 323
Manicaria, 696
Manicaria saccifera, 694–95, 721
Manihot
Manihot aesculifolia, 256
Manihot esculenta, 534
Manilkara, 294, 374, 715
Manilkara chicle, 256, 260, 273, 319, 333
Manilkara zapota, 254, 319, 712
Mantodea, 266
Marantaceae, 560
Marasmiellus, 458
Marasmius, 458
Marcgravia
Marcgravia brownei, 432
Marcgravia waferi, 211
Margarornis rubiginosus, 466
Marmosa mexicana, 354
Marpesia petreus, 270
Marsilea deflexa, 256
Matayba oppositifolia, 350
Mauria, 473
Mauria heterophylla, 351
Maxillaria, 347, 462, 718
Maxillaria adendrobium, 211
Maxillaria biolleyi, 462
Maxillaria parviflora, 211
Maytenus, 353
Mazama
Mazama americana, 354, 467, 482, 559, 686
Mazama temama, 559
Mediomastus californiensis, 110, 144, 145
Megabalanus coccopoma, 173
Megachilinae, 271
Megalops atlanticus, 546, 547, 626, 640
Megalopygidae, 323
Megaptera novaeangliae, 116, 154, 169, 170
Melaneris chagresi, 558
Melanerpes formicivorus, 466
Melanthera nivea, 255
Melastomataceae, 213, 347, 348, 354, 439, 460, 461, 462, 468, 473, 539, 561, 719
Meliaceae, 348, 468
Meliosma, 457, 718
Meliosma glabrata, 469
Meliosma vernicosa, 438
Meliponinae, 271
Mellitella stokesii, 110
Melloa, 257
Melocactus curvispinus, 256
Meloidogyne incognita, 76
Melongena, 153
Memphis, 270
Memphis forreri, 270
Menegazzia neotropica, 686
Mephitis macoura, 262
Merismopedia, 675
Merismopedia cf. *chondroidea*, 667
Merismopedia tenuissima, 667
Mesaspis monticola, 464, 511, 719
Mesembrinibus cayennensis, 630
Mesocheira, 271, 712
Mesocyclops thermocyclopoides, 660, 721
Mesoplia, 271, 713
Meteoriaceae, 463
Meteoridium, 463
Metzgeria, 510
Miconia, 347, 460, 461, 468, 469, 471, 473, 500, 512, 514, 539, 715, 718, 719
Miconia affinis, 561
Miconia argentea, 353
Miconia centrodesma, 561
Miconia dodecandra, 213, 713
Miconia kappellei, 461
Miconia lauriformis, 469
Miconia nervosa, 561
Miconia pittieri, 469
Miconia tonduzii, 472, 474
Microcerculus
Microcerculus marginatus, 552
Microcerculus philometa, 552
Microcyclops
Microcyclops ceibaensis, 667
Microcyclops varicans, 675
Microcystis, 672
Microhylidae, 551
Microrhopias quixensis, 393, 554
Microspora, 624
Microtenochira ferranti, 269
Microtia elva, 270
Microtropis, 457
Microtropis occidentalis, 469
Microtylopteryx hebardi, 559
Micrurus clarki, 392
Mikania, 461
Milnesium tardigradum, 511
Mimidae, 556
Mimosa
Mimosa aspereta, 257
Mimosa pigra, 257, 691
Mimosa trichephala, 255
Mimosa xanthocentra, 257
Mimosoideae, 274
Minquartia guianensis, 374, 715
Mionectes oleagineus, 393, 561
Mitrospingus cassinii, 554
Mixonera, 52
Mixotoxodon larensis, 51, 52
Modisimus coco, 231
Mollinedia, 468
Mollusca, 119, 144
Momotidae, 555
Monera, 6
Monimiaceae, 468
Monnina, 471, 514
Monnina crepinii, 473
Monnina xalapensis, 472
Monochaetum, 442, 461, 718
Monorhaphidium griffithii, 665, 721
Monstera, 462
Montanoa guatemalensis, 351
Montricardia, 721
Montricardia arborescens, 687, 696, 697
Mora, 696
Mora oleifera, 353, 377, 696
Moraceae, 52, 230, 254, 348, 349, 350, 396, 468
Morphnus guianensis, 355, 394
Mougeotia, 667
Mouriri, 696
Mucuma
Mucuma mutisiana, 214
Mucuma sloanei, 214
Muehlenbeckia, 469
Mugilidae, 269, 389, 546, 547, 548
Muhlenbergia, 500
Muhlenbergia flabellata, 512
Muntingia calabura, 339
Muridae, 5, 79–80, 467, 719
Musa, 335
Musa acuminata, 377
Musa velutina, 541
Musci, 211
Mustela frenata, 467
Mustelidae, 5
Mustelus
Mustelus dorsalis, 154, 711
Mustelus lunulatus, 154, 711
Myadestes melanops, 466
Mycena, 458
Mycorrhizae, 75, 87
Mycosphaerella musicola, 564
Mycteria americana, 276, 699, 702
Myiobius
Myiobius sulphureipygius, 554
Myiobius torquatus, 466
Myiodynastes maculatus, 553
Myrcia
Myrcia oerstediana, 469
Myrcia splendens, 352
Myrcianthes, 718
Myrcianthes fragrans, 352, 353, 712
Myrcianthes rhopaloides, 472
Myrcianthes storkii, 469
Myrica, 52, 440, 471, 514, 718
Myrica phanerodonta, 442
Myrica pubescens, 472
Myrmeciza
Myrmeciza exsul, 393
Myrmeciza immaculata, 554
Myrmecophaga, 5
Myrmecophaga tridactyla, 265, 277, 331, 555

Myrmecophila, 275
Myrmelachista, 433, 719
Myrmelachista lauropacifica, 387
Myrmotherula
Myrmotherula axillaris, 554
Myrmotherula fulviventris, 393
Myrmotherula quixensis, 393
Myroxylon balsamum, 351, 482
Myrrhidendron, 512
Myrrhidendron chirropoense, 513
Myrsinaceae, 217, 269, 457, 461, 467, 468
Myrsine, 461, 500, 514, 718
Myrsine coriacea, 469, 685, 721
Myrsine pittieri, 469, *501*
Myrtaceae, 217, 348
Myscelia pattenia, 325
Mytella guyanensis, 153
Myxomycetes, 208–9

Najas guadalupensis, 256
Nasua narica, 264, 466, 557
Nasutitermes, 339
Natalus stramineus, 265
Natica, 145
Natica unifasciata, 145
Natsushima sashai, 118
Navanax
Navanax aenigmaticus, 103, 594
Navanax gemmatus, 594
Navicula, 677, 721
Navicula cf. *laevissima*, 676
Navicula cryptocephala, 677
Navicula incarum, 624, 626
Navicula ingapirca, 624, 626
Navicula kohlenbachii, 624, 626
Navicula radiosa, 677
Navicula seminulum, 626
Navicula subminuscula, 626
Neckera, 463
Neckeraceae, 463
Nectandra, 81, 461, 469, 471
Nectandra aerolata, 52
Nectandra cufodontisii, 471, 472, 474
Nectandra membranacea, 260
Nectandra silicifolia, 353
Nectandra woodringi, 52
Neea laetevirens, 539
Neetroplus nematopus, 548, 549
Nematoda, 119, 144, 276
Nemertea, 119
Neoheterandria umbratilis, 547, 548, 558
Neomorphus geoffroyi, 394, 554
Neonicholsonia watsonii, 260
Neotuerta sabulosa, 323
Nephroma helveticum, 686
Nephronias tempisquensis, 693
Neptunia, 721
Neptunia natans, 257, 690
Neptunia plena, 256, 689
Nerita
Nerita funiculata, 173
Nerita scabricosta, 173–74
Neritidae, 106
Neritina latissima, 389, 635
Nesiteria concinna, 350
Nesotriccus ridgwayi, 222
Neuroptera, 231, 463
Neusticurus apodeumus, 392
Newtonia suaveolens, 353
Nitzschia
Nitzschia amphibia, 664
Nitzschia cf. *vitrea*, 676
Nitzschia frustulum, 664
Nitzschia palea, 626
Nitzschia pungens, 146
Noctilio
Noctilio albiventris, 630
Noctilio leporinus, 630, 722
Noctilionidae, 630
Noctiluca scintillans, 146
Nodilittorina modesta, 173–74
Norops, 552
Norops altae, 435
Norops aquaticus, 392
Norops intermedius, 435
Norops oxylophus, 276
Norops pentaprion, 267
Norops polylepis, 392
Norops townsendi, 220, *221*
Norops tropidolepis, 435
Notonecta, 666
Numida meleagris, 196
Nyctaginaceae, 539, 559
Nyctibiidae, 555
Nyctomys
Nyctomys nyctomys, 339
Nyctomys sumichrasti, 558
Nymphaea, 257, 721
Nymphaea amazonum, 690
Nymphaea ampla, 689
Nymphaea blanda, 687, 689
Nymphaea prolifera, 257
Nymphaea pulchella, 690
Nymphalidae, 270, 386
Nymphoides humboldtianum, 690

Ochetomyrmicini, 561
Ochroma, 381, 382, 715
Ochroma lagopus, 339
Ochroma pyramidale, 214, 261, 349, 353
Ochrophyta, 119, 509
Ocotea, 81, 347, 387, 461, 468, 469, 471, 695, 711, 718, 719
Ocotea austinii, 472
Ocotea cf. *viridiflora*, 429
Ocotea endresiana, 438, 439
Ocotea insularis, 211, 213, 216, 472, 713
Ocotea pharomachrosorum, 472, 472
Ocotea pittieri, 472, 472
Ocotea praetermissa, 474
Ocotea pseudopalmana, 472
Ocotea veraguensis, 256
Ocypode gaudichaudii, 232
Ocypodidae, 232
Odocoileus virginianus, 196, 224, 265, 559
Odonata, 269, 633
Odontomachus, 561
Odontomachus erythrocephalus, 387
Odontophoridae, 555
Oeceoclades maculata, 335
Oedipina alleni, 391
Oedogonium, 624, 720–21
Oligochaeta, 463
Olivella semistriata, 145
Olmedia aspera, 349
Ommatolampinae, 559
Oncaea venusta, 147
Oncidium, 275
Oncidium cebolleta, 275
Oncorhynchus, 336
Oncorhynchus mykiss, 463, 719
Onychophora, 78
Onychoprion fuscatus, 222
Oocystis, 667, 672
Oophaga pumilio, 551
Operculina triqueta, 257
Ophisternon aenigmaticum, 692
Ophistonema, 154
Opiliones, 231
Opisthacanthus valerioi, 227, *229*, 231
Oporornis
Oporornis formosus, 394
Oporornis philadelphia, 394
Opuntia
Opuntia cochenillifera, 256
Opuntia elatior, 257
Opuntia guatemalensis, 256
Opuntia lutea, 260
Orchidaceae, 347, 354, 462, 468, 469, 509, 510, 539, 540
Oreobolus, 512
Oreobolus goeppingeri, 513
Oreochromis, 336, 649, 692
Oreochromis aereus, 648
Oreochromis mossambicus, 648
Oreochromis niloticus, 648, 692, 722
Oreomunnea pterocarpa, 377
Oreopanax, 461, 462, 468, 469, 473, *501*, 514, 718
Oreopanax capitatus, 473, 474
Oreopanax nubigenus, 429, 432, 442, 473, 474
Oreopanax oerstedianus, 482
Ormilthidium adendrobium, 211
Ormosia macrocalyx, 260
Oropogon, 509
Orthalicidae, 511
Orthogeomys
Orthogeomys heterodus, 482
Orthogeomys underwoodii, 354
Orthomorpha coactata, 235
Orthoptera, 231, 463
Orthotrichaceae, 463
Orthrosanthus, 512
Oryctolagus cuniculus, 196
Oryza
Oryza latifolia, 623, 687, 689
Oryza sativa, 377
Oryzobus nuttingi, 553

Oryzomys, 467
  Oryzomys albigularis, 440, 466
Osaecilia osa, 391
Osa pulchra, 374, *383*
Oscillatoria, 624, 721
Ossaea
  Ossaea bracteata, 216
  Ossaea macrophylla, 216
Osteopilus septentrionalis, 551
Ostracoda, 144, 710
Ostrea iridescens, 151–52
Otoba novogranatensis, 348, 699
Ottoa oenanthiodes, 513
Ouratea, 699
  Ouratea lucens, 256
Oxalis, 458
  Oxalis frutescens, 255
Oxybelis, 552
Oxycarium cubense, 689–90
Oxydia hogei, 231
Oxyethira, 666

Pacarina, 269
Pachira
  Pachira aquatica, 697, 699
  Pachira quinata, 254, 257, 259, 260, 262, 272, 279, 377, 712
  See also *Bombacopsis*; *Pochota quinata*
Pachyarmatherium, 50
Pachycondyla, 561
  Pachycondyla bugabensis, 387
  Pachycondyla harpax, 387
  Pachycondyla theresiae, 387
Pachylia ficus, 230, 325
Paepalanthus, 721
  Paepalanthus costaricensis, 514, 686
Paguridae, 232
Pagurus californiensis, 232
Palaemon, 632
  Palaemon gracilis, 632
Palaemonidae, 233, 271
Palaeolama mirifica, 51
Palicourea, 457, 461, 468, 469, 471, 719
  Palicourea fastigiata, 696
  Palicourea guianensis, 262
  Palicourea salicifolia, 472
Palmacites berryanum, 52
Palmae, 52
Panicum, 687, 721
  Panicum maximum, 377, 623
  Panicum parvifolium, 689
Panopsis costaricensis, 350
Panterpe insignis, 469
Panthera onca, 262, 354, 467, 482, 557
Panulirus argus, 594
Panus fulvus, 257
Papilionidae, 270, 386, 539
Paracalanus parvus, 147
Parachromis
  Parachromis dovii, 269, *390*, 547, 548, 550, 640, 648, 667, 675, 722
  Parachromis loisellei, 548
  Parachromis managuense, 546, 547
Paracyclops chiltonii, 666
Paralichthyidae, 547
Paramachaerium gruberi, 374
Paraphysomonas vestita, 676
Paraponera clavata, 387–88
Paraponerinae, 388
Paraprionospio pinnata, 110, 144
Parathesis, 461
Paratrechina caeciliae, 387
Parhadjelia cairinae, 276
Parkia, 374
  Parkia pendula, 353
Parkinsonia, 701–2
  Parkinsonia aculeata, 257, *258*, 690, 691, 692, 701, 702, 721
Parmeliaceae, 459
Parmentiera valerii, 304
Parmotrema 459, 460
Parula gutturalis, 466, 511
Parulidae, 466, 556, 719
Paspalidium germinatum, 687, 690
Paspalum, 257, 347
  Paspalum pectinatum, 255
  Paspalum repens, 257, 689, 690
Passalidae, 231
Passarina cyanea, 224
Passeridae, 556
Passiflora, 347, 383, 384, *384*, 461, 469, 715, 719
  Passiflora vitifolia, 383, *383*, 715
Pavona
  Pavona chiriquensis, 179
  Pavona clavus, 113, 175, 179, 710
  Pavona gigantea, 112–13, *120*, 165, 179, 710
  Pavona maldivensis, 179
  Pavona varians, 175, 179
  Pavona xarifae, 179
Pavonia spicata, 696
Pecari tajacu, 263, 264, 354, 395, 467, 558
Pecopteris, 52
Pedilanthus nodiflorus, 256
Pelamis platurus, 221
Pelecanus occidentalis, 223
Pelliciera, 13
  Pelliciera rhizophorae, 109, 143, 353, 377
Pellobunus insularis, 231
Peltigera, 509
Peltogyne, 374
  Peltogyne purpurea, 349, 374, 715
Penaeus
  Penaeus brasiliensis, 594
  Penaeus brevirostris, 146
  Penaeus stylirostris, 146
Penelope purpurascens, 266, 466
Peniocereus hirshtianus, 256
Pennisetum, 687
  Pennisetum clandestinum, 86
Pentacalia, 512
  Pentacalia andicola, 512
  Pentacalia firmipes, 474
Pentaclethra, 532, 533, 677, 699
  Pentaclethra macroloba, 400, 540, 543, 695–98, 715, 722
Pentatomidae, 231
Peperomia, 347, 440, 461, 462, 468–69, 539, 715, 718
  Peperomia costaricensis, 440
Peponecephala electra, 154
Perenniporia, 458
Pereskia lychnidiflora, 256
Peridinium willei, 672
Peripatidae, 78
Peripatus acacioi, 78
Perissodactyla, 354, 511
Perlesta, 633
Pernettya, 458, 512, 721
  Pernettya coriacea, 474
  Pernettya prostrata, 512, 514, 686, 721
Peromyscus, 472
  Peromyscus mexicanus, 466, 467
Persea, 438, 457, 461, 471, 718
  Persea americana, 335, 461
  Persea caerulea, 350, 351
  Persea schiedeana, 461, 482
  Persea vesticula, 469
Petrolisthes armatus, 106, 146, 709
Pezopetes capitalis, 466, 511
Phaenostictus mcleannani, 339, 394, 554
Phaeocollybia, 458
Phaeophyta, *147*
Phaeothlypis fulvicauda, 222
Phaethornis superciliosus, 393
Phaeton lepturus, 223
Phalacrocoracidae, 555
Phallichthys
  Phallichthys amates, 547, 548
  Phallichthys quadripunctatus, 548
  Phallichthys tico, 547, 548
Pharomachrus
  Pharomachrus mocinno, 553, 719
  Pharomachrus mocinno ssp. *costaricensis*, 466
Phascolosoma perlucens, 146
Phaseolus, 461
  Phaseolus vulgaris, 75, 78, 85, 475
Phasmatodea, 266
Pheidole, 561
  Pheidole bicornis, 387
  Pheidole browni, 387
  Pheidole fiorii, 387
  Pheidole multispina, 387
  Pheidole pugnax, 387
Pheidolini, 561
Phellinus, 458
  Phellinus gilvus, 257, 354
  Phellinus linteus, 354
Pheucticus tibialis, 466
Philander opossum, 354
Philodendron, 347, 457, 462, 539, 715
  Philodendron thalassicum, 432
Phoebe, 471
Phoebis
  Phoebis argante, 327
  Phoebis hersilia, 327
  Phoebis philea, 327
  Phoebis sennae, 327

Phoradendron, 457
  Phoradendron tonduzii, 462
Phormidium, 624, 720
Photinus, 269
Phragmatopoma attentua, 147
Phrygionis steeleorum, 231
Phyllites costaricensis, 52
Phyllogonium viscosum, 469
Phylloporus, 458
Phyllostomus discolor, 265, 272
Phylobates vittatus, 391
Phymatolithon calcareum, 114
Physciaceae, 459
Physeteridae, 116
Physimera, 560
Picidae, 555
Pieridae, 270, 386
Pilea gomeziana, 216
Pilotrichella, 463, 719
  Pilotrichella flexilis, 469, 464
Pimelodidae, 268, 389, 547, 548, 692
Pinaroloxias inornata, 222–23, 226, 230
Pinctada mazatlanica, 151
Pinnipedia, 219
Pinnularia, 624
  Pinnularia braunii, 668, 669, 674
Pinus, 478
Piper, 273, 347, 349, 387, 440, 461, 468, 469, 539, 560, 715
  Piper amalago, 339
  Piper arieianum, 560
  Piper flavidum, 256
  Piper marginatum, 339
  Piper psilorhachis, 339
  Piper retalhuleuense, 257
  Piper reticulatum, 262
  Piper sagittifolium, 461
  Piper trigonum, 560
Piperaceae, 52, 347, 349, 440, 457, 462, 539, 560
Piperites
  Piperites cordatus, 52
  Piperites quinquecostatus, 52
Pipra
  Pipra coronata, 393
  Pipra mentalis, 393, 561
Pipridae, 556
Pistia stratiotes, 257, 377, 689, 690, 696, 697, 721
Pitcairnia, 275
  Pitcairnia calcicola, 257, 260
Pithecellobium
  Pithecellobium lanceolatus, 257
  Pithecellobium latifolium, 695
  Pithecellobium valerioi, 699
Plagiochila, 463, 510, 719
Plagiochilaceae, 463
Plagiorchioidea, 275
Plagiothecium, 463
Plantago, 458, 508, 512
Platyhelminthes, 119, 275, 634
Platymiscium
  Platymiscium curuense, 260, 262
  Platymiscium parviflorum, 259, 276, 353
  Platymiscium pinnatum, 374, 482
Platyrinchus
  Platyrinchus coronatus, 552
  Platyrinchus mystaceus, 552
Plecoptera, 633
Pleiostachya pruinosa, 560
Plethodontidae, 551
Pleuranthodendron lindenii, 260–61
Pleuromanna robusta, 147
Pleuroncodes
  Pleuroncodes monodon, 118
  Pleuroncodes planipes 118
Pleurothallis, 347, 462, 539, 715, 718
Pleurotus, 458
Plicopurpura
  Plicopurpura patula, 152
  Plicopurpura patula pansa, 152
Plina puriscalensis, 352
Pliocardia krylovata, 118
Plocosperma buxifolium, 257, 260
Plumeria rubra, 260, 262, 353, 429
Poa, 720
  Poa chirripoensis, 510
Poaceae, 52, 254, 257, 347, 350, 461, 467, 468, 499, 509, 510, 539
Pochota quinata, 272, 377
  See *Bombacopsis quinata*; *Pachira quinata*
Pocillopora, *112*, 113
  Pocillopora damicornis, 179
  Pocillopora elegans, 179
  Pocillopora eydouxi, 112, 179, 710
  Pocillopora meandrina, 113, 179
Podicipedidae, 555
Podocarpaceae, 461, 469
Podocarpus, 52, 461, 718
  Podocarpus macrostachys, 469
Poecilia, 389
  Poecilia gillii, 268, 389, 547, 548, 550, 634, 637, 692, 722
  Poecilia mexicana, 547, 550
  Poecilia sphenops, 268
Poeciliidae, 268, 389, 546, 547, 548, 558, 627, 692, 712, 715, 722
Poeciliopsis, 389
  Poeciliopsis gracilis, 269
  Poeciliopsis santaelena, 269
  Poeciliopsis turrubarensis, 389
Pogonophora, 33
Pogonopus speciosus, 260, 261
Polistes, 321
  Polistes instabilis, 327–28, *328*
  Polistes variabilis, 271
Polyarthra, 667
  Polyarthra vulgaris, 675
Polychaeta, 144
Polychrotidae, 552, 553
Polycyathus hondaensis, 179
Polydesmida, 235
Polygonaceae, 254
Polygonum, 377, 687, 721
  Polygonum acuminatum, 689
  Polygonum hispidium, 689
  Polygonum punctatum, 689
  Polygonum segetum, 668, 689
Polymesoda radiata, *151*, 153, 700
Polyommatus, 531
Polyplacophora, 153, 594
Polypodium, 347, 462
  Polypodium attenuatum, 275, 348
  Polypodium wagneri, 275
Polyporus, 458
Polytrichadelphus, 463
Pomacea costaricana, 693
Pomadasys
  Pomadasys bayanus, 389
  Pomadasys croco, 547, 548
Pompholix complanata, 675
Ponerinae, 433
Pontederia rotundifolia, 689, 690
Pontoscolex corethrurus, 78–79
Porella, 463
Porichthys, 148
Porites, *13*
  Porites evermanni, 113, 114
  Porites lobata, *112*, 113, 113–14, 169, 171, 175, 179, *180*, 710
  Porites panamensis, 113, 165
Porophyllum punctatum, 256
Porotrichodendron, 463
Porotrichum, 463
Porthidium, 552
  Porthidium nasutum, 392
  Porthidium porrasi, 392
Posoqueria
  Posoqueria grandiflora, 696
  Posoqueria latifolia, 353, 695
Potamocaranus, *634*
Potamogetonaceae, 111
Pothomorphe, 440
Potimirin, 235
Potos flavus, 557
Poulsenia armata, 374, 715
Pouteria
  Pouteria macrocarpa, 482
  Pouteria subrotata, 353
  Pouteria zapota, 260
Pradosia atroviolacea, 353
Prestoea, 461, 473, 719
  Prestoea acuminata, 429
  Prestoea decurrens, 377, 696
Priapichthys annectens, 548
Prionodon, 463
Prioria copaifera, 66, 695, 697–98, 699
Pristidae, 268, 547
Pristimantis, 531, 551
Pristis pristis, 268, 546, 547
Procnias tricarunculata, 355, 394, 553, 712, 719
Procyon
  Procyon cancrivorus, 355, *355*
  Procyon lotor, 264, 355, 466, 557
Procyonidae, 5
Proechimys semispinosus, 51
Prosapia, 437
Prosopis, 294

Protium, 374, 715
  Protium paramense, 260, 348
  Protium pittieri, 482, 540, 715
Protographium
  Protographium epidaus, 323
  Protographium philolaus, 323
Protonotoria citrea, 394
Protozoa, 6
Protura, 77
Prumnopitys, 461
  Prumnopitys standleyi, 461, 469
Prunus, 718
  Prunus annularis, 469, 474
  Prunus cornifolia, 81
  Prunus subcorymbosa, 377
Psammisia, 461, 469
  Psammisia ramiflora, 462
Psammocora
  Psammocora stellata, 179
  Psammocora superficiales, 179
Psathyrella, 458
Pselliphorus tibialis, 466
Pseudobombax septenatum, 262, 353
Pseudocolaptes lawrencii, 466
Pseudodiaptomus, 147
Pseudolmedia oxyphyllaria, 261, 350
Pseudomyrmex, 271, 274, 338, 713, 715
  Pseudomyrmex caeciliae, 387
  Pseudomyrmex ferruginea, 274
  Pseudomyrmex flavicornis, 338, 387
  Pseudomyrmex nigrocinctus, 338, 387
  Pseudomyrmex particeps, 387
  Pseudomyrmex spinicola, 338, 387
Pseudo-nitzschia pungens, 154
Pseudophallus
  Pseudophallus elcapitanensis, 268
  Pseudophallus mindii, 546, 548
  Pseudophallus starksi, 389
Pseudorca crassidens, 169, 170
Pseudostaurosira brevistriata, 664
Pseudothelphusidae, 632
Pseustes, 552
Psidium, 461
Psiguria, *384*
Psittacidae, 555
Psychotria, 347, 349, 461, 462, 469, 539, 715, 718, 719
  Psychotria chagrensis, 696, 699
  Psychotria cocoensis, 230
  Psychotria gracilenta, 217, 230
  Psychotria grandis, 539
  Psychotria poeppigiana, 699
  Psychotria turrubarensis, 350
Pteris, 458
Pterobyron, 463
Pterocarpus, 696
  Pterocarpus officinalis, 377, 598, 599, 695, 696, 697, 699, 710, 722
Pteroglossus
  Pteroglossus franzii, 356
  Pteroglossus torquatus, 562
Pterogramma cardiosomi, 232
Pteronotus, 265
Pteropsida, 52
Pterygoplichthys, 628
Ptillidae, 231
Ptychophallus paraxanthusi, 632
Ptygura, 667
Puma
  Puma concolor, 262, *263*, 467, 557
  Puma concolor ssp. *costaricensis*, 511
  Puma yagouaroundi, 262
Puya, 52, 512, 514
  Puya dasylirioides, 513, 686, *686*, 721
Pycnarthrum pseudoinsularis, 229
Pylopagurus longimanus, 232
Pyramica margaritae, 387
Pyrenulales, 509
Perginae, 275
Pyrrhobryum, 463

Quadrigula, 672
Quadrus cerealis, 560
Qualea paraensis, 374, 377, 715
Quararibea funebris, 260
Quercus, *13*, 52, 353, 458, 460, 461, 462, 467, 468, 469, 472, 475, 500, 507, 718
  Quercus cf. *seemannii*, 429
  Quercus copeyensis, 81, 463, 469, *470*, 473, 475, 478
  Quercus corrugata, 52, 349, 429
  Quercus costaricensis, 81, 467, 468, 469, 473, 475, 482
  Quercus guglielmi-treleasei, 469
  Quercus insignis, 377
  Quercus oleoides, 254, *254*, 255, 257, 272, 293, *314*, 325, 712
  Quercus oocarpa, 469
  Quercus rapurahuensis, 377, 469
  Quercus seemannii, 349, 469, 514
Quilopoda, *77*

Radula, 463, 510
Radulaceae, 211
Rallidae, 555
Ramalina, 460
Ramphastidae, 555, 561–62
Ramphastos
  Ramphastos sulfuratus, 554, 562
  Ramphastos swainsonii, 355, 554, 562
Ramphocelus passerinii, 196
Rana, 628
  Rana vaillanti, 267, 275
  Rana vibicaria, 464
  Rana warszewitschii, 628, 722
Ranidae, 551
Ranunculus, 513, 720
  Ranunculus crassirostratus, 510
Raphia, 12, 66, *368*, 377, 398, 695, 696
  Raphia taedigera, 9, 377, 598, 694–95, *694–95*, 710, 721
Rattus
  Rattus norvegicus, 219, 224, 236, 335
  Rattus rattus, 193–94, 219, 224, 236
Rauvolfia, 695
Recurvirostridae, 555
Rehdera trinervis, 257, *315*
Reithrodontomys, 467
  Reithrodontomys creper, 440, 466
  Reithrodontomys gracilis, 262
  Reithrodontomys mexicanus, 51
  Reithrodontomys paradoxus, 262
Relbunium, 462
Renealmia
  Renealmia alpinia, 561, 562
  Renealmia aromatica, 696
Rhamdia
  Rhamdia guatemalensis, 268, 389, 548
  Rhamdia nicaraguensis, 268, 547, 692
  Rhamdia rogersi, 548
Rhamnaceae, 513
Rhamnus, 458, 514, 718
  Rhamnus oreodendron, 469, 474
Rhamphichthyidae, 548
Rhapalodia gibba, 677
Rhincalanus nasutus, 147
Rhinochenus 273–74, 329–30
Rhinoclemmys, 531, 722
  Rhinoclemmys annulata, 552, 558, 636
  Rhinoclemmys funerea, 552, 558, 636
Rhinophrynidae, 551
Rhipidocladum racemiflorum, 256
Rhissomatus subcostatus, 86
Rhizobium, 75–76, 87
Rhizobium leguminosarum, 75
Rhizophora, 531
  Rhizophora harrisonii, 109, 143
  Rhizophora mangle, 9, 108, *108*, 109, 143, 353, 377, 599, *600*, 696, 710
  Rhizophora racemosa, 109
Rhizopsammia verrilli, 179
Rhizosolenia, 146
  Rhizosolenia stolterfothii, 146
Rhodobryum grandifolium, 256
Rhodocollybia 458
Rhodophyta, 119, *147*, 509, 624
Rhopalodia gibberula var. *argentina*, 624, 626
Rhus, 473
  Rhus striata, 469
Rhynchospora
  Rhynchospora corymbosa, 689, 690
  Rhynchospora polyphylla, 216
  Rhynchospora schaffneri, 514, 686
Rigodium, 463
Riodinidae, 270
Rivulidae, 268, 546, 547, 548
Rivulus, 549–50
  Rivulus fascolineatus, 268
  Rivulus isthmensis, 547, 548
Rodentia, 273, 511, 555, 558
Roeboides bouchellei, 268, 547
Roldana scandens, 461, 476
Romaleidae, 559
Rondeletia
  Rondeletia amoena, 469
  Rondeletia buddleoides, 469
  Rondeletia hameliifolia, 256
Rooseveltia franckianiana, 211

Rosaceae, 458, 461, 468, 499, 510
Rothschildia
Rothschildia erycina, 323
Rothschildia lebeau, 323
Rotylenchulus reniformis, 76
Roupala, 473
Roupala montana, 255, 260, 350, 469
Rubiaceae, 216, 217, 230, 254, 259, 347, *348*, 349, 354, 374, 439, 460, 461, 462, 468, 539, 719
Rubus, 458, 461, 468, 475, 719
Ruppia maritima, 111, *111*
Ruptiliocarpon, 374
Ruptiliocarpon caracolito, 374
Russelia sarmentosa, 255
Russula, 257, 458
Rustia occidentalis, 216, 696
Rutaceae, 349
Rynchops niger, 693

Sacciolepis myuros, 256
Sacoglottis
Sacoglottis holdridgei, 194, 213, *214*, *215*, 216, 713
Sacoglottis thrichogyne, 698
Sagitta, 148
Sagitta enflata, 148
Sagitta friderici, 148
Sagittaria guyanensis, 256, 690
Saimiri oerstedii, 265, 354, 394, 399, 712, 716
Salpidae, 592
Salvia, 458
Salvinia, 377, 721
Salvinia auriculata, 257, 689, 690, *691*
Salvinia minima, 349
Salvinia sprucei, 689
Samanea, 294
Samanea saman, 259, 261
Sapindaceae, 254, 257
Sapium, 353, 471
Sapium allenii, 374
Sapium glandulosum, 350, 353
Sapium pachystachys, 471, 472
Sapium rigidifolium, 429
Sapotaceae, 276, *348*
Sarchorachis, 440
Sarcoramphus papa, 355
Sargassum liebmanni, 105
Saturniidae, 270, 330
Satyria warszewiczii, 462
Saurauia, 438
Saurauia veraguasensis, 469
Sauria, 267
Sauvagesia pulchella, 255
Scaphiodontophis, 552
Scaphyglottis, 275, 462, 718
Sceloporus malachiticus, 464, 511, 719
Scenedesmus, 675
Scenedesmus ecornis, 667
Schausiella santarosensis, 323, 330
Scheelea, 369
Scheelea rostrata, 315, 696
Schefflera, 461, 468, 473, 514, 718
Schefflera morototoni, 260
Schefflera rodriguesiana, 429, 432, 440, 469, 475, *501*
Schizolobium parahyba, 262, 349, 402
Schlegelia, 461
Schlegelia brachyanta, 213
Schoenocaulon officinale, 255
Schwenckia americana, 255
Sciaenidae, 148, 154, 711
Scincidae, 553
Sciuridae, 79, 273
Sciurus
Sciurus granatensis, 466, 472
Sciurus variegatoides, 273
Scleractinia, 172
Scleria, 687, 721
Scleria macrophylla, 689
Scleria microcarpa, 696
Scolelepis
Scolelepis agilis, 145, 600
Scolelepis squamata, 600
Scolopacidae, 555
Scolytidae, 228–29, 231
Scorpiones, 231
Scotinomys xerampelinus, 466, 467
Scrophulariaceae, 510
Seiurus noveboracensis, 394
Selaginella
Selaginella pallescens, 275, 429
Selaginella sertata, 275
Selasphorus flammula, 511
Selenastrum, 672
Selenicereus
Selenicereus testudo, 275
Selenicereus wercklei, 275
Sellaphora pupula, 672
Sematophyllum, 463
Senecio, 458, 461, 462, 500, 512, 718, 719, 720
Senecio andicola, 512
Senecio kuhbieri, 510
Senecio oerstedianus, 512
Senna, 327
Senna atomaria, 327, 338
Senna pallida, 303, 338
Senna reticulata, 696
Serjania, 257
Serpentes, 267
Setaria geniculata, 214
Shinkai
Shinkai fontefridae, 118
Shinkai longipedata, 118
Sibon, 552
Sickingia maxonii, 695
Sicydium, 220
Sicydium adelum, 548
Sicydium altum, 548, 642
Sicydium cocoensis, 220
Sicydium salvini, 389
Sida, 257, 303, 338
Sida acuminata, 338
Sida salviifolia, 256
Siderastrea siderea, 595, 605, 607
Sideroxylon, 294
Sideroxylon capiri, 259, 260, 276, 351
Sigmodon hispidus, 51, 300
Simaba cedron, 386
Simarouba
Simarouba amara, 348, 559
Simarouba glauca, 256
Simaroubaceae, 559
Simsia santarosensis, 255
Simuliidae, 634
Siphonaria gigas, 107, 120, 153, 173–74, *175*
Siphonopora, 592
Skeletonema costatum, 146–47
Sloanea, *380*
Sloanea picapica, 353, 695
Sloanea terniflora, 254, 260, 712
Smilax, 461, 469
Smilisca
Smilisca baudinii, 392
Smilisca phaeota, 392
Smilisca sila, 391–92
Sobralia, 275
Socratea
Socratea durissima, 374, 715
Socratea exorrhiza, 540, 715
Solanaceae, 52, 254, 257, 260, *347*, 349, 439, 461, 462, 476, 559, 719
Solanum, 257, 347, 349, 458, 461, 468, 471, 500, 539, 715, 718, 719, 721
Solanum campechiense, 689
Solanum dotanum, 472
Solanum lancefolia, 687
Solanum longiconicum, 461
Solanum storkii, 469
Solenocera agassizii, 153
Solenopsidini, 561
Solenopsis altinodis, 388
Solifugae, 231
Solpugidae, 231
Sorocea cufodontisii, 374, 715
Sotalia guianensis, 595
Spathiphyllum
Spathiphyllum friedrichstallii, 687, 696, 721
Spathiphyllum laeve, 216
Sphaeradenia, 462
Sphaeroceridae, 231, 232
Sphaerodactylus pacificus, 220, *222*
Sphaeroma peruvianum, 143
Sphagneticola trilobata, 214
Sphiggurus mexicanus, 264, 466
Sphingidae, 229–30, 314, 326
Sphoeroides annulatus, 389, 635
Sphyrna lewini, 154, 170
Sphyrospermum cordifolium, 462
Spilogale putorius, 467
Spionidae, 110
Spirodela intermedia, 349
Spondias, 397
Spondias mombin, 254, 259, 273, 349, 695, 712
Spondias purpurea, 335, 338

Spondylus calcifer, 151
Squamidium, 463
Squilla
  Squilla biformes, 118
  Squilla hancocki, 154
  Squilla parva, 154
Stator
  Stator limbatus, 269
  Stator pruininus, 269
Staurastrum, 667
  Staurastrum paradoxum, 666
Staurodesmus, 667
Stauroneis, 624, 721
Stegastes partitus, 595
Stelis, 462, 718
Stellifer, 148
Stemmadenia donnell-smithii, 697
Stenella
  Stenella attenuata, 102, 116, 154, 710
  Stenella longirostris, 154
Stenocereus aragonii, 256, 259, 260, 262
Stenopterobia pumila, 624, 626
Stephaniella, 513
Sterculia apetala, 262
Sterculiaceae, 254
Stereocaulon, 509
  Stereocaulon vesuvianum, 512
Steriphoma paradoxum, 349
Stevia, 720
  Stevia westonii, 510
Sticherus remotus, 217
Sticta, 458, 459, 460, 509
Stictocardia tiliifolia, 230
Stipa, 512
  Stipa hans-meyeri, 513
Stomatopoda, 118, 146
Strabomantidae, 551
Stratiomyidae, 231
Strigidae, 555
Strombus galetus, 153
Sturnira
  Sturnira lilium, 265
  Sturnira ludovici, 467
Styrax, 457
  Styrax argenteus, 81, 469, 474
  Styrax glabrescens, 350
Sula
  Sula dactylatra, 222
  Sula leucogaster, 170, 222
  Sula sula, 170, 222, 226
Sus scrofa, 194, 224, 227
Swallenochloa, 512
Swartzia panamensis, 353
Swietenia
  Swietenia humilis, 259, 276
  Swietenia macrophylla, 256, 273, 276, 296, 332, 348, 350, 482
Sycidium salvini, 635
Sylviidae, 556
Sylvilagus, 511
  Sylvilagus brasiliensis, 559
  Sylvilagus dicei, 354, 466, 482
  Sylvilagus floridanus, 264
Symphonia, 698
  Symphonia globulifera, 9, 374, 377, 399, 696, 697, *697*, 698, 699, 715, 716, 722
Symphyla, 77
Symplocos, 457, 471, 718
  Symplocos austin-smithii, 469
  Symplocos serrulata, 471
Synbranchidae, 546, 547, 548
Synbranchus marmoratus, 268, 547, 548, 693
Syncuaria, 276
Syngnathidae, 268, 389, 546, 548
Syngonium, 462
Synurophyceae, 666, 676
Syringodium filiformis, 602, *603*, 710
Syrrhopodon, 463

Tabebuia
  Tabebuia ochracea, 254, 257, 350, 712
  Tabebuia rosea, 349, 350, 351, 353, 696, 711
Tabellaria binalis, 670
Tabernaemontana
  Tabernaemontana alba, 256, 260
  Tabernaemontana chrysocarpa, 696
Tachigali versicolor, 349, 374
Tachyphonus delattrii, 554
Tagelus, 153
Talamancalia, 509, 720
Talipariti, 225
  Talipariti tiliaceum var. *pernambucense*, 214, 215, *217*
Tamandua mexicana, 354
Tangara, 394
  Tangara gyrola, 393–94
  Tangara inornata, 553, 554
  Tangara larvata, 394
Tantilla, 552
Tapirira mexicana, 350
Tapirus, 51
  Tapirus bairdii, 262, 263, 277, *318*, *319*, 354, 397, 467, 473, 482, 511, 559, 686, 715
  Tapirus cf. *terrestris*, 51
Tardigrada, 511
Tassadia ovovata, 230
Tayassu
  Tayassu pecari, 262, 263, 277, 354, 395, 396, 559, 715
  Tayassu tajacu, 467
Tayassuidae, 5, 395
Taygetis kerea, 270
Tecoma stans, 350
Tectona grandis, 73, 279, 377, 402, 543
Teiidae, 553
Telipogon, 462, 509, 718
Terminalia, 377
  Terminalia catappa, 214, 273, 402
  Terminalia oblonga, 262, 697
Tethocyathus prahli, 179
Tetracera volubilis, 269
Tetraclita
  Tetraclita rubescens, 146, 709
  Tetraclita stalactifera, 106, 146, 173, 709
Tetragastris panamensis, 260
Tetranema floribundum, 350
Tetrathylacium, 388, *388*
  Tetrathylacium johansenii, 353
Tettigoniidae, 231
Thais brevidentata, 107, 155
Thalassia testudinum, 593, 602, *603*, 710
Thalassiosira pseudonana, 667
Thalia, 721
  Thalia geniculata, 257, 687, 690, 691, 696
Thalurania colombica, 553
Thamnophilidae, 555
Theaceae, 214
Thelypteris, 52, 347
  Thelypteris cocos, 211
  Thelypteris gomeziana, ix
  Thelypteris minor, 275
Theobroma angustifolium, 261
Theraps
  Theraps maculicauda, 546, 547
  Theraps underwoodi, *628*, 640, *645*
Thermocyclops tunuis, 666
Thioplaca cf. *chileae*, 117
Thoinia serrata, 255
Thor, 169
Thouinidium decandrum, 254, 260, 353, 712
Thraupidae, 196, 466, 556, 719
Thraupis episcopus, 196
Threskiornithidae, 555
Thryorchilus browni, 511
Thuidium, 463
Ticodendraceae, xix, 718
Ticodendron incognitum, 377, 718
Tigrisoma, 630
Tilapia, 656, 664, 699
Tillandsia, 275, 347, 462, 718
  Tillandsia punctulata, 462
  Tillandsia schiedeana, 275
  Tillandsia streptophylla, 256
Tilopus, 257
Tinamidae, 555
Tinamus major, 466
Tineidae, 231
Tipulidae, 231
Tococa guianensis, 699
Tocoyena pittieri, 482
Tolmomyias
  Tolmomyias assimilis, 554
  Tolmomyias sulphurescens, 275–76
Tomocichla
  Tomocichla tuba, 549
  Tomocichla underwoodi, 548, 549, 550
Tonina fluviatilis, 689
Topobea, 432, 462
Tortricidae, 231
Toxicodendron striatum, 350, 351
Trachelomonas, 667
  Trachelomonas volvocina, 667
  Trachelomonas volvocinopsis, 667
Trachemys, 552
Trachypenaeus byrdi, 146

- *Trachypogon plumosus*, 255, *299*, *306*, *308*, 334, 335
- *Tradescantia petricola*, 260
- *Trametes*, 458
- *Trema micrantha*, 339, 352, 353, 712
- *Triaenodon obesus*, 170–71
- *Trichechus manatus*, 265, 555, 595, 630, 696, 716, 722
- *Trichilia*, 257, 374, 473, 715, 718
  - *Trichilia havanensis*, 351, 469
  - *Trichilia pallida*, 353
  - *Trichilia pleeana*, 260
  - *Trichilia tuberculata*, 261, 696
- *Trichodesmium erythraeum*, 154
- *Trichomanes*, 462
- *Trichomorpha hyla*, 235
- Trichomycetes, 633–34
- Trichomycteridae, 391
- Trichoptera, 633
- Trigonalidae, 275
- *Trigonia rugosa*, 272
- *Trigonidium*, 275
- *Trimalaconothus novus*, 666
- *Trinectes paulistanus*, 547
- *Tripanurgus compresus*, 392
- *Triplaris melaenodendron*, 261, 348, 712
- *Tripneustes ventricosus*, 592
- *Trixis inula*, 255
- Trochilidae, 230, 466, 555, 719
- *Troglodytes ochraceus*, 466
- Troglodytidae, 556
- *Trogon*
  - *Trogon collaris*, 466
  - *Trogon melanocephalus*, 266
- Trogonidae, 555
- *Trophis racemosa*, 260
- *Tropidacris cristata*, 433
- *Tropocyclops*
  - *Tropocyclops prasinus*, 666
  - *Tropocyclops prasinus prasinus*, 665, 721
- *Tubastrea coccinea*, 179
- Turdidae, 196, 466, 556, 719
- *Turdus*
  - *Turdus albicollis*, 394
  - *Turdus grayi*, 196
  - *Turdus nigrescens*, 511, 686, 721
  - *Turdus plebejus*, 466
- *Turnera diffusa*, 255
- *Turpinia*, 457, 473
- *Tursiops truncatus*, 116, 154, 169–70, 595
- *Tylomys watsoni*, 51, 466
- *Tylopilus*, 458
- *Typha*, 249, 659, 661, 663, 691, 701–2
  - *Typha dominguensis*, 257, 280, 690, 691, 701, 721
- Tyrannidae, 466, 556, 719
- *Tyromyces*, 458
- Tytonidae, 555

- *Uca*, 145
  - *Uca panamensis*, 232
  - *Uca zacae*, 232
- *Ugni myricoides*, 514
- *Uleobryum peruvianum*, 256
- *Ulmus mexicana*, 349
- *Uncinia*, 720
  - *Uncinia hamata*, 462
  - *Uncinia koyamae*, 510
- Ungaliophidae, 553
- *Urbanus proteus*, 325
- *Urera caracasana*, 696
- *Uribea*, 374
- *Urocyon cinereoargenteus*, 265, 277, 466
- *Uroderma bilobatum*, 557
- *Urotheca*, 552
- *Urotrygon cimar*, 149, 711
- Urticaceae, 52, 216, 458, 469
- *Usnea*, 458, 459, 509, 685
- *Utricularia*, 377, 668, 721
  - *Utricularia* aff. *subulata*, 514, 686
  - *Utricularia alpina*, 432

- *Vaccinium*, 432, 458, 461, 471, 512, 514, 719
  - *Vaccinium consanguineum*, 81, 469, *472*, 474, 475, 512, 514, 685
  - *Vaccinium floribundum*, 474
- *Vachellia*, 271, 274, 713, 715
  - *Vachellia allenii*, 374, 387, *388*
  - *Vachellia collinsii*, 257, 262, 274, 338
  - *Vachellia cornigera*, 274, 338
  - *Vachellia farnesiana*, 256, 257, 335, 338
  - *Vachellia villosa*, 255
- *Valeriana*, 458
  - *Valeriana sorbifolia*, 513
- *Vampyressa thyone*, 557
- *Vampyrum spectrum*, 264
- *Vanilla*, 461
  - *Vanilla planifolia*, 275
- *Vantanea*, 374
  - *Vantanea barbourii*, 374
- Verbenaceae, 254, *315*
- *Vermivora peregrina*, 394
- Vespidae, 271, *328*
- *Viburnum*, 52, 458, 471, 514
  - *Viburnum costaricanum*, 350, 469, 471, 472
  - *Viburnum venustum*, 429
- *Viola*, 458
- Viperidae, 553
- *Vireo*
  - *Vireo flavoviridis*, 553
  - *Vireo leucophrys*, 466
  - *Vireo olivaceus*, 394
  - *Vireo philadelphicus*, 394
- Vireonidae, 556
- *Virola*, 374, 715
  - *Virola koschnyi*, 348, 353, 696
  - *Virola multiflora*, 696
  - *Virola sebifera*, 353
  - *Virola surinamensis*, 260
- Viscaceae, 462
- *Vismia*, 461
  - *Vismia baccifera*, 339
- *Vitex cooperi*, 353, 482
- *Vittaria*, 462
  - *Vittaria graminifolia*, 462
- *Vochysia*
  - *Vochysia ferruginea*, 715
  - *Vochysia guatemalensis*, 85, 543
  - *Vochysia hondurensis*, 374, 715
- *Voyria*, 699
- *Vriesea*, 462, 718
  - *Vriesea orosiensis*, 462

- *Wallinia chavarriae*, 276
- *Waltheria indica*, 255
- *Warszewiczia coccinea*, 540, 715
- *Wasmannia auropunctata*, 228
- *Websteria confervoides*, 256
- *Weinmannia*, 52, 458, 461, 473, 514, 718
  - *Weinmannia pinnata*, 81, 432, 469, 474, 475
  - *Weinmannia trianae*, 469, 474
- *Welfia*
  - *Welfia georgii*, 374, 699, 715
  - *Welfia regia*, 540, 715
- *Werauhia*, 275
- *Westoniella*, 497, 500, 509, 720
  - *Westoniella chirripoensis*, 510
  - *Westoniella eriocephala*, 510
- *Whallwachsia illuminata*, 275–76
- *Wigandia urens*, 261
- *Williamodendron glaucophyllum*, 374

- *Xanthosoma robustum*, 438
- Xantusiidae, 553
- Xenarthra, 265, 354
- *Xenops minutus*, 554
- *Xiphorhynchus*
  - *Xiphorhynchus lachrymosus*, 393
  - *Xiphorhynchus susurrans*, 394
- *Xyleborinus*
  - *Xyleborinus cocoensis*, 229
  - *Xyleborinus sparsegranularum*, 229
- *Xyleborus*
  - *Xyleborus bispinatus*, 229
  - *Xyleborus ferruginosus*, 229
- *Xylocopa*, 271
- *Xylophanes*
  - *Xylophanes anubus*, 325
  - *Xylophanes chiron*, 325
  - *Xylophanes juanita*, 323
  - *Xylophanes libya*, 326
  - *Xylophanes porcus*, 326
  - *Xylophanes tersa*, 230
  - *Xylophanes turbata*, 314, 323
- *Xylophragma seemannianum*, 257, 327
- Xylopinae, 271
- *Xylosma intermedia*, 469
- *Xyris*, 512, 721
  - *Xyris nigrescens*, 514, 686
  - *Xyris subulata*, 686

- *Zalophus*
  - *Zalophus californianus*, 170
  - *Zalophus wollebaeki*, 169, 170, 219

Zanthoxylum, 473, 718
  Zanthoxylum caribaeum, 350, 352, 712
  Zanthoxylum limoncello, 469
  Zanthoxylum melanostictum, 469
Zaretis, 270
Zea mays, 85, 475
Zeledonia coronata, 511
Zingiberales, 560
Zinowiewia
  Zinowiewia costaricensis, 352
  Zinowiewia integerrima, 350
Ziphiidae, 116
Ziphius cavirostris, 169
Zonotrichia capensis, 196, 511
Zooxanthellae, *607*
Zoraptera, 231
Zorotypidae, 231
Zygia longifolia, 260, 327
Zygothrica neolinia, 438